Petroleum Geology:
From Mature Basins to New Frontiers—
Proceedings of the 7th Petroleum Geology Conference

Volume 2

Petroleum Geology: From Mature Basins to New Frontiers— Proceedings of the 7th Petroleum Geology Conference

Held at the Queen Elizabeth II Conference Centre, London, 30 March–2 April 2009

Volume 2

Edited by

B. A. Vining
Baker Hughes, UK

and

S. C. Pickering
Schlumberger, Gatwick, UK

with

Mark Allen, Durham University, UK
John Argent, BG Group, Reading, UK
Jonathan Craig, Eni Exploration & Production, Milan, Italy
Tony Doré, Statoil, Houston, USA
Sergey Drachev, ExxonMobil International, Leatherhead, UK
Alastair Fraser, Imperial College London, UK
Scot Fraser, BHP Billiton, Houston, USA
Graham Goffey, PA Resources, London, UK

Gary Hampson, Imperial College London, UK
Howard Johnson, BGS Edinburgh, UK
Ken McCaffrey, Durham University, UK
James Maynard, ExxonMobil International, Leatherhead, UK
Jonathan Redfern, University of Manchester, UK
Bryan Ritchie, BP, Houston, USA
Robert Scott, CASP, Cambridge, UK
John Underhill, University of Edinburgh, UK

2010
Published by
The Geological Society
London

THE GEOLOGICAL SOCIETY

The Geological Society of London (GSL) was founded in 1807. It is the oldest national geological society in the world and the largest in Europe. It was incorporated under Royal Charter in 1825 and is Registered Charity 210161.

The Society is the UK national learned and professional society for geology with a worldwide Fellowship (FGS) of 10 000. The Society has the power to confer Chartered status on suitably qualified Fellows, and about 2000 of the Fellowship carry the title (CGeol). Chartered Geologists may also obtain the equivalent European title, European Geologist (EurGeol). One fifth of the Society's fellowship resides outside the UK. To find out more about the Society, log on to www.geolsoc.org.uk.

The Geological Society Publishing House (Bath, UK) produces the Society's international journals and books, and acts as European distributor for selected publications of the American Association of Petroleum Geologists (AAPG), the American Geological Institute (AGI), the Indonesian Petroleum Association (IPA), the Geological Society of America (GSA), the Society for Sedimentary Geology (SEPM) and the Geologists' Association (GA). Joint marketing agreements ensure that GSL Fellows may purchase these societies' publications at a discount. The Society's online bookshop (accessible from www.geolsoc.org.uk) offers secure book purchasing with your credit or debit card.

To find out about joining the Society and benefiting from substantial discounts on publications of GSL and other societies worldwide, consult www.geolsoc.org.uk, or contact the Fellowship Department at: The Geological Society, Burlington House, Piccadilly, London W1J 0BG: Tel. +44 (0)20 7434 9944; Fax +44 (0)20 7439 8975; E-mail: enquiries@geolsoc.org.uk.

For information about the Society's meetings, consult Events on www.geolsoc.org.uk. To find out more about the Society's Corporate Affiliates Scheme, write to enquiries@geolsoc.org.uk.

Published by The Geological Society from:
The Geological Society Publishing House
Unit 7, Brassmill Enterprise Centre
Brassmill Lane
Bath BA1 3JN, UK

(*Orders*): Tel. +44 (0)1225 445046
Fax +44 (0)1225 442836
Online bookshop: http://www.geolsoc.org.uk/bookshop

The publishers make no representation, express or implied, with regard to the accuracy of the information contained in this book and cannot accept any legal responsibility for any errors or omissions that may be made.

© Petroleum Geology Conferences Ltd 2010. All rights reserved. No reproduction, copy or transmission of this publication may be made without written permission. No paragraph of this publication may be reproduced, copied or transmitted save with the provisions of the Copyright Licensing Agency Ltd, Saffron House, 6-10 Kirby Street, London EC1N 8TS, UK. Users registered with the Copyright Clearance Center, 222 Rosewood Drive, Danvers, MA 01923, USA: the item-fee code for this publication is 978-1-86239-298-4/10/$15.00.

British Library Cataloguing in Publication Data

A catalogue record for this book is available from the British Library.

ISBN 978-1-86239-298-4

Typeset by Techset Composition Ltd, Salisbury, UK

Printed by The Charlesworth Group, Wakefield, UK

Distribution

USA
AAPG Bookstore
PO Box 979
Tulsa
OK 74101-0979
USA
(*Orders*): Tel. +1 918 584-2555
Fax +1 918 560-2652
E-mail: bookstore@aapg.org

India
Affiliated East-West Press Private Ltd
Marketing Division
G-1/16 Ansari Road, Darya Ganj
New Delhi 110 002
India
(*Orders*): Tel. +91 11 2327-9113/2326-4180
Fax +91 11 2326-0538
E-mail: affiliat@vsnl.com

Japan
Kanda Book Trading Company
Cityhouse Tama 204
Tsurumaki 1-3-10
Tama-shi, Tokyo 206-0034
Japan
(*Orders*): Tel. +81 (0)423 57-7650
Fax +81 (0)423 57-7651
E-mail: geokanda@ma.kcom.ne.jp

Contents

VOLUME 1

DVD Contents   xi

The Technical and Editorial Committee   xiii

Preface   xv

Global petroleum systems in space and time   1
S. May, R. Kleist, E. Kneller, C. Johnson & S. Creaney

The GeoControversies debates   11
J. R. Underhill

Virtual fieldtrips for petroleum geoscientists   19
K. J. W. McCaffrey, D. Hodgetts, J. Howell, D. Hunt, J. Imber, R. R. Jones, M. Tomasso, J. Thurmond & S. Viseur

Colin Oakman core workshop   27
G. J. Hampson, J. D. Collinson & P. Gutteridge

Session: Europe

Europe overview   31
G. Goffey

Exploration

North Sea hydrocarbon systems: some aspects of our evolving insights into a classic hydrocarbon province   37
D. Erratt, G. M. Thomas, N. R. Hartley, R. Musum, P. H. Nicholson & Y. Spisto

The search for a Carboniferous petroleum system beneath the Central North Sea   57
R. Milton-Worssell, K. Smith, A. McGrandle, J. Watson & D. Cameron

Channel structures formed by contour currents and fluid expulsion: significance for Late Neogene development of the central North Sea basin   77
P. C. Knutz

Source rock quality and maturity and oil types in the NW Danish Central Graben: implications for petroleum prospectivity evaluation in an Upper Jurassic sandstone play area   95
H. I. Petersen, H. P. Nytoft, H. Vosgerau, C. Andersen, J. A. Bojesen-Koefoed & A. Mathiesen

From thrust-and-fold belt to foreland: hydrocarbon occurrences in Italy   113
F. Bertello, R. Fantoni, R. Franciosi, V. Gatti, M. Ghielmi & A. Pugliese

Upper Jurassic reservoir sandstones in the Danish Central Graben: new insights on distribution and depositional environments   127
P. N. Johannessen, K. Dybkjær, C. Andersen, L. Kristensen, J. Hovikoski & H. Vosgerau

Architecture of an Upper Jurassic barrier island sandstone reservoir, Danish Central Graben: implications of a Holocene–Recent analogue from the Wadden Sea   145
P. N. Johannessen, L. H. Nielsen, L. Nielsen, I. Møller, M. Pejrup & T. J. Andersen

Sedimentology and sequence stratigraphy of the Hugin Formation, Quadrant 15, Norwegian sector, South Viking Graben   157
R. L. Kieft, C. A.-L. Jackson, G. J. Hampson & E. Larsen

Reappraisal of the sequence stratigraphy of the Humber Group of the UK Central Graben   177
P. J. Sansom

The Huntington discoveries: efficient exploration in the UK Central North Sea   213
J. M. Hollywood & R. C. Olson

The Jasmine discovery, Central North Sea, UKCS   225
S. Archer, S. Ward, S. Menad, I. Shahim, N. Grant, H. Sloan & A. Cole

Prospectivity of the T38 sequence in the Northern Judd Basin   245
J. M. Rodriguez, G. Pickering & W. J. Kirk

Can stratigraphic plays change the petroleum exploration outlook of the Netherlands?   261
F. F. N. Van Hulten

Field Development and Production

Laggan; a mature understanding of an undeveloped discovery, more than 20 years old   279
A. Gordon, T. Younis, C. Bernard-Graille, R. Gray, J.-M. Urruty, L. Ben-Brahim, J.-C. Navarre,
B. Paternoster & G. Evers

Managing the start-up of a fractured oil reservoir: development of the Clair field, West of Shetland   299
A. J. Witt, S. R. Fowler, R. M. Kjelstadli, L. F. Draper, D. Barr & J. P. McGarrity

Overcoming multiple uncertainties in a challenging gas development: Chiswick Field UK SNS   315
R. Nesbit & K. Overshott

The Ensign enigma: improving well deliverability in a tight gas reservoir   325
K. Purvis, K. E. Overshott, J. C. Madgett & T. Niven

Maximizing production and reserves from offshore heavy oil fields using seismic and drilling technology:
Alba and Captain Fields, UKNS   337
J. M. Hampson, S. F. Walden & C. Bell

Locating the remaining oil in the Nelson Field   349
C. E. Gill & M. Shepherd

The Buzzard Field: anatomy of the reservoir from appraisal to production   369
F. M. Ray, S. J. Pinnock, H. Katamish & J. B. Turnbull

The Scott Field: revitalization of a mature field   387
G. R. Brook, J. R. Wardell, S. F. Flanagan & T. P. Regan

Predicting production behaviour from deep HPHT Triassic reservoirs and the impact of sedimentary architecture
on recovery   405
S. Kape, O. Diaz De Souza, I. Bushnaq, M. Hayes & I. Turner

Sedimentology and unexpected pressure decline: the HP/HT Kristin Field   419
J. G. Quin, P. Zweigel, E. Eldholm, O. R. Hansen, K. R. Christoffersen & A. Zaostrovski

An old field in a new landscape: the renaissance of Donan   431
R. R. A. Reekie, E. L. Davies, N. J. Hart, A. T. McInally, J. R. Todd, J. R. McAteer, L. Franoux &
E. L. M. Ferguson

Techniques in Exploration and Exploitation

A road map for the identification and recovery of by-passed pay   453
P. F. Worthington

Tilting oil–water contact in the chalk of Tyra Field as interpreted from capillary pressure data   463
I. L. Fabricius & M. A. Rana

A holostratigraphic approach to the chalk of the North Sea Eldfisk Field, Norway   473
M. J. Hampton, H. W. Bailey & A. D. Jones

Role of the Chalk in development of deep overpressure in the Central North Sea   493
R. E. Swarbrick, R. W. Lahann, S. A. O'Connor & A. J. Mallon

Investigating fault-sealing potential through fault relative seismic volume analysis   509
J.-F. Dutzer, H. Basford & S. Purves

4D acquisition and processing: a North Sea case study on the relative contributions to improved 4D repeatability   517
E. C. Rushmere, M. Dyce, S. Campbell & A. J. Hill

Applying time-lapse seismic methods to reservoir management and field development planning at South Arne, Danish North Sea   523
J. V. Herwanger, C. R. Schiøtt, R. Frederiksen, F. If, O. V. Vejbæk, R. Wold, H. J. Hansen, E. Palmer & N. Koutsabeloulis

3D seismic mapping and porosity variation of intra-chalk units in the southern Danish North Sea   537
T. Abramovitz, C. Andersen, F. C. Jakobsen, L. Kristensen & E. Sheldon

Seismic imaging of variable water layer sound speed in Rockall Trough, NE Atlantic and implications for seismic surveying in deep water   549
S. M. Jones, C. Sutton, R. J. J. Hardy & D. Hardy

New aeromagnetic and gravity compilations from Norway and adjacent areas: methods and applications   559
O. Olesen, M. Brönner, J. Ebbing, J. Gellein, L. Gernigon, J. Koziel, T. Lauritsen, R. Myklebust, C. Pascal, M. Sand, D. Solheim & S. Usov

VOLUME 2

Session: Russia, Former Soviet Union and the Circum-Arctic

Russia, FSU and the Circum-Arctic: 'the final frontier'   589
J. R. Maynard, A. J. Fraser, M. B. Allen, R. A. Scott & S. Drachev

Tectonic history and petroleum geology of the Russian Arctic Shelves: an overview   591
S. S. Drachev, N. A. Malyshev & A. M. Nikishin

Assessment of undiscovered petroleum resources of the north and east margins of the Siberian craton north of the Arctic Circle
T. R. Klett, C. J. Wandrey & J. K. Pitman   621

Synchronous exhumation events around the Arctic including examples from Barents Sea and Alaska North Slope   633
P. F. Green & I. R. Duddy

Offset and curvature of the Novaya Zemlya fold-and-thrust belt, Arctic Russia   645
R. A. Scott, J. P. Howard, L. Guo, R. Schekoldin & V. Pease

Charging the giant gas fields of the NW Siberia basin   659
E. Fjellanger, A. E. Kontorovich, S. A. Barboza, L. M. Burshtein, M. J. Hardy & V. R. Livshits

Session: North Africa and Middle East

Middle East and North Africa: overview   671
J. Redfern & J. Craig

From Neoproterozoic to Early Cenozoic: exploring the potential of older and deeper hydrocarbon plays across North Africa and the Middle East   673
J. Craig, D. Grigo, A. Rebora, G. Serafini & E. Tebaldi

Palaeohighs: their influence on the North African Palaeozoic petroleum systems   707
R. Eschard, F. Braik, D. Bekkouche, M. Ben Rahuma, G. Desaubliaux, R. Deschamps & J. N. Proust

Stratigraphic trapping potential in the Carboniferous of North Africa: developing new play concepts based on integrated outcrop sedimentology and regional sequence stratigraphy (Morocco, Algeria, Libya)   725
S. Lubeseder, J. Redfern, L. Petitpierre & S. Fröhlich

Integrated petroleum systems and play fairway analysis in a complex Palaeozoic basin: Ghadames–Illizi Basin, North Africa   735
R. J. Dixon, J. K. S. Moore, M. Bourne, E. Dunn, D. B. Haig, J. Hossack, N. Roberts, T. Parsons & C. J. Simmons

Biostratigraphy, chemostratigraphy and thermal maturity of the A1-NC198 exploration well in the Kufra Basin, SE Libya   761
S. Lüning, N. Miles, T. Pearce, E. Brooker, P. Barnard, G. Johannson & S. Schäfer

Exploring subtle exploration plays in the Gulf of Suez   771
P. N. Dancer, J. Collins, A. Beckly, K. Johnson, G. Campbell, G. Mumaw & B. Hepworth

The hydrocarbon prospectivity of the Egyptian North Red Sea basin   783
G. Gordon, B. Hansen, J. Scott, C. Hirst, R. Graham, T. Grow, A. Spedding, S. Fairhead, L. Fullarton & D. Griffin

A regional overview of the exploration potential of the Middle East: a case study in the application of play fairway risk mapping techniques   791
A. J. Fraser

Appraisal and development of the Taq Taq field, Kurdistan region, Iraq   801
C. R. Garland, I. Abalioglu, L. Akca, A. Cassidy, Y. Chiffoleau, L. Godail, M. A. S. Grace, H. J. Kader, F. Khalek, H. Legarre, H. B. Nazhat & B. Sallier

Sedimentology, geochemistry and hydrocarbon potential of the Late Cretaceous Shiranish Formation in the Euphrates Graben (Syria)   811
S. Ismail, H.-M. Schulz, H. Wilkes, B. Horsfield, R. di Primo, M. Dransfield, P. Nederlof & R. Tomeh

Session: Passive Margins

Passive margins: overview   823
B. Levell, J. Argent, A. G. Doré & S. Fraser

Constraints on volcanism, igneous intrusion and stretching on the Rockall–Faroe continental margin   831
R. S. White, J. D. Eccles & A. W. Roberts

Properties and distribution of lower crustal bodies on the mid-Norwegian margin   843
R. F. Reynisson, J. Ebbing, E. Lundin & P. T. Osmundsen

The breakup of the South Atlantic Ocean: formation of failed spreading axes and blocks of thinned continental crust in the Santos Basin, Brazil and its consequences for petroleum system development   855
I. C. Scotchman, G. Gilchrist, N. J. Kusznir, A. M. Roberts & R. Fletcher

Structural architecture and nature of the continent–ocean transitional domain at the Camamu and Almada Basins (NE Brazil) within a conjugate margin setting   867
O. A. Blaich, J. I. Faleide, F. Tsikalas, R. Lilletveit, D. Chiossi, P. Brockbank & P. Cobbold

New compilation of top basement and basement thickness for the Norwegian continental shelf reveals the segmentation of the passive margin system   885
J. Ebbing & O. Olesen

Some emerging concepts in salt tectonics in the deepwater Gulf of Mexico: intrusive plumes, canopy-margin thrusts, minibasin triggers and allochthonous fragments   899
M. P. A. Jackson, M. R. Hudec & T. P. Dooley

Source-to-sink systems on passive margins: theory and practice with an example from the Norwegian continental margin   913
O. J. Martinsen, T. O. Sømme, J. B. Thurmond, W. Helland-Hansen & I. Lunt

An integrated study of Permo-Triassic basins along the North Atlantic passive margin: implication for future exploration   921
J. Redfern, P. M. Shannon, B. P. J. Williams, S. Tyrrell, S. Leleu, I. Fabuel Pérez, C. Baudon, K. Štolfová, D. Hodgetts, X. van Lanen, A. Speksnijder, P. D. W. Haughton & J. S. Daly

Sedimentology, sandstone provenance and palaeodrainage on the eastern Rockall Basin margin: evidence from the Pb isotopic composition of detrital K-feldspar   937
S. Tyrrell, A. K. Souders, P. D. W. Haughton, J. S. Daly & P. M. Shannon

Cretaceous revisited: exploring the syn-rift play of the Faroe–Shetland Basin   953
M. Larsen, T. Rasmussen & L. Hjelm

Timing, controls and consequences of compression in the Rockall–Faroe area of the NE Atlantic Margin   963
A. Tuitt, J. R. Underhill, J. D. Ritchie, H. Johnson & K. Hitchen

Episodic uplift and exhumation along North Atlantic passive margins: implications for hydrocarbon prospectivity   979
P. Japsen, P. F. Green, J. M. Bonow, E. S. Rasmussen, J. A. Chalmers & T. Kjennerud

New methods of improving seismic data to aid understanding of passive margin evolution: a series of case histories from offshore west of Ireland   1005
R. J. J. Hardy, E. Querendez, F. Biancotto, S. M. Jones, J. O'Sullivan & N. White

WATS it take to image an oil field subsalt offshore Angola?   1013
E. Ekstrand, G. Hickman, R. Thomas, I. Threadgold, D. Harrison, A. Los, T. Summers, C. Regone & M. O'Brien

Sub-basalt hydrocarbon prospectivity in the Rockall, Faroe–Shetland and Møre basins, NE Atlantic   1025
I. Davison, S. Stasiuk, P. Nuttall & P. Keane

Intra-basalt units and base of the volcanic succession east of the Faroe Islands exemplified by interpretation of offshore 3D seismic data   1033
M. Ellefsen, L. O. Boldreel & M. Larsen

Exploring for gas: the future for Angola   1043
C. A. Figueiredo, L. Binga, J. Castelhano & B. A. Vining

Session: Unconventional Hydrocarbons Resources

Unconventional oil and gas resources and the geological storage of carbon dioxide: overview   1061
H. Johnson & A. G. Doré

Bulk composition and phase behaviour of petroleum sourced by the Bakken Formation of the Williston Basin   1065
P. Kuhn, R. Di Primio & B. Horsfield

Shale gas in Europe: a regional overview and current research activities   1079
H.-M. Schulz, B. Horsfield & R. F. Sachsenhofer

UK data and analysis for shale gas prospectivity   1087
N. Smith, P. Turner & G. Williams

The Western Canada Foreland Basin: a basin-centred gas system   1099
D. J. Boettcher, M. Thomas, M. G. Hrudey, D. J. Lewis, C. O'Brien, B. Oz, D. Repol & R. Yuan

Tight gas exploration in the Pannonian Basin   1125
A. Király, K. Milota, I. Magyar & K. Kiss

Natural fractures in some US shales and their importance for gas production   1131
J. F. W. Gale & J. Holder

Athabasca oil sands: reservoir characterization and its impact on thermal and mining opportunities   1141
M. J. Peacock

Resource potential of gas hydrates: recent contributions from international research and development projects   1151
T. S. Collett

King coal: restoring the monarchy by underground gasification coupled to CCS   1155
P. L. Younger, D. J. Roddy & G. González

Geological storage of carbon dioxide: an emerging opportunity   1165
W. J. Senior, J. D. Kantorowicz & I. W. Wright

History-matching flow simulations and time-lapse seismic data from the Sleipner CO_2 plume   1171
R. A. Chadwick & D. J. Noy

Differences between flow of injected CO_2 and hydrocarbon migration   1183
C. Hermanrud, G. M. G. Teige, M. Iding, O. Eiken, L. Rennan & S. Østmo

Preparing for a carbon constrained world; overview of the United States regional carbon sequestration partnerships programme and its Southwest Regional Partnership   1189
R. Esser, R. Levey, B. McPherson, W. O'Dowd, J. Litynski & S. Plasynski

Index   1197

DVD Contents

The two proceedings volumes also include an interactive DVD. The DVD is both Mac and PC compatible and includes three main elements:

- The PDF file of the entire proceedings volumes
- 8 movies of virtual fieldtrips
- 18 posters as pdf files

The book PDF is fully searchable with hierarchical bookmarks for easy navigation. The Virtual Fieldtrips have textual links from the book PDF. Separate navigation windows allow the Posters and Fieldtrips to be selected from a separate contents listing.

Virtual Fieldtrips

Fieldtrip 1: Utilization of 3D outcrop models from Western Ireland
J. Thurmond

Fieldtrip 2: Stratigraphic architecture of an interbasinal deep-water conduit, Grand coyer sub-basin, Eocene–Oligocene Grès d'Annot formation
M. Tomasso, R. Bouroullec and D. Pyles

Fieldtrip 3: Three-dimensional characteristics of rift initiation depositional intervals using LIDAR: Abu Zenima and Nukhul formations, Suez Rift, Egypt
D. Hodgetts, P. Wilson, F. Rarity and R. Gawthorpe

Fieldtrip 4: Multi-scale, fluvial to shallow reservoir analogues, from outcrop to flow simulation: a virtual field trip to eastern Utah
J. Howell, S. Buckley, C. Carlsson, H. Enge and T. Knudsen

Fieldtrip 5: Multi-scale structural analogues: Lofoten virtual field trip
K. McCaffrey, W. Wilson and S. Bergh

Fieldtrip 6: Multi-scale structural analogues: Arkitsa Fault virtual field trip
R. Jones, K. McCaffrey and S. Kokkalas

Fieldtrip 7: Virtual outcrop models of the Moab fault to assess the impact of normal drag on across-fault juxtapositions
J. Imber, J. Long, K. McCaffrey, R. Walker, R. Holdsworth, R. Jones, F. Watson, C. Carter-Pike and R. Wightman

Fieldtrip 8: From field data to virtual outcrop models: exploring, interpreting and modelling carbonate architectures
S. Viseur, J. Borgomano, J. Gari and S. Nardon

Posters

Poster 1: South Caspian Basin opening: inference from subsidence analysis in northern Iran
M.-F. Brunet, A. Shahidi, E. Barrier, C. Muller and A. Saïdi

Poster 2: Scales, geometries and distribution of large-scale clastic intrusions in deep-water depositional systems and their impact on reservoir geometry and connectivity: a 3D seismic case study from the Paleocene of the eastern North Viking Graben
E. Dmitrieva, C. Jackson and M. Huuse

Poster 3: Basement map for the NE Atlantic margin and mainland Norway reveals influence of ancient structures on the passive margin system
J. Ebbing, O. Olesen, C. Barrère, M. Brönner, P. T. Osmundsen, C. Pascal, R. F. Reynisson and J. R. Skilbrei

Poster 4: Transfer zones: the application of new geological information from the Faroe Islands applied to the offshore of intra basalt and sub-basalt strata
D. Ellis, S. R. Passey, D. W. Jolley and B. R. Bell

Poster 5: New ways of improving and using seismic data to help the understanding of passive margin evolution: A series of case histories from West of Ireland
R. Hardy, S. Jones, E. Querendez, F. Biancotto, J. O'Sullivan and K. Fernandez

Poster 6: Sedimentology and sequence stratigraphy of the Hugin Formation, Quadrant 15, Norwegian Sector, South Viking Graben
R. Kieft, C. Jackson, G. Hampson and E. Larsen

Poster 7: Inconsistency of hydrocarbon generation potential and production data: the Bakken play of North Dakota
P. P. Kuhn, R. di Primio and B. Horsfield

Poster 8: The application of tectonostratigraphic models to the sub-basalt region of the Faroe–Shetland Basin, NE Atlantic Margin
D. J. Moy, J. Imber, R. E. Holdsworth, D. Ellis and J. W. Gallagher

Poster 9: Post-basalt deposits in the Faroese sector of the Faroe–Shetland Basin, NE Atlantic Ocean
J. Ólavsdóttir

Poster 10: Lava field thickness variations as an analogue for sedimentary dispersal patterns in the Faroe–Shetland Basin
S. R. Passey and T. Varming

Poster 11: Properties and distribution of lower crustal bodies on the mid-Norwegian Margin
R. F. Reynisson, J. Ebbing, E. Lundin, P. T. Osmundsen and O. Olesen

Poster 12: Remaining potential of a mature hydrocarbon area: integration of 3D-technology and sequence stratigraphy (Upper Jurassic, Lower Saxony Basin, Northwest Germany)
B. Seyfang

Poster 13: The importance of global consistency in resource reporting: a SPEE perspective: characterization and valuation of hydrocarbon accumulations
G. Simpson

Poster 14: Shale gas prospectivity
N. Smith and P. Turner

Poster 15: Petroleum Geology of the East Greenland Margin
H. Stendal, S. U. Hede, L. W. Møller and T. V. Rasmussen

Poster 16: Seismic imaging of variable water layer sound structure in Rockall Trough, NE Atlantic: implications for planning seismic surveys in deep water
C. Sutton, S. Jones, R. Hardy and D. Hardy

Poster 17: Insights into Palaeogene rifting in the Faroe-Shetland Basin: the Faroe Islands, NE Atlantic
R. J. Walker, R. E. Holdsworth and J. Imber

Poster 18: A road map for the identification and recovery of by-passed pay
P. Worthington

The Technical and Editorial Committee

Bernie Vining
Baker Hughes, UK
Conference Chair and
Joint Editor-in-Chief

Alastair Fraser
Imperial College London, UK
Convenor and Session Editor,
Russia, Former Soviet Union and
the Circum-Arctic

Steve Pickering
Schlumberger, Gatwick, UK
Conference Deputy Chair and
Joint Editor-in-Chief

Scot Fraser
BHP Billiton, Houston, USA
Convenor, Passive Margins

Mark Allen
Durham University, UK
Co-convenor and Session Editor,
Russia,
Former Soviet Union and
the Circum-Arctic

Graham Goffey
PA Resources, London, UK
Convenor and Session Editor
Europe

John Argent
BG Group, Reading, UK
Convenor and Session Editor,
Passive Margins

Gary Hampson
Imperial College London, UK
Convenor, Core Workshop

Jonathan Craig
Eni Exploration & Production, Milan,
Italy
Convenor and Session Editor
North Africa and Middle East

Howard Johnson
BGS Edinburgh, UK
Convenor and Session Editor,
Unconventional Hydrocarbon
Resources

Tony Dore
Statoil, Houston, USA
Convenor and Session Editor,
Passive Margins and
Unconventional Hydrocarbon
Resource

Ken McCaffrey
Durham University, UK
Convenor, Virtual field trips

James Maynard
ExxonMobil International,
Leatherhead, UK
Convenor and Session Editor,
Russia, Former Soviet Union and
the Circum-Arctic

Robert Scott
CASP, Cambridge, UK
Convenor and Session Editor, Russia,
Former Soviet Union and the Circum-
Arctic

Jonathan Redfern
University of Manchester, UK
Convenor and Session Editor,
North Africa and Middle East

John Underhill
University of Edinburgh, UK
Convenor, Geocontroversies

Bryan Ritchie
BP Houston, USA
Poster Chair and Session Editor,
North Africa and Middle East

Session editors and other technical contributors:

Matthew Allen, Stewart Clark, Richard Davies, Sergey Drachev, Steve Garrett, Jon Gluyas, Mark Lappin, Adam Law, Bruce Levell, Kevin McLachlan, Julian Rush, John Smallwood, Michael Thomas, Ian Walker

The Petroleum Geology Conferences Board Members:

Chris Bulley (Petroleum Exploration Society of Great Britain), Jerry Chessell (Petroleum Exploration Society of Great Britain), Roger Cooper (Energy Institute), Andy Fleet (The Geological Society of London), Louise Kingham (Energy Institute), John Martin (Energy Institute), Edmund Nickless (The Geological Society of London)

Organising Committee:

Marian Scutt (Petroleum Exploration Society of Great Britain), Jacqueline Warner (Energy Institute), Georgina Worrall (The Geological Society of London)

And with thanks to:

Malcolm Brown (BG Group) PGC VII Sponsorship Chairman

For help with the core workshop:

British Sedimentological Research Group (BSRG), British Geological Survey (BGS)

Grateful thanks are extended to the following companies for their generous sponsorship of the conference:

BG Group
BGS
BP
Centrica Energy
Chevron
ConocoPhillips
ExxonMobil
Hardy Oil and Gas plc
Hess
Maersk
MND Exploration & Production
Nexen
Petro-Canada
PA Resources
Schlumberger
Serica Energy
Sonangol Gas Natural
StatoilHydro
Venture Production plc

Session: Russia, Former Soviet Union and the Circum-Arctic

Russia, FSU and the Circum-Arctic: 'the final frontier'

J. R. MAYNARD,[1] A. J. FRASER,[2] M. B. ALLEN,[3] R. A. SCOTT[4] and S. DRACHEV[1]

[1]*ExxonMobil, ExxonMobil House, Ermyn Way, Leatherhead, Surrey KT22 8UX, UK*
[2]*BP Exploration, Chertsey Road, Middlesex TW166 7LN, UK (e-mail: alastair.fraser@imperial.com)*
[3]*Department of Earth Sciences, University of Durham, Durham DH1 3LE, UK*
[4]*CASP, University of Cambridge, 181A Huntingdon Road, Cambridge CB3 0DH, UK*

Sixteen papers representing the petroleum geology of the Arctic, Russia and former Soviet Union were presented over the first day and a half of PGC VII. The region is huge, diverse and has generated a great deal of excitement and outside investment in the industry over the 20 years since the collapse of the Soviet Union. The Arctic region in particular has significance as perhaps the last great frontier hydrocarbon province on Earth. The region is large, approximately 5000 km across a polar view north of the Arctic Circle (Fig. 1). Importantly, from an oil and gas exploration perspective, the Arctic Ocean has the most extensive continental shelf area of any ocean basin (*c*. 50% of offshore area). Much of this sits in the broad Russian offshore Arctic in water depths of less than 50 m. There are numerous sedimentary basins in the Arctic, some well known, but most poorly understood. Art Grantz (United States Geological Survey) and colleagues estimated resources at 114×10^9 barrels of undiscovered oil and 2000×10^{12} standard cubic feet (SCF) of natural gas. If the estimates are correct, these hydrocarbons would account for more than a fifth of the world's undiscovered resources. This great prize, in a world of diminishing reserves, has recently brought territorial issues into focus between the five countries with claims in the Arctic Ocean (Russia, Norway, Denmark, Canada and the USA). All of this is taking place against a backdrop of increasing concern for the fragile Arctic environment.

Day one was dominated by the Arctic, the first half of which focused on the various models that are considered responsible for the tectonic origin of the Arctic and its basins. That such a wide variety of models is possible emphasized how much we still need to learn about the region. The discussions prompted by this session spilled over into lively debate in the poster session. The keynote by Al Fraser (BP) on the regional context of the Arctic Frontier Basins of Canada, Russia, Norway and the USA was given by John Berry and Edith Fugelli, who had stepped in to present at short notice. They described the Arctic as comprising two major deepwater basins floored by oceanic crust. These are the Eurasia Basin and the Amerasia Basin (Fig. 1). The Eurasia Basin as the extension of the North Atlantic rift system is relatively well constrained. However, the spreading history of the Amerasia Basin is less certain and is still the subject of some debate.

This was followed by Steve Bergman (Shell) who outlined the model for the development of the Arctic Ocean. Art Grantz (USGS) spoke about the recent work undertaken by the US Survey and collaborators to assess the petroleum potential of the region. Paul Green (Geotrack International) presented the results of numerous fission track studies that show synchronous widespread Cenozoic exhumation in the region and discussed its potential impact on hydrocarbon migration and trap integrity. Sergey Drachev (ExxonMobil) discussed the tectonic and petroleum geology of the Russian Arctic sector, outlining evidence for the various theories on the opening of the Arctic Ocean and the creation of the basins in this region.

Papers in the second half of the day discussed individual Arctic basins of the Russian and Norwegian Arctic sectors. Vladimir Verzhbitsky (TGS-NOPEC) showed the results of recently acquired seismic data in the North Chukchi Basin in the Russian Arctic, which have been used to construct an interpretation of the region's geological history and contain evidence for an active petroleum system. Jan Inge Faleide (University of Oslo) presented a paper on the sedimentary basins of the Barents Sea, comparing and contrasting the Western and Eastern basins. Jorg Ebbing (Geological Survey of Norway) described a joint project with the Russian VSEGEI institute on the palaeogeographic and tectonic evolution and present-day structure of the Barents and Kara Seas. Timothy Klett (USGS) expanded on the recent Arctic assessment by highlighting the potential of the Siberian shelf and craton, which has an estimated 28 MMBOE to be discovered.

Day two was the turn of Russia's petroleum geology to be examined and the keynote delivered by Steve Creaney (ExxonMobil) gave an overview of the numerous prospective basins in this vast country, using a genetic approach. The tectonic development of the region was used as a framework to address the diverse complexity of these different basins and their hydrocarbon systems. The tectonic evolution of the prolific West Siberian Basin from the Palaeozoic to the present was described by Mark Allen (Durham University), adding to the debate on the tectonic evolution of the Arctic started on the previous day. Robert Scott (CASP) presented an explanation of the curvature of the Novaya Zemlya fold-and-thrust belt and its offset from the remainder of the Uralian Orogen, based on recently collected field observations and regional tectonic synthesis. The significance of this interpretation for adjacent hydrocarbon basins was discussed. Erik Fjellanger (ExxonMobil) described 3D modelling of the hydrocarbon system of the northern West Siberian Basin, where he characterized the plays and source rocks and calibrated the model with some of the world's giant gas fields.

The neotectonics and effects of the Fennoscandian de-glaciation on Siberia were discussed by Mark Allen (Durham University). Finally the discussion moved to the Caspian, where two Kazakh field developments were explained. Firstly the exploration and appraisal of the technically and environmentally challenging, super giant Kashagan Field was presented by Didier Terroir (ENI), and lastly Simon Beavington-Penny (BG) presented work on integrating carbonate sedimentology and cross-well seismic to optimize the phased gas injection at Karachaganak.

The discussion continued into the poster sessions which, in a change from past conferences, allocated the authors a chance to present their posters. A steady number of 15–20 people contributed to these discussions with a free flow of debate amongst the audience and presenters. Robert Scott (CASP) continued the discussion on the opening of the Arctic Ocean by considering the

Fig. 1. Sedimentary basins of the Arctic. The Arctic region is large, approximately 5000 km across this polar view north of the Arctic Circle. There are two deepwater basins floored by oceanic crust and separated by the Lomonosov Ridge. These are the Eurasia Basin and the Amerasia Basin.

various models and their significance for how sediment was dispersed in the Arctic. New seismic from offshore Sakhalin was used to explore the petroleum potential of the Deriugin Basin by Alice Little (TGS-NOPEC), an area of recent developments by western companies. A collaboration to map the basins of the Arctic was presented by two of the joint authors (Sergey Drachev, ExxonMobil and Robert Scott, CASP). This map categorizes the sedimentary accumulations of the Arctic according to their tectonic affinity. Henrik Stendal (Bureau of Minerals and Petroleum) presented his poster on the hydrocarbon potential of the East Greenland Margin. Finally, the prospectivity of the eastern Black Sea was discussed by Li Guo (CASP), who presented work on the analogue late Jurassic reefs of the western Caucasus and Crimea.

Tectonic history and petroleum geology of the Russian Arctic Shelves: an overview

S. S. DRACHEV,[1] N. A. MALYSHEV[2] and A. M. NIKISHIN[3]

[1]*ExxonMobil International Ltd, ExxonMobil House, MP44, Ermyn Way, Leatherhead, Surrey KT22 8UX, UK (e-mail: sergey.s.drachev@exxonmobil.com)*
[2]*Russian State Oil Company 'OAO NK Rosneft', 26/1 Sofiiskaya Embankment, Moscow 115035, Russia*
[3]*Geological Department of Moscow State University, 1 Vorob'evy Gory, Moscow 119899, Russia*

Abstract: The Eastern Barents, Kara, Laptev, East Siberian seas and the western Chukchi Sea occupy a large part of the Eurasian Arctic epicontinental shelf in the Russian Arctic. Recent studies have shown that this huge region consists of over 40 sedimentary basins of variable age and genesis which are thought to bear significant undiscovered hydrocarbon resources. Important tectonic events controlling the structure and petroleum geology of the basins are the Caledonian collision and orogeny followed by Late Devonian to Early Carboniferous rifting, Late Palaeozoic Baltica–Siberia collision and Uralian orogeny, Triassic and Early Jurassic rifting, Late Jurassic to Early Cretaceous Canada Basin opening accompanied by closure of the South Anyui Ocean, the Late Mesozoic Verkhoyansk–Brookian orogeny and Cenozoic opening of the Eurasia Oceanic Basin. The majority of the sedimentary basins were formed and developed in a rift and post-rift setting and later modified through a series of structural inversions. Using available regional seismic lines correlated with borehole data, onshore geology in areas with no exploration drilling, and recent Arctic-wide magnetic, bathymetry and gravity grids, we provide more confident characterization of the regional structural elements of the Russian Arctic shelf, and constrain the timing of basin formation, structural styles, lithostratigraphy and possible hydrocarbon systems and petroleum play elements in frontier areas.

Keywords: Eurasian Arctic, Barents Sea, Kara Sea, Laptev Sea, East Siberian Sea, Chukchi Sea, sedimentary basin, rift, petroleum potential, hydrocarbon system

A significant part of the Arctic is represented by the Eurasian epicontinental shelf which is the largest shelf on Earth. Its major portion (about 3.5 million km^2) is located in the Russian Arctic and is occupied by the eastern part of the Barents, Kara, Laptev, East Siberian and a western part of the Chukchi seas (Fig. 1). A systematic geological study and airborne gravity and magnetic measurements of the vast Russian Arctic shelves (RAS) was commenced soon after the end of the World War II, by the Research Institute of Arctic Geology (NIIGA, former Leningrad) and later by State Research Enterprise 'SevMorGeo' (see references below). The general results were summarized by Vol'nov *et al.* (1970), Vinogradov *et al.* (1974, 1977), Gramberg & Pogrebitskiy (1984), and recently by Suprunenko & Kos'ko (2005), Petrov *et al.* (2008), and Burlin & Stoupakova (2008).

The main exploration effort over the entire RAS was undertaken during the latest period of the Soviet era, when extensive coverage of refraction and 2D reflection seismic lines was acquired over the eastern Barents and southern Kara seas by Polar Marine Geological Expedition (PMGRE, St Petersburg), Marine Arctic Geological Expedition (MAGE, Murmansk), SevMorGeologiya (SMG, St Petersburg) and SevMorNefteGeofizika (SMNG, Murmansk) (Fig. 2). Some of the large prospects were successfully tested during the 1980s and several large discoveries were made, including the gigantic Shtokman, Rusanovskoe and Leningradskoe gas and gas condensate fields. Today the Russian Barents and southern Kara shelves represent the most explored petroleum provinces of the RAS, bearing c. 130 × 10^9 barrels of oil equivalent (BBOE) of proven resources.

The Siberian shelves, which are the most remote from the present-day markets, remain poorly explored. They represent one of the most promising petroleum frontiers worldwide. They have been explored by an irregular grid of wide-angle refraction and 2D regional multichannel seismic reflection (MCS) lines acquired mostly between 1975 and 1997 by PMGRE, MAGE, Laboratory of Regional Geodynamics (LARGE, Moscow) and SMNG in cooperation with German Federal Institute for Geosciences and Natural Resources (BGR, Hannover) in the Laptev Sea; and by LARGE and DalMorNefteGeofizika (DMNG, Sakhalin) in cooperation with Halliburton Geophysical Services, in the East Siberian and Chukchi seas. A recent seismic survey by the TGS-Nopec Geophysical Company AS provided modern high-quality data acquired with a 6 km long streamer in the Russian Chukchi Sea (Verzhbitsky *et al.* 2008).

Although the post-Soviet period did not bring new offshore discoveries due to suspended exploration activity, it was generally a time of broad regional compilation of the Soviet-era data. These became publicly available, and were incorporated into Arctic-wide digital bathymetric, gravity and magnetic grids through implementation of several international projects: International Bathymetric Chart of the Arctic Ocean (IBCAO, http://www.ibcao.org), Arctic Gravity Project (AGP, http://earth-info.nga.mil/GandG/wgs84/agp/index.html) and several compilations of the Arctic magnetic and gravity fields (Verhoef *et al.* 1996; Glebovsky *et al.* 2000; Maschenkov *et al.* 2001). Interpretation of these digital gravity and magnetic grids in combination with MCS lines allowed much more confident and accurate mapping and characterization of the Arctic regional structural elements and sedimentary basins (Ivanova *et al.* 1990; Warren *et al.* 1995; Drachev *et al.* 1998, 1999, 2001; Franke *et al.* 2000, 2001, 2004; Glebovsky *et al.* 2000; Sekretov 2000, 2001; Sherwood *et al.* 2002; Franke & Hinz 2005; Sharov *et al.* 2005; Grantz *et al.* 2009).

Sedimentary basins of the RAS are thought to bear significant volumes of undiscovered hydrocarbon (HC) resources which are still difficult to estimate due to limited geological and geophysical data. According to a recent assessment by the Russian Ministry of Natural resources, the RAS could contain as much as 700 BBOE of

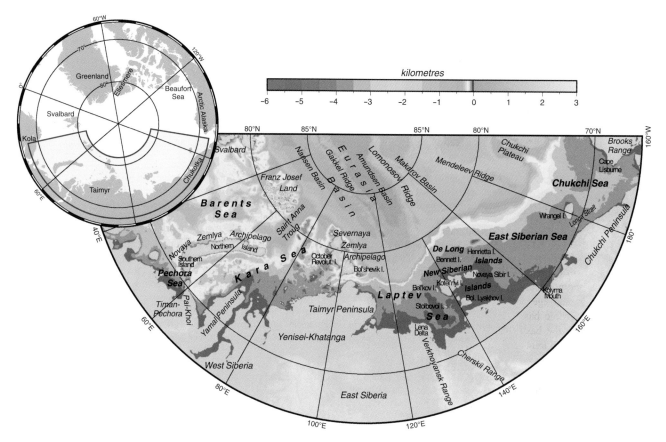

Fig. 1. Physiography of the Russian Arctic shelf. Topography is given after International Bathymetric Chart of the Arctic Ocean (IBCAO; http://www.ibcao.org). The inserted map in the upper left corner shows the location of the study area (red outline) in the Circum-Arctic.

total (discovered and undiscovered) resources. Data describing the age, composition and structural styles of the rocks composing the Arctic continental masses and islands remain a principal source of information about undrilled pre-Jurassic HC plays of the Barents–Kara region and the entire section of the Siberian Arctic shelves. This paper presents a brief overview of the RAS regional and petroleum geology. Based on available seismic, gravity/magnetic and geological data, we describe the main structural features of the East Barents, North and South Kara, Laptev, East Siberian Sea and western Chukchi provinces. The paper also summarizes the most important data on the evolution of these basins, and their known petroleum systems, and tries to apply general tectonic

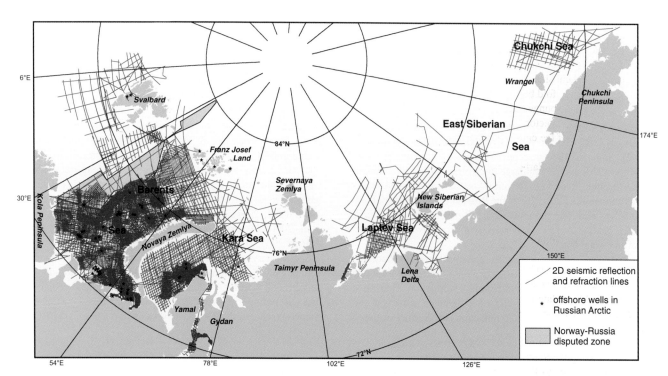

Fig. 2. Location of the 2D seismic reflection and refraction surveys and offshore wells in the Russian Arctic.

models and trans-regional correlations to draw some conclusions for those parts of the shelves where no direct observations exist to infer their geology.

Tectonic setting

As is consistent with modern plate-tectonic ideas of Arctic evolution, the structure of the consolidated continental crust underlying the Eurasian continental margin was formed during much of the Phanerozoic as a result of a series of collisions between the Laurentia, Baltica and Siberia continents and with a number of smaller microcontinents. Sedimentary basins post-dating the main Phanerozoic collisions mainly formed in response to initial rifting related to post-orogenic collapse and/or to the formation of Arctic spreading basins, for example, Eurasia and Amerasia basins. Many of these sedimentary basins were later modified through a series of intraplate structural inversions.

The formation of the RAS basins has generally been migrating over time east- and northeastward (present-day coordinates), thus basin complexity and age decrease in the same direction. The oldest Early Palaeozoic basins formed in the western sector of the RAS, and then the basin formation progressed through the Palaeozoic in the east Barents and north Kara shelves, and throughout the Early–Mid Mesozoic in the South Kara Sea and the Yenisei–Khatanga region.

The latest phase of basin formation in the RAS took place in the Laptev Sea region, where a series of rift-related basins have been evolving due to the opening of the Eurasia oceanic basin and the development of the present-day boundary between the Eurasian (EUR) and North American (NA) lithospheric plates. The Cenozoic plate-tectonic history of the Arctic is well constrained due to the decipherable set of seafloor spreading magnetic anomalies in the North Atlantic and Eurasia basins (Karasik 1968, 1974; Pitman & Talwani 1972; Vogt et al. 1979; Savostin & Karasik 1981; Karasik et al. 1983; Savostin et al. 1984a; Srivastava 1985; Cook et al. 1986; Harbert et al. 1990; Kristoffersen 1990; Glebovsky et al. 2006), and we refer to these publications when more details are required with regard to the Cenozoic plate-tectonic framework of the Arctic.

Therefore the basins of the Barents–Kara region, which rest on continental crust of the Palaeoproterozoic craton and Palaeozoic accreted crust, are mostly composed of Neoproterozoic, Palaeozoic and Mesozoic–Cenozoic carbonate and siliciclastic sequences, whereas most of the Siberian Arctic basins (the Laptev, East Siberian and Chukchi shelves) are underlain by the younger crust of the Late Mesozoic fold belts, and are filled with the Cretaceous (Aptian–Albian and younger) and Cenozoic siliciclastic sediments.

The most important tectonic events controlling the structure and petroleum geology of the entire Eurasian Arctic shelf are:

(1) Neoproterozoic to Early Cambrian Timanian orogeny;
(2) Caledonian orogeny followed by a phase of orogen collapse and crustal extension in Late Devonian–Early Carboniferous;
(3) Late Palaeozoic (Uralian) collision of the Baltica and Siberian continents;
(4) Permo-Triassic plume-related volcanic event and associated crustal extension;
(5) Jurassic rifting and subsequent opening of the Canada oceanic basin, accommodated by separation of the Arctic Alaska–Chukchi Microplate (AACM) from the Canadian margin of North America and its movement towards Siberia;
(6) Early Cretaceous closure of the Anyui Ocean due to convergence of the AACM with the Verkhoyansk-Omolon Siberian margin along the South Anyui Suture;
(7) Late Cretaceous to Paleocene opening of the Labrador Sea and Baffin Bay basins, which may also have affected the Central Arctic region between the Lomonosov and the Alpha-Mendeleev ridges;
(8) Greenland–Ellesmere and Greenland–West Barents margin convergence at 55–33 Ma and related crustal microplate re-adjustment in the Barents–Kara region;
(9) opening of the Eurasia oceanic basin at 55–0 Ma and related rifting and crustal microplate re-adjustment in the Laptev–East Siberian seas sector;
(10) the India–Eurasia collision at 40–10 Ma, causing large-scale crustal re-adjustment throughout Asia and NE Asia, which may also have reached the Eurasian Arctic continental margins.

The Cenozoic development of the RAS was controlled by continuous interaction of the NA and EUR lithospheric plates, which caused a drastic impact on the Siberian Arctic shelves. Since this region has always been in an intra-plate setting near the pole of plate rotation, even small changes in the plates' rotation have resulted in drastic changes in the basins' tectonic development and depositional environments.

Main characteristics of the RAS consolidated basement

Structurally the RAS is bordered on the south by the Baltica (also called East European, or Russian) and Siberian (also called East Siberian) cratons and adjoining Neoproterozoic, Palaeozoic and Mesozoic fold-and-thrust belts (hereinafter called fold belts). All of these first-order structures approach the Eurasian Arctic coast, and apparently extend farther offshore, where they form a tectonic basement (hereinafter also called basement) underlying sedimentary basins (Fig. 3). (By a tectonic basement we mean strongly deformed and/or metamorphosed units of rocks and their associations, compared with generally undeformed/weakly deformed and unmetamorphosed sedimentary successions existing within a sedimentary basin.)

The fact that the Eurasian Arctic onshore fold belt domains extend offshore is also supported by a number of MCS lines located close to the shoreline, as well as by gravity and magnetic maps. However, due to the lack of reliable data, there are as many points of view on possible offshore basement tectonics as there are researchers. However, many agree that a fundamental difference in tectonics and geological history exists between the western and eastern sectors of the RAS, as was earlier recognized by Vinogradov et al. (1974, 1977), Gramberg & Pogrebitskiy (1984), Savostin et al. (1984b), Zonenshain et al. (1990) and others.

Figure 3 illustrates our understanding of large-scale crustal structural pattern of the RAS based on geological data from coastal areas, published and unpublished MSC lines and publicly available gravity and magnetic grids.

Western sector of the RAS (eastern Barents and Kara shelves)

The western RAS is dominated by pre-Cambrian, Palaeozoic and Early Mesozoic crustal domains: the Neoproterozoic Timan–Varanger Fold Belt (Timanides by Gee & Pease 2005), Scandinavian Caledonides, a hypothetical Mesoproterozoic Svalbard Massif, Late Palaeozoic Uralian and Taimyr fold belts (Uralides and Taimyrides, respectively), the Early Mesozoic Novaya Zemlya and, to a small extent, South Taimyr fold belts (Fig. 3). There are still many highly disputed issues concerning possible outlines and relationships of these first-order structural domains beneath thick sedimentary cover of the Arctic seas (for the latest review see Pease 2011). Below we provide a short description of the main structural domains of the offshore basement relevant to understanding the formation and evolution of the RAS sedimentary basins.

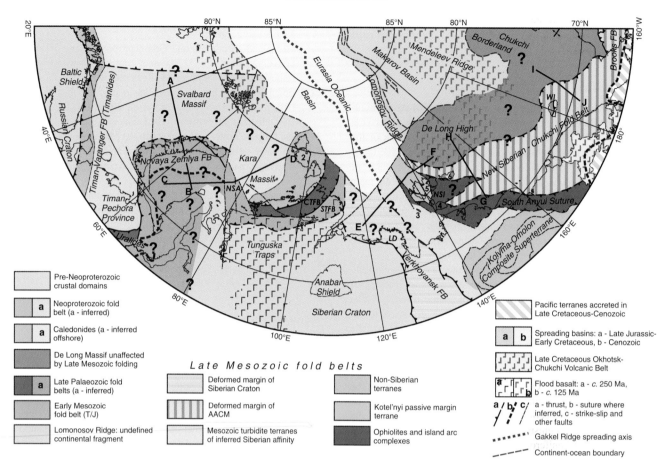

Fig. 3. Crustal tectonics of the Russian Arctic shelf and adjacent onshore and offshore regions. Bold solid lines labelled A–B, C–D, E–F, G–H and I–J show the location of crustal cross-sections given in Figures 6a, b, 11b, 14 & 16, respectively. The bold italic letters denote: NSA, North Siberian Arch; CTFB, Central Taimyr Fold Belt; STFB, South Taimyr Fold Belt; LD, Lena Delta; NSI, New Siberian Islands; WI, Wrangel Island. The bold numbers denote the following islands: 1, Bol'shevik; 2, October Revolution; 3, Stolbovoi; 4, Bol'shoi Lyakhov; 5, Kotel'nyi; 6, Novaya Sibir'. For other geographic names see Figure 1. Bold question marks denote the areas with least constrained interpretation of the basement type and age. FB denotes fold belt.

The Novaya Zemlya Fold Belt (NZFB) has a critical significance for understanding the tectonic history of the western RAS. It was formed at a very complex junction of the Baltica and Siberian cratons with the northern part of the Late Palaeozoic Uralides. It also divides two major HC provinces – East Barents and Northwest Siberian – and thus is the only area where the Palaeozoic HC systems of both provinces are exposed and thus available for direct studies.

As shown by Bondarev (1982), Lopatin *et al.* (2001), Pogrebitskiy (2004), Korago *et al.* (2004, 2009) and Vinokurov *et al.* (2009), the NZFB is mostly composed of Palaeozoic to Early Triassic successions deposited in a shelf to basin transition setting with shallow shelf facies developed along the western flank of the fold belt. The total thickness of the known section exceeds 13 km. The entire section was severely deformed at the end of the Triassic–earliest Jurassic to form a west-verging arcuate fold belt (Scott *et al.* 2010).

The basement of the NZFB is exposed locally and consists of Meso- and Neoproterozoic metaclastic and metacarbonate rocks, compressionally (east–west trending) deformed and metamorphosed in epidote–amphibolite and greenschist facies, and intruded by Neoproterozoic granite and granodiorite rocks (Korago *et al.* 2004). The recent study of metaclastic turbidites, which underlie a sharp angular unconformity at the base of unmetamorphosed variegated clastic sediments of Early Ordovician at Southern Novaya Zemlya Island, revealed the presence of Cambrian ages of detrital zircons (Pease & Scott 2009). This implies that at least some of the previously inferred Neoproterozoic rock complexes have, in fact, Cambrian age, and that the age of 'Timanian' unconformity is not older than Late Cambrian to Early Ordovician. Therefore, according to Pease (2011), the Timanian orogeny on southern Novaya Zemlya lasted until the end of Cambrian time, and the regional limit of the Timanian deformation probably extends beyond the Novaya Zemlya Archipelago into the Northern Kara region, where a contemporaneous unconformity between Cambrian and Ordovician is present on October Revolution Island (see below). A similar point of view about possible Cambrian extent of the Timanian orogeny was proposed by Bogolepova & Gee (2004).

The east–west compressed and unmetamorphosed Palaeozoic to Lower Triassic strata were deposited along the eastern margin of the Baltica (present-day orientation). The southern and central parts of the NZFB are composed of the following tectonostratigraphic rock assemblages, or complexes (Lopatin *et al.* 2001):

(1) Cambrian to Middle Devonian shallow water sandstones, dolomites, limestones, shales, siltstones, gravelites and conglomerates.
(2) Late Devonian to Early Carboniferous shallow water limestones, calcarenites, bioherms and calcareous sandstones.
(3) Middle to Late Devonian assemblage of clastic, volcanic and volcaniclastic rocks – claystones, siltstones, shales, polymictic sandstones, gravelites, conglomerates, tholeiitic basalts and tuffs. The intrusive analogues are represented by sills and dykes of gabbro-dolerites. The rock geochemistry suggest their intracontinental rift affinity.
(4) Late Devonian to Permian open marine and deepwater claystones, siltstones, rhodochrosite-bearing siliceous rocks, turbidites, olistostromes.

(5) Late Permian siliciclastic deepwater turbidites with horizons of olistostromes and calcareous sandstones.
(6) Later Permian to Lower Triassic shallow water and continental coarse-grained clastic rocks.

These rock complexes are intruded by a few small bodies of granitic rocks with Late Triassic to Early Jurassic isotopic ages (Pogrebitskiy 2004).

Baltica Timanide sources for the Palaeozoic clastic rocks are determined by detrital zircon ages (Pease & Scott 2009). Korago et al. (2009) suggested that the Upper Silurian clastic sediments, in contrast to the underlying beds, were derived predominantly from a northwesterly located Caledonian orogen. However, the Caledonian source is still highly debated (Pease 2011).

The northern part of the NZFB reveals a different type of stratigraphy. According to Lopatin et al. (2001), Pogrebitskiy (2004) and Korago et al. (2004), it is composed of a continuous c. 10–13 km thick Neoproterozoic to Lower Devonian section, which lacks any significant unconformities. The deepwater fine-grained siliciclastic metaturbidites are predominant within the Neoproterozoic to Lower Silurian successions. The overlying Upper Silurian to Lower Devonian strata are composed of shallow marine clastic and coarse clastic sediments, and the uppermost part of the section consists of Lower to Middle Devonian shallow water carbonate succession. Thin Upper Devonian and Carboniferous shallow water clastic and carbonate sediments occur sporadically and reveal a number of stratigraphic gaps. This interval could be related to the Late Palaeozoic orogenic event, which strongly affected the Taimyr Peninsula, and apparently the adjacent Kara Massif. Upper Carboniferous to Permian shallow-marine and continental coarse-grained clastic sediments coeval to main phase of the orogeny cap the section of the northern part of the fold belt.

The absence of unconformities within the Neoproterozoic to Lower Palaeozoic section of the northern block of the NZFB, which could be expected in proximity to the Timanian and possibly Caledonian (see Gee et al. 2006) deformation fronts, is one of the enigmas of the Arctic geology. One possible explanation given by Korago et al. (2004) is that this block had an independent Neoproterozoic and Palaeozoic history and became a part of the fold belt during its formation in Permian–Triassic.

The age of compressional deformations associated with the NZFB was for a long time one of the most disputed issues of the Arctic geology (see Pease 2011). The youngest strata recognized to be involved into the deformations are the Lower Triassic, which could assume a younger age of the deformations. Modern MCS data acquired in the vicinity of the western coast of Novaya Zemlya show a sharp angular unconformity at about the Triassic–Jurassic boundary, which probably corresponds to the main deformation phase (Pavlov et al. 2008).

The Kara Massif, or Microcontinent (KM) is traditionally outlined in the northern part of the Kara Sea. Structurally it is separated from the Late Palaeozoic to Early Mesozoic structural assemblage at the basement of the South Kara Basin (SKB) by a prominent linear North Siberian basement arch (or Step in the Russian literature). As depicted by the gravity field (Fig. 4, number 11), the arch strikes from the northern tip of the Novaya Zemlya Archipelago to the northwestern coast of the Taimyr Peninsula, indicating a structural relationship between the Novaya Zemlya and Taimyr fold belts. Its origin may be related to the Early Mesozoic compressional event and, given almost orthogonal orientation of the arch

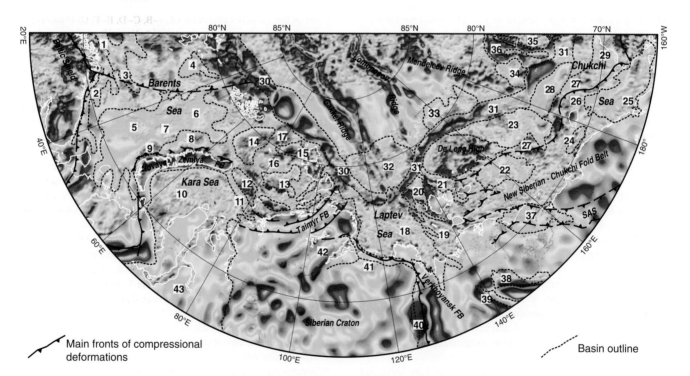

Fig. 4. Free-air gravity field over the Russian Arctic shelf and adjacent areas with outlines of the main sedimentary basins and basement highs (gravity data source is Arctic Gravity Project, http://earth-info.nga.mil/GandG/wgs84/agp). Numbers denote the following structural elements: *Barents Sea* – 1, Hammerfest Basin; 2, Varanger Trough; 3, Nordkapp Rift; 4, Olga Basin; 5–7, East Barents Megatrough (5, South Barents Basin; 6, North Barents Basin; 7, Ludlov Saddle); 8, Admiralty High; 9, Pri-Novaya Zemlya Basin. *Kara Sea:* 10, South Kara Basin; 11, North Siberian Arch; 12, Litke Trough; 13, North Kara Basin; 14, St Anna Trough; 15, Schmidt Trough; 16, Central Kara High; 17, Ushakov High. *Laptev Sea:* 18, Ust' Lena Rift; 19, Stolbovoi Horst; 20, Anisin Rift; 21, New Siberian Rift. *East Siberian Sea:* 22, East Siberian Sea Basin (East Siberian Depocentre); 23, Vil'kitskii Basin; 24, Longa Basin. *Chukchi Sea:* 25, South Chukchi (Hope) Basin; 26, Wrangel-Herald Structural Arch; 27, New Siberian-Wrangel Basin; 28, North Chukchi Basin; 29, Colville Basin. *Eurasian Continental Margin:* 30, Barents–Kara–West Laptev marginal basin; 31, East Laptev–East Siberian–Chukchi marginal basin. *Oceanic deepwater basins:* 32, South Eurasia Basin; 33, Podvodnikov Basin; 34, Chukchi Abyssal Plane Basin; 35, Northwind Basin; 36, Chukchi Plateau. *Continental realm:* 37, Lower Kolyma Basin; 38, Zyryanka Basin; 39, Moma Rift; 40, Priverkhoyansk Basin; 41, Lena-Anabar Basin; 42, Yenisei–Khatanga Basin; 43, West Siberian Basin. SAS is South Anyui Suture.

with regard to the NZFB, a considerable dextral strike-slip deformation may be expected.

The KM remains poorly studied and its geology is mainly projected from the Severnaya Zemlya Archipelago and northern Taimyr Peninsula. The first reliable isotopic data constraining the ages of magmatic, metamorphic and, therefore, tectonic events were obtained by Vernikovsky (1995, 1996). Recently new data on geology of Severnaya Zemlya were published by Metelkin et al. (2005), Lorenz et al. (2006, 2007, 2008), and Männik et al. (2009). Lorenz et al. describe the KM as the North Kara Terrane.

Tectonic basement of the KM is exposed in the northern part of the Taimyr Peninsula, and on northerly located Bol'shevik Island (Fig. 3). It is represented by a succession of Neoproterozoic siliciclastic turbidites that are commonly attributed to a passive margin of the KM (Vernikovsky 1996; Lorenz et al. 2006–2008). The detrital zircon ages provide solid argument for a Timanian Baltica source of the clastic sediments and therefore constrain Neoproterozoic setting of the KM as a part of the Baltica Continent (Lorenz et al. 2008; Pease & Scott 2009). The rocks are intensively deformed and regionally metamorphosed to lower greenschist (Bol'shevik Island) and amphibolite (Northern Taimyr) facies and intruded by Late Palaeozoic syn- and post-collisional granites (300–265 Ma, Vernikovsky 1995; Vernikovsky & Vernikovskaya 2001). While the Northern Taimyr metamorphism is related to the Late Palaeozoic Uralian orogeny (Vernikovsky 1995; Pease & Scott 2009), the data on the Neoproterozoic metaturbidites of Bol'shevik Island may suggest a Neoproterozoic (Vendian, according to Proskurnin (1999), or Riphean, according to Lorenz et al. (2008)) phase of compressional deformation and metamorphism, which may be related to a 740–600 Ma collision of the KM with an island arc terrane (currently the Central Taimyr Fold Belt; Vernikovsky & Vernikovskaya 2001).

Cambrian marine siliciclastic sediments occur on eastern October Revolution Island. The lower part of the section is represented by unfossiliferous turbidites which, according to Proskurnin (1999), reveal some similarities with the Neoproterozoic turbidites of Bolshevik Island, and therefore may have Neoproterozoic age. The fossiliferous Cambrian strata are composed of shallow marine and basinal clastic sediments with some limestone beds in the Upper Cambrian. The section is compressed in north–south trending tight folds, but the rocks are unmetamorphosed as compared with the Neoproterozoic rocks of the Bol'shevik Island (Lorenz et al. 2007, 2008).

The Ordovician shallow water clastic sediments overlie Cambrian strata with a prominent angular unconformity on eastern October Revolution Island (the Kan'on River Unconformity). Lorenz et al. (2006) consider the deformations of the Neoproterozoic turbidites and Cambrian clastics as manifestations of a Caledonian compressional phase. The Kan'on River Unconformity is a coeval analogue to the Cambrian–Ordovician unconformity on Northern Novaya Zemlya Island, and therefore may be attributed to the latest stage of the Timanian orogeny, as suggested by Pease & Scott (2009) and Pease (2011).

Multi-coloured poly-facial Early Ordovician to Late Devonian calcareous, evaporite and clastic successions form an unmetamorphosed c. 3.5–5 km thick cover of the KM. Accumulation of the Ordovician to Silurian strata was taking place in shallow-water semi-restricted basins (Männik et al. 2009). In the Early Ordovician time a magmatic event took place with emplacement of intrusive and extrusive rocks: alkaline gabbro, syenite, granite, andesite, rhyolite and trachytes. The rock geochemistry is consistent with their origin in an intracontinental rift setting (Proskurnin 1995; Gramberg & Ushakov 2000). The Mid-Ordovician section is dominated by dark shales and gypsiferous limestones. The Late Ordovician quartz-sandstones probably related to a local uplift of the KM, which was followed by deposition of carbonate rocks through most of the Silurian. Lorenz et al. (2008) correlate the Devonian shallow water and fluvial sandstone dominated clastic strata to the Old Red Sandstone Formation.

In the earliest Carboniferous, the whole Early to Middle Palaeozoic KM cover within present-day limits of the Severnaya Zemlya was affected by slight to moderate compressional or transpressional deformation, and intruded by post-orogenic granites with U–Pb zircon ages 342 ± 3.6 and 343.5 ± 4.1 Ma. This Severnaya Zemlya folding is regarded by Lorenz et al. (2007, 2008) as Caledonian related. Another coeval compressional tectonic event, which could be a potential cause of the Severnaya Zemlya deformations, is Ellesmerian, or Innuitian, folding in the Canadian Arctic (Trettin 1991).

The Carboniferous and Permian shallow marine and continental clastic sediments, post-dating compression, occur locally on the Severnaya Zemlya. These strata do not reveal any deformation, assuming this part of the KM was not affected by a compression during its collision with the Siberian continent in Late Palaeozoic, which formed a south verging Central Taimyr Fold Belt.

An Early Mesozoic phase of compression has also been reported for the South Taimyr Fold Belt (Inger et al. 1999). In the Southern Taimyr Peninsula, the Tunguska-like flood basalts with $^{40}Ar/^{39}Ar$ ages c. 229–227 Ma are folded together with Carboniferous to Lower Triassic continental clastic rocks, and are unconformably overlain by Early Jurassic strata that constrain the age of the compressional event to the Late Triassic time (Walderhaug et al. 2005). Therefore, the Early Mesozoic compression occurred across a large domain of the RAS, which is almost 2000 km in the east–west direction, from the Novaya Zemlya Archipelago to the eastern coast of the Taimyr Peninsula. The regional extent and the magnitude of the folding are comparable to other first-order tectonic events which occurred in the Arctic, like the Caledonian, Late Palaeozoic and Late Mesozoic events.

Eastern sector of the RAS (east of Taimyr Peninsula)

The eastern Siberian shelves of the RAS are considered to be mostly underlain by Late Mesozoic fold belts (Vinogradov et al. 1974; Drachev et al. 1999; Drachev 2002), which occupy a huge onshore region between the Lena River in the west and the Mackenzie River in Alaska in the east. These fold belts originated in Late Jurassic to Early Cretaceous in the course of several collisional episodes of large terranes (AACM, Omolon) with Siberian margin, and were finally consolidated in Aptian during so-called Verkhoyansk–Brookian (or Chukotka–Brookian) orogeny. The fold belts are inferred to continue offshore where the Late Mesozoic folded structural domains are exposed on New Siberian and Wrangel islands (Fig. 3).

The Verkhoyansk Fold Belt and its western branch, the Olenek fold zone, surround the Siberian Craton in the east and NE. Their sections consist of over 10 km of Upper Palaeozoic to Lower Cretaceous (Hauterivian) siliciclastic sediments, known also as a Verkhoyansk Complex, which vary from fluvial to shallow marine sediments in the proximity to the craton to deepwater turbidites in the distal zones of the fold belts. The Verkhoyansk Complex is underlain by Riphean to Middle Palaeozoic carbonate and clastic–carbonate formations, belonging to the marginal parts of the Siberian Craton. This Palaeo-Siberian passive continental margin was compressionally deformed in the Early Cretaceous (c. 130–125 Ma) in the course of collisions with Kolyma–Omolon Composite Superterrane (Kolyma Structural Loop) and the AACM. Based on available MCS and gravity data, Vinogradov & Drachev (2000) and Drachev (2002) outlined the possible offshore extent of the Verkhoyansk and Olenek fold belts beneath the Laptev Sea rifted basins post-dating the Late Mesozoic compressional deformation (Figs 3 & 10).

The Kotel'nyi Terrane is defined in the western part of the New Siberian Archipelago *c.* 350–500 km north from onshore Late Mesozoic fold belts (Fig. 3). It provides solid evidence of a major Late Mesozoic compression event affected eastern portion of the RAS. The terrane is prevailed by Middle Ordovician to Upper Devonian and Mesozoic sedimentary rocks deposited in a passive margin setting. Three main tectonostratigraphic rock assemblages were defined (Kos'ko & Nepomiluev 1975; Kos'ko *et al.* 1990; Kos'ko 1994):

(1) A 3–5 km thick Middle Ordovician to Middle Devonian succession of carbonate rocks, mainly represented by lagoonal and shallow water marine fossiliferous limestones; basinal facies of black limestones and shales are known within Lower Silurian and Lower Devonian.
(2) A 7–9 km thick Upper Devonian to lowermost Carboniferous succession of grey mudstones and siltstones with minor carbonates and sandstones occurring within a prominent NW elongated synform-like structure (the Bel'kov-Nerpalakh Trough by Kos'ko *et al.* 1990), which may represent an infill of a structurally inverted Late Devonian rift. The share of carbonates and sandstones increases in the uppermost Devonian to Carboniferous interval, where the variegated rocks appear.
(3) A 1200–1300 m thick Triassic to Jurassic succession of claystone, clayey siltstone, siltstone and sandstone. The fine-grained clastic rocks occur predominantly within the Jurassic part of the section while the Triassic interval reveals an almost total absence of clastic material and is abundant in calcite and phosphorite nodules, which may be an evidence of deepwater sedimentation (Egorov *et al.* 1987).

The Carboniferous to Permian rocks occur sporadically, and are absent over most of the terrane. Their known sections are represented by very thin (30–130 m) beds of Serpukhovian and Bashkitian shallow water fossiliferous limestones, and by *c.* 200 m thick Lower Permian black shales with siltstone and limestone interbeds. Both the Carboniferous and Permian strata overlie, with a prominent unconformity at their base, the Ordovician to Devonian strata. In several localities the unconformity is characterized as an angular unconformity (Kos'ko & Nepomiluev 1975). Therefore, these facts may suggest an occurrence of a compressional phase at earliest Carboniferous time, which may be related to Ellesmerian orogeny in the Canadian Arctic. Another possible cause for the deformation may be the tilting of blocks of lower Palaeozoic rocks during the formation of the Bel'kov–Nerpalakh Trough. At present, the lack of structural data does not allow any further conclusions to be drawn on the possible nature of the Mid Palaeozoic deformation.

The whole section of the terrane is intruded by numerous sills and dykes of gabbro-diabases that closely resemble in age and composition the Permian–Triassic Tunguska flood basalts (Kuzmichev & Pease 2007). The whole rock package of the terrane was intensively compressed in the Early Cretaceous, prior to Aptian, with clear dextral transpression component.

A narrow and highly deformed *South Anyui ophiolitic suture (SAS)* separates the New Siberian–Chukchi and Verkhoyansk–Kolyma fold belts. As shown by aeromagnetic data (Rusakov & Vinogradov 1969; Vinogradov *et al.* 1974; Spektor *et al.* 1981), it extends from the Kolyma River mouth onto the shelf where dismembered ophiolites and island-arc volcanic complexes are exposed on Bol'shoi Lyakhov Island (Drachev & Savostin 1993; Kuzmichev 2009). The further offshore continuation of the suture, though obscured by younger extensional structures, is still definable by magnetic data, which show two branches of positive anomalies: between Stolbovoi and Kotel'nyi islands and ENE of the latter (Fig. 3).

The suture formed as a result of the closure of the Anyui Ocean – a large embayment of the Late Palaeozoic–Mesozoic Pantallassa to the Pangaea II – in the course of collision between the AACM and the Siberian margin in the Late Jurassic–Early Cretaceous (prior to Aptian) time (Savostin *et al.* 1984*b*; Parfenov & Natal'in 1986; Zonenshain *et al.* 1990; Sokolov *et al.* 2002).

The Siberian portion of the *Arctic Alaska–Chukchi Microplate*, or microcontinent, is inferred to consist of two parts (Fig. 3):

(1) the New Siberian–Chukchi Fold Belt, which constitutes the southern, adjacent to the South Anyui Suture, deformed margin of the microcontinent;
(2) the De Long Massif located north of the Late Mesozoic deformation front and composed of Lower and Middle Palaeozoic complexes.

The New Siberian–Chukchi Fold Belt occupies the Chukchi Peninsula and provisionally extends offshore to include the Late Mesozoic folded complexes of Wrangel Island, and a broad area of the East Siberian Sea north of the proposed limits of the South Anyui Suture. The northern offshore limit of the fold belt has been identified on several MCS lines south of the De Long Islands and north and east of Wrangel Island, where it follows the northern flank of the Wrangel–Herald Arch (Drachev *et al.* 1999, 2001; Verzhbitsky *et al.* 2008).

Stratigraphic and structural relationships between various lithostratigraphic units of Wrangel Island remain poorly understood, although ongoing studies may illuminate its tectonic evolution. The oldest Neoproterozoic Wrangel Complex is composed of metavolcanic, metavolcanoclastic and metaclastic rocks, intruded by basic dykes and sills and small granitic bodies with isotopic ages 600–700 Ma (Kos'ko *et al.* 1993). The rocks were intensively deformed prior to deposition of thick sedimentary successions of Upper Silurian–Devonian and Carboniferous to Permian shallow marine siliciclastic rocks and shales, with some carbonate units deposited in a continental shelf setting. The uppermost part of the section is represented by an over 1–1.5 km thick unit of Triassic siliciclastic turbidites (medium to fine-grained sandstone alternating with shale and siltstone), which contrast with the older Palaeozoic shelfal strata (Kos'ko *et al.* 1993; Miller *et al.* 2010). Data on the U–Pb age of detrital zircons from Wrangel sections reveal similarities between the Upper Palaeozoic rocks of the island and the Lisburne Hills area (Western Alaska). Zircon populations from the Triassic turbidites differ significantly from both older Palaeozoic rocks of the island and from coeval rocks of the Lisburne area, although they are similar to the Triassic turbidites of the Chukchi Peninsula (Miller *et al.* 2010). All sedimentary successions were deformed in the latest Jurassic to Early Cretaceous, probably simultaneously with deformations in the Lisburne Hills area (132–115 Ma, according to Moore *et al.* 2002). On the Chukchi Peninsula and probably Wrangel Island the orogenic event was followed by an uplift and profound erosion between *c.* 117 and 95 Ma (Miller & Verzhbitsky 2009).

The De Long Massif has a very contrasting expression in gravity and magnetic fields, due to the occurrence of highly uplifted and eroded basement, and the presence of the Early Cretaceous flood basalts. The massif has repeatedly been attributed by Soviet geologists to a pre-Cambrian craton, often named the Hyperborean, or East Arctic Platform (Obruchev 1934; Shatskii 1935; Pushcharovskii 1963; Atlasov *et al.* 1970 and many others). However, today we have more evidence in favour of either Caledonian or Ellesmerian (Late Devonian) age for this feature (see below).

On Bennett Island, a 1.5 km thick succession of Middle Cambrian to Middle Ordovician fossiliferous shales and distal siliciclastic and clastic carbonate turbidites reveals rather weak deformation and a lack of metamorphism (Vol'nov et al. 1970; Drachev 1989). This section is quite unique since no similar rocks are known from the nearest Siberian Arctic. The closest occurrences of the Cambrian to Ordovician rocks to which the Bennett succession could be correlated are deepwater turbidites exposed in the Northern Greenland and the Northern Ellesmerian Island (Trettin 1991).

Henrietta Island is composed of moderately deformed clastic and volcano-clastic complexes and a unit of calc-alkaline basalts of unclear stratigraphic setting (Vinogradov et al. 1975). The section is intruded by diabase and diorite sills and dykes whose unpublished $^{40}Ar/^{39}Ar$ dates could possibly reveal a Caledonian age for the magmatism (Kaplan et al. 2001), while rock chemistry points to their island-arc affinity. The other two occurrences of the Caledonian-age structural domains in the High Arctic are at Spitsbergen and northern Ellesmere Island (Peary Terrane and adjacent area of the island). Therefore, based on the possible Caledonian age of the Henrietta magmatic rocks, we infer that a location of the De Long Massif was close to Northern Ellesmere Island prior to the Late Jurassic to Early Cretaceous opening of the Canada Basin.

On Bennett Island, Aptian coal-bearing muddy sediments, and a 200–300 m thick unit of plume-like flood basalts are separated from the underlying Lower Palaeozoic section by a sharp basal unconformity. A few K–Ar radiometric dates in the range of 119–112 ± 5 Ma constrain the age of the basalts to the Aptian to Albian (Drachev 1989; Drachev & Saunders 2006).

The above characterization of the RAS's heterogeneous folded tectonic basement points to some major uncertainties regarding the areal extent and structural relationships of the basement domains. These are:

(1) existence and extent of the Svalbard Massif (Microcontinent) and adjoining Scandinavian Caledonides and Timanides;
(2) relationships between the Kara and Svalbard massifs, and their relationship with the Baltica Continent;
(3) relationships between Late Palaeozoic Uralides and Taimyrides; extent of the Early Mesozoic Novaya Zemlya Fold Belt, and the mechanism of its formation;
(4) offshore extent of the Late Mesozoic fold belts, structural factors controlling their formation;
(5) palaeorelationships of the Kotel'nyi Terrane and the De Long Massif with Siberia, Arctic Canada, and Arctic Alaska;
(6) occurrence of Caledonian and/or Ellesmerian deformations on Kotel'nyi and Wrangel islands;
(7) offshore extent and magnitude of Permian–Triassic and Aptian–Albian flood basalt magmatic.

Presently no reliable data exist to constrain these uncertainties. However the fact that the western sector of the RAS is generally dominated by Neoproterozoic crustal domains, and the eastern sector by Late Mesozoic and, to a smaller extent, by Palaeozoic fold belts, is well supported by the modern geological and geophysical data. This has fundamental implications for characterizing the RAS sedimentary basins, history of their formations and their established and inferred HC systems.

RAS sedimentary basins and their petroleum geology

According to Grantz et al. (2009), as many as 37 large sedimentary basins of variable age and genesis exist over the entire RAS and adjoining deepwater areas. Figure 4 shows their outlines derived from MCS and gravity data. Based on the age of the basins, inferred mechanisms of their formation, composition of sedimentary infill, and known and inferred HC systems, we describe six groups of basins, or provinces, which generally fit into geographically isolated shelves: the East Barents (including offshore continuation of Timan–Pechora Basin), South Kara, North Kara, Laptev, East Siberian and Chukchi Sea (Russian sector).

East Barents Province

The structure, lithostratigraphy and petroleum geology of the East Barents and Pechora shelves are known due to the relatively dense grid of 2D seismic surveys, and a number of offshore wells (Gramberg & Pogrebitskiy 1984; Gramberg 1988; Verba et al. 1992; Bogdanov & Khain 1996; Shipilov & Tarasov 1998; Kogan et al. 2004). Shtokman, Ludlov, Ledovoe, Murmanskoe, Severo–Kildinskoe gas and gas condensate fields, Prirazlomnoe, Severo–Gulyaevskoe, Varandey–More and other oil fields are the biggest among those discovered so far.

The East Barents Province is dominated by the gigantic East Barents Megabasin (EBMB) which, due to its elongated shape, is often called a megatrough. It is bounded by the Central and North Barents platforms in the west and NW, respectively, and by the Novaya Zemlya Fold Belt in the east (Fig. 5). The southern rim of the EBMB is formed by a steep slope of the Fennoscandian Shield (Kola Monocline) and a series of NW-elongated horsts of the Timanian basement. In the north, it is limited by a high-standing block of basement of the Franz Josef Land Archipelago (Fig. 5). Some researchers also include the St Anna Trough in the EBMB. However, this basin is located in the North Kara Shelf, and is isolated from the East Barents Province by a zone of high-standing basement (the Al'banov-Gorbov Arch), and thus may have a different geological history.

The *East Barents Megabasin* extends south–north for over 1000 km, while its west–east width reaches 400–450 km. Despite a large amount of 2D MCS and refraction data and drilled wells, the EBMB is still poorly imaged beneath Jurassic strata due to the great thickness of Triassic and Upper Permian successions. Its internal structure and lithostratigraphic architecture are therefore disputable.

The EBMB is composed of two smaller depocentres: the South Barents and the North Barents basins divided by a basement high named the Ludlov Saddle (Fig. 5). Crustal velocity data show that in the central parts of the depocentres, where the total thickness of presumed post-Middle Devonian sediments reaches over 22 km, the underlying consolidated continental crust is highly attenuated or even could be completely absent (Kogan et al. 2004; Kaminsky et al. 2009). This fact is used by several researchers to propose that the EBMB is underlain by an oceanic lithosphere formed due to a failed spreading episode (Aplonov et al. 1996), or trapped during the Uralian collision event (Ustritsky 1989). As imaged by the deep seismic refraction data (Kogan et al. 2004; Ivanova et al. 2006), the EBMB generally resembles rift/post-rift basins with significant post-rift cover exceeding 20 km, for example the Pricaspian Basin.

MCS data supported by well ties both within the basin and along its southern margin in the Pechora Sea provide a good basis for understanding of the basin tectonostratigraphy. Six main seismic stratigraphic units have been identified so far (Fig. 6a). These are correlated to the siliciclastic sequences of Cretaceous (1), Jurassic (2), Triassic (3) and Late Permian (4) age, to the Late Carboniferous–Early Permian carbonate rocks (5), and to the Late Devonian to Early Carboniferous syn- and pre-rift clastic and carbonate rocks (6). The first three are well documented by numerous offshore wells that penetrated these stratigraphic intervals. Beneath the unit (6), on the flanks of the megabasin, an older pre-rift seismic stratigraphic unit is visible in seismic data (number 7 in Fig. 6a), which we correlate to Lower Palaeozoic strata known in the Timan–Pechora Basin.

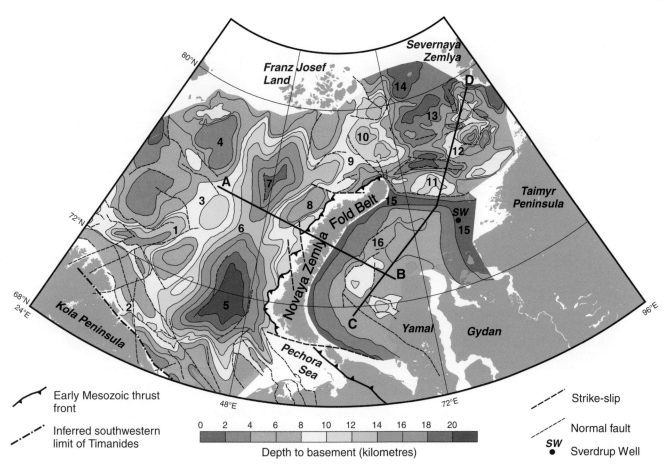

Fig. 5. Simplified depth-to-basement map of the Barents–Kara shelf based on seismic reflection and refraction data and ERS-2 gravity field (see text for the references). Numbered structures are: 1, Nordkapp Basin; 2, Varanger Basin; 3, Central Barents Platform; 4, North Barents Platform; 5, South Barents Basin; 6, Ludlov Saddle; 7, North Barents Basin; 8, Admiralty High; 9, Al'banov-Gorbov Arch. 10, St Anna Trough; 11, Litke Trough; 12, North Kara Basin; 13, Central Kara High; 14, Ushakov High; 15, North Siberian Arch; 16, South Kara Basin. Bold solid lines labelled A–B and C–D show location of cross-sections given in Figure 6.

Because of uncertainties with the stratigraphic correlation of the lower seismic horizons, the timing of basin formation is highly disputed. Proposed models differ significantly with regard to the age of the main rift event, which varies from Mezo-Neoproterozoic to Permo-Triassic (Gramberg & Pogrebitskiy 1984; Gramberg 1988; Nikishin et al. 1996; Shipilov & Tarasov 1998; Malyshev 2002; Sharov et al. 2005 and references contained therein). We support the point of view proposed in Lopatin (2000), and infer the main rifting phase to occur at Frasnian–Early Carboniferous time based on the following facts: (1) Late Devonian and Early to Mid Carboniferous rift episodes are well documented in the Norwegian Barents Shelf (Gudlaugsson et al. 1998); (2) Caledonian compressional deformation affecting the western Barents Shelf in Ordovician–Silurian terminated by Frasnian (Gee & Stephenson 2006).

The tectonic history of the EBMB can be described as a succession of the following events (Fig. 7):

(1) In Cambrian to Silurian, the basin may have been developing mainly in a shelf setting adjacent to the Uralian palaeocean. Its western (present-day) margin may have been subjected to compression and developing of a foreland basin in front of the Caledonides. Depositional environments may have varied though the EBMB from continental and fluvial–deltaic systems in the west, to carbonate platforms and deepwater conditions along its eastern flank. In the Late Silurian, clastic sediments derived from Caledonian orogen may have dominated the entire basin (Korago et al. 2009).

(2) In the latest Silurian to beginning of Devonian, large-scale strike-slip deformation and the formation of small pull-apart basins may have occurred in the southern part of the EBMB (by analogy with the Timan–Pechora Basin), followed by regional uplift and erosion during Middle to Late Devonian, prior to Frasnian. The latter could be correlated to a Svalbard Late Caledonian compressional phase, and therefore a compressional setting and basin inversion could be inferred for EBMB at the Middle to Late Devonian.

(3) Frasnian to Early Carboniferous was a time of a main rift phase accompanied by syn-rift basaltic volcanism. Continental crust was severely attenuated and formation of initial oceanic lithosphere could have taken place in the deepest parts of the megabasin. Geodynamic factors controlling the rifting are unknown. A possible association with a mantle plume event could be inferred as proposed earlier for the Russian Craton (Nikishin et al. 1996; Wilson & Lyashkevich 1996; Wilson et al. 1999). The collapse of the Caledonian orogen could also provide a possible mechanism for crustal extension. Some analogues could be drawn to the Sverdrup Basin in Canadian Arctic to explain crustal attenuation within the EBMB.

(4) The Late Carboniferous to Early Permian was a period of thermal subsidence. Carbonate platforms and buildups formed along the basin flanks while its central parts were dominated by deposition of basinal carbonates and shales.

(5) In Late Permian to Triassic, a rapid subsidence of the entire EBMB took place, accompanied by accumulation of large volume of siliciclastic sediments. Permian clinoforms show

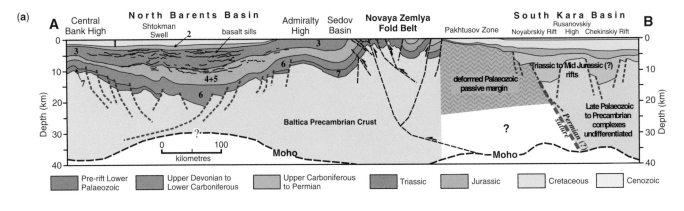

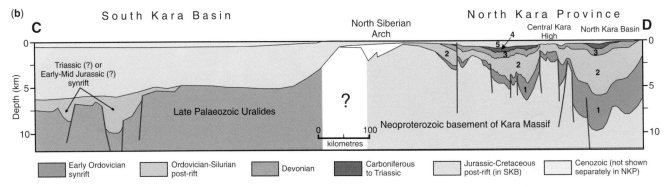

Fig. 6. Schematic geological cross-sections based on re-interpretation of deep seismic reflection and refraction data acquired and published by SMG (Sharov *et al.* 2005; Ivanova *et al.* 2006): (**a**) through the Eastern Barents and South Kara basins based on seismic transect 2-AP, (**b**) though the North Kara Province based on seismic transect 3-AP. Location is shown in Figures 3 and 5. Bold numbers from 1 to 6 in (a) and from 1 to 5 in (b) denote seismic stratigraphic units (see the text for the details).

that the main provenance areas were in the Russian Craton–Timan Pechora–Urals to the SSE of the EBMB, and to the NE within the present North Kara Shelf. Both of these regions were affected by the Late Palaeozoic Uralian orogenesis. Central and eastern parts of the EBMB remained uncompensated through the Late Permian to Early Triassic, and were probably part of a deepwater depression extending and deepening toward the present SKB. Many MCS lines approaching the Novaya Zemlya Fold Belt from the west show thickening of the Upper Permian and Lower Triassic intervals eastward, that is basinward. At the Permian–Triassic boundary, a plume-related basaltic volcanism affected some parts of the EBMB as indicated by basalts of the Pai-Khoi and Timan–Pechora (Nikishin *et al.* 2002). By the end of the Triassic the basin was probably completely filled with clastic sediments.

(6) At the Triassic–Jurassic boundary, the main Novaya Zemlya orogenic phase occurred, accompanied by inversion of the EBMB eastern flank in a foreland setting. Jurassic strata are eroded in the vicinity of the Novaya Zemlya, suggesting that the EBMB eastern flank was uplifted and subjected to erosion during most of Jurassic time.

(7) In Jurassic to Cretaceous, the central part of the EBMB continued to subside and received clastic sediments. During the latest Jurassic, the subsidence became undercompensated by sediment supply, and the EBMB became a starved basin accumulating marine organic rich sediments. By this time, the Novaya Zemlya orogen was probably completely eroded and had subsided below sea-level, uniting the EBMB with the SKB. Both of these became a part of the huge West Siberian depression and were gradually filled up in Early Cretaceous by easterly and northerly derived clastic sediments, as shown by the orientation of the Early Cretaceous clinoforms.

(8) In Aptian to Albian, there was a plume-related magmatic event that caused broad eruption of flood basalts known on Franz Josef Land and Eastern Svalbard (Amundsen *et al.* 1998).

Post-Cretaceous sediments are generally absent over most of the EBMB, which may be related to: (1) tectonic uplift during the Eocene-Oligocene and probably Early Miocene; and/or (2) recent glacial erosion. Cenozoic erosional phases were probably triggered by plate interactions in the North Atlantic and High Arctic, and by hard collision between India and Eurasia beginning around Eocene and continuing through Oligocene and Miocene. These global plate-tectonic factors caused the growth of a series of inversional swells and anticlinal structures (Fig. 8). Some of these inverted features (e.g. the Admiralty High) were formed at the Triassic–Jurassic boundary, and then re-activated in the Cenozoic time. Simultaneously, the Novaya Zemlya orogen was uplifted and became again the major divide between the Barents and Kara provinces.

The formation of the EBMB, as a whole, and especially the mechanisms of its rapid subsidence in the Triassic, remains highly debatable. Several models have been published by Nikishin *et al.* (1996), Artyushkov (2005), Levshin *et al.* (2007), Ritzmann & Faleide (2009) and others (for more comprehensive review see Gee & Stephenson 2006), and we refer readers to these publications and references contained therein for further details on the subject. In this paper we support a concept proposed by Sullivan *et al.* (2007) and by Scott *et al.* (2010), which involves the trapping of Uralian oceanic lithosphere in a pre-existing embayment of Baltica margin (present-day South Kara) during the main Late Palaeozoic collision, and following westward rollback and pulldown of the trapped lithospheric slab, accompanied by slab to cause rapid subsidence of the adjacent Barents margin (Fig. 9).

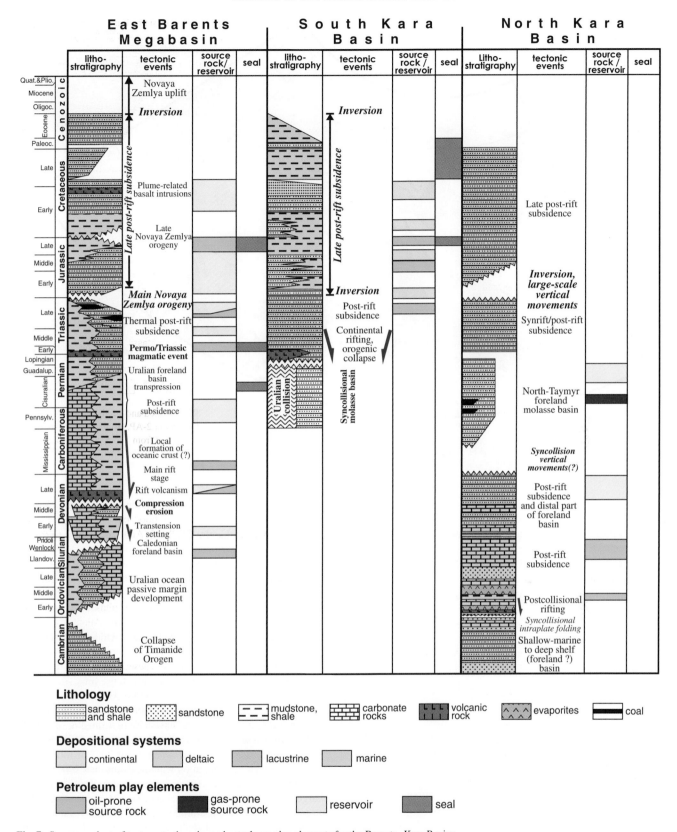

Fig. 7. Summary chart of tectonostratigraphy and petroleum play elements for the Barents–Kara Region.

Petroleum geology. HC systems and petroleum plays of the East Barents Province are known from numerous offshore wells (Fig. 2). The Palaeozoic and Lower Mesozoic plays, which occurred at greater depths throughout the basin and are currently undrilled, were tested in the Pechora Sea, and have been used to decipher the petroleum geology of the lower part of the EBMB infill. The younger Jurassic and Cretaceous plays are well studied in the EBMB where they have accumulated significant gas resources. Many data on the petroleum geology of the Barents–Pechora Shelf were summarized by Ulmishek (1982), Johansen *et al.* (1993), Ostisty & Fedorovsky (1993) and Doré (1995), and we refer readers to these publications for more detailed information. A summary of the petroleum systems and play elements of the East Barents province is given in Figure 7.

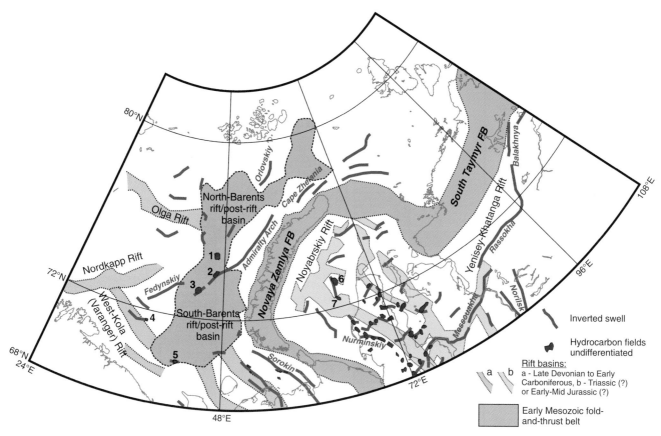

Fig. 8. Areal occurrence of main Mesozoic and Cenozoic intraplate inversional swells in the Barents–Kara Region. Bold numbers denote main gas and gas condensate fields: 1, Ludlov; 2, Ledovoe; 3, Shtokman; 4, Severo-Kil'dinskoe; 5, Murmarskoe; 6, Rusanovskoe; 7, Leningradskoe.

In the Timan–Pechora Basin, the Palaeozoic oil-prone source rocks are represented by:

(1) Upper Silurian shales;
(2) Devonian (lower Frasnian) carbonates and shales (Domanic Formation);
(3) Lower to Middle Carboniferous coaly-argillaceous sediments with mixed kerogen types.

The Lower Permian shaly sediments reveal high content of total organic carbon (TOC) in some sections, and thus could also be a local oil source.

The main reservoirs include:

(1) the Lower Devonian limestones and dolomitized limestones (encountered by a few wells at the lower levels of the Medyn–More and Prirazlomnoe fields);
(2) the Upper Carboniferous and Lower Permian carbonate rocks, especially carbonate buildups and organo-clastic limestones (the Prirazlomnoe Field);
(3) The Upper Permian to Triassic sandstones.

The properties of the Lower Devonian carbonate reservoirs are mainly controlled by highly irregular secondary porosity (7–8%) and low permeability. The Upper Carboniferous to Lower Permian carbonate rocks are the main reservoir for oil accumulations over the entire Timan–Pechora Basin. They also host the Medyn–More and Prirazlomnoe oil fields offshore. These rocks have good properties due to higher porosity, which reaches 25% in fossiliferous layers.

The main seals within the Palaeozoic and Lower Mesozoic plays of Timan–Pechora Basin are represented by:

(1) the Upper Devonian (Kynovsk–Sargaev) argillaceous succession;
(2) the Lower Carboniferous (Visean–Serpukhovian) evaporate succession;
(3) the Lower Permian (Kungurian–Upper Artinskian) carbonate–shale succession;
(4) the Lower Triassic shaly–argillaceous succession.

The Kynovsk–Sargaev and Kungurian–Upper Artinskian are the regional seals which developed throughout most of the Timan–Pechora Basin, while the others have a more limited occurrence.

The main phase of HC generation in the Timan–Pechora Basin occurred in the Permo-Triassic when about 70% of total source rocks entered the oil maturation window (Malyshev 2002).

In the EBMB, the main HC source rocks, which are inferred to charge the gigantic gas accumulations, are the Triassic organic-rich (up to 4.7–6.5% TOC) gas-prone coal-bearing shaly sediments of continental, lagoonal and shallow-marine origin. Their potential for oil generation is unknown, but cannot be ruled out, since marine facies with type II kerogen could have been deposited in the central parts of the EBMB. Marine Upper Jurassic organic rich shales (an analogue of the Bazhenov Suite of the West Siberian Basin) with TOC reaching 16% are widespread within the basin but are generally thermally immature.

The main Mesozoic petroleum plays in the EBMB are related to the Middle–Upper Jurassic (Shtokman, Ledovoe) and Lower–Middle Triassic (Severo-Kil'dinskoe and Murmansk fields) fluvial–deltaic and shallow-marine sandstones. The main pay

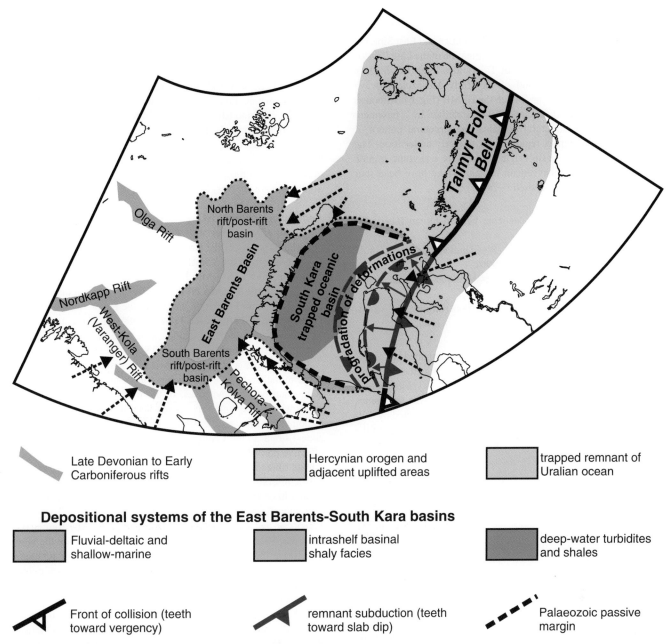

Fig. 9. Schematic chart illustrating the plate-tectonic setting and depositional environments of the East Barents and South Kara basins in Permian–Triassic and possible structural mechanism of the Novaya Zemlya Fold Belt formation. Black dashed arrows show main directions of sedimentary supply into the basins. Red arrows indicate direction of the subducted lithospheric slab roll-back.

interval of the Severo-Kil'dinskoe gas field occurs at 2440 m depth, and is hosted by Lower Triassic sandstones with porosity of c. 20%. Potential reservoirs can also be inferred within the Upper Permian siliciclastic successions forming clinoforms along the southern and northern margins of the EBMB, as well as within the Palaeozoic carbonate rocks, where these occur at drillable depths.

The only regional seal occurring throughout the EBMB is composed of the Upper Jurassic immature organic rich shales. In some parts of the basin the seal extends down-section to include the Callovian shales.

Oil and gas trap formation in the East Barents region is generally related to two inversion events:

(1) the Triassic–Jurassic orogeny;
(2) the Early Cenozoic (Eocene to Oligocene) crustal adjustment triggered by the Greenland–EUR–NA plate interactions in the North Atlantic and the Arctic.

The former has a greater impact on the eastern margin of the EBMB affected by a Novaya Zemlya foreland deformation, while the latter resulted in formation of large anticlinal quasi-isometric or elongated arches throughout the basin (Fig. 8).

The main phase of the HC generation in the EBMB started around 55–50 Ma, when the Triassic coaly source rocks became mature. Charge timing was favourable for both the traps formed in Early Cenozoic and for any older traps.

South Kara Province

The SKB is the second most petroliferous province in the RAS with established HC resources. Two large HC accumulations were discovered there at the end of the 1980s: the Leningranskoe gas and the Rusanovskoe gas condensate fields (Fig. 8). Generally the basin is fairly well explored, and many leads and prospects are known (Gramberg 1988; Shipilov

& Tarasov 1998; Kontorovich et al. 2001; Sharov et al. 2005; Vinokurov et al. 2009).

The SKB is located at a junction of the Baltica and East Siberian cratons and the intervening northern termination of the Late Palaeozoic Uralides and western flank of Taimyrides (Figs 3 & 5). It is also located at the hinterland of the Early Mesozoic Novaya Zemlya Fold Belt separating it from the EBMB (Fig. 6). Although it is generally accepted that the SKB is underlain by a basement consisting of these structural domains, the structural pattern of this tectonically complex region is not properly understood. The precise timing of the basin origin, as well as a tectonic regime causing its initiation, also remain unclear. There have been four structural mechanisms proposed so far:

(1) trapped lithosphere of Palaeozoic Uralian Ocean (Ustritsky, 1985);
(2) Late Permian to Early Triassic (Post-Uralian) abandoned small oceanic basin (Aplonov et al. 1996);
(3) Late Permian to Early/Mid Triassic intracontinental rifting (Surkov et al. 1997, and many others);
(4) Late Permian/Early Triassic subduction rollback and simultaneous crustal extension (see references below).

MCS data reveal extensional structures, dominated by half-grabens, beneath thick sedimentary cover post-dating the extension (Shipilov & Tarasov 1998; Sharov et al. 2005; Vyssotski et al. 2006). Because of the great thickness of the post-rift sediments, their lower stratigraphic intervals have not been penetrated by drilling, which results in a lack of clarity with regard to the age of the crustal extension affecting the basement beneath the basin. By analogy with the far better studied West Siberian Basin, where the age of crustal extension is considered to be Late Permian to Early Triassic (Kontorovich et al. 1975; Surkov & Zhero 1981; Nikishin et al. 2002), many researchers attribute the SKB rifts to the same Permo-Triassic extensional event. However, bearing in mind the magnitude of the Triassic–Jurassic compression, which resulted in the formation of the Novaya Zemlya and South Taimyr fold belts, we do not exclude a younger age for the South Kara rifting. Although this does not preclude the existence of the older Permo-Triassic rifts, the latter might have been affected by the Early Mesozoic compression and become a part of a folded basement underlying the SKB. The SKB grabens and half-grabens could be attributed to a crustal extension post-dating the main Triassic–Jurassic orogeny, and could therefore have originated in Early to Middle Jurassic, since the Upper Jurassic Bazhenov horizon is a well-defined seismic marker in the lower part of the post-rift cover. Indirect evidence in favour of this model is derived from the SE–NW strike of the Noyabrskiy Rift in the SKB – almost orthogonal to the dominant direction of the West Siberian rifts (Fig. 8), which suggests that these rifts could have originated independently.

The following events are considered to contribute to the origin of the SKB and its petroleum potential:

(1) the Permian Baltica–Siberia collision and possible trapping of oceanic lithosphere in the future SKB;
(2) the Permian–Triassic plume-related magmatic event;
(3) the Triassic–Jurassic compression and orogeny;
(4) the Early to Middle Jurassic rifting probably related to a collapse of the Early Mesozoic orogen.

Since the Middle Jurassic the SKB became an area of a post-rift thermal subsidence, providing room for accumulation of more than 6 km of fluvial–deltaic, shallow-marine and deepwater siliciclastic sediments. The main sedimentary supply into the basin was from the east (through the Yenisei–Khatanga Depression) and from the north.

Possible models for the SKB formation have to deal with explanation of the rapid subsidence of the EBMB in the Late Permian–Triassic, and remarkable curvature of the Novaya Zemlya Fold Belt. We believe that all these events could be integrated in the model shown in Figure 9. According to this, the slab rollback would be continuously accommodated by the westward expansion of the Uralian orogenic front into the remnant oceanic basin, which could have been completely consumed by the end of the Triassic. Modern seismic tomography data seem to support the existence of an oceanic lithospheric slab beneath the Eastern Barents and Southern Kara shelves (Levshin et al. 2007).

Petroleum geology. HC systems of the SKB are projected from the onshore West Siberian Basin. The main source rocks are the bituminous shales of the Bazhenov Formation. In addition, based on well data from the Yamal Peninsula, organic-rich beds are also present in older Early–Mid Jurassic marine shaly succession, and especially in the Tyumen Formation. Aptian–Albian deltaic coals of the Tanopchin Formation are one of the major gas sources (see Fjellanger et al. 2010). If the Triassic sediments survived the post-Triassic orogeny, they, by analogy with the East Barents Province, can also be considered as a potential source of HC.

The main gas and gas condensate accumulations of the SKB occur in the Cretaceous Tanopchin and Pokur fluvio-deltaic sandstones and pelitic sandstones. In the Rusanovskoe Field, 12 pay zones with average effective porosity c. 20% were established in Cenomanian (one zone), Albian (three zones) and Aptian (eight zones) strata. In the Leningradskoe Field, pay zones with average effective porosities of 26% were tested in Cenomanian (one zone), Albian (three zones) and Aptian (three zones) sediments. Potential clastic reservoirs may occur within the Lower Cretaceous (Neocomian) clinoforms and within the Jurassic Vasyugan Formation.

Regional seals within the SKB are represented by Turonian–Paleocene and Albian marine shaly successions. The former is a seal over 50 m thick for the Leningradskoe Field, while the latter is a 100 m thick seal in the Rusanovskoe Field. Upper Jurassic–Neocomian shales and shaly sediments form another basin-wide seal for the deeper undrilled prospects.

Numerous anticlinal structures, identified by the MCS data in the SKB, affect the Cenozoic sediments. This suggests inversion at the latest stages of the basin's formation. As shown by Vyssotski et al. (2006), the inversion in the West Siberian basin could have started as early as Campanian–Maastrichtian and culminated in the Oligocene. The existence of two systems of inverted anticlines striking to the eastnortheastern and northwestern directions, almost orthogonal one to another (Fig. 8) suggests at least two main directions of tectonic stress. Many researches refer to the far-field stresses sourced by the India–Eurasia collision to explain the West Siberia Basin inversion (e.g. Allen & Davies 2007). We believe that there could also be a more proximal source for inversion caused by convergence of the EUR–Greenland and EUR–NA plates in Early Cenozoic time.

HC generation in the SKB could have started as early as Barremian, when Tyumen Formation first entered the oil window, and persisted through the Late Cretaceous with the main HC generation from the Bazhenov Formation. One important difference between the West Siberian Basin and its offshore continuation is that all the pre-Bazhenov sources and most of the Bazhenov Formation are deeply buried within the SKB, and thus are located in the gas maturation window. Therefore, the potential for oil generation is limited to the marginal parts of the basin. According to Fjellanger et al. (2010), gas accumulations in the gigantic fields of the Yamal–Tazov Region and SKB can only be explained by the combined charge from all possible sources, including biogenic gas.

North Kara Province

North Kara Province (NKP) is the least explored part of the western RAS (Fig. 2). Current understanding of its geology is mostly based on a few published MCS lines (Shipilov & Tarasov 1998; Sharov et al. 2005; Ivanova et al. 2006), potential field data and geological observations from the Severnaya Zemlya Archipelago (Kaban'kov & Lazarenko 1982; Bogolepova et al. 2001; Metelkin et al. 2005; Lorenz et al. 2006–2008).

Tectonically the NKP is separated from the South Kara Basin by the North Siberian Arch (see above). As shown by the seismic reflection and refraction data along the 3-AP line (Sharov et al. 2005; Ivanova et al. 2006), the post-rift, especially Jurassic, sediments filling the SKB thin dramatically towards the arch. Sverdrup Well, drilled on a small same-named island in the eastern Kara Sea (Fig. 5), located just on the North Siberian Arch, penetrated about 1350 m of the Lower Cretaceous sediments and only 170 m of the Upper Jurassic sediments, which rest unconformably on metamorphic rocks of inferred pre-Cambrian age. This implies that the basement arch existed during most of the post-rift history of South Kara Province. MCS data do not show any significant faulting associated with the arch.

There are four contrasting gravity lows in the Northern Kara Sea corresponding to the main depocentres: St Anna Trough (Fig. 4, number 14; & Fig. 5, number 10), Litke Trough (Fig. 4, number 12; & Fig. 5, number 11), Schmidt Trough (Fig. 4, number 15) and North Kara Basin (Fig. 4, number 13; & Fig. 5, number 12). As shown by the MCS data, they have an extensional origin and contain c. 10 km of syn-rift and post-rift sediments (Shipilov & Vernikovsky 2010). Age calibration of the MCS data is highly controversial and can only be based on lithostratigraphic correlations with the sections described from northern part of the Novaya Zemlya and Severnaya Zemlya archipelagos.

Based on MCS data interpretation correlated with the above-described stratigraphy of the Severnaya Zemlya, we infer the following seismic stratigraphic units within the NKP offshore basins (Fig. 6b):

(1) Early Ordovician syn-rift unit consisting of shallow-marine clastic rocks and volcanic rocks;
(2) Ordovician–Silurian post-rift unit dominated by clastic rocks and shales (Ordovician), carbonate and evaporate rocks;
(3) Devonian unit dominated by syn-orogenic continental clastic molasses;
(4) Carboniferous to Triassic unit dominated by continental clastic sediments synchronous with the Taimyr orogeny;
(5) Mesozoic to Cenozoic unit composed of continental and shallow-marine sediments overlying older units with a sharp basal unconformity.

Whether the Neoproterozoic to Cambrian strata are present within the NKP sedimentary basins remains unclear. Considering data on the Northern Block of the NZFB, the pre-Cambrian siliciclastic rocks may, together with the Lower to Middle Palaeozoic strata, participate in the sedimentary cover of the KM, while the data on Severnaya Zemlya Archipelago evidences their involvement into folded Neoproterozoic basement.

Geological and MCS data suggests that the main structural inversion of the NKP basins took place in post-Triassic time, perhaps at the Triassic–Jurassic boundary, synchronous with the formation of the Novaya Zemlya Fold Belt. The inversion was accompanied by large-scale vertical movements of basement blocks, and may have generally had a transpressional character.

Petroleum geology. HC systems of the North Kara Sea and adjacent islands are not studied, and thus can only be inferred based on lithological composition and depositional environment observed in the Severnaya Zemlya sections. However, natural bitumen is reported from Silurian, Devonian and Triassic–Jurassic sediments, which may indicate the presence of source rocks generating these HCs. Oil-prone source rocks are inferred in the Middle Ordovician dark shales and Silurian carbonates, while clastic reservoirs could be expected within Devonian and Late Palaeozoic strata. The main trap formation phase may have occurred at the Triassic–Jurassic boundary, and the related structures could have been charged from Palaeozoic sources. The main exploration risks are related to seal presence, as perhaps Upper Jurassic–Lower Cretaceous (Neocomian) regional seals might not have been formed due to post-orogenic uplift of the entire NKP in Mesozoic time.

Laptev Sea Province

The Laptev Shelf is the most studied of the Siberian shelves (Fig. 2). Tectonically it represents a large, about 500 km wide and 700 km long, rift system that has been developing since Late Cretaceous time in response to the Eurasia oceanic basin breakup and consequent spreading along the Gakkel Ridge (Figs 4 & 10). Its geology has been described in detail by Drachev et al. (1998, 1999), Drachev (2000), Franke et al. (2000, 2001, 2004), Sekretov (2000) and Franke & Hinz (2005), and therefore we refer the reader to these publications for more comprehensive insight into structure and seismic stratigraphy of the Laptev shelf.

The Laptev Rift System (LRS) consists of a series of wide extensional basins and relatively narrow grabens, as shown in Figures 4, 10 and 11. These are (from west to east): the SW Laptev (Fig. 11a, number 1), Ust' Lena (Fig. 4, number 18; & Fig. 11a, number 2), Anisin (Fig. 4, number 20; & Fig. 11a, number 6), Bel'kov (Fig. 10, number 5; & Fig. 11a, number 7) and Svyatoi Nos (Fig. 11a, number 8), separated by high-standing blocks of underlying Late Mesozoic basement (East Laptev, Stolbovoi, Shiroston, and Kotel'nyi horsts, or highs; see Fig. 11a). The New Siberian Rift (Fig. 4, number 21; & Fig. 11a, number 10) occurring NE of Kotel'nyi Island is structurally isolated from the LRS, and thus is considered as a structural element of the East Siberian Shelf.

The internal structure of the LRS is controlled by a series of large-offset listric normal faults, with the main extensional detachments generally located at the eastern shoulders of the rifts (Fig. 11b). Inverted structures are widespread and occur along the listric faults; they are considered the result of compression, due to slight convergence between the EUR and NA plates in the Oligocene to Early Miocene (Savostin & Drachev 1988).

Stratigraphic correlation of the rift infill is disputable. Ivanova et al. (1990) and Sekretov (2000), based on the earlier idea of extension of the Siberian Craton far offshore (Vinogradov 1984), speculated that the rifts contain Neoproterozoic, Palaeozoic and Lower Mesozoic rocks, which represent a lithostratigraphic analogue of the craton's sedimentary cover. This concept suffers from lack of geological evidence supporting the offshore continuation of the Siberian Craton. An opposing point of view, supported in this paper, is based on a fact that the Laptev Shelf is surrounded by the Late Palaeozoic and Late Mesozoic fold belts, which apparently continue offshore and form a pre-rift basement of the rift system (Vinogradov & Drachev 2000; Drachev 2002). Therefore this concept limits the total stratigraphic range of the Laptev rift basins infill to the Upper Cretaceous and Cenozoic, and is in agreement with onshore stratigraphy of the Laptev region (Grinenko 1989; Alekseev et al. 1992).

The rift infill is composed of mainly siliciclastic non-marine, deltaic and shallow marine sediments whose total thicknesses vary from 1.5–3 to 8–10 km, and reach 13–14 km in the deepest parts of the Ust' Lena Rift (Fig. 11a, b). A clear eastward decrease of thickness and stratigraphic completeness of the syn-rift

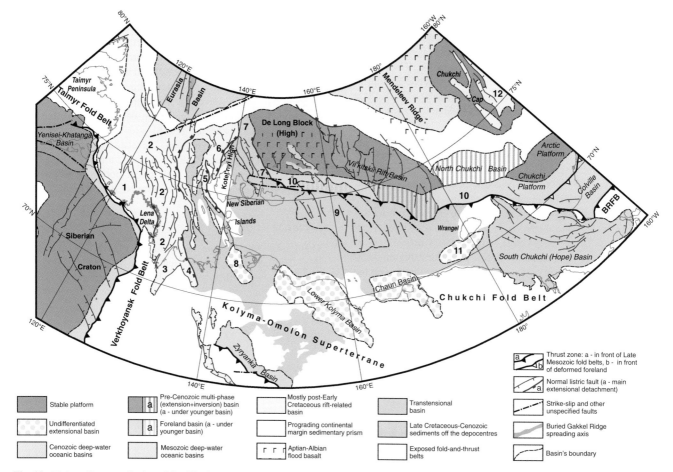

Fig. 10. Main sedimentary basins of the Siberian Arctic Shelf (modified from Grantz *et al.* 2009). Bold numbers denote the following basins: 1, Southwest Laptev Basin; 2, Ust' Lena Rift; 3, Omoloy Graben; 4, Ust' Yana Graben; 5, Bel'kov Rift; 6, Anisin Rift; 7, New Siberian Rift; 8, Tas–Takh Depression; 9, East Siberian Sea Basin; 10, New Siberian–Wrangel Basin; 11, Longa Basin; 12, Northwind Basin. BRFB, Brooks Range Fold Belt.

sequences, as well as a decrease in structural complexity of the rifts, may indicate an eastward rejuvenation of the rifts.

Based on the known history of the NA–EUR plate interaction in the Arctic, the following scenario for the evolution of the LRS is proposed (Fig. 12):

(1) In the Late Cretaceous to Paleocene, an initial rifting culminated in a breakup event and the onset of seafloor spreading in the Eurasia Basin at *c.* 55 Ma.
(2) In the Eocene, a non-rift setting existed as a result of the accommodation of the Eurasia Basin opening by the Khatanga–Lomonosov shear zone, which then became a segment of the plate boundary, and thus prevented the penetration of extensional strain onto the Laptev Shelf.
(3) During the Oligocene to Early Miocene, a non-rift or compressional setting existed, caused by global plate re-arrangement at 33 Ma.
(4) During the Late Miocene to Pleistocene, a re-activation of crustal extension occurred, representing the Second Rift Stage.

Interpreted MCS data correlated with onshore lithostratigraphy support this model and show the presence of successions deposited in a non-rift setting. These units truncate many of the earlier normal faults, and a sharp seismic stratigraphic unconformity at their base is regarded as a break-up unconformity at 55 Ma. The high reflectivity pattern on MCS data may be related to a high coaly material content. This is supported by a widespread occurrence onshore of Eocene and Oligocene strata abundant in brown coals and lignites deposited in low coastal plains during warm climatic conditions (Grinenko 1989; Stein 2008).

Petroleum geology. Data on HC systems of the Laptev Sea Province are generally absent. However, based on the lithostratigraphy of the onshore Cenozoic sections, and the inferred tectonic history of the LRS, we can speculate on possible source/reservoir/seal rock occurrences within the offshore rift basins (Fig. 12).

The main potential sources of HCs in the LRS can be attributed to:

(1) Paleocene to Eocene and, to some extent, Oligocene sediments with abundant terrestrial organic matter are potential gas-prone sources. However, Paleocene and Mid-Eocene marine organic-rich beds should not be excluded within the rift depocentres. They could be analogues of organic-rich marine shales recently encountered on the Lomonosov Ridge (Moran *et al.* 2006). Despite the uncertainty as to whether these beds could produce oil or not, their great contribution into the total HC potential of the Arctic deepwater basins and adjacent shelves cannot be ignored in future assessments.
(2) The Late Cretaceous and Paleocene syn-rift successions may contain both terrestrial and lacustrine organic-rich beds. These potential sources are presently buried at depths generally greater than 4–5 km in the main rift depocentres, and hence are likely to be mainly gas-prone.

Oligocene and particularly Lower Miocene sediments in many onshore localities are dominated by coarse-grained clastic sediments and thus can be considered as reservoir-prone successions. Little is known about the mineral composition of the sands, but considering the Palaeo-Lena River as a major supplier of the clastic sediments into the rift depocentres, especially into the Ust' Lena

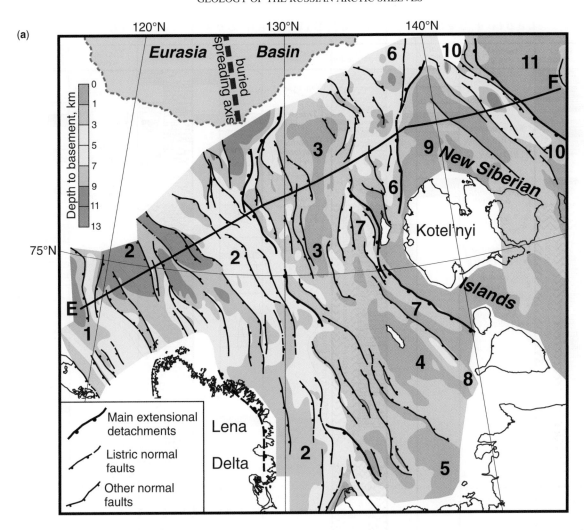

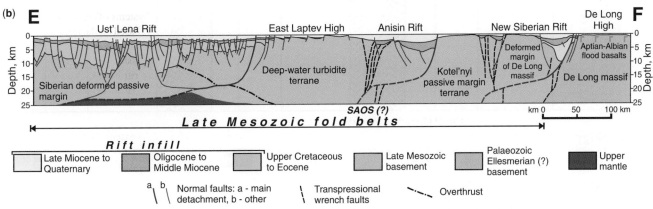

Fig. 11. Structure of the Laptev Rift System: (a) simplified depth-to-basement map of the Laptev Shelf based on 2D seismic and ERS-2 gravity data; (b) cross-section based on geological data and published BGR regional seismic lines (Franke *et al.* 2000, 2001). Location of the cross-section is shown in (a). Bold numbers in (a) denote following elements of the rift system: 1, Southwest Laptev Rift Basin; 2, Ust' Lena Rift; 3, East Laptev Horst; 4, Stolbovoi Horst; 5, Shiroston Horst; 6, Anisin Rift; 7, Bel'kov Rift; 8, Svyatoi Nos Graben; 9, Kotel'nyi High; 10, New Siberian Rift; 11, De Long High. *SAS (?)* in (b) denotes possible offshore projection of the South Anyui Suture.

Rift, we conclude that the reservoirs may potentially be of good or very good quality. Based on the broad areal extent of the Oligocene to Lower Miocene fluvial clastic sediments around the Laptev Shelf, we can predict their widespread occurrence offshore. That, in turn, makes this interval a main future HC exploration play.

In the onshore sections the Upper Miocene to Quaternary strata are dominated by clastic material against the suppressed accumulation of terrestrial organic matter. However, offshore sections are generally dominated by marine fine-grained muddy sediments, allowing the Upper Miocene and especially Pliocene to Quaternary sediments to be considered as a main regional seal.

Earliest oil generation could have started *c.* 13 Ma ago in the main rift depocentres of the western Laptev Shelf, and progressed until the end of Miocene time when the source rock is estimated to leave the oil maturation window, and the whole petroleum system became gas generating. The second phase of oil generation could have begun along the rift flanks during Late Miocene time, and has progressed up to the present.

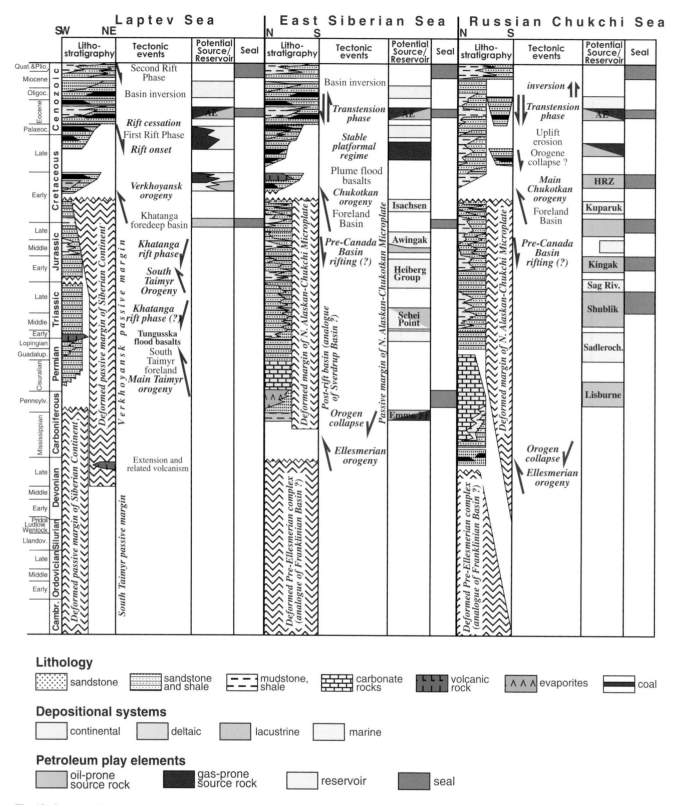

Fig. 12. Summary chart of tectonostratigraphy and petroleum play elements of the Siberian Arctic Shelf. AE denotes Azolla Event. Other bold italic words in Potential source/Reservoir columns denote names of analogue formations in the Sverdrup Basin (East Siberian Sea column) and the Alaskan Northern Slope (Chukchi Sea column). Emma FF denotes Emma Fiord Formation, and HRZ denotes Highly Radioactive Zone.

The estimated time of primary oil generation phases is quite favourable in terms of the capturing of migrating HCs in pre-existing structural traps, formed during the Oligocene to Early Miocene basin inversion, and the onset of the Second Rift Phase (Fig. 12). In most cases the traps are represented by tilted blocks and faulted rollover anticlines, as well as inverted anticlines and drapes over basement highs. The width of the structural traps is controlled by the distance between normal faults, which commonly varies around 5–7 km, while the length of the traps could be in the range of 10–20 km.

Older pre-rift HC plays may have a lesser contribution to the overall HC potential of the Laptev Shelf, as compared with the syn-rift plays. One of the pre-rift plays could be inferred in the southwestern part of the shelf, where marine organic-rich beds of latest Jurassic to earliest Cretaceous age are known to crop out along the shore in a foreland setting (Kaplan *et al.* 1973). There, a younger Neocomian succession is formed by reservoir-prone fluvial and deltaic/shallow marine facies, correlative to a set of Neocomian clinoforms that are well developed throughout the entire Yenisei–Khatanga Basin (Baldin 2001). If this basin extends offshore beneath syn-rift sedimentary cover, then the shelf area between the Lena Delta and Khatanga Bay could be favourable for both oil and gas pre-rift HC plays.

East Siberian Sea Province (ESSP)

The ESSP is the largest part of the Siberian Arctic shelf extending for over 1000 km from New Siberian Islands to Wrangel Island (Fig. 1). It is also the least studied part of the RAS (Fig. 2), and thus only general conclusions on its geology and possible HC systems can be drawn based on limited MCS, gravity and magnetic data, supported by the offshore projection of onshore geology. Earlier tectonic concepts published by Vinogradov et al. (1974, 1977), Kos'ko (1984) and Kos'ko et al. (1990) were recently reviewed with the use of regional MCS data (Roeser et al. 1995; Drachev et al. 1999, 2001; Franke et al. 2004; Franke & Hinz 2005). The following description summarizes the tectonic concepts of the ESSP history drawn from our interpretation of the available MCS data.

Two main crustal domains are recognized within the ESSP (Figs 10 & 13):

(1) the De Long Massif representing a northern (present-day) part of the AACM unaffected by the Late Mesozoic Verkhoyansk–Brookian orogeny;
(2) the New-Siberian–Chukchi Fold Belt – a southern part of the AACM involved in the Late Mesozoic compression deformation and orogeny.

Consequently, we outline two main generations of the basins whose occurrence is controlled by the above basement domains:

(1) Palaeozoic (post-Devonian?) to Mesozoic basins preserved north of the Late Mesozoic frontal thrusts (stratigraphic analogue of the US Chukchi Ellesmerian Sequence);
(2) Early Cretaceous (Aptian–Albian) to Quaternary basins, postdating the Verkhoyansk–Brookian orogeny, and evolving mainly over the New-Siberian–Chukchi Fold Belt (stratigraphic analogue of the US Chukchi Brookian Sequence).

The 70–100 km wide New Siberian Rift is clearly expressed in the gravity field (Fig. 4, number 21). It occurs between two highstanding blocks of tectonic basement – Kotel'nyi High on the west and De Long High on the east (Fig. 10). The rift sedimentary infill thickens from 5 km in the southern part of the rift, where it onlaps onto the Late Mesozoic folded basement of the Kotel'nyi High (Fig. 11b), to 10 km in the northern part of the rift, and is composed of several seismic units forming two main sets of strata:

(1) a syn-rift sequence correlates with Upper Cretaceous and Paleocene–Eocene siliciclastic terrestrial and shallow-marine sediments containing abundant coaly material;
(2) a set of post-rift seismic sequences interpreted as Oligocene to Lower Miocene, Middle to Upper Miocene and Pliocene to Quaternary sequences, deposited mainly in shallow-marine conditions.

Drachev et al. (1998, 1999) proposed that the New Siberian Rift was formed in response to a postulated divergent plate-tectonic boundary linked to the Late Cretaceous–Paleocene Labrador Sea–Baffin Bay spreading axis. Generally the rhomboid-like shape of the rift, its rather limited extent and evidence of strike-slip dislocations within its interior, are interpreted as evidence of a pull-apart origin.

The East Siberian Sea Basin (ESSB) is a NW–SE elongated 450 km long by 350 km wide depocentre, occurring in the central part of the ESSP (Fig. 4, number 22; Fig. 10, number 6; & Fig. 13). Its structural style differs significantly from the severely rifted Laptev Shelf, and is interpreted as being transtensional in origin (Franke et al. 2004; Franke & Hinz 2005). A possible cause for the transtensional regime can be inferred from the tectonic setting of the ESSP within a broad region of Early Cenozoic crustal re-adjustment between the NA and EUR lithospheric plates.

The basin is filled with siliciclastic sediments exceeding 8 km in the deepest depocentres. Their stratigraphic range is inferred to be Late Cretaceous to Quaternary in age. Based on reflection MCS data, the basin infill is subdivided into three main units (Fig. 14):

(1) a Lower Unit correlated to Upper Cretaceous (Cenomanian to Turonian), and Paleocene successions of coal-rich continental and deltaic clastic sediments exposed on Novaya Sibir' Island;
(2) a Middle Unit inferred to consist of Eocene to Middle Miocene fluvial–deltaic and shallow-marine successions, accumulated during the main stage of basin subsidence;
(3) an Upper Unit post-dating the main subsidence stage, and truncating the majority of faults identified within the Middle Unit. We correlate the Upper Unit to a set of Late Miocene to Pleistocene sequences dominated by shallow-marine shaly clastic sediments, which occur widely throughout northeastern Asia.

Considering the tectonic history of the ESSB, we can distinguish the following stages:

(1) Late Cretaceous and Paleocene – a tectonically 'quiet' platformal regime existed with accumulations of the Lower Unit, which is abundant in terrestrial organic matter.
(2) Eocene to Middle Miocene – rapid subsidence of the basin occurred in a transtensional setting controlled by dextral divergence of the NA and EUR plates. By the end of Oligocene to Early Miocene a change in the interaction of the plates caused the compression and structural inversion of some parts of the basin.
(3) Late Miocene to Pleistocene – the active tectonic movements ceased across the entire East Siberian Sea as the main zone of interaction between the EUR and NA plates moved over to the Laptev Sea region. During this sag phase the entire ESS shelf experienced thermal subsidence, and experienced broad marine transgressions.

The *New Siberian–Wrangel Foreland Basin* is recognized below the northern limb of the ESSB (Fig. 10, number 10). This narrow west–east trending basin is traced by a few lines across the ESS and north of Wrangel Island (see below) into the US Chukchi shelf, where it possibly merges into offshore prolongation of the Colville Foreland Basin (Grantz et al. 2009).

The MCS data along seismic line LARGE-8901 show the presence of two wedge-shaped packages of seismic reflectors beneath the Lower Unit north of the Late Mesozoic frontal thrust (Fig. 14). The reflectors of the upper package progressively onlap northward onto the lower wedge-shaped package, and both packages are involved into compressional deformations in the vicinity of the frontal thrust. We interpret the upper package as a siliciclastic coal-bearing infill of the foreland basin formed at terminal stages of the Verkhoyansk–Brookian orogeny (Drachev et al. 2001). The lower package may represent the Lower Mesozoic and/or Late Palaeozoic strata of the De Long Massif – analogues of the Ellesmerian Sequence of the Arctic Alaska. Alternatively, it could consist of the Lower Cretaceous flood basalts cropped

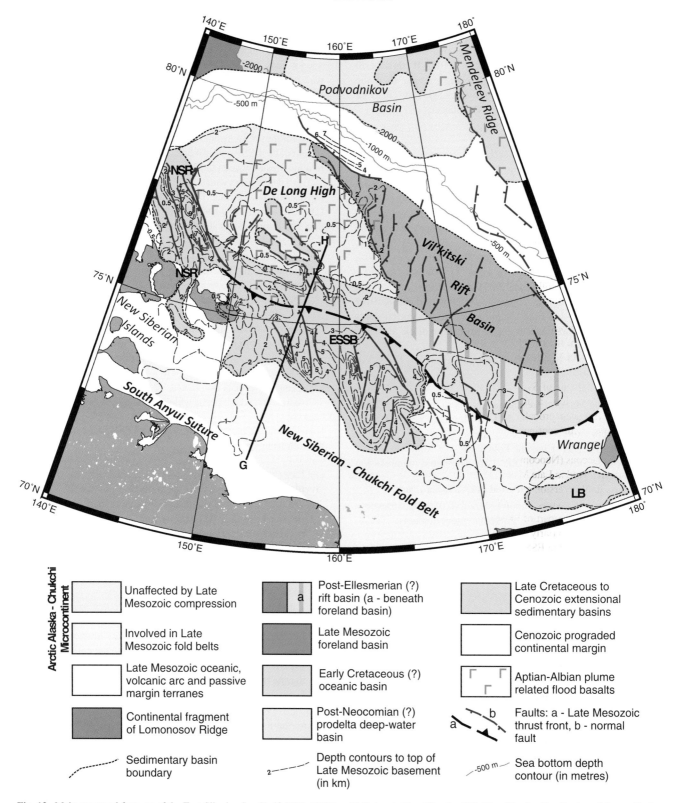

Fig. 13. Main structural features of the East Siberian Sea Shelf. NSR, ESSB and LB denote New Siberian Rift, East Siberian Sea Basin and Longa Basin accordingly. Bold line indexed G–H shows location of a cross-section given in Figure 14. The italic capital letters index the following islands: KT, Kotel'nyi; NS, Novaya Sibir'; BL, Bol'shoi Lyakhov. The accuracy of the scale bar increases towards 75°N.

out on Bennett Island. Presently we do not have data to constrain these possible interpretations.

The northeastern ESSP remains virtually unexplored, since no seismic data exist east of the De Long Archipelago (Fig. 2). The tectonic pattern of this vast area can only be inferred from the study of geophysical fields. The gravity field reveals a series of closely spaced SSE trending linear and rhomboid-shaped lows (Fig. 4, number 23) interpreted as an expression of extensional crustal features called the Vil'kitskii Trough by Kos'ko (1984), Fujita & Cook (1990) and Kos'ko et al. (1990) or the Vil'kitskii Rift System by Drachev et al. (1999). There are no data to infer the sediment age and composition in the Vil'kitskii Rift System, or to infer the scale and timing of possible crustal extension. Based on the suggested proximity of the northern ESS shelf to the Canadian Arctic prior to opening of the Canada Basin, we can only speculate as to the possible structural and stratigraphic relationships between the Vil'kitskii Rift Basin and the Sverdrup Basin. If these basins are tectonically related, then

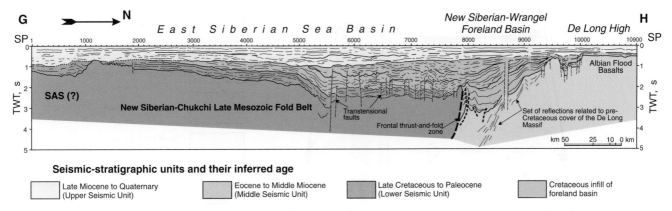

Fig. 14. Geological cross-section along seismic reflection line LARGE-89001 showing structural style and inferred seismic stratigraphy of the East Siberian Sea Basin (modified from S. Drachev *et al.* 1999). For location see Figures 3 and 13. SAS(?) denotes an offshore extent of the South-Anyui Suture.

the following tectonostratigraphic history can be proposed for the former:

(1) Early Carboniferous – the collapse of the Ellesmerian orogen, and formation of the rift system;
(2) Carboniferous to Mid Jurassic – post-rift thermal sagging and deposition of shelf carbonates, evaporates and clastic sediments, an analogue of Sverdrup Basin infill;
(3) Mid-Late Jurassic – a phase of crustal extension and initial rifting culminated probably during latest Jurassic to earliest Cretaceous, with continental breakup and spreading in the Amerasia Basin;
(4) Early Cretaceous (Neocomian) – the southern part of the basin became involved in Late Mesozoic compressional deformation and structural inversion occurred in its northern part;
(5) End Early Cretaceous – a plume-related magmatism influenced the northern part of the basin;
(6) Late Cretaceous to Early Cenozoic – followed the same development stages as the ESS Basin.

Petroleum geology. Two distinct differences exist between lithostratigraphy and, therefore, depositional environments of the Laptev and East Siberian Sea provinces:

(1) Cenomanian–Turonian terrestrial coal-bearing sediments deposited in a fluvial coastal plain setting are present on Novaya Sibir' Island, suggesting marine conditions existed northward in early Late Cretaceous time.
(2) In the Oligocene to Early Miocene, in contrast to the Laptev Province that experienced tectonic uplift and long-term regression with a prevalence of continental fluvial depositional systems, the ESSP, in turn, experienced a long period of subsidence and a dominance of marine and coastal plain environment with accumulation of mainly fine-grained sediments with abundant terrestrial organic matter and coaly material as far south as the present-day coastline (Patyk-Kara & Laukhin 1986).

This suggests that marine depositional environments were widespread through the entire Late Cretaceous and Cenozoic history of the ESSP, and therefore the corresponding sequences could be more favourable for the occurrence of marine organic rich beds – potential oil sources.

Based on this, the main potential HC sources could be expected in (Fig. 12):

(1) Upper Cretaceous to Paleocene sediments with abundant terrestrial organic matter (Lower Seismic Unit) – most probably a gas-prone source;
(2) Eocene and possibly Oligocene to Early Miocene sections of the

Based on simple analysis of basin subsidence, initiation of HC generation could have started as early as 54 Ma when potential Lower Cretaceous source rocks entered the oil maturation window and may have continued until 23 Ma. Given the mainly terrestrial composition of organic matter, most of the generated HCs are probably in the gassy fraction, although the possibility of oil generation cannot be excluded. At approximately 23 Ma, possible Oligocene source rocks may have entered the oil maturation window and the period between 23 Ma and present time could represent the main phase of HC generation.

Presently, we do not have data to constrain possible HC systems and petroleum plays of the completely unexplored Vil'kitskii Rift Basin. Speculating on possible pre-Jurassic relationships between the Canadian Arctic and ESSP, we may infer an analogue to the Sverdrup HC systems to be present in the Vil'kitskii Basin north of the Late Mesozoic deformation front (Fig. 12). However, considering the magnitude of the Cenozoic tectonically driven subsidence in better known depocentres, preservation of the oil-prone Late Palaeozoic and Early Mesozoic source rocks at the depths not exceeding the oil window becomes very questionable.

Russian sector of Chukchi Sea Province (CSP)

The Russian sector of the Chukchi Shelf has much better seismic coverage compared with the East Siberian Sea (Fig. 2), and hence its geology is better constrained. The geology and tectonic history of this region were considered by Vinogradov *et al.* (1974, 1977), Pol'kin (1984), Grantz *et al.* (1986, 1990) and recently by Verzhbitsky *et al.* (2008) and Petrovskaya *et al.* (2008), and some data on petroleum geology were summarized by Haimila *et al.* (1990) and by Warren *et al.* (1995).

Tectonically the Russian CSP is similar to the East Siberian Sea shelf (Fig. 13). Two main tectonic domains divided by a zone of frontal thrusts of the Wrangel–Herald Arch are outlined (Fig. 15):

(1) the northern part of the AACM with preserved pre-Late Cretaceous (pre-Barremian?) basins mostly filled with stratigraphic analogues of the Ellesmerian Sequence of the US Chukchi Shelf;
(2) the southern part of the AACM affected by the Late Mesozoic Chukotka–Brookian compressional event and orogeny. It comprises basins post-dating the orogeny and filled with stratigraphic analogues of the Brookian Sequence of the US Chukchi Shelf.

Three first-order basins are outlined in the Russian CSP with use of the MCS and gravity data. Two of them, the New Siberian–Wrangel and the North Chukchi basins (Fig. 4, numbers 27 and 28), occur within the older crustal domain north of the

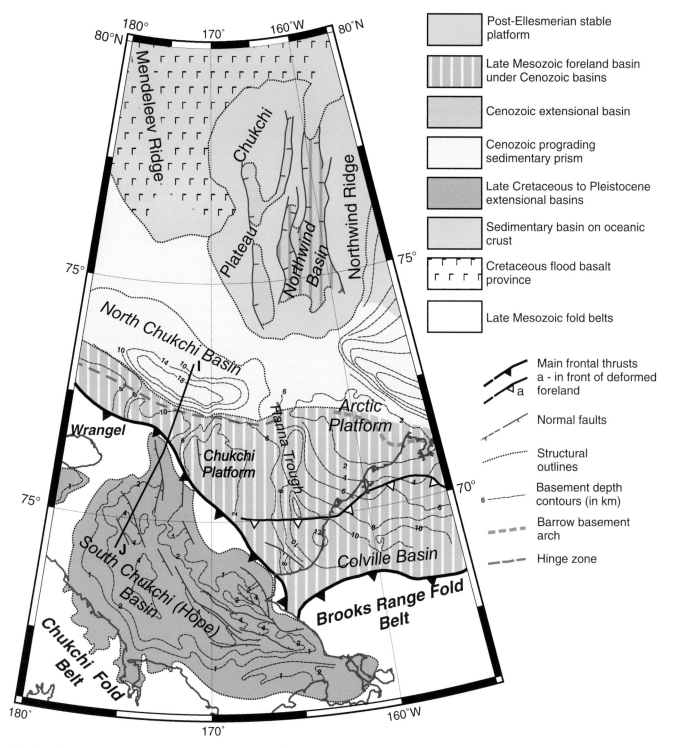

Fig. 15. Main structural elements of the Chukchi Shelf (modified from Grantz *et al.* 2009). Bold solid line labelled 'J–I' shows location of cross-section given in Figure 16. Capital italic letters denote: HT, Hanna Trough; SCB, South Chukchi Basin; KSB, Kotzebue Sound Bay. The accuracy of the scale bar increases towards 75°N.

deformational front, and the South Chukchi (Hope) Basin exists south of the Wrangel–Herald Arch (Fig. 15). A similar view on the Russian CSP tectonics was presented earlier by Grantz *et al.* (1990) and Warren *et al.* (1995).

The eastern part of the New Siberian–Wrangel Foreland Basin is well mapped by the numerous DNMG (Petrovskaya *et al.* 2008) and TGS–Nopec (Verzhbitsky *et al.* 2008) MCS lines north of Wrangel Island between the late Mesozoic deformational front and the North Chukchi Basin. From the latter it is divided by a sharp basement surface break which is assumed to be an analogue of the Hinge line by Grantz *et al.* (1990). Existence of a basin in this location was first proposed by Pol'kin (1984), who named it the North Wrangel Trough. Most of the basin infill is represented by the seismic stratigraphic units CS-2 and CS-3, which are inferred to be composed of Upper Jurassic to Lower Cretaceous clastic rocks (Fig. 16). These units are underlain by the CS-1 unit interpreted to represent Upper Palaeozoic to Lower Mesozoic sedimentary rocks, which may be analogues of the Ellesmerian Sequence of Arctic Alaska. The CS-3 forms the upper part of the basin infill and continues over the hinge line into the North Chukchi Basin. All these units are affected by moderate fold-and-thrust deformations in the vicinity of the Late Mesozoic compressional front.

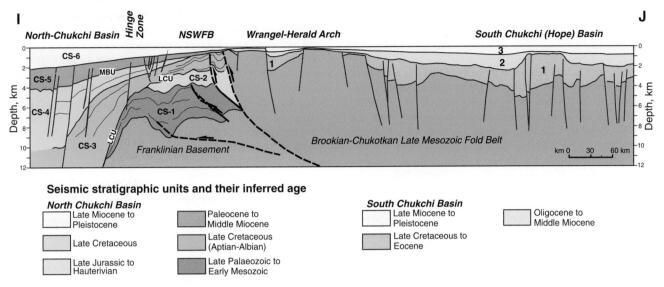

Fig. 16. Simplified cross-section illustrating internal structure and inferred stratigraphy of the South Chukchi and North Chukchi basins. Based on seismic data acquired and published by DMNG and TGS-Nopec (Petrovskaya *et al.* 2008; Verzhbitsky *et al.* 2008) and ERS-2 gravity data. Location is given in Figures 3 and 15. The bold numbers are indexes of the seismic sequences in the South Chukchi Basin after Tolson (1987). The alpha-numeric indexes CS-1 through CS-6 denote seismic sequences identified in the North Chukchi and the New Siberian–Wrangel basins (see text for further details). LCU, Lower Cretaceous Unconformity; MBU, Mid Brookian Unconformity. NSWFB denotes the New Siberian–Wrangel Foreland Basin.

The *North Chukchi Basin* extends east for more than 500 km from the 180° meridian towards the North Slope of Alaska (Figs 10 & 15). Its full extent is depicted by a prominent gravity low of −40 to −60 mGal (Fig. 4, number 28). According to interpretations of the MCS data, the basin can contain over 12 km thick sedimentary fill, which can locally reach 18 or even 20 km. Most of the basin fill is inferred to be composed of Cretaceous to Cenozoic clastic strata (Thurston & Theiss 1987; Grantz *et al.* 1990), although the presence of older sediments in the most subsided parts of the basin is highly likely.

Several seismic units have been identified within the basin and interpreted to reflect the main development stages (Fig. 16). A sharp unconformity is observed at the bottom of the seismic unit CS-3 along the steep southern slope of the basin. It deepens sharply from *c.* 3 to 12 km and more to the north, towards the basin interior. To the south, the unconformity extends into the New Siberian–Wrangel Foreland Basin, where it forms the top of seismic unit CS-2 corresponding to an early stage of the foreland deformations (Fig. 16). Therefore, we interpret this prominent seismic stratigraphic unconformity to be related to onset of the main orogenic phase which, based on geological data from the Chukchi Peninsula (Sokolov *et al.* 2002), occurred around Hauterivian to Barremian time (130–125 Ma). We further correlate this unconformity with the Lower Cretaceous Unconformity (LCU), which is one of the most distinct regional unconformities in the US Chukchi–Beaufort shelf (Craig *et al.* 1985; Thurston & Theiss 1987). A similar interpretation has been proposed by Grantz *et al.* (1990), and more recently by Verzhbitsky *et al.* (2008). The seismic stratigraphic units above the LCU form the main infill of the North Chukchi Basin, which is thus interpreted to be mostly a post-Barremian basin.

Another sharp seismic stratigraphic unconformity at the top of the seismic unit CS-4 truncates folds in the foreland basin, and extends into the central part of the North Chukchi Basin (Fig. 16). The closest analogue in the US Chukchi Sea is the well established regional Mid Brookian Unconformity (MBU) at the Cretaceous–Cenozoic boundary (Thurston & Theiss 1987). The well-laminated seismic pattern of the CS-4 unit between the LCU and MBU allows for conclusion that the post-Barremian Cretaceous infill can be dominated by marine facies, reflecting the main phase of the subsidence within the North Chukchi Basin.

The *South Chukchi (Hope) Basin* extends for over 1000 km from the Longa Strait up to the Western Alaska coast, and is limited by the transpressional Wrangel–Herald Arch in the north, and by the Chukchi Peninsula in the south (Fig. 4, number 25; Fig. 10, number 8 & Fig. 15). The basin is underlain by folded complexes of the Chukotka–Brooks Fold Belt exposed on Wrangel Island, Chukchi Peninsula and along the Alaskan rim of the basin. Two onshore exploration wells near Kotzebue in the southern US Hope Basin encountered basement of Palaeozoic schists and marbles (Thurston & Theiss 1987). In the US Chukchi Sea this basin is known as the Hope Basin, whose geology and petroleum systems were characterized by Tolson (1987) and Haimila *et al.* (1990).

MCS data reveal an asymmetric geometry of the basin. Its southern limb is formed by a gradual southward basement rise while the northern fractured limit, at the junction with the Wrangel–Herald Arch, is rather steep (Fig. 16). The internal structure of the basin is formed by a series of NW–SE trending grabens, half-grabens and dividing horsts, whose assemblage was described by Tolson (1987) as transtensional. Sediment thickness in the basin does not exceed 5–6 km in its deepest parts. Three main seismic stratigraphic units are interpreted (Fig. 16):

(1) A Lower Unit is inferred to be represented by Paleocene (perhaps, Upper Cretaceous) to Eocene continental coal-bearing and fluvial, lacustrine and shallow marine clastic sediments. In the Alaskan onshore extent of the basin, a lower stratigraphic interval penetrated by two exploration wells is known to consist of volcanic and non-marine volcaniclastic rocks, possibly of Eocene age. However, there is no direct tie of the seismic horizons with these wells, and therefore the well data may not be indicative for the offshore lithostratigraphy of the basin (Thurston & Theiss 1987).

(2) A Middle Unit is inferred to be represented by Oligocene–Middle Miocene predominantly continental clastic sediments.

(3) An Upper Unit is represented by Miocene to Quaternary mainly shallow-marine clastic and shaly sediments.

Based on the above characteristics of the Russian CSP geology, we draw the following conclusions about the tectonic history of its sedimentary basins:

(1) The North Chukchi Basin was probably initiated as a rift basin in the Early to Middle Jurassic during an extensional stage precursor to the Canada Basin opening (Grantz et al. 1998). However this remains a speculative assumption since the rocks coeval with the early stage of the North Chukchi Basin formation are buried below the depths imaged by MCS data.
(2) In latest Jurassic to Barremian, a major collision occurred between the AACM and the Siberian margin, followed by orogeny and formation of a foreland basin north of the deformation front.
(3) In Aptian to Albian, the North Chukchi Basin was affected by a drastic subsidence resulting in accumulation of a c. 10 km thick succession of clastic sediments. This event is coeval to a severe erosion which occurred at Western Alaska around 115 Ma (Moore et al. 2002).
(4) In the Late Cretaceous, formation of the South Chukchi Basin began in response to a post-orogenic collapse. The end of the stage was marked by an uplift and erosion caused by convergence of the NA and EUR plates.
(5) Paleocene to latest Eocene/earliest Oligocene – the basins evolved in an intraplate setting controlled by dextral motion between the NA and EUR plates. This is also the main phase of subsidence in the South Chukchi Basin.
(6) Oligocene to Miocene – the South Chukchi Basin was affected by NE–SW plate convergence compression causing transpressional growth of the Wrangel–Herald Arch.

Petroleum geology. Petroleum systems and HC play elements of the South Chukchi Basin can be projected from the drilled and thus to some extent better known Hope Basin. According to Haimila et al. (1990) the petroleum potential of the latter is considered to be rather limited with no identified oil-prone source and only thermally immature terrestrial organic rich beds identified in nearby onshore wells. However, offshore depocentres that subsided to significant depths may still be more prospective for generation and accumulation of HCs. During the main Eocene to Oligocene phase of tectonic subsidence, the basin remained structurally isolated from marine environment existing to the north by the Wrangel–Herald basement arch, and therefore the presence of the marine source beds in this inland basin is highly unlikely.

The older North Chukchi Basin is much more prospective for oil generation. The structural setting of the basin is quite similar to the Alaska North Slope and therefore a number of HC plays prospective for both gas and oil can be expected in this part of the Chukchi Shelf (Fig. 12).

As described above, the entire seismically imaged section of the North Chukchi Basin down to the depths of 12 km, and possibly greater, is apparently composed of Cretaceous and Cenozoic strata only. Depth to the Cenozoic base increases rapidly from a few hundred metres along the southern flank of the basin, up to 4–5 km at the basin's axial depocentre (Fig. 16). This suggests that any possible oil source rocks beneath the MBU in the central North Chukchi Basin would have already passed way below the conventional oil maturity window.

The section above the MBU has a much greater chance for the presence of lower Cenozoic marine organic rich beds – analogues of the Middle to Upper Eocene oil-prone Richards sequence in the Beaufort-Mackenzie region. However, an almost complete absence of tectonic structures within this part of the aggraded section significantly limits future exploration potential over the main part of the North Chukchi Basin axial depocentre to the stratigraphic traps.

Greater chances of finding large oil accumulations are more likely along the southern margin of the basin at the hinge zone, and to the south of the hinge zone, where pre-LCU strata occur at drillable depths for exploration wells – around 4–5 km. As illustrated in Figure 16, a basement high dividing the North Chukchi depocentre from the New Siberian–Wrangel Foreland Basin resembles the Barrow Arch. Therefore, mature HC plays existing along the Arctic coast of Alaska could be considered as the closest analogues of the potential untested petroleum plays of the southern flank of the North Chukchi Basin in the Russian CSP.

Summary and conclusions

The vast Russian Arctic Shelf is, to a large extent, sparsely explored due to its harsh environment, high cost of operations and remoteness from modern markets, and its undiscovered HC potential is still highly unconstrained. We have summarized the results of many regional geological and geophysical studies accomplished over the past few decades. Based on 2D regional seismic surveys correlated with borehole data and with onshore geological data in the areas where no exploration wells have yet been drilled, as well as on low-resolution gravity and magnetic data, we have outlined the main sedimentary basins and constrained their structural styles, lithostratigraphy and possible HC systems. Concluding this review of the RAS, we would like to highlight the following major points:

- The Barents and South Kara seas have the largest discovered gas resources in the Russian Arctic, and thus are commonly considered as gas-prone provinces. However, the chances for finding oil plays in the marginal zones of these basins are relatively high, and thus these could be of future exploration interest.
- The NKP is virtually unexplored, and its tectonostratigraphy is mainly constrained by available geological data from adjacent land areas. Several mid-sized depocentres inferred from gravity data are considered to be of mostly Early to Middle Palaeozoic age, and probably experienced inversion related to the Late Palaeozoic and Early Mesozoic orogenic events. Petroleum systems of the North Kara Shelf are unconstrained.
- The Siberian Arctic shelves were severely affected by the Late Mesozoic (Neocomian) orogeny, which largely shaped their tectonic basement and also had a dramatic impact on their potential petroleum systems.
- The Laptev rift province provides solid evidence of high HC potential. The basins are filled with thick Late Cretaceous to Cenozoic terrestrially sourced clastic sediments with abundant coaly matter, and thus are inferred to be mainly gas prone. However, if the pre-rift Bazhenov-type source rocks were not destroyed by the Late Mesozoic orogeny in the southwestern part of the Laptev shelf, or if the Early Cenozoic marine oil-prone source rocks accumulated within the rift depocentres, the Laptev shelf could also be prospective for the presence of oil plays. As the Laptev Shelf was the site of active rifting in Late Cretaceous to Paleocene and Recent times, and many normal faults reach the sea bottom, significant exploration risks may be related to the seal integrity.
- The East Siberian and Chukchi shelves are dominated by two generations of basins: (1) the Late Cretaceous to Cenozoic basins occurring south of the offshore extent of the Late Mesozoic deformational front; and (2) older (Late Palaeozoic(?) to Early Cretaceous) basins occurring north of the Late Mesozoic deformation front.
- The Late Cretaceous to Cenozoic basins are interpreted to have originated in a transtensional setting caused by dextral relative

motions of the NA and EUR lithospheric plates. Lithostratigraphy and possible HC systems of these basins are inferred from the better studied Hope Basin in the US Chukchi Sea, and are also constrained by onshore sedimentary sections. Their potential is considered to be mainly related to gas-prone terrestrial sources, although the presence of Early Cenozoic thermally mature lacustrine or shallow marine oil-prone sources cannot be disregarded.

- The inferred Late Palaeozoic to Early Mesozoic basins may be similar tectonostratigraphically to the Sverdrup Basin in the Canadian Arctic. The petroleum geology of the latter has been used to constrain possible HC systems of the Vil'kitsky Basin, which is identified on the basis of gravity and magnetic data only.
- The New Siberian–Wrangel Foreland Basin occurs in front of the Late Mesozoic compressional deformations. Its tectonostratigraphy is inferred to resemble the basins of the Alaska's North Slope and thus the petroleum geology of the latter with its rich oil potential has been projected north of Wrangel Island to highlight possible oil potential of this foreland basin.
- North Chukchi Basin dominates the Russian Chukchi Sea and is mainly filled with post-Neocomian siliciclastic sediments with a total thickness exceeding 18 km. Its petroleum plays may be similar to the known Late Cretaceous and Cenozoic plays of the Beaufort Sea.

The authors during their research career have collaborated with many geoscientists from Russia (G. E. Bondarenko, L. I. Lobkovskiy, V. E. Khain, A. V. Savitskiy, L. A. Savostin, K. O. Sobornov, S. D. Sokolov) and from Western countries (S. Creaney, A. Grantz, K. Hinz, L. Johnson, W. Jokat, N. Kaul, R. Scott, J. Thiede, G. Ulmishek, P. Ziegler). We extend our gratitude to all of them for contributing to our understanding of the Arctic geology. We are thankful to the management of ExxonMobil and Rosneft for providing the opportunity to publish this article. The whole paper was considerably improved after it had been thoroughly reviewed by N. McAllister, V. Pease and D. Thurston. E. A. Miloradovskaya (Karasik) helped with the editing of the text.

References

Alekseev, M. N., Arkhangelov, A. A. et al. 1992. Laptev and East Siberian Seas. Cenozoic. In: Palaeogeographic Atlas of the Shelf Regions of Eurasia for the Mesozoic and Cenozoic, 1. Robertson Group plc, UK, and Geological Institute, Academy of the Sciences, USSR, 1-14–1-33.

Allen, M. B. & Davies, C. E. 2007. Unstable Asia: active deformation of Siberia revealed by drainage shifts. Basin Research, 19, 379–392.

Amundsen, H., Evdokimov, A., Dibner, V. & Andresen, A. 1998. Petrogenetic significance and evolution of Mesozoic magmatism, Franz Josef Land and the Barents Sea. In: Solheim, A., Musatov, E. & Heintz, N. (eds) Geological Aspects of Franz Josef Land and the Northernmost Barents Sea (the Northern Barents Sea Geotraverse). Norsk Polarinstitutt, Meddelelser, 151.

Aplonov, S. V., Shmelev, G. B. & Krasnov, D. K. 1996. Geodynamics of the Barents–Kara Shelf: geophysical evidence. Geotectonics, 30, 309–326.

Artyushkov, E. V. 2005. The formation mechanism of the Barents basin. Russian Geology and Geophysics, 46, 700–713.

Atlasov, I. P., Dibner, V. D. et al. 1970. Explanatory Notes to Tectonic Map of the Arctic and Sub-Arctic. Scale 1:5,000,000. Nedra, Leningrad (in Russian).

Baldin, V. A. 2001. Geology and petroleum potential of Upper Jurassic to Neocomian successions of the western Yenisei–Khatanga Depression. Candidate of Sciences Thesis, Moscow (in Russian).

Bogdanov, N. A. & Khain, V. E. (eds) 1996. Explanatory Notes to the Tectonic Map of the Barents Sea and the Northern Part of European Russia. Scale 1:2,500,000. ILRAN, Moscow (in Russian).

Bogolepova, O. K. & Gee, D. 2004. Early Paleozoic unconformity across the Timanides, NW Russia. In: Gee, D. & Pease, V. (eds) The Neoproterozoic Timanide Orogen of Eastern Baltica. Geological Society, London, Memoirs, 30, 145–158.

Bogolepova, O. K., Gubanov, A. P. & Raevskaya, E. G. 2001. The Cambrian of the Severnaya Zemlya Archipelago, Russia. Newsletter Stratigraphy, 39, 73–91.

Bondarev, V. I. (ed.) 1982. Geology of the South Island of Novaya Zemlya. PGO 'Sevmorgeologiya', Leningrad.

Burlin, Yu. K. & Stoupakova, A. V. 2008. Geological conditions of petroleum potential of the Russian Arctic. Oil and Gas Geology, 4, 13–23 (in Russian).

Cook, D. B., Fujita, K. & McMullen, C. A. 1986. Present-day plate interactions in Northeast Asia: North American, Eurasian, and Okhotsk plates. Journal of Geodynamics, 6, 33–51.

Craig, J. D., Sherwood, K. W. & Johnson, P. P. 1985. Geologic Report for the Beaufort Sea Planning Area, Alaska. US Minerals Management Service OCS Report 85-0111.

Doré, A. G. 1995. Barents Sea geology, petroleum resources and commercial potential. Arctic, 48, 207–221.

Drachev, S. S. 1989. Tectonics and Mesozoic to Cenozoic geodynamics of the New Siberian Islands. PhD (Candidate of Geological and Mineralogical Sciences) thesis. Geological Department of Moscow State University.

Drachev, S. S. 2000. Tectonics of the Laptev Sea rift system. Geotectonics, 34, 467–482.

Drachev, S. S. 2002. On the basement tectonics of the Laptev Sea Shelf. Geotectonics, 36, 483–498.

Drachev, S. S. & Saunders, A. 2006. The Early Cretaceous Arctic LIP: its geodynamic setting and implications for Canada Basin opening. In: Scott, R. A. & Thurston, D. K. (eds) Proceedings of the Fourth International Conference on Arctic Margins, Dartmouth, Nova Scotia, 30 September to 3 October. U. Department of the Interior, Anchorage, Alaska, 216–223.

Drachev, S. S. & Savostin, L. A. 1993. Ophiolites of Bol'shoi Lyakhov Island (New Siberian Islands). Geotektonika, 6, 33–51 (in Russian).

Drachev, S. S., Savostin, L. A., Groshev, V. G. & Bruni, I. E. 1998. Structure and geology of the continental shelf of the Laptev Sea, Eastern Russian Arctic. Tectonophysics, 298, 357–393.

Drachev, S. S., Johnson, G. L., Laxon, S., McAdoo, D. & Kassens, H. 1999. Main structural elements of the Eastern Russian Arctic Continental Margin derived from satellite gravity and multichannel seismic reflection data. In: Kassens, H., Bauch, H. A. et al. (eds) Land–Ocean Systems in the Siberian Arctic: Dynamics and History. Springer, Berlin, 667–682.

Drachev, S. S., Elistratov, A. V. & Savostin, L. A. 2001. Structure and seismostratigraphy of the East Siberian Sea Shelf along the Indigirka Bay–Jeannette Island seismic profile. Doklady Earth Sciences, 377A, 293–297.

Egorov, A. Y., Bogomolov, Y. A., Konstantinov, A. G. & Kurushin, N. I. 1987. Stratigraphy of Triassic rocks of Kotel'nyi Island (New Siberian Islands). In: Dagys, A. S. (ed.) Boreal Triassic. Nauka, Moscow, 66–80 (in Russian).

Fjellanger, E., Kontorovich, A. E., Barboza, S. A., Burshtein, L. M., Hardy, M. J. & Livshits, V. R. 2010. Charging the giant gas fields of the NW Siberia Basin. In: Vining, B. A. & Pickering, S. C. (eds) Petroleum Geology: From Mature Basins to New Frontiers – Proceedings of the 7th Petroleum Geology Conference. Geological Society, London, 659–668; doi: 10.1144/0070659.

Franke, D. & Hinz, K. 2005. The structural style of sedimentary basins on the shelves of the Laptev Sea and western East Siberian Sea, Siberian Arctic. Journal of Petroleum Geology, 28, 269–286.

Franke, D., Hinz, K. et al. 2000. Tectonics of the Laptev Sea Region in North-Eastern Siberia. In: Roland, N. W. & Tessensohn, F. (eds) III International Conference on Arctic Margins. Polarforschung, 68, 51–58.

Franke, D., Hinz, K. & Oncken, O. 2001. The Laptev Sea rift. Marine and Petroleum Geology, 18, 1083–1127.

Franke, D., Hinz, K. & Reichert, Ch. 2004. Geology of the East Siberian Sea, Russian Arctic, from seismic images: structures, evolution, and implications for the evolution of the Arctic Ocean Basin. Journal of Geophysical Research, 109, 1–19.

Fujita, K. & Cook, D. B. 1990. The Arctic continental margin of eastern Siberia. *In*: Grantz, A., Johnson, G. L. & Sweeney, J. F. (eds) *The Arctic Ocean Region*. The Geology of North America, L. Geological Society of America, Boulder, CO, 257–288.

Gee, D. G. & Pease, V. (eds) 2005. *The Neoproterozoic Timanide Orogen of Eastern Baltica*. Geological Society, London, Memoirs, **32**.

Gee, D. G. & Stephenson, R. A. (eds) 2006. *European Lithosphere Dynamics*. Geological Society, London, Memoirs, **30**.

Gee, D. G., Bogolepova, O. K. & Lorenz, H. 2006. The Timanide, Caledonide and Uralian orogens in the Eurasian high Arctic, and relationships to the palaeo-continents Laurentia, Baltica and Siberia. *In*: Gee, D. G. & Stephenson, R. A. (eds) *European Lithosphere Dynamics*. Geological Society, London, Memoirs, **32**, 507–520.

Glebovsky, V., Kovacs, L. C., Maschenkov, S. P. & Brozena, J. M. 2000. Joint compilation of Russian and US Navy Aeromagnetic data in the Central Arctic Seas. *Polarforschung*, **68**, 35–40.

Glebovsky, V. Yu., Kaminsky, V. D., Minakov, A. N., Merkur'ev, S. A., Childers, V. A. & Brozena, J. M. 2006. Formation of the Eurasia Basin in the Arctic Ocean as inferred from geohistorical analysis of the anomalous magnetic field. *Geotectonics*, **40**, 263–281.

Gramberg, I. S. (ed.) 1988. *Barents Shelf Plate*. Nedra, Leningrad.

Gramberg, I. S. & Pogrebitskiy, Yu. E. (eds) 1984. *Geological Structure of the USSR and Trends in Distribution of Mineral Resources, 9, Soviet Arctic Seas*. Nedra, Leningrad (in Russian).

Gramberg, I. S. & Ushakov, V. I. 2000. *Severnaya Zemlya – Geology and Mineral Resources*. VNIIOkeangeologia, St Petersburg (in Russian).

Grantz, A., Mann, D. M. & May, S. D. 1986. *Multichannel Seismic-reflection Data Collected in 1978 in the Eastern Chukchi Sea*. US Geological Survey Open-File Report, **86-206**.

Grantz, A., May, S. D. & Hart, P. E. 1990. Geology of the Arctic continental margin of Alaska. *In*: Grantz, A., Johnson, G. L. & Sweeney, J. F. (eds) *The Arctic Ocean Region*. The Geology of North America, L. Geological Society of America, Boulder, CO, 257–288.

Grantz, A., Clark, D. L., Phillips, R. L. & Srivastava, S. P. 1998. Phanerozoic stratigraphy of Northwind Ridge, magnetic anomalies in the Canada basin, and the geometry and timing of rifting in the Amerasia basin, Arctic Ocean. *Geological Society of America Bulletin*, **110**, 801–820.

Grantz, A., Scott, R. A., Drachev, S. S. & Moore, T. E. 2009. *Map Showing the Sedimentary Successions of the Arctic Region ($58°–64°$ to $90°N$) that May be Prospective for HCs*. Open-File Spatial Library, World Wide Web Address: http://gisudril.aapg.org/gisdemo/.

Grinenko, O. V. (ed.) 1989. *Palaeogene and Neogene of North-East of the USSR*. Scientific Center of Siberian Department of Academy of Sciences of USSR, Yakutsk (in Russian).

Gudlaugsson, S. T., Faleide, J. I., Johansen, S. E. & Breivik, A. 1998. Late Palaeozoic structural development of the south-western Barents Sea. *Marine and Petroleum Geology*, **15**, 73–102.

Haimila, N. E., Kirschner, C. E., Nassichuk, W. W., Ulmichek, G. & Procter, R. M. 1990. Sedimentary basins and petroleum resource potential of the Arctic Ocean region. *In*: Grantz, A., Johnson, G. L. & Sweeney, J. F. (eds) *The Arctic Ocean Region*. The Geology of North America, L. Geological Society of America, Boulder, CO, 503–538.

Harbert, W., Frei, L., Jarrard, R., Halgedahl, S. & Engebretson, D. 1990. Palaeomagnetic and plate-tectonic constraints on the evolution of the Alaskan-eastern Siberian Arctic. *In*: Grantz, A., Johnson, G. L. & Sweeney, J. F. (eds) *The Arctic Ocean Region*. The Geology of North America, L. Geological Society of America, Boulder, CO, 567–592.

Inger, S., Scott, R. A. & Golionko, B. G. 1999. Tectonic evolution of the Taimyr Peninsula, northern Russia: implications for Arctic continental assembly. *Journal of the Geological Society, London*, **156**, 1069–1072.

Ivanova, N. M., Sekretov, S. B. & Shkarubo, S. I. 1990. Geological structure of the Laptev Sea shelf according to seismic studies. *Oceanology*, **29**, 600–604.

Ivanova, N. M., Sakoulina, T. S. & Roslov, Yu. V. 2006. Deep seismic investigation across the Barents–Kara region and Novozemelskiy Fold Belt (Arctic Shelf). *Tectonophysics*, **420**, 123–140.

Johansen, S. E., Ostisty, B. K. *et al.* 1993. Hydrocarbon potential in the Barents Sea region: play distribution and potential. *In*: Vorren, T. O., Bergsaker, E., Dahl-Stamnes, Ø. A., Holter, E., Johansen, B., Lie, E. & Lund, T. N. (eds) *Arctic Geology and Petroleum Potential*, Norwegian Petroleum Society Special Publication 2. Elsevier, Amsterdam, 273–320.

Kaban'kov, V. Ya. & Lazarenko, N. P. (eds) 1982. *Geology of Severnaya Zemlya Archipelago*. Sevmorgeo, Leningrad (in Russian).

Kaminsky, V. D., Glebovsky, V. Yu., Leychenkov, G. L., Golynsky, A. V., Masolov, V. N. & Gandyukhin, V. V. 2009. Role of geophysical methods in geological study of mineral resources of Arctic and Antarctic and some results. *In*: *Geology of Polar Regions of the Earth. Materials of XLII Tectonic Conference, I*. GEOS, Moscow, 253–257 (in Russian).

Kaplan, A. A., Copeland, P. *et al.* 2001. New radiometric ages of igneous and metamorphic rocks from the Russian Arctic. *In*: *Abstracts of AAPG Regional Conference 2001*. VNIGRI and AAPG, St Petersburg, 2001.

Kaplan, M. E., Yudovnyi, E. G. & Zakharov, V. A. 1973. Depositional environment of marine sediments of Paksa Peninsula transitional between Jurassic and Cretaceous (Anabar Bay). *Doklady Academii Nauk SSSR*, **209**, 691–694 (in Russian).

Karasik, A. M. 1968. Magnetic anomalies of the Gakkel Ridge and the origin of the Eurasian Subbasin of the Arctic Ocean. *In*: *Geophysical Survey Methods in the Arctic*. NIIGA, Leningrad, **5**, 8–19 (in Russian).

Karasik, A. M. 1974. The Eurasia Basin of the Arctic Ocean from the point of view of plate tectonic. *In*: *Problems in Geology of Polar Areas of the Earth*. Nauchno-Issledovatel'skii Institut Geologii Arktiki, Leningrad, 23–31 (in Russian).

Karasik, A. M., Savostin, L. A. & Zonenshain, L. P. 1983. Parameters of the lithospheric plate movements within Eurasia Basin of North Polar Ocean. *Doklady Academii Nauk SSSR, Earth Sciences Section*, **273**, 1191–1196 (in Russian).

Kogan, L. I., Malovitski, Ya. P. & Murzin, R. R. 2004. Deep structure of the East Barents Megabasin: evidence from wide-angle deep seismic profiling. *Geotectonics*, **38**, 224–238.

Kontorovich, A. E., Nesterov, I. I., Salmanov, F. K., Surkov, V. S., Trofimuk, A. A. & Ervye, Yu. G. 1975. *Geology of Oil and Gas of West Siberia*. Nedra, Moscow (in Russian).

Kontorovich, V. A., Belyaev, S. Yu., Kontorovich, A. E., Krasavchikov, V. A., Kontorovich, A. A. & Suprunenko, O. I. 2001. Tectonic structure and history of development of the East-Siberian geosineclise. *Russian Geology and Geophysics*, **42**, 1832–1845.

Korago, E. A., Kovaleva, G. N., Lopatin, B. G. & Orgo, V. V. 2004. The Precambrian rocks of Novaya Zemlya. *In*: Gee, D. G. & Pease, V. L. (eds) *The Neoproterozoic Timanide Orogen of Eastern Baltica*. Geological Society, London, Memoirs, **30**, 135–143.

Korago, E. A., Kovaleva, G. N. *et al.* 2009. To a problem of continental crust age on west of Eurasian Arctic (based on data from zircon geochronology from Ordovician and Silurian of north-western Novaya Zemlya). *In*: *Geology of Polar Regions of the Earth. Materials of XLII Tectonic Conference, I*. GEOS, Moscow, 285–289 (in Russian).

Kos'ko, M. K. 1984. East Siberian Sea. *In*: Gramberg, I. S. & Pogrebitskii, Yu. E. (eds) *Geological Structure of the USSR and Trends in Distribution of Mineral Resources, 9, Seas of the Soviet Arctic*. Nedra, Leningrad, 60–67 (in Russian).

Kos'ko, M. K. & Nepomiluev, V. F. 1975. Reconstructing Palaeozoic litho-structural zonation of Anzhu Islands. *In*: *Tectonics of the Arctic*. NIIGA, Leningrad, **1**, 26–30 (in Russian).

Kos'ko, M. K., Lopatin, B. G. & Ganelin, V. G. 1990. Major geological features of the islands of the East Siberian and Chukchi Seas and the northern coast of Chukotka. *Marine Geology*, **93**, 349–367.

Kos'ko, M. K., Cecile, M. P., Harrison, J. C., Ganelin, V. G., Khandoshko, N. V. & Lopatin, B. G. 1993. Geology of Wrangel Island, between Chukchi and East Siberian Seas, Northeastern Russia. *Geological Survey of Canada Bulletin*, **461**.

Kos'ko, M. K. 1994. Major tectonic interpretations and constraints for the new Siberian Islands region, Russian Arctic. *In*: *Proceedings International Conference on Arctic Margins*, Anchorage, AK, September 1992. US Department of the Interior, Minerals Management, 195–200.

Kristoffersen, Y. 1990. Eurasia Basin. *In*: Grantz, A., Johnson, G. L. & Sweeney, J. F. (eds) *The Arctic Ocean Region*. The Geology of

North America, L. Geological Society of America, Boulder, CO, 365–378.

Kuzmichev, A. B. 2009. Where does the South Anyui suture go in the New Siberian Islands and Laptev Sea? Implications for the Amerasia basin origin. *Tectonophysics*, **463**, 86–108.

Kuzmichev, A. B. & Pease, V. 2007. Siberian trap magmatism on the New Siberian Islands: constraints for east Arctic Mesozoic plate tectonic reconstructions. *Journal of the Geological Society, London*, **164**, 959–968.

Levshin, A. L., Schweitzer, J., Weidle, C., Shapiro, N. M. & Ritzwoller, M. H. 2007. Surface wave tomography of the Barents Sea and surrounding region. *Geophysical Journal International*, **170**, 441–459.

Lopatin, B. G. (ed.) 2000. *State Geological Map of Russian Federation*. Scale 1,000,000 (New Series). Sheet S-36, 37 – Barents Sea. Explanatory Notes. St Petersburg, P. P. Karpinsky Russian Geological Research Institute.

Lopatin, B. G., Pavlov, L. G., Orgo, V. V. & Shkarubo, S. I. 2001. Tectonic structure of Novaya Zemlya. *Polarforschung*, **69**, 131–135.

Lorenz, H., Gee, D. G. & Bogolepova, O. K. 2006. Early Palaeozoic unconformity on Severnaya Zemlya and Relationships to the Timanian margin of Baltica. *In*: Scott, R. A. & Thurston, D. K. (eds) *Proceedings of the Fourth International Conference on Arctic Margins*, Dartmouth, Nova Scotia, 30 September to 3 October 2003. US Department of the Interior, Anchorage, AK, 14–30.

Lorenz, H., Gee, D. G. & Whitehouse, M. J. 2007. New geochronological data on Palaeozoic igneous activity and deformation in the Severnaya Zemlya Archipelago, Russia, and implications for the development of the Eurasian Arctic margin. *Geological Magazine*, **144**, 105–125; doi: 10.1017/S001675680600272X.

Lorenz, H., Männik, P., Gee, D. G. & Proskurnin, V. 2008. Geology of the Severnaya Zemlya Archipelago and the North Kara Terrane in the Russian high Arctic. *International Journal of Earth Sciences (Geol. Rundsch.)*, **97**, 519–547; doi: 10.1007/s00531-007-0182-2.

Malyshev, N. A. 2002. *Tectonics, Evolution and Petroleum Potential of Sedimentary Basins of the Russian European North*. Uralian Branch of Russian Academy of Sciences, Ekaterinburg (in Russian).

Männik, P., Bogolepova, O. G., Poldvere, A. & Gubanov, A. P. 2009. New data on Ordovician–Silurian conodonts and stratigraphy from the Severnay Zemlya Archipelago, Russian Arctic. *Geological Magazine*, 497–516; doi: 10.1017/S0016756809006372.

Maschenkov, S. P., Glebovsky, V. Yu. & Zayonchek, A. V. 2001. New digital compilation of Russian aeromagnetic and gravity data over the North Eurasian Shelf. *Polarforschung*, **69**, 35–39.

Metelkin, D. V., Vernikovsky, V. A., Kazansky, A. Yu., Bogolepova, O. K. & Gubanov, A. P. 2005. Palaeozoic history of the Kara microcontinent and its relation to Siberia and Baltica: paleomagnetism, paleogeography and tectonics. *Tectonophysics*, **398**, 225–243.

Miller, E. L. & Verzhbitsky, V. 2009. Structural studies in the Pevek region, Russia: implications for the evolution of the East Siberian Shelf and Makarov Basin of the Arctic Ocean. *In*: Stone, D. B., Fujita, K., Layer, P. W., Miller, E. L., Prokopiev, A. V. & Toro, J. (eds) *Geology, Geophysics and Tectonics of Northeastern Russia: A Tribute to Leonid Parfenov*. Stephan Mueller, Special Publications Series, **4**, 223–241.

Miller, E. L., Gehrels, G. E., Pease, V. & Sokolov, S. 2010. Stratigraphy and U-Pb detrital zircon geochronology of Wrangel Island, Russia: implications for Arctic paleogeography. *AAPG Bulletin*, **94**, 665–692.

Moore, T. E., Dumitru, T. A., Adams, K. E., Witebsky, S. N. & Harris, A. G. 2002. Origin of the Lisburne Hills–Herald Arch–Wrangel Arch structural belt: stratigraphic, structural, and fission-track evidence from the Cape Lisburne area, northwestern Alaska. *In*: Miller, E. L., Grantz, A. & Klemperer, S. L. (eds) *Tectonic Evolution of the Bering Shelf–Chukchi Sea–Arctic Margin and Adjacent Landmasses*. Geological Society of America, Boulder, CO, Special Papers, **360**, 77–109.

Moran, K., Backman, J. *et al.* 2006. The Cenozoic palaeoenvironment of the Arctic Ocean. *Nature*, **441**; doi: 10.1038/nature04800.

Nikishin, A. M., Ziegler, P. A. *et al.* 1996. Late Precambrian to Triassic history of the East-European Craton: dynamics of sedimentary basin evolution. *Tectonophysics*, **268**, 23–63.

Nikishin, A. M., Ziegler, P. A., Abbott, D., Brunet, M.-F. & Cloetingh, S. 2002. Permo-Triassic intraplate magmatism and rifting in Eurasia: implications for mantle dynamics. *Tectonophysics*, **351**, 3–39.

Obruchev, S. V. 1934. Materials on tectonics of Northeastern Asia. *Aspects of Soviet Geology*, **6**, 182–200, 77–78 (in Russian).

Ostisty, B. K. & Fedorovsky, Y. F. 1993. Main results of oil and gas prospecting in the Barents and Kara Seas inspire optimism. *In*: Vorren, T. O., Bergsaker, E. *et al.* (eds) *Arctic Geology and Petroleum Potential*, Norwegian Petroleum Society Special Publication 2. Elsevier, Amsterdam, 243–252.

Parfenov, L. M. & Natal'in, B. A. 1986. Mesozoic tectonic evolution of northeastern Asia. *Tectonophysics*, **127**, 291–304.

Patyk-Kara, N. G. & Laukhin, S. A. 1986. Cenozoic evolution of the Arctic coastal relief of North-Eastern Asia. *Sovetskaya Geologiya*, **1**, 75–84 (in Russian).

Pavlov, S. P., Shlykova, V. V. & Grigoryeva, B. M. 2008. HC potential of Palaeozoic deposits of eastern flank of North Barents depression. *In*: *Proceedings of the International Conference 'Oil and Gas of the Arctic Shelf'*, Murmansk, 12–14 November (CDROM).

Pease, V. 2011. Eurasian orogens and Arctic tectonics: an overview. *In*: Spencer, A. M., Gautier, D., Stoupakova, A., Embry, A. & Sørensen, K. (eds) *Arctic Petroleum Geology*. Geological Society, London, Memoirs (in press).

Pease, V. & Scott, R. A. 2009. Crustal affinities in the Arctic Uralides, northern Russia: significance of detrital zircon ages from Neoproterozoic and Paleozoic sediments in Novaya Zemlya and Taimyr. *Journal of the Geological Society, London*, **166**, 517–527; doi: 10.1144/0016-76492008-093.

Petrov, O. V., Morozov, A. F., Strelnikov, S. I., Ivanov, V. L., Kaminsky, V. D. & Pogrebitskiy, Yu. E. (eds) 2008. *Geological Map of Russian Federation and Adjoining Seas*. Scale 1:2,500,000. Ministry of Natural Resources of the Russian Federation, Federal Agency for Mineral Resources, All-Russia Geological Research Institute, All-Russia Research Institute of Geology and Mineral Resources of World Ocean. 12 sheets.

Petrovskaya, N. A., Trishkina, C. V. & Savishkina, M. A. 2008. Main geological features of the Russian Chukchi Sea. *Oil and Gas Geology*, **6**, 20–28 (in Russian).

Pitman, W. C. III & Talwani, M. 1972. Sea-floor spreading in the North Atlantic. *Geological Society of America Bulletin*, **83**, 619–649.

Pogrebitskiy, Yu. E. (ed.) 2004. *Novaya Zemlya and Vaigach Island. Geological Structure and Miragenia*. NIIGA-VNIIOkeangeologiya, St Petersburg, **205** (in Russian).

Pol'kin, Ya. I. 1984. Chukchi Sea. *In*: Gramberg, I. S. & Pogrebitskiy, Yu. E. (eds) *Geological Structure of the USSR and Trends in Distribution of Mineral Resources, 9, Seas of the Soviet Arctic*. Nedra, Leningrad, 67–77 (in Russian).

Proskurnin, V. F. 1995. New volcano-plutonic association of Severnaya Zemlya and the features of its metallogeny. *In*: Samojlov, A. G. (ed.) *Taimyr's Subsoil. Collection of the Papers*. VSEGEI, Taimyrkomprirodresursy, Norilsk, 93–100 (in Russian).

Proskurnin, V. F. 1999. On angle unconformities in upper Pre-Cambrian and lower Palaeozoic of the Severnaya Zemlya Archipelago. *In*: Simonov, O. N. (ed.) *Taimyr's Subsoil. Collection of the Papers*. VSEGEI, Taimyrkomprirodresursy, Norilsk, 68–76 (in Russian).

Pushcharovskii, Yu. M. (ed.) 1963. *Tectonic Map of the Arctic*. Scale 1:10 000 000. Geological Institute of USSR Academy of Sciences, Moscow (in Russian).

Ritzmann, O. & Faleide, J. I. 2009. The crust and mantle lithosphere in the Barents Sea/Kara Sea region. *Tectonophysics*, **470**, 89–104.

Roeser, H. A., Block, M., Hinz, K. & Reichert, C. 1995. *Marine Geophysical Investigations in the Laptev Sea and the Western Part of the East Siberian Sea*. Alfred Wegener Institute for Polar and Marine Research, Reports on Polar Research, **176**, 367–377.

Rusakov, I. M. & Vinogradov, V. A. 1969. Eugeosynclinal and miogeosynclinal areas of North-East of USSR. *In*: *Proceedings of NIIGA, Regional Geology*, **15**, 5–27 (in Russian).

Savostin, L. A. & Drachev, S. S. 1988. The Cenozoic compression in New Siberian Islands region and its relationship with Eurasia basin opening. *Okeanologiya*, **28**, 775–782 (in Russian).

Savostin, L. A. & Karasik, A. M. 1981. Recent plate tectonics of the Arctic basin and of northeastern Asia. *Tectonophysics*, **74**, 111–145.

Savostin, L. A., Karasik, A. M. & Zonenshain, L. P. 1984*a*. The history of the opening of the Eurasia basin in the Arctic. *Doklady Earth Sciences*, **275**, 79–83.

Savostin, L. A., Natapov, L. M. & Stavsky, A. P. 1984b. Mesozoic palaeogeodynamics and palaeogeography of the Arctic region. *In*: *27th International Geological Congress (Moscow), Arctic Geology. Reports*, **4**. Nauka, Moscow, 217–237.

Scott, R. A., Howard, J. P., Guo, L., Schekoldin, R. & Pease, V. 2010. Offset and curvature of the Novaya Zemla fold-and-thrust belt, Arctic Russia. *In*: Vining, B. A. & Pickering, S. C. (eds) *Petroleum Geology: From Mature Basins to New Frontiers – Proceedings of the 7th Petroleum Geology Conference*. Geological Society, London, 645–657; doi: 10.1144/0070645.

Sekretov, S. B. 2000. Petroleum potential of the Laptev Sea basins: geological, tectonic and geodynamic factors. *In*: Roland, N. W. & Tessensohn, F. (eds) *III International Conference on Arctic Margins. Polarforschung*, **68**, 179–186.

Sekretov, S. B. 2001. Northwestern margin of the East Siberian Sea, Russian Arctic: seismic stratigraphy, structure of the sedimentary cover and some remarks on the tectonic history. *Tectonophysics*, **339**, 353–371.

Sharov, N. V., Mitrofanov, F. P., Verba, M. L. & Gillen, K. (eds) 2005. *Lithospheric Structure of the Russian Barents Sea Region*. Karelian Research Centre of Russian Academy of Sciences, Petrozavodsk (in Russian).

Shatskii, N. S. 1935. On tectonics of the Arctic. *Geology and Mineral Resources of the USSR's North*. Glavsevmorput', Leningrad, 149–165 (in Russian).

Sherwood, K. W., Johnson, P. P. *et al.* 2002. Structure and stratigraphy of the Hanna Trough, U.S. Chukchi Shelf, Alaska. *In*: Miller, E. L., Grantz, A. & Klemperer, S. L. (eds) *Tectonic Evolution of the Bering Shelf–Chukchi Sea-Arctic Margin and Adjacent Landmasses*. Geological Society of America, Boulder, CO, Special Papers, **360**, 39–66.

Shipilov, E. V. & Tarasov, G. A. 1998. *Regional Geology of Oil and Gas Bearing Sedimentary Basins of the Russian West-Arctic Shelf*. Kola Science Centre of Russian Academy of Sciences, Apatity (in Russian).

Shipilov, E. V. & Vernikovsky, V. A. 2010. The Svalbard–Kara plates junction: structure and geodynamic history. *Russian Geology and Geophysics*, **51**, 58–71.

Sokolov, S. D., Bondarenko, G. Y., Morozov, O. L., Shekhovtsov, V. A., Glotov, S. P., Ganelin, A. V. & Kravchenko-Berezhnoy, I. R. 2002. South Anyui suture, northeast Arctic Russia: facts and problems. *In*: Miller, E. L., Grantz, A. & Klemperer, S. L. (eds) *Tectonic Evolution of the Bering Shelf–Chukchi Sea-Arctic Margin and Adjacent Landmasses*. Geological Society of America, Boulder, CO, Special Papers, **360**, 209–224.

Spektor, V. B., Andrusenko, A. M., Dunko, E. A. & Kareva, N. F. 1981. Continuation of the South Anyui suture in the Primorya lowlands. *Doklady Academii Nauk SSSR, Earth Science Section*, **260**, 1447–1450 (in Russian).

Srivastava, S. P. 1985. Evolution of the Eurasian Basin and its implications to the motion of Greenland along Nares Strait. *Tectonophysics*, **113**, 29–53.

Stein, R. 2008. *Arctic Ocean Sediments: Processes, Proxies, and Palaeoenvironment. Developments in Marine Geology*, **2**. Elsevier, Amsterdam.

Sullivan, M. A., Creaney, S. *et al.* 2007. Hydrocarbon system framework of Russia: Pre-Cambrian to present-day. *In*: *International AAPG Conference Athens Greece*, 19–27 November. Abstracts.

Suprunenko, O. I. & Kos'ko, M. K. 2005. Russian Arctic shelf sedimentary basins. *Developments in Sedimentology*, **57**, 237–271.

Surkov, V. S. & Zhero, O. G. 1981. *Basement and Development of the Platform Cover of the West Siberian Platform*. Nedra, Moscow (in Russian).

Surkov, V. S., Kazakov, A. M., Devyatkov, V. P. & Smirnov, V. L. 1997. Lower-Middle Triassic rift sequence of the West Siberian basin. *Otechestvennaya Geologiya*, **3**, 31–37 (in Russian).

Thurston, D. K. & Theiss, L. A. 1987. *Geologic Report for the Chukchi Sea Planning Area, Alaska*. Minerals Management Service Outer Continental Shelf Report 87-0046.

Tolson, R. B. 1987. Structure and stratigraphy of the Hope Basin, Southern Chukchi Sea, Alaska. *In*: Scholl, D. W., Grantz, A. & Vedder, J. G. (eds) *Geology and Resource Potential of the Continental Margin of Western North America and Adjacent Ocean Basins; Beaufort Sea to Baja California*. Earth Sciences Series. Circum-Pacific Council for Energy and Mineral Resources, Houston, TX, **6**, 59–72.

Trettin, H. P. (ed.) 1991. *Geology of the Innuitian Orogen and Arctic Platform of Canada and Greenland*. The Geology of North America, E. Geological Society of America, Boulder, CO.

Ulmishek, G. 1982. *Petroleum Geology and Resource Assessment of the Timan–Pechora Basin, USSR, and the Adjacent Barents–Northern Kara Shelf*. Argonne National Laboratory, Report EES-TM-199.

Ustritsky, V. I. 1985. On the relationship between the Urals, Pay–Khoy, Novaya Zemlya and Taymyr. *Geotectonics*, **538**, 51–61.

Ustritsky, V. I. 1989. On the tectonic origin of the Barents–North Kara Megabasin. *In*: *Problems of Petroleum Geology of the World Ocean. Moscow*. Nedra, Leningrad, 182–191.

Verba, M. L., Daragan-Sushchova, L. A. & Pavlenkin, A. D. 1992. Riftogenic structures of the western Arctic shelf investigated by refraction studies. *International Geology Review*, **34**, 753–764.

Verhoef, J., Walter, R. R., Macnab, R. & Arkani-Hamed, J. Members of the Project Team 1996. *Magnetic Anomalies of the Arctic and North Atlantic Oceans and Adjacent Land Areas*. Geological Survey of Canada. Open File 3125a.

Vernikovsky, V. 1995. Riphean and Paleozoic metamorphic complexes of the Taimyr foldbelt: condition of formation. *Petrol*, **3**, 55–72.

Vernikovsky, V. A. 1996. The geodynamic evolution of the Taimyr folded area. *Geology of the Pacific Ocean*, **12**, 691–704.

Vernikovsky, V. & Vernikovskaya, A. 2001. Central Taimyr accretionary belt (Arctic Asia): Meso-Neoproterozoic tectonic evolution and Rodinia breakup. *Precambrian Research*, **110**, 127–141.

Verzhbitsky, V., Frantzen, E. *et al.* 2008. New seismic data on the South and North Chukchi sedimentary basins and the Wrangel Arch and their significance for the geology of Chukchi Sea shelf (Russian Arctic). *In*: *Proceedings of the EAGE 3rd St. Petersburg International Conference & Exhibition*, Saint Petersburg.

Vinogradov, V. A. 1984. Laptev Sea. *In*: Gramberg, I. S. & Pogrebitskiy, Yu. E. (eds) *Geological Structure of the USSR and Trends in Distribution of Mineral Resources, 9, Seas of the Soviet Arctic*. Nedra, Leningrad, 51–60 (in Russian).

Vinogradov, V. A. & Drachev, S. S. 2000. Southwestern shelf of the Laptev Sea and tectonic nature of its basement. *Doklady Earth Sciences*, **372**, 601–603.

Vinogradov, V. A., Gaponenko, G. I., Rusakov, I. M. & Shimaraev, V. N. 1974. *Tectonics of the Eastern Arctic Shelves of the USSR*. Nedra, Leningrad (in Russian).

Vinogradov, V. A., Kameneva, G. I. & Yavshits, G. P. 1975. On Hyperborean Platform in light of new data on geology of the Henrietta Island. *In*: *Tectonics of Arctic*, **1**. NIIGA, Leningrad, 21–25 (in Russian).

Vinogradov, V. A., Gaponenko, G. I., Gramberg, I. S. & Shimaraev, V. N. 1977. Structural-associational complexes of the Arctic shelf of eastern Siberia. *International Geology Review*, **19**, 1331–1343.

Vinokurov, I. Yu., Belyaev, V. I., Egorov, A. S., Kalenich, A. P., Matveev, Yu. I., Prudnikov, A. N. & Roslov, Yu. V. 2009. Deep-crustal model and evolution of Barents-Kara region. *In*: *Geology of Polar Regions of the Earth. Materials of XLII Tectonic Conference*, **I**. GEOS, Moscow, 102–106 (in Russian).

Vogt, P. R., Kovacs, L. C., Johnson, G. L. & Feden, R. H. 1979. The evolution of the Arctic Ocean with emphasis on the Eurasia Basin. *In*: *Proceedings of Norwegian Sea Symposium*. Norwegian Petroleum Society, Oslo, 1–29.

Vol'nov, D. A., Voitsekhovsky, V. N., Ivanov, O. A., Sorokov, D. S. & Yashin, D. S. 1970. New Siberian Islands. *In*: Tkachenko, B. V. & Egiazarov, B. K. (eds) *Geology of USSR*, **26**. Nedra, Moscow, 224–274 (in Russian).

Vyssotski, A. V., Vyssotski, V. N. & Nezhdanov, A. A. 2006. Evolution of the West Siberian Basin. *Marine and Petroleum Geology*, **23**, 93–126.

Walderhaug, H. J., Eide, E. A., Scott, R. A., Inger, S. & Golionko, E. G. 2005. Palaeomagnetism and $^{40}Ar/^{39}Ar$ geochronology from the South Taimyr igneous complex, Arctic Russia: a Middle–Late Triassic magmatic pulse after Siberian flood-basalt volcanism. *Geophysical Journal International*, **163**, 501–517; doi: 10.1111/j.1365-246X.2005.02741.x.

Warren, T., Sherwood, K. W. *et al.* 1995. Petroleum exploration opportunities on the US-Russia Chukchi Sea continental shelf. *In*: *Proceedings of the Fifth International Offshore and Polar Engineering Conference*, The Hague, June 11–16, 493–500.

Wilson, B. M. & Lyashkevich, Z. M. 1996. Magmatism and the geodynamics of rifting of the Pripyat-Dnieper-Donets Rift, East European Platform. *Tectonophysics*, **268**, 65–81.

Wilson, M., Wijbrans, J., Fokin, P. A., Nikishin, A. M., Gorbachev, V. I. & Nazarevich, B. P. 1999. $^{40}Ar/^{39}Ar$ dating, geochemistry and tectonic setting of Early Carboniferous dolerite sills in the Pechora basin, foreland of the Polar Urals. *Tectonophysics*, **313**, 107–118.

Zonenshain, L. P., Kuz'min, M. I. & Natapov, L. M. 1990. *Geology of the USSR: A Plate-Tectonic Synthesis*. Page, B. M. (ed.) Geodynamics Series, **21**. American Geophysical Union, Washington, DC.

Assessment of undiscovered petroleum resources of the north and east margins of the Siberian craton north of the Arctic Circle

T. R. KLETT, C. J. WANDREY and J. K. PITMAN

US Geological Survey, Denver Federal Center, MS 939, Box 25046, Denver, CO 80225 USA
(e-mail: tklett@usgs.gov)

Abstract: The Siberian craton consists of crystalline rocks and superimposed Precambrian sedimentary rocks deposited in rift basins. Palaeozoic rocks, mainly carbonates, were deposited along the margins of the craton to form an outwardly younger concentric pattern that underlies an outward-thickening Mesozoic sedimentary section. The north and east margins of the Siberian craton subsequently became foreland basins created by compressional deformation during collision with other tectonic plates. The Tunguska Basin developed as a Palaeozoic rift/sag basin over Proterozoic rifts.

The geological provinces along the north and east margins of the Siberian craton are immature with respect to exploration, so exploration-history analysis alone cannot be used for assessing undiscovered petroleum resources. Therefore, other areas from around the world having greater petroleum exploration maturity and similar geological characteristics, and which have been previously assessed, were used as analogues to aid in this assessment. The analogues included those of foreland basins and rift/sag basins that were later subjected to compression. The US Geological Survey estimated the mean undiscovered, technically recoverable conventional petroleum resources to be c. 28 billion barrels of oil equivalent, including c. 8 billion barrels of crude oil, 103 trillion cubic feet of natural gas, and 3 billion barrels of natural gas liquids.

Keywords: Siberian craton, undiscovered petroleum resources, geological analogues, assessment methodology

This study describes the method used by the US Geological Survey (USGS) to assess undiscovered, technically recoverable conventional petroleum resources of the Siberian craton. The assessed areas are north of the Arctic Circle. Areas south of the Arctic Circle were previously assessed by the USGS (Ulmishek 2001a, b) and will not be addressed in this paper. Petroleum occurrences in the Siberian craton have been studied for many years, and numerous papers have been written on the subject. This study synthesizes results of previous studies of the Siberian craton and recasts the geological models on the basis of petroleum system elements.

The Siberian craton is located in the Russian Federation between longitude 86° and 138°E and between latitude 52° and 73°N. The craton, surrounded by fold and thrust belts (Fig. 1), covers c. 4 500 000 km², extending north and south of the Arctic Circle. Most of the craton is located south of the Arctic Circle.

Geological provinces

The Siberian craton consists of continental blocks that were originally separated by ocean basins but had accreted during early Proterozoic time (Rosen *et al.* 1994; Smelov & Timofeev 2007). Various tectonic structures that developed within and among the accreted blocks and along the margins of the Siberian craton separate the area into individual geological provinces (Fig. 1). Geological provinces are defined by the USGS as spatial entities with common geological attributes. Attributes might include a single dominant structural element such as a basin or a fold belt, or a number of contiguous related elements (Klett *et al.* 1997). The geological provinces along the north and east margins of the Siberian craton and north of the Arctic Circle that were delineated for this study are (from west to east): (1) Tunguska Basin (includes Turukhansk–Norilsk Folded Zone and part of the Turukhansk–Igarka Uplift); (2) Yenisey–Khatanga Basin (includes Khatanga Saddle); (3) Lena–Anabar Basin; and (4) Lena–Vilyui Basin (northern part only) (Fig. 1). Some of these provinces extend south of the Arctic Circle.

The Siberian craton consists of crystalline rocks and superimposed Proterozoic sedimentary rocks deposited in intracratonic rifts (Ulmishek 2001a). Palaeozoic rocks, mainly carbonates, were deposited along the margins of the craton and are overlain by a progressively outward-thickening Mesozoic section (Persits *et al.* 1998). The north and east margins subsequently became foreland basins created by compressional deformation during collision with other tectonic plates during the Mesozoic (Zonenshain *et al.* 1990). The Tunguska Basin developed as a Palaeozoic rift/sag basin over Proterozoic rifts. The main styles of the basins are primarily Proterozoic to Palaeozoic extensional rift/sag, rifted passive margin and foreland basins that were later deformed by compression, uplift and tilting (Clarke 1991).

Geological models of petroleum occurrence

The sedimentary successions of the geological provinces along the north and east margins of the Siberian craton are similar because of their proximity and similar depositional histories (compare cross-sections in Fig. 3). Known and inferred petroleum sources include organic-rich mudstone in the Proterozoic, Cambrian, Permian, Triassic and Jurassic stratigraphic sections (Bakhturov 1985; Bakhturov *et al.* 1990; Kontorovich *et al.* 1991; Shenfil' 1991; Kuznetsov 1997; Shishkin & Isaev 1999). Reservoirs include Proterozoic, Permian, Triassic, Jurassic and Lower Cretaceous sandstones and seals are intraformational mudstones (Kontorovich *et al.* 1991; Kuznetsov 1997). Traps include compressional structures, updip stratigraphic pinchouts, progradational systems, and karsted platform and reef-related carbonates (Kontorovich *et al.* 1991). Based on petroleum-generation modelling, major petroleum generation and migration probably occurred during the Mesozoic.

Numerous discoveries of crude oil and natural gas were made along the margins of the craton (Fig. 1; IHS Energy 2007). In the Tunguska Basin Province, seven gas and condensate fields have

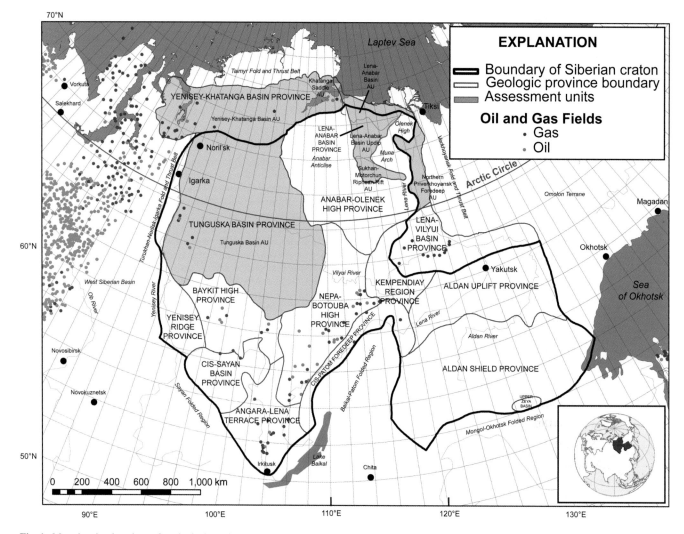

Fig. 1. Map showing locations of geological provinces and assessment units. Data from Persits & Ulmishek (2003); Persits *et al.* (1998); Environmental Systems Research Institute Inc. (1999); and IHS Energy (2007).

been discovered in Proterozoic rocks, but these fields are south of the Arctic Circle. Several gas and condensate fields were discovered north of the craton in the Yenisey–Khatanga Basin Province. Petroleum in these fields is in upper Palaeozoic and Mesozoic clastic rocks (Grebenyuk *et al.* 1988; Kontorovich *et al.* 1991). No fields have been discovered in the Lena–Anabar Basin Province, but some wells indicate the presence of petroleum (Grausman 1996; Sekretov 2003). In the Lena–Vilyui Basin Province south of the Arctic Circle, fields contain mostly natural gas and condensates in upper Palaeozoic and Mesozoic clastic rocks (Vagin *et al.* 1987; Sokolov 1989). Extensive degraded petroleum (bitumen, not assessed in this study) deposits crop out along the margins of the Anabar–Olenek High (Ivanov 1979), but no fields have been discovered on the high.

Assessment units

Assessment units (AUs), which are mappable volumes of rock that are sufficiently homogeneous so that the methodology of resource assessment is applicable (Klett *et al.* 1997), were delineated in each of the geological provinces of the northern and eastern Siberian craton (Fig. 1). One AU, Tunguska Basin, was delineated in the Tunguska Basin Province, but it was not quantitatively assessed. Two AUs, Yenisey–Khatanga Basin and Khatanga Saddle, were delineated in the Yenisey–Khatanga Basin Province. Three AUs, Lena–Anabar Basin, Lena–Anabar Basin Updip and Sukhan–Motorchun Riphean Rift, were delineated in the Lena–Anabar Basin Province; however, the Lena–Anabar Basin Updip and Sukhan–Motorchun Riphean Rift AUs were not quantitatively assessed. One AU, Northern Priverkhoyansk Foredeep, was delineated in the Lena–Vilyui Basin Province north of the Arctic Circle.

Geological analogues for assessment

The required input for the assessment of undiscovered petroleum resources is primarily estimates of the number and sizes of undiscovered fields. In areas that are mature with respect to exploration, the input can be estimated from exploration-history analysis. However, the geological provinces along the north and east margins of the Siberian craton are immature with respect to exploration. Therefore, other areas from around the world having greater petroleum exploration maturity and similar geological characteristics, and that have been previously assessed, were used as analogues to aid in this assessment.

Estimates of undiscovered petroleum resources were calculated by statistically combining the number, sizes in terms of recoverable quantities including reserve growth, and geological risk (as AU probability) of undiscovered oil and gas fields. The statistical combination uses Monte Carlo simulation with 50 000 iterations. Calculated resource quantities and the largest field size selected during each iteration are recorded. The results are given as probability distributions. The calculated largest field size distribution is used to calibrate estimates of the practical largest undiscovered field size.

Table 1. Field densities, median oil and gas field sizes, and exploration maturities of geological analogues used for the assessment of the Khatanga Saddle, Lena–Anabar Basin, and Northern Priverkhoyansk Foredeep AUs

Province name (AU number)	Area (km²)	Field density (no. discovered per 1000 km²)	Field density (no. discovered + undiscovered per 1000 km²)	Median field size, fields ≥50 MMBOE	Largest discovered field size (binned MMBOE)	Exploration maturity (percentage petroleum discovered per total assessed)
Provinces with foreland basin AUs						
Middle Caspian Basin (11090101)	36 452	0.38	1.15	116	2826	61
Amu–Darya Basin (11540101)	156 543	0.12	0.13	111	2826	94
Amu–Darya Basin (11540102)	58 199	0.17	0.29	110	11 305	97
Amu–Darya Basin (11540103)	180 786	0.08	0.09	149	2826	96
Rub Al Khali Basin (20190101)	172 931	0.09	0.18	204	22 610	90
Rub Al Khali Basin (20190102)	57 336	0.54	0.80	175	22 610	95
Rub Al Khali Basin (20190103)	40 737	0.27	0.55	109	707	54
Zagros Fold Belt (20300101)	655 783	0.20	0.57	162	45 220	79
Zagros Fold Belt (20300102)	350 545	0.19	0.59	126	22 610	83
Zagros Fold Belt (20300201)	30 409	0.30	1.55	213	5652	57
Junggar Basin (31150101)	66 347	0.11	0.17	121	5652	98
Sichuan Basin (31420101)	99 141	0.07	0.12	98	353	89
Tarim Basin (31540101)	453 119	0.02	0.13	112	353	14
North Carpathian Basin (40470101)	40 958	0.22	0.27	119	353	99
North Carpathian Basin (40470201)	57 909	0.09	0.10	77	707	98
San Jorge Basin (60580102)	14 185	0.07	0.29	124	177	99
Middle Magdalena Basin (60900101)	6178	1.13	1.25	177	707	99
Middle Magdalena Basin (60900102)	5504	0.73	0.92	115	177	91
Llanos Basin (60960101)	24 630	0.24	0.44	123	707	58
Llanos Basin (60960102)	92 074	0.09	0.22	147	1413	66
East Venezuela Basin (60980101)	35 724	0.64	1.38	187	2826	73
East Venezuela Basin (60980102)	40 319	0.07	0.46	90	353	16
Maracaibo Basin (60990102)	13 549	0.44	0.60	112	353	90
Greater Antilles Deformed Belt (61170101)	138 853	0.02	0.03	102	177	67
Median		0.18	0.37	120	874	89
Mean		0.26	0.51	132	9626	78
Provinces with AUs having rifted passive margins and foreland basin components						
Timan–Pechora Basin (11090102)	203 079	0.26	0.44	115	1413	74
Timan–Pechora Basin (11090103)	86 305	0.01	0.23	103	2826	53
Volga–Ural Basin (10150101)	471 793	0.25	0.29	114	11 305	100
Volga–Ural Basin (10150102)	95 001	0.14	0.14	116	353	100
Middle Caspian Basin (11090101)	36 452	0.38	1.15	116	2826	61
Amu–Darya Basin (11540101)	156 543	0.12	0.13	111	2826	94
Nepa–Botuba Arch (12100101)	40 737	0.02	0.07	117	11 305	76
Ma'Rib–Al Jawf/Masila Basin (20040101)	90 789	0.20	0.41	100	707	66
Fahud Salt Basin (20160201)	17 281	0.23	0.28	223	1413	92
Zagros Fold Belt (20300101)	655 783	0.20	0.57	162	45 220	79
Zagros Fold Belt (20300102)	350 545	0.19	0.59	126	22 610	83

(*Continued*)

Table 1. *Continued*

Province name (AU number)	Area (km^2)	Field density (no. discovered per 1000 km^2)	Field density (no. discovered + undiscovered per 1000 km^2)	Median field size, fields ≥50 MMBOE	Largest discovered field size (binned MMBOE)	Exploration maturity (percentage petroleum discovered per total assessed)
Pelagian Basin (20480101)	93 016	0.12	0.19	157	707	94
Pelagian Basin (20480201)	180 311	0.01	0.02	63	177	66
Tarim Basin (31540101)	453 119	0.02	0.13	112	353	14
Bombay (80430101)	180 720	0.06	0.11	106	2826	82
Median		0.14	0.23	115	1666	79
Mean		0.15	0.32	123	8007	76
Combined analogues (duplicates removed)						
Median		0.15	0.29	116	1025	86
Mean		0.22	0.43	129	5810	78
Maximum		1.13	1.55	223	45 220	100

Three elements of geological risk are incorporated in the AU probability: charge, adequate reservoir and seal rocks, and timing of maturation and trap formation. The AU probability is the probability of the existence of one field anywhere in the AU exceeding a minimum size of 50 million barrels of oil equivalent (MMBOE). AUs having a probability of 0.10 or less were not quantitatively assessed.

The numbers of undiscovered fields in an AU were estimated by comparing field densities (estimated number of undiscovered fields plus number of discovered fields, exceeding 50 MMBOE per 1000 km^2) of the analogue datasets. The density of discovered fields, which is less than the density of both discovered and undiscovered fields, was used to calibrate the densities of the undiscovered fields.

Sizes of undiscovered fields were estimated by comparing the median and maximum sizes of the analogue datasets (discovered and undiscovered fields exceeding 50 MMBOE). Minimum sizes are always 50 MMBOE, which is 50 million barrels (MMB) of crude oil or 300 billion cubic feet (BCF) of natural gas (6 BCF equals 1 MMBOE). Petroleum composition and properties of undiscovered fields estimated for an AU are based on fields both discovered in the AUs and in the analogue datasets and on global statistics.

For the assessment of the Khatanga Saddle, Lena–Anabar Basin and Northern Priverkhoyansk Foredeep AUs, two analogue datasets (Table 1) were chosen from the USGS Analogue Database (Charpentier *et al.* 2007): (1) foreland basins (24 analogues) and (2) rifted passive margins and foreland basins with mixed clastic and carbonate depositional systems (15 analogues). The analogues were used to estimate the numbers and sizes of undiscovered fields (Fig. 2). Both analogue sets have discovered fields greater than the minimum size defined for this assessment (50 MMBOE). The analogue categories include extensional and compressional structures and traps having carbonate and clastic depositional systems. Four AUs are common to both analogue sets.

Two different analogue datasets (Table 2) were used to estimate the number of undiscovered fields for the Yenisey–Khatanga Basin AU: (1) rift/sag basins that were subsequently compressed (seven analogues); and (2) basins with slope, clinoforms and turbidites depositional systems (20 analogues) (Fig. 3). The analogue sets contain primarily clastic reservoir rocks and have discovered fields of 50 MMBOE or greater.

Khatanga Saddle, Lena–Anabar Basin, and Northern Priverkhoyansk Foredeep Assessment Units

The median and maximum densities of discovered fields in the combined analogue datasets are 0.2 and 1.1 fields per 1000 km^2, respectively, whereas the median and maximum densities of discovered plus undiscovered fields are 0.3 and 1.6 fields per 1000 km^2, respectively. The median undiscovered oil field size of the combined analogue datasets is 116 MMBOE (Fig. 2) and the median of maximum undiscovered field size is 1000 MMBOE.

Khatanga Saddle Assessment Unit

The area of the Khatanga Saddle AU is *c.* 45 000 km^2 and is located entirely north of the Arctic Circle (Fig. 1). The likelihood that the Khatanga Saddle AU contains at least one field equal to or greater than the minimum size of 50 MMBOE is estimated at 50% (0.50). Input distributions and AU probabilities for this AU are shown in Figure 4.

Densities of 0.02, 0.22 and 1.1 (minimum, median and maximum, respectively) were estimated for numbers of fields per 1000 km^2 in the assessment of the Khatanga Saddle AU. The total minimum, median and maximum numbers of undiscovered fields are estimated at 1, 10 and 50, respectively (Fig. 4). The estimated number of undiscovered oil fields is 0 (minimum), 4 (median) and 40 (maximum), and the estimated numbers of undiscovered gas fields are 1 (minimum), 6 (median) and 45 (maximum), assuming oil–gas mixtures of 0.1 (minimum), 0.4 (mode) and 0.8 (maximum).

The median undiscovered field sizes in the AU are 100 MMB of crude oil and 400 BCF of natural gas. The median undiscovered oil field size (100 MMBOE) is slightly less than the median field size of the analogue dataset (116 MMBOE; Fig. 2). The low-probability maximum field size of 1000 MMBOE equals the median size of the maximum values of the combined analogue datasets (1000 MMBOE). On an oil-equivalent basis, the median and maximum (4000 BCF) sizes of undiscovered gas fields are smaller than undiscovered oil fields because of a greater risk for preservation. The mean of the largest undiscovered field sizes (called the 'expected maximum size') is chosen based on the analogue dataset and matches those provided by the Monte Carlo simulation to calculate resource quantities, *c.* 200 million barrels (MMB) for undiscovered oil fields and *c.* 900 BCF, or 150 MMBOE, for

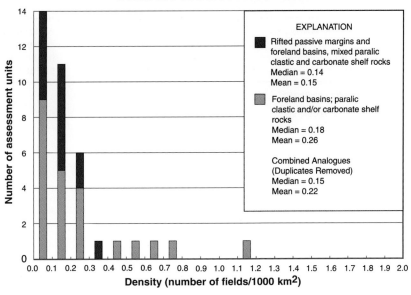
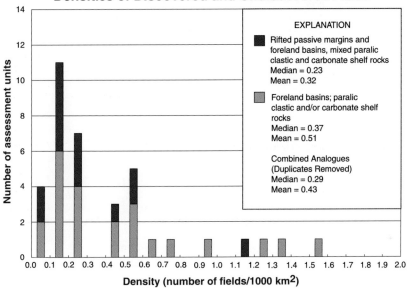
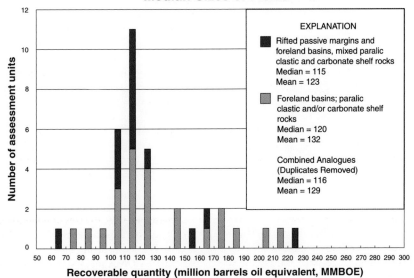

Fig. 2. Densities and median sizes of fields in the geological analogues used for the assessment of the Khatanga Saddle, Lena–Anabar Basin and Northern Priverkhoyansk Foredeep AUs. Four AUs are common to both analogue sets. Data from Charpentier *et al.* (2007).

Table 2. Field densities, median oil and gas field sizes, and exploration maturities of geological analogues used for the assessment of the Yenisey–Khatanga Basin AUs

Province name (AU number)	Area (km^2)	Field density (no. discovered per 1000 km^2)	Field density (no. discovered + undiscovered per 1000 km^2)	Median field size, fields ≥50 MMBOE	Largest discovered field size (binned MMBOE)	Exploration maturity (percentage petroleum discovered per total assessed)
Provinces with rift/sag basin AUs that were later compressed						
Middle Caspian Basin (11090201)	80 295	0.09	0.15	142	2826	95
North Ustyurt Basin (11500101)	24 898	0.16	0.29	251	1413	82
North Ustyurt Basin (11500201)	141 758	0.01	0.04	105	177	84
West Siberia Basin (11740101)	999 936	0.15	0.32	132	22 610	79
West Siberia Basin (11740301)	423 539	0.15	0.35	224	45 220	92
Malay Basin (37030101)	72 558	0.39	0.64	108	707	69
Malay Basin (37030201)	32 955	0.36	0.51	186	707	94
Median		0.15	0.32	142	1268	84
Mean		0.19	0.33	164	13 749	85
Provinces with AUs having slope, clinoforms, and turbidites depositional systems						
Baram Delta/Brunei–Sabah (37010102)	54 084	0.09	0.28	100	353	39
Bonaparte Gulf Basin (39100201)	72 634	0.03	0.14	95	353	8
Bonaparte Gulf Basin (39100201)	146 975	0.03	0.14	112	1413	68
Browse Basin (39130101)	133 150	0.02	0.14	125	1413	48
Northwest Shelf (39480101)	50 436	0.40	1.53	108	2826	56
Northwest Shelf (39480201)	309 715	0.00	0.02	111	707	41
Pannonian Basin (40480101)	99 022	0.11	0.09	102	353	98
Pannonian Basin (40480201)	54 076	0.15	0.13	95	177	99
Po Basin (40600101)	75 427	0.21	0.35	110	707	76
Sergipe–Alagoas Basin (60290102)	23 320	0.09	0.51	105	353	5
Campos Basin (60350101)	40 408	0.74	1.74	178	2826	55
Santa Cruz–Tarija Basin (60450101)	98 239	0.19	0.38	115	1413	68
Neuquen Basin (60550103)	7734	0.65	0.76	130	177	98
Tobago Trough (61030101)	23 864	0.38	0.94	128	707	42
West-Central Coastal (72030201)	258 108	0.05	0.14	114	1413	24

(Continued)

Table 2. Continued

Province name (AU number)	Area (km^2)	Field density (no. discovered per 1000 km^2)	Field density (no. discovered + undiscovered per 1000 km^2)	Median field size, fields ≥ 50 MMBOE	Largest discovered field size (binned MMBOE)	Exploration maturity (percentage petroleum discovered per total assessed)
West-Central Coastal (72030302)	199 541	0.01	0.49	140	2826	23
Orange River Coastal (73030101)	394 664	0.00	0.01	110	353	50
Bombay (80430102)	65 776	0.20	0.24	114	707	88
Ganges–Brahmaputra Delta (80470301)	176 260	0.05	0.18	116	353	41
Irrawaddy (80480102)	466 281	0.01	0.04	107	707	37
Median		0.09	0.21	112	575	49
Mean		0.17	0.41	116	1075	53
Combined analogues (duplicates removed)						
Median		0.11	0.28	114	850	68
Mean		0.17	0.39	128	4361	62
Maximum		0.74	1.74	251	45 220	99

undiscovered gas fields (Fig. 4). The sizes are reduced in this AU from the median of the largest discovered field size in the analogue dataset because of the complex tectonic history and uplift since petroleum generation.

Lena–Anabar Basin Assessment Unit

The area of the Lena–Anabar Basin AU is c. 55 000 km^2 and is located entirely north of the Arctic Circle. Input distributions and AU probabilities for the Lena–Anabar Basin AU are shown in Figure 4.

Assessment of the Lena–Anabar Basin AU was based on two scenarios because a single timing probability was not sufficient to characterize the age of petroleum generation in Proterozoic and Palaeozoic source rocks with respect to trap development. Permian and Mesozoic rocks directly overlie Cambrian rocks over much of the AU, but whether or not a complete Palaeozoic section was deposited and subsequently eroded is unknown. The scenarios are: (1) base case (having a 90% probability) – non-deposition of a middle and late Palaeozoic section followed by deposition, petroleum generation and accumulation during the Permian and Mesozoic; and (2) worst case (having a 10% probability) – deposition of a middle and upper Palaeozoic section with petroleum generation and accumulation followed by middle to late Palaeozoic erosion. The likelihood that the Lena–Anabar Basin AU contains at least one field equal to or greater than the minimum field size of 50 MMBOE is estimated to be c. 48% (0.48) for the base case scenario and 32% (0.32) for the worse case scenario.

Densities of 0.02, 0.4 and 1.8 (minimum, median, and maximum, respectively) were estimated for scenario 1 in the assessment of the Lena Anabar Basin AU because accumulations sourced by Proterozoic and Palaeozoic source rocks could exist in the Mesozoic section. For scenario 2, the median and maximum densities were estimated to be approximately half of those of scenario 1 (0.02, 0.2 and 0.9) because accumulations sourced by Proterozoic and Palaeozoic source rocks would exist only in the Palaeozoic section. For scenario 1, the total minimum, median and maximum numbers of undiscovered fields are 1, 20 and 100, respectively (Fig. 4). The estimated numbers of undiscovered oil fields are 1 (minimum), 20 (median) and 100 (maximum) and the estimated numbers of undiscovered gas fields are 0 (minimum), 2 (median) and 20 (maximum), assuming oil–gas mixtures of 0.8 (minimum), 0.9 (mode) and 1.0 (maximum). The densities and estimated numbers of undiscovered fields for scenario 2 are half of those for scenario 1 (Fig. 4). The total minimum, median and maximum numbers of undiscovered fields are 1, 10 and 50, respectively. The estimated numbers of undiscovered oil fields are 1 (minimum), 9 (median) and 50 (maximum) and the estimated numbers of undiscovered gas fields are 0 (minimum), 1 (median) and 10 (maximum).

The median sizes of undiscovered oil fields (125 and 100 MMBOE, scenarios 1 and 2, respectively) in the AU were estimated to approximate the median size of the analogue datasets (Fig. 2). The low-probability maximum oil field size, 2500 MMBOE (Fig. 4), is larger than the median of the largest discovered field size in the rifted passive margin and foreland basin analogue dataset (1700 MMBOE), whereas sizes of gas fields are lower (median sizes of 83 and 75 MMBOE and maximum sizes of 167 MMBOE) because gas-prone Mesozoic source rocks may not be fully mature with respect to generation, resulting in incomplete trap fill. The expected maximum crude oil size (c. 750 MMBOE; Fig. 4) is based on the distribution of sizes of discovered fields in the analogue dataset, particularly the median of the maximum discovered field sizes. The expected maximum gas field size (c. 600 BCF or 100 MMBOE; Fig. 4) is less because gas-prone Mesozoic source rocks may not be fully mature with respect to generation.

Northern Priverkhoyansk Foredeep Assessment Unit

The area of the Northern Priverkhoyansk Foredeep AU is c. 56 000 km^2 and 99% of the area is located north of the Arctic Circle. The likelihood that the AU contains at least one field equal to or greater than the minimum field size (50 MMBOE) is 40% (0.40). Input distributions and AU probabilities for the Northern Priverkhoyansk Foredeep AU are shown in Figure 5.

Densities of 0.02, 0.1 and 0.9 (minimum, median and maximum, respectively) were used in the assessment of the Northern Priverkhoyansk Foredeep AU. These densities are slightly less

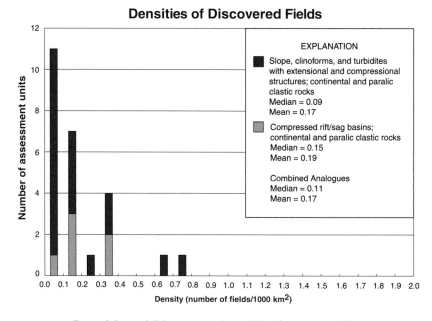
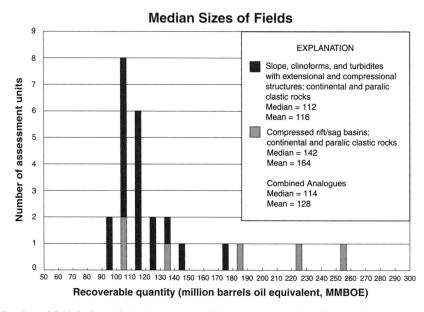

Fig. 3. Densities and median sizes of fields in the geological analogues used for the assessment of the Yenisey–Khatanga Basin Assessment Unit. Data from Charpentier *et al.* (2007).

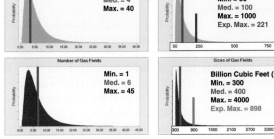

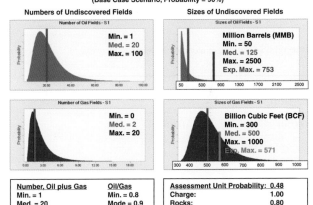

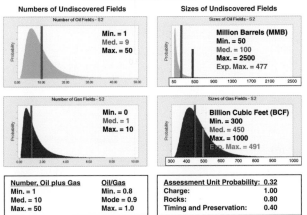

Fig. 4. Input distributions and AUs probabilities for the Khatanga Saddle and Lena–Anabar Assessment Units. Min., minimum; Med., median; Max., maximum; Exp. Max., calculated expected maximum.

than those of the analogue dataset (Fig. 2) because: (1) outlying values in the analogue set that do not represent this AU were excluded; (2) no economic discoveries were made in the AU despite exploration; and (3) a carbonate shelf might not have existed over the entire area. The total minimum, median and maximum numbers of undiscovered fields are 1, 7 and 50, respectively (Fig. 5). The estimated numbers of undiscovered oil fields are 1 (minimum), 4 (median) and 35 (maximum), and the estimated numbers of undiscovered gas fields are 1 (minimum), 4 (median)

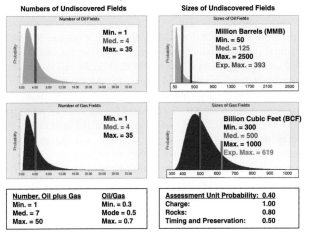

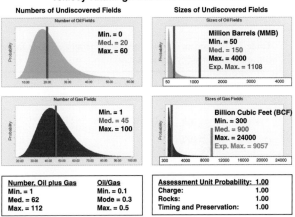

Fig. 5. Input distributions and assessment unit probabilities for the Northern Priverkhoyansk Foredeep and Yenisey–Khatanga Basin AUs. Min., minimum; Med., median; Max., maximum; Exp. Max., calculated expected maximum.

and 35 (maximum), assuming an oil–gas mixture of 0.3 (minimum), 0.5 (mode) and 0.7 (maximum).

The median undiscovered field sizes in the AU are 125 MMB of crude oil and 500 BCF of natural gas (Fig. 5). The median undiscovered oil field size is approximately equal to the mean size of the analogue dataset (c. 130 MMBOE; Fig. 2), but the median undiscovered gas field size is considerably less because traps for natural gas in this AU might not be efficient. Low-probability maximum undiscovered oil and gas field sizes in the AU are 2500 MMB and 1000 BCF, allowing for the probability of a large oil field within the Proterozoic and Palaeozoic section. The expected maximum oil and gas field sizes (c. 400 and 100 MMBOE, respectively; Fig. 5) are based on the distribution of sizes of discovered fields in the analogue dataset, particularly the median of the maximum discovered field sizes, but excluding values that do not represent this AU.

Yenisey–Khatanga Basin Assessment Unit

The area of the Yenisey–Khatanga Basin AU is c. 345 000 km² and is located entirely north of the Arctic Circle. The likelihood that the AU contains at least one field equal to or greater than the minimum field size of 50 MMBOE is estimated at 100% (1.00). Input distributions and AU probabilities for the Yenisey–Khatanga Basin Assessment Unit are shown in Figure 5.

The median and maximum densities of discovered fields in the combined analogue datasets are 0.1 and 0.7 fields per 1000 km², respectively, whereas the median and maximum densities of discovered plus undiscovered fields are 0.3 and 1.7 fields per 1000 km^2, respectively (Fig. 3). Minimum sizes of undiscovered fields defined for the AU are 50 MMB of crude oil and 300 BCF of natural gas. The median undiscovered oil field size of the combined analogue datasets is 114 MMBOE (Fig. 3), and the median of maximum undiscovered field sizes is 850 MMBOE.

Densities of <0.01, 0.2 and 0.3 (minimum, median and maximum, respectively) were estimated for numbers of fields per 1000 km^2 in this AU. The median density of 0.2 is between the median densities of the analogue datasets, and the maximum density of 0.3 approximates the density of the neighbouring West Siberian Basin AUs (0.35, included in the analogue set). The total minimum, median and maximum numbers of undiscovered fields are 1, 62 and 112, respectively (Fig. 5). The estimated numbers of undiscovered oil fields are 0 (minimum), 20 (median) and 60 (maximum), and the estimated numbers of undiscovered gas fields are 1 (minimum), 45 (median) and 100 (maximum), assuming oil–gas mixtures of 0.1 (minimum), 0.3 (mode) and 0.5 (maximum).

The median field size, 150 MMBOE, approximates the median size of the rift/sag basin analogue set (142 MMBOE), but is greater than the median field size of the combined analogue dataset (114 MMBOE; Fig. 3). This median field size is consistent with the average size of discovered fields in this AU and the neighbouring Northern West Siberian Onshore Gas AU (with a median size of c. 200–300 MMBOE, included in the analogue set). The low-probability maximum field sizes of 4000 MMB for undiscovered oil accumulations and 24 000 BCF for undiscovered gas accumulations (Fig. 5) approximates the means of the maximum field sizes in the combined analogue database. The expected maximum size of undiscovered fields is based on the analogue dataset with a smaller oil field size than gas to be consistent with the sizes of discovered fields in this AU: c. 1000 MMB for undiscovered oil fields and 1500 MMBOE for undiscovered gas fields (9000 BCF; Fig. 5). These sizes are based on a distribution of

Table 3. Assessment results of geological provinces along the north and east margins of the Siberian craton (technically recoverable, conventional, undiscovered resources)

Assessment unit	Khatanga Saddle	Yenisey–Khatanga Basin	Lena–Anabar Basin (aggregated)	Northern Priverkhoyansk Foredeep	Total (means only)
Assessment unit code	11750101	11750102	12000101	12140101	
Oil in oil fields (MMB)					
F95	0	2201	0	0	
F50	0	4847	0	0	
F05	1376	9716	7207	1732	
Mean	327	5257	1913	377	7874
Gas in oil fields (associated and dissolved gas) (BCF)					
F95	0	11 604	0	0	
F50	0	26 571	0	0	
F05	932	55 375	6022	1448	
Mean	206	29 078	1502	296	31 082
Natural gas liquids in oil fields (MMB)					
F95	0	305	0	0	
F50	0	710	0	0	
F05	25	1529	163	39	
Mean	6	786	40	8	840
Gas in gas fields (non-associated gas) (BCF)					
F95	0	38 629	0	0	
F50	0	66 089	0	0	
F05	6764	108 413	2538	4319	
Mean	1797	68 884	605	1039	72 325
Natural gas liquids in gas fields (MMB)					
F95	0	1009	0	0	
F50	0	1754	0	0	
F05	182	2929	69	116	
Mean	48	1835	16	28	1927

MMB, million barrels; BCF, billion cubic feet. Results shown are fully risked estimates. For gas fields, all liquids are included under the natural gas liquids category. F95 denotes a 95% of at least the amount tabulated; other fractiles are defined similarly. Fractiles are additive under the assumption of perfect positive correlation.

sizes of fields in this AU and neighbouring AUs, and are constrained by the discovery history.

Tunguska Basin Assessment Unit

The Tunguska Basin AU is c. 854 000 km² and 39% of the area is located north of the Arctic Circle. It was not quantitatively assessed and no analogues were selected. The likelihood that the AU contains at least one field equal to or greater than 50 MMBOE is c. 6% (0.06). AUs with probabilities less than 1% were not quantitatively assessed.

Assessment results

The USGS estimated the mean undiscovered, technically recoverable, conventional petroleum resources in the seven AUs in the Arctic portion of the Siberian craton (Fig. 1) to be c. 28 billion barrels of oil equivalent, including c. 8 billion barrels of crude oil, 103 trillion cubic feet of natural gas and 3 billion barrels of natural gas liquids (Table 3).

Conclusions

The assessment of undiscovered petroleum resources along the north and east margins of the Siberian craton was accomplished using geological analogues. Individual analogues, sets of analogues with similar geological characteristics, and combinations of analogue sets (as used in this study) facilitate the assessment of geological provinces that are immature with respect to exploration such that exploration-history analysis alone cannot be used. Number and sizes of undiscovered oil and gas fields can be obtained from analogues and, when statistically combined with AU probability, provide estimates of undiscovered petroleum resources.

The authors would like to acknowledge F. M. Persits for GIS support, and D. L. Gautier and G. F. Ulmishek for their valuable reviews and comments. The manuscript was greatly improved by the reviews and comments of A. Khudoley, A. Stoupakova, S. Drachev and S. Pickering. The authors are grateful to the USGS Library staff for their help in obtaining hard-to-find and rare geological articles from the Russian scientific literature.

References

Bakhturov, S. F. 1985. *Bituminoznye karbonatno-slantsevye formatsii Vostochnoy Sibiri (Bituminous carbonate-shale formations of eastern Siberia)*. Trudy Instituta Geologii i Geofiziki, **617**. Sibirskoye Otdelenie, Akademiya Nauk SSSR (in Russian).

Bakhturov, S. F., Yevtushenko, V. M. & Pereladov, V. S. 1990. Kuonam bituminous carbonate shale complex. *Petroleum Geology*, **24**, 124–133; translated from Trudy Instituta Geologii i Geofiziki, **671**, Sibirskoye Otdelenie, Akademiya Nauk SSSR (1988).

Charpentier, R. R., Klett, T. R. & Attanasi, E. D. 2007. *Database for Assessment Unit-scale Analogs (Exclusive of the United States)*. United States Geological Survey Open-File Report 2007-1404.

Clarke, J. W. 1991. Petroleum geology of East Siberia (abstract). American Association of Petroleum Geologists Bulletin, **75**, 554 (presented at the *1991 American Association of Petroleum Geologists Annual Convention*, Dallas, TX, 7–10 April 1991).

Environmental Systems Research Institute Inc. 1999. *ESRI Data and Maps*. Environmental Systems Research Institute Inc., Redlands, CA.

Grausman, V. V. 1996. Upper Precambrian deposits of the Olenek Uplift from deep drilling data. *Tikhookeanskaya Geologiya*, **12**, 775–781.

Grebenyuk, V. V., Gurari, F. G., Lugovtsov, A. D. & Moskvin, V. I. 1988. Specifics of oil–gas formation in the Mesozoic depressions of the Siberian Platform. *Petroleum Geology*, **19**, 602–605; translated from *Geologiya i neftegasonosnost' mesozoiskikh sedimentatsionnikh basseinov Sibiri*. Akademiia Nauk SSSR, Sibirskoe Otdelenie, Trudy Institut Geologii i Geofiziki, **532**, 132–139 (1983).

IHS Energy. 2007 [includes data current through October 2007]. *International Exploration and Production Database*. IHS Energy Group, Englewood, CO.

Ivanov, V. L. 1979. *Olenekskoe mestorozhdenie bitumov (Geologicheskoe stroenie i usloviia formirovaniia)* (Olenek bitumen field, geologic layer and conditions of formation). Nauchno-issledovatel'skii institut geologii Arktiki, Trudy (Leningrad, 'Nedra', Leningrad Branch), **182**.

Klett, T. R., Ahlbrandt, T. S., Schmoker, J. W. & Dolton, G. L. 1997. *Ranking of the World's Oil and Gas Provinces by Known Petroleum Volumes*. United States Geological Survey Open-File Report 97-463.

Kontorovich, A. Eh., Bakin, V. E., Grebenyuk, V. V., Kuznetsov, L. L. & Nakaryakov, V. D. 1991. Khatanga-Vilyui Upper Paleozoic–Mesozoic petroleum province (abstract). American Association of Petroleum Geologists Bulletin, **75**, 1414 (presented at the *American Association of Petroleum Geologists International Conference*, London, 29 September to 2 October, 1991).

Kuznetsov, V. G. 1997. Riphean hydrocarbon reservoirs of the Yurubchen–Tokhom zone, Lena–Tunguska province, NE Russia. *Journal of Petroleum Geology*, **20**, 459–474.

Persits, F. M. & Ulmishek, G. F. 2003. *Maps Showing Geology, Oil and Gas Fields, and Geologic Provinces of the Arctic*. United States Geological Survey Open-File Report 97-470-J.

Persits, F. M., Ulmishek, G. F. & Steinshouer, D. W. 1998. *Map Showing Geology, Oil and Gas Fields, and Geologic Provinces of the Former Soviet Union*. United States Geological Survey Open-File Report 97-470E.

Rosen, O. M., Condie, K. C., Natapov, L. M. & Nozhkin, A. D. 1994. Archean and early Proterozoic evolution of the Siberian Craton: a preliminary assessment. *In*: Condie, K. C. (ed.) *Archean Crustal Evolution, Developments in Precambrian Geology*, **11**. Elsevier, Amsterdam, 411–459.

Sekretov, S. B. 2003. Hydrocarbon potential of the Yenisei–Lena region of the Arctic Siberia and adjacent southwestern Laptev Sea (abstract). Abstracts of the *Fourth International Conference on Arctic Margins (ICAM IV)*, Dartmouth, Nova Scotia, 30 September–3 October, 2003).

Shenfil', V. Yu. 1991. *The Late Precambrian of the Siberian Platform*. Nauka Sibirskoye Otdelenie, Novosibirsk.

Shishkin, B. B. & Isaev, A. V. 1999. Structure of the Precambrian and Cambrian deposits in the northeast of the Siberian platform. *Russian Geology and Geophysics*, **40**, 1763–1775.

Smelov, A. P. & Timofeev, V. F. 2007. The age of the North Asian Cratonic basement: an overview. *Gondwana Research*, **12**, 279–288.

Sokolov, B. A. (ed.) 1989. History of oil–gas formation and accumulation in the east of the Siberian Craton. *Petroleum Geology*, **23**, 12–45; translated from Akademiya Nauk SSSR, Sibirskovo Otdeleniya, Yakutskii Filial, Institut Geologii, Moscow, Izdatelctvo Nauka, (1986).

Ulmishek, G. F. 2001a. *Petroleum Geology and Resources of the Nepa–Botuoba High, Angara–Lena Terrace, and Cis–Patom Foredeep, southeastern Siberian Craton, Russia*. United States Geological Survey Bulletin 2201-C.

Ulmishek, G. F. 2001b. *Petroleum Geology and Resources of the Baykit High Province, East Siberia, Russia*. United States Geological Survey Bulletin 2201-F.

Vagin, S. B., Samsonov, Yu. V., Shashin, A. V. & Dongaryan, L. Sh. 1987. Paleo-geologic and paleo-hydrogeologic bases for assessing the oil–gas prospects of the Vilyuy Basin. *Petroleum Geology*, **22**, 458–460; translated from *Neftegazovaya Geologiya i Geofizika*, **1986**, 2–4.

Zonenshain, L. P., Kuzmin, M. I. & Natapov, L. M. 1990. Geology of the USSR; a plate-tectonic synthesis. *In*: Page, B. M. (ed.) *Geodynamics Series*. American Geophysical Union, Washington, DC, **21**.

Synchronous exhumation events around the Arctic including examples from Barents Sea and Alaska North Slope

P. F. GREEN and I. R. DUDDY

Geotrack International Pty Ltd, 37 Melville Road, West Brunswick, Victoria 3055, Australia
(e-mail: mail@geotrack.com.au)

Abstract: In many areas of the Arctic, sedimentary sequences have been exhumed from significantly greater depths during the Cenozoic, with 2 km of section or more removed in some areas. Implications for exploration include enhanced maturity levels, possible loss of reservoired hydrocarbons as a result of seal breach, and phase changes due to pressure reduction. While the importance of Cenozoic exhumation to hydrocarbon prospectivity in individual basins is widely recognized, less well recognized is the regional synchroneity in the main phases of Cenozoic exhumation over wide areas of the Arctic and North Atlantic. Thermal history reconstruction studies in the Barents Sea and the Alaskan North Slope, based on application of apatite fission track analysis and vitrinite reflectance, reveal three main episodes of exhumation, in Paleocene, Eocene–Oligocene and Miocene times, and correlative exhumation episodes have been identified in a number of published studies in these and other areas. Previous attempts to explain these episodes of exhumation have been focussed on local mechanisms. However, our results reveal a pattern of regionally synchronous exhumation over a wide region, not only of the Arctic but also in many areas around the European North Atlantic margin, suggesting that events in each area are a regional response to events at plate boundaries, perhaps coupled to imbalances of crustal forces at continental boundaries. To date, no convincing mechanism has been put forward for producing such regional exhumation episodes, despite the fact that in many areas they exert critical control on regional hydrocarbon prospectivity. We suggest that serious attention should be directed to investigating the underlying mechanisms.

Keywords: AFTA, exhumation, Barents Sea, Alaskan North Slope, thermal history, burial history reconstruction

In many areas of the Arctic, sedimentary sequences have been exhumed from significantly greater depths during the Cenozoic, with up to 2 km of section or more removed in some areas. Examples include Barents Sea (Cavanagh *et al.* 2006), Svalbard (Blythe & Kleinspehn 1998), North Slope Alaska (O'Sullivan *et al.* 1993; O'Sullivan 1996, 1999), Sverdrup Basin (Arne *et al.* 2002), West Greenland (Japsen *et al.* 2005) and East Greenland (Thomson *et al.* 1999; Hansen *et al.* 2001). Implications for exploration include enhanced maturity levels (compared with assessments that underestimate former burial depths), possible loss of reservoired hydrocarbons as a result of seal breach and spillage, and phase changes due to pressure reduction (see Doré *et al.* 2002, for an extensive review of the influence of exhumation on hydrocarbon systems). In addition, the timing of generation becomes a key issue in such areas, particularly for plays involving structures formed in the initial stages of exhumation, as these structures were not available for charging during the main phase of generation, which occurred during burial prior to the onset of exhumation.

Exhumation is often considered as a negative factor for prospectivity, but exhumed basins in many parts of the world contain significant quantities of hydrocarbons, such as the East Irish Sea Basin (e.g. Jackson *et al.* 1995) and southern North Sea (Cameron *et al.* 1992) of the UK, the Reconcavo Basin of Brazil (Magnavita *et al.* 1994) and much of North Africa (e.g. Logan & Duddy 1998). In these basins, discovering hydrocarbons requires detailed knowledge of magnitude and timing of exhumation and associated effects.

As evidence accumulates regarding the relatively common occurrence of exhumed basins, it is becoming increasingly clear that episodes of exhumation display a broad regional synchroneity across wide areas. A recent synthesis of stratigraphic information from the European North Atlantic margin (Praeg *et al.* 2005; Stoker *et al.* 2005) highlighted the presence of synchronous unconformities at Paleocene, Eocene–Oligocene, Base Miocene, Middle Miocene and Early Pliocene levels from offshore Ireland (Porcupine) to offshore Norway (Lofoten). As discussed below, each of these unconformities was interpreted as representing a plate-wide response to a major tectonic episode. Holford *et al.* (2009) showed that these unconformities correlate with episodes of kilometre-scale exhumation affecting wide areas of the East Irish Sea Basin and adjacent regions of the UK, in an intra-plate setting well inboard of the continental margin. In these regions, the magnitude of these events was such that up to a kilometre or more of section has been removed and all stratigraphic evidence for these events has been erased, whereas lower amounts of erosion along the margin have resulted in the preservation of evidence as hiatuses in the incomplete stratigraphic record.

Here we show that episodes of exhumation in many areas of the Arctic, including the Barents Sea, Alaskan North Slope, Arctic Canada, West and East Greenland, also correlate with these key episodes of tectonism originally defined along the European North Atlantic Margin. While the importance of Cenozoic exhumation to hydrocarbon prospectivity is now widely recognized in many areas of the Arctic, the nature of the underlying processes is poorly understood. Previous attempts to explain exhumation in these regions have been focussed purely on local mechanisms (e.g. Cavanagh *et al.* 2006 in the Barents Sea). However the results discussed here reveal a pattern of broadly synchronous exhumation over wide regions, suggesting that events in these areas represent manifestations of a regional response to events at plate boundaries.

The work reported here is based on application of apatite fission track analysis (AFTA®) and vitrinite reflectance (VR) to provide rigorous constraints not only on the magnitude but also on the timing of exhumation. We present case studies based on individual wells from two key areas to illustrate how AFTA provides independent timing constraints on the onset of exhumation, followed by a brief review of results from other areas. We then synthesize this information to show that major phases of exhumation show a broad synchroneity across wide areas of the Arctic in three key time intervals, viz. Paleocene, Latest Eocene and Miocene–Pliocene. We then briefly discuss possible underlying mechanisms, and the resulting implications for hydrocarbon exploration. However, first we present a brief outline of the methods employed.

Thermal history reconstruction using AFTA and VR

Constraints on the magnitude and timing of exhumation episodes have been defined using thermal history reconstruction based on integrated application of AFTA and VR data (e.g. Green et al. 2004; Japsen et al. 2007). Application of AFTA in sedimentary basins allows determination of the maximum palaeotemperature attained by an individual rock sample from a sedimentary unit, and the time at which the sample cooled from that palaeotemperature. Multiple cooling episodes can often be resolved from data in individual samples. VR data allow complementary determinations of the maximum post-depositional palaeotemperatures of individual sedimentary units.

Extraction of thermal history information from AFTA and VR data begins by construction of a Default Thermal History, which is the history that can be reconstructed based on the preserved stratigraphic section and present-day thermal regime. Because the responses of both the AFTA and VR systems are dominated by maximum temperature, if the measured parameters can be explained purely by the Default Thermal History (possibly combined with the presence of shorter tracks inherited from sediment provenance regions), then no further information on the post-depositional thermal history can be extracted from the data. However, if the AFTA data show a greater degree of fission track annealing, and/or if VR values are higher than predicted from the Default Thermal History, then the sampled unit has been hotter in the past. In this case, information on the timing and magnitude of the palaeothermal maximum can be extracted from the data, by forward modelling of AFTA parameters through a range of likely thermal history scenarios, to define the range of conditions that match the measured values within 95% confidence limits. More details of factors which determine precision and accuracy of the resulting estimates are provided by Green et al. (2002).

Combination of the two methods, with AFTA data from sandstones and VR data from fine-grained units, allows data to be gathered over as wide a range of section as possible, allowing definition of the variation of palaeotemperatures with depth in key palaeothermal episodes. The slope of these palaeotemperature profiles then provides quantitative estimates of the range of allowed palaeogeothermal gradients, and extrapolation of the fitted linear profiles to an appropriate palaeosurface temperature provides definition of the corresponding amounts of additional section required to produce the observed palaeotemperatures in each episode. For further examples of this approach and practical details, see for example Green et al. (2004), Japsen et al. (2005) and Turner et al. (2008). It is emphasized that this approach is totally data-driven, and is independent of any assumptions or theoretical considerations regarding basin formation and tectonic processes.

Wherever possible, additional constraints on the magnitude of removed section are sought from methods based on sediment compaction, such as sonic velocities or porosity (e.g. Holford et al. 2005; Japsen et al. 2007). Consistency between amounts of removed section determined from palaeothermal (i.e. AFTA and VR) and palaeoburial methods (sonic velocities) provides evidence of exhumation that can be regarded as highly reliable.

Exhumation histories in the Barents Sea

The Barents Sea forms a broad offshore shelf to the north of the Norwegian mainland, containing significant thicknesses of dominantly Late Palaeozoic and Mesozoic sedimentary units in a mosaic of basins and platforms (Doré 1995). Further details of the regional geology and stratigraphy are provided by, for example, Faleide et al. (1993) and Gabrielsen et al. (1997). Cenozoic exhumation across the Barents Sea is widely recognized. Cavanagh et al. (2006) provided an extensive review of published studies, illustrating the wide diversity of opinions as to both the timing and magnitude of exhumation.

The main reason for the wide disparity of views regarding the timing of exhumation is that few studies actually employ methods which constrain the timing directly. Instead, most studies have been based on assumed timings derived from ideas regarding the nature of the underlying processes. Many studies favour a dominant phase of Late Pliocene to Pleistocene exhumation, based on the presence of marginal sedimentary wedges and perceived mechanisms of erosion related to glaciation (e.g. Nyland et al. 1992; Riis & Fjeldskaar 1992; Cavanagh et al. 2006).

Here we review AFTA and VR data from the 7120/9-2 well, located in the Hammerfest Basin (Fig. 1). The well encountered c. 60 m of Pliocene to Quaternary section, separated by an unconformity from c. 700 m of Paleocene to Eocene section, which in turn unconformably overlies c. 840 m of Hauterivian to Santonian section. Further unconformities separate c. 50 m of Kimmeridgian section, c. 130 m of Pliensbachian to Bajocian section, and c. 2970 m of Upper Permian to Hettangian section, and the well reached total depth (TD) in Upper Permian section at a depth of 5072 m (rkb). A present-day geothermal gradient of 30.1 °C/km was derived from corrected BHT data, using a sea-bed temperature of 5 °C.

Results from this well form part of a larger (unpublished) study, and are typical of those from other wells in the Hammerfest Basin. AFTA parameters from the 7120/9-2 well are plotted against depth in Figure 2a, where they are compared with trends predicted from the default burial history (derived from the preserved section in this well and the present-day thermal regime described above). These trends define the patterns of fission track age and mean track length that would be expected if units throughout this well are currently at their maximum post-depositional palaeotemperatures.

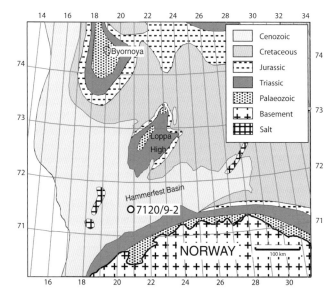

Fig. 1. Simplified geology of the Barents Sea, showing the location of the 7120/9-2 well. Generalized pre-Quaternary subcrop geology after Nøttvedt et al. (1993).

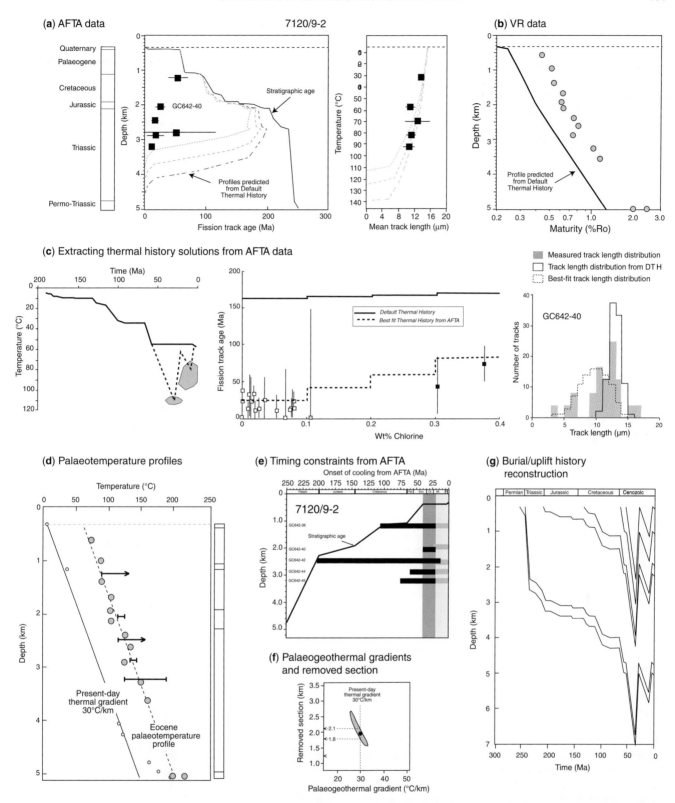

Fig. 2. AFTA and VR data in 7120/9-2, illustrating the process of burial history reconstruction. (**a**) AFTA parameters plotted against depth and compared with predictions of the Default Thermal History. A stratigraphic column is shown on the left, and the variation of stratigraphic age through the section is shown in the central panel. (**b**) VR data, also plotted against depth and compared with the profile predicted from the Default Thermal History. (**c**) Quantitative assessment of the AFTA data in sample GC642-40 shows that two episodes of heating and cooling in addition to the Default Thermal History are required to explain all aspects of the data. (**d**) Maximum post-depositional palaeotemperatures derived from AFTA and VR data in this well define a linear profile, sub-parallel to the present-day temperature profile and offset to higher temperatures by c. 60 °C. This is diagnostic of heating due to deeper burial, with little or no change in heat flow. Results from a later episode as shown in (e) have been omitted for clarity. (**e**) AFTA data in all samples define two major episodes of cooling, which began in the intervals 40–20 Ma and 20–0 Ma. (**f**) Variation of palaeotemperatures with depth as shown in (d) allow definition of the range of allowed values (within 95% confidence limits) of palaeogeothermal gradient and removed section, as shown by the hyperbolic elliptical shaded region. Constraints on the later episode are much broader and are not shown. (**g**) Integration of the palaeothermal episodes with the preserved stratigraphic section results in reconstruction of the history of post-Early Eocene burial and subsequent exhumation. The option illustrated here, with all Eocene section removed and a further 1 km or so of Oligocene to Early Miocene section deposited prior to the final phase of exhumation, represents one end-member of a range of viable solutions.

The measured fission track ages are all much younger than predicted from the Default Thermal History scenario, showing that the sampled units have been hotter than the present-day temperatures at some time since deposition.

The measured fission track ages show the characteristic pattern of a section that has been significantly hotter in the past (Green et al. 2002), with a rapid decrease at shallow depths and a break in slope (at around 2 km) denoting the transition from partial to total annealing of tracks formed prior to the onset of cooling. Fission track ages around 20 Ma below the break in slope show that these samples have only retained tracks since cooling to sufficiently low temperatures in the mid-Cenozoic. However, these ages do not directly date the onset of cooling, because tracks formed after cooling have been reduced in length (mean lengths are $c.$ 12 μm in these samples, Fig. 2a) due to the prevailing temperatures. Nevertheless, this initial, qualitative assessment of the AFTA data from the 7120/9-2 well shows quite clearly that the section in this well began to cool from palaeotemperatures considerably higher than present-day temperatures at some time in the mid-Cenozoic.

Quantitative assessment of the data, including the fission track age and track length data and their variation with chlorine content in each sample (following principles outlined by Green et al. 2002), allows extraction of thermal history solutions from the AFTA data. This process is illustrated in Figure 2c, based on AFTA data in sample GC642-40, collected from depths between 1980 and 2130 m (rkb) in Early Jurassic section. Two episodes of heating and cooling in addition to the Default Thermal History are required to explain the AFTA data in this sample. Synthesis of AFTA data in all samples from this well (Fig. 2e) results in definition of two dominant episodes of cooling, beginning in the intervals 40–20 Ma (Eocene to Early Miocene) and 20–0 Ma (Early Miocene to Recent). Note that such estimates of the onset of cooling from AFTA refer to the 95% confidence interval within which cooling *began*, and we do not imply that cooling was restricted to this interval.

VR data are plotted against depth in Figure 2b, where they are compared with the profile predicted from the Default Thermal History. Measured VR values are consistently higher than predicted from the Default Thermal History, confirming that the sampled units have been hotter at some time after deposition. Note particularly that this well contains a significant thickness of Paleocene to Early Eocene section (see stratigraphic column and variation of stratigraphic age with depth in Fig. 2), and VR values from this sequence plot well above the profile predicted from the Default Thermal History and define a consistent trend with deeper values, showing that cooling from the palaeothermal maximum post-dates deposition of the Early Eocene section.

Palaeotemperatures characterizing the earlier of the two episodes identified from AFTA, and the independent estimates of maximum post-depositional palaeotemperature derived from the VR data, are highly consistent (Fig. 2e) and define a linear depth profile, sub-parallel to the present-day temperature profile and offset to higher temperatures by $c.$ 60 °C. This is a clear signature of heating due dominantly to deeper burial by additional section (Bray et al. 1992) with cooling due mainly to subsequent removal of this additional section by uplift and erosion (exhumation). Palaeotemperatures characterizing the Early Miocene to Recent episode are less clearly defined in this well, but a similar interpretation is inferred on the basis of results in other wells (Geotrack, unpublished results). The onset of cooling in both episodes identified from AFTA, between 40 and 20 Ma and between 20 and 0 Ma, falls within the time interval represented by the unconformity in this well between Early Eocene and Quaternary sedimentary units (Fig. 2), which is also consistent with this style of interpretation.

Figure 2f shows the range of values of palaeogeothermal gradient and additional burial ('removed section') consistent with the Eocene–Early Miocene palaeotemperatures in this well within 95% confidence limits (using methods outlined e.g. by Bray et al. 1992; Green et al. 2002). The present-day thermal gradient of $c.$ 30 °C/km falls towards the centre of the allowed range of palaeogradients, and corresponds to between 1.8 and 2.1 km of additional section, assuming a constant surface temperature of 5 °C. Importantly, the constraints in Figure 2f show that palaeogeothermal gradients higher than $c.$ 33 °C/km are not compatible with the results from this well, and scenarios involving significantly elevated basal heat flow can therefore be ruled out. Constraints on the more recent episode are less well defined, but for a constant palaeogeothermal gradient are consistent with between 0.25 and 1.35 km of additional burial at the palaeothermal peak between 20 and 0 Ma. Thus, the AFTA data in this well suggest a history involving little or no significant change in heat flow with time, with around 2 km of additional section deposited on top of the youngest preserved Early Eocene section, and subsequently removed in two episodes of exhumation.

Note that the constant surface temperature adopted for this study represents a convenient simplification, and it is possible that higher or lower values may be more appropriate for each episode. The magnitude of removed section required to explain the observed palaeotemperatures can be easily adjusted by subtracting or adding the difference in depth equivalent to the change in palaeosurface temperature, for the appropriate palaeogeothermal gradient. Increasing the palaeosurface temperature by 10 °C, for a palaeogeothermal gradient of 30 °C/km, would require a reduction of 333 m in the amount of removed section needed to explain the observed palaeotemperatures.

As already noted, both episodes of cooling identified in this well began during the interval represented by the unconformity between Earliest Eocene and Quaternary units (Fig. 2). Because AFTA and VR define only the maximum palaeotemperatures in the Eocene to Early Miocene episode and the peak palaeotemperatures in the Early Miocene to Recent episode, and not the

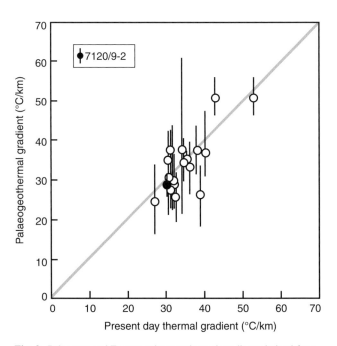

Fig. 3. Paleocene and Eocene palaeogeothermal gradients derived from AFTA and VR data in a number of Barents Sea wells, plotted against present-day thermal gradient. These results provide strong evidence for an explanation of the observed heating expressed in the AFTA and VR data in terms of deeper burial, with little or no change in heat flow.

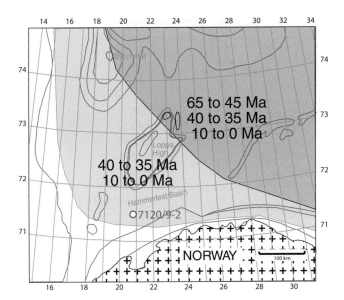

Fig. 4. Summary of areas of the Barents Sea dominated by different Cenozoic exhumation episodes.

intervening history, it is not possible to delineate the total amount of section removed in the Eocene–Early Miocene episode, because of the lack of constraint on the degree of reburial before the re-commencement of cooling (exhumation) from the Miocene to Recent palaeothermal peak. In this context, the estimates of 'removed section' shown in Figure 2 are better considered as estimates of 'additional burial' at the time that cooling in each episode began. That is, the quoted values refer to the amount by which the present unconformity horizon was more deeply buried at the onset of each cooling episode.

The final burial/uplift history reconstruction for the 7120/9-2 well is shown in Figure 2g. In this reconstruction, all of the additional section deposited prior to Latest Eocene cooling is removed and a further 1 km of Oligocene–Miocene section is deposited prior to the onset of Late Miocene exhumation. On the basis of the above discussion, this represents one end-member of a range of viable styles of Late Cenozoic uplift and erosion, with the other end-member being removal of around half of the additional Eocene section in the Latest Eocene episode, with the remainder removed in the Late Miocene episode.

Similar burial/uplift histories to that shown for the 7120/9-2 well in Figure 2g have been identified in a number of wells from the region, while other wells also show evidence of an earlier episode of Paleocene cooling (Geotrack, unpublished results). Combining evidence from all wells in the Barents Sea in which data are available shows that all results can be described in terms of three dominant episodes of Cenozoic cooling, beginning in the intervals 65–55 Ma (Paleocene), 40–30 Ma (Eocene–Oligocene) and 10–5 Ma (Late Miocene). As shown in Figure 3, palaeogeothermal gradients in each of these episodes are very close to present-day gradients, showing no evidence to suggest any significant variation of basal heat flow throughout the Cenozoic.

Figure 4 summarizes unpublished results from a number of wells across the Barents Sea region, distinguishing between those regions where Cenozoic exhumation began in the Paleocene and those characterized by an onset of exhumation in Eocene–Oligocene times. Paleocene exhumation dominates the NE of the region, where Palaeogene sediments are not present, while in the SE, Cenozoic exhumation began in the Eocene–Oligocene, following deposition of the Palaeogene sedimentary units in that region. However, Eocene–Oligocene exhumation is also recognized in the NE, and this episode together with Late Miocene exhumation are recognized across the entire region. It seems likely that the Palaeogene sediments present in the SW of the region represent the erosional products of the Paleocene erosion recognized in the AFTA data from the NE.

With data from all Barents Sea wells pointing to a most recent phase of major cooling/exhumation beginning in the interval 10–5 Ma (above), no results from the Barents Sea region show any evidence for significant amounts of cooling beginning within the last few million years that could be attributed to glacial action, as suggested in many earlier studies as discussed earlier. This is not to say that there has been no significant exhumation in this region in Late Pliocene to Recent times, and significant uplift and erosion at this time is indicated by Pliocene to Pleistocene sediment fans to the west of the shelf break (e.g. Cavanagh et al. 2006). However, any exhumation/cooling over this period is clearly negligible compared with that achieved in earlier events, and is not recorded in the AFTA data.

Exhumation histories from the North Slope Alaska

While the mountains of the Brooks Range, Alaska, display obvious contractional deformation and present-day high relief, indicating major Cenozoic denudation (e.g. O'Sullivan et al. 1997), the adjacent North Slope region is characterized by low relief at the present day. Nevertheless, numerous studies of the North Slope have provided abundant evidence of widespread Cenozoic exhumation across the region (e.g. Johnsson et al. 1993; Burns et al. 2005),

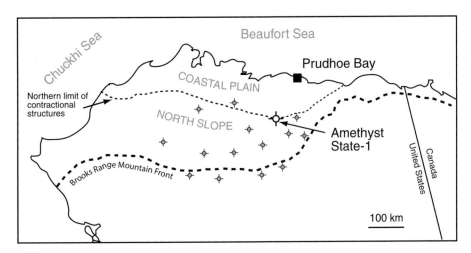

Fig. 5. Location map of the North Slope of Alaska, showing the Amethyst State-1 well, together with other wells from which results are synthesized here. Structural features are taken from figure 1 of Houseknecht & Bird (2005).

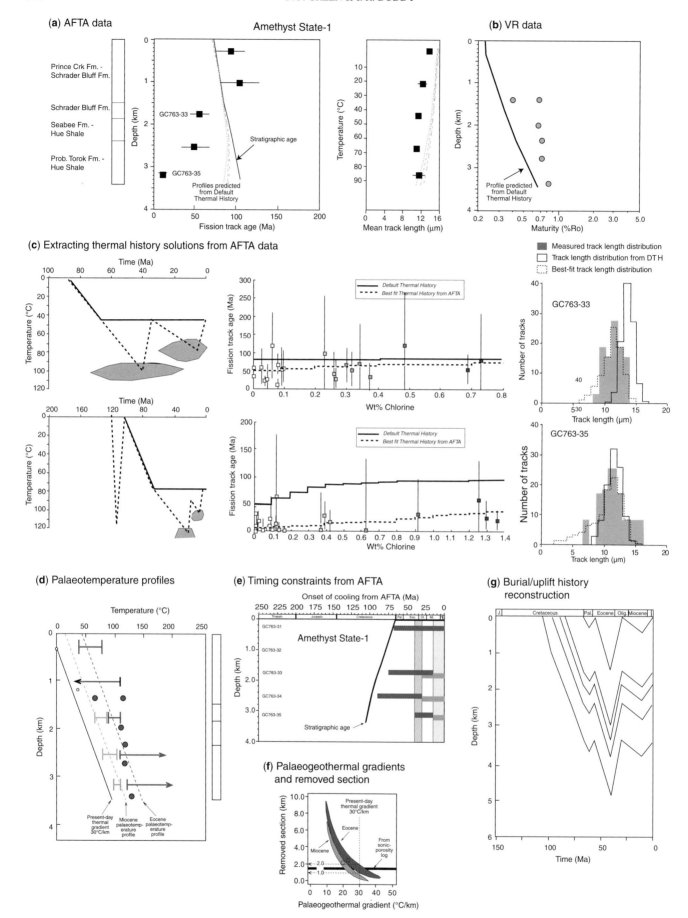

Fig. 6. AFTA and VR data in the Amethyst State-1 well, Alaska. All details are as in Figure 2. The Default Thermal History employed in construction of this figure is based on a thermal gradient below base permafrost of 30°C/km, as indicated by the AFTA data.

with amounts of section removed increasing from north to south across the region towards the Brooks Range fold-and-thrust belt. Bird & Molenaar (1992) provide a detailed review of the regional geological framework. Previous apatite fission track studies in the region (O'Sullivan et al. 1993; O'Sullivan 1996, 1999) have shown that exhumation began in Paleocene times, between 64 and 56 Ma, while also suggesting at least three later episodes of exhumation beginning at c. 45, 40–30 and 28–22 Ma (again quoting the *onset* of exhumation in each episode).

Here we discuss AFTA and VR data from the Amethyst State-1 well (Fig. 5). Again, results from this well form part of a wider (unpublished) study, and are typical of results from other wells in the immediate vicinity. The Amethyst State-1 well intersected Colville Group sediments of Cretaceous (Campanian to Albian) age, containing no recognized unconformities, and reached a total depth of 3394.5 m. A present-day geothermal gradient of 38°C/km from base permafrost at a depth of 304.8 m was estimated from corrected BHT values, but this was revised to 30°C/km on the basis of the AFTA data. Summary AFTA parameters in samples from this well are plotted against depth in Figure 6a (all details being similar to those for well 7120/9-2 in Fig. 2). Fission track ages in two shallower samples are greater than the respective depositional ages, suggesting only relatively mild post-depositional heating in these samples, but mean track lengths are significantly less than predicted from the Default Thermal History. Measured fission track ages in the three deepest samples are significantly less than predicted from the Default Thermal History. Thus, initial qualitative assessment of these data shows that all five have been hotter than present-day temperatures at some time since deposition.

Detailed analysis of the AFTA data in the five samples from this well results in definition of two major cooling episodes (Fig. 6c, e). Synthesis of data from all samples analysed from this well shows that cooling from maximum post-depositional palaeotemperatures began at some time between 40 and 30 Ma (Eocene–Oligocene), while cooling from a later palaeothermal peak began between 15 and 0 Ma (Miocene–Recent). VR data from this well, plotted against depth in Figure 6b, all plot above the profile predicted from the Default Thermal History, again confirming that the sampled units have been hotter in the past. Maximum palaeotemperatures derived from these VR data are highly consistent with the Eocene–Oligocene palaeotemperatures derived from AFTA data (Fig. 6e) and the combined constraints define a linear profile, sub-parallel to the present-day temperature profile and offset to higher temperatures by c. 40°C. Again, this is a clear signature of heating due dominantly to deeper burial by additional section (Bray et al. 1992) with cooling due mainly to subsequent removal of this additional section by uplift and erosion (exhumation).

While these palaeotemperatures allow quite a range of palaeogeothermal gradients (Fig. 6f), due largely to the relatively small number of VR values in this well, the present-day thermal gradient of 30°C/km falls well within the range of allowed palaeogradients, and corresponds to between 1000 and 2000 m of additional section, removed since the onset of cooling between 40 and 30 Ma. Miocene to Recent palaeotemperatures result in much broader constraints on the amount of additional section present at the Miocene to Recent palaeothermal peak, but again assuming a constant thermal gradient, the results suggest a value between 250 and 650 m. The resulting burial/uplift history reconstruction for the well is shown in Figure 6g.

Whereas results from the Amethyst State-1 well show that exhumation at the location of this well began between 40 and 30 Ma, results from our wider (unpublished) study in other wells across the region show that, to the west, exhumation began earlier, at around 60 Ma (consistent with earlier studies, as reviewed above), while the two episodes identified in the Amethyst State-1 well are recognized in wells across the region. Synthesis of all data defines three major episodes of exhumation which began in the intervals 60–50, 40–35 and 15–10 Ma. The magnitude of

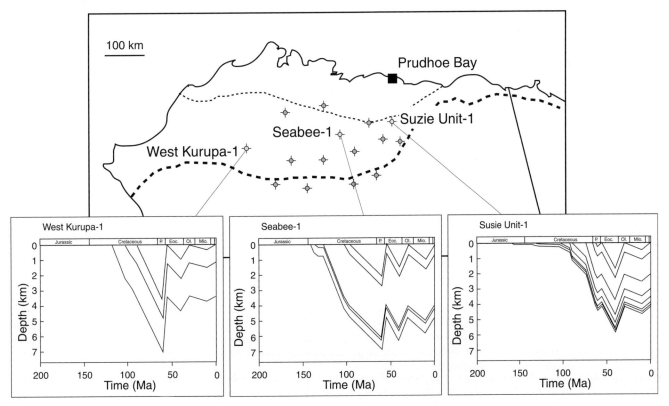

Fig. 7. Variation in burial/uplift history styles across the Alaskan North Slope. From west to east, the magnitude of Paleocene exhumation decreases, such that in wells located to the east, the onset of exhumation from maximum burial depths began in the Eocene episode. Structural elements as in Figure 5.

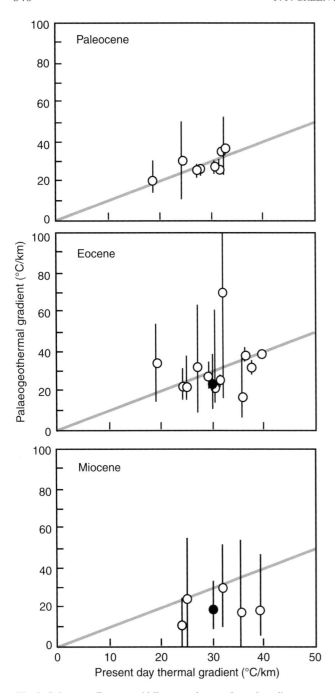

Fig. 8. Paleocene, Eocene and Miocene palaeogeothermal gradients derived from AFTA and VR data in a number of North Slope Alaska wells (locations shown in Fig. 6), plotted against present-day thermal gradient. Values for the Amethyst State-1 well are shown with black symbols. These results provide strong evidence for an explanation of the observed heating expressed in the AFTA and VR data in terms of deeper burial, with little or no change in heat flow.

Note that the estimated amount of additional section derived from AFTA and VR data in this well (between 1000 and 2000 m) is highly consistent with the value of 1486 m (4875 ft) estimated from porosity–depth relationships in this well by Burns et al. (2005), as illustrated in Figure 6f. This consistency suggests that the reconstruction provided for this well can be regarded with some confidence. Burns et al. (2005) suggested that VR data are 'demonstrably unreliable predictors of amounts of exhumation', compared with sonic-porosity methods and AFTA. We suggest that this perceived problem with VR data is related mainly to problems with data quality and analytical approach, rather than representing an inherent shortcoming of the technique. As discussed in detail elsewhere (Green et al. 2002, 2004), analysis of VR using the methods recommended by the International Committee on Coal and Organic Petrography (www.iccop.org), based on identification of indigenous vitrinite in polished thick rock sections, invariably provides results which are consistent with AFTA data from adjacent samples, and generates reliable indications of amounts of additional burial.

Other areas

A number of published studies have reported evidence of Cenozoic exhumation in Arctic regions. Arne et al. (2002) reported evidence based on apatite fission track data for Paleocene exhumation across the Sverdrup Basin of Arctic Canada, with an onset of exhumation in the range 65–60 Ma or perhaps earlier in some places.

Blythe & Kleinspehn (1998) reported evidence from apatite fission track data in Svalbard for three discrete phases of cooling, with the earliest beginning in the interval 70–50 Ma with further phases at c. 35 and 5 Ma. While their data provided no direct constraints on palaeogeothermal gradients and the mechanisms of heating and cooling, Blythe & Kleinspehn (1998) interpreted their three cooling episodes as all representing periods of exhumation, with the two earliest linked to tectonic influences – the most recent episode resulting from climate deterioration. The timing of these events is consistent with our own (unpublished) results from outcrops on Svalbard. AFTA data in samples of Mesozoic and Palaeozoic sedimentary units define cooling in the intervals 65–45 and 10–0 Ma, while samples of Paleocene sandstones began to cool from their maximum post-depositional palaeotemperatures in the 50–35 Ma interval and also show evidence of later cooling in the 10–0 Ma interval.

Green et al. (2005) reported evidence from AFTA for exhumation beginning in the interval 60–40 Ma in the Mackenzie Valley of the Northwest Territory of Canada, as well as a later phase in the last 15 Ma. Because data across much of this region are dominated by initial exhumation in the Jurassic (190–170 Ma) during a period of elevated heat flow, resolution of the detail of the Cenozoic exhumation history is less well defined than in other areas. Based on data from Alaska and the other areas discussed here, it seems likely that three Cenozoic episodes also affected this region.

AFTA and VR data from the Gro-3 well in the Nuussuaq Basin, West Greenland (Japsen et al. 2005), reveal a major phase of exhumation in this region which began in the interval 40–30 Ma, with subsequent episodes which began between 11 and 10 Ma and between 7 and 2 Ma. It is notable that this region shows no evidence of a Paleocene phase of exhumation, although there is evidence of latest Cretaceous to earliest Cenozoic uplift in the region from incised valleys that were filled by latest Maastrichtian sediments which later underwent sub-aerial erosion, prior to burial beneath overlying Palaeogene flood basalts. Japsen et al. (2009) subsequently showed using AFTA that Eocene–Oligocene exhumation also affected the basement region adjacent to the Nuussuaq basin, with the onset of exhumation refined to the interval 36–30 Ma. Again, dominance of earlier episodes in the AFTA

the Paleocene event shows a general increase to the west, as illustrated in Figure 7, towards the northward extension of the Brooks Range deformation front.

While the constant thermal gradient option adopted here for the Amethyst State-1 well is only one of a range of scenarios that would be allowed by the AFTA and VR data, results from other wells in the vicinity of the Amethyst State-1 well show a high degree of consistency between palaeogeothermal gradients and present-day thermal gradients in all three episodes (Fig. 8), and show no suggestion of elevated heat flow at any time. We therefore regard this scenario as the most realistic style of reconstruction for wells of the Alaskan North Slope.

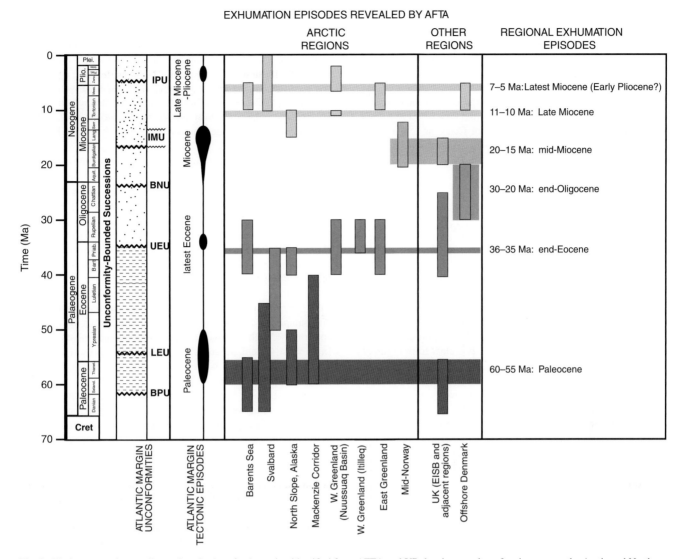

Fig. 9. Timing constraints on Cenozoic episodes of exhumation identified from AFTA and VR data in a number of regions across the Arctic and North Atlantic region, as discussed here, not only show a regional consistency between themselves (horizontal bars) but also correlate with regional unconformities defined along the Atlantic margin and to major episodes of tectonism (Praeg et al. 2005; Stoker et al. 2005). This emphasizes that these episodes of exhumation are controlled by regional processes acting on the scale of tectonic plates, rather than being due to local processes. References to results from individual areas are provided in the text. Note only results from our own studies are shown here, to ensure consistency of approach.

data from this region precludes resolution of any later cooling episodes.

A number of thermal history studies in East Greenland have provided important insights into the exhumation history of Mesozoic and Palaeozoic sediments of the Jamesonland Basin and NE Greenland, as well as adjacent basement terrains. Thomson et al. (1999) reported two episodes of Cenozoic cooling, based on AFTA data in a series of outcrop samples from Traill Island, which began between 40 and 30 Ma (Eocene–Oligocene) and between 10 and 5 Ma (Late Miocene). While this later episode was interpreted as due to uplift and erosion related to a change in the North Atlantic spreading vector, the origin of the Eocene–Oligocene cooling was not immediately clear, because the timing for the onset of cooling closely matches the timing of syenite intrusions at Kap Parry and Kap Simpson (c. 35 Ma). Thomson et al. (1999) suggested that hydrothermal circulation, deeper burial prior to exhumation (uplift and erosion) or elevated basal heat flow (or some combination of the three) provided viable alternative mechanisms for the Eocene–Oligocene palaeotemperatures.

Johnson & Gallagher (2000) reported apatite fission track data in a series of samples of Carboniferous sandstone from a vertical section on Clavering Island and found that the section began to cool from maximum post-depositional palaeotemperatures at c. 274 Ma (Early Permian), with subsequent cooling episodes at c. 206, 140 and 23 Ma. Johnson & Gallagher (2000) interpreted these cooling episodes largely in terms of denudation involving removal of several kilometres of section, but suggested that Cenozoic palaeothermal effects could be related, at least in part, to Palaeogene volcanic activity. Although the section sampled by Johnson & Gallagher (2000) is overlain by Palaeogene sediments and basalts, their analysis failed to take full account of this basic constraint on the thermal history of the sequence and this may explain the mismatch between their preferred onset of Cenozoic cooling and the other studies described above.

Hansen et al. (2001) reported results of an extensive regional apatite fission track study of the Jamesonland Basin and adjacent regions, and interpreted their results to reflect a complex interplay between deeper burial prior to Cenozoic exhumation, circulation of hot fluids associated with Mid-Cenozoic intrusive activity, and locally elevated heat flow associated with the intrusive activity. Hansen et al. (2001) favoured an onset of exhumation at c. 55 Ma in the south of the basin, and 'before 25 Ma' in the north. However, preservation of fission track ages in excess of

200 Ma in relatively close proximity to much younger ages, together with significant variations in VR levels over short distances, suggest that regional Cenozoic heating is relatively minor in this region, and that local heating effects dominate the data. We believe that all evidence suggests that any Palaeogene palaeothermal effects in this area are due either to contact heating from the widespread minor intrusions of this age or associated hydrothermal effects (or both). Based on a compilation of all available data, we conclude that the main phase of Cenozoic exhumation in East Greenland began in the interval 40–30 Ma, and continued through possibly several later episodes, some of which appear to vary in magnitude across the region.

Regional synchroneity of exhumation episodes, and underlying mechanisms

Figure 9 illustrates the timing of the exhumation episodes discussed in preceding sections (showing results from our own studies only, for consistency), within the context of the synthesis of key tectonic events along the North Atlantic margin and regional unconformities from Stoker et al. (2005) and Praeg et al. (2005), together with the timing of major episodes of intra-plate exhumation recognized in the UK by Holford et al. (2009). It is clear from Figure 9 that many of these exhumation episodes display a broad regional synchroneity (within the limits of uncertainty on the timing from AFTA across a wide region), while other events appear to be more restricted in extent.

Results from Alaska, Arctic Canada and the Barents Sea all show consistent evidence for major cooling which began in the Paleocene, with all results consistent with an onset of cooling due to exhumation in the interval 60–55 Ma (Late Paleocene). Results from a large area of the UK presented by Holford et al. (2009) also define a major period of exhumation at this time. As illustrated in Figure 9, this timing correlates with a period of pronounced tectonic activity affecting the North Atlantic region (Praeg et al.

2005), as well as the Arctic, as expressed for example by the Eurekan Orogeny (De Paor et al. 1989; Harland 1997).

Results from all the areas showing Paleocene cooling (above), as well as West and East Greenland, also show evidence for cooling which began at around 36–35 Ma (Latest Eocene). Again, the timing of this episode shows a close correlation with Latest Eocene tectonism in the North Atlantic, as expressed by the Upper Eocene Unconformity recognized by Stoker et al. (2005). Cooling at a similar time was again recognized in the UK region by Holford et al. (2009).

In contrast, the timing of the most recent exhumation episodes in Alaska, the Barents Sea and Greenland does not show the degree of consistency shown by the two earlier episodes. However, results from most areas are consistent with exhumation beginning in the 7–5 Ma interval (Fig. 9), which is very close to the timing of the 'Intra-Pliocene Unconformity' ('IPU' in Fig. 9) defined by Stoker et al. (2005). Results from a number of wells onshore and offshore Denmark (Japsen et al. 2007) also record this Late Miocene episode (Fig. 9). Japsen et al. (2007) suggested that the slight mismatch between the onset of exhumation from AFTA and the stratigraphically defined timing may be due to an anomalous result in one sample in their study, but the results shown in Figure 9 confirm the slight mismatch, which remains unresolved. Results from West Greenland and Alaska define an additional, slightly earlier episode, which is particularly well defined in the Gro-3 well from West Greenland (Japsen et al. 2005). These results appear to indicate further episodes of Miocene cooling, which either have not as yet been recognized in the stratigraphic record or else are of only local extent. Further work is required in order to resolve this uncertainty.

AFTA data from the Felicia-1 and Hans-1 wells, Offshore Denmark (Japsen et al. 2007), also define a period of Late Oligocene to Early Miocene (30–20 Ma) exhumation. Japsen et al. (2007) correlated this episode with the 'base-Neogene Unconformity' defined by Stoker et al. (2005), indicated by 'BNU' in Figure 9. Similarly, results from the UK (Holford et al.

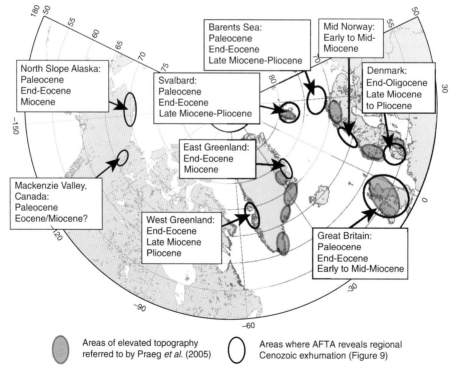

Fig. 10. Areas displaying Cenozoic exhumation with timing displayed in Figure 9 are highlighted, emphasizing the regional nature of these events. Also shown are regions of present-day elevated topography around the North Atlantic region discussed by Praeg et al. (2005). The mismatch between areas of exhumation and the modern topography emphasizes that the latter provides only limited insight into the extent of Cenozoic exhumation.

2009) define a Mid-Miocene onset of cooling (20–15 Ma) which correlates with the Intra-Miocene unconformity (IMU in Fig. 9). Cooling at this time is also identified in AFTA data from wells along the Mid-Norway margin (Green *et al.* 2007).

While it is beyond the scope of the present paper to provide a detailed synthesis of regional tectonic events across the Arctic, Figure 9 emphasizes that the timing of exhumation episodes identified across a wide area of the Arctic and North Atlantic not only shows a close correlation from region to region but also correlates with regional unconformities identified by Praeg *et al.* (2005) and Stoker *et al.* (2005) as reflecting major tectonic events affecting adjacent plate boundaries. While the true nature of the underlying processes remains as yet unknown, Figure 9 provides strong evidence that each of these exhumation episodes represents part of a regional response to plate boundary events, possibly as a result of compressive intra-plate stresses transmitted from plate boundaries. Similar observations have been made in other parts of the world, for example by Cobbold *et al.* (2001), who pointed out that episodes of Cretaceous to Cenozoic exhumation affecting the Atlantic margin of Brazil correlate closely in time with three prominent phases of Andean orogeny, representing periods of rapid convergence at the Andean margin of South America.

Regional extent of Cenozoic exhumation episodes

Figure 10 illustrates those areas where the effects of Paleocene, Eocene and Miocene exhumation have been identified to date. While data coverage is relatively sparse on the scale of Figure 10, these apparently synchronous episodes have clearly been recognized over a wide region, and it is likely that as further results become available these episodes will become increasingly familiar. The regional extent of these Cenozoic exhumation episodes suggests that the controlling processes act on a plate scale.

Note that the areas affected by these episodes show little correspondence with regions of high modern-day topography. Praeg *et al.* (2005) presented a description of the European North Atlantic margin characterized by a series of topographic highs resulting from two main episodes of Cenozoic uplift, accompanied by subsidence of offshore basins. The results of this study show that this picture is over-simplistic, and that many areas that now form offshore basins, as well as the onshore highs, have undergone multiple episodes of Cenozoic uplift and erosion, as reflected in the exhumation histories described here. Thus, the differentiation between topographic highs and basins of deposition depicted by figure 2 of Praeg *et al.* (2005) simply reflects the present-day setting, and holds only limited clues to earlier episodes of uplift and erosion that have affected wide areas through the Cenozoic (further discussion on the relationship between Cenozoic exhumation and modern topography is provided by Japsen *et al.* 2010). Again, the nature of the underlying processes remains uncertain, due in large part to the fact that in previous studies the effects of multiple episodes of uplift and erosion have not been recognized. The timing constraints on exhumation episodes established by AFTA and complementary methods, as summarized here, provide a rigorous framework within which a more complete understanding of the processes affecting these regions can be achieved.

Implications for hydrocarbon prospectivity

The exhumation episodes identified in this study, together with the corresponding prior burial, exert critical control on regional hydrocarbon prospectivity in many areas, including the Barents Sea, the Sverdrup Basin and both west and east Greenland. A fundamental issue in these and other exhumed basins is that maturity levels are considerably higher than expected on the basis of current burial depths. In these areas, the main phase of hydrocarbon generation took place during deeper burial prior to the onset of exhumation, and generation effectively ceased when exhumation began. One important outcome in relation to prospectivity is that traps formed as a result of the uplift which initiated exhumation were not present during generation and only earlier structures will have been available for charging. Subsequent episodes of uplift and erosion may have led to redistribution of any hydrocarbons trapped in those early structures, while phase changes as a result of the reduction in pressure during exhumation will favour preservation of gas and condensate over oil at the present day. Possible breaching of seals during exhumation is another key factor that must be taken into account.

The relative magnitude of the individual episodes clearly varies across the Barents Sea and the Alaskan North Slope, and in some parts of each region the Paleocene palaeothermal episode is of lesser magnitude than that of the Eocene–Oligocene episode, which represents the palaeothermal maximum (and hence the termination of hydrocarbon generation) in these areas. This implies that the later maturation in areas dominated by Eocene–Oligocene effects can produce hydrocarbons to charge structures formed during the Paleocene episode. With factors such as this crucial in the preservation of hydrocarbons to the present day, a detailed understanding of the magnitude and timing of exhumation is essential in order to ensure further discoveries in these exhumed basins.

References

Arne, D. C., Grist, A. M., Zentilli, M., Collins, M., Embry, A. & Gentzis, T. 2002. Cooling of the Sverdrup Basin during Tertiary basin inversion: implications for hydrocarbon exploration. *Basin Research*, **14**, 183–205.

Bird, K. J. & Molenaar, C. M. 1992. The North Slope foreland basin, Alaska. In: Macqueen, R. W. & Leckie, D. A. (eds) *Foreland Basins and Fold Belts*. American Association of Petroleum Geologists, Tulsa, OK, Memoirs, **55**, 363–393.

Blythe, A. & Kleinspehn, K. L. 1998. Tectonically versus climatically driven Cenozoic exhumation of the Eurasian plate margin, Svalbard: fission track analyses. *Tectonics*, **17**, 621–639.

Bray, R. J., Green, P. F. & Duddy, I. R. 1992. Thermal history reconstruction using apatite fission track analysis and vitrinite reflectance: a case study from the UK East Midlands and the Southern North Sea. In: Hardman, R. F. P. (ed.) *Exploration Britain: Into the Next Decade*. Geological Society, London, Special Publications, **67**, 3–25.

Burns, M., Hayba, D. O., Rowan, E. L. & Houseknecht, D. W. 2005. Estimating the amount of eroded section in a partially exhumed basin from geophysical well logs: an example from the North Slope. USGS Geological Survey Profession Paper **1732-D**.

Cameron, T. D. J., Crosby, A., Balson, P. S., Jeffrey, D. H., Lott, G. K., Bulat, J. & Harrison, D. J. 1992. *United Kingdom Offshore Regional Report: The Geology of the Southern North Sea*. British Geological Survey, London.

Cavanagh, A. J., Di Primio, R., Scheck-Wenderoth, M. & Horsfield, B. 2006. Severity and timing of Cenozoic exhumation in the southwestern Barents Sea. *Journal of the Geological Society, London*, **163**, 761–774.

Cobbold, P. R., Meisling, K. & Mount, V. S. 2001. Reactivation of an obliquely rifted margin, Campos and Santos basins, southeastern Brazil. *American Association of Petroleum Geologists Bulletin*, **85**, 1925–1944.

De Paor, D. G., Bradley, D. C., Eisenstadt, G. & Phillips, S. M. 1989. The Arctic Eurekan Orogeny: a most unusual fold-and-thrust belt. *Bulletin of the Geological Society of America*, **101**, 952–967.

Doré, A. G. 1995. Barents Sea geology, petroleum resources and commercial potential. *Arctic*, **48**, 207–221.

Doré, A. G., Corcoran, D. V. & Scotchman, I. C. 2002. Prediction of the hydrocarbon system in exhumed basins, and application to the NW European margin. In: Doré, A. G., Cartwright, J., Stoker, M. S.,

Turner, J. P. & White, N. (eds) *Exhumation of the North Atlantic Margin: Timing, Mechanisms and Implications for Petroleum Exploration*. Geological Society, London, Special Publications, **196**, 331–354.

Faleide, J. I., Vagnes, E. & Gudlaugsson, S. T. 1993. Late Mesozoic–Cenozoic evolution of the southwestern Barents Sea. *In*: Parker, J. R. (ed.) *Petroleum Geology of Northwest Europe: Proceedings of the 4th Conference*. Geological Society, London, 993–950; doi: 10.1144/0040993.

Gabrielsen, R. H., Grunnaleite, I. & Rasmussen, E. 1997. Cretaceous and tertiary inversion in the Bjørnøyrenna fault complex, southwestern Barents Sea. *Marine and Petroleum Geology*, **14**, 165–178.

Green, P. F., Crowhurst, P. V. & Duddy, I. R. 2004. Integration of AFTA and (U-Th)/He thermochronology to enhance the resolution and precision of thermal history reconstruction in the Anglesea-1 well, Otway Basin, SE Australia. *In*: Boult, P. J., Johns, D. R. & Lang, S. C. (eds) *Eastern Australian Basins Symposium II*. Petroleum Exploration Society of Australia, Special Publications, 117–131.

Green, P. F., Duddy, I. R. & Hegarty, K. A. 2002. Quantifying exhumation from apatite fission-track analysis and vitrinite reflectance data: precision, accuracy and latest results from the Atlantic margin of NW Europe. *In*: Doré, A. G., Cartwright, J., Stoker, M. S., Turner, J. P. & White, N. (eds) *Exhumation of the North Atlantic Margin: Timing, Mechanisms and Implications for Petroleum Exploration*. Geological Society, London, Special Publications, **196**, 331–354.

Green, P. F., Duddy, I. R., Japsen, P. & Holford, S. P. 2007. Synchronous regional kilometre-scale Late Cenozoic exhumation events around the North Atlantic region (extended abstract). *ICAM V Conference*, Tromso.

Green, P. F., Duddy, I. R., Slind, L., Calverley, F. & Rayer, F. 2005. Multiple paleo-thermal episodes in Mackenzie Corridor, NWT, Canada revealed by THR: implications for hydrocarbon prospectivity (extended abstract). *AAPG Meeting*, Calgary.

Hansen, K., Bergman, S. C. & Henk, B. 2001. The Jameson Land basin (East Greenland): a fission track study of the tectonic and thermal evolution in the Cenozoic North Atlantic spreading regime. *Tectonophysics*, **331**, 307–339.

Harland, W. B. 1997. *The Geology of Svalbard*. Geological Society, London, Memoirs, **17**.

Holford, S. P., Green, P. F., Duddy, I. R., Turner, J. P., Hillis, R. R. & Stoker, M. S. 2009. Regional intraplate exhumation episodes related to plate boundary deformation. *Bulletin of the Geological Society of America*, **121**, 1611–1628.

Holford, S. P., Green, P. F. & Turner, J. P. 2005. Palaeothermal and compaction studies in the Mochras borehole (NW Wales) reveal early Cretaceous and Neogene exhumation and argue against regional Paleogene uplift in the southern Irish Sea. *Journal of the Geological Society, London*, **162**, 829–840.

Houseknecht, D. W. & Bird, K. J. 2005. *Oil and gas resources of the Arctic Alaska petroleum province*. Geological Society of America, Boulder, CO, Professional Papers, **1732-A**.

Jackson, D. I., Jackson, A. A., Evans, D., Wingfield, R. T. R., Barnes, R. P. & Arthur, M. J. 1995. *United Kingdom Offshore Regional Report: The Geology of the Irish Sea*. British Geological Survey, London.

Japsen, P., Bonow, J. M., Green, P. F., Chalmers, J. A. & Lidmar-Bergström, K. 2009. Formation, uplift and dissection of planation surfaces at passive continental margins – a new approach. *Earth Surface Processes and Landforms*, **34**, 683–699.

Japsen, P. J., Green, P. F., Bonow, J. M., Rasmussen, E. S., Chalmers, J. A. & Kjennerud, T. 2010. Episodic uplift and exhumation along North Atlantic passive margins: implications for hydrocarbon prospectivity. *In*: Vining, B. A. & Pickering, S. C. (eds) *Petroleum Geology: From Mature Basins to New Frontiers – Proceedings of the 7th Petroleum Geology Conference*. Geological Society, London, 979–1004; doi: 10.1144/0070979.

Japsen, P., Green, P. F. & Chalmers, J. A. 2005. Separation of Palaeogene and Neogene uplift on Nuussuaq, West Greenland. *Journal of the Geological Society, London*, **162**, 299–314.

Japsen, P., Green, P. F., Nielsen, L. H., Rasmussen, E. S. & Bidstrup, T. 2007. Mesozoic-Cenozoic exhumation in the eastern North Sea Basin: a multi-disciplinary study based on palaeo-thermal, palaeo-burial, stratigraphic and seismic data. *Basin Research*, **19**, 451–490.

Johnson, K. & Gallagher, K. 2000. A preliminary Mesozoic and Cenozoic denudation history of the North East Greenland onshore margin. *Global and Planetary Change*, **24**, 303–309.

Johnsson, M. J., Howell, D. G. & Bird, K. J. 1993. Thermal maturity patterns in Alaska: implications for tectonic evolution and hydrocarbon potential. *AAPG Bulletin*, **77**, 1874–1903.

Logan, P. & Duddy, I. R. 1998. An investigation of the thermal history of the Ahnet and Reggane Basins, Central Algeria, and the consequences for hydrocarbon generation and accumulation. *In*: MacGregor, D. S., Moody, R. T. & Clark-Lowes, D. D. (eds) *Petroleum Geology of North Africa*. Geological Society, London, Special Publications, **132**, 131–155.

Magnavita, L. P., Davison, I. & Kusznir, N. J. 1994. Rifting, erosion, and uplift history of the Reconcavo-Tucano-Jatoba Rift, northeast Brazil. *Tectonics*, **13**, 367–388.

Nøttvedt, A., Cecchi, M. *et al.* 1993. Svalbard-Barents Sea correlation: a short review. *In*: Vorren, T. O., Bergsager, E., Dahl-Stammes, Ø. A., Holter, E., Johansen, B., Lie, E. & Lund, T. B. (eds) *Arctic Geology and Petroleum Potential*. Norwegian Petroleum Society, Trondheim, Special Publications, **2**, 363–375.

Nyland, B., Jensen, L. N., Skagen, J., Skarpnes, O. & Vorren, T. 1992. Tertiary uplift and erosion in the Barents Sea: magnitude, timing and consequences. *In*: Larsen, R. M., Brekke, H., Larsen, B. T. & Talleraas, E. (eds) *Structural and Tectonic Modelling and its Application to Petroleum Geology*. Norwegian Petroleum Society, Trondheim, Special Publications, **1**, 153–162.

O'Sullivan, P. B. 1996. Late Mesozoic and Cenozoic thermotectonic evolution of the Colville Basin, North Slope, Alaska. *In*: Johnsson, M. J. & Howell, D. G. (eds) *Thermal Evolution of Sedimentary Basins in Alaska*. US Geological Survey, Reston, VA, Bulletins, **2142**, 45–79.

O'Sullivan, P. B. 1999. Thermochronology, denudation and variations in paleo-surface temperature: a case study from the North Slope foreland basin, Alaska. *Basin Research*, **11**, 191–204.

O'Sullivan, P. B., Green, P. F., Bergman, S. C., Decker, J. U., Duddy, I. R., Gleadow, A. J. W. & Turner, D. L. 1993. Multiple phases of Tertiary uplift and erosion in the Arctic National Wildlife Refuge, Alaska, revealed by apatite fission track analysis. *AAPG Bulletin*, **77**, 359–385.

O'Sullivan, P. B., Murphy, J. M. & Blythe, A. E. 1997. Late Mesozoic and Cenozoic thermotectonic evolution of the central Brooks Range and adjacent North Slope foreland basin, Alaska: including fission track results from the Trans-Alaska Crustal Transect (TACT). *Journal of Geophysics Research*, **102**, 20 821–20 845.

Praeg, D., Stoker, M. S., Shannon, P. M., Ceramicola, S., Hjelstuen, B. O., Laberg, J. S. & Mathiesen, A. 2005. Episodic Cenozoic tectonism and the development of the NW European 'passive' continental margin. *Marine and Petroleum Geology*, **22**, 977–1005.

Riis, F. & Fjeldskaar, W. 1992. On the magnitude of the late Tertiary and Quaternary erosion and its significance for the uplift of Scandinavia and the Barents Sea. *In*: Larsen, R. M., Brekke, H., Larsen, B. T. & Talleraas, E. (eds) *Structural and Tectonic Modelling and its Application to Petroleum Geology*. Norwegian Petroleum Society, Trondheim, Special Publications, **1**, 163–185.

Stoker, M. S., Praeg, D., Hjelstuen, B. O., Laberg, J. S., Nielsen, T. & Shannon, P. M. 2005. Neogene stratigraphy and the sedimentary and oceanographic development of the NW European Atlantic margin. *Marine and Petroleum Geology*, **22**, 977–1005.

Thomson, K., Green, P. F., Whitham, A. G., Price, S. P. & Underhill, J. R. 1999. New constraints on the thermal history of North-East Greenland from apatite fission-track analysis. *Geological Society of America Bulletin*, **111**, 1054–1068.

Turner, J. P., Green, P. F., Holford, S. P. & Lawrence, S. R. 2008. Thermal history of the Rio-Muni (West Africa)–NE Brazil margins during continental breakup. *Earth and Planetary Science Letters*, **270**, 354–367.

Offset and curvature of the Novaya Zemlya fold-and-thrust belt, Arctic Russia

R. A. SCOTT,[1] J. P. HOWARD,[1] L. GUO,[1] R. SCHEKOLDIN[2] and V. PEASE[3]

[1]*CASP, Department of Earth Sciences, University of Cambridge, West Building, 181A Huntingdon Road, Cambridge CB3 0DH, UK (e-mail: robert.scott@casp.cam.ac.uk)*
[2]*Department of Historical and Dynamic Geology, St Petersburg State Mining Institute, House 2, 21 Line, 199106 St Petersburg, Russia*
[3]*Department of Geology & Geochemistry, Stockholm University, 104 05 Stockholm, Sweden*

Abstract: The Novaya Zemlya archipelago contains a predominantly west-vergent fold-and-thrust belt that separates two contrasting hydrocarbon basins with enigmatic subsidence histories. On the foreland side is the deep depression of the eastern Barents Shelf that hosts the Shtokman gas condensate discovery; on the hinterland side is the South Kara Basin, an offshore continuation of the gas-dominated northern West Siberian Basin. Much of the compressional deformation recorded in Novaya Zemlya appears to have been later than the onset of subsidence in adjacent basins, and may therefore be expected to have had a potentially significant influence on hydrocarbon systems within them. Two characteristics of Novaya Zemlya immediately stand out on any topographic map: the c. 600 km westward offset compared with the remainder of the Uralian Orogen and the plan-view curvature (convex towards the Barents Shelf). Any regional tectonic model developed for the Novaya Zemlya fold-and-thrust belt must be able to explain these first-order features, and a wide range of mechanisms, geometries and timings has been proposed in the literature. However, as far as we are aware, there has been no previous attempt to link the geometry of structures on the archipelago with a potential mechanism that explains their curvature in plan view. Using field observations, information on geological maps and interpretation of satellite imagery, we demonstrate the link between structural geometries in Novaya Zemlya and the basins of the adjacent eastern Barents Shelf. We find no evidence to support previous interpretations of the fold-and-thrust belt as an orocline (bending of an originally straight deformation belt) or a far-travelled thin-skinned allochthon, and conclude that the offset from the remainder of the Uralian Orogen is a primary feature that results from an original embayment on the margin of Baltica.

Keywords: tectonics, Novaya Zemlya, Barents Shelf, Russia, Arctic, fold-and-thrust belt, orogeny

A continuous belt of compressional deformation can be traced from the Aral Sea in Kazakhstan, northwards along the linear Ural mountain chain and then curving eastward through the Russian Arctic toward the edge of the Arctic Ocean in eastern Taimyr (Fig. 1). The overall cause of this deformation was the collision of the Siberia and Kazakhstan palaeocontinents with the Baltica margin of Euramerica (Laurussia) (e.g. Hamilton 1970; Zonenshain *et al.* 1984; Ziegler 1988, 1989), one of the final stages in the construction of Pangea. Most of the deformation took place in Late Palaeozoic time and is attributed to the Uralian Orogeny (Puchkov 1997), with contemporaneous granitic magmatism present in the Ural mountains (Montero *et al.* 2000) and in Taimyr (Vernikovsky *et al.* 1995). For the most part, this deformation belt does not display any abrupt changes in plan-view orientation. However, the Novaya Zemlya archipelago is the exception to this rule. When viewed on a large-scale map, two first-order characteristics of Novaya Zemlya immediately stand out: its westward offset of c. 600 km from the general trend of the orogen and its curvature (convex towards the Barents Shelf) (Fig. 1). An explanation for these features is the principal concern of this contribution.

Previous studies have adopted two contrasting hypotheses.

(1) Novaya Zemlya was originally aligned with the remainder of the Uralian orogen, such that the offset and curvature of Novaya Zemlya must result from late-stage displacement events during orogeny (i.e. Novaya Zemlya was transported c. 600 km westward during deformation). This mechanism implies significant relative movement between Novaya Zemlya and adjacent crustal blocks to the north and south.

(2) Novaya Zemlya was always offset relative to the remainder of the Uralian orogen owing to a specific characteristic of the pre-collisional continental margin, such as the presence of an embayment. This mechanism requires no relative movement between Novaya Zemlya and adjacent crustal blocks during deformation.

In the first category of model (i.e. those that explain offset and curvature to be a consequence of displacement events during orogeny), the orocline hypothesis is one of the earliest; Carey (1955) used Novaya Zemlya as an example in his development of the orocline concept, in which an initially straight orogen is subsequently bent around a vertical axis to create curvature, implying a two-stage deformation process. Hamilton (1970) proposed a variant of the orocline concept in which the c. 90° bend between Novaya Zemlya and Pai Khoi was an oroclinal fold, but the connection between northern Novaya Zemlya and Taimyr was a strike-slip fault. Mezhvilk (1995) suggested that both margins of the Novaya Zemlya block were strike-slip faults and that Novaya Zemlya had been emplaced NW relative to the remainder of the Uralides. In the same year, Otto & Bailey (1995) published an interpretation in which Novaya Zemlya was originally part of a continuous linear Uralian Orogen that was transported NW during Triassic time as a gravity-driven, thin-skinned allochthon bounded by strike-slip faults. The mechanism proposed by Otto & Bailey has remained influential in subsequent tectonic interpretations (e.g. O'Leary *et al.* 2004; Torsvik & Cocks 2004; Gee *et al.* 2006). Some authors (e.g. Zonenshain *et al.* 1990; Metelkin *et al.* 2005 and references therein) have suggested that the North Kara

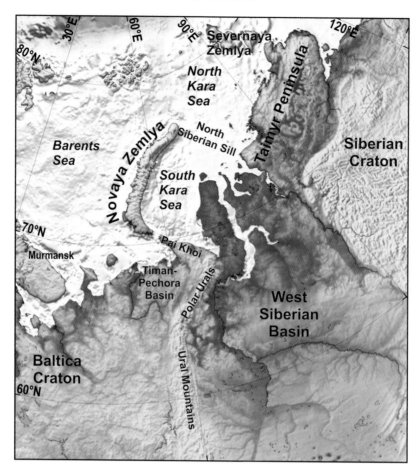

Fig. 1. Topography of northern Russia showing the curvature of the Novaya Zemlya archipelago (convex towards the Barents Shelf) and its offset with respect to the Urals and Taimyr.

region (comprising northern Taimyr, Severnaya Zemlya and the northern Kara Sea) was an independent crustal block, named the North Kara Terrane or Kara Massif, thus complicating the collisional history between Baltica and Siberia in the Arctic.

In the second category of model (i.e. those in which a precollisional embayment on the continental margin of Baltica predetermines offset and curvature), the interpretations of Ustritsky (1985) and Ziegler (1988, 1989) are the principal examples. Ustritsky (1985) proposed that a segment of oceanic crust became trapped in the South Kara Basin region, although the absence of specific tectonic information on his diagrams makes objective discussion of the subsequent deformation mechanism impossible. Ziegler (1988, 1989) depicted a pre-collisional embayment on the margin of Baltica, into which a promontory on the margin of Siberia (located in the Yamal Peninsula region) then entered during Uralian collision. This implies that a hinterland indenter was also involved in the creation of curvature. No strike-slip displacement along the margins of the embayment was depicted on Ziegler's palaeogeographic reconstructions during or after collision, although the Novaya Zemlya segment of the fold belt was depicted to propagate westward relative to the remainder of the Uralian Orogen during Permian to Early Triassic time.

In addition to identifying the mechanism(s) by which offset and curvature of Novaya Zemlya developed, we are also concerned with how the development of these features is related to the subsidence histories of adjacent basins. The fold-and-thrust belt in Novaya Zemlya is surrounded and, in part, overlain by some of the most important hydrocarbon basins in Russia (Barents Sea, Timan–Pechora, West Siberia, South Kara Sea). For the offshore basins in particular, many questions remain about the timing, magnitude and mechanism of subsidence. Furthermore, the relationship between extension and subsidence in these basins and the Novaya Zemlya compressional deformation zone that separates them is a significant issue.

In contrast to the predominantly Late Palaeozoic age of Uralian deformation elsewhere, the principal phase of compression in Novaya Zemlya is considered by the majority of authors to be Early Mesozoic (e.g. Korago et al. 1992). Stratigraphic relationships in adjacent basins and the preliminary results of our own fission track studies are consistent with this interpretation. An unresolved question in the literature, however, is whether the Early Mesozoic deformation is superimposed on earlier deformation of typical Late Palaeozoic Uralian age. Although we focus here specifically on the development of the fold-and-thrust belt, this research is part of a larger study seeking to develop a coherent tectonic model that links the Barents Shelf to the Kara Sea through Novaya Zemlya.

The South Kara Basin is generally considered to be a northward continuation of the West Siberian Basin, and began rifting in latest Permian to Early Triassic time, with a temporal, and potentially genetic, link to the eruption of the Siberian traps and/or cessation of Late Palaeozoic Uralian deformation (e.g. Hamilton 1970; Ziegler 1988, 1989; Şengör & Natal'in 1996; Allen et al. 2006; Vyssotski et al. 2006). However, this potential link with Late Palaeozoic orogenic collapse is more difficult to explain in the South Kara Basin if the main compressional deformation in Novaya Zemlya occurred in Late Triassic time (i.e. c. 50 Ma after the onset of subsidence).

The North and South Barents basins are characterized by successions approaching 20 km thick (see Fig. 6), of which around half are interpreted to be of Permo-Triassic age (e.g. Johansen et al. 1993). It is theoretically possible to model subsidence in these depocentres as the product of Permo-Triassic lithospheric extension alone (e.g. Corfield et al. 2006); however, this mechanism in isolation is considered to be an unlikely explanation of total subsidence because insufficient normal faults are visible on seismic reflection profiles to accommodate Permo-Triassic extension of the magnitude required. Additional subsidence is therefore required from pre-Permian rift events, and/or from some alternative subsidence mechanism. However, identifying any pre-Permian subsidence events on the eastern Barents Shelf, let alone constraining their timing and magnitude, is extremely difficult because the thick package of Permo-Triassic sediment masks underlying successions. It is logical to extrapolate information on pre-Permian Palaeozoic rift events from Timan–Pechora (e.g. O'Leary et al. 2004), but this inevitably relies on assumptions that are impossible to validate with current datasets. There are very few papers in which the relationship between Novaya Zemlya and the eastern Barents Shelf is directly addressed. Otto & Bailey (1995) concluded that Late Permian extension-driven subsidence in the eastern Barents Shelf produced the necessary gradient and gravitational instability to promote Late Triassic allochthonous emplacement of the Novaya Zemlya fold-and-thrust belt from an original position approximately 600 km further east. During this emplacement, the allochthonous thin-skinned thrust sheets overrode any foreland basin deposits, an argument used by O'Leary et al. (2004) to explain their modelled 1D subsidence data from the South Barents Basin.

No detailed database of structural measurements exists for Novaya Zemlya, nor is the collection of such a comprehensive dataset ever likely to be possible, considering the access restrictions to substantial portions of the archipelago, the permanent ice cap in the north and the deleterious effects of frost-shattering on many inland exposures in subdued terrain. However, using data collected during CASP expeditions to the south (2004) and north (2005) of the archipelago, interpretation of satellite images and published maps, and regional considerations, it is possible to draw some new conclusions concerning the origin of curvature and offset, and its significance for adjacent basins. As far as we aware there has been no previous attempt to link the geometry of structures on the archipelago with a potential mechanism that explains their curvature in plan view.

Curvature in fold-and-thrust belts: causes and classification schemes

Many fold-and-thrust belts have an element of curvature in plan view, and contain a distinctive pattern of structural trend lines (fault traces, fold axial plane traces and cleavage) within them – Novaya Zemlya represents a typical example. However, relatively few, if any, fold-and-thrust belts have been studied in sufficient detail for the processes involved in generating curvature to be unequivocally identified (Wezel 1986; Hindle & Burkhard 1999). It is clear that no single mechanism can explain all cases, and a number of genetic classification schemes for arcuate fold-and-thrust belts have been proposed (e.g. Marshak 1988, 2004; Ferrill & Groshong 1993; Hindle & Burkhard 1999; Macedo & Marshak 1999). Even in the most thoroughly studied examples (e.g. the Jura arc; Hindle & Burkhard 1999) where there is a comprehensive understanding of structural geometries and a substantial database of strain measurements with which to predict the precise trajectory of material during deformation, the mechanism by which the curvature developed is still controversial.

Hindle & Burkhard (1999) suggested a genetic classification scheme with three end-members identified on the basis of total strain patterns and displacement vector fields (Fig. 2). The end-members were: (1) oroclines (pure bending of an initially straight orogen, implying a two-stage process); (2) Piedmont glacier (divergent transport directions); (3) primary arcs (uniform transport direction). Identification of end-member categories based on displacement directions requires knowledge of the total strain distribution in a fold-and-thrust belt (or at least gradients in strain along- and across-strike) and such data are not available for Novaya Zemlya. However, our principal reason for citing Hindle & Burkhard (1999) is that they depicted Novaya Zemlya as an example of the 'orocline' end-member, reproducing illustrations from Carey (1955), that is, two-stage formation with bending of an originally straight Uralian Orogen about a vertical axis. This starting geometry was also proposed by Otto & Bailey (1995), although their model then invokes a different mechanism to generate offset and curvature. We therefore approach the problem of explaining the curvature of Novaya Zemlya, and its offset with respect to the remainder of the Uralian Orogen, with these previous hypotheses in mind.

The classification scheme employed by Macedo & Marshak (1999) and Marshak (2004) is similar to that proposed by Hindle & Burkhard (1999) and identifies two end-members: (1) rotational curves (equivalent to the orocline of Hindle & Burkhard 1999); and (2) non-rotational curves (equivalent to the primary arc of Hindle & Burkhard 1999) (Fig. 2). The Macedo & Marshak (1999) and Marshak (2004) classification is, however, more readily applied to Novaya Zemlya because it draws additionally on the information that can be obtained from plan-view patterns of structural trend lines (Fig. 3), without the need for detailed strain data.

The most common control on plan-view curvature in fold-and-thrust belts identified by Macedo & Marshak (1999) and Marshak (2004) was some geometric and/or mechanical property inherited from pre-existing sedimentary basins within the foreland of a fold-and-thrust belt, either acting alone or in combination. They referred to this type of curvature as 'basin controlled'. Properties of precursor sedimentary depocentres such as along-strike variations in the thickness of the deformed succession, critical-taper angle and angle of basal detachment determine the width of a fold-and-thrust belt for a given quantity of upper-crustal shortening (Fig. 3). The wider segments of the fold-and-thrust belt develop into salients (plan-view curves that are convex in the direction of tectonic transport; e.g. Aitken & Long 1978; Marshak & Wilkerson 1992; Marshak et al. 1992; Calassou et al. 1993; Boyer 1995; Mitra 1997; Macedo & Marshak 1999; Marshak 2004). Basin-controlled salients are characterized by structural trend lines that are more widely spaced at the apex than at the endpoints (Fig. 3). Thrusts are also commonly blind near the apex and emergent closer towards the endpoints, where deeper levels of the geology may be exposed. A further important characteristic of basin-controlled curves noted by Marshak (2004) is that structures are generally initiated as curves and do not involve oroclinal bending.

Of all these properties inherited from pre-existing sedimentary basins, the most significant appears to be along-strike change in the depth to detachment. In a study of fold-and-thrust belts worldwide, Macedo & Marshak (1999) noted a significant coincidence between salient apex location and pre-deformational depocentre position, implying that salients form where sediment thickness is greatest. The greater the sediment thickness, the greater the depth to detachment, and thus, for a given critical taper angle, the greater the width of the resulting fold-and-thrust belt, given the need to maintain volume balance (Fig. 3). The position of a depocentre may also coincide with a weaker basal detachment horizon compared with adjacent fold belt segments (e.g. because the lithology forming the detachment is thicker and/or more continuous in the basin). Where an underlying detachment is relatively weak, the critical taper angle decreases (Boyer 1995; Mitra 1997), forming wedges that can be transported further into the foreland than adjacent stronger rock masses for a given amount of

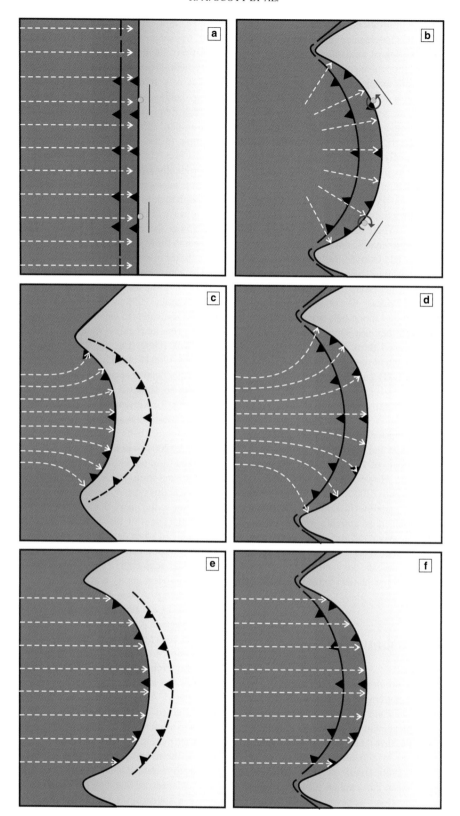

Fig. 2. The three end-member classification of arcuate fold-and-thrust belts proposed by Hindle & Burkhard (1999). In each case, the right-hand side diagram represents a later stage in development compared with left-hand side. (**a, b**) 'Orocline' (bending of an initially straight belt); (**c, d**) 'Piedmont glacier' (divergent transport directions); (**e, f**) 'primary arc' (uniform transport direction). White arrows represent displacement vectors. Red arrows in (b) denote the sense of rotation on the fold belt flanks as curvature develops. Thrusts indicated in (c) and (e) with a dashed line represent their incipient position in the foreland as the fold-and-thrust belt propagates. In the Marshak (2004) classification, (a) and (b) represent development of a rotational curve and (e) and (f) illustrate the development of a non-rotational curve. Marshak (2004) did not include the intermediate 'Piedmont glacier' curve in his classification scheme.

regional shortening (Davis & Engelder 1985; Jaumé & Lillie 1988; Marshak 2004).

Less common stimuli to the development of curvature identified by Macedo & Marshak (1999) and Marshak (2004) include irregularities on colliding margins (e.g. hinterland indenters or foreland promontories), interaction with later strike-slip faults, and distortion of the underthrust slab. The collision of a hinterland indenter into a sedimentary basin will generate a salient in the

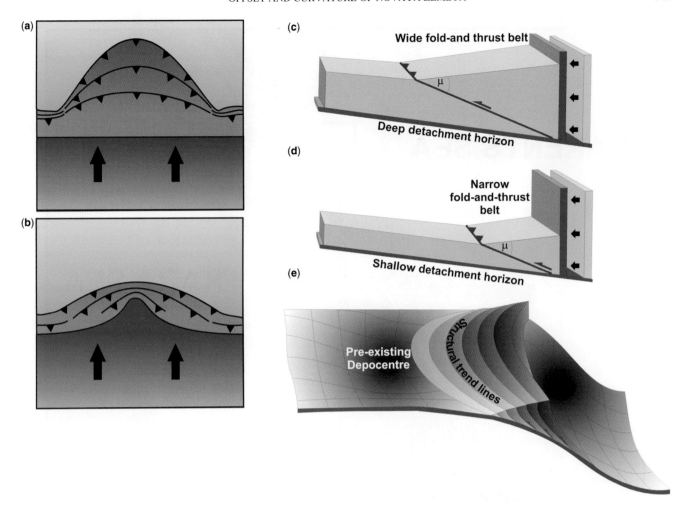

Fig. 3. Comparison of typical structural trend-line patterns in (**a**) a basin-controlled salient v. (**b**) a salient formed by collision of a hinterland indenter. (**c, d**) Schematic illustrations demonstrating how depth to detachment influences the width of a fold-and-thrust belt for a given amount of displacement. (**e**) Schematic illustration of the frontal part of a salient demonstrating how a pre-existing depocentre with gradually sloping margins produces smooth curvature. If the pre-existing depocentre has abrupt, steeply dipping margins, the salient margins are more likely to be affected by strike-slip faults. μ, critical taper angle. All diagrams adapted from illustrations in Marshak (2004).

foreland, but with precisely the opposite pattern of structural trend lines compared with a basin-controlled salient (Macedo & Marshak 1999) (Fig. 3). If the foreland to an advancing fold-and-thrust belt is irregular (i.e. it has promontories and/or embayments), a margin-controlled curve may develop as the deformation belt adapts to the shape of the margin. As Marshak (2004) pointed out, the effect of pre-deformational shape of a continental margin may be intimately linked with the effect of basin geometry on curve formation, because recesses in a continental margin may correspond to the location of wider and deeper basins. Rotation of segments of a fold-and-thrust belt around a vertical axis (i.e. oroclinal bending *sensu stricto*) is most likely to occur if curvature is induced by strike-slip faulting and/or foreland obstacles.

Structural patterns in Novaya Zemlya

The Novaya Zemlya archipelago separates the eastern Barents Shelf from the southern Kara Sea and contains a fold-and-thrust belt that exposes a Precambrian through Early Triassic succession. The fold-and-thrust belt is structurally connected to the Polar Urals through the Pai Khoi fold belt in the south, and to the Taimyr fold belt through the submerged Siberian Sill in the north (Fig. 1). The depositional setting and fauna of the Palaeozoic succession exposed on the Novaya Zemlya archipelago implies that it was located along the eastern margin of Baltica (present-day orientation) prior to deformation, with platform sediments predominant on the western side of the archipelago and slope deposits on the eastern side (e.g. Bondarev 1982) adjacent to the Uralian Ocean. The affinity with Baltica has been confirmed using detrital zircon ages from Early Palaeozoic sediments (Pease & Scott 2009). The west-to-east transition from platform to slope deposits on the margin of Baltica is also characteristic of the external parts of the main Uralian deformation belt further south (e.g. Puchkov 1997).

The Novaya Zemlya archipelago comprises two main islands separated by a narrow channel located around 73°20′ N, and forms a sweeping arc more than 900 km long that separates the Barents Sea from the Kara Sea (Figs 1 & 4). The pattern of major structures exposed on Novaya Zemlya broadly follows the trend of the archipelago. In detail, however, the plan-view shape of the archipelago is more complex, and is not a simple curve of constant curvature. The southern part of the archipelago is characterized by a relatively tight curve coincident with a broadening of the south island to *c.* 140 km wide (Fig. 4). The central part of the archipelago has virtually no curvature, with the width decreasing to <90 km in the vicinity of the channel between the north and south islands. In the northern part of the archipelago the width of the north island increases to *c.* 110 km as the curvature also increases, and towards the northernmost part of the island curvature decreases again as the width is reduced to around 60 km.

The main structural trend lines in Novaya Zemlya are summarized on Figure 5, which is based on more detailed compilations

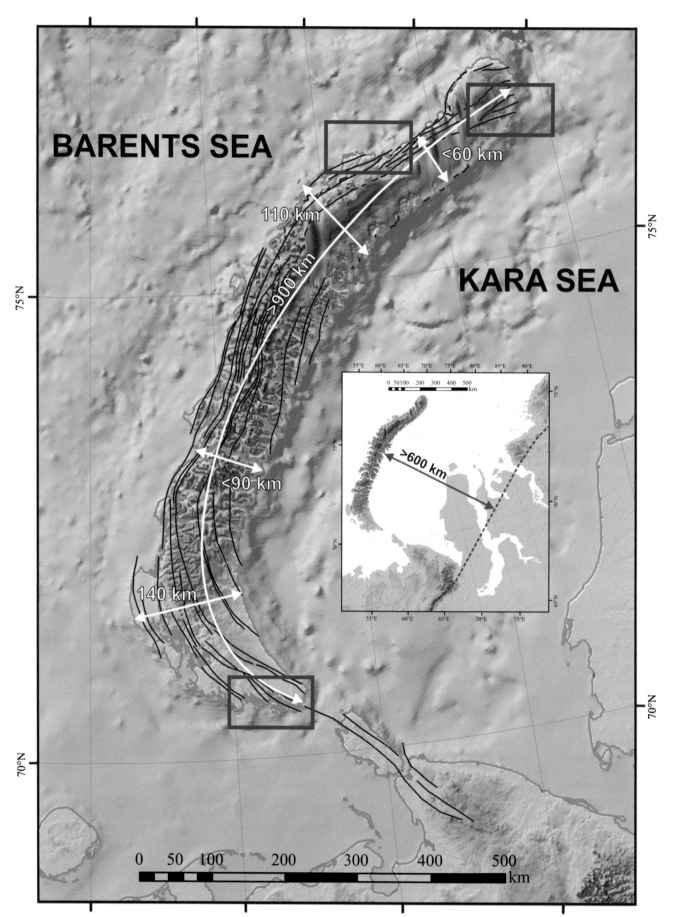

Fig. 4. Principal dimensions of the Novaya Zemlya archipelago. Structural trend lines (major faults and fold axes) are shown in black. Field work localities are indicated by the red boxes.

from geological maps and public-domain Landsat and Corona satellite imagery. Also shown is the assumed structural zonation of the fold-and-thrust belt, extrapolated from information presented in Korago et al. (1992). Korago et al. identified four north–south trending structural domains in the central portion of the archipelago (outlined by the white dashed box in Fig. 5). From west to east, these structural domains were described by Korago et al. (1992) as follows:

- Western Structural Domain – contains the most intense deformation and is characterized by an imbricate, west-directed thrust belt.
- Central Structural Domain – described as a negative structure, with less intense deformation than adjacent zones and with more continuity and better preservation of folds.
- Axial Structural Domain – structural intensity approaches that in the Western Structural Domain, but is characterized by eastward vergence.
- Eastern Structural Domain – low-lying region with limited exposure, but is characterized by less intense deformation and generally upright symmetrical or slightly east-verging folds.

The structural domains, as defined by Korago et al. (1992), occur in an area of Novaya Zemlya that is inaccessible to western geologists and therefore cannot be corroborated with our own observations. We have extrapolated these four domains along the length of the entire archipelago using the pattern of structural trend lines on the basis that our own observations in southern and northern Novaya Zemlya are broadly consistent with the Korago et al. (1992) descriptions. The overall structural style that we observed in the field, which is reported by Korago et al. (1992) and can be discerned on published geological maps and sections, is consistent with a predominantly thick-skinned style of deformation characterized by limited horizontal transport (a few tens of kilometres maximum). Only on the western side of the fold-and-thrust belt do structures exhibit a trend towards more thin-skinned geometries. For the arguments presented here, however, the trend of the domains is more important than the internal detail of their structural architecture.

Using the terminology of Marshak (2004), the entire fold-and-thrust belt exposed on Novaya Zemlya constitutes a large salient. However, it is apparent from the pattern of structural trend lines and structural domains depicted in Figure 5 that the large salient is a compound feature comprising two smaller salients: a gentle curve in the north which is convex towards the NW, and a tighter curve in the south which is convex towards the SW. Although the current geographic width of the archipelago is not completely equivalent to the width of the known fold-and-thrust belt (see Fig. 5), there is a clear systematic relationship between the pattern of structural trend lines and the width of the archipelago: (1) the northern salient coincides with the broadening of the archipelago to c. 110 km, and the southern salient coincides with the broadening of the archipelago to c. 140 km (Fig. 4); (2) trend lines are further apart near the apexes of each of these salients than at the endpoints, and trend lines are most closely spaced in the central part of the archipelago between the two salients – this is the part of the archipelago which thins to <90 km and has no significant curvature (Fig. 4).

It is also apparent that the northern and southern salients expose different elements of the structural domains. In the northern salient all four of the domains defined by Korago et al. (1992) are exposed, and this is the only part of the archipelago where the Eastern Structural Domain is exposed. Structural trend lines have to be projected beneath the permanent ice sheet in northern Novaya Zemlya, adding an element of uncertainty. Further south, structures consistently trend offshore on the eastern (Kara Sea) side of the archipelago so that the southern salient is dominated only by exposures of the Western and Central Structural Domains. We assume that the Axial and Eastern Structural Domains are present beneath the SW part of the Kara Sea shelf (Fig. 5).

The extrapolated trend lines suggest that the Central Structural Domain has a much greater outcrop width in the southern salient than in the northern salient. This may be an artefact of the inherent uncertainties involved in extrapolation, but arguments could equally be made that the change of width is real. For example, the northern salient is characterized by much higher topographic relief than the southern salient (Fig. 4), which may indicate some basement impediment to horizontal propagation of the fold-and-thrust belt in the north compared with the south. This impediment may have been the Admiralty Arch, located in the Barents Sea close to the apex of the northern salient (Fig. 5). This structural high is commonly depicted as a long-lived positive feature on palaeogeographic maps (e.g. Smelror et al. 2009).

The outcrops of Precambrian–Cambrian 'basement' rocks are also shown on Figure 5. Precambrian–Cambrian rocks crop out in only a few locations in Novaya Zemlya and are unconformably overlain by Palaeozoic strata (Korago et al. 2004; Pease & Scott 2009). Although the outcrops are relatively small, they are highly significant for understanding the evolution of the fold-and-thrust belt because they represent the areas in which the deepest geological levels are exposed. The 'basement' outcrops have a clearly defined relationship with the structural pattern: (1) they are consistently located close to the boundary between the Western Structural Domain and the Central Structural Domain; (2) they occur only at the margins of the salients and are symmetrically disposed with respect to the apex of each salient (Fig. 5).

Interpretation of structural patterns in Novaya Zemlya

Cause of curvature

The plan-view geometry of structures in Novaya Zemlya is consistent with an overall westward emplacement of thrust sheets, and the structural pattern clearly implies that the basinal geometry of the eastern Barents Shelf had a significant influence on development of the Novaya Zemlya fold-and-thrust belt. The two salients that we have identified on Novaya Zemlya were emplaced towards the two main depocentres on the eastern Barents Shelf, whereas the intervening recess lies adjacent to the Ludlov Saddle, the structural high that separates the North Barents Basin from the South Barents Basin (Fig. 6). This implies that the two salients are basin-controlled curves in the terminology of Marshak (2004), and demonstrates that the North and South Barents basins were distinct entities separated by an intervening high at the onset of deformation.

The distribution of Precambrian–Cambrian outcrops (Fig. 5) is consistent with the observation that deeper structural levels are exposed at the margins of salients (Macedo & Marshak 1999; Marshak 2004). This adds further weight to the argument that increased depth to detachment in the pre-existing eastern Barents depocentres was an important element that controlled curvature. With estimated sediment thicknesses up to 20 km in the South and North Barents basins, the detachment horizon(s) would be expected to be significantly deeper in the basin centres than the basin margins, although this pattern may have been disrupted in the north by the presence of the Admiralty Arch (this potential disruption to palaeogeographic connectivity may explain why the northern salient is less well developed than the southern salient). The expanding pattern of structural trend lines towards the apex of the salients is also consistent with basin-controlled curves, and there is no evidence from the pattern of trend lines that hinterland indenters were significant (cf. Ziegler 1988, 1989).

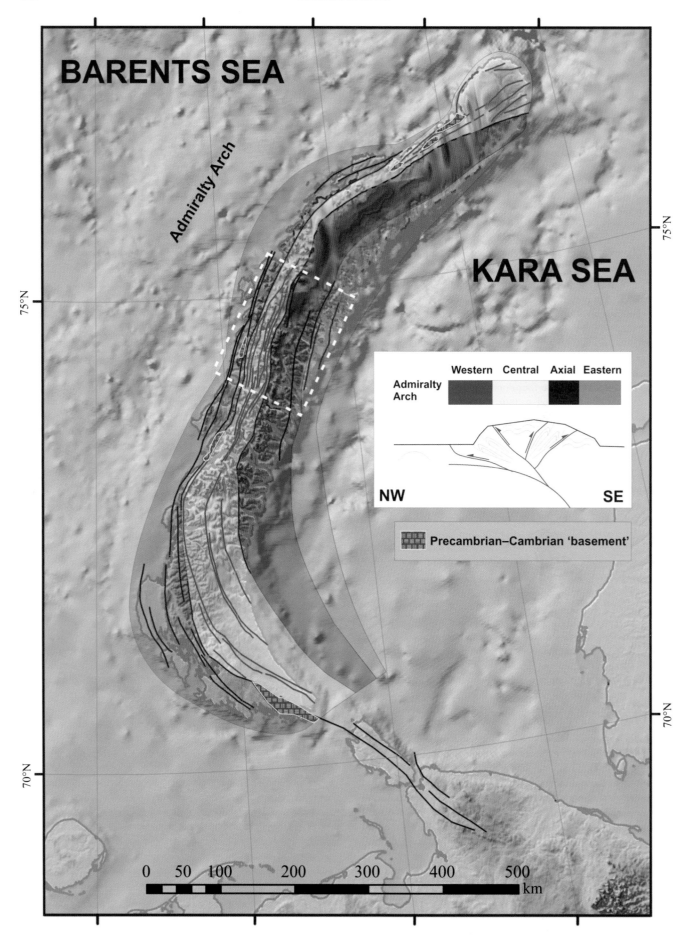

Fig. 5. Topography and structural trend lines of Novaya Zemlya from Figure 4, overlain by structural domains extrapolated from those defined by Korago *et al.* (1992) in central Novaya Zemlya (in the area outlined by the white dashed box). Inset figure shows a schematic cross-section across the northern half of Novaya Zemlya illustrating the assumed relationships between the four structural domains. Structural trend lines (major faults and fold axes) are shown in black. Precambrian–Cambrian exposures are also shown.

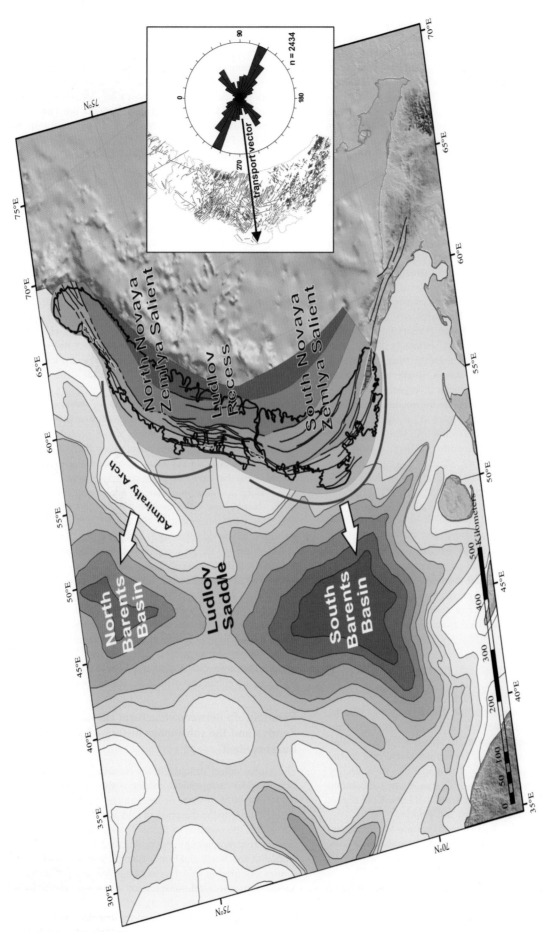

Fig. 6. Oblique view illustrating the relationship between the geometry of the Novaya Zemlya fold-and-thrust belt and structure features on the eastern Barents Shelf. Contouring on Barents Shelf represents depth to basement, with sediment thicknesses of c. 20 km in the South Barents Basin. The North Novaya Zemlya Salient and South Novaya Zemlya Salient have developed by enhanced propagation induced by a deepening of basal detachment levels in the pre-existing North Barents Basin and South Barents Basin, respectively (cf. Fig. 3e). Note that the presence of the Admiralty Arch may have inhibited the development of the northern salient compared with the southern salient. The Ludlov Recess has formed in response to the presence of a pre-existing high in the vicinity of the Ludlov Saddle. Novaya Zemlya structural domains (green shading) are named on Figure 5. Precambrian/Cambrian exposures are also shown (see also Fig. 5). The inset shows the relationship between lineament orientation taken from satellite images and the South Novaya Zemlya salient.

It is not clear from field observations at what stratigraphic level the main detachment occurs on the foreland side of the Novaya Zemlya fold-and-thrust belt, and in this respect it is difficult to predict whether the weakness of any particular detachment is enhanced in the basinal areas. In the southern and middle Urals, Brown et al. (2006) recognized foreland detachments in units of Late Riphean, Late Vendian and Ordovician age. In Timan–Pechora, Sobornov (1994) illustrated Early Palaeozoic and Artinskian (mid-Permian) detachments. Fine-grained clastic units occur at many levels in the Late Precambrian and Palaeozoic succession exposed on Novaya Zemlya, offering numerous potential décollement levels.

When seeking to explain the formation of an arcuate fold-and-thrust belt, the instinctive interpretation implied by the geometry is that thrust sheets are transported perpendicular to the thrust front at all localities, and thus there must have been radially divergent spreading of material toward the foreland. In the case of Novaya Zemlya, this line of reasoning could be used to explain significant extensional collapse in the hinterland (i.e. formation of the South Kara Basin). However, structural lineaments visible in satellite imagery of southern Novaya Zemlya are arranged in a simple conjugate pattern that is symmetrically distributed about the salient apex (Fig. 6). This symmetry with the presumed transport vector towards the South Barents Basin implies that formation of these lineaments was intimately linked with emplacement of the fold-and-thrust belt. Furthermore, the simple symmetrical pattern is consistent with a single emplacement event and would argue against significant divergent transport or oroclinal bending. We therefore conclude the southern salient in Novaya Zemlya is more likely to be closer to the 'non-rotational curve' end-member in the classification of Marshak (2004) (equivalent to the 'primary arc' end-member of Hindle & Burkhard 1999). Owing to the permanent ice sheet in northern Novaya Zemlya, insufficient lineaments can be identified to draw inferences for the northern salient with the same certainty; however, there is no evidence to suggest that it is any different to the southern salient.

We note also that Hindle & Burkhard (1999) considered their 'primary arc' end-member to be the most appropriate interpretation for the Jura arc based on extensive analysis of 2D balanced sections and other strain data. As they pointed out, severe space problems would be encountered in the hinterland after even limited radial divergence of material because so much rock mass is required to be derived from the same area in the centre of the arc.

Cause of offset

Structural patterns in Novaya Zemlya argue against significant oroclinal bending. It follows, therefore, that the only way that the fold-and-thrust belt could have been offset from a position originally in straight-line continuity with the remainder of the Uralian Orogen is by long-distance (c. 600 km) emplacement of a crustal block bounded by strike-slip faults. The model of Otto & Bailey (1995) invokes such a mechanism, with a thin-skinned allochthon bounded by major strike-slip fault systems in the Pai Khoi region and between northern Novaya Zemlya and Taimyr.

A number of observations lead us to question a long-distance, thin-skinned emplacement mechanism involving substantial horizontal thrust transport:

(1) the overall structural style of Novaya Zemlya is thick-skinned and implies limited horizontal translation.
(2) if Novaya Zemlya had originally been aligned with the remainder of the Late Palaeozoic Uralian Orogen and then emplaced westward at a later stage, the fold-and-thrust belt would be expected to preserve a complex, multi-stage deformation history, although, as we describe above, only a simple structural pattern is apparent, consistent with a single, uncomplicated emplacement event.
(3) no structural evidence was observed for significant strike-slip displacement, discontinuity of major structures or rotation of rock masses during CASP field work, even though this field work includes some of the most likely localities in which evidence would be expected according to the Otto & Bailey hypothesis (e.g. southernmost and northernmost Novaya Zemlya; Fig. 4). Significant strike-slip displacement or structural discontinuity is also not obvious on satellite images, nor can it be observed on geological maps. (We accept, however, that on its own this lack of evidence is not proof, because the main bounding fault systems may be offshore.)
(4) Even with a basal décollement that has a dip of <1°, the gravity-induced lateral translation of 600 km proposed by Otto & Bailey (1995) would require an elevation difference between each end of the décollement of >10 km. The cross-section through Novaya Zemlya and the eastern Barents Sea depicted by Otto & Bailey (their fig. 4) shows a decollement beneath Novaya Zemlya at a depth of 2–2.5 km below sea level. On this basis, the hinterland end of the décollement should have been at an elevation of c. 8 km, which is clearly unfeasible, particularly considering the fact that subsidence in the South Kara Basin may well have begun prior to emplacement of the fold-and-thrust belt.

Given these reservations about the scale of thrust-sheet transport, we consider that the offset of Novaya Zemlya with respect to the remainder of the Uralian Orogen most probably reflects the presence of an original embayment along the margin of Baltica. Such embayments are not unusual, and may reflect segments of oceanic crust that have a different spreading history to adjacent segments along a continental margin. For example, the embayment in the Norwegian–Greenland Sea adjacent to the Møre Basin coincides with a segment of oceanic crust that has a different spreading history to areas north and south (e.g. Scott et al. 2005).

Assuming that the presence of an original embayment along the margin of Baltica controlled the offset of Novaya Zemlya, this would make Novaya Zemlya a margin-controlled salient in the terminology of Marshak (2004). However, as we have demonstrated in the current study, superimposed on the broad curve of Novaya Zemlya are two smaller salients, the plan-view morphology of which must at least in part be generated in response to the presence of pre-existing depocentres on the former margin of Baltica. In this respect, therefore, Novaya Zemlya is probably best considered to represent a combination of margin- and basin-controlled salients.

Relationship between structural curvature in Novaya Zemlya and the subsidence history of the eastern Barents Shelf

The double salient structure of the Novaya Zemlya fold-and-thrust belt provides independent evidence that the North and South Barents basins were discrete depocentres, separated by an intervening high, prior to the inception of fold-and-thrust belt development. This raises the question of precisely when and how they achieved this bipartite morphology. According to Otto & Bailey (1995), separation of the North and South Barents basins into discrete depocentres occurred in mid Triassic time due to uplift of the Ludlov Saddle during sinistral transpression along a reactivated Timanide fault trend. This is only shortly before, or penecontemporaneous with, development of the fold-and-thrust belt. The timing of depocentre development on the eastern Barents Shelf is important to our understanding of the later subsidence history, the thermal history of

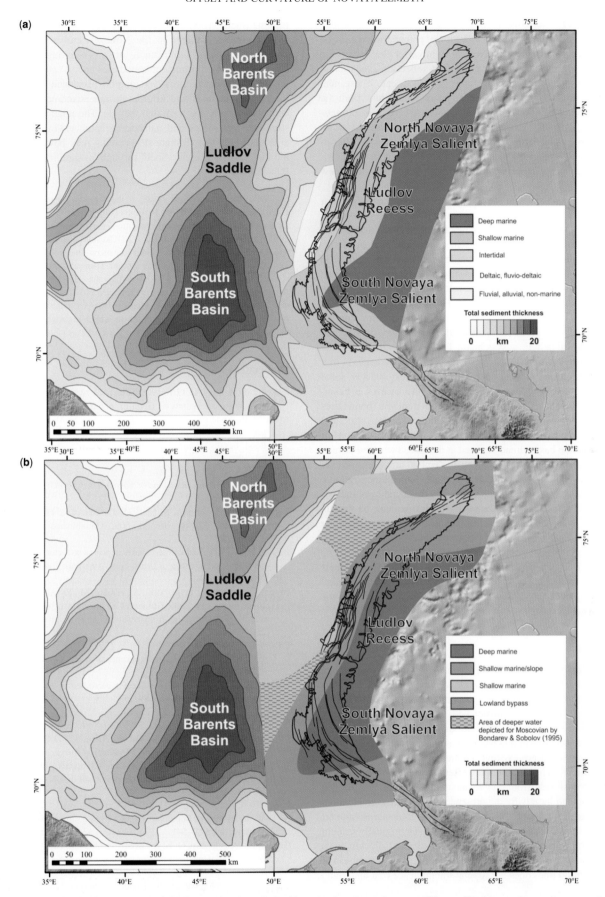

Fig. 7. (a) Lochkovian (Early Devonian) and (b) Kasimovian (Late Carboniferous) palaeogeography maps of Novaya Zemlya superimposed on a map showing total sediment thickness of the Barents Shelf and structural trend lines on Novaya Zemlya. There is a clear relationship between salient location, structural trend lines and palaeogeographic environment boundaries which implies that the North and South Barents basins, or their precursors, existed as distinct entities at these times.

the basin and the palaeogeographic evolution, which, for example, may have implications for the distribution of potential source rocks.

Palaeogeographic maps of Novaya Zemlya for the Palaeozoic consistently show deeper water to the east, reflecting a location at the margin of Baltica adjacent to the Uralian Ocean. These map sequences also show periodic deeper-water connections from the Uralian margin across Novaya Zemlya, which have been extrapolated through to the eastern Barents Shelf (e.g. Smelror et al. 2009; Guo et al. 2010; Fig. 7). Recognition of the two basin-controlled salients in Novaya Zemlya allows us to consider the significance of these possible basinal connections with new understanding.

As all the exposures in Novaya Zemlya used to constrain our palaeogeographic maps (Fig. 7) are now within an early Mesozoic fold-and-thrust belt, it can be assumed that the current distribution of palaeogeographic environments has been modified by deformation to some degree. For example, many of the maps also show a distribution of palaeo-environment boundaries that at least in part reflects the curvature of the fold-and-thrust belt, including the double salient structure (Fig. 7). However, we argue that, if any of the palaeogeographic environment boundaries can be demonstrated to cross-cut structural trend lines within the fold-and-thrust belt in a systematic fashion, it implies that the current distribution of palaeogeographic environments is not solely a function of tectonic transport but also reflects some pre-existing palaeobathymetric control. In the case of a basin-controlled salient, if palaeogeographic environment boundaries for any particular age have a more acute curvature than structural trend lines (i.e. are more acutely convex toward the foreland), it implies that the basin already existed at that age. Two pre-Permian palaeogeographic maps for Novaya Zemlya are presented as examples in Figure 7, with the structural trend lines superimposed. Inevitably, there are limitations imposed by the accuracy of palaeogeographic maps and only tentative conclusions can be drawn. However, many of the maps that we have prepared (spanning Late Ordovician to Late Carboniferous time) show palaeogeographic boundaries systematically cross-cutting structural trend lines, a phenomenon particularly apparent in the South Barents Basin. This would imply a prolonged pre-Permian history of differential subsidence on the eastern Barents Shelf, consistent with the rifting history of Timan–Pechora.

Conclusions

Novaya Zemlya contains an early Mesozoic fold-and-thrust belt characterized by an overall thick-skinned structural style and relatively low levels of exhumation compared with the Polar Urals and Taimyr, implying a 'soft' collision with limited horizontal transport of thrust sheets. Curvature of the fold-and-thrust belt was controlled by west-directed transport as a non-rotational primary arc, modified by the geometry of pre-existing sedimentary basins on the eastern Barents Shelf. The substantial offset of the Novaya Zemlya fold-and-thrust belt from the remainder of the Uralian orogen was determined largely by the presence of an original embayment on the margin of Baltica, and there is little evidence to support either development as an orocline (bending of an originally straight deformation belt) or emplacement of a thin-skinned allochthonous segment on bounding strike-slip faults. Furthermore, the pattern of structural trend lines does not support the influence of a hinterland indenter.

Compressional deformation in Novaya Zemlya occurred significantly after the onset of subsidence in adjacent basins. Palaeogeographic maps for the Palaeozoic, drawn with no attempt to compensate for later compressional deformation in the Novaya Zemlya fold-and-thrust belt, show periodic connections between the Uralian Ocean margin and depocentres on the eastern Barents Shelf, crossing Novaya Zemlya. Recognition of the two basin-controlled salients in the Novaya Zemlya fold-and-thrust belt, and of the significance of cross-cutting relationships between palaeo-environment boundaries and structural trend lines within the salients, provides tangible evidence in support of these interpretations and affords an understanding of why these connections have the form that they do. The relationship between structural trend-lines and palaeogeographic environment boundaries indicates that discrete depocentres may have been present on the eastern Barents Shelf through a significant portion of Palaeozoic time.

We thank the captains and crews of the research vessels Gorizont and Gydrolog, both operated by the Russian Hydrographic Office in Murmansk, for their logistic support during expeditions to Novaya Zemlya in 2004 and 2005, respectively. We also thank VNIIOkeangeologiya in St Petersburg and David Gee (Uppsala) for their organizational role in the Novaya Zemlya expeditions, and Henning Lorenz (Uppsala) for assistance with the satellite imagery. CASP acknowledges the industrial sponsors of its Arctic research with gratitude. Tim Kinnaird provided helpful comments on an earlier version of this manuscript. Sergey Drachev and Anatoly Nikishin provided constructive criticism which improved the quality of the manuscript.

References

Aitken, J. D. & Long, D. G. F. 1978. Mackenzie tectonic arc – reflection of early basin configuration? Geology, 6, 626–629.

Allen, M. B., Anderson, L., Searle, R. C. & Buslov, M. M. 2006. Oblique rift geometry of the West Siberian Basin: tectonic setting for the Siberian flood basalts. Journal of the Geological Society, London, 163, 901–904.

Bondarev, V. I. (ed.) 1982. Geologiya Yuzhnogo ostrova Novoy Zemli (Geology of the South Island of Novaya Zemlya). PGO 'Sevmorgeologiya', Leningrad.

Bondarev, V. I. & Sobolev, N. N. 1995. The Paleozoic of Novaya Zemla. Arctic and Russian Report 609. Unpublished report. CASP, Cambridge.

Boyer, S. E. 1995. Sedimentary basin taper as a factor controlling the geometry and advance of thrust belts. American Journal of Science, 295, 1220–1254.

Brown, D., Juhlin, C., Tryggvason, A., Friberg, M., Rybalka, A., Puchkov, V. & Petrov, G. 2006. Structural architecture of the southern and middle Urals foreland from reflection seismic profiles. Tectonics, 25, TC1002.

Calassou, S., Larroque, C. & Malavieille, J. 1993. Transfer zones of deformation in thrust wedges: an experimental study. Tectonophysics, 221, 325–344.

Carey, S. W. 1955. The orocline concept in geotectonics. Proceedings of the Royal Society of Tasmania, 89, 255–288.

Corfield, R. I., Hansen, E.-K., Fugelli, E., Milner, P., Kittilsen, J.-E., Roberts, A. & Matthews, S. 2006. Structural evolution of the Russian Barents Sea. In: Fraser, A., Allen, M., O'Callaghan, M. & Hamilton, R. (eds) Oil and Gas Habitats of Russia and Surrounding Regions. Geological Society, London, 9. (abstract volume).

Davis, D. M. & Engelder, T. 1985. The role of salt in fold-and-thrust belts. Tectonophysics, 119, 67–88.

Ferrill, D. A. & Groshong, R. H. J. 1993. Kinematic model for the curvature of the northern Subalpine Chain, France. Journal of Structural Geology, 15, 523–541.

Gee, D. G., Bogolepova, O. K. & Lorenz, H. O. 2006. The Timanide, Caledonide and Uralide Orogens in the Eurasian High Arctic, and Relationships to the Palaeo-continents Laurentia, Baltica and Siberia. Geological Society, London, Memoirs, 32, 507–520.

Guo, L., Schekoldin, R. & Scott, R. A. 2010. Sedimentology, palaeogeography and hydrocarbon significance of the Devonian succession in northern Novaya Zemlya, Arctic Russia. Journal of Petroleum Geology, 33, 105–122.

Hamilton, W. 1970. The Uralides and the motion of the Siberian and Russian platforms. Bulletin of the Geological Society of America, 81, 2553–2576.

Hindle, D. & Burkhard, M. 1999. Strain, displacement and rotation associated with the formation of curvature in fold belts; the example of the Jura arc. *Journal of Structural Geology*, **21**, 1089–1101.

Jaumé, S. C. & Lillie, R. J. 1988. Mechanics of the Salt Range–Potwar Plateau, Pakistan: a fold-and-thrust belt underlain by evaporites. *Tectonics*, **7**, 57–71.

Johansen, S. E., Ostisty, B. K. et al. 1993. Hydrocarbon potential in the Barents Sea region: play distribution and potential. *In*: Vorren, T. O., Bergsager, E., Dahl-Stamnes, Ø. A., Holter, E., Johansen, B., Lie, E. & Lund, T. B. (eds) *Arctic Geology and Petroleum Potential*. NPF Special Publication 2. Elsevier, Amsterdam, 273–320.

Korago, E. A., Kovaleva, G. N., Il'in, V. F. & Pavlov, L. G. 1992. *The Tectonics and Metallogeny of the Early Kimmerides of Novaya Zemlya*. Nedra, Leningrad.

Korago, E. A., Kovaleva, G. N., Lopatin, B. G. & Orgo, V. V. 2004. *The Precambrian Rocks of Novaya Zemlya*. Geological Society, London, Memoirs, **30**, 135–143.

Macedo, J. & Marshak, S. 1999. Controls on the geometry of fold-thrust belt salients. *Bulletin of the Geological Society of America*, **111**, 1808–1822.

Marshak, S. 1988. Kinematics of arc and orocline formation in thin-skinned orogens. *Tectonics*, **7**, 73–86.

Marshak, S. 2004. *Salients, Recesses, Arcs, Oroclines, and Syntaxes – a Review of Ideas Concerning the Formation of Map-view Curves in Fold-thrust Belts*. American Association of Petroleum Geologists, Tulsa, OK, Memoirs, **82**, 131–156.

Marshak, S. & Wilkerson, M. S. 1992. Effects of overburden thickness on thrust belt geometry and development. *Tectonics*, **11**, 560–566.

Marshak, S., Wilkerson, M. S. & Hsui, A. 1992. Generation of curved fold-thrust belts: insights from simple physical and analytical models. *In*: McClay, K. R. (ed.) *Thrust Tectonics*. Chapman & Hall, London, 83–92.

Metelkin, D. V., Vernikovsky, V. A., Kazansky, A. Y., Bogolepova, O. K. & Gubanov, A. P. 2005. Paleozoic history of the Kara microcontinent and its relation to Siberia and Baltica: paleomagnetism, paleogeography and tectonics. *Tectonophysics*, **398**, 225–243.

Mezhvilk, A. A. 1995. Thrust and strike-slip zones in northern Russia. *Geotectonics*, **28**, 298–305.

Mitra, G. 1997. Evolution of salients in a fold-and-thrust belt: the effects of sedimentary basin geometry, strain distribution and critical taper. *In*: Sengupta, S. (ed.) *Evolution of Geological Structures in Micro- and Macro-scales*. Chapman & Hall, London, 59–90.

Montero, P., Bea, F. et al. 2000. Single-zircon evaporation ages and Rb–Sr dating of four major Variscan batholiths of the Urals: a perspective on the timing of deformation and granite generation. *Tectonophysics*, **317**, 93–108.

O'Leary, N., White, N. J., Tull, S., Bashilov, V., Kuprin, V., Natapov, L. M. & Macdonald, D. 2004. Evolution of the Timan–Pechora and South Barents Sea basins. *Geological Magazine*, **141**, 141–160.

Otto, S. C. & Bailey, R. J. 1995. Tectonic evolution of the northern Ural Orogen. *Journal of the Geological Society, London*, **152**, 903–906.

Pease, V. & Scott, R. A. 2009. Crustal affinities in the Arctic Uralides, northern Russia: significance of detrital zircon ages from Neoproterozoic and Paleozoic sediments in Novaya Zemlya and Taimyr. *Journal of the Geological Society, London*, **166**, 517–527.

Puchkov, V. N. 1997. *Structure and Geodynamics of the Uralian Orogen*. Geological Society, London, Special Publications, **121**, 201–236.

Scott, R. A., Ramsey, L. A., Jones, S. M., Sinclair, S. & Pickles, C. S. 2005. *Development of the Jan Mayen Microcontinent by Linked Propagation and Retreat of Spreading Ridges*. Norsk Petroleumsforening, Trondheim, Special Publications, **12**, 69–82.

Şengör, A. M. C. & Natal'in, B. A. 1996. Paleotectonics of Asia: fragments of a synthesis. *In*: Yin, A. & Harrison, T. M. (eds) *The Tectonic Evolution of Asia*. Cambridge University Press, Cambridge, 486–640.

Smelror, M., Petrov, O. V., Larssen, G. B. & Werner, S. C. (eds) 2009. *Geological History of the Barents Sea*. Geological Survey of Norway, Trondheim.

Sobornov, K. O. 1994. *Structural Relationship of the Northern Urals and Adjacent Basins*. Canadian Society of Petroleum Geologists, Alberta, Memoirs, **17**, 145–154.

Torsvik, T. H. & Cocks, L. R. M. 2004. Earth geography from 400 to 250 Ma: a palaeomagnetic, faunal and facies review. *Journal of the Geological Society*, **161**, 555–572.

Ustritsky, V. I. 1985. On the relationship between the Urals, Pay-Khoy, Novaya Zemlya and Taymyr. *Geotectonics*, **1**, 51–61.

Vernikovsky, V. A., Neimark, L. A., Ponomarchuk, V. A., Vernikovskaya, A. E., Kireev, A. D. & Kuz'min, D. S. 1995. Geochemistry and age of collision granitoids and metamorphites of the Kara microcontinent (Northern Taimyr). *Geologiya i Geofizika*, **36**, 46–60.

Vyssotski, A. V., Vyssotski, V. N. & Nezhdanov, A. A. 2006. Evolution of the West Siberian Basin. *Marine and Petroleum Geology*, **23**, 93–126.

Wezel, F. C. (ed.) 1986. *The Origin of Arcs*. Elsevier, Amsterdam.

Ziegler, P. A. 1988. *Evolution of the Arctic–North Atlantic and the Western Tethys*. American Association of Petroleum Geologists, Tulsa, OK, Memoirs, **43**, 1–198 + 30 plates.

Ziegler, P. A. 1989. *Evolution of Laurussia: a Study in Late Palaeozoic Plate Tectonics*. Kluwer, Dordrecht.

Zonenshain, L. P., Korinevsky, V. G., Kazmin, V. G., Matveenkov, V. V. & Khain, V. V. 1984. Plate tectonic model for the development of the south Urals. *Tectonophysics*, **109**, 95–135.

Zonenshain, L. P., Kuzmin, M. I. & Natapov, L. M. 1990. *Geology of the USSR: a Plate Tectonic Synthesis*. American Geophysical Union, Washington, DC.

Charging the giant gas fields of the NW Siberia basin

E. FJELLANGER,[1] A. E. KONTOROVICH,[2] S. A. BARBOZA,[3] L. M. BURSHTEIN,[2] M. J. HARDY[4] and V. R. LIVSHITS[2]

[1]*Esso Norge AS, PO Box 60, 4064 Stavanger, Norway (e-mail: erik.fjellanger@exxonmobil.com)*
[2]*Institute of Petroleum Geology and Geophysics, Russian Academy of Sciences Siberian Branch, Koptuga 3, Novosibirsk, 630090, Russia*
[3]*ExxonMobil Upstream Research Company, PO Box 2189, Houston, TX 77008, USA*
[4]*ExxonMobil International Ltd., Ermyn Way, Leatherhead, Surrey KT22 8UX, UK*

Abstract: The West Siberia basin is the largest petroleum province in Russia, with 80% of the country's gas resources in the Cenomanian Pokur Formation. Significant undiscovered gas resources have been assessed as on trend with the giant gas fields. However, the origin of the large amounts of dry, isotopically light gas is still an enigma, albeit extensively addressed in the literature. This study aims at quantifying the gas contribution from all relevant thermal sources. The West Siberia Basin is the world's largest intracratonic basin, comprising up to 12 km of Mesozoic and Cenozoic clastic rocks. The Basement is composed of Palaeozoic accretionary crust. Northward-trending Permian–Triassic rifts were filled by fluvial–deltaic sediments from the south and east, punctuated by marine transgressions from the north. Cenozoic basin inversion formed traps for petroleum. A regional high-resolution 3D basin simulation was used to model the thermal evolution of the northern West Siberia basin. Geostatistical modelling was applied to assess source rock richness and quality. Basal heat flow was modelled by calibration to bottom-hole temperature and vitrinite measurements. Hydrocarbon generation kinetic parameters were derived from measurements performed on West Siberia rock samples. Thermal gas charge expelled from the hydrocarbon kitchen drainage areas of key fields were compared with the gas volumes accumulated in these fields. The study found that Cretaceous terrestrial sources can generate sufficient early thermal gas to charge accumulations in the South Kara Sea area, and additional Jurassic sources can charge the remaining accumulations of the study area if favourable conditions apply. Biogenic gas is likely to have contributed to the gas accumulations. Mixing of thermal and biogenic gas could explain the observed isotopic composition. Sensitivity analyses show that the timing of structuring and uplift is the most critical factor of the assessment. Variations in glaciation, heat flow and source kinetics show less effect on the hydrocarbon accumulation.

Keywords: basin modelling, source rock, gas charge, Cenomanian, Pokur Formation (Fm.), West Siberia

The West Siberia Basin is one of the world's largest petroleum provinces, estimated to contain 356 billion barrels of oil equivalent (BBOE) of petroleum resources, of which 40% is oil and 60% is gas (Ulmishek 2003). The West Siberia basin is estimated to contain more than 70% of Russia's remaining undiscovered petroleum resources (USGS 2000), and the northern West Siberia basin is assessed to hold the largest undiscovered gas potential of any Arctic basin (Bird *et al.* 2008).

While the dominant oil play is the Neocomian clastics in the central parts of the basin, the dominant gas play is the Cenomanian Pokur Fm. in the northern West Siberia basin (Figs 1–3; Kontorovich *et al.* 1975; Surkov & Zhero 1981; Peterson & Clarke 1991). The Cenomanian gas play is moderately explored onshore in the Yamal and Gydan peninsulas, and poorly explored offshore in South Kara Sea.

Several giant gas fields have been discovered in the Cenomanian play (Grace & Hart 1986; Ulmishek 2003). The gas is characterized by being very dry and isotopically light. Methane content is normally up to 99%, and the average isotope ratio is in the range -45 to -65‰ $\delta^{13}C$ (Grace & Hart 1986; Littke *et al.* 1999). The largest gas field in West Siberia is the Urengoy Field with 330×10^{12} SCF in place, the second largest gas field in the world. In this study, Urengoy, together with Bovanenkovskoye (170×10^{12} SCF) and Leningradskoye (40×10^{12} SCF), will be used as reference fields to estimate gas expelled from different sources and drainage areas to charge these fields (gas-in-place estimates are approximate numbers from the ExxonMobil internal database).

The Cenomanian reservoir section consists of fluvial–deltaic sandstone, grading to shallow marine sandstone northward into the South Kara Sea (Fig. 2b). Maastrichtian marine, transgressive mudstones form an excellent seal rock in addition to intraformational sealing mudstones causing multi-reservoir levels in many of the fields. Cenozoic compression formed valid traps as four-way dip closures or fault-bounded inverted structures. Stratigraphic traps by facies pinchout add to the petroleum potential.

Several potential source rocks are present in the northern West Siberia basin to charge the Cenomanian fields. However, the single source or combination of sources to provide the large amounts of dry and isotopically light gas is still an enigma in spite of being actively studied by many authors for years (Kortsenshteyn 1977; Vasilyev *et al.* 1979; Rice & Claypool 1981; Kontorovich 1984; Grace & Hart 1986; Galimov 1988; Prasolov 1990; Stroganov 1990; Rovenskaya & Nemchenko 1992; Surkov & Smirnov 1994; Schoell *et al.* 1997; Cramer *et al.* 1999; Kontorovich *et al.* 1999; Littke *et al.* 1999; Nemchenko *et al.* 1999; Ulmishek 2003).

The main proposed models are (Ulmishek 2003):

(1) Early thermal gas from Cretaceous terrestrial organic matter (Vasilyev *et al.* 1979; Galimov 1988; Stroganov 1990; Rovenskaya & Nemchenko 1992; Nemchenko *et al.* 1999). Coals and disseminated organic matter in the Pokur Fm. and Tanopcha Fm. can produce isotopically light gas to match the gas in Cenomanian fields. However, the formations are early to immature

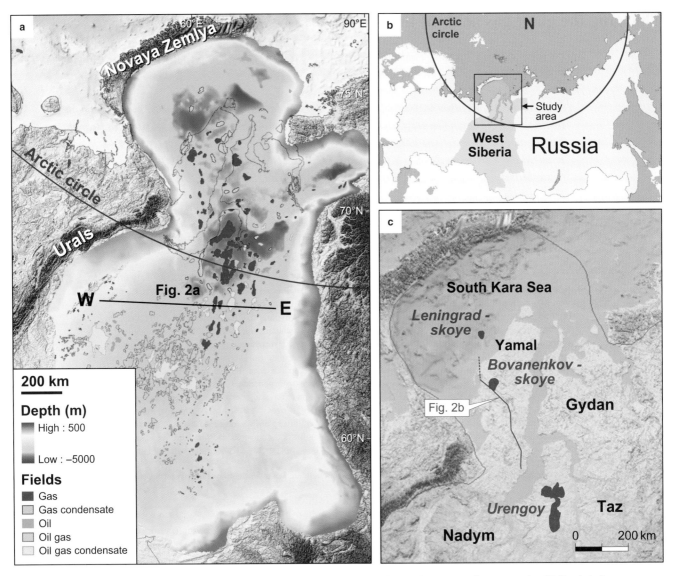

Fig. 1. (a) West Siberia basin Top Jurassic depth map including hydrocarbon fields superimposed on present-day topography. (b) Russia index map including West Siberia basin and study outline. (c) Northern West Siberia study area of 3D basin modelling project including reference fields.

for gas generation, and sufficient expulsion to charge the giant fields may not be achieved.

(2) Late thermal gas from Jurassic marine organic matter (Prasolov 1990). The prolific Jurassic Bazhenov Fm. and Tyumen Fm. oil-prone mudstones are currently in the gas window in most of the study area (Fig. 4), and gas produced from kerogen or cracked from oil may charge the Cenomanian fields. Migration pathways from the Jurassic sources to the Cenomanian fields are known to exist from source to oil correlation of remnant oil along the rims of Cenomanian fields (Kontorovich et al. 1999). However, the isotopic composition and dryness of the Cenomanian gas is not in agreement with a heavy and wet gas charge expected from late thermal gas sources.

(3) Biogenic gas is dry and isotopically light, but the volume generated is difficult to estimate. Cretaceous coals and mudstones may be in a favourable position for biogenic gas production, and methane may have been produced by biodegradation from previous hydrocarbon accumulations (Larter et al. 2005).

(4) Storage of thermal gas in Cretaceous water coupled with degassing after uplift and trap formation. This method was first discussed by Kortsenshteyn (1977) and later by Littke et al. (1999).

(5) Diffusion is discussed as a potential method by Ulmishek (2003).

The aim of this study was to apply a high-resolution 3D basin model to quantitatively assess contribution from Cretaceous and Jurassic sources to charge the Cenomanian gas fields (models 1 and 2 above) and to qualitatively assess the conditions for biogenic gas contribution from Cretaceous organic matter (model 3 above). The study was performed as a joint research project between the Institute of Petroleum Geology and Geophysics – Russian Academy of Sciences Siberian Branch (IPGG – RAS SB) and ExxonMobil.

Geology

Structure

The West Siberia basin is the world's largest intracratonic basin covering more than $3 \times 10^6 \text{ km}^2$ (Fig. 1). It extends under the South Kara Sea, and is delineated by Novaya Zemlya in the north, the Ural mountains in the west, the East Siberia craton in the east and the Altay mountains in the south (Kontorovich et al. 1975; Surkov & Zhero 1981; Peterson & Clarke 1991). The basement was formed during the Carboniferous–Earliest Permian as accretionary crust between the colliding Baltic and Siberian cratons, forming a significant part of the Pangea supercontinent (Ziegler 1989; Sengor & Natal'in 1996). The collision formed the

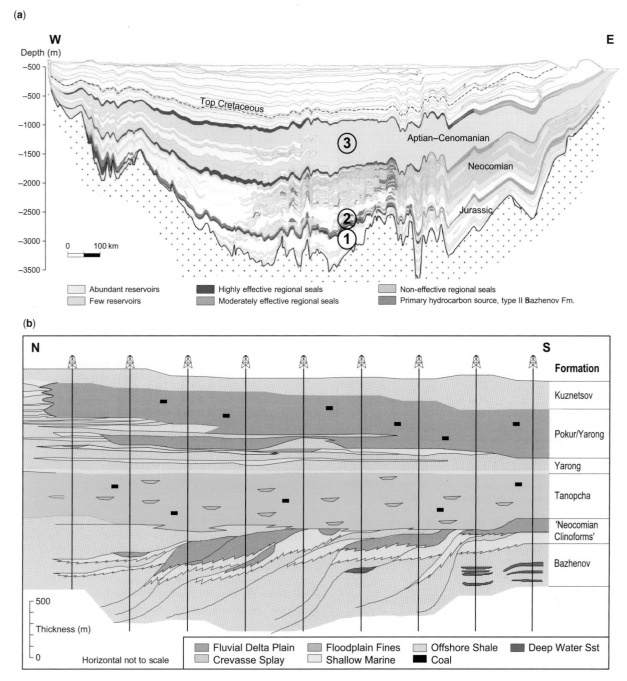

Fig. 2. (**a**) Schematic cross-section of the petroleum systems of West Siberia basin. For location see Figure 1. Key source intervals are (1) Mid Jurassic Tyumen Fm. coal and shale, (2) Late Jurassic Bazhenov Fm. marine shale and (3) Aptian–Cenomanian Pokur Fm. terrestrial coal and shale. (**b**) Schematic correlation through fields in the Kara Sea and Northern Yamal Peninsula (J. Maynard, pers. comm.).

Urals in the west by structural compression migrating from the south to the north (Nikishin *et al.* 1996; Fokin *et al.* 2001).

During the Early Permian, Pangea started breaking up, and a compressional phase formed the Novaya Zemlya Island in the north (Nikishin *et al.* 1996; Fokin *et al.* 2001). A dominantly north–south trending rift developed in northern West Siberia. Thick and extensive basalts were extruded to fill the rifts and surrounding areas, known as the East Siberian traps (Shipilov & Tarasov 1998; Vyssotski *et al.* 2006).

Clastics contributed to fill the Permian rifts, and during the subsequent thermal sag, the basin was filled with up to 12 km of Mesozoic sediments. In the Cenozoic the basin was uplifted and eroded as a result of India's collision and the opening of the arctic Eurasia basin (Vyssotski *et al.* 2006). Faults were inverted, forming traps for hydrocarbon fields.

Stratigraphy

The northern and central West Siberia basin stratigraphy is summarized in Figure 3. During the Permian continental deposits and volcanics filled the rift graben, and up to 7 km of Triassic clastics were deposited in the northern West Siberia basin.

In the Jurassic, fluvial and deltaic deposits were sourced from the Altay mountains in the south. However, marine transgressions periodically flooded the rifted basin from the north, gradually covering large parts of the basin with marine sediments. Organic-rich mudstones of Tyumen Fm. and Bazhenov Fm. were deposited and became excellent source rocks, with shallow marine sandstones of the Tyumen Fm. and Vasyugan Fm. forming good reservoirs.

During the Early Cretaceous (Neocomian) fluvial–deltaic deposits of the Megion Fm. prograded from east to west as a series of

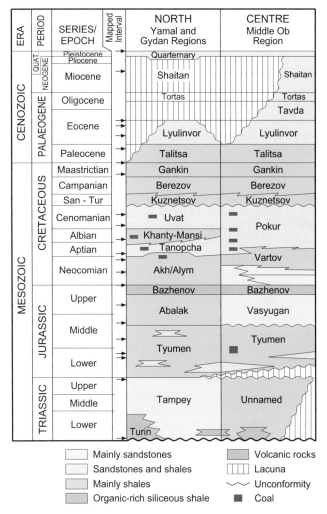

Fig. 3. Mesozoic chronostratigraphy of the northern and central West Siberia basin (modified from Ulmishek 2003). Stratigraphic levels of structure maps used for the study are marked by arrows.

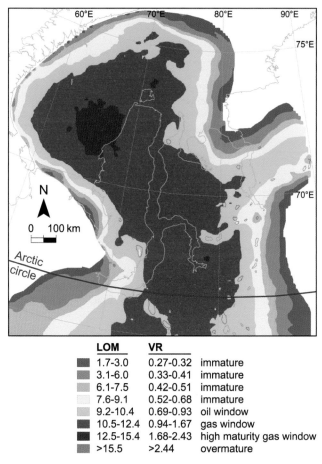

Fig. 4. Bazhenov Fm. present-day maturity map. Coastline and field outlines are annotated.

clinoforms (Fig. 2a). Deep-marine sands of the Achimov Fm. were deposited at the toe slope of the clinoforms. The Neocomian fluvial–deltaic deposits constitute the main oil play in the central West Siberia basin. An Aptian marine flooding event caused deposition of continuous mudstone (Alym Fm.) that acted as a regional seal to the Neocomian deposits.

Two renewed phases of fluvial–deltaic sediments filled the basin, firstly in Aptian (Tanopcha Fm./Pokur Fm.) and then, following an Albian marine flooding event (Yarong Fm.), Cenomanian fluvial–deltaic deposits (Pokur Fm.). These deposits grade to shallow marine sandstones in the northern part of the West Siberia basin, and deep marine mudstones in central parts of the South Kara Sea (Fig. 2b). The Pokur Fm. is interpreted as dominantly fluvial–deltaic along the South Kara Sea basin margins. These deposits form the Cenomanian gas play discussed in this paper. A Turonian flooding event (Kuznetsov Fm.) formed marine mudstone acting as a seal to the Pokur Fm. play.

In central parts of the West Siberia basin the Pokur Fm. includes the Aptian to Cenomanian fluvial–deltaic sediments. In the northern West Siberia basin, the Albian Yarong Fm. marine mudstone separates the Aptian from the Cenomanian fluvial–deltaic sediments, and the Aptian section is named Tanopcha Fm. (Fig. 3). For simplicity, in this paper the Aptian fluvial–deltaic sediments are referred to as the Tanopcha Fm. and the Albian to Cenomanian fluvial–deltaic sediments are referred to as the Pokur Fm. within the entire study area.

Cenozoic uplift brought continental conditions to the West Siberia basin, with erosion, sediment by-pass or fluvial deposits at various times (Fig. 3; Surkov & Smirnov 1994; Ulmishek 2003). During Pleistocene several glacial events caused ice cover and tills to be deposited in NW Siberia.

Today the West Siberia basin is a flat-lying, swampy basin with permafrost in the upper 300–500 m and a shallow sea (South Kara Sea) in the northernmost part of the basin.

Analyses

A high-resolution 3D basin model was constructed for the northern West Siberia basin (Fig. 1), using ExxonMobil's proprietary Stellar™ 3D basin modelling software. The model had many purposes, one of which was to assess the gas charge to the Cenomanian play. Questions raised were: how much gas is generated from thermal sources within the drainage areas of selected fields after trap formation? To what extent can this help in understanding the gas charge to the Cenomanian fields?

The high-resolution 3D basin model was constructed by establishing a stratigraphic section, boundary conditions and a source model. The stratigraphic section was built from isopach maps of 22 structural horizons covering the entire Triassic to present-day sedimentary interval (Fig. 3). The structure maps were dominantly derived from well information onshore and seismic interpretation offshore. Missing section maps were produced from estimated erosion during Cenozoic uplift. Lithology composition was entered for each isopach, split into percentage sandstone, mudstone and coal. The lithology composition was calibrated against well logs from a large number of onshore wells, and then extrapolated to the offshore and undrilled areas based on depositional environment maps.

Thermal boundary conditions were defined to enable source maturity calculations to be performed. The thermal model is bounded by the surface temperature at the top of the sedimentary section and the basal heat flow from below the sediments through time.

The palaeosurface temperatures were entered using ExxonMobil's global climate model through geological history. Palaeowater depth maps at 16 key intervals were entered to provide correct subsea temperature, palaeotopography and decompaction of the sedimentary section. Surface temperature estimates during the Quaternary glaciation required special consideration. Mapped permafrost thickness was used to calibrate the effect of glaciation on present-day subsurface temperature.

The basal heat flow was modelled as the sum of the Triassic (240 Ma) rifting heat flow anomaly and the background heat flow. The background heat flow represents the heat loss from the earth's interior together with radiogenic heat production, and is modelled as constant through time. The heat flow associated with the Triassic rifting is represented by a gamma map, which represents the thinning of the crust during rifting. The present-day regional basal heat flow map was calibrated against more than 2000 temperature and vitrinite measurements from 72 wells across the study area.

A source model was built to perform yield (expulsion) calculations. Maps of source richness (total organic carbon, TOC) and quality (hydrogen index, HI) are required as input parameters for yield analyses. An extensive rock-eval database was established and geostatistical analyses performed to determine the optimal mapped representation of the source richness and quality.

Four sources for thermogenic hydrocarbons were considered for the basin model: (1) Cenomanian deltaic coals and mudstones (Pokur Fm.); (2) Aptian deltaic coals and mudstones (Tanopcha Fm.); (3) Upper Jurassic marine mudstones (Bazhenov Fm.); and (4) Middle Jurassic marine mudstones and coals (Tyumen Fm.).

Stellar^{TM} default kinetic parameters for kerogen transformation were applied. However, with respect to the sensitivity analyses of early gas generation from Cretaceous mudstones and coals, the parameters were derived from relevant Cretaceous coals in West Siberia.

Results

The gas charge to the Cenomanian play was tested by running the Stellar 3D basin model over the entire study area, and then calculating the amounts of gas expelled within the drainage areas of three representative fields. The gas yield was then compared with the original gas-in-place volumes accumulated in the Cenomanian reservoirs of these fields. The three fields were Urengoy, Bovanenkovskoye and Leningradskoye, comprising 330, 130 and 40×10^{12} SCF gas, respectively.

First the thermal yields from the Cretaceous sources Pokur Fm. and Tanopcha Fm. were calculated within the drainage areas, applying best estimate parameters ('base case Cretaceous sources'). Secondly, key parameters were optimized to evaluate a scenario of early gas generation from the Cretaceous sources ('high case Cretaceous sources'). The parameters were also each varied to test their relative contribution to the increased volumes obtained by the early gas case scenario. Thirdly, the cumulative thermal yields from both the Cretaceous and the Jurassic sources were calculated for the same drainage areas ('base case all sources'). Finally, an early gas scenario was run to optimize the expelled amounts of gas including both Cretaceous and Jurassic sources ('high case all sources').

The yield from the Cretaceous sources applying base case parameters is represented on Figure 5. The graph on Figure 5a

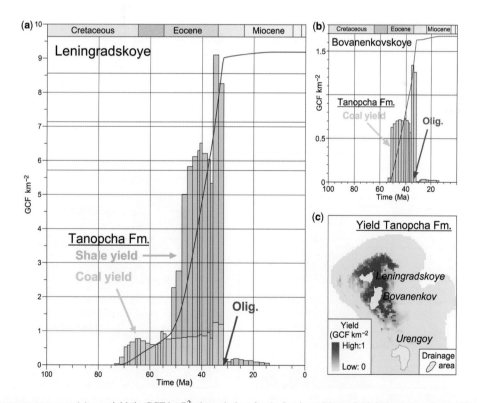

Fig. 5. Base case Cretaceous sources: (**a**) gas yield (in GCF km^{-2}) through time for the Leningradskoye field drainage area for the Tanopcha Fm. source applying base case parameters (Table 2). The Pokur Fm. was immature for gas generation. Shale yield is plotted in dark green and coal yield in bright green. Note the large volumes expelled post-Paleocene compared with post-Oligocene, and thus the importance of trap timing. (**b**) Gas yield through time for the Bovanenkovskoye field drainage area for the Tanopcha Fm. source applying base case parameters (Table 2). The Pokur Fm. was immature for gas generation. (**c**) Map of gas yield for Cretaceous sources applying base case sensitivity. Drainage areas of the selected fields are shown. All yield is from the Tanopcha Fm. The Pokur Fm. was immature for gas generation in the entire study area.

Table 1. Gas volumes accumulated in Cenomanian reservoirs and expelled from sources in the drainage areas of the Leningradskoye, Bovanenkovskoye and Urengoy fields (approximate numbers)

Field/drainage gas volumes	Leningradskoye ($\times 10^{12}$ SCF)	Bovanenkovskoye ($\times 10^{12}$ SCF)	Urengoy ($\times 10^{12}$ SCF)
Cenomanian field in-place gas volume (approximate)	40	170	330
Gas available from drainage area			
Base case Cretaceous source	2	3	0
Base case Jurassic + Cretaceous sources	8	8	7
High-case Cretaceous source	333	81	0
High-case Jurassic + Cretaceous sources	504	618	411

shows gas yield per km² within the drainage area of the Leningradskoye field plotted against time. Gas yield occurs from the Tanopcha Fm. only as the younger Pokur Fm. is not buried sufficiently deep for gas generation to be initiated. The Tanopcha Fm. yield started in late Cretaceous and peaked in late Eocene. A subsequent marked drop in yield is modelled due to basin uplift and folding. This folding event corresponds to the time of trap formation in the base case scenario, which implies that only a minor amount of gas expelled after the start of the Oligocene at 32 Ma is available for entrapment. In the Leningradskoye drainage area, 2×10^{12} SCF only were generated post-Eocene in our base case, which is far from sufficient to charge the 40×10^{12} SCF gas-in-place in Leningradskoye (Table 1).

A similar situation applies for Bovanenkovskoye, where 1×10^{12} SCF gas only is expelled from the drainage area (Fig. 5b), and gas from coals only are generated here. In the Urengoy drainage area there is no gas yield from Cretaceous sources in the base case.

The regional distribution of gas yield from the Cretaceous sources is displayed on Figure 5c. The drainage areas of the study fields Leningradskoye, Bovanenkovskoye and Urengoy are also annotated on the map. Gas expulsion from Cretaceous sources is limited to the deepest parts of the basin in South Kara Sea, and from the Tanopcha Fm. only when base case parameters apply.

Table 2. Sensitivity parameters applied to base case and high case scenarios listed in order of impact on results

Sensitivity factor	Base case	High case
Trap timing	Early Oligocene	Early Palaeocene
Eocene erosion (m)	300	0
Source thickness	Best estimate map	Best estimate +20% map
Kinetic parameters	Best estimate	Early generation
Basal heat flow	Best estimate map	Best estimate map +5 mW m^{-2}

Eocene erosion adds to the missing section maps prepared for post-Eocene isopach maps, which was applied to all sensitivity cases.

A high case scenario was defined by applying a sensitivity range on key parameters to realistically improve the yield from the Cretaceous sources. In the base case, trap timing was defined to occur in Early Oligocene at 32 Ma, representing a major tectonic folding phase in West Siberia related to the collision of India with the Asian plate. All Cenomanian traps were in place after this event. In the high case, trap timing was set to the Early

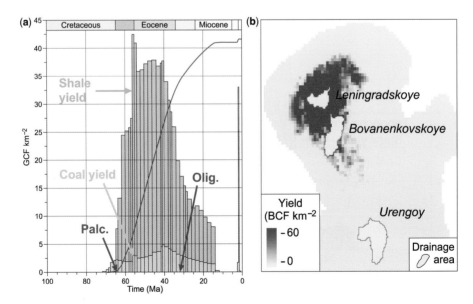

Fig. 6. High case Cretaceous sources: (**a**) Gas yield through time for the Leningradskoye field drainage area for the Tanopcha Fm. source applying high case sensitivity (Table 2). The Pokur Fm. was immature for gas generation. Note the large volumes expelled post-Paleocene compared with post-Oligocene, and thus the importance of trap timing. (**b**) Map of gas yield for Cretaceous sources applying high case sensitivity. Drainage areas of the selected fields are shown. All yield is from the Tanopcha Fm. The Pokur Fm. was immature for gas generation in the entire study area. Note the scale difference between Figures 5b and 6c.

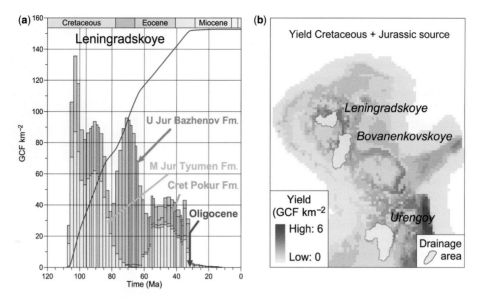

Fig. 7. Base case all sources: (**a**) gas yield through time for the Leningradskoye field drainage area for Cretaceous and Jurassic sources (Table 2 base case). Bazhenov Fm. source yield is in dark blue, Tyumen Fm. yield in light blue and Tanopcha Fm. yield in green. (**b**) Map of gas yield for Cretaceous and Jurassic sources using base case parameters.

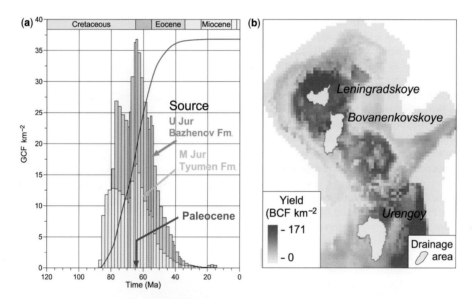

Fig. 8. High case all sources: (**a**) gas yield through time for the Urengoy field drainage area for Cretaceous and Jurassic sources (Table 2 high case). The contribution is from Jurassic sources only as the Cretaceous sources are immature at this location. (**b**) Map of gas yield for Cretaceous and Jurassic sources (high case sensitivity).

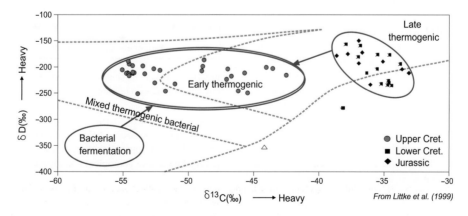

Fig. 9. Isotopic composition of methane from fields in northern West Siberia. Modified from Littke *et al.* (1999). Genetic classification after Schoell (1980) and Whiticar (1990).

Paleocene at 65 Ma. Uplift and basin inversion are described in the literature (Vyssotski *et al.* 2006) and interpreted on seismic lines to be as early as Maastrichtian in age, which means that Cenomanian traps may have started forming already in the Paleocene. The exact trap timing varies across the basin, and thus trap timing is used as a sensitivity factor. Uplift and erosion were also important factors that affected the gas yield. Cenozoic erosion was estimated based on missing section maps of Cenozoic intervals, and for simplicity interpreted to take place at Base Quaternary. However, additional erosion was set to occur in the Eocene, being varied between 0 and 500 m as a sensitivity factor. Source thickness was varied by ±20%, a range of kinetic parameters were applied, and heat flow was varied by ±5% (Table 2).

The gas yield from the Cretaceous sources, including all high case parameters within the Leningradskoye drainage area, is displayed in Figure 6a. The cumulative gas yield has increased significantly, from 9 to 42 GCF km^{-2} (red curves on Figs 5a & 6a). Generation started in the Maastrichtian (70 Ma), but significant generation did not start until the Early Paleocene (65 Ma). The yield peaked in the Eocene and then gradually decreased during the Oligocene–Miocene due to restricted burial.

Simulations were also run varying one parameter at the time. The relative contribution from each parameter in the high case is listed in Table 2. Trap timing stands out as the dominant factor. Large additional volumes are included by changing trap timing from 32 Ma in Early Oligocene to 65 Ma in Early Paleocene (Fig. 6a). The effect of reducing uplift and erosion is expressed by additional late gas yield compared with the base case (post 30 Ma). Source thickness, kinetics and heat flow as defined in the model had a minor impact on the results.

The high case parameters for charge from Cretaceous sources were also applied to the Bovanenkovskoye and Urengoy drainage areas (Fig. 6b). When applying high case yield parameters, the Bovanenkovskoye gas charge is about half that required to fill the trap, and there is still no expulsion in the Urengoy drainage area. This shows that the Cretaceous sources alone cannot charge all the Cretaceous fields.

Two Jurassic sources, the Upper Jurassic Bazhenov Fm. and the Middle Jurassic Tyumen Fm., were introduced to the 3D basin simulation, and base case sensitivity parameters were applied. Figure 7 shows the gas yield for the base case scenario from the Leningradskoye field drainage area through time for all thermal sources. In dark blue is the Bazhenov Fm. source yield and in light blue the Tyumen Fm. yield. Jurassic gas yield starts in Early Cretaceous. Gas co-generated with oil is expelled until Campanian (80 Ma). After Campanian the Bazhenov Fm. passed through the late gas generation stage with expulsion of significant amounts of cracked gas. The Tyumen Fm. expelled cracked gas during the Cenozoic, in parallel with early mature gas from the Tanopcha Fm. A marked drop in gas expulsion occurred during uplift and erosion post-Eocene.

The total volume expelled in the drainage area is 8×10^{12} SCF, which is not enough to charge the Leningradskoye field. Similar situations apply for Bovanenkovskoye and for Urengoy drainage areas, charging 8 and 7×10^{12} SCF, respectively (Table 1).

As a final test, the high case parameters were applied to all thermal sources. Significant gas volumes are expelled in all drainage areas, including the Urengoy drainage area (Fig. 8). Table 1 shows that when all high case parameters are applied, the Urengoy drainage area yields more than 400×10^{12} SCF gas. This is 25% more than the 300×10^{12} SCF gas stored in the Urengoy Cenomanian reservoir, that is, a charge multiplier of 1.25. Table 1 shows that the charge multiplier is 4 for Bovanenkovskoye and 10 for Leningradskoye. By varying the sensitivity parameters one by one, trap timing became the dominant factor and erosion was of second most importance.

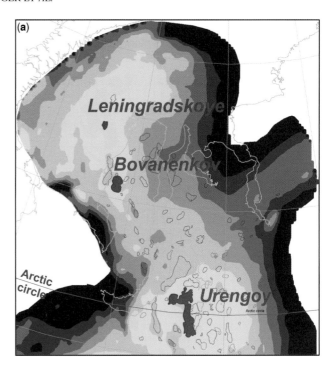

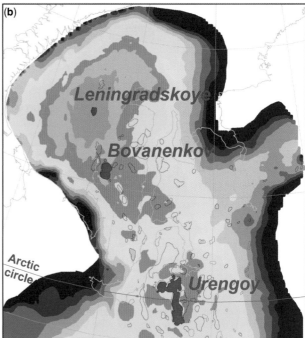

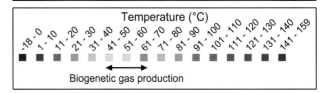

Fig. 10. (**a**) Pokur Fm. temperature map in Miocene (10 Ma). (**b**) Tanopcha Fm. temperature map in Miocene (10 Ma).

In summary, more thermal gas has been generated in the drainage areas of the Cenomanian fields in northern West Siberia than accumulated in the traps if all thermal sources are taken into account and if very favourable geological conditions apply.

Discussion

The simulation results show that, when applying high case parameters for all sources, the amounts of thermal gas expelled in the drainage areas of the selected fields exceeds the original in-place volumes in the fields. However, migration loss has not been taken into account. The Urengoy field has a charge multiplier of 1.25 and it is unlikely that sufficient gas has reached the structure from its immediate kitchen. However, fill–spill mechanisms have not been considered, and this may expand the drainage area for Urengoy substantially.

Even if additional volumes of late thermal gas from Jurassic sources are made available by expanding the drainage area to the Urengoy field, its gas charge cannot be explained by Jurassic charge alone since the isotopic composition of the gas stored in the field does not correspond to a late thermal source. This is documented by Littke et al. (1999), who plotted isotopic values of $\delta^{13}C$ against deuterium of methane from fields in northern West Siberia (Fig. 9). The graph shows the zones where early and late thermogenic methane and biogenic methane generally plots. Gases from the West Siberia Cenomanian reservoirs plot in the early thermogenic range, marked in red. This is consistent with the maturity of the Cretaceous sources.

West Siberia gases from Jurassic sources plot in the late thermogenic area, marked in black. Hence Jurassic sources cannot be the only source charging Urengoy since the isotopic ratios do not match. However, if the Jurassic gases are mixed with bacteria-produced gas, the mixture may plot in the zone of the Cenomanian fields, and thus be the origin of the gas, or at least make a contribution together with gas from the Cretaceous sources.

Hence a unique source to gas correlation is hard to establish. Furthermore, the extreme dryness of the gas (98–99%) implies that biomarkers and wet gas parameters cannot be used to discriminate the origin of the gas.

Biogenic gas is formed at shallow depths and low temperatures, between 40 °C (or even lower) and 70–80 °C, by anaerobic bacterial decomposition (biodegradation) of sedimentary organic matter and accumulated oil (e.g. Winters & Williams 1969). The Tanopcha Fm. and Pokur Fm. contain significant amounts of coals and disseminated organic matter. Figure 10 shows the temperature map of the Tanopcha Fm. and Pokur Fm. 10 Ma ago. The Stellar model shows that for the last 10 million years, the Tanopcha Fm. in the Urengoy area was between 40 and 70 °C, marked by green colours on the map. The Leningradskoye and Bovanenkovskoye drainage areas are buried deeper and have temperatures above 70 °C. In the last 10 Ma the Pokur Fm. has been exposed to temperatures in the range 40 °C–70 °C in the entire study area. It is therefore likely that biogenic gas is produced from the Pokur Fm.

Biodegraded oils from Jurassic sources are present in Cenomanian gas fields (e.g. Kontorovich 1984; Peters et al. 1994), and residual oil is normally found along the rim of the gas accumulations (Grace & Hart 1986). In biodegraded oil fields, isotopically lighter methane is commonly found mixed with thermogenic methane (Larter et al. 2005). The carbon isotopic composition of methane associated with subsurface oil biodegradation conditions in marine petroleum systems is frequently in the range of −45 to −55‰ $\delta^{13}C$, which is in agreement with the observations made by Littke et al. (1999) for the West Siberia Cenomanian gas fields (Fig. 9). Therefore, isotopically light methane in the Cenomanian fields may have been generated from biodegradation of oil charged from Jurassic sources.

In summary, there are favourable conditions for biogenic gas to supplement the gas volumes stored in the Cenomanian gas fields, as biogenic gas generated both from disseminated organic matter and from biodegraded oil pools.

Conclusions

The Cenomanian gas fields in northern West Siberia comprise the world's largest gas province. However, the gas charge to the fields is poorly understood in spite of being extensively studied by many authors.

In this study, gas yield from established Cretaceous and Jurassic sources was calculated using ExxonMobil's Stellar™ 3D basin modelling software to quantitatively estimate charge to selected Cenomanian fields. This quantitative approach eluded the most likely source intervals to charge the fields in different parts of the basin.

The study showed that Cretaceous terrestrial sources generate sufficient early thermal gas to charge fields in the deepest areas of South Kara Sea (e.g. Leningradskoye), and may contribute to the gas charge in the Yamal area (e.g. Bovanenkovskoye). No gas is generated from Cretaceous sources in the Nadym–Taz regions (e.g. Urengoy) due to shallow burial. Hence Cretaceous sources can fully or partly charge the Cenomanian gas fields in the deepest parts of South Kara Sea, and may contribute to charging gas fields on the Yamal peninsula.

Additional Jurassic sources may generate sufficient thermal gas to charge the fields in the entire study area, but only if favourable geological conditions apply. Sensitivity analyses showed that trap timing and amount of erosion were the most sensitive geological factors for thermal gas charge. Kinetics, heat flow and source thickness were of less importance in this study.

The isotopic composition of methane from Cenomanian fields shows an early gas generation signature, which is consistent with early thermal gas from the shallow Cretaceous sources, but not with the late thermal gas generated from the deeply buried Jurassic sources in the study area (Fig. 9). However, biogenic gas admixing with late thermal gas from Jurassic sources may balance the isotopic composition observed for the Cenomanian gas accumulations and plot as a gas with an early thermal generation signature. There are favourable temperature conditions for biogenic gas production from organic-rich intervals in the basin. Hence biogenic gas is likely to have contributed to the gas fields and admixed with gas from Jurassic sources.

The authors would like to thank ExxonMobil and IPGG–RAS SB for permission to publish this paper. Many thanks to the numerous geoscientists in IPGG and ExxonMobil who provided technical input during the course of the study. Special thanks to Michael Brown and James Maynard for their thorough review of the paper.

References

Bird, K. J., Charpentier, R. R. et al. 2008. Circum-Arctic Resource Appraisal: Estimates of Undiscovered Oil and Gas North of the Arctic Circle. US Geological Survey Fact Sheet 2008-3049.

Cramer, B., Poelchau, H. S., Gerling, P., Lopatin, N. V. & Littke, R. 1999. Methane released from groundwater – the source of natural gas accumulations in northern West Siberia. *Marine and Petroleum Geology*, **16**, 225–244.

Fokin, P. A., Nikishin, A. & Ziegler, P. A. 2001. Pre-Uralian and Peri-Palaeo-Tethyan Rift systems of the East European Craton. In: Ziegler, P. A., Cavazza, W., Robertson, A. H. F. & Crasquin-Soleau, S. (eds) *Peri-Tethys Memoir 6: Peri-Tethyan Rift/Wrench Basins and Passive Margins*. Muséum National d'Histoire Naturelle, Mémoires, **186**, 347–368.

Galimov, E. M. 1988. Sources and mechanisms of formation of gaseous hydrocarbons in sedimentary rocks. *Chemical Geology*, **71**, 77–95.

Grace, J. D. & Hart, G. F. 1986. Giant gas fields of northern West Siberia. *American Association of Petroleum Geologists Bulletin*, **70**, 830–852.

Kontorovich, A. E. 1984. Geochemical methods for the quantitative evaluation of the petroleum potential of sedimentary basins. In: Demaison, G. J. & Murris, R. J. (eds) *Petroleum Geochemistry and Basin Evaluation*. American Association of Petroleum Geologists, Tulsa, OK, Memoirs, **35**, 79–109.

Kontorovich, A. E., Danilova, V. P. et al. 1999. Oil-source formations of West Siberia – old and new looks at the problem. In: Kontorovich, A. E. &

Vyshemirsky, V. S. (eds) *Organic Geochemistry of Oil-source Rocks of West Siberia (Organicheskaya geokhimiya nefteproizvodyashchikh porod Zapadnoy Sibiri)*. NITS OIGGM, Novosibirsk, 10–12.

Kontorovich, A. E., Nesterov, I. I., Salmanov, F. K., Surkov, V. S., Trofimuk, A. A. & Ervye, Yu. G. 1975. *Geology of Oil and Gas of West Siberia (Geologiya nefti i gaza Zapadnoy Sibiri)*. Nedra, Moscow.

Kortsenshteyn, V. N. 1977. *Aquifer Systems of the Largest Gas and Gas Condensate Fields of the USSR (Vodonapornye sistemy krupneyshikh gazovykh i gazikondensatnykh mestorozhdeniy SSSR)*. Nedra, Moscow.

Larter, S. R., Head, I. M. *et al.* 2005. Biodegradation, gas destruction and methane generation in deep subsurface petroleum reservoirs: an overview. *In*: Doré, A. G. & Vining, B. A. (eds) *Petroleum Geology: North-West Europe and Global Perspectives: Proceedings of the 6th Petroleum Geology Conference*. Geological Society, London, 633–640, doi: 10.1144/0060633.

Littke, R., Cramer, B. *et al.* 1999. Gas generation and accumulation in the West Siberian basin. *AAPG Bulletin*, **83**, 1642–1665.

Nemchenko, N. N., Rovenskaya, A. S. & Schoell, M. 1999. Origin of natural gases in giant gas fields of northern West Siberia. *Geologiya Nefti i Gaza*, **1–2**, 45–56.

Nikishin, A. M., Ziegler, P. A. *et al.* 1996. Late Precambrian to Triassic history of the East European Craton: dynamics of sedimentary basin evolution. *Tectonophysics*, **268**, 23–63.

Peters, K. E., Kontorovich, A. E., Huizinga, B. J., Moldowan, J. M. & Lee, C. Y. 1994. Multiple oil families in the West Siberian Basin. *AAPG Bulletin*, **78**, 893–909.

Peterson, J. A. & Clarke, J. W. 1991. *Geology and Hydrocarbon Habitat of the West Siberian Basin*. AAPG Studies in Geology #32. American Association of Petroleum Geologists, Tulsa, OK.

Prasolov, E. M. 1990. *Isotopic Geochemistry and Origin of Natural Gases (Izotopnaya geokhimiya i proiskhozhdeniye prirodnykh gazov)*. Nedra, Leningrad.

Rice, D. D. & Claypool, G. E. 1981. Generation, accumulation, and resource potential of biogenic gas. *AAPG Bulletin*, **65**, 5–25.

Rovenskaya, A. S. & Nemchenko, N. N. 1992. Prediction of hydrocarbons in the West Siberian basin. *Bulletin Centre de Recherche Exploration – Production Elf Aquitaine*, **16**, 285–318.

Schoell, M. 1980. The hydrogen and carbon isotopic composition of methane from natural gases of various origins. *Geochimica et Cosmochimica Acta*, **44**, 649–661.

Schoell, M., Picha, F., Rovenskaya, A. & Nemchenko, N. 1997. Genetic characterization of natural gases in NW Siberia [abstract]. *American Association of Petroleum Geologists International Conference*, Vienna, Abstract Volume, 46–47.

Sengor, A. M. C. & Natal'in, B. A. 1996. Paleotectonics of Asia: fragments of a synthesis. *In*: Yin, A. & Harrison, T. M. (eds) *The Tectonic Evolution of Asia*. Cambridge University Press, New York, 486–641.

Shipilov, V. & Tarasov, G. A. 1998. *Regional Geology of Oil and Gas Bearing Sedimentary Basins of the Russian West-Arctic Shelf*. Apatity, Russia, 306.

Stroganov, L. V. 1990. Genetic criteria and prognosis of zones of oil accumulation on Yamal Peninsula. *Geologiya Nefti i Gaza*, **10**, 2–5.

Surkov, V. S. & Smirnov, L. V. 1994. Cenozoic tectonic events and phase differentiation of hydrocarbons in the Hauterivian–Cenomanian complex of the West Siberian basin. *Geologiya Nefti i Gaza*, **11**, 3–6.

Surkov, V. S. & Zhero, O. G. 1981. *Basement and Development of the Platform Cover of the West Siberian Platform*. Nedra, Moscow, 142.

Ulmishek, G. F. 2003. Petroleum Geology and Resources of the West Siberia Basin, Russia. *USGS Bulletin*, **2201-G**, 1–49.

USGS World Energy Assessment Team. 2000. *US Geological Survey World Petroleum Assessment 2000 – Description and Results*. US Geological Survey Digital data Series – DDS60. World Wide Web Address. http://pubs.usgs.gov/dds/dds-060.

Vasilyev, V. N., Yermakov, V. I. & Nemchenko, N. N. 1979. Origin of natural gas in fields of the northern West Siberian plain. *Geologiya Nefti i Gaza*, **4**, 7–10.

Vyssotski, A. V., Vyssotski, V. N. & Nezhdanov, A. A. 2006. Evolution of the West Siberia Basin. *Marine and Petroleum Geology*, **23**, 93–126.

Winters, J. C. & Williams, J. A. 1969. *Microbiological Alteration of Crude Oil in the Reservoir*. American Chemical Society, Division of Petroleum Chemistry, New York, Meeting Preprints, **14**, E22–E31.

Whiticar, M. J. 1990. A geochemical perspective of natural gas and atmospheric methane. *Organic Geochemistry*, **16**, 531–547.

Ziegler, P. A. 1989. *Evolution of Laurussia*. Kluwer Academic, London, **102**, 14 plts.

Session: North Africa and Middle East

Middle East and North Africa: overview

J. REDFERN[1] and J. CRAIG[2]

[1]*University of Manchester, School of Earth, Atmospheric and Environmental Sciences, Basin Analysis and Petroleum Geoscience Group, Williamson Building, Oxford Road, Manchester M13 9PL, UK*
[2]*Eni Exploration & Production Division, Via Emilia 1, 20097 San Donato Milanese, Milan, Italy*
(e-mail: jonathan.craig@eni.com)

North Africa and the Middle East hold huge reserves and resources of oil and gas and are some of the most important regions for future hydrocarbon production. Recent drilling successes onshore Algeria and offshore Egypt, and renewed industry interest in Libya have re-invigorated exploration in North Africa. The Middle East holds the majority of the world's remaining oil and gas and offers a showcase for new research into the major plays and recent exploration/production advances. The North Africa and Middle East session at the 7th Petroleum Geology Conference brought together 12 papers ranging from keynotes on regional play evaluation to detailed field studies. Ten of those papers are published in this volume.

Craig *et al.* provide a regional perspective that spans North Africa and the Middle East. The paper examines the entire stratigraphic interval and highlights the more challenging, older and deeper plays and higher risk, but more conventional, plays in under-explored frontier areas. It offers some thoughts on the potential to develop new play concepts and to extend known producing plays into new areas. Focusing on North Africa, Eschard *et al.* discuss the timing of uplift of palaeohighs and how they have influenced the Palaeozoic petroleum systems in terms of reservoir and source rock development and trapping configuration. They document how the interplay of tectonics, eustacy and subsidence rate produce complex stratigraphic wedge geometries which have important exploration significance. The paper by Lubeseder *et al.* concentrates on the early Carboniferous (Tournaisian to Visean) interval in North Africa, an alternating series of widespread shallow marine and more discrete fluvial reservoirs with interbedded offshore mudstone seals. It examines the glacio-eustatic controls of facies patterns along the stable continental margin, and identifies four potential stratigraphic trap types, using selected outcrop examples placed into a regional sequence stratigraphic context. Dixon *et al.* provide a case study of an integrated petroleum systems and play fairway analysis of the Ghadames Basin. Pulling together a large regional dataset and linking this to published work, the results offer insight into the timing of hydrocarbon charge and prospectivity of the basin. Luening *et al.* examine the frontier Kufra Basin in SE Libya. This basin is the subject of a current exploration campaign by several companies, hoping to unlock its as yet unproven potential. This paper provides information on the recent well, A1-NC198, drilled by RWE. Despite being dry and lacking hydrocarbon shows, the well still provides important data to improve the understanding of the regional petroleum play.

Moving east, two papers cover important petroleum plays in Egypt. Dancer *et al.* review the Gulf of Suez Basin, a classic extensional rift basin of Miocene age, with both syn-rift and pre-rift hydrocarbon plays. Exploration has been carried out here since the late 1800s, with over 10 billion barrels discovered to date, but poor seismic quality has hindered the identification of deep targets, such as the key pre-rift Nubia Sandstone reservoir. The paper highlights improvements in seismic data acquisition, using 3D ocean bottom cable (OBC) to enhance imaging. The authors present models of the tectonically controlled sedimentation that characterizes the syn-rift section, and show how this approach can re-invigorate exploration. Gordon *et al.* present recent work by a multi-disciplinary team evaluating the prospectivity of the North Red Sea. A new regional biostratigraphic and environmental analysis is presented from North to South through the Gulf of Suez and into the Red Sea, placing the Nubian sequences into a regional chronostratigraphic framework. Again, recent reprocessing and newly acquired seismic data has produced a step change improvement in imaging of the prospective pre-rift section. The authors suggest that all the key elements of the Gulf of Suez petroleum system exist in the North Red Sea, thereby high grading the prospectivity of the area.

Finally, three papers on the Middle East, the first by Fraser, uses an analysis of the prolific Upper Jurassic 'Arab' play fairway of the Middle East, and the application of play fairway risk mapping techniques, to demonstrate the play systems and assess remaining prospectivity. Ismail *et al.* present a short paper on analysis of the Late Cretaceous Shiranish Fm, in the Central Euphrates Graben of Syria. Detailed analyses characterize this potentially important source rock that is within the oil window in many areas. Finally, Garland *et al.* describe a detailed study of the appraisal of the Taq Taq Field, located within an anticline in the gently folded zone of the Zagros mountains, northeastern Iraq. This paper reviews the main reservoirs and the importance of characterizing the pervasive fracture system, which provides the reservoir connectivity and deliverability. The authors describe modelling of the fractures using data from core and image logs, and their critical role in a dual-media dynamic model used for field development planning.

From Neoproterozoic to Early Cenozoic: exploring the potential of older and deeper hydrocarbon plays across North Africa and the Middle East

J. CRAIG, D. GRIGO, A. REBORA, G. SERAFINI and E. TEBALDI

Eni Exploration & Production Division, Via Emilia 1, 20097 San Donato Milanese, Milan, Italy
(e-mail: jonathan.craig@eni.com)

Abstract: As the traditional exploration plays in the main productive basins of North Africa and the Middle East become more 'mature', attention is increasingly focusing on more challenging, older and deeper plays in the main producing basins and on high-risk, but more conventional, plays in under-explored frontier areas. This shift brings with it a range of technical and commercial challenges that must be addressed, if exploration in the region is to remain an attractive proposition. Exploration in North Africa and the Middle East has traditionally focused on the prolific Mesozoic- and Cenozoic-sourced petroleum systems of the Nile Delta, the Sirte Basin, the Pelagian Shelf, and the Arabian Plate and on the Palaeozoic-sourced petroleum systems of the Berkine, Ghadames, Illizi, Ahnet and Murzuq basins, the Central Arabian Basin, the Qatar Arch and the Rub Al Khali Basin. Together these form one of the most prolific petroleum provinces in the world and, as a consequence, there has been little commercial incentive to invest in exploring more challenging and riskier plays in these areas. However, as the need to find new reserves becomes imperative, attempts are increasingly being made to test new play concepts and to extend already proven plays into new areas. Key recent developments in this regard include the recognition of the hydrocarbon potential of the Neoproterozoic to Early Cambrian ('Infracambrian') sedimentary section lying below the traditionally explored Palaeozoic succession in many basins in North Africa. In some areas, particularly the Berkine Basin in Algeria, the Nile Delta in Egypt and the Rub Al Khali Basin in Saudi Arabia, attention is also increasingly being focused on developing deeper gas plays, both in new areas and beneath existing producing fields. The technical challenges associated with these deeper gas plays are immense and include difficult seismic imaging of deep prospects, low porosity and permeability, high temperature and pressure and a critical need to identify 'sweet spots' where either locally preserved primary reservoir characteristics or secondary enhancement of reservoir quality through palaeo-weathering and/or fracturing allow commercial rates of gas production to be achieved. Despite these challenges, it is clear that the future for exploration in many of the more mature basins of North Africa and the Middle East will increasingly lie in evaluating such older and more deeply-buried plays.

Keywords: North Africa, West Africa, petroleum systems, Neoproterozoic, Palaeozoic, Oligocene, Taoudenni Basin, Berkine Basin, Nile Delta

Older and deeply buried petroleum systems are usually characterized by complex geological histories, and this is certainly the case for the Neoproterozoic and Palaeozoic petroleum systems of North Africa and the Middle East. In these systems, the efficiency of the source rocks and the potential to generate, migrate and trap hydrocarbons in a time-frame that allows hydrocarbons to be retained are often the most critical risks. Hydrocarbons can usually only be trapped for a few tens of millions of years (Berg 1975; Watts 1987; Sylta 1991, 1993) because even the most perfect seals are permeable over longer periods of time. One of the most critical issues determining the efficiency of older and/or deeply buried petroleum systems is, therefore, their burial history, and specifically the existence of a 'late' burial phase that can allow hydrocarbons to be generated, expelled, migrated and trapped in a suitably recent timeframe. Exceptions, such as the Neoproterozoic petroleum system of the Amadeus and Officer basins of Australia (Ghori et al. 2009) or the Late Neoproterozoic–Early Cambrian petroleum systems of Oman (Blood 2001; Al-Siyabi 2005) and the Indian Sub-continent (Peters et al. 1995) generally occur where evaporate super-seals are present (Schoenherr et al. 2007) and/or where the post-trapping history is dominated by extreme tectonic stability.

Burial history, thermal history and the presence, quality and reactivity of source rocks are key factors in determining the efficiency of the older and deeper petroleum systems in basins across North Africa and the Middle East (Fig. 1). Thermal events are particularly important in North Africa where rift-related volcanic activity in the Early Jurassic, related to incipient Atlantic rifting (Weis et al. 1987; Liegeois et al. 1991; Logan & Duddy 1998), and the passage of a mantle plume beneath the region during the Cenozoic (Busrewil & Wadsworth 1980a, b; Woller & Fediuk 1980; Loper 1991; Nataf 1991) are locally as important, or even more important, in controlling hydrocarbon maturation than the history of burial and uplift.

The presence of Cenozoic volcanic centres in North Africa that decrease in age from the Mediterranean coast to the Hoggar and Tibesti massifs records the progressive movement of the African Plate eastwards above a mantle plume or plumes (Busrewil & Wadsworth 1980a, b; Woller & Fediuk 1980). The increased heat flow produced by the passage over this plume system seems to be an important factor in rejuvenating the Palaeozoic-sourced petroleum systems in several North African basins, particularly where the overlying Mesozoic succession is thin or absent and the Palaeozoic source rocks are primarily affected only by older, pre-Hercynian, phases of burial.

The burial history, thermal evolution and hydrocarbon prospectivity of the even older and more deeply buried Proterozoic petroleum systems in North Africa and the Middle East are even more uncertain, partly because of the long period of geological time involved and the difficulty of dating the sequences, but also because their pre-Palaeozoic tectonic evolution and their geometry and facies architecture are poorly known. For these systems to be prospective, it is seems to be essential that there is residual hydrocarbon potential remaining after the long Neoproterozoic

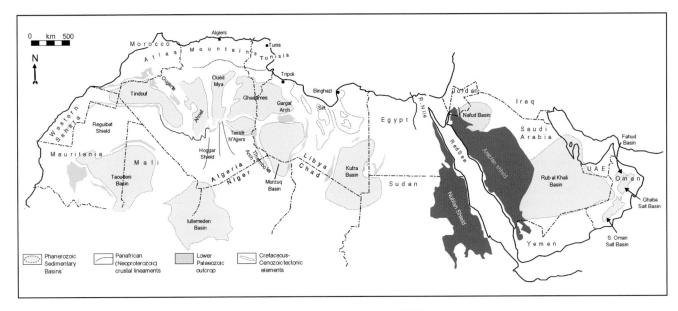

Fig. 1. Major Palaeozoic Basins of North Africa and the Middle East (after Le Heron *et al.* 2009).

and Palaeozoic history and that this residual potential is then activated during the subsequent Mesozoic and Cenozoic evolution, due either to volcanic activity and/or a phase of 'late' burial.

Neoproterozoic plays: the new frontier for exploration in North Africa and the Middle East

The Neoproterozoic Eon, which extends from 1000 Ma to the base of the Cambrian at 542 Ma, is relatively poorly known from a petroleum perspective, despite the existence of potential, proven or producing plays in many parts of the world, including Oman, Mauritania, Siberia, China, India, Pakistan, Australia and both North and South America (Bhat *et al.* 2009; Craig *et al.* 2009; Ghori *et al.* 2009).

The Neoproterozoic was a period of massive atmospheric, climatic and tectonic change. It was dominated by the 'freeze-fry' cycles of the Cryogenian (850–650 Ma) 'Snowball Earth' glaciations, when global mean surface temperature may have oscillated between −50 and +50 °C over periods of ±5–10 Ma (Hoffman *et al.* 1998; Hoffman & Schrag 2000; Halverson *et al.* 2005; Smith 2009). Evolution studies suggest that these extreme climatic variations were ultimately responsible for the emergence of the first animal life around 600 Ma during the Ediacaran (630 Ma–542 Ma) and, in turn, for the eventual explosion of life forms in the latest Neoproterozoic and Early Cambrian. In climatic, tectonic and petroleum systems terms, the Neoproterozoic to Early Cambrian period can be divided into three distinct phases: a Tonian to Early Cryogenian phase, prior to about 750 Ma, dominated by the formation, stabilization and initial breakup of the supercontinent of Rodinia; a Mid Cryogenian to Early Ediacaran phase (*c.* 750–600 Ma) including the major global-scale 'Sturtian' and 'Marinoan' glaciations and a Mid Ediacaran to Early Cambrian (*c.* post 600 Ma) phase corresponding to the formation and stabilization of the Gondwana Supercontinent (Trompette 1997; Collins & Pisarevsky 2005; Zhu *et al.* 2008; Craig *et al.* 2009; Scotese 2009; Fig. 2).

Many of the proven 'Infracambrian' (Neoproterozoic to Early Cambrian) hydrocarbon plays around the world depend on the presence of prolific 'Infracambrian' source rocks. There is increasing evidence that the deposition of many of these organic-rich units was triggered by strong post-glacial sea-level rise, on a global scale, following the major Neoproterozoic glaciations, coupled, in some areas at least, with basin development and rifting on a more local scale (e.g. Le Heron *et al.* 2009).

Interest in exploring for stratigraphically older and deeper 'Infracambrian' plays along the margin of the Gondwana supercontinent, from northern South America, across North Africa and the Middle East, into Pakistan and India (Fig. 3), is growing. This interest is rooted in the greatly improved understanding of the prolific Neoproterozoic to Early Cambrian (Huqf Supergroup), intra-salt Ara 'Stringer' carbonate and Athel silicilyte plays, and the increasingly important older 'pre-salt' plays, in the South Oman Salt Basin, and of the prolific Neoproterozoic and 'Infracambrian' plays in Siberia and southern China.

Oman and the Middle East

The Late Neoproterozoic–Early Cambrian Ara Group portion of the Huqf Supergroup in Oman (Sharland *et al.* 2001; Allen 2007) contains one of the oldest proven hydrocarbon systems in the world (Fig. 2). The reservoirs are dolomitic carbonates and silicilytes encased in salt, with complex trapping geometries, usually strongly affected by salt movements (Amthor *et al.* 2005). Predicting the nature and spatial distribution of these reservoirs in the subsurface is extremely challenging due to their complex geometry and tectonic history (e.g. Droste 1997; Blood 2001). Oil was initially, and rather unexpectedly, discovered in the 'intrasalt Ara carbonate stringers' in 1976 and the play was then explored for the next two decades with indifferent results. Commerciality was achieved in 1997 with the Harweel Deep-1 well and through this and subsequent discoveries the play now accounts for some 70% of the oil production in Oman (Al-Siyabi 2005).

The older Neoproterozoic pre-salt section, underlying the Ara Group, is called the Nafun Group. This consists of five formations, containing both siliciclastic and carbonate lithologies. Pre-salt exploration in Oman started in the 1950s, but renewed interest was triggered in the late 1990s by the discoveries in the Ara Group. The Buah Formation is the youngest and most prospective of the pre-salt formations. It consists of carbonates deposited in a ramp environment with high-quality reservoirs in shallow water peloidal-ooidal and/or stromatolitic facies. These can be laterally extensive over distances of tens of kilometres. Reservoir quality and deliverability are enhanced locally by fractures and karst phenomena. The Buah Formation petroleum system is 'self-sourced' by off-ramp basinal facies shales with total organic content (TOC) values in the range of 2.5–3.5% (Cozzi & Al Siyabi 2004; Cozzi *et al.* 2004).

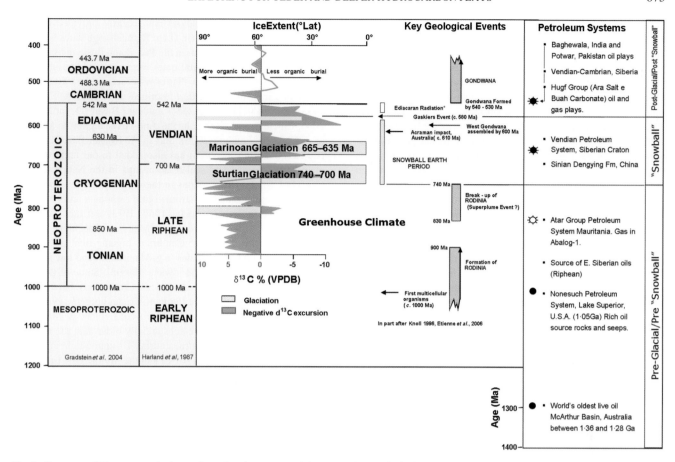

Fig. 2. Summary of Neoproterozoic timescale, carbon isotope record, ice extent, key geological events and petroleum systems (after Craig et al. 2009).

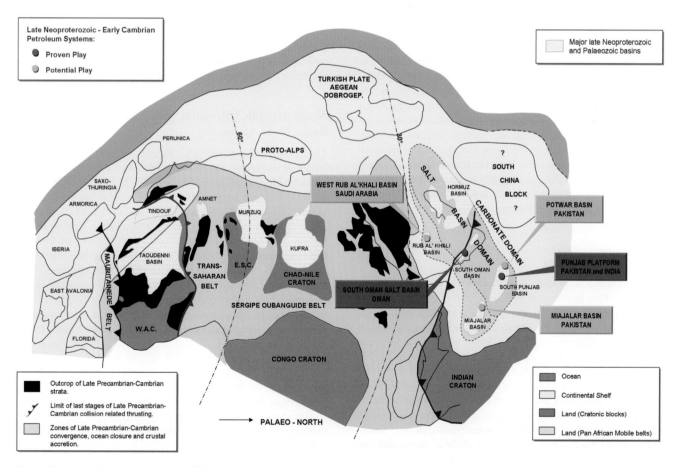

Fig. 3. Generalized Late Neoproterozoic (Ediacaran) to Early Cambrian palaeogeography of the 'Peri-Gondwana Margin' (c. 610–520 Ma) with associated petroleum systems (after Craig et al. 2009).

It is possible that Neoproterozoic plays, similar to those in Oman, exist within and beneath the Hormoz salt in neighbouring Saudi Arabia and Yemen, and perhaps also in Iran (Waltham 2009) and Iraq, but relevant data are currently very limited, with few well penetrations. Similar plays are, however, already being explored actively within the age-equivalent Marwar and Machh supergroups in the Sindh and Punjab regions of Pakistan and in Rajasthan State in western India (e.g. Sati et al. 1997; Riaz et al. 2003), and attention is increasingly turning to the potential for age-equivalent, and possibly older, Neoproterozoic plays across the rest of the Middle East. Neoproterozoic to Early Cambrian sediments occur at outcrop in Saudi Arabia (e.g. Delfour 1970), southern Jordan, Yemen and Oman (e.g. Beydoun 1991) and seismic evidence suggests the presence of Neoproterozoic–Early Cambrian Hormoz/Ara-equivalent salt within 'Infracambrian' graben beneath the Rub Al Khali Basin along the NW–SE trending Najd fault system extending from Saudi Arabia into northern Yemen. The outcropping Neoproterozoic to Early Cambrian sediments are predominantly conglomerates, sandstones and shales, locally with andesite and basalt and often with carbonates and evaporates in the upper parts, deposited under shallow marine to tidal flat conditions along the Gondwana margin. They rest unconformably on Precambrian basement rocks and are commonly overlain unconformably by Cambro–Ordovician sandstones. Interestingly, oil recovered from the Tarfyat-1 well on the North Hadramaut Arch in Yemen is geochemically similar to oil derived from Neoproterozoic to Early Cambrian Huqf Supergroup source rocks in the South Oman Salt Basin (Grosjean et al. 2008). Most of the Persian Gulf region is underlain by the Hormoz salt at depths of 4–10 km, which has been mobilized to form diapirs that reach the surface in more than 200 salt domes, mainly in the coastal region and the Zagros Mountains of Iran. The Hormoz salt was originally more than a kilometre thick in this region. It was deposited in two major units separated by a suite of volcanic rocks and shales which occur as rafts, some more than 1 km across, within the salt domes. The shales contain Cambrian trilobites which date the sequence. Rafts of granitic and gabbroic basement rocks (probably detached from basement faults and then entrained in the flowing salt), together with beds of anhydrite and sylvite, are also preserved in some of the salt domes. The depositional red banding of the salt, produced by the inclusion of some 2–10% of insoluble material (including 'earthy' hematite of syngenetic origin) in thin beds within the pure white halite, defines spectacular flow folds in many of the diapirs in Iran (Waltham 2009). In the Zagros region, salt movement largely began during the Jurassic, with the development of salt pillows, and the diapirism driven by the subsequent increase in overpressure during the Cretaceous and Cenozoic.

North and west Africa

The Neoproterozoic to Early Cambrian ('Infracambrian') successions in northern and western Africa are also emerging as hydrocarbon exploration targets with proven petroleum systems in several areas, including the Taoudenni Basin in Mauritania, Mali and southern Algeria, the Tindouf Basin in southern Morocco and NW Algeria and the Sirte–Cyrenaica rift margin in NE Libya. Large basins with excellent surface outcrops and thick sedimentary fills of Neoproterozoic and Early Palaeozoic age are widespread throughout north and west Africa (Fig. 1). Away from the outcrops, relatively little is known about these successions because they have only rarely been penetrated by wells. However, in the Taoudenni Basin, the Tindouf Basin and the Sirte–Cyrenaica rift margin, the few wells that have been drilled suggest that these successions contain tantalizing evidence of active petroleum systems with clear analogies to major producing Neoproterozoic petroleum systems in other parts of the world.

New biostratigraphic analysis has recently provided the first definitive Tonian–Cryogenian (Late Riphean) age dates for reservoir sequences containing gas in the Taoudenni Basin in Mauritania (Lottaroli et al. 2009). Similar dates were obtained for subsurface sequences in the Cyrenaica Platform bordering the eastern Sirte Basin of Libya in the 1970s (described in Arnauti & Shelmani 1988). Recent fieldwork in the Taoudenni Basin (Deynoux et al. 2006; Shields et al. 2007), in the Anti-Atlas region of Morocco (e.g. Geyer & Landing 2006), and in Al Kufrah Basin in Libya (Le Heron & Howard 2008) has also added to our understanding of reservoir, source and seal relationships in the Neoproterozoic to Early Cambrian successions in these basins and has confirmed the widespread presence of stromatolitic carbonate units of potential reservoir facies (Trompette 1969, 1973; Bertrand-Sarfati & Trompette 1976; Bertrand-Sarfati & Moussine-Pouchkine 1985).

Neoproterozoic to Early Cambrian organic-rich strata were deposited in both high latitudes (e.g. Mauritania) and low latitudes (e.g. Oman) on the 'Peri–Gondwana Margin'. Some of the black shales deposited on the West African Craton may be as old as 1000 Ma and clearly predate the Pan-African orogenic event (e.g. the Bleida–Tachdamt Group, in Morocco; Leblanc & Moussine-Pouchkine 1994). These are substantially older than the majority of the Neoproterozoic ('Infracambrian') organic-rich units that occur across much of North Africa and the Middle East (including those in Oman), which typically range from c. 850 to c. 540 Ma in age. 'Infracambrian' black 'oil shales' and laminated organic-rich dolomites, that are broadly equivalent in age to the prolific Late Neoproterozoic–Early Cambrian source rocks in Oman, are also known to generate, respectively, mature, low-sulphur, light oil (42–50° API) capable of relatively long distance migration and low-maturity, high-sulphur, heavy oil capable of only short distance migration, in northern and eastern Pakistan and western India. The giant Baghewala Oil Field in Rajasthan, India contains an estimated 628 MBBL in place of heavy oil in four separate reservoirs of Neoproterozoic, Neoproterozoic to Early Cambrian and Late Cambrian age. Biomarkers in the oil recovered in the Baghewala-1 well suggest that it is derived from a Late Neoproterozoic–Early Cambrian ('Infracambrian') source rock (Peters et al. 1995).

Neoproterozoic petroleum system in the Taoudenni Basin, West Africa

The most prospective Neoproterozoic plays in northern and western Africa appear to be in the Taoudenni Basin in Mali, Mauritania and southern Algeria, and in the Kufra Basin (Al Kufrah) in southeastern Libya. These basins have broadly similar stratigraphy and tectonic evolution, despite being more 2500 km apart.

The Taoudenni Basin lies on the West African Craton, one of the pre-Pan-African cratonic blocks that formed the core of the Gondwana Supercontinent during the Late Neoproterozoic and the Early Palaeozoic (Trompette 1973). Palaeomagnetic constraints on the position of West African Craton during the earlier Neoproterozoic, before the amalgamation of West Gondwana, are rather poor, but most modern palaeogeographic reconstructions place it as a separate continental fragment located close to the palaeo-south pole (e.g. Collins & Pisarevsky 2005; Pisarevsky et al. 2008; Scotese 2009). Today, the West African Craton and the overlying Taoudenni Basin occupy most of NW Africa and are flanked to the west and east by north–south trending Pan-African 'mobile belts', formed during the accretion of the Gondwana Supercontinent and reactivated, at least partly, during the Late Carboniferous–Early Permian Hercynian orogeny (Fig. 4).

Stratigraphy and tectonic evolution

Outcrops around the margins of the Taoudenni Basin allow a robust stratigraphic and facies framework to be developed for the

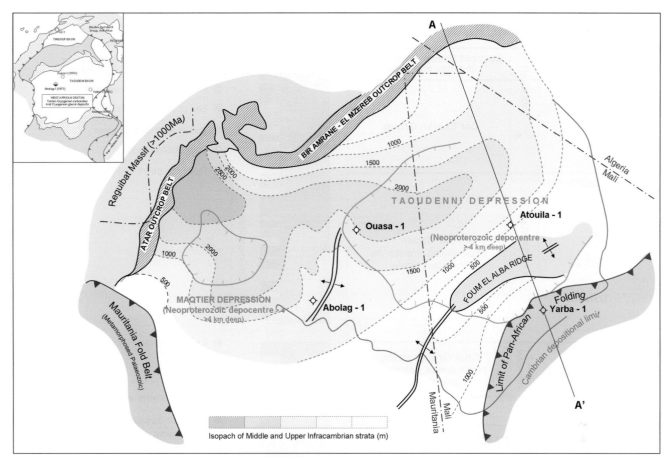

Fig. 4. Neoproterozoic geology of the Taoudenni Basin, Mauritania, Mali and southern Algeria (in part after Moussine-Pouchkine & Bertrand-Sarfati 1997, Baraka Petroleum, Mali Petroleum SA and Societé Mauritanienne des Hydrocarbures 2007).

Meso- and Neoproterozoic succession (Fig. 5). The main tectonic events are recorded by major unconformities within the 1000 m thick succession exposed along the 1100 km long outcrop belt on the northern flank of the basin through Mauritania, northern Mali and into southernmost Algeria. The outcrops in the Atar Region of Mauritania, at the SW end of this belt, are relatively well known and are documented in a series of comprehensive and accurate geological maps and sections (e.g. Moussine-Pouchkine & Bertrand-Sarfati 1997; Deynoux et al. 2006 and references therein; Alvaro et al. 2007). The Neoproterozoic succession here is well defined and radiometrically dated (e.g. Clauer 1981, 1982). An unconformity at the base of the succession marks the Eburnean Orogeny and defines the boundary between the igneous basement and the oldest sediments derived from the weathering and erosion of the Rodinia Supercontinent. The overlying Mesoproterozoic (Early Riphean) Char Group was deposited during a phase of intra-cratonic extension and consists of clastic sediments deposited in fluvial to shallow marine environments. These are overlain by a uniformly east-dipping succession of interbedded stromatolitic carbonates and black shales belonging to the Tonian to Mid-Cryogenian Atar Group that range in age from c. 1000 Ma to somewhat younger than 775 Ma. The top of the Atar Group marks the final breakup of Rodinia supercontinent and is overlain unconformably by a sub-horizontal succession of younger Neoproterozoic sediments with, at the base, a series of glacial/interglacial cycles represented by the alternating clastic and calcareous deposits of the Late Cryogenian Dar Cheikh Group. These glaciogenic sediments are radiometrically dated as younger than 630 Ma which, if correct, would suggest that they are broadly 'Marinoan' or younger in age (Fig. 2). Rapid lateral changes in thickness of the glacial units reflect inherited relief resulting from a long period of erosion and uplift associated with the initial stages of the Pan-African orogeny, combined with palaeo-relief associated with the Cryogenian glaciation. A thin glacial diamictite marks the unconformity associated with the glaciation and is overlain by the sandstones, siltstones and shales of the Late Neoproterozoic (Ediacaran) Assabet Group. A marked unconformity at the base of the Assabet Group is related to the Pan-African compression (Phase 1). Another unconformity associated with a second phase of Pan-African compression (Phase 2) seals the Assabet Group and marks the onset of the Eocambrian glaciation.

All the major unconformities observed in the outcrops surrounding the Taoudenni Basin can be identified on seismic data in the subsurface in areas of the basin where intra-cratonic sub-basins are well developed and the sedimentary successions exhibit clear geometries and internal architectures. Recently acquired seismic data suggest that much of the uplift along the southeastern flank of the Taoudenni Basin is younger than Lower Palaeozoic in age and most probably occurred during the Hercynian Orogeny (Fig. 6).

Hydrocarbon plays

The main Neoproterozoic hydrocarbon play in the Taoudenni Basin comprises Atar Group stromatolitic limestone reservoirs, with hydrocarbons sourced laterally and vertically from organic-rich shales deposited in restricted sub-basins between the stromatolic carbonate complexes, sealed by interbedded shales. The subsurface portion of the Taoudenni Basin remains extremely under-explored. There are only four exploration wells in the basin and only two of these, Abolag-1 drilled in 1973 in Mauritania and Yarba-1 drilled in 1982 in Mali, have tested the Neoproterozoic play. Abolag-1 penetrated more than 600 m of succession which, although undated at

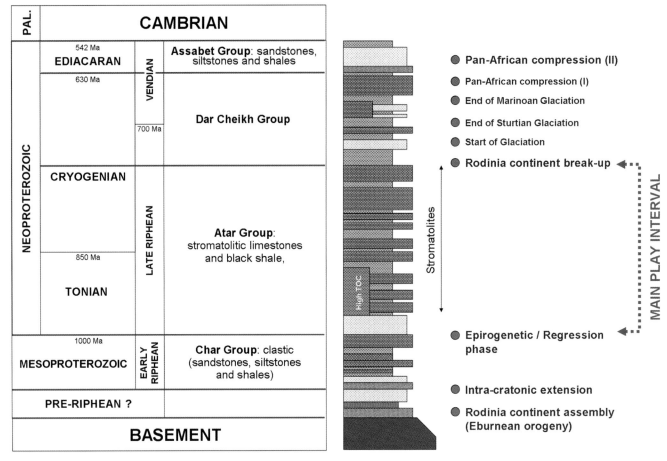

Fig. 5. Summary of the Precambrian stratigraphy, lithologies, main tectonic events and hydrocarbon play in the Taoudenni Basin.

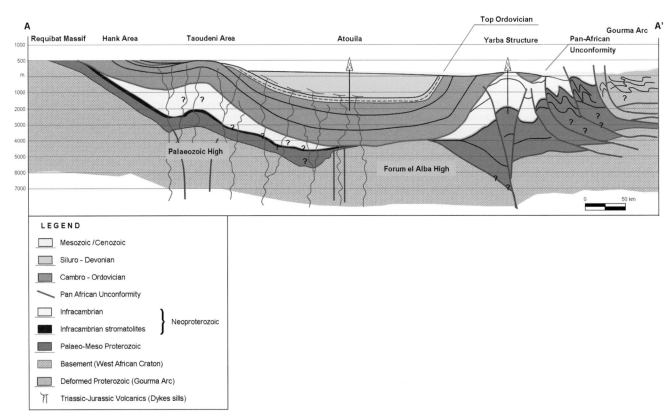

Fig. 6. Schematic geological cross-section through the Taoudenni Basin in western Mali (see Fig. 4 for location) based on seismic and well data.

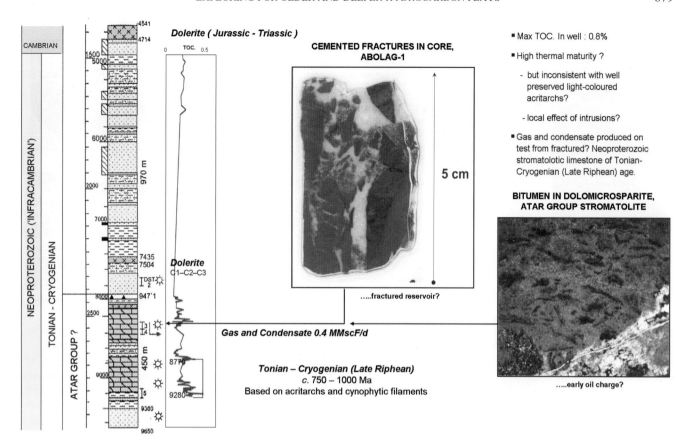

Fig. 7. Summary of the Neoproterozoic succession and the associated petroleum system encountered in the Abolag-1 well in the Taoudenni Basin, Mali.

the time, was assigned to the 'Infracambrian' (Fig. 7). The succession consists of an upper clastic sequence and a lower carbonate sequence, both of which contain hydrocarbons. Gas (and condensate?) was recovered from the carbonate sequence on test at a rate of 13 600 m^3 per day. The carbonate succession in the Abolag-1 well has subsequently been dated biostratigraphically to c. 1000–800 Ma on the basis of an assemblage of acritarchs similar to those recovered from Tonian to Early Cryogenian successions in Siberia, Australia and southern Poland (Lottaroli et al. 2009). There is, therefore, little doubt that the lower, predominantly stromatolitic carbonate succession in Abolag-1 is of Tonian to Early Cryogenian age and that the gas bearing section is the lateral equivalent of the Atar Group at outcrop. The succession in the well has a low TOC and appears to be of high thermal maturity, although this is somewhat inconsistent with the well preserved state of the acritarchs and could be partially a local thermal effect associated with the abundant dolerite intrusions. At outcrop, the equivalent black shales within the Atar Group are of low thermal maturity. The carbonates in the well are tight, heavily fractured and contain bitumen (Fig. 7; Lottaroli et al. 2009), which might suggest that there was an earlier phase of oil generation and migration in this part of the basin.

The distribution and geometry of the stromatolitic carbonates in the Taoudenni Basin can be interpreted using the depositional model of a distally steepened ramp developed by Cozzi & Al Siyabi (2004) for the probably somewhat younger Neoproterozoic stromatolitic carbonate reservoir of the Buah Formation in Oman. In this model, the inner ramp is characterized by the deposition of columnar stromatolites, with oolitic, cross-stratified, grainstones occurring in more distal locations. The latter facies, which in Oman have the best reservoir quality, appears not to have been penetrated by the wells drilled so far in the Taoudenni Basin. Recent interpretations of the depositional setting suggest that the wells are in a mid-outer ramp setting where limited, metre-scale stromatolites with an overall orientation perpendicular to the palaeocoastline are characteristic (Fig. 8). This suggests that there may be areas of the Taoudenni Basin where the reservoir quality of the Neoproterozoic carbonates may be significantly better. Predicting the location of these more prospective areas remains one of the key challenges for future exploration in the Taoudenni Basin and for the Neoproterozoic stromatolitic carbonate play elsewhere in North Africa.

Thermal evolution and hydrocarbon generation

The thermal evolution of the Neoproterozoic petroleum system in the Taoudenni Basin is dominated by two major phases of burial, the first during the Neoproterozoic and terminated by the Pan-African orogeny, and the second during the Palaeozoic and terminated by the Hercynian orogeny (Fig. 9), which was responsible for the emplacement of the internal nappes in the Mauritanide fold belt (Alvaro et al. 2007). The magnitude of the uplift and erosion associated with these two orogenic events varies across the basin. The Pan-African erosion is greatest in the eastern sub-basin, while the Hercynian erosion is more significant in the western sub-basin. Maximum burial was reached just prior to the Hercynian uplift and erosion (at c. 250 Ma) in both sub-basins, and there is little or no significant re-burial after the Hercynian orogeny. This burial history implies that there were two main phases of hydrocarbon generation and migration from any potential Early Neoproterozoic source rocks in the Taoudenni Basin, the first at c. 550 Ma during the latest Neoproterozoic and the Early Cambrian, and the second at c. 250 Ma during the Late Permian and the Early Triassic. Significant volumes of hydrocarbons were probably generated, expelled, migrated and, potentially, trapped during the earlier phase but then leaked away or were re-migrated when traps were tilted or breached during subsequent tectonic events. The lack of significant Mesozoic or Cenozoic burial

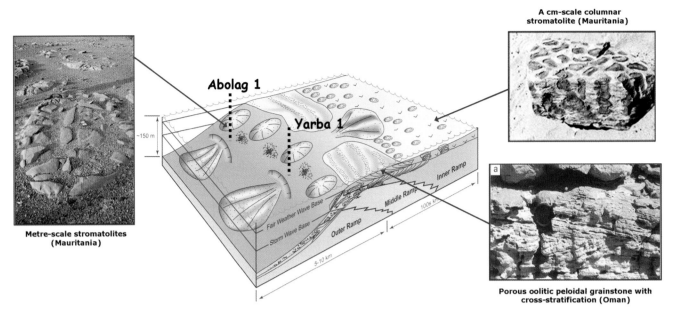

Fig. 8. Depositional model for stromatolitic carbonates in a distally-steepened ramp setting (after Cozzi & Al-Siyabi 2004) and the inferred location of the Neoproterozoic stromatolitic carbonates penetrated by the Abolag-1 and Yarba-1 wells in the Taoudenni Basin in Mauritania and Mali, respectively.

precludes more recent hydrocarbon generation. This is particularly the case around the flanks of the basin where there is remaining residual hydrocarbon potential.

The existence of a viable Neoproterozoic petroleum system in the Taoudenni Basin has yet to be proven and remains the subject of considerable debate. A new phase of exploration is currently underway in the basin and there are both negative and positive indications from the data now available. The negative factors include the low porosity (c. 2–5%) of the main stromatolitic carbonate reservoir and the low flow rate achieved in the Abolag-1 well, the general absence of thick shale seals in the wells, the complex tectonic evolution of the basin, the presence of extensive Triassic–Early Jurassic intrusions and the probability that any Neoproterozoic source rocks have reached high thermal maturities, at least in the central parts of the basin, and that 'early' generation of hydrocarbons (from Ordovician times onwards) resulted in any

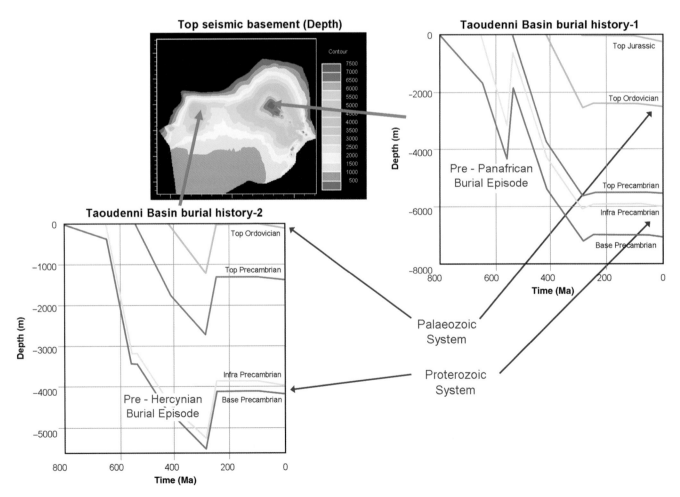

Fig. 9. Burial histories and regional tectonic events for the western and eastern sub-basins of the Taoudenni Basin, Mali and Mauritania.

liquid hydrocarbons being cracked to gas during the Triassic–Early Jurassic thermal event. These are balanced by several positive indications, including the presence of gas (and, possibly, condensate) in Abolag-1, the reported occurrence of gas shows in either Neoproterozoic or Lower Palaeozoic formations in the Ouasa, Asongo, Yarba and Atouilla wells, the existence of oil-prone black shales with a TOC of 10–20% (the so-called 'burning shales') within the Atar Group which are 'early mature' at outcrop, the presence of saline formation water, the low present-day thermal gradients (15–18°C/km), the uncertain thermal history of the basin with conflicting 'high' and 'lower' maturity indicators and, finally, the presence of numerous untested four-way dip and fault-related closures, mainly of 'Hercynian' age.

Neoproterozoic plays in other parts of North and West Africa

Exploration for Neoproterozoic or 'Infracambrian' (Late Neoproterozoic–Early Cambrian) plays in other parts of northern and western Africa is rather less well advanced. The presence of gas shows in the post-Pan-African, mixed clastic and carbonate, Neoproterozoic and Early Cambrian succession (including the Late Neoproterozoic Adoudou, Lie de Vin and Igoudine/Tislit formations) in the AZ-1 well in the Anti-Atlas of Morocco suggests that there may also be some potential both here and in the Tindouf Basin to the south. In addition, oil is produced from fractured granitic basement and overlying 'basal sandstones' on the Augila–Nafoora High along the Sirte–Cyrenaica rift margin in NE Libya. The basement in this area has been radiometrically dated and is Late Precambrian (Neoproterozoic) in age, but the 'Basal sandstones', which are preserved in downfaulted blocks along the flanks of the high, are largely undated and could be, at least in part, 'Infracambrian' in age. However, the source of the oil in this region is clearly not the 'Infracambrian' (which appears to lack any source potential in this area), but is probably the onlapping Cretaceous succession (including the Rachmat Shale), so this is not a true Neoproterozoic petroleum system. Interest has also been sparked by what is assumed to be a remnant Neoproterozoic or 'Infracambrian' basin, containing more than 1500 m of strata, observed on seismic data in the Kufra Basin (Al Kufrah) in SE Libya (Fig. 10) lying, with marked unconformity, below the relatively uniform and flat-lying Palaeozoic succession (e.g. Craig et al. 2008). This has been variously interpreted as an erosional remnant of a once much more extensive 'Infracambrian' succession, possibly preserved in a graben or even, perhaps, given the broadly rhombohedral shape and relationship with potential bounding faults, in a 'pull-apart' basin. The succession remains undrilled and, therefore, undated but, from a seismic facies perspective, it appears to consist of three distinct stratigraphic units, truncated by two well defined unconformities in the upper part. These unconformities have been interpreted as representing the two main phases of Pan-African compression that are observed in the Taoudenni Basin and elsewhere.

The very limited and fragmented nature of both the outcrops and the subsurface penetrations of the Neoproterozoic to Early Cambrian succession in North Africa make it extremely difficult to develop general models of prospectivity or to define prospective play fairways. Further investigations of the Neoproterozoic and Neoproterozoic–Early Cambrian ('Infracambrian') black shales in northern Africa are required in order to better understand their source potential and their relationship with the Cryogenian glaciations. High-resolution biostratigraphic, isotope and other palaeo-temperature proxy records need to be documented for suitable type sections and constrained, where possible, with absolute radiometric dates, so that the timing of black shale deposition

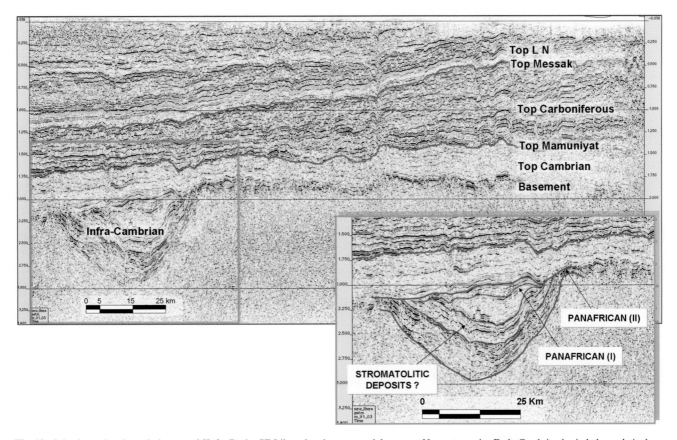

Fig. 10. Seismic section through the central Kufra Basin, SE Libya showing a potential remnant Neoproterozoic–Early Cambrian basin below relatively flat-lying Palaeozoic strata.

can be determined and placed within the Neoproterozoic to Early Cambrian climatic context. Initial observations suggest that there is a higher chance of encountering viable Neoproterozoic petroleum systems in areas underlain by the stable Precambrian cratons than in those underlain by the Pan-African mobile belts (Fig. 11). Areas such as the Taoudenni Basin on the West African craton contain, at least locally, relatively undeformed Neoproterozoic–Early Cambrian successions with Pre-Cryogenian, Cryogenian and Post-Cryogenian plays and rich Neoproterozoic source rocks (see Craig et al. 2009 for discussion). In contrast, the pre-Pan-African Neoproterozoic successions within the Pan-African mobile belts are typically heavily deformed and any 'Infracambrian' plays in theses areas are probably restricted to post-Pan-African inter-montane basins or graben which are typically filled with molasse sediments that have little or no source potential. Overall, our knowledge of the Neoproterozoic Eon and its global petroleum systems is growing rapidly and there is a widespread perception that Neoproterozoic and Late Neoproterozoic–Early Cambrian ('Infracambrian') plays will continue to be an important target for exploration in North Africa and the Middle East in the future.

Palaeozoic plays: expanding the play – the future for frontier oil and deep gas exploration in North Africa and the Middle East

The Palaeozoic hydrocarbon plays of the North African Saharan Platform and the Middle East are far better known and generally more widely explored than their Neoproterozoic counterparts. However, over large parts of the region the Palaeozoic succession is deeply buried beneath Mesozoic and Cenozoic successions (Fig. 12) that themselves contain prolific petroleum systems and, as a result, in the past there has been relatively little commercial incentive to address these more challenging, deeper and therefore, riskier, plays, particularly in the central parts of the main basins which still remain relatively under-explored (Rusk 2001). As the traditional Mesozoic and Cenozoic plays in these areas have become more mature and as the need to find new reserves has become imperative, attention has increasingly turned towards extending the Palaeozoic plays into new areas, including both 'frontier' areas on the periphery of the proven play fairway and the more deeply buried parts of the main proven fairway.

The collisional amalgamation of the African, South American, Indian, Australian and Antarctic basement terrains during the Late Precambrian, and the subsequent delamination of the underlying mantle, resulted in the uplift and subsequent peneplanation of basement rocks across the new Gondwana Supercontinent and the development of a wide, relatively stable shelf around its margins. The assembly of the Gondwana Supercontinent occurred in two main phases, c. 640–600 and c. 570–510 Ma, and was largely complete by the Mid Cambrian (e.g. Collins & Pisarevsky 2005; Pisarevsky et al. 2008; Craig et al. 2009). The peneplanation resulted in the development of an extensive and very gently inclined surface across much of the Gondwana margin encompassing North Africa, the Middle East and the Indian subcontinent. This surface was gradually buried under sediment eroded and transported broadly northwards from the remains of the Pan-African mountains to the south and was periodically flooded by marine transgressions from the north to form a broad shallow marine continental shelf throughout much of the Palaeozoic (Casati & Craig 2003). Reactivation of Pan-African structures during the Early Palaeozoic triggered the development of broad intra-cratonic sag basins that remained active depocentres throughout the Palaeozoic. Thick successions of Palaeozoic strata accumulated

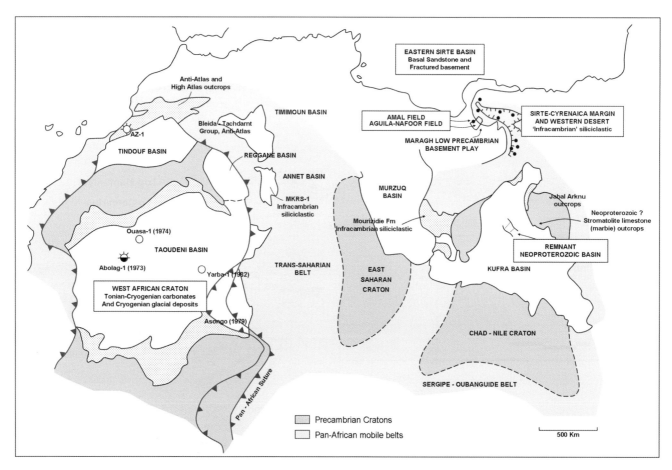

Fig. 11. Summary of Neoproterozoic and Neoproterozoic–Early Cambrian plays in North Africa.

(a) Depth to Hercynian Unconformity

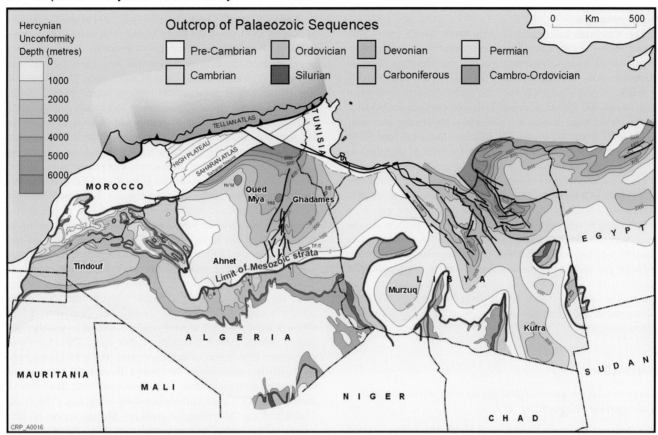

(b) Subcrop to Hercynian Unconformity

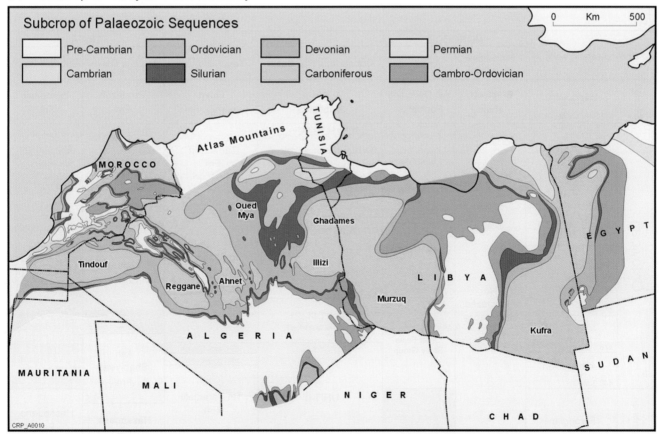

Fig. 12. Thickness of post-Hercynian succession and subcrop to the Hercynian unconformity across North Africa.

in many of these basins, but the shelf was sufficiently stable that similar and broadly correlatable sequences were deposited along the entire margin.

The Cambrian succession in both North Africa and across the Arabian plate typically rests unconformably on Precambrian crystalline or meta-sedimentary rocks or on Neoproterozoic or Neoproterozoic–Early Cambrian ('Infracambrian') sediments. It consists predominantly of fine- to coarse-grained sandstones with occasional pebble beds and conglomerates, deposited in a fluvial to marginal marine environment with primary sediment provenance from the south. The Silurian, Devonian and Early Carboniferous sequence are dominated by belts of marginal to shallow marine sandstones that migrated laterally with changing sea level and pass into deeper marine shales further offshore. The Cambro-Ordovician and Siluro-Devonian sequences are predominantly lowstand systems consisting of coarse clastic sediments deposited in a variety of continental fluvial and shallow marine environments. Major transgressions during the Rhuddanian (earliest Silurian) and the Frasnian (Late Devonian) flooded large parts of the shelf with variably anoxic waters. The associated transgressive and highstand systems contain organic-rich, radioactive ('hot') shales which form primary hydrocarbon source rocks and leaner shales which together act as regionally extensive seals.

North Africa

Cambro-Ordovician stratigraphy, tectonic evolution and hydrocarbon prospectivity

The Cambro-Ordovician succession in North Africa (Fig. 13) was deposited during a period of rifting associated with the opening of the Rheic Ocean (Stampfli & Borel 2002). The main Cambrian reservoirs are separated from the overlying Ordovician Quartzite de Hamra ('Hamra Quartzites') by a *c.* 200 m thick sequence of Argiles D'El Gassi (El Gassi Shale). This vertical barrier has probably isolated the Cambrian succession from overlying younger reservoir units and from the key Early Silurian and Devonian (Frasnian) source rocks in most areas. The charge risk that this creates is substantially reduced where there is lateral fault contact between the Silurian source rocks and Cambrian reservoirs or between Cambrian reservoirs and Ordovician reservoirs, but such occurrences appear to be relatively rare because vertical fault displacements rarely achieve the 500 m required to juxtapose the Cambrian and Silurian successions. The petrophysical properties of the Cambrian reservoirs are generally poor due to extensive quartz cementation, although some porosity is retained locally in facies where kaolinite has been transformed into dickite during burial. The porosity and permeability of the Cambrian sandstones is higher over the Hassi Messaoud High where there was less re-burial after the Hercynian uplift and fracturing is more intense.

The black and dark grey, indurated, silty and locally pyritic El Gassi Shale is sometimes considered to be a viable source rock, but the limited data available suggest that the organic content varies considerably, is usually low (less than 1.5%), although locally as high as 8.6%, and it is almost invariably overmature, although this still requires further study. There have been suggestions that the oil in the super-giant Hassi Messaoud Field (9–11 BBOE recoverable) is derived from both Silurian and Ordovician sources (Balducchi & Pommier 1970; Bacheller & Peterson 1991). Other Ordovician shale units (e.g. The Azzel Shales and the 'Microconglomeratique' Shales) appear to have rather limited source rock potential with TOC values typically less than 1%.

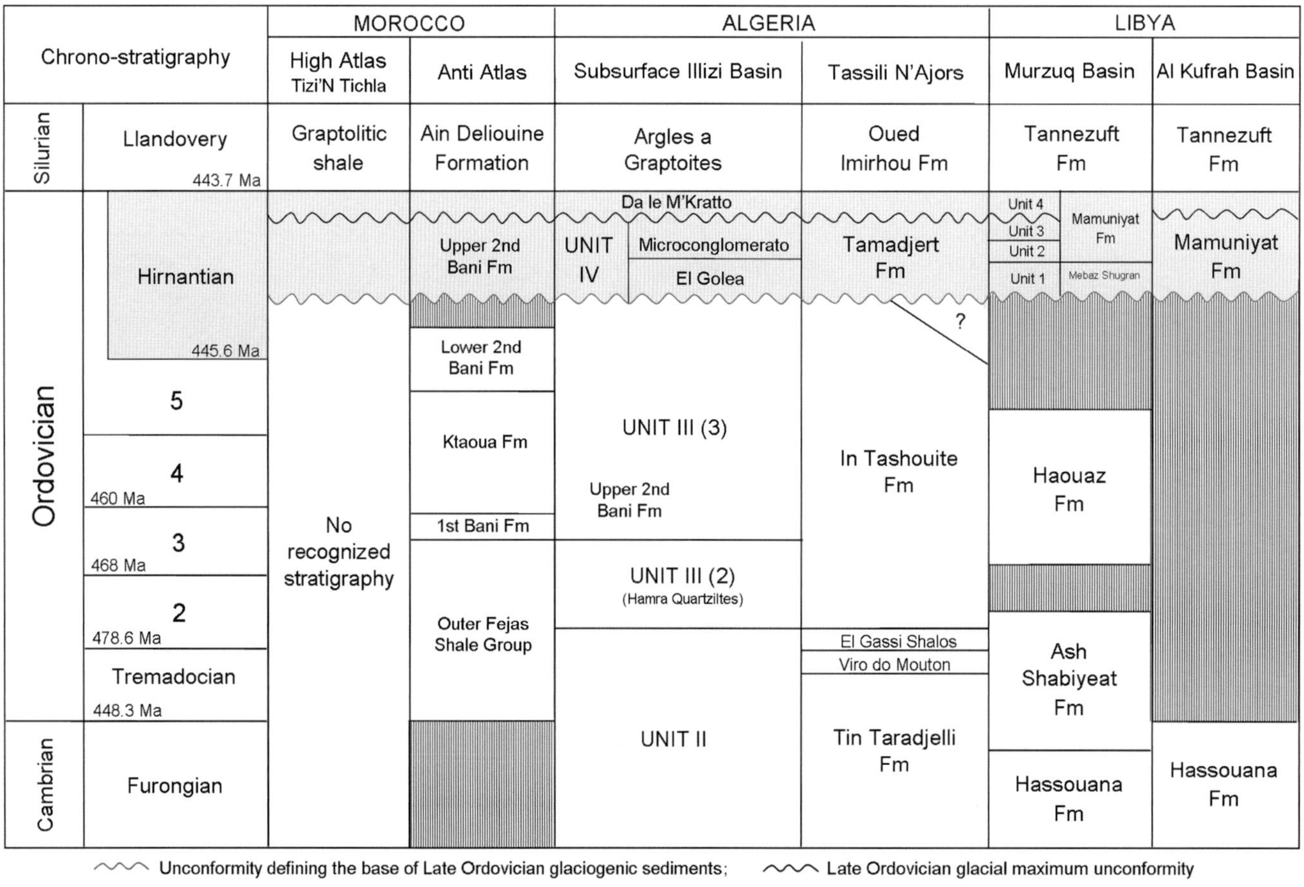

Chrono-stratigraphy		MOROCCO		ALGERIA		LIBYA	
		High Atlas Tizi'N Tichla	Anti Atlas	Subsurface Illizi Basin	Tassili N'Ajors	Murzuq Basin	Al Kufrah Basin
Silurian	Llandovery 443.7 Ma	Graptolitic shale	Ain Deliouine Formation	Argles a Graptoites	Oued Imirhou Fm	Tannezuft Fm	Tannezuft Fm
Ordovician	Hirnantian 445.6 Ma		Upper 2nd Bani Fm	Da le M'Kratto / UNIT IV / Microconglomerato / El Golea	Tamadjert Fm	Unit 4 / Unit 3 / Unit 2 / Unit 1 / Mamuniyat Fm / Mebaz Shugran	Mamuniyat Fm
	5	No recognized stratigraphy	Lower 2nd Bani Fm				
	4 460 Ma		Ktaoua Fm	UNIT III (3)	In Tashouite Fm	Haouaz Fm	
	3 468 Ma		1st Bani Fm	Upper 2nd Bani Fm			
	2 478.6 Ma		Outer Fejas Shale Group	UNIT III (2) (Hamra Quartziltes)			
	Tremadocian 448.3 Ma				El Gassi Shalos / Viro do Mouton	Ash Shabiyeat Fm	
Cambrian	Furongian			UNIT II	Tin Taradjelli Fm	Hassouana Fm	Hassouana Fm

~~~ Unconformity defining the base of Late Ordovician glaciogenic sediments;   ~~~ Late Ordovician glacial maximum unconformity

Fig. 13. Cambro-Ordovician stratigraphy of North Africa (after Le Heron *et al.* 2009).

The reservoir potential of the Ordovician Hamra Quartzites is typically relatively low as a result of extensive quartz cementation during pre- and post-Hercynian burial. The highest remaining porosity is confined to areas where bitumen coating developed during the pre-Hercynian phase of hydrocarbon generation and migration. Fracture development is also important in enhancing the reservoir potential of the Hamra Quartzites, as even in areas where the matrix porosity is very low (typically less than 1%), the presence of fractures increases the overall reservoir permeability substantially. The search for highly fractured 'sweet-spots' in the Hamra Quartzite and equivalent reservoirs is one of the key technical challenges facing future exploration across much of northern Africa.

The equivalent Cambrian to Lower Ordovician succession (the Hassauna and Haouaz formations; Fig. 13) in the Libyan portion of the Ghadames Basin consists of a thick sequence of continental to shallow marine sands and is overlain by a sequence of transgressive marine shales, assigned to the Melez Chograne Formation (Bertello et al. 2003; Craig et al. 2008; Galeazzi et al. 2010).

The Upper Ordovician reservoirs in North Africa (Unit IV, Mamouniat Formation and equivalents) contain at least 5 BBOE of reserves in more than 50 fields scattered across a broad area from the Murzuq Basin in SW Libya to the Ahnet Basin in central Algeria (Fig. 14), including the giant El Feel (more than 500 MMBBL), Tiguentourine (1.2×10^{12} SCF and 40 MMBBL) and Tin Fouye (8×10^{12} SCF and 710 MMBBL) fields (Davidson et al. 2000; Echikh & Sola 2000; Hirst et al. 2002; Ghienne 2003; Le Heron et al. 2004). Most of these reservoirs were deposited in glacially influenced, generally shallow marine settings, on the continental shelf at, or beyond, the margins of a continental ice sheet (Craig et al. 2009; Le Heron & Craig 2008; Le Heron & Dowdeswell 2009; Le Heron et al. 2009). In keeping with their ice-proximal to distal setting, they exhibit complex and rapid changes in facies and reservoir quality, and contain a wide range of glacially induced, syn-sedimentary structures. Deposition was controlled at the basin-scale by the location of fast-flowing ice streams active during glacial maxima, and by meltwater release during glacial recession. Much of the succession consists of fluvioglacial and glaciomarine sediments deposited in a series of deep subglacial incisions into the underlying pre-glacial strata at or near the ice margin (Le Heron & Craig 2009). These tunnel valleys were filled in two main phases. The initial phase was characterized by debris flow release, while during later phases of glacial retreat, glaciofluvial, shallow glaciomarine to shelf sediments were deposited dependent on the water depth at the ice front. The reservoir sand bodies deposited in these tunnel valleys have a complex and irregular distribution that is directly related to the topography created by the erosive action of the ice sheets. The reservoir distribution is closely associated with the position of 'grounding lines' where the glacial topography was enhanced and sediment accumulated in thick glacial outwash ice-contact fans (Le Heron & Craig 2009). Unequivocal evidence of the presence of ice on the continental shelf of North Africa is sparse, but includes isolated occurrences of outsized, exotic, faceted and striated (ice-rafted?) clasts in shales and the presence of locally extensive soft sediment striated 'ice-pavements' (Sutcliffe et al. 2005). The Upper Ordovician glaciogenic rocks characteristically exhibit very rapid lateral and vertical changes in facies (e.g. Sutcliffe et al. 2000; Ghienne 2003). These make it notoriously difficult to establish a sound stratigraphic framework for the sediments, or to correlate them from one area to another, even within a single oil field. However, new work has enabled the Upper Ordovician glaciogenic sediments of North Africa to be subdivided into ice-contact, glacimarine shelf,

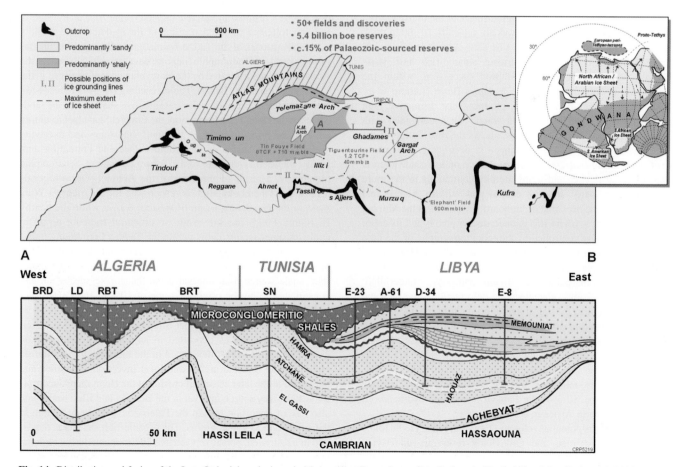

Fig. 14. Distribution and facies of the Late Ordovician glaciogenic Mamouniyat Formation and equivalents in North Africa (after Craig et al. 2008). Inset: distribution of the Late Ordovician ice sheets (after Le Heron & Dowdeswell 2009).

and rebound units, based on an analysis of the facies preserved within the Upper Second Bani Formation in Morocco, the Hassi el Hadjar Formation in Algeria, and the Melez Chograne and Memouniat Formations in Libya. Together these indicate ice-contact to distal glacimarine shelf settings on a high-latitude shelf influenced by an extensive grounded ice sheet (Le Heron & Craig 2009; Le Heron et al. 2010; Craig et al. in preparation). The different facies of the Upper Ordovician glaciogenic succession have widely differing reservoir properties. Successful exploration for, and development of these reservoirs requires a detailed understanding of sand body distribution, geometry and vertical and lateral continuity. Ice-proximal fluvioglacial deposits and high-density turbidites typically form the best-quality reservoirs, but punctuated coarsening-up shoreface deposits in the post-glacial isostatic rebound succession also have considerable potential. Appraisal and development of these glaciogenic reservoirs is further complicated by the presence of a wide range of syn-depositional, glacially induced heterogeneities, including subglacial and ice marginal fold-thrust belts, tunnel valleys, soft-sediment load structures, intra-formational shear surfaces, dewatering structures and micro-faults, all of which have the potential to act variously as barriers, baffles or conduits for fluid flow on both geological and production time-scales.

The Late Ordovician glaciation was short lived (<0.5 Ma) and glacial conditions ended abruptly near the base of the Late Hirnantian *persculptus* Zone (Vecoli 2000; Vecoli et al. 2009). The glacio-eustatic sea level rise associated with the collapse of the ice sheets produced a return to normal oceanic stratification and flooded the previously exposed continental shelf with anoxic waters, pushing the contemporary shoreline southwards by many hundreds of kilometres. Organic-rich, graptolitic pyritic, radioactive ('hot') shales were deposited in isolated topographic depressions in the former glacial landscapes in the Rhuddanian (earliest Llandovery), during the initial stages of the transgression. The organic-rich 'hot' shale intervals are typically 10–60 m thick with TOC contents ranging from 3 to 17%. A younger, less well developed and more areally restricted organic-rich 'hot' shale horizon occurs higher in the Silurian succession, within the Telychian, in some parts of the Ghadames Basin (Luning et al. 2000). The organic-rich horizons produce very distinctive high amplitude events on seismic data which, when combined with information from well penetrations, allows the distribution of the 'hot' shale source rocks to be mapped in detail (Fig. 15, Luning et al. 2000; Craig et al. 2009). The basal Silurian 'hot' shale interval appears to be thicker and more laterally continuous in the more distal or 'out-board' areas of central and western Algeria and Tunisia than in the proximal or 'in-board' areas such as the Murzuq Basin of SE Libya, where it is thinner, discontinuous and its deposition was strongly controlled by the relict topography of the Late Ordovician glacial landscape (Luning et al. 2000; Le Heron et al. 2009). The complex post-glacial topography in these areas reflects a combination of palaeo-ice stream pathways, 'underfilled' tunnel valley incisions, glaciotectonic deformation structures and re-activation of older crustal structures during post-glacial isostatic rebound. Competing hypotheses link the formation of these organic-rich source rocks to either coastal upwelling or freshening of marine water due to the influx of meltwater as the ice sheets collapse (Luning et al. 2000; Armstrong et al. 2005, 2009). These locally discontinuous basal Silurian black shales are the source of at least 80% of the Palaeozoic-derived hydrocarbons discovered in North Africa to date.

Estimates of the potential volumes of gas that remain to be found in the deep Cambrian and Ordovician reservoirs underlying the more conventional Silurian–Devonian Acacus plays in the Ghadames Basin of eastern Algeria and western Libya and the Triassic plays in the Berkine Basin in central Algeria, vary widely. Detailed 3D petroleum systems models suggest that P50 volumes in the range of $70-100 \times 10^{12}$ SCF would not be unreasonable for the Berkine Basin. About half of this volume is likely to be trapped in the Hamra Quartzite and deeper Cambrian reservoirs because the size of the structures generally increases with depth. Our models also suggest that most of this gas will be trapped in a limited number of very large structures, most of which underlie existing producing Triassic fields. The petrophysical characteristics of these deeply buried reservoirs are very uncertain, largely due to the very limited number of well penetrations at these depths, but it would be reasonable to assume that low porosities and permeabilities will be the norm and that tight gas production technologies will need to be utilized, with the consequent impact on the commercial viability. However locally, sandstones in the Late Ordovician Memouniat Formation can have porosities in the range of 8–12% with fair permeabilities (Bertello et al. 2003).

Silurian and Devonian stratigraphy, tectonic evolution and hydrocarbon prospectivity

Sea level continued to rise through the Early Silurian, flooding most of the remaining topography by the Aeronian (mid-Llandovery), before a major north to northeastward-prograding delta system deposited the sandy sediments of the Acacus Formation across much of the North African margin. The basal members of this formation are one of the primary Palaeozoic reservoir targets in the southern and eastern portions of the Ghadames Basin, particularly in northwestern Libyan and southern Tunisia, where they have porosities ranging from 10 to 20% and permeabilities that are often in excess of 100 mD (Bertello et al. 2003). Inversion-related uplift and erosion during the Early Devonian (the so-called 'Caledonian event'), associated with the separation of the 'Hun Superterrane' from African Gondwana (Stampfli & Borel 2002), resulted in variable erosion of the Acacus Formation and was followed by the deposition of fluvio-deltaic and shallow marine sandstones of the Tadrat Formation on the resulting unconformity surface. Stratigraphic traps are developed locally where the Acacus sands are truncated by the Hercynian unconformity (e.g. the Tigi Field, Ghadames Basin). The succeeding Devonian succession in the Ghadames Basin includes transgressive shallow marine sandstones of the Ouan Kasa Formation, the Emgyet Shale (representing a maximum flooding surface), the shelf sandstones and overlying shallow marine shales of the Aouinet–Ouenine Formation and, finally, the shelf sandstones of the Tahara Formation. Stratigraphic trapping as a result of pinchout of the Aouinet–Ouenine sands against the Tihemboka High is responsible for the giant El Wafa Field in the Ghadames Basin of western Libya, while the nearby El Hamra Fields are the result of structural trapping of oil in updip closures against faults within the Tadrart Formation.

The Upper Devonian (Frasnian) sequence (the upper shaly part of the Aouinet–Ouenine Formation) contains the second major 'hot' shale source rock interval n the Palaeozoic succession and is estimated to be the origin of some of 10% of the Palaeozoic-sourced hydrocarbons in North Africa. Its distribution is more restricted than that of the deeper Silurian 'hot' shales, due in part to the fact that it was deposited in an active foredeep (Dixon et al. 2010) so it is best developed in the basin centre and thin to absent on the regional arches and local inversion structures, and partly due to the later effects of erosion at the Hercynian Unconformity. It is a key source interval in the Berkine and Illizi basins and also has an important role in the Palaeozoic petroleum system in the Ahnet and Reggane basins. The organic-rich shale was deposited in a restricted, poorly oxygenated environment during regional transgression of an unconformity surface (the 'Frasnian Unconformity') with palaeo-relief of at least 200 m in the northern Berkine Basin. The 'Frasnian shale' (Frasnian–Famennian?) varies in

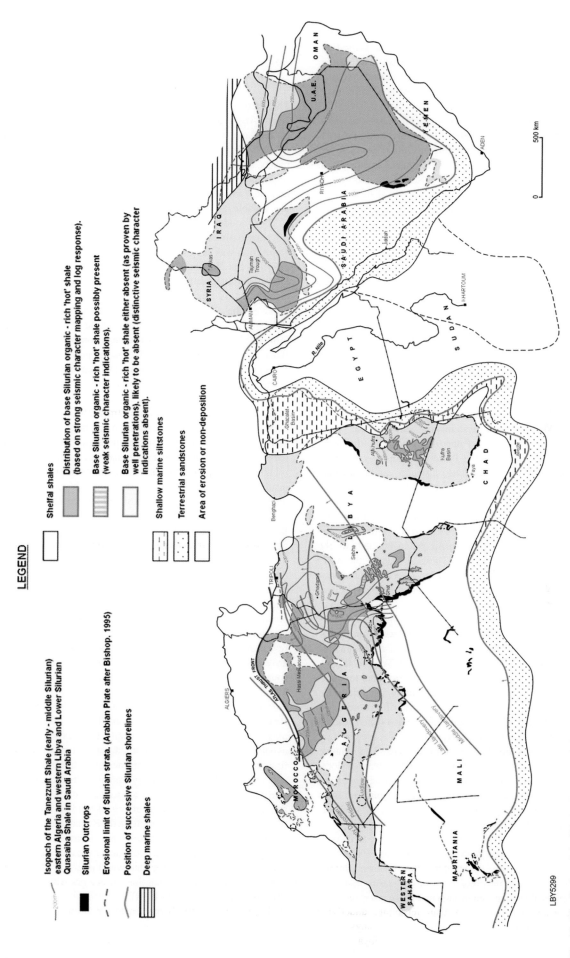

Fig. 15. Distribution of the Silurian Tannezuft Formation and equivalents across North Africa and the Middle East.

thickness from 0 to 150 m across this topography and consists of a series of high-frequency, high-amplitude, metre-scale, gamma ray cycles with TOC typically in the range of 3–5%, but locally up to 14%. As in the Silurian succession, there is a second organic-rich interval present locally, particularly in the northern parts of the Berkine Basin. Hydrocarbons expelled from the Frasnian source rock are sometimes encountered in younger Devonian and Lower Triassic reservoirs (Bertello et al. 2003).

The lateral continuity of the major stratigraphic sequences was a key characteristic of Palaeozoic deposition on the North African continental margin until Late Carboniferous times when more isolated and restricted continental basins developed as a result of tectonic movements marking the initial pulses of the Hercynian orogeny. During the Carboniferous, several mostly shaly units of shelfal and shallow marine facies (the M'rar, Assedjefar and Tiguentourine formations) were deposited before sedimentation was terminated abruptly by uplift, erosion and regional northward tilting as a result of the Hercynian orogeny, during the Late Carboniferous and Early Permian.

The Hercynian orogeny produced widespread compressional deformation in northern and western Morocco and in northern Algeria and uplift and, locally, profound erosion in intra-plate areas such as Algeria and western Libya. Petroleum systems established as a result of burial and maturation of Palaeozoic source rocks prior to the Hercynian orogeny were largely 'frozen' or destroyed by the deformation, uplift and erosion that accompanied the development of the Hercynian unconformity (Casati & Craig 2003; Craig et al. 2008). The subcrop pattern and the topography of the Hercynian unconformity surface (Fig. 12) are key elements in the present-day petroleum systems of North Africa. They control the gross distribution of Palaeozoic reservoirs, the communication pathways between the Palaeozoic source rocks and reservoirs and the pattern of long distance migration into post-Hercynian reservoirs.

Sedimentation resumed in the Early Permian with carbonate deposition in parts of southern Tunisia, but was delayed until the Early to Mid Triassic over much of North Africa with the deposition of clastic, carbonate and evaporitic sediments, that wedge strongly towards the north (Fig. 12). These include, at the base, the important Lower Triassic fluvio-deltaic reservoir sandstones of the TAGI/TAGS in the Algerian portion and the broadly equivalent Ouled Cheb and Ras Hamia formations in the Libyan and Tunisian portions of the Ghadames Basin. The early post-Hercynian sedimentation in the Oued Maya, Berkine and Ghadames basins of eastern Algeria and western Libya was characterized by the deposition of a complex system of Triassic fluvial, aeolian and sabkha sediments across the subtle erosional topography of the Hercynian unconformity surface with regional drainage towards the NNE into the developing Neo-Tethys rift (Turner et al. 2001; Turner & Sherif 2007). Sedimentation and associated volcanism during the Late Triassic and Early Jurassic was increasingly influenced by NE–SW trending extensional faults. The Triassic fluvial and aeolian sandstones are the traditional primary exploration target in these areas. They host some 50% of the Palaeozoic-sourced reserves in structural, stratigraphic and combination traps that are ultimately sealed by thick Early Jurassic (Liassic) evaporates and, locally, by laterally equivalent volcanic lavas.

It is notable that some of the younger Palaeozoic reservoirs in these areas locally retain very high porosities (up to 25%) due to the preservation of porosity by chlorite grain coatings formed during early diagenesis in the presence of mixed marine and fresh water in estuarine depositional environments. These reservoirs constitute more attractive and less risky exploration targets. Most of the Lower–Middle Devonian (Eifelian and Emsian) successions in this area were deposited in offshore or shoreface environments and typically have low reservoir porosities and permeabilities. However, Emsian and Eifelian (F4/F5) sandstones form thin, laterally discontinuous, but relatively porous reservoirs where they are embedded in offshore shales, with the preservation of reservoir quality being, at least in part, due to early overpressure development.

Thermal evolution and hydrocarbon generation

The thermal evolution of the Palaeozoic petroleum systems in North Africa and the Middle East is generally well constrained by the large well database. This is particularly the case for the maturity data necessary to reconstruct the burial history of the basins, although the maturity measurements for pre-Carboniferous sequences are not usually based on vitrinite data, but use other methods to estimate equivalent values, which are inherently less reliable.

The Taoudenni Basin was contiguous with the Tindouf and Reggane basins to the north during the Early Palaeozoic and this connection was only broken as a result of regional inversion and uplift of the Regubiat Rise during the Late Silurian and Early Devonian (Guerrak 1989). The Taoudenni Basin was uplifted and eroded at this time and the resulting unconformity completely removes the Silurian succession in some areas of the basin (Dixon et al. 2010). The thermal evolution of any potential Early Silurian Tannezuft Formation source rocks in the Taoudenni Basin is influenced strongly by burial prior to the Hercynian orogeny (Fig. 9). Uplift and erosion associated with the Hercynian events was quite significant in the western Taoudenni sub-basin, but comparatively minor in the eastern sub-basin. Basin modelling suggests that the maximum burial, reached at 250 Ma, was not sufficient to cause generation, expulsion and migration of significant amounts of hydrocarbons from any potential basal Silurian source rocks in either sub-basin, but particularly in the west, although this is highly dependent on the assumed thermal regime. There was also insufficient re-burial during the Mesozoic and Cenozoic to increase the maturity and reactivate the residual hydrocarbon potential.

The evolution of the Palaeozoic petroleum systems in the Berkine Basin is very different (Fig. 16). Here, the amount of burial during the Palaeozoic, prior to Hercynian orogeny, was comparatively minor in the present-day basin centre, but much greater on the flanks. The key difference in the evolution compared with the Taoudenni Basin is the occurrence of a second important burial phase during the Mesozoic, prior to a minor episode of uplift and erosion related to the Alpine deformation. The lack of substantial Hercynian erosion in the basin centre combined with the pre-Alpine re-burial is sufficient to produce late hydrocarbon generation and an efficient petroleum system in areas of the basin where the source rock potential was not exhausted during the pre-Hercynian burial phase. A similar evolution is observed in many other Palaeozoic basins in North Africa, including the Oued Mya Basin, the Reggane and Ahnet basins (Fig. 1), although the evolution appears to complicated locally by a pronounced heating event at around 200 Ma, recorded through apatite and zircon fission track analysis and the presence of both intrusive and extrusive volcanic rocks. This heating event is related to Late Triassic to Early Jurassic volcanic activity associated with the onset of Atlantic rifting and it overprints the effects of heating caused by simple burial before the Hercynian uplift. In the Reggane and Ahnet basins, this seems to have produced a two-stage maturation history, with an early phase of generation and expulsion of mainly liquid hydrocarbons and a later phase, associated with the c. 200 Ma 'heat spike', during which significant quantities of dry gas were generated and expelled (Logan & Duddy 1998).

In many of the Palaeozoic basins of North Africa, additional burial during Jurassic and Early Cretaceous times seems to have been sufficient to at least partially reactivate the hydrocarbon generation of the main Palaeozoic source rocks and result in further

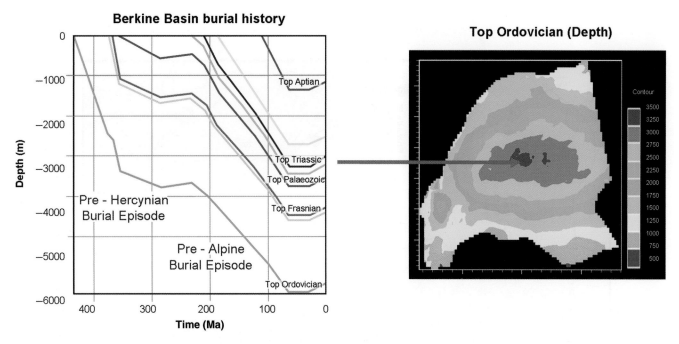

Fig. 16. Burial history and regional tectonic events in the Berkine Basin, central Algeria.

expulsion and migration of both oil and gas into Palaeozoic and basal Triassic reservoirs between the Late Cretaceous and the present day. Deformation and uplift during the Early Cretaceous 'Austrian' and Mid Cenozoic 'Alpine' events modified existing traps, and created new traps in many areas. Locally, a final phase of tilting, re-migration of hydrocarbons and freshwater flushing of reservoirs occurred with the massive Mid to Late Cenozoic uplift of the Hoggar Massif.

The presence or absence of significant Hercynian erosion is often considered to be a critical element in the maturation history of the key Silurian and Devonian source rocks in North Africa. However, detailed 3D Petroleum Systems Modelling of the Tunisian portion of the Ghadames Basin (Tonetti & Grigo 2007) suggests that this may be less important than generally envisaged. In this area, the Silurian source rocks reach approximately the same level of present-day maturity irrespective of whether the Hercynian subcrop is considered to represent palaeotopographic relief with thicker Palaeozoic sequences deposited in topographic lows and thinner sequences or depositional hiatus over the topographic highs, or the result of variously eroded thicknesses of more uniformly distributed Palaeozoic strata. The main difference between these two scenarios is in the timing of generation and expulsion, with the Silurian source rock entering the oil window earlier, just prior to the onset of Hercynian erosion, in the basin depocentre in the latter case. However, the overall quantity, and the phases, of the trapped hydrocarbons are similar in both cases, with the maximum expulsion of both oil and gas during the Middle Oligocene as a result of the combination of burial and of a peak in heat flux related to the passage of a Cenozoic hot spot beneath North Africa.

There is a strong correlation between the fluid phase trapped in the overlying Triassic reservoirs in the Berkine Basin and the maturity and lateral extent of the main basal Silurian and Frasnian source rocks. Gas fields predominate in the SW and east of the Berkine Basin, where the Frasnian source rock is eroded and the charge is probably directly from the highly mature Silurian source rock. Oil fields predominate in the north of the basin where the Frasnian source rock is early mature or in the peak oil generation window and condensates or mixed oil and gas fields characterize the central part of the basin where the Frasnian source rock is highly mature. A similar relationship is observed in the northern Oued Mya Basin (Fig. 17) where the Frasnian source has a more restricted distribution but there is a strong correlation between the phase of hydrocarbons in the fields and the maturity of the Silurian source rock within the corresponding 'fetch areas'. There is a strong differentiation between the SW portion of the Oued Mya Basin and the Mouydir Basin, where generation and migration occurred before the Hercynian orogeny and where there are no significant remaining accumulations, and the northern Oued Mya Basin where maximum maturity was achieved mainly in Albian–Aptian times and where all the major present-day oil and gas fields are located. The Hassi Messaoud, El Gassi and Rhourde El Baguel fields can all be charged from areas subject to post-Hercynian oil and wet gas generation with the southwestward increasing gas–oil ratio (GOR) of the fields along the Hassi Messaoud Arch corresponding with increasing maturity of the Silurian source rock. It is notable that in this scenario the giant Hassi R'Mel Gas Field cannot be charged from the Oued Mya Basin, but must be charged from a deep gas kitchen further to the north. This is consistent with the concept that the Hassi R'Mel Field is stratigraphically trapped along its southern edge.

There was also significant pre-Hercynian generation of oil and gas from the Silurian Tannezuft 'hot' shale source in the Ghadames and Illizi basins and probably also more minor generation of oil and gas from the Frasnian source in the deepest part of the Ghadames Basin (e.g. Dixon et al. 2010). Present day, the Silurian source rock is overmature and the Frasnian source rock is in the gas window in the central part of the Ghadames Basin (Craig et al. 2008).

The basal Silurian 'hot' shale source rock in the Murzuq Basin in SW Libya has a similar two-phase burial history (Fig. 18), with an initial pre-Hercynian burial phase and a subsequent Mesozoic phase, during which the source rock reached its maximum burial in most parts of the basin. The depth of burial in the Murzuq Basin is generally less than in the Oued Mya and Berkine basins and significant volumes of hydrocarbons were only generated and expelled during the later Mesozoic, pre-Alpine burial. Pre-Hercynian and pre-Alpine burial phases also occur in the Kufra Basin (Al Kufrah) in SE Libya, but here the pre-Hercynian phase is relatively weak and there appears to be a significant amount of Alpine uplift

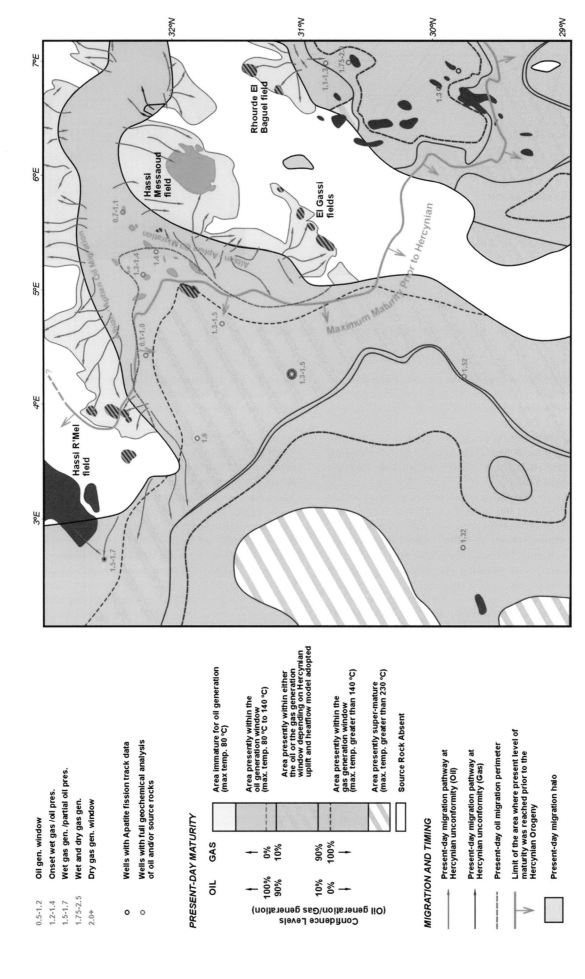

Fig. 17. Thermal maturity of the Early Silurian Tannezuft 'Hot' shale and relationship to the distribution of oil, gas and gas condensate fields in the northern Oued Mya Basin, central Algeria.

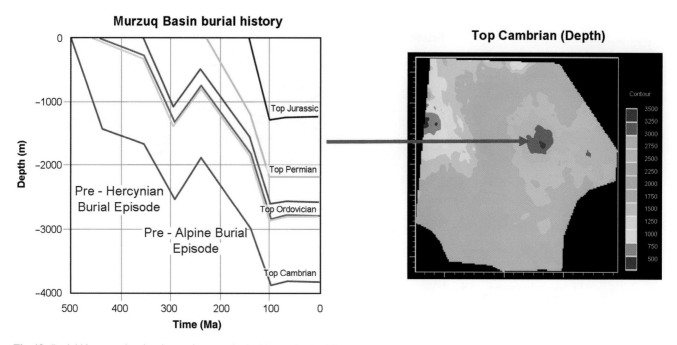

Fig. 18. Burial history and regional tectonic events in the Murzuq Basin, SW Libya.

(Fig. 19). This is likely to have resulted in a rather inefficient Palaeozoic petroleum system in this basin because any potential mature Silurian source rocks are likely to have been lifted out of the 'oil window' rather rapidly during the Alpine uplift.

The ultimate intensity of deformation and the degree of thermal maturity of the main Palaeozoic hydrocarbon source rocks in North Africa decreases broadly eastwards away from the Hercynian and Alpine collision zones (Craig et al. 2008). The present-day concentration of large oil and gas fields in eastern Algeria and southwestern Libya is a direct result of the favourable timing of hydrocarbon generation and expulsion, trap formation and, crucially, hydrocarbon preservation in these areas on the margins of the main orogenic belts. It is in these same areas that the best remaining potential for deeper accumulations below the traditional Triassic and/or Devonian objectives in North Africa is likely to be found (Fig. 20).

Saudi Arabia and the Middle East

The sedimentary basins of the Arabian Plate are universally recognized as one of the richest petroleum provinces in the world. The remaining reserves in the Kingdom of Saudi Arabia alone are estimated to be some 300×10^9 BOE, with the majority being light oil in the 'traditional' reservoirs of Jurassic and Cretaceous age (Pollastro 2003). Exploration of the deeper Palaeozoic sequences in Saudi Arabia only began in the 1980s. The first success came in 1989, with the discovery of oil in the Permian Unayzah Formation in what subsequently became the Hawtah field. Many new Palaeozoic oil and gas fields have been discovered in Saudi Arabia over the past 30 years and a significant proportion of the Saudi Arabia production now comes from Palaeozoic reservoirs. The Palaeozoic succession is well exposed in the central western

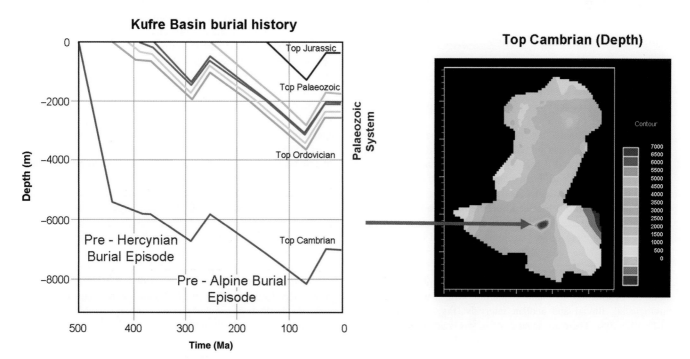

Fig. 19. Burial history and regional tectonic events in the Kufra Basin, SE Libya.

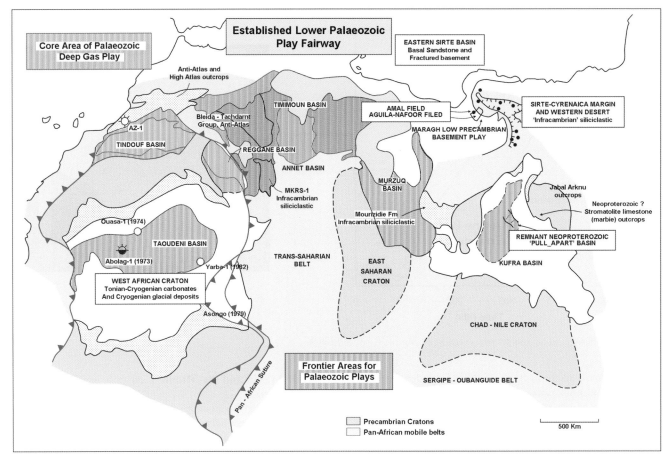

Fig. 20. Summary of Precambrian and Palaeozoic plays in North Africa.

part of the Arabian Peninsula, where it forms extensive outcrops along the eastern flank of the Arabian Shield (Fig. 21). The Palaeozoic succession in these outcrops dips gently eastwards and is soon buried under a thick sequence of mostly Mesozoic sediments which are more than 6000 m thick in the centre of the main Rub Al Khali Basin (Fig. 21).

Stratigraphy and tectonic evolution

The Late Permian Khuff Formation is the best explored and most prolific Palaeozoic reservoir in the Arabian Gulf region. It is the main reservoir in the super-giant North Field in Qatar, which was discovered in 1971, but was not brought on production until 1989. The Khuff Formation is also a reservoir for gas and condensate in Saudi Arabia (e.g. in the Ghawar, Abqaiq and Abu Safah fields) and for gas in Bahrain (Awali Field). It consists mainly of limestone and dolomite, deposited in shallow to open marine environments, with interbedded evaporites that form local intra-formational seals (Ziegler 2001).

The Palaeozoic units below the Khuff Formation in Saudi Arabia are traditionally referred to as the 'pre-Khuff section' and this is indicative of the more limited exploration for, and more limited knowledge of, these older units. The most important and best explored 'pre-Khuff' reservoir is the Late Carboniferous to Mid Permian Unayzah 'Group', a highly variable complex of clastic units of continental to shallow marine facies, with frequent fluvioglacial, fluvial and aeolian intervals (Fig. 22; Senalp & Al-Duaiji 1995). There are several major gas condensate fields in the Unayzah Group in Saudi Arabia, mainly in the so-called 'Central Area' to the south and west of the Ghawar Field. The aeolian facies sandstones typically have the best reservoir characteristics in these fields with good porosity, but locally with reduced permeability. In the tightest sandstones, effective permeability is only present where the formations are heavily fractured.

The base of the Unayzah complex in Saudi Arabia is represented by the regionally extensive Hercynian unconformity surface which cuts across the underlying Early–Middle Carboniferous to Cambrian succession. The main prospective reservoir unit in this pre-Hercynian succession is the Devonian Jauf Formation which comprises a sequence of interbedded shales and sandstones deposited in shallow marine to fluvio-deltaic environments. Trapping in the Jauf Formation is often stratigraphic and related either to the pinchout of the individual sandstone reservoirs or to their truncation by the Hercynian unconformity. Gas condensate is produced from the Jauf Formation in the Hawiyah field, NE of Ghawar (Wender *et al.* 1998). The Hawiyah Field is one of the largest gas projects in the Kingdom of Saudi Arabia and the first to be dedicated solely to the production of non-associated gas.

The underlying Silurian succession is characterized by shallow to deep marine deposition with, at its base, organic-rich radioactive ('hot') shales (Qalibah Formation, Qusaiba Member) that are age-equivalent to the basal Silurian Tannezuft 'hot' shales in North Africa. These organic-rich shales are considered to be the main hydrocarbon source rock for the entire Arabian Palaeozoic petroleum system (Abu-Ali *et al.* 1999; Konert *et al.* 2001; Abu-Ali & Littke 2005; Fig. 22). Some sandy facies within the Qalibah Formation, including the so-called 'Mid Qusaiba Sandstones' and the 'Rhuddanian Sandstones' at the base of Qusaiba Member have also recently become targets for exploration. These units usually have poor to fair reservoir properties, but their distribution and geometry are still very poorly known. Under

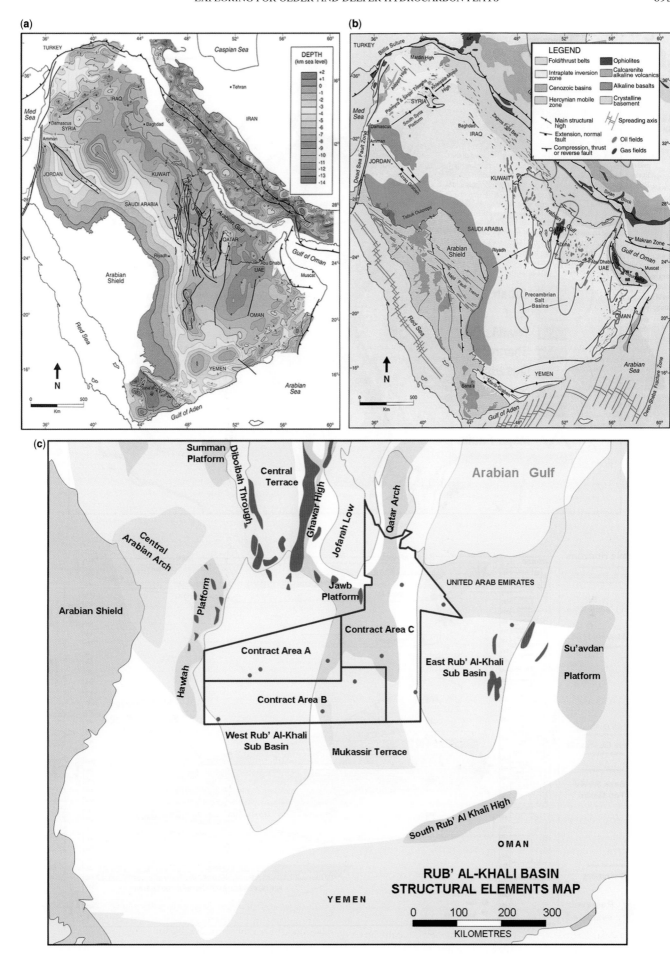

Fig. 21. Regional geology of the Middle East. (**a**) Depth to basement. (**b**) Major tectonic elements. (**c**) Structural elements of the Rub Al Khali Basin and the location of contract areas A, B and C.

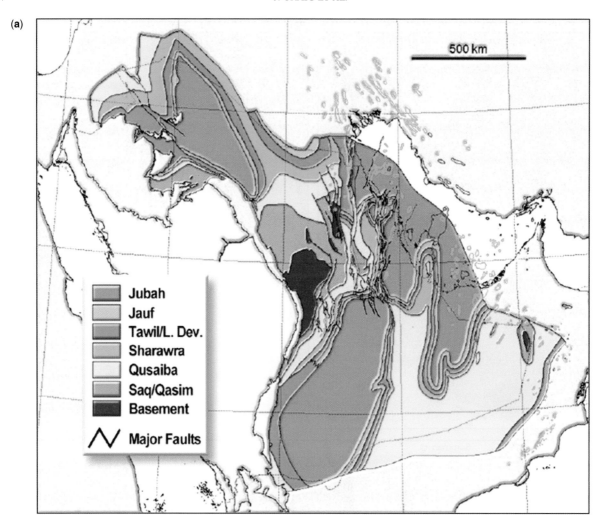

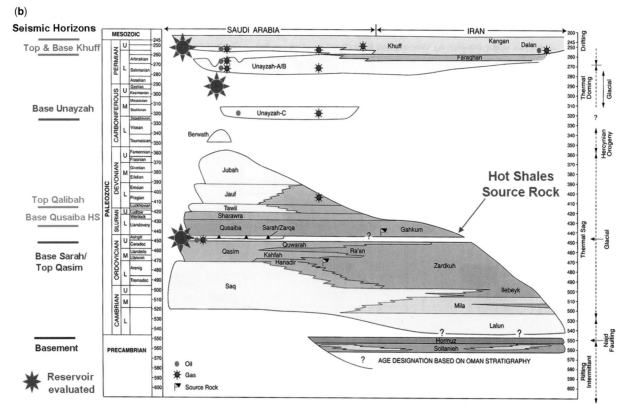

Fig. 22. Palaeozoic geology of the Middle East. (**a**) Subcrop to the Hercynian unconformity. (**b**) Chronostratigraphy, seismic horizons and distribution of source rocks, reservoirs and hydrocarbons.

Fig. 23. Outcrop of a striated glacial surface within the Late Ordovician Sarah Formation, western Saudi Arabia.

favourable conditions, the basal Silurian Qusaiba Member source rock can also charge the underlying Ordovician sandstones of the Sarah and Qasim formations, either by downward migration or through fault juxtaposition.

The latest Ordovician (Ashgillian) Sarah Formation in the Rub Al Khali Basin consists of sandstones and conglomerates deposited in braided-fluvial and fluvioglacial environments, while the underlying Caradocian Qasim Formation consists of shallow marine sandstones and open marine shales. Deposition of the Sarah Formation is strongly controlled by a well developed system of glacial valleys, and sedimentary structures of probably glacial origin are well exposed and documented at outcrop along the western flank of the basin (Fig. 23).

There are relatively few subsurface penetrations of these Ordovician reservoir targets, so information about their reservoir quality at depth is still very limited. The Palaeozoic petroleum system in the Rub Al Khali Basin is largely unaffected by major erosion episodes, with almost continuous burial until the end of the Mesozoic and only relatively minor Alpine uplift and erosion thereafter (Fig. 24). This burial history produces the optimum conditions for generation and expulsion of significant volumes of hydrocarbons (Fig. 25). The main risk seems to be the possible local absence of the basal Silurian 'hot' shale source, particularly in the eastern sub-basin.

The only commercial development of hydrocarbons in the Ordovician succession in Saudi Arabia to date has been in the Khurais region, but most penetrations of the Palaeozoic succession have been the result of deepening of wells targeted at reservoirs in the shallower Mesozoic sequences, with the Palaeozoic succession representing a relatively low-cost 'near-field' upside potential. As a consequence, most of these penetrations are concentrated in the Saudi Aramco core development areas and the Palaeozoic potential in many other areas of the Arabian plate, including the vast Rub Al Khali Basin, remains largely unexplored. It is only recently that exploration for non-associated gas in deep Palaeozoic targets has become a priority in Saudi Arabia, with the award of exploration acreage in the Rub Al Khali Basin to international companies in 2003. Four joint operating companies, Luksar (Lukoil), SRAK (Shell), SSGL (Sinopec) and EniRepSa (Eni and Repsol), each in partnership with Saudi Aramco, are now actively exploring for Palaeozoic gas plays in the Rub Al Khali Basin.

Exploration challenges

The challenges in exploring for Palaeozoic plays as 'stand alone' targets in this region are considerable. The Rub Al Khali desert is an extremely remote area, and considerable time and investment is required for the construction of skid roads and the preparation of well sites under very hostile climatic conditions. In addition, the Palaeozoic targets are extremely deep, typically in the range of 5000–6000 m, and are associated with high-pressure/high-temperature regimes that are close to, and sometimes exceed, the normal operational limits for drilling and logging tools and materials. Rates of penetration are generally extremely low because of the hard and abrasive nature of the Palaeozoic sandstones and the drilling and testing of a single well commonly takes more than one year to complete.

In total, some 22 exploration wells have been drilled (or are currently under operation) in the Rub Al Khali Basin in the past five years by the four operating companies as part of their contractual work commitments. Most of these have been drilled to the Ordovician Sarah and Qasim formations, at depths often exceeding 6000 m. Few results have been released, but several wells are reported to have been tested and flowed unspecified quantities

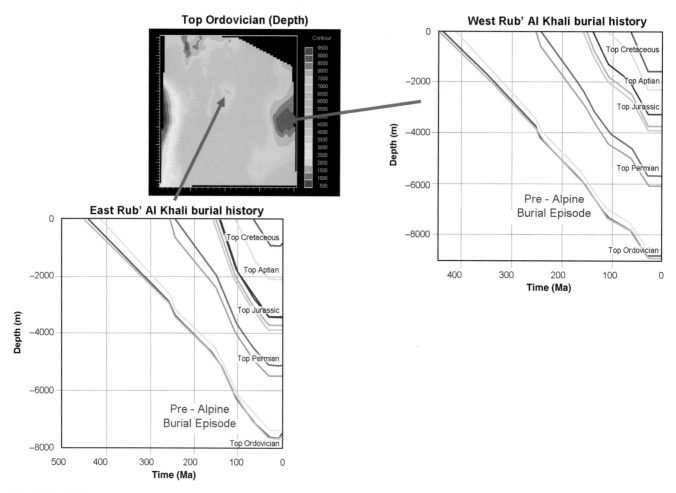

Fig. 24. Burial history and associated regional tectonic events in the Rub Al Khali Basin, Saudi Arabia.

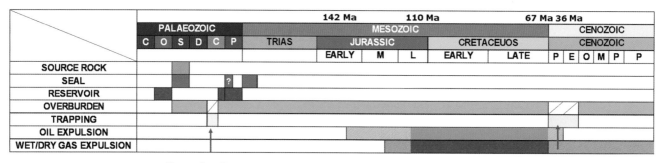

Reservoirs
- Sarah Fm. (Ordovician) sandstones
- Lower Unayzah (Permo-Caroniferous) sandstones
- Upper Unayzah (Permo-Caroniferous) sandstones

Source Rock
- Silurian Hot Shales (Qusaiba Fm.) (Kerogene type II, T.O.C. 4%, H.I. 400)

Seals
- *Qusaiba Shales (Silurian)*
- Lower Unayzah Shales
- Khuff basal clastic shales and Khuff tight limestones and anydrite

Wet/Dry Gas Expulsion Time
- End of Jurassic to recent (142-0 Ma)

Traps
- Four ways dip closure

Time of Trap Development
- Carboniferous (Sarah Fm.)
- Upper Cretaceous – Base Cenozoic

Oil Expulsion Time
- End of Jurassic to Eocene (142-52 Ma)

Fig. 25. Petroleum systems summary chart for the Rub Al Khali Basin, Saudi Arabia.

of gas and/or condensate, confirming that there is an effective Palaeozoic petroleum system in the basin, although no 'commercial' discoveries have yet been announced. Most of the exploration effort has been targeted at the Ordovician Sarah and the Permo-Carboniferous Unayzah formations. These typically exhibit fair to poor reservoir quality, particularly in terms of permeability, and constitute a 'tight gas' play, with significant technical and commercial challenges. The likely low ultimate recovery per well, coupled with the very high costs due to the difficult surface and subsurface environments, makes the economic development of any discoveries in this play extremely challenging. The current price scenario for domestic gas in Saudi Arabia also has a significant impact on project economics and the present scenario makes wet gas discoveries, with high condensate yields, far more attractive than dry gas discoveries. However, condensate production can represent an additional challenge in such low-permeability reservoirs due to 'condensate banking'. In the light of these challenges, significant effort has been devoted to locating reservoir 'sweet spots' where deliverability may be enhanced, either as a result of better primary reservoir characteristics (such as in the aeolian facies of the Unayzah Formation) or by the presence of connected fracture networks. Experience from Palaeozoic discoveries and fields elsewhere in Saudi Arabia suggests that there is likely to be a significant stratigraphic component to any hydrocarbon traps in the Rub Al Khali Basin. An overall improvement in the seismic imaging of these deep and tight formations is critical for further de-risking of prospects, with the application of appropriate surface static corrections and suppression of multiples being key steps towards achieving reliable seismic attribute analysis for reservoir quality prediction.

Palaeozoic plays in other parts of the Middle East

Palaeozoic plays remain relatively unexplored elsewhere in the Middle East with, to date, one significant discovery in western Iraq, five Palaeozoic discoveries in Syria, two Ordovician gas discoveries in Jordan, a single important Ordovician gas and condensate discovery in northern Saudi Arabia (Whaley 2004) and three producing gas fields (Barik, Saih Rawl and Saih Nihayda) and one recent giant gas discovery (Khazzan) in the Upper Cambrian to Lower Ordovician low porosity–very low permeability Barik Sandstone Member in Oman (Millson et al. 2008). There is also further deep potential in the Cambrian clastic reservoirs (Barik, Miqrat and Amin formations, Droste 1997) in Oman (e.g. Millson et al. 2008) and for Cambro-Ordovician, Devonian and Lower Permian clastic reservoirs in adjacent parts of Yemen. Reservoir quality appears to be a key issue for the Palaeozoic reservoirs in many of theses areas. In the Akkas Field in western Iraq, the Ordovician and Silurian sandstone reservoirs have porosities of 6.5–7.6% and permeabilities of 0.13–0.2 mD (Al-Hadidy 2007), but there is clearly considerable potential for further tight gas exploration in deep Palaeozoic plays throughout NW Arabia (Ramseyer et al. 2004). The reservoir quality of the Cambro-Ordovician sandstones encountered in Yemen seems to be rather better with porosities of up to 20% and permeabilities of several hundred milliDarcies recorded in some wells. Although the gas sector is relatively underdeveloped in much of the Middle East, the demand for gas is increasing steadily. For example, in Oman, the supply–demand balance suggests that without further significant gas discoveries Oman may need to import gas by 2012 and the government is actively encouraging gas exploration with an objective of discovering an additional 1×10^{12} SCF of gas per year. Such requirements should make deep gas exploration an increasingly attractive commercial proposition in many parts of the Middle East.

Mesozoic plays: the Sirte Basin, Western Desert and Levantine Basin

The Mesozoic plays in North Africa and the Middle East are, without doubt, the most studied and best explored in the region and are becoming increasingly mature from an exploration perspective, with the possible exception of the deep graben systems within the Sirte Basin of Libya, parts of the western desert of Egypt and the Levantine Basin, the easternmost portion of the Mediterranean basin. This latter, largely under-explored area has yet to be evaluated fully, but several wells in the offshore Sinai Peninsula and offshore Israel have highlighted the existence of active Mesozoic petroleum systems, below more traditional Cenozoic objectives. The major structural elements of the eastern Mediterranean offshore (Garfunkel 1998, 2004; Robertson 1998; Robertson & Comas 1998; Walley 1998; Ziegler 2001; Ben-Avraham et al. 2002; Longacre et al. 2007; Roberts & Peace 2007; Peck 2008)

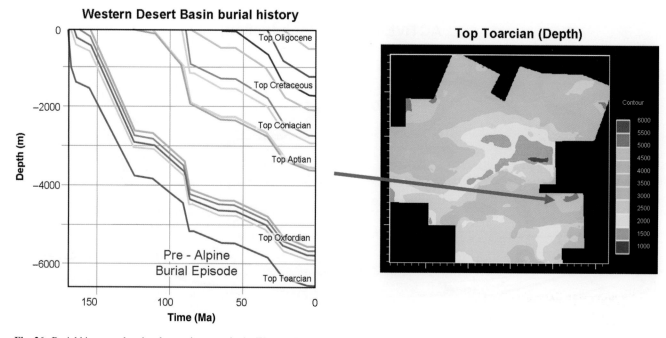

Fig. 26. Burial history and regional tectonic events in the Western Desert, Egypt.

show a marked similarity to the main Syrian Arc folds in the Egyptian Western Desert, where the availability of well data allows the main factors controlling the effectiveness of different Mesozoic petroleum systems to be better constrained. The burial history of the deep Mesozoic petroleum system in the Western Desert (cf. also Metwalli & Piggot 2005) is characterized by almost continuous burial, with some local variation in the rate of subsidence, but no major periods of uplift and erosion through the Cenozoic (Fig. 26). Critical factors controlling the effectiveness of the Mesozoic petroleum system in the Western Desert seem to include the rather indifferent quality of the Mesozoic source rock and the general lack of regional seals. In the Sirte Basin of Libya, the Upper Jurassic–Lower Cretaceous Nubian sandstone members seem to offer significant potential for future exploration below the more traditional and better explored Upper Cretaceous and Cenozoic reservoirs (e.g. Rusk 2001).

Cenozoic plays: deep Oligocene play in the offshore Nile Delta

The offshore Nile Delta of Egypt is a mature gas province with several well explored and proven plays and more than 60×10^{12} SCF of proven reserves. Most of the fields discovered to date are in the shallow portion of the delta (Fig. 27), in reservoirs ranging from Upper Miocene to Plio-Pleistocene in age (Bertello *et al.* 1996; Hart *et al.* 2002; Marten *et al.* 2004; Samuel *et al.* 2005). This Late Cenozoic interval is the most significant and well known portion of the delta sediment cone and has been the subject of numerous studies over the past 20 years (e.g. Dolson *et al.* 2000*a*, *b*; Aal *et al.* 2001; El Barkooky & Helal 2002), Recent detailed mapping of fluid released structures (pockmarks, mud volcanoes associated with gas chimney, etc.) on the seafloor across the Nile Delta cone has revealed significant new information about these shallow petroleum systems (Loncke & Mascle 2004; Loncke *et al.* 2006).

The deeper, Upper Oligocene and older portion of the Nile Delta cone (Fig. 28) is much less well known and remains largely unexplored. However, it forms a significant, if challenging, new frontier for exploration in the region. Recent high-pressure/high-temperature discoveries such as Satis and Port Foaud Marine Deep, at depths in excess of 5000 m, have confirmed the potential of these deep plays to deliver significant new gas and condensate reserves.

Stratigraphy, tectonic evolution and hydrocarbon generation

The main stratigraphic and tectonic events of the Nile Delta region are illustrated in Figure 29 (modified from Meciani *et al.* 2002). The deltaic system initially developed during the late Early Oligocene and subsequent deposition has been characterized by a general increase in terrigenous input, punctuated by several progradational and retrogradational cycles. The most dramatic event to disrupt this long-term depositional trend occurred in Messinian time, when isolation and subsequent desiccation of the Mediterranean resulted in the deposition of a laterally extensive evaporite succession that reaches a maximum thickness of 2000 m in the eastern part of the delta.

The relationship between the younger Late Cenozoic and the older and deeper Early Cenozoic portions of the Nile Delta sedimentary cone is shown in Figure 29. This regional dip geological section shows two different tectonostratigraphic regimes:

(1) a thick sedimentary succession ranging from Late Miocene to Recent in age and 0 to c. 5000 m in depth with a well defined delta progradational geometry and prominent growth faulting in the upper part;
(2) a deeper sedimentary succession of Mesozoic and Early Cenozoic age cut by transpressional features that reactivate older (Mesozoic) tectonic lineaments.

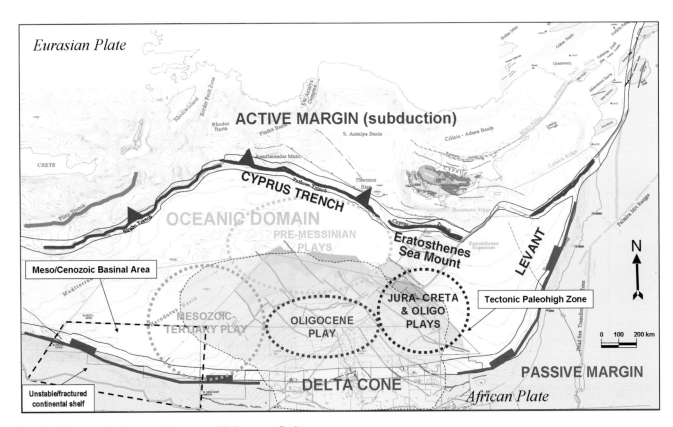

Fig. 27. Major tectonic elements of the eastern Mediterranean Basin.

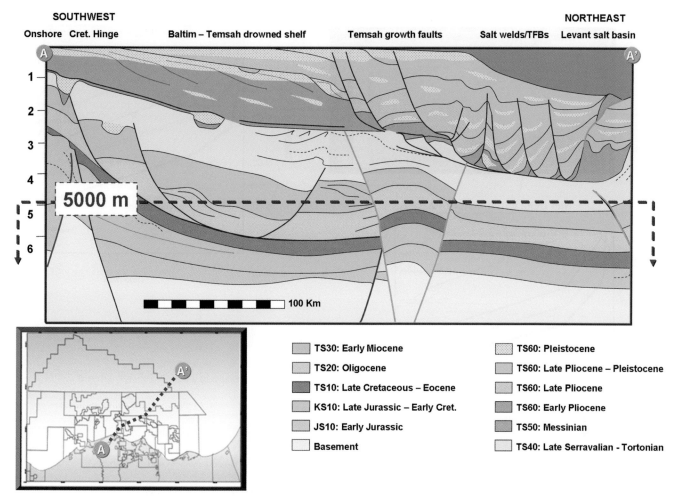

Fig. 28. Regional geological section across the Nile Delta cone, offshore Egypt (modified from Dolson 2001).

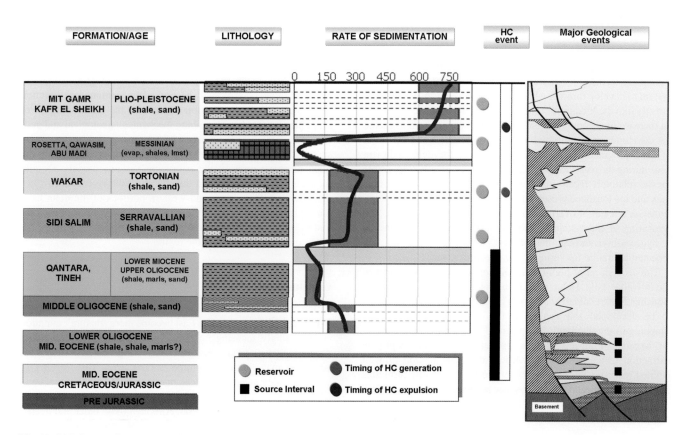

Fig. 29. Lithology, sedimentary stacking patterns, major tectonic events and effective petroleum systems of the offshore Nile Delta, Egypt.

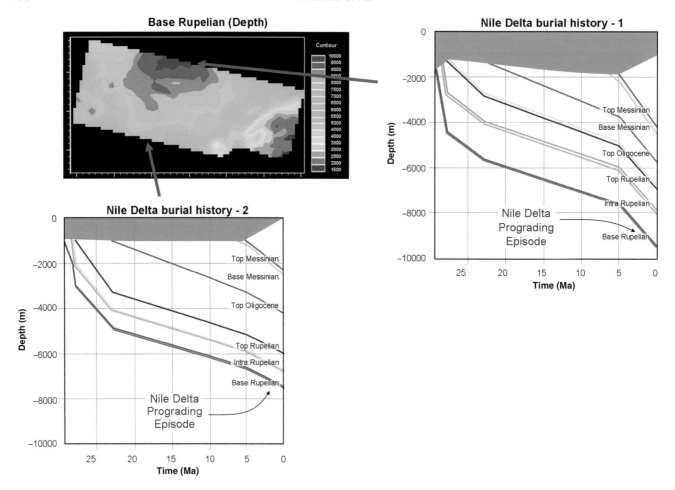

Fig. 30. Stratigraphy and sedimentary characteristics of the Oligocene sequence in the Nile Delta, offshore Egypt.

Early Cenozoic regional tectonic events were responsible for the initiation of the Nile Delta and the first pulses of clastic deposition are well dated biostratigraphically as Middle to Late Oligocene in age (Fig. 30). Detailed palaeogeographic reconstructions indicate that there were two distinct phases of deposition during the early stages of development of the Nile Delta. Initially, the lack of a significant terrigenous platform allowed a channelized turbidite system to develop through which sediment was carried northwards directly into the Mediterranean basin. Later, during a second phase, an increase in accommodation space at the basin margin triggered the development of marginal marine environments and the onset of delta progradation. The loading effect of the delta sediment cone during the past 5 million years (Fig. 31) is directly responsible for the relatively recent maturation of the Early Cenozoic source rock and the localized preservation of reservoir quality in the sandstones as a result of overpressuring of the interbedded shales. The pressuring of the system is also responsible for the re-migration of hydrocarbons through the different petroleum systems through repeated hydraulic fracturing of the intervening seals.

The acquisition of high-quality 3D seismic data across most of the offshore Nile Delta area in recent years (e.g. Wescott & Boucher 2000; Galbiati et al. 2009) has allowed new play concepts to be developed and applied to the deep petroleum systems. Mapping of the distribution of potential reservoirs has, in particular, benefited from the rigorous application of seismic amplitude extraction techniques and has led to the identification of well defined reservoir fairways characterized by sand-prone turbidites and individual meandering deepwater channel systems (Fig. 32).

Exploration challenges

The exploration of these deep targets in the offshore Nile Delta region carries significant risks from both a technical and an operational perspective. The considerable depth of the reservoirs, coupled with the high-temperature/high-pressure environment, means that reservoir quality, seal integrity, the high cost of drilling and the optimization of high-pressure/high-temperature well design are all critical issues. Understanding the reasons for the success of wells that have targeted the play, and of the failures in terms of lack of charge, top seal breach, dis-migration through faults or other reasons, will be critical to continued exploration success as will the application of 'state of the art' techniques of seal efficiency analysis (combing mechanical, geological and geochemical data), elastic inversion of 3D seismic data for overpressure and the integration of reservoir quality prediction for the main reservoir units in the petroleum systems models. Despite the technical challenges and uncertainties, the recent discoveries made in both the eastern and western portions of the delta have already proven more than 3×10^{12} SCF of reserves and indicate that the Oligocene play represents a very promising challenge for the future of exploration in the region.

Conclusions

As the traditional exploration plays in the main productive basins of North Africa and the Middle East become more 'mature', attention is increasingly focusing on more challenging, older and deeper plays in the main producing basins and on high-risk, but more conventional, plays in underexplored frontier areas. This shift brings with it a wide range of technical, operational and commercial challenges that must be addressed if exploration in the region is to remain an attractive proposition. Although these older and deeper plays have widely differing ages, from Neoproterozoic to Early Cenozoic, they share many characteristics. The technical challenges involved in their exploration include difficult seismic

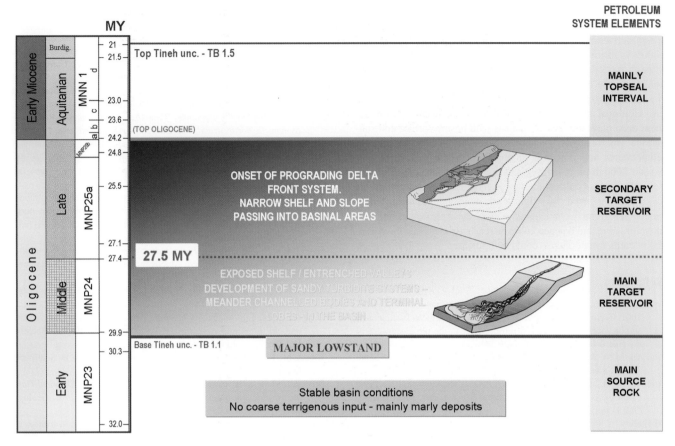

Fig. 31. Burial history and regional tectonic events of the Nile Delta depositional system, offshore Egypt.

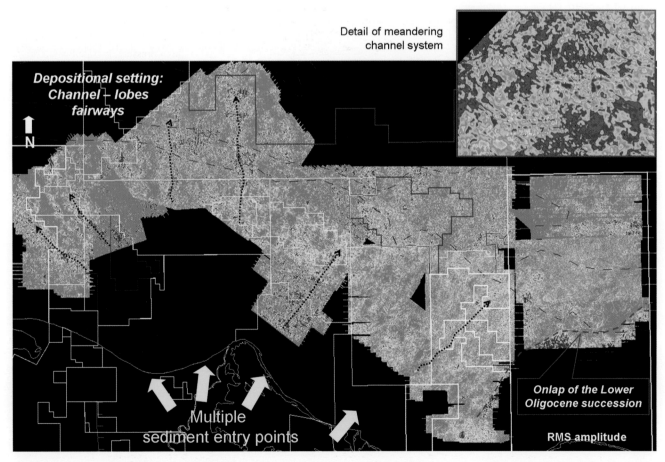

Fig. 32. Regional seismic amplitude anomaly display showing depositional features of the Oligocene succession, offshore Nile Delta, Egypt.

imaging, low porosity and permeability, high temperature and pressure and a critical need to identify reservoir 'sweet spots' where deliverability may be enhanced either through the preservation of primary reservoir quality or secondary enhancement through fracturing. In the case of the Neoproterozoic and Palaeozoic plays, the distribution and continuity of source rocks and the risk of early generation of hydrocarbons due to the deep burial are additional risks. From an operational perspective, many of these deep plays are located in areas with hostile surface and subsurface environments, in remote locations, where the extreme temperatures and pressures at objective depths that commonly exceed 5000 m place drilling and logging tools at, or sometimes beyond, their operational tolerances. The associated high drilling costs and the need for 'tight gas' development technologies renders it extremely challenging to make these projects commerciality attractive, but it is becoming increasingly clear that the potential rewards from doing so could be immense.

References

Aal, A., El Barkooky, A., Gerrits, M., Meyer, H., Schwander, M. & Zaki, H. 2001. Tectonic evolution of the eastern Mediterranean basin and its significance for hydrocarbon prospectivity in the ultradeepwater of the Nile delta. *GeoArabia*, **6**, 363–384.

Abu-Ali, M. A. & Littke, R. 2005. Paleozoic petroleum systems of Saudi Arabia: a basin modeling approach. *GeoArabia*, **10**, 131–168.

Abu-Ali, M. A., Behar, F., McGillivray, J. G. & Rudkiewicz, J. L. 1999. Paleozoic petroleum system of Central Saudi Arabia. *GeoArabia*, **4**, Gulf PetroLink, Bahrain.

Al-Hadidy, A. B. 2007. Palaeozoic stratigraphic lexicon and hydrocarbon habitat of Iraq. *GeoArabia*, **12**, 63–130.

Allen, P. A. 2007. The Huqf Supergroup of Oman: basin development and context for Neoproterozoic glaciation. *Earth Science Reviews*, **84**, 139–185.

Al-Siyabi, H. A. 2005. Exploration history of the Ara intrasalt carbonate stringers in the South Oman Salt Basin. *GeoArabia*, **10**, 39–72.

Alvaro, J. J. *et al.* 2007. The Ediacaran sedimentary architecture and carbonate productivity in the Atar cliffs, Adrar, Mauritania: palaeoenvironments, chemostratigraphy and diagenesis. *Precambrian Research*, **153**, 236–261.

Amthor, J. E., Ramseyer, K., Faulkner, T. & Lucas, P. 2005. Stratigraphy and sedimentology of a chert reservoir at the Precambrian–Cambrian Boundary: the Al Shomou Silicilyte, South Oman Salt Basin. *GeoArabia*, **10**, 89–122.

Armstrong, H. A., Turner, B. R., Makhlouf, I. M., Weedon, G. P., Williams, M., Alsmadi, A. & Abu Salah, A. 2005. Origin, sequence stratigraphy and depositional environment of an upper Ordovician (Hirnantian) deglacial black shale, Jordan. *Palaeogeography, Palaeoclimatology, Palaeoecology*, **220**, 356–360.

Armstrong, H. A., Abbott, G. D., Turner, B. R., Makhlouf, I. M., Muhammad, A. B., Pedentchouk, N. & Peters, H. 2009. Black shale deposition in an Upper Ordovician–Silurian permanently stratified, peri-glacial basin, southern Jordan. *Palaeogeography, Palaeoclimatology, Palaeoecology*, **273**, 368–377.

Arnauti, A. & Shelmani, M. 1988. A contribution to the northeast Libyan stratigraphy with emphasis on the Pre-Mesozoic. *In*: Arnauti, A., Owens, B. & Thusu, B. (eds) *Subsurface Palynostratigraphy of Northeast Libya*. Garyounis University Publications, Benghazi.

Bacheller, W. D. & Peterson, R. M. 1991. Hassi Messaoud Field; Algeria, Trias Basin, Eastern Sahara Desert. *In*: Foster, N. H. & Beaumont, E. A. (comps) *AAPG Treatise of Petroleum Geology: Atlas of Oil and Gas Fields*. American Association of Petroleum Geologists, Tulsa, OK, 211–225.

Balducchi, A. & Pommier, G. 1970. Cambrian Oil Field of Hassi Messaoud, Algeria. *In*: Halbouty, M. T. (ed.) *Geology of Giant Petroleum Fields*. American Association of Petroleum Geologists, Tulsa, OK, Memoirs, **14**, 477–488.

Ben-Avraham, Z., Ginzburg, A., Makris, J. & Eppelbaum, L. 2002. Crustal structure of the Levant Basin, eastern Mediterranean. *Tectonophysics* **346**, 23–43.

Berg, R. R. 1975. Capillary pressure in stratigraphic traps. *American Association of Petroleum Geologist Bulletin*, **59**, 939–956.

Bertello, F., Barsoum, K., Dalla, S. & Guessarian, S. 1996. Temsah discovery: a giant gas field in a deep sea turbidite environment. *In*: Youssef, M. (ed.) *Proceedings of the 13th Petroleum Conference*, Cairo, Egypt. Egyptian General Petroleum Corporation, 165–180.

Bertello, F., Visentin, C. & Ziza, W. 2003. An overview of the evolution and the petroleum systems of the Eastern Ghadames (Hamra) Basin, Libya. *AAPG Hedburg Conference 'Palaeozoic and Triassic Petroleum Systems in North Africa'*, February 2003, Algiers, Algeria, 18–20.

Bertrand-Sarfati, J. & Moussine-Pouchkine, A. 1985. Evolution and environmental conditions of the Conophyton associations in the Atar Dolomite (Upper Proterozoic, Mauritania). *Precambrian Research*, **29**, 207–234.

Bertrand-Sarfati, J. & Trompette, R. 1976. Use of stromatolites for intrabasinal correlation: example from the Late Proterozoic of the north western margin of the Taoudenni Basin. *In*: Walter, M. R. (ed.) *Stromatolites*. Elsevier, Oxford, 517–522.

Beydoun, Z. R. 1991. *Arabian Plate Hydrocarbon Geology and Potential – a Plate Tectonic Approach*. American Association of Petroleum and Geological Studies, Tulsa, OK, **33**.

Bhat, G. M., Ram, G. & Koul, S. 2009. Potential for oil and gas in the Proterozoic Carbonates (Sirban limestone) of Jammu, North India. *In*: Craig, J. *et al.* (eds) *Global Neoproterozoic Petroleum Systems and the Emerging Potential in North Africa*. Geological Society, London, Special Publications.

Blood, M. F. 2001. Exploration for a Frontier Salt Basin in Southwest Oman. *GeoArabia*, **6**, 159–176.

Busrewil, M. T. & Wadsworth, W. J. 1980a. *Preliminary Chemical Data on the Volcanic Rocks of Al Haruj Area, Central Libya*. Geology of Libya. Academic Press, London, **3**, 1077–1080.

Busrewil, M. T. & Wadsworth, W. J. 1980b. *The Basanitic Volcanoes of Gharyan Area, NW Libya*. Geology of Libya. Academic Press, London, **3**, 1095–1105.

Casati, L. & Craig, J. 2003. Petroleum Geology Overview of the North Africa Saharan Platform. AAPG Hedburg Conference '*Palaeozoic and Triassic Petroleum Systems in North Africa*', Algiers, Algeria February 2003, 18–20.

Clauer, N. 1981. Rb–Sr and K–Ar dating of Precambrian clays and glauconites. *Precambrian Research*, **15**, 331–352.

Clauer, N. 1982. Geochronology of sedimentary and metasedimentary Precambrian rocks of the West African Craton. *Precambrian Research*, **18**, 53–71.

Collins, A. S. & Pisarevsky, S. A. 2005. Amalgamating eastern Gondwana: the evolution of the Circum-Indian Orogens. *Earth Science Reviews*, **71**, 229–270.

Cozzi, A. & Al-Siyabi, H. A. 2004. Sedimentology and play potential of the late Neoproterozoic Buah Carbonates of Oman. *GeoArabia*, **9**, 11–36.

Cozzi, A., Grotzinger, J. P. & Allen, P. A. 2004. Evolution of a terminal Neoproterozoic carbonate ramp system (Buah formation, Sultanate of Oman): effects of basement paleotopography. *GSA Bulletin*, **116**, 1367–1384.

Craig, J., Rizzi, C. *et al.* 2008. Structural styles and prospectivity in the Precambrian and Palaeozoic hydrocarbon systems of North Africa. *Geology of East Libya*, **4**, 51–122.

Craig, J., Thurow, J., Thusu, B., Whitham, A. & Abutarruma, Y. 2009. *Global Neoproterozoic Petroleum Systems and the Emerging Potential in North Africa*. Geological Society, London, Special Publications, **326**.

Davidson, L., Beswetherick, S. *et al.* 2000. The structure, stratigraphy and petroleum geology of the Murzuq Basin, southwest Libya. *In*: Sola, M. A. & Worsley, D. (eds) *Geological Exploration of the Murzuq Basin*. Elsevier, Amsterdam, 295–320.

Delfour, J. 1970. Le groupe de J'Balah – une nouvelle unite du bouclier arabe. *Bullétin de BRGM (deuxieme series), Section IV*, **4**, 19–32.

Deynoux, M., Affaton, P., Trompette, R. & Villeneuve, M. 2006. Pan-African tectonic evolution and glacial events registered in Neoproterozoic to Cambrian cratonic and foreland basins of West Africa. *Journal of African Earth Sciences*, **46**, 397–426.

Dixon, R. J., Moore, J. K. S. *et al.* 2010. Integrated petroleum systems and play fairway analysis in a complex Palaeozoic basin: Ghadames/Illizi Basin, North Africa. *In*: Vining, B. A. & Pickering, S. C. (eds)

Petroleum Geology: From Mature Basins to New Frontiers – Proceedings of the 7th Petroleum Geology Conference. Geological Society, London, 735–760; doi: 10.1144/0070735.

Dolson, J. C. 2001. The Petroleum Potential of Egypt. American Association of Petroleum Geologists, Tulsa, OK, Memoirs, **74**.

Dolson, J. C., Boucher, P. J. & Shann, M. V. 2000a. Exploration potential in the offshore Mediterranean, Egypt: perspectives from the context of Egypt's future resources and business challenges. EAGE Conference on Geology and Petroleum Geology, Malta.

Dolson, J. C., Shann, M., Matbouly, S. I., Hammouda, H. & Rashed, R. M. 2000b. Egypt in the twenty-first century: petroleum potential in offshore trends. GeoArabia, **6**.

Droste, H. J. 1997. Stratigraphy of the lower Paleozoic Haima supergroup of Oman. GeoArabia, **2**, 419–472.

Echikh, K. & Sola, M. 2000. Geology and hydrocarbon occurrences in the Murzuq Basin, S.W. Libya. In: Sola, M. A. & Worsley, D. (eds) Geological Exploration in the Murzuq Basin. Elsevier, Amsterdam, 175–222.

El-Barkooky, A. N. & Helal, M. A. 2002. Some Neogene stratigraphic aspects of the Nile Delta. Mediterranean Offshore Conference, Alexandria.

Galbiati, M., Fervari, M. & Cavanna, G. 2009. Seismic evaluation of reservoir quality and gas reserves of DHI supported deep water systems in the offshore Nile Delta. First Break, **27**, 95–102.

Galeazzi, S., Point, O. N., Haddadi, N., Mather, J. & Druesne, D. 2010. Regional geology and petroleum systems of the Illizi–Berkine area of the Algerian Saharan Platform: an overview. Marine and Petroleum Geology, **27**, 143–178.

Garfunkel, Z. 1998. Constraints on the origin and history of the Eastern mediterranean Basin. Tectonophysics, **298**, 5–35.

Garfunkel, Z. 2004. Origin of the Eastern Mediterranean basin: a reevaluation. Tectonophysics, **391**, 11–34.

Geyer, G. & Landing, E. 2006. Ediacaran–Cambrian depositional environments and stratigraphy of the western Atlas Regions, Morocco. UCL Maghreb Petroleum Research Group (MPRG) Infracambrian/Early Palaeozoic Field Guide Series No. 1. Beringeria, Special Issue **6**.

Ghienne, J. F. 2003. Late Ordovician sedimentary environments, glacial cycles, and post-glacial transgression in the Taoudeni Basin, West Africa. Palaeogeography, Palaeoclimatology, Palaeoecology, **189**, 117–145.

Ghori, K. A. R., Craig, J., Thusu, B., Luning, S. & Geiger, M. 2009. Global Infracambrian petroleum systems. In: Craig, J., Thurow, J., Thusu, B., Whitham, A. & Abutarruma, Y. (eds) Global Neoproterozoic Petroleum Systems and the Emerging Potential in North Africa. Geological Society, London, Special Publications, **326**, 109–136.

Grosjean, E., Love, G. D., Stalvies, C., Fike, D. A. & Summons, R. E. 2008. Origin of petroleum in the Neoproterozoic-Cambrian South Oman Salt Basin. Organic Geochemistry, **40**, 87–110.

Guerrak, S. 1989. Time & Space distribution of Palaeozoic oolitic ironstones in the Tindouf Basin, Algerian Sahara. Journal of the Geological Society, London, **46**, 197–212.

Halverson, G. P., Hoffman, P. F., Schrag, D. P., Maloof, A. C. & Rice, A. H. 2005. Towards a Neoproterozoic composite carbon-isotope record. Geological Society of America Bulletin, **117**, 1181–1207.

Hart, S., Gerrits, M. et al. 2002. The Northeast Mediterranean deep water area – results of first drilling campaign & Regional Framework Study or ... Finding the Sweet Spots in a large under-explored block. Mediterranean Offshore Conference, Alexandria.

Hirst, J. P. P., Benbakir, A., Payne, D. F. & Westlake, I. R. 2002. Tunnel valleys and density flow processes in the upper Ordovician glacial succession, Illizi Basin, Algeria: influence on reservoir quality. Journal of Petroleum Geology, **25**, 297–324.

Hoffman, P. F. & Schrag, D. P. 2000. Snowball earth. Scientific American, **285**, 50–57.

Hoffman, P. F., Kaufman, A. J., Halverson, G. P. & Schrag, D. P. 1998. A Neoproterozoic snowball Earth. Science, **281**, 1342–1346.

Konert, G., Afifi, A. M., Al Hajri, S. A. & Droste, H. J. 2001. Paleozoic stratigraphy and hydrocarbon habitat of the Arabian Plate. GeoArabia, **6**, 407–442.

LeBlanc, M. & Moussine-Pouchkine, A. 1994. Sedimentary and volcanic evolution of a Neoproterozoic continental margin (Bleida, Anti-Atlas, Morocco). Precambrian Research, **70**, 25–44.

Le Heron, D. P. & Craig, J. 2008. First-order reconstructions of the Late Ordovician Saharan Ice Sheet. Journal of the Geological Society, **165**, 19–29.

Le Heron, D. P. & Dowdeswell, J. A. 2009. Calculating ice volumes and ice flux to constrain the dimensions of a 440 Ma North African ice sheet. Journal of the Geological Society, London, **166**, 277–281.

Le Heron, D. P. & Howard, J. 2008. Short notes and guidebook on the geology of the Jabal az-Zalmah region, Northern Al Kufrah Basin. Geology of Southern Libya, Field trip, 21–25 November 2008, 60.

Le Heron, D., Sutcliffe, O., Bourgig, K., Craig, J., Visentin, C. & Whittington, R. 2004. Sedimentary architecture of Upper Ordovician tunnel valleys, Gargaf Arch, Libya: implications for the genesis of a hydrocarbon reservoir. GeoArabia, **9**, 137–159.

Le Heron, D. P., Craig, J. & Etienne, J. L. 2009. Ancient glaciations and hydrocarbon accumulations in North Africa and the Middle East. Earth Science Reviews, **93**, 47–76.

Le Heron, D. P., Armstrong, H. A., Wilson, C., Howard, J. P. & Gindre, L. 2010. Glaciation and deglaciation of the Libyan Desert: the Late Ordovician record. Sedimentary Geology, **223**, 100–125.

Liegeois, J. P., Sauvage, J. F. & Black, R. 1991. The Permo-Jurassic alkaline province of Tadhak, Mali: geology, geochronology and tectonic significance. Lithos, **27**, 95–105.

Logan, P. & Duddy, I. 1998. An investigation of the thermal history of the Ahnet and Reggane Basins, Central Algeria, and the consequences for hydrocarbon generation and accumulation. In: MacGregor, D. S., Moody, R. T. J. & Clark-Lowes, D. D. (eds) Petroleum Geology of North Africa. Geological Society, London, Special Publications, **132**, 131–155.

Loncke, L. & Mascle, J. F. 2004. Mud volcanoes, gas chimneys, pockmarks and mounds in the Nile deep sea fan (eastern Mediterranean): geophysical evidences. Marine and Petroleum Geology, **21**, 669–689.

Loncke, L., Gaullier, V., Mascle, J., Vendeville, B. & Camera, L. 2006. The Nile deep-sea fan: an example of interacting sedimentation, salt tectonics, and inherited subsalt paleotopographic features. Marine and Petroleum Geology, **23**, 297–315.

Longacre, M., Bentham, P., Hanbal, I., Cotton, J. & Edwards, R. 2007. New crustal structure of the Eastern Mediterranean basin: detailed integration and modeling of gravity, magnetic, seismic refraction, and seismic reflection data. EGM International Workshop on Innovation in EM, Grav and Mag Methods: a new Perspective for Exploration, Capri.

Loper, D. E. 1991. Mantle plumes. Tectonophysics, **187**, 372–384.

Lottaroli, F., Craig, J. & Thusu, B. 2009. Neoproterozoic–Early Cambrian (Infracambrian) hydrocarbon prospectivity of North Africa: a synthesis. In: Craig, J., Thurow, J., Thusu, B., Whitham, A. & Abutarruma, Y. (eds) Global Neoproterozoic Petroleum Systems and the Emerging Potential in North Africa. Geological Society, London, Special Publications, **326**, 137–156.

Luning, S., Craig, J., Loydell, D. K., Storch, P. & Fitches, B. 2000. Lower Silurian 'hot shales' in North Africa and Arabia: regional distribution and depositional model. Earth Science Reviews, **49**, 121–200.

Marten, R., Shann, M., Mika, J., Rothe, S. & Quist, Y. 2004. Seismic challenges of developing the pre-Pliocene Akhen Field offshore Nile Delta. The Leading Edge, April, **23**, 314–320.

Meciani, L., Laura, S. A. & Khalil, M. 2002. Tectono-sedimentary provinces of the Oligo-Quaternary in the Offshore Nile Delta: relevance to hydrocarbon trapping and reservoirs distribution. AAPG Conference, Cairo, October 2002.

Metwalli, F. I. & Piggot, J. D. 2005. Analysis of petroleum system criticals of the Matruh-Shushan Basin, Western Desert, Egypt. Petroleum Geoscience, **11**, 157–178.

Millson, J. A., Quin, J. G., Idiz, E., Turner, P. & Al-Harthy, A. 2008. The Khazzan gas accumulation, a giant combination trap in the Cambrian Barik Sandstone Member, Sultanate of Oman: implications for Cambrian petroleum systems and reservoirs. AAPG Bulletin, **92**, 885–917.

Moussine-Pouchkine, A. & Bertrand-Sarfati, J. 1997. Tectonosedimentary subdivisions in the Neoproterozoic to Early Cambrian over of the Taoudenni Basin (Algeria, Mauritania, Mali). Journal African Earth Sciences, **24**, 425–443.

Nataf, H. C. 1991. Mantle convection, plates and hotspots. *Tectonophysics*, **187**, 361–371.

Pankhurst, R. J., Trouw, R. A. J., de Brito Neves, B. B. & de Wit, M. J. (eds) *West Gondwana: Pre-Cenozoic Correlation Across the South Atlantic Region*. Geological Society, London, Special Publications, **294**, 9–31.

Peck, J. 2008. Giant oil prospects lie in distal portion of offshore East Mediterranean Basin. *Oil & Gas Journal/Oct 6*.

Peters, K. E., Clark, M. E., Das Gupta, U., McCaffrey, A. M. & Lee, C. Y. 1995. Recognition of an Infracambrian source rock based on biomarkers in the Baghewala-1 Oil, India. *AAPG Bulletin*, **79**, 1481–1494.

Pisarevsky, S. A., Murphy, J. B., Cawood, P. A. & Collins, A. S. 2008. Late Neoproterozoic and Early Cambrian palaeogeography; models and problems. *In*: Pollastro, R. M. 2003. *Total Petroleum Systems of the Paleozoic and Jurassic, Greater Ghawar Uplift and Adjoining Provinces of Central Saudi Arabia and Northern Arabian–Persian Gulf*. US Geological Survey Bulletins, **2202-H**.

Ramseyer, K., Amthor, J. E., Spotl, C., Terken, J. M. J., Matter, A., Vroonten Hove, M. & Borgomano, J. R. F. 2004. Impact of basin evolution, depositional environment, pore water evolution and diagenesis on reservoir-quality of Lower Paleozoic Haima Supergroup sandstones, Sultanate of Oman. *GeoArabia*, **9**, 107–138.

Riaz, A. S., Jamil, A. M., McCann, J. & Saqi, M. I. 2003. Distribution of infracambrian reservoirs on Punjab Platform in Central Indus Basin of Pakistan. *ATC 2003 Conference & Oil Show*, Islamabad, 1–17.

Roberts, G. & Peace, D. 2007. Hydrocarbon plays and prospectivity of the Levantine Basin, offshore Lebanon and Syria from modern seismic data. *GeoArabia*, **12**, 3.

Robertson, A. F. H. 1998. Mesozoic–Tertiary tectonic evolution of the easternmost Mediterranean area: integration of marine and land evidence. *In*: Robertson, A. H. F., Emeis, K.-C., Richter, C. & Camerlenghi, A. (eds) *Proceedings of the Ocean Drilling Program, Scientific Results*, **160**.

Robertson, A. & Comas, M. 1998. Collision-related processes in the Mediterranean region – introduction. *Tectonophysics*, **298**, 1–4.

Rusk, D. 2001. Libya: petroleum potential of the underexplored basin centers – a twenty-first century challenge. *In*: Downey, M. W., Threet, J. C. & Morgan, W. A. (eds) *Petroleum Provinces of the Twenty-First Century*. American Association of Petroleum Geologists, Tulsa, OK, Memoirs, **74**, 429–452.

Samuel, A., Kneller, B., Raslan, S., Sharp, A. & Parsons, C. 2005. Prolific deep-marine slope channels of the Nile Delta, Egypt. *AAPG Bulletin*, **87**, 541–560.

Sati, G. C., Zutshi, P. B., Pati, P. B. & Lal, N. K. 1997. Late Proterozoic–Early Palaeozoic sequence in Southeastern Jaisalmer Basin: a new play for hydrocarbon exploration. *Indian Journal of Petroleum Geology*, **6**, 43–54.

Schoenherr, J., Ural, J. L., Kukla, P. A., Littkre, R., Schleder, Z., Larroque, J. M. & Newall, M. 2007. Salt: a case study of the Infra-cambrian Ara Salt from the South Oman Salt Basin. *AAPG Bulletin*, **91**, 1541–1557.

Scotese, C. R. 2009. Late Proterozoic plate tectonics and palaeogeography: a tale of two supercontinents, Rodinia and Pannotia. *In*: Craig, J., Thurow, J., Thusu, B., Whitham, A. & Abuturruma, Y. (eds) *Global Neoproterozoic Petroleum Systems and the Emerging Potential in North Africa*. Geological Society, London, Special Publications, **326**, 67–83.

Senalp, M. & Al-Duaiji, A. 1995. Stratigraphy and sedimentation of the Unayzah reservoir, central Saudi Arabia. *In*: Al-Husseini, M. I. (ed.) *Middle East Petroleum Geosciences, GEO'94*. *GeoArabia*, **2**, 837–847.

Sharland, P. R., Archer, R., Casey, D. M., Davies, R. B., Hall, S. H., Heward, A. P., Horbury, A. D. & Simmons, M. D. 2001. *Arabian Plate Sequence Stratigraphy*. GeoArabia, Special Publications, **2**.

Shields, G. A., Deynoux, M., Culver, S. J., Brasier, M. D., Affaton, P. & Vandamme, D. 2007. Neoproterozoic Glaciomarine and Cap Dolostone Facies of the Southwestern Taoudeni Basin (Walidiala Valley, Senegal/Guinea, NW Africa). Comptes Rendus Geoscience, Paris, **339**, 3–4, 186–199.

Smith, A. G. 2009. Neoproterozoic timescales and stratigraphy. *In*: Craig, J., Thurow, J., Thusu, B., Whitham, A. & Abuturruma, Y. (eds) *Global Neoproterozoic Petroleum Systems and the Emerging Potential in North Africa*. Geological Society, London, Special Publications, **326**, 27–54.

Societé Mauritanienne de Hydrocarbures. 2007. *Hydrocarbon Potential of Taoudeni Basin*. Promotional Brochure, December 2007.

Stampfli, G. M. & Borel, G. D. 2002. A plate tectonic model for the Palaeozoic and Mesozoic constrained by dynamic plate boundaries and restored synthetic oceanic isochrones. *Earth and Planetary Science Letters*, **196**, 17–33.

Sutcliffe, O. E., Dowdeswell, J. A., Whittington, R. J., Theron, J. N. & Craig, J. 2000. Calibrating the late Ordovician glaciation and mass extinction by the eccentricity cycles of the Earth's orbit. *Geology*, **23**, 967–970.

Sutcliffe, O. E., Craig, J. & Whittington, R. 2005. Late ordovician glacial pavements revisited: a reappraisal of the origin of striated surfaces. *Terra Nova*, **17**, 486–487.

Sylta, O. 1991. Modelling of secondary migration and entrapment of multi-component hydrocarbon mixtures using equation of state and ray-tracing modeling techniques. *In*: England, W. A. & Fleet, A. J. (eds) *Petroleum Migration*. Geological Society, London, Special Publications, **59**, 111–122.

Sylta, O. 1993. New techniques and their application in the analysis of secondary migration. *In*: Doré, A. G. *et al.* (eds) *Basin Modelling: Advances and Applications*. Elsevier, Amsterdam/Norwegian Petroleum Society, Oslo, Special Publications, **3**, 385–398.

Tonetti, M. & Grigo, D. 2007. The role of Hercynian event in Tunisian Ghadames Basin: an innovative comparison between two petroleum systems modelling studies. *Epitome*, **2**, 406–407.

Trompette, R. 1969. Les stromatolites du Précambrien supérieur de l'Adrar de Mauritanie (sahara occidental). *Sedimentology*, **13**, 123–154.

Trompette, R. 1973. *Le Precambrien superieur et le Paleozoique inferieur de l'Adrar de Mauritanie (bordure occidentale du bassin de Taoudeni, Afrique de l'Ouest). Un exemple de sedimentation de craton*. Etude stratigraphique et sedimentologique. Travaux des Laboratoires des Sciences de la Terre St-Jerome, Marseille, **B-7**.

Trompette, R. 1997. Neoproterozoic (600 Ma) aggregation of Western Gondwana: a tentative scenario. *Precambrian Research*, **82**, 101–112.

Turner, P. & Sherif, H. 2007. A giant Late Triassic–Early Jurassic evaporitic basin on the Saharan Platform, North Africa. *In*: Schreiber, B. C., Lugli, S. & Babel, M. (eds) *Evaporites through Space and Time*. Geological Society, London, Special Publications, **285**, 87–105.

Turner, P., Pilling, D., Walker, D., Exton, J., Binnie, J. & Sabaou, N. 2001. Sequence stratigraphy and sedimentology of the late Triassic TAG-I (Blocks 401/402, Berkine Basin, Algeria). *Marine & Petroleum Geology*, **18**, 959–981.

Vecoli, M. 2000. Palaeoenvironmental interpretation of microphytoplankton diversity trends in the Cambrian–Ordovician of the northern Sahara Platform. *Palaeogeography, Palaeoclimatology, Palaeoecology*, **160**, 329–346.

Vecoli, M., Riboulleau, A. & Versteegh, G. J. M. 2009. Palynology, organic geochemistry and carbon isotope analysis of a latest Ordovician through Silurian clastic succession from borehole Tt1, Ghadamis Basin, southern Tunisia, North Africa: palaeoenvironmental interpretation. *Palaeogeography, Palaeoclimatology, Palaeoecology*, **273**, 378–394.

Walley, C. D. 1998. Some outstanding issues in the geology of Lebanon and their importance in the tectonic evolution of the Levantine region. *Tectonophysics*, **298**, 36–62.

Waltham, A. 2009. Salt terrains of Iran. *Geology Today*, **24**, 188–194.

Watts, N. L. 1987. Theoretical aspects of cap-rock and fault seals for single and two-phase hydrocarbon columns. *Marine and Petroleum Geology*, **4**, 274–307.

Whaley, J. 2004. The hydrocarbon potential of the underexplored Palaeozoic and Triassic petroleum systems of Northwest Arabia. *Abstract in 6th Middle East Geosciences Conference, GEO 2004. GeoArabia*, **9**, 143.

Weis, D., Liegeois, J. P. & Black, R. 1987. Tadhak alkaline ring-complex (Mali): existence of U–Pb isochrones and 'Dupal' signature 270 Ma ago. *Earth and Planetary Science Letters*, **82**, 316–322.

Wender, L. E., Bryant, J. W., Dickens, M. F., Neville, A. S. & Al Moqbel, A. M. 1998. Paleozoic (Pre-Khuff) hydrocarbon geology of the Ghawar area, Eastern Saudi Arabia. *GeoArabia*, **3**, Gulf PetroLink, Bahrain, 273–302.

Wescott, W. A. & Boucher, P. J. 2000. Imaging submarine channels in the western Nile Delta and interpreting their paleohydraulic characteristics from 3-D seismic. *The Leading Edge*, June, **19**, 580–591.

Woller, F. & Fediuk, F. 1980. *Volcanic Rocks of Jabal as Sawda; Geology of Libya*. Academic Press, London, **3**, 1081–1093.

Zhu, W., Zhang, Z., Shu, L., Lu, H., Su, J. & Yang, W. 2008. SHRIMP U–Pb zircon geochronology of Neoproterozoic Korla mafic dykes in the northern Tarim Block, NW China: implications for the long-lasting break-up process of Rodinia. *Journal of the Geological Society*, **165**, 887–890.

Ziegler, M. A. 2001. Late Permian to Holocene paleofacies evolution of the Arabian Plate and its hydrocarbon occurrences. *GeoArabia*, **6**, Gulf PetroLink, Bahrain, 445–504.

Palaeohighs: their influence on the North African Palaeozoic petroleum systems

R. ESCHARD,[1] F. BRAIK,[2] D. BEKKOUCHE,[2] M. BEN RAHUMA,[3] G. DESAUBLIAUX,[4] R. DESCHAMPS[1] and J. N. PROUST[5]

[1]*IFP, 1, 4 Avenue de Bois Préau, 92 506 Rueil-Malmaison, France (e-mail: remi.eschard@ifp.fr)*
[2]*Sonatrach Activité Amont, Avenue du 1er Novembre, Boumerdes, 35 000, Algeria*
[3]*Libyan Petroleum Institute, Tripoli, Libya*
[4]*IFP, 1, 4 Avenue de Bois Préau, 92 506, Rueil-Malmaison, France; Present address: GDF-Suez, 93 210 Saint-Denis, France*
[5]*Rennes I University, Campus de Beaulieu, 35042 Rennes, France*

Abstract: We present new insights for the characterization of the petroleum system evolution in North Africa based on a review of the stratigraphic architecture description of some selected North African Palaeozoic basins. During Palaeozoic time, the Gondwana platform was divided into sub-basins bounded by structural highs. Most of the highs were inherited from north–south and SW–NE Pan-African crustal faults which were reactivated during the Palaeozoic and later, in the Austrian and Alpine tectonic phases. We studied the stratigraphic architecture of the Palaeozoic succession around four main highs showing a clear tectonic activity during the Palaeozoic sedimentation. The Gargaff Arch, in Libya, is a major SW–NE broad anticline which slowly grew up during the Cambrian and Ordovician and stopped rising during the Silurian. The activity resumed during Late Silurian and early Devonian and during the Late Devonian. The Tihemboka High is a north–south anticline in between Libya and Algeria. The uplift started during the Cambro-Ordovician then stopped during most of the Silurian. The activity resumed during the Late Silurian and continued until the Lower Carboniferous. The Ahara High, separating the Illizi and Berkine basins in Algeria, has continuously grown during the Cambro-Ordovician, stopped rising during the Silurian, and grew again continuously during the Devonian. The Bled El-Mass High is a part of the Azzel-Matti Ridge separating the Ahnet and Reggane basins in Algeria. The high mostly rose during the Cambro-Ordovician then subsided relatively less quickly than the surrounding basins during the Silurian and Devonian. The uplift timing and chronology of each palaeohigh partly controlled the petroleum systems of the surrounding basins. Topographic lows favoured the occurrence of anoxic conditions and the preservation of Lower Silurian and Frasnian source rocks. Complex progressive unconformities developed around the palaeohighs form potential complex tectonostratigraphic traps. Finally, hydrocarbons could have been trapped around the highs during pre-Hercynian times, preserving reservoir porosity from early silicification. Mixed stratigraphic–structural plays could then be present today around the highs.

Keywords: palaeohighs, Palaeozoic, Algeria, petroleum system, stratigraphic architecture, stratigraphic trap

The Palaeozoic basins in North Africa represent one of the most prolific petroleum systems of the world. Two major source rocks, several stacked potential reservoir levels and a succession of tectonic phases favoured the accumulation of hydrocarbons in a large variety of traps and reservoirs. Hydrocarbons were first discovered and produced in the early 1960s from giant fields, such as the Hassi Messaoud or the Hassi R'Mel fields in Algeria. Since that time, tens of fields have been discovered and produced in Algeria and Libya, the exploration strategy being still mostly driven by the identification of structural traps from seismic data. Today, exploration is facing new challenges, the new targets being the deep tight reservoir gas prospects of the Lower Palaeozoic, and Palaeozoic reservoir subtle traps. Consequently, a good understanding of the reservoir and seal distribution together with the timing of the fluid migration is key for successful exploration. The objective of this paper is to describe the stratigraphic architecture of the Palaeozoic succession around four palaeohighs in Libya and Algeria from outcrop and subsurface data: the Gargaff Arch in Libya, the Ahara and Tihemboka arches in eastern Algeria, and the Bled el Mass Arch in western Algeria. We compare the timing of the uplifts and their petroleum implications.

Geological setting of the Palaeozoic platform in North Africa

Sub-basins and arches across the Algerian and Libyan platform

During Palaeozoic times, the Saharan platform was a part of the northern passive margin of the Gondwana supercontinent. The tectonic setting of this rigid craton was remarkably calm from Cambrian to Devonian times, while deformation progressively increased during Upper Devonian and Carboniferous times, announcing the late Carboniferous Hercynian compression. Major phases of deformation nevertheless episodically affected the craton, inducing low-angle regional unconformity associated with significant erosion during the Upper Ordovician (Taconic unconformity) and the Upper Silurian–Lower Devonian (Caledonian unconformity). On the Saharan platform, fluvial sedimentation episodes alternated with shallow marine ones, depending on the competition between the relative sea-level changes and subsidence. The epicontinental seas opened to the Rheic ocean located to the NW (Scotese *et al.* 1999).

The Gondwana craton itself consisted of several terranes accreted during the Pan-African orogen (Bertrand & Caby 1978; Fabre 1988; Caby 2003; Coward & Ries 2003). The main suture zone was located between the West African craton and the Touareg shield, and is marked today by a north–south fault zone in SW Algeria turning NW in the Ougarta range and in the Anti-Atlas in Morocco (Fabre 1988). Other major north–south Pan-African faults also cut the Touareg shield into narrow strips and were reactivated several times during the geological history. Away from these mobile areas, the craton was remarkably stable and rigid. As a result, the platform showed a succession of sub-basins separated by structural highs (Fig. 1) located above the Pan-African main faults. From east to west, we can recognize the Brak bin Ghanimah uplift marking the eastern termination of the Murzuk and Ghadames basins in Libya (Massa 1988); the Tihemboka High separating the Illizi from the Murzuk basins; the Amguid–El Biod, making a major north–south ridge separating the Illizi–Berkine basins from the Mouydir/Oued-Mya basins to the west; the Arak/Foum–Belrem ridge forming the eastern flank of the Ahnet Basin; and the Bled-el-Mass/Azzel–Matti Ridge limiting the Ahnet Basin from the Reggane Basin and connected northwards to the Ougarta chain. The Reggane Basin was itself limited from the Tindouf Basin by the Bou-Bernous High. Some highs also have an east–west or NE–SW orientation, such as the Gargaff High separating the Ghadames Basin from the Murzuk Basin in Libya. Similarly, the Ahara Arch separated the Illizi and Berkine basins, this basin being itself bounded northwards by the Talemzane–Dahar High.

The cited structures influenced the Palaeozoic sedimentation. However, many other highs were formed latterly during the Hercynian orogeny or during the Austrian Mid-Cretaceous or Cenozoic Alpine compressive phases.

Summary of the stratigraphic architecture of the Cambrian to Devonian succession

The different second-order cycles of the sedimentation have been already described by Boote et al. (1998) and Eschard et al. (2005). An overall transgressive trend from Cambrian to Middle Ordovician was followed by a regressive trend during Upper Ordovician. Erosion occurred in many places during an intra-Arenig unconformity. The Upper Ordovician successions were preserved in the Ougarta trough (Ghienne et al. 2007a, b), in the Anti-Atlas succession in Morocco (Destombes et al. 1985) and Libya (Echikh 1998), but was eroded in most of the Algerian sector by the Taconic unconformity, a low-angle and low-relief tectonic unconformity associated with a large flexure of the Saharan craton that occurred during the Upper Ordovician. A major glacial event then occurred during Late Ordovician (Hirnantian times), inducing a network of glacial pathways and glacial valleys (Ghienne et al. 2007a, b; Le Heron & Craig 2007). The Taconic unconformity locally merged with the Ashgill glacial erosion surface, making the interpretation rather complex.

Silurian times were marked by a major transgression depositing a thick succession of offshore shale. It includes basal hot shale layers corresponding to the main source rock that has generated hydrocarbons in almost every basin across North Africa (Legrand 2003). The Upper Silurian was deposited during a regressive event, ended by the Caledonian unconformity, another low-angle tectonic unconformity at the Siluro-Devonian boundary. The Devonian

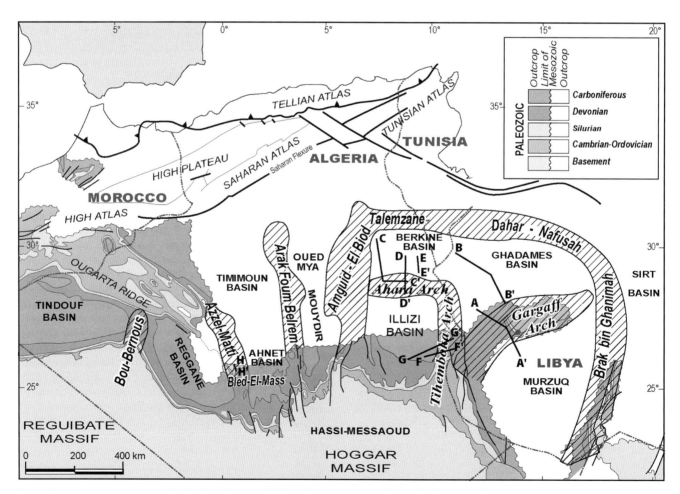

Fig. 1. Structural map showing the distribution of the main highs and basins across the Saharan Platform (partially redrawn from Boote et al. 1998). The hatched areas correspond to the main highs described in the text.

cycle initiated with low-stand deposits above the Caledonian unconformity, followed by a major transgression culminating with the deposition of the Frasnian hot-shale, forming another prolific source rock. Deformation of the craton then progressively increased during the Upper Devonian and Carboniferous, announcing the Hercynian compression.

The main characteristics of the Lower Palaeozoic sedimentation were already described by Beuf et al. (1971), summarized by Eschard et al. (2005). Figure 2 shows a summary of the stratigraphic sections observed in the Berkine, Ahnet, Ghadames and Illizi basins, and in the Tassili outcrops in Algeria. Within sub-basins, the Lower Palaeozoic stratigraphic architecture was

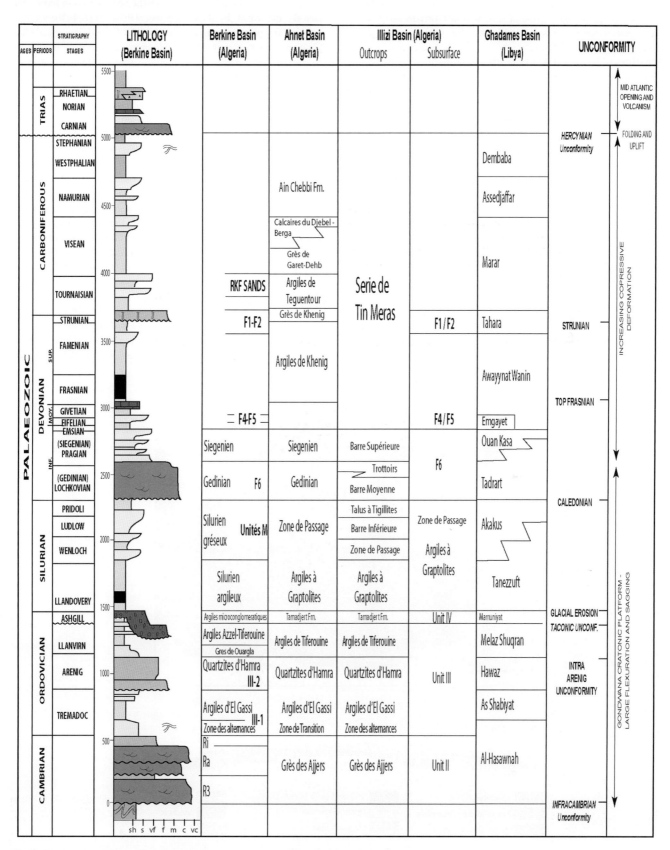

Fig. 2. Stratigraphic column of the Palaeozoic succession in the Berkine Basin, with the lithostratigraphic nomenclature of the Ahnet, Illizi, Ghadames basins and of the Ajjer Tassil outcrops.

remarkably organized in third-order sequences showing a well defined and laterally constant stacking pattern at the basin-scale. The lateral facies variations were very progressive, depositional environments being more and more distal in a NW direction. However, the stratigraphic architecture of the Palaeozoic succession abruptly changes within a few kilometres distance of the palaeohighs. The series rapidly thins out, making complex wedges with an amalgamation of several unconformities.

Stratigraphic architecture of the Palaeozoic succession around syn-tectonic highs

Gargaff Arch, Ghadames and Murzuq basins, Libya

Stratigraphic architecture. The Gargaff Arch (Fig. 3a) is a conspicuous WSW–ENE structural feature separating the Murzuq Basin from the Ghadames Basin in Libya. The arch plunges in the subsurface to the SW and probably connects to the north–south Tihemboka Arch. The Gargaff Arch continues to the NE and connects to the Tripoli–Assawada Arch. The arch forms a broad asymmetric anticline, with a relatively steep southern flank compared with the northern one (Fig. 3b).

Proterozoic granites and low-grade meta-sediments that crop out in Jabal Hasawnah were covered by Cambrian (Al-Hasawnah Formation) and Ordovician sediments (As Shabiyat, Hawaz, Melaz Shuqran and Mamuniyat formations). Lower Devonian sediments only crop out in the NW part of the high while, in the southern flank, Upper Devonian sediments directly onlap a reduced section of Silurian (Rubino & Blanpied 2000). The Hercynian unconformity then eroded the Palaeozoic sediments below, and the unconformity was itself onlapped by Cretaceous and Cenozoic sediments.

While in the subsurface to the SW of the Gargaff Arch, the Cambro-Ordovician succession thins out (Hallett 2002), in outcrops, the Cambrian sandstones remain thick (600 m, according to Massa 1988). Palaeocurrents measured in outcrops in the Cambrian and Ordovician sediments show north–NW flow directions (Vos 1981; Massa 1988; Ramos et al. 2006), which are the ones observed regionally by Beuf et al. (1971). The Ordovician succession presents a rather complex geometry in outcrops. In the western termination of the high, Lower–Middle Ordovician units (Ash Shabiyat, Hawaz and Melaz Shuqran formations) pinch out below the Late Ordovician glacial valleys (Mamouniyat Formation) from north to south.

A regional cross-section between the Ghadames Basin and the Gargaff Arch illustrates the overall geometry of the Silurian and Devonian succession (Fig. 4). In the outcrops of the northern flank of the high, only a few metres of the lower part of the Llandovery Tanezzouft shales are left above the Upper Ordovician sediments (Fig. 5), and further westwards, the Silurian is completely eroded, with Lower Devonian Tadrart Formation directly overlying the Cambro-Ordovician (Massa 1988; Dardour et al. 2004). Correlations in the Figure 4 show evidence for a discrete onlap of the lower hot-shale levels on the relief made by the Gargaff Arch; the hot shales pinch out southward. In contrast, the upper part of the Silurian succession (Akakus Formation) was progressively eroded below the base of the Lower Devonian Tadrart sandstones (Fig. 5) when approaching the palaeohigh, and is not present any more in outcrops.

The Devonian stratigraphic succession progressively thins out when approaching the high, in the southern flank of the Ghadames Basin (Fig. 4) and northern flank of Murzuk Basin (Hallett 2002). The Tadrart fluvio-estuarine sandstones are 30 m thick in outcrops,

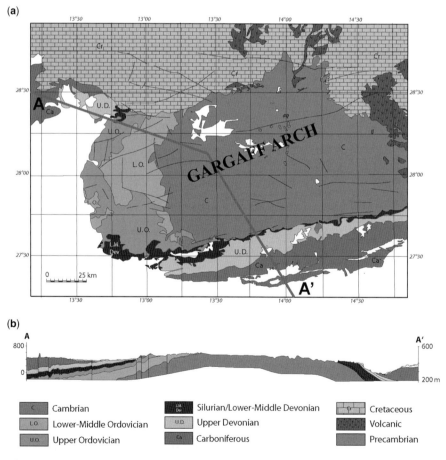

Fig. 3. (a) Simplified geological map of the Gargaf Arch. (b) Cross-section across the arch redrawn from the geological maps. Observe the asymmetry of the high and the difference in the wedge architecture between the North and South flanks.

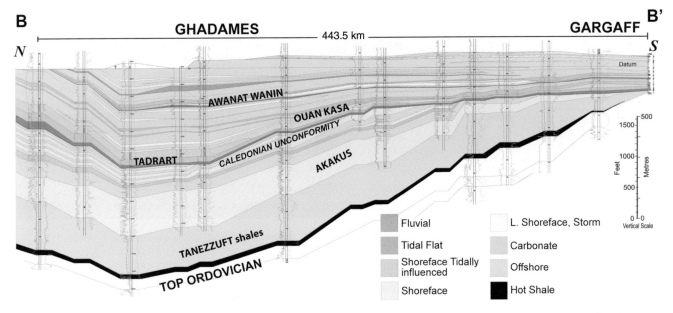

Fig. 4. Cross-section showing the stratigraphic architecture of the Siluro-Devonian succession between the Ghadames Basin and the Gargaff Arch.

thicken to 100 m in the centre of the Ghadames Basin and reach 300 m in thickness in the Berkine Basin. On the southern flank of the palaeohigh, the Lower Devonian succession is missing (Rubino & Blanpied 2000). According to Massa (1988), the Lower Devonian succession is completely pinching out over the Gargaff Arch westward in the subsurface between the Murzuk and Ghadames basins.

The Middle to Upper Devonian succession consists of fluvial, deltaic and marine sediments organized in third-order sequences which have already been described by Sutcliffe et al. (2000) for the northern flank of the Gargaff High, and by Rubino & Blanpied (2000) for its southern flank. We observed current directions and progradation northwestwards on the northern flank, and southwards on the southern flank, suggesting that the palaeohigh was expressed at that time.

In outcrops, the Devonian unconformities are of two types. First, fluvial incised valleys formed a network of narrow and deep channels filled by either fluvial or tidal sediments. An example of intra-Givetian incised valleys (Awanat Wanin Formation), formerly interpreted by Vos (1981) as a distributary channel, can be seen in Figure 6. Incised valleys, 300 m wide and 25 m thick, were incised into shelfal silty mudstones during a relative sea-level fall and filled during the subsequent relative sea-level rise by tidal deposits. Other examples of fluvial incisions can be also observed in the Eifelian sediments (Awanat Wanin/Emgayet). Second, forced regression wedges, bounded by sharp and flat erosional surfaces at their base and flooding surfaces on top, consist of laterally continuous fluvio-estuarine wedges (base of the Emsian Ouan Kasa sandstone and at the base of the Strunian Tahara sandstone) passing distally to tidal deposits. The units considerably thicken in the subsurface of the Ghadames Basin, forming lowstand prograding wedges during periods of relative sea-level fall.

Interpretation. The uplift of the Gargaff Arch started during the Cambrian, as shown by the thinning of the succession when approaching the palaeohigh. However, the palaeohigh was itself subsiding and the Cambrian fluvial system was not affected by the uplift, as proved by the constant palaeocurrent directions measured in the Cambrian fluvial channels. The Ordovician succession progressively pinched out from north to south on the western region of the high, below Upper Devonian sediments, but it is still difficult to differentiate the effects of the Taconic unconformity from Late Ordovician glacial incision and from the ones due to the local uplift.

During Lower Silurian times, the palaeohigh formed a discrete positive relief on the seafloor, which was onlapped by the basal hot shale layers. Afterwards, the palaeohigh probably remained inactive during most of the Silurian times and was covered by the Silurian succession (Fig. 4).

The major phase of uplift occurred during the Upper Silurian and Lower Devonian associated with the Caledonian unconformity. Above the high, the Upper Silurian succession was completely eroded. The thinning of the Tadrart Formation from the Ghadames Basin to the Gargaff Arch suggests a fluvial onlap configuration, and the lower part of the Devonian is probably missing in the outcrops. The Gargaff palaeohigh was emergent, and was the site of a fluvial by-pass during the Lower Devonian (Lochkovian). During the Middle and Upper Devonian, the high continually grew. Palaeocurrent orientations measured on both flanks of the high show a divergence of the palaeocurrent orientations (Rubino & Blanpied 2000), suggesting that the palaeohigh was eroded and supplied sediments both northwards and southwards.

The Hercynian orogen probably modified the shape of the arch, and the present-day Gargaff Arch was also exhumed during the Alpine compression (Underdown & Redfern 2008). The western

Fig. 5. Outcrop photo showing the Lower Devonian Tadrart formation unconformably overlying the Lower Silurian Tanezzouft Shales, Wadi Shati area, Gargaff region. Only a few metres of basal Silurian shales were preserved below the Caledonian unconformity.

Fig. 6. Incised valleys observed in the Upper Devonian, Awanat Wanin I formation, Wadi Shati area, Gargaff region. Car for scale.

part of the arch is now outcropping in the Gargaff area, while the eastern part is still buried in the subsurface separating the Murzuq and Ghadames basins. According to Hallett (2002), the highest part of the arch during the Cambro-Ordovician is now located in the subsurface between the Murzuk and the Ghadames basins. The arch has continued to influence sedimentation until the present day.

Ahara High, Illizi and Berkine basins, Algeria

Stratigraphic architecture. The Ahara High is a broad east–west anticline separating the Illizi Basin from the Berkine Basin during the Cambrian up to the Upper Devonian (Figs 7–9). The palaeohigh can be seen in the subsurface only, as it is buried under a thick series of Carboniferous and Mesozoic sediments. The structure plunges westwards and is offset by the major north–south Ramade fault complex bounding the El-Biod High. Above the El Biod High, strong uplift occurred during the Hercynian event, Cretaceous Austrian compression and Cenozoic Alpine compression, resulting in the partial erosion of the Palaeozoic succession. Eastwards, the high merges in a complex manner with the Tihemboka north–south structure. During the Mesozoic, subsidence increased in the Berkine Basin, inducing a general tilting of the Ahara High northwards (Chaouche 1992).

The pinchout of the Cambro-Ordovician succession across the palaeohigh can be seen in seismic section (Fig. 10) and in the stratigraphic correlation (Fig. 7). The thickness of the Cambro-Ordovician is still unknown in the centre of the Berkine Basin, where the succession is deeply buried. The Cambro-Ordovician sandstones onlapped a high located further eastwards than its present day position; a broad north–south trough (the Marfag trough) connected the Illizi and Berkine basins westwards at that time (Fig. 11a). The Ordovician succession in the Berkine Basin shows a series of progressive unconformities which were better expressed when approaching the palaeohigh. Hence, the 'Quartzite de Hamra' Formation successively overlies the units below (El Atchane Sandstones and El Gassi Shales), and was in turn eroded and overlain by the Ouargla Sandstones. Finally, above the palaeohigh, only a thin section of Upper Ordovician sediment, including glacial deposits (Unit IV in Algeria), was preserved below the Silurian shales. Cambrian and Lower Ordovician sediments thickened again southwards into the Illizi Basin. These successive erosions suggest that the palaeohigh was progressively growing during most of the Lower Ordovician times.

The basal Silurian succession (Argiles à Graptolithes) was deposited across the palaeohigh. However, detailed stratigraphic correlation provides evidence for an onlap of the basal Llandovery hot shales over the palaeohigh, arguing for palaeorelief at the time of the Silurian transgression. Above, the successive shoreface and tidal sequences of the Silurian prograded from south to north across the high without showing any facies or thickness change. The upper parts of the Silurian (Pridoli stage) are missing above the Ahara High, eroded below the Caledonian unconformity. The fluvial units of the lower Devonian also completely pinched out on both sides of the palaeohigh (Fig. 9). The Gedinnian unit, which is 300 m thick in the Berkine Basin first thins then pinches out both in relation with a basal fluvial onlap and with the subsequent erosion of the overlying units.

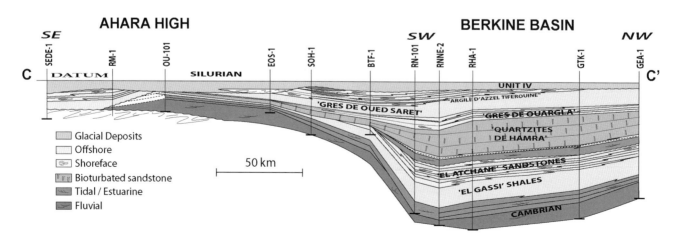

Fig. 7. Cross-section between in the Berkine Basin and the Ahara High, showing the stratigraphic pinchout of the Cambro-Ordovician succession on the northern flank of the Ahara High. The section is flattened on the Silurian flooding. Location map on Figure 1.

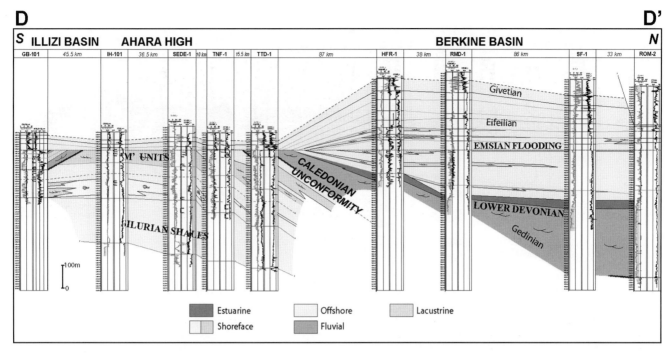

Fig. 8. Correlation of the Devonian succession on the northern flank of the Ahara High. The section is flattened on the Emsian flooding surface. Location map on Figure 1. Observe the complete pinchout of the Gedinnian fluvial unit on the flank of the high.

During Middle and Upper Devonian times, major facies and thickness variations occurred around the Ahara High. Emsian, Eifelian and Givetian sediments were onlapping on both eastern and western flanks of the palaeohigh, and sandy facies belts of shoreface and deltaic sediments extend in the southern part of the Berkine Basin (Fig. 8). In this area, the units successively pinched out southwards when onlapping the high, and passed to distal shaly facies towards the centre of the Berkine Basin. The sandstone units often show incised valleys and sharp-based shoreface sequences on logs. Finally, Upper Frasnian shales sealed the palaeohigh directly overlying the Upper Silurian sediments above the axis of the Ahara High.

Interpretation. The shape of the Ahara High has changed during the Lower Palaeozoic and the uplift rate was not constant. From Cambrian to Middle Ordovician, the uplift of the high was limited to its eastern end, and at that time (Fig. 11a), a south–north trough connected the Illizi Basin with the Berkine Basin between the El Biod and the Ahara Highs. The sediments onlapped the margin of the palaeohigh, forming wedges with a complex architecture as the constant uplift modified the available accommodation space.

The Silurian was a period during which the growth of the Ahara uplift almost completely stopped. The Silurian transgression occurred over a smoothed positive residual relief, influencing the deposition of the Lower Silurian hot shales in the lows while condensation occurred above the highs. The progradation of the Upper Silurian shorefaces and deltas was not affected at all by the high and, at that time, the Illizi Basin opened to the Berkine Basin. The uplift was reactivated during the Caledonian event, with the formation of an angular unconformity between the Upper Silurian and the Lower Devonian. The palaeohigh formed a positive relief separating the Illizi Basin from the Berkine Basin. In the Berkine Basin, an east–west trough was formed and rapidly subsided (Fig. 11b) in which a thick sedimentary succession accumulated. In contrast, the Illizi Basin formed a slowly subsiding sag. In

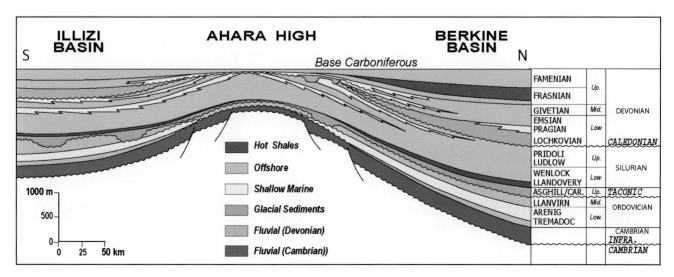

Fig. 9. Synthetic cross-section between the Illizi and Berkine basins, showing the complex stratigraphic architecture of the series around the Ahara High.

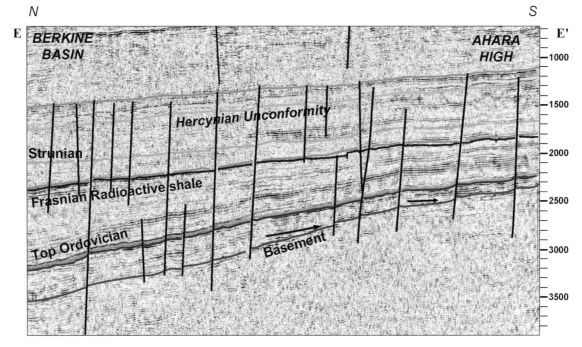

Fig. 10. Seismic section on the southern flank of the Berkine Basin illustrating the onlap configuration of the Cambro-Ordovician, and the thickness reduction of the Silurian and Devonian succession.

between, the Ahara High was the site of uplift and erosion and intermittently supplied the two basins with sediments.

The fluvial sediments of the Lower Devonian (Lochkovian and Lower Pragian) completely pinched out on the flanks of the Ahara High and the Illizi and Berkine basins were probably not connected at that time, except through narrow troughs on the eastern and western sides of the high. During Pragian, a transgression started in the Berkine Basin, and the shallow marine sediments of the Siegenien succession were deposited, while lacustrine and fluvial sediments still aggraded in the Illizi Basin during the same period. The transgression continued with the Emsian flooding, and offshore sediments were deposited all over the area, forming a major reference horizon. A regression then followed and, at that time, the Ahara High was eroded, supplying sediments towards the Berkine Basin while the high continued to rise. Periods of erosion and marine transgression then alternated during Middle and Upper Devonian times: the palaeohigh was periodically eroded during relative sea-level falls, while coastal onlaps occurred during relative sea-level rises. Forced regression wedges developed around the palaeohigh, especially in the southern part of the Berkine Basin, while less sediment arrived in the Illizi Basin, suggesting an asymmetry in the high topography (Fig. 9). Finally, the Frasnian offshore shales transgressed the palaeohigh when the uplift stopped. Frasnian shales then unconformably onlapped over the different terms of the eroded Devonian units.

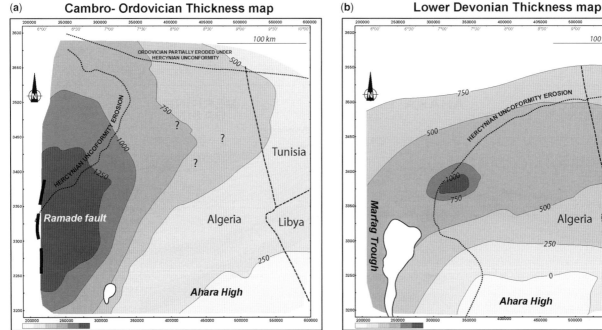

Fig. 11. (a) Isopach map of the Hamra Quartzite in the Bekine Basin; (b) Isopach map of the Lower Devonian in the Berkine Basin.

Tihemboka High, Illizi and Murzuq basins, Algeria and Libya

Stratigraphic architecture. The Tihemboka High is a major north–south structure straddling the Algerian–Libyan border. The structure extends southwards as a series of flexures and faults. Northwards, the axis of the high plunges in the subsurface of the Berkine and Ghadames basins. The Tihemboka structure may turn eastwards and possibly connect to the Gargaff Arch in Libya. The Tihemboka Arch also merges to the Ahara Arch in a complex manner.

The geometry of the Devonian stratigraphic wedges around the high was first described by Beuf *et al.* (1971) and we detail here its stratigraphic architecture by new field data and subsurface correlations in the Illizi Basin. According to the surface geology (Beuf *et al.* 1971), the Cambrian succession does not show any significant thickness variation nor palaeocurrent variation across the high in the Djanet/Ghat area. However, a stratigraphic wedge of the Cambrian and Lower Ordovician units can be observed in the subsurface in the Illizi Basin when approaching the palaeohigh from the west (Fig. 13a). The unit III-3 (Middle Ordovician) directly overlies the Precambrian in the vicinity of the Algerian–Libyan border. The Upper Ordovician glacial related valleys (Unit IV) then eroded the units below, almost reaching the Precambrian basement. Normal faults sealed by the glacial valleys may also contribute to the thickness variations of the Cambro-Ordovician units. A similar truncation of the Upper Ordovician can also be observed in the outcrops of the Djanet area in Algeria, where the Unit IV directly eroded the Cambrian sandstone (Beuf *et al.* 1971).

The Siluro-Devonian stratigraphic architecture of the Tihemboka western flank was studied in detail from outcrop and subsurface data, whereas the interpretation of the eastern flank is based on published data and map observations (Fig. 13). The high formed a broad and smooth anticline; no syn-sedimentary faults are observed in the western flank outcrops. The Lower Silurian shales (Llandovery) can be observed in outcrops and correlate on the flanks of the high without showing any significant variations (Fig. 13b). In the subsurface, the basal radioactive hot shale layers are continuous in the wells and do not onlap the palaeohigh. In contrast, the upper part of the Silurian succession was severely eroded above the palaeohigh. According to Massa (1988), the erosion reached the Middle Llandovery level (Lower Silurian) in the high Libyan flank. The upper units of the Silurian succession were progressively eroded below the Caledonian unconformity when approaching the palaeohigh, forming a low-angle regional unconformity.

The Lower Devonian succession partly pinches out over the high axis, as shown in the geological maps (Fig. 12) and well cross-sections (Fig. 13b). In the Tassili outcrops and in the Illizi Basin, the Lower Devonian succession consists of two sand-rich fluvial units separated by a fluvio-lacustrine one (see reference section in Eschard *et al.* 2005). The units progressively merged when approaching the palaeohigh, the upper one progressively truncating the lower ones, forming a progressive unconformity. In outcrops, the fluvial systems do not show any major lateral facies changes nor significant palaeocurrent rotations, even close to the high axis.

Middle and Upper Devonian units also progressively onlapped the palaeohigh. In the Emsian and Givetian sandy units, lateral facies variations can be observed, with coarser and more proximal facies towards the axis of the palaeohigh while the units thin and fine towards the Illizi Basin. The Frasnian shales onlapped the structure. The Upper Devonian units were finally eroded beneath the Carboniferous ones along the axis of the palaeohigh. In the eastern flank of the Tihemboka High in Libya, Massa (1988) observed the Frasnian shales directly overlying the Lower Silurian (Llandovery).

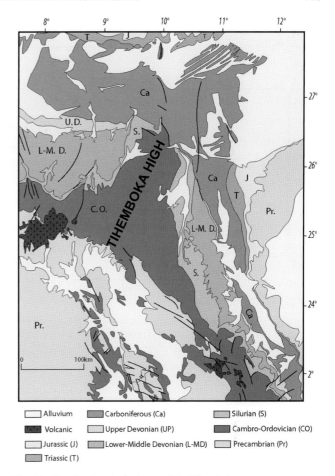

Fig. 12. Simplified geological map of the Tihemboka area.

Interpretation. The Tihemboka structure was inherited from a major north–south Pan-African fault system. The large flexure probably accommodated the motion of a deep crustal fault, forming the present-day broad anticline.

The Tihemboka area was uplifted during Cambro-Ordovician times. In the subsurface, the uplift was associated with a complex fault pattern probably forming a kind of horst which is still difficult to map precisely, as the faults were later reactivated during late compressive events. The structure continuously grew during Cambrian and Lower Ordovician times; the Cambrian units (unit II) and the lower Ordovician units (units III-3 and III-2) were progressively eroded below the Middle Ordovician units (unit III-3). The Upper Ordovician is missing because of the combined effects of the regional Taconic unconformity and of the erosion of the glacial valleys (unit IV).

The Silurian marked a period of quiescence of the palaeohigh; the Lower Silurian shales were deposited across the Upper Ordovician surface without any facies or thickness changes. Similarly, the palaeohigh did not affect the Upper Silurian shoreface progradation. Erosion resumed during Caledonian unconformity. The entire Upper Silurian shoreface succession was eroded above the high, which continued to grow almost continuously during the Devonian, forming a complex progressive unconformity. Lower Devonian fluvial units show erosive unconformities at their base while they progressively pinch out towards the axis of the palaeohigh. Each fluvial unit was first deposited on the Tihemboka High, before being eroded beneath the one above. This geometry suggests that the relative rise (i.e. the result of the tectonic uplift and of the eustasy) was not constant at that time, with base-levels falls interfering with the uplift, causing the fluvial re-incision.

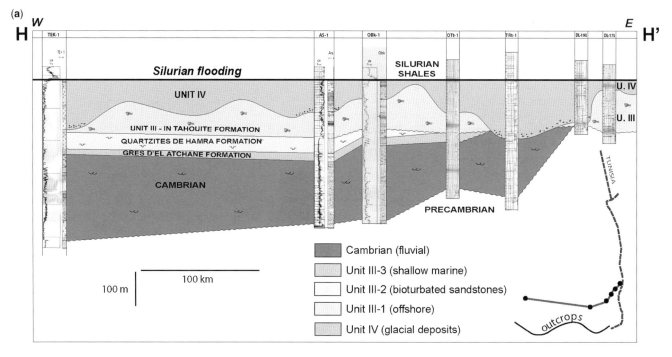

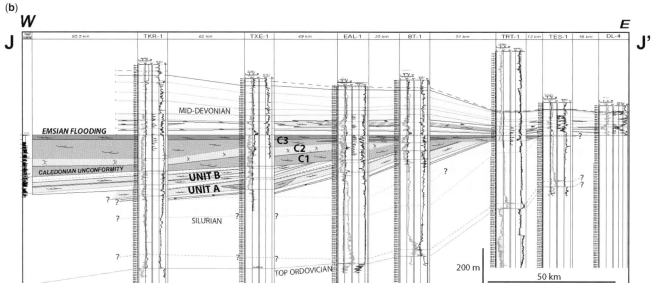

Fig. 13. East–west cross-section in the subsurface of the Illizi Basin, showing (**a**) the stratigraphic pinchout of the Cambrian and Lower Ordovician units. The section is flattened on the Silurian flooding. (**b**) The complex geometry of the Devonian stratigraphic wedge. The section was flattened on the Emsian flooding surface. Location map in Figure 1 and legend in Figure 8.

During the Upper Devonian, Emsian, Eifelian, Givetian and Frasnian shorefaces and tidal units successively onlapped the palaeohigh during transgressive events before being eroded during the following sea-level falls. At that time, the palaeohigh was eroded and supplied a clastic sediment belt around an emerged highland. Carboniferous sediments then definitively sealed the palaeohigh.

Djebel Tamamat in the Bled el Mass area

Stratigraphic architecture. The Bled el Mass area is a structurally complex zone within the Azzel Matti Ridge, a north–south high between the Ahnet and Reggane basins (Fig. 14). The Azzel Matti Ridge continues in the subsurface northwards and turns westward to connect to the Ougarta chain, finally continuing towards the Anti-Atlas area in Morocco. Southwards, the Azzel Matti Ridge can be mapped in the southern part of the Ahnet Basin. This complex structure was inherited from collision between the Reguibat and Hoggar shields during the Pan-African orogeny (Bertrand & Caby 1978). The collision implied north–south crustal transpressional faults with hundreds of kilometre offsets. From Cambrian to Devonian, a trough formed the Ougarta and Anti-Atlas areas in which accumulated a very thick stratigraphic succession (Ghienne *et al.* 2007*a, b*). In the Bled el Mass outcrops, the north–south folds (Fig. 14) at present day were formed during the Hercynian orogen, and reactivated latterly, probably during the Mid-Cretaceous Austrian event as interpreted from palaeomagnetic measurements (Haddoum *et al.* 2001; Zazoun 2001; Smith *et al.* 2006).

The studied outcrops are just a small part of the Azzel Matti and Bled El Mass ridge, as seen in the structural section in Figure 14. The outcrops of the Djebel Tamamat (Fig. 14) were recently studied, an area where Beuf *et al.* (1968) first identified a stratigraphic pinchout of the Cambro-Ordovician succession over a basement high. The north–south section shown in Figure 15 was made along the western flank of a north–south fold. The contact between the Infra-Cambrian meta-sediments of the 'Serie Pourprée' and the Cambrian sandstones was also locally offset by a reverse fault. The wedge geometry observed in the map (Fig. 14a) then results from both stratigraphic pinchout and reverse faults effect.

The Cambrian succession consists of coarse-grained fluvial sandstones which onlap a basement palaeohigh on both flanks and completely pinch out over the high. The palaeohigh formed a horst, with normal syn-sedimentary faults and tilted blocks on both sides (Fig. 15). Normal syn-sedimentary faults were sealed by the Lower Ordovician marine siltstones and shales (El Gassi Formation), which in turn onlap then pinch out over the high. Above, Lower and Middle Ordovician sediments consist of bioturbated sandstone units alternating with marine shales and siltstones (Grès d'El Atchane, Quartzites de Hamra, Argiles de Tiferouine and Grès de Oued Saret Formations). The succession significantly thins when approaching the palaeohigh, where progressive angular unconformities can be observed. Facies changes occur across the high (Fig. 15). The Quartzites de Hamra unit, which usually consists of fine-grained bioturbated sandstones, shows a transition to coarse-grained levels and gravel lags with quartzite pebbles and fragments of Precambrian schist when approaching the basement high. The material was obviously eroded from the Infra-Cambrian 'Série Pourprée', testifying that the palaeohigh was exposed during the Hamra time. Another angular unconformity overlain by conglomerates was observed at the base of the 'Argiles de Tiferouine' unit.

Ashgill aged glacial valleys can be observed flanking the palaeohigh, suggesting that the topography also influenced the glacial valley location. However, these glacial sediments sealed the palaeohigh and, just above, the Silurian shales were also deposited across the palaeohigh. The Devonian succession also regionally thins out above the high, at the scale of the Azzel Matti Ridge.

Interpretation. The 3D geometry of the palaeohigh is difficult to reconstruct only from the 2D outcrop section and map, the eastern flank of the present-day fold being also affected by a major fault. The palaeohigh probably had a NNE–SSW axis and a width of around 10 km. However, the entire Bled el Mass area was above a major mobile zone, and we observe in the outcrops only a small portion of a much larger high, the Azzel Matti Ridge.

During Cambrian times, the palaeohigh formed a horst bounded by normal faults, Cambrian fluvial sandstones being progressively eroded during the uplift as they were deposited. The palaeohigh was then flooded by a shallow sea during the Lower Ordovician. During most of the Ordovician, the palaeohigh continuously grew, as documented by progressive unconformities and the facies changes. The palaeohigh was alternatively drowned and exposed according to the sea-level variations. During the deposition of the Quartzite de Hamra Formation, the palaeohigh was sub-aerially exposed, and the products of the erosion were reworked downflank in the shoreface and tidal deposits. Then, the high was flooded while it continued to rise, marine sediments of the Argiles de Tiferouine thinning over the relief.

The activity of the high ceased before the deposition of the Late Ashgill glacial sediments. However, the Upper Ordovician sediments are missing in the area, probably due to the erosion of the Upper Ordovician Taconic unconformity, as suggested by the complex thickness variations observed at a regional scale at this level in the subsurface of the Ahnet Basin.

Silurian sediments finally sealed the high, which was inactive at that time. At the scale of the Azzel Matti Ridge, thickness variations can be observed in the Lower Devonian sections when comparing the Reggane, Bled el Mass and Ahnet areas, suggesting that the entire Bled El mass area was a still a mobile zone during the

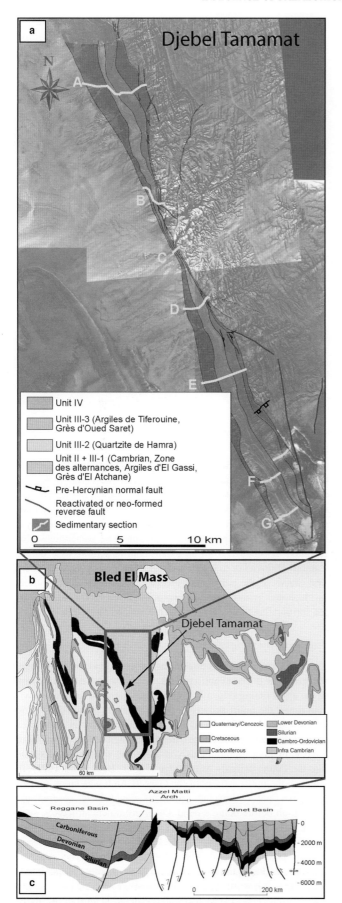

Fig. 14. (a) Photo-interpretation of the Djebel Tamamat area showing the wedge geometry. (b) Simplified geological map of the Djebal Tamamat area, showing the onlaps of the Lower Palaeozoic succession over the 'serie Pourprée'. (c) Section across the Reggane Basin, Azzel Matti Ridge and the Ahnet Basin (modified from Craig *et al.* 2006).

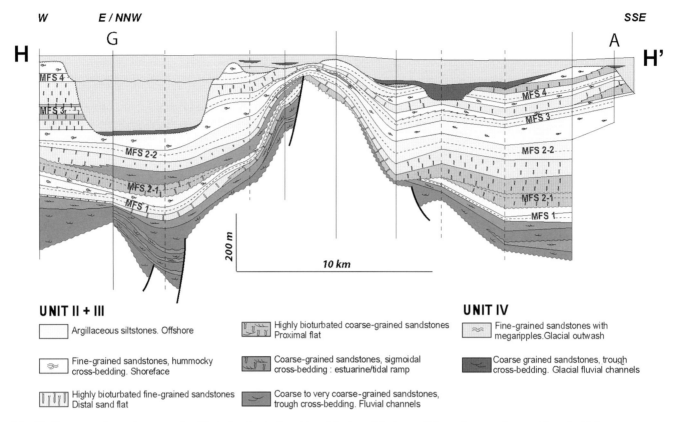

Fig. 15. Outcrop north–south cross-section illustrating the stratigraphic architecture of the Lower Palaeozoic succession apart from the high.

Devonian. The movements significantly increased during the Upper Devonian times, as proved by the major thickness variations of these series in the Ahnet Basin (Wendt et al. 2005).

Synthesis: tectonostratigraphic evolution

Structural style: the Pan-African heritage

The palaeohighs studied here were linked to the reactivation of major crustal structures. They are located at the boundary between different terranes and rigid shields accreted during the Pan-African orogeny to form the Gondwana supercontinent (Bertrand & Caby 1978). The north–south lineaments were formed during the oblique collision between the West African shield and the Hoggar shield, and were associated with major strike-slip movements (Black et al. 1979; Craig et al. 2006). The Bled el Mass/Azzel Matti Ridge was located close to the suture zone, between the Reguibat and the Hoggar shields. The Tihemboka High, the Idjerane/Arak/Foum–Belrem Ridge, the Amguid/El Biod/Hassi-Messaoud High and the Bin Ghanimah Highs (Fig. 1) are also linked to a north–south crustal fault. The crustal lineaments were then occasionally reactivated depending on the stress regime orientation and the geodynamic events which succeeded through time such as the Hercynian compression, and the Mid-Cretaceous Austrian and Cenozoic Alpine compressions. Other highs present a global east–west orientation, such as the Ahara High, the Gargaff Arch and the Talemzane–Dahar High. These structural features are more difficult to understand in the context of the Gondwana accretion.

The studied palaeohighs always present the same overall geometry: they form broad anticlines, tens of kilometres wide, with a maximum of 15° of structural dip on their flanks. The Tamamat High is apparently smaller, but outcrops are only a part of the much wider Azzel Matti Ridge.

The structural style of the palaeohighs also evolved through time (Fig. 16). During Cambro-Ordovician times, the deformation had a brittle behaviour. Several normal faults with a limited vertical throw, tilted block and half-grabens were observed along the flank of the Tamamat (Bled El Mass). In contrast, during Silurian and Devonian times, the palaeohigh formed large anticlines without any faults on their flanks. The thick Silurian shales layer then probably decoupled the deformation. The palaeohighs also recorded the change from an extensional structural style during Cambro-Ordovician to a compressive style during the Devonian. The shapes of the palaeohigh themselves changed through time, as illustrated by the geometry evolution of the Ahara High and of the surrounding basins between Cambrian and Devonian (Fig. 11).

Timing of the uplifts

The stratigraphic architecture of the series around the palaeohigh registered the different uplift phases, making possible a comparison of the uplift timing (Fig. 17). During Cambrian and Lower Ordovician times, the uplift activity of the palaeohighs varied depending on their location. The Tamamat High in the Bled el Mass area (part of the Azzel Matti Ridge), the Ahara High and the Tihemboka High grew almost continuously at that time, such as other ridges like the Foum Belrem–Arak Ridge (Fig. 1; Beuf et al. 1971). In contrast, the uplift activity of the Gargaff Arch was more limited; Cambrian and Ordovician sediments could have covered the palaeohigh before being eroded. Other prominent structures, such as the Amguid-El Biod High (Fig. 1), were not active at that time as proved by the development of the Cambro-Ordovician succession in the Hassi–Messaoud field (Boudjema 1987). The structure formed later during the Hercynian and Alpine compressions (Boudjema 1987). It is worth noting that an erosive event, the Intra-Arenig unconformity (Fig. 17), is also well registered in most of the highs.

Upper Ordovician times are complex to understand because the effects of the local uplift interfered with the Taconic unconformity, a major low-angle unconformity which occurred before the late Asghill glacial event. In many places around Algeria, the Upper Ordovician succession was not preserved and the exact nature of

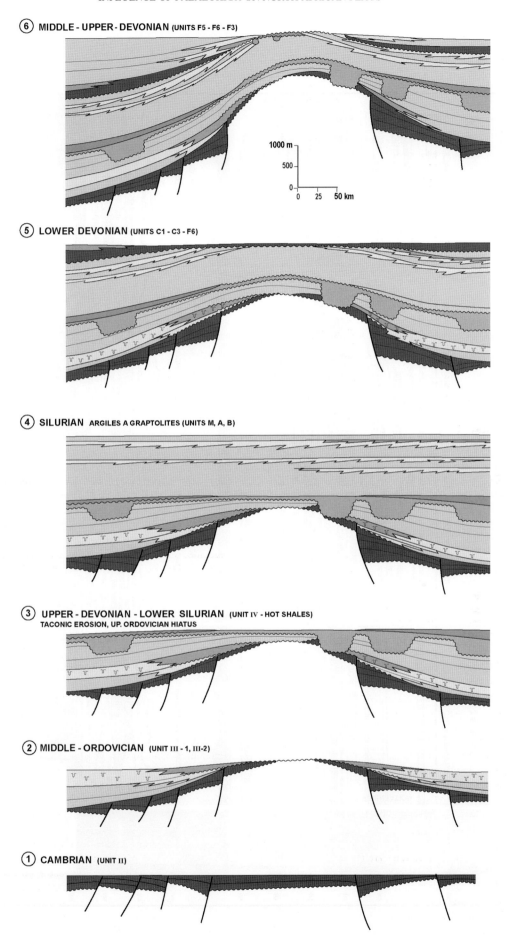

Fig. 16. Synthetic scheme showing the sedimentation associated with the deformation due to uprising palaeohigh during Palaeozoic times. Explanation in text.

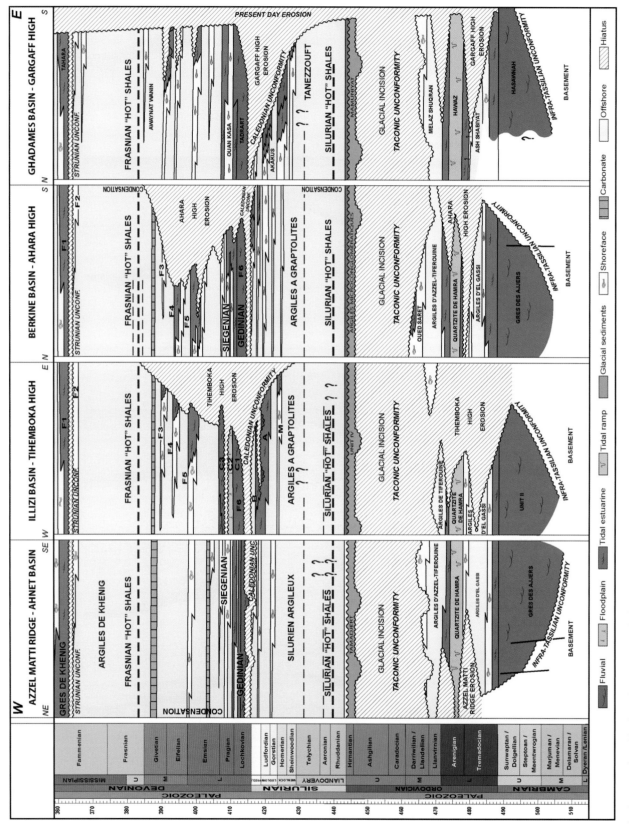

Fig. 17. Synthetic chronostratigraphic chart of the Azzel Matti Ridge–Ahnet Basin, Illzi Basin and Tihemboka Arch, Berkine Basin and Ahara High, and Ghadames Basin and Gargaff Arch. Note the very good correlation of most of the unconformities above the highs, suggesting that global eustatic variations interfered with the uplift.

the palaeohighs is poorly known for this period of time. During the Late Ordovician, the glacial pathway network itself was possibly controlled by the topography (Ghienne et al. 2007a, b; Le Heron & Craig 2007), but there is no direct evidence of tectonic activity during sedimentation except the glacio-tectonics effects. The glaciation period was probably too short (between 0.2 and 0.5 years; Brenchley et al. 1994; Sutcliffe et al. 2000) to register the effects of the long-term tectonic subsidence.

Silurian times were characterized by the stability of all palaeohighs, which were passively flooded during the Lower Silurian transgression. At that time, a residual topography remained, and was onlapped by the basal part of the transgressive Silurian shales. The Middle and Upper Silurian progradational sequences were deposited without being influenced by the palaeohigh topography.

Lower Devonian times were marked by a rejuvenation of most of the palaeohighs during the Caledonian event. As a result, the Upper Silurian succession was dramatically eroded over the Gargaff and Tihemboka arches and to a lesser extent, over the Ahara Arch. During Devonian times, significant structural growth occurred in most of the palaeohighs, with the noticeable exception of the Gargaff Arch that only slowly grew. Upper Frasnian shales generally sealed the palaeohighs, as seen over the Ahara, Gargaff and Tihemboka Highs. Frasnian shales can be then seen directly onlapping different elements of the Palaeozoic succession below. The deformation also increased significantly during Upper Devonian, especially in the western part of Algeria (Wendt et al. 2005). Compressive deformation increased again during the Carboniferous, inducing a migration of the depocentres: the Ahara High started to subside and the Tihemboka High was sealed by the Carboniferous sediments.

Controlling factors of the stratigraphic architecture around the palaeohighs

The palaeohighs were alternatively flooded or emerged depending on the eustatic sea-level variations, the uplift rates and the global deformation affecting the craton. The wedge geometry can then be rather complex when local progressive unconformities interfered with more global regional unconformities. This is especially the case for the Upper Ordovician Taconic unconformity or the Caledonian unconformity at the Silurian–Devonian boundary. These events were manifest as large flexures at the scale of the craton, inducing regional tilts and large amplitude folding. They were subjected to erosion or non-deposition of the Upper Ordovician and Upper Silurian successions, respectively. The resulting regional low-angle unconformities were then superimposed on local ones around the palaeohighs. Interpretation of the Taconic unconformity is further complicated by the Ashgill glacial valleys.

The third-order sequence architecture was controlled by the interaction between the uplift rate and the eustatic sea-level variations (Fig. 18). During transgression (or base level rise for the fluvial units), sediments were deposited above the highs, which continued slowly to rise. They were then eroded during the subsequent relative sea-level fall, inducing the creation of a network of incised valleys and forced regression wedges of shoreface and tidal sediments around the palaeohighs. The alternation of relative sea-level rise and falls explains the peculiar shape of the unconformities above the palaeohighs, which were continuously rising during some periods.

The sequence organization between the different highs is remarkably similar. This suggests that the main controlling factor of third-order sequences was related to the eustatic variations, the

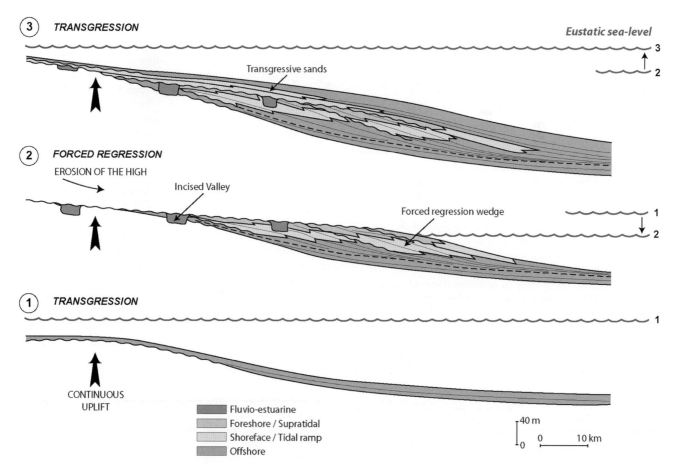

Fig. 18. Model of Devonian stratigraphic wedge developed around a high continuously rising while eustatic sea-level varied. 1, Transgression and coastal onlap during relative sea-level rise. 2, Erosion above the high during relative sea-level drop, creation of incised valley network and deposition of a forced regression wedge in the basin. 3, New transgression. Such cycles repeated several times during the Devonian.

relative sea-level changes being amplified above the highs which slowly rose during the Devonian. In such a cratonic context, the uplift rate of the highs is expected to be slow and constant. The Palaeozoic eustatic charts (Haq & Schutter 2008) show that the eustatic variations during the Devonian were relatively limited in amplitude, but sufficient to create the emersion of the palaeohighs during the eustatic sea-level falls.

Petroleum system implications

Source rock deposition

The growth of the palaeohighs had some major implications for the petroleum systems evolution. The Lower Silurian 'hot' shale layers, which constitute the oldest significant source-rock, were influenced by the topography of the Silurian flooding surface (Lüning et al. 2000, 2003a). The basal hot shale layers often onlap the residual topography above the highs. Furthermore, facies changes are also observed laterally when approaching a high within the source rock levels. Lüning et al. (2000) proposed that anoxic conditions prevailed in the topographic lows during the Silurian transgression. Above the palaeohighs, the organic-rich shales were not deposited or not preserved because of the hydrodynamics in the shallower setting (Figs 4 & 9), favouring organic matter oxidation or its erosion by marine currents (Lüning et al. 2000).

Similarly, the Frasnian source rocks deposition was controlled by the differential subsidence in the sub-basins, with anoxic conditions being created in the lows where the source rock with the best quality was preserved (Lüning et al. 2003b). In the centre of the Berkine Basin, for example, 250 m of radioactive organic-hot, Frasnian shales were deposited, sourcing the Triassic reservoirs. On the flank of the Ahara High, the source-rock layer progressively thinned out southwards and less than 5 m of organic Frasnian shales were left above the Ahara High.

Stratigraphic architecture and potential tectonostratigraphic traps

The stratigraphic relationship developed on the flanks of the palaeohighs set up potential stratigraphic trapping configurations around the palaeohighs. Toplaps of Upper Silurian shallow marine sequences can be particularly found below the Caledonian unconformity close to the palaeohighs. The Upper Silurian shoreface and tidal units were tilted and eroded during this event, especially above the palaeohighs, which were reactivated at that time. The transect between the Ghadames Basin and the Gargaff Arch (Fig. 4) illustrates this truncation. Such a geometry potentially favoured hydrocarbon migration through the Silurian drainage system to source the Devonian reservoirs.

Erosional onlaps are a common geometry observed around the palaeohighs. They formed during transgression above the rising highs, each unit onlapping the one below, which was eroded as the palaeohigh grew. These types of truncations were very frequently observed around the palaeohighs. In the Djebel Tamamat in the Bled el Mass area, the Lower Ordovician units were successively eroded onto the high as it was rising during the Ordovician transgression (Fig. 15). Above the Tihemboka, Ahara and Gargaff highs, the Lower Devonian fluvial units were also progressively eroded while onlapping the palaeohighs (Fig. 13). A major stratigraphic wedge of the Gedinnian fluvial sandstone unit can then be observed along the northern flank of the Ahara High. In this example, 250 m of fluvio-estuarine sandstones are present in the centre of the Berkine Basin, which completely pinches out on the northern flank of the Ahara High (Fig. 8). The pinchout is due to the basal fluvial onlap and erosion as the palaeohigh grew.

Lowstand wedges and incised valleys developed around the palaeohighs during the Middle and Upper Devonian times, a period during which the palaeohighs were alternatively flooded during the transgressive events and subaerially exposed during the relative sea-levels falls. At that time, the surface exposed to erosion was large enough to supply a significant amount of sediments in the adjacent basins around. A network of fluvio-estuarine incised valleys developed during the sea-level falls in Emsian, Eifelian and Givetian times around the main highs. Such valleys can be observed in the outcrops of the Gargaff Arch, and also in the subsurface of the Berkine, Illizi and Reggane basins. As in the classical models (Posamentier et al. 1992), during the incision phase, sediments transited through the valleys to accumulate in forced regression wedges during the lowstand periods (Fig. 18). Sharp-based shoreface and tidal units can be frequently observed in the clastic belts around the highs, thinning and pinching out basinwards in the offshore setting.

The geometry of the shale layers sealing the palaeohigh during the transgression is also critical to ensure the seals of the stratigraphic wedge. As the palaeohighs were uplifted, these shales were also eroded, reducing their sealing capacities, and their continuity must be carefully evaluated when chasing stratigraphic traps to ensure their closure.

Pre-Hercynian migration process and implications on reservoir quality

The petroleum systems of the North African Palaeozoic basins suffered a very long and complex geological history. The Hercynian event especially affected the geometry of the basins together with the Jurassic extensional phase, and the Austrian and Alpine compressional phases. Basin modelling approaches suggest that, in many basins, a significant amount of hydrocarbons was generated during the Palaeozoic times, from both the Silurian and the Frasnian source rocks. Understanding the Palaeozoic migration pathways and trapping is very important to predict the most porous reservoirs, as early charge inhibited porosity and permeability reduction. Different silicification phases affected the reservoir during the burial history, often significantly reducing the porosity. In the Palaeozoic traps, the oil coatings in the reservoir may have preserved porosity during cementation during later phases of diagenesis, even if most of these traps were inverted during Austrian and Alpine compressional phases. Early oil migration probably occurred in the stratigraphic wedges around the palaeohighs, explaining the unusually high porosity ranges observed in some of these reservoirs. In the Berkine Basin, after the Hercynian compression, a second phase of subsidence generated most of the hydrocarbons, charging the present-day in structures, which mostly formed during the Austrian and Cenozoic compressive phases (Underdown & Redfern 2008).

Another issue is the hydrocarbon charging of the stratigraphic wedges around the highs. As mentioned previously, the Silurian and the Frasnian source rocks are thinning out over the high. This implies a long-distance lateral migration between the oil kitchen and the stratigraphic traps to fill the stratigraphic wedges.

Conclusions

Knowledge of the timing of the uplift of the tectonic palaeohighs separating the Saharan craton into sub-basins is critical to our understanding the petroleum system evolution. The cited palaeohighs influenced sedimentation during most of the Palaeozoic times, but others may have formed during the Hercynian deformation, later during Mid-Cretaceous or Alpine deformation. In some cases (e.g. the Bled el Mass area), the same structure was reactivated during each of these major tectonic phases. Every palaeohigh formed above crustal mobile zones resulting from the terrane accretion during the Pan-African collision. Most of the crustal faults had a north–south orientation and may have branched to other NE–SW faults, and were then reactivated through geological times.

The Cambrian and Lower Ordovician was a period of active uplift, with palaeohighs forming slowly rising broad horsts. The palaeohighs were covered by the Cambro-Ordovician sediments before being eroded during the uplift phase. Although Cambrian sediments were deposited across the craton, thickness variations occurred across these palaeohighs but without major facies variations. The effect of the uplift was more pronounced during Ordovician times, during which significant erosion occurred above the palaeohighs, controlling the sequence architecture and facies distribution. Silurian times marked a period of general quiescence with limited uplift of the highs. Lower Silurian successions, including the main source rock levels, onlapped a residual topography. The subsequent progradation of the Upper Silurian from south to north across the craton was not affected by palaeorelief on the highs. The uplift resumed during latest Silurian and Lower Devonian times. Upper Silurian sediments were eroded below the Caledonian unconformity, especially at the margins of the palaeohighs. The palaeohighs then continuously grew during the Devonian, separating distinct palaeogeographic domains and controlling the local facies distribution. During Middle and Upper Devonian times, the palaeohighs were alternatively flooded and emergent depending on the relative sea-level variations. The activity of the palaeohighs almost ceased during the Frasnian times, with marine shales onlapping and sealing most of the palaeohighs. Late Devonian and Carboniferous were marked by a complete reorganization of the depocentres on the craton (Wendt *et al.* 2005), in relation to the beginning of the Hercynian compression, except for the Gargaff High, which continuously grew to the present day.

The depositional sequence architecture in such a context was controlled by the eustatic variations interfering with the palaeohigh uplift and subsidence rates, which were extremely low (a few metres per million years). In such a low accommodation context, eustatic sea-level falls were better registered above the rising palaeohighs. Progressive unconformities around the palaeohighs were superimposed onto the global tectonic unconformities, making complex stratigraphic wedge geometries. These result from the successive erosion of the reservoir levels associated with the progressive unconformities, and with the occurrence of incised valleys and forced regression wedges which developed during the relative sea-level falls.

The uplift of the palaeohigh controlled both the source rocks and reservoir rock distributions. The Lower Silurian and Frasnian source rocks were preferentially preserved in the lows where anoxic conditions prevailed, while they were less developed or absent above the palaeohighs.

The hydrocarbon prospects associated with these stratigraphic wedges must be carefully evaluated with 3D seismic surveys. If the presence of such stratigraphic wedges has been already proved and produces hydrocarbon around the palaeohighs, the lateral closure of the reservoir and the continuity of the seal above have to be investigated in detail in the prospects. The sealing offshore shales layers deposited above the palaeohighs during transgression may have been themselves eroded during the uplift, reducing their sealing capacities.

Finally, the sourcing of the wedges implies lateral hydrocarbon migration from both Silurian and Frasnian source rocks. The reconstruction of the geometry of the drainage system and its evolution during the deformation of the basin have to be taken into account to restore the migration history through time. An early pre-Hercynian migration in such traps may have prevented silicification of the reservoir, favouring relatively high-porosity ranges which make very good reservoirs during the post-Hercynian migration.

This manuscript is based on 15 years of work in Algeria performed during Triassic, Devonian, Illizi-Berkine and BerkineGas joint Sonatrach-IFP projects. Many geologists from Sonatrach and from the IFP Group were involved and they are too numerous to be individually cited, together with those in the companies who sponsored the studies (Anadarko, BP, Cepsa, ENI, GDF-Suez, Talisman Energy, Total, Repsol, Statoil). Special thanks go to D. Takherist (Alnaft), H. Chebourou, M. Malla, C. Hellal and N. Mokhtari (Sonatrach), to T. Euzen and C. Ravenne (IFP) and to T. Lorin and G. Philippe (Beicip-Franlab). Discussions with J. L. Rubino and C. Blanpied (Total) were very fruitful for improving the description of the Libyan outcrops. Special thanks are extended to the LPI management in Tripoli for their support of M. Ben Rahuma's PhD study. The contribution of F. Paris (Rennes University) and K. Boumendjel (Sonatrach) for dating the series is also invaluable. The comments from T. Patton (BP) and an anonymous reviewer improved the manuscript quality greatly.

References

Bertrand, J. & Caby, R. 1978. Geodynamic evolution of the Pan-African orogenic belt: a new interpretation of the Hoggar shield. *Geologisches Rundschau*, **67**, 357–388.

Beuf, S., Biju-Duval, B., Mauvier, A. & Legrand, Ph. 1968. Nouvelles observations sur le 'Cambro-Ordovicien' du Bled El Mass (Sahara central). *Publication du Service Géologique Algérie, Bulletin*, **38**, 39–51.

Beuf, S., Biju-duval, B., De Charpal, O., Rognon, P., Gariel, O. & Bennacef, A. 1971. *Les grès du Paléozoïque inférieur du Sahara. Sédimentologie et discontinuités. Evolution d'un craton*. Publication Institut Francais du Pétrole, Collection Science et Techniques Pétrolières, **18**. Editions Technip, Paris.

Black, R., Caby, R. *et al.* 1979. Evidence for Late Precambrian plate tectonics in West Africa. *Nature*, **278**, 223–227.

Boote, D. R., Clark-Lowes, D. D. & Traut, M. W. 1998. Paleozoic Petroleum systems of North Africa. *In*: Macgregor, D. S., Moody, R. T. J. & Clark-Lowes, D. D. (eds) *Petroleum Geology of North Africa*. Geological Society, London, Special Publications, **132**, 7–69.

Boudjema, A. 1987. Evolution Structurale Du Bassin Pétrolier (Triassique) du Sahara Nord Orientale (Algerie), Unpublished PhD thesis, Université de Paris Sud, Orsay.

Brenchley, P. J., Marshall, J. D. *et al.* 1994. Bathymetric and isotopic evidence for a short-lived late Ordovician glaciation in a greenhouse period. *Geology*, **22**, 295–298.

Caby, R. 2003. Terranes assembly and geodynamic evolution of central-western Hoggar: a synthesis. *Journal of African Earth Sciences*, **37**, 133–159.

Chaouche, A. 1992. *Genèse et mise en place des hydrocarbures dans les basins de l'Erg Oriental (Sahara Algérien)*. Thesis, Université de Bordeaux III.

Coward, M. P. & Ries, A. C. 2003. Tectonic development of North African basins. *In*: Arthur, T. J., Macgregor, D. S. & Cameron, N. R. (eds) *Petroleum Geology of Africa*. Geological Society, London, Special Publications, **207**, 61–83.

Craig, J., Rizzi, C. *et al.* 2006. Structural Styles and Prospectivity in the Precambrian and Palaeozoic Hydrocarbon Systems of North Africa. *III Symposium Geology of East Libya Binghazi*, GSPLAJ.

Dardour, A. A., Boote, D. R. D. & Baird, A. W. 2004. Stratigraphic controls on Palaeozoic petroleum systems, Ghadames Basin, Libya. *Journal of Petroleum Geology*, **27**, 141–162.

Destombes, J., Holland, H. & Willefert, S. 1985. Lower Paleozoic rocks of Morocco. *In*: Holland, C. (ed.) *Lower Paleozoic Rocks of Northwestern and Western Africa*. Wiley, New York, 91–336.

Echikh, K. 1998. Geology and hydrocarbon occurrences in the Ghadamis Basin, Algeria, Tunisia, Libya. *In*: Macgregor, D. S., Moody, R. T. J. & Clark-Lowes, D. D. (eds) *Petroleum Geology of North Africa*. Geological Society, London, Special Publications, **132**, 109–129.

Eschard, R., Abdhallah, H., Braïk, F. H. & Desaubliaux, G. 2005. The Lower Paleozoic succession in the Tasilli outcrops, Algeria: sedimentology and sequence stratigraphy. *First Break*, **23**, 27–36.

Fabre, J. 1988. Les séries du paléozoïque d'Afrique: une approche. *Journal of African Earth Sciences*, **7**, 1–40.

Ghienne, J. F., Boumendjel, K., Paris, F., Videt, B., Racheboeuf, P. & Salem, H. A. 2007a. The Cambrian-Ordovician succession in the Ougarta Range (western Algeria, North Africa) and interference of the Late Ordovician glaciation on the development of the Lower Palaeozoic transgression on northern Gondwana. *Bulletin of Geosciences*, **82**, 183–214.

Ghienne, J. F., Le Heron, D. P., Moreau, J., Denis, M. & Deynoux, M. 2007b. The Late Ordovician glacial sedimentary system of the North Gondwana platform. *In*: Hambrey, M., Christoffersen, P., Glasser, N., Janssen, P., Hubbard, B. & Siegert, M. (eds) *Glacial Sedimentary Processes and Products*. International Association of Sedimentologists, Special Publications, **39**. Blackwells, Oxford, 239.

Haddoum, H., Guiraud, R. & Moussine-Pouchkine, A. 2001. Hercynian compressional deformations of the Ahnet-Mouydir Basin, Algerian Saharan Platform: far-field stress effects of the Late Palaeozoic orogeny. *Terra Nova*, **13**, 220–226.

Hallett, D. 2002. *Petroleum Geology of Libya*. Elsevier Science, Amsterdam, 503.

Haq, B. U. & Schutter, S. R. 2008. A chronology of Paleozoic sea-level changes. *Science*, **322**, 64–68.

Legrand, P. 2003. Paléogéographie du Sahara algérien à l'Ordovicien terminal et au Silurien inférieur. *Bulletin de la Société Géologique de France*, **174**, 19–32.

Le Heron, D. P. & Craig, J. 2007. First order reconstruction of a Late Ordovician Ice sheet. *Journal of the Geological Society, London*, **93**, 47–76.

Lüning, S., Craig, J., Loydell, D. K., Storch, P. & Fitches, L. 2000. Lower Silurian 'hot shales' in North Africa: regional distribution and depositional model. *Earth Science Review*, **49**, 121–200.

Lüning, S., Archer, R., Craig, J. & Loydell, D. K. 2003a. The Lower Silurian 'Hot Shales' and 'Double Hot Shales' in North Africa and Arabia. *In*: Salem, M. J., Oun, K. M. & Seddiq, H. M. (eds) *The Geology of Northwest Libya (Ghadamis, Jifarah, Tarabulus and Sabratah Basins)*. Earth Science Society of Libya, Tripoli, **3**, 91–105.

Lüning, S., Adamson, K. & Craig, J. 2003b. Frasnian organic-rich shales in North Africa: regional distribution and depositional model. *In*: Arthur, T. J., MacGregor, D. S. & Cameron, N. R. (eds) *Petroleum Geology of North Africa: New Themes and Developing Technologies*. Geological Society, London, Special Publications, **207**, 165–184.

Massa, D. 1988. Paléozoique de Libye occidentale – Stratigraphie et Paleogéographie. Doctoral thesis, Université Nice.

Posamentier, H. W., Allen, G. P., James, D. P. & Tesson, M. 1992. Forced regressions in a sequence stratigraphic framework: concepts, examples and exploration significance. *AAPG Bulletin*, **76**, 1687–1709.

Ramos, E., Marzo, J. M., de Gibert, K. S., Tawengi, A. A., Khoja, & Bollati, N. D. 2006. Stratigraphy and sedimentology of the Middle Ordovician Hawaz Formation (Murzuq basin, Libya). *AAPG Bulletin*, **90**, 1309–1136.

Rubino, J. L. & Blanpied, C. 2000. Sedimentology and sequence stratigraphy of the Devonian to lowermost Carboniferous succession on the Gargaf Uplift (Murzuq Basin, Libya). *In*: Sola, M. & Worsley, D. (eds) *Geological Exploration in Murzuq Basin*. Elsevier Science, Amsterdam, 321–348.

Scotese, C. R., Boucot, A. J. & Mckerrow, W. S. 1999. Gondwanan palaeogeography and palaeoclimatology. *Journal of African Earth Sciences*, **28**, 99–114.

Smith, B., Derder, M. E. M. *et al.* 2006. Relative importance of the Hercynian and post-Jurassic tectonic phases in the Saharan platform: a palaeomagnetic study of Jurassic sills in the Reggane Basin (Algeria). *Geophysics Journal International*, **167**, 380–396.

Sutcliffe, O. E., Adamson, K. & Ben Rahuma, M. M. 2000. The geological evolution of the Palaeozoic rocks of western Libya: a review and field guide. Second Symposium on the Sedimentary Basins of Libya. Geology of northwestern Libya. Field Guide. *Earth Sciences Society of Libya*, **93**.

Underdown, R. & Redfern, J. 2008. Petroleum generation and migration in the Ghadames Basin, north Africa: a two-dimensional basin-modeling study. *AAPG Bulletin*, **92**, 53–76.

Vos, R. G. 1981. Deltaic sedimentation in the Devonian of Western Libya. *Sediment Geology*, **29**, 67–88.

Wendt, J., Kaufmann, B., Belka, Z., Klug, C. & Lubeseder, S. 2005. Sedimentary evolution of a Palaeozoic basin and ridge system: the Middle and Upper Devonian of the Ahnet and Mouydir (Algerian Sahara). *Geological Magazine*, **143**, 269–299.

Zazoun, R. S. 2001. La tectogenèse hercynienne dans la partie occidentale du bassin de l'Ahnet et la région de Bled El-Mass, Sahara algérien: un continuum de deformation. *Journal of African Earth Sciences*, **32**, 869–887.

Stratigraphic trapping potential in the Carboniferous of North Africa: developing new play concepts based on integrated outcrop sedimentology and regional sequence stratigraphy (Morocco, Algeria, Libya)

S. LUBESEDER, J. REDFERN, L. PETITPIERRE and S. FRÖHLICH

School of Earth, Atmospheric and Environmental Sciences, University of Manchester, Oxford Rd, Manchester M13 9PL, UK (e-mail: jonathan.redfern@manchester.ac.uk)

Abstract: The lower Carboniferous (Tournaisian to Visean) of North Africa is characterized by cycle-stacks of predominantly shelfal to marginal marine sandstones and limestones, thick shelfal mudstones and less common but important interbedded fluvio-deltaic sandstones. The cyclic sedimentation pattern continues into the Mid Carboniferous (Serpukhovian to Bashkirian), when mixed siliciclastic–carbonate sequences give way to tropical carbonates, before an abrupt return to continental deposits in the upper Carboniferous (Bashkirian to Gzhelian). The alternation of widespread shallow marine and more discrete fluvial reservoirs with interbedded offshore mudstone seals is interpreted to result from high-frequency, high-amplitude Carboniferous glacio-eustatic sea-level changes. The large base-level changes during that time, combined with climatic conditions that produced high amounts of terrigenous mud, provided favourable conditions for the development of stratigraphic traps in the clastic-prone lower Carboniferous, while the advent of tropical carbonates produced reefal buildups in the Mid Carboniferous. Four stratigraphic trapping types are recognized: (1) truncation traps in which reservoir units were eroded on subaerially exposed proximal palaeohighs and thick underlying transgressive and highstand systems tract (TST and HST) mudstones form the bottom-seal and the rapid transgression of the offshore facies forms the top-seal; (2) pinchout traps of lowstand wedges on the flanks of distal palaeohighs, which were only affected by subaqueous reworking of previous TST–HST mudstones and were buried during the subsequent transgression; (3) incised valleys of the lowstand systems tract (LST), filled with thick fluvial and tidal sandstones, cutting either into TST–HST mudstones in the lower Carboniferous, or into exposed carbonate platforms in the Mid Carboniferous; (4) Waulsortian-type reefal buildups of the Mid Carboniferous. The four trapping types are discussed using selected outcrop examples, and are placed into regional sequence stratigraphic context of the Carboniferous depositional systems and sequence development of North Africa. These concepts can be readily applied to the subsurface and offer significant potential for new plays across North Africa.

Keywords: stratigraphic traps, sequence stratigraphy, Carboniferous, Morocco, Algeria, Libya

A tentative sequence stratigraphic scheme for the Carboniferous of North Africa has been developed to guide reservoir prediction in poorly explored areas and to provide a framework for further analysis of potential stratigraphic traps. Seven important outcrop sections across North Africa were reviewed and correlated. Of these, five sections have been interpreted based on published sources and two sections (SW Anti-Atlas, Morocco; NW Murzuk Basin, Libya) were logged for this study (Figs 1 & 2). All sections are located along the southern Palaeozoic outcrop belt, comprising exposures in the Anti-Atlas, the Bechar Basin, the Reggane, Ahnet and Mouydir Basins, the southern Illizi Basin, and the northern Murzuk to southern Ghadames Basins. The presented chronostratigraphic chart has been compiled based on a comprehensive review of a large amount of published literature, of which only selected key publications are referred to in this summary paper.

North African Carboniferous sequence framework

Over many decades of studies in North Africa there have been a number of publications on the stratigraphy and sedimentology of the Carboniferous; an excellent regional summary is provided by Diaz *et al.* (1985). However, of these studies, only a few attempt any sequence-type interpretation. In this paper we present the first regional sequence stratigraphic scheme, based on a large amount of available published biostratigraphic and sedimentological data, integrated with new field work.

Good quality sedimentological and biostratigraphic published work on a regional scale is very limited, despite the vast outcrops of fossiliferous Carboniferous sediments. The sequence stratigraphic scheme and correlation presented here must be considered a preliminary working document due to the vast area covered and the relatively widely spaced dataset currently available. In certain data-poor areas the scheme remains very conjectural at present. In addition to the limited dataset, different fossil groups used for biostratigraphic dating sometimes suggest very different ages (Wendt *et al.* 2009), exemplifying the need to continue and update the research on the extensive outcrops.

The regional correlation shows the general facies patterns and palaeogeography of the Carboniferous from Morocco to western Libya. This correlation allows assessment of the broad trends, although it should be emphasized that in detail, on a basin scale, the stratigraphy and sedimentology are undoubtedly more complicated than can be shown on the generalized correlation panel.

The following sections describe each phase of Late Devonian to Late Carboniferous deposition, ending with a comparison of these phases with global sea-level cycles.

Late Devonian ('Strunian') regression

A latest Devonian ('Strunian') regression is recognized across all of NW Africa, with the deposition of shelf sandstones in the Anti-Atlas (e.g. Kaiser *et al.* 2004), Bechar Basin, Ougarta Arch, northern Hoggar region and the Illizi and Ghadames Basins (Conrad *et al.* 1986; Lubeseder 2005; Fig. 3). A regionally extensive unconformity developed in the Murzuq Basin of Libya and across surrounding palaeohighs (the Gargaf Arch and Tihemboka Arch) and presumably reached far into the Ghadames Basin. The

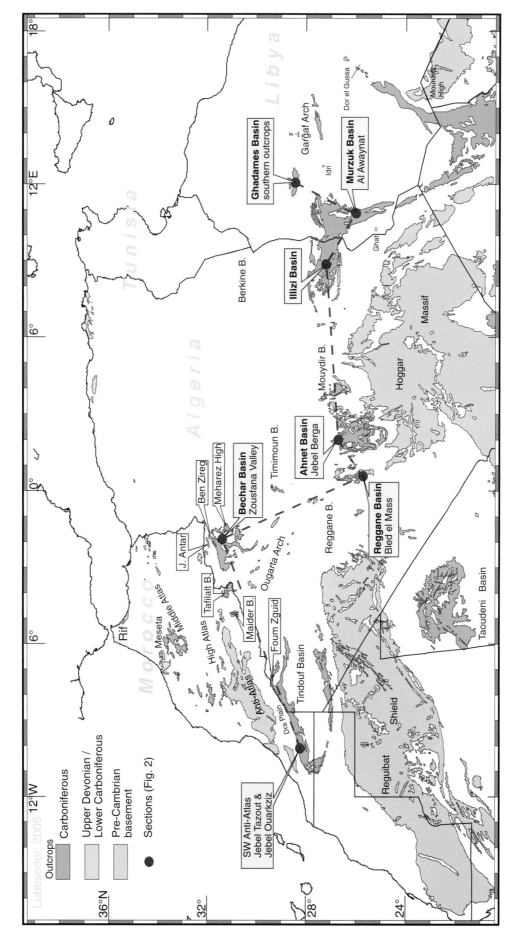

Fig. 1. Map of North Africa showing Carboniferous outcrops, location of sections (red dots), and line of correlation (shown in Fig. 2). Geological basemap after Persits *et al.* (1997).

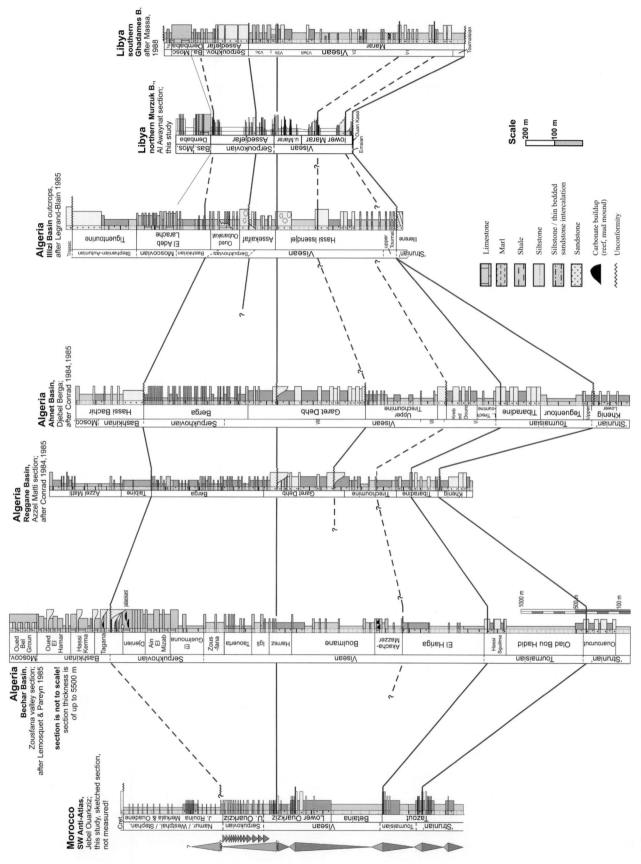

Fig. 2. Correlation of selected North African Carboniferous outcrop sections.

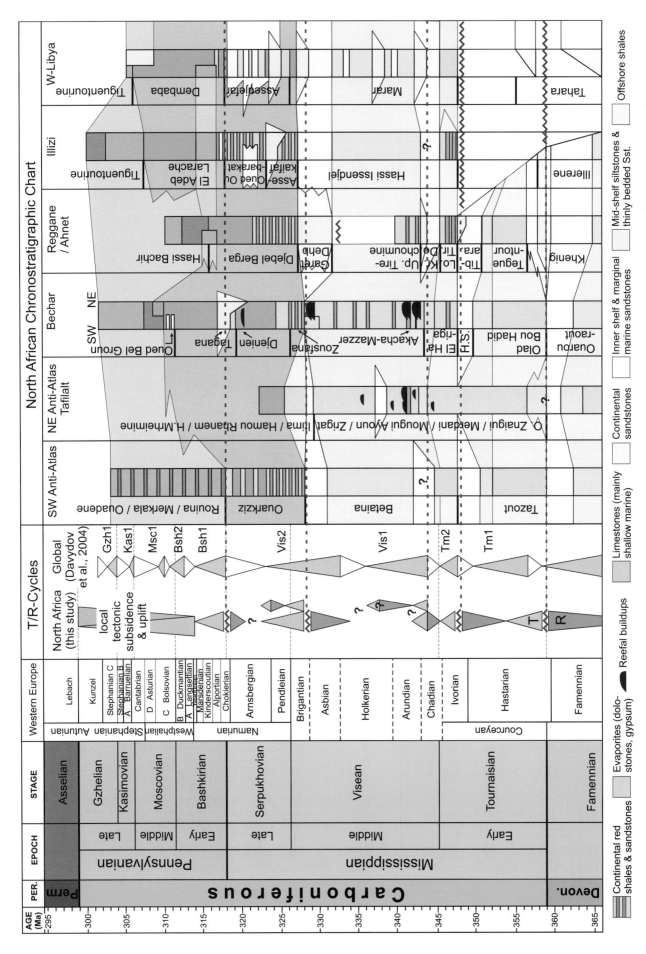

Fig. 3. Tentative chronostratigraphic correlation and transgressive-regressive (T/R) sequence framework of the Carboniferous of North Africa. Note that much more integrated biostratigraphic and sedimentological work on the Carboniferous sections of North Africa will have to be carried out before a satisfactory regional sequence framework can be achieved, and that the proposed scheme presented here is only a starting-point.

unconformity is also developed in the Reggane and Ahnet Basins of Algeria (Conrad 1984, 1985).

Early to Middle Tournaisian transgression

In the northwestern regions (the Anti-Atlas, Bechar Basin and Ougarta Arch), the subsequent transgression led to the deposition of marine shales with an Early Tournaisian fauna (the ammonoid *Gattendorfia* sp.). These shales are correlated to the shales of the Teguentour Formation in the Ahnet Basin, dated to the Mid to Late Tournaisian (Wendt *et al.* 2009). The emergence interpreted by these authors in the Ahnet Basin is interpreted here as a hiatus caused by non-deposition during maximum flooding. In the more proximal near-shore to continental dominated southwestern areas, transgressive sandstones were deposited bearing Early Tournaisian brachiopods (e.g. the upper Tahara and Ashkidah Formations of the Gargaf Arch; Mergl & Massa 2000).

Late Tournaisian regression

In the Anti-Atlas and the Bechar, Reggane and Ahnet Basins, a Late Tournaisian regression formed a shale-to-sandstone coarsening-upward hemi-cycle comparable to those of the latest Devonian. It is likely that this hemi-cycle also developed in the subsurface of several other basins (e.g. the Timimoun, Ghadames and Berkine Basins) and may have been incorporated into the 'Strunian' reservoir interval by some workers. In the southern Ahnet Basin, fluvial sandstones occur locally within the Late Tournaisian Tibaradine Formation (Conrad 1985). In the Murzuq Basin and in the southern Illizi Basin outcrops, Late Tournaisian strata were eroded or sediments bypassed these basins during this regressive phase.

Latest Tournaisian to earliest Visean transgression

The Late Tournaisian shelf sandstones are succeeded by a thick interval of outer-shelf shales in the Anti-Atlas and the Bechar Basin (Betaina Formation; El Hariga Formation), which contain latest Tournaisian faunas at their bases. In the Ahnet Basin, the shales and limestones of the Lower Tirechoumine Formation (Iridet Formation *sensu* Wendt *et al.* 2009) are dated by conodonts, ammonoids and foraminifera to the Late Tournaisian to Early Visean (Conrad 1985; Wendt *et al.* 2009). This latest Tournaisian transgression also led to the deposition of marine sediments directly above the basal Hassi Issendjel Formation unconformity (Legrand-Blain 1985) and above the basal Marar Formation unconformity in the southern Illizi and Murzuq Basins. The very bases of these two formations have been assigned to the latest Tournaisian.

Visean transgressive–regressive cycles

Most of the Visean essentially forms one large-scale regressive phase (seen in the SW Anti-Atlas Betaina Formation, more poorly developed in the Ahnet and Illizi Basins, and well developed in the Marar Formation of Libya), which ended with the onset of widespread carbonate deposition during the latest Visean. Correlation of several intra-Visean cycles across the Saharan Platform, however, is uncertain (Fig. 2). It is proposed that the Visean large-scale regression is subdivided into two, or possibly three, higher-order cycles (Fig. 3).

Two well developed Visean cycles are evident in the Betaina Formation of the Anti-Atlas and are possibly represented by the El Hariga Formation, and (partially) by the Akacha–Mazzer Formation (lower cycle), and by the Boulmane and the Harrez Formations (upper cycle), in the Bechar Basin.

In the Ahnet Basin, the regressive deltaic-marine to fluvial sandstones of the Kreb ed Douro Formation are well dated to the Early Visean by the under- and overlying marine shales and limestones with ammonoids, conodonts and foraminifera (Conrad 1985; Wendt *et al.* 2009). We suggest a different interpretation from that of Wendt *et al.* (2009), placing the subsequent fluvially dominated and undated Garet Dheb Formation into the Late Visean, not the Early Visean. Fluvial channels that are reported to have incised several tens of metres within this formation (Conrad 1985) correlate better with the Late Visean major regressive phase than with the relatively minor regression in the Early Visean. The long-lasting Visean hiatus proposed by Wendt *et al.* (2009) would, in this re-interpretation, underlie a Late Visean unconformity.

At least two higher-order cycles also occur in the Hassi Issendjel Formation of the southern Illizi Basin (Legrand-Blain 1985), but there is little detailed information available on this section. In the western Libyan outcrops, the Marar Formation shows three well developed Visean sequences, the upper two starting with incised fluvial channels at the base that grade upwards into shelf sandstones and shale (sequences LC1, LC2 and LC3 of Fröhlich *et al.* in press; biostratigraphic data mainly based on Massa & Vachard 1979 and Massa 1988).

Latest Visean to Serpukhovian transgression

In all of the reviewed sections the start of the regionally widespread carbonate deposition above the Late Visean clastics was proposed to occur in the latest Visean. In the west (the Anti-Atlas and the Bechar, Reggane and Ahnet Basins), deposition of the carbonate-rich interval continued throughout the Serpukhovian, while siliciclastics remained important in the SE (the Murzuk, Ghadames and southern Illizi Basins).

In the Anti-Atlas (Jebel Ouarkziz section) the uppermost Visean to Serpukhovian forms a well-expressed cyclic succession of shallow marine limestones, siltstones and sandstones with an overall long-term transgressive trend from lagoonal to open-shelf carbonates and a gradual disappearance of intercalated siliciclastics (Fig. 4). Similar cyclic Serpukhovian strata are recorded from the Bechar, Reggane and Ahnet Basin sections (Fig. 2). In the Libyan outcrops, oolitic and stromatolitic limestones (Collenia Beds) are part of these Serpukhovian mixed depositional systems. It appears that relative sea-level during the latest Visean and Serpukhovian rose gently, punctuated by many high-order fluctuations.

It must be noted that biostratigraphic dating in this interval is very controversial. In the Ahnet Basin, Wendt *et al.* (2009) proposed that the carbonates of the Jebel Berga Formation date entirely to the Bashkirian based on conodont data, yet foraminifera suggest a Late Visean to Serpukhovian age (Conrad 1985; Wendt *et al.* 2009). We follow the foraminifera dating since most of the other Algerian and Libyan sections were also dated using foraminifera.

Early Bashkirian regression

It appears that the gently rising Serpukhovian sea-level fell abruptly around the Serpukhovian–Bashkirian boundary, around which time a marked regression is observed that terminated the shelf carbonate deposition in the Anti-Atlas, the Reggane Basin and the Ahnet Basin. In these areas, the Bashkirian commenced with deposition of continental sandstones, marginal-marine gypsum deposits and dolomites as well as lacustrine carbonates. In the Bechar Basin, sea-level fall exposed the carbonate platform and a palaeokarst surface developed, into which sandstone-filled channels up to 150 m deep were cut (Lemosquet & Pareyn 1985). This major regression is less evident in the Libyan sections, where shelf carbonate deposition continued into the Bashkirian. Clearly, the palaeogeography and facies patterns across North Africa became more complicated across the Saharan Platform during the Bashkirian, suggesting that Variscan tectonics began

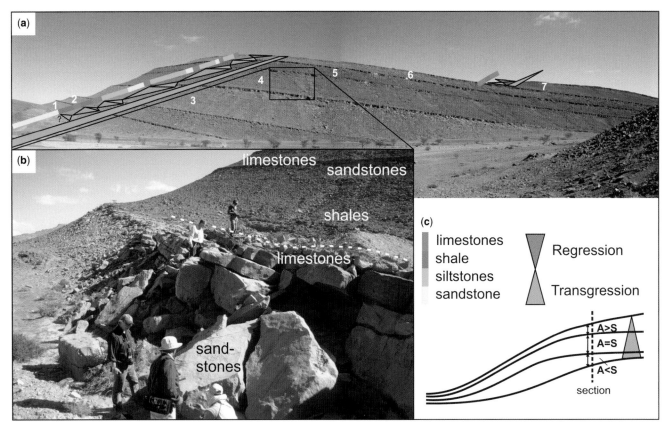

Fig. 4. (a) Serpukhovian transgressive succession of the SW Anti-Atlas (Ouarkziz Formation south of Assa) consisting of several higher-order transgressive limestone–regressive shale, siltstone, sandstone cycles (numbered 1–7). Note the overall transgressive sequence that is expressed in the upward decrease of terrigenous supply and increase of carbonate deposition as well as in the cyclical thin (cycles 1 and 2) to thick (cycles 2 and 3) to thin (cycles 4–7) stacking pattern. The section continues above the last crest with at least the same thickness of carbonate-dominated cycles. (b) Close-up view of cycle 4. (c) Legend and sketch to illustrate the interpretation of the stacking pattern as an upward passage through changing accommodation to sediment supply ratios (A/S) due to transgression (cycles 1 and 2, near-shore A < S; cycles 2 and 3, mid-shelf A = S; cycles 4–7, offshore A > S).

to destabilize the platform more effectively than the subtle tectonic movements that had prevailed since the Middle Devonian.

Late Carboniferous

Following the major end-Early Carboniferous regression, the Late Carboniferous sections vary across the region. Whilst in the Anti-Atlas (Tindouf Basin) and the Reggane and the Ahnet Basins, continental sandstones, mudstones and gypsum were deposited during the Late Carboniferous, carbonate deposition continued in the Bechar Basin, the southern Illizi Basin, and the Libyan basins during the early Late Carboniferous (Bashkirian to Moscovian). In the latter regions, continental red beds were deposited during the late Late Carboniferous (middle Moscovian to Gzhelian; middle Westphalian to Stephanian), although the full extent of their deposition is unclear, since the Hercynian Unconformity cuts down through the area. Upper Carboniferous cycles are difficult to constrain in terms of their sedimentology and biostratigraphy from the literature descriptions and are likely only to be valid locally.

Comparison of North African and global transgressive–regressive cycles

Four Early Carboniferous second-order transgressive–regressive (T/R) cycles are recognized in North Africa, which is the same as the number of cycles shown on the global scheme (Davydov et al. 2004; Fig. 3). Three out of nine North African hemi-cycle boundaries (maximum regressions and maximum transgressions) differ in age from the global scheme. Of the remaining six boundaries, four show a good correlation with the global scheme, but two are biostratigraphically too poorly constrained to allow a comparison. An additional Mid Visean cycle in North Africa is at present speculative.

The first Early Carboniferous cycle of North Africa, comprising the Devonian–Carboniferous boundary maximum regression, the Early Tournaisian transgression and the Late Tournaisian regression, compares well with the global T/R cycle Tm1 of Davydov et al. (2004) (Fig. 3). A fundamental difference between the schemes, however, exists in the preceding latest Devonian cycle. On global T/R schemes, the latest Devonian is considered to be transgressive (Johnson et al. 1985) and the following regressive phase is thought to fall entirely into the earliest Tournaisian. Yet, in North Africa, a wealth of biostratigraphic data suggest a latest Devonian ('Strunian') regression, succeeded by an Early Tournaisian transgression.

The interpreted second and third Early Carboniferous T/R cycles fit well with those of the global scheme (Tm2, Vis1; Fig. 3), although the Late Visean maximum regression appears to be younger in North Africa, constrained mainly by the age of the overlying transgressive limestones.

The fourth identified cycle, comprising the latest Visean and Serpukhovian transgression and the regression at the Serpukhovian–Bashkirian boundary, corresponds approximately to the global T/R cycle Vis2.

Although Early Hercynian tectonic activity is known to have started during the Middle Devonian and continued throughout the Late Devonian and Carboniferous in North Africa, the close match of North African and global T/R cycles suggests that Early Carboniferous sedimentary sequences were mainly controlled by eustatic sea-level fluctuations rather than tectonics. In contrast, the Late Carboniferous sequence ages, architectures and depositional systems are likely to have been controlled predominantly by regional to local tectonics and climate changes; however, the increasingly continental nature of the depositional environment means that research to date has struggled to constrain these sequences, making any conclusion regarding controls somewhat tentative.

Potential stratigraphic trapping types

Four potential stratigraphic trapping mechanisms are recognized within the Carboniferous interval. Two of the four stratigraphic trap types (truncation traps and depositional pinchout traps, Fig. 5) are controlled by the topography of palaeohighs and basins during the Carboniferous. Many of these highs are old structures that already influenced sediment distribution during the Early Palaeozoic. The structures were re-activated during the Middle to Late Devonian, and subsequently the syn-depositional relief became more and more pronounced until Hercynian uplift and peneplanation, especially in the northern basins (e.g. Bechar Basin; Malti *et al.* 2008). The other two potential trapping types are more related to depositional architecture (incised valleys, carbonate buildups, Fig. 5).

Truncation traps

Truncation, and subsequent onlap by a sealing lithology, is a proven trapping mechanism in Algeria. In several places the 'Strunian'–Tournaisian sandstones have been eroded on palaeohighs (e.g. the western Meharez High flanking the Bechar Basin, the Timimoun Basin margins and the Tihemboka Arch separating the Illizi Basin from the Ghadames Basin) either shortly after deposition (within-sequence) or later by the Hercynian Unconformity. Top-seals are either Visean shales or Triassic–Liassic shales and evaporites. The bottom-seal is provided by the Frasnian–Famennian shales. Risk of top-seal failure occurs in the case of within-sequence truncation, since laterally continuous sandstones of the following transgression may overlie the unconformity and trapping may therefore rely on sealing faults. However, the fault offsets required to give seals may only be small, since the thicknesses of the stacked transgressive sandstones are expected to be limited due to high-amplitude Carboniferous sea-level fluctuations.

Depositional pinchout of lowstand wedges on palaeohigh flanks

This stratigraphic trapping concept is probably the least-known and least-documented type in North Africa. It is similar to the truncation trap and may have been misinterpreted as such in some cases, but differs in the genetic sense and the predicted reservoir facies.

Topographic relief on the Saharan Platform influenced the facies patterns in all depositional environments, including outer shelf settings. This is well documented in the Middle to Upper Devonian carbonate facies (e.g. Wendt *et al.* 2006). In distal mid-shelf to outer-shelf dominated settings, siliciclastic supply into the basin either directly bypassed the highs or was subsequently removed by bottom-currents or storm reworking. This led to the deposition of only thinly bedded, rippled sandstones and siltstones on the highs and thickly bedded storm-induced siliciclastic turbidites and distal storm beds in the adjacent basin.

An example of this sediment partitioning is found in the 'Strunian'–Tournaisian of the Maider and Tafilalt Basins (Morocco), which are separated by a regional palaeohigh. In the centre of the Maider Basin, sandstones with large flute-marks

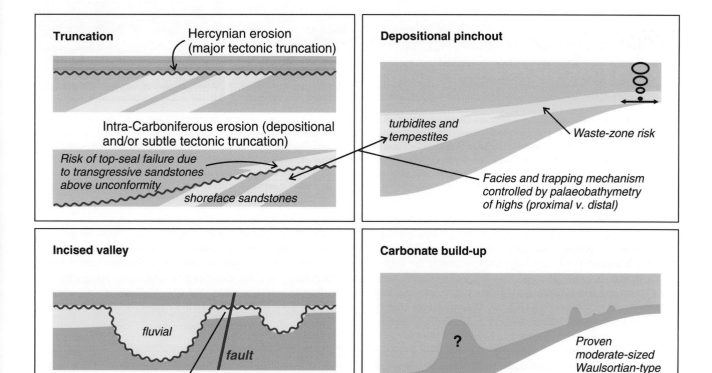

Fig. 5. The four main stratigraphic trapping types recognized in Carboniferous plays in North Africa. See text for explanations.

Fig. 6. Examples of basin-centred, distal lowstand sandstones that pinch out towards adjacent palaeo-structures thus forming potential pinchout traps on their flanks. The sandstones form 5–8 m thick bodies consisting of massive siliciclastic turbidites that pass upwards into storm deposits. The first example is the uppermost Devonian ('Strunian') sandstone exposed in the Maider Basin of Morocco (a) exhibiting large flute marks (b); note that the flute marks are filled by cross-beds and that the turbidites start at the base with the Bouma–Tc sequence. The second example is from the uppermost Givetian of the NE Dra Plain (Morocco) (c, d), where the sandstone is sandwiched between upper Givetian marls to thinly bedded calci-turbidites and Frasnian 'black' shales. Both examples are from the Middle and Upper Devonian, but very similar basin-centred lowstand sandstones are expected to occur in the NW African Carboniferous basins.

(Fig. 6) sharply overlie upper Devonian (Famennian) dark-coloured shales and are interpreted as turbiditic lowstand wedges. The sandstones pinch out onto the Tafilalt ridge and re-appear in the Tafilalt Basin. Another example of these lowstand basin-centred sandstones is the uppermost Givetian Megsem–Medersam Sandstone at Foum Zguid (SW Morocco) (Fig. 6). This sandstone sharply overlies upper Givetian thinly bedded calciturbidites, and its lower 4 m are entirely structureless. In its upper part there is a transition to hummocky cross-stratification. The transitional contact with the overlying Frasnian dark shales (related to a base Frasnian transgressive event, and elsewhere a proven source rock interval) is characterized by levels enriched in large corals and wood fragments (Meyer-Berthaud et al. 2004).

For a pinchout trap to be effective, it requires rapid pinchout of reservoir facies against the high; change from reservoir to non-reservoir facies must ideally happen over a short distance to minimize the potential for a mixed-facies 'waste-zone'. Such clear separation of reservoir-quality lowstand wedge deposits from non-reservoir deposits on the palaeohigh may require reasonable palaeo-relief, and may therefore limit this trapping style to the northwestern and northern Saharan Platform basins with, for example, known elevated subsidence rates during the Famennian.

Incised valley-fills

Fluvial channels interpreted to occur within incised valleys have been reported from several outcrop localities and stratigraphic levels of the uppermost Devonian and lower Carboniferous; and occasionally also from subsurface data. Because of the predominantly mud-rich lithologies of under- and overlying strata, these incised valley-fills may have good trapping potential, unless they pass laterally into extensive shoreline successions.

'Strunian' to Tournaisian incised valleys

Fluvial sandstones of the 'Strunian' and Tournaisian have been described (e.g. the Khenig Formation of the Ahnet and Reggane Basins, Conrad 1985; the Tazout Formation of the SW Anti-Atlas, Vos 1977), but no examples of isolated incised systems have been found in the literature. In the SW Anti-Atlas, on the northern flank of the Tindouf Basin, coarse-grained delta plain deposits occurring in up to 55 m thick and 60 km wide belts, pass laterally into extensive shoreface deposits (Vos 1977). These belts may significantly increase net reservoir thickness in structural traps, but because of the juxtaposition to the extensive shoreface deposits, they have low stratigraphic trapping potential.

Visean incised valleys

Fluvial sandstones of Visean age are widespread in the Ahnet and Reggane Basins (Garet Dehb Formation) and channels cut down several tens of metres into the underlying strata. Two main channel systems have been logged, one in the lower and one in the upper part of the formation (Azzel Matti section, Fig. 2).

Similar delta distributary channels cutting into underlying strata also occur in the southern Illizi Basin outcrops ('Gres de Champinngons Superieur'). In the northern Murzuk Basin, the Marar Formation contains recently recognized incised valleys about 0.5–3 km wide and 15–50 m thick (this study).

Serpukhovian–Bashkirian incised valleys

A different stratigraphic trapping potential may exist in the Bechar Basin. A sea-level drop towards the end of the Serpukhovian exposed the carbonate platform of the Bechar Basin, and a palaeokarst surface developed, into which cut sandstone-filled channels up to 150 m deep (Lemosquet & Pareyn 1985).

According to these authors, the fluvial sandstones were derived from the SW (Arlal Land) from the area of the future Kenadza and Abadla 'coal basins' (Colombo & Bensalah 1991).

The channels are filled with coarse, plant-bearing sandstones and occasional conglomerates. Each channel-filling episode was followed by deposition of well stratified shales and limestones containing marine faunas. The cycle of incision, fluvial sandstone deposition and open marine carbonate sedimentation suggests high-amplitude sea-level fluctuations at this time.

Carbonate buildups (Waulsortian reefs)

Visean Waulsortian-type carbonate buildups are exposed along the margins of the larger Bechar Basin (the Meharez High to the east, the Ben Zireg and Jebel Antar outcrops to the north and the Tafilalt Basin to the west). On the Meharez High, the sponge-bryozoan mounds grew vertically during times of high relative sea-level. Growth terminated when the buildups were covered by crinoidal and oolitic limestones during falling sea-level and lowstand stages (Bourque et al. 1995).

In the Tafilalt, approximately 100 Visean carbonate mud mounds are exposed (Wendt et al. 2001). However, several of these have also been interpreted as channels, filled with calcarenitic and finer-grained material, and not representing true mud mounds (Klug et al. 2006).

On the Meharez High, the exposed buildups are 50 m (locally up to 120 m) high and 2–3 km broad (Bourque et al. 1995). It may well be that even larger buildups exist in the subsurface. The size distribution of the Devonian mud mounds shows that it is inadvisable only to consider the well known examples as analogues. Most of the famous Devonian mounds in Morocco and Algeria (Brachert et al. 1992; Kaufmann 1998; Wendt & Kaufmann 1998) are rather small (few metres to tens of metres high and wide), economically unattractive buildups, but the less well-known Middle Devonian reefal mounds of the Zemour in the SW Tindouf Basin (Dumestre & Illing 1967) show that buildups at this time were also capable of accumulating 80 m thick coral-stromatoporoid-rich limestones with good reservoir potential, with aerial extents of up to 2 km.

Conclusions

A preliminary sequence stratigraphic framework is presented for the Early Carboniferous of North Africa that shows a good correlation with published global sequences and eustatic sea-level curves. However, the relatively sparse sedimentological database available regionally, and the limited biostratigraphic control, further hampered by contradictory published biostratigraphic interpretations, emphasize the need for further work in the region to develop a more robust regional synthesis of the Carboniferous sequence development in North Africa.

Four types of potential stratigraphic traps are recognized (truncation traps, depositional pinchout traps, incised valley-fills and carbonate buildups). Of these, truncation plays have the highest potential as exploration targets. Depositional pinchout traps of basin-centred sands may have been overlooked in the past and offer a higher risk but potentially volumetrically significant target. These may be imaged on future high-quality 3D seismic. Such reservoirs are believed to be a particular feature of the middle to upper Palaeozoic succession, as the basins changed from low-relief intra-cratonic sag basins (Late Cambrian to Early Devonian) to ramps with locally steepened profiles (Middle Devonian to Carboniferous). High-amplitude sea-level falls in the Carboniferous provided a mechanism to deliver clastics into the distal parts of the basins and may even have led to detached lowstand accumulations (the latter being a hypothesis that is not yet confirmed). Incised valley-fills also provide opportunities for thickened and stratigraphically discrete reservoir sections, and require a detailed knowledge of the sequence stratigraphic framework for prediction.

This study was funded by the North Africa Research Group (University of Manchester) sponsored by Anadarko, BG Group, ConocoPhillips, Hess, Maersk, Oxy, Petro-Canada, PlusPetrol, RepsolYPF, RWE, Wintershall and Woodside.

References

Bourque, P. A., Madi, A. & Mamet, B. L. 1995. Waulsortian-type bioherm development and response to sea-level fluctuations: upper Visean of Béchar Basin, western Algeria. *Journal of Sedimentary Research*, **B65**, 80–95.

Brachert, T. C., Buggisch, W., Flügel, E., Hüssner, H. M., Joachimski, M. M., Tourneur, F. & Walliser, O. H. 1992. Controls of mud mound formation: the Early Devonian Kess–Kess carbonates of the Hamar Laghdad, Antiatlas, Morocco. *Geologische Rundschau*, **81**, 15–44.

Colombo, F. & Bensalah, M. 1991. Le Westphalien continental de la région de Bechar (Algérie) Considérations sédimetologiques. *Bulletin d'Office Nationale Géologique Algérie*, **2**, 49–52.

Conrad, J. 1984. *Les séries carbonifères du Sahara central algérien. Stratigraphie, sédimentation, évolution structurale*. PhD thesis, Université d'Aix-Marseille, France.

Conrad, J. 1985. Northwestern and central Saharan areas – Ahnet–Mouydir area. *In*: Diaz, C. M., Wagner, R. H., Winkler-Prins, C. F. & Granados, L. F. (eds) *The Carboniferous of the World II: Australia, Indian Subcontinent, South Africa, South America, & North Africa*. IUGS Publication Instituto Geologico y Minero de España & Empresa Naçional Adaro de Investigaçiones Mineras, Spain, **20**, 317–322.

Conrad, J., Massa, D. & Weynant, M. 1986. Late Devonian regression and early Carboniferous transgression of the North African Platform. *Annales Société Géologique Belgique*, **109**, 113–122.

Davydov, V., Wardlaw, B. R. & Gradstein, F. M. 2004. The Carboniferous period. *In*: Gradstein, F. M., Ogg, J. O. & Smith, A. G. (eds) *A Geological Time Scale 2004*. Cambridge University Press, Cambridge, 222–248.

Diaz, C. M., Wagner, R. H., Winkler-Prins, C. F. & Granados, L. F. 1985. *The Carboniferous of the World II: Australia, Indian Subcontinent, South Africa, South America, & North Africa*. IUGS Publication Instituto Geologico y Minero de España & Empresa Naçional Adaro de Investigaçiones Mineras, Spain, **20**, 298–477.

Dumestre, A. & Illing, L. V. 1967. Middle Devonian reefs in Spanish Sahara. *In*: Oswald, D. H. (ed.) *International Symposium on the Devonian System*. Alberta Society of Petroleum Geology, Calgary, **II**, 333–350.

Fröhlich, S., Petitpierre, L., Redfern, J., Grech, P., Bodin, S. & Lang, S. in press. Sedimentological and sequence stratigraphic analysis of Carboniferous deposits in western Libya: recording the sedimentary response of the northern Gondwana margin to climate and sea-level changes. *Journal of African Earth Sciences*; doi: 10.1016/j.jafrearsci.2009.09.007.

Johnson, J. G., Klapper, G. & Sandberg, C. A. 1985. Devonian eustatic fluctuations in Euramerica. *Geological Society of America Bulletin*, **96**, 567–587.

Kaiser, S., Becker, T. R. *et al.* 2004. Sedimentary succession and neritic faunas around the Devonian-Carboniferous boundary at Kheneg Lakahal south of Assa (Dra Valley, SW Morocco). *In*: El Hassani, A. (ed.) *Devonian of the Western Anti Atlas: Correlations and Events*. Documents de l'Institut Scientifique, Université Mohammed V Agdal, Rabat, **19**, 69–74.

Kaufmann, B. 1998. Middle Devonian reef and mud mounds on a carbonate ramp: Mader Basin (eastern Anti-Atlas, Morocco). *In*: Wright, V. P. & Burchette, T. P. (eds) *Carbonate Ramps*. Geological Society, London, Special Publications, **149**, 417–435.

Klug, C., Döring, S., Korn, D. & Ebbighausen, V. 2006. The Viséan sedimentary succession at the Gara el Itima (Anti-Atlas, Morocco) and its ammonoid faunas. *Fossil Record*, **9**, 3–60.

Legrand-Blain, M. 1985. Northwestern and central Saharan areas – Illizi Basin. *In*: Diaz, C. M., Wagner, R. H., Winkler-Prins, C. F. & Granados, L. F. (eds) *The Carboniferous of the World II: Australia, Indian Subcontinent, South Africa, South America, & North Africa*. IUGS Publication Instituto Geologico y Minero de España & Empresa Naçional Adaro de Investigaçiones Mineras, Spain, **20**, 329–333.

Lemosquet, Y. & Pareyn, C. 1985. Northwestern and central Saharan areas – Bechar Basin. *In*: Diaz, C. M., Wagner, R. H., Winkler-Prins, C. F. & Granados, L. F. (eds) *The Carboniferous of the World II: Australia, Indian Subcontinent, South Africa, South America, & North Africa*. IUGS Publication Instituto Geologico y Minero de España & Empresa Naçional Adaro de Investigaçiones Mineras, Spain, **20**, 306–315.

Lubeseder, S. 2005. *Silurian and Devonian sequence stratigraphy of North Africa: regional correlation and sedimentology (Morocco, Algeria, Libya)*. PhD thesis, University of Manchester.

Malti, F. Z., Benhamou, M., Mekahli, L. & Benyoucef, M. 2008. The development of the Carboniferous Ben-Zireg-Zousfana Trough in the northern part of the Bechar Basin, Western Algeria: implications for its structural evolution, sequence stratigraphy and palaeogeography. *Geological Journal*, **43**, 337–360.

Massa, D. 1988. *Paleozoique de Libye occidentale – stratigraphie et paleogeographie*. PhD thesis, Université de Nice, France.

Massa, D. & Vachard, D. 1979. Le Carbonifère de Libye occidentale. Biostratigraphie et Micropaléontologie. Position dans le domaine téthysien d'Afrique du Nord. *Revue Institute Français du Pétrole*, **34**, 1–65.

Mergl, M. & Massa, D. 2000. A palaeontological review of the Devonian and the Carboniferous succession of the Murzuq Basin and the Djado Sub-Basin. *In*: Sola, M. A. & Worsley, D. (eds) *Geological Exploration in Murzuq Basin*. Elsevier, Amsterdam, 41–88.

Meyer-Berthaud, B., Rücklin, M., Soria, A., Belka, Z. & Lardeux, H. 2004. Fransian plants from the Dra Valley, southern Anti-Atlas, Morocco. *Geological Magazine*, **141**, 675–686.

Persits, F., Ahlbrandt, T., Tuttle, M., Charpentier, R., Brownfield, M. & Takahashi, K. 1997. Maps showing geology, oil and gas fields and geological provinces of Africa. United States Geological Survey, Open-File Report **97-470A**, 1–10.

Vos, R. G. 1977. Sedimentology of an Upper Paleozoic river, wave and tide influenced delta system in southern Morocco. *Journal of Sedimentary Petrology*, **47**, 1242–1260.

Wendt, J. & Kaufmann, B. 1998. Mud buildups on a Middle Devonian carbonate ramp (Algerian Sahara). *In*: Wright, V. P. & Burchette, T. P. (eds) *Carbonate Ramps*. Geological Society, London, Special Publications, **149**, 397–415.

Wendt, J., Kaufmann, B. & Belka, Z. 2001. An exhumed Palaeozoic underwater scenery; the Visean mud mounds of the eastern Anti-Atlas (Morocco). *Sedimentary Geology*, **145**, 215–233.

Wendt, J., Kaufmann, B., Belka, Z., Klug, C. & Lubeseder, S. 2006. Sedimentary evolution of a Palaeozoic basin and ridge system: the Middle and Upper Devonian of the Ahnet and Mouydir (Algerian Sahara). *Geological Magazine*, **143**, 269–299.

Wendt, J., Kaufmann, B., Belka, Z. & Korn, D. 2009. Carboniferous stratigraphy and depositional environments in the Ahnet Mouydir area (Algerian Sahara). *Facies*, **55**, 443–472.

Integrated petroleum systems and play fairway analysis in a complex Palaeozoic basin: Ghadames–Illizi Basin, North Africa

R. J. DIXON, J. K. S. MOORE, M. BOURNE, E. DUNN, D. B. HAIG, J. HOSSACK, N. ROBERTS, T. PARSONS and C. J. SIMMONS

BP Exploration, Chertsey Road, Sunbury-on-Thames, Middlesex TW16 7LN, UK
(e-mail: dixonr2@bp.com)

Abstract: The Ghadames–Illizi Basin system is a highly productive petroleum province with a long exploration history in Algeria, Libya and Tunisia (from the late 1950s to present day). Ongoing exploration success in all three countries suggests that it will continue to provide attractive exploration targets in the future. The basin has a long (Cambrian to Plio-Pleistocene) and complex geological evolution characterized by multiple phases of subsidence punctuated by significant regional uplift events. Two 'world-class' petroleum source rocks of different geological age are present (Lower Silurian and Upper Devonian) with similar depositional environments and geochemical characters. Both source horizons have generated significant volumes of oil and gas. Migration is strongly influenced by the stratigraphic architecture of the basin fill, notably distribution of regional seals and the complex patterns of subcrop and onlap across regional unconformities. Multiple reservoir–seal combinations are presented by Late Ordovician glaciogenic sediments and younger Silurian through to Carboniferous paralic sequences. Integrating the stratigraphic relationships with the complex burial history of the basin (timing of uplift, degree of tilting, amount of section removed by erosion) is not a trivial task, but is key to exploration success in such a complex basin. With the aid of 3D basin reconstruction and fluid flow modelling software, we can attempt to capture the stratigraphic and structural complexity and make exploration predictions. If basin modelling techniques are to be optimally applied in such settings, a fully integrated and geologically realistic approach involving sedimentologists, structural geologists, geophysicists and geochemists is required. A modelling approach, workflow and some results are presented.

Keywords: Ghadames–Illizi basin, petroleum systems, play fairway analysis

The location of the major Palaeozoic basins in North Africa and the major tectonic elements of the North African Margin are shown in Figure 1. At a crustal scale African Gondwana is dominated by two major cratonic blocks, the West African Craton in the west and the Saharan Metacraton in the east (Fig. 1). A strong, Late Proterozoic tectonic grain inherited from the Pan-African Orogeny runs between these two cratonic masses and has been reactivated many times, most notably and severely during the Hercynian collision of Gondwana and Laurussia. Great escarpments of tilted, resistant, Palaeozoic strata (Fig. 1) date from this time (LeFranc & Guiraud 1990) and define the remnant Palaeozoic basins as well as strongly influencing subsequent Mesozoic–Cenozoic deposition (LeFranc & Guiraud 1990; Schroter 1996). In the offshore Pelagian Basin, the Sirt Basin and the offshore areas of Egypt and the Levant Palaeozoic sequences are probably present at depth beneath thick sequences of Mesozoic and Cenozoic strata. The most likely position of the Tauride Block (most of southern Turkey) during the Palaeozoic was adjacent to African Gondwana (Fig. 1). The great similarities in Palaeozoic stratigraphy and lithology between the Tauride Block and African Gondwana have led to this hypothesis (Demirtash 1984; Monod *et al.* 2003; Smith 2006). The Tauride Block and other previously conjugate terrains (e.g. Apulia and the Hellenides; Smith 2006) are now separated from African Gondwana by Neotethyan oceanic crust (of probable Cretaceous age).

Ghadames–Illizi Basin

Figure 1 shows the location of the Ghadames–Illizi Basin. The basin is large, covering an area of some 400 000 km^2 and reaches a maximum depth in its northwestern part, where the top of crystalline basement is found at depths approaching 8000 m. In a regional sense the basin has a relatively simple 'saucer-like' structure. The basin was first explored by the oil industry in the 1950s and in total 32 BBOE of petroleum (oil and gas) have been discovered to date. Despite its relatively simple regional structure, the Ghadames–Illizi Basin has a long and complex history. For example, although now separated from the Ahnet Basin to the west and the Murzuq Basin to the SE by regional highs or arches (Fig. 1), it is considered likely that all of these basins were connected during the early part of the Palaeozoic (Beuf *et al.* 1971), only becoming separate depositional entities from the Middle Devonian onwards as the effects of 'Hercynian' compression became increasingly pronounced. The interaction of tectonics with sedimentation has exerted a very strong influence on the distribution of both reservoir and source facies as well as on later source rock maturation and migration. Our evaluation of the Ghadames–Illizi Basin has involved a combination of geological fieldwork on the surrounding outcrops in Algeria, Tunisia and Libya and subsurface studies of a large well and seismic database (*c.* 200 wells and *c.* 125 000 km of 2D seismic).

Megasequence framework

For the basin model to be geologically realistic the basin's burial history must be reconstructed using all available data (cores, wells, seismic and outcrops). The purpose of this section is to review the structural and stratigraphic framework of the basin. The key basin forming and uplift episodes represented in the Ghadames–Illizi Basin (and in the other North African basins illustrated in Fig. 1) are shown in Figure 2. Several megasequences are recognized.

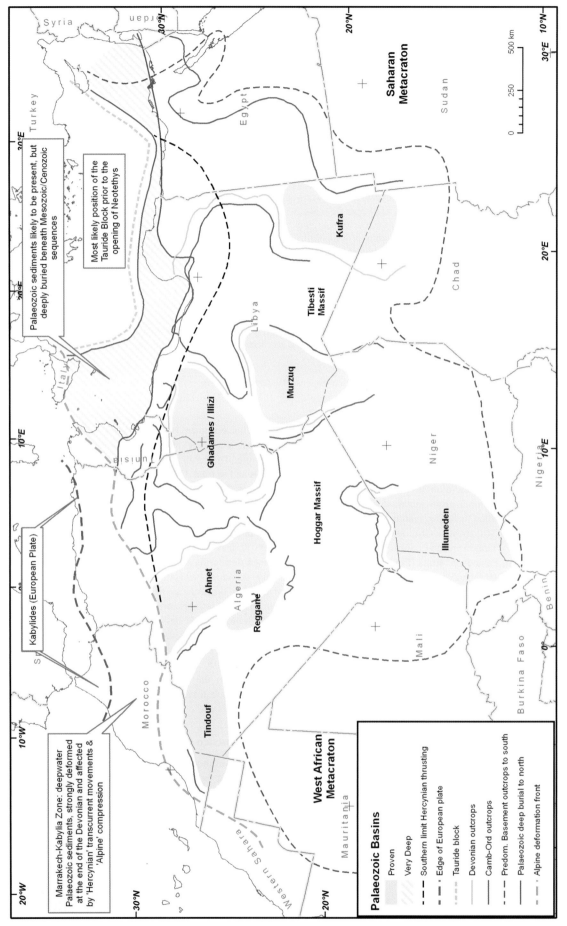

Fig. 1. Major tectonic elements and Palaeozoic basins of North Africa.

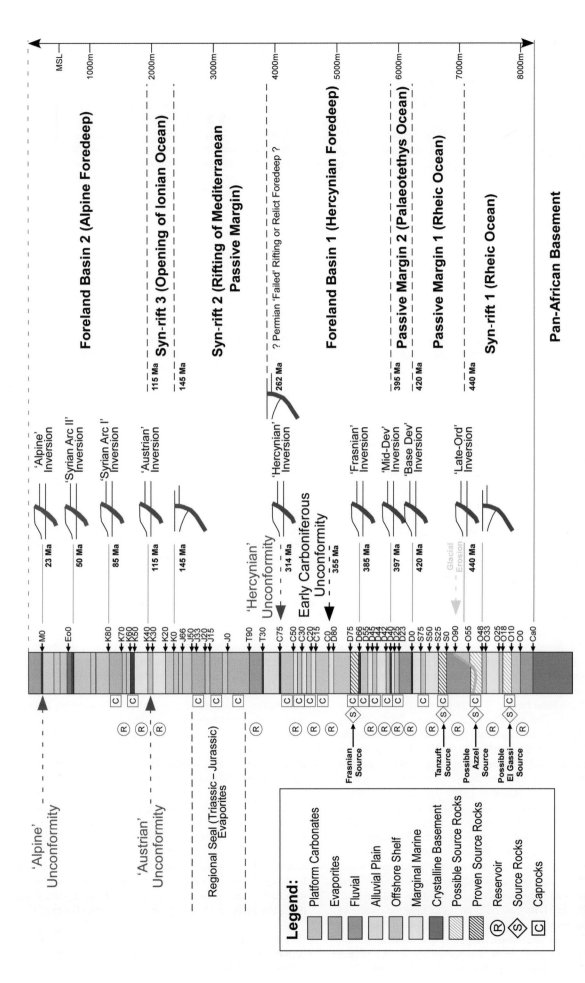

Fig. 2. Stratigraphic framework of the Ghadames–Illizi Basin. Section is schematic, but is based on the central part of the basin where up to 8000 m of sediment is preserved (arrows denote structural surfaces incorporated in the 3D basin model).

Syn-rift 1 Megasequence (Rheic Ocean rifting)

After a poorly defined Cambro-Ordovician Syn-Rift phase (opening of the Rheic Ocean; Stampfli & Borel 2002), two passive margin 'drift' megasequences were deposited, related to the Rheic Ocean and Palaeotethys Ocean, respectively, before the onset of the 'Hercynian Foredeep' (Foreland Basin 1 in Fig. 2).

Passive Margin 1 Megasequence (Rheic Ocean drift)

The first passive margin megasequence is the Rheic Ocean Drift succession, including the O90–S75 sequences of the Mamuniyat (O90–S0), Tanzuft (S0–S25) and Akakus (S25–S75). This megasequence is separated from the later Palaeotethys Ocean Drift succession by the Base Devonian Unconformity ('Caledonian'). According to Stampfli & Borel (2002), this unconformity records uplift associated with the separation of the 'Hun Superterrane' from African Gondwana ('Ridge–Push' mechanism of Stampfli & Borel 2002). The Tanzuft is a major petroleum source rock.

Passive Margin 2 Megasequence (Palaeotethys Ocean drift)

The Palaeotethys Ocean Drift succession ranges in age from D0 to D40 including the Tadrart (D0–D23), Emsian Shale (D23–D25) and Wan Kasa (D25–D40). Stampfli & Borel's (2002) 380 Ma plate reconstruction shows a wide Palaeotethys Ocean during the Late Devonian, although there are other authors who disagree with this interpretation and place Laurussia and Gondwana much closer together at this time (McKerrow et al. 2000). Both authors agree that continental collision began in the west and progressed eastwards with time, but we would agree with the McKerrow et al. interpretation and would argue that the continents started to collide in the Mid Devonian (Emsian–Eifelian). This is consistent with regional evidence of important uplift and erosion during the Middle Devonian (Emsian and Frasnian events, discussed below; Fig. 8) The Frasnian shale is a major petroleum source rock (Lüning et al. 2004b).

Foreland Basin 1 Megasequence ('Hercynian' Foredeep)

In order to reconstruct the early history of the 'Hercynian Foredeep' it is important to understand the evolution of the western Palaeozoic basins and the role of the West African Craton. The West African Craton is characterized by extremely thick crust [up to 250 km (Liegeois et al. 2005)]. Throughout the Neoproterozoic and Palaeozoic the craton remained at or close to sea-level. In the intracratonic Taoudenni Basin c. 5 km of paralic–shelfal sediment is preserved in the Gourma Trough (Villeneuve 2005). During the Lower Palaeozoic the Taoudenni Basin was connected to the Tindouf and Reggane basins (Fig. 1) and it shares a similar Cambro-Ordovician and Silurian stratigraphy. This connection was broken by regional uplift of the Regubiat Rise in the Late Silurian–Early Devonian (Guerrak 1989). The Taoudenni Basin was also uplifted and eroded at this time and a strong Base Devonian Unconformity locally completely removes the Silurian section. Deposition did not recommence in the Taoudenni area until the Emsian (Guerrak 1989; Herrera & Racheboeuf 2001).

The depositional history of the western Palaeozoic basins (Taoudenni, Tindouf, Reggane and Ahnet) starts to diverge from that of the more easterly basins (Ghadames–Illizi and Murzuq) during the Eifelian (D40–D42). Up until this time the pattern of regressions and transgressions seen in the Ghadames–Illizi Basin can be matched in the Ahnet, Reggane and Tindouf basins, suggesting that sediment supply and subsidence were balanced across the margin. During the Eifelian (D40–D42) the West African Craton appears to have stopped supplying clastic detritus into the westerly basins and they become carbonate dominated. We interpret the regional change from clastics to carbonates as the result of tectonic loading associated with the onset of the Laurussia–Gondwana collision. The West African Craton is depressed and clastic sediment supply can no longer keep pace with subsidence. The resulting clear waters allow carbonates to thrive and aggrade (e.g. Wendt et al. 1997). Accelerating tectonic subsidence also led to the establishment of deepwater conditions in the Ahnet Basin by Frasnian times (Dixon 1997c). The emergence of the deforming Marrakech–Kabylia Zone to the north is indicated by the development of a marginal offshore carbonate belt in the northernmost Ahnet Basin and in the Tindouf Basin (Fig. 5). This northern uplift (and related regional uplifts within the basins, e.g. Mole D'Ahara) generated a number of silled, anoxic depocentres where the organically rich, Frasnian source facies accumulated (Lüning et al. 2004a, b).

Up to 800 m of Fammenian sediments (D75–D80) overlie the Frasnian in the southern part of the Ahnet Basin and clinoform heights from regional 2D seismic (Barr et al. 1994) confirm progradation into deepwater (300–400 m). The clinoforms record the northward progradation of a large but 'muddy' delta, the topsets of which are the so-called 'Strunian Sands' (D80–C0), the equivalent of the 'F2' or 'Tahara' sandstones of the Ghadames–Illizi Basin. In the Tindouf Basin the Late Devonian section is even thicker (up to 1.8 km thick; Guerrak 1989), suggesting that it was also subsiding rapidly. Latest Fammenian sandy deltas fringe its northern margin (Cavaroc et al. 1976; Vos 1977; Graham 1982; Brice et al. 2007). These deltas are derived from the actively uplifting Marrakech–Kabylia Zone and imply that continental collision was well advanced by the end of the Devonian (cf. McKerrow et al. 2000).

'Pre-Hercynian' burial. Estimates of Pre-Hercynian burial are always fraught with uncertainty, because significant Post-Hercynian erosion has removed large amounts of this stratigraphy. Recent regional work on the Carboniferous has helped to constrain erosion estimates for the Ghadames–Illizi Basin. A maximum Carboniferous thickness of 1.4 km seems reasonable based on well data and biostratigraphic data (which proves a virtually complete section). In the Ghadames–Illizi Basin the transition from the Devonian to the Carboniferous is essentially conformable from a regional perspective (there is a stratigraphic hiatus – the Early Tournaisian is apparently missing or extremely condensed; Jones 1997). The depositional facies on either side of this hiatus are similar, although there is a marked shallowing of water depth into the Carboniferous. In the Ahnet Basin further west, the Early Tournaisian is also missing and Middle Tournaisian fluvial sediments rest unconformably on 'Strunian' shoreface sequences (Dixon 1997a, b, c). Visean marine sediments (clastics and carbonates) are also present in the Bechar area further north, unconformably overlying Frasnian shales (Madi et al. 2000), but no younger Carboniferous sediments are preserved. Barr et al. (1994) estimate that up to 2.0 km of Carboniferous was originally present in the Ahnet Basin. The Tindouf Basin is similar to the Ahnet in that nothing younger than Visean is preserved. Boote et al. (1988) estimate that at least 2.5 km of Carboniferous was originally present. The large thicknesses of Mid–Late Devonian and Carboniferous sediments in the western basins (over 4 km in the Tindouf Basin) led to significant maturation of both Early Silurian and Late Devonian source rocks during the Carboniferous and it seems likely that they passed through the gas window at this time (Boote et al. 1988).

'Pre-Hercynian' structures. In some cases it is possible to demonstrate that the regional arches also had a Pre-Hercynian expression. This can be shown for the Mole D'Ahara in the

Ghadames–Illizi Basin where subsurface mapping (2D seismic and wells) has illustrated that the structure was uplifted several times, in the Late Silurian–Early Devonian (Base Devonian or 'Caledonian' Unconformity), in the Mid Devonian and in the Late Devonian (Base Frasnian Unconformity). Further east, Devonian outcrops on the flanks of the Tibesti–Sirt Arch reveal a similar phase of Late Silurian–Early Devonian uplift. The cross-bedded fluvial sandstones in Dor el Gussa (western flank of the arch) show palaeocurrents toward the west, whereas those at Jebel Eighi (eastern flank) show palaeocurrents toward the NE (Clark-Lowes & Ward 1991). In the Ahnet Basin similar trends are seen at the southern end of the Ougarta Foldbelt, where the Bled el Mass basement high was exposed during the Middle Devonian and actively shedding sediment westward into the basin (Dixon 1997b). The Devonian and Early Carboniferous sequences accumulated on a very broad (hundreds of kilometres) fairly shallow shelf with a gentle ramp into the deeper waters of the Tindouf Basin toward the NW. Sand supply was from a poorly vegetated land area south of the Hoggar Massif (Fig. 1). Although this is regionally the case, it is clear that in a number of areas subtle, but, large-scale, regional uplift events punctuate the stratigraphy (Fig. 2). Each uplift event probably led to the emergence of new land areas and the erosion of earlier shelfal sequences, thus providing new sediment source areas for the syn-uplift sequences. A good example of this is the Mid Devonian 'F3' sand that fringes the northern and eastern flank of the Mole D'Ahara (Chaouchi et al. 1998). A wide range of shelfal sandstone architectures is possible in this environment (sheets, bars, spits, incised valleys, tidal estuaries, etc.). All of these locally offer excellent stratigraphic trapping potential (e.g. the giant Alrar and Al Wafa Fields; Chaouchi et al. 1998).

'Hercynian' Orogeny. The climax of the Hercynian orogenic phase in North Africa took place in the Late Carboniferous, although it is clear that significant pulses of deformation had been propagating through African Gondwana since the Mid Devonian (onset of Laurussia–Gondwana collision), including a major Early Carboniferous Unconformity at the northern margin of the basin (Fig. 2; Dixon et al. 2009). The southern limit of Hercynian thrusting (determined by seismic mapping; Dixon et al. 2009) is shown in Figure 1. For most of the Lower Palaeozoic regional drainage is from south to north (Beuf et al. 1971). The first evidence of a change in regional drainage comes from the Tindouf Basin in the Late Devonian where southerly directed drainage suggests that the Marrakech–Kabylia zone was a land area at this time (Cavaroc et al. 1976; Vos 1977). Early Carboniferous regional uplift is marked at the northwestern margin of the Ghadames–Illizi Basin by a pulse of Visean clastic sediments that prograde from the NW towards the basin centre (IFP/Sonatrach/Beicip-Franlab 2006) and by northward-prograding, marine clastic systems in Tunisia, Cyrenaica and Crete. Later in the Carboniferous there is a widespread change from marine to non-marine sedimentation across African Gondwana (C75–P5 sequence). These non-marine conditions prevail in all of the internal basins (probably represented by the basal sequences of the 'Continental Intercalaire'; Lefranc & Guiraud 1990), whereas limited data from outcrops in Tunisia (Busson & Burrollet 1973) and Crete (Robertson 2007) and subsurface data in Cyrenaica (El-Arnauti & Shelmani 1988) confirm the continuation of marine conditions into the Early Permian north of the Hercynian Deformation Front.

Following the deposition of these Early Permian sediments there was a prolonged hiatus (c. 60 Ma), before deposition resumed in Gondwana with the accumulation of Late Permian carbonates in Tunisia, the Levant and on the Tauride Block (Demirtash 1984). A marked angular unconformity records this event and can be seen both in the field and on regional seismic data (Fig. 4). This unconformity is most often called the 'Hercynian' Unconformity but in fact it formed some 60 Ma later than the 'Hercynian' event and is probably more related to the opening of the Neotethys Ocean (Dixon et al. 2009). Palaeozoic sediments were removed over large areas of the North African margin at this time leaving the current basins as erosional remnants (Fig. 1). The regional 'Hercynian Unconformity' map (Fig. 3) nicely illustrates the 'saucer-like' geometry of the Ghadames–Illizi Basin.

In the Ghadames–Illizi Basin the 'Hercynian' Unconformity is progressively onlapped from north to south by Mesozoic sediments. In southern Tunisia Late Permian carbonates rest unconformably on Late Carboniferous–Early Permian carbonates (Wahlman 1991). In the Djebel Tebaga outcrops Middle to Late Permian carbonate facies are exposed including reefal build-ups (Archaeolithoporella–Tubiphytes boundstones). A NW–SE trending reef belt is developed (Toomey 1991) bounded to the north by a 'shale' basin and to the south by mixed carbonate and clastic shoreline fringing the northern flank of the emergent Talemzane Arch (Fig. 4). In the 'shale basin' over 4000 m of Late Permian 'shelf' facies were deposited (Wahlman 1991). The existence of very thick, Middle to Late Permian sequences to the north of Djebel Tebaga has led some authors to suggest that this may be evidence of rifting (Dridi 2000) and other authors (e.g. Stampfli et al. 2001) to suggest that Permian oceanic crust might be present in the Eastern Mediterranean. This is a controversial topic as evidence from the Eastern Mediterranean ophiolites suggests that the most likely age of/oldest oceanic crust is Jurassic or Early Cretaceous (Robertson & Dixon 1984). An alternative model is that the Tunisian, Late Permian basin represents a small transtensional basin underlain by attenuated continental crust.

Syn-rift 2 Megasequence (rifting of Mediterranean passive margin)

Evidence for Triassic rifting in the Eastern Mediterranean is strong (Druckman 1984; Garfunkel & Derin 1984). Similar strong evidence for Triassic rifting is found in Morocco (e.g. The Argana Basin; Beauchamp 1988) and its counterpart in North America (Benson & Doyle 1988). Analysis of regional seismic data from the Ghadames–Illizi Basin reveals little evidence of Triassic rifting, although it has been described locally from the northwestern part of the basin in the Berkine area (Turner et al. 2001). In this area NNE to SSW trending 'Pan-African' lineaments have been reactivated as extensional faults. These faults controlled the thickness and facies of Triassic sediments and probably also localized extrusive volcanic activity (fissure fed eruptions) that in turn led to the development of extensive fields of basaltic lava of Carnian age [up to 68 000 km^2 in the northwestern part of the basin (IFP 1998)]. Regional drainage at this time was toward the NNE (Acheche et al. 2001; Turner et al. 2001) into the developing 'Neotethyan' rift and it seems likely that most, if not all, of the sediment was derived from uplifted Lower Palaeozoic sandstones and basement in the Hoggar region to the south (Fig. 1). The northern margin of the Ghadames–Illizi Basin at this time was formed by the Medenine High (Benton et al. 2000) a subtle NW–SE-trending feature broadly coincident with the old Permian reef trend and located on the northern flank of the Talemzane Arch. Triassic fluvial sediments thin over this high and have locally been removed completely by later erosion (Aptian), although it seems likely that they were originally deposited across the high, as palaeocurrents at Djebel Rehach (on the immediate southwestern flank of the structure) are uniformly toward the north (suggesting that the high was not there at the time). A similar relationship can be observed on the eastern flank of the Ghadames–Illizi Basin at Djebel Gharian in Libya. Here palaeocurrents suggest flow from east to west, suggesting the existence of a high on the eastern margin of the basin (Assereto & Benelli 1971). Further north in

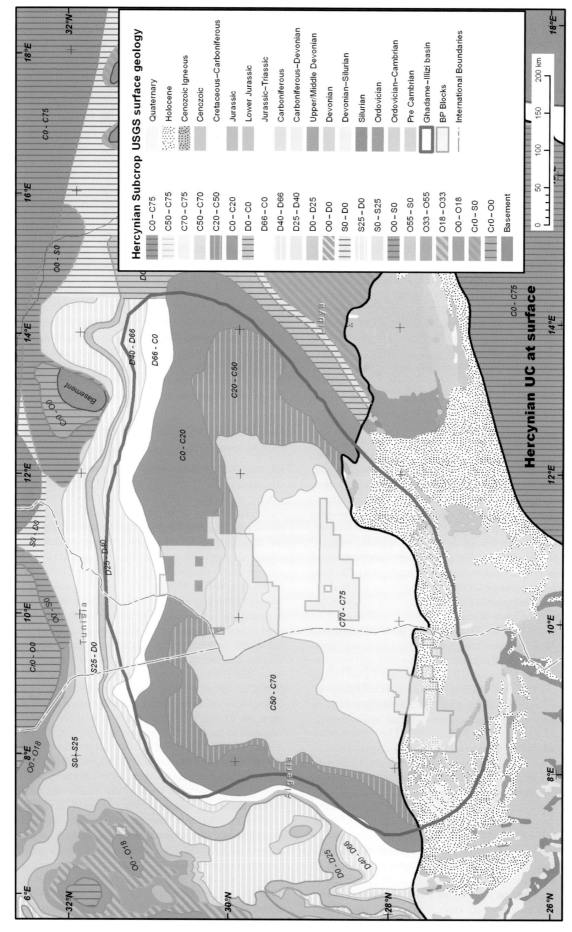

Fig. 3. Outline of the Ghadames–Illizi Basin (solid grey line) and subcrop map to the 'Hercynian' Unconformity (based on BP data).

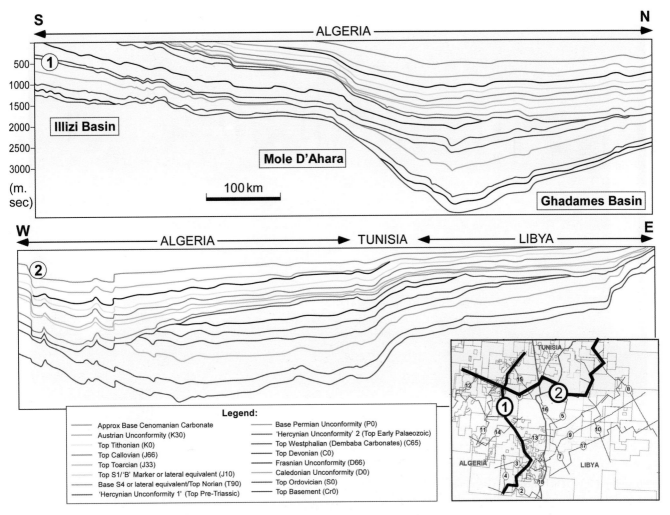

Fig. 4. Line drawings of two mega-regional seismic profiles that cross the Ghadames–Illizi Basin.

the Pelagian Basin and in Malta and Sicily, the Triassic is represented by a thick succession of platform carbonates. The Djebel Tebaga–Medenine High continued to be an important structure throughout the Jurassic, although it is unlikely ever to have been elevated above sea-level. The Jurassic sequences thin over it (Benton et al. 2000; Turner & Sherif 2007) and there are some important facies changes from south to north across it, but it seems likely that the Jurassic was only removed by later (Cretaceous–Aptian) erosion.

Post-rift 2 Megasequence (carbonates and evaporites)

Post-rift thermal subsidence after Early Triassic rifting created a broad area of Late Triassic and Jurassic evaporitic deposition in a basin that also extended northwestwards into the Moroccan Atlas (Turner & Sherif 2007). Up to 1250 m of evaporitic facies accumulated in the basin, typically with halite in the centre of the basin grading outward through anhydrites to interbedded dolomites and shales toward the edges of the basin (Turner & Sherif 2007). To the north of the Medenine High Mid Triassic evaporites are also present (Acheche et al. 2001) and are widespread across the Mediterranean, but the Jurassic is dominated by platform carbonates (Soussi & Ben Ismail 2000). The Djebel Tebaga–Medenine High seems to have acted as a 'sill' separating this marine area to the north from the evaporitic system to the south. Thinning and facies changes in wells are illustrated by Turner & Sherif (2007) and can also be observed on regional seismic lines (Fig. 4). The evaporitic sequence forms a very effective regional caprock to the underlying Triassic fluvial sandstones. These sandstones are locally juxtaposed against the Lower Palaeozoic source rocks (Early Silurian Tanzuft and the Frasnian shale) across the 'Hercynian' Unconformity (Figs 2 & 3), allowing them to be charged with petroleum. The unconformity therefore sets up a very prolific petroleum system containing the supergiant Hassi R'Mel gasfield and a number of giant oilfields in the Berkine area. As shown in Figure 4 (profile 2), the Mesozoic depocentre is offset from the Palaeozoic depocentre; this is also shown in Figure 5, a 3D representation of the main structural surfaces we mapped using our regional 2D seismic database.

Syn-rift 3 Megasequence (formation of the Ionian Ocean)

The Ghadames–Illizi Basin was on the periphery of the Early Cretaceous rifting episode so evident in the Sirt Basin (Gras & Thusu 1998), Gabes Basin (Ben Ferjani et al. 1990) and in other similar basins in Central Africa (e.g. Genik 1993) and the Sudan (e.g. Mohamed et al. 2001). No evidence of Early Cretaceous rifting has been observed on regional seismic lines from the Ghadames–Illizi Basin (Fig. 4), but Figure 2 clearly shows that the Early Cretaceous was a time of pronounced clastic input into the basin, suggesting that clastic source areas to the south of the basin (the Ougarta Chain and Hoggar Massif) were being rejuvenated at this time (Lefranc & Guiraud 1990; Benton et al. 2000). The gently tilted, Palaeozoic sandstones that form the great escarpments of the Tassili N'Ager on the northern flank of the Hoggar Massif were clearly elevated above sea-level at this time as they are onlapped by Early Cretaceous fluvio-lacustrine sediments in the Serouenout area of Hoggar in Algeria (Lefranc & Guiraud

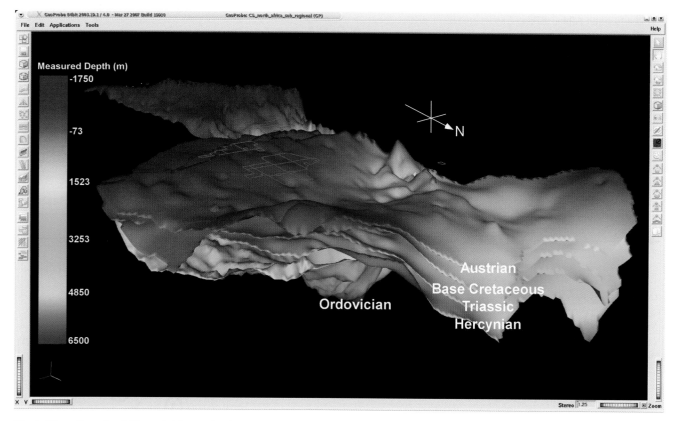

Fig. 5. Three-dimensional 'Geoprobe' image showing the offset Palaeozoic and Mesozoic depocentres and the 'Hercynian' Unconformity (Late Permian Unconformity of Dixon *et al.* 2009).

1990) and in Libya [Djebel Nafusa (Hammuda 1971) and the Sirt Basin (Schroter 1996)]. In the Ghadames–Illizi Basin the Early– Mid Cretaceous sandstones are dominated by fluvial facies (Lefranc & Guiraud 1990) and contain a rich reptilian fauna [including dinosaurs (Benton *et al.* 2000)] together with abundant flora [including ferns and very large tree trunks and stumps (Benton *et al.* 2000; Bamford *et al.* 2002)] suggestive of a stable, humid, non-seasonal, tropical environment (Lefranc & Guiraud 1990).

Slight thickening of the Early Cretaceous into the Ghadames– Illizi Basin suggests that the basin was still subsiding at this time, but the greatest variations in Early Cretaceous thickness are north of the Djebel Tebega–Medenine High, where up to 2000 m of Early Cretaceous deltaic facies are seen (Bishop 1975), and within the grabens of the Sirt Basin, where over 500 m of clastic sediment is preserved (El-Hawat *et al.* 1996; Gras & Thusu 1998). The rift shoulder of the Early Cretaceous rift was located along the northern and eastern margins of the Ghadames–Illizi Basin. The 'Neocomian' through Barremian time interval is dominated by coarse-grained, clastic sediments, but the Aptian is marked by a dramatic return to carbonate facies (in Algeria, Tunisia and Egypt), signifying a major marine transgression southward into North Africa (reaching as far south as the Sudan; Klitzch 1990). The limestone unit forms a good regional seismic horizon (Fig. 4) and is also an excellent reference point in wireline logs from the region. Detailed sedimentological and biostratigraphic work on outcrops of this stratigraphic sequence in Central Tunisia has revealed that an important stratigraphic hiatus, the 'Austrian' Unconformity, also occurs within the Aptian (Chaabani & Razgallah 2006).

The unconformity can be mapped on regional seismic and is locally a strong angular unconformity (Figs 4 & 5). Uplift, tilting and erosion are noticeable on regional seismic profiles from the southern and eastern flanks of the basin and are locally significant (Fig. 4). In the field the 'Austrian' Unconformity is locally spectacular, for example at Djebel Tebaga in southern Tunisia (Fig. 5), where gently dipping Albian sandy limestones rest with marked angular unconformity on more steeply dipping (*c.* 20°) Late Permian carbonates. Similar relationships are also observed further to the east in Jebel Gharian (Assereto & Benelli 1971) and much further to the east in the subsurface of the Western Desert of Egypt, where the karstified Early Aptian, 'Alamein' Dolomite is overlain by the Late Aptian, Dahab Shale. The origin of the 'Austrian' Unconformity in North Africa is uncertain, but it is clearly a major 'plate-scale' event generally characterized by uplift, tilting and erosion at a regional scale, that is, over major regional arches with structural wave lengths of hundreds of kilometres (see Fig. 4).The timing of this event is roughly synchronous with the maturation of and migration from the Palaeozoic source rocks in the Ghadames–Illizi Basin.

Post-rift 3 Megasequence (Neotethys and the Trans-Saharan Seaway)

To the north of the Cretaceous rift shoulder, thick sections of finegrained limestone and chalk interbedded with shales accumulated at this time. Rich petroleum source rocks of Cenomanian– Turonian and Campanian age are commonly present (Lüning *et al.* 2004*a*). Further south, the Ghadames–Illizi Basin was part of the Trans-Saharan seaway with marine connections to the South Atlantic around both sides of the Hoggar Massif. By latest Campanian the eastern seaway was apparently disconnected and only the western seaway was active, persisting until the Late Paleocene (Zaborski & Morris 1999), when the Trans-Saharan Seaway finally disappeared. The rift shoulder formed an effective 'sill' between Neotethys and the Ghadames–Illizi Basin and the Late Cretaceous sequences in the northern part of the basin are commonly evaporitic (Bracene *et al.* 2003). At the eastern end of

Djebel Nafusa the Albian Kiklah Sandstone onlaps the 'Austrian' Unconformity and the overlying Cenomanian–Turonian, Ain Tobi carbonates thin dramatically (Hammuda 1971), suggesting that this area was topographically high during the Late Cretaceous. The early phases of compression between North Africa and Europe are recorded by the Syrian Arc events (Fig. 2). Although classically expressed in Egypt and the Levant, similar aged structures may also be observed at outcrops in Libya [Jebel Akhdar (Bosworth *et al.* 2008)] and further west in Central Tunisia (e.g. the Kesra Plateau; Zaier *et al.* 1998). Typically the 'Syrian Arc' events inverted pre-existing faults. Syrian Arc 1 is a Late Cretaceous event (Santonian) whereas Syrian Arc 2 is an Eocene event.

Foreland Basin 2 Megasequence (Alpine Foredeep)

We have taken the Syrian Arc 2 event (Fig. 2) as the base of our Foreland Basin 2 Megasequence (Alpine Foredeep), although in truth this is probably a great oversimplification and the actual transition is probably more gradational and complex. To the north of the old Cretaceous rift shoulder the Cenozoic section is thick (over 3000 m) and predominantly carbonate (Swezey 2008), but to the south the Cenozoic section is thin (*c.* 300 m maximum, Swezey 2008). Analysis of well data and regional seismic profiles confirms that the Palaeogene is relatively thin (typically less than 100 m) in the Ghadames–Illizi Basin and that the Paleocene is mostly absent. The Eocene sequence is thin (rarely more than 50 m) and predominantly evaporitic, passing northwards into fully marine carbonates. It seems likely that during the Early Cenozoic the basin had a very similar geometry to that seen in the present day, but with a limited, 'silled' marine connection to the north somewhere between the Tunisian Dorsale range and the 'Djebel Tebaga–Djebel Nafusa Uplift'. In the latter area the Cenozoic was probably not deposited.

To the south of the 'sill' evaporitic facies were extensively developed whereas to the north the fully marine carbonates described by Zaier *et al.* (1998) were flourishing. The Eocene sediments are overlain unconformably by coarse-grained sandstones that are usually ascribed to the Mio-Pliocene on most oil company composite logs, although this section is very poorly dated. The main 'Alpine' Unconformity is probably of 'Miocene' age (Fig. 2) and is locally a strong angular unconformity on regional seismic profiles with locally, well defined 'channels' or canyons. The Late Cretaceous and Cenozoic sequences are also locally strongly folded (Fig. 4) and sometimes these folds may be mapped at the surface. The structural wavelength of these late folds is typically short (Fig. 4) and suggestive of reactivation of basement faults in the final phases of Alpine compression.

Petroleum source rocks

The Early Silurian and Late Devonian source rocks are very similar from a geochemical point of view, but were deposited in slightly different settings. The Early Silurian, Tanzuft source was deposited in a passive margin setting (Passive Margin 1 Megasequence) and associated with a major transgression caused by the demise of the Late Ordovician Ice Cap. The Late Devonian, Frasnian source was deposited in a Foreland Basin setting associated with widespread regional inversion (Foreland Basin 1 Megasequence). The salient points relating to the distribution of these two world-class source rocks are discussed below. Figure 2 shows the stratigraphic position of the basin's two major source rock horizons.

Early Silurian: 'Tanzuft'

In general terms the major regional control on the distribution of the Tanzuft source rock is the Saharan Metacraton. Throughout the Palaeozoic this cratonic block has been a major supplier of sandy sediment and this is particularly marked during the Silurian, Devonian and Early Carboniferous. Well data from Cyrenaica and the Western Desert of Egypt show that the Early Silurian shales are organically lean and outcrops around Djebel Uweinat show that the same stratigraphic section is dominated by sandstones (Fig. 7). To the west the source rock is generally present, although there are areas where it is not (as shown in Fig. 7). Established

Fig. 6. 'Austrian' Unconformity at Djebel Tebaga (Tunisia). Gently dipping, Albian sandy limestones overlie more steeply dipping (to the south) Late Permian carbonates. Cenomanian–Turonian carbonates cap the high escarpment in the background. Halk el Menzel locality, view to west.

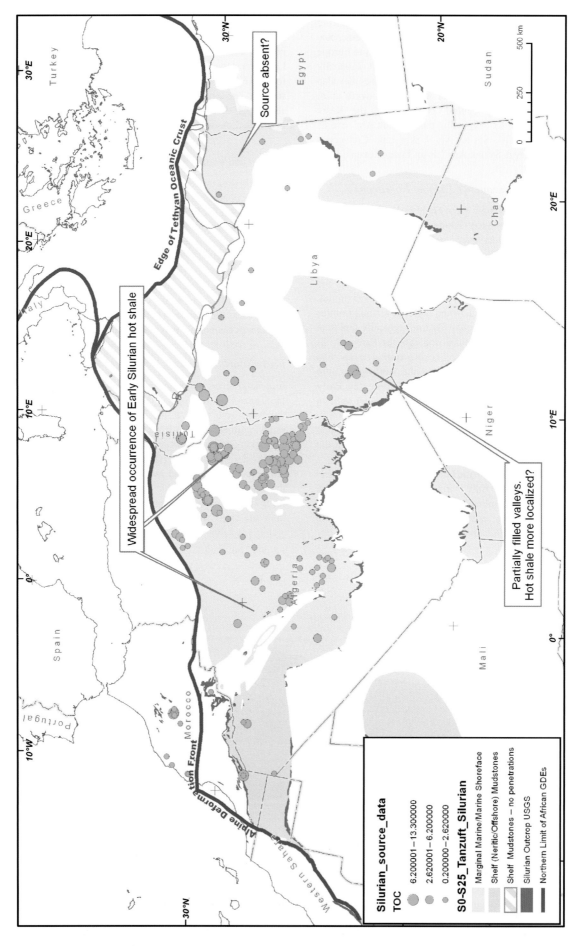

Fig. 7. Early Silurian, Tanzuft Source Rock gross depositional environment (GDE) map (green circles represent calibration points; the bigger the circles the higher the TOC content).

models for the distribution of the Tanzuft source facies (Lüning et al. 2000) invoke remnant glacial topography to provide 'silled', anoxic basins.

Late Devonian: 'Frasnian'

The distribution of the Frasnian source rock is principally controlled by the overall geometry of the evolving Hercynian Foredeep and the locally developed inversion structures that developed along its southern and eastern margins (Fig. 8). In general terms the Frasnian source facies is best developed in the basin centres and is thin to absent on the regional arches. In Cyrenaica shallow marine sandstones fringing the emergent Tibesti–Sirte Arch dilute the source facies (El-Arnauti & Shelmani 1988). In the Western Desert of Egypt the Frasnian is dominated by sandstones derived from the Saharan Metacraton (Paleoservices Report 1986). The areas covered by our gross depositional environment (GDE) maps are essentially restricted to African Gondwana. We have not attempted to extend our GDE maps into the highly deformed Marrakech–Kabylia zone, although we have extrapolated our facies belts into the deeply buried offshore zone (Fig. 1).

Reservoir–seal combinations

Figure 2 shows the main reservoir–seal combinations in the Ghadames–Illizi Basin. In our regional play fairway and 3D basin modelling we have considered 13 reservoir–seal combinations. With the exception of the glaciogenic Mamuniyat sequence, all of the other potential reservoir sequences are of a paralic character, that is, they were deposited at or close to sealevel. A wide variety of facies are represented in this category, such as fluvial, tidally influenced fluvial, shoreface. The Palaeozoic outcrops of the Tassili N'Ager and Djebel Akakus on the southern flank of the Ghadames–Illizi Basin convincingly demonstrate that a great many of these paralic sandstone units are sheet-like, regionally extensive bodies (Beuf et al. 1971) and this is confirmed in the subsurface by regional seismic mapping (Fig. 4) and evaluation/correlation of well sections.

Palaeocurrents collected from the Tassili N'Ager outcrops (Cambrian–Devonian) record gross sediment transport from south to north (Beuf et al. 1971). A similar sense of sediment transport is recorded by Dardour et al. (2004) from large-scale clinoform geometries seen the Silurian Akakus sequence. In very broad terms the Silurian and Devonian of North Africa record a second-order regressive–transgressive cycle (Fig. 8). The Lower Devonian, 'Tadrart' reservoir sequence represents the maximum northward regressive extent of paralic sandstones across the platform (Fig. 8) and the overlying Frasnian, Awaynat Wanin C shales the subsequent maximum transgression southwards across the craton. For each reservoir sequence and each seal sequence we have compiled GDE maps, examples of which are shown below (Figs 8 & 9). Each reservoir GDE shows the maximum regressive extent of the reservoir and the reservoir facies. Sand–shale ratio from wireline log data is also posted on the maps to highlight regional trends (e.g. increasing shale content northwards in Fig. 11). Each seal GDE shows the maximum transgressive extent of the seal and seal facies. Sand–shale ratios are also added to these maps to show regional trends (e.g. increasing sand content southwards in the Middle Akakus Shale; Fig. 11).

Integrated basin modelling approach and workflow

The philosophy of this work prioritizes integration across geological disciplines. A particular emphasis was placed on creating burial and uplift models that are consistent with the structural and tectonic interpretation for the basin(s) in question; it is all too easy for geochemists to settle on a model that seemingly calibrates their data with limited regard to its geological veracity. Figure 12 illustrates the multi-disciplinary approach that we have applied to Ghadames–Illizi Basin modelling. Team members collaborated and integrated their work from the start, considering palaeoclimatic change, plate reconstruction, GDE/facies mapping (including well correlation, outcrop data and field studies), structural restoration, seismic mapping (including subcrop mapping) and crustal modelling/thermal history. We used this integrated approach to utilize increasingly complex basin modelling approaches as our understanding matured, for example, 1D 'Genesis' → 2D map based 'Trinity' → 3D 'Temis' full physics.

Missing section map and erosion maps

Velocity analysis when applied to uplift requires thick shale sections, reliable logs and appropriate depth–velocity relationships. Departures from a virgin-compaction velocity–depth relationship can be used to quantify the degree of uplift (within a range of error). As such the thick, argillaceous sections recorded within the Devonian and Silurian sections of the Ghadames and Illizi basins are good candidates for this type of investigation. An example of the velocity analysis is shown in Figures 13 and 14 for the Devonian argillaceous and Silurian argillaceous shales. Organic-rich 'hot' shales were not included due to the complexity that high organic contents cause in simple stress–velocity relationships. The depth–velocity relationships used here have been derived from an extension of the work by Yang & Aplin (1999) on the practical definition of practical mudstone porosity–effective stress relationships.

The quantitative estimates of uplift obtained from shale velocity and analysis and calibration to geochemical data such as vitrinite reflectance provide insight into maximum burial depth and maximum palaeotemperature reached. However, they do not necessarily provide insight into the timing of these events, particularly in areas of long and complex geological history with multiple candidate unconformities. In order to address this issue we both tested multiple models and incorporated insights and interpretations from the structural geologists and sedimentologists. Focusing upon the Illizi Basin it was clear that significant uplift had occurred from the velocity analysis work (Figs 13 & 14), but a variety of models could account for the timing of this uplift. Three models with differing burial and uplift histories were tested and are illustrated in Figure 15. Whilst all three models can calibrate the available geochemical data and uplift estimates, not all were consistent with the observations made from the regional geology.

Observations made around the Hoggar area were particularly important. Cretaceous strata are seen to lie directly upon basement in the Amguid area and near Serouenout (Lefranc & Guiraud 1990). Reports of shoreline features suggest that this area was at or at least close to sea-level during the Early Cretaceous (Lefranc & Guiraud 1990). Given that seismic interpretation shows no evidence of thinning as the Palaeozoic approaches the surface outcrops, we conclude that over 4 km of Palaeozoic sediment which originally lay above the basement have been eroded prior to deposition of the Cretaceous sediments. The Cretaceous sediments contain vertebrate and plant fossils that indicate that they lay close to sea-level at the time of deposition (Lefranc & Guiraud 1990) but are now at 1000–1700 m a.s.l. and have been gently tilted to the north. Clearly the uplift of Hoggar was pulsed and consisted of pre- and post-Cretaceous events. In our preferred model the pre-Cretaceous uplift involved up to 4 km of exhumation (*sensu* England & Molnar 1990) whereas the post-Cretaceous comprised rock uplift (*sensu* England & Molnar 1990). The evidence that led to us to prefer this model is outlined below.

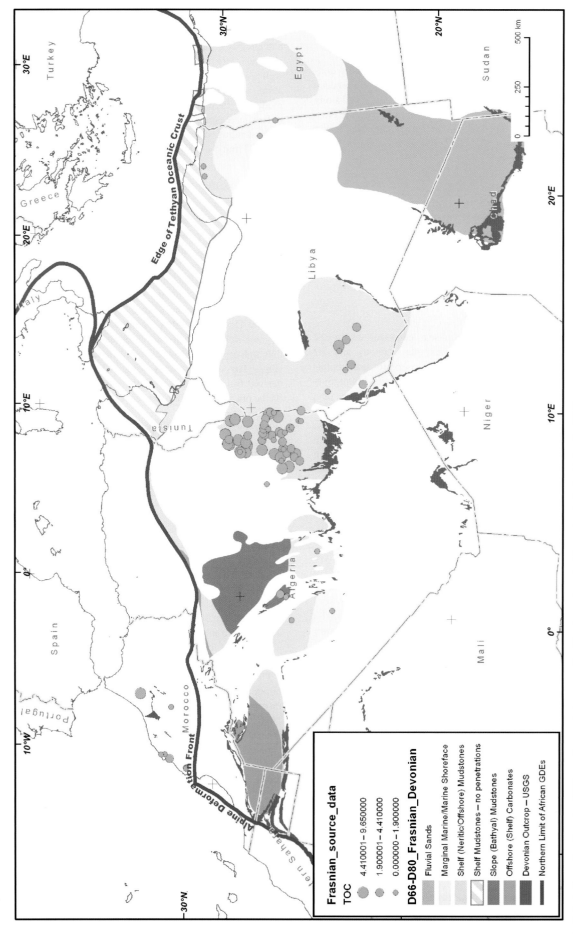

Fig. 8. Late Devonian, Frasnian GDE map (green circles are well calibration points; the bigger the circle the higher the TOC).

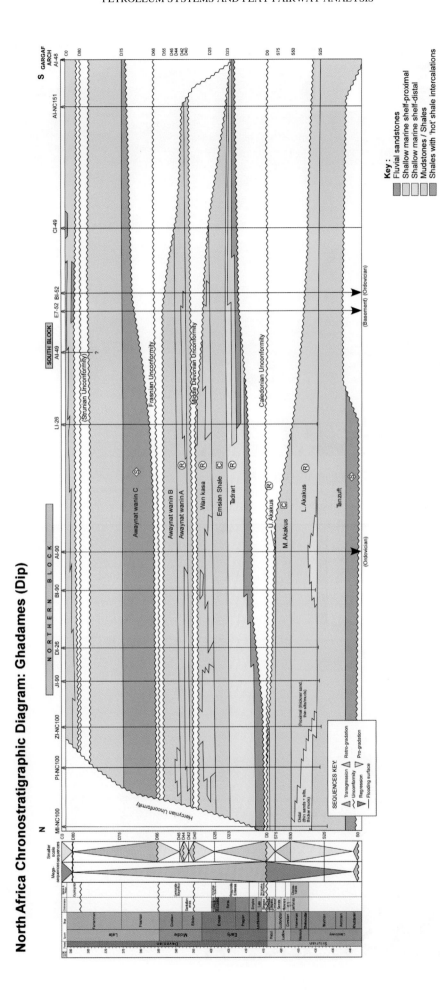

Fig. 9. North–south chronostratigraphic diagram through the Silurian and Devonian of the Ghadames–Illizi Basin.

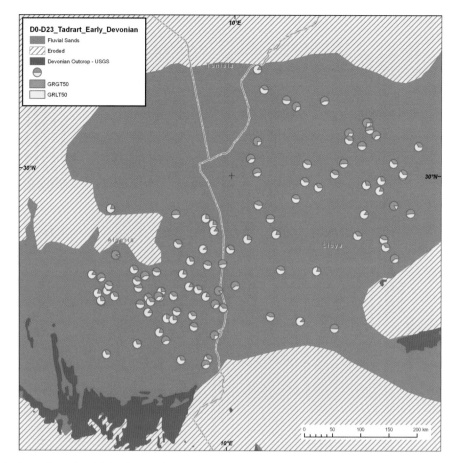

Fig. 10. GDE map for the Lower Devonian, 'Tadrart' reservoir sequence.

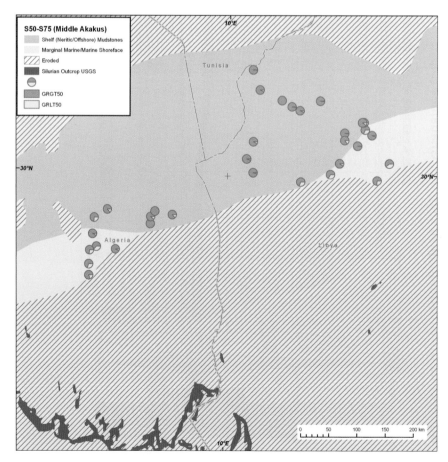

Fig. 11. GDE map for the Silurian, Middle Akakus Shale sequence.

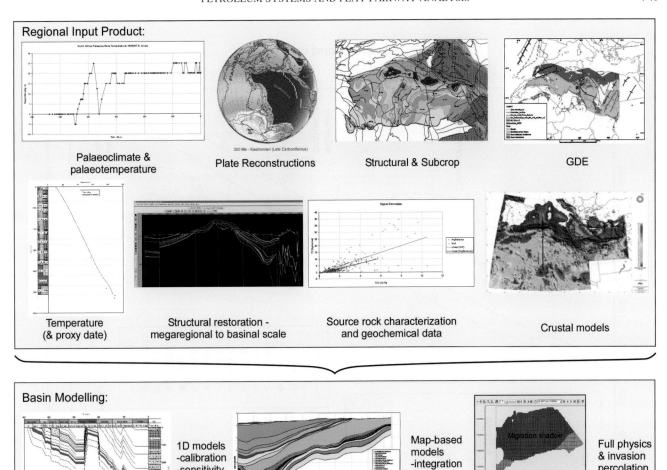

Fig. 12. Integrated basin modelling approach and workflow.

The simplest interpretation of model A would suggest that, during the Cretaceous, Hoggar (to the South of Illizi) would have been buried by a significant thickness of Palaeozoic stratigraphy, which is not consistent with the observations of Lefranc & Guiraud (1990). Similarly the observation of Early Cretaceous sediments sitting directly upon basement is also not consistent with model B. The model that most closely matches structural restorations at the mega-regional and regional scale and the apatite fission-track data is model C. This model invokes removal of most of the Palaeozoic section during the Hercynian, consistent with observation from Hoggar and the Tassili escarpment being in place at this time (cf. Lefranc & Guiraud 1990). Permian apatite fission track cooling ages (Carpena et al. 1988) from basement rocks suggest initiation of uplift post-dating the Hercynian unconformity. Outcrop observations from the In Amenas area suggest that Mesozoic sediments onlap onto the Hercynian erosion surface, potentially suggesting a gentle tilting of this surface (from south to north). There is no tectonic evidence for significant Cenozoic accommodation space being generated in the Illizi Basin; however, the observations of Cretaceous freshwater fauna around Hoggar relevant to its current elevation suggest that significant rock uplift (sensu England & Molnar 1990; whereby rock is displaced with respect to the geoid) has occurred. Assuming that much of the Cretaceous was deposited at or close to sea-level, back-interpolating the topography onto the base Cretaceous contacts and then contouring, allows estimation of the amount of post-Cretaceous uplift on the Hoggar (Fig. 15). In doing so this suggests a post-Mesozoic rock uplift of 300–700 m across the Illizi Basin.

Our model has largely viewed the basin along an effective dip-line orientated north–south from the Hoggar area to the Berkine fields of the central Ghadames Basin. This view contrasts with the approach taken by Underdown et al. (Underdown & Redfern 2007, 2008; Underdown et al. 2007), who described the basin and its evolution from a strike-line. A variety of techniques were used to estimate both gross missing section and to partition this to three main erosion events: 'Hercynian', 'Austrian' and 'Alpine' (Fig. 16). These include structural restoration, shale velocity analysis, fluid inclusions, temperature data and source maturity data. The maps at the bottom show the amount of missing section we calculated at each unconformity. Blue shows the greatest amount of erosion, whereas red indicates no erosion. From our analysis we predict that the major loss of section was during the 'Hercynian' when up to $c.$ 2000 m was eroded. Up to $c.$ 600 m was lost during the 'Austrian' and a further $c.$ 200 m at the 'Alpine' event (Fig. 17).

The recognition of the 'Austrian Unconformity' and the potentially significant erosion at this time is the major difference between this study and previous Ghadames–Illizi Basin models (e.g. Underdown & Redfern 2008 and references therein). Underdown & Redfern (2008) partition the uplift at the southern and eastern margins of the basin between the 'Hercynian' and 'Alpine'; we achieve it by having a similar loss of section between the 'Hercynian' and 'Austrian' so that maximum burial

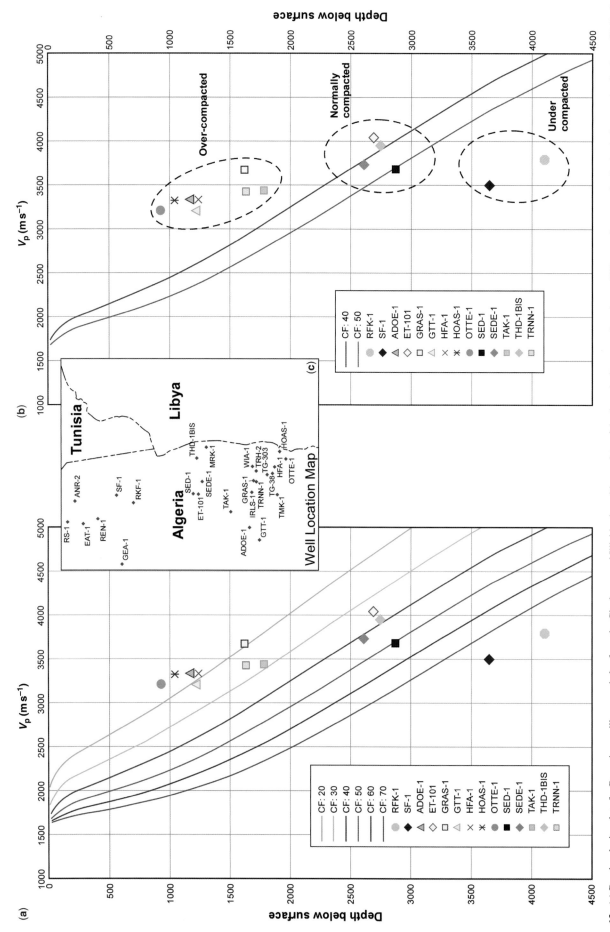

Fig. 13. (a) Depth velocity plots for Devonian argillaceous shales from the Ghadames and Illizi basins with a variety of depth–velocity curves based upon BP's proprietary mudrock physical properties database. (b) Preferred virgin compaction depth–velocity relationships and compaction state interpretation of the 14 wells analysed. (c) Location map.

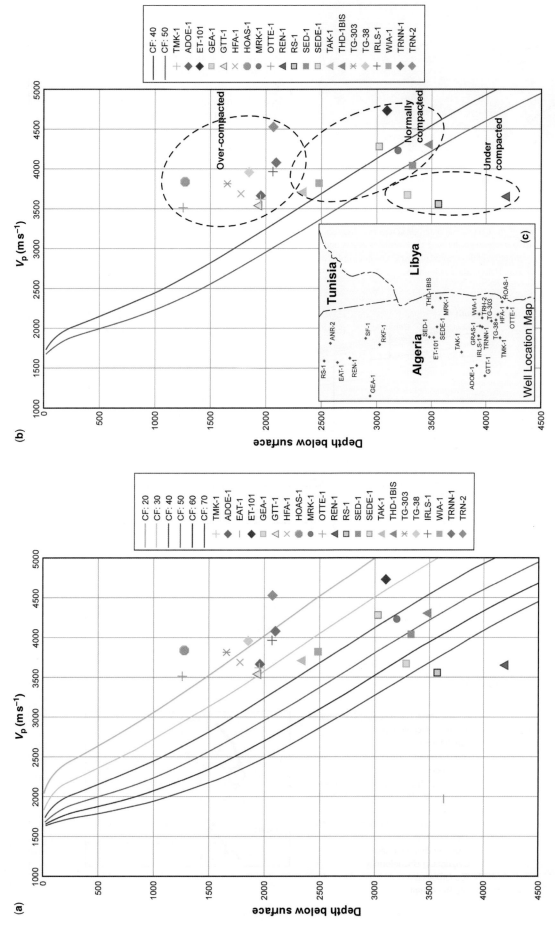

Fig. 14. (a) Depth velocity plots for Silurian argillaceous shales from the Ghadames and Illizi basins with a variety of depth–velocity curves based upon BP's proprietary mudrock physical properties database. (b) Preferred virgin compaction depth–velocity relationships and compaction state interpretation of the 22 wells analysed. (c) Location map.

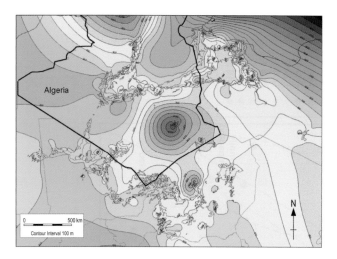

Fig. 15. Map of the base of the Cretaceous outcrops across the Hoggar; back-interpolating the topography onto the base Cretaceous contacts and then contouring allows estimation of the amount of post Cretaceous uplift on the Hoggar (courtesy of Tom Patton, BP Exploration).

is achieved in the Cretaceous rather than Miocene as in the Underdown & Redfern (2008) model in the east, but agree that maximum burial is reached in the Cenozoic in the central and western portions of the basin. Field observations (e.g. Fig. 6), well data and regional 2D seismic profiles (e.g. Fig. 4) clearly demonstrate the importance of the 'Austrian Unconformity' and also the generally thin nature of the Cenozoic section in the Ghadames–Illizi Basin. There is no geological evidence for the deposition of a thick (c. 1700 m) Cenozoic section at the eastern margin of the Ghadames–Illizi Basin and its subsequent removal during the 'Alpine' phase, rather the evidence is that this margin of the basin has been a high since the Cretaceous with little or no Cenozoic deposition (Hammuda 1971; Drake et al. 2008).

Hydrocarbon distribution

There are significant (32 BBOE) discovered resources of oil and gas in the Ghadames–Illizi Basin with the phase split of oil to gas and condensate of c. 50:50. The bulk of these resources are in Algeria in comparison to Libya and Tunisia (Fig. 18). In

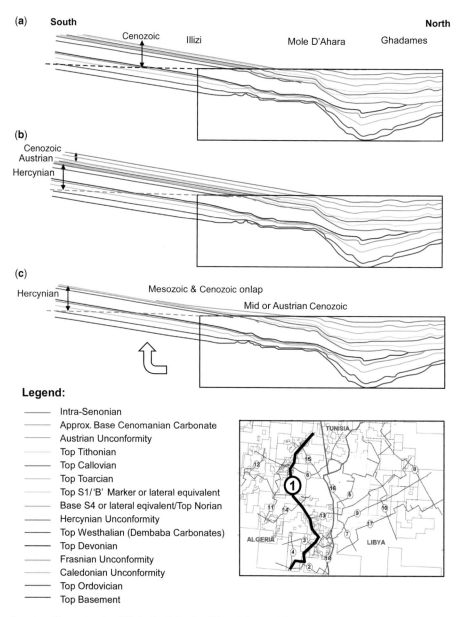

Fig. 16. Cartoon regional sections illustrating the differing burial and uplift models tested for the Illizi Basin. (**a**) Cenozoic to Hercynian section removed by the Alpine unconformity; (**b**) Cenozoic to Hercynian section removed incrementally by the Hercynian, Austrian and Alpine unconformities; (**c**) Palaeozoic section removed by the Hercynian unconformity; onlap of Mesozoic section onto this surface and erosion at the Austrian unconformity and minor Cenozoic deposition followed by uplift of rock (*sensu* England & Molnar 1990).

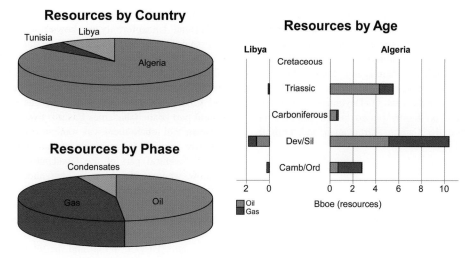

Fig. 17. Ghadames–Illizi Basin petroleum resources by country, phase and age (data from IHS and BP).

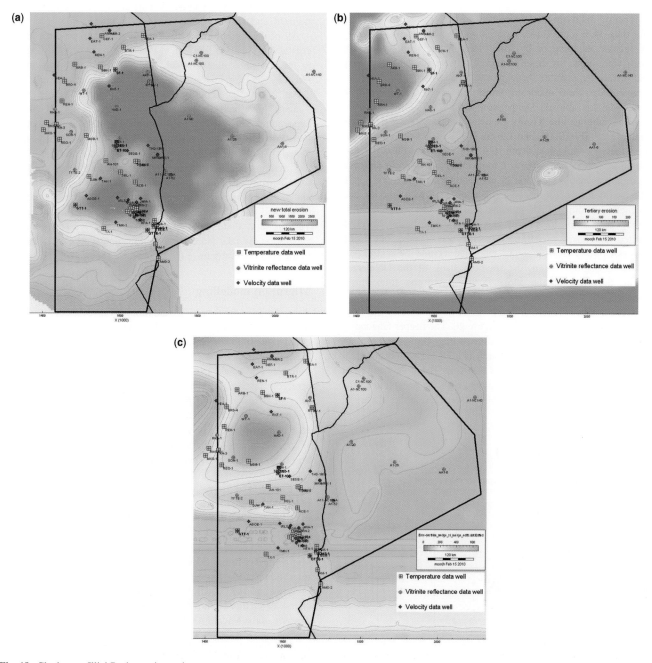

Fig. 18. Ghadames–Illizi Basin erosion estimates.

Algeria, these hydrocarbons are predominantly reservoired in the Cambro–Ordovician, Silurian–Devonian and Triassic clastic reservoirs with only moderate resources in the Carboniferous. Gas is the dominant phase in the Cambro–Ordovician, oil is the dominant phase in the Triassic and there is an c. 50:50 split of oil and gas in the Silurian–Devonian reservoirs (Fig. 18). In Libya the bulk of the resources are in the Silurian–Devonian clastic reservoirs and the phase split is slightly biased to oil. Only minor resources are present in the Cambro–Ordovician and Triassic clastic reservoirs (Fig. 18).

Source rock presence

All of the available source screening data were plotted on the GDE maps for the Frasnian (Fig. 8) and the Tanzuft 'hot' shales (Fig. 7) intervals. These data indicate that both these intervals are excellent oil-prone source rock intervals. These source intervals have TOC values up to 20 wt% and P2 values up to 100 kg/tonne.

Source access

A 'Trinity' 3D basin model was built using 35 structural surfaces and thus 34 layers. These surfaces were either derived directly by seismic mapping or by using seismic and well data. GDE maps were input for these 34 layers as were palaeo-bathymetry and surface temperature through time. In addition to the 35 structural surfaces, three major unconformities are also represented in the model (as above). Temperature data (DST and PLT data) used to calibrate the Trinity model came from three principle sources (BP, IFP (2003) and Libyan NOC). When these temperatures are plotted against depth (Fig. 19), it can be observed that the Ghadames–Illizi Basin is a moderately warm basin and that there is reasonable variation in temperature at any given depth. The warm nature of the basin is consistent with gravity and magnetics studies which indicate that the continental crust is fairly thick. Prior to running the 'Temis 3D' basin model, a limited 1D basin modelling exercise was carried out using 'Genesis'. The four 'shotpoints' modelled include the central part of the Ghadames Basin, the southern flank of the Ghadames Basin, the crest of the Mole D'Ahara and the Illizi Basin depocentre.

Representative Genesis burial plots for the Ghadames and Illizi Basin depocentres are shown in Figure 20. The deeper Ghadames Basin was separated from the shallower Illizi Basin by the long-lived Mole D'Ahara structural arch (Fig. 4). There is minimal loss of section for the Ghadames and Illizi Basin depocentre at each of the three main unconformities, with the exception of the Illizi Basin depocentre, where c. 500 m were lost at the 'Austrian' Unconformity. The timings of oil and gas expulsion relative to the 'Hercynian' Unconformity are shown for the Tanzuft and Frasnian 'hot' shales at the four modelled shot point locations in Figure 21. For the Tanzuft 'hot' shale most of the oil expulsion took place prior to the 'Hercynian' Unconformity with the exception of the crest of the Mole D'Ahara. In the present day the Mole D'Ahara and Illizi depocentres are within the gas window, whereas the central part of the Ghadames Basin is overmature for gas. For the Frasnian 'hot' shale there was some oil expulsion Pre-Hercynian, but only from the deepest part of the Ghadames Basin. Present day, the central part of the Ghadames Basin is into the gas window whereas the Mole D'hara and Illizi depocentres are within the oil window (Fig. 21).

Kitchen maps were then produced within 'Trinity' but these were then superseded by the kitchen maps produced within 'Temis 3D'. Three thermal models were run initially which varied the upper crustal ratio (UC/UC + LC) and the total basement thickness to ensure a good calibration for all wells with DST temperature data. The three models were a cooler model (crust ratio 0.4 and total basement thickness 110 km), an initial model (crust ratio 0.5 and total basement thickness 90 km) and a warmer model (crust ratio 0.6 and total basement thickness 70 km).

All of the calibration wells were then used to produce gridded crustal ratio and total basement thickness maps which were used to run the most likely model. Present day transformation ratio kitchen maps for this most likely model were produced for the Tanzuft 'hot' shale and the Frasnian 'hot' shale. Present day the Tanzuft 'hot' shale is overmature (even for gas) in the Ghadames depocentre. This kitchen map would indicate that gas would be the most likely phase in the central parts of the basin and oil would be the most likely phase in the basin margins (Fig. 22). In the present day the Frasnian 'hot' shale is in the gas window in the Ghadames depocentre where the source rock is at its richest and thickest. There was also significant 'Pre-Hercynian' generation of oil and gas throughout the Ghadames–Illizi Basin from the Tanzuft 'hot' shale. Prior to the 'Hercynian', oil and minor gas generation did take place from the Frasnian source rock, but only in the very deepest part of the Ghadames Basin.

Migration modelling

Representative cross-sections and lithology distributions within the 'Temis 3D' model are illustrated in Figure 23. The cross-sections clearly resemble the seismic lines (Fig. 4) and also illustrate the multiple reservoir–seal combinations shown in Figure 2. The Tanzuft 'hot' shale generally overlies the Cambro–Ordovician reservoir interval. Downward migration of petroleum into this interval is low risk and the phase is predominantly gas towards the basin centre and oil on the basin flanks (Fig. 22). Petroleum also migrates vertically out of the Tanzuft and into the Akakus (Fig. 24). In the northern part of the basin most of this charge will be trapped in the Lower Akakus reservoir because of the overlying Middle Akakus Shale (Fig. 11). The Lower Akakus contains a number of intra-formational seals (Dardour et al. 2004), giving potential 'stacked pay' with varying column heights. In the south the Middle Akakus Shale is absent, because it has been removed by 'Caledonian' erosion (Fig. 23). Petroleum therefore migrates through the Lower Akakus into the Lower Devonian Tadrart reservoir (Fig. 10). Initial 'Mpath' modelling predicts the presence of a migration shadow for the Tadrart reservoir in the north due to the presence of the Middle Akakus seal beneath it and low-risk access of charge to the Tadrart reservoir in the south due to the absence of the Middle Akakus seal (Fig. 25). These modelling conclusions are consistent with the field observations, which indicate that the Akakus fields are in the north and the Tadrart fields are in the southern part of the basin. The Frasnian charge will

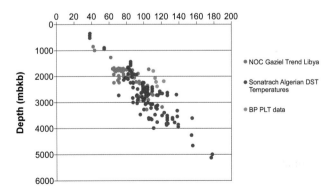

Fig. 19. Ghadames–Illizi Basin temperatures v. depth.

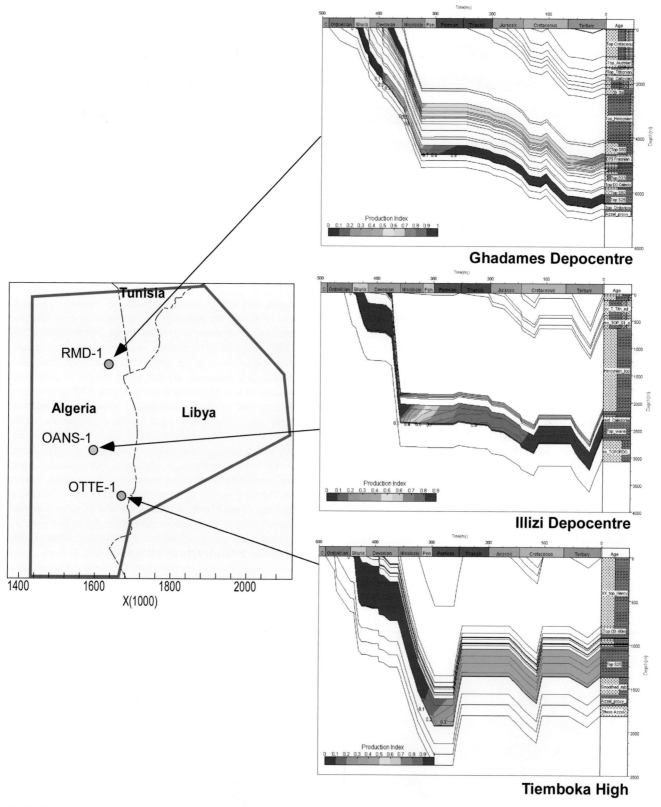

Fig. 20. Representative 'Genesis' burial plots for the Ghadames and Illizi depocentres.

migrate primarily into the Late Devonian 'F2' sand. Lateral migration will carry the charge to the 'Hercynian' subcrop where the charge will then migrate updip within the Triassic reservoir by fill and spill. Thus charge risk is low for the Triassic reservoirs close to the 'Hercynian' subcrop to the NW of the basin and this is where all the giant Triassic fields have been found. The charge risk will increase away from the Hercynian subcrop for these Triassic reservoirs as longer fill and spill chains are required. In the basin centre it is unclear what the main migration mechanism is.

Summary

The Ghadames–Illizi Basin has a long (Cambrian to Plio-Pleistocene) and complex geological evolution characterized by

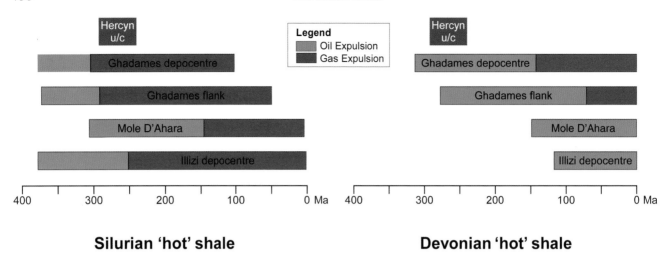

Fig. 21. Timing of oil and gas expulsion for the four modelled 'shotpoint' locations. Oil expulsion, green; gas expulsion, red.

multiple phases of subsidence punctuated by significant regional uplift events. Two major petroleum source rocks of different geological age are present (Lower Silurian and Upper Devonian) with similar depositional environment and geochemical character. Multiple reservoir–seal combinations are presented by Late Ordovician glaciogenic sediments and younger Silurian through to Carboniferous paralic sequences. The sedimentary architecture of the basin fill has been captured in a 'Temis 3D' basin model with 35 structural surfaces and thus 34 layers. Lithology calibration came from GDE maps and palaeo-bathymetry and surface temperature through time (for each layer) was also included. Three regional unconformities have had a major impact on the Ghadames–Illizi Basin petroleum system, the 'Hercynian', 'Austrian' and 'Alpine'. These events are significant, because they were characterized by uplift, tilting and erosion and their timing overlapped with petroleum maturation and migration. The 'Hercynian' event probably had the greatest impact, stopping the active 'Pre-Hercynian' petroleum system and leading to wholesale remigration of its trapped petroleum (oil and gas). Erosion is greatest in the northwestern and southern flanks of the basin.

The uplift and tilting associated with the 'Austrian' event is regional in extent, but probably an order of magnitude less than the 'Hercynian'. Erosion is greatest on the eastern flank of the basin and in the south. There is little evidence of erosion associated with the 'Alpine' event in the Ghadames–Illizi Basin. The Paleocene and Eocene sequences are thin and it is likely that the basin was an area of non-deposition for most of the Oligocene. Alpine compression led to the reactivation of a number of basement faults and the propagation of folds through to the surface. At the play fairway level the Ghadames–Illizi Basin is considered low risk for source presence and reservoir presence, although charge access for some sequences is high risk. For example, initial 'Mpath' modelling predicts the presence of a migration shadow for the Tadrart reservoir in the northern part of the basin (due to the presence of the Middle Akakus Shale seal between it and the Tanzuft source). In the southern part of the basin, however, the Tadrart has a low charge access risk, because the Middle Akakus Shale seal is absent (removed by erosion) and Tanzuft charge can access the Tadrart directly from the Lower Akakus sandstones across the 'Caledonian' Unconformity. These modelling

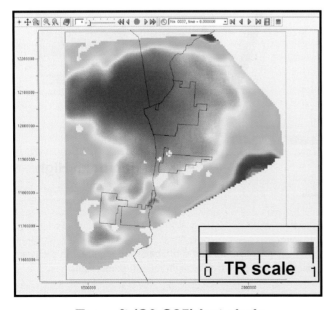

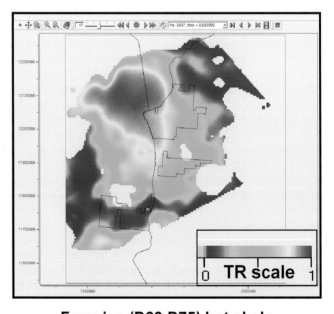

Fig. 22. Kitchen maps: present day transformation ratio for the Tanzuft and Frasnian source rocks.

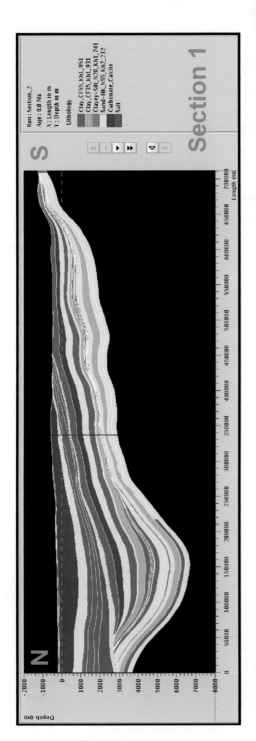
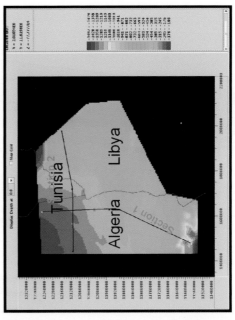
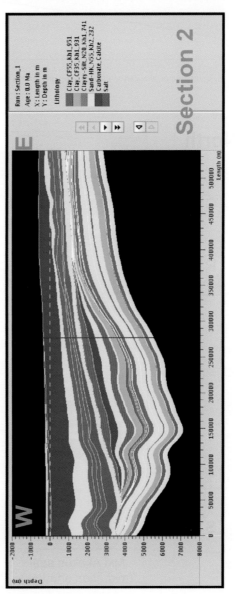

Fig. 23. 'Temis 3D' model: representative cross-sections and lithology distributions.

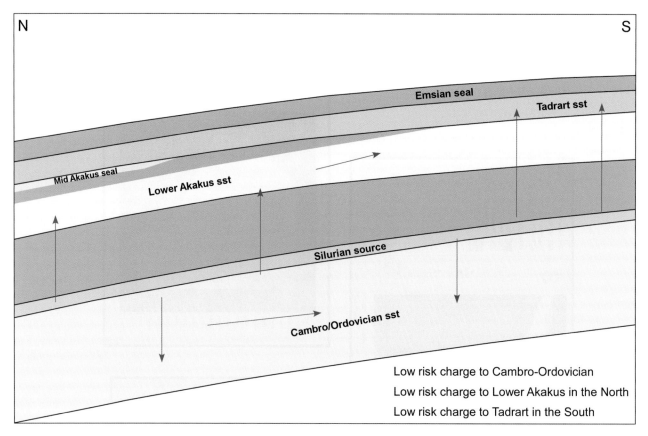

Fig. 24. Ghadames–Illizi Basin charge access cartoon for the Tanzuft source rock.

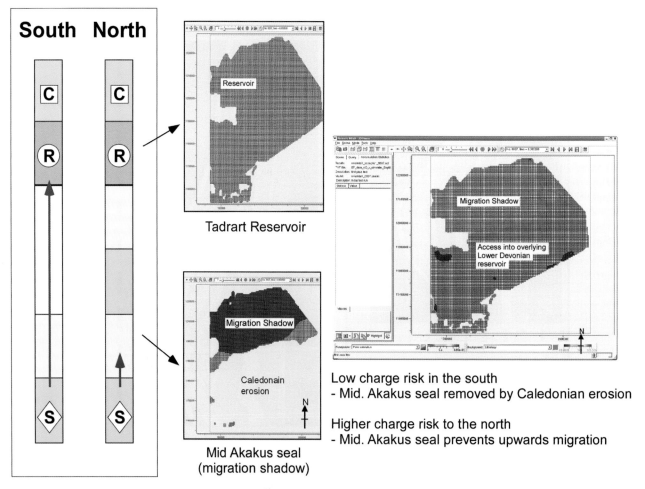

Fig. 25. 'MPath' modelling highlighting the Lower Devonian 'Tadrart play'.

conclusions are consistent with the field observations which indicate that the Akakus fields are in the north and the Tadrart fields are in the southern part of the basin.

We acknowledge BP Exploration for permission to publish. The comments of two reviewers improved the content and presentation of the manuscript.

References

Acheche, M. H., M'Rabet, A., Ghariani, H., Ouahchi, A. & Montgomery, S. L. 2001. Ghadames Basin, Southern Tunisia: a reappraisal of Triassic reservoirs and future prospectivity. *American Association of Petroleum Geologists Bulletin*, **85**, 765–780.

Assereto, R. & Benelli, F. 1971. Sedimentology of the Pre-Cenomanian Formations of the Jebel Gharian, Libya. *In*: Grey, C. (ed.) *Symposium on the Geology of Libya*. Faculty of Science, University of Libya, Tripoli, 37–87.

Bamford, M. K., Roberts, E. M., Famory, S., Mamadou, L. B. & O'Leary, M. A. 2002. An extensive deposit of fossil conifer wood from the Mesozoic of Mali, Southern Sahara. *Palaeogeography, Palaeoclimatology. Palaeoecology*, **186**, 115–126.

Barr, D., Benkali, M., Riviere, M. & Wilson, N. 1994. Algeria District 3 Convention BP/Sonatrach Joint Study. Module 2: Tectonics and Basin Evolution. BP Report EXT67024/2.

Beauchamp, J. 1988. Triassic sedimentation and rifting in the High Atlas (Morocco). *In*: Manspeizer, W. (ed.) *Triassic–Jurassic Rifting. Part A*. Elsevier, Amsterdam, 477–497.

Ben Ferjani, A., Burollet, P. F. & Mejri, F. 1990. *Petroleum Geology of Tunisia*. Entreprise Tunisienne d'Activities Petrolieres.

Benson, R. N. & Doyle, R. G. 1988. Early Mesozoic rift basins and the development of the United States middle Atlantic continental margin. *In*: Manspeizer, W. (ed.) *Triassic–Jurassic Rifting. Part A*. Elsevier, Amsterdam, 41–79.

Benton, M. J., Bouaziz, S., Buffetaut, E., Martill, M., Ouaja, M., Soussi, M. & Trueman, C. 2000. Dinosaurs and other fossil vertebrates from fluvial deposits in the Lower Cretaceous of Southern Tunisia. *Palaeogeography, Palaeoclimatology, Palaeontology*, **157**, 227–246.

Beuf, S., Biju-Duval, B., De Charpal, O., Rognon, P., Gariel, A. & Bennacef, A. 1971. *Les Gres Du Palaeozoique Inferieur Au Sahara*. L'Institut Francais Du Petrole, **18**.

Bishop, W. F. 1975. Geology of Tunisia and adjacent parts of Algeria & Libya. *American Association of Petroleum Geologists Bulletin*, **59**, 413–450.

Boote, D. R. D., Clark-Lowes, D. D. & Traut, M. W. 1988. Palaeozoic petroleum systems of North Africa. *In*: Macgregor, D. S., Moody, R. T. J. & Clark-Lowes, D. D. (eds) *Petroleum Geology of North Africa*. Geological Society, London, Special Publications, **132**, 7–68.

Bosworth, W., El-Hawat, A. S., Helgeson, D. E. & Burke, K. 2008. Cyrenaican 'shock absorber' and associated inversion strain shadow in the collision zone of north–east Africa. *Geological Society of America Bulletin*, **36**, 695–698.

Bracene, R., Patriat, M., Ellouz, N. & Gaulier, J. M. 2003. Subsidence history of basins in Northern Algeria. *Sedimentary Geology*, **156**, 213–239.

Brice, D., Legrand Blain, M. & Nicollin, J. P. 2007. Brachiopod Faunal Changes Across the Devonian/Carboniferous Boundary in North West Sahara (Morocco & Algeria). Geological Society, London, Special Publications, **278**, 261–271.

Busson, G. & Burollet, P. F. 1973. *La limite Permien-Trias sur la plateforme saharienne (Algérie, Tunisie, Libye)*. Canadian Society of Petroleum Geologists, Memoirs, **2**, 74–88.

Carpena, J., Kienast, J. R., Ouzegone, K. & Tehamo, C. 1988. Evidence of contrasted fission-track clock behaviour of the apatite from In Ouzzal Carbonates (north west Hoggar): the low-temperature thermal history of Archean basement. *Bulletin of the Geological Society of America*, **100**, 1237–1243.

Cavaroc, V. V., Padgett, G., Stephens, D. G., Kanes, W. E., Boudda, A. & Woolen, I. D. 1976. Late Palaeozoic of the Tindouf Basin – North Africa. *Journal of Sedimentary Petrology*, **46**, 77–88.

Chaabani, F. & Razgallah, S. 2006. Aptian sedimentation: an example of the interaction of tectonics and eustatics in Central Tunisia. *In*: Moratti, G. & Chalouan, A. (eds) *Tectonics of the Western Mediterranean and North Africa*. Geological Society, London, Special Publications, **262**, 55–74.

Chaouchi, R., Malla, M. S. & Kechou, F. 1998. Sedimentological evolution of the Givetian–Eifelian (F3) sandbar of West Alrar Field, Illizi Basin, Algeria. *In*: Macgregor, D. S., Moody, R. T. J. & Clark-Lowes, D. D. (eds) *Petroleum Geology of North Africa*. Geological Society, London, Special Publications, **132**, 187–200.

Clark-Lowes, D. D. & Ward, J. 1991. Palaeoenvironmental evidence from the Palaeozoic 'Nubian sandstones' of the Sahara. *In*: Salem, M. J., Sbeta, A. M. & Bakbak, M. R. (eds) *The Geology of Libya*. Elsevier, Amsterdam, **VI**, 2099–2153.

Dardour, A. M., Boote, D. R. D. & Baird, A. W. 2004. Stratigraphic controls on Palaeozoic petroleum systems, Ghadames Basin, Libya. *Journal of Petroleum Geology*, **27**, 141–162.

Demirtash, E. 1984. Stratigraphic evidence of Variscan and early Alpine tectonics in Southern Turkey. *In*: Dixon, J. E. & Robertson, A. H. F. (eds) *The Geological Evolution of the Eastern Mediterranean*. Geological Society, London, Special Publications, **17**, 129–145.

Dixon, R. J. 1997a. Depositional Facies and Key Stratal Surfaces of the Late Devonian (D70) and Early Carboniferous (C10), Ahnet Basin, District 3, Algeria. Part II – Field Geology Report: Timimoun Anticline. BP Report SDT/001/97.

Dixon, R. J. 1997b. Depositional Facies and Key Stratal Surfaces of the Lower Devonian (D51) and Middle Devonian (D65), Ahnet Basin, District 3, Algeria. BP Report SDT/005/97.

Dixon, R. J. 1997c. Depositional Facies and Key Stratal Surfaces of the Late Devonian (D70) and Early Carboniferous (C10), Ahnet Basin, District 3, Algeria. Part III – Gross Depositional Environments. BP Report SDT/013/97.

Dixon, R. J., Moore, J. K. S. *et al.* 2009. The 'Hercynian' Unconformity in North Africa: Its nature & significance; a case study from Southern Tunisia & North West Libya. Extended Abstract PESGB/HGS Africa Conference, September 2009.

Drake, N. A., El-Hawat, A. S., Turner, P., Armitage, S. J., Salem, M. J., White, K. J. & McLaren, S. 2008. Palaeohydrology of the Fazzan Basin: the last 7 million years. *Palaeogeography, Palaeoclimatology, Palaeoecology*, **263**, 131–145.

Dridi, M. 2000. Overview on the Stratigraphy, Palaeogeography and Structural Evolution in Southern Tunisia during the Mesozoic as Part of the African Margin. CIESM Workshop Series, **13**, 57–59.

Druckman, Y. 1984. Evidence for early–middle Triassic faulting and possible rifting from the Helez Deep Borehole in the coastal plain of Israel. *In*: Dixon, J. E. & Robertson, A. H. F. (eds) *The Geological Evolution of the Eastern Mediterranean*. Geological Society, London, Special Publications, **17**, 203–212.

El-Arnauti, A. & Shelmani, M. 1988. A contribution to the North East Libyan subsurface stratigraphy with emphasis on pre-mesozoic. *In*: El-Arnauti, A., Owens, B. & Thusu, B. (eds) *Subsurface Palynostratigraphy of North East Libya*. Garyounis University, Benghazi.

El-Hawat, A. S., Missallati, A. A., Bezan, A. & Taleb, T. N. 1996. The Nubian Sandstone in the Sirt Basin and its correlatives. *In*: Salem, M. J., El-Hawat, A. S. & Sbeta, A. M. (eds) *The Geology of Sirt Basin*. Elsevier, Amsterdam, 3–29.

England, P. & Molnar, P. 1990. Surface uplift, uplift of rocks and exhumation of rocks. *Geology*, **18**, 1173–1177.

Garfunkel, Z. & Derin, B. 1984. Permian–early Mesozoic tectonism and continental margin formation in Israel and its implications for the history of the Eastern Mediterranean. *In*: Dixon, J. E. & Robertson, A. H. F. (eds) *The Geological Evolution of the Eastern Mediterranean*. Geological Society, London, Special Publications, **17**, 187–201.

Genik, G. J. 1993. Petroleum geology of Cretaceous–Tertiary rift basins in Niger, Chad and Central African Republic. *American Association of Petroleum Geologists Bulletin*, **77**, 1405–1434.

Graham, J. R. 1982. Wave-dominated shallow marine sediments in the Lower Carboniferous of Morocco. *Journal of Sedimentary Petrology*, **52**, 1271–1276.

Gras, R. & Thusu, B. 1998. Trap architecture of early Cretaceous Sarir Sandstone in the Eastern Sirte Basin, Libya. *In*: Macgregor, D. S., Moody, R. T. J. & Clark-Lowes, D. D. (eds) *Petroleum Geology of*

North Africa. Geological Society, London, Special Publications, **132**, 317–334.

Guerrak, S. 1989. *Time and Space Distribution of Palaeozoic Oolitic Ironstones in the Tindouf Basin, Algerian Sahara*. Geological Society, London, **46**, 197–212.

Hammuda, O. S. 1971. Nature and significance of the lower Cretaceous Unconformity in Jebel Nafusa, North West Libya. *In*: Gray, C. (ed.) *Symposium on the Geology of Libya*. Faculty of Science, University of Libya, Tripoli, 87–99.

Herrera, Z. A. & Racheboeuf, P. R. 2001. Les Eodevonariides (*Chonetoidea, Brachiopoda*) Du Devonien des Bassins de Taoudeni et du Sud du Hoggar, Sahara. *Geobios*, **34**, 493–503.

IFP. 1998. Regional Synthesis of the Triassic Reservoir in Algeria. Oil Industry Consortium Report (4 Volumes).

IFP. 2003. Regional Basin Modelling Study of the Berkine/Illizi Basin (Algeria). Oil Industry Consortium Report (1 Volume).

IFP/Sonatrach/Beicip-Franlab. 2006. Berkine Gas Project: Evaluation of the Gas Potential of the Berkine Basin (Algeria). Oil Industry Consortium Project (3 Volumes).

Jones, R. W. 1997. Devonian–Carboniferous Stratigraphy of District 3, Algeria: Results of the 1996 Biostratigraphic Work Programme (Phase 1), March 1997. BP Report EXT 72613.

Klitzch, E. 1990. Palaeogeographical development and correlation of continental Strata (former Nubian Sandstone) in North East Africa. *Journal of African Earth Sciences*, **10**, 199–213.

Lefranc, J. P. & Guiraud, R. 1990. The continental intercalaire of North Western Sahara and its equivalents in the neighbouring regions. *Journal of African Earth Sciences*, **10**, 27–77.

Liegeois, J. P., Benhallou, A., Azzouni-Sekkal, A., Yahyao, R. & Bonnin, B. 2005. The Hoggar swell and volcanism: reactivation of the Precambrian Taureg Shield during Alpine Convergence and West African Cenozoic Volcanism. *In*: Foulger, G. R., Natland, J. H., Presnall, D. C. & Anderson, D. C. (eds) *Plates, Plumes & Paradigms*. Geological Society of America, Boulder, CO, Special Papers, **388**, 379–400.

Lüning, S., Craig, J., Loydell, D. K., Storch, P. & Fitches, B. 2000. Lower Silurian 'Hot Shales' in North Africa and Arabia: regional distribution and depositional model. *Earth Science Reviews*, **49**, 121–200.

Lüning, S., Kolonic, S., Belhadj, E. M., Cota, L., Baric, G. & Wagner, T. 2004a. Integrated depositional model for the Cenomanian–Turonian organic-rich strata in North Africa. *Earth Science Reviews*, **64**, 51–117.

Lüning, S., Wendt, J., Belka, Z. & Kaufmann, B. 2004b. Temporal/spatial reconstruction of the early Frasnian (Late Devonian) anoxia in NW Africa: new field data from the Ahnet Basin (Algeria). *Sedimentary Geology*, **163**, 237–264.

Madi, A., Savard, M. M., Bourque, P. A. & Chi, G. 2000. Hydrocarbon potential of the Mississippian Carbonate Platform, Bechar Basin, Algerian Sahara. *American Association of Petroleum Geologists Bulletin*, **84**, 256–287.

McKerrow, W. S., Mac Niocaill, C., Ahlberg, P. E., Clayton, G., Cleal, C. J. & Eager, R. M. C. 2000. The late Palaeozoic relations between Gondwana & Laurussia. *In*: Franke, W., Haak, V., Oncken, O. & Tanner, D. (eds) *Orogenic Processes: Quantification and Modelling in the Variscan Belt*. Geological Society, London, Special Publications, **179**, 9–20.

Mohamed, A. Y., Pearson, M. J., Ashcroft, W. A. & Whiteman, A. J. 2001. Structural development and crustal stretching in the Muglad Basin, Sudan. *Journal of African Earth Sciences*, **32**, 179–191.

Monod, O., Kozlu, H. *et al.* 2003. Late Ordovician glaciation in Southern Turkey. *Terra Nova*, **15**, 249–257.

Paleoservices Report. 1986. The Hydrocarbon Potential of the Palaeozoic Rocks of the Western Desert of Egypt. BP Report EXT 52158/1.

Robertson, A. H. F. 2007. Sedimentary evidence from the south Mediterranean region (Sicily, Crete, Peloponnese, Evia) used to test alternative models for the regional tectonic setting of Tethys during Late Palaeozoic–Early Mesozoic time. *In*: Robertson, A. H. F. & Mountrakis, D. (eds) *Tectonic Development of the Eastern Mediterranean Region*. Geological Society, London, Special Publications, **260**, 91–154.

Robertson, A. H. F. & Dixon, J. E. 1984. Aspects of the geological evolution of the Eastern Mediterranean. *In*: Dixon, J. E. & Robertson, A. H. F. (eds) *The Geological Evolution of the Eastern Mediterranean*. Geological Society, London, Special Publications, **17**, 1–74.

Schroter, T. 1996. Tectonic and sedimentary development of the Central Zallah Trough. *In*: Salem, M. J., Busrewil, M. T., Misallati, A. A. & Sola, M. A. (eds) *The Geology of the Sirt Basin*. Elsevier, Amsterdam, 123–137.

Smith, A. G. 2006. Tethyan ophiolite emplacement, Africa to Europe motions and Atlantic spreading. *In*: Robertson, A. H. F. & Mountrakis, D. (eds) *Tectonic Development of the Eastern Mediterranean Region*. Geological Society, London, Special Publications, **260**, 11–34.

Soussi, M. & Ben Ismail, M. H. 2000. Platform collapse and pelagic seamount facies, Jurassic development of Central Tunisia. *Sedimentary Geology*, **133**, 93–113.

Stampfli, G. M. & Borel, G. D. 2002. A plate tectonic model for the Palaeozoic and Mesozoic constrained by dynamic plate boundaries and restored synthetic oceanic isochrones. *Earth and Planetary Science Letters*, **196**, 17–33.

Stampfli, G. M., Borel, G. D., Cavazza, W., Mosar, J. & Zeigler, P. A. 2001. Palaeotectonic and palaeogeographic evolution of western Tethys and Peritethyan domain. *Episodes*, **24**, 222–228.

Swezey, C. S. 2008. Cenozoic stratigraphy of the Sahara. *Journal of African Earth Sciences*, **53**, 89–121.

Toomey, D. F. 1991. Late Permian reefs of Southern Tunisia: facies patterns and comparison with Capitan Reef, South Western United States. *Facies*, **25**, 119–146.

Turner, P. & Sherif, H. 2007. A giant late Triassic–early Jurassic evaporitic basin on the Saharan Platform, North Africa. *In*: Schreiber, B. C., Lugli, S. & Babel, M. (eds) *Evaporites through Space and Time*. Geological Society, London, Special Publications, **285**, 87–105.

Turner, P., Pilling, D., Walker, D., Exton, J., Binnie, J. & Sabaou, N. 2001. Sequence stratigraphy and sedimentology of the late Triassic TAG-I (Blocks 401/402, Berkine Basin, Algeria). *Marine & Petroleum Geology*, **18**, 959–981.

Underdown, R. & Redfern, J. 2007. The importance of constraining regional exhumation in basin modelling: a hydrocarbon maturation history of the Ghadames Basin, North Africa. *Petroleum Geoscience*, **13**, 253–270.

Underdown, R. & Redfern, J. 2008. Petroleum generation and migration in the Ghadames Basin, North Africa: a two-dimensional basin modelling study. *American Association of Petroleum Geologists Bulletin*, **92**, 53–76.

Underdown, R., Redfern, J. & Lisker, F. 2007. Constraining the burial history of the Ghadames Basin, North Africa: an integrated analysis using sonic velocities, vitrinite reflectance data and apatite fission track ages. *Basin Research*, **19**, 557–578.

Villeneuve, M. 2005. Paleozoic basins in West Africa and the Mauritanide Thrust Belt. *Journal of African Earth Sciences*, **43**, 166–195.

Vos, R. G. 1977. Sedimentology of an Upper Palaeozoic river, wave and tide influenced delta system in Southern Morocco. *Journal of Sedimentary Petrology*, **47**, 1242–1260.

Wahlman, G. P. 1991. Permian and Late Carboniferous Carbonates of South Central Tunisia: Biostratigraphy, Reefs and Associated Carbonate Facies. Amoco Internal Report.

Wendt, J., Belka, B., Kaufman, B., Kostrewa, R. & Hayer, J. 1997. The world's most spectacular carbonate mud mounds (Middle Devonian, Algerian Sahara). *Journal of Sedimentary Research*, **67**, 424–436.

Yang, Y. & Aplin, A. C. 1999. Definition and practical application of mudstone porosity–effective stress relationships. *Petroleum Geoscience*, April 2004, **10**, 153–162.

Zaborski, P. M. & Morris, N. J. 1999. The late cretaceous ammonite genus *Libyoceras* in the Lullemeden Basin (West Africa) and its palaeogeographical significance. *Cretaceous Research*, **20**, 63–79.

Zaier, A., Beji-Sassi, A., Sassi, S. & Moody, R. T. J. 1998. Basin evolution and deposition during the Early Palaeogene in Tunisia. *In*: Macgregor, D. S., Moody, R. T. J. & Clark-Lowes, D. D. (eds) *Petroleum Geology of North Africa*. Geological Society, London.

Biostratigraphy, chemostratigraphy and thermal maturity of the A1-NC198 exploration well in the Kufra Basin, SE Libya

S. LÜNING,[1] N. MILES,[2] T. PEARCE,[3] E. BROOKER,[3] P. BARNARD,[4] G. JOHANNSON[5] and S. SCHÄFER[1,5]

[1]*RWE Dea, Überseering 40, 22297 Hamburg, Germany (e-mail: Sebastian.Luening@rwe.com)*
[2]*Petrostrat Ltd, Tan-y-Graig, Parc Caer Seion, Conwy LL32 8FA, UK*
[3]*Chemostrat Ltd, Unit 1, Ravenscroft Court, Buttington Enterprise Park, Welshpool, Powys SY21 8SL, UK*
[4]*Applied Petroleum Technology (UK) Ltd, Second Floor, 14 Wynnstay Road, Colwyn Bay LL29 8NB, UK*
[5]*RWE Dea Libya, Barcelona Street, 15A Hai Andalus, Tripoli, Libya*

Abstract: The A1-NC198 exploration well was drilled in the Kufra Basin in 2007 by RWE Dea and represented only the third well in a large, 400 000 km² frontier basin. Despite being dry and lacking any hydrocarbon shows, the well provides important data to improve the understanding of the regional petroleum play. In the 1980s and 1990s the basin's prospectivity was questioned largely because of supposed (1) lack of structuration, (2) lack of source rock and (3) thermal immaturity at Silurian level. Following a series of academic and industry studies over the past 10 years, these assertions can no longer be upheld. The analysis of available seismic has proven the existence of Murzuq-style fault blocks as well as late Ordovician glacial erosional relict buried hills, potentially forming suitable structural and stratigraphic traps. The presence of hot shale in the Kufra Basin is evidenced by typical seismic onlaps of strong amplitude reflectors at base Silurian levels, shallow drilling results and outcrop spectral gamma-ray evidence. A spore colouration study of A1-NC198 cuttings indicates a deep oil window maturity for the Silurian, implying potential oil generation in the basin if suitable Silurian source rocks exist. The stratigraphy of the A1-NC198 succession was analysed by means of biostratigraphy and chemostratigraphy, which form the basis for improved well correlations within the basin.

Keywords: Kufra Basin, Libya, chemostratigraphy, petroleum geology

The Kufra Basin is a large, underexplored Palaeozoic frontier basin covering 400 000 km² in SE Libya and NE Chad, and extending into NW Sudan and SW Egypt (Fig. 1). Only four petroleum wildcat wells have been drilled in the basin to date, A1-NC43 and B1-NC43 in 1978 and 1981 by AGIP (Bellini *et al.* 1991), A1-NC198 by RWE Dea in 2007 and A1-171/4 by Statoil in 2008. All of these were dry (Fig. 2). Another four dry wells were drilled just outside the basin, two by Oasis in the 1960s (A1-71 and D1-71) and two by Repsol in 2009 (A1-NC203 and B1-NC203) (Fig. 2). Because of its large size, the basin remains significantly underexplored.

The petroleum play pursued by explorers in the Kufra Basin is similar to the neighbouring Murzuq Basin in SW Libya, which hosts a series of medium to large oil fields in a similar geological setting (Fig. 3) (Davidson *et al.* 2000; Lüning *et al.* 2000; Herzog *et al.* 2008). Most of the hydrocarbon play elements known from the Murzuq Basin also occur in the Kufra Basin. As in the Murzuq Basin, the Ordovician sedimentary package contains thick porous sandstones with potential reservoir quality (Turner 1980). Late Ordovician deposition was strongly influenced by glacial processes as evidenced by buried hill morphologies on seismic and glacial striations at outcrop (Le Heron *et al.* 2010).

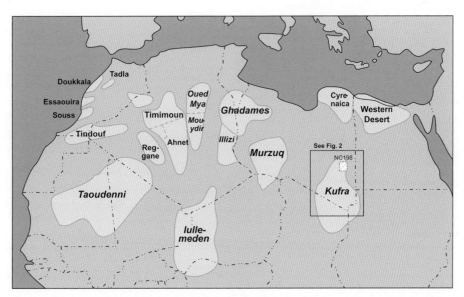

Fig. 1. Palaeozoic basins of North Africa and location of Kufra Basin and Block NC198.

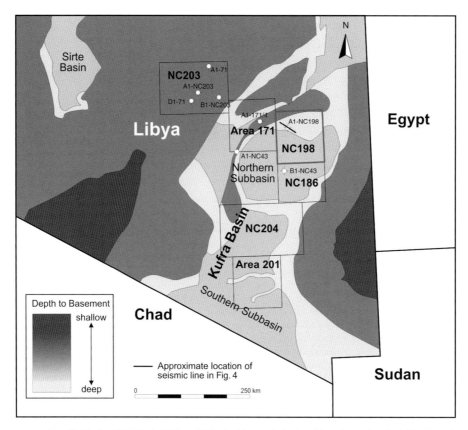

Fig. 2. Depth-to-basement map (modified after FrOGTech 2008, published with permission) and locations of wells drilled in and nearby the Kufra Basin.

In addition to buried hill palaeotopographic stratigraphic traps, a series of structural traps defined by fault blocks have also been mapped on seismic in the Kufra Basin.

The Silurian Tanezzuft Shale represents both seal and source rock for the petroleum play and is the most critical element of the play (Lüning et al. 1999). The basal Tanezzuft 'lower hot shale' has a patchy distribution and was deposited in palaeodepressions. It is the only source rock in the Murzuq Basin (Lüning et al. 2000). The hot shale also occurs in the Kufra Basin and has been identified in several outcrop sections by means of gamma-ray (GR)

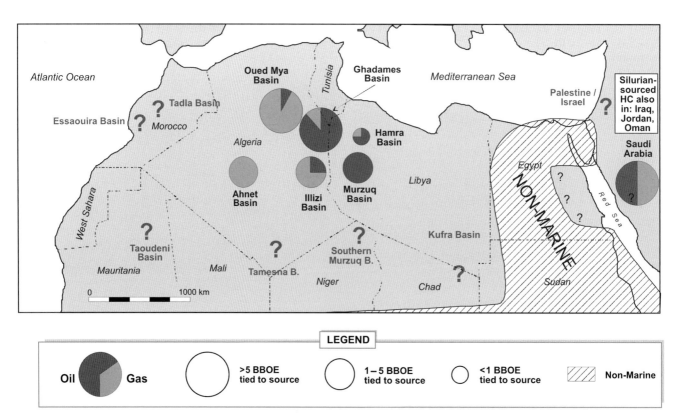

Fig. 3. Proven oil and gas occurrences in North Africa sourced by the Silurian black shales. Modified after Macgregor (1996).

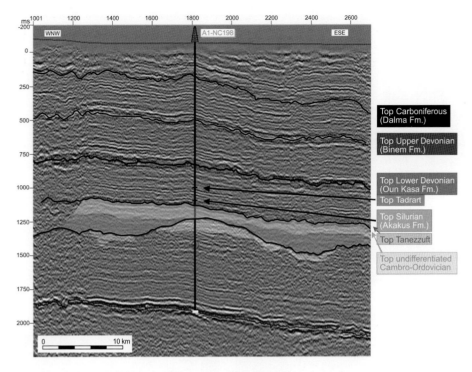

Fig. 4. Seismic cross-section through the A1-NC198 buried hill structure. Vertical scale in two-way time.

spectrometry based on its characteristic enrichment in uranium (Lüning et al. 2003; Fello et al. 2006). The source rock was also penetrated by AGIP's shallow borehole KW2 and is interpreted in the subsurface based on strong seismic amplitudes and onlap geometries. The thickness of the total Tanezzuft shale package varies greatly across the Kufra Basin (Lüning et al. 1999). In surface outcrops the Tanezzuft may be a few metres to more than 100 m thick and shale-dominated. In wells A1-NC43 and A1-NC198 the Tanezzuft is about 60 m thick; in B1-NC43 it is only 20 m (Fig. 6). The Tanezzuft shales also provide the main seal for the underlying Ordovician reservoirs. Compared with the Murzuq Basin, where Tanezzuft thicknesses of up to 500 m have been recorded, providing an excellent regional seal, the seal risk in the Kufra Basin may be higher because in some areas there is increased risk that fault offsets are greater than seal thickness. In cases where the Tanezzuft seal is breached by faults, hydrocarbons may have migrated into the Devonian–Carboniferous section, which also has reservoir potential and is characterized by a thick series of interbedded sandstones and sealing shales.

Exploration well A1-NC198 was drilled in block NC198, in the northern sub-basin (Fig. 2), between 13 April and 26 May 2007. The well targeted an upper Ordovician 'buried hill' prospect (Fig. 4), and reached basement at TD (12 193 ft, 3716 m). The well proved to be dry and did not encounter any visible hydrocarbon shows. A series of studies was carried out, including palynology, chemostratigraphy, fluid inclusions, organic geochemistry and thermal maturity, in order to better understand the petroleum play elements and implications for regional hydrocarbon prospectivity of the Kufra Basin.

Palynology

Study objective

Establishing a palyno-biostratigraphical subdivision of the section drilled by well A1-NC198 was important because the typical monotonous interbedded sandstones and shales are easily mis-correlated, especially due to the limited number of offset wells available to provide control in this frontier basin. A secondary objective was the study of depositional environments. The comprehensive biostratigraphic dataset acquired in the A1-NC198 well has enabled a more reliable correlation with AGIP's A1-NC43 and B1-NC43 wells (Fig. 5), which were palynologically studied in detail by Grigagni et al. (1991).

Material and methods

The study was based on 71 cuttings samples, mostly claystones, sampled throughout the entire well section. The analysis is based on counts of 200 palyno-specimens per sample, using 10 μm sieved preparations. The rest of the slide was then scanned for rarer taxa. In addition a 53 μm sieved preparation was analysed for most samples in order to concentrate chitinozoans, with counts of up to 100 made where possible (Fig. 6).

Results

Based on palynology, a robust biostratigraphic framework was erected for the uppermost Ordovician to Early Carboniferous encountered by well A1-NC198 (Fig. 7). Spores, chitinozoans, acritarchs and algae all proved useful in the biostratigraphic interpretation of this well section. The bulk of the Cambro-Ordovician as well as the strata overlying the Lower Carboniferous siliciclastics could not be age-dated by palynomorphs due to poor palynomorph recovery or barren samples. The majority of analysed Palaeozoic samples indicated marginal marine to inner shelf depositional environments. Only a few samples yielded evidence for non-marine settings, a scarcity which however may be related to the paucity of palynomorphs in continental sedimentary successions.

Chemostratigraphy

Chemostratigraphy involves the characterization and correlation of sedimentary strata using stratigraphic variations in inorganic geochemistry (e.g. Pearce et al. 2005; Ratcliffe et al. 2006). The technique is particularly sensitive to these variations and in many instances has proved successful in detecting and correlating them when dealing with apparently uniform lithological successions. Consequently, reservoir correlations and stratigraphic picks may

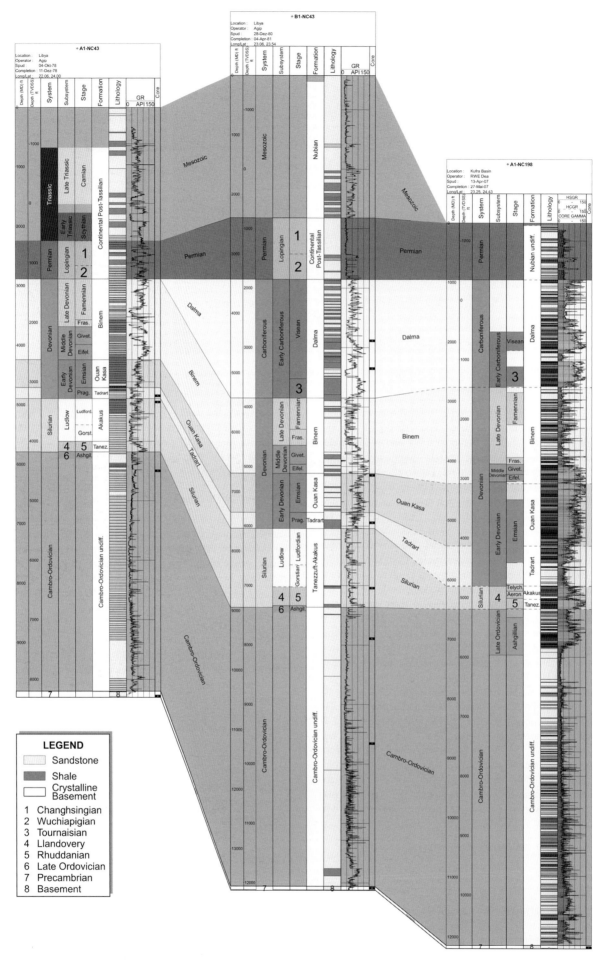

Fig. 5. Correlation of wells A1-NC43, B1-NC43 and A1-NC198 (flattened on base Permian).

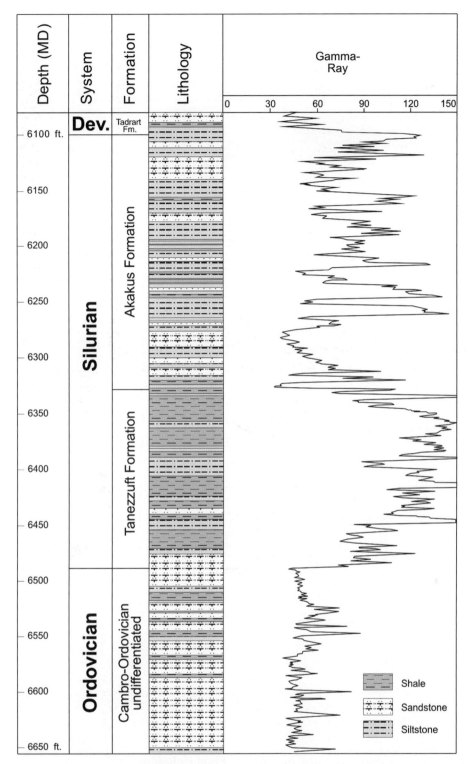

Fig. 6. Lithology and GR characteristics of the Silurian Tanezzuft Formation in well A1-NC198. GR in API units.

now be established with confidence, even for those sequences over which the E-log traces show little significant change.

Study objective

Chemostratigraphy was carried out, targeting specifically the gaps in the Cambrian to mid Ordovician and post-lower Carboniferous succession that were not resolved by palynostratigraphy.

Material and methods

A total of 69 cuttings samples were selected, distributed over the entire well, and comprised mostly sandstones but also some claystones. All samples were analysed by inductively coupled plasma–optical emission spectrometry (ICP-OES) and inductively coupled plasma–mass spectrometry (ICP-MS), with data being acquired for 50 elements.

Results

Based on the chemostratigraphic analysis, the section drilled by well A1-NC198 was subdivided into 15 discrete sedimentary packages, named P1–P15 (Fig. 8). Each package possesses a distinct set of geochemical characteristics. Because of the limited number of samples and the great thickness of the interval analysed,

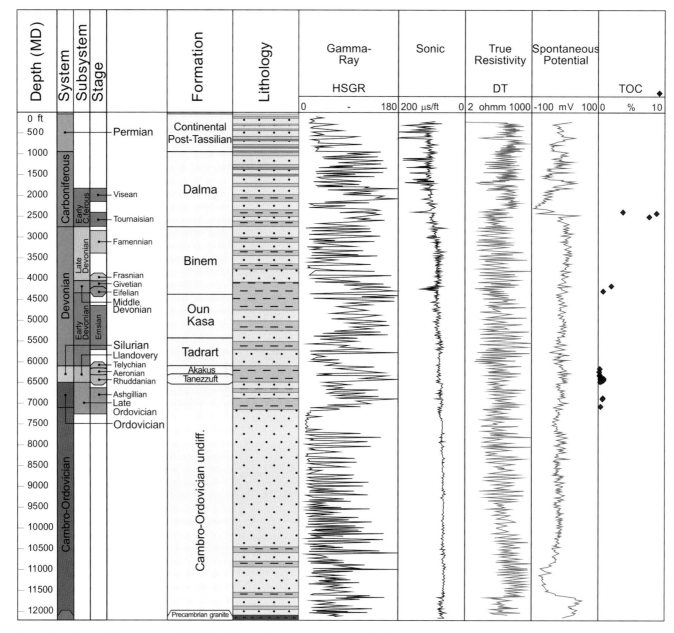

Fig. 7. Simplified well log from the A1-NC198 wildcat well and plot of measured TOC data.

chemostratigraphic distribution curves remain rather coarse for this well. However, typical trends exist both in the Cambro-Ordovician and post-lower Carboniferous that might allow chemostratigraphic correlation with neighbouring wells. In the Murzuq Basin (Kaaber et al. 2007) chemostratigraphy has been used successfully to subdivide and correlate the Cambro-Ordovician succession. Overall, the geochemical signature of the Achebayat, Hawaz and Bir Tlascin intervals identified in the Murzuq Basin are comparable to those recognized in well A1-NC198 with, for example, high Ga/Rb (attributed to high kaolinite/illite ratios) characterizing the Achebayat and moderately high K–Al (K-feldspar) occurring in the Bir Tlascin Formations. Notably, the sandstones of the Mamuniyat Formation are not identified geochemically in well A1-NC198. The Silurian/Devonian sandstones and shales of the Akakus, Tadrart and Binem Formations can each be differentiated geochemically based on variations in, amongst others, K–Al, Fe–Mn and Ti–Mg, attributed to fluctuations in feldspar, siderite and Ti-rich heavy minerals respectively. Carboniferous deposits exhibit high Na–Al associated with the occurrence of plagioclase feldspar reworked from basic volcanic provenance.

Notably, the Permian section is marked by high K–Al attributed to major influx of arkosic sandstones derived from a granitic provenance (Fig. 8).

Fluid inclusions

Fluid inclusions are representative microscopic samples of past or present-day subsurface fluids that become entrapped in rocks during the formation of diagenetic cements or healed microfractures. They are not subject to fractionation during sampling or evaporative loss during sample storage for any length of time. Fluid inclusions persist in the geological record long after the parent fluids have moved on, but are continuously formed even up to the very recent past (Hall et al. 2002).

Study objective

The study objective was the identification and quantitative analysis of volatile hydrocarbon traces within micron-sized cavities in cuttings rock samples.

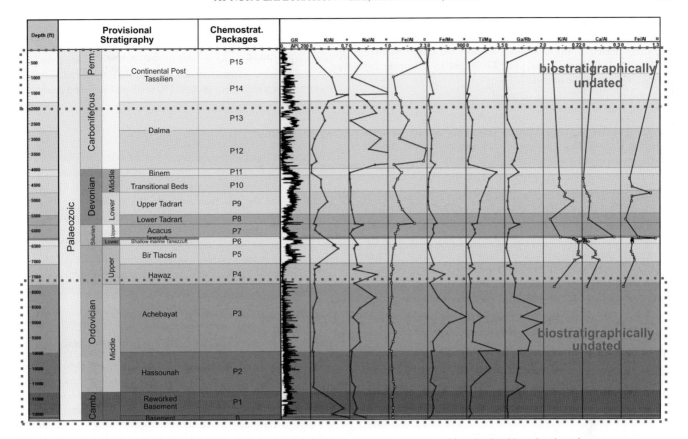

Fig. 8. Chemostratigraphy of well A1-NC198. Zones boxed with dashed lines mark intervals that could not be dated by palynology due to poor palynomorph recovery. Silurian to Carboniferous stratigraphy is based on palynomorphs.

Material and methods

The study was carried out by fluid inclusion technologies (FIT) using the fluid inclusion stratigraphy (FIS) method, a patented Amoco technology, licensed to FIT Inc. The study in well A1-NC198 was based on 78 cuttings samples, mostly sandstones. The samples were mechanically crushed and the volatile substances released instantaneously. The fluids were then pumped through mass analysers where they were ionized, separated according to their mass/charge and recorded.

Results

No liquid petroleum fluids were recorded by FIS throughout the well, indicating sub-anomalous hydrocarbon concentrations. Nevertheless, visual inspection of thin sections of Devonian and Ordovician samples resulted in the identification of rare oil inclusions (Fig. 9). In addition, a few thin zones with weak dry gas responses were recorded by FIS.

Additional evidence for the presence of liquid hydrocarbons and hence a working petroleum system in the NC198 area of the Kufra Basin also comes from a Gore^{TM} surface geochemistry microseepage survey that yielded a series of positive seepage anomalies.

Petroleum source rock evaluation

Study objective

Despite the positive evidence from high seismic amplitudes, seismically defined onlaps and radioactive Silurian shales at outcrop, an organic-rich Silurian source rock has not yet been penetrated by any of the existing exploration wells in the Kufra Basin (Lüning et al. 1999). As most wells are drilled on positive structures

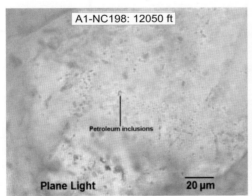
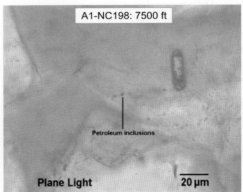

Fig. 9. Fluid inclusions containing hydrocarbon fluids in Cambro-Ordovician sandstone.

that might also have been palaeohighs during the early Silurian, the absence of organic-rich shales in these locations is not unexpected. The study objective was to evaluate whether the Lower Silurian source rock is present in well A1-NC198 and to determine the organic richness of the interval.

Material and methods

The study is based on 11 Silurian Tanezzuft samples and five Devonian-Carboniferous samples. Total organic carbon content (TOC) was analysed using a Leco system.

Results

All of the Tanezzuft samples recorded low organic richness with TOCs ranging between 0.21 and 0.84%. This is consistent with the rather low GR values of only up to 180 API, indicating the absence of the typical radioactive Silurian 'hot shale' at this location. Slightly elevated TOCs of around 2% were recorded in the Middle Devonian within the Binem Formation (Fig. 7). In contrast, high TOCs of up to 9% occur in the Carboniferous Dalma Formation (Fig. 7).

Spore colouration and vitrinite reflectivity

Study objective

The thermal maturity history of the Kufra Basin to date has been largely unknown. Therefore, a detailed thermal maturity study on the basis of spore colouration and vitrinite reflectance has been carried out on samples from the A1-NC198 well.

Material and methods

The study is based on 28 cuttings samples, mostly shales, originating from the depth range of 1995 to 7740 ft. The vitrinite reflectivity studies were carried out on isolated 10 μm sieved kerogen concentrates which were mounted in resin and polished ready for reflectivity studies.

Results

Both spore colour indices and vitrinite reflectivity data indicate that the Lower Carboniferous and the upper part of the Devonian sediments have been within the main oil window (Figs 10 & 11). Spore colour indices indicate that the Silurian and Ordovician interval is late mature for oil generation. No vitrinite-like material was observed in the Lower Palaeozoic section. Extrapolation of maturity gradients to surface intercepts indicate significant missing section, with erosion of post-Lower Carboniferous section due to uplift. Apatite fission track analyses suggest that this uplift occurred sometime during the Late Cretaceous to Cenozoic period (Lisker, pers. comm. 2009).

Implications for regional hydrocarbon prospectivity

Contrary to previous interpretations, the new results show that the potential Silurian hot shale source rock was previously deeply buried, into the oil window, as evidenced by high spore colouration indices. Unpublished apatite fission track data indicate that this oil generation phase must have terminated due to significant uplift sometime around Late Cretaceous to Early Cenozoic time.

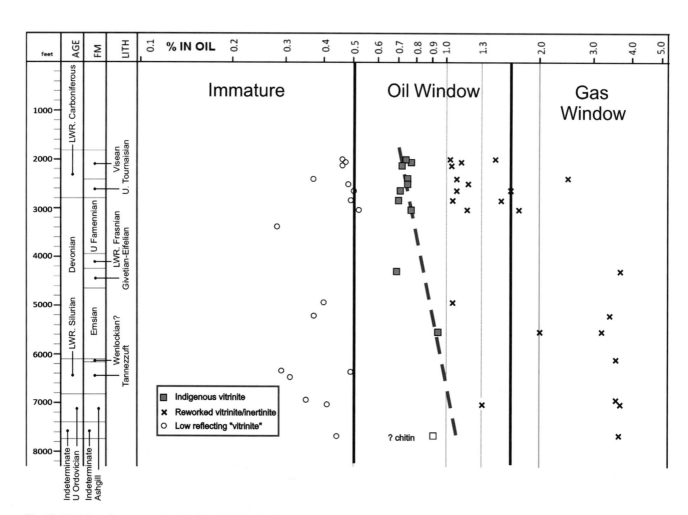

Fig. 10. Vitrinite reflectance against depth, well A1-NC198.

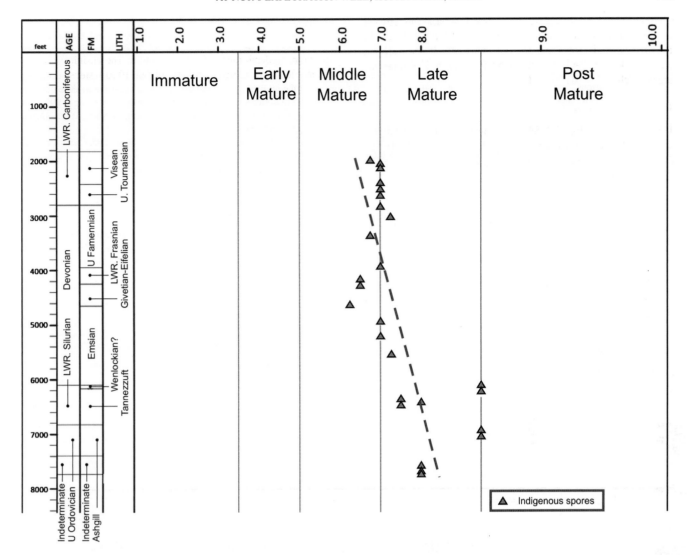

Fig. 11. Spore Colour Index against depth, well A1-NC198.

A similar structural history is known from the Murzuq and Ghadames basins (e.g. Dardour 2004; Underdown & Redfern 2008).

Based on seismic data, both glacial buried hills and tilted fault block structural traps can be identified in the Kufra Basin. However, the Tanezzuft shales which onlap these palaeohighs may not be thick enough to serve as an effective seal for the underlying Ordovician reservoirs, especially in areas where large faults exist. Over some buried hills the Tanezzuft shales may be extremely thin or even pinch out. Possible 'leakage' might have opened up migration pathways into Devonian–Carboniferous sandstones. Thick Devonian–Carboniferous shale units exist that may act as more effective seals in such cases. At the Ordovician reservoir level it may be more promising to concentrate on structural traps that developed later, such as tilted fault blocks, which may be better sealed by Tanezzuft shales than the buried hills.

In terms of source rock development, the absence of the Silurian hot shale in A1-NC198 does not challenge the current model of patchy distribution of the lower Silurian hot shale areally restricted to palaeodepressions. Evidence for the presence of the hot shale in the Kufra Basin is provided by typical seismic onlaps and strong amplitudes at base Silurian levels, and has been confirmed by shallow stratigraphic drilling results and outcrop spectral GR evidence.

Despite the two new recent dry wells in the Kufra Basin, the basin is still largely underexplored and a final determination of its hydrocarbon prospectivity is some way off. In the light of the petroliferous Murzuq Basin to the west, where a similar play exists, and which also initially had a number of negative well results before large accumulations were discovered, exploration of the Kufra basin remains a challenge well worth pursuing.

We thank RWE Dea and NOC for permission to publish this paper. The manuscript benefited greatly from helpful reviews by Professor Jonathan Redfern (University of Manchester) and an anonymous colleague.

References

Bellini, E., Giori, I., Ashuri, O. & Benelli, F. 1991. Geology of Al Kufrah Basin, Libya. In: Salem, M. J., Sbeta, A. M. & Bakbak, M. R. (eds) The Geology of Libya. Vol. 6. Elsevier, Amsterdam, 2155–2184.

Dardour, A. M. 2004. Stratigraphic controls on Palaeozoic petroleum systems, Ghadames Basin, Libya. Journal of Petroleum Geology, 27, 141–162.

Davidson, L. et al. 2000. The structure, stratigraphy and petroleum geology of the Murzuq Basin, Southwest Libya. In: Sola, M. A. & Worsley, D. (eds) Geological Exploration in the Murzuq Basin. Elsevier, Amsterdam, 295–320.

Fello, N., Lüning, S., Štorch, P. & Redfern, J. 2006. Identification of early Llandovery (Silurian) anoxic palaeo-depressions at the western margin of the Murzuq Basin (southwest-Libya) based on gamma-ray spectrometry in surface exposures. GeoArabia, 11, 101–118.

FrOGTech. 2008. Kufra Basin SE Libya Zoom Study (unpublished proprietary multiclient report).

Grigagni, D., Lanzoni, E. & Elatrash, H. 1991. Palaeozoic and mesozoic subsurface palynostratigraphy in the Al Kufrah Basin, Libya. *In*: Salem, M. J., Hammuda, O. S. & Eliagoubi, B. A. (eds) *The Geology of Libya*, Vol. IV. Elsevier, Amsterdam, 1160–1227.

Hall, D. L., Sterner, S. M., Shentwu, W. & Bigge, M. A. 2002. Applying fluid inclusions to petroleum exploration and production. AAPG Search and Discovery, www.searchanddiscovery.com, **40042**, 1–62.

Herzog, U., El-Ila, A. & Saad, S. O. 2008. Al Kufrah Basin, Libya – geological history review and refinement. *In*: Salem, A. E.-A. M. J. & El Sogher Saleh, A. (eds) *The Geology of East Libya*, Vol. 3. Earth Science Society of Libya, Tripoli, 3–18.

Kaabar, O., Algibez, J. L., Pearce, T. & Khoja, A. 2007. Chemostratigraphy of Cambro-Ordivician to Silurian sequences from Block NC186 in the Murzuq Basin, Western Libya. *Extended abstract*, 3rd North African/Mediterranean Petroleum & Geosciences Conference & Exhibition, Tripoli, Libya, 26–28 February 2007.

Le Heron, D. P., Armstrong, H. A., Wilson, C., Howard, J. P. & Gindre, L. 2010. Glaciation and deglaciation of the Libyan Desert: the Late Ordovician record. *Sedimentary Geology*, **223**, 100–125.

Lüning, S. *et al.* 1999. Re-evaluation of the petroleum potential of the Kufra Basin (SE Libya, NE Chad): does the source rock barrier fall? *Marine and Petroleum Geology*, **16**, 693–718.

Lüning, S., Craig, J., Loydell, D. K., Štorch, P. & Fitches, W. R. 2000. Lowermost Silurian 'hot shales' in North Africa and Arabia: regional distribution and depositional model. *Earth Science Reviews*, **49**, 121–200.

Lüning, S., Kolonic, S., Loydell, D. & Craig, J. 2003. Reconstruction of the original organic richness in weathered Silurian shale outcrops (Murzuq and Kufra basins, southern Libya). *GeoArabia*, **8**, 299–308.

Macgregor, D. S. 1996. The hydrocarbon systems of North Africa. *Marine and Petroleum Geology*, **13**, 329–340.

Pearce, T. J., Wray, D., Ratcliffe, K. & Wright, D. K. 2005. Chemostratigraphy of the Schooner Formation, Southern North Sea. *Proceedings of Yorkshire Geological Society, Occasional Publications*, **7**, 147–164.

Ratcliffe, K. T. *et al.* 2006. A regional chemostratigraphically-defined correlation framework for the late Triassic TAG-I Formation in Blocks 402 and 405a, Algeria. *Petroleum Geoscience*, **12**, 3–12.

Turner, B. R. 1980. Palaeozoic sedimentology of the southeastern part of Al Kufrah Basin, Libya: a model for oil exploration. *In*: Salem, M. J. & Busrewil, M. T. (eds) *The Geology of Libya*, Vol. 2. Academic Press, London, 351–374.

Underdown, R. & Redfern, J. 2008. Petroleum generation and migration in the Ghadames basin, North Africa: a two-dimensional modeling study. *AAPG Bulletin*, **92**, 53–76.

Exploring subtle exploration plays in the Gulf of Suez

P. N. DANCER,[1] J. COLLINS,[2] A. BECKLY,[2] K. JOHNSON,[3] G. CAMPBELL,[3] G. MUMAW[2] and B. HEPWORTH[3]

[1]*Dana Petroleum, Zahret El Maadi Tower (Secon Building), Corniche El Nile, Maadi, Cairo, Egypt (e-mail: nick.dancer@dana-petroleum.com)*
[2]*Senergy, 15 Bon Accord Crescent, Aberdeen AB11 6DE, UK*
[3]*Helix RDS (a Baker Hughes Incorporated Company), Peregrine Road, Westhill Business Park, Westhill, Aberdeen AB32 6JL, UK*

Abstract: The Gulf of Suez Basin is a classic extensional rift basin of Miocene age, with a number of syn- and pre-rift hydrocarbon plays. Exploration started in 1886, targeting areas around documented oil seeps, and was highly successful. During the boom in offshore exploration in the 1950s and 1960s, a combination of diligent geology and serendipity resulted in the discovery of a number of giant fields whose reserves form a major part of some 10 billion barrels discovered to date. However, the pursuit of smaller fields, both structural and stratigraphic, has been hampered by the poor quality seismic data characteristic of the basin. The poor quality of the seismic data is due to the interbedded shales and evaporites of shallow post-rift Zeit and South Gharib formations that create massive reverberation and severe attenuation of the seismic signal. Incremental progress in imaging the deeper horizons, including the key pre-rift Nubia Sandstone reservoir, has been achieved through low frequency enhancement and 3D seismic data acquisition. However it is still common for exploration and development wells to miss the Nubia objective due to the poor imaging and consequent misinterpretation of the seismic data. Exploration is now directed towards the smaller targets and subtle plays of the Gulf of Suez. To improve seismic imaging the focus has been on improved acquisition, with 3D ocean-bottom cable seismic data. This has the advantages of reduced multiple energy, higher fold and a broader bandwidth over streamer 3D data. Combined with detailed models of the tectonically controlled sedimentation that characterizes the syn-rift section, this has allowed the development of a re-invigorated exploration programme.

Keywords: Gulf of Suez, rifts, syn-rift sedimentation, petroleum systems

The Gulf of Suez has been an active exploration area since the first well (Gemsa D-1) was drilled in 1886. Oil seeps along the basin margin, especially in the Gebel Zeit area, have been historically reported, with evidence that these oils had been used during Pharonic times in the preservation of mummies (Harrell & Lewan 2002). Since the initial discovery of the Gemsa Oil Field, some 10×10^9 barrels of oil have been discovered, in approximately 132 oil fields.

The majority of the large oil fields lie offshore in the present-day Gulf, but two-thirds of the rift basin is onshore. The outcrops produced by variable erosion of the onshore areas have been the focus of much research. As a result the Gulf of Suez is often used as an analogue for other 'failed' extensional basins, including the UK and Norwegian North Sea and adjacent areas.

Gulf of Suez structural setting

The Gulf of Suez formed in Late Oligocene/Early Miocene times as the northern extension of the Red Sea Rift (Bosworth 1995; Bosworth & McClay 2001). The Red Sea Rift is the northeastern arm of the Neogene separation of the Nubia, Arabian and Somalian plates. In Mid–Late Miocene times the continued extension in the Red Sea was accommodated by sinistral strike-slip displacement along the NE–SW Gulf of Aqaba–Dead Sea transform fault system, resulting in the Gulf of Suez becoming a failed rift system. Extension within the Gulf ranges from *c.* 16 km in the northern part of the rift (extension factor $\beta = 1.33$) to *c.* 30 km in the southern part ($\beta = 1.51$) (Moustafa 2002). The primary extensional faults are commonly referred to as the clysmic faults or trend, and the movement on these faults as clysmic events (Hume 1916; Robson 1971).

The Suez Rift divides into three segments with opposite sense of fault throw and half graben dip, separated by NE–SW trending transform/accommodation zones (Fig. 1; Colletta *et al.* 1988). The segments are, from north to south, the Darag Basin (SW-dipping half graben), the Belayim Province (NE-dipping half graben) and Amal–Zeit Province (SW-dipping half graben), separated by the Galala–Abu Zenima Accommodation Zone and Morgan Accommodation Zone, respectively. The southern Amal–Zeit Province, as formed during rifting, continues south into the Red Sea to the city of Quseir, with the Quseir–Brothers–Aslam Accommodation Zone marking the southern boundary, but this is now dissected by the Late Miocene Gulf of Aqaba strike-slip system. Orthogonally to the rift access, the Suez Rift also divides into three large half graben, parallel to the clysmic trend with the central portion defining the contemporary Gulf of Suez, flanked on either side by the uplifted rotated fault blocks that form part of the rift system.

Pre-rift and syn-rift beds dip at an average of 10–15°, but due to the increased extension in the southern part of the rift, pre-rift strata locally dip up to 45–50°. Although the Gulf of Suez is considered a failed rift system, minor structural movement/uplift continues today. This is demonstrated along the coast at Gebel Zeit, where successions of Quaternary reef terraces have been uplifted on the eroded footwall. This contributes to the uplift of the rift shoulders, locally by as much as 2 km, since the cessation of extensional rifting in the Late Miocene.

Regional stratigraphy

The Gulf of Suez stratigraphy is divided into three tectonostratigraphic intervals, (1) basement and pre-rift, (2) syn-rift and (3) post-rift (Fig. 2). The basement is part of the Precambrian Pan-African craton, created by the amalgamation of several

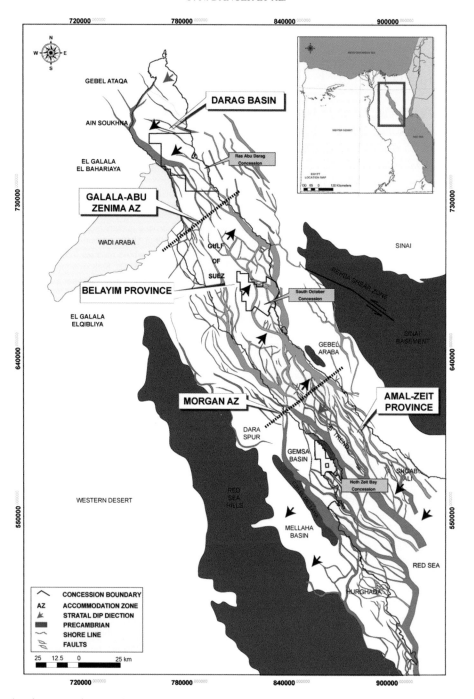

Fig. 1. Gulf of Suez – location map and structural elements.

island arcs (630–715 Ma) and modified by later transcurrent tectonics and widespread granitic intrusions. The basement is overlain by the Nubia Sandstone which comprises a number of units, predominantly of fluvial sandstones, ranging in age from Cambrian to Early Cretaceous. Overlying the Nubia are Late Cretaceous formations comprising an interbedded mixture of carbonate and clastic dominated intervals, reflecting the regional marine transgression over the continental Nubia deposits. The overlying Eocene sequence is predominantly carbonate and shale rich. The entire pre-rift stratigraphy thins dramatically southward from >2000 m at the northern tip of the Gulf to <400 m at its junction with the Red Sea. This is due to a combination of increased depositional thickness in the north due to greater subsidence, and increased erosion in the south at the onset of rifting.

The syn-rift stratigraphy encompasses the Abu Zenima and/or Nukhul Formations, overlain by the Rudeis and Kareem Formations. The Abu Zenima Formation, where present, represents the onset of the Gulf of Suez rifting and consists of continental red beds and volcanics. In southern parts of the Gulf, the formation is missing and the first representation of syn-rift sedimentation is the Nukhul basal conglomerates, resting directly on the pre-rift units. This, however, may reflect differences in the onshore exposures with the basal rift unit of the immediate hanging wall areas exposed in the north, but only footwall crests in the south.

This represents the breakup of the pre-rift structure (pre-Early Miocene) along clysmic faults, with half graben development, transgression and establishment of generally shallow marine, mixed clastic and carbonate environments. Extensive carbonate reef build-ups developed on the footwall crest palaeo-highs at this time, especially on the fault margins of the rift.

The overlying Rudeis Formation represents the regional highstand reflecting both the rift maximum and eustatic sea-level

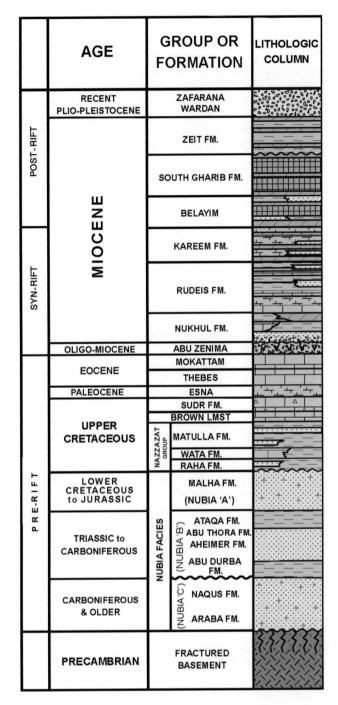

Fig. 2. Generalized stratigraphic column for the Gulf of Suez.

cross-faults. This is discussed further in the sections on specific sub-basins. The basin axes, independent of sediment input direction, are characterized by axial turbidite systems.

The overlying post-rift units are dominated by evaporites. The South Gharib and Belayim Formations consist of massive halite and anhydrite units, with interbedded carbonate, shale and clastic units. The evaporites form the regional top seal for many Gulf of Suez fields. They also pose a major challenge in acquisition and processing of seismic.

Source rocks

The primary source rocks in the basin are considered to occur in the pre-rift interval, but the sourcing of oil in the reservoirs of the Gulf of Suez is still subject to debate (Wever 2000). The major source rock is the Type II/I oil prone Brown Limestone (Duwi) Member of the Upper Cretaceous Sudr Formation. Secondary source rock intervals occur locally in the pre-rift Eocene Thebes Limestone (type II/I oil prone), and in the Miocene syn- and post-rift Rudeis, Kareem and Belayim Formations (type II oil and gas prone rocks). Oil generation began during the Late Miocene or Pliocene times, and is considered to be present day mature.

Reservoirs

Reservoir intervals in the Gulf are found throughout the entire stratigraphy.

Pre-rift reservoirs

The Nubia Formation Sandstone is the primary reservoir in the Gulf (Ras el Ush, October and East Zeit Fields), varying from 200 to 600 m in thickness and is known to be present across the whole of the Gulf of Suez. In some areas it has been subdivided into three members, A, B and C, where B is largely shale prone, but this stratification is not present across the whole of the Suez rift basin. The Nubia was deposited in a range of environments from continental to shallow marine. It is a high quality reservoir, but burial reduces the effective porosity such that, at 4000 m burial, porosities are in the range of 9–13%. In general, the other pre-rift reservoirs which have significantly less net reservoir, become economic when associated with Nubia accumulations, and include fractured Precambrian basement (Zeit Bay Field) as well as the overlying Nezzazat Group.

The overlying Nezzazat Group consists of clastic and carbonate sediments interpreted to have been deposited in a range of shelf environments. The group contains three formations with reservoir potential: Matulla, Wata and Raha Formations. The Matulla Formation is present across the Gulf, usually between 100–150 m thick, with high reservoir quality sandstones (B Trend Fields). The Wata and Raha Formations have minor sandstones with limestones, and are not generally effective reservoirs (indeed the Raha Formation is a potential source rock in the northern part of the Gulf).

The Lower to Middle Eocene Thebes Formation is a massive limestone with thin chert bands. Although generally a tight limestone (and potentially a seal), variable facies and secondary porosity development have allowed this locally to become an effective reservoir (Ras Bakr Field, Ras Sudr Field).

Syn-rift reservoirs

The Nukhul Formation covers most of the central Gulf of Suez. Areas with no Nukhul Formation present relate to erosion of fault block crests or non-deposition. Facies range from non-marine

high. Carbonaceous and marl lithologies predominate, with clastic deposits being common at the base of the formation. Carbonate reefs continued to develop on palaeo-highs throughout Rudeis times on the rift margin.

The Kareem Formation comprises similar lithologies to the Rudeis Formation with local development of evaporite units. Clastic deposits are variably developed during lowstands, in particular at the mid-Rudeis unconformity and in the lower part of the Kareem Formation. These clastic intervals reflect the exposure and erosion of footwall crests from earlier, marginal rift basins. Sediment input is both into the hanging wall of the fault scarp and down the dip-slope from the footwall crest, although the most significant sedimentation is focused through offsets in the clysmic faults, particularly at transfer zones. The direction of sediment dispersal is critical to reservoir development and is dependent on the interaction of the contemporary drainage system with sandstones to marine sandstones, shales and carbonates. The distribution of facies varies dramatically, reflecting multiple sediment input points and different hinterland lithology. Deposition is locally accompanied by syn-rift volcanism (Abu Rudeis–Sidri Field).

The Rudeis Formation is a deeper, more open marine, shale-dominated section containing significant syn-rift sandstones. The sands were deposited as a system of major basin margin alluvial outwash fans feeding deepwater submarine fans in active hanging wall depocentres. Sediment input is structurally controlled at transfer faults which breach the clysmic trend. This means that the sandstones can reach over 350 m thick (well South July-1X) and yet pinchout rapidly away from the clastic source and entry point. Distinctive productive reservoirs are the Ysur Sandstone in the Lower Rudeis (e.g. July Field) and the Asl Sandstone in the Upper Rudeis (e.g. Belayim Marine Field).

Post-rift reservoirs

Reservoir sandstones are present within the post-rift section (Zeit, South Gharib and Belayim Formations; Gemsa Field and Belayim Land Field), but they are not volumetrically significant in terms of discovered hydrocarbons. The key risk with these reservoirs is the unlikelihood of hydrocarbon charge, as they lie within the halite top seal.

Play types

The primary hydrocarbon play in the Gulf of Suez is the crest of rotated fault blocks, both as primary footwall crests (B Trend Fields, Zeit Bay Field and July Field) and downthrown terraces from these (e.g. Ras el Ush Field). The primary seal in all cases is provided by drape of post-rift evaporites. The large angle of rotation of the fault blocks allows the high-quality Nubia Sandstones to be within closure, with secondary reservoir intervals in the overlying Nezzazat Group.

Generally there is no effective seal between the Nubia Sandstones and the overlying Nezzazat Group. This results in a single oil–water contact (OWC) for multi-reservoir pre-rift fields and, to have a successful Nubia accumulation, fault throw needs to be sufficient to juxtapose the Nubia against syn- and post-rift marls and evaporites. Significant intrafield faults do not appear to form barriers and may in fact contribute to vertical connectivity through juxtaposition or open fractures. A consequence of this is that, when the Nezzazat Group has no well developed sandstone members, it can be regarded as a waste zone for exploration, occupying pre-rift closure at the expense of the Nubia Sandstone.

A secondary issue for the pre-rift play is charge, as the source rock stratigraphically overlies the main reservoirs. Effective charge is therefore generally achieved by juxtaposition.

In the syn-rift interval, there are three-way faulted dip closures (e.g. Belayim Marine Field) as well as occasional four-way dip closed anticlines (e.g. Belayim Land Field), both of which are again sealed by the overlying post-rift evaporites.

Combination structural–stratigraphic traps are also common, but contain smaller accumulations than the pure structural traps. These occur due to the rapid thickness variations associated with the syn-rift sedimentation; the thinner reservoir intervals allow stratigraphic plays and subtle hanging wall structural closures to be developed against faults. These include unconformity-related truncations, onlap pinchouts (e.g. Nukhul and Rudeis Formation sandstones) and reefal carbonates. In all cases the seal to the clastic reservoirs is provided by the overlying shales, and in contrast to the pre-rift plays, sands within the same stratigraphic unit can have different fluids and contacts.

Seismic data quality

The seismic data quality in the area ranges from poor to fair, depending on the local overburden geology and the depth or stratigraphic level of the target. The prime causes of poor seismic data are the lack of energy propagation to the target reflectors and the significant reverberation caused to the seismic signal. The water depth and seabed characteristics are benign to seismic acquisition, and the data is not heavily contaminated with seabed multiples. However, the mid-range section, from typically 300–500 ms until 1500 ms, is dominated by the Zeit and South Gharib Formations, which contain interbedded halites, anhydrites, shales and sands. This is compounded in areas by thick halite salt swells, which add seismic distortion to the already degraded signal. Traditional 2D data have limited penetration beneath the syn-rift section, and even relatively modern 3D seismic datasets struggle to image the pre-rift reservoirs. In addition to not successfully imaging structural configurations, the poor data quality precludes any amplitude analysis to relate to fluid types within the reservoir intervals.

Various attempts to improve the imaging by pre-stack depth migration have not significantly improved data quality, as the lack of signal propagation, and deep stratigraphic illumination, cannot be resolved with depth imaging. More recently, 3D ocean-bottom cable (OBC) surveys have been acquired, which show significant improvement in data quality. Initially these surveys were acquired due to shallow-water constraints at the margins of the Gulf, but the improved data quality is promoting their use in areas that were previously covered by streamer surveys. The reason why OBC data offers improved resolution is three-fold. Firstly, OBC surveys tend to be acquired with higher fold, enhancing the signal-to-noise ratio and hence the suppression of noise. Secondly, the OBC data have greater bandwidth (due to the receiver ghost effect, and also the use of geophones), especially at the lower end of the frequency spectrum, which is less prone to attenuation from thin interbedded stratigraphy. Finally the OBC acquisition layout allows for a more uniform azimuthal illumination.

Exploration targets

The dearth of high-quality seismic datasets has hampered recent exploration in the Gulf of Suez and continues to prove problematic as the remaining trap size diminishes. Initial exploration in the Gulf focused on the extrapolation of surface geological features and limited gravity data. With the advent of reflection seismology, the identification of the Upper Miocene evaporite became relatively straightforward. Significant four-way dip closures were identified at this level and most of the major fields were discovered underlying these highs. However, a number of dry holes were drilled based on the premise of a hard linkage between pre- and post-evaporite structure. The gradual understanding that there is an offset structural relationship has led to improvement in well prognosis in targeting the syn- and pre-rift fault blocks.

With the extensive drilling in the Gulf of Suez, most (if not all) of the simple fault blocks and anticlinal features have been drilled, leaving little potential for pre-rift plays requiring large fault offsets. Thus current exploration focus is on smaller structures. This is challenging even with 3D seismic data due to the limitations of imaging in this area: even over producing fields it is still possible to drill wells that do not encounter the reservoir horizons, due to the poor structural control provided by seismic data. Therefore the potential for significant discoveries is in the underexplored but riskier stratigraphic and downthrown fault block of the syn-rift suite.

These play types in the syn-rift are discussed in more detail below, with reference to selected sub-basins, which encompass the three main structural provinces of the Gulf of Suez. Firstly, in the southern Gulf, the syn-rift plays of the Amal–Zeit Province are discussed with reference to the Gemsa Basin. Secondly, the syn-rift plays of the central and northern Gulf (the Belayim and Darag Basin Provinces) are jointly discussed by reference to two concessions: the South October Concession in the Belayim Province and the Ras Abu Darag Concession in the Darag Province (Fig. 1).

Gemsa Basin footwall prospectivity

The Gemsa Basin is located onshore in the southern Gulf of Suez, Amal–Zeit Province. Immediately north of the basin lies the Morgan Accommodation Zone, while the basin is bordered to the SW by Esh el Mellaha (Fig. 1). Both areas are major hinterland sources for clastic deposition in the northwest and SW of the basin. The basin is a highly rotated half graben block, with the entire stratigraphy exposed, but thinned, in the footwall crest to the NE at Gebel Zeit. The pre-rift stratigraphy dips by as much as 45–50°SW, while the top syn-rift succession dips 15°SW (Fig. 3).

The basement exposures at the footwall crest are separated into North and South Gebel Zeit by a saddle draped by post-rift evaporites. In the larger exposure of North Gebel Zeit the erosion of the softer Nubia Formation creates a fault parallel valley, Wadi Kabrit, between the Precambrian basement and the harder Miocene deposits (Fig. 4). At South Gebel Zeit the basement complex also forms a prominent ridge, but the area is structurally complex and the majority of the pre-rift stratigraphy is absent, resulting in a more random topography.

With the exposure of the footwall, the footwall crest play is precluded in this area. The exposures at Gebel Zeit, however, have provided an excellent opportunity to examine the stratigraphy and structure of a footwall crest, which is the main play of the Gulf of Suez and as such the area has been intensely studied (Evans & Moxon 1988; Helmy 1990; Bosworth 1995; Winn et al. 2001; Aboud et al. 2003). Poor seismic imaging has also resulted in the pre-rift being considered an unattractive target in the Gemsa Basin.

The syn-rift Rudeis and Kareem Formations progressively onlap and pinchout onto the footwall crest (Fig. 3). The only tested discoveries in the updip area are in the Kareem Formation of Abydos and Gazwarina 1 wells. Each contains two vertically isolated thin (less than 6 m) sandstones and their relationship to a pair of anhydrite marker beds indicates that sandstones are developed at a minimum of three stratigraphic levels, indicating that they are not laterally continuous. The same marker beds are used to subdivide the Kareem Formation lithostratigraphically into the Shagar and Rahmi Members; an anhydrite is taken as the base of each. The lower anhydrite therefore marks the boundary between the Kareem and Rudeis Formations, although biostratigraphy suggests the formation boundary is slightly deeper. This is consistent with a repeated upward succession of lithofacies for each of the anhydrites from laminated shale, through birdcage anhydrite into banded anhydrite. The sand at the base of the Shagar Member clearly erodes into and locally removes the upper anhydrite. The development of the anhydrite indicates the isolation of the Gemsa Basin from the rest of the Gulf and therefore the emergence of the footwall area. However, whether the renewed clastic input reflects further lowering of sea-level, tectonic footwall uplift or climatic shift to wetter conditions is less clear.

The drainage system associated with such run-off is critical to the distribution of coarse clastics in the basin. The structural position, high on the dip-slope of the footwall, clearly indicates that any clastic sediments in this area will have been derived from the emergence of the footwall crest. Today, Wadi Kabrit and the comparable system of Wadi Araba to the north are characterized by axial drainage (Fig. 4) with entry onto the adjacent dip and scarp slopes controlled by cross-faults. Although it is difficult to demonstrate whether an equivalent drainage system existed during the Kareem Formation, the presence of a large incised feeder system towards the southern end of the present day Wadi Kabrit is certainly compatible with this. The incision cuts down into the pre-rift, and limited biostratigraphic evidence suggests it may be of Kareem age (Evans & Moxon 1988), a view supported by the absence of sandstones in the Rudeis Formation.

The Kareem depositional environment has been a subject of much conjecture and the sands have been regarded as turbidite deposits, largely on the evidence of deepwater bio-facies above and below. However, this view can be questioned based on a number of observations:

(1) The feeder system, previously described as a turbidite channel, is in a very proximal position with respect to the footwall crest, which has adequate hinterland to allow the erosion.
(2) The fill of the feeder contains abundant bioturbation indicative of *Thalassinoides/Ophiomorpha* ichnofacies, suggesting shallow marine conditions.
(3) The anhydrite units commonly display chicken-wire fabric, which in isolation is a poor environmental indicator, but when supported by the other observations is consistent with a shallow-water/sabkha environment (Machel & Burton 1991). The foraminifer genus *Amphistegina* has been reported from the anhydrite in one well. This shallow-marine genus has been particularly successful in invading the eastern Mediterranean through the Suez Canal (Gruber et al. 2007). The occurrence of *Amphistegina* would therefore be consistent with sporadic flooding of a restricted basin.

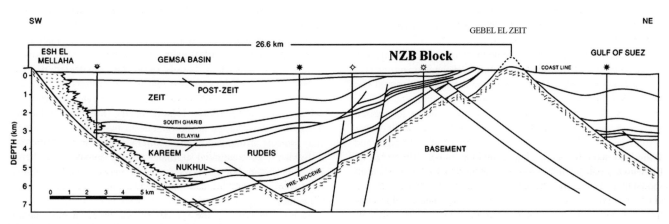

Fig. 3. Gemsa Basin half graben cross-section (after Bosworth 1995).

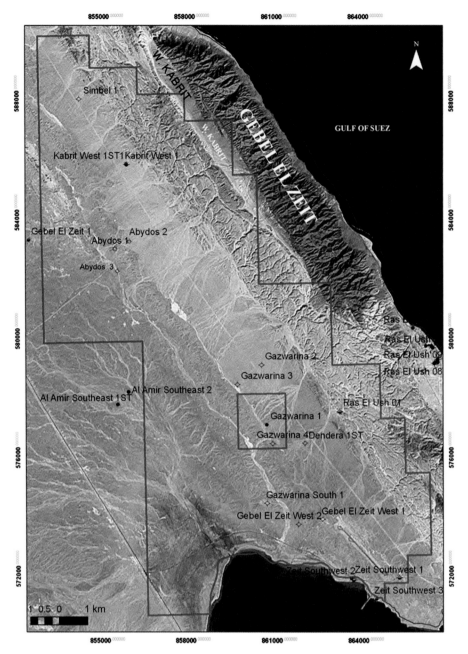

Fig. 4. Gemsa Basin, landsat 7742: geological image.

The channelized turbidite model has been used to propose sediment by-pass of the slope area, with deposition limited to isolated channels. However, an alternative interpretation of the outcrop channel is a shallow marine fill of a sub-aerially eroded incised valley. This allows potential for shallow water re-working and lateral distribution across the dip-slope.

Although it has to be tempered with the uncertainty in the quality of the seismic data, the seismic character of the Kareem interval appears to support this. On seismic data, the Kareem interval can be divided into an upper, relatively continuous unit, and a much more variable lower unit, which appears to onlap an incised surface (a possible palaeo-shoreline and cliff) on the dip-slope. This is broadly compatible with the character of the Shagar and Rahmi Formations seen on log correlation, although it has to be emphasized that the thickened interval has been very poorly constrained by well data. The alternative model suggested here is that the incised surface represents a lowstand erosive down-cut and this may have provided accommodation space for additional sands. The recent drilling of the Abydos-2X well supports this model, encountering Base Shagar sands in an erosive down-cut of the Rahmi Formation, which were not present in the Abydos-1X well some 500 m away.

The trapping mechanism remains conjectural because the individual sandstones are below seismic resolution, but is suspected to be a combination of stratigraphic pinchout and small fault offsets, given the individual units are thin. However, the recovery of 2.3×10^6 stock tank barrels (MMSTB) on depletion drive from Gazwarina 1 does suggest reasonable connectivity. In conjunction with this updip trapping there is clear evidence from the seismic of low amplitude folding of the Kareem Formation. These fold axes appear to be oblique to the cross-faults, which may indicate an element of strike-slip movement in their genesis.

Evidence of continued tectonic activity into the 'post-rift' phase in the area of Gebel Zeit is demonstrated by the majority of movement on an antithetic fault, 9 km downdip of the footwall crest, occurring during the post-rift Zeit Formation. This fault movement may also contribute to the continued uplift of Gebel Zeit. The area of Gebel Zeit therefore offers a very different perspective on the depositional processes to those derived from the exposures in the north-eastern Gulf (Young et al. 2000, 2003; Jackson et al. 2005). There the studies have focused on the deeply dissected deposits of the immediate hanging wall in areas of only moderate rotation, close to transfer zones. In Gebel Zeit, outcrop shows the deposition environment on the proximal dip-slope of the footwall crest, in an area of much greater rotation, and the position in the southern Gulf seems to have allowed continued tectonic activity into the period of Red Sea rifting, questioning the designation of the Kareem as post-rift, and the rigid syn-rift/post-rift division of the Gulf of Suez.

Upper Rudeis prospectivity of the offshore Ras Abu Darag and South October Concessions

The Ras Abu Darag and South October Concessions lie either side of Galala–Abu Zenima Accommodation Zone in the Darag Basin and Belayim Province, respectively (Fig. 1). This results in a shift from SW-dipping half graben in the Darag Basin to NE-dipping half graben in the Belayim Province. Geologically both concessions are representative of their respective sub-basin and share a common depositional history as axial basins within the Gulf of Suez rift.

Pre-rift reservoirs have been drilled in both concessions. However, deep burial and poor seismic imaging with the currently available 2D and 3D datasets make the pre-rift targets less attractive than those in the syn-rift sequence.

The syn-rift stratigraphy is dominated by the Nukhul and Rudeis Formations and, in contrast to the Gemsa Basin further south, sandstones are developed in the Rudeis Formation and not in the Kareem Formation. The major clysmic faults control the thickness of the Upper Rudeis Formation, with the maximum thickness of the Rudeis Formation within the two concessions occurring in well GS 56-1A in the Ras Abu Darag Concession, where it attains 2125 m. Active rifting appears to have ended before the deposition of the Kareem Formation evaporates, and hence the syn-rift hydrocarbon prospectivity focuses on the Upper Rudeis Formation, in which the main reservoir sandstone is informally called Asl Sandstone. The Lower and Middle Rudeis Formation are also described, as this sets the prospective interval in its depositional, sequence stratigraphical and structural context.

In contrast to the generally shallow marine to brackish depositional conditions of the underlying Nukhul Formation, planktonic foraminifera assemblages at the base of the Rudeis Formation show a sharp increase in water depth to upper bathyal conditions. The Rudeis Formation is generally conformable on the Nukhul Formation. However, the Nukhul Formation and the basal part of the Rudeis Formation are absent in wells in the southwestern part of the Ras Abu Darag Concession associated with footwall crests. This may reflect onlap of the Early Miocene clastic systems onto the evolving, rotated fault blocks, particularly on the flanks of the graben. In the east of the South October concession, structural terraces flanking the main basin margin fault show the Rudeis Formation directly overlying basement. It is unclear whether this is a fault contact or a major unconformity, possibly recording transpressional uplift along fault-bounded structural terraces before the onset of rift subsidence and onlap by the deepwater Rudeis system.

The three-fold subdivision of the Rudeis Formation is based on the 'N' zonation of planktonic foraminifer (Blow 1969), derived from 1980s vintage reports from wells in both the South October and the Ras Abu Darag Concessions:

- Upper Rudeis: *Globigerinoides sicanus – Praeorbulina transitoria* zone (lower part of N8), described in (Schütz 1994).
- Middle Rudeis: *Globigerina ciperoensis – Globigerinoides trilobus* zone (N7).
- Lower Rudeis: *Globigerinoides primordius – Globigerina ciperoensis* zone (?N6).

Well picks of the three units were adjusted to maximum flooding surfaces where possible (Fig. 5). The regional, c. 18.25 Ma T_{20} Mid-Clysmic Unconformity, as described in proprietary well and outcrop studies, could not be identified biostratigraphically in the Ras Abu Darag and South October Concessions, probably reflecting the basinal position of the wells during Rudeis times. However, it is possible that this regional structural event may be an important control on the distribution of Rudeis Formation Asl Sandstone reservoirs on the correlative unconformity (Fig. 5).

- *Lower–Middle Rudeis Formation.* In the hanging wall of the major clysmic faults, the Lower–Middle Rudeis Formation is interpreted as a rift climax, deep marine sedimentary system. The lower part of the Rudeis contains (?benthonic) foraminifera indicative of upper bathyal conditions, which pass upwards into outer shelf assemblages and finally into sub-littoral conditions at the upper part of the formation. This overall shallowing-upward profile probably reflects the progressive development of the T_{20} Mid-Clysmic Unconformity. Rapid facies changes into shallower water, more calcareous or sandy sediments occur towards the basin margins and around palaeo-highs (Schütz 1994).
- *Upper Rudeis Formation.* This unit is characterized by rift margin fan deltas feeding deep marine clastic systems in hanging-wall depocentres. The thick development of potential reservoir quality sandstones (Asl Sandstone) is limited to the Upper Rudeis Formation. These occur mainly as sharp-based sand bodies, possibly as lowstand fans associated with the T_{20} Mid-Clysmic Unconformity (Fig. 5). Deposition is either in incised channels/valleys which are back-filled during transgression, or on slope apron systems deposited in front of clysmic faults. Other Asl sands occur as progradational, submarine fan, lobe and channel complexes.

A map of Upper Rudeis sandstone distribution from well data shows the inferred lateral input and axial transport dispersal patterns controlled by the clysmic faults in the northern Gulf of Suez (Fig. 6). Three main types of Upper Rudeis sediment supply have been identified:

(1) Point sources of sediment input structurally controlled by transfer zones which offset the major clysmic faults, for example, the Belayim Transfer and the Ekma Transfer faults.
(2) Axial transport southward of coarse-grained clastics input from the Wadi Araba High. These sands were focused along the rift axis in the immediate hanging wall of normal faults.
(3) Hanging-wall sands locally derived from the erosion of intra-rift footwall highs, for example, erosion of the Belayim Marine Field crest resulted in deposition of submarine fan systems in the immediate hanging wall. Notably this contrasts with the Gemsa Basin where the focus is on sands derived from the footwall crest onto the dip-slope.

The Rudeis Formation in the South October Concession

Reservoir development is clearly controlled by the nature of the eroded hinterland. The onshore area south of the Nezzazat Fault (Fig. 6) has undergone extensive uplift with erosion down to basement in many places. This has resulted in erosion of the Nubia

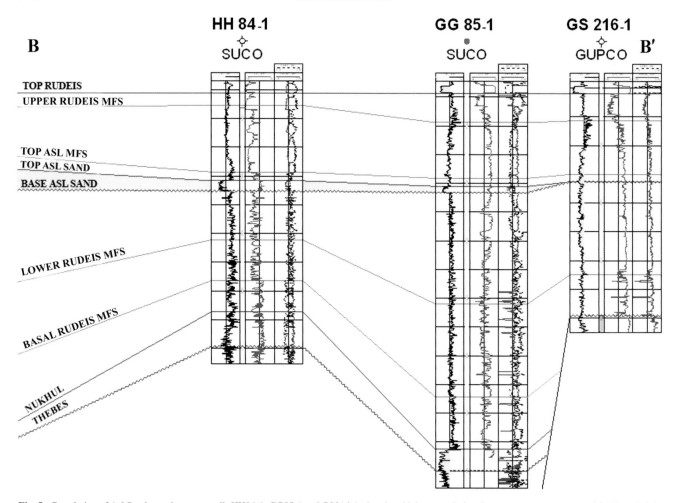

Fig. 5. Correlation of Asl Sandstone between wells HH84-1, GG85-1 and GS216-1, showing thickness variation. Log curves are gamma ray (black), resistivity (pink) and density/neutron (red/dashed black). Location of correlation shown in Figure 6. Vertical scale: 50 m per division.

Sandstone which has been reworked and deposited as syn-rift sandstones. In contrast, within the Belayim province (Fig. 6) there has been less uplift with erosion of predominantly pre-rift carbonates. This has resulted in the local development of calcarenitic syn-rift sediments along the eastern basin margins. However within the rift axis, quartzose sands are present, having been transported south from the Wadi Araba High.

In the South October Concession the thickness of the Rudeis Formation is greatest in the immediate hanging wall of clysmic faults. However the Rudeis Formation shows an overall thinning to the SE in accordance with regional sediment distribution patterns. Thickness ranges from 700 m in the hanging wall to 300 m on the dipslope. Rudeis Formation depocentres do not necessarily contain a significant thickness of sandstone, the input of which is controlled by the composition of the hinterland and location of transfer zones (Fig. 6). However the overall Rudeis Formation thickness is significantly less than that seen in the Ras Abu Darag Concession.

The Asl Sandstone is a proven reservoir in the GG85-1 discovery, which is located in the southeastern part of the South October Concession, on terraces flanking the main basin-bounding fault (Fig. 6). The Belayim Land Field lies directly south of the discovery and is separated from it by a structural low. GG85-1 occurs in the hanging wall of the major fault and is a possible transpressional feature associated with a transfer zone. It is a combined structural–stratigraphic trap, consisting of a SW-plunging anticline. From well data it is known to have lateral stratigraphic pinchout of the reservoir sandstones to the north. This is demonstrated by the gradual thinning and absence of the sands in wells GS216-1 (Fig. 5). Trapping to the east is provided by hanging wall fault closure, possibly against basement. Top seal is provided by Upper Rudeis Formation shales. Well GG 85-1 was drilled in 1983 and tested 27° API oil from an 11 m thick Asl sandstone reservoir. Understanding of the precise depositional nature (i.e. deep v. shallow marine to continental) and the distribution of the Asl sands in the discovery is uncertain. The discovery lies alongstrike from thick (>300 m of net reservoir) Asl sands of probable alluvial fan origin in the Belayim Land Field located to the south, adjacent to a major transfer zone that acts as a point source for significant sediment input. To the west of the Belayim field lies the deepwater, turbidite-prone basin of the central Gulf. One possibility, given the sharp base to the reservoir sand interval, is that the sands occur above an unconformity as a transgressive package within an erosional valley, but other interpretations (e.g. deepwater slope apron sands) are possible.

The Rudeis Formation in the Ras Abu Darag Concession

North of the Galala–Abu Zenima Accommodation Zone there is good evidence showing thickening of the Rudeis Formation in the northwestern part of the Ras Abu Darag Concession. This suggests a sediment entry point in this part of the northern Gulf of Suez, probably feeding from the west (Fig. 6). This requires connection to a large antecedent drainage network in the hinterland (possibly the Wadi El-Abyad drainage system). Widespread reservoir sands in the NE part of the Ras Abu Darag Concession suggest another structurally controlled point source fed by the Wadi Sudr drainage system.

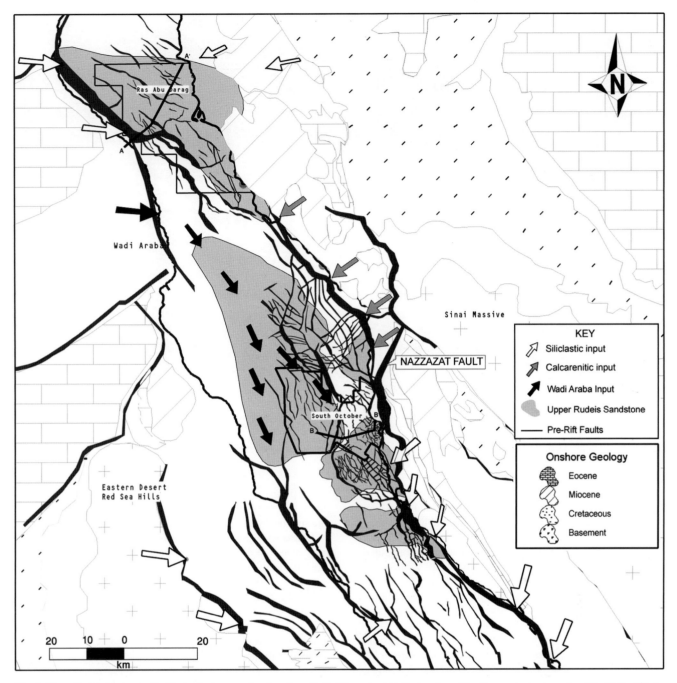

Fig. 6. Upper Rudeis sandstone distribution in the central and northern Gulf of Suez. A–A' is the location of the discussed seismic section (Fig. 7). B–B' is the location of the discussed well correlation (Fig. 5).

Increased distance from the clastic source leads to sand-poor distal portions of fan lobes near the base of the Upper Rudeis. Coarsening-upwards packages with carbonate-rich tops may reflect shallowing to shelfal conditions. Geological interpretation suggests that the Darag and Belayim Provinces behaved as separate en-echelon structurally controlled basins, each with its own sediment entry and dispersal patterns, but sharing synchronous rifting and sea-level variation.

Although the basinal Ras Abu Darag wells have encountered significant sections of Upper Rudeis formation, where thicknesses range between 460 and 1035 m, on the footwall crest of the main basin-bounding fault in the southwestern part of Ras Abu Darag, the Upper Rudeis is absent to thin (23–140 m). Possible causes of the thinning may be due to a combination of reduction in accommodation space on structural highs, onlap, faulting at footwall crests and erosion of footwall crests. The evidence from the Gemsa Basin is that all of these may occur. Any footwall-sourced sediment is likely to be more variable in character and quality and volumetrically less important. The scale of the sand distribution in the Gemsa Basin in this structural setting confirms that economic accumulations in an offshore setting are unlikely.

Given the narrowness of the deepwater Darag Basin during Upper Rudeis times, it is probable that large areas of the basin floor are sand-prone, given that the likely sediment transport processes are gravity-driven mass flows, allowing for distal sand deposition. Contemporary shelf deposits are known on the regional dip-slope offshore and onshore east of the Gulf within the Darag Basin.

Ras Abu Darag well Nebwi 81-1 encountered a 73 m thick sandstone interval which contains high proportions of net reservoir within the Upper Rudeis Formation. Non-net intervals within these gross sand intervals are either calcite-cemented sandstones or primary carbonates. Average PHIE of the Asl sands is 16%. Petrophysical work identified a unit called the Asl Sandy Limestone with thick (up to 30 m), high net-to-gross units with average

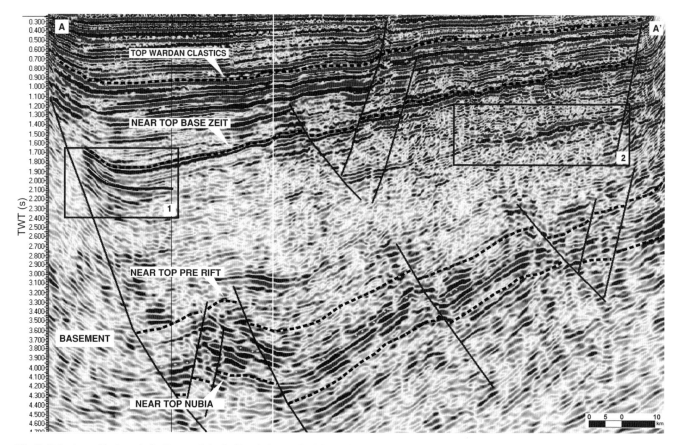

Fig. 7. Seismic profile through the Darag sub-basin. Box 1 shows a Rudeis proximal sandstone fan. Box 2 shows a shallow water carbonate. Profile location shown in Figure 6.

porosities of 15%. The top of the Upper Rudeis Formation in the Fina Z 80-1/1A well contains a 24 m thick, stacked sand body, which consists almost entirely of net reservoir and has an average porosity of 19%.

Away from the well control in Ras Abu Darag, seismic amplitude anomalies (Fig. 7) are identified as likely sandstones or calarenites, both in hanging wall and dip-slope settings, which offer potential targets in the Upper Rudeis. Hanging-wall traps offer the additional advantage of updip seal against major faults. Submarine slope apron deposits may be associated with the active major fault scarps. Dip-slope sands rely more heavily on updip stratigraphic termination of the reservoirs, as demonstrated in the Kareem Formation in the Gema Basin. This may result from either depositional pinchout of the reservoir or erosional truncation during relative sea-level fall acting on the generally uniformly dipping, ramped margin. As suggested for Gebel Zeit, such ramps in shallow marine settings could act as sites of sand deposition associated with forced regressions during the falling stage of the relative sea-level cycle. Carbonate build-ups and shoals may also be developed along the ramp margin. In deeper marine settings, lowstand submarine fans represent another possible play. However, most of these stratigraphic configurations require the addition of structural closure to form effective traps so that the acquisition of high-quality seismic data remains paramount.

Conclusion

With mean expected field sizes getting smaller, it is increasingly difficult to detect reliable traps with conventional 3D streamer seismic data. Even with good definition from the use of high-fold 3D OBC seismic data, these targets are without Direct Hydrocarbon Indicators, and therefore are significantly riskier than comparable targets in other basins. However the Gulf benefits from excellent outcrop exposure that directly relates to onshore/offshore prospectivity, and combined with geological control from adjacent wells, the structural and sedimentological risk can be managed.

Unless widespread 3D OBC seismic data is acquired to offer a step chance in deep pre-rift imaging, the lack of clearly defined pre-rift fault blocks will cause exploration to focus on more subtle plays within the syn-rift section. Outcrop geology shows clearly the rapid variation in syn-rift sedimentation and the occurrence of sand pinchouts. This play also benefits by being above the regional source rocks and closer to the regional seal. The Gemsa Basin is a clear example where significant exploration effort focused on the pre-rift has failed, but the syn-rift Kareem Formation reservoirs demonstrate successful stratigraphic trapping and downthrow fault plays. These plays are difficult to define seismically, but the combination of excellent outcrop and well control allow for moderately risked exploration targets to be created and drilled successfully. However, this is in an onshore setting and the very different economics of offshore drilling increase the challenge in the offshore areas.

Remaining potential exists for further hydrocarbon discoveries with a significant stratigraphic component in the syn-rift systems of the Gulf of Suez. Current understanding of the syn-rift systems in terms of their stratigraphy, provenance, sedimentary architecture and sequence stratigraphic and structural context should form the basis for increased recognition of these potential reservoirs.

Further application of this type of play, and the conventional fault block play within the Rudeis and the Kareem Formations, will allow additional targets in the offshore Gulf to be defined. If combination structural stratigraphic targets can be developed these should provide attractive volumes to pursue.

This paper is published with the permission of Dana Petroleum, and the authors would like to thank Dana Petroleum and EGPC for their support. The authors are grateful for the constructive discussion from various co-workers within their respective companies and would like to thank Paul Wilson and Paul Woodman for introducing the complexities of the Gulf of Suez to the authors as field trip leaders in 2008. The authors also thank Hassan Aboudy for drafting the figures. Stefan Lubesedar is gratefully thanked for his review and constructive remarks for improving the clarity of this paper. Since completion of this paper we regret to announce the death of our respected colleague Barry Hepworth who made an invaluable contribution to this paper.

References

Aboud, E., Salem, A. & Ushijima, K. 2003. Interpretation of aeromagnetic data of Gebel El-Zeit area, Gulf of Suez, Egypt using magnetic gradient techniques. *Memoirs of the Faculty of Engineering, Kyushu University*, **63**, 139–149.

Blow, W. H. 1969. Late Middle Eocene to Recent planktonic foraminiferal biostratigraphy. *In*: Bronnimann, R. & Renz, H. H. (eds) *Proceedings of the First International Conference on Planktonic Microfossils*, Geneva, 1967, **1**, 199–421.

Bosworth, W. 1995. A high-strain rift model for the southern Gulf of Suez (Egypt). *In*: Lambiase, J. J. (ed.) *Hydrocarbon Habitat in Rift Basins*. Geological Society, London, Special Publications, **80**, 72–102.

Bosworth, W. & McClay, K. R. 2001. Structural and stratigraphic evolution of the Gulf of Suez rift: a synthesis. *In*: Zeigler, P. A., Cavazza, W., Robertson, A. H. F. R. & Crasquin Soleau, S. (eds) *Peritethyian Rift/Wrench Basins and Passive Margins*. Museum National d'Historie Naturelle de Paris, Memoirs, **186**, 567–606.

Colletta, B., Le Quellec, P., Letouzy, J. & Moretti, I. 1988. Longitudinal evolution of the Suez rift structure (Egypt). *Tectonophysics*, **153**, 221–233.

Evans, A. L. & Moxon, W. 1988. Gebel Zeit chronostratigraphy: Neogene syn-rift sedimentation atop a long-lived palaeohigh. *In*: Hantar, G. (ed.) *Proceedings of the 8th Egyptian General Petroleum Corporation Exploration Seminar*, Cairo, **1**, 251–265.

Gruber, L., Lazar, S., Hyams, O., Sivan, D., Herut, B. & Almogi-Labin, A. 2007. *Amphistegina lobifera*, a larger symbiont-bearing foraminiferal migrant from the Red Sea, now dominates rocky coasts of the Israeli Mediterranean. *Geophysical Research Abstracts*, **9**, 01407.

Harrell, J. A. & Lewan, M. D. 2002. Sources of mummy bitumen in Ancient Egypt and Palestine. *Archaeometry*, **44**, 285–293.

Helmy, H. M. 1990. Southern Gulf of Suez, Egypt: structural geology of the B-Trend oil fields. *In*: Brooks, J. (ed.) *Classic Petroleum Provinces*. Geological Society, London, Special Publications, **50**, 353–363.

Hume, W. F. 1916. *Report on the Oilfield Region of Egypt, with a Geological Report on the Abu Shaar El Quibli (Black Hill) District*. Petroleum Reservoir Bulletin, **6**. Government Press, Cairo.

Jackson, C. A. L., Gawthorpe, R. L., Carr, I. D. & Sharp, I. R. 2005. Normal faulting as a control on the stratigraphic development of the shallow marine syn-rift sequences: the Nukhul and Lower Rudeis Formations, Hammam Faraun fault block, Suez rift, Egypt. *Sedimentology*, **52**, 313–338.

Machel, H. G. & Burton, E. A. 1991. Burial-diagenetic sabkha-like gypsum and anhydrite nodules. *Sedimentary Petrology*, **61**, 394–405.

Moustafa, A. R. 2002. Controls on the geometry of transfer zones in the Suez rift and northwest Red Sea: implications for the structural geometry of rift systems. *AAPG Bulletin*, **86**, 979–1002.

Robson, D. A. 1971. The structure of the Gulf of Suez (clysmic) rift, with special reference to the eastern side. *Journal of the Geological Society, London*, **127**, 247–276.

Schütz, K. K. 1994. Structure and stratigraphy of the Gulf of Suez, Egypt. *In*: Landon, S. M. (ed.) *Interior Rift Basins*. American Association of Petroleum Geologists, Tulsa, OK, Memoirs, **59**, 57–96.

Wever, H. E. 2000. Petroleum and source rock characterization based on C7 star plot results: examples from Egypt. *AAPG Bulletin*, **84**, 1041–1054.

Winn, R. D., Crevello, W. D. & Bosworth, W. 2001. Lower Miocene Nukhul Formation, Gebel el Zeit, Egypt: model for structural control on early synrift strata and reservoirs, Gulf of Suez. *AAPG Bulletin*, **85**, 1871–1890.

Young, M. J., Gawthorpe, R. L. & Sharp, I. R. 2000. Sedimentology and sequence stratigraphy of a transfer zone coarse-grained delta, Miocene Suez rift, Egypt. *Sedimentology*, **47**, 1081–1104.

Young, M. J., Gawthorpe, R. L. & Sharp, I. R. 2003. Normal fault growth and early syn-rift sedimentology and sequence stratigraphy: Thal Fault, Suez rift, Egypt. *Basin Research*, **15**, 479–502.

The hydrocarbon prospectivity of the Egyptian North Red Sea basin

G. GORDON, B. HANSEN, J. SCOTT, C. HIRST, R. GRAHAM, T. GROW, A. SPEDDING, S. FAIRHEAD, L. FULLARTON and D. GRIFFIN

Hess, Level 9, The Adelphi Building, 1–11 John Adam Street, London WC2N 6AG, UK
(e-mail: graeme.gordon@hess.com)

Abstract: Recent work by a multi-disciplinary team has led to a significantly better understanding of the prospectivity of the North Red Sea. New regional biostratigraphic and environmental analysis from north to south through the Gulf of Suez and into the Red Sea have placed the Nubian sequences into a regional chronostratigraphic framework. The Nubian Upper Cretaceous pre-rift sandstones are observed in the field on both the Egyptian and Saudi Arabian side of the North Red Sea. This regionally extensive sequence was deposited in a continental to shallow marine setting fringing the Mesozoic Tethys Ocean, which lay further north. Extensive onshore fieldwork and mapping of sediment input points, fault orientations and fault linkages have helped to develop an understanding of the expected controls on syn-rift sandstone and carbonate deposition offshore. Thick halite with interbedded evaporite and clastics in the Late Miocene sequences of the Red Sea pose seismic imaging challenges. Recent reprocessing and newly acquired seismic data have produced a step change improvement in imaging of the prospective pre-rift section. Petroleum systems modelling incorporating new information on rift timing and crustal thinning as well as onshore core analysis for source rock properties and temperature variation through time indicates that oil expulsion occurs in the inboard section of North Red Sea – Block 1. This is supported by hydrocarbon shows in the drilled offshore wells which can be typed to pre-rift source rocks from stable isotope and biomarker data. All the key elements of the Gulf of Suez petroleum system exist in the North Red Sea. An integrated exploration approach has enabled prospective areas in the North Red Sea – Block 1 to be high-graded for drilling in early 2011.

Keywords: Red Sea, Gulf of Suez, Dakhla, Duwa, Nubia, seismic imaging, subsalt, Gebel Duwi, gravity magnetics

The North Red Sea – Block 1 concession is located in the NW of the Red Sea, near the mouth of the Gulf of Suez and covers c. 9445 km^2 (Fig. 1). Esso held the block from the mid 1970s through to 1984. They acquired 2D seismic and drilled five unsuccessful wells between 1976 and 1981. Phillips drilled two dry wells to the south of the block in 1977. Failure of these wells was largely attributed to the difficulties in drilling valid structures due to the poor quality seismic, which suffered from serious multiple contamination and limited sub-salt illumination. BG was awarded the block in 1995 and acquired 1600 km^2 of 3D seismic in 1999, then relinquished it in 2000. BP took the block in 2004, then farmed it out to Hess in 2006, who acquired 2008 km^2 3D seismic in 2006 and a further 753 km^2 in 2008 (Fig. 2).

Stratigraphy and tectonic framework

Prior to the development of the Red Sea rift system, this part of North Africa was located on the southern margin of the Tethys Ocean (Stampfli & Borel 2002). An extensive passive margin succession developed showing an overall transgressive character, culminating with extensive carbonate deposition in the Late Cretaceous and Eocene, resulting in the passive margin megasequence stratigraphy of the Red Sea and Gulf of Suez (Guiraud & Bosworth 1999; Bosworth & McClay 2001; Ziegler 2001; Khalil & McClay 2009). This pre-rift sequence contains the primary reservoir target of the Upper Cretaceous Nubian sandstone and the prolific Maastrichtian age Dakhla source (equivalent to the Gulf of Suez Brown Limestone) (Fig. 3) (Robison & Engel 1993). Onshore exposure of the pre-rift shows depositional thickness variations along old structural trends, as well as some erosion from footwall highs exposed in the Miocene. South of Quseir the pre-rift section is not exposed onshore and is postulated to be absent offshore due to non-deposition and/or erosion. These onshore observations significantly upgrade the exploration potential of the pre-rift in the Northern Red Sea relative to the central area south of Quseir.

A variety of fault orientations are mapped onshore and can be interpreted on seismic. These fall into three main categories:

- Precambrian – ESE–WNW Najd shear zone features (Fig. 4);
- rift parallel SE–NW clysmic trend faults from Gulf of Suez rift extension – 25–15 Ma;
- Gulf of Aqaba parallel NNE–SSW structures due to the oblique spreading – 15–0 Ma (Lelek et al. 1992; Patton et al. 1994; Seber 2000; Ziegler 2001; Bosworth et al. 2005; Polis et al. 2005).

Source

The Cretaceous Maastrichtian age Dakhla formation, a carbonate-rich shale, overlies the leaner Campanian Duwi regionally. Dakhla data from surface samples and cores analysed from the Red Sea Hills in 2005–2008, as well as regional studies, indicate a world class source rock. From onshore unweathered cores, total organic content (TOC) values for the Dakhla range from 2 to >8%, hydrogen index (HI) from 400 to 600 mg g^{-1} and the source rock thickness is >50 m, similar to that of the Brown Limestone in the Gulf of Suez (Robison & Engel 1993). The age equivalent of the Brown Limestone is the Duwi in the Red Sea. This is a leaner shallow marine source rock with lower TOCs and HIs. From petroleum systems modelling, oil generation began at 20 Ma with peak generation at 17–15 Ma. This correlates with oil show extracts from on-block wells contain geochemical biomarkers correlating to pre-rift source rocks (Fig. 5).

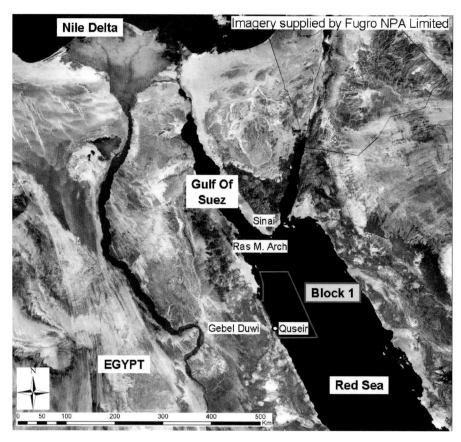

Fig. 1. Location map of Block 1, North Red Sea, Egypt.

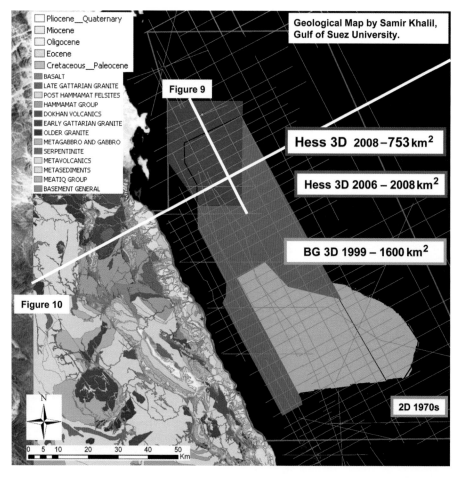

Fig. 2. Block outline showing seismic data coverage and onshore field mapping. Locations of seismic line and aeromagnetic modelling section are labelled, see Figs 9 and 10.

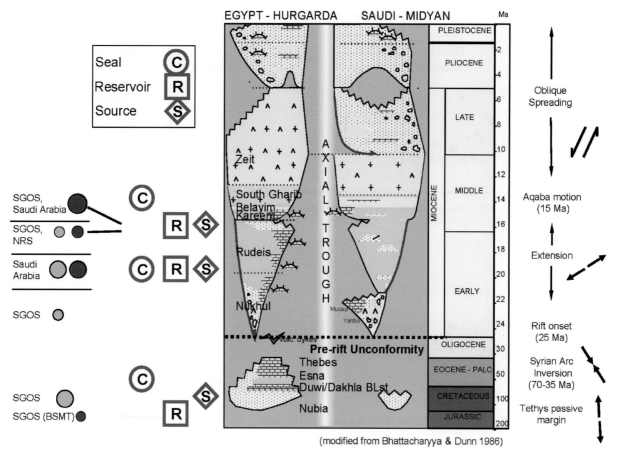

Fig. 3. Stratigraphy and tectonic framework of the North Red Sea showing ages of important tectonic events and petroleum system elements.

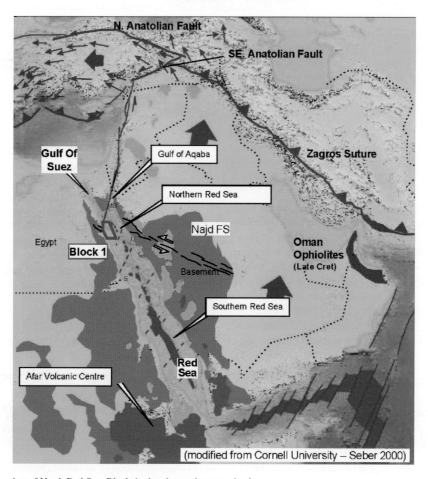

Fig. 4. Regional tectonic setting of North Red Sea, Block 1, showing major tectonic elements.

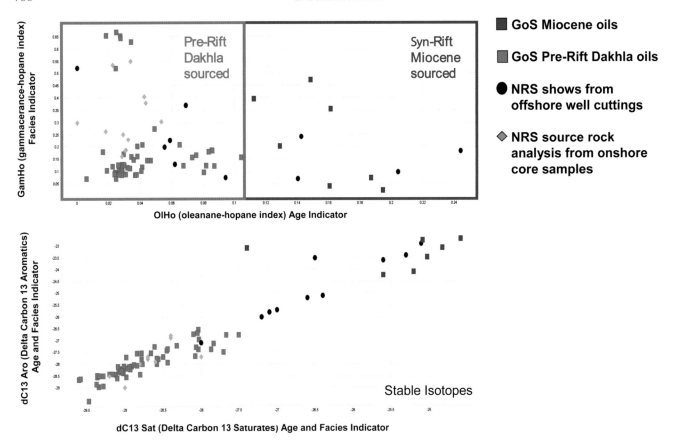

Fig. 5. Comparison of Gulf of Suez (GoS) v. North Red Sea (NRS) source rock biomarker and isotope analysis. NRS offshore shows from well cuttings and onshore core samples type to Cretaceous age GoS brown limestone.

Reservoir and seal

The principal reservoir interval within the pre-rift is the Cretaceous and older Nubian sands. The Nubia represents the initial clastic dominated continental to shallow marine fringe with the Tethyan passive margin and was deposited directly upon basement (McDonald 1979; Ward & McDonald 1979). The Nubian section thins southwards through the Gulf of Suez and is locally absent across the Kharga Arch at the southern tip of the Sinai Peninsula (Klitzch & Squyres 1990; Patton et al. 1994; Guiraud & Bosworth 1999). South of this feature, the Nubia thickens dramatically as seen in outcrop at Gebel Duwi, located onshore of the North Red Sea Block 1, where in excess of 300 m of Nubia is present (Fig. 6) (Bhattacharyya & Dunn 1986; Mitchell et al. 1992). The Nubia age-equivalent Adaffa sands are also present on the Saudi Arabian side of the North Red Sea (Hughes & Johnson 2005). As the rift evolved, sediment entry points became fixed at major fault transfer points. In the Gulf of Suez the Holocene drainage pattern is indicative of key sediment entry points into the rift basin. Likewise, in the Red Sea the main onshore wadis and drainage basin fall lines interpreted from topographic maps are considered keys to the past when evaluating point sources for Miocene sands (Fig. 7; Richardson & Arthur 1988; Gawthorpe et al. 1990; Heath et al. 1998; Younes & McClay 2002; Polis et al. 2005; Khalil & McClay 2009). In the pre-rift section, the main seal is the Maastrichtian age Dakhla and in the syn-rift, Lower Miocene intraformational shales are penetrated in the offshore wells, as well as an ultimate seal at the evaporite layers of the South Gharib, and Zeit.

Seismic acquisition and processing

Thick halite with interbedded evaporites and clastics, as well as diapiric salt, pose severe challenges to the seismic imaging of sub-salt targets. A 753 km^2 3D seismic survey was acquired in 2008 at 30° azimuth to the vintage data to provide improved subsalt illumination and a lower signal to noise multiazimuth dataset in the overlap area. Deeper streamers and sources were used than in the 2006 acquisition and the source configuration was optimized to help maximize low frequency signal penetration of the thick evaporite section. Hess has then carefully merged and reprocessed the full 3D vintage seismic datasets on-block during 2008–2009. 3D surface related multiple elimination (SRME) processing was used to reduce the impact of multiples from the rugose seabed and Top Zeit. Previous depth migrations used a flood velocity beneath the Top Zeit, but the cleaner reprocessed data allowed a laterally varying velocity field within the evaporite sequence to be generated using tomography. This replaced the flood velocity model with a more geologically consistent velocity variation (Fig. 8). Reverse-time depth migration (RTM) processing also provided a cleaner, more coherent image sub-salt where low frequencies dominate. Overall, optimized acquisition parameters and the application of the latest geophysical processing techniques have resulted in a significant uplift in the quality of subsalt imaging as compared with the 1999 BG and original 2006 3D seismic volumes. The 2D data shot in 1975 and 1988 which formed the basis for the earlier drilled wells was unable to image below the salt.

Seismic mapping

The imaging of the pre- and syn-rift sequences has markedly improved from the original time-migrated data (Fig. 9). Each of the acquisition and processing steps has provided an incremental improvement which when added together help to de-risk the sub-salt prospectivity of this new frontier basin. The latest seismic data quality at the target level has a wider frequency bandwidth, a reduction in migration noise and reduced multiple contamination. This has led to better fault definition and allows

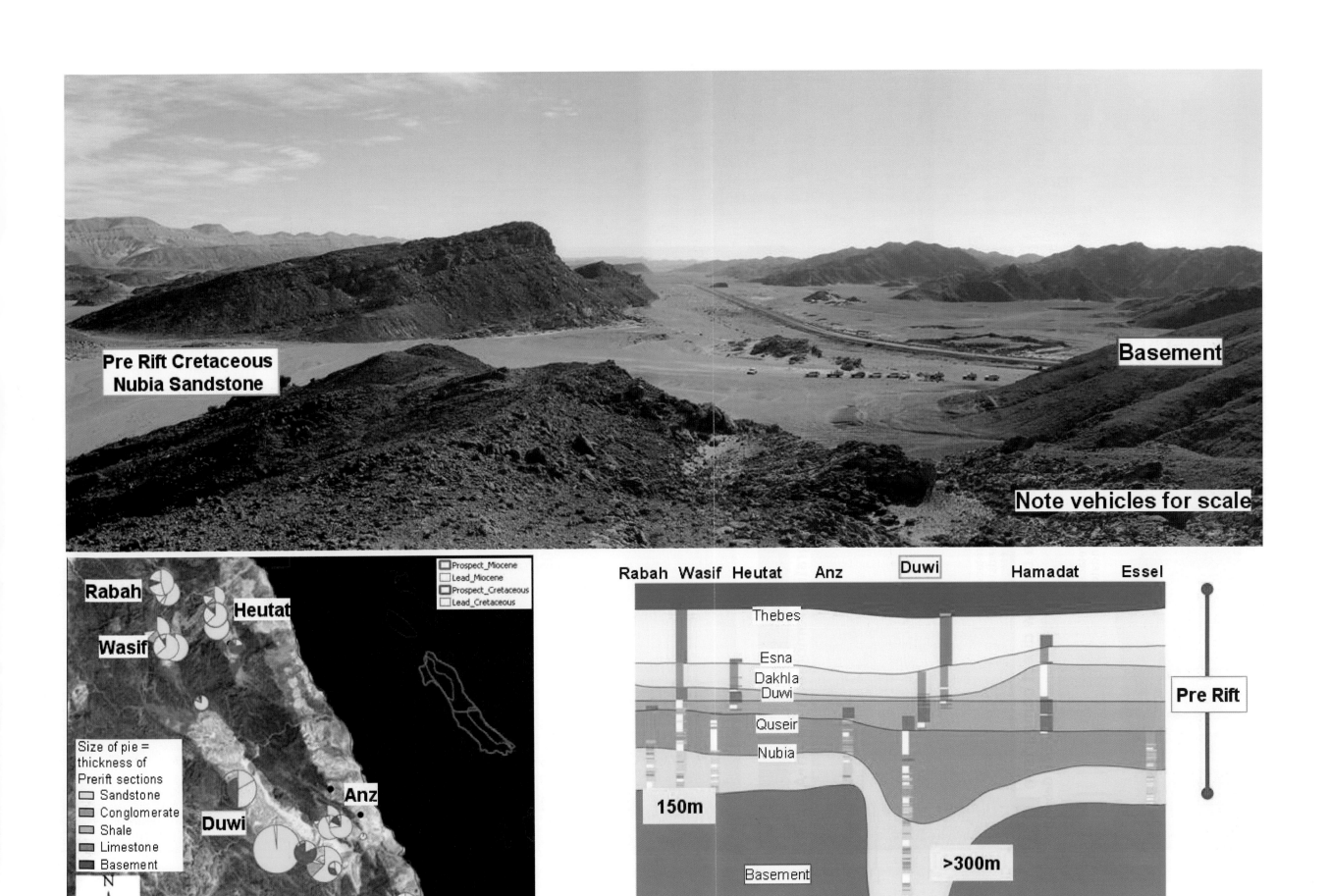

Fig. 6. Field photograph of Gebel Duwi North Red Sea Hills. Showing the geometry of a tilted pre-rift Cretaceous Nubia sandstone on basement. Map and cross section show consistent regional thickness of Nubia sandstone of 150 m with thickening at Gebel Duwi due to infilling of palaeotopography of palaeostructuring

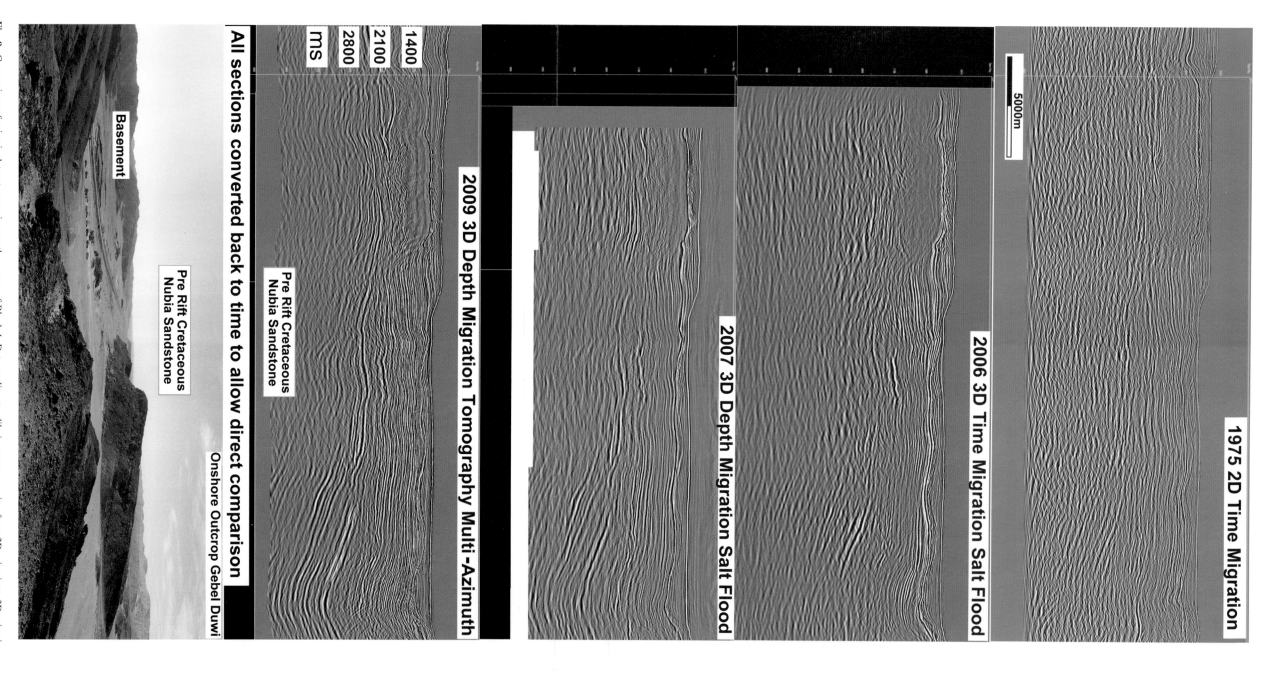

Fig. 9. Comparison of seismic datasets covering northern area of Block 1. Data quality steadily improves moving from 2D seismic to 3D seismic, from pre-stack time migration (PSTM) to pre-stack depth migration (PSDM), and form a salt flood velocity model to a laterally varying velocity model derived from tomography. With improved data quality faults are more easily delineated and the pre-rift section is imaged.

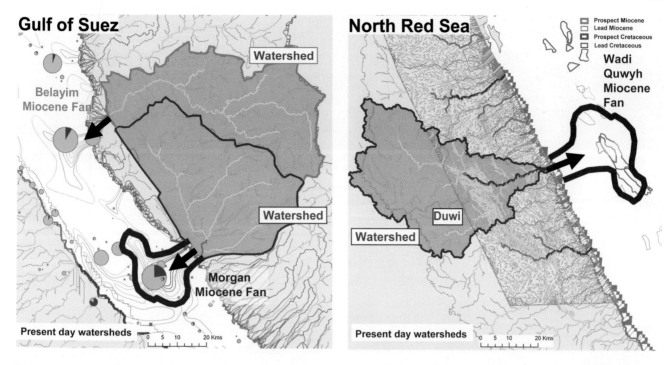

Fig. 7. Present day watersheds from digital elevation data. In the Gulf of Suez, the present day drainage system outputs correlate with Miocene syn-rift depositional thicks. This Miocene fan delta analogue was used to compare between Gulf of Suez discovered fields and North Red Sea prospects and leads.

increased confidence in the mapping of subsalt events. However, the current 3D data quality still does not lend itself to an intricate sequence-stratigraphic or AVO/amplitude mapping approach. Rather, the top and base of gross tectonic sequences have been mapped, such as the top evaporites (Top Zeit), the base evaporites (base Belayim), the rift-onset unconformity and the top of igneous basement (Fig. 9). Mapping at these levels allows detailed structural interpretation and geological models based on onshore mapping and regional understanding has identified prospectivity at both syn-rift and pre-rift levels.

Gravity/magnetics

The aeromagnetic survey coverage acquired in 2006 provides a link between the outcrops on the Red Sea margin, offshore well penetrations and seismic data. Integration with the 2008 ship-borne

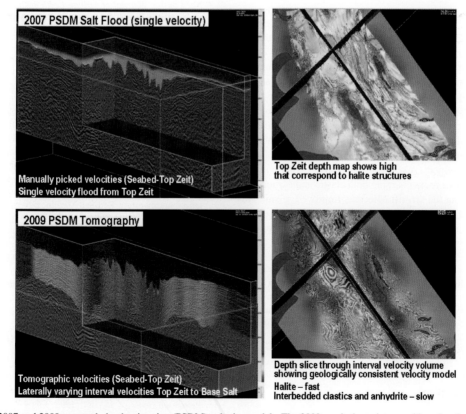

Fig. 8. 3D views of 2007 and 2009 pre-stack depth migration (PSDM) velocity models. The 2009 result shows improved imaging due to a more geologically consistent velocity model, incorporating Top Zeit rugosity and laterally varying velocities within the South Gharib and Zeit evaporite section.

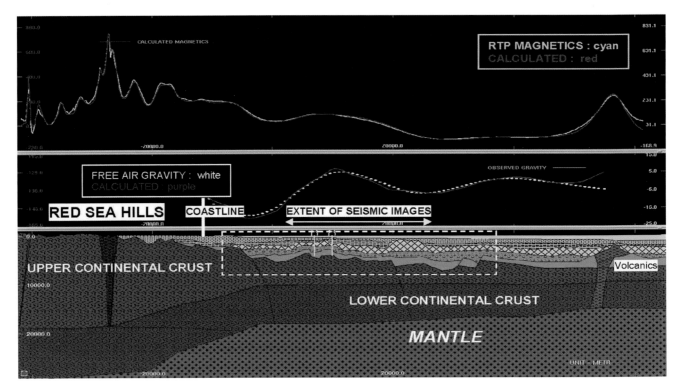

Fig. 10. Joint modelling of high frequency aeromagnetic data and regional gravity data allows salt thickness and depth to basement to be modelled for comparison with seismic interpretation.

gravity and magnetics data helps to constrain the offshore seismic interpretation (Fig. 10). In areas of imaging uncertainty, this has been particularly beneficial for placement of sub-salt faults, and to determine the thickness of salt and depth to basement where the seismic is poorly imaged.

Summary

From the extensive regional work, the northwestern margin of the North Red Sea is interpreted to be a continuation of the Gulf of Suez petroleum system. The primary play is the pre-rift Dakhla source rock charging pre-rift Nubian sands in structural traps set up by extensional tilted fault blocks. Onshore the basis for a working petroleum system can be demonstrated in the source rock exposure at outcrop and oil shows in the wells offshore have been typed to pre-rift Dakhla source rock. This confirms that the pre-rift play extends into the offshore region albeit that the wells only saw syn-rift on basement. A secondary play also exisits with syn- and/or pre-rift source rocks charging both clastic and carbonate potential reservoirs in the syn-rift Miocene. Improved seismic imaging has de-risked these plays and an integrated exploration approach summarized in this paper has enabled prospective areas to be high-graded.

The authors would like to thank EGPC for their permission to present this paper and Dr Samir Khalil, Suez Canal University, Egypt, for inclusion of his geological map. Special thanks for the contributions made by Dave Peel, Laura Lawton, Dr Paul Whitehouse, Andy Pepper, Hans Ladegaard, Dr Niall McCormack and Dr Stuart Lake from Hess, Dr Ahmed Barkooky from Cairo University, Jonathan Redfern from the University of Manchester, and Dr Nicky White from the University of Cambridge.

References

Bhattacharyya, D. P. & Dunn, L. G. 1986. Sedimentologic evidence for repeated pre-Cenozoic vertical movements along the North-East margin of the Nubian Craton. *Journal of African Earth Sciences*, **5**, 147–153.

Bosworth, W. & McClay, K. 2001. Structural and stratigraphic evolution of the Gulf of Suez Rift, Egypt, a synthesis (in Peri-Tethys memoir 6, Peri-Tethyan Rift/Wrench basins and passive margins). *Memoires du Museum National d'Histoire Naturelle*, **186**, 567–606.

Bosworth, W., Huchon, P. & McClay, K. 2005. The Red Sea and Gulf of Aden basins. *Journal of African Earth Sciences*, **43**, 334–378.

Gawthorpe, R. L., Hurst, J. M. & Sladen, C. P. 1990. Evolution of Miocene footwall-derived coarse-grained deltas, Gulf of Suez, Egypt, implications for exploration. *AAPG Bulletin*, **74**, 1077–1086.

Guiraud, R. & Bosworth, W. 1999. Phanerozoic geodynamic evolution of North-Eastern Africa and the northwestern Arabian Platform (in basin dynamics and basin fill, models and constraints; Part I). *Tectonophysics*, **315**, 73–108.

Heath, R., Vanstone, S. *et al.* 1998. Renewed exploration in the offshore North Red Sea region–Egypt. *In*: Eloui, M. (ed.) *Proceedings of the 14th Petroleum Conference.* Egyptian General Petroleum Corporation, Cairo, 16–34.

Hughes, G. & Johnson, R. S. 2005. Lithostratigraphy of the Saudi Arabian Red Sea. *GeoArabia*, **10**, 49–126.

Khalil, S. M. & McClay, K. R. 2009. Structural control on syn-rift sedimentation, northwestern Red Sea Margin, Egypt. *Marine and Petroleum Geology*, **26**, 1018–1034.

Klitzch, E. H. & Squyres, C. H. 1990. Paleozoic and mesozoic geological history of Northeastern Africa based upon new interpretation of Nubian Strata. *AAPG Bulletin*, **74**, 1203–1211.

Lelek, J. J., Shepherd, D. B., Stone, D. M. & Abdine, A. S. 1992. October Field: The Latest Giant under Development in Egypt's Gulf of Suez. *In*: *Giant Oil and Gas Fields of the Decade 1978–1988.* American Association of Petroleum Geologists, Tulsa, OK, Special Volumes, **M54**, 231–249.

McDonald, K. C. 1979. *The Nubia Formation, Quseir – Safaga Area, Eastern Desert.* University of Kansas, Egypt.

Mitchell, D. J. W., Allen, R. B., Salama, W. & Abouzakm, A. 1992. Tectonostratigraphic framework and hydrocarbon potential of the Red Sea (in The Red Sea–Gulf of Aden, II, Beydoun). *Journal of Petroleum Geology*, **15**, 187–209.

Patton, T. L., Moustafa, A. R., Nelson, R. A. & Abdine, A. S. 1994. *Tectonic Evolution and Structural Setting of the Suez Rift (in Interior Rift*

Basins, Landon). American Association of Petroleum Geologists, Tulsa, OK, Memoirs, **59**, 9–55.

Polis, S. R., Angelich, M. T. *et al.* 2005. Preferential deposition and preservation of structurally-controlled synrift reservoirs. *Northeast Red Sea and Gulf of Suez*, **10**, 97–124.

Richardson, M. & Arthur, M. A. 1988. The Gulf of Suez-Northern Red Sea Neogene Rift, a quantitive basin analysis. *Marine and Petroleum Geology*, **5**, 247–270.

Robison, V. D. & Engel, M. H. 1993. Characterization of the source horizons within the Late Cretaceous transgressive sequence of Egypt. *In*: Katz, B. J. & Pratt, L. M. (eds) *Source Rocks in a Sequence Stratigraphic Framework*. AAPG Studies in Geology, **37**, 101–117.

Seber, D., Steer, D., Sandvol, E., Sandvol, C., Brindisi, C. & Barazangi, M. 2000. Design and development of information systems for the geosciences; an application to the Middle East. *GeoArabia (Manama)*, **5**, 269–296.

Stampfli, G. M. & Borel, G. D. 2002. A plate tectonic model for the paleozoic and mesozoic constrained by dynamic plate boundaries and restored synthetic oceanic isochron. *Earth and Planetary Science Letters*, **196**, 17–33.

Ward, W. C. & McDonald, K. C. 1979. Nubia formation of central Eastern Desert, Egypt, major subdivisions and depositional setting. *AAPG Bulletin*, **63**, 975–983.

Younes, A. I. & McClay, K. 2002. Development of accommodation zones in the Gulf of Suez-Red Sea Rift, Egypt. *AAPG Bulletin*, **86**, 1003–1026.

Ziegler, M. A. 2001. Late Permian to holocene paleofacies evolution of the Arabian plate and its hydrocarbon occurrences. *GeoArabia (Manama)*, **6**, 445–504.

A regional overview of the exploration potential of the Middle East: a case study in the application of play fairway risk mapping techniques

A. J. FRASER

BP Exploration, Chertsey Road, Sunbury-on-Thames, Middlesex TW16 7LN, UK
(e-mail: alastair.fraser@imperial.ac.uk)

Abstract: The Middle East is the world's most prolific petroleum province, containing the world's top five countries in terms of oil reserves and four of the top 10 oil producers. This success is largely due to the stacked nature of what are essentially very simple play systems with multiple carbonate platform and deltaic clastic reservoirs, widespread evaporitic seals, world class source rocks and the overprint of very large compressional anticlines in both fold belt and foreland settings. As with most working petroleum systems, all areas of the Middle East are not equally endowed with petroleum resources and future potential. Using an analysis of the prolific Upper Jurassic 'Arab' play fairway of the Middle East, the application of play fairway risk mapping techniques is demonstrated.

Keywords: Middle East play fairway analysis, common risk segments, exploration triangle

In terms of future global oil and gas resources, the Middle East is of immense importance. Much has already been written about its scale and the fundamental geological reasons underlying this (Murris 1980; Beydoun 1986, 1998; Stoneley 1990; Konert *et al.* 2001; Sharland *et al.* 2001; Alsharhan & Nairn 2003; Sorkhabi 2010). This assessment suggests that the Middle East contains a significant volume of undiscovered oil and gas: that yet to be found is almost certainly well in excess of 100 BBOE (billion barrels of oil equivalent). Applying existing best in class technology to the more difficult or currently inaccessible parts of the play system suggests it is very likely that world class, super-giant fields, containing more than 1 BBOE remain to be found.

The constraints on delivering this resource are two-fold. Firstly, there is a general lack of competitive fiscal terms. These are needed if companies are to earn returns that are accretive to their financial performance. Secondly, there are very serious non-technical barriers and obstacles to foreign participation.

In this short paper, frequent reference will be made to an extensive regional study undertaken to assess the existing and future resource potential of the Arabian plate. The paper will address the principles and benefits of constructing a series of plate-scale, gross depositional environment (GDE) maps to describe basin evolution and as the fundamental building blocks of the common risk segment (CRS) mapping process. The prolific Upper Jurassic 'Arab' fairway which has estimated resources of around 300×10^9 barrels in the region will be used as an example of how these GDE maps are turned into a series of plate-wide CRS maps and subsequently used to focus exploration effort.

History

Exploration in the Middle East effectively commenced on 26 May 1908 when William Knox D'Arcy made the discovery at Masjid-i-Sulaiman in Cenozoic Asmari limestones in the Persian Zagros Mountains in what is now Iran. The photograph (Fig. 1) taken in 1911 shows one of the early wells on the field. Reserves are estimated at 1.3×10^9 barrels. The first discoveries were made on the basis of field mapping which identified surface anticlines and associated seeps. Today, we have access to modern 3D seismic data, sophisticated petroleum systems models and play fairway analysis tools to locate new oil and gas resources.

The Middle East contains estimated proven reserves of some 750×10^9 barrels of oil and 2680×10^{12} SCF of gas, representing c. 65 and 35% of global oil and gas reserves, respectively, making it, by some margin, the world's richest petroleum province (BP Statistical Review of World Energy 2009). These reserves are encountered in reservoirs of Late Palaeozoic to Cenozoic age in a NW–SE trending zone that runs from Oman to Turkey. The world's largest oil field (Ghawar) and gas field (South Pars/North Dome) are located in the region (Fig. 2). Masjid-i-Sulaiman (discovered by D'Arcy, later to form BP) is also highlighted on this map.

Saudi Arabia contains by far the majority of the oil resources of the region but this analysis is not just about Saudi Arabia (although this is the focus of the play fairway risk mapping), but also Iran, Iraq, Kuwait and Abu Dhabi, five of the top six in terms of proven oil reserves at the present day (BP Statistical Review 2009). This study utilizes a unique database of surface and subsurface information covering the region, sourced from the archives of the three heritage companies BP, Amoco and ARCO.

Fig. 1. Photograph taken in 1911 showing one of the very early BP wells drilled on Masjid-i-Sulaiman after penetrating the Cenozoic Asmari carbonate reservoir. Reserves are estimated at 1.3×10^9 barrels of oil. Photo BP archives.

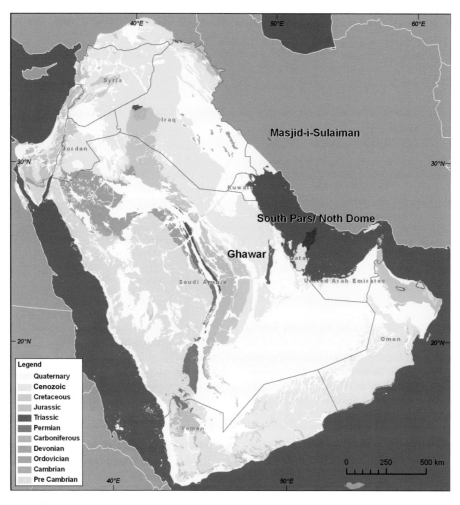

Fig. 2. Geology of the Arabian Plate.

Middle East plays

Figure 3 illustrates a tectonostratigraphic diagram to summarize the tectonic history and stratigraphic fill of the Arabian plate. It is best thought of as a horizontally compressed chronostratigraphic diagram. In essence both charts fulfil a similar function. The chronostratigraphic section is consistent along the line of section with the GDE maps, and complements them by providing a temporal interpretation of the facies associations. Additionally, the chronostratigraphic diagram illustrates the missing sections and time, something a suite of GDE maps are unable to achieve on their own. The chronostratigraphic chart also shows the spatial relationship of plays through time, such as the multiple stacked shale, carbonate and evaporite cycles observed in the Permian to Cenozoic section of the Middle East. It is used to highlight the key tectonic events recorded in the region and link these to broader plate margin processes. The stratigraphy is divided into megasequences and their constituent sequences, to identify the spatial and temporal distribution of the key reservoir, source and seal facies. A link between trap and charge timing can be observed. Ultimately, the chronostratigraphic chart highlights the major play fairways, typically based on the presence of major reservoir intervals having both a regional seal and charge system.

The most notable aspects of Middle East geology that are illustrated on this chart are as follows:

- Early north–south oriented Infra-Cambrian rifts that developed on the northern margin of Gondwanaland and contain in some places fairly thick salt – the Hormuz salt (Edgell 1996; Sharland *et al.* 2001).
- A long-lived post-rift phase through much of the Palaeozoic which ended with the onset of NW–SE trending Neo-Tethyan rifting from the early Permian (Konert *et al.* 2001; Sharland *et al.* 2001).
- A repeated set of shale (source rocks), carbonate (reservoirs) and evaporite (seal) cycles throughout the Mesozoic forming a series of stacked, essentially self-contained play fairways. For the purposes of this short paper we will concentrate on the most prolific of these, the late Jurassic 'Arab' play fairway, when discussing the principles of GDE and CRS mapping (Murris 1980; Ellis *et al.* 1996; Loosveld *et al.* 1996; Youssif & Nouman 1997; Sharland *et al.* 2001).
- Closure of Neo-Tethys and formation of the Zagros Mountains in late Cretaceous and Cenozoic times, the major trap-forming event on the Arabian plate (Sharland *et al.* 2001).

A play cartoon generated to describe the play systems of the Middle East is shown in Figure 4. The cartoon illustrates a section running roughly west–east through southern Iraq, northern Kuwait and across the Persian Gulf into Iran and is loosely based on a regional 2D seismic line from the region.

The Middle East can, in simple terms, be described as a series of inverted north–south trending Infra-Cambrian and NW–SE trending Early Permian (Neo-Tethyan) rifts. The Zagros Mountains are interpreted as a largely thick skinned fold belt with the foreland dominated by a series of associated inversion anticlines. Ghawar, the largest oilfield in the world, is effectively a large inversion anticline underpinned by an inverted north–south trending Infra-Cambrian rift.

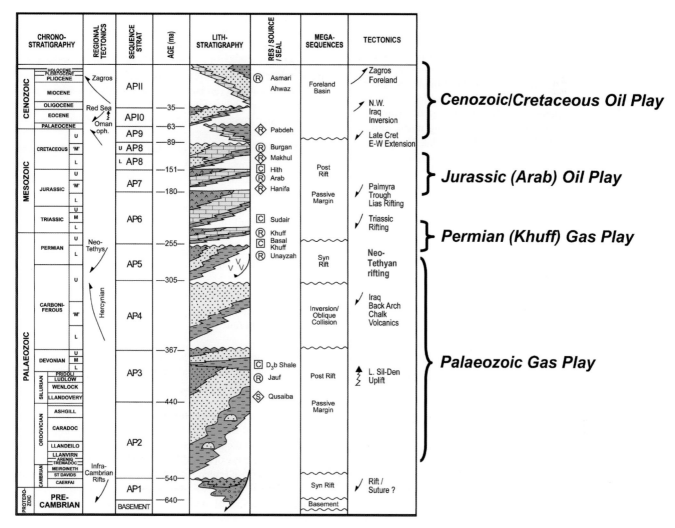

Fig. 3. Tectonostratigraphic chart for the Middle East showing key stratigraphic, tectonic and play elements. R, reservoir; S, source; C, seal. Sequence stratigraphy after Sharland *et al.* (2001).

The play cartoon also highlights the five world-class source rocks in the region. These are the Qasaiba in the Silurian, the Hanifa in the Jurassic, the Khaz Dumi and Garau in the Cretaceous and the Pabdeh in the Cenozoic. All are associated with multiple reservoir-seal pairs. In fields such as Rumaila, West Qurna and Burgan, there are up to 12 individual reservoir-seal pairs resulting in super-giant fields with reserves in excess of 40×10^9 barrels (Jassim & Al-Gailani 2006).

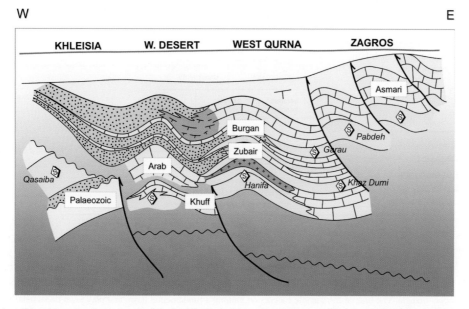

Fig. 4. Play cartoon describing the major play systems of the Middle East. Of note are the five world-class source rocks present in the region. These are the Qasaiba in the Silurian, the Hanifa in the Jurassic, the Khaz Dumi and Garau in the Cretaceous and the Pabdeh in the Cenozoic.

Exploration process

The fundamental understanding of what makes basins work has evolved over the past 15 years into a systematic approach which focuses thinking at the regional or basin scale before considering individual prospects and leads. The strength of the methodology presented in this paper, which we refer to as 'geology from the bottom up', is that it allows the early high-grading of basins and plays for further study/investment, thus negating much wasted time, effort and money in areas with little potential. The process can be elegantly summarized in the form of the 'exploration triangle' (Fig. 5). The bottom part of the triangle is about focusing on the basins/plays which are likely to work at a scale that fits with our exploration strategy of targeting basins with billion barrel potential, that is, being in the right basins and the right parts of these basins. The top part of the triangle is about shotpoint risk – locating wells and drilling 'no dry holes' using techniques such as surgical mapping, detailed charge modelling, AI/AVO analysis, etc. and ultimately building a quality prospect inventory that allows us to make better exploration investment decisions through a process of 'quality through choice'.

Jurassic of the Arabian plate

GDE maps form an essential tool in the exploration process. They fulfil two important roles. Firstly, they provide an easily comprehensible view of the depositional fill of the basin through time. Secondly, they constrain the areal distribution of source, seal and reservoir facies and thus provide the basis for defining and risking play fairways. They are used at all scales of exploration work from the regional basin scale to prospect scale analysis.

For this Middle East study, a suite of 14 individual GDE maps were generated which describe the stratigraphic evolution of the Arabian Plate from the Infra-Cambrian to the present. For the purposes of this short paper we will concentrate on the two GDE maps used to constrain the areal distribution of reservoir, source and seal for the late Jurassic (Arab) play fairway, one of the most prolific play systems in the world with over 300×10^9 barrels of oil reserves. The maps for the Upper Jurassic describe a broad carbonate shelf located on the southeastern margin of Neo-Tethys. Two isolated, intra-shelf basins that were subsiding at a faster rate than the surrounding areas are highlighted in shaded green/blue – perhaps related to underlying Infra-Cambrian salt (Fig. 6). Widespread anoxia developed in these areas, resulting in the widespread deposition of the Hanifa and equivalent type IIs source rocks. These outlines can be transferred directly to a source presence CRS map and combined with source rock maturity to produce a map combining charge presence and effectiveness for oil (Fig. 7).

A CRS is defined as an area of a play fairway which has common critical risk factors. The risk factors include charge, reservoir and seal, although the latter in some settings (e.g. conformable sequences) may be a prospect specific issue. CRS maps are typically constructed at the basin scale for each of the key risk factors. The detail of these depends entirely on the density of available data. The individual components are convolved together to produce a composite CRS description of the fairway. These maps provide the focus on the lowest risk segments of the play fairway.

The real value of CRS mapping is its simplicity. Having refined the technique over the past 15 years, we have seen it evolve to fulfil several important functions.

- Firstly, it provides a process whereby each explorer is forced to work through a systematic assessment which addresses all (stratigraphic) elements in a play fairway analysis.
- Secondly, once individual maps are combined into a composite CRS map we have a very effective means of focusing on the key 'low-risk' part of a basin or play fairway and by implication defocusing on the higher-risk 'red' areas of the map.
- Thirdly, by the use of only three 'traffic light' colours to describe relative play risk (green for low-risk, yellow for moderate-risk and red for high-risk), we as geoscientists are able to share often complex geological analyses in an easily comprehensible fashion.

As mentioned above, hydrocarbons in the Upper Jurassic 'Arab' reservoir are sourced directly from the underlying Upper Jurassic (Oxfordian), Lower Hanifa Formation. The classic oil-producing region of central Saudi Arabia is situated in an area of thick

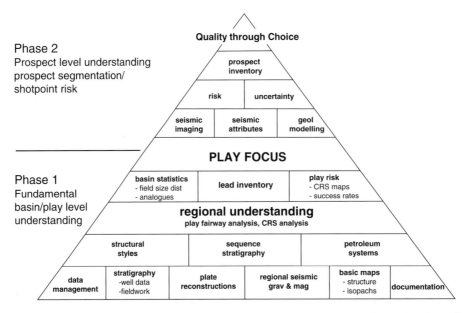

Fig. 5. The exploration process triangle. The base of the triangle focuses on gaining a fundamental basin and play level understanding using subsurface integration tools such as GDE and CRS mapping. The top part of the triangle is about shot-point risk, that is, locating wells, drilling no dry holes, using techniques such as surgical mapping, detailed charge modelling and AI/AVO analysis, and ultimately building a quality prospect inventory.

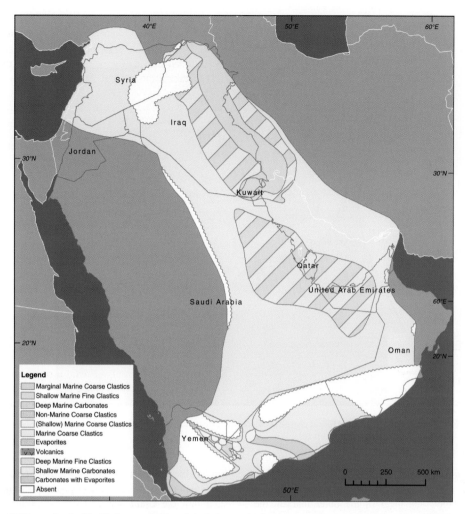

Fig. 6. Upper Jurassic GDE map for the Hanifa (source rock) and Arab (reservoir) formations. This map illustrates a broad carbonate shelf located on the southern margin of Neo-Tethys. Shaded green/blue are two intra-shelf basins that were subsiding faster than the surrounding areas. Widespread anoxia developed in these areas, resulting in the accumulation of the world-class Hanifa and equivalent class IIs source rocks. Compiled from extensive BP in-house database and reference to Murris (1980), Edgell (1996), Ellis *et al.* (1996), Loosveld *et al.* (1996), Youssif & Nouman (1997) and Sharland *et al.* (2001).

Hanifa source rock development that lies in the present-day oil window. The more deeply buried source rocks to the east in the Rub Al Khali basin are currently in the gas window. Note that on the resulting CRS oil charge map (Fig. 7), this area is coloured yellow for moderate-risk on the premise that the Hanifa oil-prone source rocks were in the oil window in the past and it is only recent gas flushing that makes gas the most likely, but not the sole, petroleum phase.

The Arab Formation carbonate grainstones are widespread across the Arabian Plate around the margins of the two intra-shelf basins, as illustrated on the Upper Jurassic (Hanifa/Arab) GDE map (Fig. 6). In the southern basin the carbonates were able to prograde out over the basin and importantly were deposited upon the Hanifa source rocks during periods of sea-level highstand. In the northern Gotnia basin continued subsidence confined the carbonates to rimmed shelves on the margins of the basin. Clearly this has important implications for reservoir prediction in the northern Gotnia depocentre situated in Kuwait, southern Iraq and Iran. This analysis from the GDE map is transferred directly to the reservoir presence and effectiveness CRS map (Fig. 8).

The GDE map for the upper part of the Upper Jurassic (Hith Formation) highlights the presence of widespread evaporite seals over the Arabian Plate overlying the Arab reservoirs particularly in the area of the southern basin in Saudi Arabia (Fig. 9). Note the presence of halites in the northern area which rest directly on Hanifa equivalent source rocks with Arab equivalent reservoir only present around the margins of the basin. There is an important facies change to the east where the Hith anhydrite pinches out into eastern Saudi Arabia and Abu Dhabi. This facies change effectively defines the eastern limit of the working Upper Jurassic 'Arab' play fairway and is represented by a rapid transition from low-risk (green) to high-risk (red) on the Seal Presence and Effectiveness CRS map (Fig. 10).

The Hith evaporites have been very effective in providing a classic regional seal for the giant Arab oil fields of the Arabian Peninsula. Additionally, intra-formational evaporitic seals exist within the Arab at the top of each of the A, B, C and D members. Beyond the eastern limit of the Hith seal, oil sourced in the Jurassic migrates vertically to charge Cretaceous reservoirs, as illustrated in Figure 11.

The stratigraphic elements of the Upper Jurassic 'Arab' play described in the GDE and CRS maps above can be summarized on a simplified play cartoon (Fig. 11). This elegantly makes the point that big oil really is simple. The more complicated the play system is required to be, the less likely it is to work. The diagram is a representation of the southern passive margin of Neo-Tethys as it existed during the Late Jurassic. The outer part of the platform is formed by a rimmed carbonate shelf. In the intra-shelf basin, silled behind this barrier, the type IIs, Hanifa source rocks were deposited. As relative sea-level fell in the intra-shelf basin, the

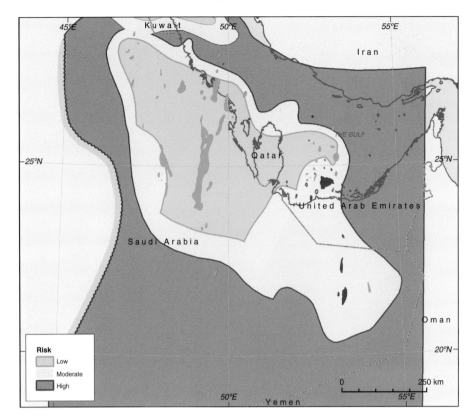

Fig. 7. Upper Jurassic 'Arab' play fairway. CRS map for oil charge presence and effectiveness. Existing oil and gas fields are shown in dark green and red, respectively.

Arab Formation grainstones were able to prograde from both the eastern and western margins to fill the intra-shelf basin. As sea-level fell further below the level of the outer margin, the basin was capped by Hith evaporites comprising mainly anhydrites with halites towards the central part of the basin. This vertical facies association repeated itself several times through the Late Jurassic and into the Cretaceous, forming multiple stacked world class source, reservoir and seal combinations.

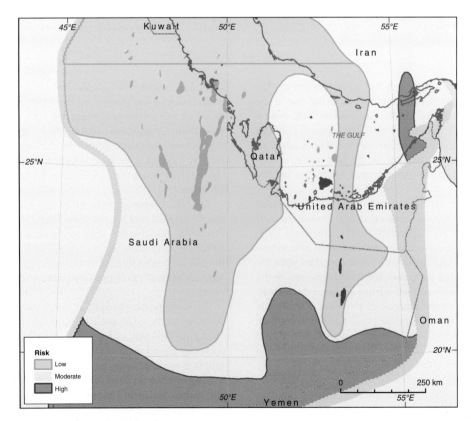

Fig. 8. Upper Jurassic 'Arab' play fairway. CRS map for reservoir presence and effectiveness. Existing oil and gas fields are shown in dark green and red, respectively.

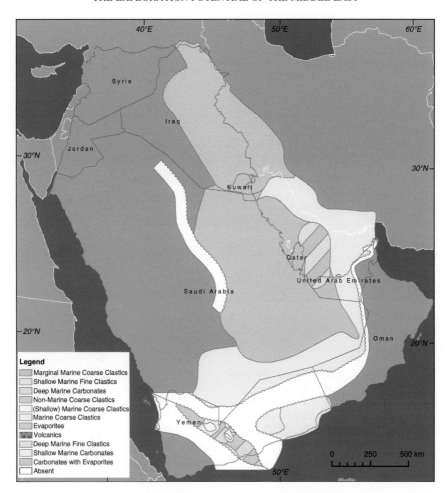

Fig. 9. Upper Jurassic GDE map for the Hith (seal) Formation. The map highlights the presence of widespread halite and anhydrite seals over the Arabian plate. Compiled from extensive BP in-house database and reference to Murris (1980), Edgell (1996), Ellis *et al.* (1996), Loosveld *et al.* (1996), Youssif & Nouman (1997) and Sharland *et al.* (2001).

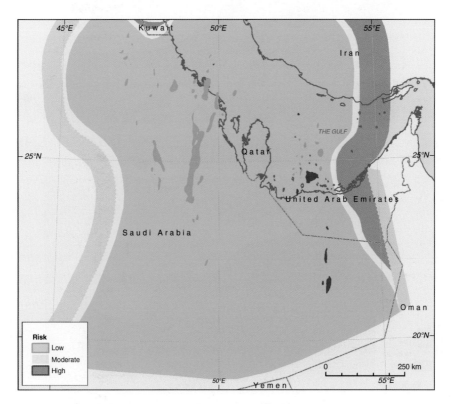

Fig. 10. Upper Jurassic 'Arab' play fairway. CRS map for seal presence and effectiveness. Existing oil and gas fields are shown in dark green and red, respectively.

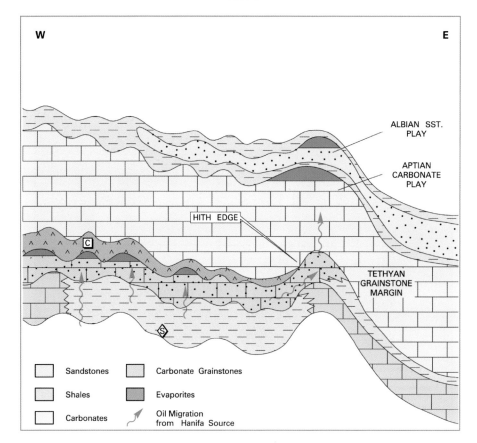

Fig. 11. Play cartoon illustrating the Upper Jurassic 'Arab' play fairway. Inherently a very simple petroleum system with the seal sitting directly above the reservoir and source. All elements of 'world class' quality. These are gently folded into large simple traps. Beyond the Hith edge, Upper Jurassic sourced hydrocarbons are free to migrate vertically into Cretaceous reservoirs.

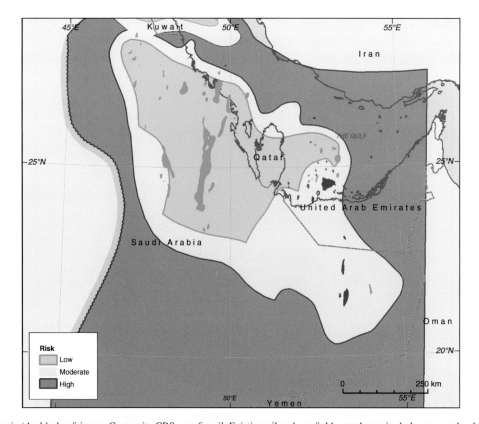

Fig. 12. Upper Jurassic 'Arab' play fairway. Composite CRS map for oil. Existing oil and gas fields are shown in dark green and red, respectively. The map presents no surprises as all the large Arab Formation oilfields such as Ghawar are located within the low-risk (green) segment.

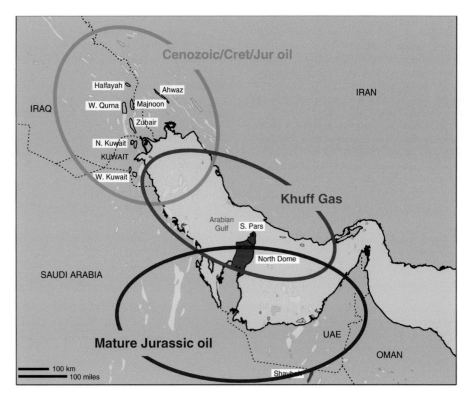

Fig. 13. Composite play fairway map of Middle East summarizing future resource potential. The map is based on a synthesis of all play fairways reviewed in this study.

To summarize, world-class Hanifa Formation source rocks are overlain by a world-class reservoir in the Arab Formation grainstones, which are in turn overlain by the Hith Formation evaporite seals. To be an effective petroleum system, it only needed sufficient late Mesozoic and Cenozoic burial to mature the source rocks and some gentle Zagros compression to form the simple, broad inversion structures which contain the oil fields.

The composite CRS map for oil for the Upper Jurassic 'Arab' play fairway is shown in Figure 12. Before discussing the results it is important to recap the process for compiling a composite CRS map. The composite CRS is assigned the risk of the most risky component CRS. Thus when the component maps are convolved, only three combined low-risk (green) segments (for reservoir, charge and seal) will generate an overall low-risk (green) composite CRS. The presence of one high-risk (red) component is sufficient to turn the composite CRS high-risk (red). Likewise, one moderate-risk (yellow) will turn an otherwise low-risk (green) segment moderate-risk (yellow).

The composite CRS map for the Upper Jurassic 'Arab' play fairway presents somewhat of a self-fulfilling prophesy with all the large Arab Formation oilfields firmly located within the low-risk (green) segment. The gas fields of the central Rub Al Khali are contained within a moderate-risk (yellow) segment as this is a map for oil. However, one oil accumulation, the North Dome field in the Persian Gulf, sits in the high-risk (red) segment. This is a classic case of long-distance lateral migration to the NE along the major Qatar Arch, revealing the only real flaw in the CRS methodology. In an essentially vertically stacked evaluation process it is has proven difficult to adequately address the impact of lateral migration of petroleum through regional carrier beds.

Summary of Middle East future potential

The future exploration potential of the Middle East is summarized in a composite play fairway map which combines the results of play fairway analyses for all the major plays in the region (Fig. 13). The already discovered and developed giant and super-giant fields of the region provide the basis for the majority of the long term reserves and are located in the Upper Jurassic of eastern Saudi Arabia and the UAE. Further significant reserves are expected to be delivered from these fields by continuing investment in improved oil recovery. The bulk of the undiscovered gas resources are expected to lie in deep Upper Palaeozoic reservoirs in an area centred on the Persian Gulf. However, the majority of the remaining undiscovered oil potential is likely to lie further north in the Cenozoic and Mesozoic trends of the Zagros fold belt and related foreland areas. However, having a technical view on the scale and location of additional hydrocarbon resources in a region the size of the Middle East is only part of the story. There are very serious non-technical barriers and obstacles to foreign participation in the region such as regional politics and the lack of competitive fiscal terms. Overcoming these and other constraints will be a major test of industry leadership in the next decade.

As with any significant piece of regional work there are many colleagues involved in locating, compiling and interpreting the large dataset required to provide a synthesis of the Arabian plate on this scale. In particular, I would like to acknowledge the work of Jeremy Goff, Bob Jones, Christoph Lehman, Steve Matthews and Ivor Simpson. For the graphics I am indebted to the efforts of Pat Randell and Dave Johnson. I also thank Jonathan Redfern of Manchester University for his comments on an earlier draft which significantly improved the text.

References

Alsharhan, A. S. & Nairn, A. E. M. 2003. *Sedimentary Basins and Petroleum Geology of the Middle East*. Elsevier, Amsterdam.

Beydoun, Z. R. 1986. The petroleum resources of the Middle East: a review. *Journal of Petroleum Geology*, **9**, 5–28.

Beydoun, Z. R. 1998. Arabian plate oil and gas: why so rich and prolific? *Episodes*, **21**, 74–81.

BP Statistical Review of World Energy 2009. BP plc, London.

Edgell, H. S. 1996. Salt tectonism in the Persian Gulf Basin. *In*: Alsop, G. I., Blundell, G. J. & Davidson, I. (eds) *Salt Tectonics*. Geological Society, London, Special Publications, **100**, 129–151.

Ellis, A. C., Kerr, H. M., Cornwell, C. P. & Williams, D. O. 1996. A tectono-stratigraphic framework for yemen and its implications for hydrocarbon potential. *Petroleum Geoscience*, **2**, 29–42.

Jassim, S. Z. & Al-Gailani, M. 2006. Hydrocarbons. *In*: Jassim, S. Z. & Goff, J. C. (eds) *Geology of Iraq*. Dolin, Prague, 232–250.

Konert, G., Afifi, A. M., Al-Hajri, S. A., de Groos, K., Al Naim, A. A. & Droste, H. J. 2001. Paleozoic stratigraphy and hydrocarbon habitat of the Arabian Plate. *In*: Downey, M. W., Threet, J. C. & Morgan, W. A. (eds) *Petroleum Provinces of the Twenty-First Century*. American Association of Petroleum Geologists, Tulsa, OK, Memoirs, **74**, 483–515.

Loosveld, R. J. H., Bell, A. & Terken, J. J. M. 1996. The tectonic evolution of interior Oman. *GeoArabia*, **1**, 28–51.

Murris, R. J. 1980. The Middle East: stratigraphic evolution and oil habitat. *American Association of Petroleum Geologists Bulletin*, **64**, 597–618.

Sharland, P. R., Archer, R. *et al*. 2001. Arabian plate sequence stratigraphy. *GeoArabia Special Publications*, **2**, 1–371.

Sorkhabi, R. 2010. Why so much oil in the Middle East? *GeoExpro*, **7**, 21–26.

Stoneley, R. 1990. The Middle East basin: a summary overview. *In*: Brooks, J. (ed.) *Classic Petroleum Provinces*. Geological Society, London, Special Publications, **50**, 293–298.

Youssif, S. & Nouman, G. 1997. Jurassic geology of Kuwait. *GeoArabia*, **2**, 91–110.

Appraisal and development of the Taq Taq field, Kurdistan region, Iraq

C. R. GARLAND,[1] I. ABALIOGLU,[2] L. AKCA,[3] A. CASSIDY,[1] Y. CHIFFOLEAU,[1] L. GODAIL,[1] M. A. S. GRACE,[1] H. J. KADER,[2] F. KHALEK,[1] H. LEGARRE,[1] H. B. NAZHAT[4] and B. SALLIER[1]

[1]*Addax Petroleum Services and Sinopec, 16 Avenue Eugène-Pittard, CH1211, Geneva, Switzerland (e-mail: chris.garland@addaxpetroleum.com)*
[2]*Taq Taq Operating Company Limited, Via Tower Bestepeler Mahallesi, Nergiz Sokak No 7/52, Sogutozu Ankara, Turkey*
[3]*Genel Enerji AS, Via Tower Bestepeler Mahallesi, Nergiz Sokak No 7/8, Sogutozu Ankara, Turkey*
[4]*Kurdistan Regional Government-IRAQ, Council of Ministers, Ministry of Natural Resources*

Abstract: The Taq Taq Field is located within an anticline in the gently folded zone of the Zagros mountains, northeastern Iraq, approximately 50 km ESE of Erbil. The main reservoirs are fractured limestones and dolomites of Late Cretaceous age, with an oil column exceeding 500 m in thickness. Eocene limestones and dolomites at shallow depth form a subsidiary reservoir. The structure is a gentle thrust-related fold which has also been affected by dextral transpression. A pervasive fracture system is present within the reservoirs, giving good connectivity and deliverability. Initial discovery and appraisal was made in 1978 when three wells were drilled. The recent appraisal programme started in 2005 and by the end of 2008 two seismic surveys had been acquired and eight additional wells had been drilled. Mapping has incorporated a seismic principal component analysis for horizon and lithology identification. Modelling of the fractures has utilized a comprehensive data set derived from core and image logs. Special core analysis has been directed towards the understanding of the pore system and its interaction with the fractures. Synthesis of all these elements is performed in a dual-media dynamic model which is currently in use for development planning.

Keywords: fracture model, Iraq, Middle East Cretaceous reservoirs, seismic facies analysis, thrust-related fold

Taq Taq was one of the first new oil accumulations to be put on production in northern Iraq for 40 years. It had been discovered in 1978 by the Northern Oil Company of Iraq, who drilled a well on the crest of a four-way dip-closed anticline 60 km NE of Kirkuk in the Zagros gently folded zone (Fig. 1). Two other wells, TT-02 and TT-03, were also drilled in this period. Production commenced in 1994 under the auspices of the Kurdistan Regional Government (KRG): the project consisted of the building of a workover rig, the testing and completion of wells TT-02 (2150 bopd from the Cenozoic Pilaspi Fm.) and TT-01 (5000 bopd from the Cretaceous Shiranish Fm.) and the construction of a refinery in Suleimaniya (Kader 2002). Delivery was by tanker from the site to the refinery and over 4×10^6 barrels of crude oil have been produced from the Cretaceous and 1×10^6 barrels from the Cenozoic up until October 2008.

Genel Enerji AS signed the first PSA in July 2002 and operation of the Taq Taq field was assigned to Genel in February 2003. In 2006 operatorship of the field was transferred to the Taq Taq Operating Company (TTOPCO), a joint venture between Genel Enerji AS and Addax Petroleum Corporation. Appraisal commenced with the acquisition of a 170 km 2D seismic survey and by the end of 2008 eight new wells had been drilled and 166 km² of new 3D seismic had been acquired. In 2009 a production facility was installed and connected to seven wells. At the time of proof-reading (March 2010), production of up to 55 000 bo had been achieved totalling 8×10^6 bo.

The anticline is clearly visible at the surface, where Miocene sandstones crop out forming ridges in an elliptical shape approximately 13 km long and 7 km wide. The stratigraphy comprises (from the top, Fig. 2): red beds, fluviatile clastics and conglomerates of the Fars Formations of Miocene–Pliocene age; limestones and dolomites of the Pila Spi Formation of Eocene age, which form a minor reservoir at the crest of the structure; shelf and shoal limestones overlying marine clastics of the Kolosh Formation of Paleocene–Eocene age, which form the main seal to the Cretaceous fractured reservoirs at a depth of approximately 1000 m below

Fig. 1. Iraq location map showing main outcrop, Taq Taq and other fields noted in the text. BH, Bai Hassan.

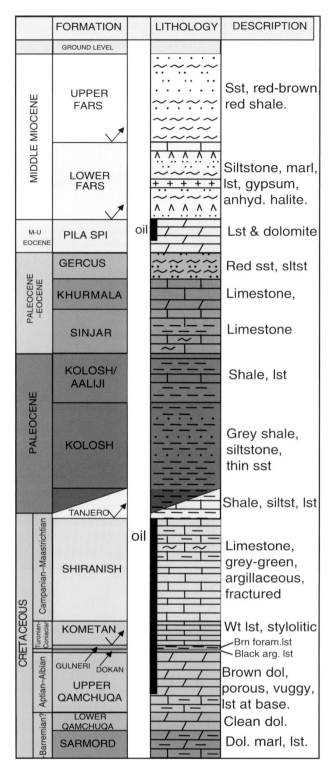

Fig. 2. General stratigraphy at Taq Taq showing reservoirs, lithology, formation names and approximate ages. Colours selected for contrast only.

Reservoirs

The main productive stratigraphic formations are described in detail in van Bellen *et al.* (1959) and Jassim & Goff (2008) and include three main reservoirs:

(1) The *Shiranish Formation* (Campanian–Maastrichtian) comprises up to 350 m of greenish-grey to dark brown foraminiferal

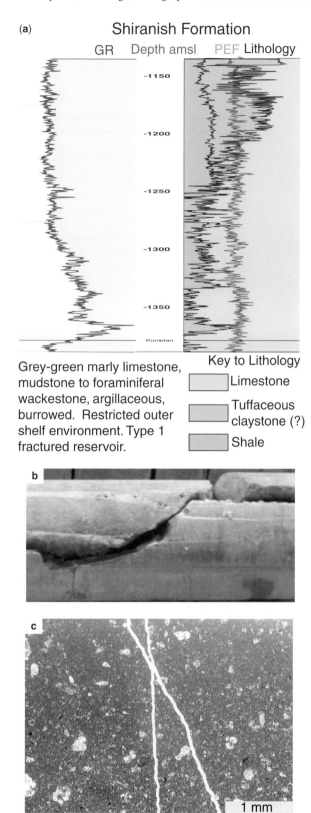

Fig. 3. Shiranish Formation. Brief description and (**a**) typical well logs; (**b**) partially cemented en-echelon open fractures in core; (**c**) thin section (plane polarized light) showing foraminifera and cemented fractures.

mean sea level (bmsl). This paper is concerned with these main Cretaceous reservoirs. Beneath these, TT-01 penetrated the entire Cretaceous and Jurassic sections, terminating in the Triassic at approximately 3300 m bmsl.

Source rocks for Taq Taq oil have been shown to be Jurassic in age: this is consistent with regional studies (Pitman *et al.* 2003) which show the Naokelekan and Sargelu Formations (which are present in Taq Taq) to be the main source rocks in northern Iraq. These began generating hydrocarbons in the Miocene prior to the main compressional phase of Zagros folding (Ameen 1991).

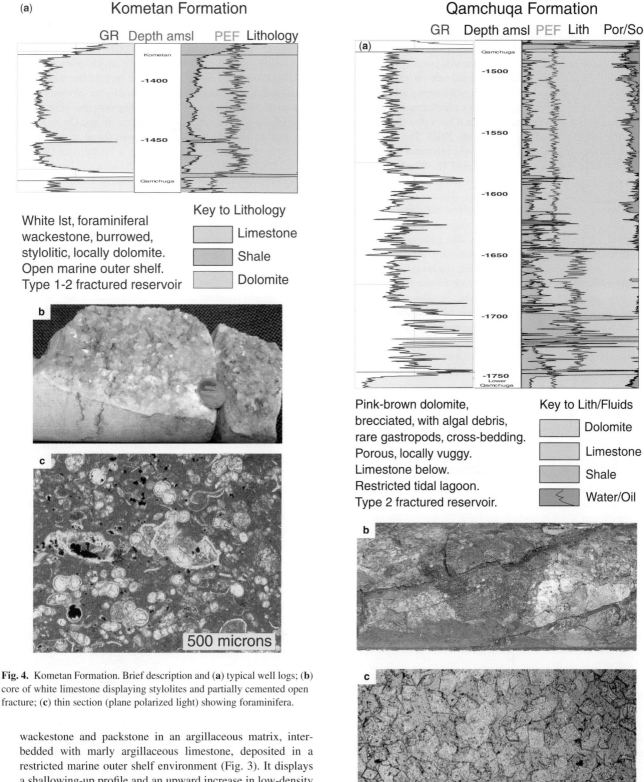

Fig. 4. Kometan Formation. Brief description and (**a**) typical well logs; (**b**) core of white limestone displaying stylolites and partially cemented open fracture; (**c**) thin section (plane polarized light) showing foraminifera.

wackestone and packstone in an argillaceous matrix, interbedded with marly argillaceous limestone, deposited in a restricted marine outer shelf environment (Fig. 3). It displays a shallowing-up profile and an upward increase in low-density clays and lithic fragments (showing green on the log) which have been attributed to wind-transported dust or volcanic material. Its thickness is reduced over the crests of the fault blocks. Despite its low matrix porosity, the Shiranish is pervasively fractured, especially in the lower part, which makes it a significant hydrocarbon reservoir with high deliverability exceeding 19 000 bopd.

(2) The *Kometan Formation* (Turonian–Santonian) comprises up to 120 m of white to light-grey, compact, burrowed, well-bedded, stylolitic, foraminiferal wackestone, similar to the Shiranish but without the major argillaceous component (Fig. 4). It was deposited in a pelagic marine environment. A disconformity is

Fig. 5. Qamchuqa Formation. Brief description of upper part of reservoir and (**a**) typical well logs with porosity and oil-saturation evaluation; (**b**) core showing brecciated lithology and steeply-dipping sub-parallel open fractures; (**c**) thin section showing dolomite crystals and blue stain in pore space.

developed at the top of the Kometan: this is interpreted from variations in thickness of the overlying Shiranish Formation and from seismic evidence that shows many faults terminating upwards at this level. There is, however, no evidence of subaerial exposure or karstification and matrix porosity is generally low. This confirms the results of regional outcrop studies (Karim et al. 2008). Locally, especially where fracturing is intense, the Kometan is dolomitized and displays a higher matrix porosity. Studies of this dolomite indicate that it has a post-depositional, hydrothermal origin (Davies & Smith 2006). The Kometan is generally more fractured than the Shiranish and flowed over 16 000 bopd on test. A thin interval (6–10 m) below the Kometan is occupied by the Gulneri and Dokan Formations.

(3) The *Qamchuqa Formation* (Aptian–Albian) comprises over 300 m of light brown microcrystalline dolomite, originally wackestones to grainstones whose original fabrics are locally preserved, including: mouldic pores after gastropods, bivalves, ostracods and gypsum crystals; laminated algal fabrics and crusts cut by burrows of various sizes; and brecciated intervals comprising angular lithified carbonate fragments, suggesting clastic gully fill or debris flow (Fig. 5). The upper part of the Qamchuqa displays a karst fabric of interconnected vugs and fracture fill, indicating subaerial exposure of the original sediment, which would have been deposited with a very high primary porosity (>50%). This indicates a lagoonal and tidal flat setting with erosion during periods of lowered sea-level. With depth, its composition becomes increasingly dominated by limestone and marl (blue on the log) down to an unconformity at its base. The connected matrix porosity of the Qamchuqa Formation makes this the main hydrocarbon reservoir. The Qamchuqa is pervasively fractured and produced over 17 000 bopd on test: the fractures are an important feature of the recovery mechanism and give access to hydrocarbon resources in the low-permeability matrix. The Qamchuqa is also a matrix reservoir in the Kirkuk area some 80 km to the SW (Al Shdidi et al. 1995). The cleaner water-bearing dolomite at the base of the log, currently termed 'Lower Qamchuqa', is probably the equivalent of the Shuaiba Formation further south.

Structure

A key seismic section from the 2D survey (Fig. 6a) shows the structure as a simple anticline with increasing complexity at depth. The interpretation (Fig. 6b) includes structural as well as stratigraphic elements. At the western end of the section the synclinal part of the fold indicates maximum depths of burial; at the base of the section the Jurassic shows reverse faulting and other complexities indicative of compressional stress; two major reversed faults emerge upward from the Jurassic on the southwestern and northeastern limbs of the fold; between them the Cretaceous reservoir shows a block-faulted style within the anticline and, in the shallow part, the Cenozoic is gently curved into a smooth parallel fold with outer-arc extension. The Cretaceous reservoirs between the two reverse faults occupy an intermediate structural position, isolated from the outer arc above by décollement within the shales and siltstones of the Kolosh Formation and separated from the deeper compressive regime by another discontinuity in the shales and limestones of the Sarmord formation.

The latest folding and faulting is related to the compressive phase of Zagros mountain-building which reached its climax in Late Miocene to Pliocene time (Ameen 1991). This late history of the structure is written in the deposition and parallel folding of the fluviatile sands and conglomerates of the Bakhtiari and Upper Fars Formations, which together attain 1500 m in thickness west of Taq Taq. The main anticline was formed as a result of compression above a thrust plane which probably lies several kilometres below the recorded seismic data. The present-day maximum horizontal stress in this area has been calculated, both from induced fractures in the Taq Taq wells and from regional earthquake studies (Reinecker et al. 2004), to be approximately NNE–SSW. This is consistent with the major orientation of the open fractures in the reservoir. Uplift of the anticline was enhanced by unloading of the crest of the structure by the erosion of at least 1500 m of clastic deposits. This unloading may also have enhanced some of the open fracturing in the Cenozoic and Cretaceous: it is only one of many factors that have been found to determine the final style of fracturing and faulting in a folded zone (Bazalgette & Petit 2007). Since the formation of the structure began subsequently to the commencement of oil generation, the open state of the fracture system may be partly due to inhibition of cementation by the presence of oil.

When the anticline is restored to its Late Cretaceous position (Fig. 7), the faults cutting it delineate a tilted block arrangement. This appears to have many similarities with that previously described for Northern Iraq in which the ramp-and-flat block-faulted style of the Cretaceous is due to one or more episodes of extension and faulting in Late Cretaceous time (Haddad & Ameen 2007). The depositional effect of this configuration in wells on the Bai Hassan and Qara Chauq structures to the SW of Taq Taq (Fig. 1) was described in terms of the sequence stratigraphy of the Gulneri and Kometan Formations and their position on the fault blocks. A similar structural control on deposition is confirmed by well data also to have taken place in the Taq Taq reservoirs, where the more shaly lower part of the Shiranish is only present in the deeper fault blocks. This phase of faulting clearly occurred following deposition of the Kometan because, whilst Top Kometan is displaced by several faults, it is not possible to trace their displacement upwards through the Shiranish. The minor variations in Kometan thickness may be related to an earlier phase of movement on these faults. Even though the two major faults have been reactivated as reverse faults during the formation of the anticline, it appears that the other faults have not been reactivated with any vertical displacement.

The depth map of Top Shiranish (Fig. 8) is derived from a principal component interpretation of the 3D seismic survey and shows details in the anticline which point to the presence of a strike-slip element in the generation of the fold. The double-plunging anticline is traversed by a set of subsidiary short folds crossing the axis, which are interpreted to be parasitic to the main anticline. These anticlines are predicted to occur in an oblique orientation to a right-lateral wrench fault (Harding 1974): in this case its orientation is NW–SE parallel to the main faults giving parasitic folds in a northeasterly to easterly direction. It has been shown from core and image-log analysis that the dominant open fracture direction is usually close to that of the nearest one of these folds and it is concluded that they are the loci of fracture corridors. Having made this connection, curvature processing has enabled them to be delineated with greater accuracy and adopted in the fracture model.

Evidence for this strike-slip element has also been found during mapping of the surface outcrop of sandstones and red shales of the Upper Fars Formation. These are extensively fractured and faulted into tilted blocks with a dominantly north–south azimuth. Detailed study of these fractures at several different localities revealed strike-slip displacements that point to dextral transpression along a NW–SE orientation. In this scenario the dominantly north–south fractures seen at the surface are synthetic Riedel shears. Strike-slip movement is also suggested by oblique-slip slickensides on cored faults. It is interpreted that, in the reservoir, the stress regime naturally exploited existing faults and fractures rather than creating new ones, and retained the Cretaceous dominant NE–SW open fracture direction during the uplift of the anticline

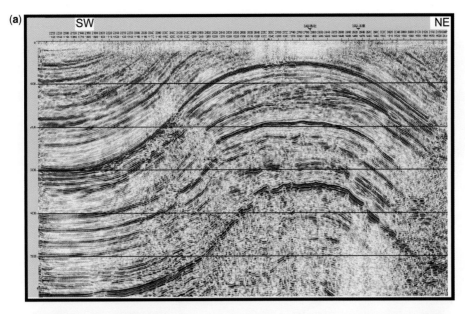

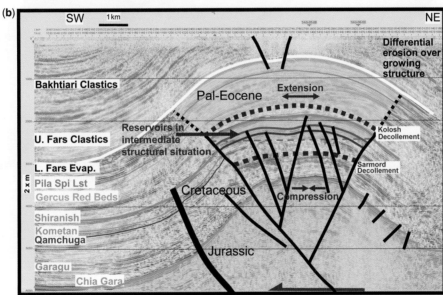

Fig. 6. (a) 2D seismic dip line across Taq Taq field. (b) The same line showing stratigraphic units and structural interpretation. The Taq Taq anticline is a response to shortening along a deep thrust plane lying below the base of the seismic section. Reverse faulting within the Jurassic indicates compression and normal faulting in the Cenozoic section indicates extension. The Cretaceous lies in an intermediate to extensional position between two neutral surfaces and is also affected by strike-slip movement.

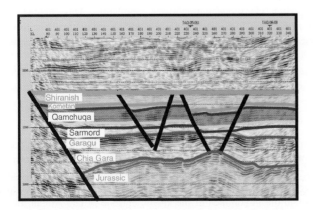

Fig. 7. 3D seismic line flattened on top Shiranish. This emphasizes fault movement prior to Late Cretaceous deposition of the Shiranish Formation.

and during filling of the growing structure with oil. Reactivation of the normal faults as strike-slip faults may also have occurred, but without any vertical movement.

Thus the Cretaceous reservoirs are in the favourable position of retaining original extensional structures as well as undergoing subsequent shear, which explains the dominant NE–SW fracture direction in the reservoir, in contrast to the north–south fractures in the Cenozoic sandstones at the surface, which formed for the first time within the later dextral shear regime. The low compressional stress within the reservoir is interpreted to partly explain the relatively open state of the fracture system.

Seismic facies analysis

Core obtained from the reservoirs and special analysis of many core plugs were a basis for the identification from logs of facies groups to be used for each well in the matrix model of the Qamchuqa and the Kometan Formations. Following a conventional system, the spectral gamma-ray log was used to differentiate clay from

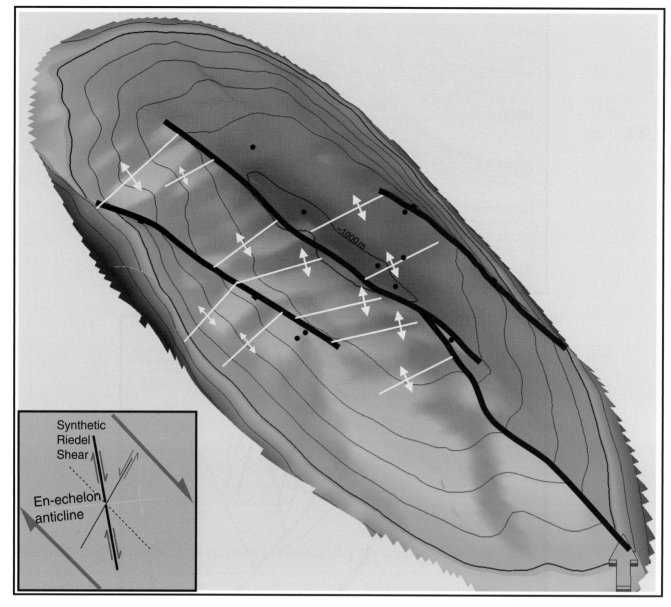

Fig. 8. Top Shiranish model depth surface, colour interval (c.i.) 100 m. Main fault displacement is strike-slip at this level. Parasitic folds, marked in yellow, perturb the smooth surface of the main anticline and are the major control on fracture corridors. Inset shows orientation of strike-slip features (after Harding 1974).

carbonate, the PEF log was used to differentiate dolomite from calcite and sonic v. density porosity was used to identify vuggy porosity from disseminated matrix porosity (Dehghani et al. 1997). These were used to define facies groups with different porosity statistics, different porosity–permeability characteristics and different water saturation response under the varying capillary pressure regime within a long transition zone. These groups were shale, limestone, low-porous dolomite, medium-porous dolomite, high-porous dolomite, medium-porous vuggy dolomite and high-porous vuggy dolomite. A study of seismic attributes was carried out to assist the mapping of these facies classes and their porosity.

Prediction of lithofacies from seismic amplitudes is difficult in carbonates where reflectivity variations are typically subtle. A statistically based modelling method was developed, consisting of sampling seismic amplitudes using a narrow sliding time window and analysing a large number of windowed samples using principal component analysis (Scheevel & Payrazyan 1999). The method made a selection of principal components to capture variance and other information within the seismic signal. These attributes of seismic samples projected into a multi-attribute space enabled a classification to be made according to the lithofacies distribution at the wells. Prediction of lithofacies was then carried out probabilistically throughout the seismic volume using a geostatistical approach via sequential gaussian simulation. This yielded the most probable facies at any location, giving a statistical, but geologically constrained, lithofacies model, as shown on the section (Fig. 9). The resultant facies volume confirmed some of the suspected trends in the data and revealed others that had not been interpreted from the limited well data alone.

Pore system

Estimation of matrix porosity in the Qamchuqa Formation has been a major target of the coring programme, partly because its character is, to some extent, masked by the exceptionally high well deliverability due to the fractures. Detailed study of the cores has confirmed the importance of the karstic episode in late Albian time. Porosity in the karstified upper zones of the Qamchuqa locally exceeds 20% with average in the order 4–12% and low permeability in the range 0.1–10 mD. In addition, late hydrothermal

- Fractured Limestone (Shiranish)
- Fractured Limestone (Kometan/Qamchuqa)
- Shale, Marl
- Low Porous Dolomite
- Medium Porous Dolomite
- High Porous Dolomite
- Vuggy Dolomite (Med-High)

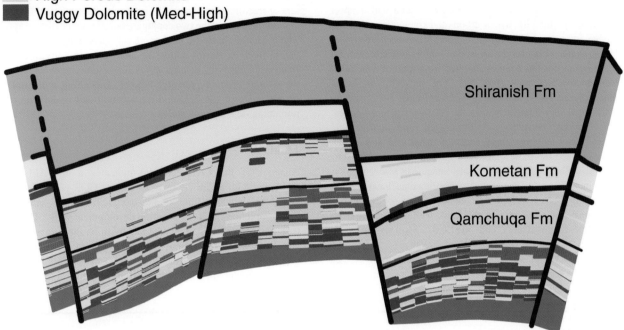

Fig. 9. Lithological model derived from seismic facies analysis (not to scale). The model shows development of vuggy porosity in the Qamchuqa formation on the crestal part of the anticline and lower porosity in the graben. The Kometan includes local dolomitization with low matrix porosity. The lithology is one of the controlling factors in the generation of fractures in each formation.

dolomitization has led to local development of matrix porosity in the Shiranish and Kometan formations, as identified by the seismic analysis. The porosity has been mapped together with facies using the seismic amplitude method described above.

A pervasive and well connected fracture system is seen as a benefit in this low-permeability matrix, which under natural or assisted water flood enables imbibition and effective oil displacement during production. The process is illustrated in Figure 10. Overall recovery will be dependent on the extent and characteristics of the fracture system, including spacing, connectivity and permeability.

A dual-media model is in use with the whole system being treated as a type 2 fractured reservoir in which connectivity within the matrix is largely provided by the fractures. To optimize recovery, the production strategy will be directed towards achieving a steady rise in water level in the fracture system so that imbibition has time to act in the matrix before the producing wells are overwhelmed by the water flood. It is believed that, with this treatment, the matrix, which contains up to 50% of the total STOIIP, will yield a high percentage of its movable oil to production. A wide range in probable volumes is carried for the porous matrix.

Fractures

Drill stem tests performed on well Taq Taq-1 in 1978 showed high productivity from the Shiranish formation, which has very low porosity and permeability. This paradox was interpreted to indicate the presence of a major fracture system in the well.

During the recent appraisal programme, high productivity has also been achieved from both the Kometan and the Qamchuqa

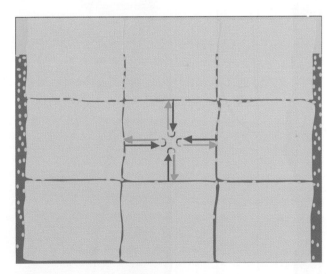

Fig. 10. Cartoon of matrix–fracture interaction under water flood. Rock matrix blocks (tan) are initially oil-bearing. Water introduced into the fractures is imbibed (blue arrows) into the matrix and oil is displaced into the fractures (green arrows). Oil droplets (green) form in the fractures, rise through the water column and merge into the oil column in the fracture network.

formations, and it was interpreted at an early stage that the Shiranish and Kometan formations could be considered to be type 1 fractured reservoirs, in which both storage and deliverability are provided only by the fractures. The Qamchuqa, however, would have to be treated as a type 2 fractured reservoir, in which storage is provided by the matrix and deliverability is provided by the fractures (Hubbert & Willis 1955). The appraisal programme was directed towards understanding both the matrix and the fractures. XRMI borehole image logs, WSTT dipole sonic logs and oriented cores were obtained from the initial appraisal wells. At the same time, surface mapping was carried out with emphasis on the fractures in the Upper Fars sandstones to gain an understanding of the stress fields related to the formation of the anticline. The cores have been studied intensively and every one of more than 1100 individual cored fractures has been described with respect to orientation, aperture, fill, hydrocarbons and kinematic indicators. The image logs across the cored intervals have been set up as standards for the whole reservoir section and a complete suite of fracture descriptions has been established for a total of over 6000 open conductive fractures (Fig. 11). The powerful statistical analysis tools available in FRACA software have been used to group fractures into families within the 'diffuse fracture system' and the 'fracture corridors'. A lithological and stratigraphic control has emerged from these analyses. Mud loss and inflow data have enabled the most significant open fractures to be identified in the wells from the appropriate image log trace. Finally, well tests, including interference and monitoring data, have verified the interpretation of a system so well connected that, for example, a pressure pulse due to testing of the flank well TT-08 could be detected almost immediately both in TT-07, 3 km to the south and in TT-06 2 km to the NW.

The trends of the axis-parallel faults interpreted from the 3D survey are also seen in those of one of the fracture families but, as noted above, the numerous small-scale anticlines with broadly ENE–WSW orientation, transverse to the main fold axis but parallel to the dominant fracture directions, are taken to be the loci of major fracture zones. This relationship has been described in many tectonic settings (e.g. Keating & Fischer 2008) and seismic curvature processing, which has improved the definition of these features, has been a major element in the interpretation. The current fracture model (Fig. 12) incorporates the interpreted clusters of fracture corridors and swarms with the statistical frequency and orientations of the two conjugate families in the diffuse system. The model has been the framework for uncertainty studies investigating likely scenarios within the assessed variability of the data. In terms of reservoir volume, a wide uncertainty remains regarding the overall fracture porosity. Fracture modelling will guide the development of the field and the location and orientation of the remaining development wells.

Fluids

Analysis of the Cretaceous oil indicates a 48° API gravity combined with a low viscosity and a low GOR of approximately 25 scf/bbl which is unusual and has facilitated the safe drilling and completion of all seven of the development wells. The source rock has been identified geochemically to be the Jurassic Sargelu and Naokelekan Formations, which have been penetrated within the field area. Another feature of the Taq Taq oil is its lower sulphur content (300 ppm) compared with the older producing

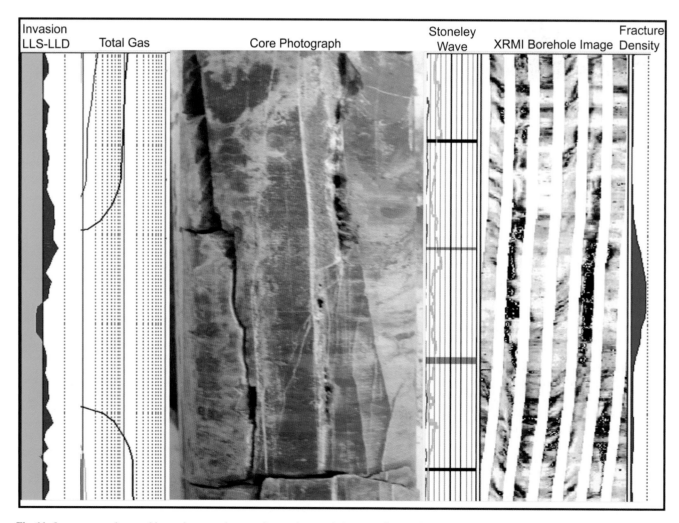

Fig. 11. Log, core, mud gas and image interpretation over fractured reservoir (not to scale). Total gas drops with severe mud loss during drilling. Core photograph of oriented whole core prior to slabbing. Stoneley wave from dipole sonic indicates zone of fractures. XRMI image detects fracture on both sides of the borehole. Fracture density calculated from core and/or from image calibrated to mean aperture.

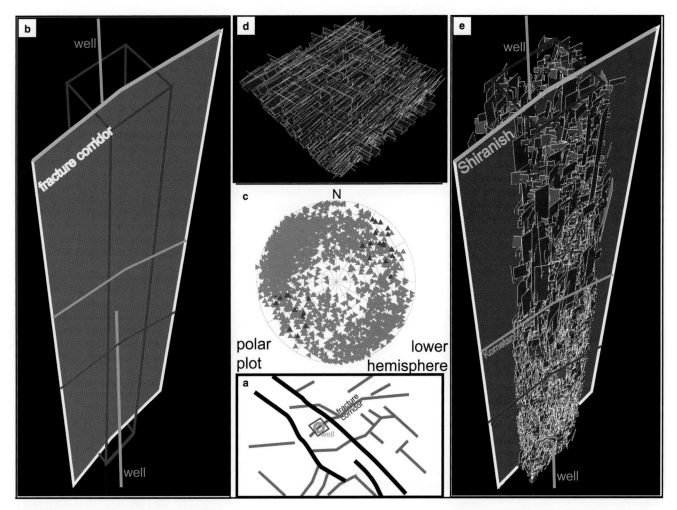

Fig. 12. Fracture model. Well located on fracture corridor (**a**) is explicitly incorporated into model (**b**). Statistics of all the fractures identified as open are analysed and split into red and green families of the diffuse fracture system (**c**). Fractures are propagated into the model for each zone based on the statistics (**d**). Zones are combined with the corridor (**e**) and the other wells in the final model.

fields in Iraq, and handling of the hydrogen sulphide has been relatively simple to incorporate into the design of the treatment facilities. Special core analysis has indicated an intermediate wettability for the Qamchuqa matrix with effective imbibition under water flood, which implies the potential for a favourable recovery factor, especially where the fracture density is high (Warren & Root 1963; Kazemi 1969). As has been found in many carbonate reservoirs, it has not been possible to define a single OWC from log interpretation. Part of the problem is due to varying water salinity where the formation water was originally saline but the active aquifer has been flushed with meteoric water since the emergence of the Zagros in Oligocene time and may still be in the process of displacing the original fluids. Also, as in many carbonate reservoirs, it has not everywhere been possible to obtain good RFT pressure data due to the difficulty of obtaining a good seal with the probe in hard and fractured dolomite. So far, the pressure gradients derived from static pressure data in both the oil column and the aquifer following production tests (of which there have been over 20) indicate a free water level implying a total potential oil column of some 525 m. It is planned to measure the true free water level in an open hole static test in a future well.

Current plans

Dynamic simulation of this hydrocarbon system, incorporating both the matrix and the fracture models, is being carried out with a dual media simulator which will enable optimization of well locations, orientations and production strategy under several possible scenarios within the uncertainty range. A production facility is being built close to the field with initial plans for tanker loading. At the time of writing (December 2008), various plans were being considered including a pipeline and possibilities for ramping up production. These facilities have been in production (March 2010) at rates up to 55 000 bopd with export by tanker. For the present, our models carry a wide uncertainty in many parameters: with production these uncertainties will narrow as the Taq Taq reservoirs progressively reveal their petrophysical secrets.

The support and cooperation of the Ministry of Natural Resources of the Kurdistan Regional Government of Iraq is acknowledged both for the writing of this paper and for the development of the Taq Taq field. Significant assistance is also acknowledged from the staff and students of the Department of Geology, University of Sulaimani, for helping to put Taq Taq subsurface geology into the context of local outcrops. Core and XRMI fracture analysis and sedimentological studies have been carried out by P. Lucas and A. Mann of Fugro-Robertson Ltd. Principal-component seismic facies analysis was carried out by J. Scheevel, of Scheevel Geotechnologies. The initial structural interpretation was carried out by B. Meier of Proseis AG, Zurich. Thanks are also due to the management of Genel Enerji (SA) and of Addax Petroleum Corporation.

References

Al Shdidi, S., Thomas, G. & Delfaud, J. 1995. Sedimentology, diagenesis and oil habitat of the qamchuqa group, N Iraq. *AAPG Bulletin*, **79**, 763–779.

Ameen, M. S. 1991. Alpine geowarpings in the Zagros-taurus range: influence on hydrocarbon generation, migration and accumulation. *Journal of Petroleum Geology*, **14**, 417–428.

Bazalgette, L. & Petit, J.-P. 2007. Fold amplification and style transition involving fractured dip-domain boundaries. *In*: Lonergan, L., Jolly, R. J. H., Rawnsley, K. & Sanderson, D. J. (eds) *Fractured Reservoirs*. Geological Society, London, Special Publications, **270**, 157–169.

Davies, G. R. & Smith, L. B. 2006. Structurally controlled hydrothermal dolomite reservoir facies: an overview. *AAPG Bulletin*, **90**, 1641–1690.

Dehghani, K., Edwards, K. A. & Harris, P. M. 1997. *Modeling of Waterflood in a Vuggy Carbonate Reservoir*. SPE Annual Technical Conference and Exhibition, San Antonio, Texas, SPE 38910.

Haddad, S. N. S. & Ameen, M. A. 2007. Mid-Turonian–Early Campanian sequence stratigraphy of northeast Iraq. *GeoArabia*, **12**, 135–176.

Harding, T. P. 1974. Petroleum traps associated with wrench faults. *Journal of Structural Geology*, **19**, 59–75.

Hubbert, M. K. & Willis, D. G. 1955. Important fractured reservoirs in the United States. *Proceedings of the 4th World Petroleum Congress*, 57–81.

Jassim, Z. J. & Goff, J. C. 2008. *The Geology of Iraq*. Dolin, Prague.

Kader, H. J. 2002. *Taqtaq Oil Field developed by Kurdistan Regional Government*. Babagurgur Centre for Kurdistan Resources Studies.

Karim, K. H., Ismail, K. M. & Ameen, B. M. 2008. Lithostratigraphic study of the contact between Kometan and Shiranish Formations (Cretaceous) from Suleimaniya Governorate, Kurdistan Region, NE Iraq. *Iraqi Bulletin of Geology and Mining*, **4**, 16–27.

Kazemi, H. 1969. Pressure transient analysis of naturally fractured reservoirs with uniform fracture distribution. *SPE Journal*, **9**, 451–462.

Keating, D. P. & Fischer, M. P. 2008. An experimental evaluation of the curvature-strain relation in fault-related folds. *AAPG Bulletin*, **92**, 869–884.

Pitman, J. K., Steinshouer, D. W. & Lewan, M. D. 2003. *Generation and Migration of Petroleum in Iraq: Modeling Study of Jurassic Source Rocks*. US Geological Survey Open-File Report.

Reinecker, J., Heidbach, O., Tingay, M., Connolly, P. & Muller, B. 2004. *World Stress Map*. Heidelberg Academy of Sciences and Humanities.

Scheevel, J. R. & Payrazyan, K. 1999. Principal component analysis applied to 3D seismic data for reservoir property estimation. *SPE Annual Technical Conference and Exhibition*, Houston, TX, SPE 56734.

van Bellen, R. C., Dunnington, H. V., Wetzel, R. & Morton, V. 1959. *Lexique Stratigraphique International Asie Iraq Vol 3c (Stratigraphic Lexicon of Iraq)*. CNRS Editions, France, reprinted by Gulf PetroLink 2005.

Warren, J. & Root, P. 1963. The behavior of naturally fractured reservoirs. *SPE Journal*, **3**, 245–255, SPE 426.

Sedimentology, geochemistry and hydrocarbon potential of the Late Cretaceous Shiranish Formation in the Euphrates Graben (Syria)

S. ISMAIL,[1] H.-M. SCHULZ,[1] H. WILKES,[1] B. HORSFIELD,[1] R. DI PRIMO,[1] M. DRANSFIELD,[2] P. NEDERLOF[3] and R. TOMEH[2]

[1]*Helmholtz Centre Potsdam-GFZ German Research Centre for Geosciences, Sect. 4.3 Organic Geochemistry, Telegrafenberg, D-14473 Potsdam, Germany (e-mail: shirin78@gfz-potsdam.de)*
[2]*Al-Furat AFPC, Damascus, Syria*
[3]*Shell International E&P, Rijswijk, the Netherlands*

Abstract: The Shiranish Formation consists of mudstones and wackestones in the central Euphrates Graben which are rich in organic carbon. Here the Shiranish Formation is more than 700 m thick with a minor increase in organic maturity with depth. The Shiranish Formation sediments are characterized by a continuously increasing hydrogen index to the top whereas the oxygen index is markedly lower in the Upper Shiranish Formation (USF). The Lower Shiranish Formation (LSF) is characterized by lower hydrogen indices and higher oxygen indices relative to the USF. These organic geochemical characteristics enable a rough subdivision into a lower and an upper part of the Shiranish Formation. Furthermore, mineralogical results enable a subdivision of the USF into two parts (USF-1, lower part; USF-2, upper part) each with individual mineralogical signatures due to a modified depositional environment and differing diagenetic history. The LSF resembles mineralogically the USF-2. Ankerite, together with higher pyrite contents in the LSF and USF-2, reflect similar diagenetic pathways which were controlled by higher clay contents. During early diagenesis, a traceable conversion of metabolizable organic matter led to mineral assemblages due to significant methanogenesis. Intervals in the USF with total organic carbon (TOC) contents up to around 4% and hydrogen indexes up to 500 mg HC/g TOC indicate the presence of very good potential source rock intervals for oil generation. Additionally, intervals of the LSF also contain gas-prone organic material. Bulk kinetic investigations show a broad activation energy of the LSF and a narrow activation energy pattern for the USF for hydrocarbon generation. Furthermore, the predicted petroleum formation temperatures are 136°C for the USF and 144°C for the LSF, respectively. This corresponds to c. 630 m difference in burial depth for petroleum formation. These differences in activation energies and corresponding depth to reach oil window maturity are controlled by facies, and less by maturity.

Keywords: organic carbon, hydrocarbon source rock, Cretaceous, graben, petroleum

Organic matter-rich sediments in modern and ancient depositional environments are the result of a complex interaction between several factors, such as primary productivity in the water column, supply of nutrients, redox–oxidation processes and sediment accumulation rates (Demaison & Moore 1980; Pedersen & Calvert 1990; Tyson 1995). An extraordinary geological test site to unravel the significance of these single factors which may have controlled organic matter richness is the marine Shiranish Formation in the Euphrates Graben (Figs 1 & 2). Sediments of the Shiranish Formation can locally reach more than 1000 m thickness in the central Graben and were deposited during the late Campanian and Maastrichtian. The thickness of the Shiranish Formation predominantly is a result of syn- and post-rift controlled subsidence which changed regionally and temporally. The Lower Shiranish Formation (LSF) was deposited during the syn-rift development of the graben, whereas the Upper Shiranish Formation (USF) is a part of the post-rift sequence (De Ruiter *et al.* 1995; Litak *et al.* 1998). The Tayarat Limestone intercalated between the Lower and Upper Shiranish Formation marks this change. The Shiranish Formation exhibits strong variations in thickness along the graben axis (e.g. up to 1800 m in the SE; De Ruiter *et al.* 1995). The remarkable thickness of the Shiranish Formation deposited within a relatively short time is the result of high sedimentation rates both of the clayey and carbonate mineral matrix, coupled with a strong subsidence, predominantly in the graben area. In terms of petroleum geology, these sediments are considered as seal and source rock having total organic carbon (TOC) contents of up to 1.7% (De Ruiter *et al.* 1995; Brew *et al.* 2001). On the other hand, the Shiranish Formation has reservoir properties outside the Euphrates Graben where shallow marine depositional environments prevailed (e.g. Oudeh Field in northern Syria and Kirkuk and Mosul area in Iraq, after Ziegler 2001).

In this manuscript we present detailed investigations of the source rock potential of the Shiranish Formation in the Euphrates Graben at a high resolution. Our investigations focus on two wells (well A and well B both drilled in 1987; Fig. 2), which are located in the central part of the Euphrates Graben. Well A was drilled with water-based mud and well B was drilled with oil-based mud. Well A recovered cuttings material from the Shiranish Formation between 2840 and 3550 m depth (well B: 2760–3525 m depth). In both wells the Shiranish Formation exhibits all features of continuous sedimentation and has a thickness of about 800 m.

The organic matter-rich sediments of the Shiranish Formation are an attractive research topic due to the high TOC contents in a thick sedimentary column deposited within a relatively short time interval during the latest Cretaceous. It is the aim of this contribution to couple data from geochemical, mineralogical and sedimentological investigations in order to evaluate the source rock properties as a consequence of depositional environment and diagenesis.

Regional and petroleum geology of Syria

Four major tectonic zones occur in Syria: the Palmyride foldbelt, the Euphrates Graben system, the Abd El Aziz/Sinjar uplifts and

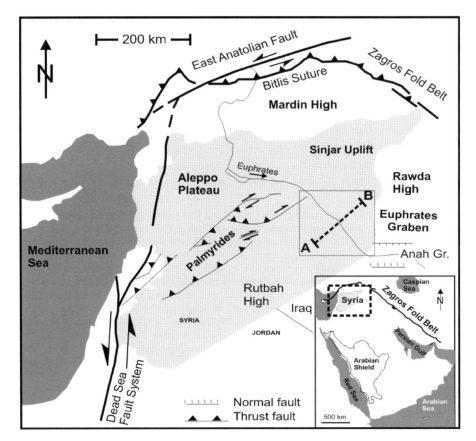

Fig. 1. Tectonic map of Syria. Geological cross-section A–B in Fig. 2.

the Dead Sea fault system (Barazangi et al. 1993; Brew et al. 2001; Fig. 1). The 160 km long Euphrates Graben system is interpreted as an aborted NW-trending intercontinental rift of Late Cretaceous age that has subsequently been hidden by sediments due to Cenozoic burial (Litak et al. 1998). It consists of a system of normal and strike-slip faults and is a result of a transtensive regime during the Late Cretaceous. The tectonic processes formed a complex of many smaller grabens and half-grabens which are less expressed in the northwestern part. The underplating of the Arabian plate in the Neo-Tethys subduction zone ended the rifting in the Late Cretaceous. It is related to convergence associated with the emplacement of amphibolites along the Arabian Plate (Litak et al. 1998). On a wider regional scale, the depositional environment during the uppermost Cretaceous in the Middle East is controlled by the northeastwards dipping Arabian Plate. This stable tectonic element enabled the deposition of carbonates with minor variation of facies (shallow and open-marine) which graded into deep marine clastics in the Levantine Basin and the Mesopotamian foreland basin (Brew et al. 2001; Ziegler 2001).

Oil and gas fields occur throughout the tectonic zones with the exception of the Dead Sea fault system. However, the majority of the oil fields are located in the Euphrates Graben system. This graben system is a relatively small but highly prolific oil province with more than 38 oil fields discovered since the mid-1980s. Reservoirs are predominantly siliciclastic and of Triassic and Early Cretaceous age (De Ruiter et al. 1995).

Potential source rock units in Syria can be subdivided into two groups on the basis of their lithofacies characteristics: (1) relatively deep and open-marine lithofacies, which include the Palaeozoic

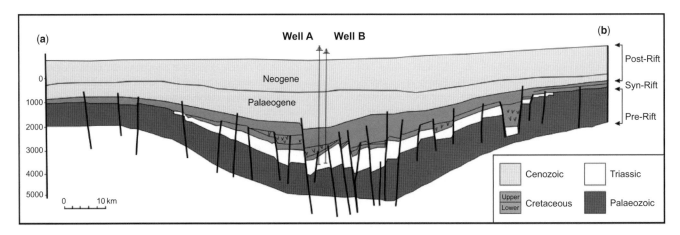

Fig. 2. Geological cross-section across the Euphrates Graben (modified after De Ruiter et al. 1995).

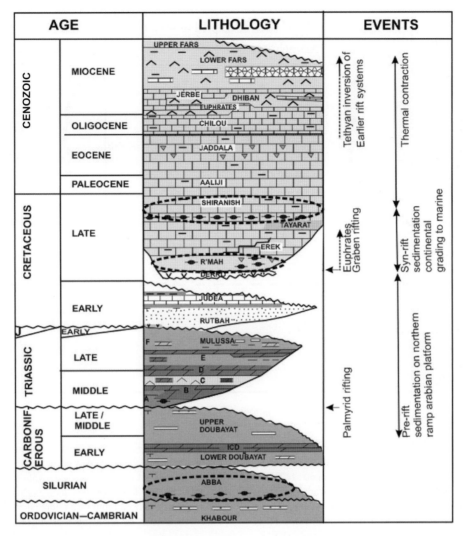

Fig. 3. General stratigraphy of the Euphrates Graben (modified after De Ruiter *et al.* 1995). Potential source rocks are presented in dotted circles.

Swab–Tanf Formation, the Late Cretaceous Shiranish Formation and the Paleocene Kermav Formation; and (2) shallow-marine and lagoonal lithofacies, including the Amanus Shale Kurra Chine Formation (Lower–Middle Triassic) and the Upper Cretaceous Soukhne Formation (Abboud *et al.* 2005).

Besides supposed Palaeozoic source rocks (e.g. the Silurian Tanf/Abba Formation) sediments predominantly of Cretaceous age are regarded as source rocks in the Euphrates Graben (De Ruiter *et al.* 1995, Fig. 3). A proven source rock is the R'mah Formation (alternatively Soukhne Formation) which is considered to be deposited as a first central grabenfill during the Early Upper Cretaceous (De Ruiter *et al.* 1995).

Samples and methods

The sample set was provided by Al-Furat (Damascus, Syria) and Shell (Rijswijk, The Netherlands). Additionally, data from Rock-Eval Pyrolysis and TOC contents of 28 samples from well A were kindly provided. Because of the limited amount of cuttings material from well A, inorganic geochemical and mineralogical analyses were carried out on cuttings material from well B. A direct comparison of data is possible, as both wells are close to each other and are similar in thickness and sedimentology.

The mineralogical and inorganic chemical composition of 30 cuttings samples from well B was investigated by XRD and XRF analysis respectively. Bulk pyrolysis was performed using 11 cuttings samples from well A. Kinetic parameters of hydrocarbon generation were determined using the experiential setup described by Schaefer *et al.* (1990).

Results and interpretation

The Shiranish Formation in well A is *c.* 700 m thick (Fig. 4). Lithologically, it consists of alternating mudstones and wackestones (terminology according Dunham 1962). A subdivision into a Lower and Upper Shiranish Formation with the Tayarat Limestone at their boundary follows internal considerations by Shell and Al-Furat, and is supported by the analysis of petrophysical properties (Ismail *et al.* 2008). This subdivision into a lower and an upper unit of the Shiranish Formation is kept in this contribution (Fig. 4) and extended to a further subdivision of the USF (Fig. 7).

Maturity and hydrocarbon source rock potential

A potential hydrocarbon source rock is characterized by its organic carbon content, the type and quality of the organic matter, and organic maturity. A compilation of different data log motifs from RockEval pyrolysis for the Shiranish Formation in well A (Fig. 4) together with a compilation of selected data in interpretive x–y plots (Fig. 5) shows the subdivision of a Lower and an Upper Shiranish Formation, the latter of marginally lower organic maturity (T_{max} averages slightly lower than 440 °C). In general, The LSF is characterized by TOC contents between 1.37 and

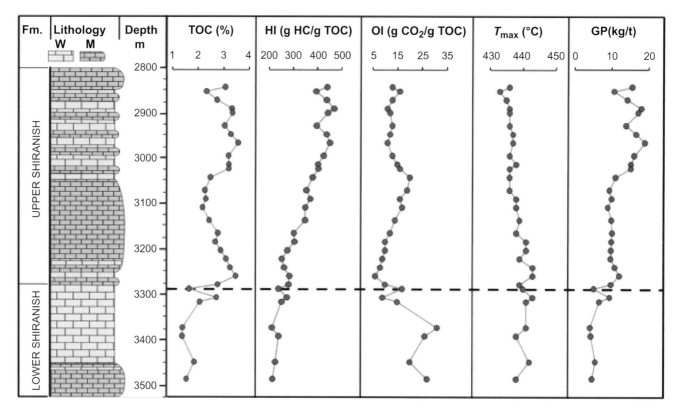

Fig. 4. Lithology and organic geochemical proxies for well A. Fm., Formation; W, wackestone; M, mudstone.

2.73%, hydrogen indices ranging from 212 to 275 mg HC/g TOC, and oxygen indices with values between 9–31 mg CO_2/g TOC. Moreover, a maximum temperature T_{max} of around 440°C is needed for the total thermal conversion into hydrocarbons. These characteristics of the LSF enable the calculation of the genetic potential (GP) of around 5 kg/t, and a production index PI of more than 0.2 (equilibrium at 0.5). In contrast, higher organic carbon contents (2.27–3.58%) occur in the USF, which has relatively higher hydrogen indices with values from 263 to 469 mg HC/g TOC, but lower oxygen indices (8–20 mg CO_2/g TOC). This organic material requires a temperature (T_{max}) slightly lower than 440°C to start releasing thermogenic hydrocarbons. As a

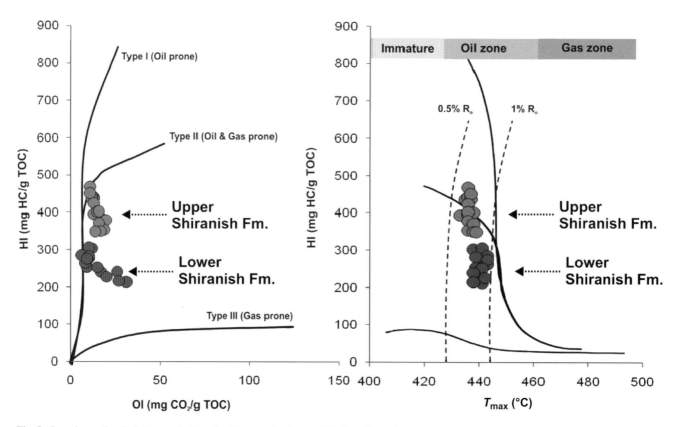

Fig. 5. Organic geochemical characterization of cutting samples from well A. Fm., Formation.

consequence, the organic material of the USF has a higher GP, due to a lower content of terrigenous organic matter, but a lower PI.

Organic facies

The organic matter of the Shiranish Formation can be classified into two groups, similar to the subdivision into the Lower and Upper Shiranish Formation. According to the van-Krevelen definition (Espitalié et al. 1980; Fig. 5), the USF contains a type II kerogen, whereas a type II–III kerogen mixture occurs in the LSF. However, TOC contents and Rock-Eval data have to be critically evaluated regarding their significance for a useful interpretation (Dembicki 2009). For example, the mineral matrix of a source rock with TOC contents less than 2.0% (as for intervals in the LSF excluding an interval at depth 3330 m) can also influence Rock-Eval kerogen typing. This phenomenon can be the result of retention on mineral grains, which cause a significant reduction in the hydrogen index (HI; Espitalié et al. 1980; Katz 1983). As the Shiranish Formation is calcareous in lithology, thermal decomposition of small amounts of carbonate minerals during the Rock-Eval analysis can contribute carbon dioxide and increase the oxygen index (OI) in low TOC sediments (Katz 1983). However, there are alternative ways to correct the interpretation of Rock-Eval data in terms of organic matter typing and to define organic matter sources. The first one is to simply deconvolve the HI into its two components, S2 and TOC (Katz & Elrod 1983; Langford & Blanc-Valleron 1990; Fig. 7). Second, higher C_1–C_5 contents generated by pyrolysis-gas chromatography from the LSF relative to the USF (Ismail et al. 2008, Fig. 6) are evidence for a correct interpretation of Rock-Eval data in terms of organic matter type. The S2:TOC ratio calculated for the corresponding material thus implies well characterized organic material with good source rock properties for the LSF and excellent source rock properties for the USF (Fig. 7). What are the factors causing variations of TOC content and the Rock-Eval parameters in the Shiranish Formation? The observed trend in the hydrogen index appears to be related to the nature of the organic matter in the Shiranish Formation. Increasing HI values can reflect either differences in the degree of organic preservation of marine organic matter or higher changes in the relative abundance of marine and terrestrial organic matter (Katz et al. 1993). According to this, higher HI values in the USF would be indicative of better preservation and/or higher relative abundance of marine organic matter. In contrast, the lower HI values in the LSF reflect terrestrial input. Moreover, a wide range of HI values indicates variations of the organic facies. In summary, a higher palaeoproductivity led to an

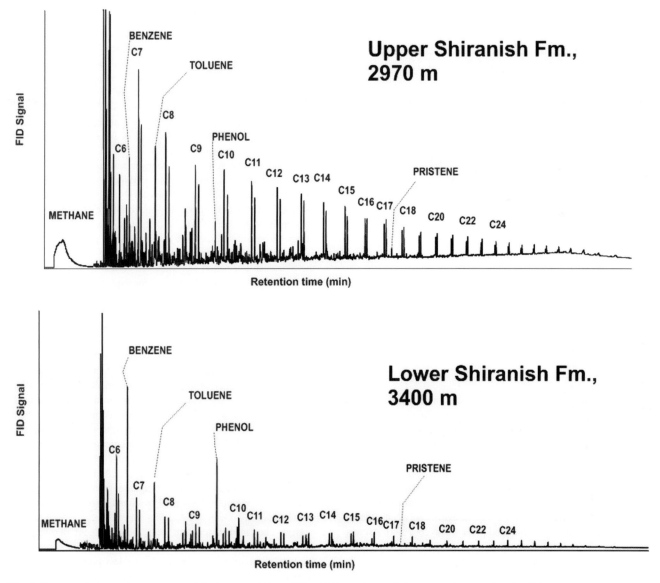

Fig. 6. Open pyrolysis gas chromatograms of the Lower and Upper Shiranish Formation carried out on cutting samples of well A.

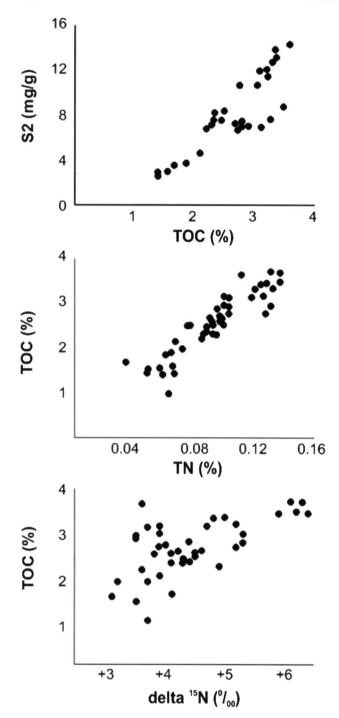

Fig. 7. TOC v. S2, total nitrogen content and $\delta^{15}N$ of cutting material from well A.

enhanced preservation potential of marine organic material due to clayey sedimentation during deposition of the USF in open marine conditions. In contrast, input of terrestrial organic material occurred during deposition of the LSF in shallow marine conditions.

TOC:N_{total} ratios remain fairly constant throughout the whole Shiranish Formation and suggest similar organic material (Fig. 7). Moreover, hydrogen indices positively correlate with $\delta^{15}N$ (Fig. 7). $\delta^{15}N$ can be used to distinguish between organic matter derived from algae and land plants (Meyers 1997). Atmospheric N_2 ($\delta^{15}N \approx 0‰$) is the nitrogen source for terrestrial plants, whereas dissolved nitrate with higher $\delta^{15}N$ values is the nitrogen source for plankton. $\delta^{15}N‰$ increases from 3 to 7‰ towards the top of the Shiranish Formation. Two scenarios can be considered as factors controlling $\delta^{15}N‰$ variability throughout the Shiranish Formation. The first one is water-column denitrification caused by relatively lower $\delta^{15}N‰$ values in the LSF due to nitrogen utilization by N-fixers. Progressive water-column denitrification may have led to relatively heavier $\delta^{15}N‰$ values in the USF due to increased productivity. Second, variations of $\delta^{15}N‰$ can be the result of mixing of isotopically light terrestrial organic matter with marine organic matter. The controversy of this interpretive approach and relevant references are presented in Calvert (2004).

Mineralogy of the Shiranish Formation and early diagenesis

The inorganic matrix of the cuttings material from well B is predominantly composed of calcite (up to c. 90%). Quartz, kaolinite and ankerite occur as well as calcite, and traces of pyrite have been found (Fig. 8). Besides the general subdivision into a Lower and Upper Shiranish Formation, the distribution patterns of the single mineral phases allow a further subdivision of the USF into a lower and upper unit, both with different mineralogical compositions. Consequently, a new classification of the Shiranish Formation into three sub-units is proposed into units A (the LSF), B (lower part of the USF) and C (upper part of the USF). In terms of mineralogical composition, unit C resembles unit A (Fig. 8). The carbonate contents are similar in both units A and C, ranging from 51 to 80%, while higher carbonate contents (72–91%) occur in unit B. High concentrations of siliciclastic minerals such as quartz and kaolinite are observed in units A and C. Tracer minerals to unravel the diagenetic pathways of the Shiranish Formation are ankerite and pyrite. The contents of both are lower in unit B.

Several mineral phases occur in the Shiranish Formation, either deposited primarily or the result of early diagenesis of organic matter-rich matrix. Calcite, quartz and kaolinite are regarded as primary minerals whereas pyrite and ankerite are interpreted as characteristic diagenetic products in organic-rich sediments (for an overview see Einsele 2000).

In general, relative sea-level changes can greatly affect carbonate sedimentation as widespread carbonate accumulations are often associated with global sea-level highstands (Tucker & Wright 1990). Thick carbonate sequences are thus deposited as transgressive systems tracts, and relative sea-level change may affect carbonate sediments by a diagenetic process (Tucker & Wright 1990). The variation in relative calcite content throughout the Shiranish Formation can therefore be considered as a general proxy for sea-level variations. As a consequence, high calcite contents in unit B could point to a highstand. Moreover, high calcite contents may reflect high palaeoproductivities due to positive correlations with relatively higher TOC contents. Periods with less carbonate flux and deposition were coupled to increased terrigenous input (high oxygen indices together with higher quartz and clay contents, here kaolinite). In general, kaolinite is associated with very low energy depositional environments (Chamley 1989; Hallam et al. 1991), but can also be formed diagenetically (Ghandour et al. 2003). Kaolinite is the predominant clay mineral in the Shiranish Formation and is at its lowest concentration in unit B (Fig. 8). This variation may be related to relative sea-level changes, which affect the distribution of clay minerals such as kaolinite and smectite (Steinke et al. 2008). As kaolinite is of sedimentary origin in the Shiranish Formation, increased contents can be related to a relative sea-level lowstand during sedimentation of unit B.

High clay mineral content decreases sediment permeability. Thus, the positive correlation between the kaolinite content and the pyrite and ankerite content indicates less permeable sediment properties which enhance sulphate reduction and minor pyrite precipitation followed by intense ankerite crystallization (Berner

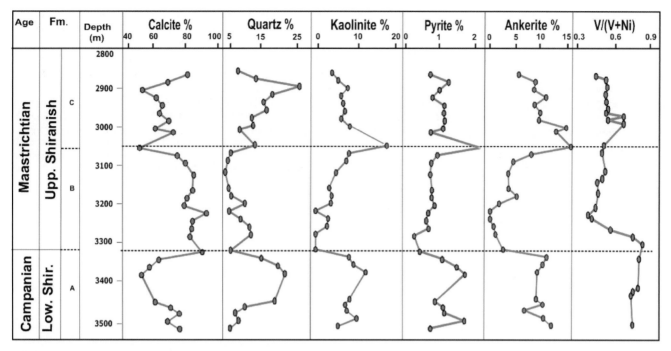

Fig. 8. Mineralogical composition of the Shiranish Formation in well B.

1982). Ankerite may indicate that intervals with relatively high kaolinite contents had methanogenic pore water conditions during early diagenesis, during which organic matter was intensively converted. Thus, intervals with significant ankerite contents originally had relatively higher TOC contents than those with low ankerite contents.

Trace element palaeoredox proxies

Trace element concentrations and ratios of significant elements have successfully been applied to evaluate the palaeo-O_2/H_2S boundary in ancient, stratified water columns (e.g. Cr/V, V/(V + Ni) etc.; Hatch & Leventhal 1992).

For example, vanadium is more effectively fixed than nickel in sediments under anoxic water conditions and in the presence of hydrogen sulphide (Lewan & Maynard 1982; Breit & Wanty 1991). The application of the ratio V/(V + Ni) suggests relative changes in oxygenation in the water column throughout deposition of the Shiranish Formation (according to Lewan & Maynard 1982; Breit & Wanty 1991). Hatch & Leventhal (1992) suggested V/(V + Ni) ratios >0.84 to indicate euxinic conditions, and lower values indicate higher oxygen contents in the bottom-water column. According to this classification, V/(V + Ni) ratios around 0.8 in the LSF could be linked to oxygen-depleted depositional conditions.

Bulk petroleum generation kinetics

The transformation of sedimentary organic matter, mainly kerogen, to petroleum during burial in a sedimentary basin can be calculated as a function of time and temperature. It is controlled by parallel and successive chemical reactions (Tissot & Welte 1984). This basic approach was applied to 11 samples from the Lower and Upper Shiranish Formation, each with a characteristic, but different organic material. The approach includes the calculation of the bulk parameters activation energy E_a and per-exponential factor A. Both will be used to determine the timing of hydrocarbon generation at a linear geological heating rate of 3.3 K/Ma. This heating rate corresponds to an average geological heating rate in sedimentary basins (Schenk et al. 1997), and is applied as a rough working hypothesis for the Euphrates Graben. Under consideration of this generalized heating rate, the Lower and Upper Shiranish Formations differ significantly in their diagnostic activation energy distributions and petroleum yields (Fig. 9). The USF is characterized by high petroleum yields at low activation energy (maximum at 54 kcal/mol) and frequency factors ranging between 1.52×10^{14} and

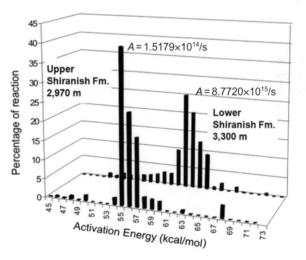

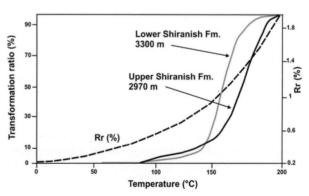

Fig. 9. Distribution of activation energy, computed transformation ratio and vitrinite reflectance in the USF and LSF.

6.3×10^{14}/s. A relatively narrow activation energy distribution indicates a kerogen type II organic matter in the USF. On the other hand, calculations for the LSF show lower petroleum yields. For the sediments of the LSF a broader activation energy distribution has been determined (54–67 kcal/mol) with frequency factors between 1.04×10^{15} and 8.77×10^{15}/s. Calculations to predict the timing of petroleum generation result in different onset and peak generation temperatures for the Lower and Upper Shiranish Formations. The generation of petroleum from the USF is predicted to start at 136°C (lowest geol. $T_{max} = 151$°C) and at 144°C (lowest geol. $T_{max} = 170$°C) in the LSF. These different temperatures would correspond to c. 630 m difference in burial depth for petroleum formation assuming a mean geothermal gradient of 33°C/km. This difference is due to the different organic facies within the Shiranish Formation, with higher contents of terrigenous organic material in the lower unit.

Summary and conclusions

Clear source rock facies variations, with significant impact of the timing of petroleum formation, have been identified for the Shiranish Formation in the central Euphrates Graben. Source rock screening of the Shiranish Formation reflects a subdivision into a Lower and an Upper Shiranish Formation. An organic matter type II to type II–III is preserved within the USF, while a kerogen type II–III is incorporated in the LSF.

The major controls on the heterogeneities of organofacies are changes in the depositional environment, and changes in the redox environment due to the deposition during a time of a sea-level change coupled with an input of terrestrial organic matter to the LSF. This terrestrial organic matter input delivered gas-prone material in contrast to the predominance of oil-prone organic matter in the USF.

A new classification of the Shiranish Formation into three sub-units is proposed according to mineralogical composition: unit A (the LSF), unit B (lower part of the USF), and C (upper part of the USF). The units A and C resemble depositional conditions. Both are characterized by more clayey sediments and ankerite as a signal for palaeomethanogenesis.

The different organic matter types of the USF and LSF cause differences in the petroleum formation kinetics. The USF characterized by HI values >350 mg HC/g TOC requires lower activation energy (53–59 kcal/mol) and frequency factors (1.52×10^{14} to 6.3×10^{14}/s). A uniform heating rate of 3.3°C/Ma can cause an onset of petroleum formation (10% TR) at c. 136°C. In contrast, the LSF has HI values <350 mg HC/g TOC and shows a higher activation energy (54–67 kcal/mol) associated with frequency factors between 1.04×10^{15} to 8.77×10^{15}/s. The onset of petroleum formation (10% transformation ratio (TR)) is predicted to be at c. 144 °C.

In summary, the more than 700 m thick Shiranish Formation in the central Euphrates Graben is a unique example of a potential hydrocarbon source rock which has efficiently stored organic carbon during the uppermost Cretaceous. The large thickness in relation to the relatively short period of deposition requires both strong subsidence and high sedimentation rates. Further investigations will show how the huge amounts of organic material could have been produced and which factors led to an efficient storage, but also how these processes affect the today's hydrocarbon potential. A complex interaction of factors like productivity, favourable incorporation in the mineral matrix and less effective diagenetic overprint offer potential for a high resolution of depositional features.

This study is part of the Ph.D. studies of S. Ismail. The authors thank Shell International E&P and Al-Furat AFPC for providing cutting material and data. We are grateful to Dr R. Naumann for his mineralogical and inorganic analyses, Dr B. Plessen for the analysis of stable nitrogen isotopes, and F. Perssen for pyrolysis experiments.

References

Abboud, M., Philp, R. P. & Allen, J. 2005. Geochemical correlation of oils and source rocks from central and NE Syria. *Petroleum Geology*, **28**, 203–218.

Barazangi, M., Serber, D., Chaimov, T., Best, J., Litak, R., Al-Saad, D. & Sawaf, T. 1993. Tectonic evolution of the northern Arabian plate in western Syria. *In*: Boschi, E., Mantonani, E. & Morelli, A. (eds) *Recent Evolution and Seismicity of the Mediterranean Region*. Kluwer Academic, Dordrecht, 117–140.

Berner, R. A. 1982. Burial of organic carbon and pyrite sulfur in the modern ocean: its geochemical and environmental significance. *American Journal of Science*, **282**, 451–473.

Breit, G. N. & Wanty, R. B. 1991. Vanadium accumulation in carbonaceous rocks: a review of geochemical controls during deposition and diagenesis. *Chemical Geology*, **91**, 83–97.

Brew, G., Barazangi, M., Al-Maleh, A. K. & Sawaf, T. 2001. Tectonic and geologic evolution of Syria. *Gulf PetroLink, Bahrain. GeoArabia*, **6**, 573–616.

Calvert, S. E. 2004. Beware intercepts: interpreting compositional ratios in multi-component sediments and sedimentary rocks. *Organic Geochemistry*, **35**, 981–987.

Chamley, H. 1989. *Clay Sedimentology*. Springer, Heidelberg, 623.

De Ruiter, R. S. C., Lovelock, P. E. R. & Nabulsi, N. 1995. The Euphrates Graben of eastern Syria: a new petroleum province in the northern Middle East. *In*: Al-Husseini, M. I. (ed.) *GEO 94: The Middle East Petroleum Geosciences*, **1**. Gulf PetroLink, Manama, Bahrain, 357–368.

Demaison, G. J. & Moore, G. T. 1980. Anoxic environments and oil source bed genesis. *American Association of Petroleum Geologists Bulletin*, **64**, 1179–1209.

Dembicki, Jr, H. 2009. Three common source rock evaluation errors made by geologists during prospect or play appraisals. *American Association of Petroleum Geologists Bulletin*, **93**, 341–356.

Dunham, R. J. 1962. Classification of carbonate rocks according to depositional texture. *In*: Ham, W. E. (ed.) *Classification of Carbonate Rocks – A Symposium*. AAPG, Tulsa, OK, Memoirs, **1**, 108–121.

Einsele, G. 2000. *Sedimentary Basins. Evolution, Facies and Sediment Budget*. Springer, Berlin, 685.

Espitalié, J., Madec, M. & Tissot, B. 1980. Role of mineral matrix in kerogen pyrolysis: influence on petroleum generation and migration. *American Association of Petroleum Geologists Bulletin*, **64**, 59–66.

Ghandour, I. M., Harue, M. & Wataru, M. 2003. Mineralogical and chemical characteristics of Bajocian-Bathonian shales, G. Al-Maghara, North Sinai, Egypt: climatic and environmental significance. *Geochemical Journal*, **37**, 87–108.

Hallam, A., Grose, J. A. & Ruffell, A. H. 1991. Paleoclimatic significance of changes in clay mineralogy across the Jurassic–Cretaceous boundary in England and France. *Palaeogeography, Palaeoclimatology, Palaeoecology*, **81**, 173–187.

Hatch, J. R. & Leventhal, J. S. 1992. Relationship between inferred redox potential of the depositional environment and geochemistry of the Upper Pennsylvanian (Missourian) Stark Shale Member of the Dennis Limestone, Wabaunsee County, Kansas, USA. *Chemical Geology*, **99**, 65–82.

Ismail, S., Schulz, H. M., Wilkes, H., Horsfield, B., Dominik, W. & Nederlof, P. 2008. Petroleum source rock characteristics of the Shiranish Formation in the Euphrates Graben, Syria. *GEO 2008 – 8th Middle East Geoscience Conference and Exhibition*. Manama, Bahrain.

Katz, B. J. 1983. Limitations of Rock-Eval pyrolysis for typing organic matter. *Organic Geochemistry*, **4**, 195–199.

Katz, B. J. & Elrod, L. W. 1983. Organic geochemistry of DSDP Site 467, offshore California, Middle Miocene to Lower Pliocene strata. *Geochimica et Cosmochimica Acta*, **47**, 389–396.

Katz, B. J., Breaux, T. M. *et al.* 1993. Implications of stratigraphic variability of source rocks. *AAPG Studies in Geology*, **37**, 5–16.

Langford, F. F. & Blanc-Valleron, M. M. 1990. Interpreting Rock-Eval Pyrolysis data using graphs of pyrolizable hydrocarbons vs. total organic carbon. *American Association of Petroleum Geologists Bulletin*, **74**, 799–804.

Lewan, M. D. & Maynard, J. B. 1982. Factors controlling enrichment of Vanadium and nickel in the bitumen of organic sedimentary rocks. *Geochimica et Cosmochimica Acta*, **46**, 2547–2560.

Litak, R. K., Barazangi, M., Brew, G., Sawaf, T., Al-Imam, A. & Al-Youssef, W. 1998. Structure and evolution of the petroliferous Euphrates graben system, southeast Syria. *American Association of Petroleum Geologists Bulletin*, **82**, 1173–1190.

Meyers, P. A. 1997. Organic geochemical proxies of paleoceanographic, paleolimnologic, and paleoclimatic processes. *Organic Geochemistry*, **27**, 213–250.

Pedersen, T. F. & Calvert, S. E. 1990. Anoxic vs. productivity: what controls the formation of organic carbon-rich sediments and sedimentary rocks? *American Association of Petroleum Geologists Bulletin*, **74**, 454–466.

Schaefer, R. G., Schenk, H. J., Hardelauf, H. & Harms, R. 1990. Determination of gross kinetic parameters for petroleum formation from Jurassic source rocks of different maturity levels by means of laboratory experiments. *Organic Geochemistry*, **16**, 115–120.

Schenk, H. J., Horsfield, B., Kroos, B., Schaefer, R. G. & Schwochau, K. 1997. Kinetics of petroleum generation and cracking. *In*: Welte, D. H., Horsfield, B. & Baker, D. R. (eds) *Petroleum and Basin Evolution*. Springer, Berlin, 231–270.

Steinke, S., Hanebuth, T. J. J., Vogt, C. & Stattegger, K. 2008. Sea level induced variations in clay mineral composition in the southwestern South China sea over the past 17 000 yr. *Marine Geology*, **250**, 199–210.

Tissot, B. P. & Welte, D. H. 1984. *Petroleum Formation and Occurrence*. Springer, Berlin, **699**.

Tucker, M. & Wright, V. 1990. *Carbonate Sedimentology*. Blackwell Science, Oxford, 28–308.

Tyson, R. V. 1995. *Sedimentary Organic Matter*. Chapman and Hall, London, **615**.

Ziegler, M. A. 2001. Late Permian to Holocene paleofacies evolution of the Arabian Plate and its hydrocarbon occurrences. Gulf PetroLink, Bahrain. *GeoArabia*, **6**, 445–504.

Session: Passive Margins

Passive margins: overview

B. LEVELL,[1] J. ARGENT,[2] A. G. DORÉ[3] and S. FRASER[4]

[1]*Shell International E and P bv., Kessler Park 1, Rijswijk, 2280AB, The Netherlands
(e-mail: bruce.levell@Shell.com)*
[2]*BG Group plc, Thames Valley Park, Reading, Berkshire RG6 1PT, UK*
[3]*Statoil USA E&P, 2103 CityWest Boulevard, Suite 800, Houston, TX 77042, USA*
[4]*BHP Billiton Petroleum Inc, 1360 Post Oak Boulevard, Suite 150, Houston, TX 77056, USA*

Abstract: Passive margins have been the reliable, accessible mainstay of exploration success worldwide for the last 25 years, and have hosted the spectacularly fast exploitation of deepwater resources (Angola, Nigeria, Brazil, Trinidad, USA Gulf of Mexico, Egypt, Australia and India). Despite, or perhaps because of this, there is still much to learn about the variety of hydrocarbon habitats they present.

For example: (1) deep seismic observations and deep sea drilling have revealed more of the diversity of passive margins geodynamics. This liberates explorationists from simple geodynamic models, with consequences not only for new views of thermal history but also for the whole tectonic and stratigraphic evolution. For example, the time significance assigned to the geometries traditionally labelled 'pre-rift, syn-rift and sag' may be misleading. This has implications for correlations, the significance assigned to unconformities and sequence boundaries, heat flow and structural history. (2) New deep imaging of the sedimentary sections has revealed mistaken assumptions about the importance of 'mobile substrate' in major deltas and allowed the detailed unravelling of salt and shale movement and its implications for reservoir and trap. (3) Depositional models for deepwater reservoirs have increased in predictive capability and modern seismic imaging supports new models for shallow water sequences. (4) Discoveries of very large amounts of dry bacterial methane in stratigraphic traps have challenged old assumptions about prospectivity based on thermally matured source rocks. (5) New engineering and development technologies are opening up the commercialization of remote frontiers. As a consequence there is legitimate scope to re-visit old 'dogmas' and to propose that each passive margin segment is best regarded as unique, with analysis and interpretation rooted in observation rather than models (at least while the newly proposed models evolve to stability). Many of these themes were visited in the Passive Margins session of the Seventh Petroleum Geology Conference, held in London in 2009. This paper outlines some of these ideas, and considers how exploration along passive margins in the next decade can use new geoscience thinking.

Keywords: passive margins, exploration technology, petroleum systems, rifting, continental breakup, deepwater plays

Despite more than 70 years of active exploration, modern passive continental margins still remain an exploration frontier. With a current aggregate length of 105 000 km (for a recent review see Bradley 2008), they represent a substantial exploration domain, and also a long-lived one: mean Phanerozoic passive margin preservational lifetime, including the modern passive margins, is about 135 Ma (Bradley 2008). Present day passive margins have a mean (incomplete) life span of 104 Ma. The passive margin sequences (post rift) are estimated to host approximately 35% of all giant field discoveries (Mann et al. 2003), which in turn represent 67% of discovered conventional hydrocarbons.

Deep sea drilling, a wide variety of geophysical methods (e.g. SCREECH, Funck et al. 2003; Van Avendonk et al. 2008; ISIMM, White et al. 2008, 2010) and geological studies have demonstrated that there is substantial variety in the history of passive margins. In this paper we set out to identify some new thoughts which can trigger critical re-examination of already partly explored areas, and add to the effectiveness of exploration in new frontiers in this basin type.

A snapshot of industry activity shows the diversity of plays currently being pursued including:

(1) major deltaic depocentres (e.g. Cenozoic of US Gulf of Mexico; Venezuela, Orinoco; Brazil Campos, Santos; Niger, Congo and Baram Deltas);
(2) smaller deltaic depocentres (e.g. Alaska, Beaufort Sea; Cameroon, Isongo; Equatorial Guinea; India, Mahanadi, Cauvery; Guyana, Essequibo; South Africa, Orange; Mozambique, Rovuma; Tanzania, Rufiji; Kenya, Lomu);
(3) slope by-pass systems (Ghana; Equatorial Guinea; India, Krishna-Godavari);
(4) carbonate platforms and their margins (Pakistan, Indus; Senegal; Mauritania; Morocco);
(5) deepwater fold belts (Mozambique, Rovuma; Nigeria, São Tomé and Príncipe);
(6) lightly explored frontier Arctic margins (USA Beaufort Sea; Canada, deepwater Mackenzie Delta);
(7) deepwater far-outboard areas with a wide variety of reservoirs and tectonic settings (Norway Outboard Voring; Brazil Santos; Australia, Outer Exmouth);
(8) syn-rift sections (Gabon sub-salt; India Krishna-Godavari; Norway, Voring Nordland);
(9) pre-rift sections on many margins.

Passive margin evolution and its implications

Better understanding of passive margins is demonstrated by the observation that it is now surprisingly hard to find an example of a 'classical' model of margin evolution (e.g. Allen & Allen 2005). Perhaps the best is the Labrador margin (Chalmers &

Pulvertaft 2001). The exciting new information on deeper structure of continental margins generated by large-scale refraction and reflection experiments as well as direct drilling supports a wide spectrum of plausible passive margin evolutionary mechanisms. This variety has triggered a wave of model-based inferences, in particular on post-rift evolution of continental margins. These insights have, to some extent at least, liberated the exploration geologist, struggling to interpret the deeper data on industry reflection seismic lines from simple models of the rift, thermal sag, rift/drift unconformity and drift phases. They include:

(1) The common occurrence of multiple rifting events with an irregular relationship to the line of final breakup (Doré et al. 1999; Ren et al. 2003). A recently identified potential example is described by Scotchman et al. (2010), who postulate a pre-breakup rift, oblique to the continental margin, in the Santos Basin, offshore Brazil, thought to have proceeded almost to oceanic status.
(2) The possibility of polyphase faulting in any given rift event. Redfern et al. (2010) suggest multiple rift events in the Permo-Triassic sequences of the North Atlantic margins, more varied than the overlying Jurassic and younger basins.
(3) A better understanding of magmatic addition to passive margins. Reynisson et al. (2010) systematically describe high-velocity, high-density lower crustal bodies on the mid-Norwegian margin and infer that, contrary to the common view that these bodies constitute magmatic additions to the base of the crust, many such features may be better explained as high-grade metamorphics remaining from the Caledonian orogeny, or as serpentinized mantle. In a similar vein, White et al. (2010) show from deep penetration and wide-angle seismic data on the Faroes–Hatton margin the probable existence of massive lower crustal intrusion, confined to the margin and with a sharp landward boundary. This again calls into question the concept of a widespread underplate, and the many phenomena assumed to result from it.
(4) The occurrence of very wide sag-type basins, relative to the syn-rift phase, possibly related to differential lower to mid-crustal stretching rather than a thermal sag related to pure shear rifting. The importance of this structural regime is attested to by recent discoveries on São Paulo Plateau of the Santos Basin (Machado et al. 2009).
(5) The possibilities of either hot/wet spot rifting with intra-crustal and supra-crustal magmatic addition or cold/dry-spot or otherwise amagmatic rifting margin. For a summary of magma-poor margins see Reston (2008). Hyperextension and mantle exhumation are currently the source of prolific literature and more examples of hyperextension are regularly being proposed. Blaich et al. (2010) discuss the nature of the conjugate magma-poor Camamu–Almada (Brazil) and Gabon margins, and in particular the nature of a prominent deep detachment surface, the M-Reflector. It is uncertain whether the reflector is underlain by exhumed mantle, but it appears to be an intraplate decoupling surface that accommodated significant pre-breakup extension.

As well as the obvious and ubiquitous thermal maturation implications, these insights also potentially impact other issues important to explorers, for example:

(1) the 3D isostatic behaviour of margins, and their consequent palaeobathymetric evolution;
(2) the correlation of poorly dated 'syn-rift units', possibly representing multiple rift phases, and the interpretation of basement penetrations or reflections as non-conformities v. rotated fault planes (and hence the significance to be attached to 'basement' penetrations in terms of completely testing a play);
(3) the nature and significance of unconformities, in particular the presence, significance and age of the 'rift-drift' unconformity (Tucholke et al. 2007) – other considerations include local footwall uplifts v. regional changes in crustal stretching and thinning and, again, unconformities due to poly-phase rifting rather than the single rift event of classic passive margin evolution.

Two-dimensional views of margin evolution are clearly only partial. Segmentation of margins can relate to pre-existing geomechanical interfaces in the basement. Many margin geometries are typically attributed to the breaking of continental crust along old lines of structural weakness. Ebbing & Olesen (2010) follow this line in documenting basement thickness variations and structural segmentation along the mid-Norwegian margin. These variations are correlated with Caledonian and Precambrian basement domains, and also with later basement detachments and normal faults mapped close to shore or onshore, and prolongated oceanwards. This paper also provides a valuable compilation of Moho depth and basement thickness for the entire mid-Norwegian shelf and western Barents Sea. It is worth noting, however, that only about 50% of the entire Gondwana supercontinent's rifted margins are even sub-parallel to pre-existing structure, based at least on the structural level of the current Gondwana surface geological map. Hence old structures are not necessarily parallel to either rift/drift tectonic strike or depositional strike. Together with intracontinental deformation, these factors can cause significant trend complexity and segmentation along margins. This can be obscured, particularly in restoring early rift units by generalized continental fits, especially those that do not accommodate intra-continental deformation (De Wit et al. 2008).

Hydrocarbon charge and maturation

Major marine source rocks and oceanic anoxic events

The explored passive margin source rocks have typically been the few, major, regionally extensive source rock-bearing intervals characteristic of whole sets of basins and seemingly preponderant in the Atlantic Oceans (e.g. Toarcian, Aptian Turonian and Lower Eocene).

In this respect passive margins are of course no different from other hydrocarbon provinces. For the world as a whole, almost 60% of discovered conventional hydrocarbons were sourced from the major source rocks of the Jurassic and Cretaceous and 75% if the Cenozoic is included (Klemme & Ulmishek 1991). Plays associated with these source rocks have largely been proven, or at least defined. Future potential in these known petroleum systems will arise from changing political/environmental or commercial circumstances. Examples of such 'political' openings include: Guyana (Upper Cretaceous petroleum system); potentially Mexico (Jurassic and Cretaceous); the Eastern seaboard of the USA; Georges Bank off Canada; east Greenland; and Nordland, off northern Norway (Upper Jurassic/Lower Cretaceous).

Local source rocks

Excitingly, much future passive margin exploration may rely on currently unknown, missed or more local or marginal source rocks. For example, driving the Niger Delta play into still deeper water will require an Oligocene or older charge system. In margins where the Middle or Upper Jurassic source is absent or too deep, and the Upper Cretaceous is immature due to lack of a depositional overburden, the Lower Cretaceous is often in the present-day maturity window. The often-marginal source rocks of the Lower Cretaceous will, however, need to deliver! Future large passive margin plays which would benefit from a prolific Lower Cretaceous source rock facies include: west Greenland; much of Africa – NW Africa from Morocco and the Canaries to Liberia, southern Africa, and east Africa; the Great Australian Bight; the South Atlantic, Uruguay and Argentina; and the Lord Howe Rise. Interestingly the geochemistry of the early Aptian oceanic anoxic event suggests that it is a deeper water anoxic event than, for example, the Toarcian, which was characterized by photic zone euxinia (Jenkyns 2003). The specific organic and inorganic geochemistry of source rock pods, reflecting their palaeo-ecology, may enable extrapolation from known occurrences into shallower sections and predict onlap or dilution.

Important clues to missed source rocks are provided by modern geochemical methods. Examples include diamondoid and C isotope analysis (Sassen & Post 2008), which when carried out on minor condensate in the Baltimore Canyon Trough on the eastern North American passive margin (Prather 1991) possibly point to extreme thermal cracking of oils from a Lower–Middle Jurassic source rock. Elsewhere dilution of a minor but effective charge system by a prolific one may obscure a secondary play. Detailed biomarker geochemistry can tease out the secondary charge system.

Hydrates and biogenic gas

In addition to conventional source rocks, biogenic gas has proven in the Nile, Krishna Godavari and Mahanadi deltas to be able to source major fields. A better understanding of biogenic gas, even a predictive understanding, may unlock many new plays.

In the Krishna Godavari (K-G) Basin a detailed model of gas generation, charge focusing and migration has been developed by workers from Reliance, following the discovery of the multi Tcf-Dhirubbai Field and its satellites (Bastia 2006; Kundu et al. 2008). The K-G play is believed to be biogenic based on uniformly high negative $\delta^{13}C > 90$ per mil in the gas. The impingement of an oxygen minimum zone on the slope has resulted in preservation of a steady 1–2% total organic carbon (TOC) through the Mio-Pleistocene section. In areas with too rapid accumulation (perhaps >1200 m/Ma) the source rock is too dispersed. In areas with too slow deposition (perhaps <500 m/Ma) there is insufficient organic preservation, or sulphate is not dispersed, or the methane itself can be oxidized. However, in zones with the right sedimentation rate, bacteria generate methane from the organic matter in the shallow subsurface. As the sedimentary section accumulates, this methanogenic zone rises, gas migrates upwards until it reaches the hydrate stability zone where it first forms hydrate, then pools beneath the impermeable layer. This layer itself rises through the sediment column as fresh insulating sediments are deposited and deeper layers dissociate releasing biogenic gas.

The hydrate layer, following the conical sea bottom of the delta, has focused migration updip and broadly towards the delta axis into the ponded slope turbidites of a terrace formed by a linked extensional/compressional fault system.

Biogenic gas systems can be driven not only by organic material in the sediment hosting the methanogenic bacteria, but also by supply of other substrates from thermally maturing source rocks. Known biogenic gas systems also provide a basis for chasing new plays.

Heat flow models

For conventional charge systems thermal maturation is clearly important. In the case of the K-G Basin biogenic system the cooling of the seafloor along the continental slope due to changes in oceanic circulation driven ultimately by glaciations, critically extended the hydrate stability field updip. More usually, heating from below and insulation by sediment are regarded as the major issues. Classical basin models rely on a relationship between observable, if local, sedimentation histories, through regional unloaded crustal subsidence, to a model of isostatically balanced crustal thinning which has heat flow consequences. The 'McKenzie' pure shear stretching model for rifts (McKenzie 1978) has stood the test of time as a robust basis for modelling in various different software suites. It provides, for example, a simple explanation of the occurrence of oil and gas in the passive margin off Congo and Angola related to the Cenomanian–Turonian Iabe Formation, as well as rift basins such as the Kimmeridge Clay of the North Sea. This may in part be due to its general nature which, despite the rigour of back-stripping, still allows matching of temperature histories with loosely constrained crustal and conductivity parameters. Recent observations however challenge the validity of classical pure shear rift models for the later stages of passive margin geodynamics (e.g. Kusznir & Karner 2007). It is plausible that thermal basin models have in some cases been right for the wrong reasons, or alternatively that thermal histories (when calibrated for example to recent temperature data) can be somewhat insensitive to the precise mechanisms of lithospheric deformation.

In addition there has been re-assessment of some of the 'constants', for example the heat-generating capacity and thermal structure of the continental crust itself (Jackson et al. 2008) and kinetic parameters used in hydrocarbon modelling (Stainforth 2009).

Seismic observations of shallow or even exposed mantle in the outer parts of passive margins, and direct drilling evidence, cannot be reconciled with pure shear extension, and imply depth-variable extension (so called 'depth-dependent'), which can be logically related to lithospheric mechanics (Reston & Perez-Gussinye 2007).

For hydrocarbon prospectivity in passive margins this has the following implications:

(1) Accommodation space creation, both in the early rift and in the late rift to drift phases, may not follow models based on pure shear. Specifically, the empirical observations are that accommodation space is lower than predicted in the early rifting phase, leading to sustained shallow water depositional environments. This is typically followed by subsidence so rapid that it may not be balanced by sediment supply leading to rapid deepening.
(2) In inboard areas, the brittle extension of the rift phase ends well prior to the onset of drifting, or may even relate to a previous phase of rifting, leading to a rift-drift transition which is younger than the level of 'top extensional faulting', sometimes picked as the rift-drift unconformity. This can lead in frontier settings to erroneous jump correlations, with implications for reservoir and source rock prediction. It may also lead to an erroneous understanding of the regional context of the older rift faulting.
(3) Conversely in outboard areas, rifting, even to an extreme degree, can continue well after true seafloor spreading has commenced (Tucholke et al. 2007).
(4) The heat flow models related to pure shear, and the attendant mantle upwelling, may not be relevant to mantle which although uplifted is apparently not uplifted on a 'normal' mantle adiabat, and moreover undergoes extensive serpentinization (Blaich et al. 2010; Reynisson et al. 2010).

These effects are currently being assessed. The ability to confidently model temperature history based on geodynamics awaits clarity on the physical mechanisms for lithospheric mantle and lower crustal extension.

New thermal history algorithms will probably fall into two domains:

(1) 'pick-and-mix' forward models which can deal with varying margin evolution scenarios and are used to classify margins into genetic types (e.g. magma-rich v. magma-poor, depth-dependent stretching or pure shear);
(2) use of observed geometric attributes (sedimentary backstripping, whatever crustal profiling data are available) and boundary temperature assumptions without explicit choice of a geodynamic model to invert to a crustal profile, followed by a forward temperature model.

An interesting approach to the latter is the 2D inversion of strain rate as defined by stratigraphic geometries (e.g. Bellingham & White 2002; White et al. 2004; Crosby et al. 2008). In its first exposition this approach was simplified, but it illustrates well the potential of inversion based on observation, rather than forward models based on a sometimes arbitrary choice of geodynamic model.

Exploration on oceanic crust

Exploration is already being pursued onto oceanic crust in areas where sedimentary overburden is thick enough for thermal blanketing to compensate for lower crustal heat production. Typically such plays are sourced by rocks related to major oceanic anoxic events. Examples include the Gulf of Mexico, the outer Niger Delta, the outer Congo Delta and the Ganges–Brahmaputra deep sea fan off Eastern India. There are suggestions that some thick passive margin sequences, such as the anomalously unstructured simple prograding wedges offshore Mozambique, may also be underlain by transitional oceanic crust (Watts 2001).

Reservoirs

Clearly a wide range of reservoir types have proven productive on passive margins: aeolian sandstones (Kudu, Namibia); hydrothermal dolomites (Deep Panuke; Wierzbicki et al. 2006); the full range of carbonate facies (Campos and Santos Basin lacustrine coquinas and microbial carbonates, reef and fore-reef limestones in Mexico–Campeche, platform sequences such as the India–Bombay High and salt-rafted limestones in Angola and Congo); shallow marine clastics (NW Australia, Gabon); deltaics (Niger, Cameroon, the Canadian Mackenzie, Nile); and last but not least slope and basin-floor turbidites (all the above deltas, Gulf of Mexico, Norway and UK Atlantic Margin, Angola, Congo, Equatorial Guinea, NW Australia).

It is probably fair to say, however, that turbidite reservoirs have stolen the show as exploration has moved into ever-deeper water. Approximately 125×10^9 BOE have been discovered in water depths of more than 400 m in the last 30 years, and with 30 plus BBOE in the last 4 years, this global play is not yet creamed. The great majority of these volumes have been from passive margins (the major exceptions being the south Caspian and NW Borneo). Reasonable estimates, based on play analysis, creamed field size distribution curves for each province or simply basin creaming curve analysis, are for a further 150–200 BBOE from passive margin deepwater plays via new discoveries and reserve growth.

Notably, recent major exploration discoveries in the deepwater Brazilian Santos Basin have taken the spotlight from turbidite reservoirs (Scotchman et al. 2010). These resources are hosted in Late Barremian and Aptian non-marine microbial carbonates in the late syn-rift and sag sequences prior to deposition of the South Atlantic salt basin (Machado et al. 2009; Wright & Racey 2009). Reservoir quality is strongly facies-controlled overprinted by a multiphase diagenetic history, culminating in late-stage corrosion enhancing the remnant primary porosity and sequent reservoir deliverability. Predicting reservoir distribution across these giant fields poses an unique challenge for development. However, these discoveries raise the question of whether similar pre-salt petroleum systems are yet to be discovered along the conjugate margins in the South Atlantic (Jones et al. 2009).

Reservoirs on outer margins

Outboard portions of passive margins have typically been dominated by exploration for deepwater reservoirs. A number of processes can result in shallow water deposits being drowned in the deep waters of the outer margins and yet still be at economically drillable depths:

(1) sediment starvation, outboard due for example to an active inboard rift-related basin, perhaps in its thermal sag phase;
(2) depth-dependent mid-crustal stretching (Kusznir & Karner 2007), for example on the outboard Exmouth Plateau;
(3) severe crustal thinning and mantle exhumation (e.g. review in Reston 2009);
(4) micro-continent isolation by rift jumping (e.g. São Paulo High, Scotchman et al. 2010).

Reservoirs and 'basement' in multi-phase rifts

Much older industry seismic data shot with 4 km streamers warrants re-analysis in the light of the realization that a classical rift then drift model of passive margin evolution is probably too simple. Apart from simply correct imaging of basement, the picking of top basement reflections and an understanding of possibly rotated early fault planes which appear to be top basement is one point to check. A second is the significance and age of 'syn-rift' or 'pre-rift' sequences. Syn- which rift? Pre- which rift? Proximal alluvial sequences may well have been dated by inference from an assumed tectonostratigraphic scheme rather than by biostratigraphic data.

Large drainage basin systems

Some 50 major river drainage systems drain most of the land surface of the present-day continents and of these about half drain to modern passive margins, giving the thermal blanketing, reservoir and seal for recent hydrocarbon generation and preservation. As exploration shifts away from the current major river deltas and/or into deeper sections, it is worth questioning past palaeogeographies to see if drainage basins differed. The widespread development, for example, of Upper and 'Middle' Cretaceous inland (epeiric) seas on all continents suggests that major drainages were busy infilling these essentially foreland basin domains, with relatively minor drainages feeding the then young (Atlantic) passive margins. Earth systems, thinking about interactions between erosion, sedimentation, palaeoclimate and palaeoceanography, may also be a stimulating source of new ideas about reservoirs in older sequences (e.g. for the Cretaceous; Skelton et al. 2003). Notable Upper Cretaceous discoveries along the equatorial African margin in the Rio Muni Basin, Equatorial Guinea and the West Tano Basin, Ghana have shown that these earlier drainage systems can provide high-quality turbidite reservoirs.

Shelf edge deltas and plays

The compelling outcrop evidence that sediment supply drives deltas to the shelf edge during high stands (e.g. Uroza & Steel 2008), demonstrates that rigorous application of 'standard' sequence stratigraphic models (which for didactic purposes have underplayed the role of variable sediment supply) may have left the possibilities of these depositional systems unexplored. High stand shelf edge deltas can be expected to be characterized by thicker progradational parasequences, in general more linear and wave-dominated shorefaces, and more hyperpycnal flow direct from river mouths. The realization that interactions between palaeoclimate and changing atmospheric conditions may have caused peaks of sediment input allows explorers to investigate these controls for specific time intervals of higher weathering.

The importance of small, localized but high volume, sediment inputs controlled by drainage systems was highlighted by Martinsen et al. (2010). They describe a holistic 'source-to-sink' approach relating onshore drainage and geomorphology to subsurface clastic reservoirs, using the Norwegian Sea Paleocene play as an example. New geomorphological insights and data highlighting old river capture patterns are another source of inspiration for undiscovered reservoir sequences. Studies of mantle-related dynamic topography imply a variability in sediment supply due to erosion of regional uplifts, which is in contrast with the often implicit assumption of uniform sediment supply rates in some sequence stratigraphic models.

Improved techniques for establishing provenance enhance precision in this regard. For example, Redfern et al. (2010) and Tyrrell et al. (2010) describe an approach based on Pb isotope analysis of detrital feldspars that has helped to identify quite surprising and diverse basement origins for the sediments in the Rockall Basin and its margins.

Remaining deepwater plays

It can be expected that deep-marine turbidite plays will continue to be important in passive margins – not only sub-salt as in Angola and the Gulf of Mexico but also in unexplored frontier deepwater areas (e.g. the outboard Mackenzie Delta). Smaller river systems, some driven to the shelf edge as described above, and older drainage systems such as those in Cretaceous depocentres are all currently being pursued (e.g. east Greenland draining to the Norwegian outer Vøring Basin, Morocco–Canaries, Casamance Delta off Senegal). Based on improved seismic imaging, Larsen et al. (2010) also proposes that the Cretaceous syn-rift play of the Faroe–Shetland Basin is worth revisiting. Their interpretation revisits the source-to-sink theme, and the concept of rift segmentation, alluded to in this paper and elsewhere in this volume.

It is also possible that, in the thickest deepwater fans with massive slope by-pass of sand such as the Congo and the Ganges–Brahmaputra, largely unstructured compactional fan lobe and channel plays may be possible even on oceanic crust (Anka et al. 2009), given of course source rocks and a thick enough thermal blanket of sediment.

Retention

Subtle migration paths

Offshore Equatorial Guinea discoveries in Miocene slope channels spectacularly demonstrate the ability of thin (10–20 m) slope and basin floor sands to form stratigraphic traps in largely unstructured sections (Stephens et al. 1996), as well as the ability of oil and gas condensate to migrate vertically through thick mudstone packages from relatively deep Upper Cretaceous source rocks. This impressive feat of migration is mediated by compactional faults related to the seafloor rugosity along transform faults at the top of the oceanic crust, but which have very limited offsets. Much remains unknown about the fault-related focusing of migration flux in such systems.

Late structural evolution

Anderson (2007) argues on the basis of the lack of strength of rocks in tension that a tectonic plate can be seen as a unit of lithosphere in a state of compressive stress: part of a self-organized system essentially held in that state by plate boundary forces. It is therefore not surprising that passive margins that are born and die at plate margins but spend their middle years in plate interiors record mid-life crises, expressed as a fair amount of compressive tectonic structuring. As is well known, the World Stress Map (Heidbach et al. 2008) shows that the current orientation of principal horizontal stress is less related to the direction of rifting than to the current direction of plate motion, giving present-day and presumably therefore palaeocompressive stresses at oblique angles to the margins. The relationship of such plate motion-induced stresses, or body-force stresses associated with plate boundaries, to observed strain phenomena is discussed in the paper by Doré et al. (2008). Broad compression-related folds in the Cretaceous–Cenozoic cover successions are observed over a wide area between the Mid-Norwegian shelf and the Rockall–Faroes area. The folds appear to be episodic and to have multiple origins, and are interesting as potential late-formed hydrocarbon traps.

Late uplift, unrelated or connected only indirectly to the actual formation of a passive margin, is now well documented from many margins (e.g. Nielsen et al. 2008). In west Greenland, Japsen et al. (2006) describe three separate phases of uplift: broadly Oligocene (36–30 Ma), Miocene (11 Ma) and Pliocene (7–2 Ma), all clearly post-dating the classical rift shoulder-related uplift in the Eocene. Such events/features may have local or regional causes (Doré et al. 2008; Cloetingh et al. 2008; Holford et al. 2008). There is also clear evidence for substantial transient rapid uplift (Rudge et al. 2008; Shaw-Champion et al. 2008). Dynamic topography related to mantle phenomena appears to be implicated. The impacts on exploration are potentially profound, with the cessation of active maturation and possibly expulsion of hydrocarbons, embrittlement of seals, creation or destruction of overpressures, and expansion of gas caps being among the more obvious on older uplifted sequences. Furthermore the new input of sediment from erosional products may trigger very recent maturation and expulsion and far-field compressive stresses and fault reactivation may trigger new trapping possibilities. Japsen et al. (2010) provide comprehensive documentation of Cenozoic uplift around the North Atlantic, including timing and implications for sedimentation and petroleum systems. They suggest that uplift may be an implicit tendency of passive margin borderlands, speculatively a function of changes in crust and lithosphere thickness over short distances.

Around Africa late continental uplift, associated by many with dynamic topography, has generated seaward dips which bedevil exploration by tilting out the traps. This late uplift is sometimes obscured by salt movement or other forms of gravitational collapse, and is often multi-phase. For example in the apparently simple and 'classical' Orange River Basin margin of NW South Africa, two phases of later-than-rift uplift (Upper Cretaceous and Mid Cenozoic), are revealed (Paton et al. 2008).

Technology

New development concepts

The full globalization of LNG is opening up gas exploration in what were previously oil basins (e.g. Vining et al. 2010), or in countries or areas without developed gas markets. Furthermore, the ability to process LNG and load tankers offshore will enable development of remote fields, or fields off difficult coastlines, with the benefit also of spreading the burden of LNG plant construction around the world. Floating LNG production is currently considered likely to first be used for developments off NW Australia, but remote passive margins worldwide could follow.

New engineering technologies are also enabling smaller oil accumulations and more difficult fluids to be developed further offshore and in yet deeper water. Recent examples include very deepwater spar developments such as that in the Perdido Fold Belt, Gulf of Mexico, where the Shell group's Great White production well in 2934 m of water holds the current record. This trend too can be expected to continue.

Exploration technology refinement, leading to new exploration models

Improvements to seismic data quality are still occurring at a rapid pace. For example, Hardy *et al.* (2010) describe potential strategies and processing flows for deepwater seismic resolution, with applications for exploration off western Ireland. Wider use of multi-azimuth seismic and pre-stack depth imaging not only for seeing below salt (Ekstrand *et al.* 2010), but also simply for higher-fidelity imaging, has helped in the development of new plays such as the highly successful Gulf of Mexico and South Atlantic sub-salt, and is likely to continue to open up new plays. As costs come down (as, e.g. shooting geometries are optimized), these techniques will become even more widespread. The vastly increased visibility of sub-salt sequences in recent years has been accompanied by significant refinement of salt models, and an increased understanding of how salt and sediment behaves in areas of high sedimentation and multiphase halokinesis. Such models are important offshore Brazil and west Africa, although the key testing ground has been the US Gulf of Mexico. Jackson *et al.* (2010) describe some of these emerging concepts, such as the movement of salt canopies in the subsurface and the transport of exotic sediment rafts over tens of kilometres seaward by migrating canopies. This paper also addresses some of the current enigmas regarding minibasin formation.

Imaging beneath a thick basalt cover is a significant concern and the key to further exploration success in the North Atlantic Faroe–Shetland and Møre Basins. In the few wells that have attempted to test sub-basalt traps, the depth to base basalt has been underestimated and the wells abandoned without reaching their objective. Two studies in this volume address this theme. Davison (2010) describes imaging using deep towed seismic and its results for Cenozoic and Cretaceous trap models, while Ellefsen *et al.* (2010) stress the need for understanding internal volcanic facies at the base of the basalt succession to better resolve the key base basalt reflector.

Use of controlled source electro-magnetic surveying is increasingly becoming routine, and can, for example, usefully distinguish low saturation gas which seismic methods may not be able to separate from higher gas saturation reservoirs. Better basin models based simply on more calibration data as well as new geodynamic insights, more refined geochemical extrapolation of source rock facies based on palaeo-ecological and palaeoclimatic insights, and a clearer understanding of late uplift through fission track geothermometry and satellite-based palaeogeomorphological studies, can all be expected.

Frontiers

In addition to new ideas for already part-explored areas, there are of course still margins which have yet to be explored. Prospective margins which have suffered commercial challenges include the difficult climatic conditions of the Arctic margins and areas subject to border disputes or moratoria, such as much of the western North Atlantic margin.

Arctic

Positions are being taken along the lightly explored margins of the greater Arctic, for example, in the Beaufort Sea–Mackenzie Delta margin, with proven plays in the Paleocene to Miocene, and in west Greenland and Labrador margins. Exploratory survey work is being conducted off east Greenland. Acreage has also been offered recently in the Laptev Sea.

Palaeo-passive margins

Although not the subject of this meeting, it has to be remembered that ancient passive margins also contain oil and gas, the Brookian sequence of Alaska being a major example. In general the conversion of the passive continental margin to an active plate margin, as is currently happening between Timor and Australia, either destroys or overwrites the petroleum system. The significance of palaeo-passive margins as petroleum systems really depends on where the cratonward, updip limit of a passive margin system is placed. Examples where this might be important include the Palaeozoic of the Western Canadian Sedimentary Basin and the Mesozoic of the northeastern (Tethyan) rim of Arabia. In general these sequences are traditionally treated as belonging to continental epeiric sea basins rather than being inboard passive margins. Perhaps new insights into passive margin geodynamics will change this too.

The views and opinions in this paper are entirely those of the authors. No representation or warranty, express or implied, is or will be made in relation to the accuracy or completeness of the information in this paper and no responsibility or liability is or will be accepted by Shell International E and P BV, BG Group plc, Statoil USA E&P, BHP Billiton Petroleum Inc. or any of their respective subsidiaries, affiliates and associated companies (or by any of their respective officers, employees or agents) in relation to it.

References

Allen, P. A. & Allen, J. R. 2005. *Basin Analysis: Principles and Applications* (2nd edn). Wiley Blackwell, Chichester.

Anderson, D. L. 2007. *New Theory of the Earth*. Cambridge University Press, Cambridge.

Anka, Z., Séranne, M., Lopez, M., Scheck-Wenderoth, M. & Savoye, B. 2009. The long-term evolution of the Congo deep-sea fan: a basin-wide view of the interaction between a giant submarine fan and a mature passive margin (ZaiAngo project). *Tectonophysics*, **470**, 42–56.

Bastia, R. 2006. An overview of Indian sedimentary basins with special focus on emerging East Coast deepwater frontiers. *The Leading Edge*, July, 818–829.

Bellingham, P. & White, N. 2002. A two-dimensional inverse model for extensional sedimentary basins. 2. Applications. *Journal of Geophysical Research*, **107**, 2260; doi: 10.1029/2001JB000174.

Blaich, O. A., Faleide, J. I., Tsikalas, F., Lilletveit, R., Chiossi, D., Brockbank, P. & Cobbold, P. 2010. Structural architecture and nature of the continent–ocean transitional domain at the Camamu and Almada Basins (NE Brazil) within a conjugate margin setting. *In*: Vining, B. A. & Pickering, S. C. (eds) *Petroleum Geology: From Mature Basins to New Frontiers – Proceedings of the 7th Petroleum Geology Conference*. Geological Society, London, 867–883; doi: 10.1144/070867.

Bradley, D. C. 2008. Passive margins through earth history. *Earth Science Reviews*, **91**, 1–26; doi: 10.1016/j.earscirev.2008.08.001.

Chalmers, J. A. & Pulvertaft, T. C. R. 2001. Development of continental margins of the Labrador Sea: a review. *In*: Wilson, R. C. L., Whitmarsh, R. B., Taylor, B. & Froitzheim, N. (eds) *Non-volcanic Rifting of Continental Margins: A Comparison of Evidence from Land and Sea*. Geological Society, London, Special Publications, **187**, 77–105.

Cloetingh, S., Beekman, F., Ziegler, P., Van Wees, J.-D. & Sokoutis, D. 2008. Post-rift compressional reactivation of passive margins and extensional basins. *In*: Johnson, H., Doré, A. G., Gatliff, R. W., Holdsworth, R., Lundin, E. R. & Ritchie, J. D. (eds) *The Nature* and *Origin of Compression in Passive Margins*. Geological Society, London, Special Publications, **306**, 27–70; doi: 10.1144/SP306.2.

Crosby, A., White, N., Edwards, G. & Shillington, D. J. 2008. Evolution of the Newfoundland–Iberia conjugate rifted margin. *Earth and Planetary Science Letters*, **273**, 214–226.

Davison, I., Stasiuk, S., Nuttall, P. & Keane, P. 2010. Sub-basalt hydrocarbon prospectivity in the Rockall, Faroe–Shetland and Møre basins, NE Atlantic. *In*: Vining, B. A. & Pickering, S. C. (eds) *Petroleum Geology: From Mature Basins to New Frontiers – Proceedings of the 7th Petroleum Geology Conference*. Geological Society, London, 1025–1032; doi: 10.1144/0071025.

De Wit, M. J., Stanciewicz, J. & Reeves, C. 2008. Restoring pan African–Brasiliano connections: more Gondwana control, less trans-Atlantic corruption. *In*: Pankhurst, R. J., Trouw, R. A. J., Brito Neves, B. B. & de Wit, M. J. (eds) *West Gondwana Pre-Cenozoic Correlations Across the South Atlantic Region*. Geological Society, London, Special Publications, **294**, 399–412.

Doré, A. G., Lundin, E. R., Jensen, L. N., Birkeland, Ø., Eliassen, P. E. & Fichler, C. 1999. Principal tectonic events in the evolution of the northwest European Atlantic margin. *In*: Fleet, A. J. & Boldy, S. A. R. (eds) *Petroleum Geology of Northwest Europe: Proceedings of the 5th Conference*. Geological Society, London, 41–61; doi: 10.1144/0050041.

Doré, A. G., Lundin, E. R., Kusznir, N. J. & Pascal, C. 2008. Potential mechanisms for the genesis of Cenozoic domal structures on the NE Atlantic margin: pros, cons and some new ideas. *In*: Johnson, H., Doré, A. G., Gatliff, R. W., Holdsworth, R., Lundin, E. R. & Ritchie, J. D. (eds) *The Nature and Origin of Compression in Passive Margins*. Geological Society, London, Special Publications, **306**, 1–26; doi: 10.1144/SP306.1.

Ebbing, J. & Olesen, O. 2010. New compilation of top basement and basement thickness for the Norwegian continental shelf reveals the segmentation of the passive margin system. *In*: Vining, B. A. & Pickering, S. C. (eds) *Petroleum Geology: From Mature Basins to New Frontiers – Proceedings of the 7th Petroleum Geology Conference*. Geological Society, London, 885–897; doi: 10.1144/0070885.

Ekstrand, E., Hickman, G., Thomas, R., Threadgold, I., Harrison, D., Los, A., Summers, T., Regone, C. & O'Brien, M. 2010. WATS it take to image an oil field subsalt offshore Angola? *In*: Vining, B. A. & Pickering, S. C. (eds) *Petroleum Geology: From Mature Basins to New Frontiers – Proceedings of the 7th Petroleum Geology Conference*. Geological Society, London, 1013–1024; doi: 10.1144/0071013.

Ellefsen, M., Boldreel, L. O. & Larsen, M. 2010. Intra-basalt units and base of the volcanic succession east of the Faroe Islands exemplified by interpretation of offshore 3D seismic data. *In*: Vining, B. A. & Pickering, S. C. (eds) *Petroleum Geology: From Mature Basins to New Frontiers – Proceedings of the 7th Petroleum Geology Conference*. Geological Society, London, 1033–1041; doi: 10.1144/0071033.

Funck, T., Hopper, J. R., Larsen, H. C., Louden, K. E., Tucholke, B. E. & Holbrook, W. S. 2003. Crustal structure of the ocean continent transition at Flemish Cap: seismic refraction results. *Journal of Geophysical Research*, **108**, 2531; doi: 10.1029/2003JB002434.

Hardy, R. J. J., Querendez, E., Biancotto, F., Jones, S. M., O'sullivan, J. & White, N. 2010. New methods of improving seismic data to aid understanding of passive margin evolution: a series of case histories from offshore west of Ireland. *In*: Vining, B. A. & Pickering, S. C. (eds) *Petroleum Geology: From Mature Basins to New Frontiers – Proceedings of the 7th Petroleum Geology Conference*. Geological Society, London, 1005–1012; doi: 10.1144/0071005.

Heidbach, O., Tingay, M., Barth, A., Reinecker, J., Kurfeß, D. & Müller, B. 2008. The 2008 release of the World Stress Map. World Wide Web Address: www.world-stress-map.org.

Holford, S. P., Green, P. F. *et al.* 2008. Evidence for kilometer-scale exhumation driven by compressional deformation in the Irish Sea basin system. *In*: Johnson, H., Doré, A. G., Gatliff, R. W., Holdsworth, R., Lundin, E. R. & Ritchie, J. D. (eds) *The Nature and Origin of Compression in Passive Margins*. Geological Society, London, Special Publications, **306**, 91–119; doi: 10.1144/SP306.4.

Jackson, M. P. A., Hudec, M. R. & Dooley, T. P. 2010. Some emerging concepts in salt tectonics in the deepwater Gulf of Mexico: intrusive plumes, canopy-margin thrusts, minibasin triggers and allochthonous fragments. *In*: Vining, B. A. & Pickering, S. C. (eds) *Petroleum Geology: From Mature Basins to New Frontiers – Proceedings of the 7th Petroleum Geology Conference*. Geological Society, London, 899–912; doi: 10.1144/0070899.

Jackson, J., McKenzie, D., Priestley, K. & Emmerson, B. 2008. New views on the structure and rheology of the lithosphere. *Journal of the Geological Society, London*, **165**, 453–465.

Japsen, P., Bonow, J. M., Green, P. F., Chalmers, J. A. & Lidmar-Bergstrom, K. 2006. Elevated passive continental margins: long term highs or Neogene uplifts? New evidence from West Greenland. *Earth and Planetary Science Letters*, **248**, 330–339; doi: 10.1016/j.epsl.2006.05.036.

Japsen, P., Green, P. F., Bonow, J. M., Rasmussen, E. S., Chalmers, J. A. & Kjennerud, T. 2010. Episodic uplift and exhumation along North Atlantic passive margins: implications for hydrocarbon prospectivity. *In*: Vining, B. A. & Pickering, S. C. (eds) *Petroleum Geology: From Mature Basins to New Frontiers – Proceedings of the 7th Petroleum Geology Conference*. Geological Society, London, 979–1004; doi: 10.1144/0070979.

Jenkyns, H. C. 2003. Evidence for rapid climate change in the Mesozoic-Palaeogene greenhouse world. *Philosophical Transactions of the Royal Society, London A*, **361**, 1885–1916.

Jones, W., Guevara, M. *et al.* 2009. The subsalt play in the lower Congo and Kwanza Basins, Angola: a seismic study. Abstract from the *American Association of Petroleum Geologists Annual Convention and Exhibition*, Denver, CO, 7–10 June.

Klemme, H. D. & Ulmishek, G. F. 1991. Effective petroleum source rocks of the world: stratigraphic distribution and controlling depositional factors. *Bulletin of the American Association of Petroleum Geologists*, **75**, 1809–1851.

Kundu, N., Pal, N., Sinha, N. & Budhiraja, I. L. 2008. Do paleo-hydrates play a major role in deepwater biogenic gas reservoirs in Krishna–Godavari Basin? *Methane Hydrate Newsletter, National Energy Technology Laboratory*, Fall 13–15.

Kusznir, N. J. & Karner, G. D. 2007. Continental lithospheric thinning and breakup in response to upwelling divergent mantle flow: application to the Woodlark, Newfoundland and Iberia margins. *In*: Karner, G. D., Manatschal, G. & Pinhero, L. M (eds) *Imaging, Mapping and Modelling Continental Lithosphere Extension and Breakup*. Geological Society, London, Special Publications, **282**, 389–419; doi: 10.1144/SP282.16.

Larsen, M., Rasmussen, T. & Hjelm, L. 2010. Cretaceous revisited: exploring the syn-rift play of the Faroe–Shetland Basin. *In*: Vining, B. A. & Pickering, S. C. (eds) *Petroleum Geology: From Mature Basins to New Frontiers – Proceedings of the 7th Petroleum Geology Conference*. Geological Society, London, 953–962; doi: 10.1144/0070953.

Machado, M. A. P., Moraes, M. F. B. *et al.* 2009. The pre-salt sequence of ultra-deep water Santos basin: geological aspects and key factors controlling the major oil accumulations. Abstract, *American Association of Petroleum Geologists International Conference and Exhibition*, Rio de Janeiro, 15–18 November.

Mann, P., Gahagan, L. & Gordon, M. B. 2003. *Tectonic Setting of the World's Giant Oil and Gas Fields*. American Association of Petroleum Geologists, Tulsa, OK, Memoirs, **78**, 15–105.

Martinsen, O. J., Sømme, T. O., Thurmond, J. B., Helland-Hansen, W. & Lunt, I. 2010. Source-to-sink systems on passive margins: theory and practice with an example from the Norwegian continental margin. *In*: Vining, B. A. & Pickering, S. C. (eds) *Petroleum Geology: From Mature Basins to New Frontiers – Proceedings of the 7th Petroleum Geology Conference*. Geological Society, London, 913–920; doi: 10.1144/0070913.

McKenzie, D. P. 1978. Some remarks on the development of sedimentary basins. *Earth and Planetary Science Letters*, **40**, 25–32.

Nielsen, S. B. *et al.* 2008. The evolution of western Scandinavian topography: a review of Neogene uplift versus the ICE (isostasy–climate–erosion) hypothesis. *Journal of Geodynamics*, **47**, 72–95; doi: 10.1016/j.jog.2008.09.001.

Paton, D. A., van der Spuy, D., di Primio, R. & Horsfield, B. 2008. Tectonically induced adjustment of passive-margin accommodation space; influence on the hydrocarbon potential of the Orange Basin, South Africa. *Bulletin of the American Association of Petroleum Geologists*, **92**, 589–609.

Prather, B. E. 1991. Petroleum geology of the Upper Jurassic and Lower Cretaceous, Baltimore Canyon Trough, Western North Atlantic Ocean. *Bulletin of the American Association of Petroleum Geologists*, **75**, 258–277.

Redfern, J. & Shannon, P. M. *et al.* 2010. An integrated study of Permo-Triassic basins along the North Atlantic passive margin: implication for future exploration. *In*: Vining, B. A. & Pickering, S. C. (eds) *Petroleum Geology: From Mature Basins to New Frontiers – Proceedings of the 7th Petroleum Geology Conference*. Geological Society, London, 921–936; doi: 10.1144/0070921.

Ren, S., Faleide, J. I., Eldholm, O., Skogseid, J. & Gradsetin, F. 2003. Late Cretaceous to Paleocene development of the NW Voring Basin. *Marine and Petroleum Geology*, **20**, 177–206.

Reston, T. J. 2009. The structure, evolution and symmetry of the magma-poor rifted margins of the North and Central Atlantic: a synthesis. *Tectonophysics*, **468**, 6–27.

Reston, T. J. & Perez-Gussinye, M. 2008. Lithospheric extension from rifting to continental breakup at magma-poor margins: rheology, serpentinization and symmetry. *In*: Bernouilli, D., Brun, J.-P. & Burg, J.-P. (eds) Special issue, Continental Extension. *International Journal of Earth Sciences*, **96**, 1033–1046.

Reynisson, R. F., Ebbing, J., Lundin, E. & Osmundsen, P. T. 2010. Properties and distribution of lower crustal bodies on the mid-Norwegian margin. *In*: Vining, B. A. & Pickering, S. C. (eds) *Petroleum Geology: From Mature Basins to New Frontiers – Proceedings of the 7th Petroleum Geology Conference*. Geological Society, London, 843–854; doi: 10.1144/0070843.

Rudge, J. F., Shaw-Champion, M. E., White, N., McKenzie, D. & Lovell, B. 2008. A plume model of transient diachronous uplift at the Earth's surface. *Earth and Planetary Science Letters*, **267**, 146–160.

Sassen, R. & Post, P. 2008. Enrichment of diamondoids and ^{13}C in condensate from Hudson Canyon, US Atlantic. *Organic Geochemistry*, **39**, 147–151; doi: 10.1016/j.orggeochem.2007.10.004.

Scotchman, I. C., Gilchrist, G., Kusznir, N. J., Roberts, A. M. & Fletcher, R. 2010. The breakup of the South Atlantic Ocean: formation of failed spreading axes and blocks of thinned continental crust in the Santos Basin, Brazil and its consequences for petroleum system development. *In*: Vining, B. A. & Pickering, S. C. (eds) *Petroleum Geology: From Mature Basins to New Frontiers – Proceedings of the 7th Petroleum Geology Conference*. Geological Society, London, 855–866; doi: 10.1144/0070855.

Shaw-Champion, M. E., White, N. J., Jones, S. M. & Lovell, J. P. B. 2008. Quantifying transient mantle convective uplift: an example from the Faroe Shetland Basin. *Tectonics*, **27**; doi: 10.1029/2007TC002106.

Skelton, P. W., Spicer, R. A., Kelley, S. P. & Gilmour, I. 2003. *In*: Skelton, P. W. (ed.) *The Cretaceous World*. The Open University/Cambridge University Press, Buckingham/Cambridge.

Stainforth, J. G. 2009. Practical kinetic modeling of petroleum generation and expulsion. *Marine and Petroleum Geology*, **26**, 552–572.

Stephens, A. R., Monson, G. D. & Reilly, J. M. 1996. *Seismic Amplitudes: Their Relevance for Hydrocarbon Exploration in the Late Miocene/Early Pliocene Sediments of the Distal Niger Delta: Examples from the Zafiro Field*. Equatorial Guinea: Offshore West Africa Conference, Libreville.

Tucholke, B. E., Sawyer, D. S. & Sibuet, J.-C. 2007. *The Iberia Newfoundland Continental Extension System (Geological and Geophysical Constraints)*. Geological Society, London, Special Publications, **282**, 9–46.

Tyrrell, S., Souders, A. K., Haughton, P. D. W., Daly, J. S. & Shannon, P. M. 2010. Sedimentology, sandstone provenance and palaeodrainage on the eastern Rockall Basin margin: evidence from the Pb isotopic composition of detrital K-feldspar. *In*: Vining, B. A. & Pickering, S. C. (eds) *Petroleum Geology: From Mature Basins to New Frontiers – Proceedings of the 7th Petroleum Geology Conference*. Geological Society, London, 937–952; doi: 10.1144/0070937.

Uroza, C. A. & Steel, R. J. 2008. A highstand shelf margin delta system from the Eocene of West Spitsbergen, Norway. *Sedimentary Geology*, **203**, 229–245.

van Avendonk, H. J. A., Lavier, L. L., Shillington, D. J. & Manatschal, G. 2008. Extension of continental crust at the margin of the eastern Grand Banks Newfoundland. *Tectonophysics*, **468**, 131–148; doi: 10.1016/j.tecto.2008.05.030.

Watts, A. B. 2001. Gravity anomalies, flexure and crustal structure at the Mozambique rifted margin. *Marine and Petroleum Geology*, **18**, 445–455.

White, R. S., Eccles, J. D. & Roberts, A. W. 2010. Constraints on volcanism, igneous intrusion and stretching on the Rockall–Faroe continental margin. *In*: Vining, B. A. & Pickering, S. C. (eds) *Petroleum Geology: From Mature Basins to New Frontiers – Proceedings of the 7th Petroleum Geology Conference*. Geological Society, London, 831–842; doi: 10.1144/0070831.

White, N., Haines, J. & Hanne, D. 2004. Towards an automated strategy for modelling extensional basins and margins in four dimensions. *In*: Davies, R. J., Cartwright, J. A., Stewart, S. A., Lappin, M. & Underhill, J. R. (eds) *3D Seismic Technology: Applications to the Exploration of Sedimentary Basins*. Geological Society, London, Memoirs, **29**, 321–331.

White, R. S., Smith, L. K., Roberts, A. W., Christie, P. A. F., Kusznir, N. J. & the rest of the ISIMM team. 2008. Lower-crustal intrusion on the North Atlantic continental margin. *Nature*, **452**, 460–464.

Wierzbicki, R., Dravis, J. J., Al-Aasm, I. & Harland, N. 2006. Burial dolomitization and dissolution of Upper Jurassic Abenaki platform carbonates, Deep Panuke reservoir, Nova Scotia, Canada. *Bulletin of the American Association of Petroleum Geologists*, **90**, 1843–1861; doi: 10.1306/03200605074.

Wright, P. V. & Racey, A. 2009. Pre-salt microbial carbonate reservoirs of the Santos Basin, offshore Brazil. Abstract from the *American Association of Petroleum Geologists Annual Convention and Exhibition*, Denver, CO, 7–10 June 2009.

Constraints on volcanism, igneous intrusion and stretching on the Rockall–Faroe continental margin

R. S. WHITE,[1] J. D. ECCLES[2] and A. W. ROBERTS[3]

[1]*Bullard Laboratories, University of Cambridge, Madingley Road, Cambridge CB3 0EZ, UK*
(e-mail: rsw1@cam.ac.uk)
[2]*The Institute of Earth Science and Engineering, University of Auckland, New Zealand (present address)*
[3]*Department of Earth Sciences, Durham University, UK (present address)*

Abstract: The northern North Atlantic margins are classic examples of 'volcanic' rifted continental margins, where breakup was accompanied by massive volcanism. We discuss strategies used to obtain good intra- and sub-basalt seismic penetration so as to map the structure and the extruded and intruded igneous volume. We recorded deep penetration reflection data using a 12 000 m long single sensor (Q-)streamer and wide-angle seismic profiles with 85 4-component ocean bottom seismometers, along two transects across the Faroe and Hatton Bank continental margins in the NE Atlantic. Tomographic inversion of both compressional (P) and shear (S) wave crustal velocities are crucial in improving the reflection image and in constraining the nature of the sub-basalt lithology and the volume of extruded and intruded melt. Beneath the basalts, which reach 5 km thickness, is a low-velocity zone with P- and S-wave velocities characteristic of sedimentary rocks intruded by basalt sills. The underlying stretched continental basement contains abundant intrusive igneous sills on the rifted margin. Near the Faroe Islands, for every 1 km along-strike, 340–420 km^3 of basalt was extruded, while 560–780 km^3 was intruded into the continent–ocean transition (COT). Lower-crustal intrusions are focussed mainly into a narrow zone less than 50 km wide on the COT, whereas extruded basalts flow >100 km from the rift. Melt on the COT is intruded into the lower crust as sills which cross-cut the stretched and tilted continental fabric, rather than as 'underplate' of 100% melt, as has often been assumed previously. Our igneous thickness and velocity observations are consistent with the dominant control on the melt production being rifting above mantle with a temperature elevated above normal. The mantle temperature anomaly was up to 150 °C above normal at the time of continental breakup, decreasing by c. 70–80 °C over the first 10 Ma of seafloor spreading.

Keywords: North Atlantic, rifted continental margin, volcanism, igneous intrusion, subsidence, crustal structure

Continental breakup of the northern North Atlantic at c. 55 Ma was accompanied by the eruption of huge volumes of basaltic lavas. The basalts flowed across the continental hinterlands on both sides of the new ocean basin. In the North Atlantic region, the volume of extrusive lavas reached more than 1×10^6 km^3 (White & McKenzie 1989; Coffin & Eldholm 1994; Eldholm & Grue 1994). The evolved composition of the extruded basalts (typically 6–8% MgO, compared with 16–18% MgO for the primitive melts from which they were derived) means that considerable volumes of residual igneous rocks must remain either in the crust or in the upper mantle. For flood basalts, Cox (1980) estimated on petrological grounds that the intrusive volume must be 'at least as large as the amount of erupted surface lava'. Just how much of the total igneous budget remains below surface is much debated, with literature estimates of the ratio of intruded to extruded igneous rock ranging from 0.25:1 to 4:1 (Coffin & Eldholm 1994). It is important to constrain the total volume of igneous rocks not just for understanding the geological history of the new ocean basin, but also for calculating the subsidence history, its heat flow and maturation history, and the temperature and properties of the underlying mantle from which the melt was generated. This is crucial both in understanding the geological history of the continental margins and adjacent shelves and for assessing their hydrocarbon potential.

In this paper we summarize the geophysical techniques used to image the igneous rocks and to constrain their volumes. A major focus of the iSIMM (integrated Seismic Imaging and Modelling of Margins) project from which seismic work is reported here was on quantifying the total amount and distribution of igneous rocks produced during the breakup of the Rockall and Faroe continental margins. Many of the detailed results are discussed in fuller publications to which we refer at the relevant points.

When volcanic continental margins were first studied in detail, it became apparent that the widespread extrusive volcanics were invariably accompanied by high-velocity lower-crust (HVLC, P-wave velocities higher than 7.0 km s^{-1}) beneath the continent–ocean transition (COT). This was generally interpreted as due to 'underplated' igneous crust (e.g. Mutter et al. 1984; LASE 1986; Vogt et al. 1998; Klingelhöfer et al. 2005; Voss & Jokat 2007). Our high-quality iSIMM seismic reflection profiles across the Faroes continental margin show the presence of numerous lower-crustal reflections beneath the COT in the region displaying HVLC. We interpret these reflections as caused by igneous sills intruding the country rock. Therefore the HVLC on the Faroes margin is better interpreted as 'intruded lower-crust' than as 'underplated' igneous crust (White et al. 2008; White & Smith 2009). It is likely that, where similar HVLC wedges are observed elsewhere on the COT, they are also caused by sill intrusions. The widespread use of the terminology of 'underplated' igneous crust rather than 'intruded lower-crust' makes a significant and, we argue, sometimes erroneous, difference to the way the cause of the widespread magmatism at the time of continental breakup is interpreted.

The locations of the two main iSIMM profiles that cross the Faroes and Hatton continental margins of NW Europe on which we report in this paper are shown in Figure 1. Both profiles had dense deployments of 85 ocean bottom seismometers (OBS) along them. These OBS provide control on the structure from wide-angle seismic data with an unprecedented density and

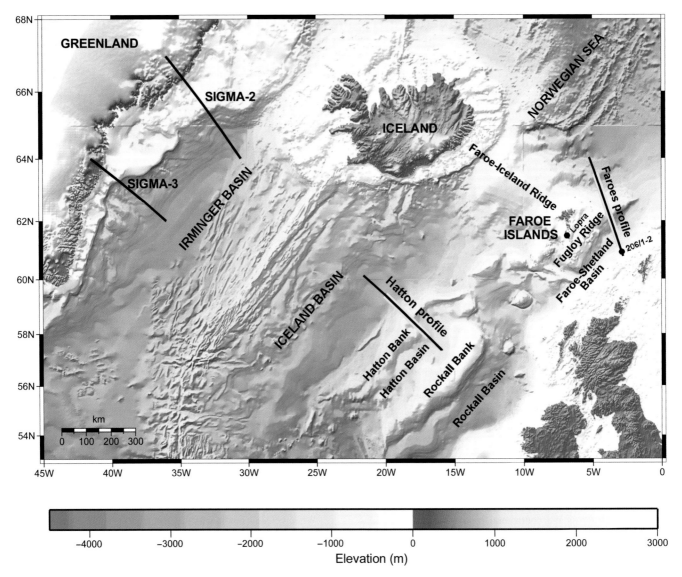

Fig. 1. Location of seismic profiles across the North Atlantic rifted continental margins. Portions of Faroes and Hatton profiles highlighted in yellow are shown in Figure 4, and results from the SIGMA profiles (Korenaga et al. 2000; Hopper et al. 2003), on the conjugate east Greenland margin, are shown in Figure 5. Lopra and 206/1-2 refer to location of wells from which data are shown on Figure 7.

number of arrivals. Only the Faroes profile had a coincident deep penetration reflection profile, although the reflection data recorded while shooting into the OBS on the Hatton profile was sufficient to constrain the sediment thickness, which was used in the analysis of the wide-angle data.

Source and receiver design

In order to achieve penetration of seismic energy through thick (c. 1 km or more) layered basalts, it is necessary to tune the seismic source to produce low frequencies. This is because the basalts strongly attenuate higher frequency energy (Ziolkowski et al. 2001; Maresh & White 2005). If the attenuation is expressed as an effective quality factor, Q_{eff}, then values of typically 25–45, but sometimes as low as 15, are measured from vertical seismic profiles in boreholes through basalts on the North Atlantic margin (Maresh et al. 2006; Shaw et al. 2008). These values of Q_{eff} are as low as those of near-surface unconsolidated sediments. The reason for the high attenuation is not the intrinsic absorption of basalts, for which Q is in the region of 200–600 (i.e. low intrinsic attenuation; Gordon & Davis 1968; Brennan & Stacey 1977; Maresh et al. 2006), but is probably due to scattering and/or, if the crack configurations and interconnectedness are right, to energy loss by water squirt between cracks as the seismic wave propagates (Shaw et al. 2008). The high-frequency scattering may arise from the fine velocity layering created by the alternation of the interiors of basalt flows with relatively high velocities, alternating with weathered and altered flow tops with much lower seismic velocities. The presence of inter-flow sediments may produce even lower velocity layers. These alternating high and low velocity layers can cause high frequency loss even if the media are flat-lying and laterally homogenous (Christie et al. 2006a; Maresh et al. 2006). Scattering of the high frequencies may also be caused by the 3D irregular undulating nature of the interfaces between individual flows within the basalt units (Maresh et al. 2006).

The iSIMM airgun source was tuned to produce peak output at 9–10 Hz. Low frequencies were generated in both the wide-angle and conventional seismic reflection sources using large airgun chambers and large total capacities (the source array for the Faroes reflection profile totalled 167 litre (10 170 cubic in.) from 48 guns). Towing the guns deeper than usual at 18 m below the sea surface also provided low-frequency enhancement due to the sea-surface ghost reflection. The choice of the source tow depth is a trade-off between enhancing the low frequencies with the reflection off the sea surface, which is better the deeper the guns, and the increase in frequency produced by an airgun of given chamber size which results from the increased ambient pressure as it is towed deeper. We also implemented an experimental configuration of shooting so as to align the bubble pulses from the individual guns rather than the conventional method of aligning the first peaks. Bubble tuning produces a lower frequency waveform than does conventional peak tuning, and also produces a more compact waveform at low frequencies (Lunnon et al. 2003; Christie et al. 2006b): a similar method termed 'single-bubble' tuning was implemented by Avedik et al. (1993, 1995). The final effect of all these factors was to produce a source for the reflection profile with a peak output of 228 dB re 1 μPa Hz^{-1} at 1 m, and a peak frequency of 9 Hz (Christie et al. 2006b). Although the bubble tuning did enhance the low-frequency output, our conclusion from comparison between the bubble-tuned and a conventional peak-tuned source with the same chamber sizes and distribution is that for most purposes the peak-tuned source is preferable because it has a smaller dependence of the waveform on the take-off angle than does the bubble-tuned source. The greatest improvement in producing low frequencies comes from the simple expedient of towing the airguns deep.

At the receiver end it is beneficial to tow the streamer at a similar depth to the source so as to also benefit from the enhancement of low frequencies by the sea surface ghost reflection. The OBS on the seafloor in deep water are far removed from the surface ghost, so the low frequencies are not enhanced in the same way, but have the additional benefit that the downward travelling water multiples can be attenuated by stacking the hydrophone and vertical geophone data because these arrivals have opposite polarities. The seabed receivers are also far removed from the acoustic noise generated by rough weather and remained quiet even when there was a Force 7 storm at the surface in the data we recorded on the iSIMM project. However, an additional source of noise to which seabed receivers may be exposed arises from strong bottom currents, which in our study were most apparent in degrading the signal-to-noise ratio of OBS at the foot of the continental slope.

Wide-angle seismic data

By deploying fixed OBS, it is possible to record out to any desired offset, limited only by the power of the source and the noise levels on the OBS. In practice we routinely recorded wide-angle arrivals to offsets of 150 km on the iSIMM profiles reported here. The importance of this is that we were able to record turning waves from throughout the lower crust, as well as wide-angle reflections from the Moho. These phases arrive outside the water wave and are normally muted when building velocity models from conventional seismic reflection data. However, they carry information on the velocity structure which can be used through tomographic inversion to create a well constrained model of the velocity variations throughout the entire crust. Furthermore, the diving waves from the deep crust appear as first arrivals for much of their trajectory, so are not contaminated by the intra-bed multiples which are such a problem in analysing conventional reflection data from shorter offset streamers. Typically the resolution of the velocities available from semblance analysis or from velocity focusing in prestack depth migration decreases with depth and, even with streamers as long as 12 000 m, provides only poor discrimination on velocities in the lower crust. This is in marked contrast to diving waves, which have maximum velocity resolution at the depths in the crust at which they are turned back towards the surface.

Two further advantages of wide-angle arrivals have been utilized in basalt covered areas. The first is the characteristic 'step-back' in the diving waves caused by a low velocity zone (LVZ), such as may occur when basalts flow across a pre-existing sedimentary system (Richardson et al. 1999; White et al. 1999). An example of such step-backs is shown in Figure 2a, from an OBS positioned on the crest of the Fugloy Ridge on the Faroes shelf, where 3.6 km thick basalt flows overlie a 1.2 km thick LVZ (Fig. 2c). The accompanying ray-trace diagram (Fig. 2b) shows rays traced through the final velocity model for the Fugloy Ridge (from Roberts et al. 2009), which produce travel times that match the observed data. The red rays are returned by the velocity gradient in the basalt layer. The maximum range to which the rays within the basalt layer propagate is governed by the thickness of the layer and the velocity gradient within it (Fliedner & White 2001): at maximum range the turning rays just graze the base of the basalt layer.

As we discuss later, the sub-basalt LVZ is probably caused by sedimentary rock intruded by igneous sills. The magnitude of the travel time step-back depends on the velocity and thickness of the LVZ and is an easily identifiable indicator of the presence of low-velocity material beneath the basalts. However, for basalt thicknesses typical of the Faroes shelf, this step-back in the diving waves usually occurs at ranges in excess of 15 km, which means that it is beyond the maximum offset of conventional streamers. Hence the need either to synthesize super-long streamers using two-ship methods, or to shoot into fixed seabed receivers. Using the presence and magnitude of the step-backs recorded by two-ship profiles in conjunction with conventional 4000 and 6000 m profiles enabled White et al. (2003) to map the regional variations in thickness of both the basalt flows and the sub-basalt LVZ from a 2D grid of profiles across the Faroes shelf.

A further advantage of using wide-angle reflections is that their amplitudes increase markedly towards the critical distance at which refractions from the underlying layer emerge (Fliedner & White 2001). This means that the wide-angle reflections produce arrivals with excellent signal-to-noise ratios, and they can be migrated to produce normal incidence reflection profiles with high amplitudes, largely free of multiples. By this means the base-basalt reflection has been identified and imaged on profiles across the Faroes shelf (Fliedner & White 2003). Conventional seismic reflection profiles still provide better images of the subsurface than do the wide-angle images, because the conventional shorter offsets exhibit less distortion from the stretching of the signal in the normal moveout correction stretch than do the far-offset arrivals, but the migrated wide-angle reflections can be used to identify which of the much weaker reflections on the conventional reflection profile are primary rather than multiple, and then the interpretation made from the conventional profile.

The use of three-component geophones in the seabed seismometers also allows the direct detection of S-waves. Airguns in water only produce acoustic waves, but significant S-wave energy may be generated by mode conversion at suitable boundaries as the energy propagates through the crust. The most efficient mode conversion occurs when the P-wave velocity on one side of an interface is similar to the S-wave velocity on the other side, provided the interface is sharp with respect to the wavelength (White & Stephen 1980). On the North Atlantic margins, such an interface is found at the top of the basalt flows where unconsolidated sediments overlie basalts, and there is indeed strong mode conversion at this interface (Eccles et al. 2007, 2009). The arrivals can be verified as S-waves from their travel times, their moveout and their particle motions.

An example of strong converted S-wave arrivals on wide-angle data is shown in Figure 3, using data recorded by OBS79, which lies near the oceanward end of the Faroes profile. The vertical component geophone of the OBS (Fig. 3a) records strong P-waves from the crustal diving wave (Pg) and from the Moho reflection off the base of the crust (PmP). A linear reduction

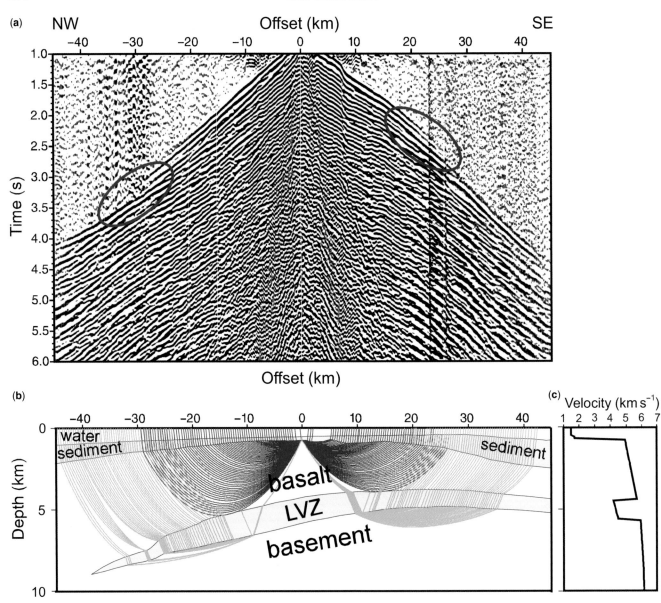

Fig. 2. (a) Example of seismic data from the vertical component of OBS53 on the Fugloy Ridge, Faroes profile to illustrate the effect of the LVZ on the travel times of diving waves. The location of 'step-backs' in the first arrival diving waves, which are characteristic of the presence of LVZs, are marked by red ovals. Note that the step-back occurs later in travel time and at a greater offset on the northwestern, seaward side (negative offsets) than on the southeastern, landward side (positive offsets) of OBS53 due to the deeper location of the LVZ on the seaward side. (b) Raytracing for first arrivals recorded at OBS53 through the final velocity model including the LVZ. Only one-third of the rays are shown for clarity. Red rays are from diving waves returned from within the extrusive basalt layer, and blue rays are from rays which have penetrated through the LVZ to the crust of the basement beneath. Note the termination of basalt diving waves (red) at the offset where they penetrate to the base of the basalt layer. (c) P-wave velocity-depth profile from Roberts et al. (2009) at the location of OBS53 (0 km offset on this figure).

velocity of 7.0 km s^{-1} has been applied to the travel times, which causes phases turning at 7.0 km s^{-1} to appear as horizontal arrivals on the time–offset plot.

The radial component of the OBS (Fig. 3b) shows the converted arrivals which are recorded as S-waves at the seafloor. The travel times have a linear velocity moveout of 3.9 km s^{-1} applied, so arrivals with a phase velocity of 3.9 km s^{-1} appear as horizontal on this plot. The ratio of the reduction velocities between Figure 3a, b is c. 1.8, which approximates the V_p/V_s ratio of the lower oceanic crust. Thus the Pg and PmP arrivals on the vertical component are sub-horizontal on Figure 3a, while the equivalent Sg and SmS arrivals on the radial component are sub-horizontal on Figure 3b. Sg and SmS are phases which were converted to S-waves at the interface between the sediment and the top of the basalt on the path down from the shot, and then travelled through the remainder of their paths as S-waves. Note too that the Pg phases are much weaker on the radial than on the vertical component, as is to be expected since they produce particle motions parallel to the ray path, which is sub-vertical at the seafloor. A clear example of this is shown by the PPS phase, which is converted on the way up at the sediment-basalt interface, and the P-wave water multiple generated near the OBS receiver (Fig. 3a, b). Both repeat the travel time v. offset behaviour of the primary crustal diving wave Pg and the Moho reflection PmP, but the PPS phase is delayed an additional 2.5 s by the path through the sediments as a S-wave and the water multiple phase is delayed 3.6 s by the additional two paths through the water layer. On the vertical component (Fig. 3a) the PPS phase is weak but the water multiple Pg phase is strong. On the radial component (Fig. 3b) the relative amplitudes of the PPS and water multiple arrivals are reversed: the PPS phase is strong because it arrives as a S-wave while the water multiple P-phase is weak. The delay between the

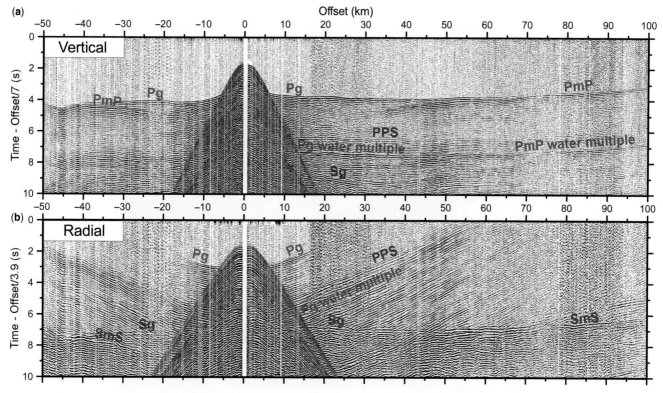

Fig. 3. Example of wide-angle data from OBS79 at the oceanic end of the Faroes profile (location shown in Fig. 1). Arrival phases labelled in red are direct P-wave arrivals: Pg are diving waves returned from the crust; PmP are reflections off the Moho and the Pg and PmP water multiples are delayed by a water multiple bounce in the water at the OBS end of the path. Arrivals labelled in blue are recorded as S-waves at the OBS: Sg and SmS are diving waves returned from the crust and reflections off the Moho, respectively, with conversion from P- to S-wave at the top basalt interface on the down-going path. The PPS phase has travelled as a P-wave through the crust and converted to S-wave at the top-basalt interface beneath the OBS on the upward travelling path. The labels of all the phases on this figure sit immediately above the phase to which they refer. (**a**) Vertical geophone component receiver gather displayed with a linear reduction velocity of 7.0 km s^{-1}. (**b**) Rotated radial geophone component displayed with a linear reduction velocity of 3.9 km s^{-1} to emphasize the converted shear wave arrivals. Figure modified from Eccles *et al.* (2009).

primary P-wave arrival and the corresponding PPS arrival allows the V_p/V_s ratio in the sediments beneath the OBS to be constrained and thus these strong converted S-waves recorded on the horizontal components of the OBS allow us to model the S-wave structure of the entire crust, which adds considerable value to the conventional P-wave velocity structure. As we discuss later, the additional information from S-wave velocities can be of crucial importance in discriminating between possible lithologies.

Crustal velocity models

The crustal velocity structure was modelled by picking the travel times of individual wide-angle arrivals recorded by each OBS from each shot and then performing a tomographic velocity analysis using the arrival times. Over 103 000 wide-angle P-wave arrivals and more than 60 000 converted S-wave arrivals were picked from the two wide-angle profiles discussed here (see Table 1). Sediment velocities were calculated from semblance analysis of wide-angle reflections on the 12 000 m long streamer data. For the sediments, this gave better velocity control than the sediment diving waves, which were largely obscured beneath the water wave and its coda. Furthermore, the diving waves are only available beyond the critical distance, whereas the reflections can be traced back to zero-offset arrivals.

The crustal velocity structure was derived first by a raytracing inversion, which used a layered model to match the modelled travel times to the observed arrival times, with the model iterated one layer at a time until the observed and modelled travel times matched to within their estimated uncertainties (Zelt & Smith 1992). This was followed by a grid-based joint reflection and refraction tomographic inversion (Korenaga *et al.* 2000). The grid-based inversion is more objective because it does not assume any initial layered boundaries, but it is poor at handling LVZs because it attempts to fit a smooth velocity field through them, whereas in reality they may have sharp interfaces with discontinuous velocities. In the extracts of P-wave velocity models shown here in Figure 4, we therefore used the grid-based result for the Hatton profile (Fig. 4a), and the raytraced model for the Faroes profile with its prominent LVZ (Fig. 4b). Uncertainties in the velocities were estimated by making multiple inversions with

Table 1. *Number of wide-angle arrival times used in tomographic inversions*

Phase	No. of arrival picks
Faroes Profile	
Pg	52 050
PmP	25 800
Sg	26 900
SmS	14 900
Hatton Profile	
Pg	17 650
PmP	7850
Sg	9500
SmS	9000

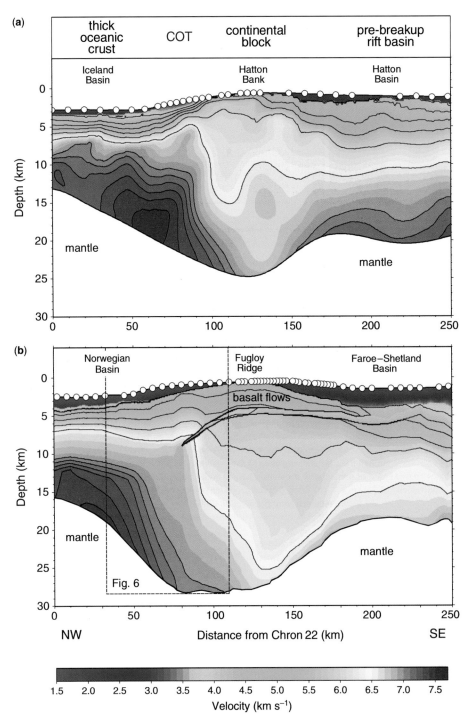

Fig. 4. Seismic velocity structure of crust across the COT near (**a**) Hatton Bank and (**b**) the Faroe Islands. For line locations see Figure 1. Profiles are aligned with 0 km distance at seafloor spreading magnetic anomaly chron 22. The main structural regions, which are found consistently on both profiles, are labelled along the top of the figure. Velocities of post-rift Cenozoic sediments (purple) are constrained by semblance moveout analysis of reflections recorded on the 12 km long hydrophone streamer and sub-sediment crustal velocities are constrained by tomography using wide-angle reflections and diving waves from OBS data (OBS locations shown by circles at seafloor). Colour bands are at 0.1 km s^{-1} intervals, with contours above 7.0 km s^{-1} spaced every 0.1 km s^{-1} to highlight the lower-crustal velocity. Diagram modified from White *et al.* (2008).

100 randomized starting velocity models: velocities are generally constrained to better than 0.1 km s^{-1} and Moho depths to within 1 km. The P-wave structure of the oceanic sections of both profiles was modelled by Craig Parkin (Parkin & White 2008), the Hatton P-wave profile by Lindsey Smith (White & Smith 2009), the Faroes P-wave profile by Alan Roberts (Roberts *et al.* 2009) and the S-wave structure across both profiles by Jennifer Eccles (Eccles *et al.* 2009): the reader is referred to these papers for full details of the modelling, the resolution testing and the uncertainty tests.

The P-wave velocity structure across both the Hatton and the Faroes rifted continental margins shows striking similarities, as marked by the summary of crustal type along the top of Figure 4, though the two profiles are some 800 km apart along strike and are at different distances from the centre of the thermal anomaly in the mantle at the time of continental breakup. Both profiles are aligned with 0 km distance at the prominent seafloor spreading magnetic anomaly 22, which is visible on the oceanic sections. Both profiles are bounded on the landward side by older Mesozoic failed rifts with thinned crust, the Hatton–Rockall Basin in

Figure 4a and the Faroe–Shetland Basin in Figure 4b. Immediately adjacent to the COT on the landward side are blocks of continental crust some 24–27 km thick, called the Hatton Bank on the southern profile and the Fugloy Ridge on the northern profile near the Faroe Islands. These continental blocks are covered in both cases by basalts which flowed landward from the continental breakup rift. The continental blocks are thinned a little from the presumed original continental crustal thickness of 30–32 km found beneath nearby Ireland and northern Britain (e.g. Barton 1992; Landes et al. 2005; Tomlinson et al. 2006). The seismic velocity of the Hatton Bank and Fugloy Ridge continental blocks averages 6.7 km s^{-1} in the lower crust, which is typical of the velocity of continental crust beneath the UK well away from the oceanic rift. This suggests that there has been only limited lateral intrusion of igneous rocks away from the COT during the breakup phase, because otherwise the igneous intrusions would have caused the seismic velocity of the lower crust beneath Hatton Bank and the Fugloy Ridge to be increased.

The transition from this continental crust to fully igneous oceanic crust is surprisingly narrow, with the COT being less than 50 km wide. Over this interval the seismic velocity of the lower crust increases by approximately 0.6 km s^{-1}, from an average of 6.7 km s^{-1} beneath the Hatton Bank to 7.3 km s^{-1} beneath the first-formed oceanic crust (White & Smith 2009), and with a similar increase across the Faroes margin. As we show later, the increase in seismic velocity of the lower crust is due to an increasing percentage of igneous sill intrusion as the oceanic rift is approached.

The first formed (i.e. the oldest) oceanic crust which, apart from the subsequently deposited sediments, is 100% igneous, exhibits the highest seismic velocities on the profiles, with the average lower-crustal (oceanic layer 3) velocity reaching 7.3 km s^{-1}: this velocity is markedly higher than the average layer 3 velocity for normal oceanic crust of 6.95 km s^{-1} (White et al. 1992). The oceanic crustal thickness of 8–17 km is also considerably greater than the normal oceanic thickness of 6.5 km (White et al. 1992). Both observations, of increased seismic velocity and increased thickness, can be explained by the generation of the melt which forms the oceanic igneous crust from mantle with its temperature elevated as much as 150 °C above normal (White & Smith 2009). Over the subsequent 10 million years following the onset of sea-floor spreading in the North Atlantic both the average lower-crustal velocity and the crustal thickness of the oceanic crust measured along our profiles decreased (Fig. 5), suggesting that the mantle plume temperature decreased by about 70–80 °C from its peak at the time of continental breakup (Parkin & White 2008; White & Smith 2009). Besides controlling the formation of melt, this change in mantle temperature would have exerted a strong regional control on the subsidence history. The rifted margin and its hinterland was elevated above, or near, sea-level during breakup in the Paleocene and then subsided as the mantle thermal anomaly decreased (Barton & White 1997; White & Lovell 1997). It is important to include such changes in the regional base-level in studies of subsidence relevant to hydrocarbon exploration.

In modelling subsidence, changes in crustal thickness are the dominant control on the uplift or subsidence history. Stretching and crustal thinning causes subsidence. The addition of mass to the crust causes uplift. On the continent–ocean boundaries the thinning caused by stretching at the rift is counteracted by the thickening caused by igneous intrusion and extrusion. This often results in the surface remaining at about the same elevation. Thus the extrusive volcanics on the COT were erupted close to sea-level throughout the rifting history, despite the original crust having been thinned by more than a factor of five. Landward of the COT, there is much less stretching, but still some crustal thickening by the injection of igneous rocks and the flow of basalts across the surface. The thickening of the crust by magmatism causes permanent uplift,

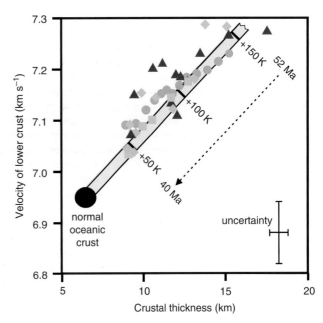

Fig. 5. Average crustal thickness and lower-crustal (oceanic seismic layer 3) P-wave velocity after correction to a standard reference pressure of 230 MPa and a reference temperature of 150 °C, calculated every 10 km along the oceanic portions of the Faroes profile (red triangles, data from Roberts et al. 2009), the Hatton profile (blue diamonds, data from White & Smith 2009), and the east Greenland SIGMA-3 profile approximately conjugate to the Hatton profile (green circles, data from Hopper et al. 2003), and the east Greenland SIGMA-2 profile (green squares, data from Korenaga et al. 2000), which is a similar distance from the centre of the mantle thermal anomaly as is the Faroes profile. Locations of profiles are shown in Figure 1. Uncertainty ranges typical of the data points are shown in the bottom right corner. Yellow arrow shows representative trend of changes in total igneous thickness and lower-crustal P-wave velocity for passive decompression beneath an oceanic spreading centre of mantle of increasing temperature, with tick marks approximately every 50 K above the normal mantle temperature (White & Smith 2008).

whilst temperature increases in the underlying mantle cause transient uplift which decreases as the mantle cools back to its normal temperature. So again there can be a delicate balance between subsidence and uplift that is governed largely by the magmatism and changes in underlying mantle temperature. These factors are frequently ignored in modelling the subsidence history, but are clearly of prime importance.

In the Rockall–Faroes region, local compressional structures that post-date the tectonics associated with the continental breakup are often superimposed on them (Tuitt et al. 2010). Thus there is evidence that Hatton Bank, which was already a structurally relatively high region at the end of continental breakup, was subsequently further uplifted during the Mid to Late Eocene by compressional forces unrelated to the breakup and that the Fugloy Ridge on the Faroes shelf was formed by inversion after the rift-related basalts had flowed across the shelf.

Geological interpretation from combined seismic reflection and wide-angle data

The combination of good velocity control from wide-angle seismic data with a deep penetration seismic reflection profile not only enables improvements to be made to the processing of the seismic reflection data in order to enhance the image, but also allows better interpretation to be made of the crustal structure. This is shown well by the section of the reflection profile across the Faroes COT and first-formed oceanic crust shown in Figure 6 (for the location see the box on Fig. 4b). The seismic P-wave velocity structure from Roberts et al. (2009) has been superimposed in colour on the reflection profile, using the same colour scale as in Figure 4. One of the most striking results on Figure 6 is that the region of high-velocity (6.7–7.3 km s^{-1}) lower crust under the COT between 55 and 90 km distance exhibits marked sub-horizontal reflectivity. This is interpreted as caused by lower-crustal igneous intrusions. They underlie the thickest section of extrusive basalts, which on the COT produce striking upward-convex seaward-dipping reflectors. Both the intrusions and the seaward dipping reflectors were produced in the continental rift zone at the time of continental breakup. The intrusions cross-cut the local dipping fabric of the continental basement visible between 75–110 km distance on Figure 6.

The region of HVLC containing the intrusions has velocities intermediate between that of the continental crust on the landward side and the fully oceanic crust on the seaward side. If we take the velocities of the continental and oceanic crust as the two end-members, then the percentage of intrusion at any position across the COT can be estimated by a simple mixing law. By summing the percentage of intrusion at all positions and depths across the COT, it is then possible to estimate the total area on the 2D profile of intruded igneous rock emplaced during continental breakup. For a 1 km deep slice centred on the Faroes profile this yields a total intruded igneous volume of 560–780 km^3 (Roberts et al. 2009). The total area in cross-section of extruded basalt which accompanied this intrusion can be measured directly off the seismic reflection profile, remembering that it flowed more than 100 km landward (Fig. 4b), so has to be summed across that region. On the Faroes profile, the area can be measured easily because the base of the basalt is defined well by the underlying LVZ (Figs 2, 4b & 6). On the Hatton profile the basalts lie directly over higher velocity crust, and their base cannot be so easily delineated. For a 1 km deep slice centred on the Faroes profile the total volume of extruded basalts is 340–420 km^3 (Roberts et al. 2009).

Although we do not know the age of the lower-crustal intruded rocks, because they have never been sampled directly, it is known in the North Atlantic flood basalt province, as is true of others around the world, that the main phase of the extrusive igneous activity was very short-lived, lasting typically less than 1 million years (White & McKenzie 1989, 1995; Eldhom & Grue 1994). Therefore it is likely that the entire volume of igneous melt including the intrusions, totalling 900–1200 km^3 for every 1 km alongstrike in the Faroes region, was produced in a geologically short time beneath the continental rift as it opened to form the new ocean basin. A similar volume was emplaced at the same time on the conjugate margin (White & Smith 2009). This is an astonishingly large volume of melt, which is most plausibly explained by decompression melting of abnormally hot mantle some 150 °C hotter than normal as the rift opened. As we have seen from Figure 5, the mantle temperature subsequently decreased by 79–80 °C over the next 10 Ma as the ocean opened. It is a feature of thermal boundary layer instabilities such as those

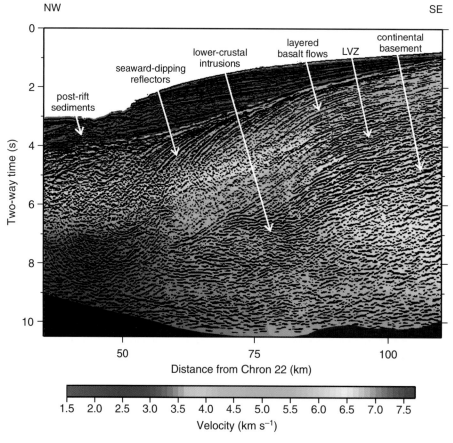

Fig. 6. Section of iSIMM seismic profile crossing the COT north of the Faroe Islands (see Fig. 1 for location) with superimposed P-wave velocity field from tomographic inversion of wide-angle arrivals on OBS (from White et al. 2008 and Roberts et al. 2009). Purple section beneath seafloor is post-rift Cenozoic sediments. Underlying extrusive lava flows are imaged as seaward-dipping reflectors between 50 and 90 km along profile and as sub-horizontal layered basalts at distances >90 km. LVZ marks the low-velocity zone immediately beneath the basalt flows. Lower-crustal layering coincident with high (>7.0 km s^{-1}) velocities, caused by igneous intrusions, lies beneath the basalts on the COT, with a termination at c. 90 km against continental crust with lower velocities. Moho reflection at base of crust shallows markedly from continental (SE) to oceanic (NW). Processing of the reflection profile included source designature, multiple suppression and post-stack time migration.

postulated to form mantle plumes that they are hotter at the time of initiation and then cool as convection continues, so this behaviour fits well the observations in the North Atlantic, with the anomalously hot mantle of the Iceland plume arriving shortly before rifting. The oldest volcanic rocks in the North Atlantic province date from 62 Ma, so this age probably marks the arrival of the mantle plume.

Lithological constraints from P- and S-wave velocities

Measurements of the *in-situ* S-wave velocity (V_s) in addition to the P-wave velocity (V_p) provide useful constraints on the composition of subsurface rocks that are not available from V_p measurements alone. We demonstrate this by considering two crucial regions that have been the subject of differing interpretations in the past: the LVZ that immediately underlies the basalts on the Faroes shelf, and the region of HVLC that is found beneath the COT.

The presence of a LVZ under the basalt flows on the Faroes shelf has been well established by wide-angle seismic constraints from first-arriving P-waves (Figs 2 & 4b), as discussed above, and is visible by a change of reflection character on the seismic reflection profile (Fig. 6). The main question concerning the LVZ under the basalt flows is whether it comprises sedimentary rock or igneous hyaloclastite material. This makes a large difference to its hydrocarbon prospectivity. One of the deepest penetrating wells through the basalts in the Faroes region is the Lopra 1/1A borehole (Fig. 1). The original 2178 m deep well was deepened to 3565 m on the basis of strong sub-bottom reflectors that were interpreted to be sedimentary. However, drilling showed that the deepest section was hyaloclastites (with lower P-wave velocity than the overlying basalt flows) and that the strong reflectors previously interpreted as sedimentary in origin were caused by basalt sills (Christie *et al.* 2006*a*). Sedimentary units were not reached before the Lopra well was abandoned. Elsewhere, on the eastern flank of the Fugloy Ridge, Spitzer *et al.* (2005) inferred on the basis of P-wave velocities derived from the 12 km offset seismic reflection data that there was a layer of hyaloclastite immediately beneath the base of the basalts, but that this was underlain by lower velocity material that was interpreted as sedimentary in origin.

Wide-angle seismic constraints from the Faroes profile constrain the average P-wave velocity in the LVZ as 4.5 km s^{-1}, with a possible range from 4.0–4.8 km s^{-1} and an average V_p/V_s ratio of 1.81 ± 0.1. Plotted on a graph of V_p v. V_p/V_s ratio (Fig. 7), it is clear that these velocities fall outside the main region of hyaloclastite properties measured from the nearby Faroes Lopra 1/1A well. The P-wave velocity of the material in the LVZ is lower than most hyaloclastites and the V_p/V_s ratio is also towards the lower limit of the bulk of hyaloclastite values. However, the P-wave velocity is somewhat higher than expected for the sedimentary rocks that are likely to lie under the basalts. On Figure 7 we show the average velocities of Paleocene sedimentary rocks recovered from well 206/1-2 located towards the southeastern end of our Faroes profile (see Fig. 1 for location), beyond the feather edge of the basalts. These sedimentary units may be similar to those found under the basalts. The deeper section of the Paleocene sediments, which lie at the same depth as the LVZ, has an average P-wave velocity of 3.8 km s^{-1}, with an inferred V_p/V_s ratio similar to that of the LVZ (Fig. 7). For comparison, we show on Figure 7 the equivalent values for Brent sandstone, which have similar seismic properties to the well 206/1-2 sediments. The P- and S-wave velocities of the material in the LVZ therefore indicate the presence of neither pure sedimentary rock nor hyaloclastites. The most likely explanation is that the LVZ primarily comprises sedimentary units intruded by basaltic sills, since the LVZ V_p and V_p/V_s values lie on a mixing line intermediate between these two end-members

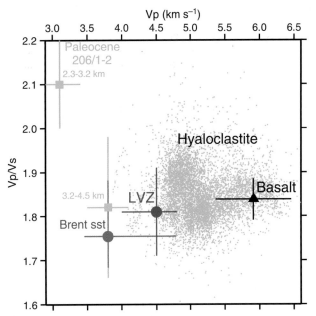

Fig. 7. V_p v. V_p/V_s ratio for the sub-basalt LVZ on the Faroes shelf and literature comparisons. The average result determined for the LVZ from analysis of the wide-angle seismic data is shown by the red circle. Grey dots represent well log measurements of hyaloclastites from the Faroes Islands Lopra-1/1A well (see Fig. 1 for location); the black triangle is an average literature value for basalt at 200 MPa (Christensen 1996). The properties of the Paleocene sedimentary section from the 206/1-2 well (see Fig. 1 for location), subdivided into two sections based on a major change in geophysical character, are shown in green, with V_s calculated using the mudrock relation of Castagna *et al.* (1985). The properties of Jurassic sandstone from the Brent field (Strandenes 1991) are shown by the blue circle for comparison. Data from Eccles *et al.* (2009).

(Fig. 7). There are sills throughout the sedimentary section in the Faroe–Shetland Trough, as well as in the hyaloclastites penetrated by the Lopra well (Christie *et al.* 2006*a*), and it would be surprising if sills were not also present in the sedimentary section beneath the basalt flows on the Faroes Shelf.

The S-wave measurements also assist the interpretation of the material which forms the HVLC under the continent–ocean boundary. We have already noted the lateral change in the average P-wave velocity of the lower crust from the continental end-member to the fully igneous oceanic end-member across the COT: the average P-wave lower crustal velocity increases from *c*. 6.7 to *c*. 7.3 km s^{-1} (Figs 4 & 6). Incorporation of S-wave measurements shows that there is also a linked change in the V_p/V_s ratio across the same COT region, with an increase from *c*. 1.75 at the continental end to *c*. 1.80 at the oceanic end (Fig. 8). The same trend is visible on both the Hatton profile data (blue diamonds) and the Faroes profile data (red triangles). We have already suggested that the gradient in P-wave velocity across the COT is caused by an increasingly dense intrusion of high-velocity igneous sills into the lower-velocity continental crust, an interpretation supported by the imaging of sub-horizontal reflectors in this part of the crust, inferred to be sills (Fig. 6). The trend of the V_p/V_s ratio confirms this interpretation. The continental lower crust in the region is likely to comprise Lewisian Gneiss, and *in-situ* seismic measurements from beneath northern Britain show that the lower crust exhibits not only relatively low P-wave velocities, but also a low V_p/V_s ratio (black square on Fig. 8, from Assumpção & Bamford 1978). The mafic intrusions by contrast have high P-wave velocities and higher V_p/V_s ratios. The systematic trends of increasing V_p and increasing V_p/V_s ratio across the COT are thus entirely compatible with a linear mixing trend between the

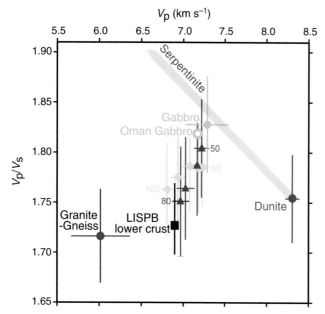

Fig. 8. V_p v. V_p/V_s ratio of the lower crust across the Faroes (red triangles) and Hatton (blue diamonds) COTs, corrected to a standard pressure of 230 MPa and a temperature of 200 °C. Values are averaged over a 10 km distance window. Small numbers show distance in km from chron 22 along the Faroes (red numbers) and the Hatton (blue numbers) profiles, matching the distance scales on Figures 4 and 6. The *in-situ* value for the lower crust under Scotland from the LISPB seismic experiment (black square) is from Assumpção & Bamford (1978). The properties of dunite, gabbro and granite-gneiss from laboratory experiments, corrected to 230 MPa, are from Christensen (1996). The properties of an Oman olivine gabbro with 10% olivine (Browning 1984) are shown by an open green circle. The properties of serpentinite with variable degrees of serpentinization (purple band) are from Horen *et al.* (1996) and Carlson & Miller (1997). The trend of increasing V_p/V_s ratio with decreasing V_p as the degree of serpentinization increases is orthogonal to the trend of increasing V_p/V_s ratio with increasing V_p measured for the HVLC on the continental margins, ruling out serpentinization as the cause of the HVLC. The trend of values observed for the HVLC on the Faroes and Hatton profiles is consistent with increasing intrusion of mafic material as sills into the continental lower crust, since they lie on a mixing line between the end-points for the lower crust (from LISPB) and high magnesium gabbro.

end members of Lewisian gneiss at the continental end and high magnesium gabbros at the oceanic end (Fig. 8). An interpretation which is ruled out by the V_p/V_s constraints is that the wedge of HVLC on the COT might be serpentinite, as has been postulated for parts of the Møre continental margin off Norway by Reynisson *et al.* (2010). Although serpentinite can certainly exhibit the range of P-wave velocities of 6.7–7.3 km s^{-1} that we observe, the accompanying V_p/V_s ratios of *c.* 1.89–1.84 for these degrees of serpentinization are much higher than are observed on the North Atlantic margins. Furthermore, the trend of increasing V_p/V_s ratio with decreasing V_p as the degree of serpentinization increases (purple shading on Fig. 8) is orthogonal to the trend of increasing V_p/V_s ratio with increasing V_p measured for the HVLC on the continental margins.

Conclusions

Layered basalts pose problems for deep sub-basalt imaging for two main reasons: high-frequency seismic energy is scattered and the high impedance contrasts at the margins of basalt sequences produce strong interbed multiples which make the interpretation of sub-basalt structure difficult. The iSIMM project has shown that sub-basalt imaging can be improved greatly by careful attention to the source and receiver design, to ensure that the source spectrum contains sufficient energy at low frequencies to make it capable of penetrating through layered basalt sequences. In order to address the problem of interbed multiples, the use of long streamers helps discriminate between primary and multiple arrivals during processing.

In addition to improvements to the conventional seismic reflection imaging mentioned above, the deployment of OBS along a profile adds another dimension by making it possible to construct a well constrained P-wave velocity model of the crust. This is useful in the mechanics of improving the reflection image through pre-stack depth migration and other processing, but it is also useful in a further way because by combining the velocity information with the reflection image it allows greatly improved geological interpretations to be made, as we have discussed with specific reference to the layered intrusions in the HVLC on the COT. Further constraints on the S-wave velocities obtained from mode-converted waves recorded directly at the OBS have the potential to add even more value to the interpretation by allowing lithological discrimination in a way which may remain ambiguous if only P-wave velocities are available. Both these improvements are of importance in hydrocarbon exploration.

Combination of tomographic velocity results with reflection imaging allows us to constrain quantitatively the total volume of melt produced during continental breakup at the rift underlying the COT. The profile across the Faroes margin, which is close to the centre of the mantle thermal anomaly, shows that a total of approximately 1000 km^3 of melt is generated from the mantle for every kilometre along strike. This is a huge volume, but of course is only half of the total produced at the continental rift near the Faroe Islands, because there is a similar volume on the east Greenland conjugate margin. Furthermore we have shown that more than half the melt is intruded in the lower crust on the COT, and never reaches the surface. This is consistent with earlier suggestions by Cox (1980, 1993) based on petrological arguments. We are able to show, at least near the Faroe Islands, that the ratio of intruded to extruded melt is about 1.5:1, thus considerably tightening the possible range of 0.25–4:1 reported by Coffin & Eldholm (1994). This is of crucial importance in modelling the subsidence and heat flow history of the continental margin and adjacent shelf.

Furthermore our conclusion that large quantities of melt are intruded in the lower crust at the rift zone is consistent with recent results from the northern volcanic rift in Iceland reported by Maclennan *et al.* (2003). They show that melt inclusions carried in extruded basalts had stalled in magma chambers near the base of the crust or in the uppermost mantle, and that some fractionation occurred before they were subsequently carried to shallow magma chambers near the surface and eventually erupted. The current northern volcanic rift of Iceland has many similarities to the Faroes margin at the time of early Cenozoic breakup, because it is a place where rifting is presently occurring through older crust between 20–30 km thick, above a thermal anomaly in the mantle caused by a mantle plume (Darbyshire *et al.* 2000a, b). Studies of the seismicity currently occurring under the Icelandic rift show considerable numbers of small earthquakes at depths of 15–30 km beneath the active Askja volcano, which are interpreted as caused by melt intrusion happening now in the lower crust (Soosalu *et al.* 2010). So it seems that we have consistent information from the seismic profiles across the Cenozoic rifted continental margin, from petrological and geochemical arguments and from seismic activity currently occurring beneath an active volcanic rift zone, that large quantities of melt are intruded in the lower crust before fractionating, moving upward towards the surface and finally erupting.

The presence of these enormous quantities of igneous rock intruded and extruded at volcanic rifts are important not only for understanding the geological history of such rifts, but also for modelling the thermal and subsidence history. Most subsidence modelling programmes assume that mass is conserved in the crust: if in fact large volumes of melt are added as the results presented here make clear, this makes an enormous difference to the subsidence pattern and the thermal history of the rifts (Barton & White 1997). Coupled to this are large base-level changes caused by changes in the temperature of the underlying mantle. Unless the mass and crustal thickness changes caused by magmatism and the elevation changes caused by the thermal history of the mantle are included in the subsidence analysis, the predicted subsidence history may be greatly in error.

We are grateful for support of the iSIMM project by Liverpool and Cambridge Universities, Schlumberger Cambridge Research Ltd, Badley Geoscience Ltd, WesternGeco, Amerada Hess, Anadarko, BP, ConocoPhillips, ENI UK, Statoil, Shell, the NERC and the DTI. The Q-streamer acquisition was undertaken by WesternGeco. The full iSIMM team comprises A. Chappell, P. A. F. Christie, J. D. Eccles, R. Fletcher, D. Healy, N. Hurst, N. J. Kusznir, H. Lau, Z. Lunnon, C. J. Parkin, A. M. Roberts, A. W. Roberts, L. K. Smith, R. Spitzer, V. J. Tymms and R. S. White. Department of Earth Sciences, Cambridge contribution number ESC1284.

References

Assumpção, M. & Bamford, D. 1978. LISPB – V. Studies of crustal shear waves. *Geophysical Journal of the Royal Astronomical Society*, **54**, 61–73.

Avedik, F., Renard, V., Allenou, J. P. & Morvan, B. 1993. 'Single-bubble' air-gun array for deep exploration. *Geophysics*, **58**, 366–382.

Avedik, F., Nicolich, R., Hirn, A., Maltezou, F., McBride, J., Cernobori, L. & the Streamers/Profiles Group. 1995. Appraisal of a new, high-energy low-frequency seismic pulse generating method on a deep seismic reflection profile in the Central Mediterranean Sea. *First Break*, **13** (July), 277–290.

Barton, P. J. 1992. LISPB revisited: a new look under the Caledonides of northern Britain. *Geophysical Journal International*, **110**, 371–391.

Barton, A. J. & White, R. S. 1997. Crustal structure of the Edoras Bank continental margin and mantle thermal anomalies beneath the North Atlantic. *Journal of Geophysical Research*, **102**, 3109–3129.

Brennan, B. J. & Stacey, F. D. 1977. Frequency dependence of elasticity of rock – tests of seismic velocity dispersion. *Nature*, **268**, 220–222.

Browning, P. 1984. Cryptic variation within the cumulate sequence of the Oman ophiolite: magma chamber depth and petrological implications. *In*: Gass, I. G., Lippard, S. J. & Shelton, A. W. (eds) *Ophiolites and Oceanic Lithosphere*. Geological Society, London, Special Publications, **13**, 71–82.

Carlson, R. L. & Miller, D. J. 1997. A new assessment of the abundance of serpentinite in the oceanic crust. *Geophysical Research Letters*, **24**, 457–460.

Castagna, J. P., Batzle, M. L. & Eastwood, R. L. 1985. Relationships between compressional-wave and shear-wave velocities in clastic silicate rocks. *Geophysics*, **50**, 571–581.

Christensen, N. I. 1996. Poisson's ratio and crustal seismology. *Journal of Geophysical Research*, **101**, 3129–3156.

Christie, P., Gollifer, I. & Cowper, D. 2006a. Borehole seismic studies of a volcanic succession from the Lopra-1/1A borehole in the Faroe Islands, NE Atlantic. *Geology of Denmark Survey*, **9**, 23–40.

Christie, P. A. F., Lunnon, Z. C., White, R. S. & iSIMM Team. 2006b. iSIMM experience with peak- and bubble-tuned sources for generating low frequencies. *68th Meeting of the European Association of Geoscientists and Engineers, Vienna, expanded abstract* A034.

Coffin, M. F. & Eldholm, O. 1994. Large igneous provinces: crustal structure, dimensions, and external consequences. *Reviews of Geophysics*, **32**, 1–36.

Cox, K. G. 1980. A model for flood basalt vulcanism. *Journal of Petrology*, **21**, 629–650.

Cox, K. G. 1993. Continental magmatic underplating. *Philosophical Transactions of the Royal Society, London, A*, **342**, 155–166.

Darbyshire, F. A., White, R. S. & Priestley, K. P. 2000a. Structure of the crust and uppermost mantle of Iceland from a combined seismic and gravity study. *Earth and Planetary Science Letters*, **181**, 409–428.

Darbyshire, F. A., Priestley, K. P., White, R. S., Stefánsson, R., Gudmundsson, G. B. & Jakobsdóttir, S. S. 2000b. Crustal structure of central and northern Iceland from analysis of teleseismic receiver functions. *Geophysical Journal International*, **143**, 163–184.

Eccles, J. D., White, R. S., Roberts, A. W., Christie, P. A. F. & iSIMM Team. 2007. Wide angle converted shear wave analysis of a North Atlantic volcanic rifted continental margin: constraint on sub-basalt lithology. *First Break*, **25** (October 2007), 63–70.

Eccles, J. D., White, R. S. & Christie, P. A. F. 2009. Identification and inversion of converted shear waves: case studies from the European North Atlantic continental margins. *Geophysical Journal International*, **179**, 381–400; doi: 10.1111/j.1365-246X.2009.04290.x.

Eldholm, O. & Grue, K. 1994. North Atlantic volcanic margins: dimensions and production rates. *Journal of Geophysical Research*, **99**, 2955–2988.

Fliedner, M. M. & White, R. S. 2001. Seismic structure of basalt flows from surface seismic data, borehole measurements and synthetic seismogram modeling. *Geophysics*, **66**, 1925–1936.

Fliedner, M. M. & White, R. S. 2003. Depth imaging basalt flows in the Faeroe-Shetland Basin. *Geophysical Journal International*, **152**, 353–371.

Gordon, R. B. & Davis, L. A. 1968. Velocity and attenuation of seismic waves in imperfectly elastic rock. *Journal of Geophysical Research*, **73**, 3917–3935.

Hopper, J. R., Dahl-Jensen, T. *et al.* 2003. Structure of the SE Greenland margin from seismic reflection and refraction data: implications for nascent spreading center subsidence and asymmetric crustal accretion during North Atlantic opening. *Journal of Geophysical Research*, **108**, 2261–2291; doi: 10.1029/2002JB001996.

Horen, H., Zamora, M. & Dubuisson, G. 1996. Seismic wave velocities and anisotropy in serpentinized peridotites from Xigase ophiolite: abundance of serpentinite in slow spreading ridges. *Geophysical Research Letters*, **23**, 9–12.

Klingelhöfer, F., Edwards, R. A., Hobbs, R. W. & England, R. W. 2005. Crustal structure of the NE Rockall Trough from wide-angle seismic data modelling. *Journal of Geophysical Research*, **110**, B11105; doi: 10.1029/2005JB003763.

Korenaga, J., Holbrook, W. S., Kent, G. M., Kelemen, P. B., Detrick, R. S., Hopper, J. R. & Dahl-Jensen, T. 2000. Crustal structure of the southeast Greenland margin from joint refraction and reflection seismic tomography. *Journal of Geophysical Research*, **105**, 21591–21614.

Landes, M., Ritter, J. R. R., Readman, P. W. & O'Reilly, B. 2005. A review of the Irish crustal structure and signatures from the Caledonian and Variscan orogenies. *Terra Nova*, **17**, 111–120.

LASE Study Group. 1986. Deep structure of the US East Coast passive margin from large aperture seismic experiments (LASE). *Marine Geophysical Research*, **3**, 234–242.

Lunnon, Z. C., Christie, P. A. F. & White, R. S. 2003. An evaluation of peak and bubble tuning in sub-basalt seismology: modelling and results from OBS data. *First Break*, **21** (December), 51–56.

Maclennan, J., McKenzie, D., Grönvold, K., Shimizu, N., Eiler, J. M. & Kitchen, N. 2003. Melt mixing and crystallization under Theistareykir, northeast Iceland. *G3 Geochemistry Geophysics Geosystems*, **4**, 8624; doi: 10.1029/2003GC000558.

Maresh, J. & White, R. S. 2005. Seeing through a glass, darkly: strategies for imaging through basalt. *First Break*, **23** (May), 27–32.

Maresh, J., White, R. S., Hobbs, R. W. & Smallwood, J. R. 2006. Seismic attenuation of Atlantic margin basalts: observations and modeling. *Geophysics*, **71**, B211–B221; doi: 10.1190/1.2335875.

Mutter, J. C., Talwani, M. & Stoffa, P. L. 1984. Evidence for a thick oceanic crust adjacent to the Norwegian Margin. *Journal of Geophysical Research*, **89**, 483–502.

Parkin, C. J. & White, R. S. 2008. Influence of the Iceland mantle plume on oceanic crust generation in the North Atlantic. *Geophysical Journal International*, **173**, 168–188; doi: 10.1111/j.1365-246X.2007.03689.x.

Reynisson, R. F., Ebbing, J., Lundin, E. & Osmundsen, P. T. 2010. Properties and distribution of lower crustal bodies on the mid-Norwegian margin. In: Vining, B. A. & Pickering, S. C. (eds) *Petroleum Geology: From Mature Basins to New Frontiers – Proceedings of the 7th Petroleum Geology Conference*, Geological Society, London, 843–854; doi: 10.1144/0070843.

Richardson, K. R., White, R. S., England, R. W. & Fruehn, J. 1999. Crustal structure east of the Faroe Islands. *Petroleum Geoscience*, **5**, 161–172.

Roberts, A. W., White, R. S. & Christie, P. A. F. 2009. Imaging igneous rocks on the North Atlantic rifted continental margin. *Geophysical Journal International*, **179**, 1024–1038; doi: 10.1111/j.1365-246X.2009.04306.x.

Shaw, F., Worthington, M. H., White, R. S., Andersen, M. S., Petersen, U. K. & the SeiFaba Group. 2008. Seismic attenuation in Faroe Islands basalts. *Geophysical Prospecting*, **56**, 5–20; doi: 10.1111/j.1365-2478.2007.00665.x.

Soosalu, H., Key, J., White, R. S., Knox, C., Einarsson, P. & Jakobsdóttir, S. S. 2010. Lower-crustal earthquakes caused by magma movement beneath Askja volcano on the north Iceland rift. *Bulletin of Volcanology*, **72**, 55–62; doi: 10.1007/s00445-009-0297-3.

Spitzer, R., White, R. S. & iSIMM Team. 2005. Advances in seismic imaging through basalts: a case study from the Faroe-Shetland Basin. *Petroleum Geoscience*, **11**, 147–156.

Strandenes, S. 1991. Rock physics analysis of the Brent Group reservoir in the Oseberg Field. In: *Stanford Rockphysics and Borehole Geophysics Project*, Special Volumes. Stanford University, Stanford, CA.

Tomlinson, J. P., Denton, P., Maguire, P. K. H. & Booth, D. C. 2006. Analysis of the crustal velocity structure of the British Isles using teleseismic receiver functions. *Geophysical Journal International*, **167**, 223–237.

Tuitt, A., Underhill, J. R., Ritchie, J. D., Johnson, H. & Hitchen, K. 2010. Timing, controls and consequences of compression in the Rockall–Faroe area of the NE Atlantic margin. In: Vining, B. A. & Pickering, S. C. (eds) *Petroleum Geology: From Mature Basins to New Frontiers – Proceedings of the 7th Petroleum Geology Conference*. Geological Society, London. 963–977; doi: 10.1144/0070963.

Vogt, U., Makris, J., O'Reilly, B. M., Hauser, F., Readman, P. W., Jacob, A. W. B. & Shannon, P. M. 1998. The Hatton Basin and continental margin: crustal structure from wide-angle seismic and gravity data. *Journal of Geophysical Research*, **103**, 12545–12566.

Voss, M. & Jokat, W. 2007. Continent–ocean transition and voluminous magmatic underplating derived from P-wave velocity modelling of the East Greenland continental margin. *Geophysical Journal International*, **170**, 580–604.

White, N. & Lovell, B. 1997. Measuring the pulse of a plume with the sedimentary record. *Nature*, **387**, 888–891.

White, R. & McKenzie, D. 1989. Magmatism at rift zones: the generation of volcanic continental margins and flood basalts. *Journal of Geophysical Research*, **94**, 7685–7729.

White, R. S. & McKenzie, D. 1995. Mantle plumes and flood basalts. *Journal of Geophysical Research*, **100**, 17543–17585.

White, R. S. & Smith, L. K. 2009. Crustal structure of the Hatton and the conjugate east Greenland rifted volcanic continental margins, NE Atlantic. *Journal of Geophysical Research*, **114**, B02305; doi: 10.1029/2008JB005856.

White, R. S. & Stephen, R. A. 1980. Compressional to shear wave conversion in oceanic crust. *Geophysical Journal of the Royal Astronomical Society*, **63**, 547–566.

White, R. S., McKenzie, D. & O'Nions, R. K. 1992. Oceanic crustal thickness from seismic measurements and rare earth element inversions. *Journal of Geophysical Research*, **97**, 19683–19715.

White, R. S., Fruehn, J., Richardson, K. R., Cullen, E., Kirk, W., Smallwood, J. R. & Latkiewicz, C. 1999. Faroes Large Aperture Research Experiment (FLARE): imaging through basalts. In: Fleet, A. J. & Boldy, S. A. R. (eds) *Petroleum Geology of Northwest Europe: Proceedings of the 5th Conference*. Geological Society, London, 1243–1252; doi: 10.1144/0051243.

White, R. S., Smallwood, J. R., Fliedner, M. M., Boslaugh, B., Maresh, J. & Fruehn, J. 2003. Imaging and regional distribution of basalt flows in the Faroe-Shetland Basin. *Geophysical Prospecting*, **51**, 215–231.

White, R. S., Smith, L. K., Roberts, A. W., Christie, P. A. F., Kusznir, N. J. & iSIMM Team. 2008. Lower-crustal intrusion on the North Atlantic continental margin. *Nature*, **452**, 460–464.

Zelt, C. A. & Smith, R. B. 1992. Seismic travel time inversion for 2-D crustal velocity structure. *Geophysical Journal International*, **108**, 16–24.

Ziolkowski, A., Hanssen, P., Gatliff, R., Li, X. & Jakubowicz, H. 2001. The use of low frequencies for sub–basalt imaging. In: *71st Annual International Meeting of the Society of Exploration Geophysicists*, 74–77.

Properties and distribution of lower crustal bodies on the mid-Norwegian margin

R. F. REYNISSON,[1,2,3] J. EBBING,[1,2] E. LUNDIN[3] and P. T. OSMUNDSEN[2]

[1]*Norwegian University of Science and Technology, Department of Petroleum Engineering and Applied Geophysics, S.P. Andersens vei 15A, 7491 Trondheim, Norway (e-mail: rrey@statoil.com)*
[2]*Geological Survey of Norway, Leiv Eirikssons vei 39, 7040 Trondheim, Norway*
[3]*Statoil, Rotvoll, Arkitekt Ebbels veg 10, 7005 Trondheim, Norway*

Abstract: Anomalously high velocity and high density bodies have been detected in the lower crust on the mid-Norwegian margin. The lower crustal bodies (LCB) are pronounced on the Møre and Vøring margins segments and have mainly been interpreted as either magmatic or high-grade metamorphic in origin. Evolutionary models of the whole margin are heavily affected by the interpretation of the LCB and so are estimates of vertical movements and thermal structure in the area. A 3D gravity and magnetic model of the mid-Norwegian margin was constructed to map the main geological features of the margin and acquire the distribution of the LCB. The model utilizes the most recent potential field compilations on the margin and is constrained by extensive reflection seismic data and published refraction profiles. Further constraints on the model were attained from studying the isostatic state of the lithosphere. We present a map showing the distribution of the different LCB and discuss the implications for the structural and thermal evolution of the margin. The properties of the LCB vary across the margin and at least three different processes may be responsible for their existence. The LCB is commonly interpreted as igneous rock either intruded into the lower crust or underplated beneath it. The distribution of the LCB along the Vøring margin has an apparent correlation with the offshore prolongations of major onshore detachments stemming from Late Caledonian orogenic collapse. This may point towards some relation between the LCB and these old zones of weakness and that the LCB represents high-grade metamorphic rocks. Detailed modelling on the Møre margin shows a spatial link between parts of the LCB and extremely thin crustal thickness, suggesting a serpentinized exhumed mantle origin.

Keywords: Norwegian continental shelf, lower crustal body, magmatic underplating, exhumed mantle, passive continental margins, flexure and isostasy, potential fields

This study focuses on the mid-Norwegian continental margin (Fig. 1), a part of the NE Atlantic margin, and addresses the origin of a high-velocity, high-density layer at the base of the crust, which is detected by ocean bottom seismometer (OBS) seismic experiments with a seismic P-wave velocity of 7.0–7.9 km s^{-1} (e.g. Skogseid *et al.* 1992; Raum *et al.* 2002; Mjelde *et al.* 2005, 2009*a*, *b*; Tsikalas *et al.* 2008). We refer to this layer as a lower crustal body (LCB) to follow convention, but suggest that it can be part of the mantle as well as the crust. The LCB is often referred to as magmatic underplating (e.g. Skogseid *et al.* 1992; van Wijk *et al.* 2004; Mjelde *et al.* 2005). However, while the high velocity/high density of the LCB may be considered as an objective observation, its origin as a layer of igneous rock either intruded into the lower crust or underplated beneath it is an interpretation that dates back to work by, for example, White *et al.* (1987). To our knowledge, underplated material has never been observed in outcrop nor sampled by drilling and the usage as a descriptive term should be abandoned; it is clearly an interpretation. In recent years, there has been a renewed discussion about the interpretation of the LCB (e.g. Gernigon *et al.* 2004; Mjelde *et al.* 2005, 2009*a*; Ebbing *et al.* 2006), and of the distinction between igneous underplating below stretched continental crust and igneous intrusion into the lower crust (White *et al.* 2008; White & Smith 2009).

The mid-Norwegian margin was formed by episodic extensional events during Late Palaeozoic–Triassic, Late Jurassic–Early Cretaceous and Late Cretaceous–Paleocene times (Ziegler 1988; Blystad *et al.* 1995; Doré *et al.* 1999; Brekke 2000). Early Cenozoic continental breakup and initial seafloor spreading between Eurasia and Greenland was characterized by emplacement of significant volumes of magmatic rocks (e.g. Eldholm & Grue 1994). The magmatic rocks were partially extruded on the surface as flood basalts and tuffs and partially intruded as central complexes, sills and dykes into the sedimentary rocks and the crystalline crust. In addition, a lower crustal high-velocity body has been recognized along many parts of the margin, and is commonly interpreted to represent magmatic material added beneath the crust (Eldholm & Grue 1994; Mjelde *et al.* 2001), or intruded in the lower crust (e.g. White *et al.* 1987, 2008; White & Smith 2009). Notably, the interpreted magmatic body has been proposed to constitute between 60 and 80% of the total magmatic rock volume in the NAIP (White *et al.* 1987, 2008; Eldholm & Grue 1994). In the southernmost Vøring basin, however, the lower crustal layer shows anomalously high P-wave velocities (8.4 km s^{-1}) and has been interpreted as eclogite by Raum *et al.* (2006) and Mjelde *et al.* (2009*b*). In the remainder of the basin the layer is interpreted as mafic intrusions emplaced during the last phase of rifting, but it cannot be excluded that the body consists of older (Caledonian?) mafic rocks (Ebbing *et al* 2006; Mjelde *et al.* 2009*a*). Determining the nature of the LCB clearly is relevant for the thermal history of volcanic margins, and arguably also for the entire concept of the development of such margins.

We present a deep crustal configuration based on the results of a 3D model of the Møre margin compiled with studies by Ebbing *et al.* (2006), Osmundsen & Ebbing (2008), Tsikalas *et al.* (2008) and Mjelde *et al.* (2009*a*). These studies provide means to address the properties of the LCB on the mid-Norwegian margin by incorporating flexural isostasy considerations.

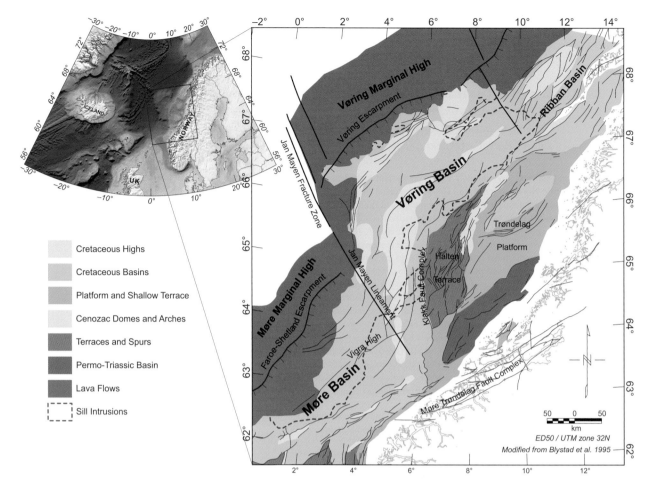

Fig. 1. Study area. The study area is the mid-Norwegian margin, a part of the NE Atlantic margin, and comprises the Møre, Vøring and Lofoten-Vesterålen margin segments. The figure shows the main geological and structural features of the margin based on Blystad *et al.* (1995).

Compilation of 3D models

The geometries of the deeper crust and the upper mantle on the mid-Norwegian margin are reasonably well defined in two dimensions by OBS data. From 3D potential field models, which have been integrated in the current study, different horizons can be mapped, which allows discussing regional changes in the geometry of the study area. The spatial extension of the horizons, with the exception of the Moho, is limited to the NW by the main escarpments and to the SE by the coast of Norway. Here, we utilize top crystalline basement, top LCB and Moho horizons in alliance with seismic interpretation and isostatic considerations to shed light on the properties of the LCB. See Ebbing *et al.* (2006) and Omsundsen & Ebbing (2008) for more details on the integration of potential field data, well information and seismic data in the construction of 3D models of the mid-Norwegian margin.

Figure 2 shows the thickness of the high-density LCB as defined by the integrated 3D models and seismic data. The LCB is primarily modelled on the Møre and Vøring margins and extends from the escarpments well into the basins. The overall structural trend of the LCB follows the general strike of the margin as indicated by seismically mapped faults. Two distinct LCB ridges that are up to 10 km thick exist on the Møre margin. The Vøring margin contains a LCB that is somewhat more evenly distributed, varying between 3 and 8 km thickness. The Lofoten margin has only a veneer of LCB that is little more than 1 km thick and is not present north of the Jennega Transfer Zone (Tsikalas *et al.* 2008).

By subtracting the Moho from the top basement horizon a crustal thickness map was acquired (Fig. 3). The map shows a very thin crystalline crust (less than 5 km) below the centre of the Møre Basin. Thin crust (less than 10 km) is noted along the whole mid-Norwegian margin and coincides with the extent of the Cretaceous depocentres, suggested to relate to Early Cretaceous rifting (Lundin & Doré 1997). Comparison of the crustal thickness with the LCB distribution reveals a strong first-order correlation along the entire mid-Norwegian margin. Locally, there is not always a one-to-one correlation between thickest modelled LCB and thinnest modelled crust. There are several possible explanations for this, such as different origin of the LCB (magmatic, metamorphic, serpentinized). In particular, densities of serpentinized mantle rocks may overlap with densities of crustal rocks and may therefore not be recognized by gravity modelling.

The distribution of volcanics in the study area was compiled from a seismic study and from previously published maps (Blystad *et al.* 1995; Olesen *et al.* 2002; Planke *et al.* 2005). Comparison of distributions of the LCB and the limit of lavas and sills (Fig. 4) reveals that all the LCB on the Vøring margin is overlain by volcanics and so is a considerable part of the Lofoten margin. On the Møre margin the LCB is overlain by volcanics, apart from the landward LCB ridge that coincides with the very thin crust.

Isostatic considerations

The Moho boundary is assumed to reflect the isostatic equilibrium surface at the base of the lithosphere (Braitenberg *et al.* 2002; Wienecke *et al.* 2007). Because the Moho is shallower than the base lithosphere and has a high density contrast, it is the main contributor of isostatic compensation to the gravity field. Comparison

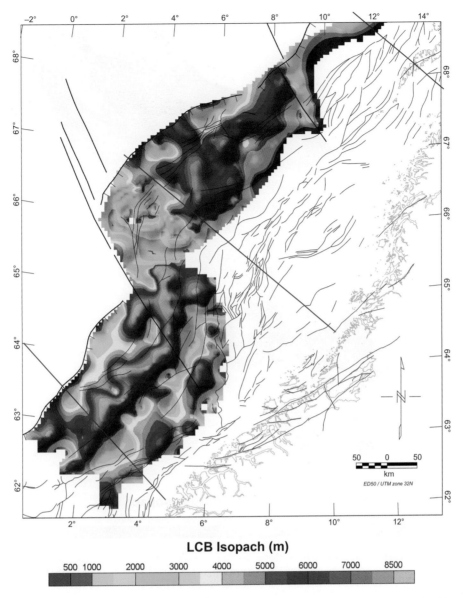

Fig. 2. LCB on the mid-Norwegian margin. LCB isopach resulting from integrated 3D models of the mid-Norwegian margin. The structural features are the same as in Figure 1. The red lines indicate the location of profiles in Figure 8.

of the Moho from the integrated models with an isostatic compensation surface provides insight into the whole crust characteristics of the study area. By comparing the two differently acquired Mohos, it is proposed that a distinction can be made between whether the LCB should be assigned to the crust or the mantle. This distinction has a profound influence on the genetic origin of the LCB. Crust-originated LCB indicates underplated or intruded igneous material in the lower crust and mantle-originated LCB reflects a low-density mantle. It is feasible to interpret the LCB associated with the mantle as serpentinized mantle and further that this body spatially coincides with very thin crust.

Assuming that the mid-Norwegian margin is in isostatic equilibrium, we calculated the isostatic compensation for the de-loading of the margin by the relatively low-density sedimentary infill and water. For the isostatic calculation, we used the LithoFlex software (Braitenberg et al. 2006; Wienecke et al. 2007). The loading of the sediments was calculated by applying a linear depth-dependent density function from 2200 kg m^{-3} at sea surface to 2700 kg m^{-3} at 10 km depth and constant below this level. The surrounding basement had density of 2750 kg m^{-3}. The simplified density distribution leads to de-loading of the lithosphere, which in an isostatic concept requires crustal thinning to balance the load. A relatively low value for the strength of the lithosphere ($T_e = 10$ km) was applied. The reference depth was 30 km and the density contrast at the isostatic flexural base was 400 kg m^{-3}, which corresponds to a lower crust density of 2900 kg m^{-3} and a mantle density of 3300 kg m^{-3}.

The resulting isostatic Moho (Fig. 5) shows a similar regional trend to the model Moho (Fig. 6) with depth decreasing from the coast to the axial part of the basins and increasing again towards the marginal high. The difference map (Fig. 7) highlights substantial differences between the isostatic and modelled Moho and gives valuable insight into the margin. A positive difference reflects that the model Moho is shallower than the isostatic Moho and negative difference indicates locations where the model Moho is deeper. In the southern Vøring, the isostatic and model Moho are at similar depths, while in the central and northern Vøring margin the isostatic Moho is more than 4 km shallower than the model Moho, corresponding to the thickness of the LCB. The same observation is made for the outer Møre margin. In the inner Møre margin and the Lofoten margin, the isostatic Moho is deeper than the model Moho. Three profiles representative for each margin segment further illustrate the difference between the isostatic and model Moho (Fig. 8).

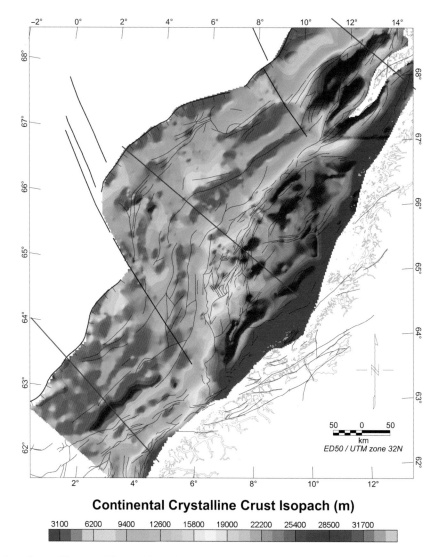

Fig. 3. Thickness of continental crystalline crust. The map shows that a very thin crystalline crust (less than 5 km) exists below the centre of the Møre Basin. Thin crust (less than 10 km) is noted locally on the whole mid-Norwegian margin and coincides with the extent of the Cretaceous depocentres, suggested to relate to Cretaceous rifting (Lundin & Doré 1997). The structural features are the same as in Figure 1. The red lines indicate the location of profiles in Figure 8.

The lateral volcanic distribution coincides mainly with negative Moho difference but the location of thin crust is most often associated with positive difference or a modelled Moho shallower than isostatic Moho (compare Figs 4 & 7). Most of the LCB has a distribution that follows the trend of the negative Moho difference, although the most inboard part of the LCB on the Møre and Vøring margins and the whole LCB on the Lofoten margin are spatially associated with positive Moho difference and thin crust. The positive difference coinciding with LCB introduces a problem to a static crustal model if the LCB is assumed part of the crust. This is because the high density body is included in the model but not in the isostatic calculations. From isostatic considerations a high-density body of LCB in the crust should suppress the Moho. It is therefore more feasible to assign the LCB in these scenarios to the mantle than the crust. Kimbell *et al.* (2004) compared Moho depth based on isostatic consideration with Moho depth observed in seismic data and noted discrepancies where estimated Moho was both shallower and deeper than observed Moho. Amongst several possibilities to explain the isostatic Moho being deeper than the seismic Moho, Kimbell *et al.* (2004) favoured a negative upper mantle density anomaly. This means that the modelled high-density LCB is not a part of the crust but a low-density mantle and the model Moho is defined at the top of the LCB, which results in an even larger positive Moho difference.

Process orientated approach

A profile from the Møre margin demonstrates the Moho configuration and the discrepancy between model Moho and isostatic Moho (Fig. 8). Accepting that LCB coinciding with a positive Moho difference represents a low-density mantle, the Moho is defined at the top of the LCB in the centre of the basin but at the base of the LCB at the flanks of the basin. In order to reconstruct this Moho configuration with isostatic and flexural considerations a process orientated approach (Watts 2001) is needed. This approach allows us to take into account different crustal strength regimes during rifting and subsequent sedimentation. We used elastic thickness values $T_e = 5$ km during rifting and $T_e = 25$ km during sedimentation to reconstruct the Moho geometry below the basin. The high Moho relief is preserved from the rifting period when the crust was weak. During sedimentation the crust is more rigid and therefore distributes the load caused by the sedimentary fill over larger areas of the crust. The resulting Moho configuration resembles the model Moho configuration in a very shallow Moho in the centre of the basin. This approach demonstrates how the positive difference between the model Moho and the isostatic Moho can exist and still be isostatically reasonable.

From the above, it possible to exclude that the LCB coinciding with positive Moho difference is magmatic underplate or highly

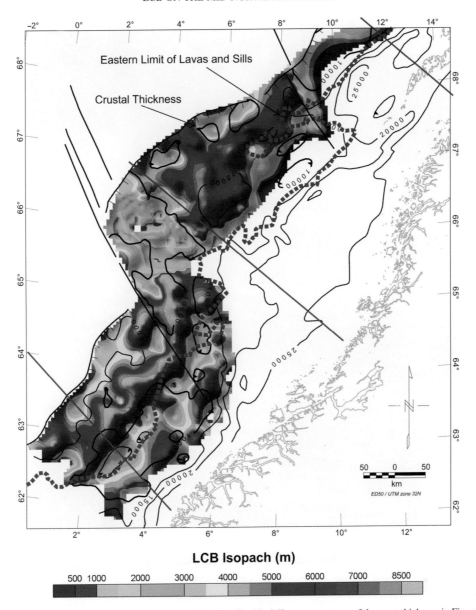

Fig. 4. LCB distribution compared to magmatic rocks and crustal thickness. The black lines are contours of the crust thickness in Figure 3 and are drawn with 5000 m interval. The red dotted line presents the eastern limit of lavas and sills on the margin. Comparison of LCB distribution and volcanic distribution reveals that all the LCB on the Vøring margin is overlain by volcanics and considerable part on the Lofoten margin. On the Møre margin the LCB is overlain by volcanics apart from the landward ridge that coincides with the very thin crust. Comparing the crustal thickness to the LCB distribution shows that in the Møre Basin the thinnest crust spatially coincides with the most landward ridge of the LCB. The landward edge of the LCB in the Vøring Basin coincides with underlying thin crust. On the Lofoten margin the whole LCB coincides with thin crust. Main lineaments are the same as in Figure 1. The red lines indicate the location of profiles in Figure 8.

intruded lower crust. An underplated or intruded lower crust would be related to the Early Cenozoic continental breakup (e.g. White et al. 1987; Eldholm & Grue 1994), which is much later (in the order of 80 Ma) than the main rifting period on the margin in the Late Jurassic–Early Cretaceous (Lundin & Doré 1997) and likewise after most of the sediment infill had taken place. If the lower crust is underplated or greatly intruded below the axial part of the basins the crust would reflect a more local isostatic architecture because high heat flows and consequently weak crust are the effects of magmatic underplating, which would be in disagreement with a strong crust ($T_e = 25$) during the whole sedimentation period. The spatial coincidence of very thin crust, positive Moho difference and LCB further point towards a mantle origin of the LCB. From this it is more feasible to assign the origin of the LCB to unroofed mantle rocks that have undergone serpentinization. Such settings are prone to develop décollements at the crust–serpentinized mantle interface (e.g. Reston 2009), which in turn can lead to core complex-like structural geometries. A modelled Moho deeper than isostatic Moho can on the other hand be explained by isostatic readjustments of weak crust caused by igneous underplate or intrusions in the lower crust.

Regional considerations

Geological history predating separation of the mid-Norwegian margin from that of east Greenland included two events that are particularly important with respect to the discussion of LCB origin. The first event was the Late Silurian–Early Devonian Caledonian Orogeny, which was succeeded by Devonian–Early Carboniferous extension and unroofing (e.g. Osmundsen et al. 2006). High-grade metamorphic rocks formed during this orogeny are well preserved in outcrop on both conjugate margins and it can readily be concluded that the deeper crust of the original

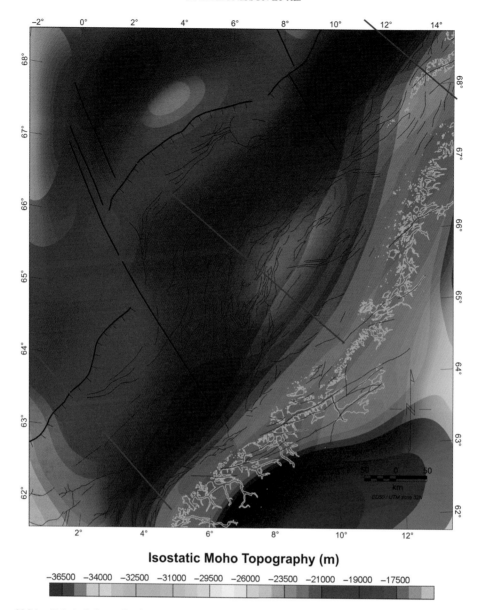

Fig. 5. Isostatic flexural Moho. Relatively low value for the strength of the lithosphere ($T_e = 10$ km) was applied to calculate the isostatic flexural Moho. The reference depth was chosen as 30 km and the density contrast at the isostatic flexural base with 400 kg m^{-3}, which correlates to a lower crust density of 2900 kg m^{-3} and a mantle density of 3300 kg m^{-3}. The structural features are the same as in Figure 1. The red lines indicate the location of profiles in Figure 8.

Caledonian root also consisted of high-grade rocks. The protoliths of this orogenic root probably varied considerably. Even so, it appears very likely that remnants from the orogenic root likely would have achieved P-wave velocities in the 7–8 km s^{-1} range (cf. Hürich et al. 2001).

The second important tectonic event was the Late Jurassic–Early Cretaceous rifting, which established the precursor to our current NE Atlantic margins (Doré et al. 1999). Following the Mid Jurassic opening of the Central Atlantic, Early Cretaceous rifting followed by plate separation that propagated northward from Iberia–Newfoundland into the Bay of Biscay, Rockall Trough and Labrador Sea (e.g. Johnston et al. 2001). Severe Early Cretaceous rifting also extended from the Rockall Trough via the Faroe–Shetland Basin, Møre Basin, Vøring Basin, Lofoten–NE Greenland margins into the Tromsø and Bjørnøya Basins of the SW Barents margin (e.g. Shannon 1991; Faleide et al. 1993; Lundin & Doré 1997). Of the mentioned areas the Iberia and Newfoundland margins have been sampled via several academic drilling campaigns, proving the presence of exhumed and serpentinized mantle (e.g. Tucholke et al. 2007 and references therein). For quite some time it has been clear that the Labrador Sea conjugate margins are characterized by belts of exhumed mantle (Louden & Chian 1999; Srivastava & Roest 1999). The Rockall Trough has also been interpreted to be underlain by exhumed and serpentinized mantle (O'Reilly et al. 1996; Morewood et al. 2005). The highly stretched southern axial part of the Porcupine Basin, where the so-called Median Volcanic Ridge is located, is now considered to be highly thinned continental crust underlain by serpentinized mantle (Reston 2009). A review of the mentioned magma-poor margins is provided by Reston (2009). Our contribution in this paper illustrates that also the Møre Basin is probably underlain by serpentinized mantle, and we argue that the Vøring basin is probably also underlain by serpentinized mantle, at least in part. Like the Møre Basin, the Vøring Basin is characterized by a thick Cretaceous sedimentary fill, and the basin lacks significant Cenozoic extension except in the outermost part (e.g. Lundin & Doré 1997; Roberts et al. 1997).

While mantle exhumation in the Porcupine Basin probably relates to Late Jurassic rifting, the other mentioned areas were thinned during Cretaceous time. Notably, the plate separation rates were slow during the Cretaceous, consistent with the development of the mentioned magma-poor margins (Bown & White

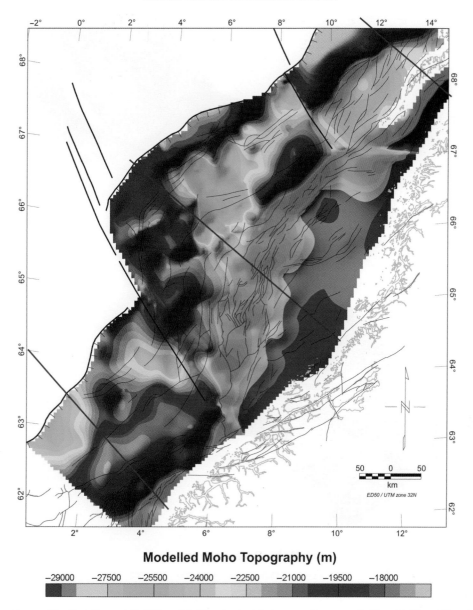

Fig. 6. Moho configuration resulting from the compiled 3D models. The structural features are the same as in Figure 1. The red lines indicate the location of profiles in Figure 8.

1995). During the Early Eocene opening of the magma-rich NE Atlantic, however, the plate separation rates increased dramatically for a short period. Commonly, this increase in plate separation rate and the magma-rich opening of the NE Atlantic is associated with the arrival of a mantle plume (e.g. White *et al.* 1987). High P-wave velocities of a LCB along the continent–ocean transition has been interpreted to represent magmatic intrusion in the lower crust or underplating, stemming from rapid decompressional melting of anomalously hot mantle related to the mantle plume (e.g. White & McKenzie 1989; Eldholm & Grue 1994). Numerous OBS studies by Mjelde and co-workers along the mid-Norwegian margin (see Mjelde *et al.* 2009*a* and references therein) have outlined an extensive LCB under the mid-Norwegian margin, and this LCB has generally been interpreted to represent Early Cenozoic magmatic underplating.

Ascribing the LCB along the Edoras–Hatton margin and the conjugate SE Greenland margin to Early Cenozoic magmatic intrusion or underplating (White *et al.* 1987; White & Smith 1989) appears reasonable. In this area breakup of Pangaea occurred within the Laurentian plate and the presence of Caledonian high-grade metamorphic rocks is unlikely. As mentioned, the path of Early Cretaceous rifting and related mantle serpentinization went via the Rockall Trough into the mid-Norwegian margin. Lack of evidence for major Cretaceous extension along the SE Greenland/Edoras–Hatton Bank part of the NE Atlantic also makes mantle serpentinization an unlikely explanation for the LCB here. However, the situation is quite different along the mid-Norwegian/East Greenland margins. Breakup of this part of the NE Atlantic certainly reopened the Caledonian Orogen, and both the mid-Norwegian Basins and Thetis Basin off NE Greenland were severely thinned during the Early Cretaceous rift event. Thus, the LCB beneath the mid-Norwegian margins in all likelihood relates to at least three different processes: (1) magmatic underplating or lower-crustal intrusion during Early Eocene breakup; (2) high-grade metamorphism during the Caledonian orogeny; and (3) mantle serpentinization during Late Jurassic–Early Cretaceous rifting. Of these processes, it appears that Early Cretaceous mantle serpentinization may be the dominant process. Recent work along the Hatton–Edoras margin reveal that the LCB there is confined to a relatively narrow body, *c.* 40 km in width (White *et al.* 2008; White & Smith 2009). We have no problem accepting that this LCB relates to Early Cenozoic intrusions of the lower crust

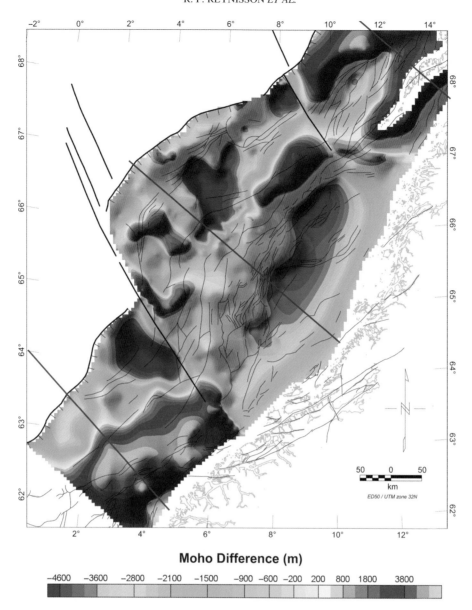

Fig. 7. Moho difference. Positive difference reflects scenarios where the model Moho is shallower than the isostatic Moho and negative difference indicates locations where the model Moho is deeper. The difference map highlights substantial differences between the isostatic and model Moho. In the southern Vøring, the isostatic and model Moho are at similar depths, while on the central and northern Vøring margin the isostatic Moho is more than 4 km shallower than the model Moho. The same observation can be made on the outer Møre margin. On the inner Møre margin and the Lofoten margin, the isostatic Moho is deeper than the model Moho. The structural features are the same as in Figure 1. The red lines indicate the location of profiles in Figure 8.

as proposed. However, the much wider LCB, observed on the mid-Norwegian margin coincides closely with the deep Cretaceous basins and we propose a genetic relationship between the severe Cretaceous extension and the LCB. Notably, V_p/V_s ratios along regional transects across the mid-Norwegian margin reveal ratios in the 1.8–1.9 range for large areas. Such V_p/V_s ratios are consistent with serpentinized mantle, but also with gabbro (Miller & Christensen 1997; Escartin et al. 2001). The weakness in interpreting this part of the mid-Norwegian LCB as underplated material (Mjelde et al. 2009a) lies foremost in the absence of Cenozoic extension.

Discussion and conclusions

The origin of the LCB is still under discussion and different interpretations exist for the body and the strong deep crustal reflectors, which are often associated with it (e.g. Gernigon et al. 2003). The early proposal that the anomalously high velocity may reflect the concentration of MgO within magmatically underplated or intruded material, which in turn was suggested to reflect mantle temperature (White & McKenzie 1989), has subsequently been challenged (Eldholm et al. 1995; Gernigon et al. 2004). Nevertheless, it is still common for workers (e.g. Mjelde et al. 2005) to propose that the LCB represents magmatically underplated material, and that the density and velocity distribution may signify differences in the magma composition, in turn possibly reflecting asthenospheric temperatures or compositional inhomogeneities in the asthenospheric source. OBS studies indicate large variations in the thickness and velocity within the LCB, and these have been proposed to relate to the distribution of mantle melts via feeder dykes (Mjelde et al. 2002). The velocities of this layer vary between 7.0 and 7.7 km s^{-1}, and have been reinterpreted as functions of differences in the magma composition due to inhomogeneities in the asthenospheric source (Mjelde et al. 2005) or as a function of mafic differentiation inside the LCB (Gernigon et al. 2005). White et al. (2008) suggest that variations in seismic velocity in the LCB may reflect variable percentages of intruded igneous rock into pre-existing continental crust.

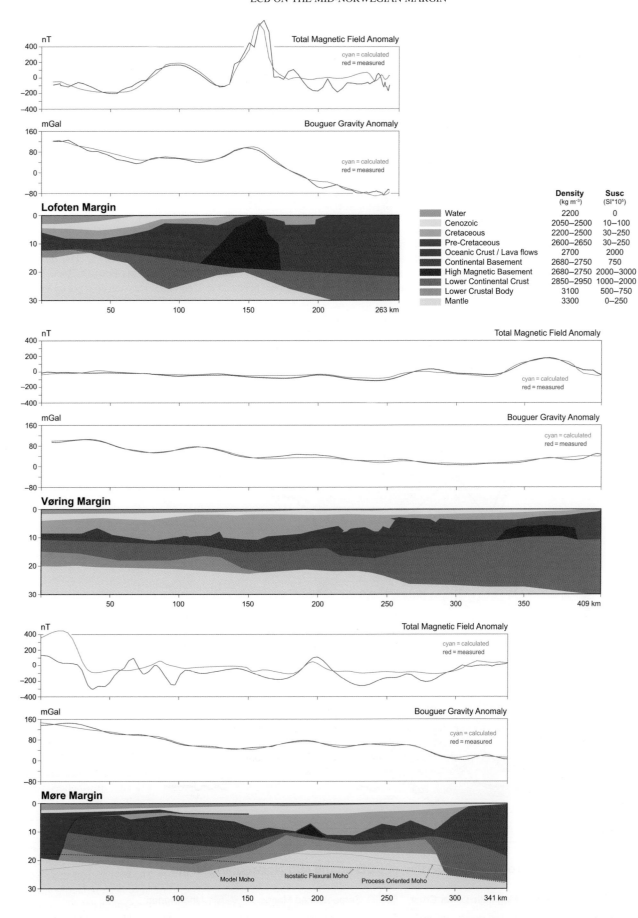

Fig. 8. Representative profiles of the mid-Norwegian margin segments. Three profiles representative for each margin segment illustrate the main geological bodies of the mid-Norwegian margin. The properties used in the potential field models are listed in the legend. See the Møre profile for comparison of different Moho types. The location of the profiles is shown with thick red lines in Figures 2–7.

Gernigon et al. (2003, 2004) discussed a range of possible origins for the LCB, where ultra high-pressure metamorphic rocks were the preferred alternative. Upper amphibolite to eclogite facies metamorphic rocks of a mafic composition have the same velocity range as those of the LCB (Hürich et al. 2001). The crust of the Horda Platform (Christiansson et al. 2000) and the Møre coastal area (Ólafsson et al. 1992) are interpreted to be underlain by eclogite. Raum et al. (2006) and Mjelde et al. (2009b) interpreted the lower crust in the southwestern corner of the Vøring Basin as in situ eclogite based on anomalously high P-wave velocities (8.4 km s^{-1}). The apparent correlation between the boundaries of the LCB and the interpreted offshore extensions of the post-Caledonian detachments support this interpretation (Ebbing et al. 2006), but it is unlikely that the entire region has the composition of eclogite. After all, only parts of the western gneiss region, onshore Norway, consist of these rocks and the bulk composition is a granitic or granodioritic gneiss. Depending on the protolith, metamorphism to upper amphibolite–eclogite facies can result in the same velocity range (c. 7–8 km s^{-1}) as the rocks that traditionally are interpreted as magmatically underplated material (Hürich et al. 2001). Thus, based on P-wave velocity alone it is not possible to distinguish between these alternatives and magmatic underplating. The V_p/V_s ratio should be distinctive (Eccles et al., 2009; White et al. 2010) but V_p/V_s ratios of LCB on the mid-Norwegian margin have been reported between 1.8 and 1.9 (Mjelde et al. 2009a, b). These V_p/V_s ratios are compatible with mafic rocks of both igneous and metamorphic origin. Such V_p/V_s ratios are also consistent with serpentinized mantle (Miller & Christensen 1997; Escartin et al. 2001).

In order to address the possible genetic relationship between the LCB and interpretations of magmatic underplating, it is important to consider the tectonic setting of the layer. For instance, the LCB extends far inboard of the eastern limit of significant Cenozoic extension, while it apparently stops abruptly against the Bivrost Lineament. The abrupt termination of the LCB against the Bivrost Lineament is difficult to explain since it is clear that the passive margin continues north to the Senja Fracture Zone. Seen from a regional perspective, Cenozoic extension does not terminate

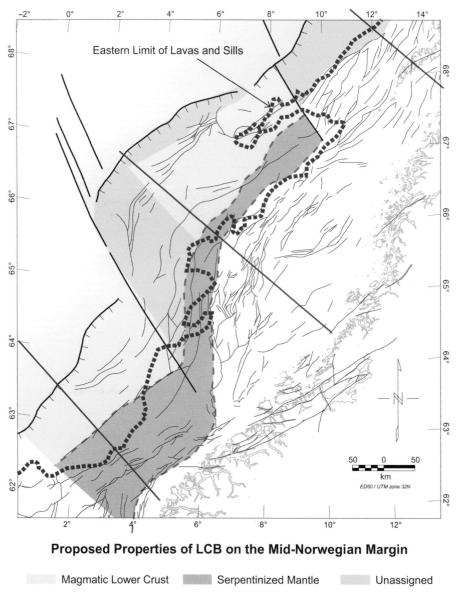

Fig. 9. Proposed properties of LCB on the mid-Norwegian margin. The main conclusions are summarized in this figure which shows the distribution of LCB on the margin and its varying properties. The yellow colour indicates LCB most likely of magmatic intrusive or underplate origin. The violet colour defines the distribution of LCB that is presumably serpentinized mantle. LCB that has indecisive origin is shown in orange. Here, the origin could be any of the proposed origins: (1) high-grade metamorphic, (2) mantle serpentinization or (3) magmatic underplating. The red dotted line presents the eastern limit of lavas and sills on the margin. The structural features are the same as in Figure 1. The red lines indicate the location of profiles in Figure 8.

against the Bivrost Lineament. Thus, it is difficult to attribute the lack of a LCB body north of the Bivrost Lineament to the lack of Cenozoic extension since it is obvious that the Lofoten margin segment also went to breakup.

Some authors (Eldholm & Grue 1994) have proposed that magmatic underplating will preferentially be located at pre-existing Moho relief locations. This concept could possibly explain how Early Cenozoic underplated material could be preferentially located under the Mesozoic Møre and Vøring basins. However, the explanation appears unlikely; the Rockall Trough is an example of a highly thinned basin that ought to have been ideally situated to preferentially receive Early Cenozoic magmatic underplating – yet it did not. It appears far easier to propose a mechanism that links the LCB in the central Møre and Vøring basins in time to the Mesozoic crustal thinning, that is, suggesting a serpentinized mantle origin. From isostatic considerations it is also unlikely, as it is not reasonable to sustain a Moho relief with material added to the base of a rifted crust without assuming unrealistically rigid crust.

On the mid-Norwegian margin, the presence of the LCB is linked to the differences between the isostatic and seismic Moho. The occurrence of LCB is either a response of the subsidence and rift history of the margin or the LCB has influenced the rift and subsidence history of the margin. Our proposal that the mid-Norwegian margin LCB can be related to at least three different geological events (Fig. 9) and processes influences several issues. The magmatic rock volume of the North Atlantic large igneous province may be significantly less than previously thought, considering that 60–80% of the volume of this LIP has been estimated to reside in the LCB (White et al. 1987; Eldholm & Grue 1994; White et al. 2008). An origin other than magmatic underplating for the LCB will change the heat flow history of the margin significantly (to a cooler scenario). The concept of ponding of magmatic underplate beneath pre-existing Moho relief is questioned.

In order to further enhance our understanding of the LCB observed on passive margins worldwide, it is important not only to integrate different geophysical methods but also to integrate different geological scale studies. The regional Moho configuration acquired by isostatic and modelling studies compared with seismically obtained Moho can shed light on the properties of LCB on all passive margins and improve understanding of the evolution of the margins.

The Geological Survey of Norway and TGS-NOPEC provided the gravity, magnetic and bathymetry data. The 3D modelling was performed using the software package IGMAS. This study is part of a PhD project titled 'Subbasalt exploration using integrated gravimetric-, magnetometric- and seismic- (velocity) models, with isostatic considerations' financed by Shell. Odleiv Olesen is thanked for constructive feedback that led to modifications of the manuscript. We are also grateful to Robert S. White and Tony Doré, who provided helpful reviews that improved the manuscript considerably.

References

Blystad, P., Brekke, H., Færseth, R. B., Larsen, B. T., Skogseid, J. & Tørudbakken, B. 1995. *Structural Elements of the Norwegian Continental Shelf. Part II: The Norwegian Sea Region*. Norwegian Petroleum Directorate, Bulletins, **8**.

Bown, J. W. & White, R. S. 1995. Effect of finite extension rate on melt generation at rifted continental margins. *Journal of Geophysical Research*, **100**, 18011–18029.

Braitenberg, C., Ebbing, J. & Götze, H.-J. 2002. Inverse modelling of elastic thickness by convolution method – the eastern Alps as a case example. *Earth and Planetary Science Letters*, **202**, 387–404.

Braitenberg, C., Wienecke, S. & Wang, Y. 2006. Basement structures from satellite-derived gravity field: South China Sea ridge. *Journal of Geophysical Research*, **111**, B05407; doi: 10.1029/2005JB003938.

Brekke, H. 2000. The tectonic evolution of the Norwegian Sea Continental Margin with emphasis on the Vøring and Møre Basin. *In*: Nøttvedt, A. (ed.) *Dynamics of the Norwegian Margin*. Geological Society, London, Special Publications, **167**, 327–378.

Christiansson, P., Faleide, J. I. & Berge, A. M. 2000. Crustal structure in the northern North Sea; an integrated geophysical study. *In*: Nøttvedt, A. (ed.) *Dynamics of the Norwegian Margin*. Geological Society, London, Special Publications, **167**, 15–40.

Doré, A. G., Lundin, E. R., Jensen, L. N., Birkeland, Ø., Eliassen, P. E. & Fichler, C. 1999. Principal tectonic events in the evolution of the northwest European Atlantic margin. *In*: Fleet, A. J. & Boldy, S. A. R. (eds) *Petroleum Geology of Northwest Europe: Proceedings of the 5th Conference*. Geological Society, London, 41–61; doi: 10.1144/0050041.

Ebbing, J., Braitenberg, C., Bjørnseth, H. M., Fichler, C. & Skilbrei, J. R. 2006. *Basement Characterisation by Regional Isostatic Methods in the Barents Sea*. Extended Abstracts for EAGE Meeting, St Petersburg.

Eccles, J. D., White, R. S. & Christie, P. A. F. 2009. Identification and inversion of converted shear waves: case studies from the European North Atlantic continental margins. *Geophysical Journal International*, **179**, 381–400.

Eldholm, O. & Grue, K. 1994. North-Atlantic volcanic margins: dimensions and production-rates. *Journal of Geophysical Research – Solid Earth*, **99**, 2955–2968.

Eldholm, O., Skogseid, J., Planke, S. & Gladczenko, T. P. 1995. Volcanic margin concepts. *In*: Banda, E. (ed.) *Rifted Ocean–Continent Boundaries*. NATO ASI, **5**. Kluwer Academic, Norwell, MA, 1–16.

Escartín, J., Hirth, G. & Evans, B. 2001. Strength of slightly serpentinized peridotites: implications for the tectonics of oceanic lithosphere. *Geology*, **29**, 1023–1026.

Faleide, J. I., Vågnes, E. & Gudlaugsen, S. T. 1993. Late Mesozoic-Cenozoic evolution of the south-western Barents Sea in a regional rift-shear setting. *Marine and Petroleum Geology*, **10**, 186–214.

Gernigon, L., Ringenbach, J. C., Planke, S., Le Gall, B. & Jonquet-Kolstø, H. 2003. Extension, crustal structure and magmatism at the outer Voring Basin, Norwegian margin. *Journal of the Geological Society*, **160**, 197–208.

Gernigon, L., Ringenbach, J. C., Planke, S. & Le Gall, B. 2004. Deep structures and breakup along volcanic rifted margins: insights from integrated studies along the outer Voring Basin (Norway). *Marine and Petroleum Geology*, **21**, 363–372.

Gernigon, L., Ringenbach, J.-C., Planke, S. & Le Gall, B. 2005. *Tectonic and Deep Crustal Structures along the Norwegian Volcanic Margin: Implications for the 'Mantle Plume or Not?' Debate*. World Wide Web Address: http://www.mantleplumes.org/VM_Norway.html.

Hürich, C. A., Deemer, S. J., Indares, A. & Salisbury, M. 2001. Compositional and metamorphic controls on velocity and reflectivity in the continental crust: an example from the Grenville Province of eastern Quebec. *Journal of Geophysical Research*, **106**, 665–682.

Johnston, S., Doré, A. G. & Spencer, A. M. 2001. The Mesozoic evolution of the southern North Atlantic region and its relationship to basin development in the south Porcupine Basin, offshore Ireland. *In*: Shannon, P. M., Haughton, P. D. W. & Corcoran, F. V. (eds) *The Petroleum Exploration of Ireland's Offshore Basins*. Geological Society, London, Special Publications, **188**, 237–263.

Kimbell, G. S., Gatliff, R. W., Ritchie, J. D., Walker, A. S. D. & Williamson, J. P. 2004. Regional three-dimensional gravity modelling of the NE Atlantic margin. *Basin Research*, **16**, 259–278.

Louden, K. E. & Chian, D. 1999. The deep structure of non-volcanic rifted continental margin. *Philosophical Transactions of the Royal Society, London*, **357**, 767–804.

Lundin, E. & Doré, A. G. 1997. A tectonic model for the Norwegian passive margin with implications for the NE Atlantic: Early Cretaceous to breakup. *Journal of the Geological Society*, **154**, 545–550.

Miller, D. J. & Christensen, N. I. 1997. Seismic velocities of lower crustal and upper mantle rocks from the slow-spreading Mid-Atlantic Ridge, south of the Kane Fracture Transform Zone (MARK). *In*: Karson, J. A., Cannat, M., Miller, D. J. & Elthon, D. (eds) *Proceedings of ODP, Sciences Results*, **153**. Ocean Drilling Program, College Station, TX, 437–454.

Mjelde, R., Digranes, P., van Schaack, M., Shimamura, H., Shiobara, H., Kodaira, S. & Næss, O. 2001. Crustal structure of the outer Vøring Plateau, offshore Norway, from ocean bottom seismic and gravity data. *Journal of Geophysical Research*, **106**, 6769–6791.

Mjelde, R., Kasahara, J. *et al.* 2002. Lower crustal seismic velocity-anomalies; magmatic underplating or serpentinized peridotite? Evidence from the Vøring Margin, NE Atlantic. *Marine Geophysical Research*, **23**, 169–183.

Mjelde, R., Raum, T., Digranes, P., Shimamura, H. & Kodaira, S. 2003. Vp/Vs-ratio along the Vøring Margin, NE Atlantic, derived from OBS-data; implications on lithology and stress-field. *Tectonophysics*, **369**, 175–197.

Mjelde, R., Raum, T., Breivik, A., Shimamura, H., Murai, Y., Takanami, T. & Faleide, J. I. 2005. Crustal structure of the Vøring Margin, NE Atlantic: a review of geological implications based on recent OBS data. *In*: Doré, A. G. & Vining, B. A. (eds) *Petroleum Geology: North-West Europe and Global Perspectives: Proceedings of the 6th Petroleum Geology Conference*. Geological Society, London, 804–813; doi: 10.1144/0060804.

Mjelde, R., Faleide, J. I., Breivik, A. J. & Raum, T. 2009a. Lower crustal composition and crustal lineaments on the Vøring Margin, NE Atlantic: a review. *Tectonophysics*, **472**, 183–193.

Mjelde, R., Raum, T., Kandilarov, A., Murai, Y. & Takanami, T. 2009b. Crustal structure and evolution of the outer Møre Margin, NE Atlantic. *Tectonophysics*, **468**, 224–243.

Morewood, N. C., Mackenzie, G. D., Shannon, P. M., O'Reilly, B. M., Readman, P. W. & Makris, J. 2005. The crustal structure and regional development of the Irish Atlantic margin region. *In*: Doré, A. G. & Vining, B. A. (eds) *Petroleum Geology: North-West Europe and Global Perspectives: Proceedings of the 6th Petroleum Geology Conference*. Geological Society, London, 1023–1033; doi: 10.1144/0061023.

Ólafsson, I., Sundvor, E., Eldholm, O. & Grue, K. 1992. Møre margin – crustal structure from analysis of expanded spread profiles. *Marine Geophysical Researches*, **14**, 137–162.

Olesen, O., Lundin, E. *et al.* 2002. Bridging the gap between the onshore and offshore geology in Nordland, northern Norway. *Norwegian Journal of Geology*, **82**, 243–262.

O'Reilly, B. M., Hauser, F., Jacob, A. W. B. & Shannon, P. M. 1996. The lithosphere below the Rockall Trough: wide-angle seismic evidence for extensive serpentinisation. *Tectonophysics*, **255**, 1–23.

Osmundsen, P. T. & Ebbing, J. 2008. Styles of extension offshore mid-Norway and implications for mechanisms of crustal thinning at passive margins. *Tectonics*, **27**, TC6016; doi: 10.1029/2007TC002242.

Osmundsen, P. T., Eide, E. *et al.* 2006. Kinematics of the Høybakken detachment zone and the Møre-Trøndelag Fault Complex, central Norway. *Journal of the Geological Society, London*, **163**, 303–318.

Planke, S., Rasmussen, T. E., Rey, S. S. & Myklebust, R. 2005. Seismic characteristics and distribution of volcanic intrusions and hydrothermal vent complexes in the Vøring and Møre basins. *In*: Doré, A. G. & Vining, B. A. (eds) *Petroleum Geology: North-West Europe and Global Perspectives: Proceedings of the 6th Petroleum Geology Conference*. Geological Society, London, 1–12; doi: 10.1144/0060001.

Raum, T., Mjelde, R. *et al.* 2002. Crustal structure of the southern part of the Voring Basin, mid-Norway margin, from wide-angle seismic and gravity data. *Tectonophysics*, **355**, 99–126.

Raum, T., Mjelde, R. *et al.* 2006. Crustal structure and evolution of the southern Vøring Basin and Vøring Transform Margin, NE Atlantic. *Tectonophysics*, **415**, 167–202.

Reston, T. J. 2009. The structure, evolution and symmetry of the magma-poor rifted margins of the North and Central Atlantic: a synthesis. *Tectonophysics*, **468**, 6–27.

Roberts, A. M., Lundin, E. R. & Kusznir, N. J. 1997. Subsidence of the Vøring Basin and influence of the Atlantic continental margin. *Journal of the Geological Society, London*, **154**, 551–557.

Shannon, P. M. 1991. The development of Irish offshore sedimentary basins. *Journal of the Geological Society, London*, **148**, 181–189.

Skogseid, J., Pedersen, T., Eldholm, O. & Larsen, B. T. 1992. Tectonism and magmatism during NE Atlantic continental break-up: the Vøring Margin. *In*: Storey, B. C., Alabaster, T. & Pankhurst, R. J. (eds) *Magmatism and the Causes of Continental Break-up*. Geological Society, London, Special Publications, **68**, 305–320.

Srivastava, S. P. & Roest, W. R. 1999. Extent of oceanic crust in the Labrador Sea. *Marine and Petroleum Geology*, **16**, 65–84.

Tsikalas, F., Faleide, J. I. & Kusznir, N. J. 2008. Along-strike variations in rifted margin crustal architecture and lithosphere thinning between northern Vøring and Lofoten margin segments off mid-Norway. *Tectonophysics*, **458**, 68–81.

Tucholke, B. E., Sawyer, D. S. & Sibuet, J.-C. 2007. Breakup of the Newfoundland–Iberia rift. *In*: Karner, G. D., Manatchal, G. & Pinheiro, L. M. (eds) *Imaging, Mapping and Modelling Continental Lithosphere Extension and Breakup*. Geological Society, London, Special Publications, **282**, 9–46.

van Wijk, J. W., van der Meer, R. & Cloetingh, S. A. P. L. 2004. Crustal thickening in an extensional regime: application to the mid-Norwegian Vøring margin. *Tectonophysics*, **387**, 217–228.

Watts, A. B. 2001. *Isostasy and Flexure of the Lithosphere*. Cambridge University Press, Cambridge.

White, R. S. & McKenzie, D. 1989. Magmatism at rift zones: the generation of volcanic continental margins and flood basalts. *Journal of Geophysical Research – Solid Earth and Planets*, **94**, 7685–7729.

White, R. S. & Smith, L. K. 2009. Crustal structure of the Hatton and the conjugate east Greenland rifted volcanic continental margins, NE Atlantic. *Journal of Geophysical Research*, **114**, B02305; doi: 10.1029/2008JB005856.

White, R. S., Spence, G. D., Fowler, S. R., McKenzie, D. P., Westbrook, G. K. & Bowen, N. 1987. Magmatism at rifted continental margins. *Nature*, **330**, 439–444.

White, R. S., Smith, L. K., Roberts, A. W., Christie, P. A. F., Kusznir, N. J. & iSIMM Team. 2008. Lower-crustal intrusion on the North Atlantic continental margin. *Nature*, **452**, 460–464; plus supplementary information at www.nature.com,doi:10.1038/nature06687.

White, R. S., Eccles, J. D. & Roberts, A. W. 2010. Constraints on volcanism, igneous intrusion and stretching on the Rockall–Faroe continental margin. *In*: Vining, B. A. & Pickering, S. C. (eds) *Petroleum Geology: From Mature Basins to New Frontiers – Proceedings of the 7th Petroleum Geology Conference*, Geological Society, London, 831–842; doi: 10.1144/0070831.

Wienecke, S., Braitenberg, C. & Götze, H. J. 2007. A new analytical solution estimating the flexural rigidity in the Central Andes. *Geophysical Journal International*, **169**, 789–794.

Ziegler, P. A. 1988. *Evolution of the Arctic–North Atlantic and the Western Tethys*. American Association of Petroleum Geologists, Tulsa, OK, Memoirs, **43**.

The breakup of the South Atlantic Ocean: formation of failed spreading axes and blocks of thinned continental crust in the Santos Basin, Brazil and its consequences for petroleum system development

I. C. SCOTCHMAN,[1] G. GILCHRIST,[2] N. J. KUSZNIR,[3] A. M. ROBERTS[4] and R. FLETCHER[1,3]

[1]*Statoil (UK) Ltd, 1 Kingdom Street, London W2 6BD, UK (e-mail: isco@statoil.com)*
[2]*Consultant, Statoil do Brasil, Rio de Janeiro, Brazil*
[3]*Department of Earth and Ocean Sciences, University of Liverpool, Liverpool L69 3BX, UK*
[4]*Badley Geoscience Ltd, North Beck House, North Beck Lane, Spilsby, Lincolnshire PE23 5NB, UK*

Abstract: The occurrence of failed breakup basins and deepwater blocks of thinned continental crust is commonplace in the rifting and breakup of continents, as part of passive margin development. This paper examines the rifting of Pangaea–Gondwanaland and subsequent breakup to form the South Atlantic Ocean, with development of a failed breakup basin and seafloor spreading axis (the deepwater Santos Basin) and an adjacent deepwater block of thinned continental crust (the Sao Paulo Plateau) using a combination of 2D flexural backstripping and gravity inversion modelling. The effects of the varying amounts of continental crustal thinning on the contrasting depositional and petroleum systems in the Santos Basin and on the São Paulo Plateau are discussed, the former having a predominant post-breakup petroleum system compared with a pre-breakup system in the latter. An analogy is also made to a potentially similar failed breakup basin/thinned continental crustal block pairing in the Faroes region in the NE Atlantic Ocean.

Keywords: Brazilian rifted margin, continental breakup, Santos Basin, Sao Paulo Plateau, Faroes, Atlantic margin, subsidence, gravity inversion

Continental lithospheric thinning and rifted margin formation is a poorly understood process, the kinematics of which can be affected by many factors, including pre-existing lithospheric heterogeneities, variations in plate kinematics and the presence of mantle features such as plumes (e.g. Dunbar & Sawyer 1989; Corti *et al.* 2003; Ziegler & Cloetingh 2004). Evidence of complex kinematics of rifted margins is seen at many margins, such as the South Atlantic Ocean off Brazil and the NE Atlantic Ocean margin close to the Faroe Islands, both being discussed in this paper.

The Santos Basin

The Santos Basin is the southernmost of the petroliferous chain of basins along the western margin of the South Atlantic Ocean in Brazil (Fig. 1). These basins, the Santos, Campos and Espirito Santo Basins, resulted from rifting of the Gondwanaland 'supercontinent' in the earliest Cretaceous with breakup and subsequent seafloor spreading. The Santos Basin is a NE–SW-trending basin that covers about 200 000 km² of the Brazilian continental margin, bounded by the Cabo Frio Arch to the north and by the Florianopolis Platform to the south, both features being related to magmatic activity associated with the westward prolongation of the ocean fracture zones (Cainelli & Mohriak 1998). The western limit of the basin is defined by the uplifted Precambrian rocks of the Serra do Mar, a coastal range reaching up to 2000 m high, while to the east the basin is flanked by the Sao Paulo Plateau (SPP). The Florianopolis Fracture Zone (FFZ) is a major transform feature which defines the southern limit of the Santos Basin and marks a major break in South Atlantic Ocean breakup history. To the south, seafloor spreading is constrained by the M4 and M0 magnetic anomalies (Mueller *et al.* 1997), dated at 125.7 and 120.6 Ma, respectively. To the north of the FFZ, the age of breakup is poorly constrained as it took place during the Cretaceous normal polarity superchron between the M0 and C34 (83.5 Ma) magnetic anomalies.

The Santos Basin and the Sao Paulo Plateau

Regional geology

Rifting and breakup of the Santos Basin and SPP areas appear to have been very complex with several apparent attempts to extend seafloor spreading north of the FFZ. This resulted in various intrusive and volcanic features located in the southwestern Santos Basin and to the east of the SPP, where the Avedis volcanic chain was interpreted by several authors as a failed spreading centre (Cobbold *et al.* 2001; Meisling *et al.* 2001; Gomes *et al.* 2002). However, recent exploration and drilling indicate these features to be probably of pre-rift origin and not the result of failed Early Cretaceous-aged breakup. They form large regional structural highs (the Tupi and Sugar Loaf structures of Gomes *et al.* 2009) which drilling indicates to have trapped extremely large volumes of hydrocarbons in overlying syn-rift and sag phase reservoirs. However, data presented in this paper suggest that an earlier attempt at breakup took place in the centre of Santos Basin, extending northeastwards into the basin centre from the area of likely oceanic crust emplaced in the southwestern Santos Basin (Meisling *et al.* 2001; Gomes *et al.* 2002). The formation of the SPP and the resultant development of the prolific pre-salt hydrocarbon province recently discovered in the area appear intimately related to this failed breakup event.

Rifting in the Santos Basin began around 140 Ma in the Neocomian (Karner & Driscoll 1999), contemporaneous with eruption of the Parana volcanics (Renne *et al.* 1996; Fig. 2). Syn-rift deposition, overlying and interfingering with late stage basalts

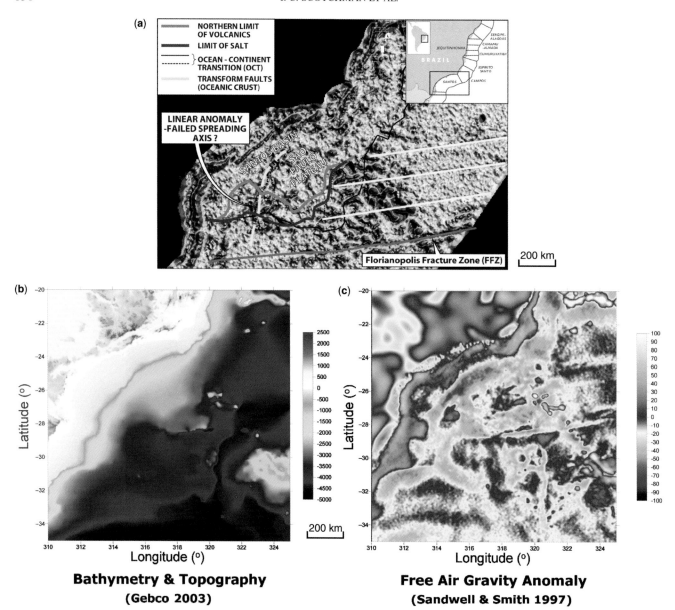

Fig. 1. (a) Bouguer gravity anomaly map (200 km high-pass filter) over the central area of Santos Basin showing the linear feature with strong negative gravity anomaly adjacent to the SPP. (b) Bathymetry (metres) and (c) free air gravity (mgal) data for the Santos and Campos and adjacent segments of the Brazilian rifted margin.

(Chang et al. 1992), comprises fluvial–lacustrine siliciclastics which infilled much of the early rift topography. Subsequent deposition, during the late rift sag phase, comprised basinal lacustrine, organic-rich shales lapping onto coquinas of the Lagoa Feia Formation, which were deposited on structural highs and the basin flanks (Pereira & Feijo 1994). The sag phase is overlain, generally unconformably (Karner & Gamboa 2007), by evaporites comprising intercalations of anhydrite and halite, reflecting development of the post-rift South Atlantic Salt Basin (Fig. 3). These evaporites were deposited in the remnant sag basin system at or at most several hundred metres below ambient sea-level (Karner & Gamboa 2007) by marine incursions from the north (Davison & Bate 2004). While Davison (2007) favours formation in a deep rift basin, evaporite formation close to ambient sea-level is indicated by several lines of evidence: the microbilitic limestones immediately underlying the salt, forming the reservoir in hydrocarbon discoveries such as Tupi, probably accumulated in less than 3 m of water (J. Lukasik, pers. comm.), the layered evaporite sequence overlying the basal halite section on the SPP was probably formed under shallow water sabkha-type conditions with periodic emergence and, finally, the large rift block bounding faults have a substantial post-salt displacement (e.g. Fig. 3, Section 2 inset). Thick evaporitic sections up to 2 km thick developed in the adjacent Santos Basin where there was more accommodation space, greatly influencing basin fill during the later evolution of the basin with extensive halokinesis producing large salt diapirs and walls within the deepwater basin.

In contrast, recent drilling on the SPP has found the structure to have thicker, well developed syn-rift and sag phase sections (Fig. 4; e.g. Gomes et al. 2009) which are capped by porous lacustrine limestones of algal, stromatolitic or thrombolitic origin which form the reservoir rocks in the recent large hydrocarbon discoveries in the area, such as Tupi (Mello et al. 2009). In contrast to the great thicknesses of evaporites seen in the Santos Basin to the west and Campos Basin to the north (Davison 2007), on the SPP a generally thinner layer of mobile salt overlies the syn-rift section; the rest of the evaporitic section of up to 2 km in thickness appears from seismic to be well bedded but is also tightly folded with diapiric structures (Fig. 3). This well bedded section above the mobile salt, previously identified as an extension of the Late Cretaceous

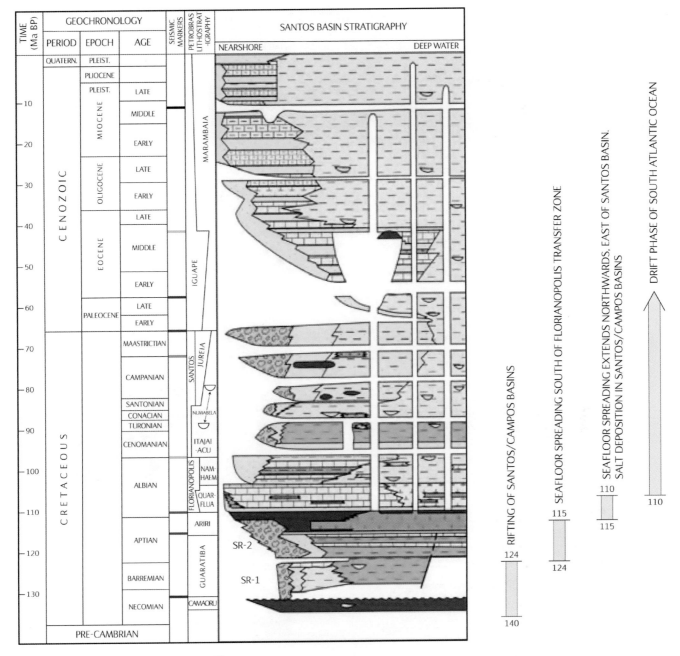

Fig. 2. Santos Basin stratigraphy and breakup history.

turbidite play north of the Santos Fault, has been found by recent drilling to consist of bedded evaporites, comprising halite and anhydrite as well as complex evaporites such as carnallite, bischofite, sylvite and tachyhydrite (Poiate et al. 2006). These complex minerals are 'end members' of the evaporitic system, indicating development of extreme conditions.

Karner & Gamboa (2007) suggest a date for evaporite deposition in the Santos Basin as 110–113 Ma, after which period major marine flooding of the basin took place, probably linked to breakup of the margin to the north adjacent to the Campos Basin. This was marked in the Santos Basin by deposition of Albian-age shallow marine carbonates, grey shales and sandstones which were followed by coarse turbiditic sandstones and shales deposited during progressive deepening of the basin, with the maximum flooding conditions being marked by dark grey-black shales of Cenomanian–Turonian age Itajai–Acu Formation. In the proximal areas, a thick conglomerate package and shallow marine sandstones were deposited during the Santonian to Maastrichtian in response to the first phase of Serra do Mar uplift (dated at 100–80 Ma by Lelarge 1993).

The age of the bedded evaporite deposition on the SPP is equivocal as no biostratigraphic data from recent wells has been published. Because of its extreme distal location, effectively isolated between the African and Brazilian margins, evaporite deposition under sabkha-type conditions may have continued up into the Albian as no equivalent shallow marine carbonate platform development is present, the evaporites being overlain by possibly Cenomanian or younger Late Cretaceous turbidites and shales. However, the age-equivalent section to the Albian carbonate platform may instead be present as shales at the base of the marine section overlying the evaporites.

By the Late Cretaceous and through the Cenozoic, deepwater turbidite and shale deposition was predominant across both the Santos Basin and the SPP, characterized by a basinward progradation of siliciclastics over platform/slope shales and marls. However, the SPP remained relatively sediment-poor with only

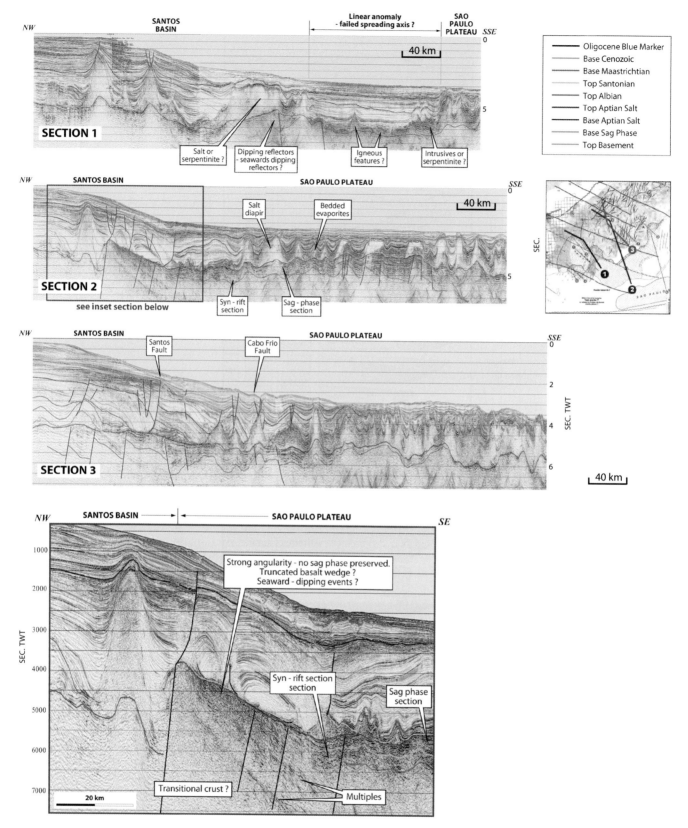

Fig. 3. Seismic cross-sections across the deepwater Santos Basin and SPP.

around 2 km drift section compared with 4–5 km or more equivalent thickness inboard in the Santos Basin.

Regional structure

Regional evaluation of the Santos Basin and SPP, using an approximate 4 × 4 km grid of 2D seismic data combined with both satellite and ship-borne gravity and magnetic data, has shown that the structure of the basin is complex with a linear negative gravity anomaly stretching from the outer southwestern part of the basin north-northeastwards into the basin centre, with a large area to the ESE comprising the SPP (Fig. 1). Regional seismic across the linear gravity anomaly in the southwestern–central Santos Basin (Fig. 3) shows a linear anomaly with volcanic features of oceanic

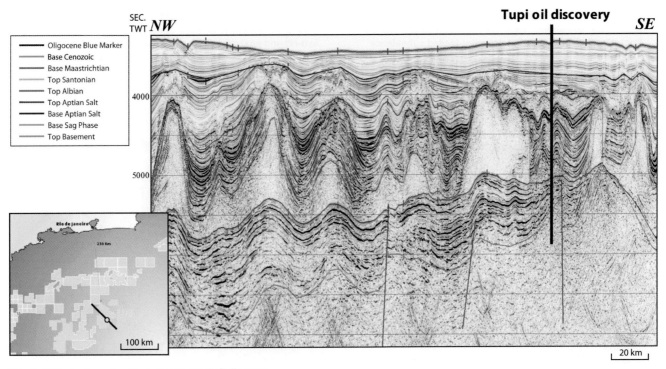

Fig. 4. SPP seismic section illustrating the Tupi oil discovery.

affinity extending from a crustal igneous feature in the southwestern part of the basin northeastwards into the basin centre where it terminates in a major fault, the Santos Fault. The feature forms the northwestern margin of the SPP, giving rise to the observation that these two features are connected. The contrasting structural and stratigraphic characteristics of the SPP and the Santos Basin to the north are illustrated by seismic line 3 of Figure 3, showing the thick, well developed syn-rift section capped by over 2 km of largely bedded evaporites in the former, while the latter shows a thin syn-rift section capped by diapiric salt buried by thick Cretaceous–Cenozoic turbidites. In contrast, the equivalent turbidite section on the SPP is only thinly developed, unlike either the Santos Basin or the other basins to the north (Campos and Espirito Santo Basins).

In order to investigate the linear anomaly within Santos Basin and its relationship to the SPP, along with the anomalous structural and sedimentary development of the latter, crustal modelling was undertaken. This used subsidence analysis from 2D backstripping, based on a regional grid of eight seismic lines, in conjunction with determination of crustal thickness from gravity inversion based on the bathymetry and free air gravity anomaly (Fig. 1) as detailed below.

Continental lithosphere thinning and crustal thickness of the Santos Basin and Sao Paulo plateau

Crustal thickness, continental lithosphere thinning and Moho depth for the Santos Basin and SPP areas of the Brazilian rifted margin were studied by subsidence analysis using flexural backstripping (Kusznir *et al.* 1995; Roberts *et al.* 1998) and gravity inversion performed in the 3D spectral domain (Parker 1972), the latter using a new method incorporating a lithosphere thermal gravity anomaly correction (Greenhalgh & Kusznir 2007; Chappell & Kusznir 2008).

Subsidence analysis

Subsidence analysis using flexural backstripping to produce water-loaded subsidence, and gravity inversion using a new method incorporating a lithosphere thermal gravity anomaly correction, were used to determine continental lithosphere thinning, Moho depth and continental crustal thickness for the Santos Basin and SPP areas of the Brazilian rifted margin. The results of the subsidence analysis using flexural backstripping are described in Scotchman *et al.* (2006) and only a brief summary is given below.

Water-loaded subsidence was determined using the 2D flexural backstripping (Roberts *et al.* 1998) of profiles across the Santos Basin and SPP. The assumption was made of the palaeobathymetric constraint of ambient sea-level for the base Aptian salt at the time of deposition, as discussed above. The water-loaded subsidence was then processed using the extensional basin formation model of McKenzie (1978) to determine continental lithosphere stretching and thinning for (i) the assumption that base salt subsidence from Aptian to present is due to both syn-rift and post-rift subsidence, and (ii) alternatively that base salt subsidence from Aptian to present is due to post-rift thermal subsidence only. An 'average' age of 120 Ma for rift development across the basin was used in the modelling, an age older than the 110–113 Ma estimates for that of the Aptian salt (Karner & Gamboa 2007), to acknowledge the existence of syn-rift deposition beneath the salt.

It is very likely that the base salt horizon within the Santos Basin experienced syn-rift tectonic subsidence, followed by continued thermal subsidence to the present day. In this likely scenario, treating the base-case salt water-loaded subsidence as only post-rift leads to an overestimate of the β stretching factors required to model the subsidence and the prediction of oceanic crust (with infinite thinning) over much of the Santos Basin (Fig. 5a). This is clearly in conflict with the seismic and gravity data as any ocean–continent transition derived from this model is too far inboard.

Conversely, the salt does not represent the whole of the syn-rift sequence and, as a consequence, treating the base salt water-loaded subsidence as representing the whole of the syn-rift and post-rift thermal subsidence underestimates continental lithosphere stretching and thinning (Fig. 5b). This assumption predicts finite (non-infinite) thinning factors for the SPP, implying that this region is underlain by thinned continental crust. A region of highly stretched and thinned continental crust is predicted in the deepwater Santos Basin to the NW, separating the SPP from the Brazilian margin.

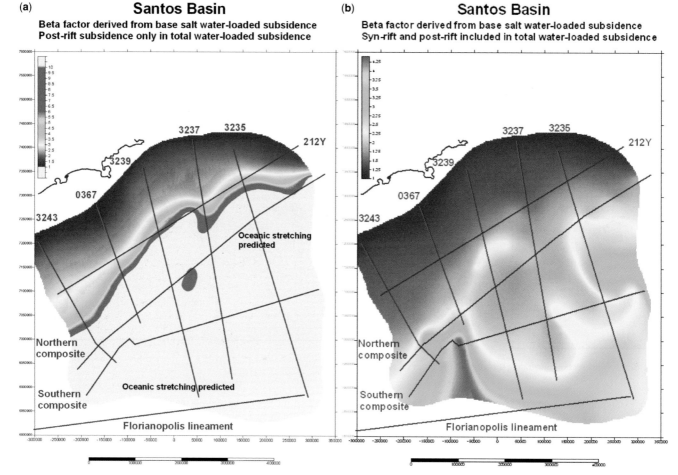

Fig. 5. Beta factor maps for the Santos Basin and SPP from 2D backstripping from Scotchman *et al.* (2006): (**a**) assuming subsidence of base salt is post-breakup only; (**b**) assuming subsidence of base salt is both syn-rift and post-breakup.

Crustal thickness determination

The input data used in the Santos Basin and SPP gravity inversion study were satellite free air gravity (Sandwell & Smith 1997), digital bathymetry (Gebco 2003), sediment thickness to base salt derived from 2D/3D seismic reflection mapping and ocean age isochrons to define unequivocal oceanic crust (Mueller *et al.* 1997). Crustal thicknesses produced by the gravity inversion applied to the Santos Basin and SPP are shown in Figure 6. Gravity inversion results are shown both omitting sediment thickness information (Fig. 6a) and including sediment thickness (Fig. 6b) to give a more accurate depth to Moho where such data exists. Crustal cross-sections with Moho depth determined using the gravity inversion method are shown in Figure 7 and indicate that crustal basement thickness in the Santos Basin decreases southwards. The low crustal thicknesses in the south of the deepwater Santos Basin, shown on cross-section 3 of Figure 7, and located to the north of the FFZ, suggest that the SW Santos Basin is underlain by oceanic crust. In contrast the SPP is underlain by crust between 12 and 16 km thick which is interpreted as thinned continental crust (Figs 6 & 7).

Both the gravity inversion results and the flexural backstripping subsidence analysis indicate that (i) the SPP is underlain by thinned continental crust and (ii) that a 'tongue' of oceanic crust extends north of the FFZ into the deepwater Santos Basin.

Discussion and conclusions

The failed breakup model for the Santos Basin

The results of the subsidence modelling from flexural backstripping indicate a zone of high subsidence in the southwestern Santos Basin, extending northeastwards into the central part of the basin, albeit with decreasing subsidence. Estimates of crustal thinning based on McKenzie (1978), backed by the regional geological reasoning, indicates that this represents both syn- and post-rift subsidence of the base salt layer, suggesting the outer part of the southwestern Santos Basin is underlain by either oceanic crust or by extremely thinned continental crust, which forms a tongue extending northeastwards into the basin centre. Seismic sections across this area confirm this finding (Fig. 3), with associated igneous features interpreted as seawards dipping reflectors observed within the basin. The results also indicate thinned continental crust to the SE of the feature beneath the SPP. The results of the gravity inversion work (Fig. 6) lead to similar conclusions with thin potentially oceanic crust in the outer part of the Santos Basin with decreasingly thinned crust northeastwards along the feature and less thinned crust beneath the SPP.

These results indicate that the feature is likely to be a failed seafloor spreading centre, representing an early attempt at breakup and initiation of seafloor spreading through the centre of the Santos Basin north of the FFZ in the early Aptian. The results suggest that extreme thinning occurred in the southern part of the feature, which probably represents incipient oceanic crust. However, the breakup and seafloor spreading event appears to have been short lived, probably due to an adjustment of plate kinematics.

Rifting and breakup history of the Santos Basin/Sao Paulo Plateau

By 140 Ma in the Neocomian, rifting followed by breakup of the southernmost South Atlantic Ocean took place south of the crustal

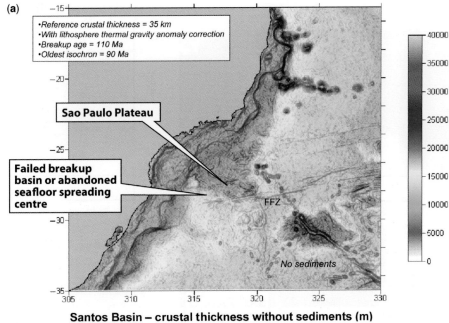

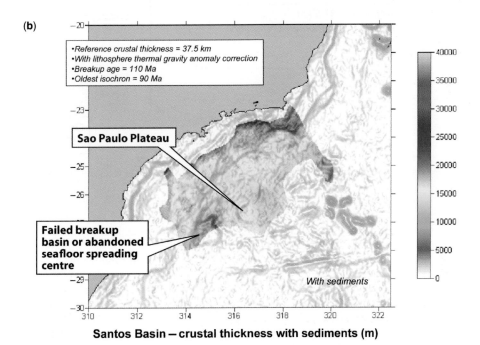

Fig. 6. (a) Crustal thickness map of the Santos Basin derived from gravity inversion assuming zero sediment thickness showing thick continental crust underlying the SPP and thinned crust beneath the linear anomaly in the SW Santos Basin. (b) Crustal basement thickness map derived from gravity inversion incorporating sediment thickness data from seismic reflection grid showing thinned continental crust underlying the SPP and oceanic crust beneath the SW Santos Basin.

lineament subsequently re-activated as the Albian-aged FFZ. North of this lineament, rifting began around 140 Ma (Fig. 8a), probably contemporaneous with the intrusion of the Ponta Grossa dyke swarm and extrusion of the Parana volcanics, associated with the Tristao da Cunha hotspot. Initiation of seafloor spreading occurred south of the FFZ by 126 Ma, while to the north sag-sequence deposition took place in the rift basin system with deposition of the lacustrine source rocks and sandstones. During this time the Tristao da Cunha hotspot appears to have moved into the rift area north of the FFZ with intrusion of igneous bodies. Associated with the hotspot, the spreading centre south of the FFZ may have 'tipped-out' into the rift basin on the northern side where it developed into incipient breakup and seafloor spreading on an ENE–SSW trend towards the Brazilian coastline around 120 Ma.

Crustal thinning appears to have affected the area with associated thick development of sag-phase sedimentation, particularly over what is now the SPP. Extreme crustal thinning took place along the line of the incipient breakup north of the FFZ, with likely emplacement of oceanic crust in its southern part, north of the FFZ, in what is now the southwestern part of the Santos Basin. However, this phase of incipient breakup and seafloor spreading appears to have failed in the early Aptian and resulted in the formation of a failed breakup basin/seafloor spreading axis in what is now the central part of the Santos Basin with an adjacent area of thinned continental crust which became the present-day SPP.

During the late Aptian, deposition of thick halite took place in the subsiding sag basins along the whole rifted margin, which had been flooded intermittently by marine water, most likely from the

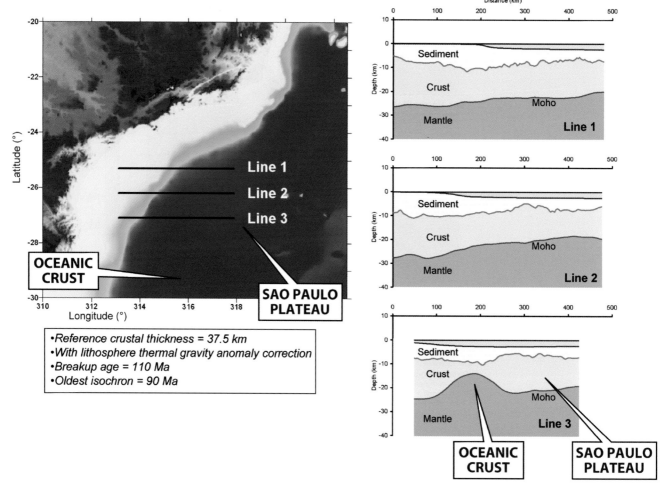

Fig. 7. Crustal cross-sections showing crustal basement thickness and Moho depth determined from gravity inversion across the Santos Basin and SPP.

north via the early rifts in the central part of the Atlantic Ocean, as the Rio Grande High appears to have prevented access by marine waters to the basin system from the south (Scotchman et al. 2008). The SPP area was at this time still on the African side of the rift system and remained relatively high, forming the eastern flank of the Santos failed breakup basin. Here the lack of large-scale regional subsidence appears to have resulted in the deposition of very shallow water algal/thrombolitic/stromatolitic lacustrine/brackish water limestones capping the syn-rift shales, forming the reservoir in recent discoveries such as Tupi. The area remained a positive feature during the subsequent deposition of evaporites with shallow, sabkha-like deposition of bedded evaporites. These largely comprise halite and anhydrite, but occasionally with complete evaporation leading to Mg- and K-salt precipitation.

Breakup finally appears to have taken place to the east of the SPP in the late Aptian–Albian, with evidence that this took place from the north (Scotchman et al. 2008), with formation of the oceanic fracture zones such as the FFZ, which as noted above appears to have re-activated an older crustal lineament. The Santos Basin to the west was flooded by shallow seas with deposition of carbonate platforms along the Brazilian flank and development of early turbiditic sandstones and shales in the deepest parts of the basin. The Albian carbonates appear to be not present or below seismic resolution on the adjacent SPP, although a lateral facies change to deep marine shales is also possible, making it difficult, without biostratigraphic data, to resolve if deposition of the marginal bedded evaporite sequence ended in the Aptian or continued into the Albian. Regional evidence perhaps favours the latter hypothesis as the Albian section in the deepwater Santos Basin flanking the northern side of the SPP comprises anoxic organic-rich black shales which form the main post-salt hydrocarbon source in the basin (Katz & Mello 2000). Marine organic shale deposition generally requires a restricted basinal setting, for example, Demaison & Moore (1980), and a narrow seaway between the coastal carbonate platform and the SPP located over the thermally subsiding failed spreading centre could easily fulfil such a role, providing additional excellent hydrocarbon source potential.

When seafloor spreading finally occurred in the area, perhaps as late as the end Albian–Cenomanian, the split between Africa and South America was completed and the Rio Grande High breached. The whole area then underwent rapid subsidence with the spread of deepwater turbidite deposition which had flooded both the Santos Basin and the SPP. Figure 8 illustrates the simplified kinematic model for the breakup of the Brazilian–African margin, showing development of the failed breakup basin (the Santos Basin) and the SPP.

Petroleum systems

The syn-rift Lagoa Feia lacustrine facies shales are the main hydrocarbon source rock in the basin system along the Brazilian Atlantic Ocean margin (Katz & Mello 2000), where they have charged post-salt turbidite reservoirs of Late Cretaceous to Cenozoic age to form a very prolific petroleum system with giant oil fields in basins such as the Campos, Espirito Santo and Santos (Guardado et al. 2000). Post-salt source rocks, particularly of Albian and Cenomanian–

BRAZILIAN MARGIN

(a) Mid to Late Aptian

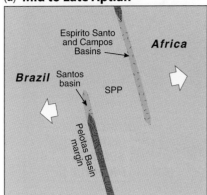

Northward propagation of sea-floor spreading into deep-water Santos Basin (SB) region. Simultaneous over-lapping southward propagation of sea-floor spreading into Espirito Santo Basin and Campos Basin (CB) deep-water regions. Two over-lapping sea-floor spreading systems separated by thinned continental crust which includes the Sao Paulo Plateau (SPP).

(b) Mid to Late Albian

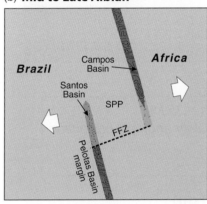

Development of Florianopolis Fracture Zone (FFZ) and "hard-linkage" plate separation between sea-floor spreading on Campos and Pelotas Basin margins. Cessation of lithospheric thinning in Santos Basin.

(c) Late Albian – Cenomanian

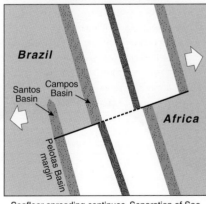

Seafloor spreading continues. Separation of Sao Paulo Plateau from African margin by sinistral motion on Florianopolis FZ.

FAROES MARGIN

(d) Late Paleocene

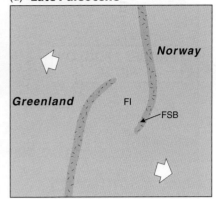

Continental extension and thinning between Norway and Greenland at 2 (or more) linked segments around Faroe Islands (FI).

(e) End Paleocene

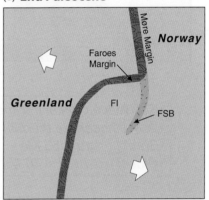

Continental lithospheric rupture. Faroes Margin acts as a transfer zone linking northern and southern segments.

(f) Early Eocene

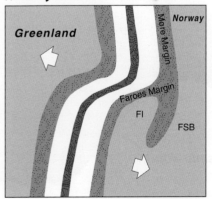

Seafloor spreading ensues. Extension and thinning in the Faroe-Shetland basin has ceased, leaving a failed breakup basin.

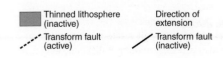

Fig. 8. Conceptual breakup models for (**a–c**) the South Atlantic in the Santos Basin and SPP and (**d–f**) the North Atlantic in the Faroes/Faroe–Shetland Basin (modified from Fletcher 2009).

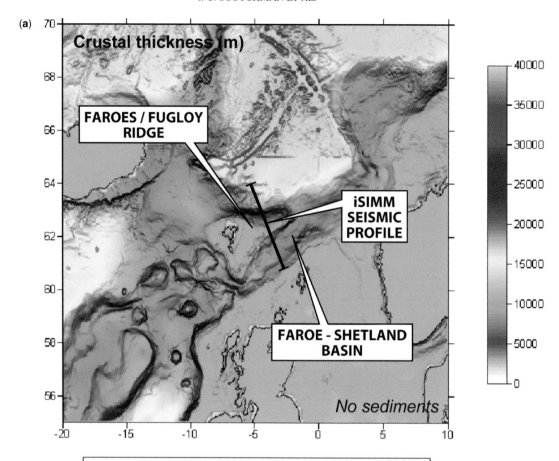
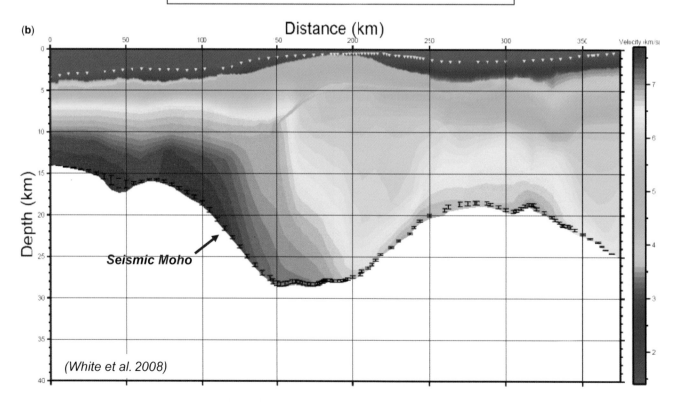

Fig. 9. (a) Crustal thickness map for the NE Atlantic Ocean showing thinned continental crust beneath the Faroe–Shetland Basin while the Faroes continental block comprises relatively un-thinned continental crust. (b) The iSIMM Deep Seismic line illustrating the deep structure of the Faroes/Faroe–Shetland Basin (from White et al. 2008).

Turonian, also provide important charging to these drift-phase deepwater reservoirs in the Santos and Espirito Santo Basins.

The syn-rift sourced petroleum system occurs in both the Santos Basin and the SPP; in the former the main reservoirs are in the post-salt turbidites while in the SPP the main reservoir is the pre-salt carbonate section capping the thick syn-rift/sag phase section. Importantly, due to the greatly different heatflow and burial depth between the Santos Basin and the SPP, the hydrocarbon phase is different. Within the Santos Basin, the subsidence associated with the failed seafloor spreading ridge resulted in the deep burial of the syn-rift source rock beneath a thick Late Cretaceous–Cenozoic turbidite section of 4–5 km (e.g. Fig. 3 seismic line 3), up to twice that of the equivalent section on the SPP, which remained a relatively positive feature. It is therefore not surprising that light hydrocarbons and gas, such as the giant Merluza gas field, predominate in this part of the Santos Basin. On the SPP, the large volumes of oil (28–32°API) have been discovered, reportedly in the syn-rift carbonate reservoir with estimated resources of $20-50 \times 10^9$ barrels, and a large gas discovery, Jupiter, discovered on the flank of the SPP. As well as greatly increased burial depths, the thermal regime in the Santos Basin differs from that on the SPP: at a depth of 3 km below mud-line, the typical temperature in the former is around 100 °C compared with 60 °C on the SPP (Poiate et al. 2006). The temperature differences appear to reflect both a lower heatflow on the SPP and the effects of the thick evaporites blanketing the structure.

Therefore the development of the prolific pre-salt oil province on the SPP appears related to its unique structural development compared to the other basins along the Brazil Atlantic Margin: the thinned crust allowed development of a thick, well developed syn-rift/sag phase section with resultant excellent source and reservoir rocks, relatively low subsidence compared with the adjacent Santos Basin and relatively low temperatures due to lower heatflows and the thick evaporites, allowing the source rocks to remain in the oil window compared with the Santos Basin, where the equivalent source rock kitchens are in the gas-window or are burnt out.

Comparison of continental lithospheric thinning and rifted margin formation in the Santos Basin/Sao Paulo Plateau with the Faroe–Shetland Basin and Faroes margin, NE Atlantic

The kinematics of continental lithosphere thinning and breakup at the Brazilian margin can be compared with those at the NE Atlantic Ocean in the Faroe–Shetland area (Fig. 9). The Faroe–Shetland Basin (FSB) and Fugloy Ridge appear to be analogous features to the Santos Basin and the SPP, respectively. The FSB, a major Cretaceous–Cenozoic depocentre, is located between the West Shetland Platform north of Scotland and the Faroes (Doré et al. 1999) and contains a series of NE–SW trending sub-basins formed by a complex tectonic history involving multiple phases of extension and volcanism (Carr & Scotchman 2003). The FSB underwent several periods of rifting, accommodated by extensional faulting from Devonian to Cretaceous times (e.g. Doré et al. 1999; Roberts et al. 1999). The basin also experienced lithospheric thinning in the Late Paleocene synchronously with crustal rupture and the onset of Atlantic Ocean seafloor spreading to the west of the Faroe Islands, at the Faroes margin (Fletcher 2009). This was accompanied by a massive outpouring of basalt which covers part of the FSB. The crustal thickness map of the Faroese region (Fig. 9a), derived using the method of Greenhalgh & Kusznir (2007) and the iSIMM refraction line (Fig. 9b) shows that the FSB is underlain by very thinned crust. Thin crust beneath the FSB is coaxial with thin crust at the Møre margin, and the FSB is postulated to be a failed breakup basin at the palaeopropagating tip of the Atlantic. A schematic diagram of the kinematics of breakup at the Faroes margin is shown in Figure 8b.

Both the Santos Basin and the FSB appear to be failed breakup basins associated with attempted propagation of seafloor spreading and consequent thinning of continental crust, while the adjacent structural highs, the SPP and Fugloy Ridge respectively, represent areas of relatively thick continental crust, albeit with greatly differing amounts of thinning. The SPP underwent considerable thinning and subsidence which accommodated a thick syn-rift section containing both source and reservoir rocks capped by a thick post-rift evaporitic section, resulting in a prolific petroleum province. The Fugloy Ridge also experienced considerable crustal thinning but comparison with the SPP is difficult as the crust has probably been re-thickened by the addition of both igneous intrusions and extrusive volcanics after it was thinned. Thick volcanics on the Fugloy Ridge mean that the syn-rift section is hard to image, although recent seismic data indicate the development of a sedimentary sequence beneath the post-rift sequence (Roberts 2007), indicating the potential for the development of a petroleum province.

Conclusions

The modelling suggests that the linear gravity and magnetic feature identified in the Santos Basin represent abandoned seafloor spreading propagation with the formation of oceanic or proto-oceanic crust at its southwestern end. This represents an early attempt at seafloor spreading initiation north of the FFZ during the early Aptian. The breakup process also resulted in the formation of an adjacent area of thinned continental crust, the SPP, with subsequent deposition of a thick syn-rift/sag phase section containing both hydrocarbon source and reservoir rocks capped by a thick post-rift bedded evaporitic section. A prolific pre-salt petroleum province developed on this feature due to preservation of the syn-rift/sag source rocks and reservoirs by the relatively low heatflows and subsidence compared with the adjacent Santos Basin. Breakup and seafloor spreading to the east of the SPP, marking the final breakup of the South Atlantic Ocean, appear to have been initiated from the north (Scotchman et al. 2008) and occurred in the late Albian–Cenomanian.

Similar analysis of the FSB area of the northeast Atlantic Ocean suggests development of a similar failed breakup/seafloor spreading basin. Both the Brazilian and Faroes margins exhibit evidence for complex breakup kinematics, where lithospheric thinning originally occurred at two or more overlapping segments before becoming linked when the lithosphere ruptured. Sedimentary basins on the regions of thinned crust on rifted continental margins have potential to be hydrocarbon provinces.

The authors would like to thank Statoil for permission to publish this work and John Kipps and Michael McCambridge for their draughting of the figures and the many revisions.

References

Cainelli, C. & Mohriak, W. U. 1998. Brazilian geology part II: geology of Atlantic Eastern Brazilian Basins. *Rio '98 AAPG International Conference and Exhibition, American Association of Petroleum Geologists/Associação Brasileira dos Geólogos de Petróleo, Short Courses*, Rio de Janeiro.

Carr, A. D. & Scotchman, I. C. 2003. Thermal history modelling in the southern Faroe–Shetland Basin. *Petroleum Geoscience*, **9**, 333–345.

Chang, K. H., Kowsmann, R. O., Figueiredo, A. M. F. & Bender, A. 1992. Tectonics and stratigraphy of the east Brazil rift system: an overview. *Tectonophysics*, **213**, 97–138.

Chappell, A. R. & Kusznir, N. J. 2008. Three-dimensional gravity inversion for Moho depth at rifted continental margins incorporating a lithosphere thermal gravity anomaly correction. *Geophysics Journal International*, **246**, 1–17; doi: 10.1111/j.1365-246X.2008.03803.x.

Cobbold, P. R., Meisling, K. & Mount, V. S. 2001. Reactivation of an obliquely rifted margin, Campos and Santos Basins, southeastern Brazil. *American Association of Petroleum Geologists Bulletin*, **85**, 1925–1944.

Corti, G., Van Wijk, J. et al. 2003. Transition from continental break-up to punctiform seafloor spreading: how fast, symmetric and magmatic. *Geophysical Research Letters*, **30**,1604; doi: 10.1029/2003GL017374.

Davison, I. 2007. Geology and tectonics of the south Atlantic Brazilian salt basins. *In*: Ries, A. C., Butler, R. W. H. & Graham, R. H. (eds) *Deformation of the Continental Crust: The Legacy of Mike Coward.* Geological Society, London, Special Publications, **272**, 345–359.

Davison, I. & Bate, R. 2004. Early opening of the South Atlantic: Berriasian rifting to Aptian salt deposition. *PESGB/HGS 3rd International Joint Meeting Africa: The Continent of Challenge and Opportunity*, 7–8 September.

Demaison, G. J. & Moore, G. T. 1980. Anoxic environments and oil source bed genesis. *Organic Geochemistry*, **2**, 9–31.

Doré, A. G., Lundin, E. R., Birkeland, Ø., Eliassen, P. E. & Jensen, L. N. 1999. The NE Atlantic Margin: implications for late Mesozoic and Cenozoic events for hydrocarbon prospectivity. *Petroleum Geoscience*, **3**, 117–131.

Dunbar, J. A. & Sawyer, D. S. 1989. How pre-existing weaknesses control the style of continental breakup. *Journal of Geophysical Research – Solid Earth and Planets*, **94**, 7278–7292.

Fletcher, R. J. 2009. *Mechanisms of continental lithosphere thinning and rifted margin formation.* PhD thesis, University of Liverpool.

GEBCO (IOC, IHO, and BODC). 2003. *Centenary Edition of the GEBCO Digital Atlas.* Published on CD-ROM on behalf of the Intergovernmental Oceanographic Commission and the International Hydrographic Organization as part of the General Bathymetric Chart of the Oceans. British Oceanographic Data Centre, Liverpool.

Gomes, P. O., Parry, J. & Martins, W. 2002. The outer high of the Santos Basin, southern Sao Paulo Plateau, Brazil: tectonic setting, relation to volcanic events and some comments on hydrocarbon potential. Extended Abstract. *American Association of Petroleum Geologists Hedberg Conference 'Hydrocarbon Habitat of Volcanic Rifted Passive Margins'*, 8–11 September, Stavanger. *Search and Discovery* Article **90022**.

Gomes, P. O., Kilsdonk, B., Minken, J., Grow, T. & Barragan, R. 2009. The outer high of the Santos Basin, southern Sao Paulo Plateau, Brazil: pre-salt exploration outbreak, paleogeographic setting and evolution of the syn-rift structures. *Search and Discovery* Article **10193**.

Greenhalgh, E. E. & Kusznir, N. J. 2007. Evidence for thin oceanic crust on the extinct Aegir Ridge, Norwegian Basin, NE Atlantic derived from satellite gravity inversion. *Geophysical Research Letters*, **34**, L06305, 1–5; doi: 10.1029/2007GL029440.

Guardado, L. R., Spadini, A. R., Brandao, J. S. L. & Mello, M. R. 2000. Petroleum system of the Campos Basin, Brazil. *In*: Mello, M. R. & Katz, B. J. (eds) *Petroleum Systems of South Atlantic Margins.* American Association of Petroleum Geologists, Tulsa, OK, Memoirs, **73**, 317–324.

Karner, G. D. & Driscoll, N. W. 1999. Tectonic and stratigraphic development of the West African and eastern Brazilian Margins: insights from quantitative basin modelling. *In*: Cameron, N. R., Bate, R. H. & Clure, V. S. (eds) *The Oil and Gas Habitats of the South Atlantic.* Geological Society, London, Special Publications, **153**, 11–40.

Karner, G. D. & Gamboa, L. A. P. 2007. Timing and origin of the South Atlantic pre-salt sag basins and their capping evaporites. *In*: Schreiber, B. C., Lugli, S. & Babel, M. (eds) *Evaporites Through Space and Time.* Geological Society, London, Special Publications, **285**, 15–35.

Katz, B. J. & Mello, M. R. 2000. Petroleum systems of South Atlantic marginal basins – an overview. *In*: Mello, M. R. & Katz, B. J. (eds) *Petroleum Systems of South Atlantic Margins.* American Association of Petroleum Geologists, Tulsa, OK, Memoirs, **73**, 1–13.

Kusznir, N. J., Roberts, A. M. & Morley, C. 1995. Forward and reverse modelling of rift basin formation. *In*: Lambiase, J. (ed.) *Hydrocarbon Habitat in Rift Basins.* Geological Society, London, Special Publications, **80**, 33–56.

Lelarge, M. L. M. V. 1993. *Thermochronologie par la methode des traces de fission d'une marge passive (dome de Ponta Grossa, SE Brásil) et au sein d'une chaine de collision (zone externe de lárc Alpin, France).* PhD thesis, Université Joseph Fourier, Grenoble.

McKenzie, D. 1978. Some remarks on the development of sedimentary basins. *Earth and Planetary Science Letters*, **40**, 25–32.

Meisling, K., Cobbold, P. R. & Mount, V. S. 2001. Segmentation of an obliquely rifted margin, Campos and Santos Basins, southeastern Brazil. *American Association of Petroleum Geologists Bulletin*, **85**, 1903–1924.

Mello, M. R., Azambula Filho, N. C., de Mio, E., Bender, A. A., de Jesus, C. L. C. & Schmitt, P. 2009. 3D modelling illuminates Brazil's presalt geology. *Offshore Magazine*, **69**, PennEnergy.

Mueller, R. D., Roest, W. R., Royer, J.-Y., Gahagan, L. M. & Sclater, J. G. 1997. Digital isochrons of the world's ocean floor. *Journal of Geophysical Research*, **102**, 3211–3214.

Parker, R. L. 1972. The rapid calculation of potential anomalies. *Geophysics Journal of the Royal Astronomy Society*, **31**, 447–455.

Pereira, J. L. & Feijo, F. J. 1994. Bacia de Santos. *Boletim de Geociências da Petrobrás*, **8**, 219–234.

Poiate, E., Maia da Costa, A. & Falcão, J. L. 2006. Drilling Brazilian salt: Petrobras studies salt creep and well closure. *Oil & Gas Journal*, **104**, 88–96. World Wide Web Address: http://www.ogj.com/articles/article_display.cfm?ARTICLE_ID=256731&p=7.

Renne, P., Deckart, K., Ernesto, M., Feraud, G. & Piccirillo, E. 1996. Age of the Ponta Grossa Dyke swarm (Brazil) and implications to the Parana flood volcanism. *Earth and Planetary Science Letters*, **144**, 199–211.

Roberts, A. 2007. *Crustal structure of the Faroes North Atlantic margin from wide-angle seismic data.* PhD thesis, University of Cambridge.

Roberts, A. M., Kusznir, N. J., Yielding, G. & Styles, P. 1998. 2D flexural backstripping of extensional basins; the need for a sideways glance. *Petroleum Geoscience*, **4**, 327–338.

Roberts, D. G., Thompson, M., Mitchener, B., Hossack, J., Carmichael, S. & Bjornset, H.-M. 1999. Palaeozoic to Tertiary rift and basin dynamics: mid-Norway to the bay of Biscay – a new context for hydrocarbon prospectivity in the deep water frontier. *In*: Fleet, A. J. & Boldy, S. A. R. (eds) *Petroleum Geology of Northwest Europe: Proceedings of the 5th Conference.* Geological Society, London, 7–40; doi: 10.1144/0050007.

Sandwell, D. T. & Smith, W. H. F. 1997. Marine gravity anomaly from Geosat and ERS1 satellite altimetry. *Journal of Geophysical Research*, **102**, 10039–10054.

Scotchman, I. C., Marais-Gilchrist, G. et al. 2006. A failed sea-floor spreading centre, Santos Basin, Brasil. *Rio Oil and Gas Conference, IBP*, Rio de Janeiro.

Scotchman, I. C., Brockbank, P., Gilchrist, G., Hunsdale, R., Roberts, A. & Kusznir, N. 2008. The formation of the South Atlantic between Brazil and West Africa: did it un-zip South to North or North to South? Abstract, *American Association of Petroleum Geologists International Conference and Exhibition*, 26–29 October, Cape Town.

White, R. S., Smith, L. K., Roberts, A. W., Christie, P. A. F. & Kusznir, N. J. 2008. Lower-crustal intrusion on the North Atlantic continental margin. *Nature*, **452**, 460–464.

Ziegler, P. A. & Cloetingh, S. 2004. Dynamic processes controlling evolution of rifted basins. *Earth-Science Reviews*, **64**, 1–50.

Structural architecture and nature of the continent–ocean transitional domain at the Camamu and Almada Basins (NE Brazil) within a conjugate margin setting

O. A. BLAICH,[1] J. I. FALEIDE,[1] F. TSIKALAS,[1,2] R. LILLETVEIT,[3] D. CHIOSSI,[3] P. BROCKBANK[3] and P. COBBOLD[4]

[1]*Department of Geosciences, University of Oslo, PO Box 1047 Blindern, NO-0316 Oslo, Norway (e-mail: o.a.blaich@geo.uio.no)*
[2]*Eni E&P, EASS, Via Emilia 1, I-20097 San Donato Milanese, Italy (present address)*
[3]*Global Exploration, Statoil, Forusbeen 50, N-4035 Stavanger, Norway*
[4]*Géosciences-Rennes, 35042 Rennes Cedex, France*

Abstract: Regional seismic reflection profiles and potential field data across the conjugate magma-poor Camamu/Almada–Gabon margins, complemented by crustal-scale gravity modelling and plate reconstructions, are used to reveal and illustrate the relationship of crustal structure to along-margin variation of potential field anomalies, to refine and constrain the continent–ocean boundary, as well as to study the structural architecture and nature of the continent–ocean transitional domain. The analysis reveals that the prominent conjugate Salvador–N'Komi transfer system appears to be a first-order structural element, governing the margin segmentation and evolution, and may have acted as an intraplate decoupling zone. The continent–ocean transitional domain, offshore northeastern Brazil, is characterized by rotated fault-blocks and wedge-shaped syn-rift sedimentary sequences overlying a prominent and undulated reflector ('M-reflector'), which in turn characterizes the boundary between an extremely thinned, possibly magmatically intruded, continental crust and normal lithospheric mantle. The 'M-reflector' in the northeastern Brazilian margin shows remarkable similarities to the S-reflector at the West Iberia margin. In the same way, the 'M-reflector' is interpreted as a detachment surface that was active during rifting. Unlike the well studied central and northern segments of the West Iberia margin, however, the present study of the northeastern Brazilian margin does not clearly reveal evidence of an exhumation phase. The latter predicts exhumation of middle and lower crust followed by mantle exhumation. Increase in volcanic activity during the late stages of rifting may have 'interrupted' the extensional system, implying a failed exhumation phase. In this setting, the break-up and drift phase may have replaced the exhumation phase. Nevertheless, the available observations cannot discount the possibility that the 'M-reflector' is underlain by partially serpentinized mantle. Our study further leads to the development of a detailed conceptual model, accounting for the complex tectonomagmatic evolution of the conjugate northeastern Brazilian–Gabon margins. This model substantiates a polyphase rifting evolution mode, which is associated with a complex time-dependent thermal structure of the lithosphere. In the conjugate margin setting, asymmetrical lithospheric extension resulted in the formation of the thinned continental crust domain prior to the formation of the approximately symmetrical transitional domain.

Keywords: northeastern Brazilian margin, gravity modelling, continent–ocean transition, crustal extension, South Atlantic conjugate margins

The extensional margin evolution from rift through break-up rupture of the continental lithosphere to progressive oceanic crust formation remains controversial (e.g. Rosendahl *et al.* 2005; Reston & Pérez-Gussinyé 2007; Sawyer *et al.* 2007). This is due to the fact that deep crustal structures along several margins are partially or totally masked by evaporite deposits and/or by magmatic materials. There are only limited amounts of high-quality geophysical data acquired at deep and ultra-deep waters. As petroleum exploration advances towards new frontiers, there is an economic need to better define the architecture and nature of the continent–ocean boundary/transition. This also has obvious implications for the reconstruction of rifted continental margins as this boundary represents the line of final continental separation.

Drilling results combined with recent studies at the Iberia and Newfoundland margins (e.g. Sawyer *et al.* 1994; Whitmarsh *et al.* 1998; Lavier & Manatschal 2006; Péron-Pinvidic *et al.* 2007; Reston 2009) and the acquisition of high-quality geophysical data at several margins have recently cast doubt on classical models of rifted margins. The latter viewed both the concurrence of continental and oceanic crust through a well-defined sharp boundary, and the thinning of the continental crust by simple or pure shear mechanisms, or a combination of the two (McKenzie 1978; Wernicke 1985; Lister *et al.* 1986; Buck *et al.* 1988). For instance, observations from the central South Atlantic margins are not consistent with classical rifted margin model predictions (Karner *et al.* 2003; Contrucci *et al.* 2004; Moulin *et al.* 2005; Karner & Gambôa 2007). For these distinctive margins, a wide region of extremely thin continental crust is observed, overlain by relatively thin and undeformed late syn-rift sediments, deposited under shallow water conditions in 'sag' basins, and being more compatible with depth-dependent extension models (e.g. Driscoll & Karner 1998; Davis & Kusznir 2004; Karner & Gambôa 2007; Kusznir & Karner 2007). Deeply subsided and possibly anomalous crust has been recognized along several margin segments in South Atlantic and described variously as 'proto-oceanic crust,' 'attenuated continental crust' and 'transitional crust' (e.g. Meyers *et al.* 1996a, b; Rosendahl & Groschel-Becker 1999; Wilson *et al.* 2003; Contrucci *et al.* 2004; Moulin *et al.* 2005; Rosendahl *et al.* 2005). The term continent–ocean 'transitional domain' is adopted in the current study and defined as the part of the lithosphere which is located between the thinned crystalline crust domain and the first appearance of oceanic crust formed by seafloor spreading (e.g. Whitmarsh & Miles 1995; Dean et al. 2000; Srivastava et al. 2000; Afilhado et al. 2008).

In this study of the Camamu and Almada basins in the northeastern Brazilian margin (Figs 1 & 2), we analyse an available grid of regional multichannel seismic (MCS) reflection profiles. We also analyse an available grid of regional MCS profiles together with published seismic reflection/refraction profiles along the conjugate Gabon margin. In this conjugate margin context, the integration of seismic, potential field data and conducted 2D gravity modelling, together with potential field plate reconstructions, provide adequate means to study the magma-poor Camamu/Almada–Gabon margins, to elucidate structural elements and features that reflect the processes that rupture the continental crust, as well as to refine and constrain the structural architecture and nature of the continent–ocean transitional (COT) domain.

Data and methodology

Data

Together with the gridded bathymetry and potential field data along the northeastern Brazilian margin and its conjugate off Gabon, a set of deep seismic reflection profiles was at our disposal (Fig. 2). The dataset was provided courtesy of Statoil and comprises several surveys acquired between 1999 and 2001. The dataset comprises 9500 km^2 of 2D high-resolution MCS reflection profiles, acquired by TGS-NOPEC with 37.5 m shotpoint interval and 2 ms sampling interval. Published seismic reflection profiles (Meyers et al. 1996a, b; Rosendahl et al. 2005) together with industrial seismic reflection and wide-angle refraction profiles (Wannesson et al. 1991) on the conjugate Gabon margin have been analysed, revised and incorporated into our interpretations.

Potential field data used in this study consist of a 1×1 min. elevation grid (GEBCO, General Bathymetric Chart of the Oceans 2003), a 1×1 min. satellite radar-altimeter gravity grid (version 16.1; Sandwell & Smith 1997; Fig. 3a), a 3×3 min. gridded aeromagnetic anomaly field map (WDMAM, World Digital Magnetic Anomaly Map 2007; Fig. 3c), as well as publicly available ship-borne bathymetry and potential field anomaly data (LDEO, Lamont–Doherty Earth Observatory; Wessel & Watts 1988; Smith 1993). A comparison between the satellite and shipborne gravity anomalies shows that there is reasonably good agreement in wavelength and amplitude of the two datasets (Marks 1996; Sandwell & Smith 1997, 2009). These results support crustal-scale modelling with the use of satellite free-air gravity. Using the broad-coverage satellite gravity data also allowed us greater flexibility in the selection of the constructed transects, being able to project the gridded data along the selected MCS profiles. Finally, a Bouguer-corrected gravity anomaly field map was constructed by utilizing the detailed bathymetry and free-air gravity anomalies and a density of 2670 kg m^{-3} as infill to the bathymetric relief (Fig. 3b).

Transect construction

The MCS profiles within the study area were compiled and extended to the ultra-deep water oceanic province, in order to construct representative regional transects used in crustal-scale gravity modelling. The regional transects are based on the seismic reflection profiles, extracted bathymetry and gravity anomaly data along the extended parts, and on initial estimates of

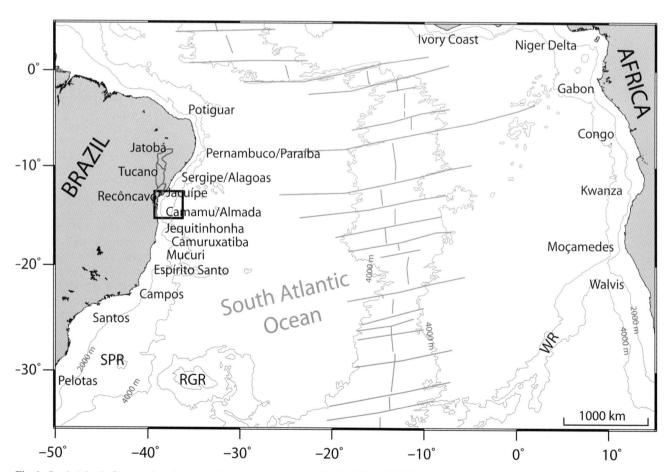

Fig. 1. South Atlantic Ocean and conjugate continental margins, outlined by the 2000 and 4000 m bathymetry contours. Rectangle outlines the study area. The location of major sedimentary basins along the East Brazilian and West Africa margins and major fracture zones are also indicated. WR, Walvis Ridge; RGR, Rio Grande Rise; SPR, São Paulo Ridge.

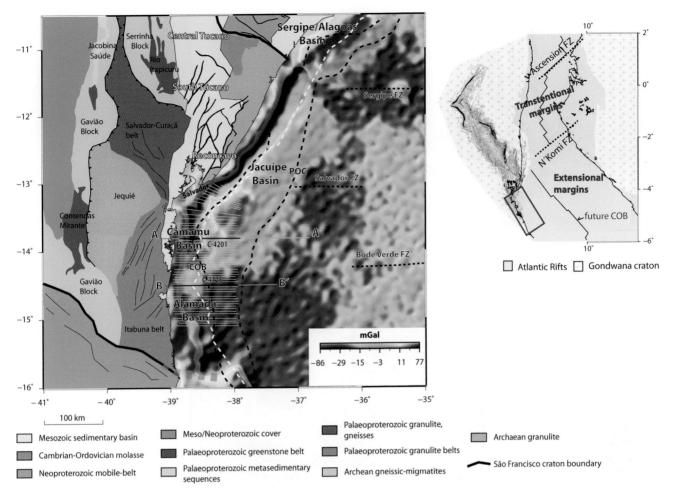

Fig. 2. Offshore: 1 × 1 min. gridded satellite radar-altimeter free-air gravity anomaly field (Sandwell & Smith 1997; version 16.1). Onshore: simplified geological and structural map of the northeastern Brazilian margin generalized from Barbosa & Sabaté (2003), Destro et al. (2003), Alkmim et al. (2006) and Leite et al. (2009). The constructed regional transects (A–A′ and B–B′) are also indicated together with the selected MCS profiles and the denser grid of MCS that was at our disposal. Interpreted oceanic fracture zones are based on observations made on the potential field data and on Gomes et al. (2000), Davison (1999) and Mohriak et al. (2000). COB, continent–ocean boundary (Karner & Driscoll 1999, white dashed line; Rosendahl et al. 2005, black dashed line); POC, proto-oceanic crust (Rosendahl et al. 2005). Inset: reconstruction of the rift system between northeastern Brazil and western Africa at the time of oceanic crust inception, showing major depocentres and bifurcation of the spreading axis in two branches (modified from Mohriak et al. 2000).

the Moho relief using both isostatic balancing and inverse gravity modelling. The latter method used in this study is described in detail by Blaich et al. (2008, 2009). The potential field plate reconstructions (Fig. 3) were performed using the PLATES software (Institute for Geophysics, University of Texas at Austin) and rotation poles of Torsvik et al. (2009). Plate reconstructions were used in this study in order to select conjugate profiles and to evaluate the tectonic evolution of the northeastern Brazilian/Gabon margins expressed on potential field data.

Margin setting

The opening of the South Atlantic Ocean and the subsequent passive margin formation resulted from lithospheric extension followed by break-up of the Palaeozoic Gondwana supercontinent and seafloor spreading during Mesozoic times (e.g. Rabinowitz & LaBrecque 1979; Chang et al. 1992). The opening of the South Atlantic, which probably started in the southern portion and propagated towards the north, resulted in considerable diachronic deformation in the Equatorial and eastern Brazilian margins (Matos 2000). The Equatorial margin evolved in response to transform motion between Brazil and Africa, resulting in complex shear-dominated basins. In contrast, the Eastern Brazilian margin is characterized by development of extensional basins due to divergent and semi-orthogonal crustal extension (Chang et al. 1992). A triple-junction is proposed near Salvador city, where the rift axis in the offshore domain changes to a more northerly direction. In this setting, the western branch of the rift forms the aborted onshore Recôncavo–Tucano–Jatobá Basins, whereas the eastern branch evolved to form the northeastern Brazilian–Gabon/Equatorial Guinea passive margin (Fig. 2; Mohriak et al. 2000).

Structural inheritance

Several studies of the northeastern Brazilian basin system postulate a pattern of structural inheritance, indicating that pre-existing continental lineaments, transfer zones and various basement terranes exerted important influence and rheological control on the structural development of the Mesozoic rifting and break-up (Fig. 2) (e.g. Milani & Davison 1988; Matos 1992, 1999; Szatmari & Milani 1999; Jacques 2003; Blaich et al. 2008). These zones of weakness played an important role in continental break-up from the earliest stage of rifting and have been correlated with failed arms of triple-junction rifts (Jacques 2003). In this setting, the pattern of extension (symmetrical v. asymmetrical), the sequence of rift phases and rift pattern leading to break-up, and the physiography of the resulting passive margin segments are all to a large extent controlled by the inherited fabric of continental lithosphere (Dunbar & Sawyer 1988, 1989).

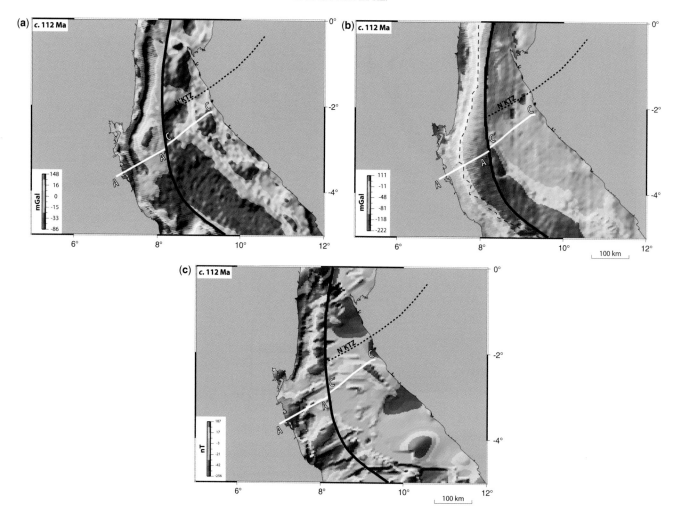

Fig. 3. Plate reconstructions at forced break-up (c. 112 Ma) utilizing the rotation poles of Torsvik et al. (2009); COB revised from Torsvik et al. (2009) and Karner & Driscoll (1999) fitting observations made in this study. (**a**) 1 × 1 min. satellite radar-altimeter free-air gravity anomaly field (Sandwell & Smith 1997; version 16.1). (**b**) Low-pass filtered (>200 km) Bouguer-corrected gravity anomaly field (utilizing a density of 2670 kg m^{-3} as infill to the bathymetric relief). (**c**) 3 × 3 min. gridded aeromagnetic anomaly field (WDMAM, World Digital Magnetic Anomaly Map 2007). The locations of conjugate transects A–A′ and C–C′ are indicated. Black dashed line indicates the COB as proposed by Karner & Driscoll (1999). N'KT, N'Komi Transfer Zone.

In the offshore domain, structural inheritance is reflected by the trend of the gravity anomalies. The gravity anomaly map (Fig. 2) shows that the anomaly low that flanks the 'edge-effect' anomaly high extends for 350–380 km parallel to the coastline. This anomaly low is abruptly terminated by a belt of positive anomalies to the NE, at latitude c. 11.2°S, along the equivalent offshore extent of the northern boundary of the São Francisco craton. The landward continuation of the Sergipe oceanic fracture zone, as proposed by Mohriak et al. (1995), is also closely related to this abrupt change on the gravity anomaly trend (Fig. 2). In this setting, inherited transfer zones appear to determine the location of recent oceanic fracture zones. Oceanwards, the prominent gravity anomaly high located at c. 13.5°S/−38°W and referred to as the Jacuípe Volcanic Complex (Davison 1999) terminates at the equivalent offshore extent of the boundary between the Salvador–Curaçá and the Itabuna Palaeoproterozoic belts (Fig. 2). Structural inheritance is also indicated by the trend of the magnetic anomalies (Fig. 3c).

Similarly, along the conjugate margin off Gabon, a prominent gravity anomaly low that flanks the 'edge-effect' gravity anomaly high abruptly terminates to the north at the N'Komi Transfer Zone (Fig. 3a). In the immediate vicinity of the prominent N'Komi Transfer Zone, abrupt changes in the deep crustal structure have been observed and the transfer zone appears to be a first-order structural element, governing the segmentation and evolution of the Gabon margin (Wannesson et al. 1991; Blaich et al. 2008). These observations suggest a distinct and probably inherited segmentation of the South Atlantic margins, where major changes in lithospheric strength may have occurred (e.g. Watts & Stewart 1998).

Crustal architecture

Northeast Brazil. The representative crustal transects obtained from the 2D gravity modelling outline the crustal structure of the individual segments along the Camamu and Almada basins. The distinct along-margin structural changes are highlighted by the modelled transects A–A′ and B–B′, which illustrate the structural complexity of the study area. Transects A–A′ and B–B′ are oriented in a west–east direction and extend c. 258 and c. 190 km eastwards along the Camamu and Almada basins, respectively (Figs 4 & 5). As the aim of this study is to illustrate the structural architecture at the transitional domain along the Camamu and Almada basins, seismic interpretation of the sedimentary units only accounts for pre-rift/syn-rift and post-rift sequences.

Interpretation of all available MCS profiles along the study area (Fig. 2) does not suggest a typical rifted continental crust, with progressive crustal thinning over an extensive distance towards oceanic crust formed due to break-up and seafloor spreading. On the contrary, although seismic interpretation combined with gravity modelling suggest extreme crustal thinning, the thick syn-rift sedimentary basins overlying the thinned crystalline

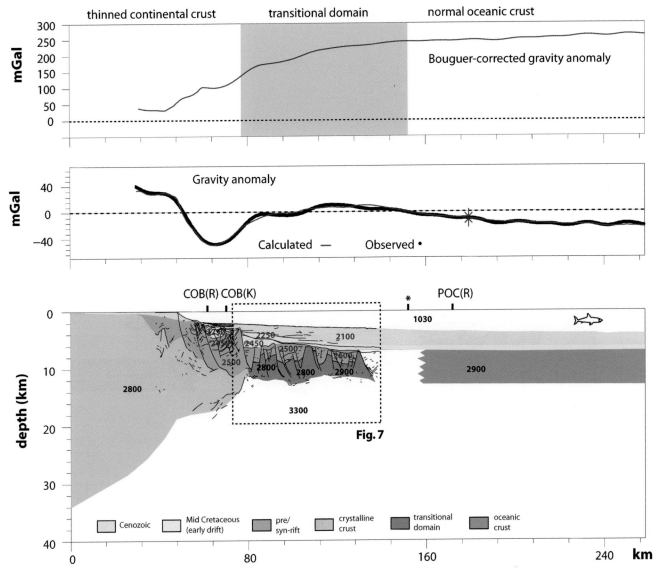

Fig. 4. Gravity modelled transect A–A′ (location in Fig. 2) utilizing the satellite radar-altimeter free-air gravity anomaly field (Sandwell & Smith 1997; version 16.1). Bouguer-corrected gravity anomaly is also shown. COB(K), continent–ocean boundary (Karner & Driscoll 1999); COB(R), continent–ocean boundary (Rosendahl et al. 2005); the asterisk indicates 'line of fit' used for plate reconstructions; POC(R), proto-oceanic crust (Rosendahl et al. 2005).

continental crust domain appear not to be the result of intense upper crust brittle deformation (Fig. 4). For instance, the data show that the continental crust along the Camamu Basin thins abruptly, from c. 30 to c. 4 km over a distance of only c. 80 km without prominent brittle deformation of the upper crust (Fig. 4).

Along-margin variation on the thickness of the pre/syn-rift sedimentary package is observed on the MCS profiles, and corresponds to prominent along-margin changes in the potential field data. In particular, a thick pre/syn-rift sedimentary package is characterized by a prominent and elongated gravity anomaly low along transect A–A′ (Figs 2 & 4). Further southwards, the thickness of the pre/syn-rift sedimentary package decreases considerably at places where the prominent and elongated gravity low is replaced by higher values (transect B–B′; Figs 2 & 5). A band of high-amplitude seismic reflections is observed along the thinned continental crust domain of transect B–B′ and at a depth of c. 14 km (Fig. 5). Such reflections are not present along transect A–A′. These reflections may represent the top of a lower crust or indicate a detachment surface; however, gravity modelling alone cannot unambiguously resolve this issue.

Although the northeastern Brazilian margin (Fig. 1) is proposed to be associated with well defined seaward-dipping reflections (Mohriak et al. 1995, 1998; Jackson et al. 2000), seismic images from this margin segment are very different from the images obtained on the well studied volcanic margins worldwide (e.g. Eldholm & Grue 1994; Gladczenko et al. 1998; Hinz et al. 1999; Franke et al. 2007; Tsikalas et al. 2008), implying that magmatic products along the northeastern Brazilian margin were not sufficiently voluminous to form seaward-dipping reflections (Sawyer et al. 2007). Seaward-dipping reflections were not recognized in the Camamu and Almada basins within the extensive seismic reflection dataset at our disposal (Fig. 2). Higher densities are required for the sedimentary package located adjacent to the normal oceanic crust (c. 110–140 km; Fig. 4), where an increase in seismic amplitude reflectivity is observed, indicating a possible increase in magmatic products characterized by volcanic flows and interbedded sediments with volcanic material.

Salt diapirs are clearly depicted along the Almada Basin, where a typical halokinetic structure is observed at the top of a well defined fault (Fig. 5). Although interpretation of the MCS profiles along the Camamu Basin indicates that prominent salt tectonism is not present (Fig. 4), previous studies have identified salt deposits along the Camamu (Karner & Gambôa 2007; Menezes & Milhomem 2009), Jacuípe (Davison 2007) and Sergipe–Alagoas basins (Castro 1988; Mohriak et al. 1995, 2000; Davison 2007; Souza-Lima 2009).

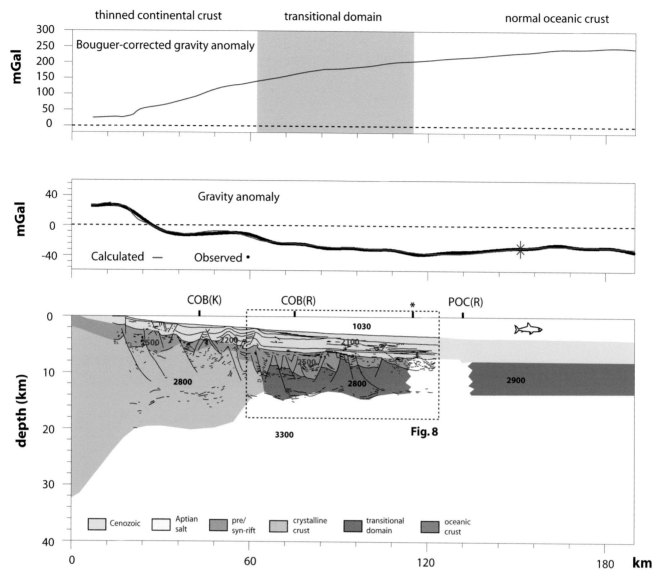

Fig. 5. Gravity modelled transect B–B′ (location in Fig. 2) utilizing the satellite radar-altimeter free-air gravity anomaly field (Sandwell & Smith 1997; version 16.1). Bouguer-corrected gravity anomaly is also shown. COB(K), continent–ocean boundary (Karner & Driscoll 1999); COB(R), continent–ocean boundary (Rosendahl et al. 2005); the asterisk indicates 'line of fit' used for plate reconstructions; POC(R), proto-oceanic crust (Rosendahl et al. 2005).

Gabon. In order to understand the total rift evolution, the conjugate margin transect C–C′ was constructed, located in the Gabon margin segment and extending for c. 204 km (Figs 3 & 6). The MCS profile is characterized by a wide salt basin that considerably masks the seismic reflection energy, implying that interpretation of the top basement reflection and rift structures are disputed and not easily recognized. A sub-horizontal, high-amplitude band of seismic reflections, located at a depth of c. 16–18 km and a distance of c. 0–80 km from the eastern end of transect C–C′, appears to be a good Moho candidate (Fig. 6). Several remarkable landward-dipping reflections located beneath the band of high-amplitude reflectivity are observed, indicating a different dip-pattern than the seaward-dipping structures related to the Mesozoic rifting. The Moho relief obtained by gravity modelling does not coincide with the high-amplitude band of seismic reflections discussed above (Fig. 6).

Based on seismic refraction and reflection analyses along the Gabon margin, a high-amplitude band of seismic reflections located at a depth of c. 16 km (depth-converted SPOG-2 profile from Wannesson et al. 1991) that exhibits a velocity increase from 6.1 to 6.9 km s^{-1} was earlier revealed. In addition, the gravity modelled SPOG-2 profile using densities derived from the wide-angle data of Wannesson et al. (1991) required a lower crustal body in order to accomplish a reliable fit in the gravity modelling. Similarly, an enigmatic high-velocity lower crustal body has also been suggested further south (Contrucci et al. 2004; Moulin et al. 2005). The analysis performed in this study indicates that the high-amplitude and sub-horizontal band of seismic reflections located at a depth of c. 16–18 km (Fig. 6) corresponds to the top of a lower crustal body rather than Moho. Such a lower crustal body is not necessarily magmatic in nature, but may be related to older tectonic episodes and, possibly, orogenies; for example, inheritance from Transamazonian and Pan-African collision belts (e.g. Meyers et al. 1996b).

Continent–ocean boundary

The location of the continent–ocean boundary (COB) along the northeastern Brazilian margin has been outlined in several studies based mostly on observations made on potential field anomalies and plate reconstructions (e.g. Karner & Driscoll 1999; Rosendahl et al. 2005). In this setting, the COB location was placed at a prominent negative–positive gradient on the Bouguer-corrected gravity field (Fig. 3b; Karner & Driscoll 1999). This location is

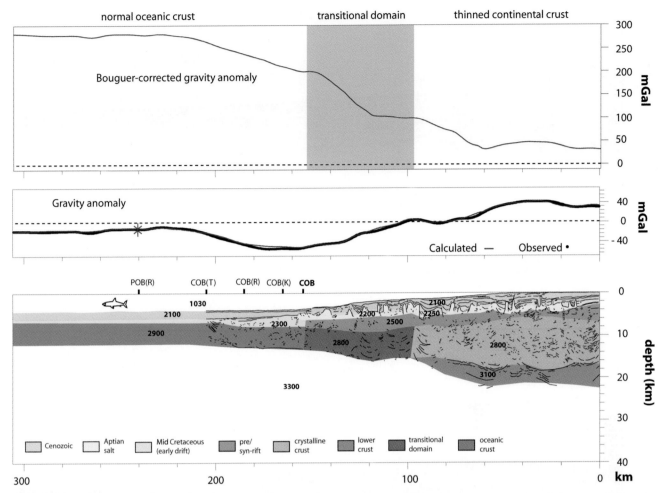

Fig. 6. Gravity modelled transect C–C′ (location in Fig. 3) utilizing the satellite radar-altimeter free-air gravity anomaly field (Sandwell & Smith 1997; version 16.1). Bouguer-corrected gravity anomaly is also shown. COB(K), continent–ocean boundary (Karner & Driscoll 1999); COB(R), continent–ocean boundary (Rosendahl *et al.* 2005); COB(T), continent–ocean boundary (Torsvik *et al.* 2009); COB, this study; POC(R), proto-oceanic crust (Rosendahl *et al.* 2005).

not consistent with current observations made on the MCS profiles, where the presence of thinned continental crust is inferred and modelled oceanwards of this boundary (e.g. along transect A–A′, Fig. 4). The COB location, as proposed earlier, seems to represent a crustal boundary; however, seismic observations and gravity modelling in the current study indicate that the crust oceanwards of the earlier proposed COB does not represent normal oceanic crust (Figs 4 & 5).

We suggest that the COB as defined by Karner & Driscoll (1999) and Blaich *et al.* (2008) is closely related to the transition from the thinned continental domain to the transitional domain. A prominent and steep fault appears to be associated with this boundary (Figs 4 & 5). The oceanwards limit of the transitional domain is, however, poorly known as the seismic reflection profiles acquired along the northeastern Brazilian margin do not extend into the normal oceanic crust domain. High-amplitude reflections within the wedge-shaped syn-rift sedimentary sequences are observed at the oceanward edge of the MCS profiles (Figs 7 & 8) and are interpreted as igneous material interbedded with sediments. The observed increased portion of volcanic features may indicate proximity to normal oceanic crust.

Along the conjugate margin off Gabon, on the other hand, a set of industrial and published deep seismic reflection profiles extends into the presumed normal oceanic crust. In this setting, the boundary between thinned continental crust and transitional domain is defined by the oceanwards edge of the anomalous lower crust, indicated by landward-dipping reflections and is characterized by shallowing of the Moho discontinuity (Fig. 6). The COB along this margin is placed approximately at the seaward limit of the salt basin, where the crust is clearly oceanic in seismic reflection character (e.g. Mohriak *et al.* 1995; Torsvik *et al.* 2009). The oceanic crust observed along the Gabon margin is highly reflective, rough and characterized by relatively flat Moho. The oceanic crust is *c*. 6 km thick and its base is imaged reasonably clear (Fig. 6). The interpreted COB along the Gabon margin is mirrored along the Camamu/Almada Basin and is defined as the 'line of fit' used for potential field plate reconstructions (Fig. 3). The 'line of fit' along the Camamu/Almada Basin is located within the portion of the transitional domain associated with increased observations of volcanic features (Figs 4 & 5).

The transitional domain along the northeastern Brazilian margin is characterized by rotated fault-blocks and wedge-shaped syn-rift sedimentary sequences (Figs 4 & 5). Below the syn-rift sediments a top basement reflector is poorly imaged; however, the transition from an area with well defined and stratified seismic reflection character to an area with more transparent character is interpreted as the top basement level. Gravity modelling of transects A–A′ and B–B′ indicates densities of 2800 kg m^{-3} for the rotated fault-blocks along the transitional domain. Somewhat higher densities (2900 kg m^{-3}) are required for the rotated fault-block located closer to the 'line of fit' along transect A–A′ (Figs 4 & 5). A high-amplitude and undulated reflector ('M-reflector') is observed and underlies the rotated fault-blocks at the transitional domain (Figs 4 & 5). These observed undulations are mainly due to a pullup effect below the tilted blocks, similar to that observed in the West Iberia margin (e.g. Reston *et al.* 2007). The performed

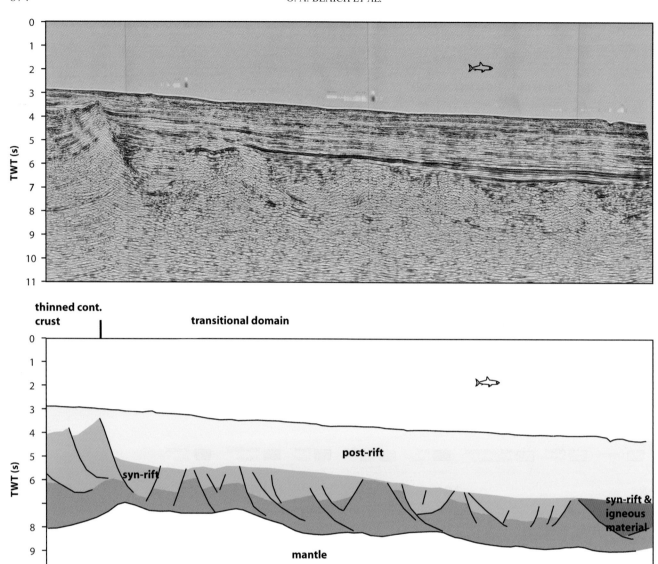

Fig. 7. Seismic example and interpretation of the transitional domain along MCS profile C-4201 (location in Fig. 2).

gravity modelling indicates that this prominent reflector probably corresponds to the Moho discontinuity that appears to be extremely shallow along transects A–A' and B–B' (Figs 4 & 5).

Along the Gabon conjugate margin, the evaporitic deposits considerably mask the seismic signal at the transitional domain, obscuring the potential identification of possible similar structures as in the Camamu/Almada margin, that is, rotated fault-blocks and wedge-shape syn-rift sedimentary sequences (Fig. 6). Due to the non-unique interpretation of gravity anomalies, it is difficult or even impossible to exclude such a possibility. Gravity modelling of transect C–C' also indicates densities of 2800 kg m^{-3} for the crust at the transitional domain (Fig. 6).

Nature of the crust at the transitional domain

Our analysis indicates the existence of a transitional domain between the thinned continental crust and the oceanic crust along the northeastern Brazilian margin and its conjugate off Gabon (Figs 4–6). In most of the passive margins worldwide the nature and the precise evolution of the transitional domain from early rifting to break-up of the continental lithosphere is still unclear and subject to intense debate (Whitmarsh et al. 1996; Wilson et al. 2003; Rosendahl et al. 2005; Afilhado et al. 2008; Reston 2009). In an equivalent setting to both margins off northeastern Brazil and Gabon, the well studied magma-poor West Iberia passive continental margin also shows evidence of a transitional domain, where the crust is regarded as: (1) highly extended and intruded continental crust (Whitmarsh & Miles 1995; Whitmarsh et al. 1996); (2) oceanic crust derived by ultra-slow or slow seafloor spreading (Whitmarsh & Sawyer 1996; Srivastava et al. 2000); and (3) mantle exhumed during continental rifting by shear tectonics (Lavier & Manatschal 2006; Péron-Pinvidic et al. 2007; Reston et al. 2007; Sibuet et al. 2007; Reston 2009). Recent studies along this margin have indicated along-margin structural changes. While there is strong evidence of mantle exhumation along the central and northern segment of the West Iberia margin (Lavier & Manatschal 2006; Péron-Pinvidic et al. 2007), there is no clear evidence of an exhumation phase in the southern segment of the same margin (Afilhado et al. 2008).

The term 'proto-oceanic crust' (POC) was adopted in previous studies of the conjugate West Africa and Brazilian margins and defined as a possible anomalous crust emplaced prior to normal oceanic crust formation (Mohriak & Rosendahl 2003; Rosendahl et al. 2005). These studies assumed the 'proto-oceanic crust' to be fundamentally oceanic in nature, where POC is coupled to and subsides with the oceanic crust, exhibiting similar thermal

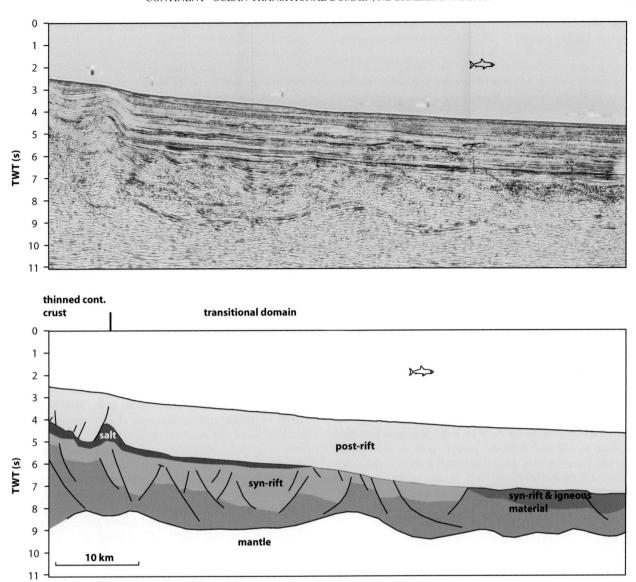

Fig. 8. Seismic example and interpretation of the transitional domain along MCS profile C-4125 (location in Fig. 2).

subsidence histories. Our analysis of seismic reflection data integrated with gravity modelling, however, casts doubt on the previous interpretation as we believe that the transitional domain observed along the conjugate northeastern Brazilian and Gabon margins is not necessarily oceanic in nature.

At several magma-dominated margins the tectonic structures at the COT zone cannot be properly resolved as they are buried beneath thick piles of lava. The deep crustal structures along the Camamu/Almada basins are not being masked by salt and/or by magmatic materials, providing a unique setting for the transitional domain to be imaged and for resolving structures related to the progressive development of a rift system through continental break-up to seafloor spreading. In this way, our analysis provides adequate means to study and refine the structural architecture and nature of the COT domain.

Exhumed mantle?

The transitional domain along the offshore northeastern Brazilian margin is characterized by rotated fault-blocks and wedge-shaped syn-rift sedimentary sequences, and occupies a zone of approximately 60 km in width. Within the transitional domain, the high-amplitude and undulated 'M-reflector' observed, at a depth of 7–9 s (TWT) shows along-margin amplitude variations. In particular, the 'M-reflector' along the Camamu Basin (Fig. 7) exhibits lower reflection coefficient than the corresponding reflector along the Almada Basin (Fig. 8). This observation has obvious implications for the composition of the crust at the transitional domain, although due to the non-unique interpretation of gravity anomalies it is not possible to account for detailed crustal heterogeneities. Along-margin variations are also observed on the gravity anomaly map (Figs 2 & 3), where a discernible gravity high characterizes the transitional domain along transect A–A′. This gravity high is related to the geometry of the Moho discontinuity and does not appear to be associated with magmatism, as earlier proposed by Davison et al. (1999).

The outcome of the seismic interpretation and gravity modelling of transects A–A′ and B–B′ (Figs 4 & 5) shows similarities to the results proposed in a similar study along the Rio Muni transform margin (Equatorial Guinea), West Africa, by Wilson et al. (2003). Similar to previous studies, however, this study also preferred the term 'proto-oceanic crust', which in this case is believed to comprise serpentinized peridotite ridges confined to discrete segments by several fracture zones. At the Rio Muni margin, serpentinized mantle was proposed to be able to move into the space generated by the oblique opening (Wilson et al. 2003). In contrast, the constructed transects A–A', B–B' and C–C' in this study are oriented parallel to the proposed fracture zones and thus are not segmented in the same way.

In accordance with the relationship between density and serpentinized mantle fraction (Miller & Christensen 1997), the 2800 and 2900 kg m^{-3} density values required by gravity modelling for the rotated fault-blocks at the transitional domain (Figs 4–6) correspond to a serpentinization fraction of c. 0.64 and 0.52, respectively. With such extreme crustal thinning as observed along the study area (Figs 4 & 5), local outcrop of serpentinized peridotite ridges across the remaining thin veil of crust may not be unusual (Sibuet 1987). If this is the case, serpentinized upper lithospheric mantle has become unroofed and emplaced at the basement surface (e.g. Whitmarsh et al. 2001; Wilson et al. 2003; Lavier & Manatschal 2006; Reston & Pérez-Gussinyé 2007; Reston 2009). In this setting, the variable reflection coefficient observed on the seismic data (Figs 7 & 8) and the variable densities required for the rotated fault-blocks for transects A–A' (Fig. 4) may be explained by diverse fractions of serpentinization and magmatic modification of the continental mantle.

Unlike the central and northern segment of the West Iberia margin, however, the present study of the Camamu/Almada Basin does not clearly reveal evidence of an exhumation phase (Figs 7 & 8). The Moho discontinuity shallows considerably at the thinned-transitional domain boundary, yet it does not cut through the crystalline crust at the transitional domain. On the contrary, the Moho discontinuity flattens out and continues beneath the rotated fault-blocks at the transitional domain (Figs 4 & 5). Finally, the rifting model for the West Iberia margin as proposed by Lavier & Manatschal (2006) predicts both an exhumation phase and a considerably large transitional domain (>100 km), as observed in the Iberia Abyssal Plain. In contrast, the transitional domain along the study area occupies only a zone of approximately 60 km in width (Figs 4 & 5).

'S-reflector' resemblance?

The high-amplitude and undulated 'M-reflector' observed at a depth of 7–9 s (TWT) along the northeastern Brazilian margin (Figs 7 & 8) shows remarkable resemblances to the prominent and undulated 'S-reflector' observed along the Galicia Bank continental margin (e.g. De Charpal et al. 1978; Sibuet 1987; Hoffmann & Reston 1992; Whitmarsh et al. 1996; Reston & Pérez-Gussinyé 2007; Reston 2009). At this margin, seismic velocities indicate that parts of the 'S-reflector' correspond to the boundary between highly thinned, fractured continental crust and the underlying zone of partially serpentinized mantle (De Charpal et al. 1978; Hoffmann & Reston 1992). The 'S-reflector' is therefore interpreted as the crust–mantle boundary at the time of margin formation (Boillot et al. 1987), characterizing a major detachment fault that was active during rifting and on which the overlying fault-blocks ride (Hoffmann & Reston 1992; Sibuet 1992; Reston et al. 2007). Only at its eastern end does the 'S-reflector' cut through the crust and is interpreted as a mid-crustal reflector (Whitmarsh et al. 1996). Similarly, the analyses performed in this study suggest that the observed undulated 'M-reflector' characterizes a major detachment surface that cuts through the crust at its western end and on which the overlying fault-blocks ride (Fig. 5).

Unlike the 'S-reflector', the 'M-reflector' described in this study does not appear to characterize a continental crust/serpentinized mantle boundary, but rather the boundary between an extremely thinned crust and normal lithospheric mantle. The undulation and discontinuity of this reflector may be distortions and imaging problems due to the passage of the seismic energy through fault-blocks characterized by a laterally varying velocity structure; equivalent observations are present in the West Iberia margin (Reston et al. 2007). A gravity model that accounts for serpentinized mantle beneath the 'M-reflector' is also feasible as serpentinized mantle may have a wide range of density values; however, gravity modelling alone cannot resolve this issue. Other possible explanations for the crust at the transitional domain that are applicable for the investigated conjugate margins in this study can be summarized as: slow spreading (atypic) oceanic crust (e.g. Srivastava & Keen 1995; Srivastava & Roest 1999); thinned and magmatically intruded continental crust (e.g. Whitmarsh & Miles 1995; Russell & Whitmarsh 2003); and/or exhumed lower continental crust material (e.g. Rosenbaum et al. 2005).

Crustal extension mode and rate

Typical rotated fault-blocks capped by sedimentary layers which parallel the sloping surfaces of the blocks are usually seen across passive continental margins that were subjected to large extension and rifting. Several studies have shown that oceanic crust formed at the early stages of break-up opening may apparently exhibit evidence of brittle deformation. In particular, excessively slow seafloor spreading (3–10 mm/year) and ridge axis propagation play an important role in thinning the oceanic crust, thereby faulting and rotating the oceanic crust (Srivastava & Keen 1995; Srivastava & Roest 1999). In the absence of identifiable magnetic anomalies, best estimates of initial half-spreading rate of c. 20 mm/year have been proposed for the plate separation at the northeastern Brazilian margin (Greenroyd et al. 2008; Müller et al. 2008; Torsvik et al. 2009), suggesting that the thin crust at the transitional domain is unlikely to be the result of ultra-slow spreading. Finally, we believe we have identified syn-rift deposits at the transitional domain along the northeastern Brazilian margin, implying that the crust there is fundamentally continental in nature (Figs 7 & 8).

Reconstructions of West Gondwana suggest that the São Francisco and Congo cratons were connected into a single continental block prior to the opening of the South Atlantic (Daziel 1997; Alkmim & Marshak 1998; Alkmim et al. 2006; De Wit et al. 2008). The Palaeoproterozoic orogeny resulted in crustal thickening within the São Francisco Craton, leading to high-grade metamorphism of the different terranes (Leite et al. 2008). Extension of overthickened continental crust is commonly characterized by an early core complex stage of extension, by movement along low-angle normal detachment faults, which exhume high-grade metamorphic lower crustal rocks (Lister & Davis 1989; Buck 1991; Hopper & Buck 1996; Rosenbaum et al. 2005). The core complex stage of extension is followed by a later stage of crustal-scale rigid block faulting (Rosenbaum et al. 2005). The model of core complex extension can explain the nature and character of the crust at the transitional domain, where the difference in densities for this domain required by gravity modelling for transects A–A' may be attributed to a variable grade of metamorphism and/or a variable amount of mafic intrusives in the exhumed lower crust. In this way, the weak and intruded crust at the transitional domain may deform differently from the unintruded crust (Fig. 4). It can be definitely postulated that there is no evidence of an exhumation phase along the study area. As 'M-reflector' characterizes a major detachment surface, the eventually exhumed lower crust or exhumed mantle cannot be situated above this detachment. This implies that the rotated fault-blocks at the transitional domain cannot be composed of exhumed material. In this setting, the rotated fault-blocks observed at the transitional domain probably characterize thinned and magmatically intruded continental crust. In the initial stage of the propagating rift model (Martin 1984; Whitmarsh & Miles 1995), rifting of the continental crust leading to extension is characterized by the formation of tilted fault-blocks and half-grabens. The rising asthenosphere enables magma generated by decompression melting to begin penetrating the thinnest continental crust (transitional domain) in a diffuse manner (Whitmarsh & Miles 1995). The diffuse magmatic intrusion stage affects mainly the continent transition zone, increasing its proper densities. Consequently, the weak and magmatically intruded crust may isostatically subside relative to the unintruded crust as a consequence of the increased mean density.

Apparently, the observed seismic reflection signature and the gravity modelling results indicate that the crust at the transitional domain can be neither interpreted as normal oceanic crust, nor as exhumed lower crust/mantle. The observed 'M-reflector' is interpreted as a major detachment surface active during rifting that defines the Moho discontinuity at the transitional domain. The absence of clearly defined magnetic anomalies, as this margin was formed during the Cretaceous magnetically quite superchron, and the lack of wide-angle seismic velocity data pose an additional difficulty for the evaluation of the alternative hypotheses mentioned above. Nevertheless, the results obtained in this study favour a model in which the rotated fault-blocks located at the transitional domain are continental in nature, overlying a high-amplitude and undulated 'M-reflector'. As serpentinized mantle may have a wide range of density values, the hypothesis that the 'M-reflector' overlies a serpentinized mantle is also viable.

Total rift setting

The process of lithospheric extension that rifts and thins continental margins, resulting in symmetrical or asymmetrical lithospheric crustal structure, is much debated and focuses on the pure shear and simple shear end-member models (McKenzie 1978; Wernicke 1981; Buck et al. 1988). For instance, upper crust decoupling from the lower crust and mantle by strain localization into detachment zones allows the dislocation of deformation (Lavier & Manatschal 2006; Regenauer-Lieb et al. 2006; Weinberg et al. 2007). In such a setting, and when extension is slow, the lithosphere cools as it extends, thereby continuously renewing its frictional plastic layer and promoting asymmetrical behaviour and widening of the rift zone (Kusznir & Park 1987; Bassi et al. 1993; Huismans & Beaumont 2002).

Asymmetrical behaviour is evident on the gravity model of the conjugate pairing transect A–A' (northeastern Brazilian margin) and transect C–C' (Gabon margin) (Fig. 9), where a wide salt basin along the Gabon margin is conjugate to minor salt deposits along the northeastern Brazilian margin. Rifting and crustal thinning across the conjugate margins also indicate striking asymmetry. The Gabon margin is characterized by a wide area where major crustal thinning occurs, by a corresponding gradual Moho shallowing, and by relatively thin syn-rift sediments overlying the attenuated crust (Figs 9 & 10). On the other hand, the northeastern Brazilian margin is characterized by abrupt Moho shallowing and by a narrow zone of crustal thinning, where the extremely attenuated crust is overlain by relatively thick syn-rift sediments (Figs 9 & 10). This striking asymmetry is also evident on the Bouguer-corrected gravity anomaly field, where a wider area of negative–positive gradients on the Bouguer-corrected gravity anomaly map is observed along the Gabon margin, indicating a wider area of crustal thinning (Fig. 3b). The study indicates a narrow and sharp crustal tapper along the northeastern Brazilian margin that is conjugate to a wide and gentle crustal tapper along the Gabon margin.

The asymmetry observed on the conjugate northeastern Brazilian–Gabon margins may reflect an asymmetry in the process of extension (Reston & Pérez-Gussinyé 2007) and/or it may be related to the tectonic heritage of this segment (e.g. Lavier & Manatschal 2006; Aslanian et al. 2009). In this way, asymmetrical behaviour may be controlled by the existence of an inherited strong lower crust along the Gabon margin (Fig. 6), as well as by detachment faulting (e.g. Lavier & Manatschal 2006). Plate reconstructions indicate that the margin segment located south of the Salvador–N'Komi transfer zones has a narrow–wide conjugate margin configuration (Figs 9 & 10). In

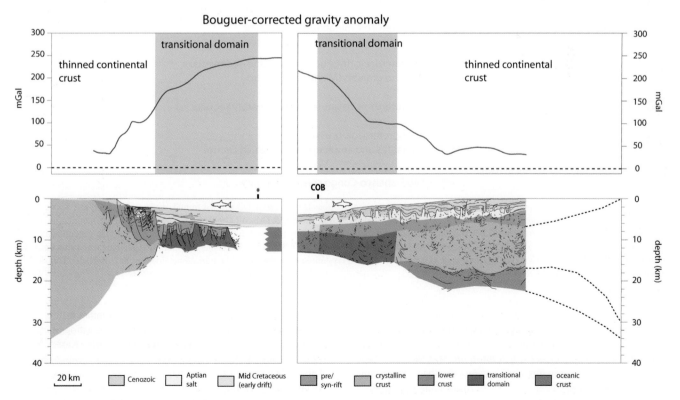

Fig. 9. (a) Conjugate transects A–A' and extended transect C–C'. COB(K), continent–ocean boundary (Karner & Driscoll 1999); COB(R), continent–ocean boundary (Rosendahl et al. 2005); COB(T), continent–ocean boundary (Torsvik et al. 2009); COB, this study; the asterisk indicates 'line of fit' used for plate reconstructions; POC(R), proto-oceanic crust (Rosendahl et al. 2005).

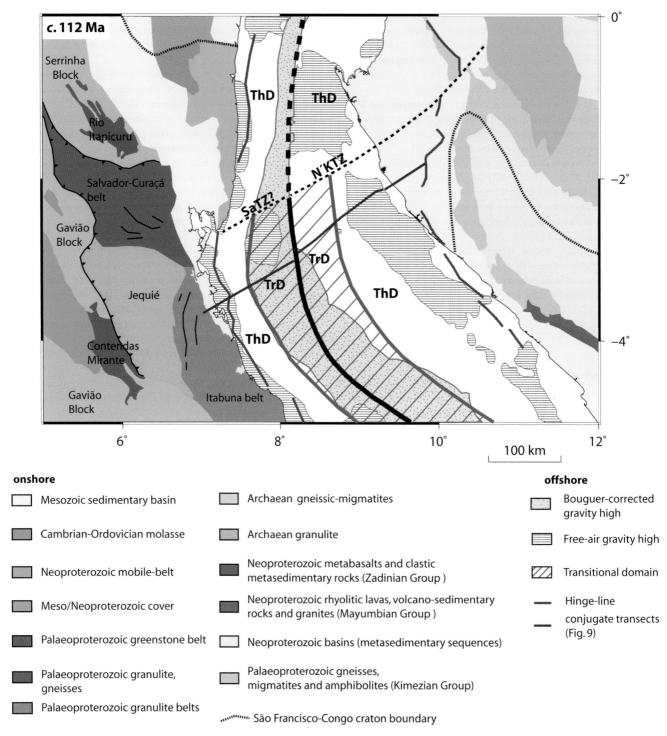

Fig. 10. Offshore: summary of prominent features observed on plate reconstructions (Fig. 3a, b); COB revised from Torsvik et al. (2009) and Karner & Driscoll (1999) fitting observations made in this study. Hinge-lines are derived from Karner & Driscoll (1999) for the northeastern Brazilian margin and Mounguengui & Guiraud (2009) for the Gabon margin. Onshore northeastern Brazilian margin: simplified geological and structural map generalized from Barbosa & Sabaté (2003), Destro et al. (2003), Alkmim et al. (2006) and Leite et al. (2009). Onshore Gabon margin: simplified geological and structural map generalized from Tack et al. (2001) and Alkmim et al. (2006). ThD, thinned crystalline crust domain; TrD, transitional domain; SaT, Salvador Transfer Zone; N'KT, N'Komi Transfer Zone.

this setting, asymmetrical lithospheric extension resulted in the formation of the thinned continental crust domain prior to the formation of the approximately symmetrical transitional domain (Fig. 10). This is further accounted for in the next section, where the analysis performed in this study suggests that the tectonic evolution of the central segment of the South Atlantic margin reflects a polyphase rifting evolution mode.

The tectono-sedimentary evolution along the central segment of the South Atlantic sedimentary basins can be summarized as indicating: (1) early Neocomian to early Barremian brittle deformation that resulted in block rotation and the deposition of characteristic syn-rift sediment wedges (early syn-rift) (Karner et al. 2003; Karner & Gambôa 2007); (2) early Barremian to late Aptian deposition of thick and tectonically undeformed 'sag' basins, including the evaporites, deposited under shallow water conditions (late syn-rift) (Moulin et al. 2005; Karner & Gambôa 2007); and (3) Albian seafloor spreading followed by significant post-rift accommodation in shallow water conditions (Karner &

Gambôa 2007). Our analysis also indicates that vertical motion prevailed in comparison to horizontal motion along the thinned continental crust domain in the northeastern Brazilian margin. In particular, the creation of syn-rift accommodation space along this domain is not characterized by horst and graben morphologies as expected for uniform extension. The observed reflections within the syn-rift basins are mostly parallel to the basement and are thus not affected by deformation that would imply significant horizontal motions (Fig. 9). On the other hand, the transitional domain suggests block rotation and the deposition of characteristic syn-rift sediment wedges (Figs 7 & 8).

The above observations are generally incompatible with uniform extension and are in better agreement with depth-dependent extension, in which stretching of the lower crust and lithospheric mantle greatly exceeds that of the upper crust (Driscoll & Karner 1998; Davis & Kusznir 2004; Karner & Gambôa 2007; Kusznir & Karner 2007). In this setting, depth-dependent thinning results in the development of relatively non-deformed upper crust (i.e. the upper plate) and a ductile-deformed lower crust and lithospheric mantle (i.e. lower plate) (Driscoll & Karner 1998).

Conjugate margin evolution

The boundary between the thinned continental crust domain and the transitional domain along the northeastern Brazilian margin coincides with a prominent Bouguer-corrected gravity high, implying abrupt crustal thinning. In contrast, the same boundary along the Gabon margin deviates gradually northwards from the prominent Bouguer-corrected gravity high (Fig. 10). This observation indicates that the Moho discontinuity along the Gabon margin shallows in a more abrupt manner in the south and progressively more gradual towards the north (Figs 9 & 10).

A similar transitional domain is also observed northwards of the prominent conjugate Salvador–N'Komi transfer zone (e.g. Rosendahl & Groschel-Becker 1999). However, the analyses performed in this study suggest that the transitional domain terminates at the conjugate Salvador–N'Komi transfer zone at break-up time (c. 112 Ma; Torsvik et al. 2009). Therefore, the conjugate Salvador–N'Komi transfer zone may have acted as an intra-plate decoupling zone where a different tectono-sedimentary evolution is observed and may be associated with differential extension between the northern and the southern segments of the Gabon margin (Figs 3, 9 & 10). A similar character for the conjugate Salvador–N'Komi transfer system has also been indicated by previous studies (e.g. Teisserence & Villemin 1989; Wannesson et al. 1991).

The tectonomagmatic evolution of the northeastern Brazilian margin and its conjugate off Gabon is complex and may reflect a polyphase rifting evolution mode, which is associated with a complex time-dependent thermal structure of the lithosphere (e.g. Lavier & Manatschal 2006; Péron-Pinvidic et al. 2007; Huismans & Beaumont 2008; Aslanian et al. 2009). We have developed a conceptual model for the tectono-sedimentary evolution of the conjugate northeastern Brazilian–Gabon (Fig. 11). According to Lavier & Manatschal (2006), the initial extension phase is characterized by distributed listric-normal faults and symmetrical extension, resulting in block rotation and the deposition of characteristic syn-rift sedimentary wedges along the conjugate margins. The analysis performed in this study does not recognize an initial symmetrical extension phase.

The transition from a broadly distributed and symmetrical extension to asymmetrical and localized rifting is directly controlled by the existence of a strong gabbroic lower crust. In this setting, the listric-normal faulting is overprinted by localized detachment faulting (Lavier & Manatschal 2006). The thinned continental crust domain suggested in this study (Figs 9 & 10) would document the onset of the thinning phase as proposed by Lavier & Manatschal (2006), where the asymmetry between the conjugate margins is evident and probably controlled by a strong lower crust (Figs 9–11a). Since the distribution and style of subsidence

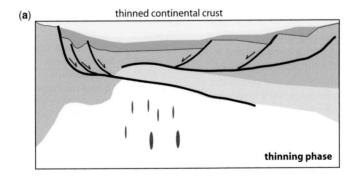

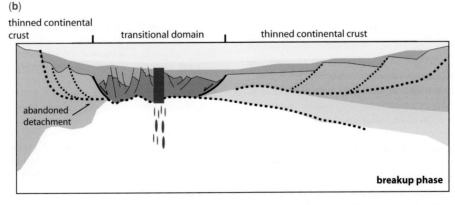

Fig. 11. Conceptual model showing the tectono-sedimentary and magmatic polyphase evolution of the conjugate northeastern Brazilian and Gabon margins. (**a**) Thinning phase, characterized by detachment faulting and depth-dependent stretching, which resulted in the deposition of thick and tectonically undeformed 'sag' basins. (**b**) Break-up phase, the increase in volcanic activity followed by the abandonment of the detachment fault may have 'interrupted' the extensional system, implying a failed exhumation phase that is replaced instead by continental break-up.

observed on both conjugate margins suggest depth-dependent stretching and upper plate subsidence patterns, detachments that separate the brittle and ductile deformation in the crust are proposed for both conjugate margins (Fig. 11a). During the thinning phase, the thick and tectonically undeformed 'sag' basin may develop under shallow water conditions (Moulin et al. 2005; Karner & Gambôa 2007). The subsequent exhumation phase, which predicts exhumation of middle and lower crust followed by mantle exhumation (Lavier & Manatschal 2006), is not evident along the conjugate margins off northeastern Brazil/Gabon. Even though the margins in the central segment of the South Atlantic show magma-poor affinity (Contrucci et al. 2004; Moulin et al. 2005; Sawyer et al. 2007), these margins also experienced magmatism during the formation of the transitional domain and break-up (Figs 7 & 8). In this setting, at the final stage of extension, magmatism increases oceanward, affecting mainly the crust at the transitional domain. The magma supply allows the release of stress by dyking, which in turn may effectively cut off the detachment fault (Ebinger & Casey 2001). The increase of volcanic activity followed by the abandoning of the detachment fault may have 'interrupted' the extensional system, implying a failed exhumation phase that is replaced instead by continental break-up.

Conclusions

An integrated analysis of regional seismic reflection profiles and potential field data across the magma-poor Camamu/Almada margin and its conjugate off Gabon, complemented by crustal-scale gravity modelling and plate reconstructions has been conducted. The analyses were used to reveal and illustrate the relationship of crustal structure to regional variation of potential field anomalies, to refine and constrain the COB, as well as to study the structural architecture and nature of the COT domain; the latter is defined as the part of the lithosphere which is located between the thinned crystalline crust domain and the first appearance of oceanic crust formed by seafloor spreading.

The study indicates the presence of along-margin tectono-sedimentary changes along the northeastern Brazilian margin and its conjugate off Gabon, and that pre-existing continental lineaments, transfer zones and various basement terranes exerted important influence and rheologic control on the structural development of the Mesozoic rifting and break-up. In particular, the prominent conjugate Salvador–N'Komi transfer zones appear to be first-order structural elements, governing the margin segmentation and evolution and may have acted as an intraplate decoupling zone.

Our analysis indicates the existence of a transitional domain between the thinned continental crust and the oceanic crust along the northeastern Brazilian margin and its conjugate off Gabon. Offshore northeastern Brazil, the transitional domain is characterized by rotated fault-blocks and wedge-shaped syn-rift sedimentary sequences overlying a high-amplitude and undulated reflector ('M-reflector') which, in turn, shows remarkable resemblances to the prominent and undulated 'S-reflector' observed along the Galicia Bank continental margin. However, unlike the 'S-reflector', the undulated reflector described in this study does not appear to characterize a continental crust/serpentinized mantle boundary, but rather the boundary between an extremely thinned, continental crust and normal lithospheric mantle. The 'M-reflector' is interpreted as a major detachment surface active during rifting that defines the Moho discontinuity at the transitional domain. Our study favours a model in which the rotated fault-blocks located at the transitional domain are continental in nature, overlying the high-amplitude and undulated 'M-reflector'. However, as serpentinized mantle may have a wide range of density values, a model that accounts for serpentinized mantle beneath the 'M-reflector' is also feasible.

Striking asymmetry is observed south of the Salvador–N'Komi transfer zones, characterized by narrow–wide thinned continental crust conjugate pair configurations. The conjugate transitional domain is approximately symmetrical. In this setting, asymmetrical lithospheric extension resulted in the formation of the thinned continental crust domain prior to the formation of the approximately symmetrical transitional domain. Vertical motion prevailed in comparison to horizontal motion along the thinned continental crust domain across the northeastern Brazilian margin reflecting depth-dependent extension. On the other hand, the transitional domain suggests block rotation and the deposition of characteristic syn-rift sedimentary wedges. A conceptual model is developed based on all our results and integrates, in a pragmatic way, several tectonic models at similar settings and regional-scale observations, providing a highly constrained polyphase model for the tectono-magmatic evolution of the conjugate Camamu/Almada–Gabon margins. This study indicates that an exhumation phase, which predicts exhumation of middle and lower crust followed by mantle exhumation, is not evident along the conjugate margins off northeastern Brazil/Gabon. The increase of volcanic activity followed by the abandoning of the detachment fault may have 'interrupted' the extensional system, implying a failed exhumation phase that is replaced instead by continental break-up. The polyphase rifting evolution mode is associated with a complex time-dependent thermal structure of the lithosphere.

We thank Statoil, TGS-NOPEC and ION for providing the MCS profiles used in this study. Eni E&P is acknowledged for providing the time and resources (to Filippos Tsikalas) to fulfil the study. We also thank the two reviewers Erik Lundin and John Argent for their constructive comments and improvements to the manuscript. The study is part of the University of Oslo Conjugate South Atlantic Margin Studies project funded by Statoil.

References

Afilhado, A., Matias, L., Shiobara, H., Hirn, A., Mendes-Victor, L. & Shimamura, H. 2008. From unthinned continent to ocean: the deep structure of the West Iberia passive continental margin at 38°N. *Tectonophysics*, **458**, 9–50.

Alkmim, F. F. & Marshak, S. 1998. Transamazonian Orogeny in the Southern São Francisco Craton Region, Minas Gerais, Brazil: evidence for Paleoproterozoic collision and collapse in the Quadrilátero Ferrífero. *Precambrian Research*, **90**, 29–58.

Alkmim, F. F., Marshak, S., Pedrosa-Soares, C., Peres, G. G., Cruz, S. C. P. & Whittington, A. 2006. Kinematic evolution of the Araçuaí–West Congo orogen in Brazil and Africa: nutcracker tectonics during the Neoproterozoic assembly of Gondwana. *Precambrian Research*, **149**, 43–64.

Aslanian, D., Moulin, M. et al. 2009. Brazilian and African passive margins of the Central Segment of the South Atlantic Ocean: kinematic constraints. *Tectonophysics*, **468**, 98–112.

Barbosa, J. S. F. & Sabaté, P. 2003. Colagem Paleoproterozóica de places Arqueanas do Cráton do São Francisco na Bahia. *Revista Brasileira de Geociências*, **33**, 7–14.

Bassi, G., Keen, C. E. & Potter, P. 1993. Contrasting styles of rifting: models and examples from the eastern Canadian margin. *Tectonics*, **12**, 639–655.

Blaich, O. A., Tsikalas, F. & Faleide, J. I. 2008. Northeastern Brazilian margin: regional tectonic evolution based on integrated analysis of seismic reflection and potential field data and modeling. *Tectonophysics*, **458**, 51–67.

Blaich, O. A., Faleide, J. I., Tsikalas, F., Franke, D. & León, E. 2009. Crustal-scale architecture and segmentation of the Argentine margin and its conjugate off South Africa. *Geophysical Journal International*, **178**, 85–105.

Boillot, G., Winterer, E. L. et al. 1987. *Proceedings of the Ocean Drilling Program, Initial Reports*, **103**. Ocean Drilling Program, College Station, TX.

Buck, W. R. 1991. Models of continental lithospheric extension. *Journal of Geophysical Research*, **96**, 20161–10178.

Buck, W. R., Martinez, F., Steckler, M. & Cochran, J. 1988. Thermal consequences of lithospheric extension: pure and simple. *Tectonics*, **7**, 213–234.

Castro, A. C. M. 1988. *Structural evolution of the Sergipe-Alagoas Basin, Brazil*. Doctorate Thesis, Rice University, TX.

Chang, H. K., Kowsmann, R. O., Figueiredo, A. M. F. & Bender, A. A. 1992. Tectonics and stratigraphy of the East Brazil Rift System; an overview. *Tectonophysics*, **213**, 97–138.

Contrucci, I., Matias, L. *et al.* 1988. Thermal consequences of lithospheric extension: pure and simple. *Tectonics*, **7**, 213–234.

Contrucci, I., Matias, L. *et al.* 2004. Deep structure of the West African continental margin (Congo, Zaire, Angola), between 5 degrees S and 8 degrees S, from reflection/refraction seismic and gravity data. *Geophysical Journal International*, **158**, 529–553.

Davis, M. & Kusznir, N. J. 2004. Depth-dependent lithospheric stretching at rifted continental margins. *In*: Karner, G. D. (ed.) *Proceedings of NSF Rifted Margins Theoretical Institute*. Columbia University Press, New York, 92–136.

Davison, I. 1999. Tectonics and hydrocarbon distribution along the Brazilian South Atlantic margin. *In*: Cameron, N. R., Bate, R. H. & Clure, V. S. (eds) *The Oil and Gas Habitats of the South Atlantic*. Geological Society, London, Special Publications, **153**, 133–151.

Davison, I. 2007. Geology and tectonics of the South Atlantic Brazilian salt basins. *In*: Ries, A. C., Butler, R. W. H. & Graham, R. H. (eds) *Deformation of the Continental Crust: The Legacy of Mike Coward*. Geological Society, London, Special Publications, **272**, 345–359.

Daziel, I. W. D. 1997. Neoproterozoic–Paleozoic geography and tectonics: review, hypothesis, environmental speculation. *Geological Society of America Bulletin*, **109**, 16–42.

Dean, S. M., Minshull, T. A., Whitmarsh, R. B. & Louden, K. E. 2000. Deep structure of the ocean–continent transition in the Southern Iberia Abyssal Plain from seismic refraction profiles. The IAM-9 transect at 40°20′N. *Journal of Geophysical Research*, **105**, 5859–5885.

De Charpal, O., Guennoc, P., Montadert, L. & Roberts, D. G. 1978. Rifting and subsidence of the northern continental margin of the Bay of Biscay. *Nature*, **275**, 706–711.

De Wit, M. J., Brito Neves, B. B., Trouw, R. A. J. & Pankhurst, R. J. 2008. Pre-Cenozoic correlations across the South Atlantic region: the ties that bind. *In*: Pankhurst, R. J., Trouw, R. A. J., Brito Neves, B. B. & De Wit, M. J. (eds) *West Gondwana: Pre-Cenozoic Correlations Across the South Atlantic Region*. Geological Society, London, Special Publications, **294**, 1–8.

Destro, N., Alkmim, F. F., Magnavita, L. P. & Szatmari, P. 2003. The Jeremoabo transpressional transfer fault, Recôncavo-Tucano Rift, NE Brazil. *Journal of Structural Geology*, **25**, 1263–1279.

Driscoll, N. W. & Karner, G. D. 1998. Lower crustal extension across the northern Carnarvon basin, Australia: evidence for an eastward dipping detachment. *Journal of Geophysical Research*, **103**, 4975–4992.

Dunbar, J. A. & Sawyer, D. S. 1988. Continental rifting at pre-existing lithospheric weaknesses. *Nature*, **242**, 565–571.

Dunbar, J. A. & Sawyer, D. S. 1989. How pre-existing weaknesses control the style of continental break-up. *Journal of Geophysical Research*, **94**, 7278–7292.

Ebinger, C. J. & Casey, M. 2001. Continental breakup in magmatic provinces: an Ethiopian example. *Geology*, **29**, 527–530.

Eldholm, O. & Grue, K. 1994. North Atlantic volcanic margins: dimensions and production rates. *Journal of Geophysical Research*, **99**, 2955–2968.

Franke, D., Neben, S., Ladage, S., Schreckenberger, B. & Hinz, K. 2007. Margin segmentation and volcano-tectonic architecture along the volcanic margin off Argentina/Uruguay, South Atlantic. *Marine Geology*, **244**, 46–67.

Gladczenko, T. P., Skogseid, J. & Eldholm, O. 1998. Namibia volcanic margin. *Marine Geophysical Researches*, **20**, 313–341.

Gomes, P. O., Gomes, B. S., Palma, J. J. C., Jinno, K. & de Souza, J. M. 2000. Ocean–continental transition and tectonic framework of the oceanic crust at the continental margin off NE Brazil: results of LEPLAC project. *In*: Mohriak, W. & Talwani, M. (eds) *Atlantic Rifts and Continental Margins*. American Geophysical Union, Washington, Geophysical Monographs **115**, 261–288.

Greenroyd, C. J., Peirce, C., Rodger, M., Watts, A. B. & Hobbs, R. W. 2008. Demerara Plateau – the structure and evolution of a transform passive margin. *Geophysical Journal International*, **172**, 549–564.

Hinz, K., Neben, S., Schreckenberger, B., Roeser, H. A., Block, M., de Souza, G. K. & Meyer, H. 1999. The Argentine continental margin north of 48°S: sedimentary successions, volcanic activity during breakup. *Marine and Petroleum Geology*, **16**, 1–25.

Hoffmann, H. J. & Reston, T. J. 1992. The nature of the S reflector beneath the Galicia Bank rifted margin: preliminary results from pre-stack depth migration. *Geology*, **20**, 1091–1094.

Hopper, J. R. & Buck, W. R. 1996. The effect of lower crustal flow on continental extension and passive margin formation. *Journal of Geophysical Research*, **101**, 20175–20194.

Huismans, R. S. & Beaumont, C. 2002. Asymmetric lithospheric extension; the role of frictional plastic strain softening inferred from numerical experiments. *Geology*, **30**, 211–214.

Huismans, R. S. & Beaumont, C. 2008. Complex rifted continental margins explained by dynamical models of depth-dependent lithospheric extension. *Geology*, **36**, 163–166.

Jackson, M. P. A., Cramez, C. & Fonck, J. M. 2000. Role of subaerial volcanic rocks and mantle plumes in creation of South Atlantic margins: implications for salt tectonics and source rocks. *Marine and Petroleum Geology*, **17**, 477–498.

Jacques, J. M. 2003. A tectonostratigraphic synthesis of the sub-Andean basins; inferences on the position of South American intraplate accommodation zones and their control on South Atlantic opening. *Journal of the Geological Society, London*, **160**, 703–717.

Karner, G. D. & Driscoll, N. W. 1999. Tectonic and stratigraphic development of the West African and eastern Brazilian margins; insights from quantitative basin modeling. *In*: Cameron, N. R., Bate, R. H. & Clure, V. S. (eds) *The Oil and Gas Habitats of the South Atlantic*. Geological Society, London, Special Publications, **153**, 11–40.

Karner, G. D. & Gambôa, L. A. P. 2007. Timing and origin of the South Atlantic pre-salt sag basins and their capping evaporites. *In*: Schreiber, B. C., Lugli, S. & Babel, M. (eds) *Evaporites Through Space and Time*. Geological Society, London, Special Publications, **285**, 15–35.

Karner, G. D., Driscoll, N. W. & Barker, D. H. N. 2003. Syn-rift regional subsidence across the West African continental margin: the role of lower plate ductile extension. *In*: Arthur, T. J., MacGregor, D. S. & Cameron, N. R. (eds) *Petroleum Geology of Africa: New Themes and Developing Technologies*. Geological Society, London, Special Publications, **207**, 105–129.

Kusznir, N. J. & Karner, G. D. 2007. Continental lithospheric thinning and breakup in response to upwelling divergent mantle flow: application to the Woodlark, Newfoundland and Iberia margins. *In*: Karner, G. D., Manatschal, G. & Pinheiro, L. M. (eds) *Imaging, Mapping and Modelling Continental Lithosphere Extension and Breakup*. Geological Society, London, Special Publications, **282**, 389–419.

Kusznir, N. J. & Park, R. G. 1987. The extensional strength of the continental lithosphere: its dependence on geothermal gradient, crustal composition and thickness. *In*: Coward, M. P., Dewey, J. F. & Hancock, P. L. (eds) *Continental Extensional Tectonics*. Geological Society, London, Special Publications, **28**, 35–42.

Lavier, L. L. & Manatschal, G. 2006. A mechanism to thin the continental lithosphere at magma-poor margins. *Nature*, **440**, 324–328.

Leite, C. De M. M., Barbosa, J. S. F., Goncalves, P., Nicollet, C. & Sabaté, P. 2009. Petrological evolution of silica-undersaturated sapphirine-bearing granulite in the Paleoproterozoic Salvador–Curaçá Belt, Bahia, Brazil. *Gondwana Research*, **15**, 49–70.

Lister, G. S. & Davis, G. A. 1989. The origin of metamorphic core complexes and detachment faults formed during Tertiary continental extension in the northern Colorado River region, U.S. *Journal of Structural Geology*, **11**, 65–94.

Lister, G. S., Etheridge, M. A. & Symonds, P. A. 1986. Detachment faulting and the evolution of passive continental margins. *Geology*, **14**, 246–250.

Marks, K. M. 1996. Resolution of the Scripps/NOAA marine gravity field from satellite altimetry. *Geophysical Research Letters*, **23**, 2069–2072.

Martin, A. K. 1984. Propagating rifts: crustal extension during continental rifting. *Tectonics*, **3**, 611–617.

Matos, R. M. D. 1992. The northeast Brazilian rift system. *Tectonics*, **11**, 766–791.

Matos, R. M. D. 1999. History of the northeast Brazilian rift system; kinematic implications for the break-up between Brazil and West Africa. *In*: Cameron, N. R., Bate, R. H. & Clure, V. S. (eds) *The Oil and Gas Habitats of the South Atlantic*. Geological Society, London, Special Publications, **153**, 55–73.

Matos, R. M. D. 2000. Tectonic evolution of the Equatorial South Atlantic. *In*: Mohriak, W. & Talwani, M. (eds) *Atlantic Rifts and Continental Margins*. American Geophysical Union, Washington, Geophysical Monographs, **115**, 331–354.

McKenzie, D. P. 1978. Some remarks on the development of sedimentary basins. *Earth and Planetary Science Letters*, **40**, 25–32.

Menezes, P. E de L. & Milhomem, P da S. 2009. Tectônica de sal nas bacias de Cumuruxatiba, do Almada e de Camamu. *In*: Mohriak, W., Szatmari, P. & Couto Anjos, S. M. (eds) *Sal: Geologia e Tectônica*. São Paulo, Brasil, 232–251.

Meyers, J. B., Rosendahl, B. R., Groschel-Becker, H., Austin, J. A., Jr. & Rona, P. A. 1996a. Deep penetrating MCS imaging of the rift-to-drift transition, offshore Douala and North Gabon basins, West Africa. *Marine and Petroleum Geology*, **13**, 791–835.

Meyers, J. B., Rosendahl, B. R. & Austin, J. A., Jr. 1996b. Deep-penetrating MCS images of the South Gabon Basin: implications for rift tectonics and post-breakup salt remobilization. *Basin Research*, **8**, 65–84.

Milani, E. J. & Davison, I. 1988. Basement control and transfer tectonics in the Reconcavo-Tucano-Jatoba Rift, Northeast Brazil. *Tectonophysics*, **154**, 41–70.

Miller, D. J. & Christensen, N. I. 1997. Seismic velocities of lower and upper mantle rocks from slow-spreading Mid-Atlantic Ridge, South of the Kane Transform (MARK). *Proceedings of the Ocean Drilling Program, Scientific Results*, **153**, 437–454.

Mohriak, W. U. & Rosendahl, B. R. 2003. Transform zones in the South Atlantic rifted continental margins. *In*: Storti, F., Holdsworth, R. E. & Salvini, F. (eds) *Intraplate Strike-Slip Deformation Belts*. Geological Society, London, Special Publications, **210**, 211–228.

Mohriak, W. U., Rabelo, J. H. L., Matos, R. D. & Barros, M. C. 1995. Deep seismic reflection profiling of sedimentary basins offshore Brazil; geological objectives and preliminary results in the Sergipe Basin. *Journal of Geodynamics*, **20**, 515–539.

Mohriak, W. U., Bassetto, M. & Vieira, I. S. 1998. Crustal architecture and tectonic evolution of the Sergipe-Alagoas and Jacuípe basins, offshore northeastern Brazil. *Tectonophysics*, **288**, 199–220.

Mohriak, W. U., Bassetto, M. & Vieira, I. S. 2000. Tectonic evolution of the rift basins in the northeastern Brazilian region. *In*: Mohriak, W. & Talwani, M. (eds) *Atlantic Rifts and Continental Margins*. American Geophysical Union, Washington, Geophysical Monographs, **115**, 293–315.

Moulin, M. B., Aslanian, D. *et al.* 2005. Geological constraints on the evolution of the Angolan margin based on reflection and refraction seismic data (Zaïango Project). *Geophysical Journal International*, **162**, 793–810.

Mounguengui, M. M. & Guiraud, M. 2009. Neocomian to early Aptian syn-rift evolution of the normal to oblique-rifted North Gabon Margin (Interior and ŃKomi Basins). *Marine and Petroleum Geology*, **26**, 1000–1017.

Müller, R. D., Sdrolias, M., Gaina, C. & Roest, W. R. 2008. Age, spreading rates, and spreading asymmetry of the world's ocean crust. *Geochemistry Geophysics Geosystems*, **9**, Q04006; doi: 10.1029/2007GC001743.

Péron-Pinvidic, G., Manatschal, G., Minshull, T. A. & Sawyer, D. S. 2007. Tectonosedimentary evolution of the deep Iberia-Newfoundland margins: evidence for a complex breakup history. *Tectonics*, **26**, TC2011; doi: 10.1029/2006TC001970.

Rabinowitz, P. D. & LaBrecque, J. L. 1979. The Mesozoic South Atlantic Ocean and evolution of its continental margins. *Journal of Geophysical Research*, **84**, 5973–6002.

Regenauer-Lieb, K., Weinberg, R. F. & Rosenbaum, G. 2006. The effect of energy feedbacks on continental strength. *Nature*, **442**, 67–70.

Reston, T. J. 2009. The structure, evolution and symmetry of the magma-poor rifted margins of the North and Central Atlantic: a synthesis. *Tectonophysics*, **468**, 6–27.

Reston, T. J. & Pérez-Gussinyé, M. 2007. Lithospheric extension from rifting to continental breakup at magma-poor margins: rheology, serpentinisation and symmetry. *International Journal of Earth Sciences*, **96**, 1033–1046.

Reston, T. J., Leythaeuser, T., Booth-Rea, G., Sawyer, D., Klaeschen, D. & Long, C. 2007. Movement along a low-angle normal fault: the S reflector west of Spain. *Geochemistry Geophysics Geosystems*, **8**, Q06002; doi: 10.1029/2006GC001437.

Rosenbaum, G., Regenauer-Lieb, K. & Weinberg, R. F. 2005. Continental extension: from core complexes to rigid block faulting. *Geology*, **33**, 609–612.

Rosendahl, B. R. & Groschel-Becker, H. 1999. Deep seismic structure of the continental margin in the Gulf of Guinea: a summary report. *In*: Cameron, N. R., Bate, R. H. & Clure, V. S. (eds) *The Oil and Gas Habitats of the South Atlantic*. Geological Society, London, Special Publications, **153**, 75–83.

Rosendahl, B. R., Mohriak, W. U., Odegard, M. E., Turner, J. P. & Dickson, W. G. 2005. West Africa and Brazilian conjugate margins: crustal types, architecture, and plate configurations. *In*: Post, P. *et al.* (eds) *Petroleum Systems of Divergent Continental Margin Basins: Bob F. Perkins Research Conference*. 25th Gulf Coast Section, Society of Sedimentary Geology, Houston, TX, 4–7 December, CD-ROM.

Russell, S. M. & Whitmarsh, R. B. 2003. Magmatism at the west Iberia non-volcanic rifted continental margin: evidence from analyses of magnetic anomalies. *Geophysical Journal International*, **154**, 706–730.

Sandwell, D. T. & Smith, W. H. F. 1997. Marine gravity anomaly from Geosat and ERS 1 satellite altimetry. *Journal of Geophysical Research*, **102**, 10039–10054.

Sandwell, D. T. & Smith, W. H. F. 2009. Global marine gravity from retracked Geosat and ERS-1 altimetry: ridge segmentation versus spreading rate. *Journal of Geophysical Research*, **114**, BO1411.

Sawyer, D. S., Whitmarsh, R. B. & Shipboard Scientific Party. 1994. *Proceedings of the Ocean Drilling Program, Initial Reports*, **149**. Ocean Drilling Program, College Station, TX.

Sawyer, D. S., Coffin, M. F., Reston, T. J., Stock, J. M. & Hopper, J. R. 2007. COBBOOM: The Continental Breakup and Birth of Oceans Mission. *Scientific Drilling*, **5**, September.

Sibuet, J. C. 1987. *Contribution à l'étude des mécanismes de formation des marges continentals passives*. Doctorat d'Etat thesis. Université de Bretagne Occidentale, Brest.

Sibuet, J. C. 1992. New constraints on the formation of the nonvolcanic continental Galicia-Flemish Cap conjugate margins. *Journal of Geological Society, London*, **149**, 829–840.

Sibuet, J. C., Srivastava, S. & Manatschal, G. 2007. Exhumed mantle-forming transitional crust in the Newfoundland-Iberia rift and associated magnetic anomalies. *Journal of Geophysical Research*, **112**, B06105; doi: 10.1029/2005JB003856.

Smith, W. H. F. 1993. On the accuracy of digital bathymetric data. *Journal of Geophysical Research*, **98**, 9591–9603.

Souza-Lima, W. 2009. Seqüencias evaporíticas da bacia de Sergipe-Alagoas. *In*: Mohriak, W., Szatmari, P. & Couto Anjos, S. M. (eds) *Sal: Geologia e Tectônica*. Beca, São Paulo, 232–251.

Srivastava, S. & Keen, C. E. 1995. A deep seismic reflection profile across the extinct Mid-Labrador Sea spreading center. *Tectonics*, **14**, 372–389.

Srivastava, S. & Roest, W. R. 1999. Extent of oceanic crust in the Labrador Sea. *Marine and Petroleum Geology*, **16**, 65–84.

Srivastava, S., Sibuet, J. C., Cande, S., Roest, W. R. & Reid, I. R. 2000. Magnetic evidence for slow seafloor spreading during the formation of the Newfoundland and Iberia margins. *Earth and Planetary Science Letters*, **182**, 61–76.

Szatmari, P. & Milani, E. J. 1999. Microplate rotation in Northeast Brazil during South Atlantic rifting; analogies with the Sinai Microplate. *Geology*, **27**, 1115–1118.

Tack, L., Wingate, M. T. D., Liégeois, J. P., Fernandez-Alonso, M. & Deblond, A. 2001. Early Neoproterozoic magmatism (1000–910 Ma) of the Zadinian and Mayumbian Groups (Bas–Congo): onset of Rodinia rifting at the western edge of the Congo craton. *Precambrian Research*, **110**, 277–306.

Teisserence, P. & Villemin, J. 1989. Sedimentary basin of Gabon – geology and oil systems. *In*: Edwards, J. D. & Santogrossi, P. A. (eds) *Divergent Passive Margin Basins*. American Association of Petroleum Geologists, Memoirs, **48**, 117–199.

Torsvik, T. H., Rousse, S., Labails, C. & Smethurst, M. A. 2009. A new scheme for the opening of the South Atlantic Ocean and the dissection of an Aptian salt basin. *Geophysical Journal International*, **177**, 817–833.

Tsikalas, F., Faleide, J. I. & Kusznir, N. J. 2008. Along-strike variations in rifted margin crustal architecture and lithosphere thinning between northern Vøring and Lofoten margin segments off mid-Norway. *Tectonophysics*, **458**, 68–81.

Wannesson, J., Icart, J. C. & Ravat, J. 1991. Structure and evolution of adjoining segments of the west African margin determined from deep seismic profiling. *Geodynamics*, **22**, 275–289.

Watts, A. B. & Stewart, J. 1998. Gravity anomalies and segmentation of the continental margin offshore West Africa. *Earth and Planetary Sciences Letters*, **56**, 239–252.

Weinberg, R. F., Regenauer-Lieb, K. & Rosenbaum, G. 2007. Mantle detachments and the break-up of cold continental crust. *Geology*, **35**, 1035–1038.

Wernicke, B. 1981. Low-angle normal faults in the Basin and Range province: nappe tectonics in an extending orogen. *Nature*, **291**, 645–648.

Wernicke, B. 1985. Uniform-sense normal simple shear of the continental lithosphere. *Canadian Journal of Earth Sciences*, **22**, 331–339.

Wessel, P. & Watts, A. B. 1988. On the accuracy of marine gravity measurements. *Journal of Geophysical Research*, **93**, 393–413.

Whitmarsh, R. B. & Miles, P. R. 1995. Models of the development of the west Iberia rifted continental margin at 40°30′N deduced from surface and deep-tow magnetic anomalies. *Journal of Geophysical Research*, **100**, 3789–3806.

Whitmarsh, R. B. & Sawyer, D. S. 1996. The ocean/continent transition beneath the Iberia Abyssal Plain and continental-rifting to seafloor-spreading processes. *Proceedings of the Ocean Drilling Program, Scientific Results*, **149**. Ocean Drilling Program, College Station, TX, 713–736.

Whitmarsh, R. B., White, R. S., Horsefield, S. J., Sibuet, J. C., Recq, M. & Louvel, V. 1996. The ocean–continent boundary off the western continental margin of Iberia: crustal structure west of Galicia Bank. *Journal of Geophysical Research*, **101**, 28291–28314.

Whitmarsh, R. B., Beslier, M. O., Wallace, P. J. & Shipboard Scientific Party. 1998. *Proceedings of the Ocean Drilling Program, Initial Results*, **173**. Ocean Drilling Program, College Station, TX, 1–294.

Whitmarsh, R. B., Manatschal, G. & Minshull, T. A. 2001. Evolution of magma-poor continental margins from rifting to seafloor spreading. *Nature*, **413**, 150–154.

Wilson, P. G., Turner, J. P. & Westbrook, G. K. 2003. Structural architecture of the ocean–continent boundary at an oblique transform margin through deep-imaging seismic interpretation and gravity modelling; Equatorial Guinea, West Africa. *Tectonophysics*, **374**, 19–40.

New compilation of top basement and basement thickness for the Norwegian continental shelf reveals the segmentation of the passive margin system

J. EBBING[1,2] and O. OLESEN[1]

[1]*Geological Survey of Norway, 7491 Trondheim, Norway (e-mail: Joerg.Ebbing@ngu.no)*
[2]*Department for Petroleum Engineering and Applied Geophysics, NTNU Trondheim, 7491 Trondheim, Norway*

Abstract: We present the first complete top basement map for the passive margin system of the Norwegian continental shelf, which covers the Northern North Sea, the mid-Norwegian margin system, and the western Barents Sea. This compilation is based largely on a review and synthesis of previously published portions on detailed depth to basement and 3D modelling studies, which in turn are based on a wealth of seismic profiles, commercial and scientific drilling on the shelf and mainland Norway, petrophysical sampling and a dense coverage of gravity and aeromagnetic data. The top basement defines the base of the sedimentary strata. Sedimentary thickness is larger than 14 km on the mid-Norwegian margin and the Barents Sea, but varies strongly along the different segments of the margin. In the northeastern North Sea, the Danish–Norwegian Basin, and over the Trøndelag Platform the top basement does not exceed 8 km in depth. The top basement map highlights the transition between the different segments of the Norwegian margin, between different basement domains (e.g. Caledonian and Precambrian basement), and shows a clear correlation with the normal faults mapped on the margin and prolongated from onshore Norway into the offshore realm. The distribution of basement domains provides a new regional understanding of the evolution of the entire passive margin system. We also present an updated Moho depth compilation, which allows calculation of a thickness of crystalline basement map for the entire margin. The basement thickness map enhances differences in crustal architecture for the different margin segments and shows extremely thin crust (less than 12.5 km) for large areas on the mid-Norwegian margin and westernmost Barents Sea, intermediate thin crust (15–20 km) below the Viking Graben, but moderately thin crust (20–30 km) below the Norwegian–Danish Basin and the remaining part of the Barents Sea.

Keywords: basement, continental shelf, potential fields, onshore–offshore, Caledonian, Precambrian

The Norwegian continental shelf is of major interest for the petroleum industry. Since the first realization of the economic potential of the oil and gas prospects in the North Sea in the 1960s, a vast amount of geophysical data at all scales has been collected to enhance our understanding of the passive margin system (e.g. Blystad *et al.* 1995; Fleet & Boldy 1999; Nøttvedt 2000, and references therein). Large parts of the Norwegian shelf are covered by 2D and 3D seismic reflection surveys, which allow detailed mapping of key horizons within the sedimentary section. The base of the sedimentary strata, the top basement and the configuration of the crust below are less clearly imaged due to the large sedimentary thickness. Acquisition of OBS profiles has helped to enhance the imaging of the deep crustal configuration, as well as the top basement below the sedimentary basins, but has a less dense aerial coverage of the margin (e.g. Mjelde *et al.* 2005). The top basement is, however, of major interest for understanding the evolution of the sedimentary basins and possible migration of hydrocarbon (e.g. Doré & Vining 2005, and references therein), and in basin modelling detailed knowledge of the basin geometry is an essential input parameter (e.g. Nøttvedt 2000). Normal faults observed in seismic lines often leave their imprint in the top basement configuration, and open pathways for fluid migration, which affects potential sources and even the thermal state of the sedimentary basin. Basement structures offshore have been indirectly investigated through the combination of potential field data with seismic reflection and refraction data, as wells penetrating the top basement are only sparsely available.

In this paper, we will provide a review of the different basement studies carried out in the past and present as a synthesis of the first complete top basement map of the Norwegian continental shelf and adjacent regions. Our definition of the basement thickness regards the entire basement as a metamorphic complex. We will briefly present the individual contributions to the compilation and discuss the main structural domains as expressed in the geometry and structure of the basement. A crustal thickness map illustrates further the structure of the Norwegian margin. In the final step, we will bring our maps into the context of the evolution of the Norwegian continental shelf and its link to the onshore geology.

Geological setting

The Norwegian continental shelf is subject to the same geological processes observed onshore. Figure 1 illustrates the onshore–offshore structural links by the distribution of normal faults, which can be traced from land into the offshore basins. On land in Norway and Sweden the remnants of the Caledonides dominate the surface geology; this mountain belt formed as a result of the convergence of the North American plate with the westward subducting margin of the Baltica plate that culminated in continent–continent collision. The stacked nappes comprise allochthonous continental and oceanic crust. They were the result of thrusting and tectonic underplating of the exotic terranes from Laurentia/Iapetus (microcontinents and island arcs) and imbrications of the Palaeozoic passive margin of Baltica (e.g. Roberts 2003 and references therein). The collisional climax was followed by a general collapse of the mountain belt (e.g. Fossen 1992; Andersen 1998) and the development of Devonian basins (e.g. Osmundsen *et al.* 1998). The processes causing the mountain chain collapse and the transition in the late Devonian to Carboniferous to the purely rift-related mechanisms are still debated.

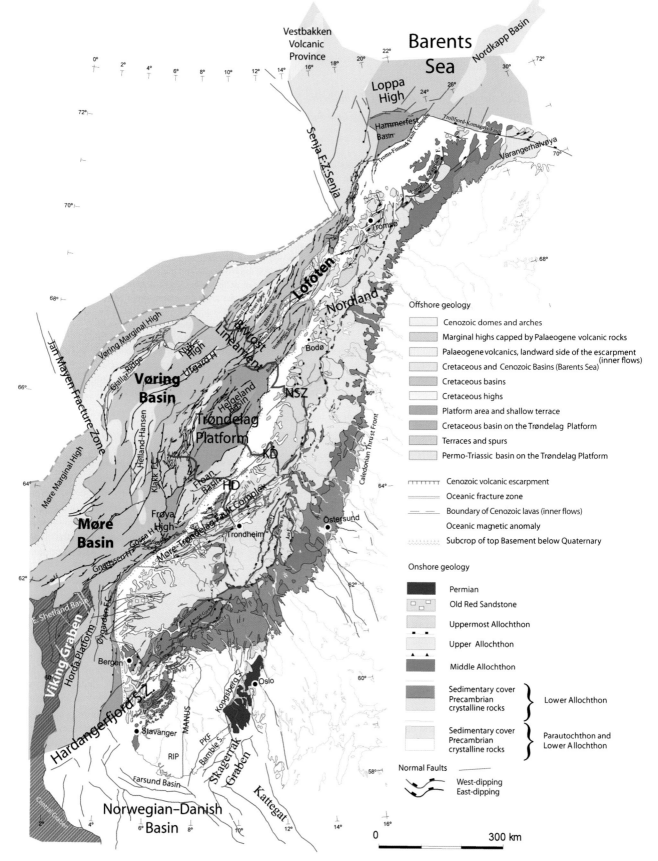

Fig. 1. Simplified onshore–offshore geological map of the Scandinavian North Atlantic passive margin (modified from Mosar 2003). In the offshore domain the different major tectono-sedimentary events are indicated (adapted and modified from Blystad et al. 1995; Brekke et al. 1999; Gabrielsen et al. 1999; Mosar 2000; Smethurst 2000). The dip directions of some of the major normal faults in the offshore, such as in the Nyk High, the Utgard High and the Gjallar Ridge, are shown according to interpretation of deep seismic surveys (Osmundsen et al. 2002). The faults have been colour-coded according to dip direction: red for west-dipping and black for east-dipping. The onshore tectonostratigraphic map is a simplified and modified version of the map by Gee et al. (1985). MANUS, Mandal–Ustaoset Fault; RIP, Rogaland Igneous Province; PKF, Porsgrunn–Kristiansand Fault; HD, Høybakken detachment; KD, Kollstraumen detachment; NSZ, Nesna Shear Zone.

Physiographically, the mid-Norwegian margin consists of a continental shelf and slope that vary considerably in width and steepness. The two adjacent shallow seas, the North Sea and the Barents Sea were, prior to the formation of the deep NE Atlantic ocean in early Cenozoic time, part of a much larger epicontinental sea between the continental masses of Fennoscandia, Svalbard and Greenland (Faleide et al. 2008).

The mid-Norwegian margin was formed by episodic extensional events during Late Palaeozoic–Triassic, Late Jurassic–Early Cretaceous and Late Cretaceous–Paleocene times (Ziegler 1988; Blystad et al. 1995; Doré et al. 1997; Brekke 2000). The Norwegian Atlantic margin is a classic passive margin that finally went from rifting to drifting in the early Cenozoic. Prior to, and during, the rift–drift transition, the margin witnessed extensive volcanic activity, which changed to normal accretionary magma volumes with subsequent continental margin subsidence and maturation from the middle Eocene to the present (Faleide et al. 2008).

Models for the development of the Norwegian Atlantic passive margin have in the past often focused on the sedimentary cover sequence (e.g. Doré 1992; Doré et al. 1997; Brekke 2000; Brekke et al. 2001), but in recent years the importance of the underlying basement and crustal structure on the development of the margin has been more and more recognized (e.g. Olesen et al. 2002; Osmundsen et al. 2002; Breivik et al. 2005).

The present structure of the passive margin reflects the cumulative effect of several consecutive rifting events that controlled basin development, and climaxed in continental breakup, the opening of the North Atlantic and formation of new oceanic crust (Mosar 2003). These rifting events were interspersed with Mesozoic uplift periods and preceded the proposed Neogene uplift that produced a significant contribution to the present-day mountain topography (e.g. Gabrielsen et al. 2005).

All these processes left an imprint on the structure of the basement offshore and on the topography observed onshore Norway. On the Norwegian passive margin different domains can be identified. In the north of Norway, we find the stable continental shelf of the Barents Sea, which is divided by the transform margin of the Barents Sea to the mid-Norwegian margin. On the mid-Norwegian margin three domains can be identified: (1) the Lofoten margin; (2) the Vøring Basin-Trøndelag Platform; and (3) the Møre Basin (Fig. 1). Further to the south we find the Viking and Central Graben system. The Norwegian–Danish Basin continues around the southern tip of Norway into the Skagerrak, Jutland and Kattegat.

Presentation of database

For the Norwegian continental margin and onshore Norway and Sweden a large amount of geophysical data is available. Figure 2 shows the gravity and magnetic field data and the regional seismic profiles available for the study area. Aeromagnetic and gravity data provide an important tool for mapping crustal structures both on mainland Norway and in the adjacent offshore regions.

Magnetic anomaly map

The magnetic anomaly map (Fig. 2a) is based on integration of multiple surveys at different scales and heights (Olesen et al. 2007, 2010). Over the last 50 years a vast number of offshore aeromagnetic surveys have been acquired with typical flight heights of 200–300 m and line spacing of 2–5 km over the continental

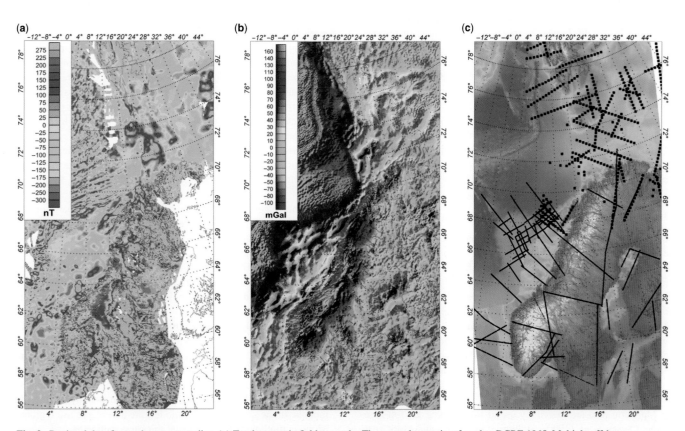

Fig. 2. Regional data for top basement studies. (a) Total magnetic field anomaly. The anomaly map is referred to DGRF-1965. Multiple offshore aeromagnetic surveys have been processed and merged to produce the displayed map (Olesen et al. 2010). (b) Bouguer anomaly. The map is calculated with a reduction density of 2670 kg m^{-3} onshore, and 2200 kg m^{-3} offshore (Olesen et al. 2010). Offshore measurements of approximately 59 000 km of marine gravity profiles have been acquired by the Norwegian Petroleum Directorate, oil companies, TGS NOPEC Geophysical Company and the Norwegian Mapping Authorities. (c) Regional wide-angle seismic profiles as presented in Kinck et al. (1993) for the onshore part, Christiansson et al. (2000) for the northeastern North Sea, Mjelde et al. (2005, 2009) for the mid-Norwegian margin, and Ritzmann et al. (2007) for the Barents Sea.

margin (Olesen et al. 2010). Outstanding features in the magnetic anomaly map are the anomalies associated with seafloor spreading, the Loppa High in the southwestern Barents Sea, the prominent high along the Lofoten islands, the Frøya High at the southern edge of the Trøndelag Platform, but also onshore structures like the magnetic high of the Oslo Graben or the north–south trending band of the Transscandinavian Igneous Belt. The anomalies are partly continuous from the Baltic Shield under the Scandinavian mountains into the continental shelf, which indicate continuous bedrock structures (e.g. Åm 1975; Olesen et al. 2002). The magnetic data provide important constraints on the interpretation of the regional basement configuration and distribution of volcanic rocks on the shelf (e.g. Olesen et al. 2002; Skilbrei et al. 2002).

Gravity anomaly map

The Bouguer anomaly map (Fig. 2b) is based on the integration of land measurements, ship-borne gravity surveys and satellite data (Olesen et al. 2010). The Bouguer anomaly is calculated with a rock density of 2670 kg m^{-3} onshore and 2200 kg m^{-3} offshore. The Bouguer anomaly map is dominated by the low along the Scandes mountain chain, and the high anomalies crossing from the continental shelf into the oceanic plate. The gravity low of the Scandes mountain belt indicates isostatic compensation, and the transition from the continental to oceanic plate reflects the crustal thinning and increase of crustal densities. On the continental shelf, anomalies are less pronounced than in the magnetic anomaly map. On the mid-Norwegian margin especially, the Trøndelag Platform stands out as a relative gravity low surrounded by higher gravity anomalies.

Regional seismic data

Seismic data are always important to constrain interpretations of potential field data. Figure 2c shows the distribution of regional seismic profiles for onshore Norway and the continental shelf. The figure does not show all available seismic lines on the shelf as many commercial 2D and 3D surveys have been and continue to be acquired. The figure features academic OBS profiles (Mjelde et al. 2005, 2009), the dataset available in a recent compilation for the Barents Sea (Ritzmann et al. 2007) and from a compilation of Moho depths for the Fennoscandian Shield (Kinck et al. 1993). All these published profiles have been used to image the base of the crust, and provide an indication of the offshore top basement, but with variable accuracy and resolution.

Importance of potential field data and data integration

Mapping of the top basement is based on different methods for combining the available geophysical data. The basement configuration is particularly visible in the potential fields, where it lies close to the surface, and the interpretation of potential field data helps to trace structures from the onshore to offshore realms, which can then be observed from the surface to depth. For example, the mid-Norwegian margin was studied by 3D modelling integrating a wealth of geophysical data: seismic profiles, commercial and scientific drilling on the shelf and mainland Norway, petrophysical sampling and a dense coverage of gravity and aeromagnetic data (e.g. Olesen et al. 2002; Skilbrei & Olesen 2005; Ebbing et al. 2006, 2009).

Magnetic data are extremely useful for mapping the top basement. Olesen et al. (1990), Skilbrei et al. (1991) and Mørk et al. (2002) show that the susceptibilities of the basement can range between 0.0005 and 0.3 SI while the susceptibilities of the overlying sediments are only in the order of 0.0003 SI, up to three orders of magnitude lower. The range of susceptibilities for the basement depends on composition and varies typically from 0.005–0.01 SI for Caledonian basement, 0.01–0.035 for Precambrian basement, to even higher values for mafic intruded basement (e.g. Barrère et al. 2009). Because of the large contrast between sedimentary rocks and underlying basement, magnetic depth estimates provide a good starting point for a genuine structural interpretation. Magnetic depth estimates calculated by applying, for example, Euler Deconvolution or Peter's Slope Method have been used effectively on the mid-Norwegian margin (e.g. Åm 1975; Skilbrei & Olesen 2005) and Barents Sea shelf (e.g. Skilbrei 1991, 1995), as well as in the Viking Graben and Norwegian Danish Basin (e.g. Hospers & Ediriweera 1991; Smethurst 2000; Olesen et al. 2004).

Skilbrei & Olesen (2005) studied the accuracy and the geological meaning of the 'magnetic basement' on the mid-Norwegian margin. They found generally good agreement between estimates made from magnetic anomalies and the depth to the Precambrian basement. In some areas non-magnetic Devonian basins may exist, and non-magnetic Caledonian nappes can overly the Precambrian basement. In the latter case, the true crystalline basement would lie above the 'magnetic basement'. Comparison of magnetic depth estimates and seismic, borehole and petrophysical data yields errors that generally vary between 5 and 15% (Skilbrei et al. 2002).

Gravity data are useful to a limited extent in top basement mapping as, owing to sedimentary compaction, the density contrast between sedimentary rocks and top basement becomes relatively small at depths greater than 5 km (e.g. Ramberg & Smithson 1975). The crystalline basement is also often difficult to recognize on seismic sections (e.g. Hospers & Ediriweera 1991). This is largely a result of a decrease in the contrast in acoustic impedance between sedimentary rocks and basement at greater depths. Another well known problem in the areas affected by volcanic activity is sub-basalt imaging, which makes an estimate of the top basement from seismic data very difficult (e.g. Reynisson et al. 2009; Reynisson 2010).

Using 3D modelling decreases the uncertainty as seismic, borehole and petrophysical data can be used directly to constrain the top basement structures, and to distinguish between different basement units. For the 3D modelling we used in most instances the Interactive Gravity and Magnetic Application System (IGMAS), which calculates the potential field effect of the model by triangulating between parallel, vertical planes (Götze & Lahmeyer 1988). The vertical planes defining the 3D geometry have typically a distance of 10–15 km depending on the geological structures and available constraining data (Ebbing et al. 2006, 2009; Reynisson 2010).

In the integration of datasets from the different parts of the margin we put emphasis on the consistency between the models, which was assured by careful integration of the different datasets. For example, the model of the mid-Norwegian margin after Ebbing et al. (2006, 2009) is an extension of the 3D model on the Nordland–Lofoten margin by Olesen et al. (2002) and therefore a continuous transition between the two areas on the margin is provided.

The individual 3D models also provide information about the base of the crust, which allows the calculation of total basement thickness maps, and linking of basement geometry with the overall crustal configuration. This minimizes the trade-off between sources at different crustal levels on the observed anomalies, and the effect of 3D geometries, which can heavily distort 2D models. In the areas where 3D models are absent, we use regional Moho compilations for the crustal thickness estimates.

In the following we will present the top basement compilation by giving an overview of the individual regional contributions to the top basement map of the Norwegian continental shelf and adjacent regions. Afterwards we will present the Moho depth for the same area and a basement thickness map.

Individual contributions to the basement depth map

The top basement map of the Norwegian shelf was compiled by integrating individual regional studies, shown in Figure 3 and summarized in Table 1. For the onshore part, our compilation basically corresponds to the topography. The Caledonian basement dominates the surface (Fig. 1) and is divided by thrusts, normal faults and detachments. Almost no sedimentary cover is observed. For some areas the thickness of these nappe units has been mapped, as well as the base of the Precambrian units in northern Norway (Olesen et al. 1990, 2002). Here we concentrate, however, on the depth to the top of the crystalline basement.

The character of the top basement surface is clearly related to different segments of the margin. For example, between the Møre margin and northeastern North Sea the grain of the top basement topography changes from SW–NE on the Møre margin to a north–south orientation, paralleling the coastline of Norway. At the same time the maximum depth of the top basement is typically less than 8 km south of the Møre margin, while on the mid-Norwegian margin the top basement is typically located at a depth of less than 10 km. Other prominent changes occur between the Vøring and Lofoten margins and in the North Sea at the offshore extension of the Hardangerfjord Shear Zone.

Barents Sea

The northernmost part of the compilation is the area of the Barents Sea. For the western Barents Sea, top basement maps have been presented by Johansen et al. (1992) and Skilbrei (1991, 1995). Johansen et al. (1992) used magnetic depth estimates integrated with seismic profiles and integrated further data from the eastern Barents Sea. The compilation by Skilbrei (1991, 1995) is more enhanced as new methods for the magnetic depth estimates have been used systematically (Skilbrei 1995). He integrated seismic profiles and well data to verify his top basement estimates. The resulting map was defined by hand contouring between the magnetic depth estimates.

The western limit of the top basement map is defined by the transition to North Atlantic oceanic crust, as indicated by the Senja Fracture Zone and Vestbakken Volcanic Province. The top basement depth reflects the deep Cenozoic basins of the Western Barents Sea with up to 14 km thickness (e.g. Nordkapp Basin). In the Loppa High, the basement rises to depths of less than 3 km and is associated with a Bouguer anomaly high. A magnetic high extends eastwards from this basement feature; this is due in part to the basement geometry, but also reflects compositional changes between different basement units. Barrère et al. (2009) showed in a recent study the influence of the different basement domains on magnetic anomalies, and provided a tentative map of their distribution. The basement is very inhomogeneous in the Barents Sea, reflecting rocks of both Caledonian and Precambrian age, as well as of younger volcanic affinity.

We also provide a top basement for the transition zone towards the eastern Barents Sea. The transition zone is associated with a basement depth of around 6–8 km before the top basement deepens rapidly below the Eastern Barents megabasins, where the top basement exceeds a depth of 20 km, just outside the map area. The data for the eastern Barents Sea are based on a study of Gramberg et al. (2001). According to these authors the interpretation is based on a few thousand kilometres of reflection and refraction lines, together with aeromagnetic surveys and gravity data. Even if the study by Gramberg et al. (2001) relies on an extended database, an unambiguous evaluation of the accuracy of their interpretation is not possible, as detailed description of the methodology is not available.

Mid-Norwegian margin

The top basement and the basement structure of the mid-Norwegian margin have been extensively studied. The first complete maps were constructed by combining regional seismic profiles and aeromagnetic depth estimates as well as onshore petrophysical studies (e.g. Olesen et al. 2002; Skilbrei et al. 2002; Skilbrei & Olesen 2005). These studies also addressed the basement type, as shown by petrophysical studies of rocks exposed in onshore areas that flank the margin. Two types of basement can be distinguished: a highly magnetic part corresponding to granulite facies Precambrian gneisses and igneous rocks and a less magnetic part corresponding to Caledonian nappes and amphibolite facies Precambrian basement (Olesen et al. 1991; Skilbrei et al. 1991).

Lofoten margin. On the Lofoten margin, the data used are from Olesen et al. (2002) with modifications in the transition to the Vøring margin by Ebbing et al. (2006). The offshore top basement estimates have been compiled from studies based on magnetic

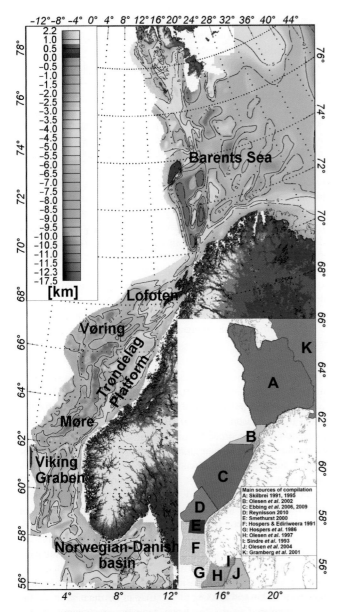

Fig. 3. Top basement map of the Norwegian continental shelf and surrounding areas. The inset map shows the distribution of the individual regional contributions. See Table 1 and text description of methodology and accuracy. Onshore top basement is represented by the topography after Dehls et al. (2000).

Table 1. Overview of individual contributions to the top basement compilation

	Area	Source	Compilation type
A	Western Barents Sea	Skilbrei (1991, 1995)	Magnetic depth
B	Nordland, Lofoten margin	Olesen et al. (2002)	3D model
C	Vøring–Lofoten	Ebbing et al. (2006, 2009)	3D model
D	Møre	Reynisson (2010)	3D model
E	Northern Viking Graben	Smethurst (2000)	Magnetic depth
F	Northern North Sea	Hospers & Ediriweera (1991)	Magnetic depth
G	Central-eastern North Sea	Hospers et al. (1986)	Magnetic depth
H	Norwegian–Danish Basin	Olesen et al. (1997)	Magnetic depth, 2D modelling
I	Skagerrak	Sindre et al. (1993)	Magnetic depth
J	Jutland/Kattegat	Olesen et al. (2004)	Magnetic depth
K	Barents Sea	Gramberg et al. (2001)	Magnetic depth

depth estimates, and validated by forward modelling using seismic information on the sedimentary succession and petrophysical data (Olesen et al. 2002). Also, the thickness of the mainland Caledonian nappes has been included in the model. The top basement on this narrow portion of the margin does not exceed 10 km in depth and locally reaches the surface, for example at the Lofoten Ridge, where the Precambrian basement crops out. The Bivrost Lineament which represents the offshore extension of the Devonian Nesna Shear Zone constitutes the border with the Vøring margin to the south.

Vøring margin. Top basement on the Vøring margin has been mapped by Skilbrei & Olesen (2005), and integrated and updated with a complete model of the crust by Ebbing et al. (2006). The latter was well constrained on this margin, but did not address the magnetic field variations. Most recently Osmundsen & Ebbing (2008) and Ebbing et al. (2009) presented an updated model of the basement, where they differentiate the distribution of different basement types on the margin. Basically, a less magnetic 'Caledonian' basement overlies the highly-magnetic 'Precambrian' part.

The resulting top basement model is based on a wealth of seismic, borehole and petrophysical information (Ebbing et al. 2006, 2009), and the top basement coincides over the major part of the margin with the top Caledonian basement. In general, as for the entire mid-Norwegian margin, the top basement deepens from the coastal area towards the continental edge. Below the Trøndelag Platform the top basement is located at depths between 3 and 10 km, while in the Vøring Basin, top basement is generally encountered between 11 and 15 km. The top basement shows a variety of local features, for example basement highs at the Nyk and Utgard highs.

The Precambrian basement below the Caledonian cover is generally located at depths of more than 12 km, but it forms extended structural highs below the Trøndelag Platform where Precambrian rocks lie at depths of less than 3 km and the Caledonian basement is absent (Ebbing et al. 2009). These highs are aligned and can appear to be a continuous high from the Frøya High along the western border of the Froan Basin. However the Frøya High is very broad and is terminated in the north by the offshore prolongation of the late-Caledonian Høybakken detachment (Osmundsen & Ebbing 2008). Below the Vøring Basin the boundary between the low- and high-magnetic basement is almost horizontal and located 2–4 km below the top basement (Ebbing et al. 2009). This geometry explains the transition from pronounced magnetic anomalies on the mid-Norwegian margin to low-amplitude magnetic anomalies over the Vøring Basin.

Below the Helgeland Basin, the top basement in our compilation is at a depth of 3–8 km. Ebbing et al. (2009) suggest a deeper top basement below the Helgeland Basin with depths up to 12 km. The latter interpretation, which was based on industrial seismic data, however, differentiated between the non-magnetic Devonian and the underlying Caledonian basement. A significant portion of the basement below the Helgeland Basin is non-magnetic, most likely caused by downfaulting of the Helgeland Nappe Complex along the offshore extensions of the Devonian Nesna shear zone and the Kollstraumen detachment (Olesen et al. 2002). The presence of Devonian sandstones is an alternative interpretation (Olesen et al. 2002). The main large-scale structural changes in this map, compared with pre-existing maps, occur in the southern Vøring area, where the structural grain of the top basement bends from around the Trøndelag Platform and continues into the Møre margin. The strike direction on the southern Vøring area is north–south, which differs from the NE–SW direction dominant on the Lofoten, central Vøring and Møre margins to the south. Although, the crustal structure indicates no compositional variation along the transition between the Møre and Vøring margin, the extension of the Jan Mayen Fracture Zone on the margin appears to have controlled the changes in strike direction (e.g. Ebbing et al. 2006).

Møre margin. The top basement has been mapped on the Møre margin by Skilbrei et al. (2002), and was updated first by Osmundsen & Ebbing (2008), and more recently by Reynisson (2010). The model by Reynisson et al. (2010) is based on a 3D gravity and magnetic model incorporating industrial seismic horizons for the sedimentary succession, regional seismic profiles and magnetic depth estimates. Scheck-Wenderoth et al. (2007) presented a similar top basement map for the Møre margin as part of a 3D model, but the crustal model did not address the magnetic anomalies over the Møre margin, which are quite prominent, for example the magnetic high along the Gossa and Gnaussen highs, which is related to some extent to the crustal geometry and in some part to the compositional variation between different basement units (Reynisson 2010).

Below the central Møre Basin, the top basement is at a depth of more than 12 km and links up to the strike of the deep basins on the Vøring margin. Clearly, the basin geometry continues from the Vøring into the Møre Basin. Towards the ocean–continent transition the top basement depth increases to more than 8 km, while on the Vøring margin the top basement is located at a depth of greater than 10 km close to the continent–ocean transition. The outer margin is highly volcanic and large volumes of sills impede imaging the sedimentary successions (e.g. Planke *et al.* 2005). Only the combined interpretation of gravity and magnetic data allows a reasonable estimate of the sub-basalt sedimentary thickness for the outer part of the margin (Reynisson *et al.* 2009).

Towards the land the Møre margin is limited against the Møre–Trøndelag Fault Complex, which extends over a distance of more than 300 km from its inferred northeastwards termination in central mid-Norway to its poorly mapped offshore extension in the Norwegian Sea. Here, the Møre–Trøndelag Fault Complex separates the deep, northern Cretaceous Møre and Vøring basins from the shallower Jurassic–Cretaceous–Cenozoic basin systems to the south in the northern North Sea (Gabrielsen *et al.* 1999).

Northeastern North Sea, Skagerrak and Kattegat

Many attempts have been made to constrain the depth to top crystalline basement in the northern North Sea, but comparatively few maps have been published. Åm (1973) reported a first interpretation based on both magnetic and gravity data. Hospers & Ediriweera (1991) published a refined depth to basement map, which significantly differed from Åm's (1973) earlier work, presumably due to the greater availability of geological information. The depth to crystalline basement of the southernmost part of the maps has been compiled from six different sources: (1) Smethurst (2000), (2) Hospers & Ediriweera (1991), (3) Hospers *et al.* (1986), (4) Olesen *et al.* (1997), (5) Sindre (1993), and (6) Olesen *et al.* (2004).

Viking Graben. Our compilation for this area is based on Hospers & Ediriweera (1991) and Smethurst (2000). The Viking Graben and the Norwegian–Danish Basin constitute the main structural elements within the study area of the 'North Sea Rift Zone'. The basement depth ranges from 1–6 km on platforms and structural highs, to 7–9 km on terraces and basins and more than 10 km below the grabens. Conventionally, the Viking and Sogn Graben are regarded as first-order features, but our basement map indicates that the dominant structural feature is a north–south trending zone, approximately 150 km wide, containing a variety of structures (graben, basins and structural highs), some of which are oblique to the zone boundaries. To the west this zone is limited by the eastern boundary faults of the East Shetland Platform. To the east the zone is limited by the Øygarden Fault Zone. The Horda Platform is not a major shallow basement platform, like the Trøndelag Platform.

The study by Hospers & Ediriweera (1991) covers the entire Viking Graben and Horda Platform. The top basement estimates are based on aeromagnetic and marine magnetic data (Hospers & Ediriweera 1988), conventional seismic reflection profiles and exploration wells. Magnetic depth estimates at the flank of the anomalies were used to construct the initial top basement map. Gravity data did not contribute independent, non-redundant information on the depth and configuration of the crystalline basement, but helped to link the overall crustal structure to the sedimentary succession. The crystalline basement is often also difficult to recognize on seismic data (cf. Hospers & Ediriweera 1991). In such instances, basement depths have been extrapolated from or interpolated between basin flanks, using whatever seismic reflection evidence is available. On the flanks of the basins, where the basement lies at relatively shallow depths, the basement reflection is usually strong and can be identified by the use of well data, by seismic correlations along or between seismic sections, or by means of the earlier map based on magnetic and well data (Hospers & Ediriweera 1991).

For the northeasternmost part of the North Sea, following the acquisition of a new aeromagnetic dataset, Smethurst (1994) produced a depth to basement map based solely on the new magnetic data. The uncertainty of this map was quite high, and the map was later refined by the integration of gravity and magnetic data (Smethurst 2000). The compilation by Smethurst (2000) is based on clusters of consistent magnetic and gravity depth estimates, which were identified on the flanks of anomalies. These were then contoured by hand with a 1 km depth interval and interpolated onto a regular grid using the minimum curvature method. Incorporation of wide-angle lines (Fig. 2c; Christiansson *et al.* 2000) further constrained the top basement interpretation. The uncertainty of this top basement depth is estimated to be relatively high away from the seismic profiles.

Norwegian–Danish Basin. Further to the south the compilation is based on the interpretations by Hospers *et al.* (1986), Sindre (1993) and Olesen *et al.* (1997, 2004). They used magnetic depth estimates for an initial top basement interpretation. Two-dimensional gravity and magnetic modelling along regional seismic profiles constrained the compilation further. The top Zechstein faults interpretation of Vejbæk & Britze (1994) and exploration wells further constrain the map. Five of the wells penetrate the Precambrian basement and the other well data represent minimum depths to crystalline basement (Olesen *et al.* 2004). The top basement estimates would be more precise if derived from a complete 3D model based on the available seismic and borehole data, and potential field anomalies, and addressing the complete crustal structure. Yegorova *et al.* (2007) present a comprehensive study for the Danish–German basin and the Norwegian–Danish basin. They present a top Zechstein map, which can be used in the future to extend the model towards the south. However, they do not provide a new top basement map and do not address the magnetic anomalies of the area (Yegorova *et al.* 2007).

In our compilation the top basement below the Norwegian–Danish Basin constitutes a saucer-shaped surface with no major faulting along the basin margins. The maximum depth is more than 10 km and the shallower estimates are in general consistent with what is known from exploration drilling (Hospers *et al.* 1986; Olesen *et al.* 2004). Most of the regional aeromagnetic anomalies in the Norwegian–Danish Basin are continuous with aeromagnetic anomalies on mainland Norway and Sweden, suggesting that the offshore anomalies to a large extent are caused by the Sveconorwegian intra-basement rocks observed at the surface on land. Lyngsie *et al.* (2005) show that these anomalies in southwestern Norway coincide with the onshore Rogaland Igneous Province (RIP), Mandal–Ustaoset Fault Zone (MAGNUS) and Porsgrunn–Kristiansand Shear Zone (PKSZ), and extend for at least 200 km into the Norwegian–Danish Basin and across the Cretaceous–Cenozoic Sorgenfrei–Tornquist zone. Here, the continental scale fault zones dominate the magnetic and gravity pattern in the northern part of the Norwegian–Danish Basin.

The geometry of the top basement in the Skagerrak is influenced by the processes forming the Oslo Graben (Lyngsie *et al.* 2005), which is bound by dense Sveconorwegian complexes (Bamble and Kongsberg) to the west (Ebbing *et al.* 2007). An extension of the high-grade gneisses within the Bamble Complex also produces magnetic anomalies in the northern Skagerrak (Sindre 1993). The Kattegat–Skagerrak Platform is generally characterized by a shallow gently dipping crystalline basement surface, which is covered by an eastward thinning Meso-Cenozoic sedimentary sequence, unconformably underlain by downfaulted Palaeozoic sediments (Lassen & Thybo 2004).

Some uncertainties are also present with respect to the Moho depth in this area. The Moho depth (see below) is shallowest below the central part of the Norwegian–Danish Basin (29 km) and the Skagerrak Graben (26–28 km), and successively deepens towards the Farsund Basin and the mainland of Norway to the north (31–32 km) and towards the south (32 km). The positive gravity effect from the shallow Moho below the Norwegian–Danish Basin is, however, not sufficient to compensate for the low-gravity effect from the low-density sediments within the basin. Olesen et al. (2004) added a body of mafic rocks immediately above the Moho. The thickness varied between 3 and 9 km and the width between 80 and 100 km. A similar deep-seated dense mafic rock unit and shallow Moho are included in the gravity model of the Skagerrak Graben by Ramberg & Smithson (1975). A similar coinciding shallow Moho and dense mafic body in the upper part of the deep crust were also applied to model the Silkeborg gravity high in Denmark by Thybo & Schönharting (1991). However, a complete 3D model incorporating the available seismic and borehole data, in addition to the potential field anomalies, would significantly enhance the accuracy of both the top basement estimates and the base of the crust.

Moho depth and basement thickness maps

The top basement map is often associated with the top of the crystalline crust. To understand the evolution of the passive margin systems, it is of interest also to look at the base of the crust as defined by the Moho interface. Figure 4a shows the Moho depths for the Norwegian shelf, mainland Norway and surrounding areas. This Moho map has a similar database to the recent compilation for the European Moho depth by Grad et al. (2009), but includes more local studies on the mid-Norwegian margin.

In the depth to the Moho map the different characteristic of the Barents Sea and the southern part of the Norwegian continental margin stand out. On the mid-Norwegian margin the Moho generally deepens from the centre of the Fennoscandian Shield towards the continent–ocean transition. Typical values at the coastline are 30–35 km and the depth decreases to values around 10–15 km at the continental edge. Over the stable shelf of the Barents Sea the crustal base is for most of the area located at a depth larger than 30 km. While on the mid-Norwegian margin and the North Sea thick sedimentary sequences correlate with a shallow Moho as expected for classical extensional, pure shear, rift basins; the deep basins of the Barents Sea do not reveal a similar expression in the Moho geometry.

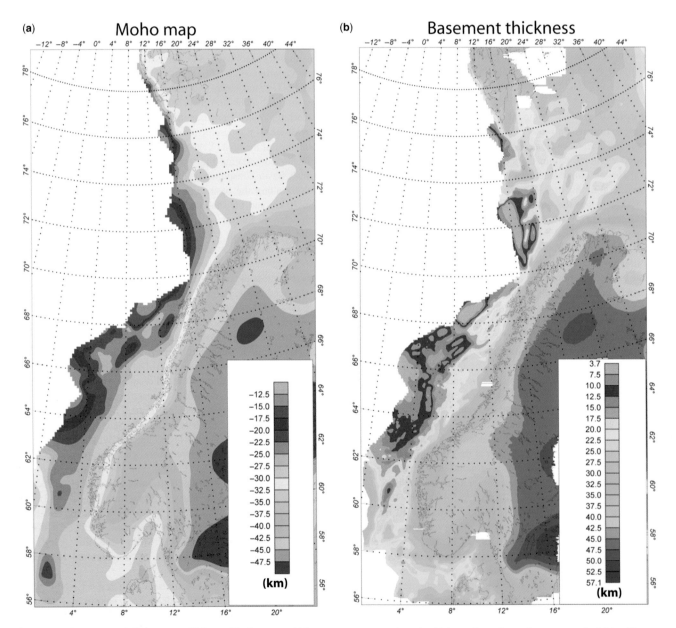

Fig. 4. (**a**) Moho map and (**b**) basement thickness. The basement thickness map corresponds to the thickness from the top basement to the Moho. The depth to Moho map is based on the regional profiles and compilations presented in Figure 2c, with modifications by Osmundsen & Ebbing (2008) for the Møre and Vøring margins, and Tsikalas et al. (2008) and Olesen et al. (2002) for the Lofoten margin.

The basement thickness map enhances this observation even more (Fig. 4b). This map is defined as the thickness of the crust between the top basement and the base of the crust (Moho). In the Barents Sea the basement thickness is typically larger than 20 km with local exceptions (e.g. the Nordkapp Basin), and despite the large sedimentary overburden, the crustal thinning is moderate. Over the Viking Graben, Møre and Vøring margins, the basement thickness is however less than 20 km, and for most areas even less than 12.5 km. Compared with the continental crustal thickness at the Norwegian coast (c. 32 km), this shows an enormous thinning of the crust, which reflects the influence of multiple rift episodes before the final opening of the Atlantic. An exception on the mid-Norwegian margin is the Trøndelag Platform, which is underlain by more than 20 km of crystalline crust. Similar basement thicknesses can be observed south of the Hardangerfjord Shear Zone. Clearly, the crustal thickness map reflects the distance to the continental edge, and indicates a direct link between the location of the margin segments and the opening history of the North Atlantic.

Discussion

Our definition of the basement thickness regards the entire basement as a metamorphic complex. Both low-magnetic Devonian basins and lower crustal bodies have been observed on the margin, which strictly are not part of the crystalline basement. Because of the age and often large depth, the density of the Devonian basement is likely to be similar to that of the underlying basement, and the base is often difficult to image seismically. Therefore, these Devonian basins are here regarded as part of the basement.

In our crustal thickness map, we include the entire thickness from the top basement to the crust–mantle transition (Moho). However, in the areas with extremely thin crust on the Vøring and Møre margins, a high-velocity, high-density body at the base of the crust has been mapped (e.g. Mjelde et al. 2005, 2009). This lower crustal body (LCB) has been mapped to have a thickness of more than 6 km, and to consist of a combination of magmatic underplated material and partially eclogitized rocks (e.g. Mjelde et al. 2005, 2009). This interpretation would imply that the LCB does not entirely correspond to crystalline crust and below parts of the Vøring Basin almost no crystalline crust might exist. Similar observations have been made on the southern Vøring margin and the Møre margin, where crustal-scale detachments can be observed culminating in the lower crust and the crustal thickness tapers out to less than 4 km (Osmundsen & Ebbing 2008). Below the Horda Platform a similar high-velocity body of eclogitic origin has been imaged on the landward side of the graben (Christiansson et al. 2000). Also below the Trøndelag Platform, increased seismic velocities and densities can be observed for the lower crust, but more in agreement with a gradual increasing velocity in normal continental crust. Hence, here the lower crust can be related to the crystalline crust.

In general, rifting that led to the formation of the Norwegian North Atlantic passive margin is associated with important detachments at depth and high-angle extensional faults near the surface (Smethurst 2000; Olesen et al. 2002; Osmundsen et al. 2002). The main structural domains of the passive margin are separated by several major normal fault systems (Fig. 1). While discussion concerning the dip directions of some major faults is ongoing, these fault systems and related structural highs and half-graben features are also reflected in the top basement map. Figure 5 shows the extension of the normal faults on the margin and their correlation with the top basement structure. The entire margin segmentation is controlled by an offshore extension of these low-angle faults and shear zones. The Hardangerfjord Shear Zone is a key example for the division of the Viking Graben to the north and the Norwegian–Danish Basin to the south. Other examples are the offshore extension of the Møre–Trøndelag Fault Complex, which divides the mid-Norwegian margin and northern North Sea, and the Nesna Shear Zone, which culminates into the Bivrost Lineament that separates the Lofoten from the Vøring margin (Blystad et al. 1995). The Kollstraumen and Høybakken detachments furthermore control the structural style of the Trøndelag Platform, and the latter bounds the Trøndelag Platform to the south (Osmundsen & Ebbing 2008). These major detachments formed during orogen-parallel extension, that is at a high angle to the orogenic front (Braathen et al. 2000, 2002; Mosar 2003).

Extrapolating the onshore structures to the offshore realm, it can be deduced that NE–SW trending (i.e. orogen-parallel) late Caledonian gravity collapse affected the entire mid-Norwegian margin, which led to the intimate link between the crust below the Trøndelag Platform and the onshore domain (Osmundsen & Ebbing 2008). Interpretation of magnetic anomalies helps to link onshore and offshore structures and correlation between detachment zones onshore and structural highs offshore can be made (e.g. Olesen et al. 2002; Osmundsen et al. 2002; Skilbrei et al. 2002). Of particular interest is the recognition of the structurally denuded basement culminations onshore Norway, which are bound by the detachments. These structural highs contain high-density and highly magnetized rocks of assumed lower crustal origin, which can be interpreted to represent reactivated core complexes (Osmundsen & Ebbing 2008; Ebbing et al. 2009).

Osmundsen & Ebbing (2008) advocate a structural model for the Jurassic–Early Cretaceous thinning of crust under the southeastern mid-Norway rifted margin, which gives an important role to the detachments. Following widespread Late Palaeozoic–Early Mesozoic half-graben formation, basin flank detachment faults developed. These faults cut the previous configuration of Palaeozoic faults and basins, and may have reactivated deep-seated older structures. At an early stage the middle and/or lower crust were still able to deform in a ductile manner, and provided a level of delamination for the early detachments, which could explain thinning of crust over the >100 km wide area that would later become the Møre and southern Vøring basins. Redistribution of ductile material led to depletion of such material (north)west of the basin flank detachments and to preferential storage of middle/lower crust under the platform and sub-basins. Isostatic denudation of the footwall and/or rotation on the Møre-Trøndelag Fault Complex led to the warping and deactivation of the Jurassic detachment, and to incision by the more planar, basin-bounding detachment fault. The main Møre fault evolved into the top basement detachment fault that, in combination with the rotated fault block arrays, controlled the basinward taper of the crystalline crust (Osmundsen & Ebbing 2008).

Basin floor detachment faults did become localized in the area of highly thinned crust 50–100 km NW of the basin flanks and the crustal thinning accommodated by the basin floor detachments resulted in extensional basins that juxtaposed rotated pre-rift and syn-rift sediments with rocks exhumed from the middle/deep crust or upper mantle (Osmundsen & Ebbing 2008).

A valuable application of our compilation will be for studies of the thermal state of the margin. Basement highs often are associated with pathways for upward fluid migration, which can lead to an anomalously high heat flow (Ritter et al. 2004), while the presence of granitoidal rocks can lead to a significantly increased heat-flow, as shown for the Transscandinavian Igneous Belt in the Central Scandes (Pascal et al. 2007). The top basement and crustal thickness map allow the influence of the crustal structure on the heat flow into the base of the sedimentary basins to be estimated.

Another step towards a better understanding of the different basement units is mapping of the base of the dominantly Caledonian or Precambrian basement. Ebbing et al. (2009) mapped the

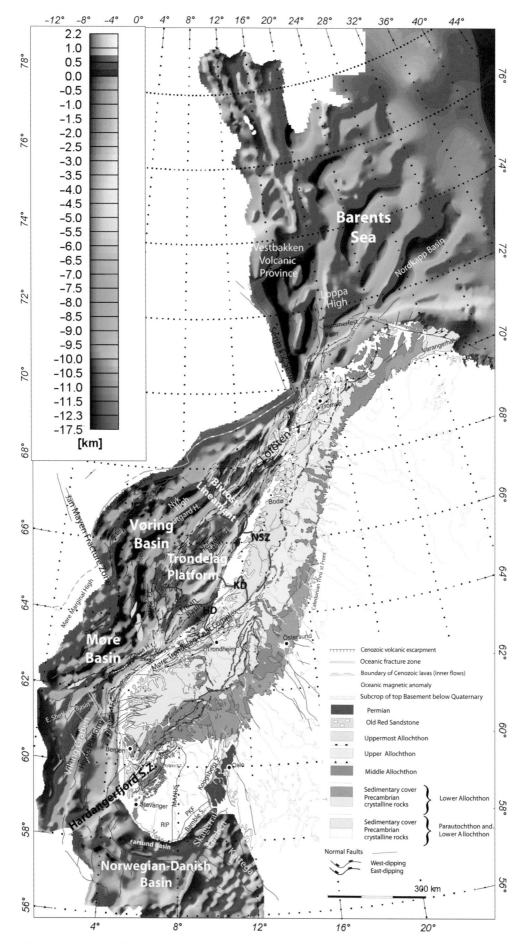

Fig. 5. Offshore top basement map and onshore tectonostratigraphic map. The map illustrates the continuation of faults from the onshore to offshore realm and illustrates the link between margin segmentation and onshore geology. Thick black lines over the Trøndelag Platform show offshore continuation of Late Caledonian detachments mapped onshore Norway.

base of the upper part of the basement on the mid-Norwegian margin, which corresponds mainly to the Caledonian basement, but also provided a tentative interpretation of the top Precambrian basement. Such studies are not yet available for the entire margin, but hopefully they will be available in the future, when the Norwegian–Danish Basin and North Sea are revisited.

Conclusions

We present for the first time a top basement map for the entire Norwegian shelf. The top basement map is based for a large part on integration of magnetic, gravity, seismic and borehole data. The quality of the compilation varies along the margin depending on the methodology used for compiling the individual contributions. The mid-Norwegian margin is the area with the largest amount of data used for the top basement mapping, partly because here the most recent studies exist. We also present a Moho compilation, which combines onshore data with the high-resolution crustal models of the margin, and which allows us to calculate the crustal thickness for the entire margin and onshore Norway. Both the top basement map and the crustal thickness map highlight the different margin segments. Comparison with the geological structure mapped onshore Norway shows the influence of normal faults on the segmentation of the margin. Previous studies mostly concentrated on different individual segments of the margin, but the compiled data will in the future allow refined studies of the evolution of the Norwegian continental shelf as a whole, and to study in more depth the influence of onshore–offshore structures on the structural style of the entire Norwegian continental shelf.

The main part of the study was done as part of the project KONTIKI (Continental Crust and Heat Generation in 3D) supported by StatoilHydro. Individual projects have been funded by different industry partners and the Research Council of Norway. We thank all authors of the individual studies for providing their results and cooperation in the individual studies, and especially appreciate the input from R. F. Reynisson, J. R. Skilbrei and M. Smethurst. We also thank the editor S. C. Pickering, an anonymous reviewer and G. S. Kimbell for their valuable comments and suggestions.

References

Åm, K. 1973. *Tolkning av magnetiske og gravimetriske data: Kontinentalsokkelen. Stavanger/Stad (59°–62°N)*. Geological Survey of Norway Report, **1175**.

Åm, K. 1975. Aeromagnetic basement complex mapping north of latitude 62 N, Norway. *Norges geologiske undersøkelse*, **316**, 351–374.

Andersen, T. B. 1998. Extensional tectonics in the Caledonides of southern Norway, an overview. *Tectonophysics*, **285**, 333–351.

Barrère, C., Ebbing, J. & Gernigon, L. 2009. Offshore prolongation of Caledonian structures and basement characterisation in the western Barents Sea from geophysical modeling. *Tectonophysics*, **470**, 71–88; doi: 10.1016/j.tecto.2008.07.012.

Blystad, P., Brekke, H., Færseth, R. B., Larsen, B. T., Skogseid, J. & Tørudbakken, B. 1995. *Structural Elements of the Norwegian Continental Shelf, Part II. The Norwegian Sea Region*. Norwegian Petroleum Directorate Bulletins, **8**.

Braathen, A., Nordgulen, Ø., Osmundsen, P. T., Andersen, T. B., Solli, A. & Roberts, D. 2000. Orogen-parallel, opposed extension in the Central Norwegian Caledonides. *Geology*, **28**, 615–618.

Braathen, A., Osmundsen, P. T., Nordgulen, Ø., Roberts, D. & Meyer, G. B. 2002. Orogen-parallel extension of the Caledonides in northern Central Norway: an overview. *Norwegian Journal of Geology*, **82**, 225–241.

Breivik, A. J., Mjelde, R., Grogan, P., Shimamura, H., Murai, Y. & Nishimura, Y. 2005. Caledonide development offshore-onshore Svalbard based on ocean bottom seismometer, conventional seismic, and potential field data. *Tectonophysics*, **401**, 79–117.

Brekke, H. 2000. The tectonic evolution of the Norwegian Sea continental margin with emphasis on the Vøring and Møre Basins. *In*: Nottvedt, A. (ed.) *Dynamics of the Norwegian Margin*. Geological Society, London, Special Publications, **167**, 327–378.

Brekke, H., Dahlgren, S., Nyland, B. & Magnus, C. 1999. The prospectivity of the Vøring and Møre basins on the Norwegian Sea continental margin. *In*: Fleet, A. J. & Boldy, S. A. R. (eds) *Petroleum Geology of Northwest Europe: Proceedings of the 5th Conference*, London, 26–29 October 1997. Geological Society, London, 261–274; doi: 10.1144/0050261.

Brekke, H., Sjulstad, H. I., Magnus, C. & Williams, R. W. 2001. Sedimentary environments offshore Norway – an overview. *In*: Martinsen, O. J. & Dreyer, T. (eds) *Sedimentary Environments Offshore Norway: Palaeozoic to Recent*. Elsevier Science, New York, 7–37.

Christiansson, P., Faleide, J. I. & Berge, A. M. 2000. Crustal structure in the northern North Sea from deep seismic reflection and refraction data. *In*: Nøttvedt, A. *et al.* (eds) *Dynamics of the Norwegian Margin*. Geological Society, London, Special Publications, **167**, 15–40.

Dehls, J. F., Olesen, O., Bungum, H., Hicks, E., Lindholm, C. & Riis, F. 2000. *Neotectonic Map: Norway and Surrounding Areas*. Geological Survey of Norway, Trondheim.

Doré, A. G. 1992. Synoptic paleogeography of the Northeast Atlantic Seaway: Late Permian to Cretaceous. *In*: Parnell, J. (ed.) *Basins on the Atlantic Seaboard: Petroleum Geology, Sedimentology and Basin Evolution*. Geological Society, London, Special Publications, **62**, 421–446.

Doré, A. G. & Vining, B. A. (eds) 2005. *Petroleum Geology: North-West Europe and Global Perspectives – Proceedings of the 6th Petroleum Geology Conference*. Geological Society, London; doi: 10.1144/0060803.

Doré, A. G., Lundin, E. R., Birkeland, Ø., Eliassen, P. E. & Jensen, L. N. 1997. The NE Atlantic Margin: implications of late Mesozoic and Cenozoic events for hydrocarbon prospectivity. *Petroleum Geosciences*, **3**, 117–131.

Ebbing, J., Lundin, E., Olesen, O. & Hansen, E. K. 2006. The mid-Norwegian margin: a discussion of crustal lineaments, mafic intrusions, and remnants of the Caledonian root by 3D density modelling and structural interpretation. *Journal of the Geological Society, London*, **163**, 47–60.

Ebbing, J., Skilbrei, J. R. & Olesen, O. 2007. Insights into the magmatic architecture of the Oslo Graben by petrophysically constrained analysis of the gravity and magnetic field. *Journal of Geophysical Research*, **112**, B4404.

Ebbing, J., Gernigon, L., Pascal, C., Olesen, O. & Osmundsen, P. T. 2009. A discussion of structural and thermal control of magnetic anomalies on the mid-Norwegian margin. *Geophysical Prospecting*, **57**, 665–681.

Faleide, J. I., Tsikalas, F. *et al.* 2008. Structure and evolution of the continental margin off Norway and the Barents Sea. *Episodes*, **31**(1), 1–10.

Fleet, A. J. & Boldy, S. A. R. (eds) 1999. *Petroleum Geology of Northwest Europe: Proceedings of the 5th Conference*, London, 26–29 October 1997. Geological Society, London.

Fossen, H. 1992. The role of extensional tectonics in the Caledonides of south Norway. *Journal of Structural Geology*, **14**, 1033–1046.

Gabrielsen, R. H., Odinsen, T. & Grunnaleite, I. 1999. Structuring of the northern Viking Graben and the Møre Basin: the influence of basement structural grain, and the particular role of the Møre–Trøndelag Fault Complex. *Marine and Petroleum Geology*, **16**, 443–465.

Gabrielsen, R. H., Braathen, A., Olesen, O., Faleide, J. I., Kyrkjebø, R. & Redfield, T. F. 2005. Vertical movements in south-western Fennoscandia: a discussion of regions and processes from the present to the Devonian. *In*: Wandås, B. T. G., Eide, E. A., Gradstein, F. & Nystuen, J. P. (eds) *Onshore–Offshore Relationships on the North Atlantic Margin*. Norwegian Petroleum Society, Oslo, Special Publications, **12**, 1–28.

Gee, D. G., Guezou, J.-C., Roberts, D. & Wolff, F. C. 1985. The central-southern part of the Scandinavian Caledonides. *In*: Gee, D. G. & Sturt, B. A. (eds) *The Caledonide Orogen – Scandinavia and Related Areas*. Wiley, New York, 109–133.

Götze, H.-J. & Lahmeyer, B. 1988. Application of three-dimensional interactive modeling in gravity and magnetic. *Geophysics*, **53**, 1096–1108.

Grad, M., Tiira, T. & ESC Working Group 2009. The Moho depth map of the European Plate. *Geophysical Journal International*, **176**, 279–292.

Gramberg, I. S. *et al.* 2001. Eurasian Arctic Margin: earth science problems and research challenges. *Polarforschung*, **69**, 3–15.

Hospers, J. & Ediriweera, K. K. 1988. *Mapping the Top of the Crystalline Continental Crust in the Viking Graben Area, North Sea.* Norwegian Geological Survey, Trondheim, Special Publications, **3**, 21–28.

Hospers, J. & Ediriweera, K. K. 1991. Depth and configuration of the crystalline basement in the Viking Graben area, Northern North Sea. *Journal of the Geological Society, London*, **148**, 261–265.

Hospers, J., Rathore, J. S., Jianhua, F. & Finnstrøm, E. G. 1986. Thickness of pre-Zechstein-salt Palaeozoic sediments in the southern part of the Norwegian sector of the North Sea. *Norsk Geologisk Tidsskrift*, **66**, 295–304.

Johansen, S. E., Ostisty, B. K. *et al.* 1992. Hydrocarbon potential in the Barents Sea region: play distribution and potential. *In*: Vorren, T. O., Bergsager, E. *et al.* (eds) *Arctic Geology and Petroleum Potential*. Elsevier, Amsterdam, NPF Special Publications, **2**, 273–320.

Kinck, J. J., Husebye, E. S. & Larsson, F. R. 1993. The Moho depth distribution in Fennoscandia and the regional tectonic evolution from Archean to Permian times. *Precambrian Research*, **64**, 23–51.

Lassen, A. & Thybo, H. 2004. Seismic evidence for Late Proterozoic orogenic structures below the Phanerozoic sedimentary cover in the Kattegat area, SW Scandinavia. *Tectonics*, **23**, TC2017; doi: 10.1029/2003TC001499.

Lyngsie, S. B., Thybo, H. & Rasmussen, T. M. 2005. Regional geological and tectonic structures of the North Sea area from potential field modeling. *Tectonophysics*, **413**, 147–170.

Mjelde, R., Raum, T. *et al.* 2005. Crustal structure of the Vøring Margin, NE Atlantic: a review of geological implications based on recent OBS data. *In*: Doré, A. G. & Vining, B. A. (eds) *Petroleum Geology: North-West Europe and Global Perspectives: Proceedings of the 6th Petroleum Geology Conference*. Geological Society, London, 803–813; doi: 10.1144/0060803.

Mjelde, R., Faleide, J. I., Breivik, A. J. & Raum, T. 2009. Lower crustal composition and crustal lineaments on the Vøring Margin, NE Atlantic: a review. *Tectonophysics*, **472**, 183–193.

Mørk, M. B. E., McEnroe, S. A. & Olesen, O. 2002. Magnetic susceptibility of Mesozoic and Cenozoic sediments off Mid Norway and the role of siderite: implications for interpretation of high-resolution aeromagnetic anomalies. *Marine and Petroleum Geology*, **19**, 1115–1126.

Mosar, J. 2000. Depth of the extensional faulting on the mid-Norway Atlantic passive margin. *Norges Geologiske Undersokelse Bulletin*, **437**, 33–41.

Mosar, J. 2003. Scandinavia's North Atlantic passive margin. *Journal of Geophysical Research*, **108**, 2360.

Nøttvedt, A. (ed.) 2000. *Dynamics of the Norwegian Margin*. Geological Society, London, Special Publications, **167**.

Olesen, O., Roberts, D., Henkel, H., Lile, O. B. & Torsvik, T. H. 1990. Aeromagnetic and gravimetric interpretation of regional structural features in the Caledonides of West Finnmark and North Troms, northern Norway. *Bulletin of the Geological Survey of Norway*, **419**, 1–24.

Olesen, O., Henkel, H., Kaada, K. & Tveten, E. 1991. Petrophysical properties of a prograde amphibolite granulite facies transition zone at Sigerfjord, Vesterålen, Northern Norway. *Tectonophysics*, **192**, 33–39.

Olesen, O., Smethurst, M., Beard, L., Torsvik, T., Bidstrup, T. & Egeland, B. 1997. *SAS-96 Part II, Skagerrak Aeromagnetic Survey 1996, Interpretation Report*. NGU Report **97.022**.

Olesen, O., Lundin, E. *et al.* 2002. Bridging the gap between the onshore and offshore geology in Nordland, northern Norway. *Norwegian Journal of Geology*, **82**, 243–262.

Olesen, O., Smethurst, M. A., Torsvik, T. H. & Bidstrup, T. 2004. Sveconorwegian igneous complexes beneath the Norwegian–Danish Basin. *Tectonophysics*, **387**, 105–130.

Olesen, O., Ebbing, J. *et al.* 2007. An improved tectonic model for the Eocene opening of the Norwegian–Greenland Sea: use of modern magnetic data. *Marine and Petroleum Geology*, **24**, 53–66.

Olesen, O., Brönner, M. *et al.* 2010. New aeromagnetic and gravity compilations from Norway and adjacent areas: methods and applications. *In*: Vining, B. A. & Pickering, S. C. (eds) *Petroleum Geology: From Mature Basins to New Frontiers – Proceedings of 7th Petroleum Geology Conference*. Geological Society, London, 559–586; doi: 10.1144/0070559.

Osmundsen, P. T. & Ebbing, J. 2008. Styles of extension offshore mid Norway and implications for mechanisms of Late Jurassic–Early Cretaceous rifting. Basement configuration of the mid-Norwegian margin. *Tectonics*, **27**, TC6016; doi: 10.1029/2007TC002242.

Osmundsen, P. T., Andersen, T. B., Markussen, S. & Svendby, A. K. 1998. Tectonics and sedimentation in the hanging wall of a major extensional detachment: the Devonian Kvamshesten Basin, western Norway. *Basin Research*, **10**, 213–224.

Osmundsen, P. T., Sommaruga, A., Skilbrei, J. R. & Olesen, O. 2002. Deep structure of the Mid Norway rifted margin. *Norwegian Journal of Geology*, **82**, 205–224.

Pascal, C., Ebbing, J. & Skilbrei, J. R. 2007. Interplay between the Scandes and the Transscandinavian Igneous Belt: integrated thermal and potential field modelling of the Central Scandes profile. *Norwegian Journal of Geology*, **87**, 3–12.

Planke, S., Rasmussen, R., Rey, S. S. & Myklebust, R. 2005. Seismic characteristics and distribution of volcanic intrusions and hydrothermal vent complexes in the Vøring and Møre basins. *In*: Doré, A. G. & Vining, B. A. (eds) *Petroleum Geology: North-West Europe and Global Perspectives: Proceedings of the 6th Petroleum Geology Conference*. Geological Society, London, 833–844; doi: 10.1144/0060833.

Ramberg, I. B. & Smithson, S. B. 1975. Geophysical interpretation of crustal structure along the southeastern coast of Norway and Skagerrak. *GSA Bulletin*, **86**, 769–774.

Reynisson, R. F. 2010. *Deep structure of the mid-Norwegian margin with emphasis on the Møre margin and sub-basalt exploration using integrated gravimetric-, and magnetometric models, with isostatic considerations*. PhD thesis, Norwegian University of Science and Technology, Trondheim.

Reynisson, R. F., Ebbing, J. & Skilbrei, J. R. 2009. The use of potential field data in revealing the basement structure in sub-basaltic settings: an example from the Møre margin, offshore Norway. *Geophysical Prospecting*, **57**, 753–771.

Ritter, U., Zielinski, G. W., Weiss, H. M., Zielinski, R. L. B. & Sættem, J. 2004. Heat flow in the Vøring Basin, Mid-Norwegian shelf. *Petroleum Geoscience*, **10**, 353–365.

Ritzmann, O., Maercklin, N., Faleide, J. I., Bungum, H., Mooney, W. D. & Detweiler, S. T. 2007. A three-dimensional geophysical model of the crust in the Barents Sea region: model construction and basement characterization. *Geophysical Journal International*, **170**, 417–435.

Roberts, D. 2003. The Scandinavian Caledonides: event chronology, palaeogeographic settings and likely modern analogues. *Tectonophysics*, **365**, 283–299.

Scheck-Wenderoth, M., Faleide, J. I., Raum, T., Mjelde, T. & Horsfield, B. 2007. The transition from the continent to the ocean: a deeper view on the Norwegian margin. *Journal of the Geological Society, London*, **164**, 855–868.

Sindre, A. 1993. *Regional tolkning av geofysiske data, kartblad Arendal*. NGU Report **92.213**.

Skilbrei, J. R. 1991. Interpretation of depth to the magnetic basement in the Northern Barents Sea (South of Svalbard). *Tectonophysics*, **200**, 127–141.

Skilbrei, J. R. 1995. Aspects of the geology of the southwestern Barents Sea from aeromagnetic data. *NGU Bulletin*, **427**, 64–67.

Skilbrei, J. R. & Olesen, O. 2005. Deep structure of the Mid-Norwegian shelf and onshore–offshore correlations: insight from potential field data. *In*: Wandås, B. T. G., Eide, E. A., Gradstein, F. & Nystuen, J. P. (eds) *Onshore–Offshore Relationships on the North Atlantic Margin*. Norwegian Petroleum Society, Oslo, Special Publications, **12**, 43–68.

Skilbrei, J. R., Skyseth, T. & Olesen, O. 1991. Petrophysical data and opaque mineralogy of high-grade and retrogressed lithologies – implications for the interpretation of aeromagnetic anomalies in Northern Vestranden, Central Norway. *Tectonophysics*, **192**, 21–31.

Skilbrei, J. R., Olesen, O., Osmundsen, P. T., Kihle, O., Aaro, S. & Fjellanger, E. 2002. A study of basement structures and onshore–offshore correlations in Central Norway. *Norwegian Journal of Geology*, **82**, 263–279.

Smethurst, M. A. 1994. *Viking-93 Aeromagnetic Survey: Interpretation Report*. NGU Report **94.011**.

Smethurst, M. A. 2000. Land–offshore tectonic links in western Norway and the northern North Sea. *Journal of the Geological Society, London*, **157**, 769–781.

Thybo, H. & Schönharting, G. 1991. Geophysical evidence for Early Permian igneous activity in a transtensional environment, Denmark. *Tectonophysics*, **189**, 193–208.

Tsikalas, F., Faleide, J. I. & Kusznir, N. J. 2008. Along-strike variations in rifted margin crustal architecture and lithosphere thinning between northern Vøring and Lofoten margin segments off mid-Norway. *Tectonophysics*, **458**, 68–81.

Vejbæk, O. V. & Britze, P. 1994. *Geological Map of Denmark 1:750 000*. Danmarks Geologiske Undersøgelse **45**.

Yegorova, T., Bayer, U., Thybo, H., Maystrenko, Y., Scheck-Wenderoth, M. & Lyngsie, S. B. 2007. Gravity signal from the lithosphere in the Central European Basin System. *Tectonophysics*, **429**, 133–163.

Some emerging concepts in salt tectonics in the deepwater Gulf of Mexico: intrusive plumes, canopy-margin thrusts, minibasin triggers and allochthonous fragments

M. P. A. JACKSON, M. R. HUDEC and T. P. DOOLEY

Bureau of Economic Geology, Jackson School of Geosciences, The University of Texas at Austin, University Station Box X, Austin, Texas 78713, USA (e-mail: martin.jackson@beg.utexas.edu)

Abstract: We summarize four emerging concepts in salt tectonics in the deepwater Gulf of Mexico, selected from a longer list of concepts that have advanced significantly in the last decade. Squeezed salt stocks are common in orogenic forelands, in inverted basins and at the toe of salt-bearing passive margins. Modelling suggests that during early shortening, an inward salt plume from the source layer inflates the diapir and arches its roof. After further shortening, diapiric salt is expelled as an outward plume back into the source layer. Salt canopies are conventionally thought to advance by glacial extrusion. However, almost all modern salt canopies are now buried and can only advance by frontal thrusting. Thrusting allows the salt canopy and its protective roof to advance together, minimizing salt dissolution. Advance is by a roof-edge thrust rooted in the leading tip of salt or by thrust imbricates forming accretionary wedges. Minibasins can sink into salt if the average density of the overburden exceeds that of salt. This requires 2–3 km of burial of siliciclastic fill, yet most minibasins first sink when much thinner. Three alternative mechanisms to negative buoyancy in the deepwater Gulf of Mexico address this paradox of initiation. First, squeezed diapirs inflate, leaving the intervening minibasins as depressions. Second, when a diapir's salt supply wanes, the overlying dynamic salt bulge subsides, allowing a minibasin to form. Third, differential loading causes the thick end of a sedimentary wedge to sink faster into the salt, creating a sag. Spreading salt canopies can transport their dismembered roof fragments tens of kilometres basinward. These exotic fragments are up to 25 km in breadth and comprise anomalously old Mesozoic through Miocene sequences. Strata of the same age underlie the salt canopy or its welded equivalent, signalling lateral transport by thick salt.

Keywords: salt tectonics, diapir, thrust, allochthonous salt, Gulf of Mexico, Louann, structural geology

Knowledge of salt tectonics has expanded rapidly on three main fronts. Petroleum exploration of passive continental margins has yielded new 3D seismic datasets, improved seismic imaging and well control, all of which reveal the large-scale geometry of salt structures. Mapping in orogenically unroofed salt basins clarifies smaller-scale field relations. Increasingly sophisticated structural modelling elucidates the kinematics and dynamics of salt tectonics.

These advances in knowledge bring new ideas and new terminology. The last attempt to summarize the conceptual growth of salt tectonics was in 1993 at the Hedberg International Research Conference on Salt Tectonics, held in Bath (UK) (Jackson 1995). That review summarized progress in understanding since 1856, when the first salt diapir was described in the geological literature.

The present paper is more restricted in scope than the Bath review. Major advances in understanding salt tectonics at seismic scale during the last decade include, for example (Fig. 1): (a) halokinetic sequences around passive diapirs (Giles & Lawton 2002; Rowan *et al.* 2003); (b) inflation and deflation of autochthonous salt layers (Hall 2002); (c) multiwavelength compressional folding in thickening strata (Peel 2003); (d) epeirogenic crustal uplift rejuvenating salt tectonics on passive margins (Hudec & Jackson 2004; Jackson *et al.* 2005); (e) emplacement of deep, old allochthonous fringes (Peel 2001); (f) directly measured rates of subaerial salt extrusion (Talbot 1998; Talbot *et al.* 2000; Talbot & Aftabi 2004); (g) salt sutures (Kadri *et al.* 2000; Liro *et al.* 2004); (h) surging salt canopies overriding small stocks (Hall 2002; Liro *et al.* 2004; Mount *et al.* 2006); and (i) strike-slip enhancing or retarding salt diapirism (Talbot & Alavi 1996; Letouzey & Sherkati 2004; Dooley *et al.* 2007). Some of these ideas have been published only as abstracts and have not been thoroughly explored.

Space allows only a brief summary of all these processes, so instead we focus on the deepwater Gulf of Mexico to highlight four advances of which we have first-hand knowledge. Despite this parochial approach, these advances could apply beyond the Gulf of Mexico, especially to equivalent deepwater settings on the West African and Brazilian continental margins or other basins where thick salt is mobilized. None of these new ideas is yet in wide currency. Some are controversial. Some may not last. However, they all seem novel and interesting. The four ideas showcased are intrusive plumes, salt-sheet thrusts, minibasin triggers and allochthonous fragments. A cross-sectional sketch shows where these processes might occur in a deepwater passive margin (Fig. 2).

Intrusive salt plumes

Within every growing diapir, salt flows upwards from its source layer. This is a classic and deeply entrenched idea (Escher & Kuenen 1929) for salt diapirs growing halokinetically and for diapirs squeezed by lateral compression. When diapirs are squeezed, salt is forced upwards, like an uncapped toothpaste tube squeezed by hand pressure. The rising salt either arches its sedimentary roof or escapes from the ruptured roof as extrusive salt.

Salt sags within model diapirs that are laterally stretched faster than they can be supplied by salt (Vendeville & Jackson 1992), and entrained dense layers partly sink inside rising model diapirs (Chemia & Koyi 2008). However, in these examples, the salt flows downward only within the diapir and does not challenge the dogma that diapiric salt does not intrude the source layer. Yet downward flow back into the source layer is indicated by modelling squeezed diapirs. Figure 3 shows this process in vertical

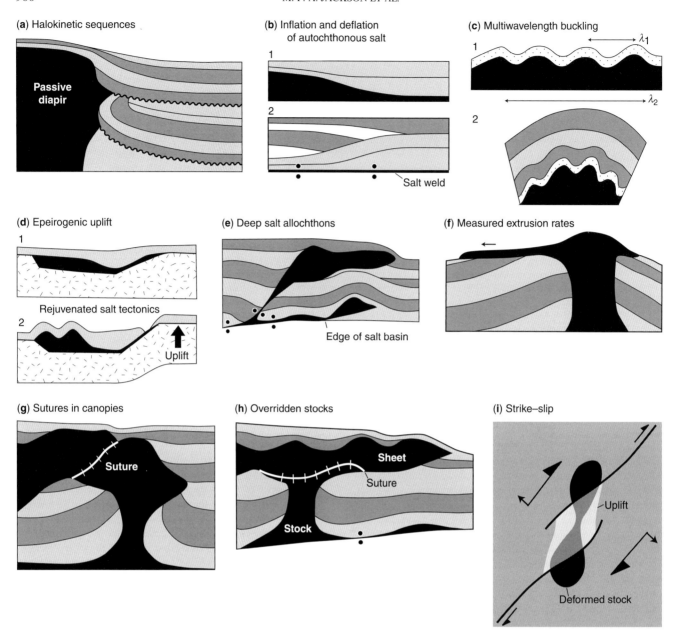

Fig. 1. Some concepts in salt tectonics recently discovered or improved: (**a**) halokinetic sequences around passive diapirs; (**b**) inflation and deflation of autochthonous salt layers; (**c**) multiwavelength (λ_1 and λ_2) compressional folding in thickening strata; (**d**) epeirogenic crustal uplift rejuvenating salt tectonics on passive margins; (**e**) emplacement of deep, old, allochthonous fringes; (**f**) directly measured rates of subaerial salt extrusion; (**g**) salt sutures; (**h**) surging salt canopies overriding small stocks; and (**i**) strike-slip (here transtensional) enhancing or retarding salt diapirism.

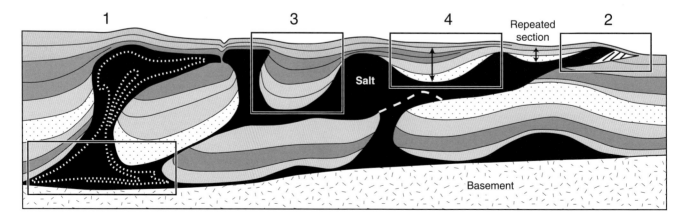

Fig. 2. Schematic cross-section of the Gulf of Mexico lower continental slope, showing the setting of the four conceptual processes highlighted in this paper. (1) Squeezed diapirs intrude plumes of salt into the source layer. (2) Buried salt canopies advance by overthrusting of allochthonous salt and its roof. (3) Three mechanisms explain why even low-density, uncompacted sediments sink into salt and start minibasins. (4) Spreading salt canopies can carry exotic fragments of the canopy roof great distances and duplicate stratigraphy above and below the canopy.

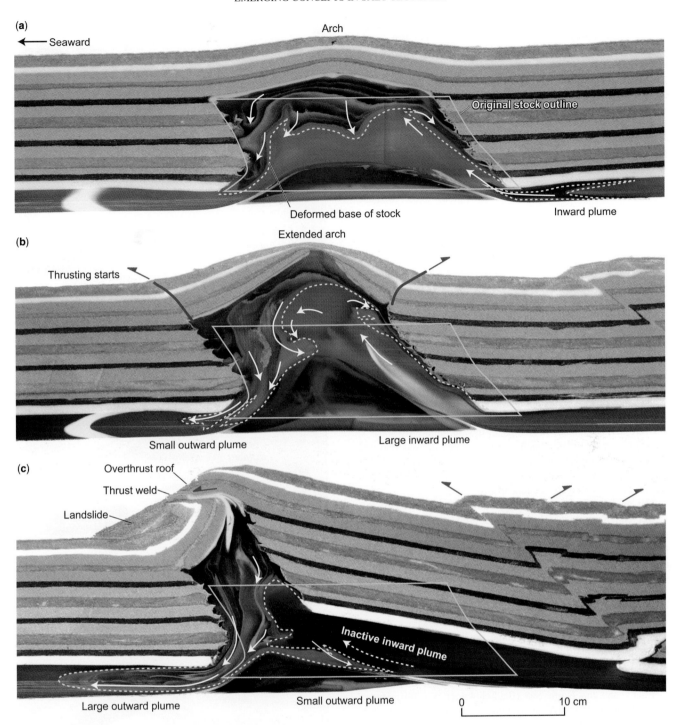

Fig. 3. Cross-sections from three physical models showing stages of squeezing of inclined stocks, whose original shapes are outlined by a polygon. (**a**) Mild shortening (7 cm) drives salt seawards. An inward-intrusive plume inflates the stock's cross-sectional area to 111% and gently arches its roof. (**b**) After moderate shortening (15 cm), the inward plume enlarges and inflates the stock area to 117%; the stock also expels a small outward plume seawards into the source layer. (**c**) After strong shortening (30 cm), two outward plumes shrink the diapir area to 42% of original area and inflate the source layer.

sections through three dynamically scaled physical models (Dooley et al. 2009). The models were sliced at different stages to yield snapshots of how they evolved. Apart from different durations of deformation, the three models were identical. Each model simulated the effects of shortening on seaward-inclined, cylindrical salt stocks having flat crests buried by thick roofs. The models show the effects of shortening on passive stocks that have evolved through a specific kinematic history. The resulting pre-shortening geometry modelled here is common in the Gulf of Mexico but not universal. As a moving endwall shortened the stocks at a constant rate, the squeezed diapir evolved through three stages.

In Stage 1, shortening thickened the source layer of salt and drove it seawards on both sides of the diapir (Fig. 3). Static pressures in the source layer were higher than those in the stock because salt was displaced from the thrust belt growing in the hinterland (to the right of the frame in Fig. 3a). Pressures in the source layer were also high because the overburden was 10% denser than the salt. The difference in hydraulic head injected a salt plume into the diapir from its landward side. The squeezed diapir expanded in vertical section to 111% of its original area, even though its width decreased by 2% because of shortening. The diapir dilated rather than shrank. The inward plume of salt arched the diapir roof ahead of the advancing thrust front (out of the frame in Fig. 3a).

In Stage 2, the thrust front jumped seawards of the salt stock (Fig. 3b), and the arched stock roof began to shorten. The inward plume continued to inflate the diapir to 117% of its original area in vertical section. Salt was pumped above regional because the diapir walls converged and because salt was imported from the source layer. The salt volume displaced above regional in the diapir was c. 1.8 times greater than the volume displaced by 11% convergence of the diapir walls. Thus salt import still dominated. However, salt export began as an outward plume of diapiric salt and began to intrude the source layer on the seaward side of the diapir.

In Stage 3, the compressed stock shrank to 42% of its original area in vertical cross-section as the inward plume of salt died and outward plumes were expelled (Fig. 3c). Serial sections show that no salt escaped along-strike. Overthrusting thickened the arched roof, which prevented salt from escaping as an extrusion. Salt flowed downwards because its roof was too thick for salt to escape upwards. The diapir grew taller because of salt inflation and lateral squeezing. Increased height of the diapir reversed the earlier hydraulic-head gradient and forced diapiric salt into the source layer, thickening it 4-fold on the landward side of the diapir. Salt intruded the source layer in outward plumes both seawards and landwards.

The three stages of shortening are shown in underside views up through the colourless, transparent silicone simulating salt (Fig. 4). Shearing of coloured marker plugs into streaked-out ribbons tracked salt flow. After minor shortening (stage 1), the blue base of the stock remained elliptical. During stage 2, salt converged seaward into the stock and rose as an inward plume. By stage 3, outward plumes of salt expelled and crumpled the blue base of the diapir and a green marker layer from half-way up the stock. Together the blue and green markers merged into the two outward plumes seawards and landwards.

The models shed light on the dynamics of squeezing of salt stocks. Gravity provides a load, as in all salt tectonics. In addition, regional compression forces together the sides of the stock. Initially the pressurized source layer intruded and inflated the stock. Then as shortening intensified, more diapiric salt was expelled as plumes. Shortening structures in the overburden rose and made space for the salt plumes.

Four other factors promote outward plumes intruding the source layer but are not essential (Dooley et al. 2009). First, a thick diapiric roof is strong enough to resist arching, so diapiric salt is forced to flow downwards into the source layer. Passive diapirs are interpreted to have emplaced wings of salt into thin post-Zechstein salt beds delaminated in compressional detachment folds in Germany and the North Sea (Baldschuhn et al. 2001; Hudec 2004; Stewart 2007). A small minority of these diapirs (e.g. Wolthausen diapir and Niendorf II diapir; Baldschuhn et al. 2001; Hudec 2004) appear to have been emergent during wing intrusion, suggesting that a thick roof is not necessary for outward plumes. Second, a wide diapir provides a large volume and broad conduit for salt to intrude the source layer. Third, an initially thick source layer provides a wide channel for salt flow in which boundary drag is proportionally low. Fourth, a seaward dip of the base of salt helps salt to flow seaward and entrains exported diapiric salt. The thick-roofed models in Figure 3 did not require a seaward-dipping source layer to form outward plumes, but our models of thin-roofed diapirs required a dipping source layer to expel plumes (Dooley et al. 2006). All four factors are more effective when combined. In contrast, although a dense overburden encourages inward salt plumes, the effect of density contrast seems small for outward plumes.

Outward plumes should be considered in restorations of cross-sections of squeezed diapirs. In a closed system the source layer thickens as salt laterally shortens. Downward expulsion of salt

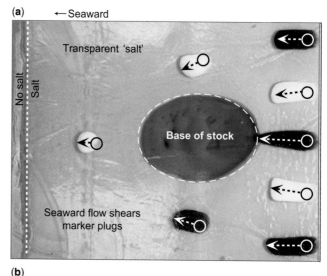
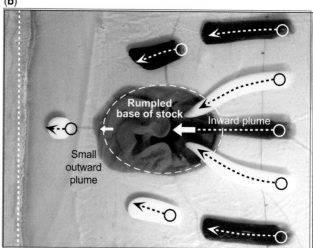
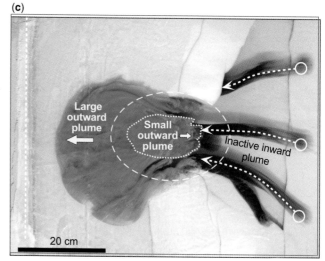

Fig. 4. Underside view upwards through the base of the physical models showing the same three stages of shortening (**a**–**c**) as in Figure 3. The salt source layer is transparent and colourless, except for the brown and yellow marker plugs embedded within it. The pale background colour is the base of the overburden. The dashed ellipse marks the shape and position of the stock before shortening.

plumes from squeezed diapirs further thickens the source layer. Neglecting downward expulsion causes an interpreter to overestimate the original thickness of the source layer. For example, diapirs such as Hildesheimer Wald and Niendorf II (Germany)

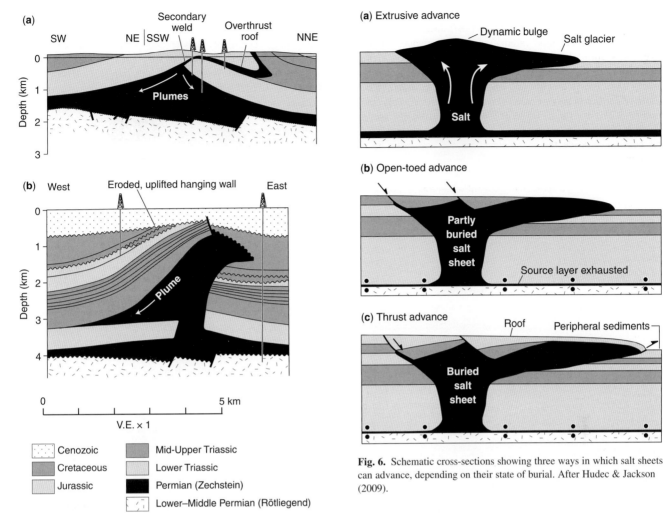

Fig. 5. Two German squeezed diapirs, showing thick adjoining salt. We attribute the thickness to plumes expelled from the diapirs by regional shortening. (**a**) Intense squeezing of Hildesheimer Wald diapir expelled an extrusive salt sheet upwards and probable plumes of diapiric salt downwards, which thickened the Permian Zechstein source layer to nearly 2 km. (**b**) Niendorf II diapir expelled Zechstein-derived diapiric salt outwards into the buckling, delaminating layer of Triassic Röt halite. Cross-sections after Baldschuhn *et al.* (2001).

Fig. 6. Schematic cross-sections showing three ways in which salt sheets can advance, depending on their state of burial. After Hudec & Jackson (2009).

are locally flanked by an unusually thick salt next to the diapir, which could have been thickened by salt expelled from the squeezed diapirs or inherited from a structurally thickened core of an anticline (Fig. 5). In addition the geometry of Niendorf II suggests that salt plumes may intrude even thin evaporite layers such as the Röt halite. Regional shortening peeled the overburden upwards, providing space for salt to escape from the squeezed diapir. Incorrectly estimating the original thickness of a salt layer also affects estimates of palaeo-heat flow and timing of welding.

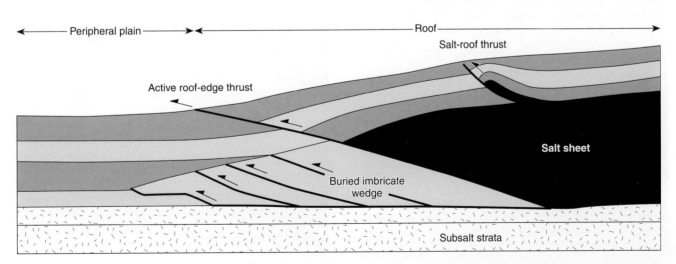

Fig. 7. Schematic cross-section of three types of thrusts in front of advancing salt canopies. (1) Roof-edge thrusts cut upsection from the salt tip and separate the condensed roof of the salt from the adjoining peripheral plain. (2) Imbricate wedges of thrusted peripheral-plain strata detach on a sole thrust rooted in the tip or base of the salt canopy. (3) Salt-roof thrusts, which shorten the roof of the canopy, are much rarer than the other two types of thrusts so are not discussed here. After Hudec & Jackson (2009).

The internal structures of most diapirs are complex (e.g. Balk 1949; Dixon 1975; Talbot & Jackson 1987; Jackson & Talbot 1989; Talbot & Aftabi 2004; Talbot *et al.* 2009). Outward intrusive plumes may further complicate the internal structure of a diapir. The models in Figure 3 show intricately refolded folds as the source layer mingled with salt originally in the diapir. Both deformed further as they intruded as an outward plume. This complexity makes it difficult to predict where even closely spaced drilling might intersect repeated anomalous layers in, for example, a potash mine. Such layers could be highly ductile potash beds (commonly sylvite and carnallite), which increase the risk of a borehole closing, or they could be shale inclusions, as in Belle Isle and Weeks Island salt mines (Gulf of Mexico; Kupfer 1974, 1980). Inclusions might increase the risk of abrupt changes in pore pressures, especially where the inclusions are continuous.

Canopy-margin thrust systems

Since the early 1990s, geoscientists have inferred that salt canopies advance extrusively as submarine salt glaciers (McGuinness & Hossack 1993; Fletcher *et al.* 1995). Extrusive advance applies to unburied and partly buried salt sheets. However, almost all salt canopies in modern marine settings are almost completely buried by sedimentary roofs (Fig. 2). A completely buried canopy can advance only if its roof overthrusts sediments in front (Fig. 6). The canopy and its roof advance together, which minimizes dissolution (Jackson & Hudec 2004; Hudec & Jackson 2009) as exemplified by the Sigsbee Escarpment in the northern Gulf of Mexico. The escarpment has a great-circle length of 560 km and a buried, sinuous length of *c.* 1100 km, making it the longest near-surface salt structure in the world. The escarpment follows the inflated leading edge of salt canopies. Roughly 60% of the escarpment is advancing, almost entirely by thrusting. Salt is exposed along less than 1% of the escarpment length, so extrusive advance is minor today.

Along the Sigsbee Escarpment, frontal thrusts can be divided into three parts (Fig. 7). A single roof-edge thrust commonly links the tip of the salt canopy to the seafloor (Fig. 8a). The hanging wall of the roof-edge thrust comprises the canopy and its sedimentary roof, which is a stratigraphically condensed carapace. The overridden footwall comprises a peripheral plain of uncondensed strata. An imbricate wedge of subsidiary thrusts can shorten the peripheral strata. Onlapping peripheral strata bury the imbricate wedge once the wedge is no longer active (Fig. 8b, c). If so, the canopy continues to advance on a large roof-edge thrust, which overrides any imbricate wedge. Frontal thrusts also sporadically fringe salt canopies landwards of the Sigsbee Escarpment.

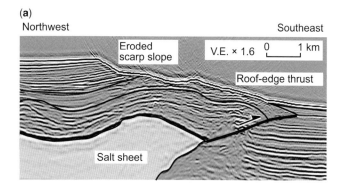

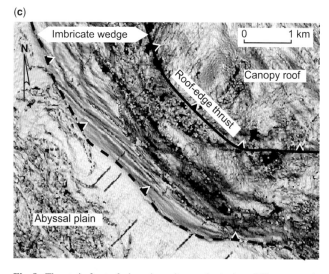

Fig. 8. Thrusts in front of advancing salt canopies in three different areas in the northern Gulf of Mexico. (**a**) A roof-edge thrust, Green Canyon area. (**b**) An imbricate wedge offset by a roof-edge thrust, Alaminos Canyon area. Horizontal lines are depth contours. (**c**) Dip map of a Pleistocene surface mapping the top of a buried imbricate wedge; backthrusts, which are rare in this setting, corrugate the top of the wedge into parallel ridges trending NW. Pre-stack depth-migrated seismic images (a) courtesy of BP, Unocal, and BHP Billiton; (b) and (c) courtesy of CGGVeritas. After Heyn *et al.* (2008).

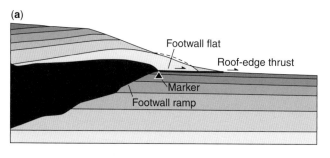

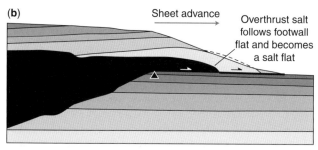

Fig. 9. Schematic cross-sections showing how a thrust flat (**a**) can later become a salt flat (**b**) as a salt canopy advances along the thrust. This change creates ambiguities in interpreting a stepped base of salt.

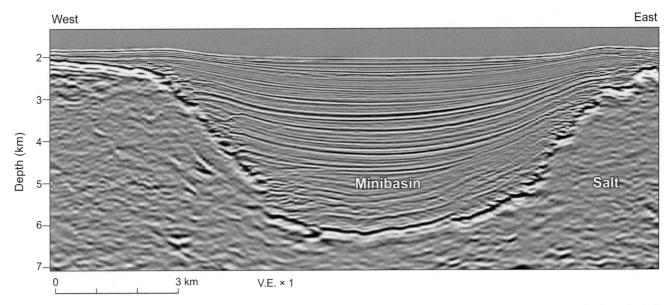

Fig. 10. A simple U-shaped minibasin sunk 4 km into a shallow salt canopy in Walker Ridge area, northern Gulf of Mexico. Pre-stack depth-migrated seismic image from Hudec *et al.* (2009), courtesy of CGGVeritas.

Evidence suggests that roof-edge thrusts continue along the base of the salt sheet for at least a few kilometres landwards (Hudec *et al.* 2009). A top-salt detachment would require a hanging-wall flat above the roof-edge thrust, which is rare. A short hanging-wall flat may be present in Figure 8a, so detachment along the top of salt might have contributed slip to the roof-edge thrust. However, Figure 8b lacks a hanging-wall flat and is far more typical. More fundamentally, the condensed hanging wall has age-equivalent cutoffs against the base of the salt sheet. Thus both the salt and its roof are in the hanging wall, and the detachment must follow the base of salt.

The preponderance of thrusting along the Sigsbee Escarpment affects how we interpret the origin of the stepped base of salt canopies. Segments of the stepped base of salt concordant with underlying strata are known as 'salt flats'. Conversely, markedly discordant segments are called 'salt ramps'. Since the early 1990s, salt ramps and flats have been attributed to variable rates of aggradation. A salt flat is inferred to mark a hiatus when salt extrusion was unhindered by sedimentation. Conversely, a salt ramp records when sedimentation accelerated, forcing extrusive salt to climb seawards over accumulating sediments.

This conventional interpretation of salt ramps and flats is invalid when a salt canopy is completely buried. Thrusting can create ramps and flats independent of aggradation rate during thrusting. As the thrust cuts through strata, variations in sediment strength create ramps and flats in the footwall (Fig. 9). As the canopy advances along the thrust, the base of the advancing canopy conforms to the stepped shape of the thrust footwall. As salt advances, footwall flats become salt flats; footwall ramps become salt ramps. Thus, both extrusion and thrusting form steps in the base of canopies. The geometry of the base is ambiguous and does not reveal the mechanism.

The ambiguity grows because most buried salt canopies advance first by extrusion and then by thrusting if they advance after burial. How can we infer when thrust advance began? The key is the age of the oldest roof strata. This age is rarely known in thin roofs because most subsalt wells in the deepwater Gulf of Mexico are drilled riserless for the first 600–900 m depth through suprasalt strata (R. Pilcher, pers. comm. 2009). However, where known, the age of the oldest roof strata records when a salt canopy was finally buried and extrusion changed to thrusting. At that stage, ramps and flats take on a different significance.

Minibasin triggers

Dish-shaped minibasins sinking into a broad salt canopy are deceptively simple (Figs 2 & 10), but within their synclinal form lies the puzzle of how they are initiated. Weakly compacted siliciclastic sediments are less dense than salt, so a sedimentary lens floats like an iceberg in a sea of salt (Fig. 11). Added sediments tend to be diverted off the structural high and pile up elsewhere – an unpromising start for a minibasin. How then do minibasins start to form and how do they deepen enough for sediments to compact and sink into the salt by negative buoyancy? This problem has been known for decades but has been largely ignored as just another mystery of salt tectonics, although some authors have invoked lateral variations in thickness and density to explain why buoyant sediments can sink into salt (e.g. Jackson & Talbot 1986; Kehle 1988; Cohen & Hardy 1996; Ge *et al.* 1997; Waltham 1997; Stewart & Clark 1999).

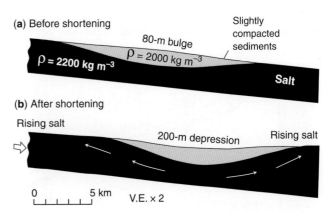

Fig. 11. Finite-element model showing the effects of gravity and regional shortening on a 1 km-thick lens of slightly compacted sediment floating on a salt canopy. (**a**) Because of its low density, the sedimentary lens floats 80 m above the surrounding salt. Extra sediments would be deflected off the bathymetric bulge and stacked on each side, so the lens does not become a minibasin. (**b**) After regional shortening, a similar lens of sediment becomes a local depression because passive diapirs rise on each side. Diapiric rise creates space for more sediments to form a minibasin. The minibasin was part of a 50 km-long model tilted 3° to the right. After Hudec *et al.* (2009).

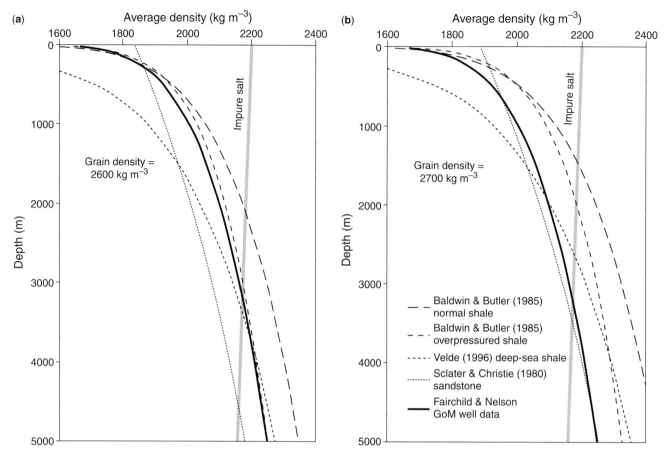

Fig. 12. Best-fit curves of depth v. average density for the entire sedimentary column in siliciclastic rocks. Compacting siliciclastic sediments must be at least 2–3 km thick before they become denser than salt. The thick black line is based on density measurements from sidewall cores and density logs (L. Fairchild & T. Nelson, pers. comm. 1989). The other curves are based on solidity measurements, assuming two different average densities of grains; an assumed grain density of 2600 kg m^{-3} (**a**) fits the measured bulk densities better than does a grain density of 2700 kg m^{-3} (**b**). The density of halite (thick grey line) at room temperature is $c.$ 2160 kg m^{-3}. Typical rock salt containing anhydrite impurities has a density rounded off to 2200 kg m^{-3}, which decreases slightly on burial because of the high thermal expansion coefficient of halite. Curves compiled by Hudec *et al.* (2009), except that of salt, which is from Jackson & Talbot (1986).

Density profiles suggest that the problem is extreme. Impure rock salt is slightly denser than halite, $c.$ 2200 km m^{-3}. During burial, salt is virtually incompressible (Heard *et al.* 1975) and its density decreases slightly as the crystalline rock salt warms and expands (Fig. 12). Siliciclastic sediments compact to densities equal to salt at burial depths of 1000–2000 m. However, only the deepest sediments are this dense. For the average density of the entire overburden to equal the density of salt, siliciclastic sediments must be 2000–4000 m thick, especially in overpressured basins (Fig. 12) (Hudec *et al.* 2009). Only at depths of several kilometres will

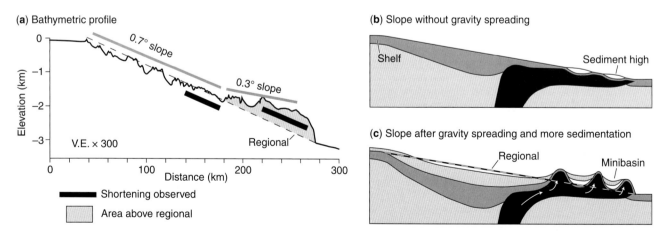

Fig. 13. Regional bathymetry can control where shortening initiates minibasins. (**a**) Bathymetric profile of the continental slope along longitude 92°35′W in the northern Gulf of Mexico; bathymetric data from Divins & Metzger (2007). The concept of an elevated lower slope trapping minibasins is from Rowan (2002). (**b**) and (**c**) Schematic cross-sections showing that in the compressed, uplifted area of the lower slope, both salt and sediment rise, but the more mobile salt diapirs rise above the adjoining minibasins.

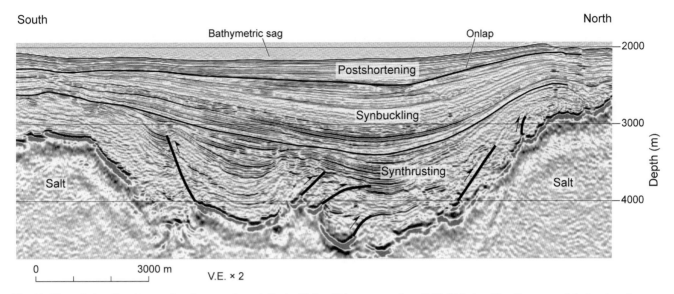

Fig. 14. Thrust faults offset the top of a salt canopy in a minibasin, Walker Ridge area, northern Gulf of Mexico. Thrusting occurred during stage 1, as shortening depressed the low-density sedimentary roof into the salt. After stage 1 thrusting ended, continued mild shortening depressed the minibasin further during stage 2. After shortening ended and stage 3 began, the minibasin compacted enough to sink by negative buoyancy. Pre-stack depth-migrated seismic data courtesy of CGGVeritas, after Hudec *et al.* (2009).

density inversion allow negative buoyancy to drive minibasin subsidence. If thick, dense carbonate or anhydrite layers overlie the salt, the density inversion will be shallower.

If buoyancy cannot initiate siliciclastic minibasins, what does? Five other mechanisms may explain minibasin subsidence when basin fill is less dense than salt (Hudec *et al.* 2009). Three of these appear common in the deepwater Gulf of Mexico: downdip shortening, decay of dynamic bulges and topographic loading. The north-central Gulf of Mexico has two zones of Pleistocene shortening above salt canopies. A landward zone of shortening in the midslope area (Fig. 13a) coincides with its probable cause: a regional break in slope from 0.7° dip to 0.3° dip. A

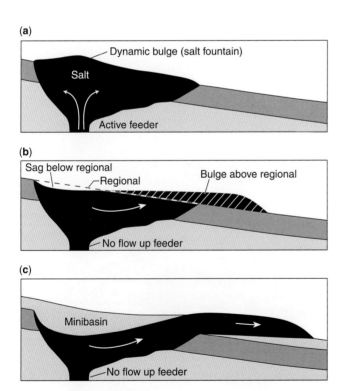

Fig. 15. Schematic cross-sections showing how decay of a dynamic bulge of salt can create a minibasin. (**a**) Above an active feeder, the dynamic pressure of rising salt supports a salt fountain. (**b**) After salt supply wanes, the unsupported salt bulge sags by gravity spreading to form a flat-topped droplet of salt. The updip part of the salt surface sags below regional, creating space for a local depocentre. (**c**) Sediments can gather in the depocentre and further deepen it to form a minibasin. After Hudec *et al.* (2009).

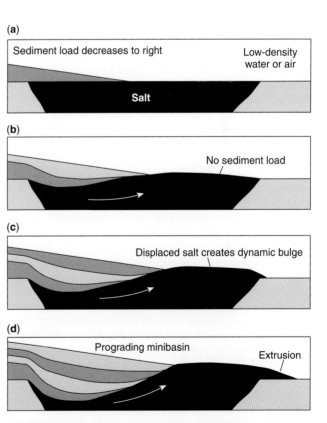

Fig. 16. Schematic cross-sections showing how the differential load of a prograding sedimentary wedge can initiate a minibasin. (**a**) A wedge of uncompacted sediments imposes a differential load because it is much denser than adjoining air or water. (**b**) The differential load causes the thick end of the sedimentary wedge to sink into salt, displacing salt seawards (**c**) and (**d**). After Hudec *et al.* (2009).

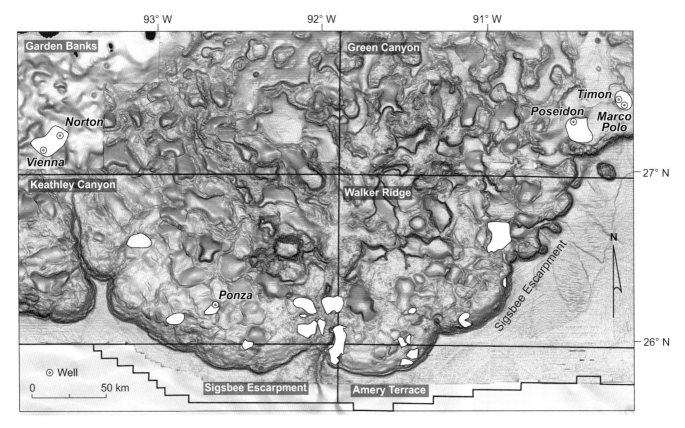

Fig. 17. Allochthonous fragments (white) interpreted to contain Mesozoic to Miocene strata at anomalously shallow burial depths in the northern Gulf of Mexico. Fragments are recognized by combinations of well penetrations from six wells, seismic character and high seismic velocities. Salt carried the exotic fragments seawards to collect near the front of the spreading Sigsbee salt canopy, which is marked by the Sigsbee Escarpment. Fragment map courtesy of Shell E&P Company, after Kilby *et al.* (2008), except for addition of Marco Polo (Mount *et al.* 2006) and Poseidon minibasins. Base map shows first derivative of bathymetric data from Bryant & Liu (2000).

seaward zone of shortening in the toe of the gravity-spreading system is raised above regional (Fig. 13a). As the canopy shortens and thickens, both salt and the floating lenses of overburden rise above regional. However, the laterally squeezed diapirs rise more than the intervening sediment lenses, leaving the minibasins as synclinal depressions perched on the inflated canopy (Fig. 13c). A compressional origin of minibasins was proposed on the basis of geometric reasoning by Rowan (2002) and later simulated by numerical modelling (Fig. 11). As further evidence, the deepest sediments in some minibasins near the Sigsbee Escarpment have been thrust (Fig. 14). Thrusting was the first of three evolutionary stages of minibasin subsidence, so a compressional trigger is plausible.

Many minibasins lack evidence of early thrusting. This lack of evidence does not preclude mild shortening that affected only intervening salt diapirs but may indicate that other mechanisms initiated minibasins. One mechanism could be the decay of salt topography above an ebbing salt feeder (Fig. 15). An emergent diapir fed by strong salt flow has a smoothly rounded summit, like a slow-moving fountain (Bailey 1931; Talbot & Jarvis 1984). The dynamic pressure of rising salt supports the bulging summit. Once upward flow ebbs, the dynamic bulge decays as the salt settles under gravity towards a flat-topped profile (Talbot 1998). The updip part of the flattening salt may sag below the sediment surface, accommodating sediments in a deepening minibasin.

Another way to trigger a minibasin without a density inversion is by the well recognized mechanism of topographic loading (Fig. 16). A wedge of sediment prograding across a flat-topped, emergent stock applies a differential load that increases toward the thick end of the wedge, which sinks fastest, regardless of density. The thick end sags as long as topography exists. Each

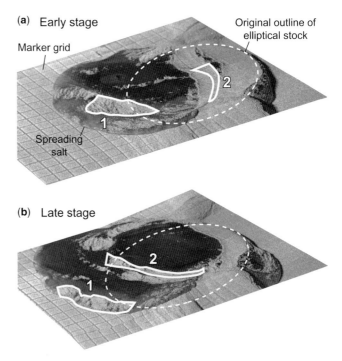

Fig. 18. Disruption of a thin roof above an extruding salt stock, shown in oblique views of a physical model. Lateral shortening forced salt to pierce and disperse a once-continuous sedimentary roof. (**a**) Extrusive salt carried roof fragment 1 almost to the salt front; fragment 2 had only begun to fracture. (**b**) Fragment 1 became stranded at the salt front, like a terminal moraine; fragment 2, now separated, was still being carried by spreading salt.

increment of sagging creates space for more sediment in the deepening minibasin.

Eroding a canyon into salt and filling it with sediments could initiate a minibasin, but only if the canyon was several kilometres deep and filled in a geological instant before the salt could respond to unroofing. More realistically, salt in the floor of an eroded canyon would well up faster than sediments could fill it, thereby inverting and destroying the canyon, as along most of Bryant Canyon where it crossed the Sigsbee salt canopy (Lee et al. 1996).

The mechanisms of compression, decay of salt topography and topographic loading are independent of density, thereby explaining a long-standing puzzle of how minibasins first form when loaded by siliciclastic sediments. The problem disappears in compacted minibasins more than 2–3 km thick, where negative buoyancy becomes the main force for subsidence.

Allochthonous fragments

Marine salt canopies are eventually dissolved or buried by a sedimentary carapace. Subparallel carapace strata tend to be stratigraphically condensed because they collect hemipelagically over a salt-induced bulge in the seafloor (Hart et al. 2004). As the canopy spreads further it may dismember a carapace into pieces that can be carried by flowing salt. The pieces are likely to be carried most effectively by freely extruding salt. A carapace fragment can be tilted or folded or come to rest subconformably on a salt weld. However, because of its distinctive character, a carapace fragment may still be recognizable as a piece of disrupted roof even when deformed or buried.

Salt flow may carry pieces of its roof basinward like glacial erratics for up to tens of kilometres, although the distance of transport is still speculative (Fig. 2). Many Mesozoic to Miocene fragments have been interpreted as salt-borne exotic pieces (Fig. 17) (Hart et al. 2004; Kilby et al. 2008). The fragmented slabs are up to 25 km in breadth and are a few kilometres thick. Two stages of a physical model (Fig. 18) show how diapiric salt can break a thin roof and widely disperse the pieces. The spreading salt strands the carapace pieces near the front of a waning flow or overrides them in a vigorous flow.

Exotic pieces are typically perched anomalously high above salt rather than buried many kilometres deeper with coeval strata. A notable example of a perched fragment was in the Norton discovery well in GB-754 (Hart et al. 2004; Fig. 17). The well intersected a moderately condensed Jurassic-to-Palaeogene interval overlain by condensed Neogene strata. The Mesozoic kerogen is thermally immature, so the strata were never deeply buried. The lack of

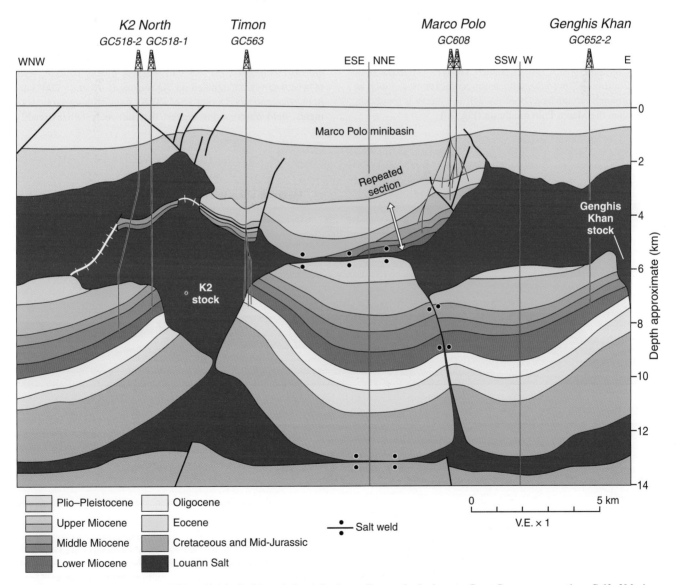

Fig. 19. Suprasalt Miocene strata c. 1500 m thick in the Marco Polo minibasin overlie coeval subsalt strata, Green Canyon area, northern Gulf of Mexico. A few kilometres NW of Marco Polo minibasin, condensed Oligocene and Lower Miocene strata are encased within the canopy. Encased strata mark a suture where a salt sheet merged with the K2 stock. After Mount et al. (2006).

repeated section suggests the Norton fragment is in place. The carapace fragment is flanked by deeper Pleistocene minibasins.

Besides their unusually shallow depth, some exotic fragments are stratigraphic duplicates. A prime example is the Marco Polo minibasin (Figs 17 & 19). About 1500 m of Oligo-Miocene strata in the minibasin overlie a thinned or welded salt canopy (Mount *et al.* 2006). At the base of the minibasin, condensed Lower Miocene strata represent the formerly elevated roof strata of the canopy. Below the salt canopy are repeated coeval strata. This stratigraphic duplication suggests that from the latest Miocene onwards, basinward-flowing salt carried the exotic fragment laterally over coeval strata.

Such a large exotic fragment can only be carried by a thick salt canopy. The thickness of repeated section in the Marco Polo minibasin is *c.* 1800 m, equivalent to *c.* 2100 m when uncompacted. The spreading salt must have been at least as thick as *c.* 2100 m; perhaps even thicker to overcome basal drag of the entrained minibasin. Is this thickness realistic for spreading salt? Analogy with salt glaciers spreading subaerially in the Zagros Mountains (Talbot & Jarvis 1984; Talbot 1998; Talbot *et al.* 2000) and buried below the Sigsbee Escarpment suggests that an unimpeded salt glacier flows across a flat surface with a thickness of 1000 m or less. That is too thin to carry a 2100 m-thick minibasin, even if its average density was less than that of salt. However, a spreading salt canopy can thicken by ponding against structural and sedimentary buttresses (Fig. 20). In this schematic restoration, by the end of the Pliocene an anticline of the Atwater fold belt formed a structural buttress 2600 m high. Then rapid early Pleistocene aggradation built a sedimentary buttress, which further ponded salt flow. Together both types of buttress allowed the salt canopy to thicken to as much as 5200 m, which would have been enough to entrain a minibasin thicker than the Marco Polo minibasin (Fig. 20).

Where do these exotic fragments originate? Imagine a possible sequence of events, partly shown in Figure 20. Initially, a minibasin sank into a broad diapir, which flared upward into a salt canopy. Next the canopy inflated and carried the minibasin basinward by viscous traction across subsalt strata coeval with the minibasin. As a result, Miocene and Palaeogene stratigraphy became repeated. Left behind was a stratigraphic gap from which the exotic fragment was plucked. The gap must have been at least as wide and as deep as the minibasin that originally nestled within the evacuated diapir. After salt evacuation Plio-Pleistocene strata could fill that Miocene gap.

Where kilometre-scale exotic pieces travel far from their source, stratigraphic prediction in a complexly deformed basin is even more difficult. These exotic pieces create odd stratigraphic repetitions. Stratigraphic gaps remain from where they were plucked, which could be many kilometres away. Despite the challenge, linking the source and destination of an exotic fragment would greatly improve structural restoration. Knowing the breadth and thickness of an exotic fragment provides clues to the size of the source gap and the thickness of the salt canopy that carried it. Knowing the thickness, a restoration can include enough structural and stratigraphic buttresses behind which the canopy can thicken.

Exotic fragments also provide stark evidence of high strains in salt canopies. As a canopy carries exotic slabs, it shears internally. This inherent shear is shown by deformed grids of numerical models or by streaked-out markers in physical models. During shear, former roof strata trapped as sutures within a salt canopy are extremely stretched into thin wisps or blocky boudins, some of which can be recognized on 3D seismic (Liro *et al.* 2004) and in physical models (Dooley *et al.* 2008). Even without exotic fragments, field observations suggest that salt recrystallizes readily,

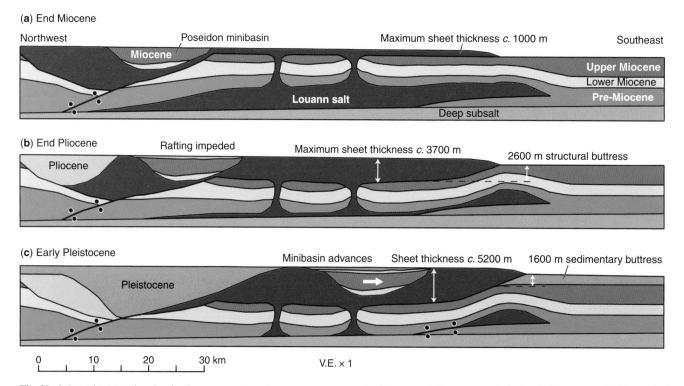

Fig. 20. Schematic restoration showing how a spreading salt canopy can carry an allochthonous minibasin seawards during the Pleistocene. (**a**) The original Miocene location of the minibasin is unknown, but the restoration shows a stratigraphic gap into which the Miocene strata may have fitted. (**b**) The salt canopy thickened by ponding against a late Miocene anticline, but this thickening was still not enough to carry the minibasin. (**c**) Rapid aggradation of lower Pleistocene sediments further thickened the salt canopy enough to transport the minibasin seawards. Vertical scale is only approximate but shows the initial thickness of Louann salt as *c.* 4 km, a thickness used in other restorations (Peel *et al.* 1995). However, the autochthonous salt layer has been variably inflated and deflated, expelled to several canopy levels, and lost to the world ocean, so estimates of the original thickness are highly speculative.

losing its strain memory (Talbot 1979). Even so, shear is likely to create an anisotropic fabric in the salt. This fabric anisotropy may cause an equivalent seismic anisotropy. Seismic anisotropy is poorly understood, but it could affect how accurately the shape and depth of the base of salt is predicted from seismic data.

State of knowledge

The abstract summarizes our current understanding of how diapiric salt can intrude downwards into a source layer, how salt canopies can advance despite being completely buried, how minibasins can form even when loaded by low-density sediments, and how the fragmented remnants of a canopy roof travel laterally to leave gaps and duplicate stratigraphy. Despite these significant advances, it is well to reflect on what we do not understand about these processes.

How much must natural diapirs shorten before they intrude their source layer? Does the outward plume from a squeezed diapir thicken only the nearby source layer? How does thickening affect the temperature of the source-layer salt and influence maturation?

Where buried salt canopies advance by thrusting, does the thrust root into the basal contact of the salt, or into a shear zone within the salt, or into an overpressured rubble zone below the salt? Can we use thrust style to predict overpressure below the salt canopy? Do imbricate wedges form rapidly or slowly? How widespread are thrust structures at the front of spreading canopies other than along the Sigsbee Escarpment?

If minibasins form when loaded by uncompacted sediments, which is the most important mechanism: compression, decay of dynamic bulge, or topographic loading? Are there regional zones where one mechanism dominates? How well do we understand the dynamics of shortening in salt canopies on the lower continental slope?

If exotic fragments of a salt-canopy roof can travel far, what proportion of these pieces have we been able to recognize? In what setting do the pieces originate? What determines whether a fragment is carried? Where and why are allochthonous fragments stranded? Will we be able to restore them all to their original positions in three dimensions? Once restored, can we infer trends and patterns in the distribution of source rocks and reservoirs?

To take a step further, what surprising or puzzling new processes will emerge in another 10 years?

The project was funded by the Applied Geodynamics Laboratory consortium, comprising the following companies: Anadarko, BHP Billiton, BP, CGGVeritas, Chevron, Cobalt, ConocoPhillips, Devon, ENI, ExxonMobil, Fugro, GX Technology, Hess, IMP, Maersk, Marathon, Mariner, Maxus, Murphy, Nexen, Noble, Pemex, Petrobras, PGS, Samson, Saudi Aramco, Shell, StatoilHydro, TGS-NOPEC, Total and Woodside. The authors received additional support from the Jackson School of Geosciences, The University of Texas at Austin. F. Diegel kindly provided his unpublished map of exotic fragments. Robin Pilcher and an anonymous reviewer provided insightful, helpful suggestions to improve the paper. We thank CGGVeritas and BP, BHPBilliton and Unocal for allowing us to reproduce seismic data. GeoLogic provided *LithoTect* software for the restoration in Figure 20. Publication was authorized by the Director, Bureau of Economic Geology, Jackson School of Geosciences, The University of Texas at Austin.

References

Bailey, E. B. 1931. Salt plugs. *Geological Magazine*, **68**, 335–336.

Baldschuhn, R., Binot, F., Frisch, U. & Kockel, F. 2001. Geotektonischer Atlas von Nordwest-Deutschland und dem deutschen Nordsee-Sektor. *Geologisches Jahrbuch*, **A 153**, 3 CD-ROM.

Baldwin, B. & Butler, C. O. 1985. Compaction curves. *American Association of Petroleum Geologists Bulletin*, **69**, 622–626.

Balk, R. 1949. Structure of Grand Saline salt dome, Van Zandt County, Texas. *American Association of Petroleum Geologists Bulletin*, **33**, 1791–1829.

Bryant, W. R. & Liu, J. Y. 2000. *Bathymetry of the Gulf of Mexico (TAMU-SG-00-606)*. Texas Sea Grant, College Station, TX, CD-ROM.

Chemia, Z. & Koyi, H. 2008. The control of salt supply on entrainment of an anhydrite layer within a salt diaper. *Journal of Structural Geology*, **30**, 1192–1200.

Cohen, H. A. & Hardy, S. 1996. Numerical modeling of stratal architecture resulting from differential loading of a mobile substrate. *In*: Alsop, G. I., Blundell, D. J. & Davison, I. (eds) *Salt Tectonics*. Geological Society, London, Special Publications, **100**, 265–273.

Divins, D. L. & Metzger, D. 2007. *NGDC Coastal Relief Model*. World Wide Web Address: http://www.ngdc.noaa.gov/mgg/coastal/coastal.html.

Dixon, J. M. 1975. Finite strain and progressive deformation in models of diapiric structures. *Tectonophysics*, **28**, 89–124.

Dooley, T. P., Jackson, M. P. A. & Hudec, M. R. 2006. Allochthonous salt extrusion, roof dispersion, and intrusive import and export of salt in squeezed stocks (abstract). *American Association of Petroleum Geologists Annual Convention and Exhibition Abstracts Volume*, **15**, 27.

Dooley, T., Monastero, F. C. & McClay, K. R. 2007. Effects of a weak crustal layer in a transtensional pull-apart basin: results from a scaled physical modeling study. *EOS*, **88**, Abstract V53F-04.

Dooley, T., Hudec, M. R. & Jackson, M. P. A. 2008. Dismembered sutures formed during asymmetric salt-sheet collision (abstract). *American Association of Petroleum Geologists 2008 Annual Convention and Exhibition Abstracts Volume*, **17**, 46.

Dooley, T. P., Jackson, M. P. A. & Hudec, M. R. 2009. Inflation and deflation of deeply buried stocks during lateral shortening. *Journal of Structural Geology*, **31**, 582–600.

Escher, B. G. & Kuenen, P. H. 1929. Experiments in connection with salt domes. *Leidsche Geologiese Meddelanden*, **3**, 151–182.

Fletcher, R. C., Hudec, M. R. & Watson, I. A. 1995. Salt glacier and composite sediment-salt glacier models for the emplacement and early burial of allochthonous salt sheets. *In*: Jackson, M. P. A., Roberts, D. G. & Snelson, S. (eds) *Salt Tectonics: A Global Perspective*. American Association of Petroleum Geologists, Tulsa, OK, Memoirs, **65**, 77–108.

Ge, H., Jackson, M. P. A. & Vendeville, B. C. 1997. Kinematics and dynamics of salt tectonics driven by progradation. *American Association of Petroleum Geologists Bulletin*, **81**, 398–423.

Giles, K. A. & Lawton, T. F. 2002. Halokinetic sequence stratigraphy adjacent to the El Papalote diapir, northeastern Mexico. *American Association of Petroleum Geologists Bulletin*, **86**, 823–840.

Hall, S. H. 2002. The role of autochthonous salt inflation and deflation in the northern Gulf of Mexico. *Marine and Petroleum Geology*, **19**, 649–682.

Hart, W. H., Jaminski, J. & Albertin, M. 2004. Recognition and exploration significance of supra-salt stratal carapaces. *24th Bob F. Perkins Research Conference*, Gulf Coast Section, Society for Sedimentary Geology (SEPM), 166–199.

Heard, H. C., Abey, A. E., Bonner, B. P. & Duba, A. 1975. *Stress–strain behavior of polycrystalline NaCl to 3.2 GPa*. Lawrence Livermore Laboratory Report, **UCRL-51743**, 16.

Heyn, T., Jackson, M. P. A. *et al*. 2008. Accretionary-wedge shortening caused by advance of the Sigsbee Escarpment, Alaminos Canyon, Gulf of Mexico (abstract). *American Association of Petroleum Geologists Annual Convention and Exhibition Abstracts Volume*, **17**, 89–90.

Hudec, M. R. 2004. Salt intrusion: time for a comeback? *24th Bob F. Perkins Research Conference*, Gulf Coast Section, Society for Sedimentary Geology (SEPM), 120–132.

Hudec, M. R. & Jackson, M. P. A. 2004. Regional restoration across the Kwanza Basin, Angola: salt tectonics triggered by repeated uplift of a metastable passive margin. *American Association of Petroleum Geologists Bulletin*, **88**, 971–990.

Hudec, M. R. & Jackson, M. P. A. 2009. Interaction between spreading salt canopies and their peripheral thrust systems. *Journal of Structural Geology*, **31**, 1114–1129.

Hudec, M. R., Jackson, M. P. A. & Schultz-Ela, D. D. 2009. The paradox of minibasin subsidence into salt: clues to the evolution of crustal basins. *Geological Society of America Bulletin*, **121**, 201–221.

Jackson, M. P. A. 1995. Retrospective salt tectonics. *In*: Jackson, M. P. A., Roberts, D. G. & Snelson, S. (eds) *Salt Tectonics: a Global Perspective*. American Association of Petroleum Geologists, Tulsa, OK, Memoirs, **65**, 1–28.

Jackson, M. P. A. & Hudec, M. R. 2004. A new mechanism for advance of allochthonous salt sheets. *24th Annual Bob F. Perkins Research Conference*, Gulf Coast Section, Society for Sedimentary Geology (SEPM), 220–242.

Jackson, M. P. A. & Talbot, C. J. 1986. External shapes, strain rates, and dynamics of salt structures. *Geological Society of America Bulletin*, **97**, 305–323.

Jackson, M. P. A. & Talbot, C. J. 1989. Anatomy of mushroom-shaped diapirs. *Journal of Structural Geology*, **11**, 211–230.

Jackson, M. P. A., Hudec, M. R. & Hegarty, K. A. 2005. The great West African Tertiary coastal uplift: fact or fiction? A perspective from the Angolan divergent margin. *Tectonics*, **24**, TC6014; doi: 10.1029/2005TC001836.

Kadri, M. S., Liro, L. M., Lahr, M. J., Hobbs, R. & Montecchi, P. 2000. Occurrence and distribution of seismic reflectors internal to allochthonous salt, Walker Ridge lease area, deepwater Gulf of Mexico (abstract). *American Association of Petroleum Geologists Annual Convention & Exhibition Abstracts Volume*, 75–76.

Kehle, R. O. 1988. The origin of salt structures. *In*: Schreiber, B. C. (ed.) *Evaporites and Hydrocarbons*. Columbia University Press, New York, 345–404.

Kilby, R. E., Diegel, F. A. & Styzen, M. J. 2008. Age of sediments encasing allochthonous salt in the Gulf of Mexico; clues to emplacement history. *American Association of Petroleum Geologists Search and Discovery Article*, **90078**.

Kupfer, D. H. 1974. Boundary shear zones in salt stocks. Fourth Symposium on Salt. *Northern Ohio Geological Society*, **1**, 215–225.

Kupfer, D. H. 1980. Problems associated with anomalous zones in Louisiana salt stocks, USA. Fifth Symposium on Salt. *Northern Ohio Geological Society*, **1**, 119–134.

Lee, G. H., Watkins, J. S. & Bryant, W. R. 1996. Bryant Canyon fan system: an unconfined, large river-sourced system in the northwestern Gulf of Mexico. *American Association of Petroleum Geologists Bulletin*, **80**, 340–358.

Letouzey, J. & Sherkati, S. 2004. Salt movement, tectonic events, and structural style in the central Zagros fold and thrust Belt (Iran). *24th Annual Bob F. Perkins Research Conference*, Gulf Coast Section, Society for Sedimentary Geology (SEPM), 753–778.

Liro, L. M., Murillas, J., Villalobos, L., Gatenby, G. & Mathur, V. 2004. Salt sutures in single- and multi-tiered allochthons, Green Canyon and Walker Ridge areas, deepwater Gulf of Mexico. *24th Annual Bob F. Perkins Research Conference*, Gulf Coast Section, Society for Sedimentary Geology (SEPM), 200–219.

McGuinness, D. B. & Hossack, J. R. 1993. The development of allochthonous salt sheets as controlled by rates of extension, sedimentation, and salt supply. *In*: Armentrout, J. M., Bloch, R., Olson, H. C. & Perkins, B. F. (eds) *Rates of Geological Processes. Fourteenth Annual Research Conference*, Gulf Coast Section, SEPM Foundation, Program with Papers, 127–139.

Mount, V. S., Rodriguez, A., Chauche, A., Crews, S. G., Gamwell, P. & Montoya, P. 2006. Petroleum system observations and interpretation in the vicinity of the K2/K2-North, Genghis Khan, and Marco Polo fields, Green Canyon, Gulf of Mexico. *Gulf Coast Association of Geological Societies Transactions*, **56**, 613–625.

Nelson, T. H. & Fairchild, L. H. 1989. Emplacement and evolution of salt sills in northern Gulf of Mexico. *American Association of Petroleum Geologists Bulletin*, **73**, 395.

Peel, F. J. 2001. Emplacement, inflation and folding of an extensive allochthonous salt sheet in the Late Mesozoic (ultra-deepwater Gulf of Mexico) (abstract). *American Association of Petroleum Geologists Annual Convention Official Program*, **10**, A155.

Peel, F. J. 2003. Styles, mechanisms and hydrocarbon implications of syndepositional folds in deepwater fold belts; examples from Angola and the Gulf of Mexico. Houston Geological Society: HGS International and HGS Dinner Meeting, December 15, 2003.

Peel, F. J., Travis, C. J. & Hossack, J. R. 1995. Genetic structural provinces and salt tectonics of the Cenozoic offshore U.S. Gulf of Mexico: a preliminary analysis. *In*: Jackson, M. P. A., Roberts, D. G. & Snelson, S. (eds) *Salt Tectonics: a Global Perspective*. Association of Petroleum Geologists, Tulsa, OK, Memoirs, **65**, 153–175.

Rowan, M. G. 2002. Salt-related accommodation in the Gulf of Mexico deepwater: withdrawal or inflation, autochthonous or allochthonous? *Gulf Coast Association of Geological Societies Transactions*, **52**, 861–869.

Rowan, M. G., Lawton, T. F., Giles, K. A. & Ratliff, R. A. 2003. Near-salt deformation in La Popa basin, Mexico, and the northern Gulf of Mexico: a general model for passive diapirism. *American Association of Petroleum Geologists Bulletin*, **87**, 733–756.

Sclater, J. G. & Christie, P. A. F. 1980. Continental stretching: an explanation of the post-mid-Cretaceous subsidence of the central North Sea basin. *Journal of Geophysical Research*, **85**, 3711–3739.

Stewart, S. A. 2007. Salt tectonics in the North Sea basin: a structural style template for seismic interpreters. *In*: Ries, A. C., Butler, R. W. H. & Graham, R. H. (eds) *Deformation of the Continental Crust: the Legacy of Mike Coward*. Geological Society, London, Special Publications, **272**, 361–396.

Stewart, S. A. & Clark, A. J. 1999. Impact of salt on the structure of the Central North Sea hydrocarbon fairways. *In*: Fleet, A. J. & Boldy, S. A. R. (eds) *Petroleum Geology of Northwest Europe: Proceedings of the 5th Conference*. Geological Society, London, 179–200; doi: 10.1144/0050179.

Talbot, C. J. 1979. Fold trains in a glacier of salt in southern Iran. *Journal of Structural Geology*, **1**, 5–18.

Talbot, C. J. 1998. Extrusions of Hormuz salt in Iran. *In*: Blundell, D. J. & Scott, A. C. (eds) *Lyell: The Past is the Key to the Present*. Geological Society, London, Special Publications, **143**, 315–334.

Talbot, C. J. & Aftabi, P. 2004. Geology and models of salt extrusion at Qum Kuh, central Iran. *Journal of the Geological Society London*, **161**, 1–14.

Talbot, C. J. & Alavi, M. 1996. The past of a future syntaxis across the Zagros. *In*: Alsop, G. I., Blundell, D. J. & Davison, I. (eds) *Salt Tectonics*. Geological Society, London, Special Publications, **100**, 89–110.

Talbot, C. J. & Jackson, M. P. A. 1987. Internal kinematics of salt diapirs. *American Association of Petroleum Geologists Bulletin*, **71**, 1068–1093.

Talbot, C. J. & Jarvis, R. J. 1984. Age, budget and dynamics of an active salt extrusion in Iran. *Journal of Structural Geology*, **6**, 521–533.

Talbot, C. J., Medvedev, S., Alavi, M., Shahrivar, H. & Heidari, E. 2000. Salt extrusion at Kuh-e-Jahani, Iran, from June 1994 to November 1997. *In*: Vendeville, B., Mart, Y. & Vigneresse, J.-L. (eds) *Salt, Shale and Igneous Diapirs in and around Europe*. Geological Society, London, Special Publications, **174**, 93–110.

Talbot, C. J., Aftabi, P. & Chemia, Z. 2009. Potash in a salt mushroom at Hormoz Island, Hormoz Strait, Iran. *Ore Geology Reviews*, **35**, 317–332.

Velde, B. 1996. Compaction trends of clay-rich deep sea sediments. *Marine Geology*, **133**, 193–201.

Vendeville, B. C. & Jackson, M. P. A. 1992. The fall of diapirs during thin-skinned extension. *Marine and Petroleum Geology*, **9**, 354–371.

Waltham, D. 1997. Why does salt start to move? *Tectonophysics*, **182**, 117–128.

Source-to-sink systems on passive margins: theory and practice with an example from the Norwegian continental margin

O. J. MARTINSEN,[1] T. O. SØMME,[2] J. B. THURMOND,[1] W. HELLAND-HANSEN[2] and I. LUNT[1]

[1]*Statoil Research, PO Box 7200, 5020 Bergen, Norway (e-mail: ojma@statoil.com)*
[2]*Department of Earth Sciences, University of Bergen, 5007 Bergen, Norway*

Abstract: Source-to-sink system analysis involves a complete, earth systems model approach from the ultimate onshore drainage point to the toe of related active deepwater sedimentary systems. Several methods and techniques have evolved in recent years, from experimental and numerical modelling through analysis of modern and recent systems, to analysis of ancient systems. A novel method has been developed, bridging between the previous approaches and dividing and analysing source-to-sink systems based on linked geomorphic segments along the source-to-sink profile. This approach builds on uniformitarian principles. The method is driven by the need to understand ancient, subsurface systems and still has high uncertainty but is an original, first-order approach to source-to-sink system analysis. In modern systems, entire onshore-to-offshore systems can be analysed with a higher degree of confidence than in ancient systems and semi-quantitative relationships can be established. Application in ancient systems is much more challenging but, in some cases, antecedent morphologies have been preserved onshore that can be matched with offshore known occurrences of, for instance, sandy submarine fan systems. Along the Norwegian North Sea and Norwegian Sea margins the Paleocene deep-marine reservoir of the giant Ormen Lange gas field is such an example. There, antecedent onshore drainage patterns which formed the feeder system to the offshore, deepwater fan system can be interpreted and aligned with onshore palaeogeomorphological evidence. Understanding the palaeogeomorphic development of basement regions such as the Fennoscandian shield is of high importance for understanding the offshore presence of deepwater sandstones.

Keywords: source-to-sink, passive margins, Norwegian margin, geomorphology

Source-to-sink analysis is a novel method in sedimentary geology that includes investigating the entire sedimentary system from its ultimate upstream source in the continental realm to the ultimate sediment sink, most commonly on deep basin plains (Fig. 1; Sømme *et al.* 2009*a, b*). While the method currently is best applied on margins with a clear division between catchment, shelf, slope and basin plain geomorphic segments, such as on passive margins, there is considerable further potential for application in other basins such as foreland regions.

The concept of source-to-sink analysis was first introduced qualitatively by Meade (1972, 1982) and the relationships between sediment source and basin and the ability to integrate various data sources and geo-disciplines were discussed by Dickinson (1997) and Hovius & Leeder (1998). However, in recent years, utilization of the concept has expanded, not least given the ability provided by new tools and digital elevation data to work with onshore and offshore data sources such as geomorphologic and seismic data in fully integrated, quantitative ways (Sømme *et al.* 2009*a*). Furthermore, the conceptual understanding of the role of onshore drainage systems on sediment sink or basin filling patterns has been widely developed (e.g. Allen 2008). Also both numerical and experimental modelling studies have experienced great development and compare more realistically with natural systems (e.g. Granjeon & Joseph 1999; Paola 2000; Syvitski *et al.* 2003; National Science Foundation Science Plans 2004).

A whole series of methods and techniques can be used to investigate the sedimentary systems, such as palaeoclimatic (deep-time) studies, continental and subaqueous geomorphology, sequence stratigraphy, classic sedimentary analysis, tectonic studies and various basin analysis techniques. Several schools have developed (see Sømme *et al.* 2009*a* for an overview) and these will be briefly reviewed below, from those that are fully quantitative and numerical to experimental flume studies and on to semi-quantitative, geomorphologically based studies. The main advantage for the petroleum industry is that source-to-sink studies potentially allow for prediction of reservoir, source and sealing rocks in basins in deep time from a host of methods and data. Thus, predictive, reliable methods that can reduce uncertainty in basin and interbasin scale exploration can be developed.

The aim of this paper is to provide a brief overview of available source-to-sink methods and demonstrate their potential by reviewing the study of Sømme *et al.* (2009*b*). This study applies source-to-sink methods to an ancient passive margin, the Paleocene Møre margin of the Mid-Norway Møre Basin, where the giant Ormen Lange gas discovery was made in 1997 (cf. Gjelberg *et al.* 2001).

Significance of the source-to-sink approach

For hydrocarbon exploration, a major challenge is to select the correct basins, plays and sedimentary systems within which to explore. The correct selection of play and complete understanding of reservoir, seal and source rocks depends on the best possible analysis of geological data from complete sedimentary systems. A favoured method to achieve this understanding is to analyse and compare with recent systems, but transfer of this knowledge to ancient systems in which petroleum systems occur is challenging since controls cannot be observed directly in past systems.

The United States National Science Foundation (NSF) has had source-to-sink studies on its science plan since the initiation of the Margins Programme in 2004. The plan states that: 'Erosion sculpts the landscape, and redistribution of the sediment creates the alluvial plains, coasts, deltas, and continental shelves upon which most of the world's population lives and derives such of

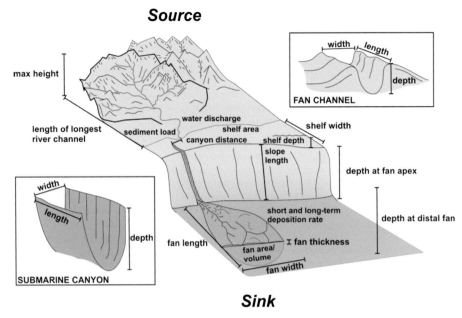

Fig. 1. Overview of key elements of a source-to-sink system developed on a margin with a distinct drainage basin–shelf–slope–basin floor profile. The key elements have quantitative characteristics that can be measured and compared between segments and between individual source-to-sink systems. The quantitative data base forms an empirical data base for application of source-to-sink analysis to ancient systems where segments (see Fig. 2) or elements of segments are preserved or can be measured. Modified from Sømme et al. (2009a).

its energy and water.' This transfer of sediment and solute mass from source to sink plays a key role in the cycling of elements such as carbon, in ecosystem change caused by global change and sea-level rise, and in resource management of soils, wetlands, groundwater and hydrocarbons. Although the source-to-sink system has been studied in its isolated component parts for more than 100 years, significant advances in our predictive capability require physical and numerical modelling of fluxes and feedbacks based on data from integrated field studies. According to the Margins programme (National Science Foundation Science Plans 2004), the Source-to-Sink Initiative is an attempt to quantify the mass fluxes of sediments and solutes across the Earth's continental margins by answering the following questions:

- How do tectonics, climate, sea-level fluctuations, and other forcing parameters regulate the production, transfer, and storage of sediments and solutes from their sources to their sinks?
- What processes initiate erosion and transfer, and how are these processes linked through feedbacks?
- How do variations in sedimentary processes and fluxes and longer-term variations such as tectonics and sea-level build the stratigraphic record to create a history of global change?

Various oil and energy companies have invested significant effort and research in understanding components of and complete source-to-sink systems, such as recent shallow-marine and deep-water systems in the Gulf of Mexico (e.g. Badalini et al. 2000). The competition for new attractive acreage globally will be based not only on access to the best technology and data, but also on access to the best geological methods and minds with the ability to master them. There is now growing recognition that new software tools and access to new quantitative datasets and new ideas facilitate analysis of complete source-to-sink systems and not just components. The depositional realm, where hydrocarbons are generated, migrate and are trapped, is significantly controlled by source area characteristics, climate and tectonics (cf. Sømme et al. 2009a, b). Thus there is a need to step out from the traditional approach of analysing only elements of the complete erosional–depositional system and describe and interpret in a source-to-sink manner.

Methods and technology

In general, four types of approaches have developed which all require different insight and technology to deliver results:

- modern and recent systems analysis;
- ancient systems analysis;
- numerical and experimental modelling;
- source-to-sink segment analysis.

Modern and recent systems analysis

The coupled process–response models of source areas and sediment sinks in basins can best be investigated along margins with recently or presently active source-to-sink systems. Numerous examples exist where particular segments have been analysed, but relatively few workers have investigated the entire source-to-sink systems. An exception is the incipient work of Leeder & Gawthorpe (1987; but followed by numerous other publications) on rift basins, years or even decades ahead in its in-depth understanding of the role of catchment development on resulting deposition. Other examples include the summary by Weaver et al. (2006) of the STRATAFORM and EUROSTRATAFORM projects, and the work of Einsele et al. (1996) and Goodbred (2003), who both investigated the giant Ganges–Brahmaputra system and showed the relationship of the sink sediments to Himalayan source area tectonic and climatic change. Other workers, such as Blum & Törnqvist (2000) and Blum (2008) have focused mainly on the drainage or source part of the system. Nevertheless, this focus is crucial because the majority of earlier work has been sink focused and much more empirical data needs to be collected and models built on the source system to create predictive models for basin-fill stratigraphy. In a recent paper, Sømme et al. (2009a; see also below) collected data and showed relationships from 29 modern and recent source-to-sink systems as an empirical data base for application to ancient systems. In general, empirical data is lacking from source-to-sink systems. Using data from

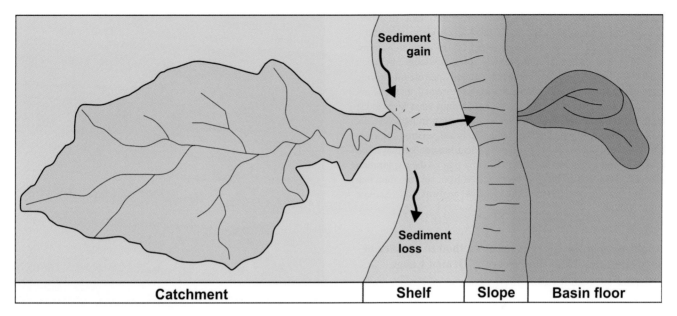

Fig. 2. Overview of the four segments of source-to-sink systems developed on margins with a distinct drainage basin–shelf–slope–basin floor profile. While the size of the individual segments is schematic and not to scale, the relative sizes are realistic, so that catchments are usually several times the area of, for example, basin floor fans (see Sømme et al. 2009a, b for statistics).

modern and recent systems to analyse ancient systems further highlights the principle of uniformitarianism and to what extent this principle can be used in source-to-sink analysis needs to be tested.

Ancient systems analysis

Very few examples of application of source-to-sink analysis (sensu the definition applied here; see above) to ancient systems have been published. There are many challenges with such application, not least that in most ancient systems single or several parts of the entire source-to-sink system are not preserved or data cannot be collected from them. In addition, while controlling factors such as climate, slope and sediment supply can be interpreted in modern and recent systems, these factors must be estimated with care in ancient systems (see also Leeder et al. 1998). Nevertheless, herein lies the power and potential of the source-to-sink method in its ability to predict the characteristics of segments which cannot be directly observed, and this is illustrated by the Ormen Lange case study (Sømme et al. 2009b; see below).

Commonly the case is that, the older the prospective source-to-sink system, the higher the uncertainty of data and interpretations, meaning that apparently the application of the method in deep time has severe limitations. Nonetheless, if uniformitarian principles apply and more empirical data are collected, the method has high potential. This will have potential significant impact on exploration in virgin basins and stratigraphy.

Numerical and experimental modelling

Over the last decade, various workers have developed software which deals with sediment delivery from drainage areas and deposition in adjacent basins (e.g. Granjeon & Joseph 1999; Paola 2000; Syvitski & Bahr 2001; Syvitski et al. 2003; National Science Foundation Science Plans 2004; Syvitski & Milliman 2007). Syvtiski's group has been working with developing a quantitative approach to source-to-sink studies using focused computer modelling and field investigation of landscape and seascape evolution, sediment transport and accumulation in two primary field areas (Fly River, SE Asia and Waipaoa system, New Zealand) where the complete source-to-sink system can be analysed. To what degree the mathematical and physical approach is fully useful from a petroleum exploration perspective, where there are large uncertainties, is a topic for future discussion.

Forward computer modelling of sedimentary basin filling has been focused on by several workers, including the fundamental work of Granjeon & Joseph (1999) and many later workers. This type of modelling is particularly useful to gain understanding of the physical processes that control basin filling.

Another experimental approach is large flume tank experiments (e.g. Paola 2000). In these experiments, controlling factors such as overall basin shape, basin gradient and sediment supply can be controlled quantitatively and scaled to each other. However, a critical point is whether the flume processes are scaled to natural systems both spatially and temporally. Nevertheless, flume modelling is an integral part of source-to-sink studies to build a quantitative understanding of sedimentary processes and sediment flux.

The linked segment-style approach

A fourth approach to analysing source-to-sink systems is driven by prediction of ancient systems and truly is a synthesis of the other approaches described above. This method (Sømme et al. 2009a), which relies on dividing the source-to-sink systems into segments (Figs 1 & 2), utilizes a uniformitarian approach where complete modern and recent source-to-sink systems are the key to predicting characteristics of ancient systems. By investigating a series of recent source-to-sink systems, an empirical framework of controlling factors which govern the delivery and distribution of sediments to basins is built. The method relies on a fundamental principle that processes and characteristics in one segment (Fig. 2) are dependent on updip segments and influence processes and characteristics in downdip segments.

In more detail, the method involves carrying out classic geomorphological studies of onshore drainage systems, collection of quantitative parameters of precipitation and river discharge to understand sediment supply and sedimentary processes, investigation of slope gradient across the terrestrial, shallow marine and deepwater parts of depositional systems, and comparison with sediment volumes deposited at the ultimate sink location of the sedimentary systems (Sømme et al. 2009a). This approach concentrates on the identification and actual measurement of the critical controls for sediment delivery to and deposition in basins. Thus, the method to a large extent avoids the scaling problems of experimental and numerical methods. Its focus is on identifying those parameters which can be used for the analysis of ancient systems, a critical consideration for transfer of knowledge between modern–recent systems and ancient, potentially petroleum-bearing systems. In many ways, numerical and experimental methods are less useful for prediction as it is often uncertain what knowledge is directly transferable from model to application in natural-scale systems. For instance, methods which rely on seismic data are likely to be much more useful for the identification of plays and prospects since this data is such an integral part of the exploration process. Quantitative data can be extracted from seismic data that can be applied directly in source-to-sink analysis (see Sømme et al. 2009b).

Application of the source-to-sink method to an ancient passive margin case: the Paleocene Ormen Lange system, Norwegian continental shelf

That the onshore geomorphology of particularly the south and central parts of Norway possesses antecedent characteristics formed prior to Neogene glaciations and which has genetic relationships to the offshore stratigraphy is widely accepted (Fig. 3; e.g. Riis 1996; Martinsen et al. 1999, 2005, and references therein). Nevertheless, the quantitative relationship between pre-glacial onshore geomorphology and offshore stratigraphy has not been published properly previously. Recently, Sømme et al. (2009b) investigated in detail the relationship between the latest Maastrichtian–Early Paleocene sediments of the Ormen Lange field and their antecedent source area, partly preserved onshore mainland Norway (Figs 4 & 5). In contrast to forward modelling using the drainage area size and characteristics to predict basinal deposits, inversion was used to test the segment-style approach of Sømme et al. (2009a). As mentioned previously, the segment-style approach resulted from analysis of 29 modern–recent source-to-sink systems. As such, the application of this database in analysing the Ormen Lange dataset is also a test of the usefulness of uniformitarianism in working with source-to-sink systems.

Critical quantitative aspects addressed by Sømme et al. (2009b) include: (i) can the volume of the Ormen Lange fan confidently predict the size of the onshore source area using the modern–recent database developed by Sømme et al. (2009b); (ii) what is the shape of the source and catchment area honouring onshore antecedent geomorphology and tectonics; and (iii) when defined, what other characteristics of the onshore drainage system can be interpreted which are of importance for the supply of reservoir sands to the deepwater basin? The Ormen Lange system has been studied extensively over the last decade, and details of the reservoir units and their formation can be found in Gjelberg et al. (2001, 2005) and references therein.

Fan volume v. onshore source area size

Calculations from published subsurface data (Gjelberg et al. 2005) show that the fan (Figs 4 & 5) reaches a maximum thickness of c. 110 m in the proximal part, from where it gradually thins into heterolithic lobe fringe deposits at its termination, giving a total fan volume of c. 104 km^3. This fan volume, compared with the global dataset of Sømme et al. (2009a), yields an estimated source area of c. 22×10^3 km^2, with minimum and maximum values of 1 and 560×10^3 km^2. The large uncertainty, possibly in part related to uncertainty regarding depositional rates, is discussed in detail by Sømme et al. 2009a, b), but the mean estimate is of great interest because it corresponds very closely to the proposed source area inferred purely from onshore geomorphologic and tectonic data (Figs 4 & 5; see also below; cf. Martinsen et al. 2002).

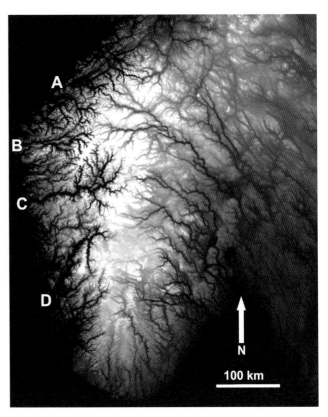

Fig. 3. Satellite image of the south of Norway highlighting major fjords and river valleys. Four fjord systems along the west coast are highlighted: (**a**) fjords of the Møre region, (**b**) Nordfjord, (**c**) Sognefjord and (**d**) Hardangerfjord. It is generally accepted that the fjords developed initially as Cenozoic fluvial drainage systems enhanced by Pleistocene glaciations (cf. Nesje & Whillans 1994 for a review). In terms of palaeomorphic shape, the Møre region fjords take on a rectangular pattern steered by the major faults of the Møre–Trøndelag Fault Complex (cf. Martinsen & Nøttvedt 2008). In contrast, the Nordfjord and Sognefjord examples have generally classic dendritic shapes uninfluenced by major tectonics. The last example, Hardangerfjord, shows a first-order, SW–NE linear shape controlled by a major tectonic lineament, the Hardangerfjord Shear Zone, widely believed to be the easternmost expression of Caledonian extensional collapse in the latest Silurian to early Devonian (cf. Martinsen & Nøttvedt 2008). The four examples, all interpreted to be major palaeodrainage systems to the Norwegian continental shelf, clearly show the importance of basement structure and geomorphology in transporting sediments from catchment to offshore basins.

This provides confidence that the linked segment-style method of Sømme et al. (2009a) is useful, but the large uncertainty needs to be handled in further work. The likely causes for the uncertainties are considered to be a general lack of empirical data from source-to-sink systems globally coupled with challenges in calculating fan volume correctly.

Shape and size of source area

The source-to-sink modelling inversion of the Ormen Lange fan system yields quantitative values of the size of the source area and catchment. Nevertheless, to be fully valuable, local geological data must be incorporated. The antecedent geomorphology and tectonics of the source area region are strongly influenced by the presence of the Palaeozoic-age Møre–Trøndelag Fault Zone and Cenozoic uplift of southern Scandes (Fennoscandia) (Figs 4 & 5; Riis 1996; Martinsen et al. 1999, 2005; Martinsen & Nøttvedt 2008; Sømme et al. 2009b). The source area did not develop a classic dendritic leaf-shape, as expected on flat, gently inclined surfaces with no tectonic influence. Instead, a rectangular pattern

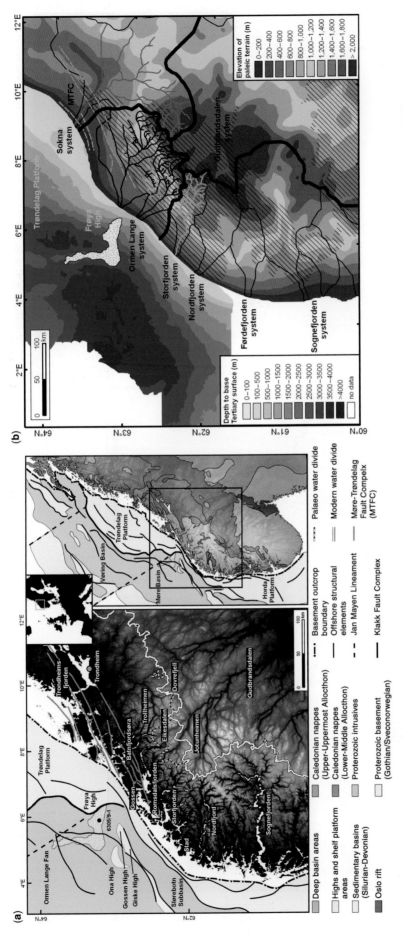

Fig. 4. (a) Overview of major geological features in the mid-Norway region and their spatial relation to the Ormen Lange deepwater fan system of latest Maastrichtian–Paleocene age. The white solid line along the highest mountains and oriented south–north–NE is the current drainage divide in Norway, generally interpreted as also roughly representing the pre-glacial Cenozoic drainage divide (Riis 1996; Martinsen et al. 2002; Martinsen & Nøttvedt 2008). A significant deviation occurs in the Møre region, shown by the dashed white line, where the palaeodrainage divide occurred further to the NW, and Neogene stream piracy moved the current divide further to the SE (cf. Sømme et al. 2009b for a discussion). This drainage divide formed an important hinterland constraint on the development of the Ormen Lange catchment, in addition to the Møre–Trøndelag Fault Zone structuring (shown by grey lines oriented in an ENE–WSW direction obliquely to the current coastline). From Sømme et al. (2009b). (b) Semi-quantitative interpretation of the pre-glacial Palaeic landscape and drainage basins in the Møre region. The Ormen Lange catchment is shown NW of the thick black line and to the north of the Gudbrandsdalen system. Note the deviation of drainage towards the SW along the Møre–Trøndelag Fault Zone and skewness towards the ENE. Modified from Sømme et al. (2009b).

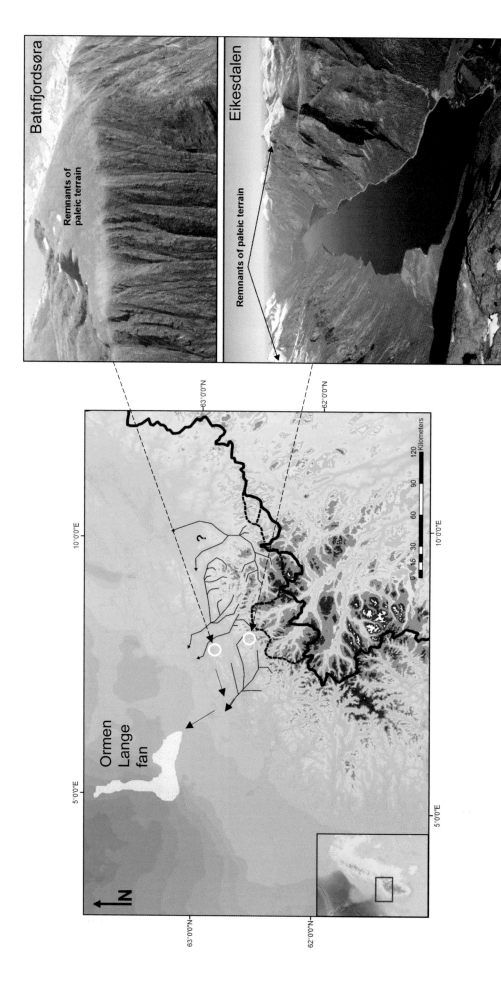

Fig. 5. Remnants of pre-glacial terrain can be observed at many places in the south of Norway. Two examples from the catchment of the Ormen Lange system are shown. The upper example from Batnfjordsøra shows a flat, slightly inclined surface on top of a mountain surrounded by glacial and post-glacial fluvial valleys and recent, steep avalanche-prone valley walls. The lower example from Eikesdalen shows gently inclined high surfaces on the margins of a major glacially cut valley modified by recent rock falls. The high surfaces are interpreted to have formed by fluvial erosion prior to Pleistocene glaciations (modified from Martinsen & Nøttvedt 2008; Sømme *et al.* 2009*b*).

developed, with a preferred drainage axis obliquely oriented in an ENE–WSW direction to the current coastline and steered by the major faults of the Møre–Trøndelag Fault Zone (Figs 3 & 4). The main sub-drainages meet at the apex of the feeder system to the Ormen Lange fan (Fig. 5), thus also explaining the general lack of sand of Paleocene age north of the Ormen Lange fan system, because no major drainage system reached these areas. It is generally believed that the southeastern boundary of this Paleocene drainage system lay at the culmination of the Palaeic surface (see Sømme et al. 2009a, b for a discussion). This culmination corresponds generally with a south–north–NE line drawn through the axis of the highest mountains in Norway (Figs 3–5), and is considered with minor modifications to be inherited from the Palaeogene (Martinsen & Nøttvedt 2008; cf. also Martinsen et al. 2002). Therefore, as a result, a drainage area skewed to the ENE developed (Fig. 4).

The skewness of the drainage area is also supported by provenance data from U–Pb isotopes in zircons, showing that the bedrock to the east and NE of the Ormen Lange fan supplied a higher proportion of sediment than the area to the SW, which is immediately inland from the fan system (Fonneland et al. 2004).

Other drainage system characteristics

The segment-style source-to-sink method developed by Sømme et al. (2009a) allows for the prediction of a range of parameters due to its basis in linkage and genetic relationships between segments. Thus, based on fan volume and catchment size calculations, other characteristics of the Ormen Lange system can be inferred. Coupled with the grain size characteristics of the depositional system (variable coarse-to-fine fractions but generally fining in a proximal-distal direction; Gjelberg et al. 2001, 2005; Sømme et al. 2009b), a relatively steep system is predicted with a peak/average discharge ratio of 100–1000, indicative of the frequency-magnitude characteristics of the system, and thus the timing and mode of sediment delivery to the fan.

The number and configurations of catchments feeding the shallow-marine area can also be interpreted, with implications for the spacing of sediment sources in the region (Fig. 4b). Sømme et al. (2009b) shows that, along the Møre margin, drainage basins were closely spaced (c. 40–80 km), implying that several smaller catchments may have fed the Ormen Lange fan.

It is also suggested that the palaeotopography and bathymetry in source-to-sink systems with a distinct shelf–slope–basin floor configuration are generically related. Thus, flattening and uplift of the catchment may occur in conjunction with flattening and deepening of the slope and basin floor as a system develops. For the Ormen Lange system, a catchment height of c. 2500 m (ranging between 900–4200 m) and water depths of c. 1800 m (ranging between 600–3200 m) are suggested for the Maastrichtian–Danian period (Sømme et al. 2009b).

A final point concerns to what extent onshore geomorphic evidence in a drainage basin can be used as an indicator and predictor of offshore deposition. Obviously, the deeper the time and age of a particular system, the more the uncertainty in application of source-to-sink analysis increases. Nevertheless, contrary to many people's expectation, many continental regions landward of prospective offshore basins contain strong evidence of the characteristics of the drainage areas that supplied the older, subsurface sediments (see e.g. Leturmy et al. 2003 for a thorough deep-time analysis of the West African margin). In the south of Norway, fjords (Fig. 3), perceived by many as generated during the Pleistocene glaciations (cf. discussion in Nesje & Whillans 1994), show a lot of evidence for a pre-glacial origin controlled by fluvial discharge and steered in some regions by major tectonic lineaments (Nesje & Whillans 1994; Martinsen & Nøttvedt 2008). There are numerous remnants of pre-glacial surfaces that relate to the supply of sediments to offshore basins (Fig. 5). These sediments are most commonly reservoirs, but the drainage history in deep time reveals important aspects of source and seal rock formation as well.

Uncertainties for applying source-to-sink methodology in exploration

In general, the older the ancient/subsurface system, the larger the uncertainty is in applying source-to-sink concepts on a particular exploration margin. In the Ormen Lange case, the inversion (i.e. estimation of the catchment area from fan volume) shows the dependency on a correctly interpreted fan volume from seismic. Thus, obviously, the quality of seismic data and the seismic interpretation is a significant factor for the confidence of the interpretation. In the Ormen Lange case, the confidence is relatively high, thus providing estimates of catchment size, shelf width and slope length that are reliable. Nevertheless, using this data in an inversion exercise also warrants calibration towards other regional and independent data such as palaeogeomorphical data (see above; Sømme et al. 2009b). In regions where such independent data do not exist, the uncertainty of the method grows significantly larger given the still small empirical database that exists for source-to-sink systems (cf. Sømme et al. 2009a).

The real value for exploration of the source-to-sink approach naturally lies in using it as a predictive tool in regions where little data exist, but also as a test and uncertainty-reduction tool where some data are present. In prospective regions with very low data quality and resolution, such as in sub-salt regions, the value is very high. There is no lower limit to how much quantitative data should be present from, for example, one segment to apply the method as a predictive tool to understand another segment, as the method is of immense value as an idea-generation tool. However, the more data is available, the better the constraints on geological models can be established. The revelation that there is a genetic link, for example, between catchment size, slope length and fan volume as established by Sømme et al. (2009a) and shown by the Ormen Lange case study, is a major breakthrough. The quest now is to collect more empirical data to reduce the uncertainty, and apply the concept to other basins.

Although the current empirical data base is established for margins with a distinct stepped profile with four segments (i.e. catchment, shelf, slope and basin floor), there is no reason to believe that similar relationships exist in other basin types such as rift and foreland basins. Conceptual relationships can be established in these and other settings, but the time-consuming and tedious work to collect the data must be done to establish similar quantitative relationships to those established by Sømme et al. (2009a).

Summary

This paper has briefly attempted to illustrate the potential of source-to-sink analysis on continental margins for predicting modern, recent and ancient, subsurface basin fill stratigraphy. Source-to-sink analysis has a wide potential, and the technique is based on experimental and numerical modelling, as well as analysis of modern to ancient systems. Previous methods for predicting basin filling have either been source or sink focused, but now new technology (software, computer processing), new data (such as coupling of high-resolution digital elevation data with seismic) and new insights provide the capability for a potential paradigm shift in how analysis of sedimentary systems and their source areas are analysed.

An approach based on linked segments along the source-to-sink profile is preferred (Sømme et al. 2009a), which builds on uniformitarian principles. Nevertheless, the validity of a uniformitarian approach needs further testing and more empirical data, not least from ancient systems.

The application of the linked segment-style analysis to the Ormen Lange system offshore mid-Norway is one valid test of the source-to-sink method. In addition, the analysis in this region has shed new light and added credibility to the use of onshore palaeogeomorphic data as predictors of offshore deposition, even in regions traditionally perceived as possessing or having preserved little evidence of antecedent supply systems to the offshore, subsurface stratigraphic record.

References

Allen, P. A. 2008. From landscapes into geological history. *Nature*, **451**, 274–276.

Badalini, G., Kneller, B. & Winker, C. D. 2000. Architecture and processes in Late Pleistocene Brazos-Trinity turbidite system, Gulf of Mexico. *In*: Weimer, P. *et al.* (eds) *Turbidite Reservoirs of the World. 20th Annual Gulf Coast Section SEPM Bob F. Perkins Research Conference Proceedings*, Gulf Coast Section, Society for Sedimentary Geology (SEPM), Houston, TX, 16–34.

Blum, M. 2008. Continental shelves as the lowstand fluvial long profile: possible implications for icehouse vs. greenhouse stratigraphic records. *American Association of Petroleum Geologists Annual Convention Abstracts*, **17**, 20.

Blum, M. D. & Törnqvist, T. E. 2000. Fluvial responses to climate and sea-level change: a review and look forward. *Sedimentology*, **47**, 2–48.

Dickinson, W. 1997. Panel throws down gauntlet. *GSA Today*, **7**, 25.

Einsele, G., Ratschbacher, L. & Wetzel, A. 1996. The Himalaya–Bengal Fan denudation–accumulation system during the last 20 Ma. *Journal of Geology*, **104**, 163–184.

Fonneland, H. C., Lien, T., Martinsen, O. J., Pedersen, R. B. & Kosler, J. 2004. Detrital zircon ages: a key to understanding the deposition of deep marine sandstones in the Norwegian Sea. *Sedimentary Geology*, **164**, 147–159.

Gjelberg, J. G., Enoksen, T., Kjærnes, P., Mangerud, G., Martinsen, O. J., Roe, E. & Vågnes, E. 2001. The Maastrichtian and Danian depositional setting, along the eastern margin of the Møre Basin (mid-Norwegian Shelf): implications for reservoir development of the Ormen Lange Field. *In*: Martinsen, O. J. & Dreyer, T. (eds) *Sedimentary Environments Offshore Norway – Paleozoic to Recent*. Norwegian Petroleum Society, Oslo, Special Publications, **10**, 421–440.

Gjelberg, J. G., Martinsen, O. J., Charnock, M., Møller, N. & Antonsen, P. 2005. The reservoir development of the Late Maastrichtian–Early Paleocene Omen Lange gas field, Møre Basin, Mid-Norwegian Shelf. *In*: Doré, A. G. & Vining, B. A. (eds) *Petroleum Geology: North-West Europe and Global Perspectives: Proceedings of the 6th Petroleum Geology Conference*. Geological Society, London, 1165–1184; doi: 10.1144/0061165.

Goodbred, S. L. 2003. Response of the Ganges dispersal system to climate change: a source-to-sink view since the last interstade. *Sedimentary Geology*, **162**, 83–104.

Granjeon, D. & Joseph, P. 1999. Concepts and applications of a 3-D multiple lithology, diffusive model in stratigraphic modeling. *In*: Harbaugh, J. W., Watney, W. L. *et al.* (eds) *Numerical Experiments in Stratigraphy: Recent Advances in Stratigraphic and Computer Simulations*. Society for Sedimentary Geology, Special Publications, **62**, 197–210.

Hovius, N. & Leeder, M. R. 1998. Clastic sediment supply to basins. *Basin Research*, **10**, 1–5.

Leeder, M. R. & Gawthorpe, R. L. 1987. Sedimentary models for extension in tilt-block/half-graben basins. *In*: Coward, M. P., Dewey, J. F. & Hancock, P. L. (eds) *Extensional Tectonics*. Geological Society, London, Special Publications, **28**, 139–152.

Leeder, M. R., Harris, T. & Kirkby, M. J. 1998. Sediment supply and climate change: implications for basin stratigraphy. *Basin Research*, **10**, 7–18.

Leturmy, P., Lucazeau, F. & Brigaud, F. 2003. Dynamic interactions between the Gulf of Guinea passive margin and the Congo River drainage basin 1: morphology and mass balance. *Journal of Geophysical Research – Solid Earth*, **108**, 1–13.

Martinsen, O. J. & Nøttvedt, A. 2008. Norway rises from the sea. *In*: Ramberg, I., Bryhni, I., Nøttvedt, A. & Rangnes, K. (eds) *The Making of a Land – Geology of Norway*. Norwegian Geological Society, Trondheim, 441–477.

Martinsen, O. J., Bøen, F., Charnock, M. A., Mangerud, G. & Nøttvedt, A. 1999. Cenozoic development of the Norwegian margin 60–64 degrees N: sequences and sedimentary response to variable basin physiography and tectonic setting. *In*: Fleet, A. J., Boldy, S. A. R. & Burley, S. D. (eds) *Petroleum Geology of North-West Europe: Proceedings of the 5th Conference*. Geological Society, London, 293–304; doi: 10.1144/0050293.

Martinsen, O. J., Fonneland, H. C., Lien, T. & Gjelberg, J. G. 2002. On latest Cretaceous–Palaeocene drainage and deep-water sedimentary systems on- and offshore Norway. *In*: Thorsnes, T. (ed.) *Onshore–Offshore Relationships on North Atlantic Margins*. Norwegian Geological Society, Trondheim, Abstracts and Proceedings, 123–124.

Martinsen, O. J., Lien, T. & Jackson, C. 2005. Cretaceous and Palaeogene turbidite systems in the North Sea and Norwegian Sea basins: source, staging area and basin physiography controls on reservoir development. *In*: Doré, A. G. & Vining, B. A. (eds) *Petroleum Geology: North-West Europe and Global Perspectives: Proceedings of the 6th Petroleum Geology Conference*. Geological Society, London, 1147–1164; doi: 10.1144/0061147.

Meade, R. H. 1972. Sources and sinks of suspended matter on continental shelves. *In*: Swift, D. J. P., Duane, D. B. & Pilkey, O. H. (eds) *Shelf Sediment Transport*. Dowden, Hutchinson & Ross, Stroudsburg, 249–262.

Meade, R. H. 1982. Sources, sinks, and storage of river sediment in the Atlantic drainage of the United States. *Journal of Geology*, **90**, 235–252.

National Science Foundation Science Plans. 2004. *Margins*. National Science Foundation, Washington, DC.

Nesje, A. & Whillans, I. M. 1994. Erosion of Sognefjord, Norway. *Geomorphology*, **9**, 33–45.

Paola, C. 2000. Quantitative models of sedimentary basin filling. *Sedimentology*, **47**, 121–178.

Riis, F. 1996. Quantification of Cenozoic vertical movements of Scandinavia by correlation of morphological surfaces with offshore data. *Global and Planetary Change*, **12**, 331–357.

Sømme, T., Helland-Hansen, W., Martinsen, O. J. & Thurmond, J. B. 2009a. Relationships between morphological and sedimentological parameters in source-to-sink systems: a basis for predicting semi-quantitative characteristics in subsurface systems. *Basin Research*, **21**, 361–387.

Sømme, T., Martinsen, O. J. & Thurmond, J. B. 2009b. Reconstructing morphological and depositional characteristics in subsurface sedimentary systems: an example from the Maastrichtian-Danian Ormen Lange system, Møre Basin, Norwegian Sea. *American Association of Petroleum Geologists Bulletin*, **93**, 1347–1377.

Syvitski, J. P. M. & Bahr, D. B. 2001. Numerical models of marine sediment transport and deposition. *Computers and Geosciences*, **27**, 617–753.

Syvitski, J. P. M. & Milliman, J. D. 2007. Geology, geography and humans battle for dominance over the delivery of fluvial sediment to the coastal ocean. *Journal of Geology*, **115**, 1–19.

Syvitski, J. P. M., Peckham, S. D., Hilberman, R. & Mulder, T. 2003. Predicting the terrestrial flux of sediment to the global ocean: a planetary perspective. *Sedimentary Geology*, **162**, 5–24.

Weaver, P. P. E., Canals, M. & Trincardi, F. (eds) 2006. Source-to-sink sedimentation on the European margin. *Marine Geology*, **234**, 1–292.

An integrated study of Permo-Triassic basins along the North Atlantic passive margin: implication for future exploration

J. REDFERN,[1] P. M. SHANNON,[2] B. P. J. WILLIAMS,[2] S. TYRRELL,[2] S. LELEU,[3] I. FABUEL PEREZ,[1] C. BAUDON,[1] K. ŠTOLFOVÁ,[2] D. HODGETTS,[1] X. VAN LANEN,[1] A. SPEKSNIJDER,[4] P. D. W. HAUGHTON[2] and J. S. DALY[2]

[1]*University of Manchester, North Africa Research Group, School of Earth, Atmospheric and Environmental Sciences, Williamson Building, Oxford Road, Manchester M13 9PL, UK (e-mail: jonathan.redfern@manchester.ac.uk)*
[2]*UCD School of Geological Sciences, University College Dublin, Belfield, Dublin 4, Ireland*
[3]*University of Aberdeen, Geology & Petroleum Geology, School of Geosciences, Meston Building, King's College, Aberdeen AB24 3UE, Scotland, UK*
[4]*Shell International Exploration and Production BV, Kessler Park 1, 2288GS, Rijswijk, Netherlands*

Abstract: Permo-Triassic rift basins offer important hydrocarbon targets along the Atlantic margins. Their fill is dominated by continental red beds, comprising braided fluvial, alluvial fan, aeolian, floodplain and lacustrine facies. These relatively lightly explored basins span both the Atlantic and Tethyan domains and developed above a complex basement with inherited structural fabrics. Sparse data in offshore regions constrain understanding of depositional geometries and sedimentary architecture, further impeded by their deep burial beneath younger strata, combined with the effects of later deformation during continental breakup. This paper provides results from a multidisciplinary analysis of basins along the Atlantic margin. Regional seismic and well data, combined with geochemical provenance analysis from the European North Atlantic margins, are integrated with detailed outcrop studies in Morocco and Nova Scotia. The research provides new insights into regional basin tectonostratigraphic evolution, sediment fill, and reservoir distribution, architecture and quality at a range of scales. Regional seismic profiles, supported by key well data, indicate the presence of post-orogenic collapse basins, focused narrow rifts and low-magnitude multiple extensional depocentres. Significantly, Permo-Triassic basin geometries are different and more varied than the overlying Jurassic and younger basins. Provenance analysis using Pb isotopic composition of detrital K-feldspar yields new and robust controls on the sediment dispersal patterns of Triassic sandstones in the NE Atlantic margin. The evolving sedimentary architecture is characterized by detailed sedimentological studies of key outcrops of age equivalent Permian–Triassic rifts in Morocco and Nova Scotia. The interplay of tectonics and climate is observed to influence sedimentation, which has significant implications for reservoir distribution in analogue basins. New digital outcrop techniques are providing improved reservoir models, and identification of key marker horizons and sequence boundaries offers a potential subsurface correlation tool. Future work will address source and seal distribution within the potentially petroliferous basins.

Keywords: Permian Triassic Atlantic Borderland basins, Morocco, Atlas, Fundy, continental red beds, provenance, rifting

Integrated regional study

The Late Mesozoic and Cenozoic development of basins along the Atlantic margin (Fig. 1) has been documented in a number of regional syntheses (e.g. Blystad *et al.* 1995; Doré *et al.* 1999; Naylor *et al.* 1999; Stoker *et al.* 2005). Seismic and well data from the Northern Atlantic margin show the extensive presence of Permo-Triassic strata, preserved in a suite of elongate, largely fault-bounded basins (Fig. 2). Their depositional extent and regional architecture are generally poorly constrained. This is due to overprinting of later rifting and continental breakup, together with deformation in the thick overlying Jurassic to Cenozoic strata and the effects of early Cenozoic igneous activity in the North Atlantic Igneous Province. However, the frequently sandy nature of the lower part of the Permo-Triassic succession makes it a potentially important reservoir target in petroleum exploration.

Previous studies have focused either on individual basins or plays. This study extends from Norway to Morocco (Fig. 1) and offers a broad perspective that examines the influence of regional as well as local and small-scale controls, investigating the complete spectrum of scales from regional crustal, through outcrop and to the pore scale. The aims of this paper are to (a) present initial results from an interdisciplinary basin analysis of Atlantic Margin Permo-Triassic successions and (b) illustrate how regional basin analysis, using a range of techniques and at scales through orders of magnitude from the basin to the pore throat, can provide an improved understanding of basin development and reservoir architecture, with implications for petroleum prospectivity.

Geological framework

The North Atlantic passive continental margin has a complex history, influenced by the inherited basement framework (Fig. 2). The Caledonian Orogeny, in latest Silurian to earliest Devonian times, resulted in the accretion of terranes with different crustal composition and structural fabrics (Coward 1995). Following the Variscan Orogeny in latest Carboniferous times, further structural fabrics were added to the basement of the newly formed Pangaean Supercontinent. In the region west of mid-Norway through to western Ireland, comprising the Caledonian domain, the major structures strike NE–SW to north–south. Further south in the

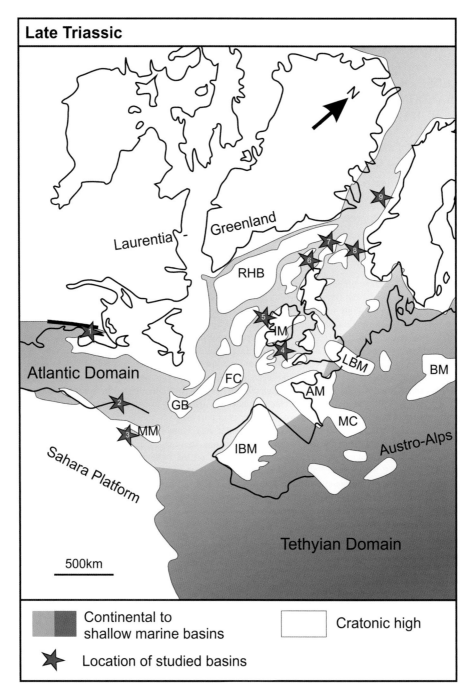

Fig. 1. Location map showing the main Permian–Triassic basins on the northern and central Atlantic borderland. Basins reviewed in this paper are denoted with numbered stars. 1, Fundy Basin; 2, Argana Basin; 3, Central High Atlas Basin; 4, Celtic Sea basins; 5, Slyne and Rockall basins; 6, Solan Bank High; 7, Magnus–East Shetland basins; 8, Horda Platform; 9, Froan Basin. MM, Moroccan Massif; IBM, Iberian Massif; FC, Flemish Cap; AM, Armorican Massif; LBM, London Brabant Massif; GB, Grand Banks; BM, Belgian Massif; MS, Massif Central; IM, Irish Massif; RHB, Rockall High Bank.

Variscan region the orientation of the structures was predominantly east–west. The inherited Caledonian and Variscan fabric provided the structural template that controlled the location, orientation and development of the Permo-Triassic basins (Naylor & Shannon 2009; Štolfová & Shannon 2009). The accreted terranes that constitute the Pangaean Supercontinent have a range of compositions and these exerted a major control on the composition of the Permo-Triassic sediments, and determined their reservoir quality. The basins examined in this study extend from the North Atlantic offshore mid-Norway, west of the major Caledonian Iapetus suture and fold belt, across the Variscan orogenic fold belt and core complex to the Moroccan basins of the central Atlantic. They also include outcrop studies from the conjugate margin basins of offshore Canada (Nova Scotia).

Structural style and evolution: regional seismic analysis

The shapes of the preserved remnant Permo-Triassic basins of the Atlantic margin show a clear association with the underlying structural fabrics (Doré *et al.* 1999). West of Norway, the UK and western Ireland basins developed in a broadly NE–SW to north–south belt, parallel to the Caledonian domain. Further south, basin orientation is dominated by an east–west trend parallel to the Variscan structures, with a similar trend documented in the region of the Innuitian fold belt north of Greenland.

The basins west of mid-Norway, exemplified by the Froan Basin on the Trøndelag Platform (Fig. 3) contain several kilometres of Permo-Triassic strata and the basins are often controlled by eastward-dipping master growth faults. The overall basin trend

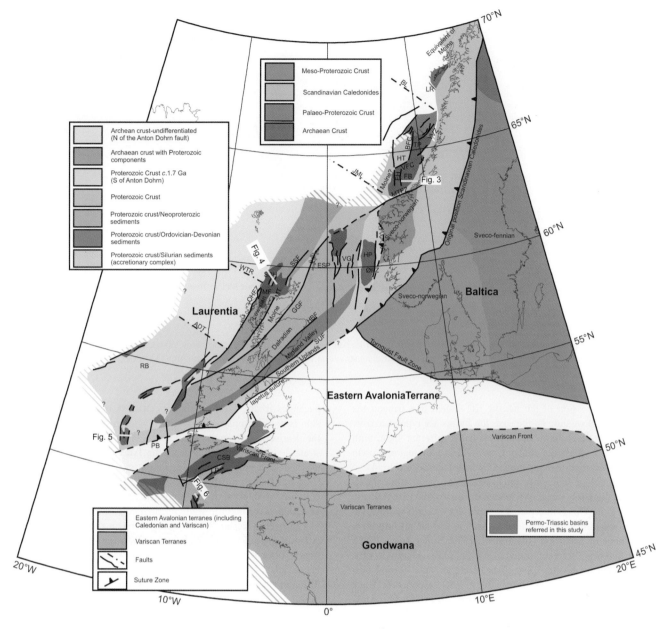

Fig. 2. Simplified structural and terrane map of the Northwest European Atlantic margin. Superimposed on this is the location of the various Permo-Triassic basins involved in the present study. The location of Figures 3–6 is also shown. Abbreviations: ADT, Anton Dohrn Transfer; BFC, Bremstein Fault Complex; BL, Bivrost Lineament; CSB, Celtic Sea basins; ESP, East Shetland Platform; FB, Froan Basin; FH, Frøya High; GGF, Great Glen Fault; HBF, Highland Boundary Fault; HP, Horda Platform; HT, Halten Terrace; JML, Jan Mayen Lineament; LP, Labadie Bank–Pembrokeshire Ridge; LR, Lofoten Ridge; MF, Minch Fault; MTFZ, Møre, Trøndelag Fault Zone; MT, Moine Thrust; ØFZ, Øygarden Fault Zone; OHFZ, Outer Hebrides Fault Zone; PB, Porcupine Basin; RB, Rockall Basin; SBH, Solan Bank High; SSF, Shetland Spine Fault; SUF, Southern Upland Fault; TP, Trøndelag Platform; VFC, Vingleia Fault Complex; VG, Viking Graben; WTR, Wyville–Thompson Ridge; YVZ, Ylvingen Fault Zone. Dashed areas represent uncertain terrane occurrence or boundary. From Štolfová & Shannon (2009).

appears to be influenced by a Caledonian structural fabric (e.g. the Ylvingen Fault Zone). However, no evidence is seen of direct reactivation of individual major Caledonian thrust structures. The basins are therefore interpreted as the product of narrow 'classical' rifting (McKenzie 1978; Buck et al. 1999).

The Permo-Triassic basins immediately west of the UK show similar pronounced half-graben geometries. The North Minch Basin is interpreted as having developed by strain localization in a narrow rift system, with a slight northwestward migration of strain through time resulting in the progressive movement of active fault systems (Štolfová & Shannon 2009). In the Magnus Basin, where the Permo-Triassic is preserved in a series of rotated fault blocks, initial palaeotopographic infill was followed by late Triassic and younger growth faulting, which also has the characteristic geometry of a narrow rift system (Štolfová & Shannon 2009).

In a number of areas a series of small, Permo-Triassic half-graben basins are controlled directly by growth faults developed along pre-existing Caledonian structures. The West Orkney Basin is probably associated with reactivation along the Caledonian and Precambrian Moine Thrust and Outer Hebrides Fault Zone complex. The Slyne and Erris basins west of Ireland (Naylor et al. 1999) and the Magnus Basin NE of Shetlands lie close to the projected offshore extensions of the Great Glen Fault complex, a major strike-slip Caledonian fault system that was reactivated initially in Devonian times and later in Permo-Triassic times.

The Solan Bank High, an elongate fault-bounded basement block running along a Caledonian structural fabric, acted as a barrier to

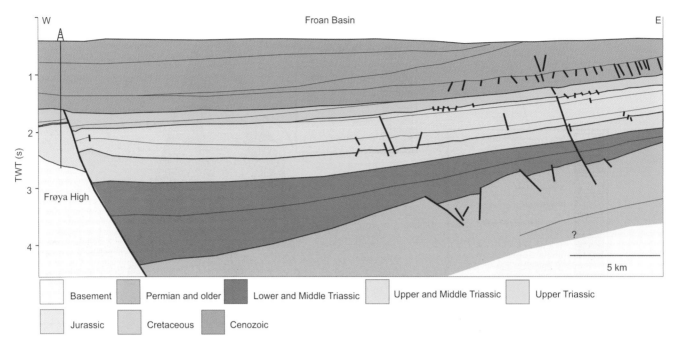

Fig. 3. Geoseismic profile across the Froan Basin showing typical half-graben geometry characteristic of a narrow 'classical' rift. The thick Permo-Triassic succession is subdivided into three seismic packages indicating different structural styles and demonstrating clear growth faulting, especially in the central, second package.

westward progression of the narrow rift system that fringed the west coast of Scotland. To the east, the basins are typical narrow rifts while the large basin to the west has a different style, with thick uniform sediments showing no evidence of major growth faulting, albeit modified and eroded by later tectonism (Fig. 4).

This pattern of uniformly thick sediments, with subtle fault-controlled thickness variations, is seen in much of the Irish Atlantic margin region. This is interpreted to represent wide rift basin development resulting from strain delocalization and lateral migration of the locus of extension (Buck et al. 1999; Buiter et al. 2008). Wide rift zones, often composed of multiple low-magnitude rifts, are predicted to develop in regions of higher than normal heat flow with relatively thick crust (Buck 1991; Hopper & Buck 1996). The margins of the Irish Rockall Basin exhibit uniformly thick interpreted Permo-Triassic successions, with subtly different internal stratigraphy between adjacent basins, inset by younger faults as a string of north–south trending 'perched' basins (Fig. 5). The overall interpretation is wide basin extension producing a broad region of Permo-Triassic deposition.

The Triassic successions over much of the Celtic Sea region, south of Ireland, are characterized by uniform sediment thickness across wide areas. Subtle internal wedging of individual seismic packages occurs, with fault polarity changes, but the overall succession is not markedly asymmetric. A wide rift basin system above the Variscan basement domain is suggested. Inferred Permian strata lie unconformably beneath the Triassic succession in small, isolated half-grabens with common syn-sedimentary wedging (Fig. 6). These probably developed in response to post-orogenic collapse of thickened, hot Variscan crust. Similar features are observed in the outcrop study of the marginal Moroccan basins.

The overall Permo-Triassic succession in the European North Atlantic margin region thickens northwards from the Celtic Sea region offshore Ireland to the mid-Norwegian basins. In most basins distinct seismic sequences can be identified, reflecting different phases of early post-Pangaean rift development. While these can be mapped through individual basins, correlation between the various basins is only tentative at this stage due largely to limited well control. Nonetheless, regional patterns can be recognized. Permian strata are only locally developed in the south but have a significant thickness in the northern basins. In the south (e.g. Celtic Sea region) interpreted Permian strata occur within small, fault-controlled half-graben basins that are hinged by extensionally reactivated Variscan structures. These are interpreted as early Variscan collapse intermontane basins. Further

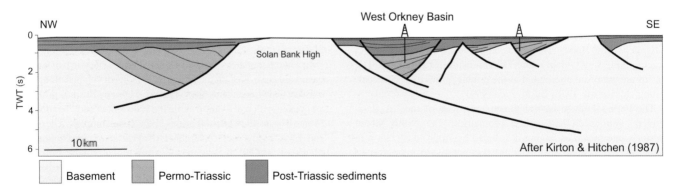

Fig. 4. Geoseismic profile across the Solan Bank High. The basement high compartmentalized the rifting style, with a series of narrow rift half-grabens developed to the SE while the remnants of an eroded and tilted wide-rift system occurs to the NW. This lacks evidence of major growth faulting and probably had a more extensive thick, uniform depositional geometry.

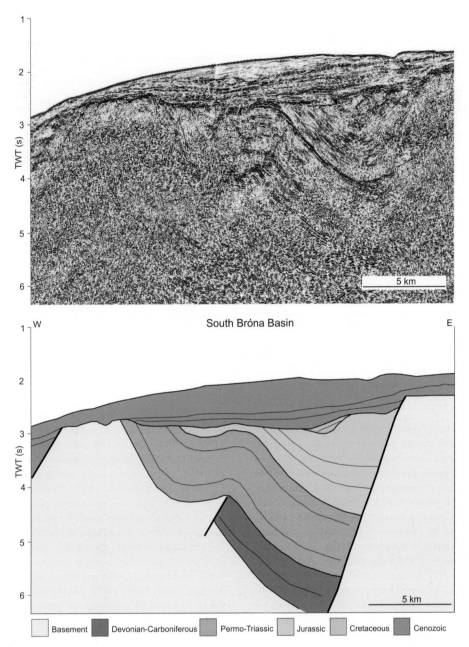

Fig. 5. Seismic profile and geoseismic profile across the South Bróna Basin on the SE flank of the large Rockall Basin. The preserved basin, incised by later faulting and erosion, had a larger depositional extent as evidenced by the absence of any evidence of thickness variations towards the bounding fault to the east or the basement block to the west. It is interpreted as an example of a wide-rift basin.

north, as typified by the Slyne Basin, basins NW of Scotland, and the Froan Basin on the Trøndelag Platform, the Permian succession is significantly thicker. The thin succession resting unconformably upon the interpreted Permian strata in the south of the region is suggested to be of Late Triassic age, while in the Norwegian region a complete Triassic succession is commonly proven by well data. This suggests that rifting commenced in the Norwegian region in Permian times, synchronous with intermontane localized Variscan collapse basins further south. Regional Triassic sedimentation did not take place in the south until later in the Triassic, when deposition was of regional extent, with wide rift processes dominating in this region.

Structural style/evolution: evidence from outcrop: Morocco

The Permo-Triassic Central High Atlas and Argana basins in Morocco (Fig. 7) are 100 km apart, separated by the Palaeozoic and Pre-Cambrian rocks of the 'Massif Ancien'. The exceptional exposures provide information on contrasting structural style, and the geodynamic response to North Atlantic and Tethyan tectonics. The basin types are similar to those documented from regional seismic, and outcrop analysis provides an analogue study to detail the fault geometries and impact of tectonics on local sedimentation and reservoir development. The outcrop evidence builds upon the regional picture and elucidates in more detail the semi-regional to reservoir scale tectonostratigraphic features of the successions.

The Oukaimeden–Ourika Valley, part of the broader Central High Atlas Basin, comprises an ENE-elongated rift basin bounded by extensionally reactivated Variscan structures. Although the present-day outcrop pattern reflects inversion due to the Alpine compression that led to major uplift in the High Atlas and subsequent erosion, it is possible to reconstruct the basin's earlier structural history. Both the ENE–WSW and NNE–SSW striking sets of normal faults show evidence for syn-sedimentary

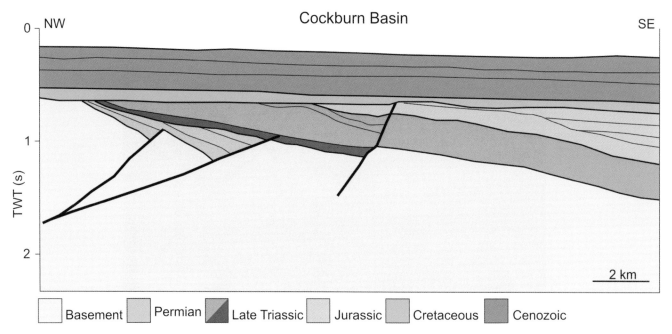

Fig. 6. Geoseismic profile from the Cockburn Basin. Interpreted Permian strata are developed in small half-graben controlled by early extensional collapse of the thick Variscan crust along reactivated thrusts. These sediments are unconformably overlain by late Triassic strata showing as relatively uniform, with subtle thickness variations, interpreted as the product of wide-rift extension. The planar nature of the unconformity, with no evidence of residual topography, suggests a significant hiatus between the two sedimentary sequences.

movement, displaying stratigraphic thickening into growth faults and associated progressive change in dip of the bedding (Fig. 8). The ENE–WSW faults are deep rooted and controlled the geographical extent of the basin, providing the accommodation for sediment deposition within a typical narrow half-graben basin geometry following NW–SE extension during the Late Triassic. The direction of extension is nearly parallel to the dip on NE to NNE striking faults, illustrated by striations. ENE trending structures display left-lateral movement, with a strong normal component.

Close to the bounding fault the main sandstone units have a higher net to gross ratio (N:G), consistent with a tectonic control on reservoir development. Increased accommodation on the downthrown side of the main faults resulted in river capture, and detailed sedimentary logging (Fabuel-Perez & Redfern 2009) shows that the main fluvial system flowed parallel to the controlling faults. More mudstone-rich overbank deposits become the dominant facies as the section thins away from the main fault-controlled depocentres (Fig. 8). The presence of thick locally derived conglomerates adjacent to the main bounding faults, with palaeocurrents oblique to the dominant axial trends for the main fluvial system (Figs 7 & 8), record alluvial fans derived from erosion of the main fault scarp during tectonic activity (Fabuel-Perez & Redfern 2009).

Smaller NE to NNE oriented faults are interpreted to be contemporaneous to the main fault set and are compatible with NW–SE extension. However, these faults have a small throw and are interpreted to have a limited influence on sediment deposition.

The second area of study in the Argana Basin, on the western margin of the High Atlas, contains exceptionally exposed Permian and Triassic red beds. This is the SE extension of the Essaouira Basin (Fig. 7). Permian and Upper Triassic beds strike approximately parallel to the NNE trend of the valley and dip 5–30° towards the NW. The Upper Permian sediments consist predominantly of conglomerates interpreted to be deposited within alluvial fans, grading vertically and laterally into sandstones deposited in fluvial channels and as floodplain deposits. A major unconformity separates the Permian from the overlying Upper Triassic sediments that mainly consist of siltstones and mudstones deposited in floodplain or playa environments, alternating with coarse alluvial conglomerates and continental fluvial or aeolian sandstones (Tixeront 1973; Brown 1980; Mader 2005).

Several models have been proposed to explain the structural evolution of the basin and the structural control on sedimentation (Tixeront 1973; Brown 1980; Medina 1988, 1991; Hofmann et al. 2000). Episodic movement of east–west trending fault blocks tilted towards the north was initially suggested to control the deposition of the whole Permo-Triassic sequence (Brown 1980). Other studies recognized two main phases of syn-sedimentary extension characterized by the two sets of normal-faults, striking east–west and NE to NNE (Laville & Petit 1984; Medina 1988, 1991, 1995). Both these interpretations attribute the evolution of Mesozoic basins in the High Atlas of Morocco to pull-apart extensional mechanism along the ENE–WSW trending fractures (Manspeizer 1982; Laville & Petit 1984).

Recently analysed field data as part of this study are modifying this interpretation. It is suggested that the large east–west striking-faults do not significantly influence Triassic deposition. The thick Permian conglomerates, deposited as large alluvial fans, were sourced mainly from the uplifted Ancien Massif (Baudon et al. 2009). These conglomeratic units are cut by, but not sourced from, the east–west faults. The Upper Permian sequence was tilted towards the NW prior to Upper Triassic sedimentation producing a marked angular inconformity. Upper Triassic sedimentation in the Argana Valley does not show significant lateral variation in thickness or facies, which suggests the east–west and smaller NNE trending normal faults were not significantly active. The main basin-bounding fault is speculated to be a sub-surface extensional fault located to the west of the valley, now masked by later Jurassic and Cretaceous cover. This is parallel to faults producing similar half-graben structures identified further NW in the Essaouira Basin (Hafid 2000; Le Roy & Pique 2001). This fault probably controlled the regional tilt of the beds. In contrast to the Ourika Basin, deposition in the Argana Basin is interpreted to have been part of the much broader Essaouira Basin, with limited local fault control on sedimentation. This pattern of uniformly thick sediments, with subtle fault-controlled thickness variations, is comparable to that observed in the seismic study from the North Atlantic,

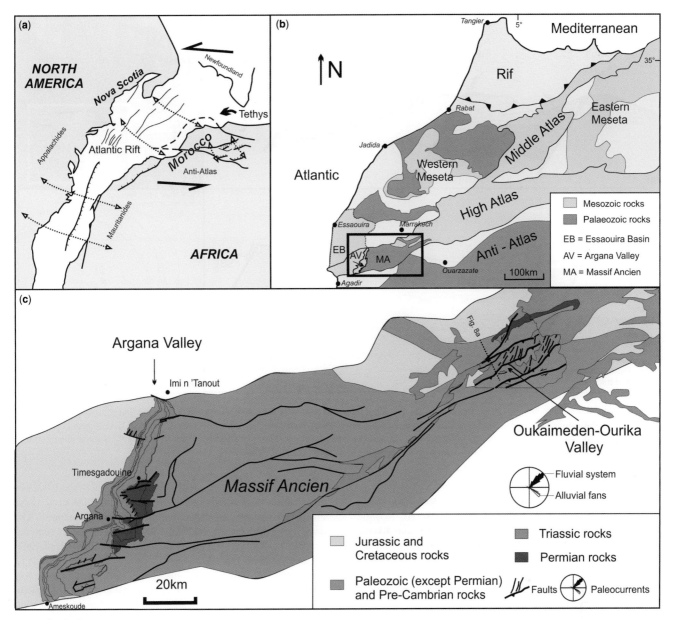

Fig. 7. (a) Palaeogeographic map of Morocco at Triassic time (modified from Laville & Piqué 1991). (b) Present-day structure summary map of Morocco (modified from Ellouz et al. 2003). The box denotes the location of the study area. (c) Close-up of study area showing the main structures. The Permian and Triassic outcrops in the Oukaimeden–Ourika Valley (modified after Taj-Eddine & Pignone 2005) and Argana Valley (modified after Tixeront 1973) are highlighted with respect to the Massif Ancien. Palaeocurrents in the Ourika Valley for the F5 Unit from Fabuel-Perez & Redfern (2009).

suggesting this Moroccan section provides a good outcrop analogue for subsurface exploration in the Atlantic borderland area.

Provenance studies: tracking Triassic sand dispersal on the NW European margin

The Moroccan outcrops described in the previous section illustrate the interplay of tectonics and sedimentation at regional and local scale and the resulting complexity of the depositional system. To better understand reservoir sand dispersal within such complex systems a novel provenance tool was utilized as part of the integrated analysis, illustrated by results from the Atlantic Margin basins west of Ireland and the UK.

Recent studies have demonstrated the utility of the Pb isotopic composition of detrital K-feldspar as a sand provenance tool, particularly when applied on a regional scale (Tyrrell et al. 2006, 2007; Clift et al. 2008). Common Pb isotopes vary in the crust on a sub-orogenic scale and it has been shown that detrital K-feldspar can retain the signature of its source despite erosion, transport and diagenesis (Tyrrell et al. 2006). The Pb isotopic signature of individual K-feldspar sand grains can be analysed in situ using laser ablation multicollector inductively-coupled plasma mass spectrometry (LA-MC-ICPMS). Imaging prior to analysis highlights heterogeneities which then can be avoided during laser ablation. The use of ion counters to measure Pb ion beams means that data can be retrieved at a spatial resolution (c. 20 μm laser spot sizes) similar to that achievable using ion microprobe techniques but with reduced analytical uncertainty (Tyrrell et al. 2009).

One of the major advantages of the Pb K-feldspar tool is that, in contrast to provenance approaches that utilize signals in robust grains (e.g. U–Pb zircon), it provides a means of assessing first-cycle sand-grain provenance. As detrital K-feldspar is unlikely to survive more than one sedimentary cycle, these grains can be tracked back directly to their basement source, allowing the scale and geometry of the drainage system to be constrained. These types of insights can aid in more accurate prediction of reservoir sandstone distribution and quality in the subsurface.

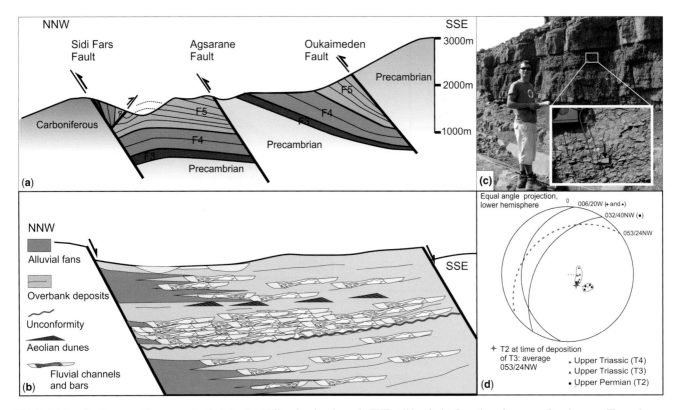

Fig. 8. (**a**) Structural cross-section across the Oukaimeden Valley showing the main ENE striking faults. Location of cross-section shows on Figure 6c. (**b**) Schematic model of the structural control on sedimentation, showing alluvial fans shed from the footwall highs, and the axial river system preferentially captured close to the main fault. (**c**) Photograph of the alluvial fan breccias, with close up of clasts. (**d**) Stereonet plot illustrating the angular unconformity between the Permian and Triassic bedding in the Argana Valley.

The Pb K-feldspar provenance tool is particularly appropriate in an investigation of Triassic sand dispersal in the NE Atlantic margin basins, given the abundance of arkosic and sub-arkosic sandstones in these successions. Recently published provenance data have shed new light on the nature and origin of these sandstones. Data from K-feldspar from Lower Triassic sandstones in the Slyne Basin (including the Corrib Gasfield), offshore west of Ireland, define two distinct Pb populations (Fig. 9), which are likely to have been sourced from Archaean and Proterozoic crust to the north and west (Tyrrell et al. 2007; McKie & Williams 2009).

Pb K-feldspar data from Permo-Triassic sandstones in the Rockall Basin (Dooish gas condensate discovery, Well 12/2-1z; Fig. 9) appear to have been derived from the north and NE, probably from elements of the Lewisian Complex (Fig. 9; Tyrrell et al. 2010). In basins west of Shetland (Fig. 9), new Pb K-feldspar data (Fig. 10; McKie & Williams 2009; Tyrrell et al. 2009) from Middle–Upper Triassic Foula Formation sandstones (part of the Strathmore Field) indicate derivation from Archaean–Palaeoproterozoic rocks on the margins of the rift basin (Nagssugtoqidian Mobile Belt of East Greenland and/or the Lewisian Complex of NW Scotland and equivalents), although whether the derivation is from the east- or west-rift margins, or from a northern axial source, cannot be distinguished. Here, the Pb K-feldspar data are in broad agreement with U–Pb detrital zircon geochronology (Morton et al. 2007).

The Pb K-feldspar data from Atlantic margin basins consistently preclude Irish Massif, the UK Mainland south of the Moine Thrust and the remnant Variscan Uplands to the far south as sources for Triassic sand (Fig. 10). This indicates no linkage between the drainage systems supplying these Atlantic margin basins and those delivering sand to onshore UK, the East Irish Sea Basin and the Central and Southern North Sea. The sedimentary contribution of non-radiogenic Archaean and Palaeoproterozoic Pb sources appears to increase in basins further to the north. The data indicate that palaeodrainage evolution in these marginal Triassic basins was strongly influenced by uplifted Archaean–Palaeoproterozoic basement highs, and not Variscan Uplands to the south, with consequent implications for potential reservoir sandstone distribution in these areas (e.g. Tyrrell et al. 2010). These studies reveal the value and effectiveness of the Pb K-feldspar provenance tool in investigations of both ancient and modern broad-scale drainage systems.

Developing analogue depositional and reservoir models

Building upon the work defining the basin architecture and provenance, detailed sedimentological analysis is improving our understanding of basin development, basin-scale depositional systems and producing analogue reservoir models. Correlations within the Central Atlantic domain have been attempted by Olsen (1997) and Olsen et al. (2000) from the Fundy Basin to Morocco. This is based on limited biostratigraphic data and often poorly defined unconformities. Because of the low resolution of biostratigraphy with red bed deposits, the only synchronous marker beds that can confidently be picked are the radiometrically dated CAMP basalt and the palynological turnover of the Rhaetian/Jurassic boundary located a few metres below the basalt. Biostratigraphical correlations across the Triassic basins of Morocco have been assessed recently by El Arabi et al. (2006) while the biostratigraphy in the Fundy Basin is still relatively poorly constrained. The stratigraphy from the Fundy Basin, Canada and the Argana and Central High Atlas basins in Morocco is compared in Figure 11. Triassic rift basins along the Central and North Atlantic margins display a similar sedimentary evolution characterized by an initial phase of

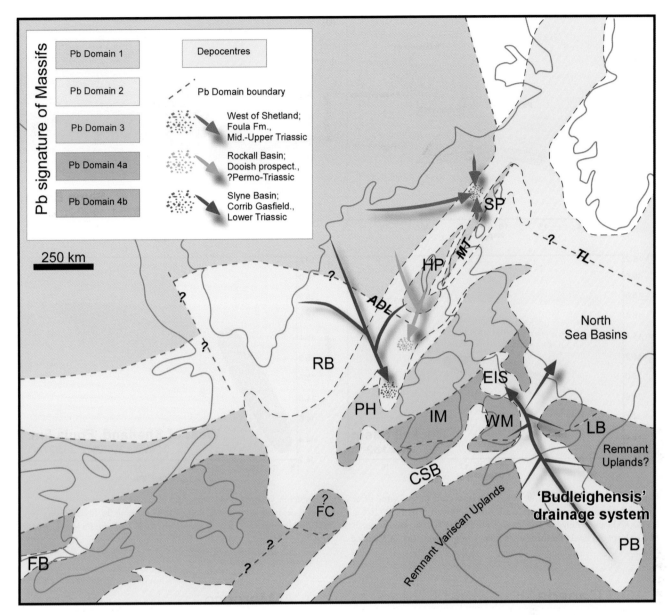

Fig. 9. Schematic Triassic palaeogeographic reconstruction of the circum-North Atlantic region (after Torsvik et al. 2001; Eide 2002; Scotese 2002) showing the configuration of massifs and depocentres. The massifs are colour-coded to reflect their broad Pb isotopic signature (after Tyrrell et al. 2007 and references therein). The 'Budleighensis' drainage system is also shown (blue). Directional arrows (red, green, brown) show likely derivation directions for sandstones from the marginal basins, based on the Pb isotopic composition of individual K-feldspar sand grains. ADL, Anton Dorhn Lineament; CSB, Celtic Sea basins; EIS, East Irish Sea Basin; FB, Fundy Basin; FC, Flemish Cap; HP, Hebridean Platform; IM, Irish Massif; LB, London–Brabant Massif; MT, Moine Thrust; PB, Paris Basin; PH, Porcupine High; RB, Rockall Bank; SP, Shetlands Platform; TL, Tornqvist Line; WM, Welsh Massif.

alluvial fan and fluvial sedimentation and a later playa/lacustrine dominated phase. Although the sedimentological evolution of these basins is similar, the timing and duration of the phases of fluvial and playa deposition vary within and between the basins. In order to understand the basin-fill evolution, and in particular the depositional systems, the sedimentology of two broadly co-eval Permo-Triassic basins is reviewed in this paper: the Minas Basin (Nova Scotia) and the Central High Atlas Basin (Morocco). These formed at different palaeolatitudes and now lie on different sides of the Atlantic passive margin.

In the Fundy Basin, detailed analysis of laterally extensive coastal outcrops allows definition of basin-scale sedimentary architecture and assessment of basin development. A comparable Moroccan section provides equally extensive sections, offering another analogue for the subsurface basins, and using innovative LiDaR (Laser Detection and Ranging) analysis, detailed sedimentary and reservoir models have been produced (Fabuel-Perez et al. 2009).

Fundy Basin, Nova Scotia, Canada

The Fundy Basin, one of a series of early Mesozoic rift basins developed along the NW Atlantic margin, contains 6–12 km of Anisian to basal Hettangian non-marine clastic sediments (Olsen et al. 1989; Wade et al. 1996; Leleu et al. 2009). It represents a large complex Triassic half-graben (Wade et al. 1996) with a strike-slip component of movement (Olsen & Schlische 1990; Withjack et al. 1995, 2009) and is subdivided into three sub-basins; amongst them the easternmost Minas sub-basin shows the most extensive outcrops and preservation of sequences (Fig. 12).

The Triassic succession in the Minas sub-basin comprises the Wolfville (<800 m) and the overlying Blomidon (<250 m) formations (Fig. 12). The Wolfville Formation is Carnian in age (Olsen et al. 1989), and lies unconformably on Carboniferous and older rocks, forming the earliest syn-rift unit in the basin (Wade et al. 1996). The Wolfville Formation comprises coarse- and fine-grained fluvial sandstones (Klein 1962; Hubert & Forlenza 1988;

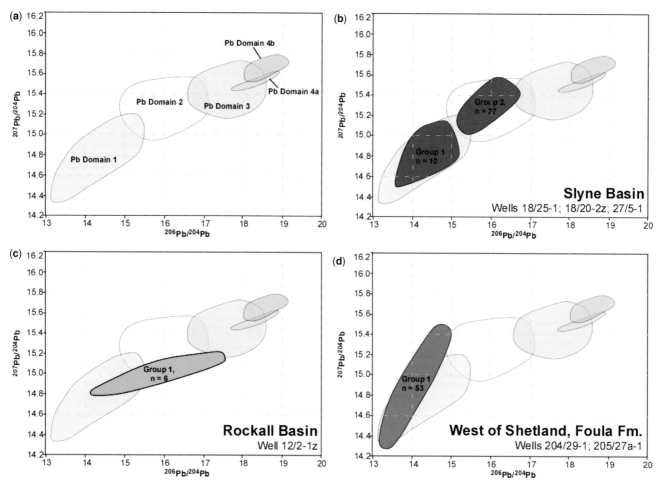

Fig. 10. Plot of $^{206}Pb/^{204}Pb - ^{207}Pb/^{204}Pb$ lead space (after Tyrrell *et al.* 2007 and references therein) illustrating (**a**) the isotopic composition of Pb basement domains in the circum-North Atlantic (Fig. 8); (**b**) the Pb isotopic range of K-feldspar from Lower Triassic sandstones in the Slyne Basin; (**c**) the Pb isotopic range of K-feldspar from Permo-Triassic sandstones in the Rockall Basin (Dooish gas condensate discovery); (**d**) the Pb isotopic range of K-feldspar in Foula Formation sandstones from basins west of Shetland.

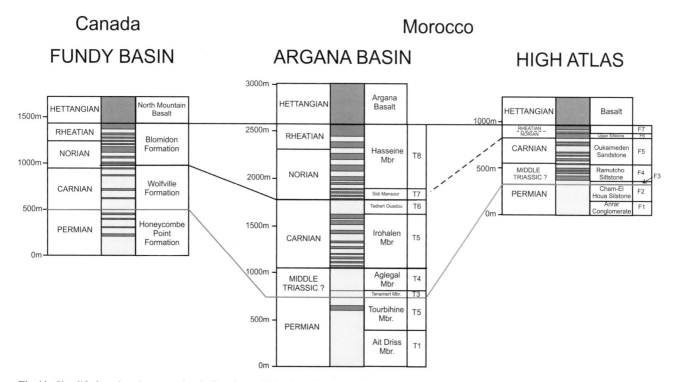

Fig. 11. Simplified stratigraphy comparing the Permian and Triassic sections in the Fundy Basin, Nova Scotia, and the Argana and Central High Atlas basins in Morocco. Modified from Olsen *et al.* (2000).

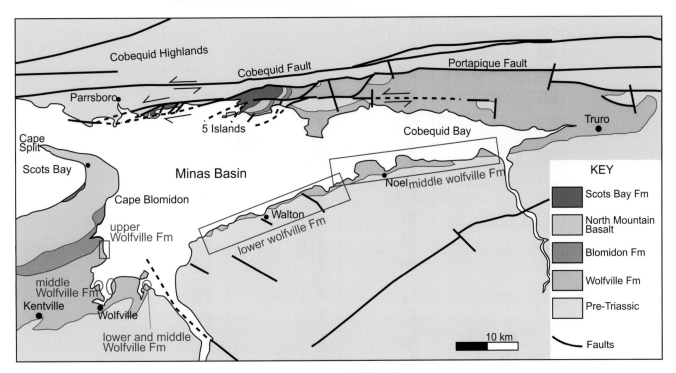

Fig. 12. Location of the Minas Basin in Nova Scotia, Canada, showing the studied Triassic outcrops.

Leleu et al. 2009) and subordinate aeolian dune deposits (Hubert & Mertz 1980, 1984) and alluvial fan sediments (Hubert & Mertz 1984). The contact between the formations corresponds to a major change in facies architecture to mud-rich playa margin deposits containing evaporites and ephemeral fluvial sheet-sandstones (Hubert & Hyde 1982; Mertz & Hubert 1990; Gould 2001). The extensive North Mountain Basalt, part of the Central Atlantic Magmatic Province event (McHone 2000), overlies the Blomidon Formation, and straddles the Triassic–Jurassic boundary (Olsen et al. 1989; Olsen 1997).

Within the Wolfville fluvial succession three distinguishable packages with very different characteristics of grain size and stacking patterns have been observed. The lower Wolfville Formation (110 m thick) is dominated by coarse-grained fluvial deposits (Fig. 12) which filled remnant palaeotopography. It comprises four mega-units and 13 smaller scale high-resolution fining-upwards cycles of pebbly conglomerates to sandstones (Fig. 13). The bounding surfaces can be correlated regionally over a distance of between 10 and 27 km (Leleu et al. 2009), thereby providing effective basin-scale stratigraphic correlation. The overall succession has a sheet-like geometry comprising stacked, erosion-dominated, multi-storey channel bodies with a drainage system that evolves to the north and NE. An individual high-resolution cycle records braid-plain development and is interpreted as a multi-storey channel belt. However, the width of the active channel belt is unknown. Recognition of key surfaces across a 10–20 km section (Fig. 14), provides criteria for correlation of these continental facies, barren of biostratigraphic markers (Leleu et al. 2009).

The middle Wolfville Formation abruptly overlies the lower Wolfville Formation (Fig. 12). It is a well sorted sandy bedload fluvial system (< 350 m thick) and consists of 15 repetitive units. These comprise stacked channel bodies forming channel belt complexes that are intercalated with upward thickening floodplain deposits. Application of LiDaR digital outcrop analysis aids the definition of detailed architecture (van Lanen et al. 2009). Large-scale sequences have been defined based on grain size evolution that encompasses several channel belt complexes. The grain size variation reveals a climatic control which influences the bedload transport capacity of the river, while smaller scale repetitive channels and channel belt migration/avulsion possibly result from autocyclic processes. The larger scale architectural evolution including the development of the channel belt architecture, channel body dimensions and the increased upwards preservation of greater floodplain units result from allocyclic processes which may reflect a tectonic control producing increased accommodation.

The upper Wolfville Formation (240 m thick) occurs beneath the lacustrine/playa deposits of the Blomidon Formation (Fig. 12). It displays repetitive packages of channelized and unconfined fluvial facies and playa deposits together with occasional aeolian dune deposits. Grain size varies from pebbly coarse sand and medium to coarse sand in the channelized facies to well sorted medium to fine sand in the unconfined facies and aeolian deposits. Development of the playa margin environment in the basin during Blomidon Formation deposition records retrogradation of the fluvial profile within the drainage basin, and the extension of the playa and playa margin towards the catchment. Retrogradation suggests a major decrease in water and sediment supply. This major shift could be interpreted as the result of climate change. However, although all Triassic basins along the Atlantic margins show a similar sedimentological evolution from fluvial to playa depositional phases, the transition occurs at different times and with variable duration. This suggests that basin development governs the large-scale basin-fill and the shift from fluvial to playa depositional phase (Schlische & Olsen 1990; Smoot 1991). In particular, because the transition from fluvial to playa/lacustrine conditions takes place at different times, it indicates that climate may not be the critical controlling factor. The transition is interpreted to be associated with a progressive decrease in source area relief related to a decline in fault-generated topography towards the end of the syn-rift phase of basin development.

The Central High Atlas Basin, Morocco

Previous work in Morocco (Petit & Beauchamp 1986; Mattis 1977; Benaouiss et al. 1996; Tourani et al. 2000) provided an overview

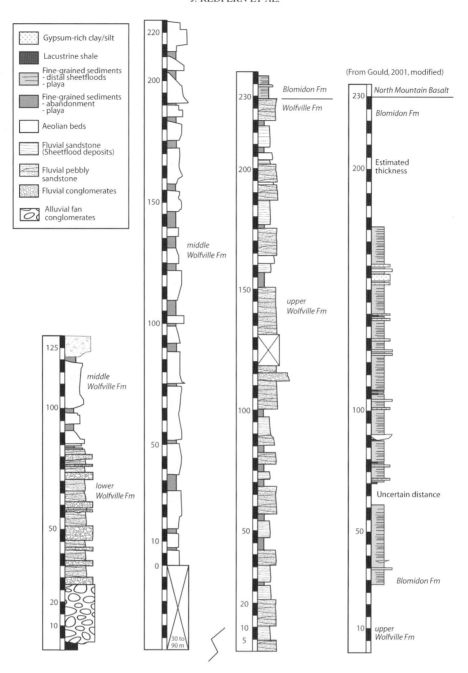

Fig. 13. Two summary sedimentary logs through the Triassic succession in the Minas sub-basin comprising the Wolfville (<800 m) and the overlying Blomidon (<250 m) formations. The Wolfville Formation is Carnian in age (Olsen *et al.* 1989), and lies unconformably on Carboniferous and older rocks.

of the broad depositional setting and stratigraphy for the Permian and Triassic section in the Central High Atlas (Fig. 11). Fabuel-Perez & Redfern (2009), as part of this research project, recently revised the sedimentological interpretation of the Oukaimeden Sandstone Formation in the Oukaimeden–Ourika valley area (Fig. 7). Detailed sedimentary models were produced using LiDaR for digital outcrop analysis, and a reservoir model was developed (Fabuel-Perez & Redfern 2009; Fabuel-Perez *et al.* 2009).

Within the Oukaimeden Sandstone Formation five major facies associations have been identified (Fig. 15). Changes in the distribution of facies associations, identification of key boundaries, surfaces and variation in the architectural style through time have been used to subdivide the Oukaimeden Sandstone Formation into three major members (Fig. 15): (a) a lower member comprising channels and bars (Facies Association 1) alternating with floodplain mudstone units (Facies Association 2); (b) a middle member, characterized by vertically stacked amalgamated channels and bars and a significant decrease in the amount of preserved mudstone units; and (c) an upper member showing a similar style to the one observed in the lower member with the first occurrence of aeolian facies (Facies Association 3) and tidally influenced sandstones (Facies Association 4).

Using a process-based depositional facies model based on genetically related packages (Fabuel-Perez & Redfern 2009), the Lower Oukaimeden Member is interpreted to record deposition from large axial ephemeral river systems. The Middle Oukaimeden displays a change in depositional style to perennial fluvial conditions. This is interpreted to record the rejuvenation of the fluvial regime related to a change in climate to more humid conditions and subsequent increased run-off. The Upper Oukaimeden is interpreted to record a change in fluvial conditions from perennial back to ephemeral combined with the presence of aeolian dunes (Facies Association 3) which reflect climatic variations shifting

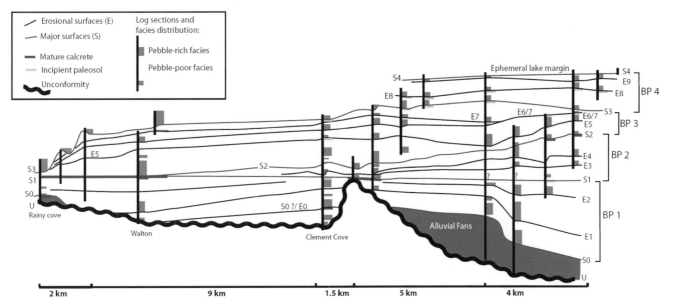

Fig. 14. Summary correlation of the lower Wolfville Formation (110 m thick). It comprises multi-storey channel bodies arranged as four mega-units bounded by 'S' surfaces and 13 smaller scale high-resolution fining-upwards cycles of pebbly conglomerate to sandstone, bounded by 'E' (or 'S') surfaces which can be correlated regionally between 10 and 27 km (Leleu et al. 2009).

towards more arid conditions. The top of the Oukaimeden Sandstone is characterized by the presence of tidally influenced facies recording the first marine incursion into the basin.

In the lower and upper member, alluvial fans were deposited in response to footwall uplift erosion of the controlling Sidi Fars fault. These alluvial fans, which pinchout towards the SW, are interpreted to force the axial rivers away from the controlling fault. The lack of alluvial fan deposits in the middle member could be due to 'toe cutting' of the alluvial fan by the axial rivers due to their increased size and switch to perennial conditions of the river system as a consequence of a climatic change (Leeder & Mack 2001).

Evidence is seen of a tectonic control on deposition and changes in architectural style. Preferred orientation of palaeocurrents in Facies Association 1 is towards the ENE. This is interpreted to be related to the palaeogeography of the basin (ENE–WSW orientated), which was tectonically controlled by the orientation of the syn-sedimentary Sidi Fars fault. Alluvial fans have a preferred palaeocurrent orientation towards the SE, normal to the ENE–WSW orientation of the Sidi Fars fault, indicating a sediment source from the uplifted footwall. N:G ratio decreases away from the fault, indicative of capture of the axial river system close in the depocentre. The morphology of the contacts also varies with distance from the controlling fault, with a sharp erosive contact

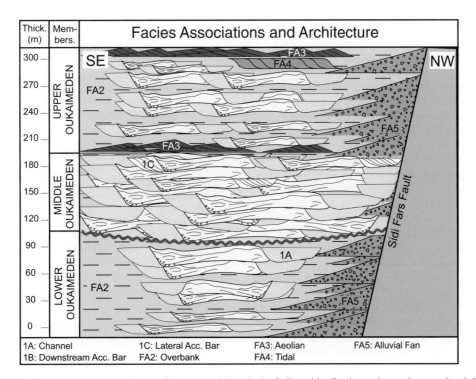

Fig. 15. Composite log summarizing the gross distribution of facies associations in the Ouikamaiden Sandstone, key surfaces used to define the three members and the changing architectural style through time.

near the fault due to successive erosional events by downcutting channels, and more gradation contact on the basin margin.

The sections in Morocco record both the tectonic and climatic influence on depositional architecture and reservoir distribution. The contact between Lower and Middle Oukaimeden is interpreted to record possible tectonic reactivation in the hinterland magnified by a concurrent change in climate towards more humid conditions. Increasing run-off and input of coarse sediment into the basin produced a change in architectural style, and a switch from ephemeral to perennial conditions. In the Upper Oukaimeden, a return to ephemeral conditions, combined with the presence of aeolian facies, suggests a climatic change back towards more arid conditions. Similar gross facies shifts, also interpreted as a response to climatic change, are observed in the Argana Basin. Improved dating and more detailed correlation are required to corroborate these as regionally correlatable events.

Conclusions

(1) An improved understanding of basin and sedimentary evolution, provided by multidisciplinary analysis, is illustrated by this regional study of the Atlantic margin Permo-Triassic basins. This addresses basin and outcrop to pore scales, across three orders of magnitude. Regional seismic scale allows analysis of a number of structurally linked basins, imaging the large-scale structure, in order to understand overall basin architecture and identify mega-sequences. Outcrop analysis provides ground-truth information to refine the seismic-scale structural interpretations and the resolution to model reservoir architecture at sequence scale, with definition of depositional systems and key correlation surfaces. Provenance analysis has the potential to test regional structural models and predict source to sink fairways, improving reservoir characterization by identifying provenance sources and sediment transport directions.

(2) A range of basin geometries are identified in the Permo-Triassic Atlantic margin basins, using both seismic (offshore, NE Atlantic) and outcrop (onshore, Morocco) studies. These include post-orogenic collapse basins, focused narrow rifts and wide, low-magnitude multiple extension depocentres. The shape and evolution of the basins were influenced by both Atlantic and Tethyan tectonic regimes, with significant control from inherited basement structures and structural fabrics.

(3) The basin fill, characterized in detail by examining the onshore Permo-Triassic outcrops in both Morocco and Nova Scotia, is dominated by continental red beds. They comprise braided fluvial, alluvial fan, aeolian, floodplain and lacustrine facies, and display a cyclicity in fluvial deposits from major ephemeral to perennial fluvial systems. Detailed outcrop analysis has identified key marker horizons and sequence boundaries that can be used to provide constraints for the delineation of laterally correlatable cycles. The depositional architecture at high-resolution outcrop scale is complex, controlled by the interplay of local and regional tectonic and climatic factors, as well as autocyclic processes. At basin scale, sedimentation records climatic changes, but the low-resolution basin fill evolution is mainly controlled by tectonically induced basin development. This has implications for reservoir distribution and quality on a regional and field scale.

(4) Application of a new Pb in K-feldspar provenance technique to determine the source of Triassic sandstones offers new and robust controls on sediment dispersal patterns in the North Atlantic. Results indicate that the regional palaeodrainage evolution of the basins was predominantly controlled by uplifted Archaean–Palaeoproterozoic basement highs to the north and NW and not, as hitherto assumed, by the Variscan uplands to the south. In addition to providing information on likely reservoir quality, the technique also has applications as a test for the robustness of large-scale structural basin tectonics and timing.

Funding is acknowledged from a Griffith Geoscience Research Award (Irish Department of Communications, Energy and Natural Resources) under the National Geoscience Programme 2007–2013. Funding from a Science Foundation Ireland Research Frontiers Programme grant (RFP06/GEO029) is also acknowledged. Thanks to K. Souders, P. Sylvester and M. Tubrett (Microanalysis Facility, Memorial University, Newfoundland, Canada) for assistance with ICPMS and A. Kronz (Electron Microprobe Laboratory, Geowissenschaftliches Zentrum, Göttingen, Germany) and to the Petroleum Affairs Division (Ireland), the Petroleum Infrastructure Programme (Ireland) and D. Sutherland (DTI core store, Edinburgh, UK) for facilitating access to core. Shell, EPE Stavanger, Shell EPX, EPTS and the North Africa Research Group (NARG) Consortium – Hess, Conoco, Repsol, Petrocanada, Maersk, Woodside, BG Group, Wintershall, RWE, ConocoPhilips, Pluspetrol, and ONHYM in Morocco – are all gratefully acknowledged for funding and support.

References

Baudon, C., Fabuel-Perez, I. & Redfern, J. 2009. Structural style and evolution of a Late Triassic rift basin in the Central High Atlas, Morocco: controls on sediment deposition. *Geological Journal Special Issue*, **44**, 677–691.

Benaouiss, N., Courel, L. & Beauchamp, J. 1996. Rift-controlled fluvial/tidal transitional series in the Oukaimeden Sandstones, High Atlas of Marrakesh (Morocco). *Sedimentary Geology*, **107**, 21–36.

Blystad, P., Brekke, H., Færseth, R. B., Larsen, B. T., Skogseid, J. & Tørudbakken, B. 1995. *Structural Elements of the Norwegian Continental Shelf, Part II: the Norwegian Sea Region*. Norwegian Petroleum Directorate, Bulletin, **8**.

Brown, R. H. 1980. Triassic rocks of Argana Valley, Southern Morocco, and their regional structural implications. *American Association of Petroleum Geologists Bulletin*, **64**, 988–1003.

Buck, W. R. 1991. Modes of Continental Lithospheric Extension. *Journal of Geophysical Research*, **96**, 20161–20178.

Buck, W. R., Lavier, L. L. & Poliakov, A. N. B. 1999. How to make a rift wide. *Philosophical Transactions of the Royal Society of London*, **357**, 671–693.

Buiter, S. J. H., Huismans, R. S. & Beaumont, Ch. 2008. Dissipation analysis as a guide to mode selection during crustal extension and implications for the styles of sedimentary basins. *Journal of Geophysical Research*, **113**, 1–20.

Clift, P. D., Van Long, H. *et al.* 2008. Evolving east Asian river systems reconstructed by trace element and Pb and Nd isotope variations in modern and ancient Red River–Song Hong sediments. *Geochemistry Geophysics Geosystems*, **9**, Q04039; doi: 10.1029/2007GC001867.

Coward, M. P. 1995. Structural and tectonic setting of the Permo-Triassic basins of northwest Europe. In: Boldy, S. A. R. (ed.) *Permian and Triassic Rifting in Northwest Europe*. Geological Society, London, Special Publications, **91**, 7–39.

Doré, A. G., Lundin, E. R., Jensen, L. N., Birkeland, Ø., Eliassen, P. E. & Fichler, C. 1999. Principal tectonic events in the evolution of the northwest European Atlantic margin. In: Fleet, A. J. & Boldy, S. A. R. (eds) *Petroleum Geology of Northwest Europe: Proceedings of the 5th Conference*. Geological Society, London, 41–61; doi: 10.1144/0050041.

Eide, E. A. 2002. *BATLAS – Mid Norway Plate Reconstruction Atlas with Global and North Atlantic Perspectives*. Geological Survey of Norway.

El Arabi, E. H., Diez, J. B., Broutin, J. & Essamoud, R. 2006. First palynological characterization of the Middle Triassic; implications for the first Tethysian rifting phase in Morocco. *Comptes Rendus Geosciences*, **338**, 641–649.

Ellouz, N., Patriat, M., Gaulier, J. M., Bouatmani, R. & Sabounji, S. 2003. From rifting to Alpine inversion: Mesozoic and Cenozoic subsidence some Moroccan basins. *Sedimentary Geology*, **156**, 185–212.

Fabuel-Perez, I. & Redfern, J. 2009. Sedimentology of an intra-montane rift-controlled fluvial dominated succession: the Upper Triassic Oukaimeden Sandstone Formation, Central High Atlas, Morocco. *Sedimentary Geology*, **218**, 103–140.

Fabuel-Perez, I., Hodgetts, D. & Redfern, J. 2009. A new approach for outcrop characterization and geostatistical analysis of a low-sinuosity fluvial-dominated succession using digital outcrop models: Upper Triassic Oukaimeden Sandstone Formation, central High Atlas, Morocco. *America Association of Petroleum Geoscience, Bulletin*, **93**, 795–827; doi: 0.1306/02230908102.

Gould, S. 2001. *Integrated sedimentological and whole-rock trace element geochemical correlation of alluvial red-bed sequences at outcrop and in the subsurface*. Unpublished PhD thesis, University of Aberdeen.

Hafid, M. 2000. Triassic–early Liassic extensional systems and their Tertiary inversion, Essaouira Basin (Morocco). *Marine and Petroleum Geology*, **17**, 409–429.

Hofmann, A., Tourani, A. & Gaupp, R. 2000. Cyclicity of Triassic to Lower Jurassic continental red beds of the Argana Valley, Morocco: implications for palaeoclimate and basin evolution. *Palaeogeography, Palaeoclimatology, Palaeoecology*, **161**, 229–266.

Hopper, J. R. & Buck, W. R. 1996. The effect of lower crustal flow on continental extension and passive margin formation. *Journal of Geophysical Research*, **101**, 20175–20194.

Hubert, J. F. & Forlenza, M. F. 1988. Sedimentology of braided-river deposits in Upper Triassic Wolfville redbeds, southern shore of Cobequid Bay, Nova Scotia. *Developments in Geotectonics*, **22**, 231–237.

Hubert, J. F. & Hyde, M. G. 1982. Sheet-flow deposits of graded beds and mudstones on an alluvial sandflat–playa system: Upper Triassic Blomidon redbeds, St. Mary's Bay, Nova Scotia. *Sedimentology*, **29**, 457–474.

Hubert, J. F. & Mertz, K. A. 1980. Eolian dune field of Late Triassic age, Fundy Basin, Nova Scotia. *Geology*, **8**, 516–519.

Hubert, J. F. & Mertz, K. A., Jr. 1984. Eolian sandstones in Upper Triassic–Lower Jurassic red beds of the Fundy Basin, Nova Scotia. *Journal of Sedimentary Petrology*, **54**, 798–810.

Kirton, S. R. & Hitchen, K. 1987. *Timing and Style of Crustal Extension N of the Scottish Mainland*. Geological Society, London, Special Publications, **28**, 501–510; doi: 10.1144/GSL.SP.1987.028.01.32.

Klein, G. D. 1962. Triassic sedimentation, Maritime provinces, Canada. *Geological Society of America Bulletin*, **73**, 1127–1145.

Laville, E. & Petit, J.-P. 1984. Role of synsedimentary strike-slip faults in the formation of Moroccan Triassic basins. *Geology*, **12**, 424–427.

Laville, E. & Piqué, A. 1991. La distension crustale atlantique et atlasique au Maroc au debut du Mésozoïque: le rejeu des structures hercyniennes. *Bulletin Societe Géologique France*, **162**, 1161–1171.

Leeder, M. R. & Mack, G. H. 2001. Lateral erosion ('toe-cutting') of alluvial fans by axial rivers: implications for basin analysis and architecture. *Journal of the Geological Society of London*, **158**, 885–893.

Leleu, S., Hartley, A. J. & Williams, B. P. J. 2009. Large-scale alluvial architecture and correlation in a Triassic pebbly braided river system, lower Wolfville Formation (Fundy Basin, Nova Scotia, Canada). *Journal of Sedimentary Research*, **79**, 265–286.

Le Roy, P. & Pique, A. 2001. Triassic-Liassic Western Moroccan synrift basins in relation to the Central Atlantic opening. *Marine Geology*, **172**, 359–381.

Mader, N. K. 2005. *Sedimentology and sediment distribution of Upper Triassic fluvio-aeolian reservoirs on a regional scale (Central Algeria, SW Morocco, NE Canada): an integrated approach unravelling the influence of climate v. tectonics on reservoir architecture*. PhD Thesis, University of Manchester.

Manspeizer, W. 1982. Triassic–Liassic basins and climate of the Atlantic passive margins. *Geologische Rundschau*, **71**, 895–917.

Mattis, A. F. 1977. Nonmarine Triassic sedimentation, central High Atlas Mountains, Morocco. *Journal of Sedimentary Research*, **47**, 107–119.

McHone, J. G. 2000. Non-plume magmatism and rifting during the opening of the central Atlantic Ocean. *Tectonophysics*, **316**, 287–296.

McKenzie, D. 1978. Some remarks on the development of sedimentary basins. *Earth and Planetary Science Letters*, **40**, 25–32.

McKie, P. & Williams, B. P. J. 2009. Triassic fluvial dispersal across northwest European basins – a north–south divide. *Geological Journal Special Issue*, **44**, 711–741.

Medina, F. 1988. Tilted-blocks pattern, paleostress orientation and amount of extension, related to Triassic early rifting of the central Atlantic in the Amzri area (Argana Basin, Morocco). *Tectonophysics*, **148**, 229–233.

Medina, F. 1991. Superimposed extensional tectonics in the Argana Triassic formations (Morocco), related to the early rifting of the central Atlantic. *Geological Magazine*, **128**, 525–536.

Medina, F. 1995. Syn- and postrift evolution of the El Jadida-Agadir basin (Morocco): constraints for the rifting models of the central Atlantic. *Canadian Journal of Earth Sciences*, **32**, 1273–1291.

Mertz, K. A. & Hubert, J. F. 1990. Cycles of sand-flat sandstone and playa-mudstone in the Triassic–Jurassic Blomidon redbeds, Fundy rift basin, Nova Scotia: implications for tectonic and climatic controls. *Canadian Journal of Earth Sciences*, **27**, 442–451.

Morton, A. C., Herries, R. & Fanning, C. M. 2007. Correlation of Triassic sandstones in the Strathmore Field, west of Shetland, using heavy mineral provenance signatures. *In*: Mange, M. & Wright, D. K. (eds) *Heavy Minerals in Use, Developments in Sedimentology*, **58**, Elsevier, Amsterdam, 1037–1072.

Naylor, D. & Shannon, P. M. 2009. Geology of offshore Ireland. *In*: Holland, C. H. & Sanders, I. S. (eds) *The Geology of Ireland*. Dunedin Academic Press, Edinburgh, 405–460.

Naylor, D., Shannon, P. M. & Murphy, N. 1999. *Irish Rockall Basin Region – a Standard Structural Nomenclature System*. Petroleum Affairs Division, Dublin, Special Publications, **1/99**.

Olsen, P. E. 1997. Stratigraphic record of the early Mesozoic breakup of Pangea in the Laurasia–Gondwana rift system. *Annual Review of Earth and Planetary Sciences*, **25**, 337–401.

Olsen, P. E. & Schlische, R. W. 1990. Transtensional arm of the early Mesozoic Fundy rift basin; penecontemporaneous faulting and sedimentation. *Geology*, **18**, 695–698.

Olsen, P. E., Schlische, R. W. & Gore, P. J. W. 1989. *Newark Basin, Pennsylvania and New Jersey; Stratigraphy (Field Guide): Field Trips for the 28th International Geological Congress*. American Geophysical Union, Washington, DC, 69–152. Seilacher, 1967.

Olsen, P. E., Kent, D. V., Fowell, S. J., Schlische, R. W., Withjack, M. O. & LeTourneau, P. M. 2000. Implications of a comparison of the stratigraphy and depositional environments of the Argana (Morocco) and Fundy (Nova Scotia, Canada) Permian-Jurassic basins. *In*: Oujidi, M. & Et-Touhami, M. (eds) *Le Permien et le Trias du Maroc, Actes de la Premièr Réunion su Groupe Marocain du Permien et du Trias*. Hilal Impression, Oujda, 165–183.

Petit, J. P. & Beauchamp, J. 1986. Synsedimentary faulting and palaeocurrent patterns in the Triassic sandstones of the High-Atlas (Morocco). *Sedimentology*, **33**, 817–829.

Schlische, R. W. & Olsen, P. E. 1990. Quantitative filling models for continental extensional basins with applications to the early Mesozoic rifts of eastern North America. *Journal of Geology*, **98**, 135–155.

Scotese, C. R. 2002. http://www.scotese.com (PALEOMAP website).

Smoot, J. P. 1991. Sedimentary facies and depositional environments of early Mesozoic Newark Supergroup basins, eastern North America. *Palaeogeography, Palaeoclimatology, Palaeoecology*, **84**, 369–423.

Stoker, M. S., Praeg, D. et al. 2005. Neogene evolution of the Atlantic continental margin of NW Europe (Lofoten Islands to SW Ireland): anything but passive. *In*: Doré, A. G. & Vining, B. A. (eds) *Petroleum Geology: North-West Europe and Global Perspectives: Proceedings of the 6th Petroleum Geology Conference*. Geological Society, London, 1057–1076; doi: 10.1144/0061057.

Štolfová, K. & Shannon, P. M. 2009. Permo-Triassic development from Ireland to Norway: basin architecture and regional controls. *Geological Journal Special Issue*, **44**, 652–676.

Taj-Eddine, K. & Pignone, R. 2005. *L'Ourika: Haut Atlas-Haouz de Marrakech, Maroc. Un patrimoine geologique, biologique et culturel exceptionnel*. SELCA, Florence, scale 1:60.000, 1 sheet.

Tixeront, M. 1973. Lithostratigraphie et minéralisations cuprifères et uranifères stratiformes syngénétiques et familières des formations détritiques permo-triasiques du couloir d'Argana (Haut-Atlas occidental, Maroc). *Notes et Memoires du Sérvice Géologique du Maroc*, **33**, 147–177.

Torsvik, T. H., Van der Voo, R., Meert, J. G., Mosar, J. & Walderhaug, H. J. 2001. Reconstructions of the continents around the north Atlantic at about the 60th parallel. *Earth and Planetary Science Letters*, **197**, 55–69; doi: 10.1016/S0012-821X(01)00284-9.

Tourani, A., Lund, J. J., Benaouiss, N. & Gaupp, R. 2000. Stratigraphy of Triassic syn-rift deposition in Western Morocco. *Zentralblatt fuer Geologie und Palaeontologie*, **1**, 1193–1215.

Tyrrell, S., Haughton, P. D. W., Daly, J. S., Kokfelt, T. F. & Gagnevin, D. 2006. The use of the common Pb isotope composition of detrital K-feldspar grains as a provenance tool and its application to Upper Carboniferous palaeodrainage, Northern England. *Journal of Sedimentary Research*, **76**, 324–345; doi: 10.2110/jsr.2006.023.

Tyrrell, S., Haughton, P. D. W. & Daly, J. S. 2007. Drainage re-organization during break-up of Pangea revealed by in-situ Pb isotopic analysis of detrital K-feldspar. *Geology*, **35**, 971–974; doi: 10.1130/G4123A.1.

Tyrrell, S., Leleu, S., Souders, A. K., Haughton, P. D. W. & Daly, J. S. 2009. K-feldspar sand-grain provenance in the Triassic, west of Shetland: distinguishing first-cycle and recycled sediment sources? *Geological Journal Special Issue*, **44**, 692–710.

Tyrrell, S., Souders, A. K., Haughton, P. D. W., Daly, J. S. & Shannon, P. M. 2010. Sedimentology, sandstone provenance and palaeodrainage on the eastern Rockall Basin margin: evidence from the Pb isotopic composition of detrital K-feldspar. *In*: Vining, B. A. & Pickering, S. C. (eds) *Petroleum Geology: From Mature Basins to New Frontiers – Proceedings of the 7th Petroleum Geology Conference*. Geological Society, London, 937–952; doi: 10.1144/0070937.

Van Lanen, X. M. T., Hodgetts, D., Redfern, J. & Fabuel-Perez, I. 2009. Applications of digital outcrop models: two fluvial case studies from the Triassic Wolfville Fm., Canada and Oukaimeden Sandstone Fm., Morocco. *Geological Journal*, **44**, 742–760.

Wade, J. A., Brown, D. E., Traverse, A. & Fensome, R. A. 1996. The Triassic–Jurassic Fundy Basin, eastern Canada: regional setting, stratigraphy and hydrocarbon potential. *Atlantic Geology*, **32**, 189–231.

Withjack, M. O., Olsen, P. E. & Schlische, R. W. 1995. Tectonic evolution of the Fundy Rift Basin, Canada – evidence of extension and shortening during passive margin development. *Tectonics*, **14**, 390–405.

Withjack, M. O., Schlische, R. W. & Baum, M. S. 2009. Extensional development of the Fundy rift basin, southeastern Canada. *Geological Journal*, **44**, 631–651.

Sedimentology, sandstone provenance and palaeodrainage on the eastern Rockall Basin margin: evidence from the Pb isotopic composition of detrital K-feldspar

S. TYRRELL,[1] A. K. SOUDERS,[2] P. D. W. HAUGHTON,[1] J. S. DALY[1] and P. M. SHANNON[1]

[1]*Sand Provenance Centre, UCD School of Geological Sciences, University College Dublin, Belfield, Dublin 4, Ireland (e-mail: shane.tyrrell@ucd.ie)*
[2]*MicroAnalysis Facility, INCO Innovation Centre and Department of Earth Sciences, Memorial University, St John's, NL A1B 3X5, Canada*

Abstract: The Rockall Basin, west of Ireland, is a frontier area for hydrocarbon exploration, but currently the age and location of sand fairways through the basin are poorly known. A recently developed provenance approach based on *in-situ* Pb isotopic analysis of single K-feldspar grains by laser ablation multi-collector inductively coupled mass spectrometry (LA-MC-ICPMS) offers advantages over other provenance techniques, particularly when applied to regional palaeodrainage issues. K-feldspar is a relatively common, usually first-cycle framework mineral in sandstones and its origin is typically linked to that of the quartz grains in arkosic and sub-arkosic rocks. Consequently, in contrast to other techniques, the Pb-in-K-feldspar tool characterizes a significant proportion of the framework grains. New Pb isotopic data from K-feldspars in putative Permo-Triassic and Middle Jurassic sandstones in Well 12/2-1z (the Dooish gas condensate discovery) on the eastern margin of the Irish Rockall Basin are reported. These data suggest that three isotopically distinct basement sources supplied the bulk of the K-feldspar in the reservoir sandstones and that the relative contribution of these sources varied through time. Archaean and early Proterozoic rocks (including elements of the Lewisian Complex and its offshore equivalents), to the immediate east, NE and north of the eastern Rockall Margin, are the likely sources. More distal sourcelands to the NW cannot be ruled out but there was no significant input from southern sources, such as the Irish Massif. These data, together with previously published regional Pb isotopic data, highlight the important role played by old, near and far-field Archaean–Proterozoic basement highs in contributing sediment to NE Atlantic margin basins. The Irish Massif appears to have acted as a significant, but inert, drainage divide from the Permo-Triassic to the Late Jurassic and hence younger, Avalonian and Variscan sand sources appear to have been less important on the Irish Atlantic Margin.

Keywords: sedimentology, palaeodrainage, provenance, Pb isotopes, K-feldspar, Mesozoic, Rockall Basin, Dooish

Understanding the provenance of sandstones helps in the reconstruction of palaeogeography and in constraining drainage scales and sediment pathways in sedimentary basins. This helps to predict the source and distribution of sandstones, with implications for reservoir quality in the subsurface. However, there are some inherent shortcomings in provenance analysis. Provenance studies are limited by the ability to geochemically distinguish the wide range of potential sourcelands. Additionally, their specific use in the reconstruction of robust palaeodrainage models is problematic, as there are difficulties in recognizing and quantifying the extent of sedimentary recycling and mixing (Tyrrell *et al.* 2009).

Provenance techniques fall into two broad categories – those that utilize the bulk composition of the sediment or sedimentary rock and those which exploit signals in individual sand grains. Bulk compositional approaches (e.g. petrological techniques, Sm–Nd whole rock) are particularly hampered by problems such as framework grain modification, failure to recognize mixing and recycling, inadequate source characterization and non-unique signals. These issues cannot be easily overcome with a bulk compositional approach. With the increased availability of *in-situ* micro-analytical techniques, provenance studies increasingly utilize geochemical or isotopic signals in single grains of a specific mineral (e.g. Cawood *et al.* 2004; Clift *et al.* 2008). However, although these techniques enable sourcelands to be distinguished, the full transport history experienced by individual framework components often remains unconstrained. Many provenance tools use 'trace' components that may be difficult to relate to the framework grains that determine reservoir characteristics. Robust mineral phases (e.g. zircon) may have been recycled through one or more intermediary sediments or sedimentary rock. In contrast, unstable mineral grains (e.g. K-feldspar) are less likely to survive sedimentary recycling and, therefore, if present, can be regarded as first-cycle components derived directly from source. The identification of first-cycle detritus is key to resolving drainage scales and sediment transport pathways in reservoir sandstone intervals. Recent studies have demonstrated the utility of the Pb isotopic composition of detrital K-feldspar as a sand provenance tool that can be applied on a regional scale (Tyrrell *et al.* 2006, 2007; Clift *et al.* 2008). As detrital K-feldspar is unlikely to survive more than one sedimentary cycle, grains can be tracked back directly to their basement source.

In recent years Pb K-feldspar data have been collected from NW European Margin basins as part of an ongoing study of Permo-Triassic to Early Cretaceous sediment dispersal (Tyrrell *et al.* 2007, 2009). The aims of this regional work are to understand better the large-scale controls on drainage development throughout the Mesozoic, to examine the evolving drainage patterns prior to and during the breakup of Pangaea, to identify potential linkages between basins on the conjugate Atlantic margins, and to consider the implications for regional sand distribution within and across the lightly explored basins west of Ireland.

Detailed investigations of the *c.* 6 km of Mesozoic and Cenozoic stratigraphy in the Rockall Basin have been hampered by the lack of wells drilled in the basin. Until recently, stratigraphic control points were limited to a single DSDP borehole (610) in the SW of the basin, which penetrated Miocene and younger sediments, and a small number of boreholes in the NE Rockall Basin within the UK sector. However, a programme of shallow borehole drilling carried out in 1999 (Haughton et al. 2005) allowed the Mesozoic stratigraphy of the eastern basin margin to be directly constrained for the first time. The oldest definitively dated rocks penetrated by these boreholes are Late Jurassic (Early Kimmeridgian). More recently, exploration drilling on the eastern Rockall Basin margin, offshore NW Ireland (well 12/2-1 and sidetrack 12/2-1z; Fig. 1), led to the Dooish gas condensate discovery which is hosted in sandstones of Permo-Triassic and Middle Jurassic age (see below). The cores from well 12/2-1z present a unique opportunity to expand our knowledge of evolving drainage in the region. In this paper, new provenance data obtained from both the Permo-Triassic and Middle Jurassic reservoir sandstone intervals in this well are discussed in detail. In addition, these data are incorporated with recently published data from Triassic, Upper Jurassic and Cretaceous sandstones from the Slyne, Porcupine and eastern Rockall basins, respectively, in order to understand better regional sediment dispersal patterns and controls on palaeodrainage development. These impact directly on the prospectivity of the Irish Atlantic margin basins.

The Pb-in-K-feldspar provenance tool: rationale and approach

The Pb-in-K-feldspar provenance tool can provide valuable information on regional sand dispersal in sedimentary basins, especially, as in this case, where the reservoir intervals are feldspathic. The common Pb isotopic signature of K-feldspar has proved to be a powerful method of mapping gross crustal structure (e.g. Connelly & Thrane 2005) and hence potential sand source areas. Systematic regional variations in common Pb composition are seen at sub-orogenic scale (<100 km) in both active mountain belts and ancient orogens (Vitrac et al. 1981) reflecting the assembly of basement domains with different ages and U–Th–Pb fractionation histories. Therefore, broad regional patterns in Pb composition can be identified by characterizing a relatively small number of granite feldspars (DeWolf & Mezger 1994; Connelly & Thrane 2005). K-feldspar typically contains between 5 and 150 ppm Pb, and has low U and Th contents (typically <1 ppm), and hence the corrections to the common Pb signature for radioactive decay are negligible.

K-feldspar is a common mineral in many sandstones and a key constituent of arkose, and it has been shown that detrital grains retain the Pb isotopic signature of their source (Tyrrell et al. 2006). Importantly, the source of K-feldspar is very likely linked to the source of the majority of the quartz grains. Therefore determining K-feldspar provenance ultimately can constrain the source of the majority of the framework grains. With the advent of *in-situ* techniques (laser ablation inductively coupled plasma mass spectrometry, LA-ICPMS, and secondary ion mass spectrometry, SIMS), it is now possible to measure variations in the Pb composition within individual K-feldspar sand grains so that the extent of intra-grain variation and mixing of different grain populations can be isolated (Tyrrell et al. 2006). The rapid, non-destructive, *in-situ* analysis of large numbers of individual grains by laser ablation multi-collector inductively coupled plasma mass spectrometry (LA-MC-ICPMS) within an arkose represents a significant advance in provenance characterization (Tyrrell et al. 2006, 2007). The use of ion counter instead of Faraday collectors (see below) allows for the analysis of low Pb and finer grain sizes (Tyrrell et al. 2009).

Five principal Pb basement source domains have been distinguished in the North Atlantic region using published Pb data (Fig. 1; Tyrrell et al. 2007 and references therein). Spatially, these zones strike NE–SW and correspond to the basement terranes involved in the assembly of Laurentia and Rodinia, the Caledonian collision of Laurentia with Avalonia, and the Variscan Orogen. Although there are variations within each of these zones, there is a broad shift towards more radiogenic Pb values towards the SE, reflecting the history of crustal growth. Pb domain 1 comprises Archaean and Palaeoproterozoic basement and has the least radiogenic Pb. Pb domain 2 corresponds mainly to older basement reworked during late Meso- and Palaeoproterozoic. Pb domain 3 largely comprises Mesoproterozoic basement with Neoproterozoic metasedimentary rocks and younger Caledonian granites. Pb domain 4a corresponds to Avalonian basement and has a large overlap with Pb domain 4b, which corresponds with the Variscan and contains radiogenic Pb remobilized from Avalonian basement during end-Palaeozoic closure of the Rheic Ocean.

Mesozoic sediment dispersal into NE Atlantic margin basins

During the Permian and Triassic, NW Europe lay at low latitudes close to the margins of the Pangaean Supercontinent. Late Permian and Early Triassic sedimentary basins in the region are characterized by non-marine deposits, comprising local alluvial fans, extensive fluvial deposits, playa lakes and aeolian sands in successions in excess of 1 km thick (Warrington & Ivimey-Cooke 1992). The Early Triassic sandstones form reservoirs for hydrocarbons offshore west of Ireland (the Corrib gas field in the Slyne Basin), in the East Irish Sea Basin (the Morecombe gas field), in the Wessex Basin of southern England (the Wytch Farm oil field) and in the North Sea. The Lower Triassic sandy deposits (Sherwood Sandstone and equivalents) are overlain by the Middle–Upper Triassic dominantly mud-prone (but locally sand-rich) successions (Mercia Mudstone). The 'Budleighensis' palaeo-river system drained northward from the Variscan Uplands into the Wessex Basin (Audley-Charles 1970) and extended further north into the Cheshire Basin, and probably as far as the East Irish Sea Basin. However, until recently the extent of drainage systems with this geometry in basins further north and west remained unclear.

In the Slyne Basin west of Ireland (Fig. 1), Sherwood sandstone equivalent Triassic sandstones host the Corrib gas field and comprise fine- to medium-grained arkosic fluvial and alluvial sandstones with subordinate sand-flat and playa mudstone deposits (Dancer et al. 2005). Previous palaeodrainage models for these sandstones, based on borehole dipmeter logs and whole-rock geochemistry, suggested derivation from the south and SW (the Variscides) with some input from the uplifted Irish mainland (Dancer et al. 2005). Recently published Pb isotopic data for the K-feldspar component suggest a contrasting interpretation: two distinct Pb populations appear to have been sourced from Archaean and Proterozoic crust, suggesting derivation from a potentially wide area to the north and west (Tyrrell et al. 2007). Significantly, grains appear to have been dispersed across the subsequent site of the Rockall Basin, with drainage scales likely to have been in excess of 500 km. These data indicate that very few (if any) grains could have been derived from the Irish Massif or, significantly, from the Variscan Uplands. Furthermore, in basins west of Shetland, new Pb K-feldspar data (Tyrrell et al. 2009) from Middle–Upper Triassic Foula Formation sandstones (Strathmore Field) indicate local derivation from Archaean–Palaeoproterozoic rocks on the margins of the rift basin (Nagssugtoqidian Mobile Belt of East Greenland and/or the Lewisian Complex of NW Scotland), although an eastern or western source cannot be distinguished. Here, the Pb K-feldspar data is in broad agreement with U–Pb detrital zircon geochronology (Morton et al. 2007). Regionally, data from Triassic sandstones suggest that sediment dispersal was at this time controlled by uplifted Archaean and Proterozoic

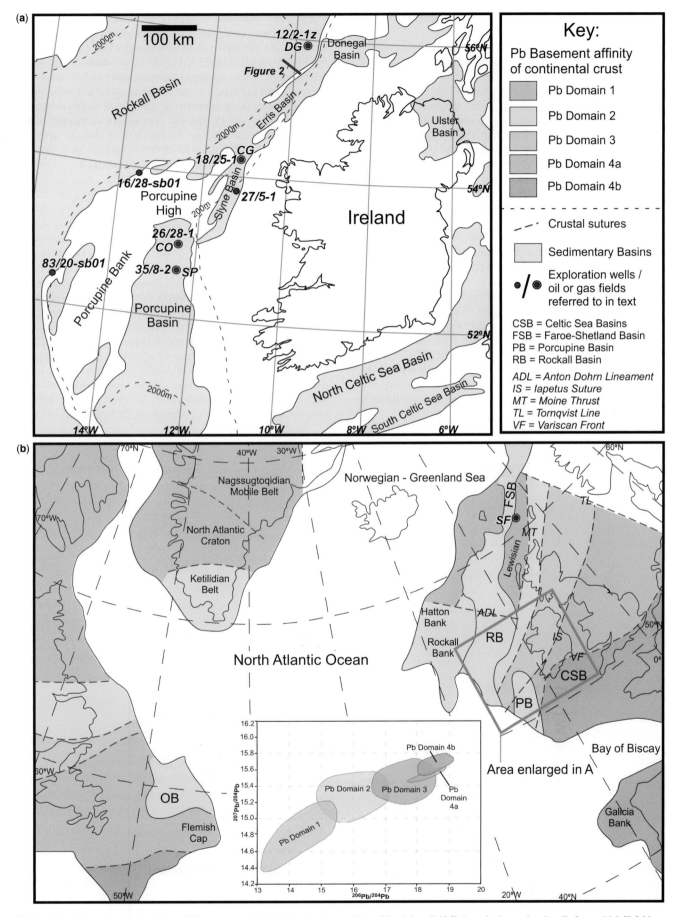

Fig. 1. Maps illustrating (a) the basins offshore western Ireland, showing the position of Dooish well 12/2-1z and other regional wells from which K-feldspar provenance data has been retrieved; and (b) the broader North Atlantic region showing the distribution of Pb domains (adapted from Tyrrell *et al.* 2007) with a plot showing Pb isotopic composition of these domains. Also shown are the approximate location of the geoseismic section in Figure 2 and the position of hydrocarbon fields and discoveries referred to in text. CG, Corrib gas field; CO, Connemara oil accumulation; DG, Dooish gas condensate discovery; SP, Spanish Point gas condensate discovery; SF, Strathmore field.

basement, with the Irish Massif acting as a drainage divide for south-derived Variscan sediment.

The Porcupine Basin, SW of the Slyne Basin (Fig. 1), includes a Jurassic sequence deposited during a phase of rifting (Croker & Shannon 1987; Naylor & Shannon 2005). In the northern part of the basin, an Upper Jurassic (Kimmeridgian–Tithonian) sequence of low-energy fluvial (meandering river) and marginal marine facies is present. These sandstones are the reservoirs for the Connemara oil accumulation in well 26/28-1 (Fig. 1). These facies are replaced southwards by marine sandstones and deep-water fans (Butterworth et al. 1999; Williams et al. 1999), which host the Spanish Point gas condensate discovery (35/8-2). Previous work (Butterworth et al. 1999) suggested north to south basin-axial flowing drainage in the Upper Jurassic section in the northern Porcupine Basin. Published Pb data from K-feldspars from the Upper Jurassic of the Northern Porcupine basin (Tyrrell et al. 2007) define two Pb isotopic populations. Both occur in sandstones from wells 26/28-1 and 35/8-2. These data have been interpreted (Tyrrell et al. 2007) to represent axial derivation from a northern source, probably from the uplifted Porcupine High, with possible transverse input from the basin margins. Dispersal distances are envisaged as being c. 100 km. These data also indicate that the sandstones in quadrants 26 and 35 share the same source and that the northern fluvial systems may have been linked with the turbiditic fans further south. Grains derived from Archaean and Palaeoproterozoic sources are very rare in this dataset, suggesting that drainage did not extend across the developing Rockall Basin at this time.

Previously published Pb data from detrital K-feldspar grains from Cretaceous sands and sandstones sampled from the western flanks of the Porcupine Bank (eastern Rockall Basin margins; 83/20-sb01 and 16/28-sb01; Fig. 1) show relatively radiogenic compositions. These sediments, sampled from shallow boreholes, are calcareous and appear to have been deposited in a high-energy shelf environment (Haughton et al. 2005). The sedimentology of these sands and the Pb isotopic composition of the K-feldspar component suggest they are locally derived directly from the Porcupine High (Tyrrell et al. 2007). Transport of K-feldspar sand grains across the Rockall Basin is not indicated.

The Dooish gas condensate discovery, Eastern Rockall Basin

The Dooish gas condensate discovery, on the eastern margins of the Rockall Basin, comprises pre-rift sandstones in a 40 km² tilted fault block sealed by post-rift sediments of Late Cretaceous age (Fig. 2). Well 12/2-1, some 125 km NW of Donegal, was drilled in 2002 in order to test the prospect. The target sandstones were presumed to be Middle–Late Jurassic in age, based on seismic jump correlation with the Slyne–Erris Basin to the south, although it was recognized that they could be as old as Permo-Triassic. Well 12/2-1 successfully demonstrated the presence of a working petroleum system in the eastern Rockall Basin, but insufficient biostratigraphic information was retrieved to date precisely the reservoir intervals. Well 12/2-1z, a deviated sidetrack of 12/2-1, was drilled in 2003 in order to determine the vertical extent of the hydrocarbon column

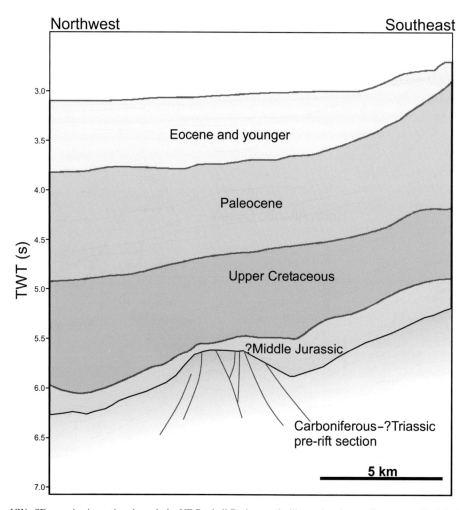

Fig. 2. Representative NW–SE geoseismic section through the NE Rockall Basin margin illustrating the pre-Cretaceous tilted fault block structure of the Dooish trend, offset to the south of well 12/2-1z. Figure is reproduced courtesy of the Petroleum Affairs Division (Ireland) and PGS Geophysical.

penetrated by the earlier well and to investigate the potential of the previously undrilled deeper section.

Well 12/2-1z terminated in Late Carboniferous interbedded sandstones and shales at a depth of c. 4.5 km (Fig. 3). Gas condensate was recorded within younger sand-prone intervals that span ages between Permian and Lower Cretaceous (see below). A total of three cores were retrieved from the well. This study utilizes new sedimentological logging of the cored intervals,

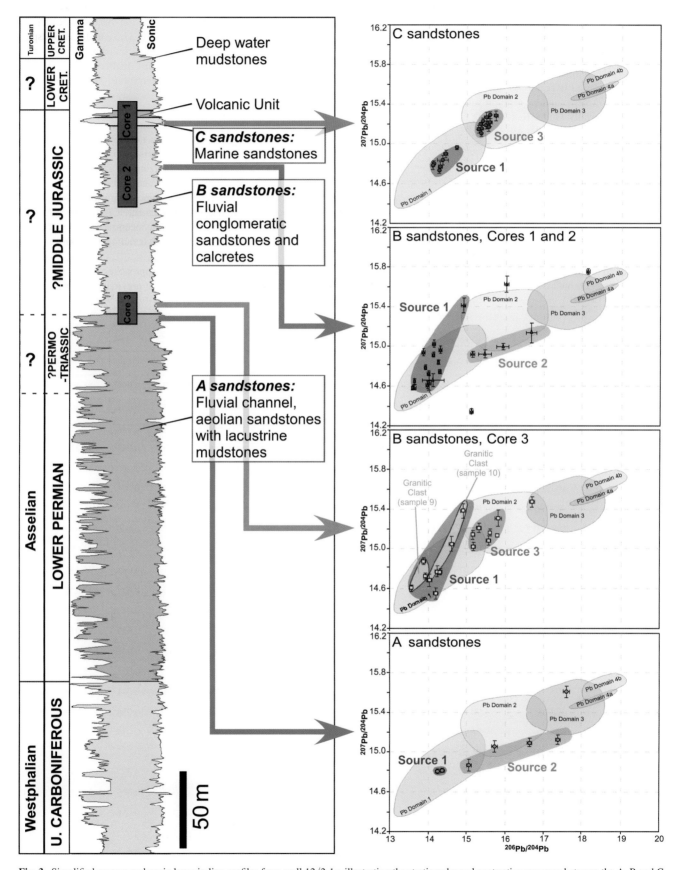

Fig. 3. Simplified gamma and sonic log wireline profiles from well 12/2-1z, illustrating the stratigraphy and contrasting response between the A, B and C sandstones (left) with $^{206}Pb/^{204}Pb$–$^{207}Pb/^{204}Pb$ lead space plots showing the variation of Pb K-feldspar populations with stratigraphy.

together with well reports and completion logs to constrain depositional and stratigraphic context for the provenance analysis.

Methodology

Core logging and sampling

The three cored intervals in well 12/2-1z, comprising a total of 90 m of section, were logged at a scale of 1:50. The cores were sampled at intervals of c. 2 m, with sand-rich beds and distinct facies intervals preferentially targeted. Wherever possible, core plug trims were used in preference to sampling the core directly. Thin sections of these samples were prepared, and their petrography assessed optically. On the basis of bulk facies and petrographic assessment, a subset of samples was selected for provenance analysis (see below).

Provenance analysis

Imaging and EMPA analysis. Thin sections of c. 300 μm thickness were prepared from which K-feldspar grains were imaged using backscatter electron (BSE) and cathodoluminescence (CL) at the Electron Microprobe Laboratory, Geowissenschaftliches Zentrum, Göttingen, Germany. The majority of the imaged grains were analysed using electron microprobe analysis at the above facility, to constrain bulk composition of the feldspar grains.

Pb analysis. The analytical techniques used are detailed in Tyrrell *et al.* (2009). Data were acquired using a Neptune MC-ICPMS linked to a Geolas 193 nm Excimer laser, at the Microanalysis Facility, InCo Innovation Centre, Memorial University, Newfoundland (MUN). Initially, a Faraday cup array was used to collect the isotopic data. However, grains with low Pb concentration (<30 ppm) produced weak Pb ion beams, resulting in unacceptably poor errors (2 SE >1%) in Pb isotopic ratios. More recently, the technique has been refined such that reliable Pb isotopic measurements can be made on grains smaller than 50 μm long axis or with low (<10 ppm) Pb concentrations. This advancement utilizes ion counters, in preference to Faraday cups, for more accurate measurement of weak Pb ion beams. This technique requires *in-situ* measurement of the Pb concentration in the K-feldspar grains prior to isotopic analysis. This enables the laser spot size to be adjusted during isotopic analysis such that sufficient, but not excessive, Pb is released. Optimizing the Pb ion beam allows for the reduction of analytical error and prevents damage to the ion counter array which can result from strong beams. Pb concentrations were measured using an Element ICPMS in the Microanalysis Facility at MUN. U and Th concentrations of individual K-feldspars, used to constrain potential radiogenic growth, were measured at the same time (Table 1).

In the case of all the analytical approaches, the data can be acquired rapidly and require only a previously imaged, thick

Table 1. Pb concentration, Pb isotopic and bulk feldspar composition data of detrital K-feldspars from sandstones in the Dooish well (12/2-1z)

Sample	Technique	Depth (m)	Grain	Or:Ab:An	Package	Pb (ppm)	^{206}Pb/^{204}Pb	2SE	^{207}Pb/^{204}Pb	2SE	^{208}Pb/^{204}Pb	2SE
1	F	4096.90	6	97.15:2.58:0.26	A	–	15.077	0.063	14.867	0.059	34.303	0.137
	F	4096.90	8	96.53:3.36:0.11	A	–	14.252	0.052	14.813	0.051	34.387	0.122
3	IC	4093.55	1	97.55:2.43:0.02	A	5	14.344	0.039	14.806	0.031	34.255	0.153
	IC	4093.55	2	97.00:3.00:0	A	7	17.360	0.062	15.116	0.043	37.214	0.257
	IC	4093.55	3	97.05:2.95:0	A	19	17.576	0.084	15.605	0.060	38.046	0.353
	IC	4093.55	5	98.31:1.69:0	A	12	16.632	0.057	15.084	0.041	36.520	0.235
	IC	4093.55	7	97.69:2.28:0.03	A	10	15.710	0.076	15.055	0.058	35.417	0.313
4	IC	4091.00	1	98.34:1.64:0.03	B1	34	14.598	0.029	15.042	0.029	34.839	0.162
	IC	4091.00	2	96.23:3.77:0	B1	11	15.314	0.048	15.203	0.047	34.981	0.250
	IC	4091.00	4	97.41:2.59:0	B1	9	15.816	0.083	15.301	0.081	35.442	0.408
	IC	4091.00	5	96.85:3.15:0	B1	8	14.197	0.057	14.540	0.057	33.653	0.297
	IC	4091.00	6	97.88:2.06:0.6	B1	17	15.163	0.042	15.132	0.043	34.880	0.228
5	F	4089.00	2	95.19:4.81:0	B1	–	14.308	0.030	14.760	0.032	34.079	0.072
6	IC	4087.50	2	87.76:11.91:0.33	B1	25	15.575	0.059	15.073	0.048	34.896	0.277
	IC	4087.50	3	83.38:15.18:1.45	B1	16	16.708	0.069	15.467	0.053	35.534	0.300
	IC	4087.50	5	96.15:3.85:0	B1	11	15.175	0.051	15.014	0.041	34.550	0.241
8	F	4084.85	1	90.85:9.15:0	B1	–	15.582	0.036	15.161	0.034	34.980	0.080
9*	IC	4084.00	2	96.44:3.02:0.54	B1	24	13.876	0.029	14.867	0.033	36.160	0.203
	IC	4084.00	3	94.63:5.37:0	B1	13	13.566	0.029	14.592	0.033	35.375	0.206
	IC	4084.00	4	95.34:4.64:0.02	B1	8	13.927	0.030	14.718	0.032	35.858	0.211
10*	IC	4083.10	1	96.21:3.79:0	B1	9	14.598	0.071	15.044	0.075	35.242	0.403
	IC	4083.10	2	95.82:4.18:0	B1	22	14.230	0.055	14.764	0.058	34.653	0.319
	IC	4083.10	3	96.91:3.09:0	B1	10	14.909	0.072	15.373	0.074	36.263	0.408
	IC	4083.10	4	96.04:3.06:0	B1	36	14.028	0.057	14.672	0.061	34.127	0.331

(*Continued*)

Table 1. *Continued*

Sample	Technique	Depth (m)	Grain	Or:Ab:An	Package	Pb (ppm)	^{206}Pb/^{204}Pb	2SE	^{207}Pb/^{204}Pb	2SE	^{208}Pb/^{204}Pb	2SE
12	F	4021.10	1	97.81:2.19:0	B2	–	14.236	0.086	13.903	0.060	32.142	0.161
	F	4021.10	6	96.44:3.50:0.05	B2	–	13.536	0.088	13.020	0.096	30.252	0.211
19	F	4006.00	1	97.03:2.97:0	B2	–	15.159	0.052	14.920	0.029	34.562	0.080
	F	4006.00	2a	96.40:3.60:0	B2	–	15.463	0.164	14.924	0.040	34.672	0.160
	F	4006.00	2b	96.40:3.60:0	B2	–	15.929	0.158	14.998	0.033	35.006	0.142
20	IC	4004.00	2	97.01:2.99:0	B2		14.140	0.043	15.026	0.036	35.102	0.274
	IC	4004.00	3	–	B2		13.871	0.042	14.942	0.036	34.568	0.265
	IC	4004.00	5	97.33:2.67:0	B2		13.977	0.042	14.611	0.035	33.878	0.264
	IC	4004.00	6	97.24:2.72:0.04	B2		14.306	0.043	14.969	0.036	34.600	0.267
21	IC	3999.70	1	97.38:2.62:0	B2		15.111	0.036	14.347	0.024	33.744	0.095
	IC	3999.70	2	96.06:3.94:0	B2		14.003	0.033	14.726	0.024	34.806	0.096
	IC	3999.70	3	97.47:2.51:0.02	B2		14.296	0.034	14.744	0.024	35.295	0.104
	IC	3999.70	4	97.64:2.36:0	B2		18.142	0.046	15.755	0.027	35.426	0.112
	IC	3999.70	5	96.23:3.77:0	B2		13.923	0.033	14.792	0.024	34.287	0.098
	IC	3999.70	6	97.36:2.64:0	B2		14.127	0.034	14.917	0.024	34.677	0.094
22	IC	3998.50	1	97.53:2.47:0	B2		16.018	0.063	15.627	0.079	35.728	0.323
	IC	3998.50	2	97.21:2.79:0	B2		14.018	0.055	14.630	0.074	33.986	0.297
	IC	3998.50	3	93.75:6.25:0	B2		14.909	0.059	15.412	0.078	36.387	0.322
23	F	3997.25	6	–	B2	–	14.254	0.027	14.840	0.018	34.465	0.044
	F	3997.25	3a	–	B2	–	13.581	0.019	14.578	0.019	33.818	0.046
	F	3997.25	3b	–	B2	–	13.626	0.052	14.585	0.023	33.811	0.066
27	F	3987.60	1	96.96:3.04:0	B2	–	16.683	0.131	15.137	0.100	35.670	0.248
	F	3987.60	5a	95.30:4.68:0.02	B2	–	14.122	0.282	14.656	0.065	34.406	0.222
	F	3987.60	5b	95.30:4.68:0.02	B2	–	13.645	0.030	14.645	0.029	34.163	0.072
30	F	3967.65	1	96.03:3.97:0	C	–	14.295	0.041	14.780	0.044	34.412	0.103
	F	3967.65	6	97.19:2.77:0.05	C	–	14.376	0.140	14.839	0.046	34.636	0.153
	F	3967.65	7	96.20:3.76:0.04	C	–	14.734	0.019	14.966	0.020	34.588	0.047
31	IC	3965.25	1	–	C	38	15.353	0.031	15.193	0.028	34.963	0.154
	IC	3965.25	2	–	C	46	15.527	0.031	15.204	0.028	35.371	0.152
	IC	3965.25	3	–	C	13	15.387	0.031	15.156	0.028	35.160	0.153
	IC	3965.25	4	–	C	25	14.290	0.031	14.733	0.030	34.156	0.161
	IC	3965.25	5	–	C	27	14.132	0.028	14.805	0.027	34.553	0.150
	IC	3965.25	6	–	C	3	15.496	0.031	15.182	0.027	35.187	0.149
	IC	3965.25	7	–	C	36	15.380	0.034	15.110	0.031	34.863	0.166
	IC	3965.25	8	–	C	11	15.593	0.036	15.292	0.034	35.350	0.181
33	IC	3963.83	1	–	C	41	14.443	0.042	14.904	0.041	34.765	0.197
	IC	3963.83	2	–	C	5	15.307	0.041	15.151	0.038	35.067	0.186
	IC	3963.83	3	–	C	38	14.121	0.037	14.783	0.036	34.484	0.174
	IC	3963.83	4	–	C	48	15.594	0.043	15.220	0.040	35.175	0.193
	IC	3963.83	5	–	C	51	15.412	0.040	15.227	0.036	35.219	0.175
	IC	3963.83	6	–	C	39	15.548	0.042	15.167	0.038	35.122	0.185
	IC	3963.83	7	–	C	14	15.509	0.058	15.265	0.056	34.956	0.265
	IC	3963.83	8	–	C	6	15.746	0.050	15.285	0.047	35.121	0.236

Note: Data have been collected using two techniques (F, Faraday collector configuration; IC, ion counter collector configuration), detailed in the text.
*Analysis of K-feldspar crystals in granitic clasts.

(c. 300 μm) polished thin section, thereby retaining the grain context within the sedimentary rock sample. The approach also uses minimal core material and can be undertaken on small core chips or plug trims. Previous workers have used multiple- or single-grain leaching techniques coupled with TIMS analysis (Hemming et al. 1998), but this approach does not allow for potential intra-grain heterogeneities to be characterized prior to and avoided during analysis. The use of ion microprobe techniques (Clift et al. 2001, 2008) allows for in-situ high spatial resolution analysis (c. 20 μm spot sizes), but produces consistently larger errors on ^{204}Pb dependent ratios than LA-MC-ICPMS analysis. In previous LA-MC-ICPMS Pb K-feldspar work (Tyrrell et al. 2006, 2007) data could only be retrieved from grains larger than c. 300 μm long axis, but the refinement of the technique has allowed for Pb analysis of very fine-grained sand and even silt grains – with similar levels of spatial resolution to those in ion microprobe analysis now possible (c. 20 μm laser spot sizes). Analytical uncertainties (2 SE on ^{206}Pb/^{204}Pb) are typically <0.5% for the Faraday configuration and typically <0.25% with the ion counter collection array. In all cases, ^{204}Pb, ^{206}Pb, ^{207}Pb and ^{208}Pb were measured. ^{202}Hg was also measured during analysis in order to correct for isobaric interference of ^{204}Hg on ^{204}Pb. Standard-sample bracketing, using NIST standard glass 612 (^{206}Pb/^{204}Pb = 17.099), was carried out in order to correct for mass bias fractionation (Tyrrell et al. 2009).

Results

Well stratigraphy

Wireline logs from 12/2-1z were used to constrain the well stratigraphy. On the basis of gamma and sonic velocity response, three post-Carboniferous sand-prone packages have been identified and targeted for further investigation (Fig. 3). These form a continuous sand-rich sequence which hosts the hydrocarbon column. The oldest of the three intervals, referred to here as the A sandstones, is c. 225 m thick and shows typically low gamma response in sandstones, intercalated with thin high gamma mudstones. The A sandstones are overlain directly by the B sandstones (c. 128 m thick). The boundary is marked by a distinct change in the character of the gamma log profile (Fig. 3), with much lower variations, but with an overall higher gamma response within the sandstones. This change is attributed to an increase in K-feldspar modal abundance. The C sandstones are very thin (c. 4.5 m) and directly overlie the B sandstones. The C sandstones have a similar low gamma response to the A sandstones.

The deepest core (core 3), from 4098 to 4078 m driller depth (d.d.) RKB, comprises a 20 m interval of well-sorted, medium-coarse-grained, red-buff sandstones with minor mudstones, and includes the boundary between the A and B sandstones (Fig. 3). Cores 1 and 2 represent a near-continuous c. 70 m section some 57 m above the top of core 3 (4021.1–3952 m d.d. RKB). These cores, corresponding to the uppermost part of the B sandstones and the entire C sandstone, comprise coarse pebbly to fine-grained sandstones, calcareous sandstones, siltstones and minor mudstones, with a distinctive volcanic horizon toward the top of the core 1.

Given the unfossiliferous nature of the cored intervals, their age has yet to be firmly established. Biostratigraphical results from released well reports suggest an early Permian (Asselian) age for the lowermost, uncored part of the A sandstones and a Middle Jurassic to Earliest Cretaceous age for the B and C sandstones. In the absence of definitive biostratigraphic data, released well reports have assigned a broad Permo-Triassic age for the uppermost and cored part of A sandstones, and B and C sandstones are assigned a tentative Middle Jurassic age. For the purposes of this study, the important issues are the compositional variations between packages and the broad correspondence between this well and others in the region at the level of Permo-Triassic compared to Jurassic–Cretaceous.

Core description

The short section of cored A sandstones (c. 4.5 m) comprises well sorted medium- to coarse-grained buff-coloured sandstones with subordinate siltstones and mudstones (Fig. 4). The sandstones are characterized by common low-angle cross-stratification and minor ripple cross-lamination. The top of the package is marked by muddier horizons, with a distinctive red mudstone bed just beneath the contact with the A package. The boundary between the A and B sandstones occurs as an erosion surface at 4092.5 m d.d. RKB. This is directly overlain by a pebbly based, medium-grained buff-coloured sandstone. B sandstones in core 3 are distinguished from the underlying A sandstones by their more abundant coarser sandstones beds, frequently with distinct erosive bases and mud rip-up clasts. The gamma log confirms an abrupt change in character at the A/B contact.

B sandstones in core 2 comprise coarse to pebbly sandstones (Fig. 4) in graded beds with common erosive bases. They tend to be cross-stratified, particularly towards bed tops (Fig. 4). The sandstones contain extraclasts (pebbles of vein quartz, granite and granitic gneiss) and rounded fragments of intraclastic calcareous sandstones. Sandstone beds frequently appear to have been internally remobilized or disrupted during deposition, and there are abundant post-depositional siderite-cemented fractures. Coarser beds become less abundant at higher levels (above c. 3991 m d.d. RKB, toward the top of core 1 and base core 2), where there are abundant calcareous bands, marls and siltstone beds. Here, the logged interval is marked by a change from buff to dominantly grey colour. There are several distinct coarsening-upward beds (0.5–1 m scale) which comprise mud-dominated bases with sandy lenses, becoming increasingly sandy toward their tops which contain mud-lenses (e.g. at depth 3976.5–3975.5 m). These heterolithic beds are frequently associated with in-situ calcareous horizons.

The boundary between the B and C sandstones occurs toward the top of core 1 (3968 m d.d. RKB) and appears conformable. It is marked by a distinct, green, pebble-based sandstone bed overlying B package muddy siltstones. The C sandstones are c. 4.5 m thick and comprise coarse- to medium-grained sandstones with distinct cm-scale pebble-rich bands. There are abundant buff, rounded, intraclastic calcareous sandstone fragments (Fig. 4), which appear to be enveloped by green remobilized sandstones. The green colour is attributed to the presence of fine-grained glauconite. The top of the C sandstone (3963.5 m d.d. RKB) is marked by a dark green, vesicular, volcanic rock.

Depositional environments

A sandstones. In cored A sandstones, the sedimentary structures, the well sorted nature of the sandstones, lack of marine indicators (e.g. fossils, glauconite) and the presence of rounded grain-types are suggestive of a non-marine aeolian or fluvially reworked aeolian facies with possible fluvial unconfined channel sandstones.

B sandstones. The coarse nature of the B sandstones in core 3 and the presence of both trough cross-bedding and low-angle cross-stratification suggest intermittent fluvial shallow channel sandstones and sandflat deposition. The calcareous layers are interpreted as calcretes, indicating a semi-arid environment. The remobilized or disrupted appearance of some beds may be due to modification during pedogenesis.

The coarse sandstone packages toward the base of core 2 are interpreted to be the deposits of a relatively high-energy alluvial/fluvial system, with channel lag deposits and sandy bar-forms interbedded with unconfined sheet sand deposits. The energy of this

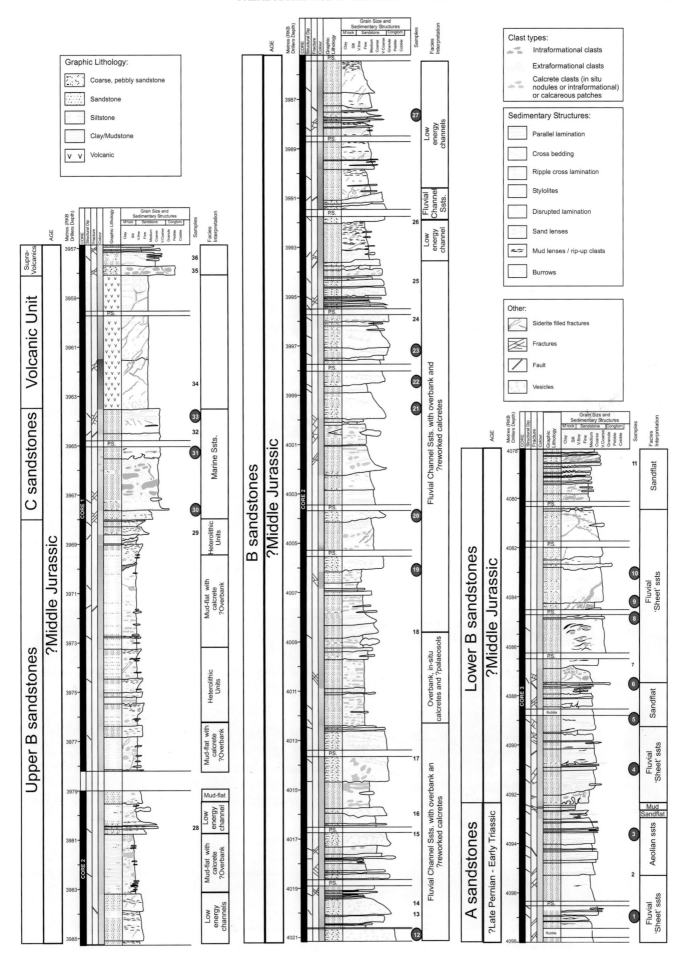

Fig. 4. Sedimentological logs of cores 1–3 in well 12/2-1z, illustrating the character and distribution of facies. Samples from which the Pb composition of K-feldspar have been analysed are highlighted with a red circle.

system wanes upward with more frequent unconfined sheet sand deposits, lower energy channels, and abundant mudstone-lenses.

The finer grained, grey, muddy lithologies observed from the upper part of core 2 appear likely to represent overbank deposits. Calcareous horizons are interpreted as palaeosols, and occur as *in-situ* continuous mature calcrete horizons, *in situ*-nodules (immature calcretes) and possibly reworked intraformational clasts. Disruption of lamination is likely attributed to soil formation processes. The heterolithic beds in core 1 and 2 suggest a tidal influence. The structures in the upper part of core 2 and lower part of core 1 are indicative of the transition from a low-energy floodplain/lacustrine to an estuarine environment with possible tidal influence.

C sandstones. The presence of glauconite in C package sandstones suggests a marine environment, possibly a shoreface setting (Fig. 4). The disrupted appearance of C package sandstones may be caused by de-watering and/or related to deposition on an unstable slope. The volcanic horizon occurring above package C is interpreted as a lava flow as its upper boundary appears blocky and weathered.

Sandstone petrography

A sandstones. Sampled sandstones are well sorted sub-arkosic arenites, with sub-angular to rounded grain shapes (Fig. 5a). Rounded grains may represent an aeolian grain population which has been reworked by fluvial action. K-feldspar is common (c. 15% modal abundance), and it has a very fresh appearance, commonly displaying perthitic textures (Fig. 5d). Traces of plagioclase, muscovite mica and detrital clay occur. Minor lithic fragments, dominantly of plutonic igneous origin, also occur. Macroporosities vary between 5 and 15%.

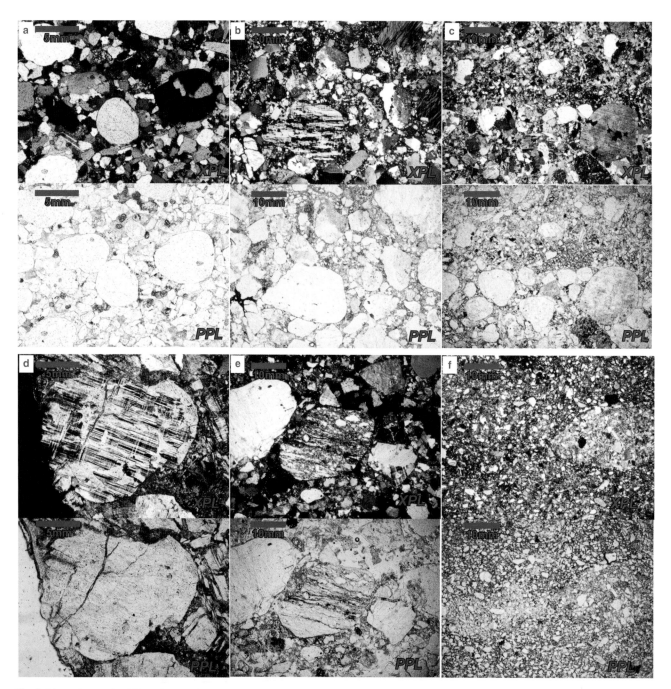

Fig. 5. Photomicrographs (XPL and PPL) of (**a**) A sandstones with typical bimodal grain distribution of rounded and finer, angular grains; (**b**) arkosic, lithic, B sandstones; (**c**) C sandstones with bimodal grain size distribution; (**d**) perthitic microcline K-feldspar grains from A sandstones; (**e**) quartzo-feldspathic gneissic lithic from B sandstones and (**f**) fine-grained calcareous sandstones associated with soil-forming processes from upper B sandstones.

B sandstones. Sampled sandstones are generally texturally more immature than A sandstones and are lithic arkoses. They comprise abundant K-feldspar (up to c. 30%) with a significantly higher lithic component than A sandstones. Lithic fragments are dominantly quartzo-feldspathic aggregates of likely plutonic igneous origin, but also including minor gneissic (Fig. 5e) and schistose fragments. Some fine-grained B sandstones are calcareous and contain micritic intraclasts. Trace accessory minerals include albite, muscovite mica and detrital clay. Macroporosities in these sandstones range between 5 and 20%. Calcite occurs as a patchy poikilotopic pore-filling cement in all the sandstones. Minor authigenic illitic clays occur, usually as a grain-rimming cement, and pyrite is present as a late phase.

C sandstones. Sampled sandstones are relatively well sorted and invariably display a bimodal grain size distribution comprising a fine angular grain component and a medium-coarse rounded grain component (Fig. 5c). The sandstones are arkosic, with K-feldspar dominant, and sublithic, with plutonic igneous fragments most common. The sandstones contain glauconite. Macroporosity is less than c. 15%.

Analytical results

Sixty-seven Pb isotopic analyses of 64 K-feldspar grains/crystals from sandstone intervals in well 12/2-1z K-feldspar were carried out and the data are presented in Table 1. Of these, 18 were obtained using a Faraday collector configuration with the remaining analysis carried out using the ion counter collector configuration (see above).

Seven grains were analysed from two samples of the A sandstones in core 3, 17 grains from six samples of B package sandstones in core 3, 21 grains from seven samples of B sandstones from cores 1 and 2, and 19 grains from three samples of C sandstones from core 1. The bulk composition of 45 of these grains was constrained using electron microprobe analysis. Results show that feldspars are dominantly orthoclase, with an average composition of Or:Ab:An = 96.08:3.83:0.08. There appears to be no link between the Pb isotopic and the bulk feldspar composition.

Overall, the feldspars largely exhibit relatively unradiogenic Pb isotopic compositions, with $^{206}Pb/^{204}Pb$ ratios varying between 13.53 and 16.70, averaging at 14.92, with only three more radiogenic grains with $^{206}Pb/^{204}Pb$ in excess of 17. Overall, the Pb data form three distinct groupings or trends. These groupings are considered to represent three distinct sources, although it should be noted that geographically different sources may have a shared basement history and thus have the same Pb isotopic composition. Source 1 corresponds to a grouping of the most unradiogenic Pb compositions plotting within Pb domain 1 (see above); source 2 corresponds to a group of data with an apparent linear trend, while source 3 forms a tight grouping of grains, more radiogenic than source 1, and plotting within Pb domain 2 (Fig. 6).

Samples 9 and 10 from B package sandstones in core 3 are granitic clasts from which several K-feldspar crystals were analysed. Pb isotopic analyses of 3 K-feldspar crystals from sample 9 exhibit an unradiogenic composition plotting within a restricted range (Fig. 6), but isotopic analyses of 4 K-feldspars from sample 10 show a significant spread in $^{207}Pb/^{204}Pb$. This illustrates that K-feldspars from the same source rock can, at least in terms of the ^{207}Pb, can have a range of Pb isotopic compositions. This

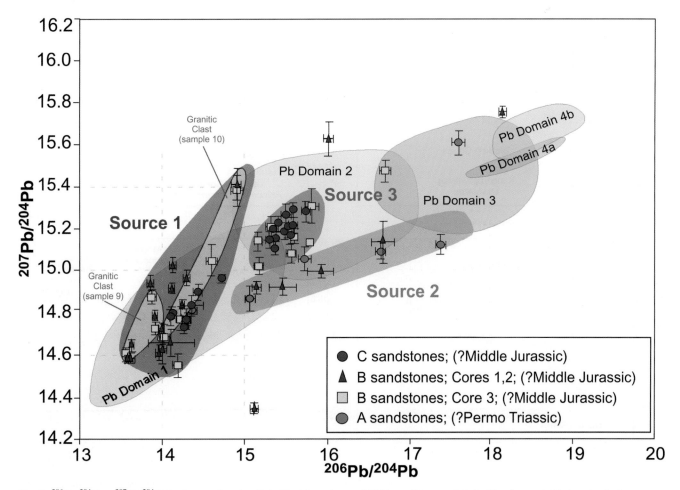

Fig. 6. $^{206}Pb/^{204}Pb$ – $^{207}Pb/^{204}Pb$ lead space plots for all the Dooish sandstones K-feldspar. Data are plotted with reference to Pb domains 1–4 (after Tyrrell *et al.* 2007, Fig. 1b). Three groupings, corresponding to sources 1–3, are highlighted.

characteristic could help explain some of the outlying Pb isotopic compositions, and could be due to small initial variations in Pb, U and Th, either between K-feldspars in the same rock or within individual zoned crystals. However, the differences appear insignificant in terms of the regional variation in Pb, for example, despite the Pb isotopic variation in sample 10, all data vary little in terms of ^{206}Pb/^{204}Pb and plot along a steep trend line.

Provenance interpretation

The majority of K-feldspars from A sandstones show a Pb isotopic composition that falls within the range defined by source 2. There is a relatively minor component (two grains) which falls within source 1 (Fig. 3). K-feldspars from lower B sandstones (i.e. those from core 3) show bimodal Pb isotopic populations, attributed to sources 1 and 3, but with no source 2 grains (Fig. 3). K-feldspars from upper B sandstones (i.e. those from cores 1 and 2) also have a bimodal Pb isotopic composition, corresponding to source 1 and 2, but with no source 3 grains. K-feldspars from C sandstones plot within sources 1 and 3, with no source 2 grains, a similar distribution pattern to that seen in lower B sandstones (Fig. 3).

In order to constrain the sourcelands, the three 'source' ranges have been plotted with Pb isotopic data for basement in the region (Fig. 7). There is good correspondence between source 1 and elements of the Lewisian Complex, in particular the offshore Stanton Banks area, but grains with this composition could also have been derived from elements of the Nagssugtoqidian of SE Greenland. Source 2 appears to correspond with the linear data arrays that are manifest in some of the blocks within the Lewisian Complex (e.g. Uist Block).

Source 3 has a more radiogenic Pb composition than source 1, but a less radiogenic composition than other sampled regional basement units (e.g. the Rhinns Complex) and offshore highs (e.g. Rockall Bank, Porcupine High). Reconstructions of the North Atlantic region suggest continuation of orogenic belts across the Atlantic (Tyrrell *et al.* 2007), yet much of the offshore basement remains poorly characterized. Source 3 may correspond to an area of as yet unsampled basement in the region. On the basis of its Pb isotopic composition, this source seems unlikely have an affinity with the oldest basement in the region (Lewisian, Nagssugtoqidian), but could feasibly represent basement of Ketilidian affinity. Ketilidian-affinity crust in the region is likely to display a broader range in Pb isotopic composition than that shown by the limited samples analysed from the Rockall Bank. Such a source could have been available at times during deposition of the Dooish sandstones, but may presently be buried. Naturally, there are problems speculating on the past geographical occurrence and areal extent of such basement but, given its Pb isotopic composition (plotting in the Pb domain 2) it could occur to the immediate east of the Dooish area or, alternatively, within the northern Rockall Bank.

Palaeogeographic reconstructions for the intervals of deposition of the Dooish sandstones and the position of likely source areas are shown in Figure 8. A sandstones have probably been derived directly from the north or from the NW, with drainage scales in the order of 300 km envisaged. More local sources, directly from

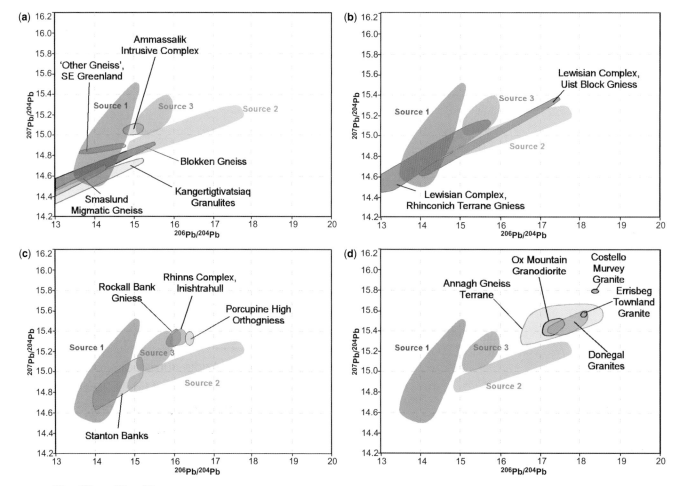

Fig. 7. ^{206}Pb/^{204}Pb – ^{207}Pb/^{204}Pb lead space plots for the three sources, plotted with Pb data from (**a**) the Nagssugtoqidian Mobile Belt of SE Greenland; (**b**) Lewisian Complex Rocks from NW Scotland; (**c**) offshore crystalline basement blocks from the NW Atlantic and (**d**) crystalline basement from onshore western Ireland. Pb basement data from Tyrrell *et al.* (2007, Fig. 1b and references therein).

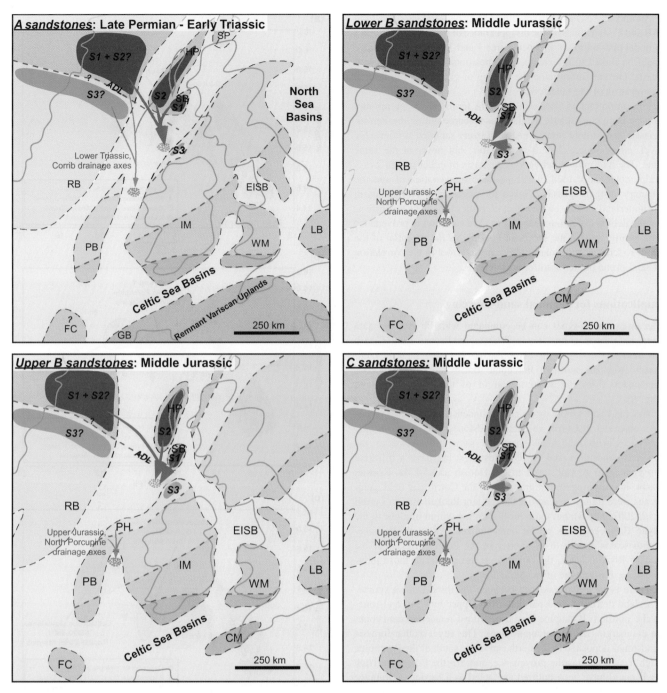

Fig. 8. Palaeogeographic reconstructions of the circum North Atlantic region during the Late Permian–Early Triassic and Middle Jurassic (after Williams et al. 1999; Ziegler 1999; Torsvik et al. 2001; Eide 2002; Scotese 2002), showing the palaeodrainage directions (arrows) during the deposition of the Dooish sandstones. The land masses are colour coded so as to reflect their bulk Pb signature (Tyrrell et al. 2007) with darker colours used to highlight areas that are potential sources. Also shown, for reference, are the orientation of drainage axes supplying the Corrib gas field sandstones (Lower Triassic, Slyne Basin) and Upper Jurassic of the northern Porcupine Basin (from Tyrrell et al. 2007). ADL, Anton Dohrn Lineament; CM, Cornubia Massif; EISB, East Irish Sea Basin; FC, Flemish Cap; GB, Galicia Bank; HP, Hebridean Platform; IM, Irish Massif; LH, Labadie Bank High; LB, London–Brabant Massif; PB, Porcupine Bank; PH, Porcupine High; RB, Rockall Bank; SP, Shetland Platform; WM, Welsh Massif.

the east, NE and north are suggested for the lower B sandstones. Data from upper B sandstones suggest dominantly north and NE derivation, while C package sandstones could have been derived from more local sources to the north, NE and east.

There is an abrupt change in source area between A sandstones and overlying lower B sandstones as sampled from core 3, manifesting as a switching off of source 2 and an influx on source 1 and 3 grains. This change of source area corresponds to an abrupt change in the wireline character at the boundary of these units, and is consistent with the apparent prolonged period of non-deposition between the Permo-Triassic and Middle Jurassic. During this hiatus it is possible the palaeo-erosion surface and hinterland geological composition changed considerably.

Upper B sandstones (from cores 1 and 2) have a different provenance character from the lower B sandstones (core 3), with the loss of source 3 grains in the upper B sandstones and the presence of a source 2 population, while source 1 continues to be important. This may reflect a more gradual evolution of drainage/erosion in the hinterland than that suggested by the abrupt change between the A and B sandstones. However, analysis of the intervening B sandstones between the top of core 3 and base of core 2 is required in order to test this hypothesis.

Another rapid change in sourcing occurs between the deposition of B and C sandstones, to the disappearance of source 2 grains and a bimodal distribution of distinct source 1 and 3 grains. This change could be linked to the change in depositional environment, with marine currents influencing the routing of grains and the detrital composition of the sandstones.

It is important to note that, in bulk terms, source 1 appears to have been consistently available as a source during the Permo-Triassic and Middle Jurassic. Also, at every sampled stratigraphic level, either source 2 or 3 is available, but never at the same time. The intermittent delivery to the basin of either source 2 or 3 grains could be caused by the periodic rejuvenation of specific tributary systems, possibly linked to variations in uplift rates in the hinterland, or may be the effect of subtle climatic factors affecting the delivery of grains to the basin. The variation of provenance with stratigraphy in the succession highlights the potential of the Pb-in-K-feldspar provenance technique as a tool in the correlation of sand-prone but unfossiliferous intervals.

Implications for regional sand-sourcing

The three source-types can be compared with Pb isotopic data from K-feldspar grains in (1) Triassic sandstones from Corrib gas field sandstones in the Slyne Basin, (2) Upper Jurassic sandstones from the northern Porcupine Basin and (3) Cretaceous sandstones from the western margin of the Porcupine Bank (Fig. 9).

Sources 1 and 3 show strong correspondence with the two groupings observed in Corrib gas field sandstones. This suggests that Triassic Corrib sandstones share the same provenance as the lower B sandstones and the C sandstones, but have a different provenance to A sandstones in the Dooish area of the Rockall Basin. It is envisaged that grains from the Corrib gas field sandstones were transported across the nascent Rockall Basin (Tyrrell et al. 2007; Fig. 8), but more local sourcing within or close to the Slyne Basin could as yet be unrecognized. Although drainage-length scales appear to be shorter in the Permo-Triassic of the eastern Rockall basin, the orientations are similar to those interpreted for the Triassic of the Slyne Basin.

There is no correspondence between the three Dooish source-types and detrital K-feldspar data from Upper Jurassic sandstones in the northern Porcupine Basin, or from Cretaceous sandstones on the margins of the Porcupine Bank. This suggests that Jurassic sandstones deposited to the north and to the south of the Porcupine High have a contrasting provenance, and that the Porcupine High itself could have been uplifted and acted as a barrier to drainage during the Middle to Late Jurassic (Fig. 8). It appears that local sand sourcing dominated during the Middle and Late Jurassic in both the Rockall and Porcupine basins, with drainage scales in both basins interpreted to be c. 100 km.

Additionally, in the Upper Jurassic of the Porcupine Basin, recycled Carboniferous sedimentary rocks are suggested to have been an important source (Robinson & Canham 2001). These seem to play a far less important role in the Dooish sandstones, where petrography clearly shows crystalline plutonic igneous and metamorphic basement as the predominant source.

In overall terms, the pattern of detrital Pb isotopic compositions of K-feldspars in Permo-Triassic to Cretaceous sandstones in basins on the NW European margins highlights the important role played by local (often poorly characterized) Archaean and Proterozoic basement blocks as sand sources and drainage divides and, consequently, in controlling reservoir sandstone distribution in the region (Fig. 8). Perhaps unsurprisingly, grains with an Archaean and Palaeoproterozoic signature (Pb Domain 1) are more abundant in basins further north, which is also consistent with provenance data from Triassic sandstones in the Faroe–Shetland

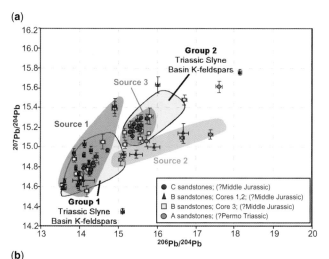

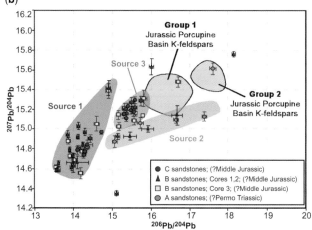

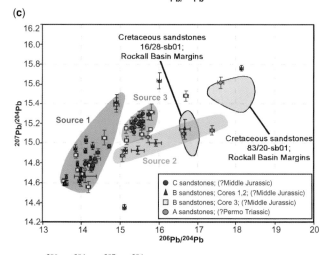

Fig. 9. $^{206}Pb/^{204}Pb$–$^{207}Pb/^{204}Pb$ lead space plots for (**a**) Dooish K-feldspars compared with populations from Triassic sandstones in the Slyne Basin, including those from the Corrib gas field; (**b**) Dooish K-feldspars compared with populations from Upper Jurassic sandstones in the northern Porcupine Basin (the Connemara field and the Spanish Point gas condensate); and (**c**) Dooish K-feldspars compared with populations from Cretaceous sandstones from shallow boreholes on the eastern Rockall Basin margin. Previously published data is from Tyrrell et al. (2007).

region (Tyrrell et al. 2009), but become increasingly less so in basins further south.

Conclusions

(1) The Pb isotopic composition of detrital K-feldspar offers a powerful method of determining the provenance of arkosic

and sub-arkosic sandstones, and prospective source areas are also relatively easily characterized. When applied to Mesozoic sandstones in basins west of Ireland, the technique can place constraints on the scale and pattern of palaeodrainage and constrain the relative contribution of different source areas.

(2) The bulk of K-feldspars in Permo-Triassic and Middle Jurassic sandstones from the eastern Rockall Basin (Dooish; 12/2-1z) appear to have been derived from three isotopically distinct source areas. Elements of the Lewisian Complex, similar to the Stanton Bank, formed an important and persistent source for all the sampled sandstones (source 1), but other Lewisian Complex rocks (such as the Uist Block and its equivalents; source 2) and a currently unidentified, likely Ketilidian, source (source 3) variably provided additional sediment during deposition of the these sandstones. Intriguingly, sources 2 and 3 do not appear to be available at the same time, which may suggest that the grains are being delivered through different tributary systems which are active at different times. There are very few grains that could have been derived from southern sources, such as the Irish Massif and the Variscan Uplands. Although sources from the far north and west (e.g. SE Greenland) cannot be entirely ruled out, there are more proximal terranes that can supply feldspar with the requisite Pb isotopic composition. Similarly, it is difficult to clearly ascertain if grains were transported across the nascent Rockall Basin, although the overall textural immaturity of the sampled sandstones supports local sourcing. The simplest model for palaeodrainage reconcilable with all the Pb data suggests derivation directions from the NW, north, NE or directly from the eastern margin of the basin. North-derived dispersal directions appear to have persisted during deposition of the Permo-Triassic and the Middle Jurassic sequences, but more northeastern and, potentially local, eastern sources appear to have supplied detritus intermittently.

(3) There are difficulties constraining transport distances both during the Permo-Triassic and the Jurassic in the Dooish area, given the uncertainty associated with the nature and palaeo-position of all the potential sources, but palaeogeographic reconstructions suggest they are unlikely to be in excess of 300 km. It is probable that local sources (transport <100 km) played an important role, especially during the Jurassic. The Irish Massif appears to have acted consistently as a significant and persistent drainage divide or barrier during the Permo-Triassic and into the Late Jurassic. Although this massif prevented the transport of Variscan sand grains into the basins west of Ireland, it does not appear to have supplied significant sediment itself.

(4) Determining the provenance of sandstones provides a first-order control on the distribution of reservoir intervals. The directions of sediment transport derived from the provenance data suggest that more distal equivalents of the sedimentary rocks in 12/2-1z may have been deposited to the south and SW, implying that sandstones with similar bulk characteristics may occur at depth within the Rockall Basin.

(5) The overall dispersal pattern of K-feldspar into basins west of Ireland during the Mesozoic suggests that Archaean and Proterozoic basement blocks played a crucial role in controlling drainage during this prolonged period. It is therefore important to understand the nature, palaeo-position and late uplift history of these crustal blocks.

This work is based on research funded by the Department of Communications, Energy and Natural Resources under the National Geoscience Programme 2007–2013 (Griffiths Geoscience Awards) awarded to the Marine and Petroleum Geology Research Group at UCD, and by a Science Foundation Ireland Research Frontiers Programme grant (RFP06/GEO029) awarded to PDWH. ST acknowledges P. Sylvester and M. Tubrett (Microanalysis Facility, Memorial University, Newfoundland, Canada) for assistance with ICPMS and A. Kronz (Electron Microprobe Laboratory, Geowissenschaftliches Zentrum, Göttingen, Germany). ST thanks Shell International and the Petroleum Affairs Division (Ireland) for facilitating access to core and providing well data. David MacDonald and Tony Doré are thanked for detailed reviews which improved the manuscript.

References

Audley-Charles, M. G. 1970. Triassic palaeogeography of the British Isles. *Quarterly Journal of the Geological Society, London*, **126**, 49–89.

Butterworth, P., Holba, A., Hertig, S., Hughes, W. & Atkinson, C. 1999. Jurassic non-marine source rocks and oils of the Porcupine Basin and other North Atlantic margin basins. *In*: Fleet, A. J. & Boldy, S. A. R. (eds) *Petroleum Geology of Northwest Europe: Proceedings of the 5th Conference*. Geological Society, London, 471–486; doi: 10.1144/0050471.

Cawood, P. A., Nemchin, A. A., Strachan, R. A., Kinny, P. D. & Loewy, S. 2004. Laurentian provenance and an intracratonic tectonic setting for the Moine Supergroup, Scotland, constrained by detrital zircons from the Loch Eil and Glen Urquhart successions. *Journal of the Geological Society, London*, **161**, 861–874.

Connelly, J. N. & Thrane, K. 2005. Rapid determination of Pb isotopes to define Precambrian allochthonous domains: an example from West Greenland. *Geology*, **33**, 953–956.

Clift, P. D., Shimizu, N., Layne, G. D. & Blusztajn, J. 2001. Tracing patterns of erosion and drainage in the Paleogene Himalaya through ion probe Pb isotope analysis of detrital K-feldspars in the Indus Molasse, India. *Earth and Planetary Science Letters*, **188**, 475–491.

Clift, P. D., Van Long, H. *et al.* 2008. Evolving east Asian river systems reconstructed by trace element and Pb and Nd isotope variations in modern and ancient Red River-Song Hong sediments. *Geochemistry Geophysics Geosystems*, **9**, Q04039.

Croker, P. F. & Shannon, P. M. 1987. The evolution and hydrocarbon prospectivity of the Porcupine Basin, Offshore Ireland. *In*: Brooks, J. & Glennie, K. W. (eds) *Petroleum Geology of North West Europe*. Graham and Trotman, London, 633–642.

Dancer, P. N., Kenyon-Roberts, S. M., Downey, J. W., Baillie, J. M., Meadows, N. S. & Maguire, K. 2005. The Corrib gas field, offshore west of Ireland. *In*: Doré, A. G. & Vining, B. A. (eds) *Petroleum Geology: North-West Europe and Global Perspectives: Proceedings of the 6th Petroleum Geology Conference*. Geological Society, London, 1035–1046; doi: 10.1144/0061035.

DeWolf, C. P. & Mezger, K. 1994. Lead isotope analyses of leached feldspars: constraints on the early crustal history of the Grenville Orogen. *Geochimica et Cosmochimica Acta*, **58**, 5537–5550.

Eide, E. A. 2002. *BATLAS – Mid Norway Plate Reconstruction Atlas with Global and North Atlantic Perspectives*. Geological Survey of Norway.

Haughton, P., Praeg, D. *et al.* 2005. First results from shallow stratigraphic boreholes on the eastern flank of the Rockall Basin, offshore western Ireland. *In*: Doré, A. G. & Vining, B. A. (eds) *Petroleum Geology: North-West Europe and Global Perspectives: Proceedings of the 6th Petroleum Geology Conference*. Geological Society, London, 1077–1094; doi: 10.1144/0061077.

Hemming, S. R., Broecker, W. S. *et al.* 1998. Provenance of Heinrich layers in core V28–82, northeastern Atlantic: $^{40}Ar/^{39}Ar$ ages of ice-rafted hornblende, Pb isotopes in feldspar grains, and Nd–Sr–Pb isotopes in the fine sediment fraction. *Earth and Planetary Science Letters*, **164**, 317–333.

Moore, T. E., Hemming, S. R. & Sharp, W. D. 1997. Provenance of the Carboniferous Nuka Formation, Brooks Range Alaska: a multicomponent isotope provenance study with implications for age of cryptic crystalline basement. *In*: Dumoulin, J. A. & Gray, J. E. (eds) *Geologic Studies in Alaska by the U.S Geological Survey, 1995*. US Geological Survey, Professional Papers, **1574**, 173–194.

Morton, A. C., Herries, R. & Fanning, C. M. 2007. Correlation of Triassic sandstones in the Strathmore Field, west of Shetland, using heavy mineral provenance signatures. *In*: Mange, M. & Wright, D. K.

(eds) *Heavy Minerals In Use*. Developments in Sedimentology, Elsevier, Amsterdam, **58**, 1037–1072.

Naylor, D. & Shannon, P. M. 2005. The structural framework of the Irish Atlantic Margin. *In*: Doré, A. G. & Vining, B. A. (eds) *Petroleum Geology: North-West Europe and Global Perspectives: Proceedings of the 6th Petroleum Geology Conference*. Geological Society, London, 1009–1021; doi: 10.1144/0061009.

Robinson, A. J. & Canham, A. C. 2001. Reservoir characteristics of the Upper Jurassic sequence in the 35/8-2 discovery, Porcupine Basin. *In*: Shannon, P. M., Haughton, P. D. W. & Corcoran, D. V. (eds) *The Petroleum Exploration of Ireland's Offshore Basins*. Geological Society, London, Special Publications, **188**, 301–321.

Scotese, C. R. 2002. PALEOMAP website; http://www.scotese.com.

Torsvik, T. H., Van der Voo, R., Meert, J. G., Mosar, J. & Walderhaug, H. J. 2001. Reconstructions of the continents around the north Atlantic at about the 60th parallel. *Earth and Planetary Science Letters*, **197**, 55–69.

Tyrrell, S., Haughton, P. D. W., Daly, J. S., Kokfelt, T. F. & Gagnevin, D. 2006. The use of the common Pb isotope composition of detrital K-feldspar grains as a provenance tool and its application to Upper Carboniferous palaeodrainage, Northern England. *Journal of Sedimentary Research*, **76**, 324–345.

Tyrrell, S., Haughton, P. D. W. & Daly, J. S. 2007. Drainage re-organization during break-up of Pangea revealed by *in-situ* Pb isotopic analysis of detrital K-feldspar. *Geology*, **35**, 971–974.

Tyrrell, S., Leleu, S., Souders, A. K., Haughton, P. D. W. & Daly, J. S. 2009. K-feldspar sand-grain provenance in the Triassic, west of Shetland: a first-cycle sedimentary fingerprint? *Geological Journal*, **44**, 692–710.

Vitrac, A. M., Albarede, F. & Allégre, C. J. 1981. Lead isotopic composition of Hercynian granite K-feldspars constrains continental genesis. *Nature*, **291**, 460–464.

Warrington, G. & Ivimey-Cooke, H. C. 1992. Triassic. *In*: Cope, J. C. W., Ingham, J. K. & Rawson, P. F. (eds) *Atlas of Palaeogeography and Lithofacies*. Geological Society, London, Memoirs, **13**, 97–106.

Williams, B. P. J., Shannon, P. M. & Sinclair, I. K. 1999. Comparative Jurassic and Cretaceous tectono-stratigraphy and reservoir development in the Jeanne d'Arc and Porcupine basins. *In*: Fleet, A. J. & Boldy, S. A. R. (eds) *Petroleum Geology of Northwest Europe: Proceedings of the 5th Petroleum Geology Conference*. Geological Society, London, 487–499; doi: 10.1144/0050487.

Ziegler, P. A. 1991. *Geological Atlas of Western and Central Europe*. Geological Society, London/Shell International Petroleum Maatschappij B.V.

Cretaceous revisited: exploring the syn-rift play of the Faroe–Shetland Basin

M. LARSEN, T. RASMUSSEN and L. HJELM

DONG Energy, E&P, Agern Allé 24-26, DK-2970 Hørsholm, Denmark (e-mail: micla@dongenergy.dk)

Abstract: Improved seismic imaging of the deep structures in the Faroe–Shetland Basin has revealed a complex Mesozoic rift system with shifting block polarity along the West Shetland Platform. Newly acquired seismic data has led to the focus of hydrocarbon exploration on structurally defined Mesozoic traps and has re-opened exploration in the deeper stratigraphic sections beyond the stratigraphic, Paleocene deep-water play. In the study area, rift geometry changes from symmetrical (south) to asymmetrical (north), the latter creating a large-scale seaward-dipping flexure. The polarity shift may link up with deep-seated basement structures (rift-oblique lineaments) segmenting the rift zone. The initial rifting along the West Shetland Platform strongly influenced the depositional setting and lateral distribution of the Lower Cretaceous sediments. During rift initiation in the Early Cretaceous faulting took place along numerous small faults, which eventually linked up, creating a set of basin master faults in the main rift phase. Sand derived from rivers and longshore currents on the West Shetland Platform was transported down the axis of relay ramps and filled the juvenile rift basins. These sediments formed thick onlapping wedges, reflecting the continuous creation of accommodation space and the overall transgressive nature of the syn-rift and early post-rift succession. In this period, rift basins were elongated, which to some extent hindered cross-rift transport of coarse material except at relay ramps and rift-oblique lineaments. As fault movements ceased, the rift topography was levelled out and allowed gravity-driven systems to reach further into the basin, overstepping former cross-rift barriers. Lower Cretaceous syn-rift sediments are well exposed at several localities along the margins of the northern North Atlantic including onshore NE Greenland. The close analogy to the syn-rift structural setting imaged in the west of Shetland seismic succession may provide valuable information on structurally controlled depositional systems, reservoir architecture and properties.

Keywords: Faroe–Shetland Basin, NE Greenland, petroleum geology, syn-rift, Cretaceous, sediments, extensional faulting, relay ramp

Faroe–Shetland exploration history

The presence of deep sedimentary basins on the continental slope west of the Shetland Islands was first recognized by academic and government-sponsored geophysical surveys in the early 1960s and was confirmed by the first petroleum industry deep reflection seismic surveys in the late 1960s. Exploration drilling began in 1972 and more than 30 wells were drilled during the following decade. Early exploration activity primarily targeted potential Mesozoic reservoir sandstones associated with tilted fault block structures, using the highly successful oil plays of the Viking Graben as analogues. Oil and gas shows were encountered in several wells, and in 1977 the giant Clair Field was discovered (Fig. 1). Oil in place was estimated at up to 4.0×10^9 barrels (Lamers & Carmichael 1999), but was retained in poor-quality reservoirs comprising fractured basement and Devonian–Carboniferous sandstones.

Exploration continued at a slower pace through the 1980s, chasing the Clair play concept, but without significant commercial success. Oil was discovered in 1990 in Triassic sandstones (Strathmore) and in 1991 in Jurassic sandstones (Solan), but volumes were relatively small and the fields were considered sub-economic. A number of wells encountered gas in the Paleocene and Cretaceous reservoirs (e.g. Laggan, Laxford, Torridon and Victory), but again field sizes were only small to moderate, and gas development was not considered economically feasible at the time.

The discovery of large volumes of oil in good quality Paleocene sandstone reservoirs in the Foinaven, Schiehallion and nearby fields in the early 1990s represented an important breakthrough in the west of Shetlands area and the new fields were on stream by 1998. Further significant oil and gas discoveries (Tobermory, Cambo, Rosebank, Laggan/Tormore and Glenlivet) have been made in the recent years in the Faroe–Shetland Basin, supporting the future development of a gas infrastructure and the possibilities for commercialization of minor gas discoveries as well.

The initial exploration was focused on the Palaeozoic and Mesozoic pre- and syn-rift plays, whereas the late exploration successes are mostly related to combined structural and stratigraphic traps in the Palaeogene (Lamers & Carmichael 1999). Numerous exploration wells, targeting the deeper structural play, were drilled on the crest of tilted fault blocks identified on poor-quality 2D seismic and encountered fractured crystalline basement, often with oil shows. However, the continued search for the Clair play concept to a large extent neglected the downdip opportunities of a Cretaceous syn-rift play. In this study, we discuss the Cretaceous syn-rift play, using high-quality reprocessed 2D seismic data and play concepts developed on the basis of onshore analogues. The depositional systems are linked to the structural evolution along the West Shetland Platform and it is demonstrated how improved seismic data, structural analysis and extensive use of analogue data may re-open the syn-rift play of the Faroe–Shetland Basin.

Methods

Reflection seismic data

Vintage 2D seismic data in the study area (Fig. 1) are from the 1970s and 1990s and are generally of poor quality. The data are characterized by multiple energy and noise, which heavily mask primary energy and hamper interpretation of the geology and structures below the base Palaeogene horizon (Fig. 2). Most of these vintage data are thus of limited use in imaging of the Mesozoic rift basins along the western edge of the West Shetland Platform.

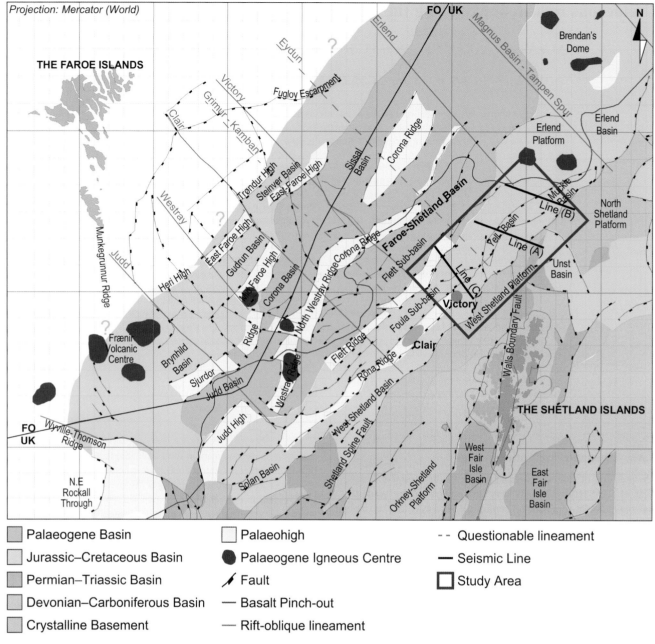

Fig. 1. Generalized structural map of the Faroe–Shetland region with the study area outline. Major Cretaceous depocentres are present along the West Shetland Platform and potentially in the sub-basalt area SE of the Faroes, whereas the Palaeogene depocentres were shifted further towards the rift axis. A series of NE–SW striking half-grabens formed along the margins of the rift in the Early Cretaceous. The sub-basins and intervening ridges (e.g. Rona Ridge) form the key structural elements in the syn-rift play of the basin.

A new 2D data set (long offset) was acquired by TGS-NOPEC in 2005 and 2006 (NSR05 and -06) in a regular grid with approximately 5 km spacing suitable for regional screening and imaging of the deeper targets. These data were successfully reprocessed in 2006 with focus on careful velocity picking and improved imaging of the fault cuts.

The seismic examples (Fig. 2) show the overall improvement from old v. new 2D seismic acquisition, modern processing techniques and the gain in resolution that can be obtained through careful reprocessing of 2D data. The reprocessed data give confidence in the interpretation of the platform–transition–outer basin trend and allow identification of deeper, tilted fault blocks. This may form the key to unlocking the remaining exploration potential along the West Shetland Platform. Regional gravimetric maps and filtered gravity data from the seismic surveys have been utilized for basement trend analysis and act as support for evaluation of the large-scale structural setting, including Caledonian and older structural grains (Kimbell et al. 2005).

Wells

The Mesozoic–Palaeozoic structural play of the Faroe–Shetland Basin was targeted during the early exploration phases and most wells of interest are thus released. Control of sedimentary facies and depositional models along the margin of the West Shetland Platform has been achieved from well analysis and onshore analogue studies. Although most of the older wells are situated at the crest of the tilted fault blocks, drilling updip from the syn-rift succession, valuable information on the syn-rift sediments was obtained from the Victory gas discovery situated downdip from the crest of a hanging wall block, east of the Rona Ridge (Goodchild et al. 1999; Fig. 1).

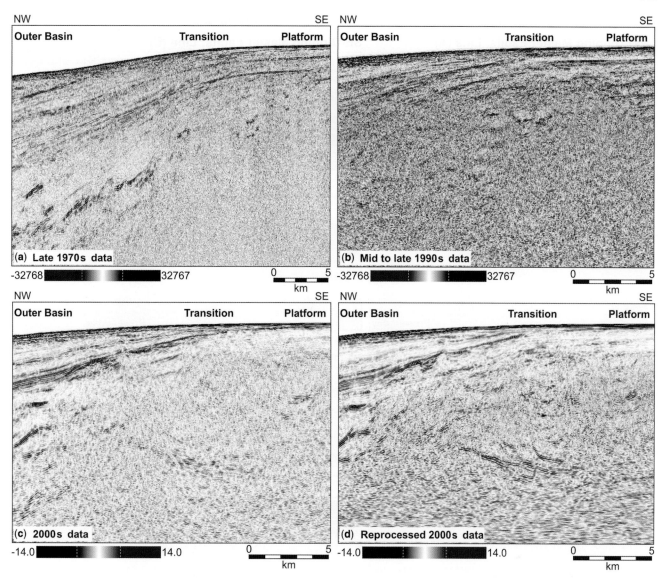

Fig. 2. Four generations of seismic acquisitions/processing showing approximately the same section across the 'Yell' Basin at the western edge of the West Shetland Platform. The differences in the seismic character are related to the acquisition parameters, especially cable length (TGS long offset data) and the processing and re-processing techniques, which have been improved over recent decades (e.g. careful velocity picking and multiple removal).

The key wells used in the present study are 206/3-01, 206/4-01, 207/1-02, 207/1-03, 207/1a-05, 208/24-1A, 208/26-01 and 209/12-01. The data comprise biostratigraphic analysis, petrophysical evaluation, and description and interpretation of relevant core.

Onshore analogues

Lower Cretaceous sandstone-dominated successions are widespread in the northern North Atlantic and excellent outcrop examples occur in southern England (Ruffell & Wach 1998), Svalbard (Gjelberg & Steel 1995) and in East Greenland, for example, Kangerlussuaq (Larsen *et al.* 1999) and Hold with Hope (Kelly *et al.* 1998; Larsen *et al.* 2001). In this study, we compare structural observations and depositional models derived from the seismic interpretation of the Cretaceous rift basins west of Shetlands with a Barremian–Aptian syn-rift succession in northern Hold with Hope, NE Greenland (Larsen *et al.* 2001). It consists of a sandstone-dominated, coarse clastic wedge onlapping a series of rotated Triassic–Jurassic fault blocks. The Cretaceous succession is up to 300 m thick and exposed in coastal cliffs for approximately 18 km.

Structural observations

The West Shetland Platform consists of a deformed high with a core of crystalline basement and pre-Jurassic sediments. The NW margin of the high is defined by a few prominent, NE–SW-striking master faults. These faults are characterized by large offsets and bound a series of asymmetrical half-grabens, approximately 10 km wide. The characteristic expression of rift structures in the Faroe–Shetland Basin is illustrated by the NE–SW-oriented Shetland Spine Fault, with a series of westward downstepping basement highs and basins (Fig. 1). Important structural trends inherited from the basement fabric are a general NE–SW 'Caledonian' trend although with a marked north–south component in the study area (probably controlled by the Walls Boundary Fault Zone) and a possibly older NW–SE 'Tornquist' trend (Fig. 1). The NW–SE structural trend is also mimicked in the transfer zones (Rump *et al.* 1993) or rift-oblique lineaments (see discussion by Moy & Imber 2009), which may correspond to zones of major changes in structural style along the West Shetland Platform. The basement fabric has frequently been reactivated and appears to have exerted important influence on the Cretaceous rifting and segmentation of the West Shetland

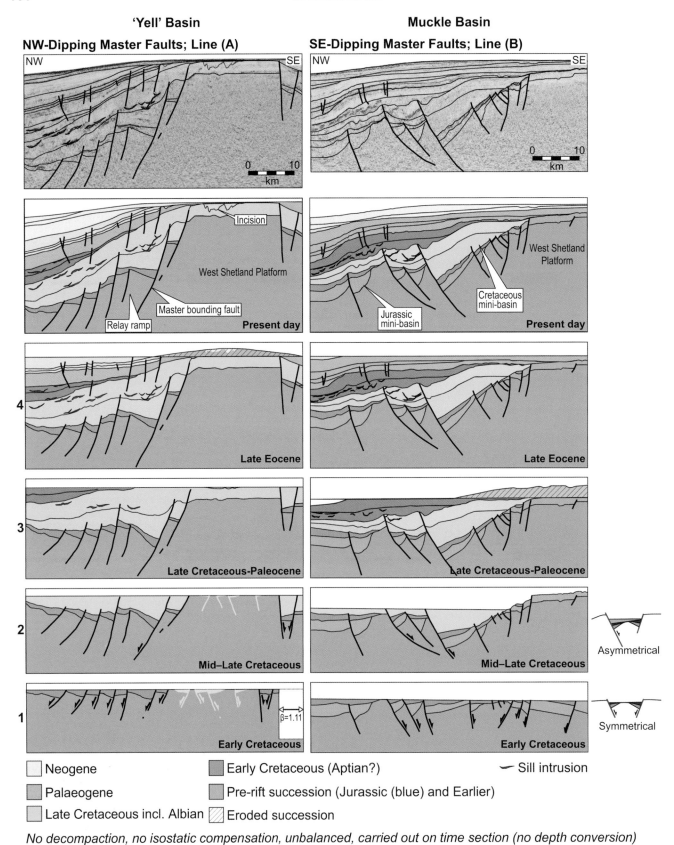

Fig. 3. Structural reconstruction of the 'Yell' and Muckle Basins along the West Shetland Platform based on interpretation of 2D seismic lines and gravimetric maps. (1) The Early Cretaceous was dominated by small, isolated to semi-connected basins formed during the rift initiation. (2) Displacement became focused on a few master faults during the early syn-rift. Because of segmentation of the West Shetland Platform, markedly different basin developments are recognized in the study area. Line A, cross-section of the 'Yell' Basin showing NW-dipping master faults. Line B, cross-section of the Muckle Basin showing SE-dipping master faults giving rise to a marginal rollover. (3–4) The rift climax and late syn-rift were controlled by thermal contraction and subsidence. See Figure 1 for location of seismic lines A and B.

Platform margin and may also have influenced sedimentation in the Palaeogene.

The varied structural configuration along the West Shetland Platform has fundamental importance for the trap configuration of the leads and plays in the area as some may be located in downfaulted, breached relay ramps and other in rotated and tilted fault blocks. The present-day structural setting indicates a highly dynamic structural evolution and the structural configuration varies over small distances in the study area. However, the overall structural evolution in the area is believed to have similar dynamics in spite of the changes in structural configuration observed along the margin. Structural analysis, simple 2D sectional back-stripping and flattening of the geometries are illustrated in Figure 3, seismic lines A and B. Rift terminology and subdivision into rift stages are adapted from Ravnås & Steel (1998).

Although the main structural elements are widely recognized in the Faroe–Shetland Basin, there is currently no standard nomenclature and some confusion exists in the published literature. DONG Energy's current usage is illustrated in Figure 1. In the quadrant 208 area we refer to the basin situated NE of the Unst Basin as the 'Yell' Basin (Fig. 1). The name is taken from one of the northern Shetland Islands. In the quadrant 209 area, the narrow, asymmetrical sub-basin lying between the volcanic extrusives of the Erlend Platform and the shallow basement of the North Shetland Platform is referred to as the Muckle Basin (Fig. 1). This name is taken from the end of well report produced after the drilling of well 209/12-01.

Pre-Early Cretaceous. The pre-Early Cretaceous structural evolution in the 'Yell' and Muckle Basin is poorly documented as seismic data below the BCU only allow identification of poorly defined, wedge-shaped geometries, suggesting the presence of older basins of probable Late Jurassic age. This is in line with regional observations indicating a general onset of the extensional pulses in the Late Jurassic–Early Cretaceous. The Jurassic rifting event decreased in intensity in the earliest Cretaceous and was followed by a period of general uplift with only minor structural modifications due to gravity-driven collapse of over-steepened faults scarps.

Early Cretaceous (rift initiation). The main episode of Cretaceous rifting began in the Valanginian as dated by coarse, clastic sediments in wells from the West Shetland and Solan Basins (Booth *et al.* 1993). Although the presence of older faults is acknowledged in the study area, the recognition of NE–SW oriented faults of Aptian–Albian age is interpreted as a significant and dominating overprint. Back-stripping suggests that several faults were present in scattered arrays across broad parts of the Faroe–Shetland Basin and that growth-faulting initiated the development of several isolated mini-basins. The scatted fault array created accommodation space in minor isolated to semi-connected basins and continental to shallow marine environments were established, leading to erosion of structural highs and deposition in the intervening fault-controlled basins (e.g. Victory Formation). Initially the graben system was probably symmetrical along the entire margin, with half-grabens developing across a wide area. This early rift stage may not have seen great subsidence rates due to the dominance of minor, isolated faults (Fig. 3, stage 1).

Mid-Cretaceous (early syn-rift). During the latest Early Cretaceous–early Late Cretaceous the scattered fault array began to link up. This transition from the initial syn-rift stage to the early syn-rift stage is marked by several faults terminating near the Aptian–Albian horizon and only a few major faults continuing into the Upper Cretaceous succession (Fig. 3, stage 2).

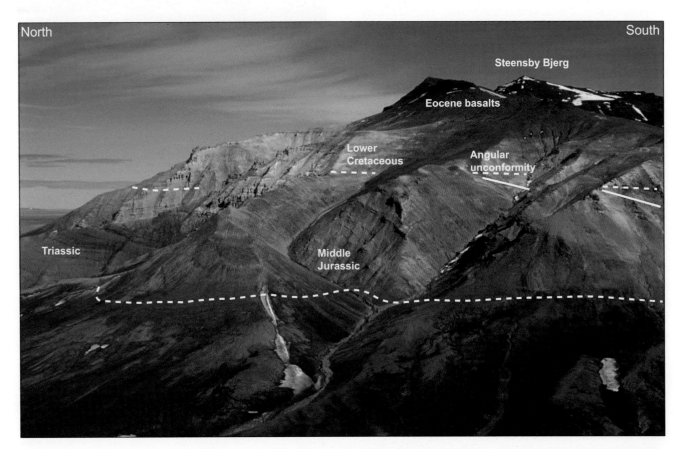

Fig. 4. Triassic/Jurassic pre-rift succession dipping towards SW unconformably overlain by Lower Cretaceous early syn-rift sediments, Hold with Hope, NE Greenland. The succession is overlain by Eocene plateau basalts forming the mountain peaks in the back. Height of succession approximately 800 m.

Displacement was focused on a few major faults and subsidence rates changed dramatically during this rift-stage transition, from scattered fault-related subsidence to a more focused subsidence related to displacement along a few linked master faults. The focused fault displacement resulted in development of major half-grabens and uplifted highs of varied geometry (Fig. 3, stage 2). Fault growth and interaction became increasingly important as faults progressively linked up into major fault segments, thus creating greater subsidence and increased marine influence. Relay ramps between major fault zones may have acted as conduits for sedimentary input, feeding mature sediments into depocentres along the platform–basin margin.

Mid–Late Cretaceous (rift climax)

Maximum subsidence occurred along a few master faults during the rift climax stage and deposition was dominated by thick mud-prone successions. The transfer of fault displacement in this stage also led to segmentation of the West Shetland Platform. In some of the segments, basin geometries changed from being symmetrical full-grabens to asymmetrical half-grabens associated with formation of a gentle marginal rollover dipping towards the Northwest (Fig. 3, stage 2). The observed segmentation and change in fault polarity between the southern 'Yell' Basin and the northern Muckle Basin may relate to interaction between Caledonian north–south structural grain (extension of the Walls Boundary Fault Zone) and Mesozoic NE–SW rifting along the West Shetland Platform (Fig. 1). Both the syn-rift and early post-rift plays are strongly affected by the asymmetrical graben geometry and formation of the marginal rollover, thus experiencing a higher risk on reservoir and trap.

Late Cretaceous (late syn-rift–early post-rift)

Continued rifting resulted in completely linked master faults and increased rotation of the associated half-grabens. This led to a dramatic increase in subsidence of the basins and transgression of the slopes. The marine transgression may have caused reworking of previously deposited sands across structural highs as they subsided and became submerged. Finally, the rift topography was filled in and the coarse-grained, syn-rift successions became capped by Upper Cretaceous marine mudstones (Fig. 3, stages 3 & 4). Post-rift thermal cooling continued in the latest Cretaceous and led to differential subsidence from West Shetland Platform towards the basin and as a result the Palaeogene depocentres became translated towards the rift axis (Fig. 1). The post-rift succession is dominated by thick mudstones, providing an excellent seal for the syn- and early post-rift plays of the Faroe–Shetland Basin. Although the general post-rift tectonic setting was affected by thermal subsidence in the basins, the West Shetland Platform itself experienced relative uplift of 300–800 m during several Eocene and Neogene pulses (e.g. Smallwood 2008).

Cretaceous reservoir formations

The Lower Cretaceous succession in the West Shetland Basin includes fan delta and shallow marine sandstones deposited in narrow half-grabens east of the Rona Ridge (Victory Formation) and submarine fan conglomerates and mass-flow sandstones and mudstones (Royal Sovereign and Cruiser Formations) deposited on the basin-ward side of the platform bounding fault system (Grant et al. 1999). Stacked shoreface sandstones, inner shelf mudstones and fan-delta sandstones were ponded in the hanging wall 'back-basin' of the Rona Ridge and other rotated fault blocks, and form the reservoir in the Victory gas discovery (Fig. 1).

The Victory wells (207/1-02, -03 and 207/1a-05) penetrated a succession of Aptian–Albian sandstones more than 200 m thick. Sandstone-dominated, Lower Cretaceous sediments were also penetrated in four neighbouring wells, 207/2-01, 208/23-01, 208/24-01 and 208/27-01. The sands in 208/23-01, 208/24-01 and 208/27-01 were of variable reservoir quality with relatively low porosity and permeability, whereas those in 207/1-02, -03 and 207/1a-05 and 207/2-01 were quite clean, with a high N/G and excellent porosity and permeability. Sedimentological interpretation based on cores from wells 207/1-03 and 207/1a-05 suggests that the conglomerates and sandstones of the Victory area were deposited as fan delta and shoreface sandstones in a shallow marine depositional environment (Goodchild et al. 1999). The younger Whiting sandstone described in well 207/1-03 may represent a second phase of sediment input or re-deposition of older sediments in the Victory area.

Greenland analogues

Lower Cretaceous sandstone-dominated successions are widespread in the northern North Atlantic and in East Greenland. In this study we refer to a Barremian–Aptian syn-rift succession in northern Hold with Hope (Larsen et al. 2001). The Lower Cretaceous succession, up to 300 m thick, rests with an angular unconformity on Lower Triassic and Middle–Upper Jurassic strata (Fig. 4). The unconformity formed after block faulting and rotation in Late Jurassic–earliest Cretaceous times and it is represented by an erosional surface overlain by a transgressive lag conglomerate.

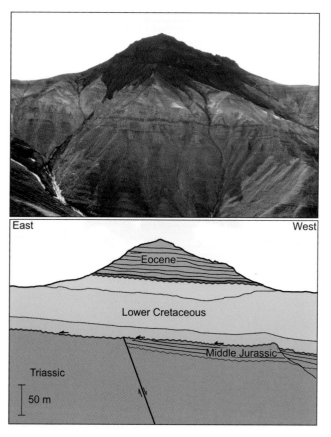

Fig. 5. Close-up photograph of the Lower Cretaceous sandstone-dominated syn-rift succession in Hold with Hope, NE Greenland. Note the minor antithetic fault terminating against the base of the Cretaceous succession corresponding to a change in fault activity, from rift initiation to early syn-rift stage, where extension is concentrated on a few major basin-forming faults and smaller faults become inactive. The Middle Jurassic sandstones form a potential pre-rift reservoir, but are only preserved in the small graben, downdip from the fault crest. The main reservoir unit would be the onlapping Lower Cretaceous syn-rift succession consisting of paralic and shallow marine sandstones.

The succession is interpreted as paralic, tidal and shallow marine, reflecting an overall transgression (Larsen et al. 2001).

The structural setting is comparable to the West Shetland Platform/Faroe–Shetland Basin transition and consists of a series of landward rotated half-grabens 10–30 km wide providing accommodation space for coarse-grained syn-rift sediments (Vischer 1943; Surlyk 1990). Hanging wall fan conglomerates are exposed in Wollaston Forland (Surlyk 1978, 1984), whereas sand-rich systems onlapping the more gentle slopes of the footwall blocks are found in eastern Wollaston Forland and Hold with Hope (Larsen et al. 2001). Minor faults truncated by the base Cretaceous unconformity (Fig. 5) indicate that the initial faulting along many smaller faults had ceased and that faulting at the time of deposition was focused along a few major faults.

The rift topography had fundamental control of the depositional systems and sediment transport direction feeding sand-rich sediments into the inherited rift topography and cross-rift along major relay ramps. The coarse-grained sediments, which seem to dominate the Lower Cretaceous successions, are a combination of the half-graben morphology, easy access to Jurassic sand-rich sections being exposed in the core of tilted footwall blocks and an overall transgressive regime piling shallow marine sands up against the exhumed fault blocks along the basin margins (Larsen et al. 1999).

Based on the outcrops, a number of potential plays can be defined: (1) pre-rift Middle Jurassic sandstones preserved in down-faulted grabens in the hanging wall block (this study, Surlyk 1977); (2) syn-rift coarse-grained submarine fan deposits along footwall blocks (Surlyk 1978, 1984); (3) Lower Cretaceous early syn-rift shallow marine sandstones deposited along the crest of the hanging wall block (Larsen et al. 2001); and (4) mass flow and mature turbiditic fans deposited after infilling of the rift topography in the late rift stage (Whitham et al. 1999; Surlyk & Noe-Nygaard 2001).

Structural control on sedimentation

Rift basins have strongly variable syn-rift sedimentary geometries because of the accommodation creation and variations in the sediment supply throughout the rift cycle (Ravnås & Steel 1998). Accommodation changes are mainly controlled by local basin-floor

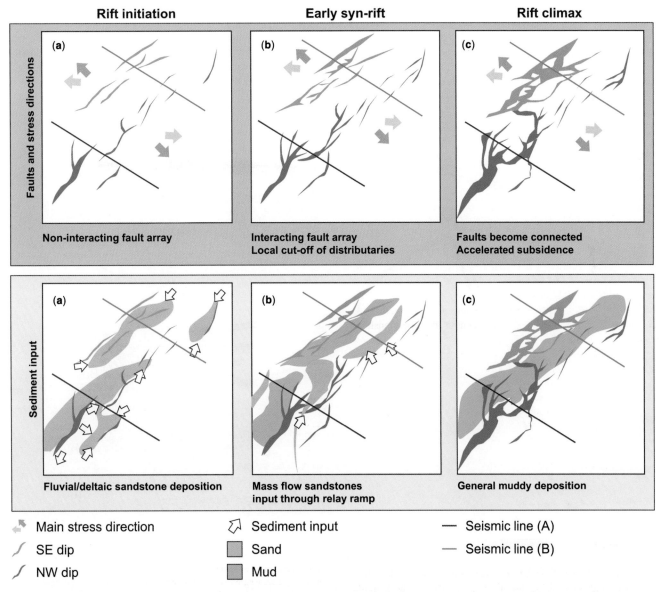

Fig. 6. Conceptual model for the structural evolution, depositional setting and resulting sand/mud distribution. Sedimentation in the Cretaceous syn-rift play is believed to be strongly controlled by structural evolution along the West Shetland Platform. The rift basin is divided into three stages: initial, early syn-rift and rift climax. (**a**) The rift initiation is characterized by non-interacting faults and deposition of fluvial/deltaic and shallow marine sandstones. (**b**) The early syn-rift is characterized by interacting faults and mass-flow sandstones sourced through relay ramp systems. (**c**) The rift climax is characterized by fully connected faults and maximum subsidence rates. Sediments deposited are generally mudstone. (See Fig. 3 for seismic and back-stripped cross-sections.)

rotation and background subsidence. Basically, the sediment supply determines how much of the accommodation is filled and in what manner. It is also controlled by the distance to the main hinterland areas and the size and sediment yield potential of any local fault block source area. Marine siliciclastic syn-rift successions are classified in terms of sediment supply as overfilled, balanced, underfilled and starved (Ravnås & Steel 1998). Sediment-overfilled and sediment-balanced infill types are characterized by a sandstone–mudstone–sandstone infill motif. The sediment-underfilled type is represented by a conglomerate–sandstone–mudstone motif and finally the sediment-starved type is commonly represented by a mudstone motif (Surlyk 1978). The sequential development, linked depositional systems, and stratigraphic signatures of the early syn-rift, the rift climax and the late syn-rift to early post-rift stages vary significantly between the rift basin infill types (Ravnås & Steel 1998).

A model for the structural evolution and sediment dispersal systems in the study area, 'Yell' Basin and Muckle Basin, offshore West Shetlands, has been proposed based on the seismic data and outcrop analogues described in the sections above. The structural evolution and the rift stages are summarized in Figure 6.

Jurassic rift stage. The seismic data suggest that the Middle to Late Jurassic sediments were deposited during an earlier rift episode. However, little is known of the Jurassic sediments as the later main rifting event obscures the seismic imaging of the succession and there is poor well control. In the Muckle and 'Yell' Basins Jurassic sediments are penetrated in wells 209/12-01 (30.5 m Upper Jurassic Rona Sandstone) and 208/24-01A (64.6 m sandstone assigned to the Rona Sandstone, but still undated). The Upper Jurassic succession unconformably overlies crystalline basement or pre-Jurassic sediments.

Rift initiation. During the rift initiation the margin was dominated by minor, scattered and unconnected faults. These faults created accommodation space in minor isolated to more connected basins along the West Shetland Platform. In these fault-bounded basins, shallow marine environments were established and stacked shallow marine shoreface sandstones, inner shelf deposits and fan-delta sandstones became deposited (Fig. 6a). The coarse sediments apparently became trapped in the hanging wall 'back-basin' of the Rona Ridge, being unable to cross the NE–SW elongated rift topography, leading to development of sediment overfilled/balanced basin fill. The onlapping, stacked sandstones form the reservoir in the Victory gas discovery (Goodchild *et al.* 1999). In contrast, immature coarse-grained submarine fans are found adjacent to the major rift controlling faults along the NW-dipping Rona Fault and represent classic syn-rift fan deposits (e.g. Royal Sovereign Formation in wells 206/3-01 and 208/26-01). These fault-related deposits were deposited in completely submerged basins and probably correspond to a sediment-underfilled outer basin fill during the rift initiation stage. This proximal to distal change in basin fill is also recognized in the syn-rift sediment fill model of Ravnås & Steel (1998).

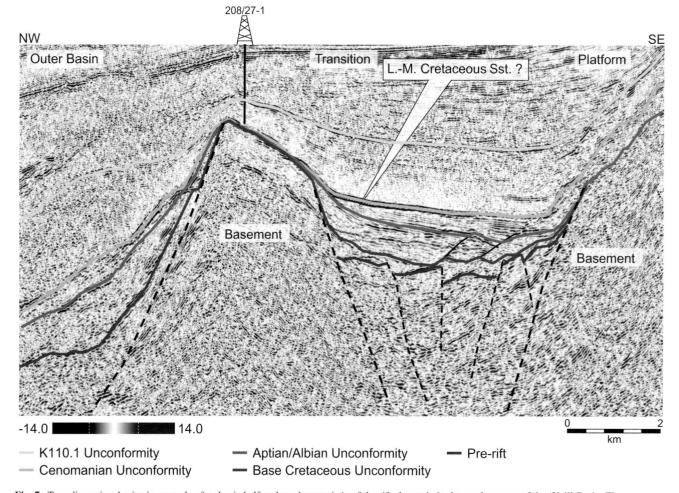

Fig. 7. Two-dimensional seismic example of a classic half-graben characteristic of the rifted margin in the southern part of the 'Yell' Basin. The exploration well targeted the Clair play and was drilled on the crest of the rotated fault block, penetrating crystalline basement and missing out potential Cretaceous syn-rift reservoirs. Despite the poor seismic quality, the data indicate the possibility of a thick Cretaceous succession (with embedded sandstones) downdip on the hanging wall block. See Figure 1 for location of seismic line C.

Lower Cretaceous shallow marine and deltaic sandstones are widespread in the North Atlantic and reflect a phase of overall transgression. The sandstones are associated with the basin margins and intra-basinal highs and are commonly dominated by locally derived material. Reworking of crystalline basement and older sediments (Devonian–Jurassic) provided large quantities of coarse material. Because of the recycled nature of the sands, they commonly show excellent reservoir properties, although weathered crystalline basement may locally source immature and clay-rich (kaolinitic) sandstones and conglomerates as seen in the Victory area (well 207/1-01).

Early syn-rift. In the early syn-rift stage, fault growth and interaction became increasingly important as faults progressively linked up to form major fault zones, thus creating greater subsidence and increased marine influence. Relay ramps connecting major fault zones and rift oblique lineaments segmenting the rift margin may have acted as conduits for sedimentary inputs to the basins (Fig. 6b). The sediments were mainly mass-flow deposited, mature sandstones (e.g. Commodore Formation) as a stable feeding system became established on the West Shetland Platform. Episodic deposition of mass-flow sandstones in the more distally located half-grabens resulted in sediment balanced/underfilled basin fill.

Rift climax. In the rift climax stage a completely linked master fault existed and resulted in a dramatic increase in subsidence of the basin with consequent transgression of the slope (Fig. 6c). Locally the marine transgression led to reworking of earlier sediments and on structural highs a second phase of shallow marine sand deposition became established (e.g. Whiting Sandstone).

Late syn-rift and early post-rift. Rifting waned during the Late Cretaceous and the entire Faroe–Shetland basin experienced thermally induced basin-wide subsidence and a thick succession of homogeneous mudstones with subordinate limestones (e.g. Kyrre Formation) was deposited.

The syn-rift plays

In spite of decades of industry activity, well control of the syn-rift succession is limited as early exploration in the Faroe–Shetland Basin was focused on the successful North Sea structural play with the reservoir forming part of the pre-rift system. Many wells are therefore situated at the crest of the tilted fault blocks, drilling updip from the syn-rift succession and failing due to lack of reservoir (Fig. 7). Improved seismic imaging of the deep structures in the 'Yell' and Muckle Basins along the West Shetland Platform has revealed the presence of a complex Mesozoic rift system with shifting fault polarity. The new seismic data led to a focus of exploration on structurally defined traps and re-opened exploration in the deeper stratigraphic sections. Based on the seismic observations, core data and onshore analogue studies, a revised syn-rift play model for the Faroe–Shetland Basin is proposed.

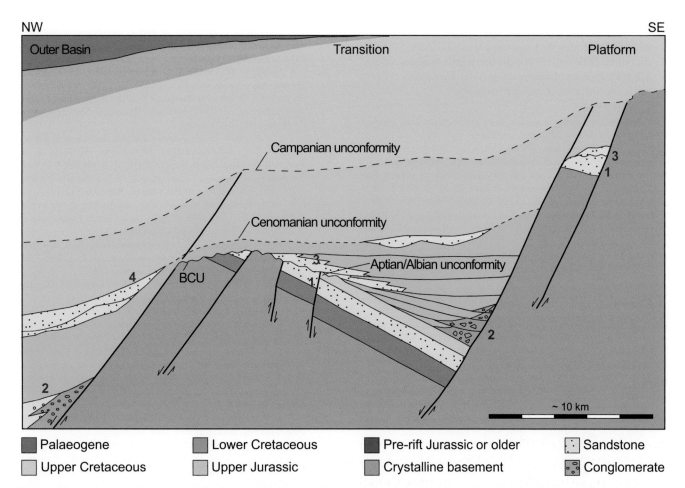

Fig. 8. Conceptual play models along the West Shetland Platform margin showing tectonic stages and possible related deposits. (1) Middle Jurassic pre-rift sandstones preserved in down-faulted grabens. (2) Lower Cretaceous rift initiation (e.g. Royal Sovereign Formation) deposited along major basin-forming faults. (3) Lower Cretaceous early syn-rift shallow marine sandstones (e.g. Victory Formation) onlapping structural highs (fault crests). (4) Mid to Upper Cretaceous early syn-rift, mature, mass-flow sandstones sourced through relay ramps (e.g. Commodore Formation). Play models inspired by Surlyk (1978, 1984), Grant *et al.* (1999) and Ravnås & Steel (1998).

The conceptual play models (Fig. 8) include:

(1) Middle Jurassic pre-rift sandstones preserved in down-faulted grabens. The play may form an interesting target in already drilled tilted fault blocks in quadrant 207 and 208. The play concept is supported by analogues in NE Greenland.
(2) Lower Cretaceous rift initiation and early syn-rift stage conglomerates and sandstones (e.g. Royal Sovereign Formation) deposited along major basin-forming faults. The immature sandstones are proven in quadrant 206 and may form an untested exploration target in the marginal basins.
(3) Lower Cretaceous early syn-rift stage sandstones deposited in shallow marine settings along the crest of structural highs during initial marine transgression (e.g. Victory Formation). The play is proven in the Victory gas discovery in quadrant 207 and probably has a large untested potential in basins along the margin of the West Shetland Platform.
(4) Mid-Cretaceous early syn-rift stage mature, mass-flow deposited sandstones sourced into the marginal basins along relay ramps and rift oblique lineaments (e.g. Commodore Formation). The reservoir unit is proven in quadrant 206, but the play apparently failed due to lack of updip sealing (Grant *et al.* 1999). It may form an attractive exploration target if sediment entry points can be identified along the West Shetland Platform margin.

The key elements for future exploration along the margins of the Faroe–Shetland Basin will be:

- careful reprocessing of existing 2D seismic data (fair to good quality) and acquisition of fit-for-purpose 3D seismic surveys;
- structural reconstructions and improved understanding of the interaction between the Caledonian structural grain, Mesozoic rift related faults and rift oblique lineaments (transfer zones);
- acknowledgement of structural control on depositional systems;
- full implementation of onshore analogues.

The Cretaceous syn-rift play models along the West Shetland Platform evolved from technical input and discussions with numerous persons; however, the presented views and conclusions are the sole responsibility of the authors. The manuscript benefited from critical reviews by Martyn Stoker and Finn Surlyk. We are especially grateful for the support from our licence partners Dana Petroleum Ltd and GDF SUEZ E&P UK Ltd. TGS long offset data are reproduced with the permission of TGS-NOPEC. Field work in East Greenland by Michael Larsen formed part of the project 'Resources of the sedimentary basins of North and East Greenland' supported by the Danish Research Councils, and additional mapping was supported by the Geological Survey of Denmark and Greenland (GEUS).

References

Booth, J., Sweicicki, T. & Wilcockson, P. 1993. The tectonostratigraphy of the Solan Basin, West of Shetlands. *In*: Parker, J. R. (ed.) *The Petroleum Geology of Northwest Europe: Proceedings of the 4th Conference*. Geological Society, London, 987–998; doi: 10.1144/0040987.

Gjelberg, J. & Steel, R. J. 1995. Helvetiafjellet Formation (Barremian–Aptian), Spitsbergen: characteristics of a transgressive succession. *In*: Steel, R. J., Felt, V. L., Johannessen, E. P. & Mathieu, C. (eds) *Sequence Stratigraphy on the Northwest European Margin*. Norwegian Petroleum Foundation, Oslo, Special Publications, 5, 571–593.

Goodchild, M. W., Henry, K. L., Hinkley, R. J. & Imbus, S. W. 1999. The Victory gas field, West of Shetland. *In*: Fleet, A. J. & Boldy, S. A. R. (eds) *Petroleum Geology of Northwest Europe: Proceedings of the 5th Conference*. Geological Society, London, 713–724; doi: 10.1144/0050713.

Grant, N., Bouma, A. & McIntyre, A. 1999. The Turonian play in the Faeroe–Shetland Basin. *In*: Fleet, A. J. & Boldy, S. A. R. (eds) *Petroleum Geology of Northwest Europe: Proceedings of the 5th Conference*. Geological Society, London, 661–673; doi: 10.1144/0050661.

Kelly, S. R. A., Whitham, A. G., Koraini, A. M. & Price, S. M. 1998. Lithostratigraphy of the Cretaceous (Barremian–Santonian) Hold with Hope Group, NE Greenland. *Journal of the Geological Society, London*, 155, 993–1008.

Kimbell, G. S., Ritchie, J. D., Johnson, H. & Gatliff, R. W. 2005. Controls on the structure and evolution of the NE Atlantic margin revealed by regional potential field imaging and 3D modelling. *In*: Doré, A. G. & Vining, B. (eds) *Petroleum Geology: North-West Europe and Global Perspectives – Proceedings of the 6th Conference*. Geological Society, London, 933–945; doi: 10.1144/0060933.

Lamers, E. & Carmichael, S. M. M. 1999. The Paleocene deepwater sandstone play West of Shetland. *In*: Fleet, A. J. & Boldy, S. A. R. (eds) *Petroleum Geology of Northwest Europe: Proceedings of the 5th Conference*. Geological Society, London, 645–659; doi: 10.1144/0050645.

Larsen, M., Hamberg, L., Olaussen, S., Nørgaard-Pedersen, N. & Stemmerik, L. 1999. Basin evolution in southern East Greenland: an outcrop analogue for Cretaceous–Paleogene basins on the North Atlantic volcanic margins. *Bulletin of the American Association of Petroleum Geologists*, 83, 1236–1261.

Larsen, M., Nedkvitne, T. & Olaussen, S. 2001. Lower Cretaceous (Barremian–Albian) deltaic and shallow marine sandstones in North-East Greenland – sedimentology, sequence stratigraphy and regional implications. *In*: Martinsen, O. J. & Dreyer, T. (eds) *Sedimentary Environments Offshore Norway – Palaeozoic to Recent*. Norwegian Petroleum Foundation, Oslo, Special Publications, 10, 259–278.

Moy, D. J. & Imber, J. 2009. A critical analysis of the structure and tectonic significance of rift-oblique lineaments ('transfer zones') in the Mesozoic–Cenozoic succession of the Faroe–Shetland Basin, NE Atlantic margin. *Journal of the Geological Society, London*, 166, 831–844.

Ravnås, R. & Steel, R. J. 1998. Architecture of marine rift-basin successions. *Bulletin of the American Association of Petroleum Geologists*, 82, 110–146.

Ruffell, A. & Wach, G. 1998. Estuarine/offshore depositional sequences of the Cretaceous Aptian-Albian boundary, England. *In*: de Graciansky, P.-C., Hardenbol, J., Jacquin, T. & Vail, P. (eds) *Mesozoic and Cenozoic Sequence Stratigraphy of European Basins*. SEPM, Special Publications, 60, 411–421.

Rump, B., Reaves, C. M., Orange, V. G. & Robinson, D. L. 1993. Structuring and transfer zones in the Faroe Basin in a regional tectonic context. *In*: Parker, J. R. (ed.) *The Petroleum Geology of Northwest Europe: Proceedings of the 4th Conference*. Geological Society, London, 7–40; doi: 10.1144/0040007.

Smallwood, J. R. 2008. Uplift, compression and the Cenozoic Faroe–Shetland Basin sediment budget. *In*: Johnson, H., Doré, A. G., Gatliff, R. W., Holdsworth, R., Lundin, E. R. & Ritchie, J. D. (eds) *The Nature and Origin of Compression in Passive Margins*. Geological Society, London, Special Publications, 306, 137–152.

Surlyk, F. 1977. Stratigraphy, tectonics and palaeography of the Jurassic sediments of the areas north of Kong Oscars Fjord, East Greenland. *Bulletin Grønlands Geologiske Undersøgelse*, 123.

Surlyk, F. 1978. Submarine fan sedimentation along fault scarps on tilted fault blocks (Jurassic–Cretaceous boundary, East Greenland). *Bulletin Grønlands Geologiske Undersøgelse*, 128.

Surlyk, F. 1984. Fan delta to submarine fan conglomerates of the Volgian–Valanginian Wollaston Forland Group, East Greenland. *In*: Koster, E. H. & Steel, R. J. (eds) *Sedimentology of Gravels and Conglomerates*. Canadian Society of Petroleum Geologists, Calgary, Memoirs, 10, 359–382.

Surlyk, F. 1990. Timing, style and sedimentary evolution of Late Palaeozoic–Mesozoic extensional basins of East Greenland. *In*: Hardman, R. F. P. & Brooks, J. (eds) *Tectonic Events Responsible for Britain's Oil and Gas Reserves*. Geological Society, London, Special Publications, 55, 107–125.

Surlyk, F. & Noe-Nygaard, N. 2001. Cretaceous faulting and associated coarse-grained marine gravity flow sedimentation, Traill Ø, East Greenland. *In*: Martinsen, O. J. & Dreyer, T. (eds) *Sedimentary Environments Offshore Norway – Palaeozoic to Recent*. Norwegian Petroleum Foundation, Oslo, Special Publications, 10, 293–319.

Vischer, A. 1943. Die Post-Devonische tektonik von Ostgrönland zwischen 74° to 75°N.Br., Kuhn Ø, Wollaston Forland, Clavering Ø und angrenzende gebiete. *Meddelelser om Grønland*, 133, 1–194.

Whitham, A. G., Price, S. P., Koraini, A. M. & Kelly, S. R. A. 1999. Cretaceous (post-Valanginian) sedimentation and rift events in NE Greenland (71–77°N). *In*: Fleet, A. J. & Boldy, S. A. R. (eds) *Petroleum Geology of Northwest Europe: Proceedings of the 5th Conference*. Geological Society, London, 325–336; doi: 10.1144/0050325.

Timing, controls and consequences of compression in the Rockall–Faroe area of the NE Atlantic Margin

A. TUITT,[1,2] J. R. UNDERHILL,[1] J. D. RITCHIE,[2] H. JOHNSON[2] and K. HITCHEN[2]

[1]*School of GeoSciences, University of Edinburgh, Grant Institute of Earth Science, King's Buildings, West Mains Road, Edinburgh EH9 3JW, UK*
[2]*British Geological Survey, Murchison House, West Mains Road, Edinburgh EH9 3LA, UK*
(e-mail: jdri@bgs.ac.uk)

Abstract: The simplest models of passive margins would suggest that they are characterized by tectonic quiescence as they experienced gentle thermal subsidence following the extensional events that originally formed them. Analysis of newly acquired and pre-existing 2D seismic data from the Rockall Plateau to the Faroe Shelf, however, has confirmed that the NE Atlantic Margin was the site of significant active deformation. Seismic data have revealed the presence of numerous compression-related Cenozoic folds, such as the Hatton Bank, Alpin, Ymir Ridge and Wyville–Thomson Ridge Anticlines. The distribution, timing of formation and nature of these structures have provided new insights into the controls and effects of contractional deformation in the region. Growth of these compressional features occurred in five main phases: Thanetian, late Ypresian, late Lutetian, Late Eocene (C30) and Early Oligocene. Compression has been linked to hotspot-influenced ridge push, far-field Alpine and Pyrenean compression, asthenospheric upwelling and associated depth-dependent stretching. Regional studies make it clear that compression can have a profound effect on seabed bathymetry and consequent bottom-water current activity. Bottom-water currents have directly formed the early Late Oligocene, late Early Miocene (C20), Late Miocene–Early Pliocene, and late Early Pliocene (C10) unconformities. The present-day Norwegian Sea Overflow (NSO) from the Faroe–Shetland Channel into the Rockall Trough is restricted by the Wyville–Ymir Ridge Complex, and takes place via the syncline (Auðhumla Basin) between the two ridges. The Auðhumla Basin Syncline is now thought to have controlled the path of the NSO into the Rockall Trough and the resulting unconformity formation and sedimentation therein, no later than the Mid Miocene.

Keywords: NE Atlantic continental margin, Cenozoic, compression, inversion, sedimentary basin, bottom-water currents

Over the past two decades there has been an increasing realization of the general role that compression may have played in the tectonic development of the NE Atlantic passive continental margin (Boldreel & Andersen 1993; Doré & Lundin 1996; Ritchie *et al.* 2003; Johnson *et al.* 2005). Previous studies have revealed Cenozoic compressional structures in the Norwegian Margin (Brekke 2000; Lundin & Doré 2002; Løseth & Henriksen 2005), Faroe–Shetland Basin (Ritchie *et al.* 2003; Smallwood 2004; Ritchie *et al.* 2008) and the Rockall–Faroe area (Boldreel & Andersen 1993, 1994, 1998; Hitchen 2004; Johnson *et al.* 2005; Ritchie *et al.* 2008). These compressional features have been attributed to normal ridge push (Boldreel & Andersen 1993), hotspot-influenced ridge push (Doré & Lundin 1996; Lundin & Doré 2002), Alpine compression (Roberts 1989; Doré & Lundin 1996; Vågnes *et al.* 1998; Brekke 2000), Pyrenean compression (Knott *et al.* 1993), plate reorganization (Boldreel & Andersen 1993; Lundin & Doré 2002) and the Iceland Plateau Body force (Doré *et al.* 2008).

The ages of compression in the Norwegian Margin (Lundin & Doré 2002) and the Faroe–Shetland Basin (Ritchie *et al.* 2003) have been derived from dating unconformities associated with compression and calibrated using well data. The ages of compression in the Rockall–Faroe area (Boldreel & Andersen 1993; Johnson *et al.* 2005), however, were not well constrained hitherto. In addition, there has been relatively little emphasis on the study of compressional structures in the Rockall–Faroe area compared with other areas of the NE Atlantic Margin.

The aim of this paper is to date Cenozoic compressional structures in the Rockall–Faroe area and to link the ages of these structures to regional events. The study area is located on the continental margin to the SE of Iceland (Fig. 1). This study utilizes 40 000 km of 2D seismic data (line-length), acquired over the past 20 years, extending from the north Rockall Plateau to the south Faroe Shelf and covering an area of *c.* 300 000 km^2 (Fig. 2). Well data have been used to date unconformities within the area to constrain the timing of both compressional and non-compressional events. With the quality of seismic data and the number of wells and boreholes available, it is now possible to calibrate and define the specific effects of compression and bottom-water current activity in the Rockall–Faroe area. The better constrained ages for these events have facilitated the linking of compression and bottom-water current activity to regional events affecting the NE Atlantic Margin.

Unconformities

Unconformities are stratigraphic surfaces which represent an interruption in the deposition of sediment (McQuillin *et al.* 1984). Eustasy (changes in sea-level) has long been thought to control unconformity formation in quiescent basins. However, it has been previously recognized in the Rockall Basin that some unconformities and progradational features are not produced by eustatic sea-level changes (Shannon *et al.* 2005). The erosional nature of some unconformities in the Rockall–Faroe area has been previously attributed to bottom-water current activity (Andersen & Boldreel 1995). In this study post-Oligocene unconformities, such as the late Early Miocene (C20) and Late Miocene–Early Pliocene unconformities, have also been ascribed to the actions of bottom-water currents.

Unconformities formed by the onlap of sediments onto tilted strata of anticlines are considered to be the result of folding

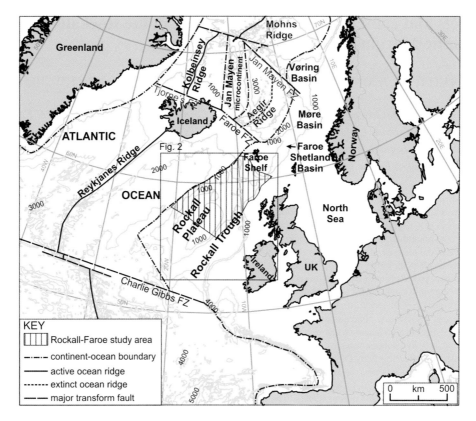

Fig. 1. Bathymetric map of the NE Atlantic Margin and surroundings showing the location of the Rockall–Faroe study area. The bathymetry is reproduced from the GEBCO Digital Atlas published by the British Oceanographic Data Centre on behalf of the IOC, IHO & BODC (2003; www.gebco.net).

(Stoker *et al.* 2005) and can provide important markers for the initiation of compressional events. In the Rockall–Faroe area, the Thanetian, late Ypresian, late Lutetian, Late Eocene (C30) and Early Oligocene unconformities are dated using well information and define distinct compressional episodes. The characteristics of these unconformities and other stratigraphic surfaces are shown in Table 1.

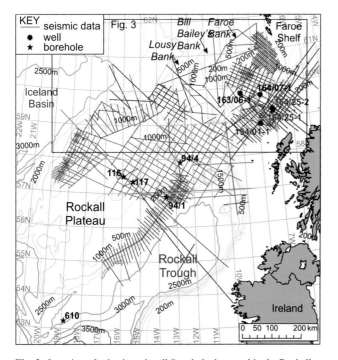

Fig. 2. Location of seismic and well/borehole data used in the Rockall–Faroe study area. Seismic data courtesy of BGS, Fugro Multi Client Services, CGG Veritas and GEUS.

Compressional structures

The NE Atlantic Margin has been affected by horizontal compressional forces during the Cenozoic (Boldreel & Andersen 1993; Doré & Lundin 1996; Lundin & Doré 2002; Doré *et al.* 2008), giving rise to the growth of anticlines, synclines and reverse faults in the Rockall–Faroe area. These anticlines and synclines have defined fold axes (Fig. 3) and have been mapped using reflection seismic data. The compressional structures vary in size, orientation and shape. The folds in the study area have long wavelengths and relatively small amplitudes representing <1% shortening.

There are three dominant orientations of folds and reverse faults: NE, NW and east–west trends. The axes of the Hatton Bank, Lousy Bank and South Faroe Bank anticlines are parallel to the adjacent NE-trending continent–ocean boundary. The West Lewis reverse faults and the Bridge Anticline are also NE-trending. The NW-trending Bill Bailey's Bank and North Faroe Bank anticlines lie adjacent to the continent–ocean boundary. Other NW-trending structures include the Faroe Bank Channel Syncline, the Vine reverse faults, and the Dawn, Wyville–Thomson Ridge, Ymir Ridge (North, Central and South) and Viera anticlines. The Hatton Bank reverse faults and the Alpin and Judd anticlines exhibit east to west trends.

Hatton Bank Anticline

Hatton Bank Anticline has a positive bathymetric expression (Fig. 3) and is also well expressed by the top-basalt surface (Fig. 4). The axial trace of this NE-trending anticline is *c.* 210 km in length, running parallel to the continent–ocean boundary. The Hatton Bank Anticline (Fig. 5) has an amplitude of 1.1 km (measured from the top-basalt surface) and a wavelength of *c.* 40 km. Marked onlap onto the Intra-Eocene 1, Intra-Eocene 2, C30 (Late Eocene) and Early Oligocene unconformities (Fig. 6) suggests that major phases of compression took place at these

Table 1. The age and characteristics of mapped seismic horizons within the Rockall–Faroe area

Age and name of mapped horizon	Seismic character and nature of mapped horizon	Source of stratigraphic correlation
C10 (late Early Pliocene)	Strong, continuous seismic reflector characterized by erosional truncation in the NE Rockall Trough and the Hatton–Rockall Basin	This unconformity was dated and constrained using wells 154/01-1, 164/25-1 and 164/25-2 in the NE Rockall Trough and borehole 116 in the Hatton–Rockall Basin
Late Miocene–Early Pliocene	Strong seismic reflector characterized by erosional truncation in the North Rockall Trough (see Fig. 10) and the NE Rockall Trough (see Fig. 17)	The age of this unconformity was inferred based on its stratigraphic position between the C20 and C10 horizons. A Pliocene unconformity of this age was mapped in the Faroe–Shetland Basin (Davies & Cartwright 2002)
C20 (late Early Miocene)	Moderately strong seismic reflector marked by onlap and erosional truncation in the North Rockall Trough (see Fig. 10) and the Hatton–Rockall Basin. It is also characterized by an onlapping layer of smectite in the South Rockall Trough (Dolan 1986)	This unconformity was dated using borehole 610 in the South Rockall Trough and borehole 116 in the Hatton–Rockall Basin. Wells 164/25-1 and 164/25-2 were used to constrain the position of the C20 stratigraphic surface in the NE Rockall Trough
Early Late Oligocene	Strong seismic reflector marked by onlap in the NE and North Rockall Trough and onlap and erosional truncation in the Hatton–Rockall Basin (see Fig. 6)	This unconformity was dated by well 164/25-2 and further constrained using well 164/25-1 in the NE Rockall Trough. The early Late Oligocene stratigraphic surface was also constrained by borehole 116 in the Hatton–Rockall Basin. This unconformity is inferred as representing the Top Palaeogene unconformity (TPU) mapped previously in south Faroe–Shetland Channel and the Faroe Bank Channel (Smallwood 2004)
Early Oligocene	Strong seismic reflector marked by onlap and defining folding in the Hatton Bank, Alpin, Mordor, Ymir and Wyville–Thomson Ridge anticlines	The Early Oligocene unconformity was dated using wells 164/25-1 and 164/25-2 in the NE Rockall Trough
C30 (Late Eocene)	Strong seismic reflector marked by onlap. Unconformity represents extensive compression across the Rockall–Faroe area and also coincides with a period of dramatic subsidence in the Rockall Trough	The C30 unconformity was dated using wells 154/01-1, 163/06-1, 164/07-1, 164/25-1 and 164/25-2 in the NE Rockall Trough and borehole 94/01 on the east margin of the Rockall Bank. The unconformity has also been constrained using boreholes 117 and 94/04 in the Hatton–Rockall Basin and the NE margin of the Rockall Bank, respectively
Late Lutetian	Moderately strong seismic reflector dated in the NE Rockall Trough and mapped from this area to the Faroe Bank Channel by jump correlation. The unconformity marks a significant phase of compression in the Mordor, Ymir Ridge (North, Central and South), Wyville–Thomson Ridge, Bridge and Onika anticlines	The late Lutetian stratigraphic surface was dated using wells 154/01-1, 163/06-1, 164/07-1, 164/25-1 and 164/25-2. This unconformity has also been inferred, in this study, in the Hatton–Rockall Basin (Intra-Eocene 2; see Fig. 6) and the South Rockall Trough
Late Ypresian	Strong seismic reflector dated in the NE Rockall Trough and mapped from this area to the Faroe Bank Channel. This unconformity marks compression in the Dawn, Mordor, North Ymir Ridge and Wyville–Thomson Ridge anticlines	The late Ypresian stratigraphic surface was dated using wells 154/01-1, 163/06-1, 164/07-1, 164/25-1 and 164/25-2. This unconformity has also been inferred, in this study, in the Hatton–Rockall Basin (Intra-Eocene 1; see Fig. 6)
Top-basalt	Strong seismic reflector marking the cessation of Palaeogene lava flow extrusion in the Late Paleocene–Early Eocene. It should be noted that the lava flows consist of other volcanic rocks, such as dacite, in addition to basalt	Seismic reflector easily identified throughout the Rockall–Faroe area and has been calibrated using wells 154/01-1, 163/06-1, 164/07-1, 164/25-1 and 164/25-2 in the NE Rockall Trough and borehole 117 in the Hatton–Rockall Basin
Thanetian	Moderately strong seismic reflector marked by onlap. The Thanetian unconformity defines inversion of the West Lewis Basin. Unconformity inferred within basalt lava flows representing initial growth of the North Ymir Ridge and Wyville–Thomson Ridge anticlines and the Faroe Bank Channel Syncline	Unconformity dated by well 164/25-1 over the West Lewis Basin. The unconformity may represent the Base Balder unconformity dated in the south Faroe–Shetland Basin (Smallwood & Gill 2002)

times. There is also some evidence of onlap of Early Eocene sediment onto the top-basalt surface. This suggests that a structural high existed prior to the Early Eocene. Eocene sediment on Hatton Bank Anticline southern limb is cut by east–west-trending reverse faults 15–25 km in length, which dip towards the north and south (Fig. 6). Reverse faulting is inferred as occurring at C30 (Late Eocene) time supported by the onlap of Early Oligocene sediment onto the C30 unconformity on the hanging wall of one of the reverse faults (Fig. 6). A compressional phase in the Late Eocene for the Hatton Bank was also proposed in previous work (Hitchen 2004; Johnson *et al.* 2005). Boldreel & Andersen (1993), however, inferred Mid Miocene reverse faults for the Hatton Bank. The early Late Oligocene, C20 (late Early Miocene) and C10 (late Early Pliocene) unconformities in the Hatton–Rockall Basin, overlying the reverse faults, are marked by onlap and erosional truncation and have been inferred as being related to bottom-water current activity.

Lousy Bank

The Lousy Bank Anticline has a positive expression on the seabed bathymetry (Fig. 3) and the top-basalt surface (Fig. 4). The axial

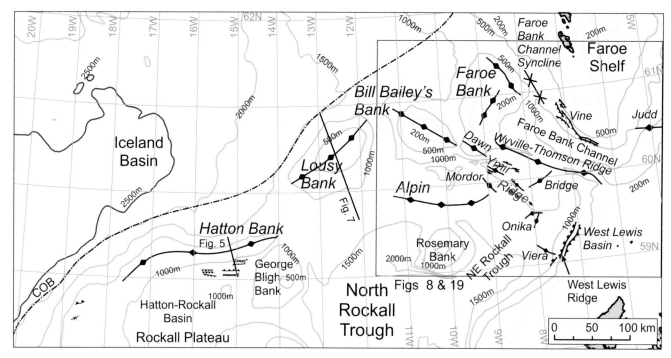

Fig. 3. Location of compressional structures in the Rockall–Faroe study area. Compressional structures are labelled in italics. The locations of seismic illustrations are shown by solid lines.

trace of this NE-trending anticline extends for 125 km, parallel to the continent–ocean boundary and was previously mapped by Boldreel & Andersen (1993, 1994). The anticline has an amplitude of up to 1.9 km (measured from the top-basalt surface) and a wavelength of c. 95 km (Fig. 7). On the southeastern limb of the anticline the C30 (Late Eocene) unconformity is marked by onlap and defines a compressional phase for the growth of the Lousy Bank. The lack of deep penetrating and high-resolution seismic data, across the Lousy Bank Anticline, precludes the mapping of any possible earlier compression-related unconformities.

Bill Bailey's Bank

The Bill Bailey's Bank Anticline, like the Lousy Bank Anticline, has a positive expression on the seabed bathymetry (Fig. 3) and the top-basalt surface (Fig. 4). The axial trace of this anticline is NW-trending and extends for 100 km. The anticline has an amplitude of up to 1.7 km (measured from the top-basalt surface) and a wavelength of c. 85 km, and was first mapped as a compressional ridge by Boldreel & Andersen (1994). The C30 (Late Eocene) unconformity, in this study, defines a compressional phase of

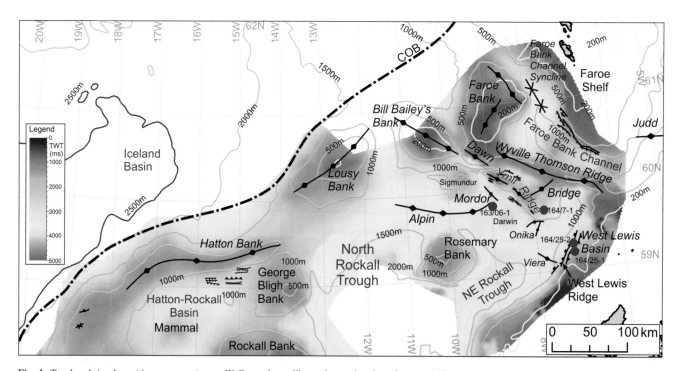

Fig. 4. Top-basalt isochron (time structure) map. Wells used to calibrate the top-basalt surface are delineated by purple circles. The edges of the top-basalt isochron indicate the limit of the data. Compressional structures are labelled in italics.

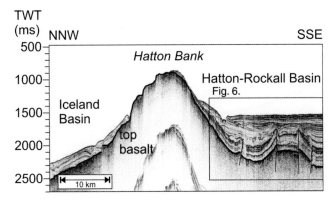

Fig. 5. NNW-trending seismic profile across the Hatton Bank Anticline. For the location of the seismic illustration, see Figure 3. Seismic profile, BGS © NERC 2009. All rights reserved.

growth of the Bill Bailey's Bank Anticline (seismic profile not illustrated).

Faroe Bank Anticlines

The Faroe Bank is well delineated on seabed bathymetry (Fig. 3) and the top-basalt surface (Fig. 4). The bank is interpreted as comprising two anticlines with orthogonal trends. The South Faroe Bank Anticline is NE-trending and has a length of 60 km. The Thanetian, late Ypresian, late Lutetian, C30 (Late Eocene) and Early Oligocene unconformities, on the southeastern limb of the fold, are inferred to mark compressional growth phases of the South Faroe Bank Anticline (seismic profile not illustrated).

The North Faroe Bank Anticline is *c.* 65 km in length and was previously mapped by Boldreel & Andersen (1994). On its northern limb the interpreted late Lutetian and C30 (Late Eocene) unconformities are marked by onlap and are interpreted to have formed as a result of compression. The C10 (late Early Pliocene) unconformity is marked by erosional truncation as a result of bottom-water current activity (seismic profile not illustrated).

Faroe Bank Channel Syncline

The Faroe Bank Channel Syncline is NW-trending and lies between the Faroe Bank and the Faroe Shelf (Fig. 8). It is characterized by the folding of Palaeogene basalt lava flows and Cenozoic sediments (Fig. 9). The syncline has an amplitude of up to 2.2 km (measured at the top-basalt surface) and is 50 km in length. Thinning and onlap onto the Thanetian, late Ypresian, late Lutetian and C30 (Late Eocene) unconformities suggest compression at these times.

Alpin Anticline

The Alpin Anticline is east–west trending and lies between Bill Bailey's Bank and Rosemary Bank (Fig. 8). The anticline (Fig. 10) has an amplitude of up to 700 m (measured from the C30 unconformity), a length of *c.* 150 km and a wavelength of 30 km. Thinning and onlap onto the C30 (Late Eocene) and Early Oligocene unconformities suggest compression at these times. Previous work ascribed Late Eocene, Late Oligocene and Mid Miocene compression in the formation of the Alpin Anticline (Stoker *et al.* 2005; Ritchie *et al.* 2008).

The C20 (late Early Miocene) and Late Miocene–Early Pliocene unconformities, over the northern limb of the anticline, are defined by onlap and erosional truncation and are interpreted as being the result of bottom-water current activity. Poor seismic imaging between the top-basalt surface and the C30 (Late Eocene) unconformity, on the western section of the fold, precludes the mapping of any possible Early Eocene unconformities within the Alpin Anticline. The Alpin Anticline comprises relatively thick Eocene sediment bounded by the Sigmundur and Darwin seamounts and the Rosemary Bank (Fig. 11). This suggests that the Alpin Anticline developed from the compression of an Eocene depocentre.

Mordor Anticline

The Mordor Anticline (Fig. 12) is a symmetrical NW-trending fold with an amplitude of 500 m (measured from the late Lutetian unconformity), and is *c.* 25 km in length. The fold has no expression on the seabed as the crest of the anticline is covered by a veneer of Late Pliocene–Recent sediment, 100 m thick. The Mordor Anticline is surrounded by the Sigmundur and Darwin seamounts and the North Ymir Ridge Anticline, and lies partly within a basin bounded by basalt escarpments (Fig. 12). The basin facilitated the accumulation of a relatively thick Early Eocene basin fill (Fig. 13) which has been compressed to form the Mordor Anticline. The Mordor Anticline is characterized by both the folding of basalt lava flows and overlying Cenozoic sediments. Folding of the Mordor Anticline is defined by the late Ypresian, late Lutetian, C30 (Late Eocene) and Early Oligocene unconformities as represented by onlap and thinning onto these stratigraphic surfaces (Fig. 12). The Late Miocene–Early Pliocene unconformity has truncated the Early Oligocene and early Late Oligocene unconformities at the crest of the fold and the C20 (late Early Miocene) unconformity located more distally above the South Ymir Ridge Anticline (Fig. 12). The erosional nature of the Late Miocene–Early Pliocene unconformity is interpreted as being the result of bottom-water current activity.

Dawn Anticline

The Dawn Anticline is a NW-trending linear anticline located on the southern end of the Faroe Bank and is parallel to the Wyville–Thomson Ridge Anticline and the North Ymir Ridge Anticline (Fig. 8). The anticline was previously mapped as part of the Ymir Ridge by Ziska & Varming (2008). In this study, however, a bathymetric gorge separates the Dawn Anticline from the North Ymir Ridge Anticline (Fig. 8). The fold is 30 km in length and has a wavelength of 18 km. The amplitude of the Dawn Anticline (measured from the top-basalt surface) is 2.1 km off the southern limb and 0.34 km off the northern limb of the

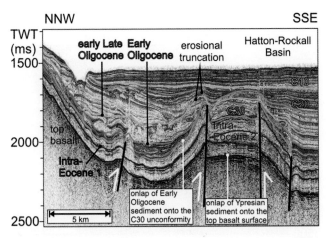

Fig. 6. NNW-trending seismic profile of unconformities and reverse faults on the southern limb of the Hatton Bank Anticline. For the position of the seismic illustration, see Figure 5. Seismic profile, BGS © NERC 2009. All rights reserved.

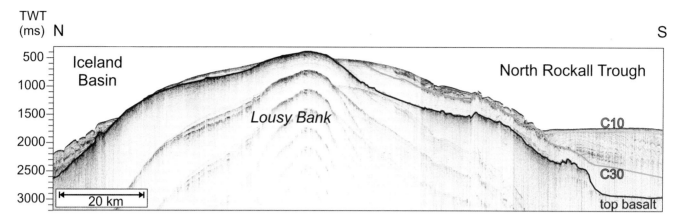

Fig. 7. North–south seismic profile of the Lousy Bank Anticline. For the location of the seismic illustration, see Figure 3. Seismic profile, BGS © NERC 2009. All rights reserved.

fold. The late Ypresian, late Lutetian, C30 (Late Eocene) and the Early Oligocene unconformities are marked by onlap and are interpreted as representing compressional phases forming the Dawn Anticline (seismic profile not illustrated).

Ymir Ridge Complex

The Ymir Ridge is a bathymetric high (Fig. 8) comprising a series of three anticlines – the North, Central and South Ymir Ridge anticlines. Ziska & Varming (2008) also segmented the Ymir Ridge into North, Central and South sections, but these differ from the sections defined in this study. The Ymir Ridge was also previously mapped by Boldreel & Andersen (1993), Johnson et al. (2005) and Ritchie et al. (2008). However, in these studies there was no attempt to differentiate the Ymir Ridge into separate anticlines.

The North Ymir Ridge Anticline is NW-trending and is 25 km in length. The fold (Fig. 14) has an amplitude of 1.4 km (measured from the top-basalt surface) and a wavelength of up to 28 km. A Thanetian unconformity is defined by thinning and onlap within the basalt lava flows on the southern limb of the North Ymir Ridge Anticline, suggesting folding at this time. The late Ypresian, late Lutetian and C30 (Late Eocene) compression-related unconformities are also marked by onlap.

The Central Ymir Ridge Anticline is NW-trending, 23 km in length and has a wavelength of c. 30 km. The fold (Fig. 15) is asymmetrical with the southern limb having a steeper gradient than the northern limb. The amplitude of the fold (measured from the top-basalt surface) is up to 1.24 km. The northern limb of the fold is cut by north-dipping reverse faults. The late Lutetian, C30 (Late Eocene) and the Early Oligocene unconformities areas are marked by onlap on both limbs of the fold and represent the ages of compression for the Central Ymir Ridge Anticline.

The South Ymir Ridge Anticline is NW-trending, and is c. 7 km in length. There are no seismic data available showing the full wavelength of the fold, but the fold is assumed to have a wavelength c. 26 km (Fig. 12). The late Lutetian, C30 (Late Eocene) and Early Oligocene unconformities area are marked by onlap on the southern limb of the fold and are inferred to represent the ages of compression of the South Ymir Ridge Anticline.

Wyville–Thomson Ridge Anticline

The Wyville–Thomson Ridge Anticline has a positive expression on seabed bathymetry (Fig. 8) and the top-basalt surface (Fig. 4). The axial trace of the fold is NW-trending and 200 km in length. The anticline extends from the Faroe Bank to the Hebrides/West Shetland Shelf, separating the Rockall Trough from the Faroe Shetland Channel. The Wyville–Thomson Ridge Anticline (Figs 14 & 15) has a wavelength of c. 28 km and an amplitude of up to 1.7 km (measured from the top-basalt surface). On the northern limb of the

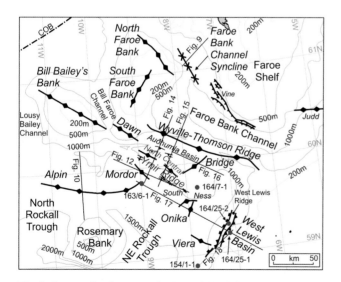

Fig. 8. Compressional structures in the north Rockall–Faroe study area. The locations of seismic illustrations are shown by solid black lines. Compressional structures are labelled in italics. The position of the continent–ocean boundary is based on Kimbell et al. (2005). For the location of the area, see Figure 3.

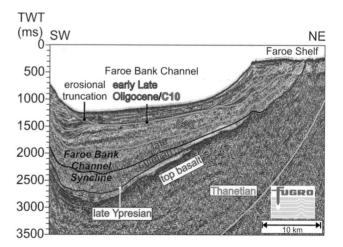

Fig. 9. North–south seismic profile of the Faroe Bank Channel Syncline. Seismic illustration courtesy of Fugro Multi Client Services. For the location of the seismic illustration, see Figure 8.

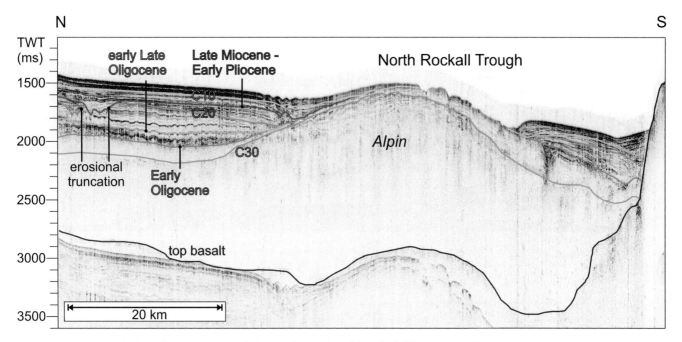

Fig. 10. North–south seismic profile of the Alpin Anticline. For the location of the seismic illustration, see Figure 8. Seismic profile, BGS © NERC 2009. All rights reserved.

Wyville–Thomson Ridge Anticline, the unconformities inferred as being the result of fold growth – Thanetian, late Ypresian, late Lutetian, C30 (Late Eocene) and the Early Oligocene – are marked by thinning and onlap of basaltic or sedimentary successions onto these surfaces. Previous work ascribed Late Paleocene–Early Eocene, Late Eocene, Oligocene and Miocene compression to the growth of the anticline (Boldreel & Andersen 1993; Johnson et al. 2005).

The Early Oligocene stratigraphic surface on the northern limb of the Wyville–Thomson Ridge is inferred as forming a composite unconformity with the early Late Oligocene unconformity. The early Late Oligocene unconformity is interpreted as the Top Palaeogene unconformity (TPU) previously mapped in the south Faroe Shetland Channel (Smallwood 2004). The C20 (late Early Miocene) unconformity is marked by onlap on the northern limb of the Wyville–Thomson Ridge. Here, the C10 (late Early Pliocene) unconformity, formed by bottom-water current activity,

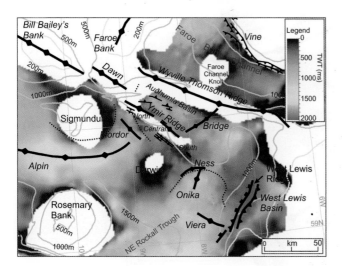

Fig. 11. Top-basalt–C30 (Late Eocene) isochore. Basalt escarpments, representing palaeo-shorelines, are delineated by dotted black lines. Compressional structures are labelled in italics.

truncates the late Ypresian, late Lutetian, C30, Early Oligocene/early Late Oligocene and C20 unconformities.

Auðhumla Basin Syncline

A syncline, the Auðhumla Basin, is located between the Ymir Ridge (North and Central) and the Wyville–Thomson Ridge anticlines (Figs 14 & 15). Within the Auðhumla Basin there are fault propagation folds (Fig. 15) which run in a NW direction, are 6–12 km in length and their associated reverse faults are northward dipping.

The development of the Auðhumla Basin Syncline is governed by the growth of both the Ymir and Wyville–Thomson Ridge anticlines. The earliest growth phase is inferred as Thanetian based on this unconformity defining growth of both the North Ymir Ridge and the Wyville–Thomson Ridge anticlines. On the northern limb of the Central Ymir Ridge anticline, the late Lutetian unconformity defines the earliest compressional phase, suggesting that the Auðhumla Basin Syncline formed at this time between the Wyville–Thomson Ridge and the Central Ymir Ridge anticlines.

Bridge Anticline

The Bridge Anticline is well expressed on the top-basalt surface (Fig. 4). The anticline is a NE-trending symmetrical fold between the Wyville–Thomson Ridge Anticline and the southern end of the Central Ymir Ridge Anticline and separates the Auðhumla Basin from the NE Rockall Trough (Fig. 8). The Bridge Anticline (Fig. 16) has an amplitude of up to 1.2 km (measured from the top-basalt surface) and a wavelength of 40 km. The fold is characterized by both the folding of basalt lava flows and Cenozoic sediments. On the crest and on the eastern limb of the fold, Middle Eocene sediment, up to 800 m thick, onlaps onto the late Ypresian unconformity. The Middle Eocene succession is interpreted as being sourced from the Wyville–Thomson Ridge to the north. These sediments are themselves onlapped by a wedge of Late Eocene sediment (c. 500 m thick). The late Lutetian unconformity, separating the Middle and Late Eocene sediment packages and the C30 (Late Eocene) unconformity on the eastern limb of the fold, are marked by strong onlap. Initial growth of the Bridge Anticline is

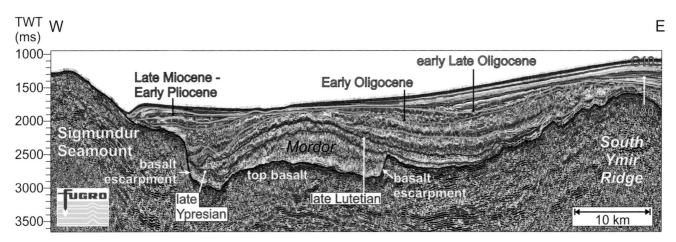

Fig. 12. East–west seismic profile of the Mordor and South Ymir Ridge Anticlines. Seismic illustration courtesy of Fugro Multi Client Services. For the location of the seismic illustration, see Figure 8.

interpreted to have occurred during the late Lutetian. Onlap and thinning onto the top-basalt surface on the eastern limb of the Bridge Anticline is interpreted as being associated with earlier growth of the proximal Wyville–Thomson Ridge Anticline.

Onika Anticline

The Onika Anticline lies between the Darwin seamount and the West Lewis Ridge (Fig. 17). The anticline is NNE-trending, has an amplitude of up to 800 m [measured from the C30 (Late Eocene) unconformity] and is not expressed at the seabed as its crest is covered with *c.* 500 m of post-Early Oligocene sediment. The Onika Anticline has been previously recognized, but not named, by Tate *et al.* (1999), and it appears to be a southeastern extension of the South Ymir Ridge Anticline. However, in this study, the Onika Anticline is considered to be a separate feature which is not linked to the Ymir Ridge. The anticline is located within a basin that is partly bounded by basalt escarpments, which has facilitated the accumulation of relatively thick Early Eocene sediment (Fig. 13). Sediment thickness between the top-basalt surface and the late Lutetian unconformity increases towards the crest of the fold and is consistent with folding of basin sediment. The thinning and onlap onto the late C30 (Late Eocene) unconformity suggest compression at this time.

West Lewis Ridge reverse faults

Reverse faults partially bound the margins of the NE-trending West Lewis Ridge, which is located on the east margin of the NE Rockall Trough (Fig. 8). These reverse faults (Figs 17 & 18) are termed the West Lewis reverse faults and are interpreted as normal faults which have undergone inversion. Along the eastern side of the ridge the reverse fault is 40 km in length, with the West Lewis Basin occurring within the hanging wall of the fault. This reverse fault cuts the late Ypresian and late Lutetian unconformities in the north (Fig. 17), but towards the south, it forms a fault-propagation fold (Fig. 18). The Thanetian unconformity, defined by onlap within the Palaeogene lava flow succession, is located

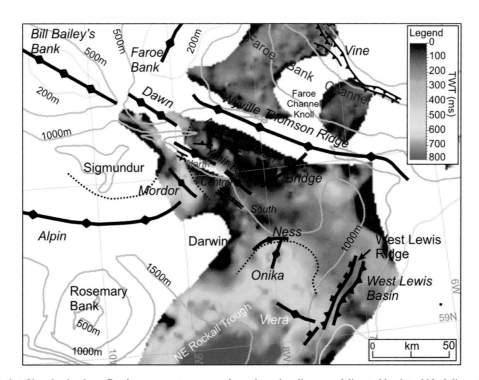

Fig. 13. Top-basalt–late Ypresian isochore. Basalt escarpments, representing palaeo-shorelines, are delineated by dotted black lines. Compressional structures are labelled in italics.

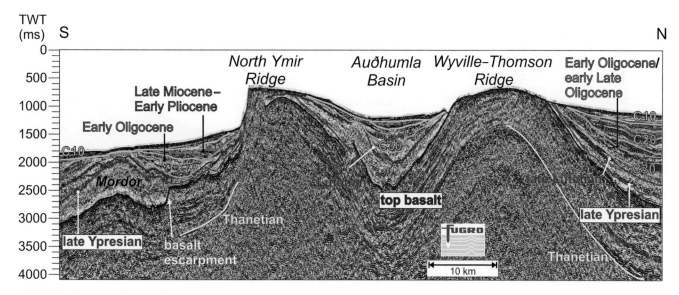

Fig. 14. North–south seismic profile of the Mordor, North Ymir Ridge and Wyville–Thomson Ridge anticlines. Seismic illustration courtesy of Fugro Multi Client Services. For the location of the seismic illustration, see Figure 8.

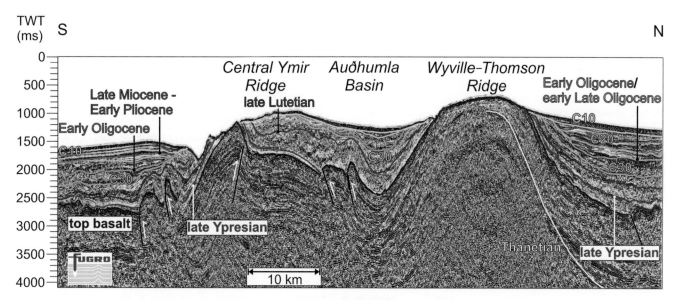

Fig. 15. North–south seismic profile of the Central Ymir Ridge and Wyville–Thomson Ridge anticlines. Seismic illustration courtesy of Fugro Multi Client Services. For the location of the seismic illustration, see Figure 8.

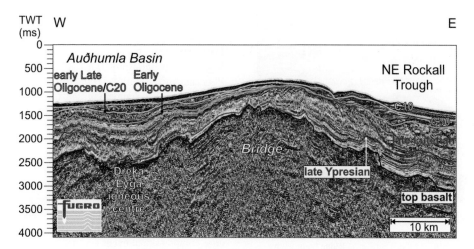

Fig. 16. East–west seismic profile of the Bridge Anticline. Seismic illustration courtesy of Fugro Multi Client Services. For the location of the seismic illustration, see Figure 8.

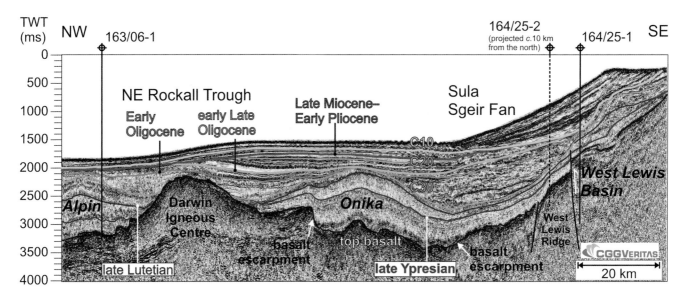

Fig. 17. NW–SE seismic profile of the Alpin Anticline, Onika Anticline and the inverted West Lewis Basin. Seismic illustration courtesy of CGG Veritas. For the location of the seismic illustration, see Figure 8.

in the south West Lewis Basin and defines the initiation of reverse faulting.

Viera Anticline

The Viera Anticline is a NW-trending fold, 30 km in length and located to the SW of the West Lewis Ridge in the NE Rockall Trough (Fig. 8). The amplitude of the anticline (measured from the late Ypresian unconformity) is up to c. 700 m (seismic profile not illustrated). The C30 (Late Eocene) unconformity, which is interpreted as marking the initial age of folding, is defined by onlap of Early Oligocene sediment. The Viera Anticline represents the folding of relatively thick Early Eocene sediment (Fig. 13).

Basin control on the orientation of compressional structures

The effects of basin inversion have been described in other areas of the NE Atlantic Margin. On the Norwegian Margin, for example,

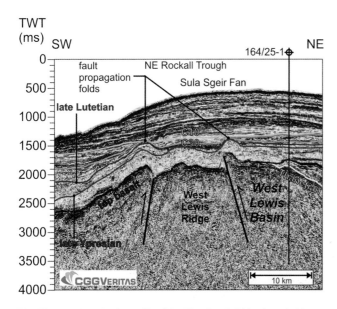

Fig. 18. SW–NE seismic profile of the West Lewis Ridge bounded by reverse faults. Seismic illustration courtesy of CGG Veritas. For the location of the seismic illustration, see Figure 8.

the Helland–Hansen Arch and other large domes are thought to be due to compressional reactivation of normal faults developed during Cretaceous rifting (Grunnaleite & Gabrielsen 1995; Vågnes et al. 1998). The Naglfar Dome, between the Vøring and Møre Basins (Fig. 1), is located within the Hel Graben (Lundin & Doré 2002). In addition, the Balder Graben in the NE Rockall Trough has been mapped beneath the Viera Anticline (Waddams & Cordingley 1999). Lundin & Doré (2002) proposed that Mid to Late Cenozoic compression resulted in basin inversion affecting deep Cretaceous depocentres on the Norwegian–Greenland Sea margins.

Results presented here suggest inversion of the West Lewis Basin from Thanetian to C30 (Late Eocene) time (Figs 17 & 18). In addition, the development of the Alpin, Mordor, Onika and Viera anticlines is considered to be due to the compression of Early Cenozoic depocentres. It has been previously recognized that basins within the NE Atlantic Margin follow the Caledonide (NE–SW) and the Lewisian (NW–SE, north–south and east–west) structural trends (Tate et al. 1999). All compressional structures in the Rockall–Faroe area have orientations which fit these trends. The orientation of the folds in the Rockall–Faroe area could, thus, be determined by their underlying basin morphologies. However, seismic masking by Palaeogene lavas largely precludes direct observation of such a relationship.

Bottom-water current activity

Post Oligocene unconformities, such as the C20 (late Early Miocene) unconformity (Fig. 10) and the Late Miocene–Early Pliocene unconformities (Figs 10 & 17), are characterized by strong erosional channelling and truncation that are most likely the result of bottom-water current activity rather than onlap onto growth folds. This is in contrast to the well documented Miocene unconformities which mark compression in the Vøring Basin (Blystad et al. 1995; Lundin & Doré 2002) and the Faroe–Shetland Basin (Ritchie et al. 2003; Johnson et al. 2005). In the Rockall–Faroe area, however, well defined reverse faults are observed which do not cut Early Miocene sediment (Fig. 6). This characteristic, and the lack of tighter folding immediately below the C20 (late Early Miocene) unconformity (Fig. 17), suggests that a Miocene compressional event was not as significant in this area.

It has been suggested that compression could alter seabed bathymetry to result in changes in bottom-water current activity

(Laberg et al. 2005). The study assesses the possible role of compression in influencing, directly or indirectly, bottom-water current activity.

Compression and C30 subsidence

The Rockall Trough is a basin which formed by rifting in the Early Cretaceous (Musgrove & Mitchener 1996). Sharp subsidence of this basin coincides with a major compressional phase at C30 (Late Eocene) time (Stoker 1997; Stoker et al. 2001; Praeg et al. 2005). Praeg et al. (2005) proposed that the C30 subsidence was the result of a loss of dynamic support by convection in the underlying mantle. However, the onset of this subsidence at the time of compression suggests that these events could be related. Flexural downwarping due to compression has been postulated for the Late Pliocene subsidence in the southern North Sea basin (Kooi et al. 1991). Compression-induced downwarping occurs as a result of folding of weak lithosphere of low flexural rigidity (Kooi et al. 1991). The Rockall Trough is characterized by weak lithosphere with an elastic thickness of approximately 6 km on the eastern margin near the Porcupine Bank (Daly et al. 2004). This is thinner than the 25 km elastic thickness of continental lithosphere (Fowler 2001). The Rockall Trough, having a relatively low elastic thickness, could have been prone to the effects of compression-induced flexural downwarping at C30 time. The deepening of the Rockall Trough facilitated the flow of bottom-water currents and the subsequent deposition of contourites (Stoker 1997; Stoker et al. 2001).

Compression and the Norwegian Sea Overflow

The present-day bottom-water currents in the North Rockall Trough stem partly from the Norwegian Sea Overflow (NSO) (Fig. 19; Stow & Holbrook 1984; Stoker 1997; Stoker et al. 2001). The NSO flows from the Faroe Bank Channel to the North Rockall Trough via a bathymetric gorge at the west end of the Auðhumla Basin Syncline (Figs 19 & 20). Previous work has proposed that the initiation of the NSO was in the Mid Miocene (Blanc et al. 1980; Bohrmann et al. 1990; Eldholm 1990; Ramsay et al. 1998; Stoker et al. 2005). The Auðhumla Basin

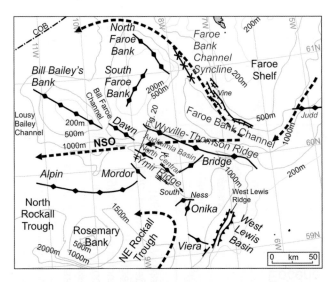

Fig. 19. Present-day circulation of bottom-water currents (dashed lines) in the north Rockall–Faroe study area (Stow & Holbrook 1984; Stoker 1997; Stoker et al. 2001). Note the bathymetric gorge between the Dawn and North Ymir Ridge anticlines, within the Auðhumla Basin, through which the Norwegian Sea Overflow passes. For the location of the area, see Figure 3.

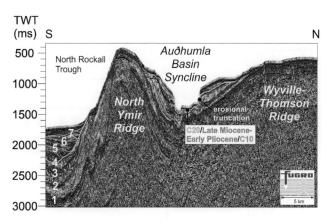

Fig. 20. North–south seismic profile across the west end of the Auðhumla Basin Syncline. Note the erosional truncation of Cenozoic sediment within the syncline. Seismic horizons: 1, top basalt; 2, late Ypresian; 3, late Lutetian; 4, C30 (Late Eocene); 5, Early Oligocene; 6, Late Miocene–Early Pliocene; 7, C10 (late Early Pliocene). Seismic illustration courtesy of Fugro Multi Client Services. For the location of the seismic illustration, see Figure 19.

Syncline would have thus been a major conduit for the passage of the NSO into the North Rockall Trough no later than Mid Miocene times. The location of this compressional feature would have influenced the direction of the NSO, affecting the subsequent distribution of contourites and unconformities within the North Rockall Trough.

Driving mechanisms

The formation of compressional features within the NE Atlantic Margin has been attributed to normal ridge push (Boldreel & Andersen 1993; Doré & Lundin 1996), hotspot-influenced ridge push (Lundin & Doré 2002), Alpine compression (Roberts 1989; Doré & Lundin 1996; Vågnes et al. 1998; Brekke 2000), Pyrenean compression (Knott et al. 1993), plate reorganization (Boldreel & Andersen 1993; Lundin & Doré 2002) and the Iceland Plateau Body force (Doré et al. 2008). Uplift and exhumation in the nearby Irish Sea during the early Palaeogene and the Late Palaeogene–Neogene has been partly attributed to Alpine compression and Atlantic continental break-up (Holford et al. 2005).

In this study, we attempt to link compressional events in the Rockall–Faroe area to ridge push (normal and hotspot-influenced), Alpine and Pyrenean compression, depth-dependent stretching/ asthenospheric upwelling and the Iceland body force (Fig. 21).

The Thanetian compressional event would coincide with the timing of postulated depth-dependent stretching which is thought to occur just prior to continental break-up as a result of asthenospheric upwelling which pushes the adjacent lithosphere towards the continental margin (Withjack et al. 1998; Kusznir et al. 2005). Depth-dependent stretching is modelled to affect a marginal zone 75–150 km from the continent–ocean boundary (Kusznir et al. 2005). The Hatton Bank, Lousy Bank, Bill Bailey's Bank, Faroe Bank anticlines and the Faroe Bank Channel Syncline are all located within such a marginal zone in the Rockall–Faroe area. The Thanetian unconformity defines the Faroe Bank Channel Syncline. Furthermore, onlap of Eocene sediment onto the top basalt lava flows of Hatton Bank suggests a structural high existed in this area prior to the Early Eocene. The North Ymir Ridge Anticline, Wyville–Thomson Ridge Anticline and the inverted West Lewis Basin also underwent Thanetian compressional growth. These structures, however, are located beyond the range of the marginal zone modelled to be affected by depth-dependent stretching. It is, therefore, postulated in this study that

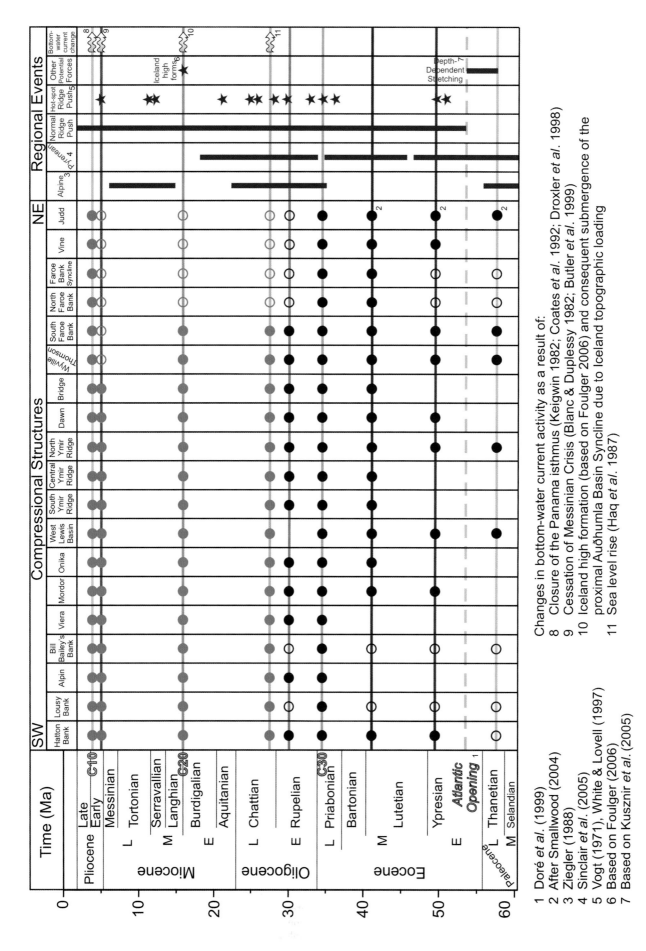

Fig. 21. Cenozoic unconformities within the Rockall–Faroe area and regional events affecting the NE Atlantic Margin. Red circles define the ages of compressional unconformities while green circles define the ages of non-compressional unconformities in the Rockall–Faroe area. Filled circles represent unconformities which have been mapped or dated. Circle outlines represent unconformities inferred to be present but have not been mapped.

these structures are more likely to be the result of compression on the periphery of a region uplifted by asthenospheric upwelling. A hotspot was present prior to NE Atlantic rifting which resulted in considerable uplift as a result of extensive asthenospheric upwelling (Skogseid & Eldholm 1988; White & McKenzie 1989). This uplift produces significant gravitational potential (White & McKenzie 1989) which can release horizontal stresses towards the rim of the uplifted area. Oligocene–Holocene compression of the Kwanza Basin, located in the continental margin off Angola, for example, has been attributed to mantle-driven uplift of the African superswell (Hudec & Jackson 2002). The Base Balder unconformity in the southern Faroe–Shetland Basin is attributed to uplift as a result of 'the introduction of asthenospheric mantle with an anomalously high potential temperature' (Smallwood & Gill 2002). The age of the Base Balder unconformity is modelled at Mid to Late Paleocene (60.5–55.2 Ma) (Jolley et al. 2002; cited in Smallwood & Gill 2002). It is the view, in this study, that the Base Balder unconformity in the southern Faroe–Shetland Channel is contemporaneous with the Thanetian (Late Paleocene) unconformity in the Rockall–Faroe area.

The late Ypresian, C30 (Late Eocene) and Early Oligocene compressional events coincide with the timings of hotspot-influenced or plume enhanced ridge push (Fig. 21). These compressional events also take place during times of Alpine and/or Pyrenean compression. Lundin & Doré (2002) negate the effects of Alpine and Pyrenean compression due to the presence of Cenozoic compressional features present on the East Greenland continental margin, indicating that the structures formed due to hotspot-influenced ridge push. The Traill Ø region on the East Greenland margin, for example, experienced compression during the Cenozoic (post-54 Ma) based on the folding of sills c. 54 Ma in age (Price et al. 1997). Kjelstad et al. (2003), however, cast doubt on any far-field stresses resulting in the compression and inversion of underlying basins. According to these authors the horizontal stress required to reverse a basement fault of a basin at 5 km depth, with a dip of 45° and an internal angle of friction of 30°, is 376 MPa based on the Mohr–Coulomb failure criterion. The force exerted onto the continental margin from hotspot-influenced ridge push is lower, modelled at 100 MPa (Bott 1993). However, the effects of hotspot-influenced ridge push could have potentially been amplified by Alpine and Pyrenean compression to result in the inversion of underlying basins in the Rockall–Faroe area. The East Greenland continental margin contains smaller scale deformation in comparison with deformation present in the NE Atlantic Margin, where structures are greater in length, width and amplitude (Tsikalas et al. 2005). This is consistent with the lack of contributing forces from the Alpine and Pyrenean compression.

However, the late Lutetian compressional event, which occurs at a time of Pyrenean compression, does not correlate with the timings of hotspot-influenced ridge push (Fig. 21). It remains uncertain in this study whether other factors could have contributed to the growth of compressional structures in the late Lutetian.

It is interesting to note that the uplifts bordering the ocean–continent boundary such as Lousy and Hatton Bank lie in a region shown to be underlain by highly intruded lower crust (White et al. 2010) and, where present, this 'underplate' is likely to have resulted in rapid isostatic uplift. However, major growth that we can observe on the anticlines bordering the ocean–continent boundary, such as Hatton Bank, occurred in Eocene and Oligocene times. The compressive strain is not restricted to the flank of the structure, but does mark its overall growth. Consequently, it would appear that growth of the Hatton Bank Anticline occurred millions of years after the 'underplate' is generally considered to have been intruded (Late Paleocene–Early Eocene times). On the basis of this timing of fold development, we prefer to explain the folds bordering the ocean–continent boundary such as Lousy and Hatton Bank as primarily compressional in origin. Of course, this would not preclude the possibility of an earlier phase of uplift associated with underplating.

Implications for prospectivity

Continental margins have often provided ideal settings for petroleum plays. Shale deposited in anoxic conditions during young ocean basin development has the potential to become an ideal source for hydrocarbons (Boillot 1981). As the ocean basin deepens, turbidite sands intercalated with hemipelagic sediments are deposited and can form reservoirs and seals (Boillot 1981). Normal faults formed as a result of rifting can act as conduits for migrating hydrocarbons. The presence of anticlines within continental margins completes the play setting by forming ideal structural traps of four-way dip closure (Doré & Lundin 1996; Doré et al. 1997). In the NE Atlantic Margin, gas accumulation in the Ormen Lange Dome in the Vøring Basin is located within such a structural trap (Lundin & Doré 2002; Smith & Møller 2003). The timing of growth of compressional structures is critical in determining their presence at the time of hydrocarbon migration. Uplift as a result of compression can remove potential source rocks from the gas/oil generation window. Thus, there is often a mismatch between the timing of petroleum migration from source rocks and later trap-forming events in structurally inverted basins. Secondary migration, however, can supply these later traps with hydrocarbons. Further understanding of the area's prospective potential will rely upon determining whether there were ever any mature source rocks and if the compressional structures described herein lay on suitable migration pathways as, or after, they formed.

Conclusions

Horizontal forces have resulted in compression within the Rockall–Faroe area to form structures such as the Lousy Bank and Alpin anticlines during Cenozoic times. The orientation and extent of compressional structures were most likely governed by their underlying basin morphologies. The growth of structures in the Thanetian could be linked to depth-dependent stretching and asthenospheric upwelling. Late Ypresian, C30 (Late Eocene) and Early Oligocene compressional phases correlate well with hotspot-influenced ridge push, but were most likely due to the combined effects of this force and the Alpine and Pyrenean stresses. The study suggests evidence for the role of compression in changing seabed bathymetry to result in changes in bottom-water current activity. Subsidence of the Rockall Trough at C30 (Late Eocene), which increased bottom-water current activity in the basin, could have resulted from compression-induced flexural downwarping. The west end of the Auðhumla Basin Syncline is a major conduit for the transport of the NSO from the Faroe Bank Channel to the North Rockall Trough. The growth of the syncline has controlled the distribution of sediments and unconformities within the North Rockall Trough by influencing the direction and existence of the NSO. A full understanding of the timing and distribution of the structures and basin subsidence histories will help unlock the petroleum prospectivity in the area.

This work was carried out jointly between the British Geological Survey and The University of Edinburgh under the auspices of a British University Funding Initiative (BUFI) Grant, which is gratefully acknowledged. The work was carried out at the British Geological Survey. We thank Fugro Multo Client Services and CGG Veritas for kind permission to publish their seismic data. We also thank M. Stoker for helpful discussions. The contributions from J. D. Ritchie, H. Johnson and K. Hitchen are published with permission of the Executive Director, British Geological Survey (NERC).

References

Andersen, M. S. & Boldreel, L. O. 1995. Effect of Eocene–Miocene compression structures on bottom-water currents in the Faeroe-Rockall area. In: Scrutton, R. A., Stoker, M. S., Shimmield, G. B. & Tudhope, A. W. (eds) *The Tectonics, Sedimentation and Palaeoceanography of the North Atlantic Region*. Geological Society, London, Special Publications, **90**, 141–143.

Blanc, P. & Duplessy, J. 1982. The deep-water circulation during the Neogene and the impact of the Messinian salinity Crisis. *Deep-Sea Research*, **29**, 12A, 1391–1414.

Blanc, P.-L., Rabussier, D., Vernaud-Grazzini, C. & Duplessy, J.-C. 1980. North Atlantic Deep Water formed by the later middle Miocene. *Nature*, **283**, 553–555.

Blystad, P., Brekke, H., Faerseth, R. B., Larsen, B. T., Skogseid, J. & Tørudbakken, B. 1995. *Structural Elements of the Norwegian Continental Shelf. Part II: The Norwegian Sea Region*. Norwegian Petroleum Directorate Bulletins, **8**.

Bohrmann, G., Henrich, R. & Thiede, J. 1990. Miocene to Quaternary Paleoceanography in the northern North Atlantic: variability in carbonate and biogenic opal accumulation. In: Bleie, U. & Thiede, J. (eds) *Geological History of the Polar Ocean*. Kluwer, Dordrecht, 647–675.

Boillot, G. 1981. *Geology of the Continental Margins*. Longman Higher Education, Harlow.

Boldreel, L. O. & Andersen, M. S. 1993. Late Paleocene to Miocene compression in the Faeroe-Rockall area. In: Parker, J. R. (ed.) *Petroleum Geology of Northwest Europe: Proceedings of the 4th Conference*. Geological Society, London, 1025–1034; doi: 10.1144/0041025.

Boldreel, L. O. & Andersen, M. S. 1994. Tertiary development of the Faeroe-Rockall Plateau based on reflection seismic data. *Bulletin of the Geological Society of Denmark*, **41**, 162–180.

Boldreel, L. O. & Andersen, M. S. 1998. Tertiary compressional structures on the Faroe–Rockall Plateau in relation to northeast Atlantic ridge-push and Alpine foreland stresses. *Tectonophysics*, **300**, 13–28.

Bott, M. H. P. 1993. Modelling the plate-driving mechanism. *Journal of the Geological Society, London*, **150**, 941–951.

Brekke, H. 2000. The tectonic evolution of the Norwegian Sea Continental Margin with emphasis on the Vøring and Møre Basins. In: Nøttvedt, A. et al. (eds) *Dynamics of the Norwegian Margin*. Geological Society, London, Special Publications, **167**, 327–378.

Butler, R. W. H., McClelland, E. & Jones, R. E. 1999. Calibrating the duration and timing of the Messinian salinity crisis in the Mediterranean: linked tectonoclimatic signals in thrust-top basins of Sicily. *Journal of the Geological Society, London*, **156**, 827–835.

Coates, A. G., Jackson, J. B. C., Collins, L. S., Cronin, T. M., Dowsett, H. J., Bybell, L. M., Jung, P. & Obando, J. A. 1992. Closure of the Isthmus of Panama: the nearshore marine record of Costa Rica and western Panama. *Geological Society of America Bulletin*, **104**, 814–828.

Daly, E., Brown, C., Stark, C. P. & Ebinger, C. J. 2004. Wavelet and multitaper coherence methods for assessing the elastic thickness of the Irish Atlantic margin. *Geophysical Journal International*, **159**, 445–459.

Davies, R. J. & Cartwright, J. 2002. A fossilized Opal A to Opal C/T transformation on the northeast Atlantic margin: support for a significantly elevated Palaeogeothermal gradient during the Neogene? *Basin Research*, **14**, 467–486.

Dolan, J. F. 1986. The relationship between 'R2' seismic reflector and a zone of abundant detrital and authigenic smectites, Deep Sea Drilling Project Hole 610. Rockall Plateau region, North Atlantic. In: Ruddiman, W. F., Kidd, R. B., Thomas, E. et al. (eds) *Initial Reports of the Deep Sea Drilling Project*, **94**, 1109–1115.

Doré, A. G. & Lundin, E. R. 1996. Cenozoic compressional structures on the NE Atlantic margin: nature, origin and potential significance for hydrocarbon exploration. *Petroleum Geoscience*, **2**, 299–311.

Doré, A. G., Lundin, E. R., Birkeland, Ø., Eliassen, P. E. & Jensen, L. N. 1997. The NE Atlantic Margin: implications of late Mesozoic and Cenozoic events for hydrocarbon prospectivity. *Petroleum Geoscience*, **3**, 117–131.

Doré, A. G., Lundin, E. R., Jensen, L. N., Birkeland, O., Eliassen, P. E. & Fichler, C. 1999. Principal tectonic events in the evolution of the northwest European Atlantic margin. In: Fleet, A. J. & Boldy, S. A. R. (eds) *Petroleum Geology of Northwest Europe: Proceedings of the 5th Conference*. Geological Society, London, 41–61; doi: 10.1144/0050041.

Doré, A. G., Lundin, E. R., Kusznir, N. J. & Pascal, C. 2008. Potential mechanisms for the genesis of Cenozoic domal structures on the NE Atlantic margin: pros, cons and some new ideas. In: Johnson, H., Doré, A. G., Gatliff, R. W., Holdsworth, R., Lundin, E. R. & Ritchie, J. D. (eds) *The Nature and Origin of Compression in Passive Margins*. Geological Society, London, Special Publications, **306**, 1–26.

Droxler, A. W., Burke, K. C. et al. 1998. Caribbean constraints on circulation between Atlantic and Pacific Oceans over the past 40 million years. In: Crowley, T. J. & Burke, K. C. (eds) *Tectonic Boundary Conditions for Climate Reconstructions*. Oxford University Press, New York, 169–191.

Eldholm, O. 1990. Paleogene North Atlantic magmatic–tectonic events: environmental implications. *Memorie Societa Geologica Italiana*, **44**, 13–28.

Foulger, G. R. 2006. Older crust underlies Iceland. *Geophysical Journal International*, **165**, 672–676.

Fowler, C. M. R. 2001. *The Solid Earth; An Introduction to Global Geophysics*. Cambridge University Press, Cambridge.

Grunnaleite, I. & Gabrielsen, R. H. 1995. Structure of the Møre Basin, mid-Norway continental margin. *Tectonophysics*, **252**, 221–251.

Haq, B. U., Hardenbol, J. & Vail, P. R. 1987. Chronology of fluctuating sea levels since the Triassic. *Science*, **235**, 1156–1167.

Hitchen, K. 2004. The geology of the UK Hatton–Rockall margin. *Marine and Petroleum Geology*, **21**, 993–1012.

Holford, S. P., Turner, J. P. & Green, P. F. 2005. Reconstructing the Mesozoic–Cenozoic exhumation history of the Irish Sea basin system using apatite fission track analysis and vitrinite reflectance data. In: Doré, A. G. & Vining, B. A. (eds) *Petroleum Geology: North-West Europe and Global Perspectives: Proceedings of the 6th Petroleum Geology Conference*. Geological Society, London, 1095–1107; doi: 10.1144/0061095.

Hudec, M. R. & Jackson, M. P. A. 2002. Structural segmentation, inversion, and salt tectonics on a passive margin: evolution of the Inner Kwanza Basin, Angola. *Geological Society of America Bulletin*, **114**, 1222–1244.

IOC, IHO & BODC. 2003. *Centenary Edition of the GEBCO Digital Atlas*. British Oceanographic Centre, Liverpool.

Johnson, H., Ritchie, J. D., Hitchen, K., McInroy, D. B. & Kimbell, G. S. 2005. Aspects of the Cenozoic deformational history of the Northeast Faroe–Shetland Basin, Wyville–Thomson Ridge and Hatton Bank areas. In: Doré, A. G. & Vining, B. A. (eds) *Petroleum Geology: North-West Europe and Global Perspectives: Proceedings of the 6th Petroleum Geology Conference*. Geological Society, London, 993–1007; doi: 10.1144/0060993.

Jolley, D. W., Clarke, B. & Kelley, S. 2002. Paleogene time scale miscalibration: evidence from the dating of the North Atlantic Igneous Province. *Geology*, **30**, 7–10.

Keigwin, L. 1982. Isotopic paleoceanography of the Caribbean and East Pacific: role of Panama uplift in late Neogene time. *Science*, **217**, 350–353.

Kimbell, G. S., Ritchie, J. D., Johnson, H. & Gatliff, R. W. 2005. Controls on the structure and evolution of the NE Atlantic margin revealed by regional potential field imaging and 3D modelling. In: Doré, A. G. & Vining, B. A. (eds) *Petroleum Geology: North-West Europe and Global Perspectives: Proceedings of the 6th Petroleum Geology Conference*. Geological Society, London, 933–945; doi: 10.1144/0060933.

Kjelstad, A., Skogseid, J., Langtangen, H. P., Bjørlykke, K. & Høeg, K. 2003. Differential loading by prograding sedimentary wedges on continental margins: an arch-forming mechanism. *Journal of Geophysical Research*, **108**, 2036; doi: 10.1029/2001JB001145.

Knott, S. D., Burchell, M. T., Jolley, E. J. & Fraser, A. J. 1993. Mesozoic to Cenozoic plate reconstructions of the North Atlantic and hydrocarbon plays of the Atlantic margins. In: Parker, J. R. (ed.) *Petroleum Geology of Northwest Europe: Proceedings of the 4th Conference*. Geological Society, London, 953–974; doi: 10.1144/0040953.

Kooi, H., Hettema, M. & Cloetingh, S. 1991. Lithospheric dynamics and the rapid Pliocene-Quaternary subsidence phase in the southern North Sea Basin. *Tectonophysics*, **192**, 245–259.

Kusznir, N. J., Hunsdale, R., Roberts, A. M. & iSIMM Team, 2005. Timing and magnitude of depth-dependent lithosphere stretching on the southern Lofoten and northern Vøring continental margins offshore mid-Norway: implication for subsidence and hydrocarbon maturation at volcanic rifted margins. *In*: Doré, A. G. & Vining, B. A. (eds) *Petroleum Geology: North-West Europe and Global Perspectives: Proceedings of the 6th Petroleum Geology Conference*. Geological Society, London, 767–783; doi: 10.1144/0060767.

Laberg, J. S., Stoker, M. S. *et al.* 2005. Cenozoic alongslope processes and sedimentation on the NW European Atlantic Margin. *Marine and Petroleum Geology*, **22**, 1069–1088.

Løseth, H. & Henriksen, S. 2005. A Middle to Late Miocene compression phase along the Norwegian passive margin. *In*: Doré, A. G. & Vining, B. A. (eds) *Petroleum Geology: North-West Europe and Global Perspectives: Proceedings of the 6th Petroleum Geology Conference*. Geological Society, London, 845–859; doi: 10.1144/0060845.

Lundin, E. & Doré, A. G. 2002. Mid-Cenozoic post-breakup deformation in the 'passive' margins bordering the Norwegian-Greenland Sea. *Marine and Petroleum Geology*, **19**, 79–93.

McQuillin, R., Bacon, M. & Barclay, W. 1984. *An Introduction to Seismic Interpretation*. Graham and Trotman, London.

Musgrove, F. W. & Mitchener, B. 1996. Analysis of the pre-Tertiary rifting history of the Rockall Trough. *Petroleum Geoscience*, **2**, 353–360.

Praeg, D., Stoker, M. S., Shannon, P. M., Ceramicola, S., Hjelstuen, B., Laberg, J. S. & Mathiesen, A. 2005. Episodic Cenozoic tectonism and the development of the NW European 'passive' continental margin. *Marine and Petroleum Geology*, **22**, 1007–1030.

Price, S., Brodie, J., Whitham, A. & Kent, R. 1997. Mid-Tertiary rifting and magmatism in the Traill Ø region, East Greenland. *Journal of the Geological Society, London*, **54**, 419–434.

Ramsay, A. T. S., Smart, C. W. & Zachos, J. C. 1998. A model of early to middle Miocene deep ocean circulation for the Atlantic and Indian Oceans. *In*: Cramp, A., MacLeod, C. J., Lee, S. V. & Jones, E. J. W. (eds) *Geological Evolution of Ocean Basins: Results From the Ocean Drilling Program*. Geological Society, London, Special Publications, **131**, 55–70.

Ritchie, J. D., Johnson, H. & Kimbell, G. S. 2003. The nature and age of Cenozoic contractional deformation within the NE Faroe–Shetland Basin. *Marine and Petroleum Geology*, **20**, 399–409.

Ritchie, J. D., Johnson, H., Quinn, M. F. & Gatliff, R. W. 2008. The effects of Cenozoic compression within the Faroe–Shetland Basin and adjacent areas. *In*: Johnson, H., Doré, A. G., Gatliff, R. W., Holdsworth, R., Lundin, E. R. & Ritchie, J. D. (eds) *The Nature and Origin of Compression in Passive Margins*. Geological Society, London, Special Publications, **306**, 121–136.

Roberts, D. G. 1989. Basin inversion in and around the British Isles. *In*: Cooper, M. A. & Williams, G. D. (eds) *Inversion Tectonics*. Geological Society, London, Special Publications, **44**, 131–150.

Shannon, P. M, Stoker, M. S. *et al.* 2005. Sequence stratigraphic analysis in deep-water, underfilled NW European passive margin basins. *Marine and Petroleum Geology*, **22**, 1185–1200.

Sinclair, H. D., Gibson, M., Naylor, M. & Morris, R. G. 2005. Asymmetric growth of the Pyrenees revealed through measurement and modelling of orogenic fluxes. *American Journal of Sciences*, **305**, 369–406.

Skogseid, J. & Eldholm, O. 1988. Early Cainozoic evolution of the Norwegian volcanic passive margin and the formation of marginal highs. *In*: Morton, A. C. & Parson, L. M. (eds) *Early Tertiary Volcanism and the Opening of the NE Atlantic*. Geological Society, London, Special Publications, **39**, 49–56.

Smallwood, J. R. 2004. Tertiary inversion in Faroe–Shetland Channel and the development of major erosional scarps. *In*: Davies, R. J., Stewart, S. A., Cartwright, J. A., Lappin, M. & Underhill, J. R. (eds) *3D Seismic Technology: Application to the Exploration of Sedimentary Basins*. Geological Society, London, Memoirs, **29**, 187–198.

Smallwood, J. R. & Gill, C. E. 2002. The rise and fall of the Faroe–Shetland Basin: evidence from seismic mapping of the Balder Formation. *Journal of the Geological Society, London*, **159**, 627–630.

Smith, R. & Møller, N. 2003. Sedimentology and reservoir modelling of the Ormen Lange field, mid Norway. *Marine and Petroleum Geology*, **20**, 601–613.

Stoker, M. S. 1997. Mid- to late Cenozoic sedimentation on the continental margin off NW Britain. *Journal of the Geological Society, London*, **154**, 509–515.

Stoker, M. S., Van Weering, T. C. E. & Svaerdborg, T. 2001. A Mid- to Late Cenozoic tectonostratigraphic framework for the Rockall Trough. *In*: Shannon, P. M., Haughton, P. D. W. & Cororan, D. V. (eds) *The Petroleum Exploration of Ireland's Offshore Basins*. Geological Society, London, Special Publications, **188**, 411–438.

Stoker, M. S., Praeg, D. *et al.* 2005. Neogene evolution of the Atlantic continental margin of NW Europe (Lofoten Islands to SW Ireland): anything but passive. *In*: Doré, A. G. & Vining, B. A. (eds) *Petroleum Geology: North-West Europe and Global Perspectives: Proceedings of the 6th Petroleum Geology Conference*. Geological Society, London, 1057–1076; doi: 10.1144/0061057.

Stow, D. A. V. & Holbrook, J. A. 1984. North Atlantic contourites: an overview. *In*: Piper, D. J. & Stow, D. A. V. (eds) *Fine Grained Sediments: Deep Water Processes and Fabrics*. Geological Society of America, Boulder, CO, Special Publications, **15**, 245–256.

Tate, M. P., Dodd, C. D. & Grant, N. T. 1999. The Northeast Rockall Basin and its significance in the evolution of the Rockall-Faeroes/East Greenland rift system. *In*: Fleet, A. J. & Boldy, S. A. R. (eds) *Petroleum Geology of Northwest Europe: Proceedings of the 5th Conference*. Geological Society, London, 391–406; doi: 10.1144/0050391.

Tsikalas, F., Faleide, J. I., Eldholm, O. & Wilson, J. 2005. Late Mesozoic–Cenozoic structural and stratigraphic correlations between the conjugate mid-Norway and NE Greenland continental margins. *In*: Doré, A. G. & Vining, B. A. (eds) *Petroleum Geology: North-West Europe and Global Perspectives: Proceedings of the 6th Petroleum Geology Conference*. Geological Society, London, 785–801; doi: 10.1144/0060785.

Vågnes, E., Gabrielsen, R. H. & Haremo, P. 1998. Late Cretaceous–Cenozoic intraplate contractional deformation at the Norwegian continental shelf: timing, magnitude and regional implications. *Tectonophysics*, **300**, 29–46.

Vogt, P. R. 1971. Asthenosphere motion recorded by the ocean floor south of Iceland. *Earth and Planetary Science Letters*, **13**, 153–160.

Waddams, P. & Cordingley, T. 1999. The regional geology and exploration potential of NE Rockall Basin. *In*: Fleet, A. J. & Boldy, S. A. R. (eds) *Petroleum Geology of Northwest Europe: Proceedings of the 5th Conference*. Geological Society, London, 379–390; doi: 10.1144/0050379.

White, R. & McKenzie, D. 1989. Magmatism at rift zone: the generation of volcanic continental margins and flood basalts. *Journal of Geophysical Research*, **94**, 7685–7729.

White, N. & Lovell, B. 1997. Measuring the pulse of a plume with the sedimentary record. *Nature*, **387**, 888–891.

White, R. S., Eccles, J. D. & Roberts, A. W. 2010. Constraints on volcanism, igneous intrusion and stretching on the Rockall Faroe–continental margin. *In*: Vining, B. A. & Pickering, S. C. (eds) *Petroleum Geology: From Mature Basins to New Frontiers – Proceedings of the 7th Petroleum Geology Conference*. Geological Society, London. 831–842; doi: 10.1144/0070831.

Withjack, M. O., Schlische, R. W. & Olsen, P. E. 1998. Diachronous rifting, drifting, and inversion on the passive margin of Central Eastern North America: An analogue for other passive margins. *AAPG Bulletin*, **82**, (5A), 817–835.

Ziegler, P. A. 1988. *Evolution of the Arctic-North Atlantic and the Western Tethys*. AAPG Memoir, **43**.

Ziska, H. & Varming, T. 2008. Palaeogene evolution of the Ymir and Wyville Thomson ridges, European North Atlantic margin. *In*: Johnson, H., Doré, A. G., Gatliff, R. W., Holdsworth, R., Lundin, E. R. & Ritchie, J. D. (eds) *The Nature and Origin of Compression in Passive Margins*. Geological Society, London, Special Publications, **306**, 153–168.

Episodic uplift and exhumation along North Atlantic passive margins: implications for hydrocarbon prospectivity

P. JAPSEN,[1] P. F. GREEN,[2] J. M. BONOW,[1] E. S. RASMUSSEN,[1] J. A. CHALMERS[1] and T. KJENNERUD[3]

[1]*Geological Survey of Denmark and Greenland (GEUS), Øster Voldgade 10, DK-1350 Copenhagen, Denmark (e-mail: pj@geus.dk)*
[2]*Geotrack International, 37 Melville Road, Brunswick West, Victoria 3055, Australia*
[3]*Exploro Geoservices, Stiklestadveien 1, 7041 Trondheim, Norway*

Abstract: We present observations that demonstrate that the elevated passive margins around the North Atlantic were formed by episodic, post-rift uplift movements that are manifest in the high-lying peneplains that characterize the coastal mountains, in the unconformities in the adjacent sedimentary basins and in accelerated subsidence in the basin centres. Results from West Greenland show that subsidence of the rifted margin took place for c. 25 Myr after rifting and breakup in the Paleocene, as predicted by classical rift theory, but that this development was reversed by a series of uplift movements (starting at c. 35, 10 and 5 Ma) that remain unexplained. East Greenland and Scandinavia seem to have had a similar evolution of post-rift subsidence followed by uplift starting at c. 35 Ma. There was no notable fall in sea-level at this time, so the subsiding basins must have been inverted by tectonic forces. We speculate that the forces causing this phase were related to the plate boundary reorganization in the North Atlantic around Chron 13 time. One feature that these areas have in common is that uplift took place along the edges of cratons where the thickness of the crust and lithosphere changes substantially over a short distance. It may be that the lateral contrasts in the properties of the stretched and unstretched lithosphere make the margins of the cratons unstable long after rifting. These vertical movements have profound influence on hydrocarbon systems, not only in frontier areas such as West and East Greenland, where Mesozoic basins are deeply truncated and exposed onshore, but also for the understanding of near-shore hydrocarbon deposits in mature areas such as the North Sea Basin, where low-angular unconformities may represent episodes of deposition and removal of significant sedimentary sections.

Keywords: uplift, landscape, hydrocarbons, Greenland, Scandinavia, North Atlantic, exhumation, erosion

There are mountains along the passive continental margins on both sides of the North Atlantic (north of Charlie–Gibbs Fracture zone; Figs 1 & 2), as along many passive margins around the world. Characteristic features of these mountains are elevated plateaux that are cut by deep river valleys that may have been enlarged by glaciers (e.g. Lidmar-Bergström et al. 2000; Bonow 2006a, b). We have previously shown that the regional peneplain in West Greenland (Fig. 2) was formed as an erosion surface after removal of up to 1 km of additional section starting at c. 35 Ma, following kilometre-scale post-rift subsidence after breakup in the Paleocene, and that the present-day mountains are the result of late Neogene uplift and dissection of this erosion surface (Japsen et al. 2006, 2009). Peneplains have also been identified at high elevations in Norway, and it has been suggested that these surfaces were formed during the Cenozoic (Lidmar-Bergström et al. 2000, 2007). Neogene uplift of the land masses along North Atlantic rift systems has been documented by numerous studies (see Japsen & Chalmers 2000 for a review and also Bonow et al. 2007), but it is less widely recognized that such elevated plateaux are evidence of at least two major phases of uplift: one phase to produce the peneplain as an erosion surface that cuts unconformably across rocks of different ages and resistance, and a second phase of uplift to bring the peneplain to its present elevation above sea-level (a.s.l.).

The sedimentary successions in many basins around the North Atlantic also provide evidence for significant vertical movements and deep exhumation during the Neogene and earlier (e.g. Green & Duddy 2010). Pre-Cenozoic rocks are commonly exposed onshore and the pre-Quaternary sediments offshore are generally of Neogene age. Inclined Palaeogene and older beds are truncated by erosional unconformities along many coastlines such as along the western and eastern margins of the North Sea Basin and along Norway (Fig. 1; Japsen & Chalmers 2000). This configuration is consistent with Neogene uplift of the continents, and of the adjacent shelf, which significantly post-dates Palaeogene breakup both in the North Atlantic and between Greenland and Canada. That most of this uplift is unrelated to late Cenozoic glaciations has been documented by several studies, notably Stoker et al. (2005), who found that early Pliocene tilting of the Atlantic continental margin of NW Europe pre-dates the onset of widespread glaciation. The sedimentary succession, however, may also include low-angle unconformities that represent deep, regional exhumation that is not easily recognized.

These observations suggest strongly that episodic, post-rift uplift along passive continental margins has formed the present high mountains and affected the adjacent sedimentary basins. Such vertical movements cannot easily be understood in terms of the classical theory of continental stretching that predicts deposition of a thick post-rift sequence overlying both the rift and its margins (e.g. McKenzie 1978; White & McKenzie 1988). The ability to understand and model geological and physical processes in the past is, however, fundamental for our understanding of how hydrocarbon systems have evolved. It is thus essential to have an open-minded approach to such observations that indicates that theories which have proven useful for decades may not be applicable under all circumstances or may not tell the complete story. Recognition of such evidence is essential in order for new theories to be

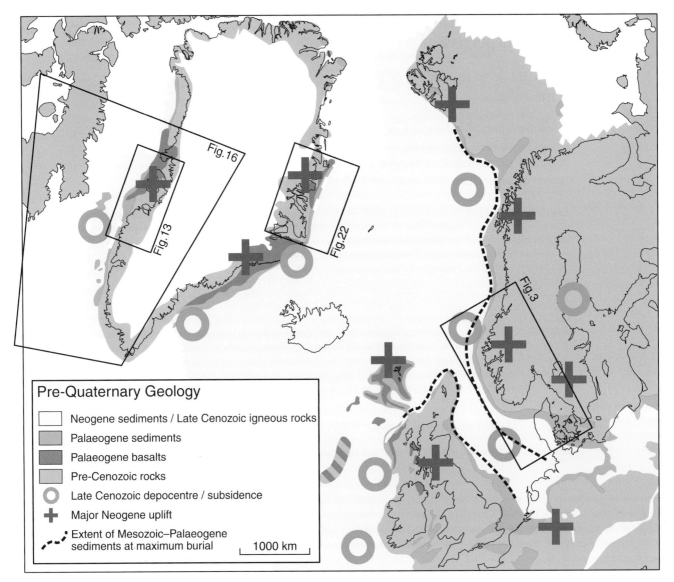

Fig. 1. Pre-Quaternary geology around the North Atlantic, showing centres of Neogene/late Cenozoic uplift and of accelerated subsidence/deposition. Pre-Cenozoic rocks are generally exposed onshore and the pre-Quaternary sediments offshore are generally of Neogene age. Between the two, inclined Palaeogene and older beds are truncated by erosional unconformities along many coastlines. Cenozoic exhumation is not limited to the areas of high topography, but has also affected much if not all of the area where Neogene cover is absent, as well as some regions where Neogene sediments are present. Along the Atlantic margin of NW Europe and Scandinavia, the extent of Mesozoic–Palaeogene sediments at maximum burial is traced (dashed line) based on numerous investigations (e.g. Japsen & Chalmers 2000; Japsen et al. 2007a; Green & Duddy 2010). Study areas discussed in this paper are indicated. The evidence presented here points to major Neogene uplift of the continents and the nearby basins which significantly post-dates the Palaeogene breakup. Modified after Japsen & Chalmers (2000).

formulated which may allow us to understand the nature of the unknown processes that are apparently driving vertical movements along passive continental margins.

It is thus a fundamental question as to whether a hiatus in the stratigraphic record represents an episode of stability and non-deposition or an event involving deposition and removal of rocks. A hiatus represents not only a gap in the stratigraphic record, but also a gap in our understanding of the geological history. The interpretation of what happened during the formation of a hiatus can thus have a profound influence on our understanding of the hydrocarbon system in an area (cf. Doré et al. 2002): when did maximum burial occur? How much section has been removed? How have migration paths been affected? Has a significant amount of hydrocarbon charge been lost from a breached reservoir during uplift? Such questions become very important not only in frontier areas such as West and East Greenland where Mesozoic basins are deeply truncated and exposed onshore, but also for the understanding of near-shore hydrocarbon deposits in mature areas such as the North Sea Basin (e.g. Japsen et al. 2007a).

Here we summarize results from recent and ongoing studies of burial, uplift and exhumation that are based on the integration of palaeothermal (apatite fission-track analysis, AFTA, and vitrinite reflectance, VR) data, palaeoburial (sonic) data and analysis of large-scale landforms constrained by the geological record. We will thus provide and review evidence for tectonic control of the palaeogeography of the eastern North Sea Basin (Case A), Neogene exhumation offshore south Norway (Case B), formation of the West Greenland mountains long after rifting and breakup (Case C), removal during the Oligocene of a significant cover offshore West Greenland (Case D), deposition and subsequent removal of a thick Ordovician–Triassic cover onshore West Greenland (Case E) and for post-rift burial of the East Greenland margin followed by uplift and exhumation (Case F).

Fig. 2. Photos of the uplifted, regional peneplain in West Greenland (the upper planation surface, UPS) on (**a**) Disko and (**b**) northeast of Sukkertoppen Iskappe, in both cases around 900 m a.s.l. The UPS cuts across Precambrian basement (b), but the erosion surface can be traced northwards, to where it cuts across Palaeogene basalt on Nuussuaq and Disko (a). The peneplain is thus of Cenozoic age across central West Greenland and interpretation of AFTA and VR data has shown that the peneplain is an erosional surface formed after removal of a rock column up to 1 km thick after c. 35 Ma (see Fig. 15; Japsen et al. 2005b, 2009). Photo locations on Figure 13. Photo a, Niels Nielsen; photo b, Karsten Secher. Modified from Japsen et al. 2006, 2009.

Methods for estimating timing and magnitude of uplift and exhumation

Corcoran & Doré (2005) presented a review of techniques for estimating the magnitude and timing of exhumation in sedimentary basins and found that they could be categorized into four reference frames: tectonic, thermal, compactional (or rather effective stress) and stratigraphic. To this list we want to add geomorphological analysis that investigates the formation, preservation and destruction of large-scale landforms relative to changes in base level.

Tectonic-based techniques

A framework for the assessment of exhumation in offshore basins is provided by the lithospheric stretching model of McKenzie (1978), which predicts an initial syn-rift (fault related) subsidence followed by an exponentially decreasing post-rift (thermal) subsidence. White & McKenzie (1988) extended this analysis into two dimensions and showed that the post-rift subsidence cover should extend beyond the margins of the actual rift. Deviations of the observed from the predicted subsidence curves may be used to estimate the magnitude and timing of exhumation (e.g. Rowley & White 1998).

Palaeothermal techniques

Since temperature increases with depth within a sedimentary basin, determining the thermal history of a sedimentary unit provides the basis for estimating the depth to which it has been buried. A variety of methods are available which essentially constrain the maximum temperature experienced by a sedimentary unit, the most widely adopted of which is VR (e.g. Tissot & Welte 1984). In contrast, AFTA (Green et al. 1989, 2002) not only allows determination of the maximum temperature attained by a sedimentary unit, but also the time at which the sample began to cool from that temperature. In tandem, AFTA and VR in a vertical sequence of samples allow determination of the palaeogeothermal gradient at the palaeothermal maximum and extrapolation to an appropriate surface temperature allows determination of the amount of section that has been removed as a result of uplift and erosion (e.g. Bray et al. 1992; Japsen et al. 2005b, 2007a, 2009).

Palaeoburial (effective-stress) techniques

Reduction in sediment porosity takes place during burial due to the increased effective stress (the load of the overburden minus the pore pressure) that causes mechanical compaction and redistribution of mass due to pressure solution at grain contacts (e.g. Sclater & Christie 1980). As porosity reduction is largely irreversible, porosity becomes a proxy for the maximum burial (effective stress) of the sedimentary rock. Sonic velocity of a sedimentary rock may also be used as a proxy for maximum burial. The sonic-velocity method is robust for estimating maximum burial of sedimentary rocks of specific lithological composition because sonic velocities are measured directly in boreholes, they represent bulk properties of the sediment and they are sensitive to changes in burial as porosity is reduced and grain contacts become firmer during diagenesis (cf. Japsen et al. 2007b).

Palaeostructure techniques

The stratigraphic approach to the quantification of exhumation at an unconformity involves the identification of the maximum preserved section beneath the unconformity (the reference section). A minimum estimate of exhumation is then determined for a particular location by comparing the thickness of the equivalent stratigraphic unit encountered at that location with the reference section (e.g. Holliday 1993). Moreover, the method can be utilized even if the unconformity is at its maximum burial depth today. The technique is, however, difficult to apply over long distances and if an area has been deeply eroded.

Palaeosurface techniques

Extensive low-relief erosion surfaces (peneplains) defined by their geometric basal plane cutting across bedrock of different age and resistance form as erosion surfaces by river systems grading to the prevailing base level (Fig. 2) (cf. Ahlmann 1941; King 1967; Lidmar-Bergström 1996; Bonow et al. 2006a, b; Japsen et al. 2006, 2009). Unless the local base level is represented by a resistant rock surface, sea-level is the most likely base level, particularly for locations at or close to continental margins (e.g. in the cases of the coastal areas in Greenland and Scandinavia during the Cenozoic). Where such a surface has risen relative to base level, valleys will be incised into it as a new phase of erosion grades the landscape to the new base level. If a low-relief surface with incised valleys is found at a substantial height above present sea-level, the parsimonious explanation is that such a surface represents an uplifted peneplain. The height difference between the valley floor and the overlying peneplain is thus a proxy for the relative fall in base level (amount of uplift). Over time the valleys below the uplifted peneplain will expand to form a new peneplain graded to the new, lower base level. Erosion surfaces that have been buried and protected by sedimentary deposits and then re-exposed provide evidence of repeated episodes of uplift and erosion, subsidence and burial and renewed uplift. Analysis of the relationship between mapped erosion surfaces (re-exposed and those never buried; horizontal, inclined or in steps) and valleys incised below the surfaces provides a relative chronology for the formation of surfaces and for uplift/incision or subsidence/covering events as well as tilting and offsets by faulting of the surfaces.

Integrated studies of uplift and exhumation

The history of burial and exhumation of a sedimentary basin can be reconstructed most confidently by integrating a variety of independent methods, particularly involving methods based on thermal data (e.g. AFTA and VR) with methods based on palaeoburial (or effective stress, e.g. sonic velocities) and those based on landscape studies. Analysis of large-scale landforms (base-level governed peneplains) is the only method that provides an absolute estimate of the amount of surface uplift, whereas studies of palaeothermal and palaeoburial indicators provide estimates of exhumation (removal of rock column). By combining such studies, it is thus possible to evaluate if exhumation was accompanied by surface uplift (e.g. Japsen et al. 2006). Uplift and subsidence may be driven tectonically, but can also be caused by isostatic compensation due to erosional unloading and depositional loading, respectively (e.g. Medvedev et al. 2008) and the trigger for the erosion and deposition may be eustatic changes in sea-level. The driving forces behind such vertical movements must thus be assessed in each and every case (see Case A and the discussion at the end of this paper). It is additionally possible to limit the likely range of the palaeothermal gradients during maximum burial by estimating the maximum burial depth independently from sonic data (Case B; cf. Japsen et al. 2007a). Studies of the Cenozoic thermal and burial history in West Greenland have shown that the regional, elevated peneplain there provides a more useful reference level for evaluating palaeotemperature profiles than present-day sea-level (Cases C and E). Integration of the results from thermochronological studies with even very simple constraints from the geological record will often highlight the episodic nature of the burial and exhumation history along a margin (Case F). Effects related to climate change can also cause cooling in the subsurface and hence potentially cause effects detectable in AFTA data. In a study in the eastern North Sea Basin, Japsen et al. (2007a) found that the effects of Eocene climate change could be resolved confidently from those of late Neogene burial only by integrating palaeothermal methods with temperature-independent estimates from sonic data.

Case A: tectonic control on Cenozoic stratigraphy and palaeogeography in the eastern North Sea Basin

The eastern North Sea Basin is a key region for studies of the Cenozoic development of Scandinavia because Paleocene to upper Miocene sediments crop out onshore Denmark and because it is possible to correlate these occurrences with the up to 3 km thick Cenozoic succession of the North Sea Basin (cf. Fig. 1). However, the Cenozoic deposits onshore Denmark occur close to the surface only because they have been exhumed from beneath their cover of younger rocks. This implies that a reconstruction of the Cenozoic development in the eastern North Sea Basin, involving both burial and exhumation, must rely on sedimentological and seismostratigraphic studies of preserved sediments as well as on physical parameters that may yield evidence of the post-depositional history of the sediments now at the surface. Only if the burial and exhumation history of the basins can be deciphered is it possible to infer the geological development in the Scandinavian hinterland where Cenozoic sediments are rarely preserved.

In a study using AFTA and VR data from eight Danish wells, Japsen et al. (2007a, 2008) identified four Mesozoic–Cenozoic palaeothermal events, that is, episodes of heating and subsequent cooling that are not reflected in the preserved rock record. These episodes were interpreted as being caused by burial and subsequent exhumation and one phase reflecting climate change during the Eocene. The study combined thermal history reconstruction with exhumation studies based on palaeoburial (sonic velocity), stratigraphic and seismic data (cf. Japsen & Bidstrup 1999; Green et al. 2002; Nielsen 2003; Rasmussen 2004; Japsen et al. 2007b). Two of the exhumation phases occurred during the mid-Jurassic and the mid-Cretaceous, but here we focus on the Cenozoic development and on the early and late Neogene exhumation phases during which up to 1 km of sediments was removed across much of the eastern North Sea Basin (Fig. 3). The inferred palaeogeographical development of the area during the Cenozoic reflects a complex interaction of subsidence, progradation after tectonic pulses in the Scandinavian hinterland and episodes of uplift and exhumation reaching far into the sedimentary basin (see Fig. 4).

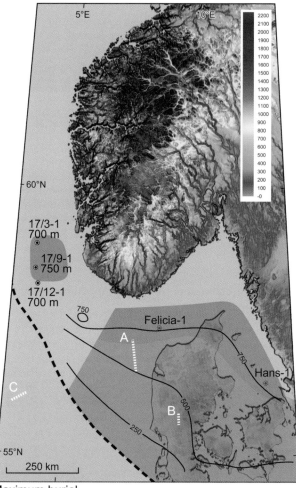

Fig. 3. The eastern North Sea Basin and southern Norway (topographic colour code). Contours of estimated thickness of the section removed during Cenozoic exhumation (modified after Japsen et al. 2007a). Maximum burial of the Mesozoic sediments took place during the late Neogene in most of the study area. Near the present coasts of Norway and Sweden, however, maximum burial occurred subsequent to deposition of the Oligocene wedges and prior to early Neogene exhumation. Only near the Hans-1 well did maximum burial occur during the Mesozoic. Modified after Globe Task Team (1999) and Japsen et al. (2007a). Location of map shown in Fig. 1.

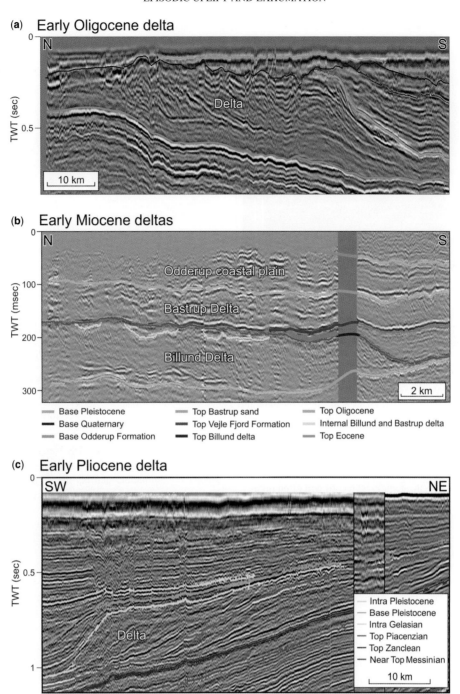

Fig. 4. Seismic sections illustrating deltas prograding in directions away from present-day Norway in response to uplift. (**a**) Early Oligocene delta prograding towards the south. This progradation represents a major change from the deposition of fine-grained hemipelagic marine sediments prior to the progradation. (**b**) Early Miocene deltas prograding southwards (modified from Rasmussen 2009*a*). The Billund delta was characterized by a braided fluvial system while the Bastrup delta was dominantly characterized by a meandering system. This indicates a tectonic influence on the sedimentary pattern in the earliest early Miocene. (**c**) Pliocene delta prograding towards SW (modified from Japsen *et al.* 2007*a*). The progradation after the Zanclean (*c*. 3.6 Ma, red reflector) is followed by renewed subsidence in the central North Sea. Locations in Figure 3.

Early and late Neogene exhumation

A major phase of exhumation of the parts of the eastern North Sea Basin adjacent to the presently exposed basement areas in Norway and Sweden began between 30 and 20 Ma ago according to the AFTA data from wells Felicia-1 and Hans-1 (Fig. 3). Japsen *et al.* (2007*a*) suggested that this phase corresponds to a regional early Neogene unconformity (*c*. 23 Ma at the Oligocene–Miocene boundary, but for simplicity we will refer to this event as early Neogene; see Rasmussen 2009*b*). Prior to this exhumation phase, the Mesozoic sediments in the Felicia-1 well were at maximum burial. The maximum burial of the Mesozoic sediments in the Hans-1 well occurred during the mid-Cretaceous, prior to inversion along the Sorgenfrei–Tornquist Zone, while renewed Cenozoic exhumation began in the early Neogene phase.

AFTA data from several wells (e.g. Felicia-1) suggest that the most recent episode of Neogene exhumation began between 10 and 5 Ma, but Japsen *et al.* (2007*a*) suggested that exhumation probably began in the early Pliocene at *c*. 4 Ma, as suggested by the prominent unconformity of this age (with only one AFTA sample limiting the onset to 5 Ma). A section of 450–850 m was removed over much of Denmark in association with this reshaping of the North Sea Basin (Fig. 3). The exhumation affected extensive regions where Neogene strata are truncated, and a sub-Cretaceous

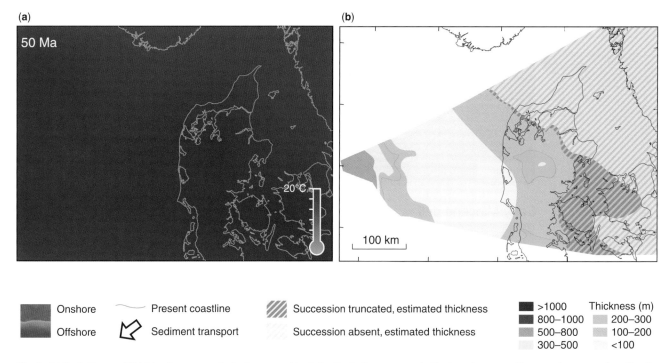

Fig. 5. (a) Early Eocene (50 Ma) palaeogeography in the eastern North Sea Basin. Deep sea, no indication of any coastline near the present Scandinavian basement areas. (b) Upper Paleocene–Eocene isochore prior to Neogene exhumation. Temperatures indicated are for the coldest month along the southern margin of the North Sea Basin (Utescher *et al.* 2009). Modified after Japsen *et al.* (2007a, 2008).

etch surface (ES) was re-exposed along the coasts of southern Norway and Sweden during the late Neogene phase (Lidmar-Bergström *et al.* 2000; Japsen *et al.* 2002).

Cenozoic palaeogeography of the eastern North Sea Basin

We have compiled a series of maps to illustrate the Cenozoic development of the eastern North Sea Basin (Figs 5–9). The isochore maps are based on present thicknesses and estimates of the total amount of section removed by Cenozoic exhumation. The estimates of the removed sections of specific ages are furthermore constrained by the known geology and our interpretation of the depositional pattern. Each isochore map shows the thicknesses prior to the first phase of exhumation that affected the distribution of the unit, that is, prior to early Neogene exhumation (starting at *c*. 23 Ma) in the case of the Palaeogene units and prior to late Neogene exhumation (starting at *c*. 4 Ma) for the Neogene units. The palaeogeographic maps show the distribution of onshore and offshore areas separated by the coastline based on the known geology. In the areas where the deposits of a given age have been removed, the maps have been drawn from an estimate based on our interpretation of the depositional system and on the inferred burial and exhumation history.

The presence of hemipelagic, deep-marine deposits of Eocene age in the eastern North Sea Basin suggests that much of Scandinavia was covered by a deep sea from the early Eocene until latest Eocene times (Fig. 5a; Heilmann-Clausen *et al.* 1985; Michelsen *et al.* 1998): Eocene deposits less than 100 km from basement currently exposed in Sweden contain a deep-marine fauna with no

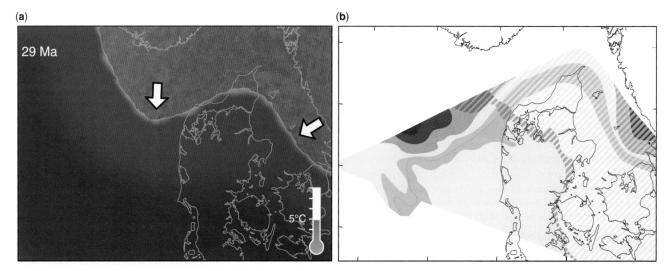

Fig. 6. (a) Mid-Oligocene (29 Ma) palaeogeography in the eastern North Sea Basin. Progradation of sedimentary wedges after first Scandinavian uplift phase (*c*. 33 Ma). (b) Oligocene isochore prior to Neogene exhumation. Legend in Figure 5. Modified after Japsen *et al.* (2007a, 2008), Schiøler *et al.* (2007).

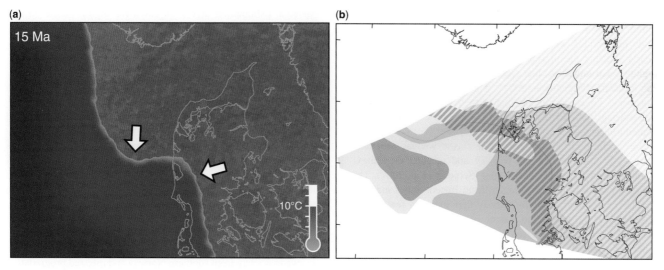

Fig. 7. (**a**) Middle Miocene (15 Ma) palaeogeography in the eastern North Sea Basin. Progradation of sedimentary wedges after second Scandinavian uplift phase (c. 23 Ma). (**b**) Lower to lower, middle Miocene isochore prior to late Neogene exhumation. Legend in Figure 5. Modified after Japsen *et al.* (2007*a*, 2008).

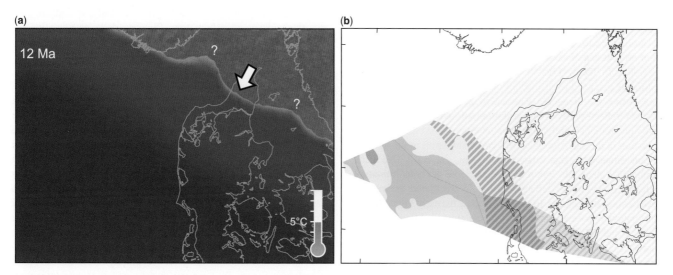

Fig. 8. (**a**) Late Miocene (12 Ma) palaeogeography in the eastern North Sea Basin. Transgression, despite colder climate, due to tectonic subsidence. (**b**) Upper, middle to lower, upper Miocene isochore prior to late Neogene exhumation. Legend in Figure 5. Modified after Japsen *et al.* (2007*a*, 2008).

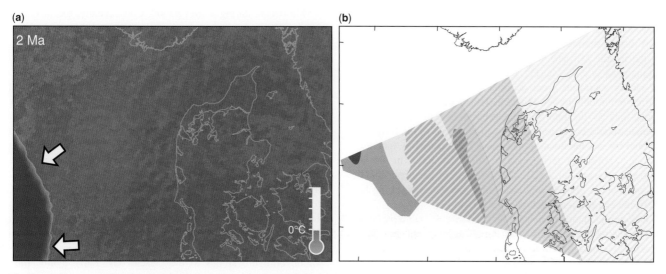

Fig. 9. (**a**) Pliocene (2 Ma) palaeogeography in the eastern North Sea Basin. Progradation and subsidence after third Scandinavian uplift phase (c. 4 Ma). (**b**) Upper, upper Miocene–Pliocene isochore prior to late Neogene exhumation. Legend in Figure 5. Modified after Japsen *et al.* (2007*a*, 2008).

signs of a near-shore fauna (Fig. 5a) (C. Heilmann-Clausen, pers. comm. 2005). Reworked Eocene dinocysts and clasts of Eocene muds interbedded with Miocene deltaic sediments in Denmark clearly indicate the presence of marine Eocene deposits in the Scandinavian hinterland that were eroded and redeposited during the Miocene (Rasmussen 2004). The thickness of the upper Paleocene–Eocene sediments that were removed from the study area during the Neogene rise of Scandinavia was probably limited to less than about 200 m (Fig. 5b).

During the early Oligocene, major clastic wedges prograded southwards from present-day Norway and reached their maximum extent at 29 Ma (Figs 4a & 6a) (Schiøler et al. 2007). This change in depositional environment from Eocene times indicates a phase of uplift of southern Norway that started a few million years before the maximum extent of the prograding wedges (i.e. c. 33 Ma) (e.g. Michelsen et al. 1998; Clausen et al. 1999; Lidmar-Bergström et al. 2000; Faleide et al. 2002). Miller et al. (2005) show no large fall in eustatic sea-level at any time within the uncertainty of these dates, so the cause of the uplift is presumably tectonic. We suggest that a thick succession of Oligocene shelf and delta sediments that prograded southwards from southern Norway and westwards from southern Sweden made up a substantial part of the section that was removed during the Neogene exhumation beginning at c. 23 Ma at the locations of the Felicia-1 and Hans-1 wells (>1 km; Fig. 6b).

During the early Miocene, deltas prograded to the south and SW from present-day Norway and southern Sweden, reaching a thickness of up to 300 m (Figs 4b & 7a). Deposition of coarse-grained sediments extended over much of the present-day Danish onshore area during the Neogene (Rasmussen 2004). These prograding systems reflect partly a redistribution of older sediments and partly erosion of strongly weathered basement. The basinward progradation in the early Neogene was caused by uplift in the Scandinavian hinterland that also resulted in the early Neogene exhumation near the craton. The occurrence of both immature and mature sediments reflects erosion of weathered as well as newly exposed basement. In the Danish area, the early Miocene tectonic activity is demonstrated by coarse-grained, braided fluvial systems and later by a sudden increase in the heavy mineral content at c. 17 Ma (Rasmussen 2004). The thickness of the lower to lower-middle Miocene sediments that was removed during late Neogene exhumation was substantial (<500 m) (Fig. 7b). The lowering of relative sea-level implied by this event does not correlate with cooler climate, because the climate was warm temperate and subtropical during both the late Oligocene and early Miocene (Utescher et al. 2009), so the lowering of sea-level must have been due to basin-wide inversion tectonism (Rasmussen 2009b).

A distinct marine flooding of the area took place after c. 15 Ma in the middle Miocene, with maximum transgression of the sea taking place at c. 12 Ma (Fig. 8a; Rasmussen 2005). Up to 150 m of clayey sediments were deposited in southwestern Denmark during the middle to late Miocene. This flooding occurred at the same time as the mid-Miocene deterioration in climate. Cool climate should correlate with low sea-level due to the formation of land ice, so this flooding must have been caused by increased tectonic subsidence in the eastern North Sea Basin in the late Miocene (Rasmussen 2004). The thickness of the upper-middle to lower-upper Miocene sediments that was removed during late Neogene exhumation was limited (<200 m; Fig. 8b).

Delta progradation resumed from the northeast and in particular the east at the end of the Miocene (Figs 4c & 9a; Rasmussen 2005). The infilling of the North Sea Basin continued during the Pliocene when up to 500 m of sediment were deposited in the central North Sea Basin. Substantial thicknesses (<500 m) of Upper–Upper Miocene–Pliocene sediments were removed during late Neogene exhumation across an extensive region of the eastern North Sea (Fig. 9b).

Phases of post-breakup uplift of southern Scandinavia

These results emphasize that the tectonic development of a region cannot be reconstructed solely on evidence from the preserved sedimentary record. Such reconstructions must also integrate evidence provided by palaeothermal and palaeoburial techniques of geological units that were once present but have now been removed. As described above, we find evidence for three phases of uplift that have affected southern Scandinavia after North Atlantic breakup at the Paleocene–Eocene transition at c. 55 Ma (e.g. Eldholm et al. 1989; Ziegler 1990; Skogseid et al. 2000):

(1) a phase that began at the Eocene–Oligocene transition at c. 33 Ma as indicated by the onset of progradation of clastic wedges away from southern Norway and inferred progradation away from southern Sweden;
(2) an early Neogene phase that began at c. 23 Ma, as indicated by exhumation of the basins adjacent to the presently exposed basement areas and by early Miocene coarse-grained braided fluvial systems south of Scandinavia;
(3) a late Neogene phase that began at c. 4 Ma, as indicated by the widespread exhumation, Pliocene delta progradation off Scandinavia and subsequent tilting of the Neogene succession in the eastern North Sea.

These phases are consistent with the stratigraphy in the NE Atlantic (Stoker et al. 2005) and around southern Norway (e.g. Michelsen et al. 1998; Faleide et al. 2002; Rasmussen 2004). The observations thus suggest that southern Norway, with peaks higher than 2 km a.s.l., emerged during several phases since Eocene times, when a deep sea covered much of the region. This deep Eocene sea may have developed in response to post-rift subsidence after breakup in the North Atlantic. However, the subsequent uplifts are clearly not in agreement with classical theory for the post-rift development of a rifted margin (e.g. McKenzie 1978; White & McKenzie 1988). We discuss this further in the Chapter 'On the origin of post-rift uplift along passive margins'.

The reconstruction of the palaeogeography of the eastern North Sea Basin (Figs 5–9) illustrates that the variations in relative sea-level cannot be explained by climate variations alone: for each palaeogeographic episode, the mean temperature of the coldest month is indicated and this shows that, whereas the high sea-level in the Eocene could reflect the warm climate, the transgression at 12 Ma occurs during a cold period. Consequently, the increased sea-level at that time must be due to tectonic subsidence.

As we have shown here, these events of uplift and erosion affected not only the exposed basement areas onshore, but also adjacent basins that are mainly offshore today. An accurate reconstruction of these events, both with respect to timing and magnitude, is important for understanding the petroleum systems of the marginal basins fringing southern Norway. For example, failure to account for the additional burial offshore can lead to underestimation of likely maturity levels. Similarly, understanding the timing of maximum burial and subsequent exhumation in relation to formation of potential trapping structures is essential to predict migration pathways and possible loss of charge due, for example, to breach of seals and phase changes during uplift. In this connection, the Neogene maximum burial off southern Norway and the resulting late hydrocarbon generation increases the likelihood of charging early-formed structures and the preservation of hydrocarbons to the present day. Early hydrocarbon expulsion can become a critical factor because it involves a long preservation phase for the hydrocarbons in the traps.

Case B: early Neogene exhumation offshore southern Norway

Early Neogene uplift caused the regional inversion of the Norwegian–Danish Basin and of the southern Central Graben (Japsen *et al.* 2007a; Rasmussen 2009b). A regional unconformity was formed, especially over elevated areas and on top of major prograding Oligocene successions. The style of the early Neogene tectonism contrasts with the pronounced tilting of the sedimentary succession produced during the late Neogene uplift, which resulted in truncated Neogene and older units overlain by flat-lying Quaternary units which makes the late Neogene event easily recognizable on seismic sections. The regional nature of early Neogene uplift and erosion results in a very low-angle unconformity over wide areas and can thus be difficult to identify on seismic sections. In addition, stratigraphic data for the Oligocene–Miocene strata are often sparse, as this section is often drilled prior to the first casing being set and is not sampled, so any base-Neogene unconformity may be difficult to identify. However, the exhumation caused by the event may, as we show here, be identified with palaeoburial (sonic) and palaeothermal (AFTA and VR) methods.

Magnitude of exhumation estimated from palaeoburial data

Basic concepts. Exhumation can be estimated from sonic velocities of sediments measured in boreholes. The basic parameter is the burial anomaly, dZ_B, which is the difference between the present-day burial depth of a rock, z, and the depth, $z_N(V)$, corresponding to normal compaction for the measured velocity, V, and the lithology in question (Japsen 1998):

$$dZ_B = z - z_N(V)$$

The depth corresponding to normal compaction (the effective depth; Japsen *et al.* 2005a) may be determined from a normal velocity–depth trend (baseline), which is a function that describes how sonic velocity increases with depth in a relatively homogeneous, brine-saturated sedimentary formation when porosity is reduced during normal compaction (mechanical or by pressure solution). Compaction is 'normal' when the fluid pressure of the formation is hydrostatic, and the formation is at maximum burial depth. Normal velocity–depth trends are specific for given lithologies such as marine shale (Japsen 1999, 2000), sandstone (Japsen *et al.* 2007b) and chalk (Japsen 1998, 2000).

The burial anomaly is zero for normally compacted sediments, whereas high velocities relative to depth give negative burial anomalies which may be caused by a reduction in overburden thickness when lithology is relatively homogeneous over the study area, and if lateral variations of horizontal stress are minor. A positive burial anomaly may indicate undercompaction due to overpressure. Any post-exhumational burial, B_E, will mask the magnitude of the missing (removed) overburden section, Δz_{miss}:

$$\Delta z_{miss} = -dZ_B + B_E \quad (1)$$

where the minus indicates that erosion reduces depth (Hillis 1995; Japsen 1998). The timing of the exhumation events is thus important for understanding the succession of events, their true magnitude and for identifying the age of the eroded succession.

Estimation of burial anomalies in three Norwegian wells. Burial anomalies have been estimated for the Mesozoic–Palaeogene succession in three Norwegian wells from quadrant 17 (Figs 3 & 10). The estimates are primarily based on a comparison between (1) the sonic data for the chalk of the Upper Cretaceous–Danian Shetland Group and the chalk baseline (Japsen 2000), (2) the sonic data for the Palaeogene shales and the shale baseline (Japsen 2000) and (3) the sonic data from below the chalk and the shale baseline. There are ample variations of the sonic data over the latter interval where shale units typically dominate above the Upper Jurassic Tau Formation, which is a local source rock with a characteristic, low sonic velocity that reflects high organic content. Very low velocities in the upper part of the Lower Cretaceous Åsgård Formation and the overlying Sola Formation may have a similar explanation.

Anomalously low velocities are found in the uppermost part of the chalk below high-velocity carbonates in two wells, 17/9-1 and 17/12-1R. There is no correlation between the gamma response and the low velocities in these chalk intervals. A possible explanation is that these low velocities (high porosities) are caused by the presence of hydrocarbons in the chalk at an early stage during the compaction of the chalk. Since 17/12-1R was the discovery well of the Bream Field, leaking from the Jurassic reservoir rocks into the chalk in the overburden is not unlikely and hence the presence of hydrocarbons at some stage during the Cenozoic burial history. The presence of hydrocarbons is known to have a porosity-preserving effect in chalk where the coating of the grains with hydrocarbons prevents pressure solution of the carbonates (e.g. Fabricius *et al.* 2008).

The estimation of the burial anomaly is based on depth-shifting shale and chalk baselines by the same amount until a match is obtained with the data for the intervals of the identified lithologies that are considered most reliable – typically the Upper Cretaceous–Danian chalk and the Palaeogene shale. The match is primarily based on the absolute level of the sonic velocities in these intervals, but also on identification of velocity–depth trends for these intervals. The benefit of this integrated approach is that, whereas the velocity–depth trend for one of the above mentioned intervals may be difficult to identify, the comparison of sonic data for two or more lithologies with their respective baselines effectively limits the range of the possible burial anomalies and also guards against the possible effects of local variations in lithology (Fig. 10).

The timing of the exhumation can only be broadly estimated from the available sonic and stratigraphic data (as being younger than the overcompacted Palaeogene shales), but the available AFTA data for these wells provide good constraints on when maximum burial took place.

Timing and magnitude of exhumation from palaeothermal and palaeoburial data

Thermal history reconstruction, based on AFTA and VR data, has been applied in the three wells to define the onset of cooling, as well as providing independent corroboration of the magnitude of the exhumation revealed from the sonic velocity analysis. The methods employed are described in detail by, for example, Japsen *et al.* (2007a). Results from all three wells provide consistent evidence that the pre-Cenozoic section has been hotter in the past. AFTA and VR data in each well provide consistent indications of maximum post-depositional palaeotemperatures, and the combined AFTA dataset from all three wells shows that cooling from the palaeothermal maximum began some time in the interval 30–15 Ma, that is, late Oligocene to middle Miocene.

Palaeotemperatures derived from AFTA and VR data in the three wells are plotted as a function of depth in Figure 11. Results from each well define consistent linear profiles, sub-parallel to the present-day temperature profile derived from corrected borehole temperature (BHT) values in each well, and offset to higher temperatures by around 20–30 °C. The form of these profiles suggests that deeper burial is the most likely explanation of the

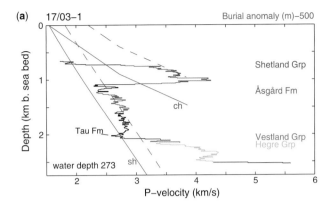

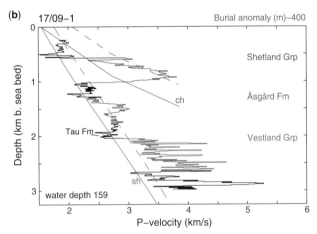

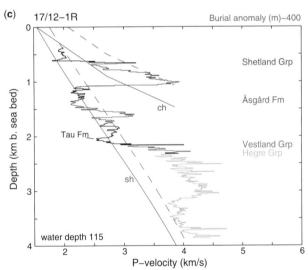

Fig. 10. Plot of sonic logs for three Norwegian wells compared with (1) the baselines for chalk and shale and (2) these baselines shifted by the burial anomaly estimated for the pre-Neogene section in each of the wells: (**a**) 17/3-1 (**b**) 17/9-1 and (**c**) 17/12-1R (location in Fig. 3). The chosen value of the burial anomaly is based on the agreement between the sonic data for the Palaeogene shale, for the chalk of the Shetland Group and for the Lower Cretaceous–Jurassic shale and the depth-shifted baselines for chalk and shale. The match is primarily based on the absolute level of the sonic velocities in these intervals, but also on identification of velocity–depth trends for these intervals. The comparison of sonic data for two lithologies with their respective baselines effectively limits the range of the possible burial anomalies. See Table 1 for details concerning the estimation of the burial anomaly in each well. Selected lithostratigraphic units are indicated in the diagram with a colour corresponding to the colouring of that part of the sonic log (cf. Well Data Summary Sheets at www.npd.no). Legend: ch, baseline for chalk; full blue line (only for velocities less than 3.9 km/s; see Japsen et al. 2005a); sh, baseline for shale; full red line (Japsen 2000); dashed line, baseline for respective lithology shifted vertically by the burial anomaly to fit the data trend. Burial anomaly <0 indicative of exhumation.

palaeotemperatures derived from AFTA and VR in these wells, supporting the conclusion derived from the sonic velocity analysis that the Palaeogene and older section in these wells has been more deeply buried and subsequently exhumed. The profile of palaeotemperatures derived from VR data in the 17/12-1 well is less well defined than in the other two wells, resulting in broader constraints on allowed palaeogeothermal gradients in this well.

The timing of cooling indicated by AFTA data in these three wells, between 30 and 15 Ma, correlates closely with the 30–20 Ma onset of cooling obtained from AFTA in the Felicia-1 well by Japsen *et al.* (2007*a*), where this cooling was correlated with exhumation corresponding to formation of a regional early Neogene unconformity (as discussed under Case A). We therefore infer that the cooling revealed by AFTA in the three wells from Quad 17 also represents exhumation at this time. Given these timing constraints, we have selected the top-Oligocene horizon identified in these wells as the reference for determining amounts of additional section (Table 1).

Using the methods outlined by Bray *et al.* (1992) and Green *et al.* (2002), we have determined the range of palaeogeothermal gradients allowed by the palaeotemperature constraints derived from AFTA and VR data in each well. Extrapolating the fitted palaeogeothermal gradients from the corresponding unconformity surface to an appropriate palaeosurface temperature allows determination of the amount of additional section that is required to explain the observed palaeotemperatures. This analysis explicitly assumes that the palaeogeothermal gradient was linear at the palaeothermal maximum. Comparison of the amounts of additional section determined from this approach with independent values determined from the sonic velocity analysis provides a check on the validity of this assumption.

Results of this analysis for the three wells are shown in Figure 12. For each well, a hyperbolic region defines the range of correlated values of palaeogeothermal gradient and removed section that can explain the observed palaeotemperatures with 95% confidence limits (see e.g. Green *et al.* 2002, 2004; Japsen *et al.* 2007*a* for further details). Also shown for each well is the range of values of additional burial defined by the sonic velocity analysis, obtained by adding the amounts of post-Oligocene re-burial in each well to the appropriate burial anomaly from Figure 10 [Equation (1)]. In two wells (17/3-1 and 17/9-1), values of additional burial defined from the palaeoburial approach show excellent agreement with the range of values indicated from palaeothermal techniques for a constant heat-flow scenario (i.e. for palaeogeothermal gradients equal to the present-day value in each well). In well 17/12-1, the palaeothermal constraints are wider than in the two other wells (Fig. 11) and comparison with the results of the sonic analysis shows a discrepancy. Based on the consistent results of the sonic data in all three wells, the sonic estimate of exhumation in well 17/12-1 is preferred. Thus, the combination of the two approaches in these wells provides strong evidence for deposition and subsequent erosional removal of around 700 m of additional section on an early Neogene unconformity in the eastern North Sea off southern Norway (Fig. 3).

Multiple episodes of uplift onshore and offshore Norway

The evidence presented here for a major exhumation event between 30 and 15 Ma in Norwegian quadrant 17, which correlates with the early Neogene event (*c.* 23 Ma) identified in Danish wells, supports the conclusion that a regional uplift event affected southern Norway and the adjacent basins at that time, as suggested in Case A. The geological response to this event is the lower Miocene deltas that prograded to the south and SW from mainland Norway and central Sweden with thicknesses of up to 300 m (cf. Figs 4b & 7a).

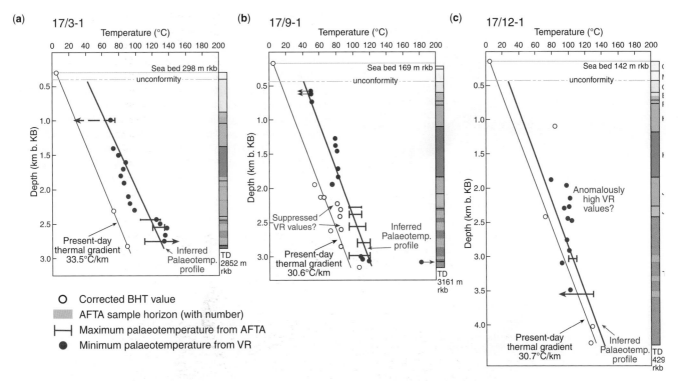

Fig. 11. Thermochronological constraints on the amounts of section removed in three Norwegian wells (location in Fig. 3). AFTA and VR data have been used in each well to determine maximum post-depositional palaeotemperatures, while AFTA have also been used to determine the time at which samples began to cool from those palaeotemperatures. As shown here, palaeotemperatures from both techniques define consistent linear profiles in the 17/3-1 and 17/9-1 wells, although some results from VR data in the 17/9-1 well appear to be anomalously low, probably due to suppression in Jurassic units. Palaeotemperatures derived from VR data in the 17/12-1 well also show excessive scatter, with some values appearing to be much higher than present-day temperatures compared with others. The highest values have not been used in determining palaeogradients and removed section. This uncertainty produces a less consistent match with results from sonic velocity analysis in Figure 12.

As with the example of the Norwegian–Danish Basin in Case A, recognition of the exhumed nature of the sedimentary section offshore southern Norway has major implications not only for hydrocarbon exploration offshore but also for understanding the nature of the tectonic processes involved. In terms of hydrocarbon systems, failure to account for the greater depths of burial prior to exhumation can lead to serious underestimation of maturity levels, while the effects of exhumation on the timing of hydrocarbon generation, changes in migration routes and on any reservoired hydrocarbons also require assessment (e.g. Doré et al. 2002). In terms of tectonics, the results presented here show that not only southern Norway but also major portions of the offshore shelf underwent exhumation in the early Neogene, and any attempt to understand the underlying processes must take into account the linkage

Table 1. Estimation of the burial anomaly for the pre-Neogene section based on sonic data (Fig. 10), the post-Oligocene reburial (cf. www.ndp.no) and the amount of exhumation during the early Neogene phase in three Norwegian wells (Eq. 1). Depths below sea bed (bsb)

Well	Burial anomaly (1) Re-burial (2) Exhumation (3) (m)	Estimate based on	Shale	Chalk
17/03-01	−500 ± 100 (1) 100 (2)* 600 (3)	Good agreement between results from chalk and shale data	Shale trend defined below the chalk and above Tau Fm., 1.1–1.9 km bsb, with some scatter. No data above the chalk	Chalk trend well defined by upper bound, 0.7–1.0 km bsb
17/09-01	−400 ± 150 (1) 300 (2) 700 (3)	Good agreement between results from chalk and shale data	Shale trend loosely defined by upper bound for the Palaeogene shales, 0.2–0.5 km bsb, and for the section between the chalk and the Tau Fm., 1.0–2.1 km bsb (disregarding peak velocity for Åsgård Fm.)	Chalk trend broadly defined by upper bound below data for Ekofisk Fm. and high-velocity chalk, 0.7–0.9 km bsb
17/12-1R	−400 ± 150 (1) 300 (2) 700 (3)	Agreement between results from chalk and shale data	Shale trend defined by limited data for the Palaeogene section, 0.4–0.6 km bsb, and by a loosely defined upper bound below the chalk and above the Tau Fm. (disregarding peak velocity for Åsgård Fm.)	Chalk trend defined by upper bound over the deeper part of the chalk, 0.8–1.0 km bsb

*Approximated by the thickness of the Nordland Group.
The burial anomaly is estimated by comparing the normal velocity–depth trends for shale and chalk and sonic log data for the relevant intervals (Fig. 10, Eq. 1). The timing of maximum burial between 30 and 20 Ma is based on AFTA data and this event is interpreted as corresponding to a regional early Neogene unconformity at c. 23 Ma (cf. Japsen et al. 2007a).

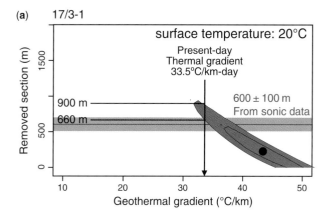

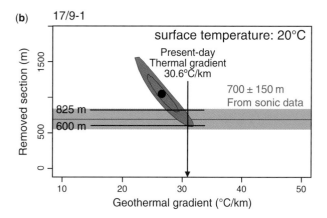

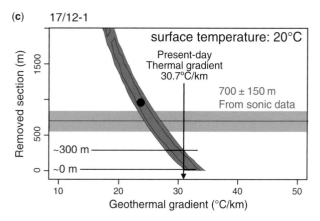

Fig. 12. The range of values of palaeogeothermal gradient and removed section that are allowed by the palaeotemperature constraints from AFTA and VR data in each of the Norwegian wells shown in Figure 11 are shown by the hyperbolic ellipsoid regions (Bray et al. 1992; Green et al. 2002). These are compared with the estimates of removed section derived from sonic velocities in each well (Table 1). In the 17/3-1 and 17/9-1 wells, the two approaches give highly consistent results for a scenario based on a constant heat flow (palaeogeothermal gradients close to present-day values). In the 17/12-1 well, the agreement is less good, due most likely to the excessive scatter in VR values. Black dots mark maximum likelihood solutions.

between onshore and offshore demonstrated by these results. Similar results from West Greenland are presented below, suggesting that this may be a common phenomenon.

Case C: West Greenland's mountains are surprisingly young

In West Greenland, where rifting occurred during mid-Cretaceous and earliest Paleocene times and seafloor spreading started in the mid-Paleocene, kilometre-thick Palaeogene post-rift volcanic sequences with incursions of marine sediments are exposed in mountains reaching to above 2 km a.s.l. (Figs 2 & 13; Chalmers et al. 1999). Japsen et al. (2006) analysed these landscapes by combining the cooling history from AFTA data with the denudation history from landscape analysis (Bonow et al. 2006a, b) and the stratigraphic record. These results demonstrate that post-rift subsidence and burial continued until the end of the Eocene when seafloor spreading ceased between Canada and Greenland (Chalmers & Pulvertaft 2001). Maximum burial of the post-rift sequence occurred at the end of the Eocene and was followed by uplift in which up to 1 km of the post-rift sequence was removed, forming a regional, low-relief erosion (or planation) surface (the upper planation surface, UPS) that was graded to sea-level during the Oligocene–Miocene. This erosion surface was uplifted to present-day altitudes of up to 2 km (Fig. 13) and offset by reactivated faults. Broad and uplifted palaeovalleys (the valley floor of these defining the lower planation surface, LPS) incised below the surface demonstrate that the uplift took place in two phases and AFTA data define the onset of these phases to around 10 and 5 Ma.

Episodes of cooling defined by AFTA data

The geomorphological and geological observations in West Greenland have been complemented by AFTA data supported by VR data (Japsen et al. 2005b, 2006, 2009). Thermal history constraints have been extracted from our AFTA data following the principles established in Green et al. (2002). Data in individual samples of both Cretaceous sediments and basement define typically two (rarely three) discrete palaeothermal episodes. Synthesis of the timing constraints for the cooling episodes identified in all samples suggests that at least six discrete episodes of cooling are required to explain all of the AFTA results.

The events identified date back to the latest Proterozoic, but here we focus on the Cenozoic (Cases C and D) and Mesozoic (Case E) episodes. The AFTA data define five cooling episodes that began in the intervals 230–220, 160–150, 36–30, 11–10 and 7–2 Ma (cooling episodes C0–C4; see Fig. 14). Note that these time intervals represent the ranges of uncertainty for the time of onset of cooling in each episode, not the intervals during which all of the cooling took place. The onset of the C1 event is recognized from AFTA in most samples of basement rocks across the area and the C2 event in almost every sample. In contrast, the palaeotemperatures from which most basement samples cooled in the C3 and C4 events are too low in most of the basement outcrop samples for these events to be detected. These late Neogene events are mainly defined from samples of sedimentary rocks from the Nuussuaq Basin because of the higher palaeogeothermal gradients in this region, and they are particularly well defined in AFTA data from the 3 km deep Gro-3 well (Japsen et al. 2005b).

The C0 cooling event that is defined by many samples from the Precambrian basement areas began in the late Triassic. Palaeotemperatures for this episode estimated from rocks at present-day outcrop increase westwards to above 100 °C near the present-day coast. Rocks of Triassic age are not known from West Greenland, either onshore or offshore, but the increase in palaeotemperatures towards the present-day coast may indicate a relation with the later Mesozoic rift events.

Integration of thermochronological and geomorphological results

A Mesozoic etch or weathering surface, formed subsequent to uplift initiated in the Late Jurassic (C1), has been re-exposed from below Cretaceous–Paleocene cover rocks after Neogene uplift (Bonow 2005). This surface is characterized by an undulating topography and has been mapped on Nuussuaq, Disko and in the

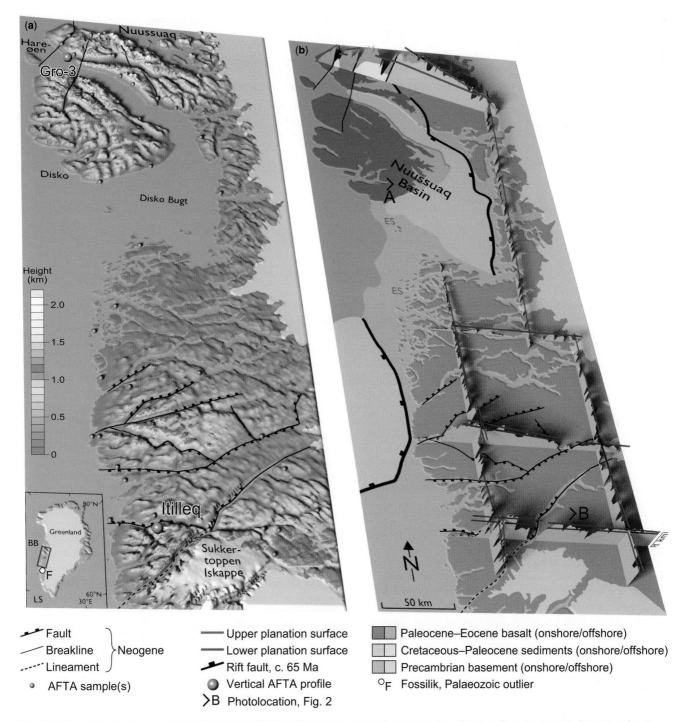

Fig. 13. Three-dimensional maps of central West Greenland. (**a**) Topography. (**b**) Geology with topographical profiles. An elevated plateau is defined by the regional upper and the partially developed lower planation surface that cut across Precambrian basement and Palaeogene volcanic rocks. The formation of the planation surfaces was uniform across the study area and they must be younger than mid-Eocene basalts. The planation surfaces now dip in different directions and are offset by faults that displace them. An ES has been re-exposed from below Cretaceous–Paleocene cover rocks on Nuussuaq, Disko and south of Disko Bugt. Note the Palaeozoic outlier at Fossilik ('F' on the key map, lower left). LS, Labrador Sea; BB, Baffin Bay. Modified from Japsen *et al.* (2006, 2009); based on Chalmers *et al.* (1999); Bonow (2005); Bonow *et al.* (2006a, b). Location of map shown in Fig. 1.

coastal regions south of Disko Bugt. The ES is cut off by a younger surface in the area south of Disko Bugt. The younger surface is tilted and rises towards the south, where it gradually splits into the UPS and LPS.

The C2 cooling episode that began at the Eocene–Oligocene transition is recognized in AFTA results from almost every sample, both within the Nuussuaq Basin and in the basement areas. The samples recording the C2 event are distributed across an area similar in extent to the UPS that has been dated independently to be post-middle Eocene because it cuts across basalts of that age (Bonow *et al.* 2006a; Fig. 13). We therefore consider that both the formation of the UPS and the C2 cooling event are evidence of the same episode of exhumation (Japsen *et al.* 2006). AFTA and VR data from the Gro-3 well show that up to 1 km of rocks were removed during Oligocene–Miocene times before the final formation of the UPS across Cenozoic, Mesozoic and basement rocks (Japsen *et al.* 2005b). Similarly, up to 1 km of rocks (most likely including Cretaceous–Palaeogene sediments) were removed before the final formation of the UPS in the basement area north of Sukkertoppen as we shall discuss in Case E (Japsen

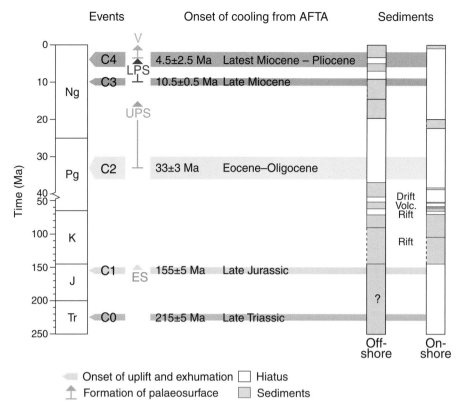

Fig. 14. Chronology of events in West Greenland. Correlation between discrete episodes during which AFTA data indicate onset of cooling (C0–C4), the formation of surfaces and valleys and the age of sediments and main tectonic events. At least six discrete episodes of cooling are required to explain all the AFTA results, but only the five youngest are shown on the figure. Dotted lines: maximum age range of sediments. Renewed development of the ES probably took place prior to Paleocene volcanism. ES, etch surface; UPS, upper planation surface; LPS, lower planation surface; V, deeply incised valleys; Tr, Triassic; J, Jurassic; K, Cretaceous; Pg, Palaeogene; Ng, Neogene; Subs, thermal subsidence; Volc, volcanism. Stratigraphy from Storey *et al.* (1998); Chalmers *et al.* (1999); Nøhr-Hansen (2003); Piasecki (2003); Schmidt *et al.* (2005); Pedersen *et al.* (2006); Sørensen (2006). Modified after Japsen *et al.* (2009).

et al. 2009). Figure 15 shows a possible reconstruction of the landscape development near Sukkertoppen Iskappe.

Uplift to present-day altitudes of up to 2 km took place in two late Neogene events (C3 and C4), resulting in dissection and tilting of the UPS into megablocks. Evidence from AFTA for these cooling events is found in samples from the bottom of valleys that have been incised by rivers subsequent to the uplift (and in subsurface samples below the valley floor, for example, the Gro-3 well). The UPS is well preserved as a low-relief surface in resistant rocks while it has started to be dissected in less resistant rocks. However, the faults offset the UPS so that it now dips in different directions in different fault blocks.

The late Neogene uplift phases post-date the final episode of rifting (in the Maastrichtian–early Danian) by *c.* 50 million years and seafloor spreading west of Greenland by *c.* 30 million years (Chalmers *et al.* 1999; Chalmers & Pulvertaft 2001), after which both the basins and their surrounds continued subsiding until the end of the Eocene. Our analysis shows that uplift started at this time, leading to erosion that led ultimately to formation of a low-relief erosion surface close to sea-level by the late Miocene. Remnants of this low-relief surface are now preserved as plateaux of large extent at elevations up to 2 km a.s.l., forming the UPS (Bonow *et al.* 2006*a*, *b*). The present-day elevated passive continental margin of West Greenland is thus in no way a reflection of the processes acting during the rifting or drifting events but developed much later.

As we will see in the following discussion of Case D, the events identified from the onshore margin discussed here are also expressed in the offshore basins (Fig. 14), where they have important implications for hydrocarbon prospectivity. Thus, integration of onshore and offshore events forms an important component of fully understanding the processes which control hydrocarbon systems in offshore basins.

Case D: the nature of the Oligocene hiatus offshore southern West Greenland

Oligocene sediments are absent on the West Greenland shelf (Sørensen 2006) and, north of about 66°N, upper Eocene sediments are separated from lower Eocene sediments by another pronounced hiatus (Fig. 16; Chalmers 2000). South of about 66°N, the two hiatuses merge into one and upper Miocene sediments lie directly on lower Eocene sediments (cf. Chalmers & Pulvertaft 2001; Nøhr-Hansen 2003; Piasecki 2003; Sønderholm *et al.* 2003; see the offshore stratigraphic column in Fig. 14). The earlier hiatus coincides with the substantial decrease in the speed of seafloor spreading in the Labrador Sea, and the latter hiatus with the final cessation of seafloor spreading between Canada and Greenland (Chalmers & Pulvertaft 2001).

It is not possible to use only reflection seismic data to evaluate whether a hiatus represents a period of non-deposition or if sediments have been deposited and then removed. The Oligocene hiatus is seen as a very low-angle unconformity on seismic data from much of the Greenland shelf and it can thus be difficult to judge if it is a truncational unconformity (Fig. 17). Such low-angle unconformities are conventionally regarded as representing intervals of stability. In the area penetrated by the Qulleq-1 exploration well, however, the middle Miocene hiatus is clearly angular and truncates lower Eocene, Paleocene and Upper Cretaceous strata (Fig. 18; Christiansen *et al.* 2001).

VR data from the Qulleq-1 well suggest that the pre-Neogene succession in the well has been more deeply buried in the past.

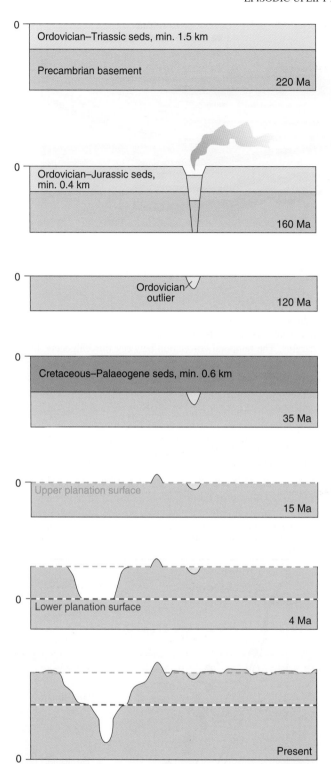

Fig. 15. Idealized profile illustrating the landscape development across the present basement area near Sukkertoppen Iskappe, West Greenland, based on AFTA data from the Itilleq vertical section (Figs 13 & 20). Note the UPS and two levels of valleys incised in the UPS: the wide valleys of the LPS and the present valleys incised in the LPS. We correlate the incision of these valley systems with the two uplift phases that formed the present topography. 220 Ma, Triassic maximum burial before C0 uplift (cooling episode C0 from AFTA data, see Fig. 14); 160 Ma, Jurassic maximum burial before C1 uplift and volcanism preserving Ordovician sediments as fall-back material; 120 Ma, most likely removal of sedimentary cover after C1 uplift and volcanism; 35 Ma, Palaeogene maximum burial after rifting and Palaeogene volcanism, before C2 uplift; 15 Ma, Oligocene–Miocene UPS; 4 Ma, late Neogene LPS after C3 and before C4 uplift; Present, deeply incised valleys after final uplift. Based on Stouge & Peel (1979); Bonow et al. (2006a, b); Japsen et al. (2006, 2009); Larsen (2006).

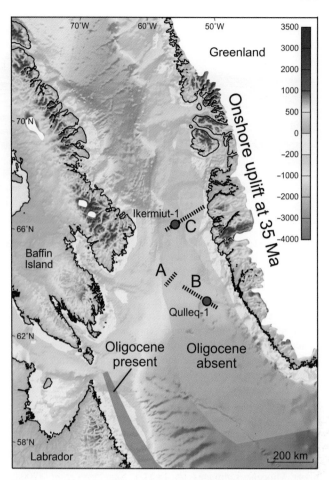

Fig. 16. Outline of the Oligocene hiatus offshore West Greenland based on seismic and well data, for example, profiles A and B, and well Qulleq-1 (Figs 18–20). The area offshore from which Oligocene deposits are known to be absent is marked. Oligocene strata have been encountered east of Labrador, but the presence of these strata in the remaining parts of the offshore area is unknown. Lower Miocene strata are also absent west of Greenland (Sørensen 2006; see the offshore stratigraphic column in Fig. 14). The Oligocene hiatus thus represents the time interval during which the Oligocene–Miocene peneplain, UPS, was graded to sea-level onshore after the C2 uplift event that began c. 35 Ma. Bathymetry and bedrock topography in metres. Icecaps indicated as white. Profile C is shown in Figure 21. Map location shown in Fig. 1.

In Figure 19, VR data from the well is seen to plot above curves of predicted VR trends based on the 'default history' derived from the preserved stratigraphy and based on a likely range of geothermal gradients between 20 and 30 °C/km (in this history, hiatuses are assumed to represent periods of non-deposition, not erosion). The discrepancy between observed and predicted VR values indicates that the drilled section below the unconformity has been hotter in the past, most likely due to deeper burial below a section that has been removed. This possibility is strengthened by the observation that the VR value for the only Neogene sample matches the default history, indicating that this sample is now at its maximum burial depth.

Further studies, based on palaeothermal (AFTA and VR) data and palaeoburial (sonic) data, constrained by the preserved stratigraphy, are needed to evaluate the timing, magnitude and extent of the exhumation related to the Oligocene hiatus offshore West Greenland. Such studies would provide firm constraints on both the thermal and the burial history of the preserved sedimentary section, allowing definition of those areas where maximum burial was reached during mid-Cenozoic times. This, in turn, would allow identification of areas where hydrocarbon generation ceased prior to exhumation, and conversely those areas where hydrocarbon

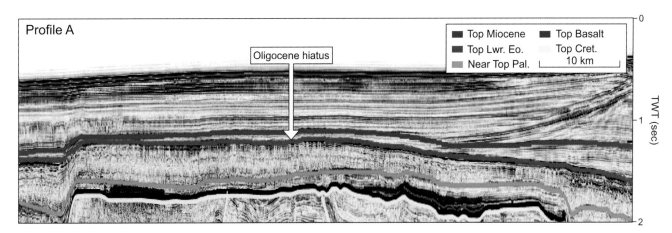

Fig. 17. Seismic profile A showing a low-angle Oligocene unconformity offshore West Greenland with lower Eocene strata below and Miocene strata above the unconformity. It is not possible from the seismic profile to evaluate whether the hiatus represents a period of no deposition or whether sediments have been deposited and then removed during the time interval represented by the unconformity. TWT, seismic two-way-time. Location in Figure 16. Modified after Sørensen (2006).

generation proceeded until the present day, allowing charging of structures formed during the mid-Cenozoic uplift and erosion.

Such studies will also be important for understanding the tectonic evolution of the Arctic during the Cenozoic within a wider context: a hiatus similar to that encountered on the West Greenland shelf is also present in the Sverdrup Basin in the eastern Canadian Arctic (Harrison et al. 1999) and a similar stratigraphic break was unexpectedly penetrated on the Lomonosov Ridge (central Arctic Ocean) by the Integrated Ocean Drilling Program (IODP) Expedition 302 (Backman et al. 2008). The drilled core documented a 26 Myr hiatus, separating middle Eocene (c. 44 Ma) from lower Miocene sediments (c. 18 Ma). Sangiorgi et al. (2008) were unable to determine whether that hiatus was generated by sediment erosion or by non-deposition, but emphasized that it conflicts with classical post-rift thermal subsidence models for passive margins. The temporal correlation between this Oligocene hiatus and the time interval during which the Oligocene–Miocene peneplain, UPS, was graded to sea-level onshore West Greenland, emphasizes the regional controls on such processes, as was also highlighted by Green & Duddy (2010).

Case E: are there Ordovician source rocks offshore West Greenland?

The thermal and burial history of the Precambrian basement south of the rifted Nuussuaq Basin and east of the rifted basins now offshore (Chalmers & Pulvertaft 2001) has been determined by analysing AFTA data from five samples taken at different heights on an

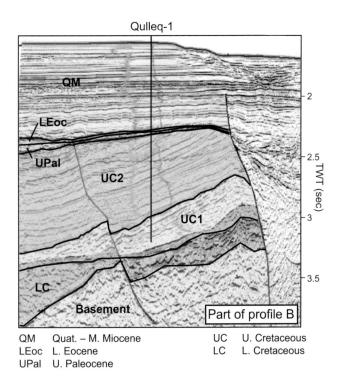

QM Quat. – M. Miocene
LEoc L. Eocene
UPal U. Paleocene
UC U. Cretaceous
LC L. Cretaceous

Fig. 18. Seismic line through the Qulleq-1 well, offshore West Greenland. Note the pronounced angular unconformity below the Neogene succession (QM) that truncates lower Eocene, Paleocene and Upper Cretaceous strata. The gamma-ray log from the Qulleq-1 well is shown in green to the left of the well and the acoustic impedance log is shown in red to the right. TWT, seismic two-way-time. Location in Figure 16. Modified after Christiansen et al. (2001).

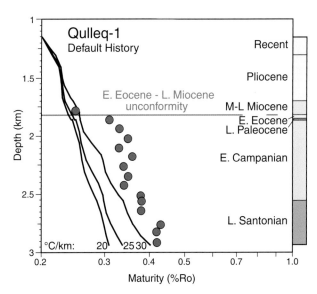

Fig. 19. Plot of VR data v. depth for the Qulleq-1 well, offshore West Greenland. Drilled succession shown on the right-hand side of the plot. Black curves indicate predicted VR trends based on the default history (preserved stratigraphy and geothermal gradients ranging from 20 to 30 °C/km). The discrepancy between observed and predicted VR values indicates that the section has been hotter in the past, possibly due to deeper burial below a section that was subsequently removed. Only the VR value for the sample above the base–Neogene unconformity matches the default history. This suggests that the early Eocene–middle Miocene hiatus represents the removal of sediments of that age. Location in Fig. 16. Modified after Christiansen et al. (2001).

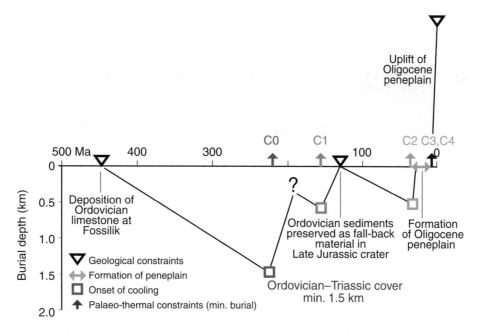

Fig. 20. Diagram illustrating the burial and uplift history of the summit at Sukkertoppen Iskapppe, West Greenland (c. 2 km a.s.l., location in Fig. 13). The interpretation is based on AFTA data from the Itilleq vertical section (minimum burial depths prior to exhumational events), interpretation of the landscape development (formation and uplift of the Oligocene peneplain, UPS) and geological constraints (presence of Ordovician sediments as fallback material in Late Jurassic volcanic breccia). The presence of Ordovician sediment at Fossilik within Upper Jurassic rocks shows that the section removed during the Triassic (C0) event must have consisted of Phanerozoic sedimentary rocks and thus that a minimum of 1500 m of Ordovician to Triassic sedimentary rocks must have been present above the basement prior to the C0 event (cf. Fig. 15).

elevation section up to a summit of 1848 m in the Itilleq fjord, NW of Sukkertoppen Iskappe (Fig. 13). The AFTA data, together with the thermal history solutions extracted from individual samples, are presented by Japsen et al. (2009), whereas the full details of these analyses and a regional dataset are provided in the online data repository associated with Japsen et al. (2006).

Three palaeothermal episodes can be defined from the AFTA data in the Itilleq profile over a range of altitudes and these events correspond well with regional cooling events (C0, C1, C2; see Fig. 14). The range of palaeotemperatures for all events is wide, but the palaeogeothermal gradients must have been less than 40 °C/km during these three events and much of the heating reflected in the observed palaeotemperatures must represent greater depth of burial; that is, the present summit must have been buried below a section with a minimum thickness of 1500 m prior to the Triassic C0 event, a minimum of 400 m prior to the Late Jurassic C1 event and a minimum of 600 m prior to the Cenozoic C2 event corresponding to the maximum allowed palaeogeothermal gradients of 40, 35 and 22.5 °C/km, respectively, during these events (Fig. 20). Lower palaeogeothermal gradients would require larger amounts of additional section.

Implications of the presence of an Ordovician outlier

An Ordovician outlier is known from a Late Jurassic explosion breccia at Fossilik, south of Sukkertoppen Iskappe (Stouge & Peel 1979; see Fig. 13, lower left). These rocks were probably preserved as fall-back deposits into a crater formed by a violent volcanic explosion that shattered the overlying rocks in Late Jurassic times (Larsen 2006). The presence of Palaeozoic sediment at Fossilik in the Late Jurassic shows that the section removed during the Triassic (C0) event must have consisted of Ordovician to Triassic sedimentary rocks because Ordovician sediments were still present after the Triassic erosional event. The AFTA data show that the minimum thickness of the Ordovician–Triassic cover was 1500 m (Fig. 15).

It is likely that the basement around Sukkertoppen Iskappe was exposed following Late Jurassic (C1) exhumation and prior to the Cretaceous rifting. Regional AFTA data show extensive cooling across central West Greenland following the onset of the C1 event and this is likely to have been caused by exhumation because all sedimentary cover around and within the Nuussuaq Basin was stripped at this time prior to the development of the Cretaceous ES (Bonow 2005). This is further supported by the evidence for Late Jurassic removal of cover rocks at Fossilik (Larsen 2006; Steenfelt et al. 2006). These observations imply that the C2 palaeothermal peak recognized in AFTA samples from Itilleq represents the culmination of Cretaceous–Palaeogene reburial in this area.

Dalhoff et al. (2006) reported that several samples of marine carbonates of late Ordovician age were dredged by seabed sampling west of Greenland where dipping reflectors of unknown age extend to near the seabed. One Ordovician carbonate sample had oil staining. The authors suggested a correlation between these samples and lower Palaeozoic carbonate successions of the SE Arctic Platform and the Hudson Platform in eastern and northeastern Canada (Sanford & Grant 2000) and North Greenland, and those at Fossilik, onshore West Greenland (Stouge & Peel 1979).

Indications of thick successions of pre-Cretaceous strata offshore Greenland are provided by the presence of the so-called 'Deep Sequence' (Chalmers & Pulvertaft 2001), a thick succession of reflections below lower to mid-Cretaceous sequences (Fig. 21). What these reflections represent is unknown, but Ordovician carbonates are not unlikely, although Triassic or even Jurassic sediments cannot be ruled out. If the successions are Ordovician, the oil-stained sample recovered by Dalhoff et al. (2006) leads us to suggest that this sequence may contain a source rock horizon which has generated oil. The possibility of thick Palaeozoic deposits being present offshore West Greenland is further strengthened by the documentation here that a significant Ordovician–Triassic cover has been deposited even in the onshore areas.

Case F: post-rift burial of the East Greenland margin

Published thermal history studies in East Greenland have provided important insights into the thermal and tectonic history of both

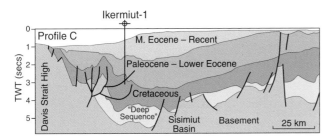

Fig. 21. Interpretation of seismic profile C, offshore West Greenland. The 'Deep Sequence' is defined by sub-horizontal reflections below and truncated by reflections that come from sediments that can be dated as Cretaceous by the Ikermiut-1 well (Chalmers & Pulvertaft 2001). The age and lithology of the 'Deep Sequence' is unknown, but Ordovician carbonates are not unlikely, although Triassic or even Jurassic sediments cannot be ruled out. If these successions are Ordovician, the oil-stained sample recovered by Dalhoff et al. (2006) suggests that they could act as a regional source rock. The possibility of thick Palaeozoic deposits being present offshore West Greenland is further strengthened by the documentation that a significant Ordovician–Triassic cover has been deposited onshore (see Figs 15 & 20). TWT, seismic two-way-time. Location in Figure 16.

Palaeozoic–Mesozoic sediments and basement terrains. Early studies revealed considerable Cenozoic cooling of country rock along the margin, usually interpreted in terms of kilometre-scale denudation related in some way to the process of rifting and separation of the North Atlantic Ocean (e.g. Gleadow & Brooks 1979; Hansen 1988, 1992, 1996). Here we will show that the later studies, despite the many uncertainties remaining, point to a quite different interpretation.

Thermal effects of Palaeogene intrusive bodies and of Cenozoic burial and exhumation revealed by AFT data

Thomson et al. (1999) published the first results from the sedimentary sequences to the north of Jameson Land (Fig. 22). They reported that AFTA data in a series of outcrop samples from Traill Ø and adjacent regions showed two dominant episodes of Cenozoic cooling, plus an earlier episode of cooling beginning between 225 and 165 Ma, recognized only in samples from westernmost locations. Cooling in the earlier of the two Cenozoic episodes began between 40 and 30 Ma (late Eocene–Oligocene), closely matching the timing of syenite intrusions (c. 35 Ma, e.g. Price et al. 1997) at the western extremity of Traill Ø. Because of the proximity of many of the samples analysed by Thomson et al. (1999) to these intrusions, the origin of the Eocene–Oligocene palaeothermal effects identified in these samples was not clear. The more recent Cenozoic episode, in which cooling began in the interval 10–5 Ma, was interpreted as due to uplift and erosion related to a change in the North Atlantic spreading vector. Several samples cooled from palaeotemperatures as high as 90–100 °C in this late Cenozoic episode, and measured AFT ages less than 20 Ma in several samples testify to the pronounced degree of Neogene cooling that has occurred in this region. Thomson et al. (1999) commented specifically that 'No paleo-thermal effects have been identified related to the onset of rifting in the Early Tertiary'.

Subsequently, Johnson & Gallagher (2000) reported AFT data from samples of Carboniferous sandstones from a vertical section on Clavering Ø (Fig. 22). Although this location lies near two samples analysed by Thomson et al. (1999), Johnson & Gallagher (2000) made no reference to the earlier study. This is made more surprising by the general consistency in the results of the two studies. Johnson & Gallagher (2000) found that the section began to cool from maximum post-depositional palaeotemperatures at c. 274 Ma (early Permian), with subsequent cooling episodes at c. 206, 140 and 23 Ma. The c. 206 Ma episode correlates closely with the earliest episode identified by Thomson et al. (1999), while the c. 23 Ma episode falls between the two Cenozoic episodes identified by Thomson et al. (1999) and the single Cenozoic episode defined by Johnson & Gallagher (2000) may well represent the unresolved effects of these two. The c. 140 Ma cooling episode identified by Johnson & Gallagher (2000) was not recognized in the Thomson et al. (1999) study, possibly because of the pronounced effect of Cenozoic events on Traill Ø, which would have masked any Mesozoic effects.

Johnson & Gallagher (2000) interpreted these cooling episodes largely in terms of denudation involving removal of several kilometres of section, while also suggesting that Cenozoic palaeothermal effects could be related to Palaeogene volcanic activity. One notable shortcoming of the Johnson & Gallagher (2000) study is that, despite their sampled section being overlain by Palaeogene sediments and basalts (their Fig. 1, cf. Henriksen et al. 2000), their analysis took no note of this basic constraint on the thermal history of the sequence. The thermal history that Johnson & Gallagher (2000) reported for their sampled section shows temperatures of c. 60 °C or more during Palaeogene times, and only a minor reheating to the Cenozoic palaeothermal peak around 70 °C at 23 Ma (Fig. 23). The presence of the Palaeogene deposits shows, however, that their uppermost sample must have been at near-surface temperature in the Palaeogene. This implies a major degree of re-burial by Palaeogene volcanics and/or younger sediments prior to the onset of cooling from the mid-Cenozoic palaeothermal peak. Note also that only the palaeothermal peaks are constrained by AFT data, and that it thus remains unknown what degree of cooling and reheating (e.g. exhumation and reburial) may have taken place between the peaks. The Carboniferous rocks studied by Johnson & Gallagher (2000) may thus have been near the surface several times after initial burial in the Carboniferous. As with the Thomson et al. (1999) study, Johnson & Gallagher (2000) reported no evidence of Palaeogene palaeothermal effects.

Hansen et al. (2001) reported results from the Jameson Land Basin and adjacent regions, based largely on AFT data (Fig. 22b). The results presented by Hansen et al. (2001) show a very consistent regional pattern, with youngest ages less than 25 Ma restricted to a region around the north and south coasts of Kong Oscar Fjord (Traill Ø and northern Jameson Land, respectively), surrounded by a wider region characterized by ages between 25 and 45 Ma, while older ages are largely restricted to higher elevations and/or locations to the south and west of Jameson Land. Hansen et al. (2001) interpreted their results as reflecting a complex interplay between deeper burial prior to Cenozoic exhumation, circulation of hot fluids associated with mid-Cenozoic intrusive activity around the eastern end of Kong Oscar Fjord and also possibly elevated basal heat flow in eastern Traill Ø. Hansen et al. (2001) favoured an onset of exhumation at c. 55 Ma in the south of the basin, and 'before 25 Ma' in the north, suggesting removal of as much as 4–5 km for palaeogeothermal gradients around 25 °C/km.

A notable feature of the data from Jameson Land presented by Hansen et al. (2001) is the erratic variation of fission track age with elevation. Preservation of fission track ages in excess of 200 Ma in SW Jameson Land, in relatively close proximity to much younger ages (Fig. 22b), suggests that regional Cenozoic heating was relatively minor in this region, and that local heating effects dominate the data. In view of these observations, any evidence for Palaeogene exhumation in this dataset is tentative at best, with all evidence suggesting that any Palaeogene palaeothermal effects are localized, due either to contact heating from the widespread minor intrusions of this age or associated hydrothermal effects (or both). In contrast to the erratic variation in heating

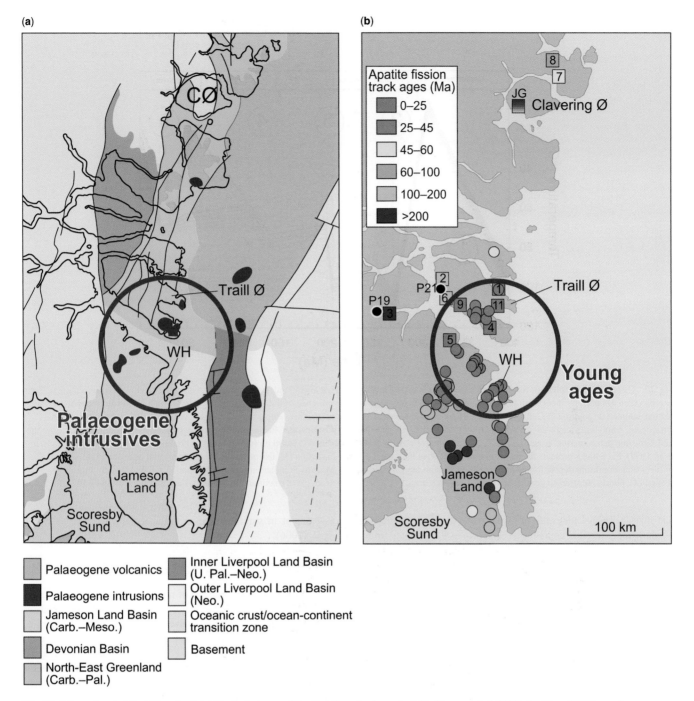

Fig. 22. Maps of central East Greenland. (**a**) Geological map of the East Greenland margin (after Hamann *et al.* 2005). (**b**) Map of AFT ages from Thomson *et al.* (1999; squares with sample numbers) and Hansen *et al.* (2001; circles). The coincidence of the area with Palaeogene intrusives (map A) and the area with young AFT ages (map B) is highlighted with red circles. Vertical profile of samples analysed by Johnson & Gallagher (2000, marked JG) and Swift *et al.* (2008, black circles P19 and P21) is also indicated. JL, Jameson Land; WH, Wegener Halvø. Location of map shown in Fig. 1.

effects in the south of Jameson Land, the pattern of fission track ages in Figure 22b shows that Neogene palaeothermal effects in eastern Traill Ø and NE Jameson Land around Wegener Halvø are more uniform. The correspondence of these younger ages with the presence of late Eocene–Oligocene intrusive bodies suggests a causal relationship.

A number of basin modelling and maturation studies of the Jameson Land Basin have used AFT data to constrain the history of uplift and denudation (Christiansen *et al.* 1990, 1992; Mathiesen *et al.* 1995, 2000). Mathiesen *et al.* (2000) assumed that generation of hydrocarbons was dominated by deeper burial after deposition of the thick volcanic pile during the early Paleocene and prior to denudation beginning in the late Paleocene, with a total of 2–3 km of section removed since that time. Christiansen *et al.* (1992) drew attention to the presence of hydrothermal effects in upper Permian carbonates in Wegener Halvø, which correlates closely with the AFT age 'hot spot' in Figure 22b.

Mismatch between apatite (U–Th)/He dating and AFT results

Swift *et al.* (2008) presented apatite (U–Th)/He dating results from two vertical sections where one profile (number 21) is located close to samples 2 and 6 from the Thomson *et al.* (1999) study (Fig. 22). Apatite (U–Th)/He ages reported from Profile 21 by Swift *et al.* (2008) show erratic variation with elevation, with youngest ages between 23 and 29 Ma in three samples (with uncertainties of ±1 or 2 Ma) while other samples give ages between 56

998   P. JAPSEN ET AL.

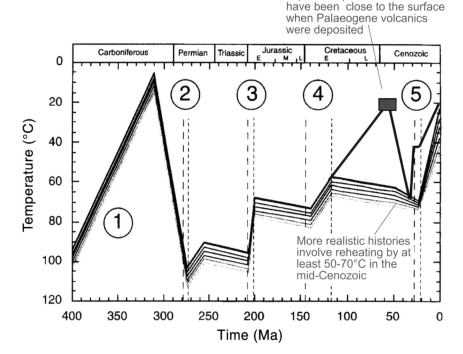

Fig. 23. Thermal history solutions for a vertical section of rocks on Clavering Ø, East Greenland, from Johnson & Gallagher (2000), showing a more realistic interpretation obtained by incorporating basic geological constraints based on the presence of the Palaeogene deposits at the highest elevations above the section sampled by Johnson & Gallagher (2000). These constraints show that the uppermost sample must have been at near surface temperature in the Palaeogene, implying a major degree of re-burial by additional volcanics and/or post-volcanic Palaeogene sediments prior to the onset of cooling from the mid-Cenozoic palaeothermal peak of c. 70 °C. The alternative cooling history indicated is based on two episodes of cooling at c. 35 and 10 Ma (Thomson et al. 1999). Each line represents the temperature history of an individual sample. In section (1), fission tracks are produced and annealed in the source region of the sediment prior to deposition; in section (2), maximum temperatures at 274 Ma mid-Permian; in section (3), cooling at 206 Ma early Jurassic; in section (4), poorly constrained cooling at 140 Ma; and in section (5), Cenozoic reheating and maximum temperatures at 23 Ma. Location on Fig. 22.

and 70 Ma with one older value of 137 ± 8 Ma. Results from Profile 19 show more internal consistency, with two of the samples from low elevations giving reproducible (U–Th)/He-ages around 70–80 Ma. Swift et al. (2008) interpreted these results as indicating a phase of rapid denudation which began at 74 ± 15 Ma. This conclusion stands in stark contrast to previous studies in this region which, as reviewed above, showed no evidence of late Cretaceous/Palaeogene cooling (Thomson et al. 1999; Johnson & Gallagher 2000). Swift et al. (2008) also claimed that 'AFTA data published by Thomson et al. (1999) for sea level bedrock samples located near to profiles 19 and 21 are consistent with rapid cooling from temperatures of c. 70 °C at c. 70 Ma.' This is not correct. For example, Thomson et al. (1999) reported that their sample 2 cooled from between 85 and 95 °C in the interval 40–5 Ma.

Swift et al. (2008) stated that He ages can be interpreted qualitatively as reflecting the time at which samples cool through 75–35 °C. We should then expect He ages in sample 2 from Thomson et al. (1999) of perhaps 30 Ma or younger, in contrast to most of the measured He ages, which are c. 60 Ma or older. This mismatch can be understood in terms of the now widely recognized problems with the (U–Th)/He system, in which the He retentivity is enhanced over the 'normal' systematics as a result of the accumulation of radiation damage within the apatite lattice (Green et al. 2006; Shuster et al. 2006). The apatite (U–Th)/He ages cannot therefore be interpreted simply in terms of cooling below 35–75 °C as stated by Swift et al. (2008). Comparison of their (U–Th)/He ages with the interpretations provided by Thomson et al. (1999) suggests that the three youngest ages from Profile 21 probably provide a more reliable assessment of the low-temperature history of the sampled rock section, while ages around 60 Ma or older are likely to be spuriously old as a result of the factors discussed above. Unfortunately, the results provided by Swift et al. (2008) do not contain sufficient detail to allow assessment in similar terms to the analysis presented by Green et al. (2006).

Development of the East Greenland margin

In summary, despite a number of regional, thermal-history studies of the East Greenland margin, resulting in a consistent regional pattern of variation, significant uncertainty remains with regard to the precise time at which individual rock units reached their palaeothermal maximum, how this timing varies across the region, and about the nature of the mechanisms responsible for heating and cooling, including palaeogeothermal gradients and amounts of section removed by uplift and erosion.

Remarkably, no convincing evidence for regional palaeothermal effects related to Palaeogene breakup has been reported; Cenozoic maximum temperatures appear to have occurred in the mid- and late Cenozoic. In particular, results from Clavering Ø show that Palaeogene basalts preserved on the summits there have been heated to peak temperatures of c. 70 °C prior to cooling in the late Cenozoic. As there is no Palaeogene intrusive activity reported from this area, a major cause of the post-basalt heating must be due to burial below rocks that have been removed, presumably by uplift and erosion in the Neogene.

The lack of evidence for palaeothermal cooling (which might reflect uplift) related to breakup is supported by evidence from the geological record. The Scoresby Sund area (Fig. 22) was characterized by net subsidence during the eruption of the basalts at the Paleocene–Eocene transition with no evidence for crustal upwarping (Brooks 1985; see also Larsen & Tegner 2006). There are marine incursions in the youngest plateau lavas (see Larsen et al. 1989), and outliers of Eocene marine sediments overlie the basalt just south of Scoresby Sund (see Larsen et al. 2005). The margin therefore subsided following rifting and breakup, and deposition of this post-rift sequence during thermal subsidence is just what is predicted from the theory of continental stretching (McKenzie 1978; White & McKenzie 1988).

Medvedev et al. (2008) investigated the effects of erosional unloading based on geological and topographical observations from the Scoresby Sund area where peaks reach 2.5–3.7 km a.s.l. According to their calculations, erosional unloading due to incision below the highest peaks in the area can account for 1.1 km of uplift, which is comparable with the elevation of Mesozoic marine sediments that are now at an elevation of 1.2 km a.s.l. The authors concluded that 'the active erosion of the region is the first-order feature responsible for significant uplift of the marine sediments and adjacent Fjord Mountains'. Medvedev et al. (2008) did not, however, consider that these marine sediments may have been buried below a pile of Upper Cretaceous–Palaeogene sediments, in which case the true amount of uplift of these marine sediments since their maximum burial would be much larger than 1.2 km. Much of the high ground north of Scoresby Sund appears to consist of a dissected plateau similar to those in Norway and West Greenland (Ahlmann 1941; see Henriksen 2008, pp. 22–23). As noted above, formation of such a plateau requires at least two episodes of uplift; a first phase during which an erosion surface is formed near sea-level and a second to uplift it to its present elevations. Medvedev et al.'s (2008) calculations may show that a large part of the uplift of the erosion surface in the Scoresby Sund area is due to isostatic response to erosion, but none of the calculations address the reasons for any earlier episode.

In East Greenland, uplift events have led to exhumation of remnants of hydrocarbon accumulations hosted in Jurassic sandstones now at elevations of around 1 km on Traill Ø (Price et al 1997; Price & Whitham 1997). The generation of hydrocarbons in the vicinity of eastern Traill Ø from Upper Jurassic source rocks (Price & Whitham 1997; Price et al. 1997) was most likely enhanced due to the increased heat flow in this region associated with the intrusions in the late Eocene (combined with the palaeoburial at that time). This suggestion agrees well with the very low organic carbon content of the reservoir rocks (GEUS unpublished data). According to J. Bojesen-Koefoed (pers. comm. 2009) the organic content is much lower than would be expected if the petroleum was degraded through biogeochemical processes upon exhumation of the reservoir, and it is thus more likely explained by destruction of the hydrocarbons by high temperature rather than by leakage and biodegradation. The implication of this is that the presence of these accumulations is likely to be closely associated with the local thermal perturbation caused by the intrusions. Further afield, in areas where heat flow was not enhanced, any source rocks present would have remained at lower maturity levels at the end of the Eocene. Particularly in offshore regions that were not affected by mid-Cenozoic exhumation, such source rocks may then have generated hydrocarbons during later burial in areas of Neogene subsidence. Therefore the presence of these exhumed oil accumulations may be seen as a positive feature for exploration in the region, instead of the more negative impact generally attributed to exhumation. Moreover, uplift of the margin may have created potential traps for these hydrocarbons, suggesting that hydrocarbons may be present in the offshore region. In exploring for these hydrocarbons, detailed mapping of the key uplift events will be crucial in identifying suitable targets in more prospective regions where the timing relationships are most favourable.

Are elevated rift margins formed by processes related to rifting and breakup?

Evidence from West Greenland clearly shows that the present-day elevated continental margin there is not a remnant of the original rifting process, as rifting terminated in the Paleocene (Dam et al. 1998), after which kilometre-scale subsidence occurred (Pedersen et al. 2002; Japsen et al. 2006). Instead the uplift occurred in Oligocene to Recent times (Japsen et al. 2005b, 2006; Bonow et al. 2006a, b). Observations presented here from the sedimentary basins adjacent to southern Norway similarly indicate that the mountains in that region emerged during several phases of uplift long after North Atlantic breakup at the Paleocene–Eocene transition. The mountains along these passive continental margins are characterized by elevated plains and deeply incised valleys which is also the case for margins in other parts of the world, such as southern Africa and southeastern Australia (see King 1967; Partridge & Maud 1987; Lidmar-Bergström et al. 2000). The results from West Greenland and the eastern North Sea Basin therefore lead us to suggest that other elevated rift margins may also have been formed by processes that are related neither to rifting nor to continental breakup (Japsen et al. 2009).

Both southern Africa and southeastern Australia have been well studied by both thermochronological and geomorphological standpoints. Partridge & Maud (1987) concluded after a thorough morphological analysis of the South African margin that the regional peneplain that defines the elevated topography in the area (the African Surface; King 1967) was uplifted in the Neogene, thereby agreeing with one of King's (e.g. 1967) major conclusions. In contrast to these results, many of the more recent studies of that area are predicated on the basic assumption that the morphology reflects the processes of rifting (e.g. Ollier 1985; Brown et al. 2002; van der Beek et al. 2002). However, AFTA ages from outcrop samples across a large area of southern Africa (Gallagher & Brown 1999) post-date rifting by many tens of millions of years, showing that rocks forming the extensive low-relief surface there were at temperatures around 100 °C well after the rift episode. Similar results were reported by Raab et al. (2005) from Namibia, where young AFT ages extend several hundreds of kilometres inland, as they also do in southeastern Australia (e.g. O'Sullivan et al. 1995; Kohn et al. 1999). Along the Brazilian margin, AFTA ages are younger than rifting by up to c. 50 Ma (Gallagher et al. 1996). None of these observations are consistent with simple models of rift margin development as summarized by, for example, Persano et al. (2002), and it seems far from straightforward to infer that these results reflect processes related directly to rifting.

In West Greenland, integration of stratigraphic evidence provided by the sediments and erupted volcanic rocks of the Nuussuaq Basin, objective interpretation of the denudational history from study of the landscape and a detailed thermochronological dataset have provided unique constraints on the development of the margin. In contrast to this setting, the majority of previous thermochronological studies in other regions have been focused on basement terrains and the data have been interpreted by assuming monotonic cooling, presumably because it is assumed that such areas never had a sedimentary cover and have undergone continued emergence. Our results show that these assumptions are false in West Greenland. We have shown that even the basement terrain in this region was buried repeatedly during the Phanerozoic by sedimentary covers that were subsequently removed during episodes of burial and exhumation (Japsen et al. 2009). We suggest that similar histories are likely to be appropriate to many other passive continental margins (see Bonow et al. 2007). We further suggest that studies designed to investigate the development of passive continental margins should be focused on areas where remnants of the syn- and post-rift sequence are preserved onshore (e.g. Brazil; Cobbold et al. 2001) since these areas contain far more information on the history of vertical displacements than those where basement is exposed at the surface. Unfortunately, the high terrain in Norway is an example of a continental margin where only rocks that predate rifting and breakup by hundreds of millions of years are preserved, limiting severely the information that can be obtained on the evolution of this region.

The sedimentary record in the basins adjacent to the elevated passive continental margins of West Greenland and southern Norway also provide evidence for the episodic post-rift uplift and erosion of the offshore shelves along these margins. This evidence comes partly from seismic and stratigraphic data that reveal sedimentary wedges prograding away from the centres of uplift in the present highlands and partly from palaeothermal and palaeoburial data from wells that make it possible to show when a hiatus represents a time interval of burial followed by exhumation. The application of this approach has demonstrated that the Neogene uplift of Scandinavia was accompanied by exhumation that reaches far into the North Sea Basin. The exhumation thus reaches beyond the truncated Palaeogene strata in Figure 1 and also affected areas with a Miocene cover, far from the centre of uplift in the Scandinavian hinterland.

In East Greenland, a major cause of the post-basalt heating reported in several studies (e.g. Thomson et al. 1999; Johnson & Gallagher 2000) must be due to burial below rocks that have been removed presumably by uplift and erosion in the Neogene. This interpretation resembles the burial and exhumation history established for West Greenland by Japsen et al. (2006) where rifting and breakup was followed by kilometre-scale subsidence and burial over an interval of many millions of years until this trend was reversed by uplift and exhumation during the mid- and late Cenozoic. These interpretations imply that post-rift subsidence of both the West and East Greenland margins, together with burial of the pre- and syn-rift sections, has taken place after rifting and breakup, just as predicted by classical rift theory. Subsequently, this subsidence was reversed by a series of uplift movements (in the West Greenland case at c. 35, 10 and 5 Ma) which as yet remain unexplained.

On the origin of post-rift uplift along passive margins

The regional nature of the post-rift uplift movements along passive, continental margins is now well documented (e.g. Japsen & Chalmers 2000; Green & Duddy 2010) and it is instructive to compare these events in the North Atlantic domain.

Post-breakup subsidence seems to have continued on most passive margins around the North Atlantic until the Eocene–Oligocene transition (c. 35 Ma), after which it was interrupted. There was no particularly notable fall in sea-level at this time (Miller et al. 2005), so it is difficult to see how the change could be due to a fall in eustatic sea-level. It seems instead that the subsiding basins must have been inverted by tectonic forces. We speculate that the forces causing this phase may have been related to the major plate boundary reorganization in the North Atlantic around Chron 13 time at c. 33 Ma (Gaina et al. 2009). We would also point out that this is also the approximate time at which the Iceland plume started to pulse, as revealed by gravity anomalies and the V-shaped ridges (Poore et al. 2009).

The next events of regional exhumation occur at very different times in Greenland and Scandinavia; in the late Miocene in West Greenland at c. 10 Ma (most likely at about the same time in East Greenland; cf. Thomson et al. 1999) and in the early Neogene at c. 23 Ma in southern Scandinavia. The time difference alone rules out the possibility that these events can be explained by eustasy; differential tectonics are needed to explain these movements. West Greenland provides the example where the chronology of events is best understood, and here the fully developed peneplain prior to 10 Ma provided no relief to be eroded. Thus, even if erosion was increased due to climate deterioration, no valley incision could have occurred to initiate late Neogene uplift and, as there was no significant change in eustatic sea-level at this time (Miller et al. 2005), a tectonic trigger was required. The early Neogene phase of regional uplift in Scandinavia correlates with the Savian phase of Alpine tectonics in central Europe (Rasmussen 2009b).

The third event of post-rift uplift in the region seems to have affected almost all margins beginning around the early Pliocene and, within the accuracy of the methods applied, almost simultaneously (e.g. Green & Duddy 2010): West Greenland between 7 and 2 Ma, Norwegian–Danish Basin c. 4 Ma, NE Atlantic margin c. 4 Ma and Barents Sea 10–5 Ma. The effects of this event are probably more widespread, but they cannot be resolved in many areas because of the relatively low palaeotemperatures (<60 °C that could equate to deeper burial by up to 1.5 km) of outcrop samples prior to this final phase of Cenozoic exhumation. These movements overlap with the early Pliocene warm period in NE Greenland, and predate the late Pliocene glacial maximum (Funder 1989). In fact this event caused significant surface uplift across the region and this relief, combined with climate deterioration, may thus have had a triggering role in the formation of the Greenland and Scandinavian ice sheets (Letréguilly et al. 1991; Japsen et al. 2006). Miller et al. (2005) show that lowstands of sea-level fell by about 25 m starting at 6 Ma and that there was significant fall in average sea-level after c. 2 Ma, during the latest Pliocene and in the Pleistocene. The latter fall must have reduced base levels significantly, leading to increased erosion and consequent enhanced uplift, but we question whether the earlier fall in lowstands alone is enough to trigger the final phase of uplift (cf. Stoker et al. 2005). That some tectonic trigger is needed is suggested by there being large negative gravity lows over all of the uplifted areas around the North Atlantic (Sandwell & Smith 1992; Laxon & McAdoo 1998). Such lows would not be produced by only an isostatic response to lowered base level, but need some form of low density or vertical motions in either the lower crust or upper mantle. The early Pliocene phase does not seem to be related to any of the plate reorganizations in the NE Atlantic found by Gaina et al. (2009).

The regional synchronicity between these vertical movements (and to some extent also the lack of synchronicity) and the emerging correlation between these events and episodes of plate reorganization (e.g. Gaina et al. 2009) and magmatic activity in Iceland (Doré et al. 2008; Poore et al. 2009) indicates that there have been tectonic triggers to the uplift movements, but does not, however, indicate a direct causal relationship between uplift movements and plate reorganization. The forces produced by the moving plates do not seem to be sufficient alone to produce the regional uplifts with kilometre-scale magnitudes along passive margins discussed in this paper. Doré et al. (2008) thus argued that the mild compressional structures that formed mainly between 20 and 15 Ma on the passive margins along NW Europe were generated by horizontal stress from the significant bathymetric–topographic high around Iceland. Doré et al. (2008) found that ridge push, generally thought to be the dominant body force acting on passive margins, was able to generate enough stress to cause only mild deformation. However, the regional uplift movements as well as the events of plate reorganization may both be expressions of deep-rooted changes in, for example, the lithosphere–asthenosphere system. The considerable distance between uplift movements along passive margins and active plate boundaries suggests that the causal mechanisms must be located in the deep crust or the upper mantle where the thickness of the crust and lithosphere changes substantially over a short distance (cf. King & Anderson 1998; Praeg et al. 2005; Japsen et al. 2006). The response to such tectonic movements is then amplified by the isostatic response to resulting erosion and redeposition of mass in the basins (cf. Medvedev et al. 2008).

Conclusions

The study of the burial and exhumation history of the eastern North Sea Basin emphasizes that the tectonic development of a region cannot be reconstructed solely on evidence from the preserved sedimentary record. Such reconstructions also need to be based on the record of physical indicators of palaeothermal and palaeoburial events related to the former presence of geological units that are now removed. Many studies of the sedimentary record across Denmark and offshore Norway have been published, but unconformities, particularly the low-angle unconformities, have received less attention and gaps in the sedimentary record have often been considered as periods of non-deposition (e.g. Faleide et al. 2002). As we have shown, such unconformities often represent intervals during which significant amounts of previously deposited section have been eroded. Only by considering episodic exhumation as an inherent aspect of the sedimentary record can the tectonic evolution in general, and the history of petroleum systems in particular, be reconstructed.

The study of the eastern North Sea Basin provides evidence for three major uplift episodes that have affected southern Scandinavia during the Cenozoic: starting at the Eocene–Oligocene transition (c. 33 Ma), in the early Neogene (c. 23 Ma) and in the late Neogene (c. 4 Ma). These phases, of which the two youngest involve exhumation of the offshore basin as well as the onshore region, are consistent with the sedimentary record in the NE Atlantic and around southern Norway. The observations indicate that the high mountains of southern Norway have emerged since Eocene times when a deep sea covered much of the area.

The West Greenland case underlines the importance of investigating the development of passive continental margins where sedimentary and volcanic sequences are preserved, since these areas contain far more information on the history of vertical displacements than those where only basement is exposed at the surface. The preservation of evidence of kilometre-scale subsidence within the Palaeogene volcanic–sedimentary succession makes it possible to rule out the assumption that the landscape was already high subsequent to rifting in West Greenland. However, AFTA data from areas where only basement is exposed have allowed us to infer that, even there, Cenozoic maximum burial and subsequent uplift to present elevations occurred well after rifting. We have shown that even basement terrains were covered by a Phanerozoic sedimentary cover that was subsequently removed during several episodes of burial and exhumation and in particular that a cover of at least 1500 m of Ordovician to Triassic sediments must have been deposited and then removed over the present basement high near Sukkertoppen Iskappe. This increases the possibility that the thick 'Deep Sequence' identified on seismic data from west of Greenland below deposits of Cretaceous age may include Palaeozoic sediments and possibly intervals that may act as source rocks for hydrocarbon systems.

Oligocene strata are absent from offshore West Greenland, but seismic data show that over parts of the shelf this hiatus is an erosional unconformity where middle Miocene strata truncate lower Eocene, Paleocene and Upper Cretaceous deposits. Interpretation of VR data from the Qulleq-1 well suggests that the pre-Neogene sediments have been hotter in the past and that it is likely that the Oligocene hiatus represents a period when previously deposited Palaeogene and locally older sediments were removed by erosion. Mesozoic and Cenozoic source rocks in this region are likely to have reached their maximum post-depositional palaeotemperatures as a result of burial by these eroded sediments, and hydrocarbon generation in these areas will therefore have terminated at the onset of uplift and exhumation in mid-Cenozoic times. Any structures formed during this uplift were therefore not available for charging during the main phase of generation and only structures formed earlier are likely to be viable exploration targets. Further definition of areas where timing relationships may be more favourable must await more detailed definition of regional variation in the timing and magnitude of exhumation, based on palaeothermal and palaeoburial indicators together with the regional stratigraphy.

In East Greenland, Cenozoic maximum palaeotemperatures occurred in the mid- and late Cenozoic. In particular, results from Clavering Ø show that Palaeogene basalts preserved on the summits there have been heated to peak temperatures of c. 70 °C prior to cooling in the late Cenozoic. As there is no Palaeogene intrusive activity reported from this area, a major cause of the post-basalt heating must be due to burial below rocks that have been removed, presumably by uplift and erosion in the Neogene. The lack of evidence for palaeothermal cooling (which might reflect uplift) related to breakup is supported by evidence from the geological record that indicates net subsidence during the eruption of the basalts at the Paleocene–Eocene transition. The margin therefore subsided following rifting and breakup, and deposition of this post-rift sequence during thermal subsidence is just what is predicted from the theory of continental stretching, whereas uplift in the mid- and late Cenozoic led to the formation of the present-day landscape. The uplift caused exhumation of remnants of hydrocarbon accumulations hosted in Jurassic sandstones on Traill Ø (Price et al., 1997). The generation of hydrocarbons here was most likely enhanced due to the increased heat flow in this region associated with the intrusions in the late Eocene (combined with the palaeo-burial at that time). Further afield, in areas where heat flow was not enhanced, any source rocks present would have remained at lower maturity levels at the end of the Eocene, and Neogene burial may have resulted in generation of significant amounts of hydrocarbons that remain to be discovered.

Our observations demonstrate that elevated passive margins around the North Atlantic were formed by episodic, post-rift uplift movements that are manifest in the high-lying peneplains that characterize the coastal mountains and in the unconformities in the adjacent sedimentary basins. Moreover, the results from West Greenland, supported by observations from East Greenland and Scandinavia, show that post-rift subsidence took place for c. 25 Myr (in West Greenland between c. 60 and c. 35 Ma) after rifting and breakup, as predicted by classical rift theory, but that this development was reversed by a series of uplift movements (in West Greenland starting at c. 35, 10 and 5 Ma). These post-rift uplift movements are clearly unexplained by the existing theory of continental stretching and rifting (e.g. McKenzie 1978; White & McKenzie 1988).

The tectonic forces driving the phases of post-rift uplift along passive continental margins are unknown, but the regional nature of these movements, their considerable distance from active plate boundaries and, in some cases, their correlation with phases of plate reorganization (e.g. Gaina et al. 2009) and magmatic activity around Iceland (Doré et al. 2008) suggest that similar causal mechanisms may be active along many so-called passive continental margins. One feature that all of these areas have in common is that uplift has taken place along the edges of cratons where the thickness of the crust and lithosphere changes substantially over a short distance. It may be that the lateral contrasts in the properties of the stretched and unstretched lithosphere make the margins of the cratons unstable long after rifting, causing uplift movements that form both the present-day mountains, the regional unconformities in the adjacent basins and accelerated subsidence in the basin centres.

We thank Det norske oljeselskap ASA for permission to publish results from a study of exhumation based on sonic data in the southern Norwegian sector and ConocoPhillips for permission to publish AFTA results from three Norwegian wells. We also thank Tony Doré for a complementary, instructive and thoughtful review that improved the paper. The paper is published with permission from the Geological Survey of Denmark and Greenland.

References

Ahlmann, H. W. 1941. Studies in North–East Greenland 1939–1940. *Geografiska Annaler*, **23**, 145–209.
Backman, J., Jakobsson, M. et al. 2008. Age model and core-seismic integration for the Cenozoic Arctic Coring Expedition sediments from

the Lomonosov Ridge. *Paleoceanography*, **23**, 1–15; doi: 10.1029/2007PA001476.

Bonow, J. M. 2005. Re-exposed basement landforms in the Disko region, West Greenland – disregarded data for estimation of glacial erosion and uplift modelling. *Geomorphology*, **72**, 106–127.

Bonow, J. M., Japsen, P., Lidmar-Bergström, K., Chalmers, J. A. & Pedersen, A. K. 2006a. Cenozoic uplift of Nuussuaq and Disko, West Greenland – elevated erosion surfaces as uplift markers of a passive margin. *Geomorphology*, **80**, 325–337.

Bonow, J. M., Lidmar-Bergström, K. & Japsen, P. 2006b. Palaeosurfaces in central West Greenland as reference for identification of tectonic movements and estimation of erosion. *Global and Planetary Change*, **50**, 161–183.

Bonow, J. M., Lidmar-Bergström, K., Japsen, P., Chalmers, J. A. & Green, P. F. 2007. Elevated erosion surfaces in southern Norway and West Greenland: their significance in studies of uplift and tectonic events. *Norwegian Journal of Geology*, **87**, 181–196.

Bray, R., Green, P. F. & Duddy, I. R. 1992. Thermal history reconstruction in sedimentary basins using apatite fission track analysis and vitrinite reflectance: a case study from the east Midlands of England and the Southern North Sea. *In*: Hardman, R. F. P. (ed.) *Exploration Britain: Into the Next Decade*. Geological Society, London, Special Publications, **67**, 3–25.

Brooks, C. K. 1985. Vertical crustal movements in the Tertiary of central East Greenland: a continental margin at a hot-spot. *Zeitschrift für Geomorphologie*, **54**, 101–117.

Brown, R., Summerfield, M. A. & Gleadow, A. J. W. 2002. Denudational history along a transect across the Drakensberg Escarpment of southern Africa derived from apatite fission track analysis. *Journal of Geophysical Research*, **107**, 2350–2367.

Chalmers, J. A. 2000. Offshore evidence for Neogene uplift in central West Greenland. *Global and Planetary Change*, **24**, 311–318.

Chalmers, J. A. & Pulvertaft, T. C. R. 2001. Development of the continental margins of the Labrador Sea: a review. *In*: Wilson, R. C. L., Withmarsh, R. B., Taylor, B. & Froitzheim, N. (eds) *Non-volcanic Rifting of Continental Margins: A Comparison of Evidence from Land and Sea*. Geological Society, London, Special Publications, **187**, 77–105.

Chalmers, J. A., Pulvertaft, T. C. R., Marcussen, C. & Pedersen, A. K. 1999. New insight into the structure of the Nuussuaq Basin, central West Greenland. *Marine and Petroleum Geology*, **16**, 197–224.

Christiansen, F. G., Piasecki, S. & Stemmerik, L. 1990. Thermal maturity history of the Upper Permian succession in the Wegener Halvø area, East Greenland. *Rapport Grønlands Geologiske Undersøgelse*, **148**, 109–114.

Christiansen, F. G., Larsen, H. C. *et al.* 1992. Uplift study of the Jameson Land Basin, East Greenland. *Norsk Geologisk Tidsskrift*, **72**, 291–294.

Christiansen, F. G., Bojesen-Koefoed, J. A. *et al.* 2001. Petroleum geological activities in West Greenland in 2000. *Geology of Greenland Survey Bulletin*, **189**, 24–33.

Clausen, O. R., Gregersen, U., Michelsen, O. & Sørensen, J. C. 1999. Factors controlling the Cenozoic sequence development in the eastern parts of the North Sea. *Journal of the Geological Society, London*, **156**, 809–816.

Cobbold, P. R., Meisling, K. E. & Mount, V. S. 2001. Reactivation of an obliquely rifted margin, Campos and Santos basins, southeastern Brazil. *American Association of Petroleum Geologists Bulletin*, **85**, 1925–1944.

Corcoran, D. V. & Doré, A. G. 2005. A review of techniques for the estimation of magnitude and timing of exhumation in offshore basins. *Earth Science Reviews*, **72**, 129–168.

Dalhoff, F., Larsen, L. M. *et al.* 2006. Continental crust in the Davis Strait: new evidence from seabed sampling. *Geological Survey of Denmark and Greenland Survey Bulletin*, **10**, 33–36.

Dam, G., Larsen, M. & Sørensen, J. C. 1998. Sedimentary response to mantle plumes: implications from Paleocene onshore successions, West and East Greenland. *Geology*, **26**, 207–210.

Doré, A. G., Corcoran, D. V. & Scotchman, I. C. 2002. Prediction of the hydrocarbon system in exhumed basins, and application to the NW European margin. *In*: Doré, A. G., Cartwright, J. A., Stoker, M. S., Turner, J. & White, N. (eds) *Exhumation of the North Atlantic Margin: Timing, Mechanisms and Implications for Petroleum Explorations*. Geological Society, London, Special Publications, **196**, 401–429.

Doré, A. G., Lundin, E., Kusznir, N. J. & Pascal, C. 2008. Potential mechanisms for the genesis of Cenozoic domal structures on the NE Atlantic margin: pros and cons and some new ideas. *In*: Johnson, H., Doré, A. G., Gatliff, R. W., Holdsworth, R. E., Lundin, E. & Ritchie, J. D. (eds) *The Nature and Origin of Compression in Passive Margins*. Geological Society, London, Special Publications, **306**, 1–26.

Eldholm, O., Thiede, J. & Taylor, E. 1989. Evolution of the Vøring volcanic margin. *In*: Eldholm, O., Thiede, J. & Taylor, E. (eds) *Proceedings of the Ocean Drilling Program (ODP), Scientific Results*, College Station, TX (Ocean Drilling Program), **104**, 39–68.

Fabricius, I. L., Gommesen, L., Krogsbøll, A. & Olsen, D. 2008. Chalk porosity and sonic velocity versus burial depth: influence of fluid pressure, hydrocarbons, and mineralogy. *American Association of Petroleum Geologists Bulletin*, **92**, 201–223.

Faleide, J. I., Kyrkjebø, R. *et al.* 2002. Tectonic impact on sedimentary processes during Cenozoic evolution of the northern North Sea and surrounding areas. *In*: Doré, A. G., Cartwright, J. A., Stoker, M. S., Turner, J. P. & Whites, N. (eds) *Exhumation of the North Atlantic Margin: Timing, Mechanisms and Implications for Petroleum Exploration*. Geological Society, London, Special Publications, **196**, 235–269.

Funder, S. 1989. Quaternary geology of the icefree areas and adjacent shelves of Greenland. *In*: Fulton, J. R. (ed.) *Quaternary Geology of Canada and Greenland*. The Geology of North America. Geological Society of America, Boulder, CO, 741–792.

Gaina, C., Gernigon, L. & Ball, P. 2009. Palaeocene–Recent plate boundaries in the NE Atlantic and the formation of the Jan Mayen microcontinent. *Journal of the Geological Society, London*, **166**, 601–616.

Gallagher, K. & Brown, R. 1999. The Mesozoic denudation history of the Atlantic margins of southern Africa and southeast Brazil and the relationship to offshore sedimentation. *In*: Cameron, N. R., Bate, R. H. & Clure, V. S. (eds) *The Oil and Gas Habitats of the South Atlantic*. Geological Society, London, Special Publications, **153**, 41–53.

Gallagher, K., Hawkesworth, C. J. & Mantovani, M. 1996. The denudation history of the onshore continental margin of SE Brazil inferred from apatite fission track data. *Journal of Geophysical Research*, **99**, 18117–18145.

Gleadow, A. J. W. & Brooks, C. K. 1979. Fission-track dating, thermal histories and tectonics of igneous intrusions in East Greenland. *Contributions to Mineralogy and Petrology*, **71**, 45–60.

GLOBE Task Team and others (Hastings, D. A., Dunbar, P. K. *et al.*, eds) 1999. *The Global Land One-kilometer Base Elevation (GLOBE) Digital Elevation Model, Version 1.0*. National Oceanic and Atmospheric Administration, National Geophysical Data Center, Boulder, CO. Digital database on the World Wide Web http://www.ngdc.noaa.gov/mgg/topo/globe.html and CD-ROMs.

Green, P. F. & Duddy, I. 2010. Synchronous exhumation events around the Arctic including examples from Barents Sea, Alaska North Slope. *In*: Vining, B. A. & Pickering, S. C. (eds) *Petroleum Geology: From Mature Basins to New Frontiers – Proceedings of the 7th Petroleum Geology Conference*. Geological Society, London, Special Publications, 633–644; doi: 10.1144/0070633.

Green, P. F., Duddy, I. R., Gleadow, A. J. W. & Lovering, J. F. 1989. Apatite fission track analysis as a palaeotemperature indicator for hydrocarbon exploration. *In*: Naeser, N. D. & McCulloh, T. (eds) *Thermal History of Sedimentary Basins – Methods and Case Histories*. Springer, New York, 181–195.

Green, P. F., Duddy, I. R. & Hegarty, K. A. 2002. Quantifying exhumation from apatite fission-track analysis and vitrinite reflectance data: precision, accuracy and latest results from the Atlantic margin of NW Europe. *In*: Doré, A. G., Cartwright, J. A., Stoker, M. S., Turner, J. P. & Whites, N. (eds) *Exhumation of the North Atlantic Margin: Timing, Mechanisms and Implications for Petroleum Exploration*. Geological Society, London, Special Publications, **196**, 331–354.

Green, P. F., Crowhurst, P. V. & Duddy, I. R. 2004. Integration of AFTA and (U–Th)/He thermochronology to enhance the resolution and precision of thermal history reconstruction in the Anglesea-1 well, Otway Basin, SE Australia. *In*: Boult, P. J., Johns, D. R. & Lang, S. C. (eds) *Eastern Australian Basins Symposium II*. Petroleum Exploration Society of Australia, 117–131.

Green, P. F., Crowhurst, P. V., Duddy, I. R., Japsen, P. & Holford, S. P. 2006. Conflicting (U–Th)/He and fission track ages in apatite: enhanced He retention, not anomalous annealing behaviour. *Earth and Planetary Science Letters*, **250**, 407–427.

Hamann, N. E., Whittaker, R. C. & Stemmerik, L. 2005. Geological development of the Northeast Greenland Shelf. *In*: Doré, A. G. & Vining, B. A. (eds) *Petroleum Geology: North-West Europe and Global Perspectives: Proceedings of the 6th Petroleum Geology*. Geological Society, London, 887–902; doi: 10.1144/0060887.

Hansen, K. 1988. Preliminary report of fission track studies in the Jameson Land basin, East Greenland. *Rapport Grønlands Geologiske Undersøgelse*, **140**, 85–89.

Hansen, K. 1992. Post-orogenic tectonic and thermal history of a rifted continental: the Scoresby Sund area, east Greenland. *Tectonophysics*, **216**, 309–326.

Hansen, K. 1996. Thermo-tectonic evolution of a rifted continental margin: fission track evidence from the Kangerlussuaq area, SE Greenland. *Terra Nova*, **8**, 458–469.

Hansen, K., Bergman, S. C. & Henk, B. 2001. The Jameson Land basin (east Greenland): a fission track study of the tectonic and thermal evolution in the Cenozoic North Atlantic spreading regime. *Tectonophysics*, **331**, 307–339.

Harrison, J. C., Mayr, U. et al. 1999. Correlation of Cenozoic sequences of the Canadian Arctic region and Greenland; implications for the tectonic history of North America. *Bulletin of Canadian Petroleum Geology*, **47**, 223–254.

Heilmann-Clausen, C., Nielsen, O. B. & Gersner, F. 1985. Lithostratigraphy and depositional environments in the upper Palaeocene and Eocene of Denmark. *Bulletin of the Geological Society of Denmark*, **33**, 287–323.

Henriksen, N. 2008. *Geological History of Greenland*. Geological Survey of Denmark and Greenland (GEUS), Copenhagen.

Henriksen, N., Higgins, A. K., Kalsbeek, F. & Pulvertaft, T. C. R. 2000. *Greenland from Archaean to Quaternary. Descriptive Text to the Geological Map of Greenland 1:2 500 000*. Geology of Greenland Survey Bulletins, **185**.

Hillis, R. R. 1995. Quantification of tertiary exhumation in the United Kingdom southern North Sea using sonic velocity data. *American Association of Petroleum Geologists Bulletin*, **79**, 130–152.

Holliday, D. W. 1993. Mesozoic cover over northern England: interpretation of apatite fissiontrack data. *Journal of the Geological Society, London*, **150**, 657–660.

Japsen, P. 1998. Regional velocity–depth anomalies, North Sea Chalk: a record of overpressure and Neogene uplift and erosion. *American Association of Petroleum Geologists Bulletin*, **82**, 2031–2074.

Japsen, P. 1999. Overpressured Cenozoic shale mapped from velocity anomalies relative to a baseline for marine shale, North Sea. *Petroleum Geoscience*, **5**, 321–336.

Japsen, P. 2000. Investigation of multi-phase erosion using reconstructed shale trends based on sonic data. Sole Pit axis, North Sea. *Global and Planetary Change*, **24**, 189–210.

Japsen, P. & Bidstrup, T. 1999. Quantification of late Cenozoic erosion in Denmark based on sonic data and basin modelling. *Bulletin of the Geological Society of Denmark*, **46**, 79–99.

Japsen, P. & Chalmers, J. A. 2000. Neogene uplift and tectonics around the North Atlantic: overview. *Global and Planetary Change*, **24**, 165–173.

Japsen, P., Bidstrup, T. & Lidmar-Bergström, K. 2002. Neogene uplift and erosion of southern Scandinavia induced by the rise of the South Swedish Dome. *In*: Doré, A. G., Cartwright, J., Stoker, M. S., Turner, J. P. & White, N. (eds) *Exhumation of the North Atlantic Margin: Timing, Mechanisms and Implications for Petroleum Exploration*. Geological Society, London, Special Publications, **196**, 183–207.

Japsen, P., Mavko, G. et al. 2005a. Chalk background velocity: influence of effective stress and texture. *EAGE 67th Conference & Exhibition*, Madrid, 13–16 June.

Japsen, P., Green, P. F. & Chalmers, J. A. 2005b. Separation of Palaeogene and Neogene uplift on Nuussuaq, West Greenland. *Journal of the Geological Society, London*, **162**, 299–314.

Japsen, P., Bonow, J. M., Green, P. F., Chalmers, J. A. & Lidmar-Bergström, K. 2006. Elevated, passive continental margins: long-term highs or Neogene uplifts. New evidence from West Greenland. *Earth and Planetary Science Letters*, **248**, 315–324.

Japsen, P., Green, P. F., Nielsen, L. H., Rasmussen, E. S. & Bidstrup, T. 2007a. Mesozoic–Cenozoic exhumation in the eastern North Sea Basin: a multi-disciplinary study based on palaeo-thermal, palaeo-burial, stratigraphic and seismic data. *Basin Research*, **19**, 451–490.

Japsen, P., Mukerji, T. & Mavko, G. 2007b. Constraints on velocity–depth trends from rock physics models. *Geophysical Prospecting*, **55**, 135–154.

Japsen, P., Rasmussen, E. S., Green, P. F., Nielsen, L. H. & Bidstrup, T. 2008. Cenozoic palaeogeography and isochores before Neogene exhumation in the eastern North Sea Basin. *Geological Survey of Denmark and Greenland Bulletin*, **15**, 25–28.

Japsen, P., Bonow, J. M., Green, P. F., Chalmers, J. A. & Lidmar-Bergström, K. 2009. Formation, uplift and destruction of planation surfaces at passive continental margins. *Earth Surface Processes and Landforms*, **34**, 683–699.

Johnson, K. & Gallagher, K. 2000. A preliminary Mesozoic and Cenozoic denudation history of the North East Greenland onshore margin. *Global and Planetary Change*, **24**, 303–309.

King, L. C. 1967. *The Morphology of the Earth* (2nd edn). Oliver & Boyd, Edinburgh.

King, S. D. & Anderson, D. L. 1998. Edge-driven convection. *Earth Planetary Science Letters*, **160**, 289–296.

Kohn, B. P., Gleadow, A. J. W. & Cox, S. J. D. 1999. Denudation history of the Snowy Mountains: constraints from apatite fission track analysis. *Australian Journal of Earth Science*, **46**, 181–198.

Larsen, L. M. 2006. *Mesozoic to Palaeogene Dyke Swarms in West Greenland and their Significance for the Formation of the Labrador Sea and the Davis Strait*. Danmarks og Grønlands Geologiske Undersøgelse Rapport **2006/34**.

Larsen, R. B. & Tegner, C. 2006. Pressure conditions for the solidification of the Skaergaard intrusion: eruption of East Greenland flood basalts in less than 300 000 years. *Lithos*, **92**, 181–19.

Larsen, L. M., Watt, W. S. & Watt, M. 1989. Geology and petrology of the Lower Tertiary plateau basalts of the Scoresby Sund region, East Greenland. *Grønlands Geologiske Undersøgelse Bulletin*, **157**, 164.

Larsen, M., Heilmann-Clausen, C., Piasecki, S. & Stemmerik, L. 2005. At the edge of a new ocean: post-volcanic evolution of the Palaeogene Kap Dalton Group, East Greenland. *In*: Doré, A. G. & Vining, B. A. (eds) *Petroleum Geology: North-West Europe and Global Perspectives: Proceedings of the 6th Petroleum Geology Conference*. Geological Society, London, 923–932; doi: 10.1144/0060923.

Laxon, S. & McAdoo, D. 1998. Satellites provide new insights into polar geophysics. *EOS Transactions, American Geophysical Union*, **79**, 72–73.

Letréguilly, A., Huybrechts, P. & Reeh, N. 1991. Steady-state characteristics of the Greenland ice sheet under different climates. *Journal of Glaciology*, **37**, 149–157.

Lidmar-Bergström, K. 1996. Long term morphotectonic evolution in Sweden. *Geomorphology*, **16**, 33–59.

Lidmar-Bergström, K., Ollier, C. D. & Sulebak, J. C. 2000. Landforms and uplift history of southern Norway. *Global and Planetary Change*, **24**, 211–231.

Lidmar-Bergström, K., Näslund, J.-O., Ebert, K., Neubeck, T. & Bonow, J. M. 2007. Cenozoic landscape development on the passive margin of northern Scandinavia. *Norwegian Journal of Geology*, **87**, 181–196.

Mathiesen, A., Christiansen, F. G., Bidstrup, T., Marcussen, C., Dam, G., Piasecki, S. & Stemmerik, L. 1995. Modelling of hydrocarbon generation in the Jameson Land Basin, East Greenland. *First Break*, **13**, 329–341.

Mathiesen, A., Bidstrup, T. & Christiansen, F. G. 2000. Denudation and uplift history of the Jameson Land Basin, East Greenland – constrained from maturity and apatite fission track data. *Global and Planetary Change*, **24**, 303–309.

McKenzie, D. 1978. Some remarks on the development of sedimentary basins. *Earth and Planetary Science Letters*, **40**, 25–32.

Medvedev, S., Hartz, E. H. & Podladchikov, Y. Y. 2008. Vertical motions of the fjord regions of central East Greenland: impact of glacial erosion, deposition, and isostasy. *Geology*, **36**, 539–542.

Michelsen, O., Thomsen, E., Danielsen, M., Heilmann-Clausen, C., Jordt, H. & Laursen, G. 1998. Cenozoic sequence stratigraphy in the eastern North Sea. In: de Graciansky, P.-C. et al. (eds) *Mesozoic and Cenozoic Sequence Stratigraphy of European Basins.* Society of Economic Paleontologists and Mineralogists, Special Publications, **60**, 91–118.

Miller, K. G., Kominz, M. A. et al. 2005. The phanerozoic record of global sea-level change. *Science*, **310**, 1293–1298.

Nielsen, L. H. 2003. Late Triassic–Jurassic development of the Danish Basin and Fennoscandian Border Zone, southern Scandinavia. In: Ineson, J. & Surlyk, F. (eds) *The Jurassic of Denmark and Greenland. Geological Survey of Denmark and Greenland Bulletin*, **1**, 459–526.

Nøhr-Hansen, H. 2003. Dinoflagellate cyst stratigraphy of the Palaeogene strata from the Hellefisk-1, Ikermiut-1, Kangâmiut-1, Nukik-1, Nukik-2 and Qulleq-1 wells, offshore West Greenland. *Marine and Petroleum Geology*, **20**, 987–1016.

Ollier, C. D. 1985. Morphotectonics of passive continental margins: introduction. *Zeitschrift für Geomorphologie Neue Folge, Suppl.*, **54**, 1–9.

O'Sullivan, P. B., Kohn, B. P., Foster, D. A. & Gleadow, A. J. W. 1995. Fission track data from the Bathurst batholith: evidence for rapid mid-Cretaceous uplift and erosion within the eastern highlands of Australia. *Australian Journal of Earth Sciences*, **42**, 597–607.

Partridge, T. C. & Maud, R. R. 1987. Geomorphic evolution of southern Africa since the Mesozoic. *South African Journal of Geology*, **90**, 179–208.

Pedersen, A. K., Larsen, L. M., Riisager, P. & Dueholm, K. S. 2002. Rates of volcanic deposition, facies changes and movements in a dynamic basin: the Nuussuaq Basin, West Greenland, around the C27n-C26r transition. In: Jolley, D. W. & Bell, B. R. (eds) *The North Atlantic Igneous Province. Stratigraphy, Tectonic, Volcanic and Magmatic Processes.* Geological Society, London, Special Publications, **197**, 157–181.

Pedersen, A. K., Pedersen, G. K. & Larsen, L. M. 2006. Geologien. In: Bruun, L., Nielsen, N., Pedersen, G. K. & Pedersen, P. M. (eds) *Arktisk Station, 1906–2006.* Arktisk Station, Københavns Universitet with Forlaget Rhodos, Copenhagen, 17–32.

Persano, C., Stuart, F. M., Bishop, P. & Barfod, D. N. 2002. Apatite (U–Th)/He age constraints on the development of the Great Escarpment on the southeastern Australian passive margin. *Earth and Planetary Science Letters*, **219**, 1–12.

Piasecki, S. 2003. Neogene dinoflagellate cysts from Davis Strait, offshore West Greenland. *Marine and Petroleum Geology*, **20**, 1075–1088.

Poore, H. R. N., White, N. & Jones, S. 2009. A Neogene chronology of Iceland plume activity from V-shaped ridges. *Earth & Planetary Science Letters*, **283**, 1–13; doi: 10.1016/j.epsl.2009.02.028.

Praeg, D., Stoker, M. S., Shannon, P. M., Ceramicola, S., Hjelstuen, B. O., Laberg, J. S. & Mathiesen, A. 2005. Episodic Cenozoic tectonism and the development of the NW European 'passive' continental margin. *Marine and Petroleum Geology*, **22**, 977–1005.

Price, S. P. & Whitham, A. G. 1997. Exhumed hydrocarbon traps in east Greenland: analogs for the lower–middle Jurassic play of northwest Europe. *American Association of Petroleum Geologists Bulletin*, **81**, 196–221.

Price, S. P., Brodie, J. A., Whitham, A. G. & Kent, R. 1997. Mid-Tertiary rifting and magmatism in the Traill Ø region, East Greenland. *Journal of the Geological Society, London*, **154**, 419–434.

Raab, M. J., Brown, R. W., Gallagher, K., Webber, K. & Gleadow, A. J. W. 2005. Denudational and thermal history of the Early Cretaceous Brandberg and Okenyenya igneous complexes on Namibia's passive margin. *Tectonics*, **24**, TC3006.

Rasmussen, E. S. 2004. The interplay between true eustatic sea-level changes, tectonics, and climatical changes: what is the dominating factor in sequence formation of the Upper Oligocene–Miocene succession in the eastern North Sea Basin, Denmark? *Global and Planetary Change*, **41**, 15–30.

Rasmussen, E. S. 2005. The geology of the upper Middle–Upper Miocene Gram Formation in the Danish area. *Palaeontos*, **7**, 5–18.

Rasmussen, E. S. 2009a. Detailed mapping of marine erosional surfaces and the geometry of clinoforms on seismic data: a tool to identify the thickest reservoir sand. *Basin Research*, **21**, 721–737.

Rasmussen, E. S. 2009b. Neogene inversion of the Central Graben and Ringkøbing-Fyn High, Denmark. *Tectonophysics*, **465**, 84–97.

Rowley, R. & White, N. 1998. Inverse modeling of extension and denudation in the East Irish Sea and surrounding areas. *Earth and Planetary Science Letters*, **161**, 57–71.

Sandwell, D. T. & Smith, W. H. E. 1992. *Global Marine Gravity Field from ERS-1, Geosat and Seasat.* Scripps Institute of Oceanography, La Jolla, CA.

Sanford, B. V. & Grant, A. C. 2000. Geological framework of the Ordovician system in the Southeast Arctic Platform, Nunavut. In: McCracken, A. D. & Bolton, T. E. (eds) *Geology and Paleontology of the Southeast Arctic Platform and Southern Baffin Island, Nunavut.* Geological Survey of Canada, Bulletins, **557**, 13–38.

Sangiorgi, F., Brumsack, H. J. et al. 2008. A 26 million year gap in the central Arctic record at the greenhouse-icehouse transition: looking for clues. *Paleoceanography*, **23**, 1–13; doi: 10.1029/2007PA001477.

Schiøler, P., Andsbjerg, J. et al. 2007. *Lithostratigraphy of the Palaeogene – Lower Neogene Succession of the Danish North Sea.* Geological Survey of Denmark and Greenland, Bulletins, **12**.

Schmidt, A. G., Riisager, P., Abrahamsen, N., Riisager, J., Pedersen, A. K. & van der Voe, R. 2005. Palaeomagnetism of Eocene Talerua Member lavas on Hareøen, West Greenland. *Bulletin of the Geological Society of Denmark*, **52**, 27–39.

Sclater, J. G. & Christie, P. A. F. 1980. Continental stretching; an explanation of the post-Mid-Cretaceous subsidence of the central North Sea basin. *Journal of Geophysical Research*, **85**, 3711–3739.

Shuster, D. L., Flowers, R. M. & Farley, K. A. 2006. The influence of natural radiation damage on helium diffusion kinetics in apatite. *Earth and Planetary Science Letters*, **249**, 148–161.

Skogseid, J., Planke, S., Faleide, J. I., Pedersen, T., Eldholm, O. & Neverdal, F. 2000. NE Atlantic continental rifting and volcanic margin formation. In: Nøttvedt, A. (ed.) *Dynamics of the Norwegian Margin.* Geological Society, London, Special Publications, **167**, 295–326.

Sønderholm, M., Nøhr-Hansen, H., Bojesen-Koefoed, J. A., Dalhoff, F. & Rasmussen, J. A. 2003. *Regional Correlation of Mesozoic–Palaeogene Sequences across the Greenland–Canada Boundary.* Danmarks og Grønlands Geologiske Undersøgelse Rapport **2003/25**.

Sørensen, A. B. 2006. Stratigraphy, structure and petroleum potential of the Lady Franklin and Maniitsoq Basins, offshore southern West Greenland. *Petroleum Geoscience*, **12**, 221–234.

Steenfelt, A., Hollis, J. A. & Secher, K. 2006. The Tikiusaaq carbonatite: a new Mesozoic intrusive complex in southern West Greenland. *Geological Survey of Denmark and Greenland Bulletin*, **10**, 41–44.

Stoker, M. S., Praeg, D. et al. 2005. Neogene evolution of the Atlantic continental margin of NW Europe Lofoten Islands to SW Ireland: anything but passive. In: Doré, A. G. & Vining, B. A. (eds) *Petroleum Geology: North-West Europe and Global Perspectives: Proceedings of the 6th Petroleum Geology Conference.* Geological Society, London, 1057–1076; doi: 10.1144/0061057.

Storey, M., Duncan, R. A., Pedersen, A. K., Larsen, L. M. & Larsen, H. C. 1998. $^{40}Ar/^{39}Ar$ geochronology of the West Greenland Tertiary volcanic province. *Earth and Planetary Science Letters*, **160**, 569–586.

Stouge, S. & Peel, J. S. 1979. Ordovician conodonts from the Precambrian shield of southern West Greenland. *Rapport Grønlands Geologiske Undersøgelse*, **91**, 105–109.

Swift, D. A., Persano, C., Stuart, F. M., Gallagher, K. & Whitham, A. 2008. A reassessment of the role of ice sheet glaciation in the long-term evolution of the East Greenland fjord region. *Geomorphology*, **97**, 109–125.

Thomson, K., Green, P. F., Whitham, A. G., Price, S. P. & Underhill, J. R. 1999. New constraints on the thermal history of North-East Greenland from apatite fission-track analysis. *GSA Bulletin*, **111**, 1054–1068.

Tissot, B. P. & Welte, D. H. 1984. *Petroleum Formation and Occurrence.* Springer, Berlin.

Utescher, T., Mosbrugger, V., Ivanov, D. & Dilcher, D. L. 2009. Present-day climatic equivalents of European Cenozoic climates. *Earth and Planetary Science Letters*, **284**, 544–552.

van der Beek, P., Summerfield, M. A., Braun, J., Brown, R. W. & Fleming, A. 2002. Modelling postbreakup landscape development and denudational history across the southeast African (Drakensberg Escarpment) margin. *Journal of Geophysical Research*, **107**, 2351; doi: 10.1029/2001JB000744.

White, N. & McKenzie, D. 1988. Formation of the steers head geometry of sedimentary basins by differential stretching of the crust and the mantle. *Geology*, **16**, 250–253.

Ziegler, P. A. 1990. *Geological Atlas of Western and Central Europe* (2nd edn). Elsevier for Shell Internationale Petroleum Maatschappij, Amsterdam.

New methods of improving seismic data to aid understanding of passive margin evolution: a series of case histories from offshore west of Ireland

R. J. J. HARDY,[1] E. QUERENDEZ,[1] F. BIANCOTTO,[1] S. M. JONES,[1] J. O'SULLIVAN[2] and N. WHITE[3]

[1]*Trinity College Dublin, School of Natural Science, Department of Geology, Dublin 2, Ireland (e-mail: hardyr@tcd.ie)*
[2]*Providence Resources Plc Airfield House, Airfield Park, Donnybrook, Dublin 4, Ireland*
[3]*Bullard Laboratories, Department of Earth Sciences, Madingley Road, Cambridge CB3 0EZ, UK*

Abstract: Deepwater sedimentary basins are attracting increased attention from the hydrocarbon industry and academia as they remain one of the last geological frontiers still to be fully explored and understood. As part of a regional project to understand basin development offshore Ireland, we have developed and tested new ways of improving seismic images, and of incorporating these improvements into geological interpretations. Here we illustrate this methodology using three case histories from the Porcupine, Slyne and Erris Basins. In each case, the first stage of the workflow consists of reprocessing a selection of key seismic data. Processing includes relative amplitude preservation and advanced demultiple and interpretation-driven pre-stack depth imaging. Interpretation of the data is assisted by incorporating products such as multiple models, pre-stack gathers, velocity models and attributes. Finally, we show how a velocity model can be inverted to exhumation estimates. The results and approach developed here can be applied to other deepwater exploration areas.

Keywords: seismic data, seismic data processing, seismic data interpretation, exhumation

Recent technological and modelling advances provide an excellent opportunity to constrain the quantitative temporal and spatial evolution of Irish sedimentary basins and passive margins. Margin formation and growth play an integral role in the Mesozoic and Cenozoic geological history of Ireland (Naylor & Shannon 2005). The deeper reaches of margins are of considerable interest to the hydrocarbon industry since they constitute an important exploration frontier. However, methods are required to reduce exploration risk since drilling costs are high. Despite the large volumes of seismic data acquired offshore Ireland over the last 20 years, considerable improvements in imaging are possible. Here we apply state-of-the-art processing and interpretation technology to legacy seismic reflection datasets and well-log information from the Irish Continental Shelf, concentrating on the Porcupine, Slyne and Erris Basins (Fig. 1).

The first phase of this project was to understand and improve the quality of seismic reflection imaging throughout the sedimentary pile and basement. In collaboration with the project sponsors, we selected regional profiles for reprocessing from field and stacked records for each of the three basins of interest. The purpose of the reprocessing phase was to produce uniform quality depth images for a range of analyses at basin and prospect scales. We used novel pre-stack processing techniques such as deterministic deconvolution, advanced multiple suppression and interpretation driven pre-stack depth imaging. For the Porcupine Basin study area, the new depth images have been calibrated against well-log information to produce uniform quality reinterpretations. Reprocessed and existing legacy datasets have been integrated and carefully interpreted using other available geological and geophysical information in order to constrain basin configuration. The main improvements are in the imaging of syn-rift stratigraphy, but post-rift stratigraphic imaging has also benefited from our approach. Integrated use of pre-stack data, gravity and seismically derived velocity fields has been the key to better mapping. The Cenozoic exhumation history of the Slyne Basin has also been studied using reprocessed depth sections and derived velocity fields utilizing a suite of in-house tools that exploit thermal data (e.g. fission track, vitrinite reflectance), compaction data (e.g. sonic velocities, stacking velocities) and remnant subsidence data.

Porcupine Basin

The Porcupine Basin is generally believed to have formed by lithospheric stretching. However, several aspects of basin evolution are controversial. There is a significant discrepancy between the amount of extension measured across normal faults and that estimated from analysis of subsidence and crustal thinning. Early studies interpreted significant syn-rift igneous activity (Tate *et al.* 1993) while later work cited the basin as a natural laboratory to study the pre-breakup evolution of amagmatic passive margins (Reston *et al.* 2004). Good quality seismic imaging at depth is clearly required in order to resolve these controversies.

Database

Prior to commencement of our study in 2003, no academic digital database existed containing all released well and seismic data for the region. Our initial project requirement was therefore to establish and maintain a local digital database of seismic surveys (field, stacked and migrated 2D and 3D data), well logs (e.g. sonic, density, gamma), seabed mapping (e.g. sidescan sonar surveys) together with potential field information (gravity, magnetic). Our focus was on pre-stack seismic data. Seismic field tapes were required to enable pre-stack analysis of prospect-scale targets, which can be quantitatively compared with similar analysis worldwide. Stacked data, where available, formed an additional resource for this project. For example, investigation of stacked data can facilitate identification of high impedance scatterers (e.g. volcanic packages or salt features) within the stratigraphic section. Furthermore, interpretation of migrated 2D regional lines can be considerably improved by mapping and compensating for out-of-plane effects at line ties using the stack data. Available datasets ranged in vintage from 5 to 20 years and little had been reprocessed using pre-stack imaging or advanced demultiple techniques. No prestack gathers (either time or depth migrated) or depth migrated

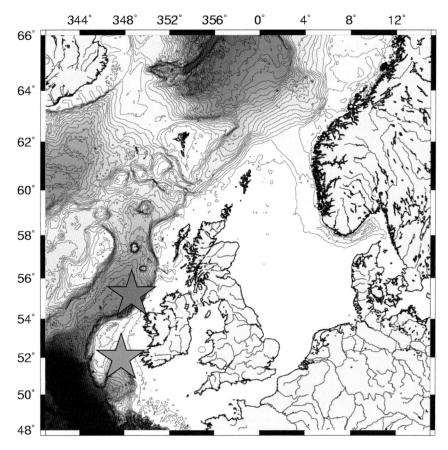

Fig. 1. Location map showing basins of interest. Porcupine Basin (lower star), Slyne–Erris basins (upper star).

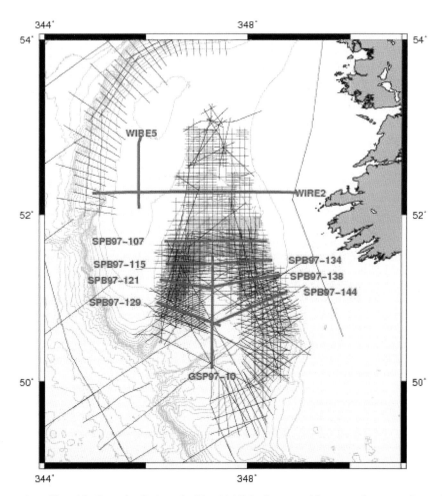

Fig. 2. Basemap of interpretation grid used for Porcupine Basin study. Lines highlighted were used for pre-stack reprocessing studies.

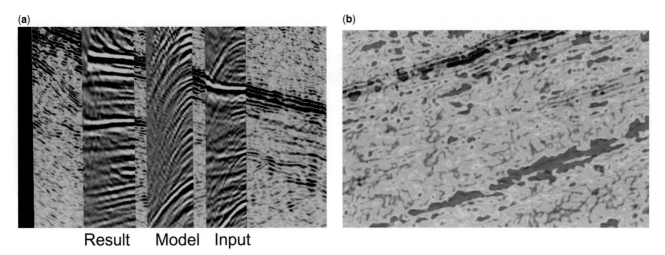

Fig. 3. (a) Visualization quality control of surface-related multiple elimination showing pre-stack gathers used during interpretation of deep water offshore seismic data. (b) Colour overlay of multiple energy and seismic data for quality control and reduction of interpretation risk.

sections were available. Recently there have been several programmes to acquire long-offset data which can benefit both multiple suppression and pre-stack imaging. However, these data are not yet generally released.

Seismic data reprocessing

An optimal grid of eight lines of released seismic data was selected for test reprocessed from field and stacked records (Fig. 2). Our philosophy is that novel seismic processing is an integral component of seismic interpretation and basin modelling and that data reprocessing (particularly depth imaging) can reduce uncertainty in the basin modelling phases.

In the last 10 years, there has been a revolution in the quality of multiple suppression and seismic imaging. One example is the surface-related multiple elimination (SRME) or inverse scattering approach of Weglein *et al.* (1997). Another example is pre-stack wave-field extrapolation imaging in place of the more limited traditional Kirchhoff approach (O'Brien *et al.* 2001). Both approaches have become economically feasible thanks to the upsurge of commodity PC equipment which can be linked together to provide supercomputer power at minimal cost (Bednar & Bednar 2001). Results of the new processing algorithms are commonly shown for sub-salt imaging of synthetic data, and in these cases the increase in signal-to-noise ratio and image quality, even on 2D data, is often remarkable. We anticipated similar improvements for the Irish Continental Shelf, where the main cause of image degradation is basaltic igneous intrusions (Martini & Bean 2002).

Our processing flow incorporated 2D SRME to suppress multiples. Initially we were disappointed by the SRME results. In particular, considerable residual multiple energy remained following the adaptive subtraction phase. Recent industry demonstrations have indicated that 3D SRME is dramatically better than 2D SRME, and in part this perhaps explains our poor initial 2D SRME results for some areas such as the Slyne Basin. However, we were able to use the 2D SRME multiple prediction to aid geological interpretation. An example is shown in Figure 3, which depicts a modern visualization-based development of methods originally reported by Hardy *et al.* (1990). We then developed an alternative adaptive subtraction technique (Hardy & Mongan

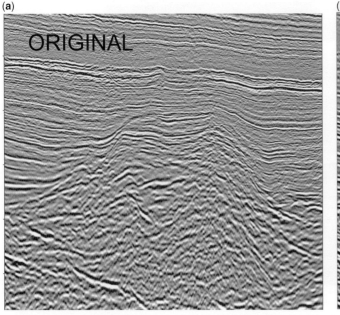 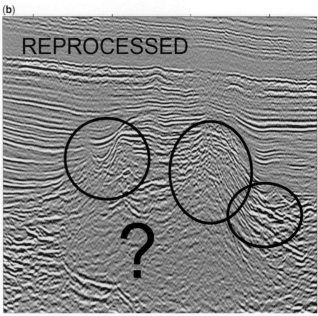

Fig. 4. (a) Original and (b) reprocessed section from the Porcupine Median Ridge. Circles indicate areas of improvement. '?' represents uncertainty in structure in the core of the ridge.

2006) which is a multichannel approach that preserves primary reflections and can be combined with residual radon transform-based multiple suppression. Additionally we used predictive techniques based on 3D side-scan sonar data to isolate areas where 2D SRME should be effective. Once multiple suppression was resolved, we developed an iterative, interpreter-driven workflow for producing high quality depth images. Each image required around eight iterations of velocity picking and residual moveout analysis. We also incorporated horizon-guided smoothing in order to increase the geological bias of the resulting velocity field. This procedure should improve results when extrapolating interpretation away from existing sparse well locations.

Figure 4 shows a comparison of initial time migration processing (left) with our radon transform multiple suppression and pre-stack time migration processing (right). This structure, part of the Porcupine Median Ridge (PMR) complex, has been interpreted as being of volcanic or serpentinitic origin (Reston et al. 2004). The PMR is thought to be capped by a commercially viable carbonate reservoir. Our new imaging of this capping structure shows several areas of improvement (circled), particularly on the steep flanks. The core of the PMR is seismically transparent in the new imaging. We attribute this change to more effective multiple suppression, that is, some of the deep reflections in the original images were residual multiples. More recent long offset data and a revised structural interpretation of this enigmatic structure are currently underway.

In general we have found that many existing vintages of post-stack migrated data from offshore Ireland are under-migrated

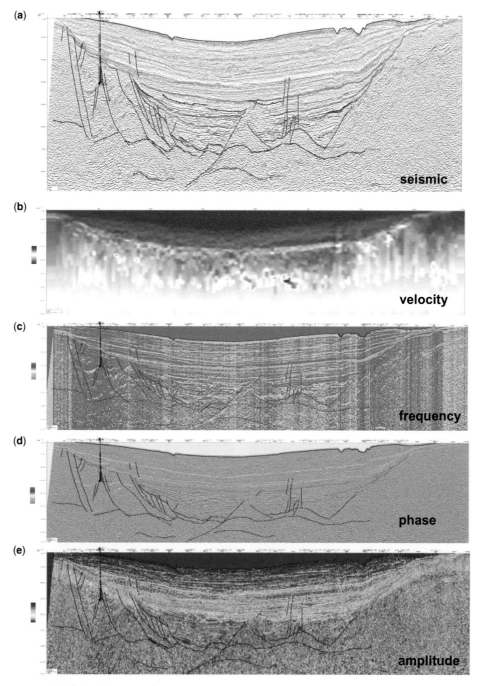

Fig. 5. (**a**) Example of interpretation showing extrapolation of horizons to basin centre from flank well. (**b**) Extrapolation was performed with conventional interpretation aided by interval velocity models; (**c**) instantaneous frequency; (**d**) instantaneous phase; and (**e**) amplitude envelope. Horizons shown (in order of increasing age): Base Oligocene unconformity [brown]; Near Top Danian (Top Chalk) [light green]; Intra-Cenomanian unconformity; Base Albian unconformity; Intra-Aptian unconformity; Base Cretaceous unconformity [dark green]; Top Crystalline Basement [purple]; Seismic Moho [pink].

with residual diffraction energy present. For some surveys where we have not been able to obtain field data, we have been able to demigrate (or diffract) the data using a background velocity field and then re-image it with a modern steep-dip algorithm. This procedure provides a quicker method of producing improved images than full pre-stack reprocessing.

Seismic data interpretation: Porcupine Basin

Over 17 000 line km of digital 2D seismic data, representing seven major surveys and five minor surveys, were incorporated into our interpretation together with all released wells. We used sonic and density logs where available to attempt to extrapolate stratigraphy from the basin margin highs (where most of the wells are drilled) to the deeper undrilled parts of the basin to the south. Standard seismic attributes such as instantaneous amplitude, frequency and phase were used to help extrapolate horizons (Fig. 5). The seismic stacking velocity model was interpolated and converted to SEGY, then loaded as an attribute. In this way the velocities could also be used to extrapolate lithology, where suitable well ties were available.

Figure 6 shows examples of depth maps for two horizons in the post-rift stratigraphy. These are the first published maps to use all released digital data. Depth data and interpretations are currently being utilized in basin modelling and studies of basin formation.

Erris Basin

Data reprocessing

Figure 7 illustrates results of data reprocessing from the Erris Basin. In the main basin itself we improved multiple suppression (using standard Radon demultiple), imaging around and beneath sills, and imaging of fault blocks utilizing the pre-stack imaging techniques described above.

In the deeper water sections we found that residual diffracted multiple energy hampered imaging of key targets. Radon demultiple (applied during the original and reprocessed version) can remove the simple deep water multiples but leaves residual parts of the multiple wave field which subsequently 'overmigrate' into the image and degrade it. We therefore found that better deep imaging was possible when the multiple wave field was retained in the data (Fig. 8) which compares original processing versus a reprocessed line in which the multiples were 'left in' prior to pre-stack imaging. Note the pre-stack imaging has not only improved images of the sill complex, but has also improved the deeper parts of the section, potentially dykes feeding the sill complex. Therefore in deep water it may not be necessary or even optimum to suppress the multiples prior to imaging. Possibilities exist for multiple suppression post-pre-stack imaging; however in this case these are not needed as the interpretation is fairly clear.

Slyne Basin

In this section, we focus on a study from the Slyne Basin, offshore Ireland, which has been affected by rift flank and epeirogenic uplift (Biancotto *et al.* 2007). Measuring uplift directly is generally impossible because reference levels are usually destroyed or at least modified by the processes of erosion. A way to address this problem is to evaluate the magnitude and distribution of exhumation at regional unconformities. The most common methods used exploit the thermal or mechanical properties of rocks, such as apatite fission track, vitrinite reflectance and sonic velocity modelling, but they all have the disadvantage of being restricted to the location of boreholes. In addition, there is often a large scatter in these sparsely distributed measurements. We show in

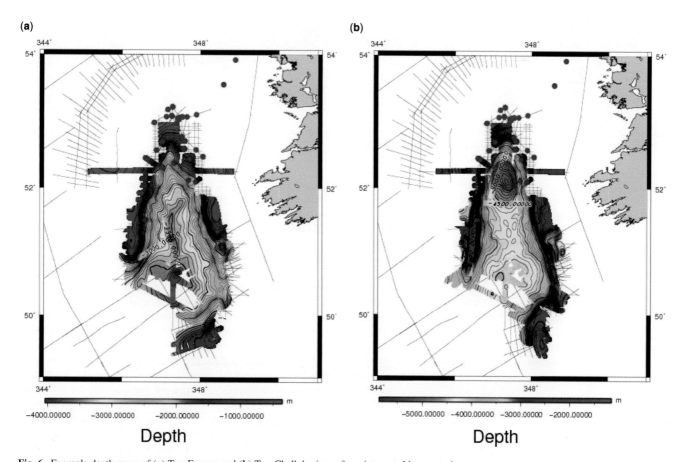

Fig. 6. Example depth maps of (**a**) Top Eocene and (**b**) Top Chalk horizons from integrated interpretation.

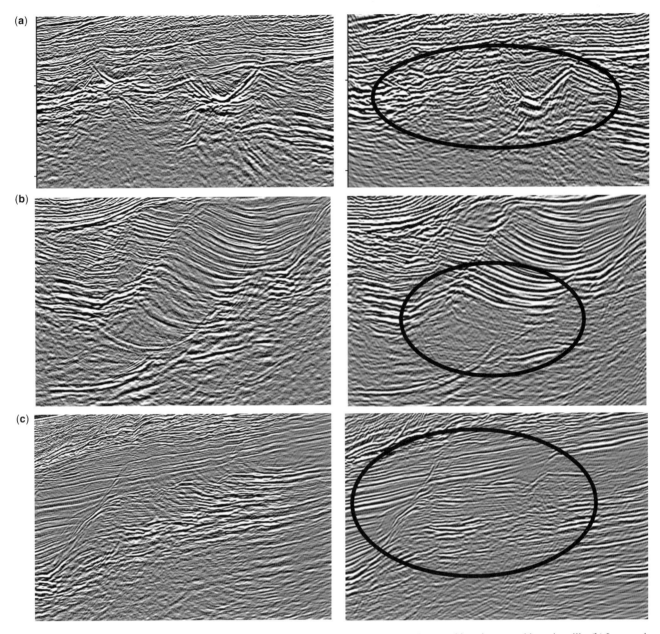

Fig. 7. Examples of reprocessing from the Erris Basin: left (original) and right (reprocessed). (**a**) Improved imaging around intrusive sills. (**b**) Improved multiple suppression and fault plane imaging. (**c**) Improved fault block imaging.

this study that inversion of seismic velocity profiles from seismic reflection datasets can be a useful tool to spatially constrain the distribution and the magnitude of exhumation (Walford & White 2005). The subject is important to oil exploration in the region in order to calibrate maturity models. This is the first such study which attempts to integrate 2D and 3D seismic-based measurements.

The Slyne Basin lies about 60 km off the NW coast of Ireland, with water depth ranging between 150 and 500 m. As part of a series of extensional basins trending NNE–SSW, it is a half-graben cut by transfer faults, dividing it into northern, central and southern parts (Scotchman & Thomas 1995). Dancer *et al.* (1999) have documented the sediment fill of the Slyne Basin from regional well data.

Exhumation estimates from seismic velocity data

Seismic velocity is controlled by the mineralogy, porosity, density, pore fluid composition and properties, pressure and depth of the imaged succession. However, the porosity of the medium is often the primary control factor on the seismic velocity because the sonic velocity of the pore fluid is much lower than that of solid sediment grains. Therefore, by measuring the sonic velocity of a sedimentary rock, we can evaluate its porosity. Similar methods are routinely used by the hydrocarbon industry, for example inverting seismic amplitude to porosity in chalk reservoirs (e.g. Dvorkin & Alkhater 2004). As sediment is buried, it is subjected to higher pressure and higher temperature and cementation begins. These processes of cementation and compaction reduce porosity. The decrease in porosity of sediments with depth of burial is largely irreversible. If a sedimentary rock is subsequently uplifted, it will keep the signature of the porosity reached at the maximum depth of burial.

Thus, by measuring the present-day porosity of a sedimentary rock and estimating the porosity decay trend with the depth of an unexhumed sedimentary column, we can calculate the maximum depth of burial of that sedimentary rock. The difference between present-day and maximum depth indicates the magnitude of exhumation. Here we are ignoring the secondary effects of cementation and diagenesis together with complications such as the presence of gas hydrates and hydrocarbons (Walford & White 2005). Thus,

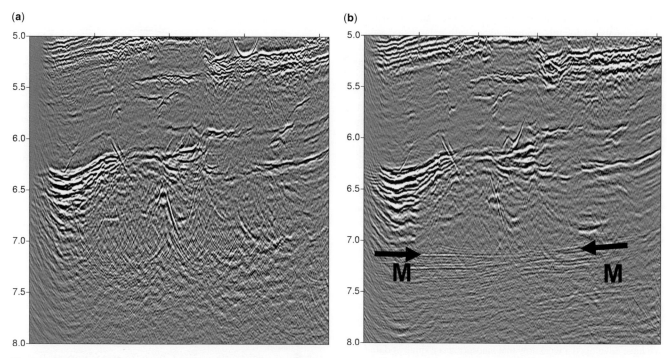

Fig. 8. Example of reprocessing from deeper water areas where multiple suppression is inadequate, particularly peg-leg multiples from instrusions. (**a**) Original processing where residual multiple energy contaminates deeper image. (**b**) Reprocessed pre-stack migration with no multiple suppression. Image is clearer beneath multiple (shown with arrows and 'M' text).

following these assumptions, we can use seismic velocity to estimate the amount of denudation of a sedimentary region. Walford & White (2005) developed an inverse model to determine the magnitude of denudation at the seabed that best fits stacking velocity measurements obtained from individual common midpoint gathers during seismic reflection processing. By building a simple, theoretical, 1D velocity profile and by minimizing the misfit between this simple synthetic profile and the observed profile, the amount of denudation can be constrained.

We have automated the inversion technique to make the method suitable for 2D lines, and by extension, 3D surveys (Biancotto *et al.* 2007). Generally 2D lines are inexpensive to reprocess and provide good quality control of velocities as long as the structures are not too complex and vary strongly out of the 2D plane.

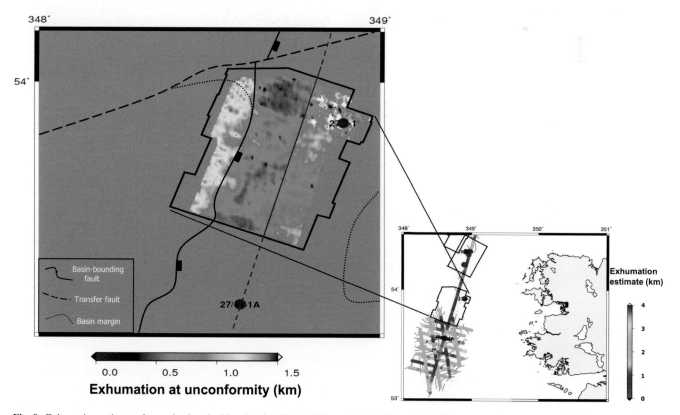

Fig. 9. Exhumation estimates from seismic velocities showing 3D and 2D results in the Slyne Basin. Estimates are in agreement with other 'well-based' methods and can be extrapolated away from wells.

Figure 9 shows the location of a 3D survey and 2D seismic profiles analysed for this study together with exhumation values calculated from seismic velocity around the Oligo-Miocene unconformity. Our estimates at the well location vary from 0.9 to 1.7 km exhumation. Scotchman & Thomas (1995) used several more classic methods of measuring the denudation, on well 27/13-1. They found magnitude of uplift ranging from 0.7 to 1.9 km. With apatite fission track analysis, uplift and erosion of 1.6 km were obtained. Shale sonic log interval velocities suggest a magnitude of denudation between 0.7 and 1.05 km. Biomarker analyses show results indicating denudation between 0.7 and 1.06 km. On these biomarker data, vitrinite reflectance studies and sterane analysis were conducted and respectively indicate 1.9 km of erosion and 1.06 km of denudation. The seismic method therefore gives results which are in good agreement with traditional methods and additionally can be extrapolated away from the well locations.

Conclusions

In this paper we have summarized the results of recent research into Irish offshore basins which utilizes, for the first time in an academic study, all available well and seismic data in a digital interpretation. We have developed a workflow that incorporates a spine of data for seismic data reprocessing and we have shown some reprocessed data comparisons. Even if images are not spectacularly improved, integration of the reprocessed depth images, together with prestack gathers and velocity model, can improve the reliability of a regional interpretation. Finally we demonstrated results of a specific application which can invert seismic velocity to exhumation and compared results from 2D and 3D seismic data with those previously obtained at a well location in the Slyne Basin.

This publication uses data and survey results acquired during a project undertaken on behalf of the Irish Shelf Petroleum Studies Group (ISPSG) of the Irish Petroleum Infrastructure Programme (PIP) Group 4. The ISPSG comprises Chevron Upstream Europe, ENI Ireland BV, ExxonMobil International plc, Island Oil & Gas plc, OMV (Ireland) Exploration GmbH, Providence Resources plc, Serica Energy plc, Shell E&P Ireland Ltd, Sosina Exploration Ltd, Statoil Exploration (Ireland) Ltd, Total E&P UK plc and the Petroleum Affairs Division of the Department of Communications, Energy and Natural Resources. The EOSG (PIP) is thanked for provision of computer equipment. Schlumberger are thanked for donation of software licences (Petrel, Geoframe, Omega). Fugro are thanked for permission to use and show data in Figures 4 and 5. Statoil are thanked for permission to use and show data in Figures 7 and 8. TGS and Spectrum and WesternGeco are thanked for data contributions to this study. Enterprise Ireland are thanked for funding of FB and the exhumation study.

References

Bednar, C. J. & Bednar, J. B. 2001. Whither geophysical technology: a true paradigm shift? *The Leading Edge*, **20**, 550–551.

Biancotto, F., Hardy, R., Jones, S. M., Brennan, D. & White, N. J. 2007. Estimating denudation from seismic velocities offshore northwest Ireland. *SEG Expanded Abstracts*, **26**, 407; doi: 10.1190/1.2792452.

Dancer, P. N., Algar, S. T. & Wilson, I. R. 1999. Structural evolution of the Slyne trough. *In*: Fleet, A. J. & Boldy, S. A. R. (eds) *Petroleum Geology of Northwest Europe: Proceedings of the 5th Conference*. Geological Society, London, 445–453; doi: 10.1144/0050445.

Dvorkin, J. & Alkhater, S. 2004. Pore fluid and porosity mapping from seismic. *First Break*, **22**, 53–57.

Hardy, R. J. J. & Mongan, J. 2006. *Identification, Interpretation and Attenuation of Multiples from West Africa*. 68th EAGE Conference and Exhibition.

Hardy, R. J. J., Hobbs, R. W., Warner, M. R. & Jones, I. F. 1990. The reliability of multiple suppression. *First Break*, **08**, 297–304.

Martini, F. & Bean, C. J. 2002. Interface scattering versus body scattering in subbasalt imaging and application of prestack wave equation datuming. *Geophysics*, **67**, 1593; doi: 10.1190/1.1512750.

Naylor, D. & Shannon, P. M. 2005. The structural framework of the Irish Atlantic margin. *In*: Doré, A. G. & Vining, B. A. (eds) *Petroleum Geology: North-West Europe and Global Perspectives: Proceedings of the 6th Petroleum Geology Conference*. Geological Society, London.

O'Brien, J., Addis, D., Drummond, J., Raney, G., Walraven, D., Weigant, J., Stein, J. & Key, S. 2001. 3-D subsalt wave-equation depth imaging: a case study from the Hickory Field. *SEG Expanded Extracts*, **20**, 957; doi: 10.1190/1.1816798.

Reston, T. J., Gaw, V., Oennell, J., Klaeschen, D., Stubenrauch, A. & Walker, I. 2004. Extreme crustal thinning in the south Porcupine Basin and the nature of the Porcupine Median High: implications for the formation of non-volcanic rifted margins. *Journal of the Geological Society, London*, **161**, 783–798.

Scotchman, I. C. & Thomas, J. R. W. 1995. Maturity and hydrocarbon generation in the Slyne Trough, northwest Ireland. *In*: Croker, P. F. & Shannon, P. M. (eds) *The Petroleum Geology of Ireland's Offshore Basins*. Geological Society, London, Special Publications, **93**, 385–411.

Tate, M., White, N. & Conroy, J. J. 1993. Lithospheric extension and magmatism in the Porcupine Basin, West of Ireland. *Journal of Geophysical Research*, **98**, 13905–13923.

Walford, H. L. & White, N. J. 2005. Constraining uplift and denudation of west African continental margin by inversion of stacking velocity data. *Journal of Geophysics Research*, **110**, B04403; doi: 10.1029/2003JB002893.

Weglein, A. B., Gasparotto, F. A. F., Carvalho, P. M. & Stolt, R. H. 1997. An inverse-scattering series method for attenuating multiples in seismic reflection data. *Geophysics*, **62**, 1975; doi: 10.1190/1.1444298.

WATS it take to image an oil field subsalt offshore Angola?

E. EKSTRAND,[1] G. HICKMAN,[2] R. THOMAS,[3] I. THREADGOLD,[4] D. HARRISON,[2] A. LOS,[2] T. SUMMERS,[2] C. REGONE[5] and M. O'BRIEN[6]

[1]*BP America Inc, 200 Westlake Park Boulevard, Houston, TX 77079, USA (e-mail: ekstraej@bp.com)*
[2]*BP Exploration Operating Company Ltd, Chertsey Road, Sunbury-on-Thames TW16 7LN, UK*
[3]*RT Geophysics Ltd, 108 Kenilworth Avenue, Wimbledon, London SW19 7LR, UK*
[4]*BP America Inc, 501 Westlake Park Boulevard, Houston, TX 77079, USA*
[5]*Consulting Geophysicist, 1650 Shillington Drive, Katy, TX 77450, USA*
[6]*Allied Geophysics Inc, 27891 Man O War Trail, Evergreen, CO 80439, USA*

Abstract: BP's exploration success in deepwater Block 31, offshore Angola, has been driven by conventional narrow-azimuth 3D seismic data coupled with the latest available imaging algorithms. However limitations in these data are now apparent and the data is deemed insufficient for the appraisal and development of the subsalt discoveries in the western part of the block. 3D acoustic finite-difference modelling was applied to Block 31 to evaluate the potential data quality uplift from a wide-azimuth towed streamer (WATS) survey. Results showed that a significant improvement in data quality is possible. The modelling also investigated key acquisition variables (acquisition direction, sail line separation, number of tiles, cable length) to arrive at a solution that optimized both data quality and cost. Acquisition of this survey began in December 2008 and it is expected to complete in August 2009. This is the first WATS seismic survey outside the Gulf of Mexico and the first in Angola.

Keywords: wide-azimuth, WATS, subsalt, 3D finite-difference modelling, deepwater, Angola

Block 31 is located approximately 150 km offshore Angola in 1200–2500 m of water (Fig. 1). The block covers part of the Lower Congo Basin and lies to the south and east of the modern day Zaire fan. The Block 31 licence group consists of BP as operator, Exxon-Mobil, Sonangol P&P, Statoil, Marathon and Total.

The petroleum geology of Block 31 involves sandstone reservoirs of Oligocene to earliest Miocene age, sealed by overlying deep water mudstones, and sourced from Late Cretaceous to Palaeogene organic-rich marine sediments. The reservoirs are typically deposited in erosional channel complexes within a lower slope depositional environment. The structural evolution of the block is complex and dominated by salt-induced tectonics. Traps are commonly formed through the uplift of turbidite channels on the flanks and crests of salt-cored ridges.

Block 31 exploration activities to date have resulted in 19 discoveries that are currently the focus of appraisal and development activities. Following the acquisition of a block-wide conventional exploration 3D seismic survey in 1999, BP had early success in the block outside and adjacent to salt (referred to as extrasalt), with initial exploration focus in the NE open salt basins (Fig. 1). The majority of the reservoirs in this area are well imaged on seismic data and direct hydrocarbon indicators are present. Exploration activities then moved to the southeastern area, which contains isolated salt canopies and diapirs. Imaging in this part of the block is more complex, consisting of steeply dipping salt flank structures with portions of the trap obscured by salt.

Exploration efforts were then focused in the western area of Block 31, where reservoirs are completely covered by an extensive allochthonous salt canopy (Fig. 1). Seismic imaging in this area is extremely poor compared with the extrasalt regions of the block. A deep-tow, long cable 3D seismic survey was acquired in 2004 prior to the subsalt exploration drilling campaign to optimize the imaging on conventional narrow-azimuth seismic data. The first subsalt discovery in Block 31 and in Angola was made in 2006. There are currently five subsalt discoveries in Block 31, shown as red stars in Figure 1. Approximately 60% of Block 31 is covered by salt, which is characterized by complex 3D geometries (Fig. 2). Subsequent complexities in the velocity model result in significant subsalt imaging challenges for both appraisal and development.

Subsalt imaging challenges

Deepwater Angola is typically associated with excellent seismic imaging, based on shallow-tow, high-resolution 3D seismic surveys which have become the standard for extrasalt developments in Angola over the last decade. The root mean square (RMS) amplitude extraction shown in Figure 3a illustrates the detailed channel complex geometries which are resolvable with high-resolution 3D seismic in the extrasalt areas in Block 31. The RMS amplitude extraction in Figure 3b is from conventional 3D data over a subsalt reservoir interval in Block 31, where only a gross outline of the channel trend is visible. Although the frequency content between shallow and conventional tow-depth data is very different, the key imaging difference between Figure 3a, b is associated with poor illumination, low signal-to-noise and significant residual multiple energy, which is characteristic of the subsalt environment.

Sections through 3D pre-stack depth-migrated seismic data over two Block 31 subsalt discoveries are shown in Figure 4 to illustrate the subsalt imaging challenges for appraisal and development. Imaging is generally good over Well A (Fig. 4a); however the high amplitude reflectors targeted by the well were found to be associated with wet sands. The well encountered hydrocarbons in the updip side-track; therefore the field is actually located in an area of extremely poor imaging. Overall imaging is somewhat better at Well B (Fig. 4b); however poor imaging in the crestal area results in high levels of uncertainty in reservoir extent, container dimension and resource size. Although adequate for exploration, seismic imaging is clearly not sufficient for appraisal and

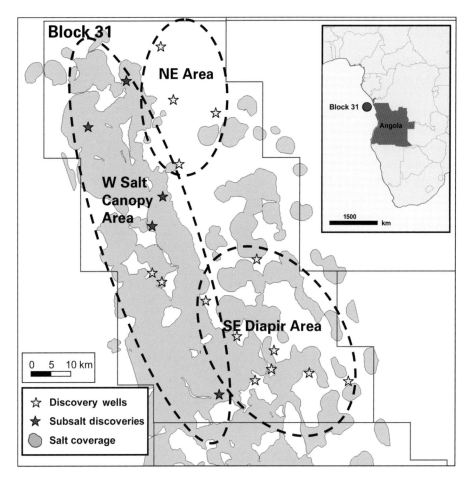

Fig. 1. Location of Block 31, offshore Angola, illustrating discovery wells, salt coverage and exploration focus areas.

development of these subsalt discoveries and requires a step change in quality.

As illustrated in Figure 4, complex salt bodies and associated velocity variations distort the seismic wave field. Historical attempts to image subsalt data in deepwater Angola and the Gulf of Mexico have used conventional narrow-azimuth towed streamer (NATS) seismic technology (shown in Fig. 4). This approach to a very complex 3D problem (using conventional 3D seismic surveys consisting of relatively narrow swathes of data) results in significant gaps in the seismic wave field. BP's subsalt experience in the Gulf of Mexico has shown that processing alone will not overcome limitations in conventional narrow-azimuth seismic acquisition. In order to address these acquisition limitations, BP has developed a portfolio of wide-azimuth solutions over the last decade that is

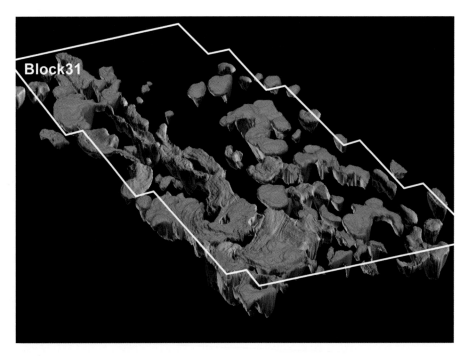

Fig. 2. Block 31 salt geometries.

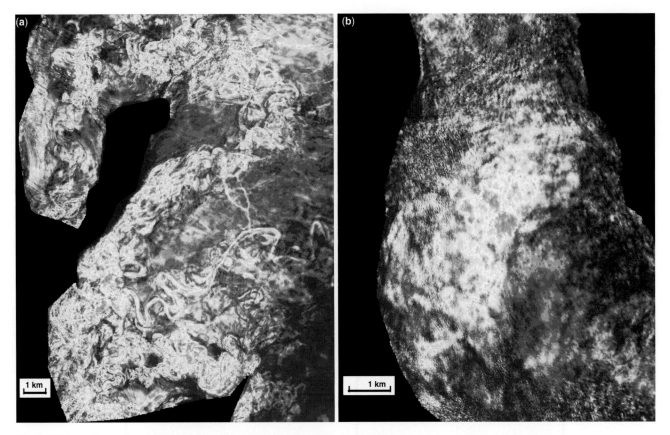

Fig. 3. RMS amplitude extractions over (a) extrasalt and (b) subsalt areas in Block 31.

showing significant quality improvement over conventional narrow-azimuth data (Barley & Summers 2007). The wide-azimuth towed streamer (WATS) seismic approach is considered by BP as the preferred option for imaging under large salt bodies (characterized by long-wavelength velocity anomalies) across the value chain from exploration through appraisal and development (Barley & Summers 2007).

The wide-azimuth solution: WATS

Michell et al. (2004) illustrated the benefits of variable illumination from dual-azimuth imaging over the subsalt Mad Dog Field in the Gulf of Mexico. Analysis in this area, however, highlighted the ineffectiveness of combining a finite number of conventional narrow-azimuth surveys (acquired in different orientations), to sufficiently address imaging challenges below large salt bodies, such as poor illumination, low signal-to-noise and significant multiple energy (Barley & Summers 2007). BP conducted a field trial in 2004 over the Mad Dog Field to acquire a wide range in azimuths with a single towed streamer acquisition (Michell et al. 2006; Threadgold et al. 2006), utilizing full 3D finite-difference (3DFD) modelling to optimize survey design (Regone 2006). The field trial confirmed the 3DFD modelling and demonstrated that this type of wide-azimuth survey can provide vastly improved imaging below salt. This acquisition method is called wide-azimuth towed streamer, in contrast with conventional NATS acquisition.

A comparison of NATS and WATS acquisition design is illustrated in Figure 5. Conventional NATS acquisition utilizes a single vessel towing both the source and receiver arrays, acquiring

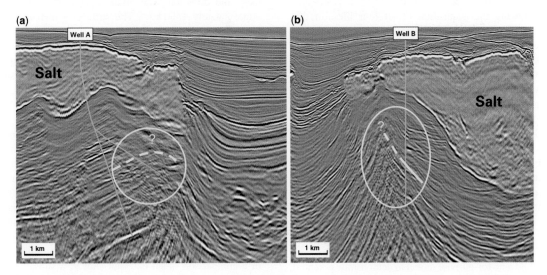

Fig. 4. Pre-stack depth-migrated data (common-azimuth wave equation algorithm) over subsalt discovery: (a) Well A and (b) Well B.

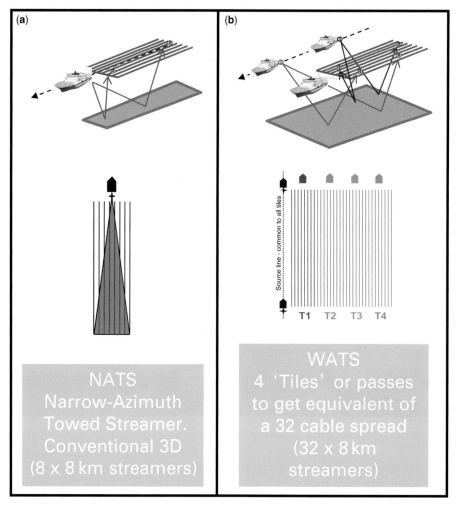

Fig. 5. Acquisition configuration for (**a**) NATS (narrow-azimuth towed streamer) and (**b**) WATS (wide-azimuth towed streamer) data.

a narrow range in azimuths (Fig. 5a). WATS acquisition utilizes two source vessels offset at the head and the tail of the streamer array, which is towed by a separate, third vessel (Fig. 5b). The source/receiver crossline offset is varied by acquiring successive, overlapping passes (Michell *et al.* 2006; Threadgold *et al.* 2006). In the WATS configuration shown in Figure 5b, four passes, or 'tiles' are acquired to build up a 4×8 km receiver patch, or the equivalent of a 32-cable spread (32×8 km). The basic vessel layout for WATS acquisition is illustrated in Figure 6a. The source and receiver array positions are shown for a four-tile WATS acquisition in Figure 6b. Following the acquisition of all four tiles, in this example, the source line position is offset by 500 m (source line or sail line separation) to acquire the next set of tiles (Fig. 6c). This process is then repeated to cover the entire survey area.

As each tile is an independent 3D survey, the entire WATS survey area is effectively acquired multiple times depending on the number of tiles used in the design. With three vessels required plus support, costs are high for WATS compared with conventional NATS acquisition. WATS acquisition parameters in general order of relative cost impact are shown in Table 1. The challenge in WATS survey design is to optimize these parameters for the specific survey objectives, particularly sail line separation and number of tiles, since they have a linear relationship with cost – in addition to building confidence that imaging uplift will actually be obtained.

3D finite-difference modelling

The key is to build a fit-for-purpose model to demonstrate potential WATS imaging uplift and to determine optimized acquisition parameters. The reservoirs of interest in Block 31 are relatively shallow subsalt channel complexes; therefore modelling needs to contain appropriate detail to demonstrate the potential WATS data uplift in both stratigraphic as well as structural definition. One of the earth models generated in western Block 31 is shown in Figure 7, illustrating the extent of the Miocene channel complexes that were investigated, comprising five individual channels. The geometries modelled were selected to investigate variation in channel wavelength, confinement width, channel body width, sinuosity and direction. The input area of this particular model is approximately 1600 km^2. The surfaces which define the model are based on interpretation of actual NATS data from Block 31 (Fig. 8a). The model is then populated with appropriate densities (Fig. 8b) in order to generate sufficient acoustic impedance contrasts since the velocity field is smooth apart from salt (Fig. 8c). An example of the band-limited reflectivity volume is shown in Figure 8d to illustrate the 'ideal' model image.

A 3DFD model was then generated based on the earth model. This methodology is discussed in detail by Regone (2006), in which 3DFD modelling was used in the Gulf of Mexico to design BP's first WATS survey over the Mad Dog Field in 2004 and an ocean bottom seismic (OBS) nodes survey over the Atlantis Field in 2005. The Block 31 input model area (yellow) is illustrated in Figure 9 with salt coverage shown in pink. Propagating wave fields of individual shots are modelled with a finite-difference (FD) algorithm, generating FD shot records (Fig. 10a). The significant impact of salt on the seismic wave field is illustrated in a comparison of an extrasalt shot v. a subsalt shot shown in Figure 10b, c. Referring back to Figure 9, data are recorded for

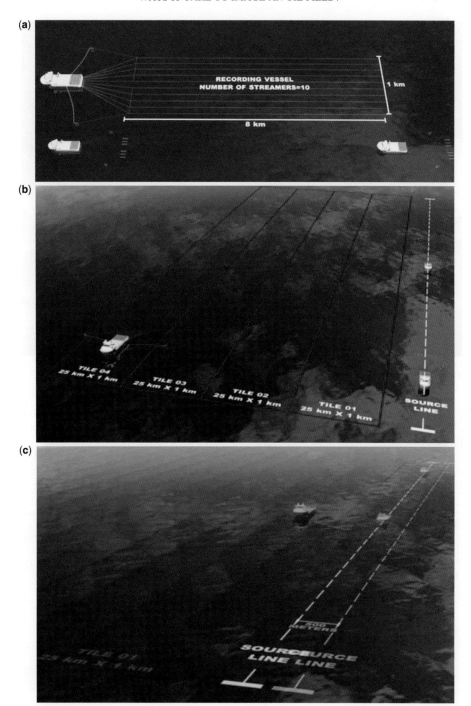

Fig. 6. WATS acquisition details: (**a**) basic vessel layout; (**b**) four-tile acquisition summary; (**c**) 500 m source line (sail line) spacing example.

Table 1. Relative cost impact of WATS acquisition parameters

Parameters	Relative cost impact
Sail line separation (source line)	High
Number of tiles	↕
Migration aperture and survey area	↕
Acquisition direction (rectangular surveys)	↕
Acquisition configuration and streamer length	↕
Record length and shotpoint interval	↓
Source and receiver tow depths	Low

each shot in a sub-model receiver spread of 16 × 16 km (blue); the image area (orange) is covered by shots on a 250 m grid. Data is then sorted and migrated with wave-field shot migration with the exact velocity model to create different acquisition geometries (Fig. 11). 3DFD modelling produces realistic simulations of 3D seismic data. Since the velocity model is known, modelling allows the imaging impact of velocity errors, free-surface multiple effects and acquisition geometry to be isolated (Regone 2006). 3DFD modelling is also very computer-intensive, with runtime being proportional to the fourth power of frequency (f^4). Modelled datasets were consequently generated with a maximum frequency of 15 Hz (which is sufficient given modelling parameters).

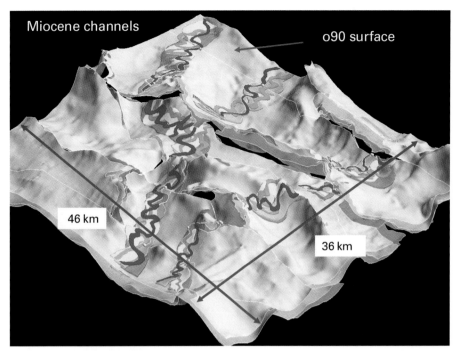

Fig. 7. Earth model incorporating channel complex details.

Model evaluation

A representative cross-section from 3DFD model data is shown in Figure 12 to illustrate the expected imaging uplift from WATS data in western Block 31. Both the NATS (Fig. 12a) and WATS (Fig. 12b) datasets are modelled with free-surface multiples. The density model for this cross-section is shown in Figure 12c to illustrate the structural surfaces and channel body geometries contained in the input model. This specific WATS section is based on 250 m sail line separation, 8 km streamers and four tiles. Model WATS data (Fig. 12b) shows significant improvements in imaging,

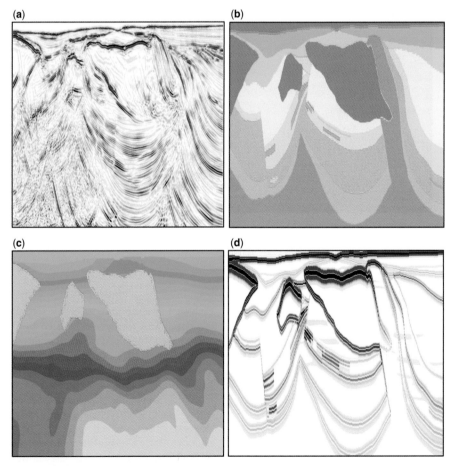

Fig. 8. Earth model construction: (**a**) structural horizons based on actual pre-stack depth-migrated NATS data; (**b**) density model; (**c**) velocity model; (**d**) model-based band-limited reflectivity.

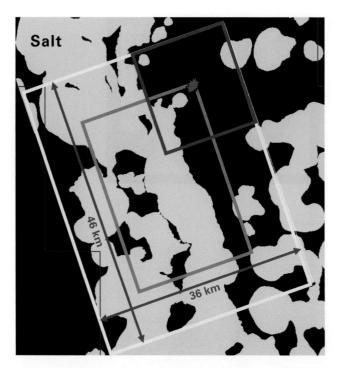

Fig. 9. 3DFD model area – input area, yellow; output area, orange; data area recorded for each shot (sub-model receiver spread), blue.

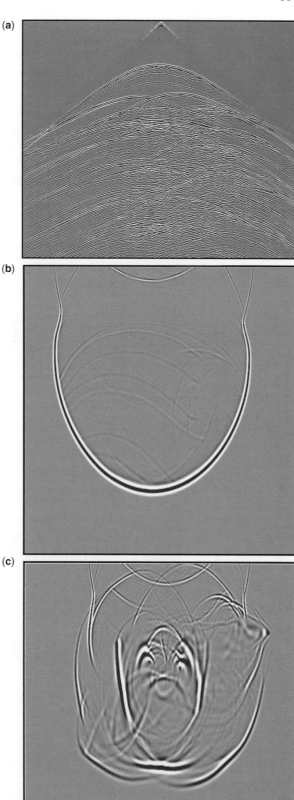

Fig. 10. 3DFD model generation: (**a**) typical shot record; (**b**) extrasalt shot; (**c**) subsalt shot.

reflector continuity and signal-to-noise compared with the model NATS section (Fig. 12a). It is clear on the model WATS data that reflectors either extend up to salt or are offset by faulting prior to reaching the salt interface. Model WATS data also clearly allow more confident horizon correlation across the section, particularly in the deeper parts of the cross-section where multiple energy is very strong on the NATS data. As supported by actual WATS datasets in the Gulf of Mexico, one of the benefits of WATS acquisition is that, owing to acquisition design, specular multiples are significantly reduced simply through the stacking process (the details of this effect are modelled and discussed by VerWest & Lin 2007).

To illustrate the impact of the number of WATS tiles (or the width of the acquisition receiver patch) on imaging improvements, Figure 13 shows RMS amplitude extractions over the deepest, isolated channel interval from Figure 12c. The channel is associated with the red, north–south trending feature on the amplitude maps in Figure 13. A substantial improvement in continuity and signal-to-noise can be seen simply going from NATS to a two-tile WATS case, with progressive improvements up to six-tile WATS. All of the model WATS data in Figure 13 are based on a 500 m sail line separation and 8 km streamer length.

In addition to comparing RMS amplitude extractions and cross-sections through focused areas of interest, a more quantitative analysis of data quality uplift was also done. One approach utilized is based on a coherency attribute which is the first principle component of the data, extracted over appropriate modelled intervals. The area of coherent data from different modelled acquisition volumes can be measured to compare the progressive uplift in data quality with, for example, increasing tile number. The graph shown in Figure 14a plots the normalized area or percentage of 'good' data for the NATS case and increasing tile number WATS cases. As illustrated in the RMS amplitude extraction maps in Figure 13, the most significant uplift in data quality occurs between the NATS and two-tile WATS cases, with smaller incremental improvements in quality with increasing tile number. The curve appears to be flattening around the five-tile case. Although this coherency attribute extraction is from one modelled interval over a large subsalt basin, the trends are considered representative of WATS data uplift overall. The curve can be translated into generic costs using a 2008 cost reference, and acquisition assumptions of 8 × 8 km streamers, 8 km data aperture, 500 m sail line

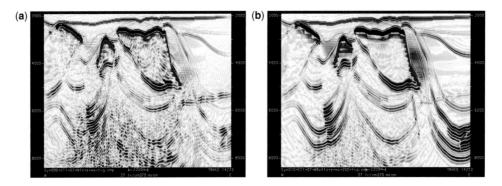

Fig. 11. 3DFD model output examples: (a) model NATS data; (b) model WATS data.

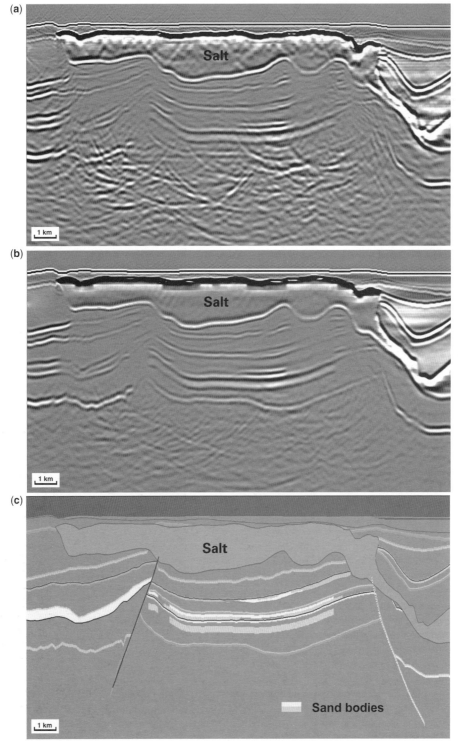

Fig. 12. 3DFD model NATS v. WATS comparisons: (a) model NATS data; (b) model WATS data; (c) input density model.

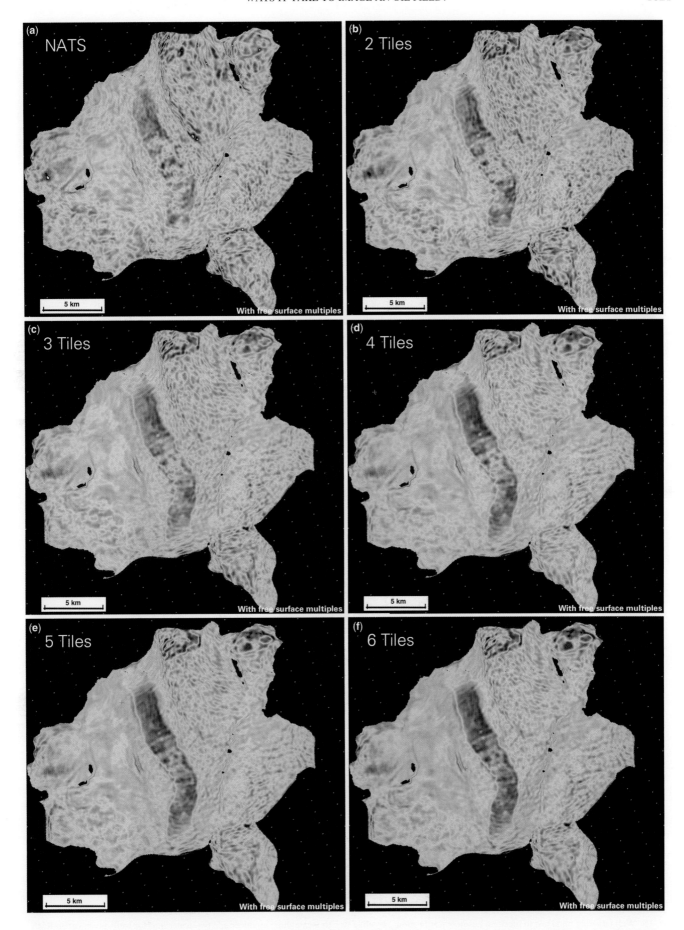

Fig. 13. Model NATS v. WATS tile number comparison: (a) NATS; (b) two-tile WATS; (c) three-tile WATS; (d) four-tile WATS; (e) five-tile WATS; (f) six-tile WATS.

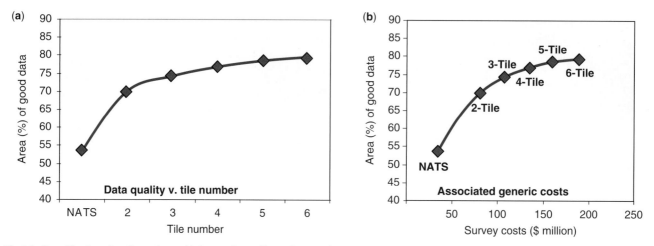

Fig. 14. Quantification of quality and cost: (**a**) data quality v. tile number: area/percentage of 'good' (coherent) model data from different acquisition designs; (**b**) associated generic costs.

separation and full-fold survey area of 1000 km². This approach allows both quality and costs to be optimized, consistent with survey objectives. These types of graphs were utilized in conjunction with analysis of targeted cross-sections and RMS amplitude extractions to determine the final recommended survey parameters. As we can see, there are significantly higher costs for WATS acquisition v. conventional NATS. However, the costs of not having better quality data will be significant and probably higher.

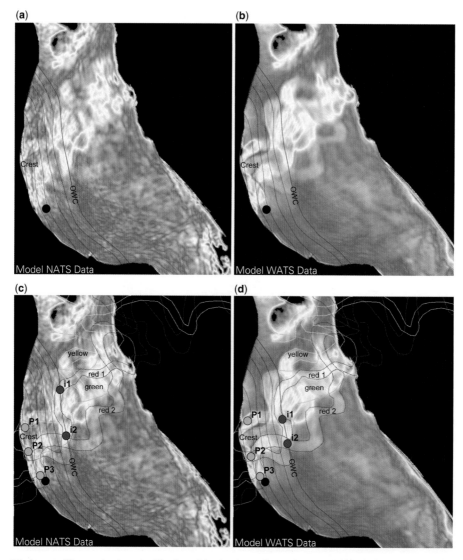

Fig. 15. Well targeting blind test to illustrate business justification: objective was to locate three producer and two water injector wells given (**a**) an RMS attribute map from model NATS data, location of a discovery well and the reservoir container geometry; (**b**) an RMS attribute map from model WATS data was then provided with an opportunity to revise the initial well locations. Input model channel geometries (not provided in the blind test) are shown on the (**c**) RMS attribute map from model NATS data with a representative example of initial NATS-based well locations and on the (**d**) RMS attribute map from model WATS data with the WATS-based revised well locations.

Business case support

Gulf of Mexico WATS experience suggests that benefits in improved imaging of a WATS survey will result in: increased confidence in top reservoir mapping subsalt; definition of reservoir compartments and better description of internal reservoir architecture leading to optimized well locations, improved recovery and potentially reduced well count; greater understanding of potential risks during drilling; and associated cost benefits. Block 31 3DFD model data results were used to test some of these potential business case benefits.

A well targeting blind test was constructed to simulate the development implications of WATS imaging improvements. This blind test was given to a number of subsurface staff, with a representative example illustrated in Figure 15. The objective of the blind test was to locate three producer and two water injector wells given an RMS attribute map from model NATS data, the location of a discovery well and the reservoir container geometry (Fig. 15a). An RMS attribute map from model WATS data (Fig. 15b) was then provided with an opportunity to revise the initial well locations. Input model channel geometries (not provided in the blind test) are shown on the RMS attribute map from model NATS data with a representative example of initial NATS-based well locations (Fig. 15c) and on the RMS attribute map from model WATS data with the WATS-based revised well locations (Fig. 15d). The model WATS data shows significant imaging improvements compared with the model NATS data which is clearly reflected in the updated well locations shown in Figure 15d.

By making some basic assumptions linking recovery factor directly to well targeting efficiency, the example shown in Figure 15 resulted in a 20% increase in recovery based on improved imaging from the model WATS dataset. Although a 20% increase in recovery cannot be guaranteed from actual WATS data, this blind test example demonstrates that better imaging from WATS data will lead to better decision-making in actual appraisal and development.

Recommended WATS survey

The actual WATS survey acquisition for Block 31 based on the 3DFD modelling discussed in this paper is shown in Figure 16. The survey consists of four tiles, acquired in a NW–SE direction, using an 8 × 8 km cable configuration. The survey area is targeted over key subsalt discoveries and consists of two parts: a 1735 km^2 appraisal quality survey shown in green, and a 295 km^2 development quality over a focused area shown in red. Appraisal quality is defined by a sail line separation of 500 m over the entire survey area (green). The red area will be infilled to a 250 m sail line separation, providing a development-quality WATS dataset around a focused, priority area. Because of the scalability of the WATS acquisition design, surveys can be optimized across the value chain. Should further development opportunities be identified, it will be straightforward to subsequently upgrade the appraisal quality WATS seismic survey to development quality by acquiring additional infill data. Acquisition of this survey began in December 2008 and is expected to complete in August 2009. This is the first WATS seismic survey outside the Gulf of Mexico and the first in Angola.

Summary

Subsalt imaging in Block 31 is significantly poor compared with the extrasalt environment of current Angola developments. Existing narrow-azimuth data is clearly not sufficient for subsalt appraisal and development; experience in the Gulf of Mexico supports the case that WATS will provide a step change in Angola subsalt imaging. 3DFD modelling has been key in developing the technical justification for WATS, optimizing survey design (quality and costs) and building business case support. Modelling is high effort, in both cost and time, but is also high value. Over 100 scenarios (volumes) were generated in the Block 31 modelling project discussed in this paper. However this commitment is small

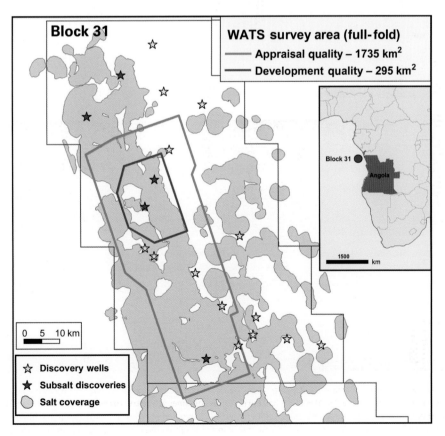

Fig. 16. Block 31 WATS survey acquisition area.

compared with the cost of WATS acquisition and the future cost benefits for appraisal and development.

The authors would like to thank Sonangol, BP, ExxonMobil, Sonangol P&P, Statoil, Marathon and Total for permission to present the data in this paper.

References

Barley, B. & Summers, T. 2007. Multi-azimuth and wide-azimuth seismic: shallow to deep water, exploration to production. *The Leading Edge*, **26**, 450–458.

Michell, S., Billette, F. J., Sharp, J. & Turner, J. 2004. Taking advantage of dual-azimuth analysis for model building and imaging over Mad Dog. *SEG International Exposition and 74th Annual Meeting, Expanded Abstracts*, 410–413.

Michell, S., Shoshitaishvili, E., Chergotis, D., Sharp, J. & Etgen, J. 2006. Wide azimuth streamer imaging of Mad Dog; have we solved the subsalt imaging problem? *SEG International Exposition and 76th Annual Meeting, Expanded Abstracts*, 2905–2909.

Regone, 2006. A modeling approach to wide-azimuth design for subsalt imaging. *The Leading Edge*, **25**, 1467–1475.

Threadgold, I. M., Zembeck-England, K., Aas, P. G., Fontana, P. M., Hite, D. & Boone, W. E. 2006. Implementing a wide azimuth towed streamer field trial: the what, why and mostly how of WATS in Southern Green Canyon. *SEG International Expostion and 76th Annual Meeting, Expanded Abstracts*, 2901–2904.

VerWest, B. J. & Lin, D. 2007. Modeling the impact of wide-azimuth acquisition on subsalt imaging. *Geophysics*, **72**, SM241–SM250.

Sub-basalt hydrocarbon prospectivity in the Rockall, Faroe–Shetland and Møre basins, NE Atlantic

I. DAVISON,[1] S. STASIUK,[2] P. NUTTALL[3] and P. KEANE[3]

[1]*GEO International Ltd, 38–42 Upper Park Road, Camberley, Surrey GU15 2EF, UK (e-mail: i.davison@earthmoves.co.uk)*
[2]*International Geoscience Ltd., Burnham, Slough, Berks SL18AS, UK*
[3]*Ion/GXT, 2105 City West Boulevard, Building III, Suite 900, Houston, TX 77042 USA*

Abstract: The seismic imaging below the basalts in the NE Atlantic basins is generally poor and the thickness of the basalts is difficult to predict. Two recent wells (William, 6005/13-1 and Brugdan, 6104/21-1) drilled in Faroes waters were suspended before reaching the main sub-basalt reservoir targets, because the basalts were thicker than expected. New deep-tow (18 m) seismic reflection data has now allowed more detailed imaging of the sub-basalt Mesozoic strata. The base of the basalt is not always reflective and is probably a transitional contact, with low acoustic impedance. The underlying Mesozoic strata produce coherent reflectivity. Large structural closures produced by Cretaceous rifting and Cenozoic folding have been imaged in the Rockall and Faroe–Shetland basins and these may contain Cretaceous and Paleocene clastic reservoirs at drillable depths of about 4 km. The Rockall Trough contains 4–7 km thickness of Mesozoic strata and appears to be floored by stretched continental crust throughout. The lower section may contain Jurassic source rocks similar to the areas to the south, east and north of the basin. Large Cenozoic age folds are present which produce potential traps at Late Cretaceous level, where sandstones equivalent to the Upper Cretaceous Nise Formation may occur. These folds represent some of the largest undrilled structures in NW Europe.

Keywords: sub-basalt prospecting, sub-basalt imaging, NE Atlantic margin

Basalts of Palaeogene age extend over much of the Faroe–Shetland, Rockall and Møre basins and reach up to approximately 6 km thickness around the Faroe Islands (White *et al.* 2005). The existing seismic data quality is generally very poor below the basalt. Promising oil and gas field discoveries are present in the shallower eastern portions of both the Rockall and Faroes–Shetlands, but no wells drilled through the thick basalts have encountered Cretaceous reservoir targets. The new deep-tow seismic data presented here have imaged the deeper Mesozoic strata, which enhances new play possibilities for oil exploration across this area.

Seismic data acquisition and processing

Various seismic acquisition techniques have been used to try to image below the volcanic rocks in the Faroes–Shetlands area: over–under technique; ocean-bottom seismometer surveys; wide-angle reflection and seismic refraction (e.g. White *et al.* 1999, 2003, 2005; Hobbs & Jakubowicz 2000; England *et al.* 2005; Gallagher & Dromgoole 2007, 2008; Christie & White 2008). However, most of these surveys have been limited to a few lines due to the high cost of acquisition.

New seismic data presented in this paper was shot and processed by ION/GXT in 2008 (see Fig. 1 for line locations). The seismic line acquisition and Pre-Stack Depth Migration (PSDM) processing are all designed to provide optimum seismic imaging below thick (>1 km) Palaeogene basalts, which cover much of the NE Atlantic margin (Fig. 1). This was achieved with:

(1) long receiver offset (10.2 km);
(2) large source size (7440 cu. in.) with 173 barm peak output measure through system filters;
(3) deep-towed airgun (17.5 m);
(4) deep-towed geophone streamer (18 m); and
(5) minimum bubble interference.

The airgun and streamer deep tow provided a maximum energy input at 20–30 Hz frequency. This low-frequency energy penetrates through the basalts more effectively. Another significant data quality benefit of deep tow is the reduced noise caused by near-surface current and wave action.

This acquisition produced more coherent reflections below the basalts which have a high seismic attenuation factor. Data was recorded to 18 s TWT, to image the complete crustal structure (Fig. 2). All the PSDM seismic data was migrated using velocities derived from iterative tomographic velocity modelling. The velocity-modelled depths are within 5% of the depths from several calibration wells in the Faroe–Shetland Basin.

Faroe–Shetland and Møre basins

The Faroe–Shetland and Møre basins were proved to be important new hydrocarbon provinces when the Foinaven, Scheihallion and Ormen Lange fields were discovered in the 1990s. However, subsequent discoveries have been smaller and more difficult to find. Large areas of the UK and Faroes sector are covered by thick Palaeogene volcanics from the Iceland plume eruptions. The Lopra-1 well on the Faroe Islands drilled 3565 m of the lower basalt series (Ziska & Andersen 2005). Their total thickness is not known but seismic data suggest perhaps >6 km of basalt may be present below the Faroe Islands (White *et al.* 2005). Only two wells (6104/21-1 and 6005/13-1) have been drilled in the Faroes continental shelf into thick basalt and the results have not yet been released into the public domain. The volcanics have a high seismic absorption (Q factor, Maresh *et al.* 2006; Shaw *et al.* 2008) caused by interbedded lower velocity sediments, coals and palaeo-sol horizons, which produce a large number of internal reflections. Several seismic lines have been shot with wide aperture arrays or sea-bottom seismometers that have produced encouraging results with sub-basalt imaging (Richardson *et al.* 1999; White *et al.* 2003, 2005). However, until now there have been no

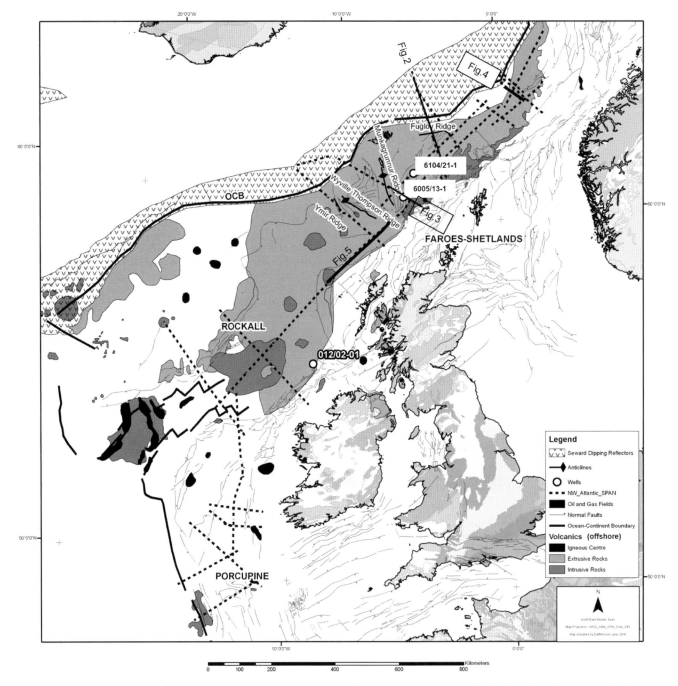

Fig. 1. Map of Faroes and Rockall basins showing location of seismic lines used in this study as well as the distribution of the Palaeogene volcanics. The locations of key wells mentioned in the text are also shown.

basin-wide seismic data sets available which image the whole crustal structure.

The base basalt is interpreted to be a transitional zone of thinly bedded tuffs, lavas and sediments as there is usually no clear seismic event associated with the base (Fig. 3). However, deeper stratal reflections are clearly imaged in many areas below a weakly reflective zone. Large-scale Cenozoic age folds (Munkagrunnur Ridge and Fugloy Ridge; Ritchie et al. 2008) and Mesozoic rotated fault blocks constitute major potential traps with Mesozoic reservoirs and Jurassic source rocks present (Lamers & Carmichael 1999; Figs 2–4). The Cenozoic folds are cored by Late Cretaceous strata (well 6004/16-1; Smallwood 2009) and are laterally extensive. More closely spaced seismic lines will be required to properly define structural closures on these long ridges. To our knowledge many of these deeper sub-basalt structures were not visible on previous seismic data. They are located in water depths up to 2 km, at drillable depths, and in the oil window (4–6 km), making them viable exploration plays which have yet to be tested.

The onset of volcanism started at 61–62 Ma (Saunders et al. 1997), with the main period of extrusion following at 56–53 Ma, which accompanied seafloor spreading of the North East Atlantic (Chron 24). Magmatism along the North East Atlantic produced a thick seaward dipping reflector (SDR) zone which is oriented parallel to the ocean–continental crust boundary (Spitzer et al. 2008; Fig. 1). The individual SDRs extend for some 20 km horizontally in a NW–SE direction orthogonal to the spreading centre. This length of lava run-out is thought to be indicative of sub-aerial volcanism (Fig. 2). This suggests that the mid-ocean ridge was sub-aerial at this time due to the anomalously high mantle potential temperature created by the Iceland plume. The ocean–continent crust boundary is here defined as the seaward end of the last

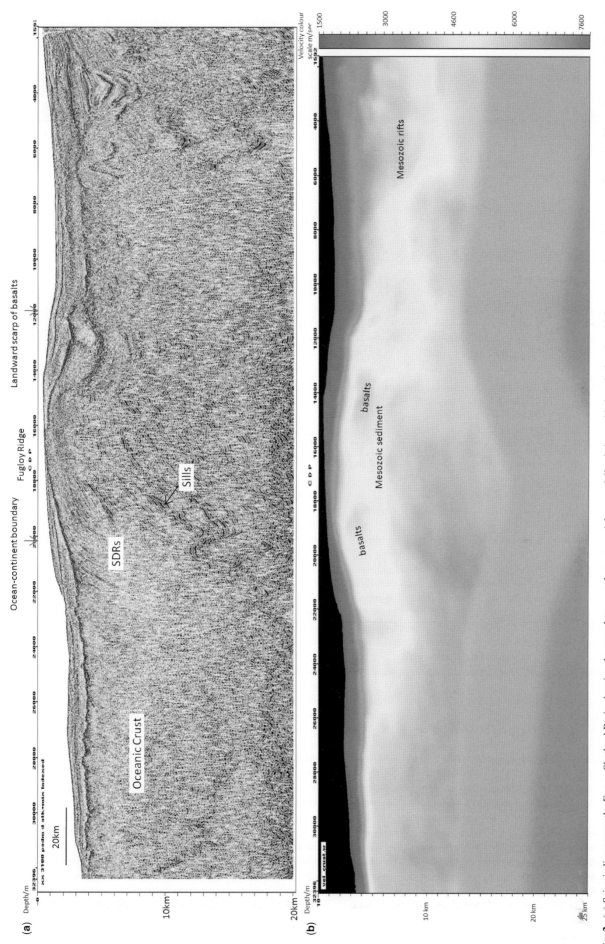

Fig. 2. (a) Seismic line across the Faroe–Shetland Basin showing the complete crustal structure. A large anticline is imaged below the basalts which may constitute a large potential trap in Mesozoic strata. This line is shot along the same location as the iSIMM line (White et al. 2005). (b) Velocity model used to produce the PSDM image in (a).

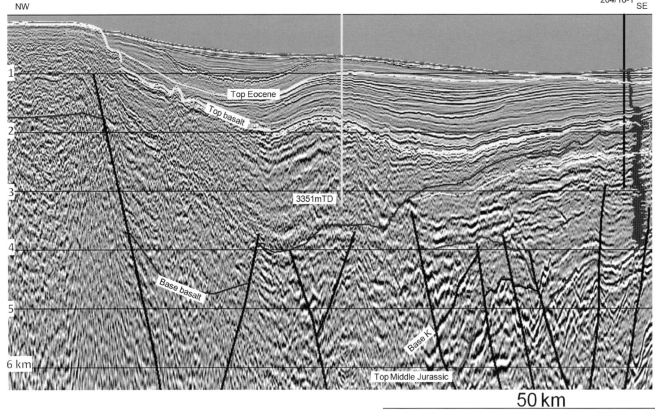

Fig. 3. Seismic section through the 6005/13-1 well illustrating the problem with definition of the base basalt.. The underlying rotated fault blocks are well imaged and may be viable exploration targets situated at >4 km depth. The well log for 204/18-1 is the sonic curve.

visible rotated fault block below the thick basalts (>2 km), which lies close to the landward edge of the main SDR sequence (Fig. 2).

The SDRs reach up to approximately 6–7 km maximum thickness. We interpret this to be new oceanic crust erupted above sea-level when they reach this thickness. The basalts flowed landward from the spreading centre to produce lava deltas, with eastward prograding foresets. These reach up to 1 km in vertical relief, indicating the probable palaeowater depths (Fig. 4). Below the landward (eastern) edge of the basalts, rifted sedimentary strata are clearly imaged which reach up to 5–7 km in thickness. These can be traced for a short distance below the basalts (Figs 2 & 4). Stratal reflections are imaged, even under the thickest sequence of SDRs; however, these are apparently injected by a large number of magmatic sills (Fig. 4). Strong irregular reflections occur at depths of 12–16 km below the SDRs, and are interpreted to be basic intrusions injected into Lewisian basement (Fig. 2). Prominent intrusions are occasionally imaged where there is a strong landward (eastward)-dipping reflection imaged at depths of 10–15 km (Fig. 2).

Many igneous sills were injected into the crust during the Palaeogene (Fig. 4) and 70 such sills have been intersected by the well 164/7-1; these sills range in thickness from 2 to 150 m in thickness (Linnard & Nelson 2005). The sills are mainly injected around 1–3 km below the level of the Paleocene basalts, where the depth of sill injection was determined by the depth at which the magmatic pressure exceeded the prevailing lithostatic pressure. Below this depth, the magmatic pressure was not great enough to overcome the lithostatic pressure and only dyke injection would have occurred. Hence, the deeper Jurassic source rocks may have escaped pervasive heating from sill injection as they are situated at greater depths than 3 km below base basalt.

Rockall

The nature of the crust below Rockall has been suggested to be either oceanic or continental (e.g. England & Hobbs 1995, 1997; O'Reilly et al. 1996; Shannon et al. 1999, Morewood et al. 2004; Chappell & Kusznir 2005; Klingelhoer et al. 2005). The presence of rifted continental crust would significantly improve the prospectivity of the basin with the possibility of the occurrence of Jurassic source rocks in deep half-graben. The new data presented here support the presence of continental rifted blocks throughout the basin (Fig. 5).

The Rockall Trough has proven hydrocarbons on the eastern perched rifted margin with gas condensate discovered in the Dooish well. However, the central part of the basin is virtually unexplored with no wells drilled below the basalt and only a sparse 2D seismic grid is available. Basalts cover over two-thirds of this basin (Fig. 1) and a large number of igneous sills are also injected into the Late Cretaceous strata that inhibit seismic penetration in the deepest parts of the basin (Archer et al. 2005).

The new seismic data indicate that between 2 and 7 km of sedimentary strata are present throughout the basin, of which up to 4 km may lie below the basalts (Fig. 5). If a Jurassic source rock were present in the area, it is predicted to be in the oil generation window at the present time.

Structural inversion and uplift occurred in the Middle to Upper Eocene when the Wyville–Thompson and Ymir Ridges were formed. These structures were also reactivated in the Miocene (Ritchie et al. 2003). These ridges are capped by basalts and it has never been possible to image the sedimentary packages below the basalt. The new data presented here indicate that underlying layered Mesozoic? strata are present and that rotated fault

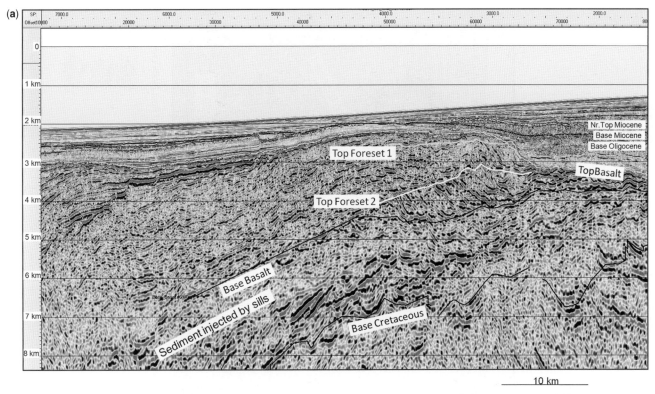

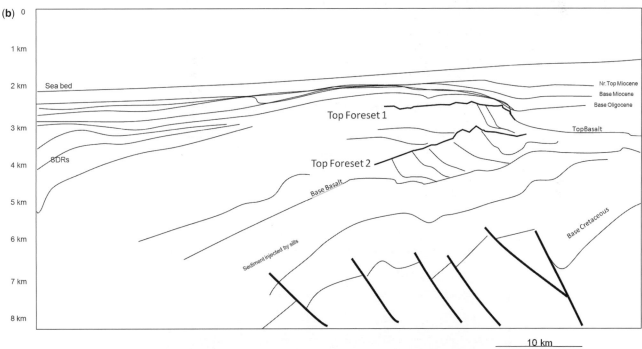

Fig. 4. Seismic section through the Møre Basin showing a lava pile with a landward-dipping scarp edge. Two foreset sequences of lavas are piled one on top of one another. The lavas overlie rifted fault blocks which have been injected by sills (bright reflections). These rifted basins are buried to depths of 5–6 km and reservoir quality is predicted to be poor at these depths.

block geometries are present in the anticlines. Consequently, large structural closures may be expected below the basalts.

The age of rifting in the Rockall Trough is not known, as the deeper section has never been drilled and it is debatable whether a Jurassic source rock will be present within the basin (Shannon et al. 1999; Tate et al. 1999). However, Jurassic source rocks are present in the Porcupine Basin to the south, in the Slyne–Erris Trough to the east and in the Faroe–Shetland Basin to the north, suggesting the presence of a regional Jurassic source (see Scotchman 2001; Scotchman et al. 2006). The Rockall Basin is situated 100–400 km from the British mainland but Late Cretaceous turbidite sandstones may have also been sourced from the Greenland, similar to the Nise Formation in the Vøring Basin farther north (Fig. 6). The large rotated fault blocks in Rockall are located at depths of 3–4 km in water depths of 1000 m, making these viable exploration targets.

Conclusions

New PSDM seismic data along the NE Atlantic margin have successfully imaged the sub-basalt strata and indicate that there are many potentially large structural closures located below

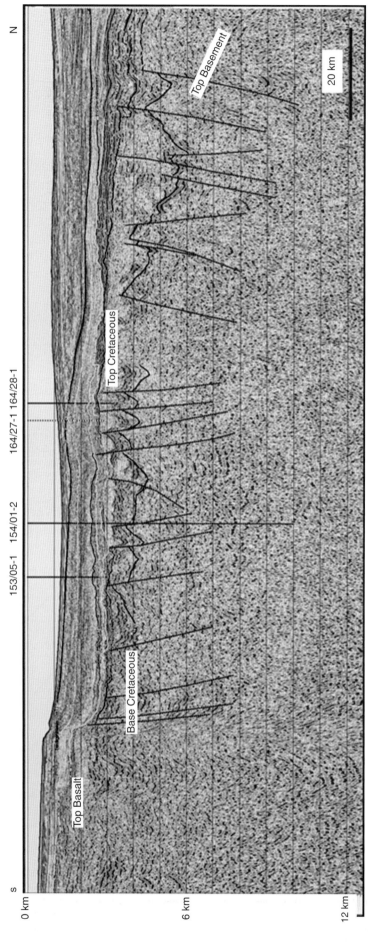

Fig. 5. NE–SW strike line in the northern part of the Rockall Basin showing rotated faults (blocks?) of presumed Jurassic/Cretaceous age with up to 7 km of sedimentary strata in the basin. Rotated fault block crests are potential targets at depths of 4–5 km.

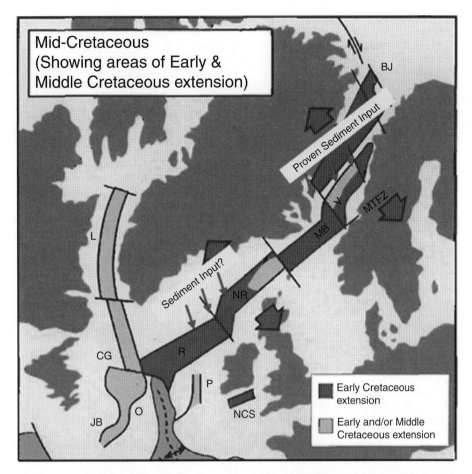

Fig. 6. Map of the NE Atlantic showing a pre-drift reconstruction (Doré *et al.* 1999). Cretaceous strata are derived from Greenland in the Vøring Basin and Paleocene strata were derived from Greenland in the Shetlands area. We suggest here that Late Cretaceous sandstones in the Rockall could also be sourced from west of the basin.

the Palaeogene basalts and sills in the Faroe–Shetland, Rockall and Møre basins. The base of the Palaeogene volcanics is not clearly imaged on the seismic data, probably because this is a transitional contact with little acoustic impedance contrast; the thickness of basalts is therefore very difficult to predict. This increases the risk of exploration targets but this does not preclude exploration as the deeper Mesozoic half-grabens are well-imaged 1–2 km below the presumed base of the volcanics.

The Rockall Trough contains on average 4–7 km of Mesozoic strata which were deposited on rifted continental crust with rotated fault blocks imaged throughout the basin. The presence of Jurassic source rocks and good quality Cretaceous reservoirs is not yet proven but is considered to be the key to successful exploration in the Rockall. The new seismic data presented here have helped to identify many potential structures below the basalts, which are some of the largest undrilled structures in NW Europe. Deep-towed long streamer seismic data have successfully imaged below the basalt province.

We would like to thank ION/GXT BasinSPAN programs for permission to show the seismic data used in this paper. John Smallwood is thanked for a very useful review of this paper.

References

Archer, S. G., Bergman, S. C., Lliffe, J., Murphy, C. M. & Thornton, M. 2005. Palaeogene igneous rocks reveal new insights into the geodynamic evolution and petroleum potential of the Rockall Trough, NE Atlantic Margin. *Basin Research*, **17**, 171–201.

Chappell, A. R. & Kusznir, N. J. 2005. Evidence for slow spreading ocean ridge in the southern Rockall Trough from satellite inversion and seismic data. *EOS Transactions, AGU*, **86**, Abstract T53A-05.

Christie, P. A. F. & White, R. S. 2008. Imaging through Atlantic margin basalts: an introduction to the sub-basalt mini-set. *Geophysical Prospecting*, **56**, 1–4.

Doré, A. G., Lundin, E. R., Birkeland, Ø., Eliassen, P. E. & Fischler, C. 1999. Principal tectonic events in the evolution of the northwest European Atlantic margin. *In*: Fleet, A. J. & Boldy, S. A. R. (eds) *Petroleum Geology of Northwest Europe: Proceedings of the 5th Conference*. Geological Society, London, 41–61; doi: 10.1144/0050041.

England, R. W. & Hobbs, R. W. 1995. Westline: a deep near normal incidence reflection profile across the Rockall Trough. *In*: Croker, P. F. & Shannon, P. M. (eds) *The Petroleum Geology of Ireland's Offshore Basins*. Geological Society, London, Special Publications, **93**, 423–427.

England, R. W. & Hobbs, R. W. 1997. The structure of the Rockall Trough imaged by deep seismic reflection profiling. *Journal Geological Society London*, **154**, 497–502.

England, R. W., McBride, J. H. & Hobbs, R. W. 2005. The role of Mesozoic rifting in the opening of the NE Atlantic: evidence from deep seismic profiling across the Faroe-Shetland Trough. *Journal Geological Society London*, **162**, 661–673.

Gallagher, J. W. & Dromgoole, P. W. 2007. Exploring below the basalt, offshore Faroes: a case history of sub-basalt imaging. *Petroleum Geoscience*, **13**, 213–225.

Gallagher, J. W. & Dromgoole, P. W. 2008. Seeing below the basalt – offshore Faroes. *Geophysical Prospecting*, **56**, 33–45.

Hobbs, R. & Jakubowicz, W. 2000. Sub-basalt imaging using low frequencies. *Geophysics*, **56**, 190–201.

Klingelhoer, F., Edwards, R. A., Hobbs, R. W. & England, R.W. 2005. Crustal structure of the NE Rockall Trough from wide-angle seismic data modelling. *Journal of Geophysical Research*, **110**, B11105; doi: 10.1029/2005JB003763.

Lamers, E. & Carmichael, S. M. M. 1999. The Palaeocene deepwater sandstone play west of Shetlands. *In*: Fleet, A. J. & Boldy, S. A. R. (eds) *Petroleum Geology of Northwest Europe: Proceedings of the 5th Conference*. Geological Society, London, 635–644.

Linnard, S. & Nelson, R. 2005. Effects of Tertiary volcanism and later events upon the Faroese hydrocarbon system. *Faroe Islands Exploration Conference Proceedings of the 1st Conference*. Annales Societatis Scientarum Faeroensis, Supplementum, **43**, Torshavn, 44–53.

Maresh, J., White, R. S., Hobbs, R. W. & Smallwood, J. R. 2006. Seismic attenuation of Atlantic margin basalts. *Geophysics*, **56**, 645–653.

Morewood, N. C., Shannon, P. M. & Mackenzie, G. D. 2004. Seismic stratigraphy of the southern Rockall Basin: a comparison between wide-angle seismic and normal incidence reflection data. *Marine and Petroleum Geology*, **21**, 1149–1163.

O'Reilly, B. M., Hauser, F., Jacob, A. W. B. & Shannon, P. M. 1996. The lithosphere below the Rockall Trough: wide-angle seismic evidence for extensive serpentinisation. *Tectonophysics*, **255**, 1–23.

Richardson, K. R., White, R. S., England, R. W. & Fruehn, J. 1999. Crustal structure east of the Faroe Islands: mapping sub-basalt sediments using wide angle seismic data. *Petroleum Geoscience*, **5**, 161–172.

Ritchie, J. D., Johnson, H. & Kimbell, G. S. 2003. The nature and age of Cenozoic contractional deformation within the NE Faroe–Shetland Basin. *Marine and Petroleum Geology*, **20**, 399–409.

Ritchie, J. D., Johnson, H., Quinn, M. F. & Gatliff, R. W. 2008. The effects of Cenozoic compression and the Cenozoic Faroe-Shetland Basin and adjacent areas. *In*: Johnson, H., Dore, T. G., Gatliff, R. W., Holdsworth, R. W., Lundin, E. R. & Ritchie, J. D. (eds) *The Nature and Origin of Compression in Passive Margins*. Geological Society, London, Special Publications, **306**, 121–136.

Saunders, A. D., Fitton, J. G., Kerr, A. C., Norry, M. J. & Kent, R. W. 1997. *In*: Mahoney, J. J. & Coffin, M. F. (eds) *The North Atlantic Igneous Province*. American Geophysical Union, Washington, DC, Geophysical Monographs, **100**, 45–93.

Scotchman, I. C. 2001. *Petroleum Geochemistry of the Lower and Middle Jurassic in Atlantic Margin Basins of Ireland and the UK*. Geological Society, London, Special Publications, **188**, 31–60.

Scotchman, I. C., Carr, A. D. & Parnell, J. 2006. Hydrocarbon generation modelling in a multiple rifted and volcanic basin: a case study in the Foinaven Sub-basin, Faroe–Shetland Basin, UK Atlantic margin. *Scottish Journal of Geology*, **42**, 1–19.

Shannon, P. M., Jacob, A. W. B., O'Reilly, B. M., Hauser, F., Readman, P. W. & Makris, J. 1999. Structural setting, geological development and basin modelling in the Rockall Trough. *In*: Fleet, A. J. & Boldy, S. A. R. (eds) *Petroleum Geology of Northwest Europe: Proceedings of the 5th Conference*. Geological Society, London, 42–43; doi: 10.1144/0050042.

Shaw, F., Worthington, M. H., White, R. S., Andersen, M. S. & Petersen, U. K. 2008. Seismic attenuation in Faroe Islands basalts. *Geophysical Prospecting*, **56**, 5–20.

Smallwood, J. R. 2009. Back-stripped 3D seismic data: a new tool applied to testing sill emplacement models. *Petroleum Geoscience*, **15**, 259–268.

Spitzer, R., White, R. S. & Christie, P. A. F. 2008. Seismic characterisation of basalt flows from the Faroes margin and the Faroe–Shetland Basins. *Geophysical Prospecting*, **56**, 21–31.

Tate, M. P., Dodd, C. D. & Grant, N. T. 1999. The Northeast Rockall Basin and its significance in the evolution of the Rockall–Faeroes/East Greenland rift system. *In*: Fleet, A. J. & Boldy, S. A. R. (eds) *Petroleum Geology of Northwest Europe: Proceedings of the 5th Conference*. Geological Society, London, 391–406; doi: 10.1144/0050391.

White, R. S., Fruehn, J., Richardson, K. R., Cullen, E., Kirk, W., Smallwood, J. R. & Latkiewicz, C. 1999. Faroes Large Aperture Research Experiment (FLARE): imaging through basalt. *In*: Fleet, A. J. & Boldy, S. A. R. (eds) *Petroleum Geology of Northwest Europe: Proceedings of the 5th Conference*, 1243–1252; doi: 10.1144/0051243.

White, R. S., Smallwood, J. R., Fliedner, M. M., Boslaugh, B., Maresh, J. & Fruehn, J. 2003. Imaging and regional distribution of basalt flows in the Faeroe–Shetland Basin. *Geophysical Prospecting*, **51**, 215–231.

White, R. S., Spitzer, R., Christie, P. A. F., Roberts, A., Lunnon, Z., Maresh, J. & iSIMM. 2005. Seismic imaging through basalt flows on the Faroes Shelf. *Proceedings of the 1st Faroe Islands Exploration Conference*. Annales Societatis Scientiarum Færoensis, Supplementum, **43**, Tórshavn, 11–31.

Ziska, A. H. & Andersen, C. 2005. Exploration opportunities in the Faroe Islands. Faroe Tórshavn. *Proceedings of the 1 Faroe Islands Exploration Conference*. Annales Societatis Scientiarum Færoensis, Supplementum, **43**, Tórshavn, 146–162.

Intra-basalt units and base of the volcanic succession east of the Faroe Islands exemplified by interpretation of offshore 3D seismic data

M. ELLEFSEN,[1,2] L. O. BOLDREEL[1] and M. LARSEN[2]

[1]*Department of Geography and Geology, Copenhagen University, Øster Voldgade 10, DK-1250 Copenhagen, Denmark (e-mail: melle@dongenergy.dk)*
[2]*DONG Energy, Exploration, Agern Alle 24-26, DK-2970 Hørsholm, Denmark*

Abstract: Recent exploration activities in the basalt covered part of the offshore Faroe Island region has stressed the need for a better understanding of the volcanic succession, especially the lowermost part and the transition to the underlying non-volcanic succession. Using the proposed interpretation methods, which include careful identification of inclined reflector segments and flattening to various reflectors representing pronounced continuous markers embedded horizontally and 3D visualization, six volcanic units have been identified. The upper three units consist of plane-parallel reflector packages and probably represent subaerially extruded lava flows. The three lower units are characterized by irregular hummocky and inclined reflectors and are interpreted as lava deltas composed mainly of hyaloclastites. Within the data coverage area, the direction of extrusion is inferred from the seismic interpretation and four units originate from the NW and two from the SE. It is believed that the base of the volcanic succession is situated at a deeper level than the base of the parallel bedded units. At the base of the basalt succession and in the lower part of the volcanic succession pronounced reflector segments are ascribed to saucer-shaped sills. Analysing the seismic data using principles of seismic stratigraphy and observing them in a 3D environment, it is possible to obtain a broader understanding of the volcanic succession in the area and thereby divide the volcanic succession into characteristic facies units and separate it from the non-volcanic lithology and sills. The study shows that applying the proposed interpretation technique allows an evolution model to be put forward and it is suggested that the approach used in the study can be of use for hydrocarbon exploration in other basalt-covered areas.

Keywords: Faroe Islands, volcanics, basalts, hyaloclastites, facies interpretation, 3D seismic data

The Faroe Islands consist of 18 islands situated in the North Atlantic Ocean. The islands are mainly composed of layered basaltic rocks of the FIBG (Faroe Island Basalt Group) (Passey 2009). The FIBG is on a regional-scale part of the North Atlantic Palaeogene flood basalt province that is characterized by widespread lava successions extruded subaerially during continental breakup (e.g. Saunders *et al.* 1997; Larsen *et al.* 1999). The province covers an area of more than 2000 km in diameter and the subaerially extruded lavas can be observed onshore in East and West Greenland (e.g. Pedersen & Dueholm 1992; Larsen *et al.* 1999; Hopper *et al.* 2003; Peate *et al.* 2003; Pedersen *et al.* 2006), on the Faroe Islands (Rasmussen & Noe-Nygaard 1970, 1990; Waagstein 1988, 1996; Passey 2007), Baffin Island in northeastern Canada, and in parts of northern Scotland (Ritchie *et al.* 1999). In the Faroe region the extrusion and evolution of the volcanic succession has been treated from a non-geochemical aspect by, for example, Rasmussen & Noe-Nygaard (1969, 1970), Boldreel & Andersen (1994), Ellis *et al.* (2002), Passey (2004, 2009), Passey *et al.* (2006), Passey & Bell (2007) and Ellis *et al.* (2009). There seems to be an agreement that the volcanics spread eastwards into basins and covered the former subaerially exposed landscapes. A regional Paleocene correlation in the Faroe–Shetland area, which includes this idea, was published by Ellis *et al.* (2009).

The thickness of the volcanic successions and whether siliciclastic sediments are present below the basalts and volcanic rocks in the Faroes area have been debated for many years. Results of refraction seismic experiments give the impression that sediments are to be found and that the volcanic succession is 2–7 km thick (Eccles *et al.* 2009). Details of the volcanic and pre-volcanic succession based on reflection seismic methods are difficult to obtain as sub-basalt imaging is difficult. In order to enhance sub-basalt imaging by reflection seismic methods, a large number of studies have been carried out related to acquisition and processing (e.g. Eccles *et al.* 2009; Gallagher & Dromgoole 2009; Leathard *et al.* 2009) and pronounced progress has been obtained. Exploration wells offshore Faroe Islands and west of Shetland have been drilled at the distal edge of the basalt cover where the thickness of the basalt succession is limited and, based on released wells offshore the Faroe Islands, it is clear that the Paleocene/Eocene basalts have been drilled through (Varming 2009). Onshore the Faroe Islands a deep scientific well drilled through the lowermost part of the plane-parallel bedded basaltic succession (Beinisvørð Formation) and had total depth (TD) in a succession of hyaloclastites (The Lopra Formation; Ellis *et al.* 2009). Geological observations and studies have resulted in a better understanding of the architecture of flood basalts and the complexity of volcanic succession (e.g. Nelson & Jones, 1970; Heinesen 1987; Pedersen *et al.* 2006; Passey 2007; Nelson *et al.* 2009). Results from earlier studies have in this study been used in the interpretation of 3D reflection seismics (e.g. Boldreel & Andersen 1994; Planke & Eldholm 1994; Andersen *et al.* 2007).

The aim of our study is to use seismic stratigraphic principles based on facies architecture of volcanic successions (e.g. Pedersen *et al.* 2006; Andersen *et al.* 2007; Nelson *et al.* 2009) and lithostratigraphic interpretation of wireline logs in flood basalt (e.g. Boldreel 2006; Andersen & Boldreel 2009) to carry out interpretation of a sub-basalt 3D digital reflection seismic survey. The interpretation methodology is applied in order to enhance sub-basalt interpretation and investigate the internal structure of the volcanic succession and its base on the scale of resolution based on conventional reflection seismic data. In this paper we describe the strategy and methods used in the offshore study area east of the Faroe Islands.

The background of the study

Results from fieldwork on outcrops in West Greenland and the interpretation of the wireline logs in the Lopra-1 well illustrate the problem of locating the base of the volcanic succession. In West Greenland, it is found that the lower part of the volcanic succession consists of lavas extruded into a large lake forming a lava delta showing foreset-bedded hyaloclastites. As the depression of the terrain was filled up, the plane-parallel bedded basalts became embedded (Pedersen et al. 2006). A similar scenario is outlined for the transition of hyaloclastites to massive lava flows in the Lopra-1 well (Boldreel 2006). These two examples illustrate that the base of the plane-parallel bedded basalts does not constitute the base of the volcanic succession if the lavas were extruded into water-filled lows where lava deltas developed. Only in the case that the lavas were extruded onto dry land can it be expected that the base of the parallel bedded basalt will actually form the substratum for the volcanic succession. Even under these circumstances it should be expected that the lowest part of the volcanic succession may not be an even surface as the lava filled up topographic depressions and locally developed pyroclastic deposits and lava deltas.

At the distal edge of the basalt-covered region east of the Faroe Islands it seems from the seismic data that steeply dipping foresets indicating lava deltas (hyaloclastites) are absent below the plane-parallel bedded basalts whereas a thick pile of hyaloclastites is found in the Lopra-1 well on the Faroe Islands. This raises the question of how far out from the Faroe Islands the lava deltas extend; that is, where does the base of the plane-parallel bedded basalt form the base of the volcanic succession? This has proved to be a difficult task using 2D reflection seismic data and therefore it was decided to carry out a study in order to investigate if a recently acquired 3D survey of high quality could support the ideas from West Greenland and the Lopra-1 well regarding the base of the volcanic succession. The idea of the project was to investigate the volcanic succession, especially the part found below the parallel bedded basaltic units, by detailed interpretation of the 3D survey using seismic stratigraphy and modern interpretation – viewing – and mapping software.

Data and methods

Multichannel reflection seismic data was acquired in the 1970s by Western Geophysical in the North Atlantic and, based on part of their data, it was found that the surface of the basalt was represented by a pronounced strong positive reflector, that the basalt succession was characterized by parallel bedded reflections and that the eastern limit of the basalt cover is located in the Faroe–Shetland Channel. In the northern part of the Faroe–Shetland Channel an escarpment indicates that the upper part of the plane-parallel bedded unit was extruded into a water-covered basin. In the southern part, the escarpments is absent, thus suggesting that the upper part of the basaltic unit was extruded onto dry land. The interpretation of the data also showed that sediments are present below the outer parts of the basalt covered area and that these sediments extend westwards towards the Faroe Islands for some distance before sub-basalt seismic imaging can no longer be obtained. Western Geophysical was awarded the obligation to acquire new multichannel reflection seismic data using the state of the art in the period 1994–1995 and the seismic data showed pronounced improvement of subsurface and sub-basalt imaging and thus the zone of sub-basalt imaging moved further to the west towards the Faroe Islands. Especially in the eastern part of the Faroes part of the Faroe–Shetland Channel the sub-basalt imaging was improved and pronounced inclined (foresets) and likely basin floor fan configurations were seen below a parallel bedded basalt unit (Boldreel et al. 1996). In addition, distinct seismic units could be identified below the foreset bedded units (Boldreel et al. 1996). Since 1996 a large number of conventional reflection seismic surveys, both 2D and 3D, have been carried out offshore the Faroe Islands in addition to experiments aimed at improving sub-basalt imaging. This has resulted in improvement of the sub-basalt imaging, but it is still a difficult task to carry out interpretation below the parallel bedded flood basalts in order to locate the base of the volcanic succession.

The data used in this study is 3D offshore reflection seismic data acquired in 2003 in license area 006 on the East Faroe High SE of the Faroe Islands. The survey covers an area of approximately 12×24 km (Fig. 1). A bubble-tuned source on a depth of 15 m was used, together with 8 km-long cables, placed in a depth of 20 m below the sea surface (Gallagher & Dromgoole 2007, 2009). The data was processed as zero phase data and frequencies higher than 38 Hz were removed. Great effort was put into processing the data, as discussed by Gallagher & Dromgoole (2007, 2009), and this resulted in a pronounced improvement in the subsurface imaging of the basaltic unit. The vertical resolution of the data is estimated to be of the order of 50–60 m using an average frequency of 20–25 Hz and 5000 m s^{-1} for the velocity of subaerially extruded basalt as based on the interpretation of subaerially extruded flood basalt in the Lopra-1 well (Boldreel 2006). The thickness of individual basaltic beds (lava flows) on the Faroe Islands differs according to the stratigraphic position (Series) they are part of, but the average in the Beinisvørd Formation (former Lower Series) is 5–25 m (Hald & Waagstein 1984; Boldreel 2006) and similar thicknesses of basalt flows are noticed by interpreting wireline logs from exploration wells located in the West of Shetland area (Andersen & Boldreel 2009). Similar thickness variations of simple flows from various regions are published by Nelson et al. (2009). Thus, it is expected that most individual lava flows in the surveyed area are below seismic resolution, as is also illustrated by Nelson et al. (2009). This implies that the reflections observed in the basaltic unit may represent a number of lava flows. A

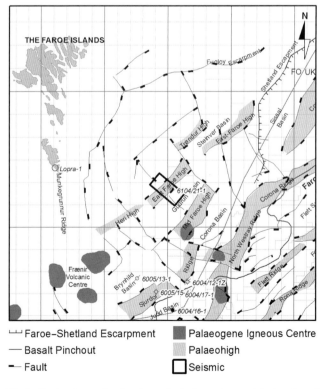

Fig. 1. Regional map showing the geological structures and the location of the study area.

similar problem of resolution has been addressed in a study of flood basalts from West of Shetland area, where it was shown that the reflectors interpreted as lava flows actually are made up of a number of lava flows (Hansen 2007). In another study addressing chalk mounds it was shown that the size of the individual mounds was below seismic resolution but the overall form of the mound and directions of growth could still be determined (Nielsen *et al.* 2004).

In order to estimate the thickness of the basaltic units in metres, a velocity has to be ascribed to the basaltic succession. Several authors have dealt with this subject by measuring the velocity of rock samples, from velocity picking in reflection seismic data, refraction seismic studies and interpreting acoustic wireline logs. It was shown that the velocity of basaltic rock varies according to whether it represents a compound or simple flow, hyaloclastites or intrusions. In addition the internal composition of, for example, a simple flow shows a large variation of velocity in the massive core and porous top and base (Planke 1994; Boldreel 2006; Andersen *et al.* 2007; Andersen & Boldreel 2009). In this study the findings by Planke (1994), Boldreel (2006), Andersen *et al.* (2007) and Andersen & Boldreel (2009) are applied with a velocity of 5000 m s^{-1} for units 1–3 (parallel bedded basalt) and a velocity of 3000 m s^{-1} for units 4–6 (hyaloclastites). Stacking velocities from reflection seismic surveys east of the Faroe Islands reveal a velocity of 5000–6000 m s^{-1} for the parallel bedded basalt unit and a drop in velocity leaving the unit and entering the unit below the parallel bedded basalt is seen.

Seismic interpretation

The basalt has been drilled in various places offshore UK and Faroese and by use of long-distance seismic ties it is believed that the succession which is the focus of the present study in the upper part consists of basalt. This has recently been confirmed by released information from the exploration well 6104/21–1 situated within the study area (Ellis pers. comm. 2009). From the interpretation of 2D seismic profiles in the East Faroe High area and in a westwards direction it is found that pronounced escarpments are located below the parallel bedded basalt, indicating that the palaeo-environment was probably water covered (Boldreel *et al.* 1996). The Lopra-1/1A well encountered the base of the Beinisvørð Formation at some 2489 m depth before entering the Lopra Formation, which consists of hyaloclastites (Boldreel 2006; Ellis *et al.* 2009). Thus it is expected that a hyaloclastite succession could be present below the base of the parallel bedded basalts in the study area.

Previous work has shown that the surface of the subaerial extruded basalt is easily mapped and that the principles of seismic stratigraphy are applicable for interpreting basalt successions (e.g. Gatliff *et al.* 1984; Boldreel & Andersen 1994; Andersen *et al.* 2007). Various studies have dealt with the architecture of flood basalts and their seismic expression on 2D reflection seismic data, for example, Planke *et al.* (1999, 2000), Jerram (2002), Andersen *et al.* (2007), Nelson *et al.* (2009). In this study we have compared the volcanic succession with five seismic facies types (cf. Andersen *et al.* 2007) that are also recognized in the subsurface of the British and Faroe Island continental shelves. The five types are

- The *parallel bedded platform* facies represent subaerially erupted volcanics and have been sampled by drilling or surface sampling.
- The *oblique progradational* facies represent basaltic hyaloclastic rocks tentatively interpreted as foreset breccias of basaltic flows that flowed into a subaqueous environment, and this type has been sampled by drilling or surface sampling.
- The *sigmoid aggradational* facies are believed to represent basaltic lavas that flowed into water. In contrast to the *sigmoid aggradational* facies, the *oblique progradational* facies are characterized by a disconformity above the foresets. It is proposed that this may reflect a relatively larger lava supply during formation of the *oblique progradational* than during formation of the *sigmoidal aggradational* facies. This type has been interpreted by comparison with well exposed volcanic provinces in Antarctica, West Greenland, Hawaii and Iceland.
- The *chaotic hummocky mounded* facies are believed to represent submarine volcanic products representing volcanic material deposited around a single feeder dyke (or stock). This type is inferred by comparison to well exposed volcanic provinces in Antarctica, West Greenland, Hawaii and Iceland.
- The chaotic *hummocky basin floor* facies probably represents submarine volcanic products that have been deposited over wider areas and this type was interpreted by comparison to well exposed volcanic provinces in Antarctica, West Greenland, Hawaii and Iceland.

In the basalt covered region around the Faroe Islands, West of Shetlands and in Norwegian waters, irregular, wavy, very pronounced high-amplitude reflectors of limited extent and often crossing bedding planes are observed. These saucer-shaped reflectors have been interpreted as magmatic intrusions (sills) by Hansen *et al.* (2004), Planke *et al.* (2005), Hansen & Cartwright (2006) and Smallwood & Harding (2009).

The interpretation method used in the present study was first to interpret and map the surface of the basalt followed by interpretation of an intrabasaltic reflector that forms the base of unit 1. In the study area both the surface of the basalt and the lower boundary of unit 1 are regional surfaces that were almost horizontal at the time of formation. Therefore the seismic flattening technique on these boundaries was applied to eliminate the effects of later tectonic events and morphological expression on the lower lying units (Fig. 2). The top basalt reflector was flattened and the interpretation of the base unit 1 reflector was refined. Following this, the base unit 1 reflector was flattened. The flattening technique was used for all of the plane-parallel bedded basalt units. Below the plane-parallel bedded basalt units, the flattening technique was applied to the part of the units that was originally embedded horizontal. In this manner a total of six units were interpreted.

In places it was difficult to identify the base of the lower seismic units from later intruded sills that make up a large proportion of the reflections around the base of the succession (Naylor *et al.* 1999; Ellis *et al.* 2002). Applying the principles of saucer-shaped sills and the facility of working with 3D data, it was possible to carry out a detailed interpretation of the individual sills and thus to separate the intrusions from the background facies (Planke *et al.* 2005; Hansen 2006). Furthermore, working with 3D data it was possible to select optimal orientated seismic profiles for interpreting the individual seismic facies units. The detailed study of the facies units and their reflectors made it possible to establish the direction of foreset dip and thus interpret the local migration direction of lava-flows within the lowermost units, and to recognize and reconstruct the palaeocoastline in the southeastern part of the study area. Finally, the interpreted seismic reflectors were displayed as surfaces in a 3D visualization environment to increase the understanding of the geological evolution of the area.

Description of facies units

A typical seismic profile through the volcanic succession in the study area is found in Figure 2. The succession is divided into seismic facies units numbered from top to base. The thickness and outer shape of the units are illustrated in Figure 3, which shows the outline of the basalt facies units in a 3D perspective.

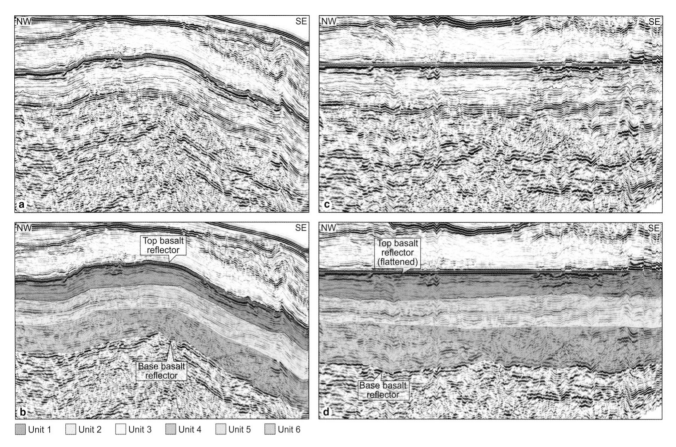

Fig. 2. Representative seismic crossline profile orientated NW–SE. (**a**) Seismic profile. (**b**) Seismic profile with interpreted units and top basalt and base basalt reflector indicated. (**c**) Seismic profile flattened to the surface of the basalt. (**d**) Seismic profile flattened to the surface of the basalt with interpreted units and top basalt and base basalt reflector indicated.

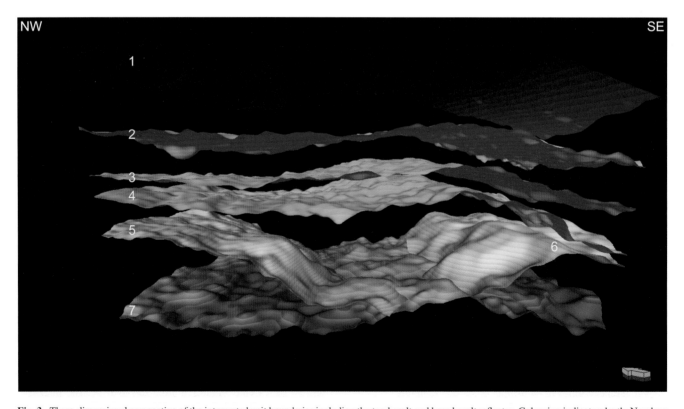

Fig. 3. Three-dimensional perspective of the interpreted unit boundaries including the top basalt and base basalt reflector. Colouring indicates depth. Numbers indicate top surfaces of units and number 7 indicates base basalt surface.

Facies unit 1. The upper boundary of unit 1 is a very distinct continuous reflector characterized by pronounced strong positive (hard kick) high amplitude. The reflector is smooth in appearance and very slightly undulating in the survey area, although the continuity of the reflector is broken at minor geographical places. At the base, unit 1 is defined by a continuous rather smooth, slightly undulating reflector with positive medium amplitude. The reflector parallels the top reflector, although slightly more irregular. The reflector and the internal bedding of unit 1 are broken at the same geographical locations as the top reflector. The internal reflection pattern is continuous with sub-parallel to partly irregular parallel reflections of medium amplitude. The external form of the unit seems to be sheet-like, but the unit extends outside the study area.

The upper boundary of the unit is interpreted as top basalt reflector (Ellis pers. comm. 2009; Fig. 2). Facies unit 1 experiences a rather constant thickness of approximately 800 m (Fig. 3). The unit is interpreted as subaerially parallel bedded platform facies erupted on an almost horizontal surface. In limited areas small changes in the internal pattern may indicate individual lava flows, or groups of lava flows, that stand out. The unit is widespread with the source area situated outside the study area.

Facies unit 2. The base of unit 2 is a positive (hard kick), partly broken reflector of medium to low amplitude (Fig. 2). In the southeastern part of the study area the reflector is smoother and more regular than in the northwestern part of the study area. In limited locations the base of unit 2 is not a pronounced reflector as the lowermost part of the units is almost reflection-free. Here the base of the unit is defined by the change in reflection pattern to unit 3. The internal reflector configuration is somewhat thicker and more faded as compared with unit 1, which could be a result of thinning of layers below seismic resolution. The internal reflection pattern of unit 2 consists of fairly continuous, partly irregular parallel to sub-parallel reflections of medium amplitude. The external form of the unit is a sheet that continues outside of the study area. The thickness is fairly constant at approximately 500 m (Fig. 3). Facies unit 2 belongs to the subaerially erupted parallel bedded platform facies that is emplaced over a large flat-lying area with the source area outside the study area. Within the upper part of unit 2 individual younger lava flows or groups of flows are seen to build out in limited locations. There is a slight change of the internal reflections pattern of unit 2 as compared with unit 1 that may indicate that unit 2 represents a change in style of eruption and thus there might have been a halt in lava production between units 1 and 2.

Facies unit 3. The base of unit 3 is defined by a low-amplitude negative (soft kick) reflector that in places is non-continuous (Fig. 2). In the southeastern part of the study area the reflector is easily followed. The lowermost part of the unit is nearly reflection-free, especially in the northwestern part of the study area, and here the base of unit 3 is defined by the change in reflection configuration between units 3 and 4. The internal reflection pattern consists of pronounced continuous, closely spaced plane-parallel reflections of medium amplitude in the northeastern part of the area. Gradually the pattern becomes less continuous towards the SW and furthest to the SW the pattern consists of irregular and broken sub-parallel reflections. The unit is cut by a small number of faults, most of which are located in the northwestern part of the study area. The external form of the unit is sheet-like and the unit extends out of the study area. The thickness is approximately 300 m but differs slightly, being thicker where the unit fills small depressions in the underlying surface (Fig. 3). Facies unit 3 belongs to the subaerially erupted parallel bedded platform facies and was extruded onto an almost flat land area. The unit spread out in thinner lava flows as compared with units 1 and 2.

Facies unit 4. The base of unit 4 is made up of three different reflectors as the unit fills in a basin (Fig. 2). In the NW the base of unit 4 is defined by a positive (hard kick) continuous reflector of medium to high amplitude that gradually dips towards the basin. The internal reflector pattern shows downlap onto the base reflector. In the SE the lower boundary of unit 4 is a positive (hard kick) fairly continuous reflector of medium to high amplitude that gradually dips towards the basin. The internal reflector pattern shows downlap towards the lower boundary. In the central part of the study area the lower boundary of unit 4 is defined by a very complex wavy negative (soft kick) reflector. In some locations the reflector is continuous with a high amplitude whereas in other locations the unit is nearly reflection-free and the lower boundary difficult to identify. The internal reflection configuration in unit 4 is mostly characterized by medium to low amplitude reflector segments that have a wavy chaotic pattern with little continuity. Analysing the unit in more detail, a number of smaller internal units are identified. The unit is easily distinguished from the underlying units 5 and 6 except for a few locations. The external form is a basin fill-form. The extent of the upper part of the unit goes beyond the study area. The thickness of unit 4 differs greatly throughout the study area and the maximum thickness is found in the central part of the study area, where it reaches approximately 800 m (Fig. 3). Facies unit 4 is ascribed to a hummocky basin-floor facies. It is suggested that unit 4 is subaquatically extruded and consists of a mixture of lava flows from various directions that have flowed out into a basin and here formed lava deltas (hyaloclastites). The basin extends out of the study area both to the north and south. After emplacement, unit 4 has been heavily influenced by intrusions.

Facies unit 5. Unit 5 is located in the northwestern part of the study area and its lower boundary is a very complex wavy negative (soft kick) reflector (Fig. 2). In places it is a high-amplitude continuous reflector whereas in other places the reflector is hard to distinguish as the unit here is nearly reflection-free. The internal reflection configuration is very complex as in most places the pattern shows downlap toward the lower boundary, whereas in minor areas it appears chaotic or reflection-free. Unit 5 mainly consists of a mixture of minor prograding sigmoid and complex sigmoid-oblique clinoforms, with reflections of medium to high amplitude, that dip from NW towards the SE (Figs 2 & 4). The external form of the unit is a sigmoid/complex sigmoid-oblique prograding clinoform where the northwestern part of the unit extends out of the study area (Fig. 3). The thickness of unit 5 differs between 650 and 1150 m as the unit is thickest in the SW and thins towards the NE. Unit 5 is made up of a large number of smaller units consisting of prograding clinoforms with a low inclination.

Facies unit 6. This unit is found beneath unit 4 in the southeastern part of the study area. The lower boundary is a very complex wavy negative (soft kick) reflector (Fig. 2). In places the reflector is continuous with a high amplitude whereas in other places the unit is nearly reflection-free and the boundary difficult to identify. The internal reflection configuration in unit 6 consists of medium-amplitude S-shaped prograding clinoforms prograding from the SE towards the NW. In some places the configuration is rather chaotic and the S-shaped clinoforms can be difficult to distinguish from one another and it is difficult to see whether the clinoforms are sigmoid, complex sigmoid-oblique or oblique. A careful analysis of the internal reflector configuration reveals that a large number of smaller units can be outlined that mostly dip to the NW, although at the outer parts of unit 6 the flows show various directions (Fig. 4). The external form of unit 6 is a sigmoid-oblique prograding clinoform and the southeastern part of the unit extends beyond

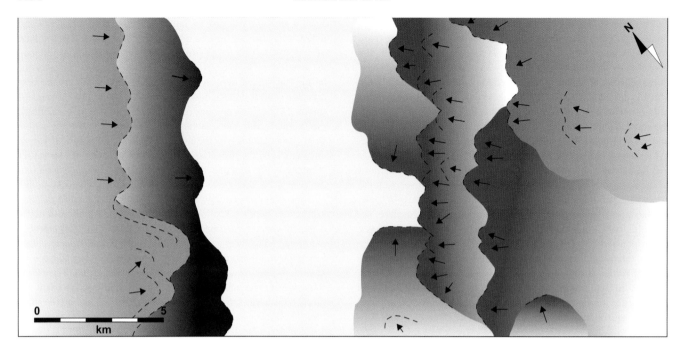

Fig. 4. Facies map of units 5 (purple colour) and 6 (grey colour) showing the locations of main prograding subunits and direction of movement.

the study area. The thickness of unit 6 varies but is largest in the southeastern part of the study area. The internal reflection pattern of the unit is more regular than unit 5 and the maximum thickness of the unit is approximately 800 m (Fig. 3). Breakpoints on the prograding clinoforms are mapped and thus it is possible to construct the extent of a palaeocoastline. Unit 6 shows almost ideal sigmoid-oblique clinoforms as opposed to unit 5 and it is suggested that unit 6 consists of a mixture of lava and sedimentary deposits that alternately prograded across an existing coastline. The unit has been intruded by younger intrusions, for example, sills or dykes.

'Base volcanic reflector' and subvolcanic unit. The 'base volcanic reflector' is interpreted as a composed surface corresponding to the base of volcanic units 4–6. It is a very complex wavy negative (soft kick) reflector that is disrupted by intrusions and faults (Figs 2 & 3). The reflector is generally characterized by downlap of younger reflectors and reflector segments, as illustrated by flattening of the surface of the parallel bedded basalt units (Fig. 2). The reflector is in places distinct and continuous with high amplitude, whereas it is hard to identify the reflector in other locations as the overlying unit is nearly reflection-free. Flattening to the surface of basalt unit 1 reveals that the 'base volcanic reflector' in places undulates as it drapes minor highs and lows. Beneath the 'base volcanic reflector' the reflection pattern changes are pronounced and the internal reflector pattern of the underlying units resembles what is usually interpreted as sedimentary, faulted units (Fig. 2). In the northwestern part of the investigated area below unit 5 the internal reflector pattern consists of reflectors of medium amplitude that show dip to the SE. The reflector pattern between the uppermost of these inclined reflectors and the base basalt onlaps towards an inclined reflector. Beneath unit 6 in the southwestern part of the study area pronounced strong diverging reflectors of slightly wavy appearance are found that originate in the SE and dip to the NW. Below unit 4 the seismic signal is incoherent and this, in addition to the large amount of sills, makes it hard to interpret the geological succession below the basin fill.

Intrusions. On the majority of the seismic profiles intrusions are mainly identified in the level of the base of the volcanic succession. The intrusions are seen as pronounced high-amplitude continuous positive (hard kick) reflectors that on the individual inlines and crosslines are arc shaped and exhibit a limited extent. Using 3D display facility it is seen that the intrusions are individual reflectors that are saucer-shaped and have a characteristic behaviour similar to what was reported by Planke *et al.* (2005) and Hansen (2006). Analysing the intrusions in detail it is found that in places the intrusions are connected and this probably indicates that they have been fed from the same source, similar to the findings of Smallwood & Harding (2009).

Analogues

Onshore analogues to the parallel bedded basalt, inclined basalt flows and hyaloclastites are found at various volcanic provinces around the world. To present an analogue to the seismic study, a comparison with the nearby geographical area of the Faroe Islands and West Greenland, which was extruded within the same time frame, is suggested (Rasmussen & Noe-Nygaard 1970; Pedersen *et al.* 2006; Passey 2007). Based on photogrammetry of the south coast of Nuussuuaq, west Greenland, five detailed geological maps have been constructed (Pedersen *et al.* 2006) that show a similar setting as revealed from the present seismic study. The volcanic succession is more than 2 km thick and consists of plane-parallel bedded basalt flows, hyaloclastites and foreset beds exposed on a scale that is similar to seismic resolution. At Nuussuuaq, plane-parallel basaltic lava-flows in association with sediments can be shown to have prograded into a more than 700 m deep marine basin, creating a basin-fill mixture of hyaloclastites and sediments (Pedersen & Dueholm 1992). Above this succession plane-parallel bedded lava flows are found. On the Faroe Islands the Lopra-1/1A well encountered the transition from subaerial extruded plane-parallel bedded lava flows to the hyaloclastites, and downhole wireline logs measured in the entire hole form an important dataset for studying the plane-parallel bedded basalt in the Beinisvørð Formation and the transition to the hyaloclastites in the Lopra Formation. From the interpretation of the wireline logs it is found that the petrophysical properties of the basalt succession in the borehole change drastically from hyaloclastites to plane-parallel bedded lava flows, especially in relation to P-wave velocity, density and porosity, which make an important contribution to the seismic expression of basalts belonging to the hyaloclastites and plane-parallel bedded facies (Boldreel 2006). This means that the transition from plane-parallel bedded basalt to hyaloclastites is characterized by a negative reflection coefficient (soft kick) which was also seen from the interpretation of the VSP survey of the Lopra-1 well (Kiørboe & Petersen 1995). From the UK part of the Faroe–Shetland Channel, a seismic modelling study was carried out in which results from the interpretation of wireline logs in released wells were used as input and released 2D reflection seismic profiles were used for comparison. The study showed that the overall seismic expression of hyaloclastites and parallel bedded basalt could be produced (Hansen 2007).

Geological development of the area

Findings from the detailed interpretation of a 3D digital reflection seismic survey using principles of seismic stratigraphy, flattening technique to selected reflectors and 3D visualization technique combined with 2D seismic reflection studies in the region and released wells and analogy studies to the Faroe Islands and West Greenland, have proposed the following geological evolution of the study area (Fig. 5a–f).

(a) The oldest identified unit in this study shows a negative reflection coefficient (soft kick) at the transition to the overlying volcanic succession (units 4–6), which means a unit characterized by a predominately lower velocity than the oldest volcanic succession. The unit shows a strong resemblance to a sedimentary succession with indications of highs and basins offset by faults. It is known that sedimentary successions are present below the flood basalts at the edge of the basalt covered region in the Faroe–Shetland Channel and thus it could be concluded that a sedimentary succession is present beneath the volcanic unit in the study area.

(b–d) In the northwestern part of the survey area reflectors show onlap towards older inclined reflectors in the lowermost part of the volcanic succession (unit 5). Further to the SE, the unit that contains the dipping reflectors diminishes and becomes less inclined.

Unit 5 is made up of minor prograding sigmoid and complex sigmoid-oblique clinoforms that extend from NW to SE. Unit 5 seems not to be truncated at the upper boundary. The fronts of a large number of prograding subunits within unit 5 have been mapped (Fig. 4) and suggest that the lava flows of unit 5 prograded in a southeastern direction. It is suggested that unit 5 consists of minor basalt flows of northwestern source that gradually build out into a water-covered area. The outer parts of unit 5 gradually build out as a lava delta (hyaloclastite) in the basin and interfinger laterally and vertically with unit 6. The geographic change in the inclination may indicate the outer basinward part of subaerial extruded basalt, and this implies that subaerial extruded plane-parallel bedded basalt or units of basalt foresets are present in the direction towards the Faroe Islands, as suggested by Ellis et al. (2002). The lack of truncation at the upper level suggests that the relative sea-level rose following the eruption of unit 5.

In the southeastern part of the study area, part of a lava delta named unit 6 is found. The delta builds out from the SE towards the NW, crossing an existing coastline. The detailed interpretation of the lava delta reveals that unit 6 consists of an alternating sigmoid and oblique prograding facies constructed of subaerially extruded basalt that entered an aquatic environment. It is not possible to judge if unit 6 is truncated at the upper level or whether the uppermost part consists of thin basalt flows below seismic resolution. If truncation has occurred, this would imply that erosion took place before a relative sea-level rise prior to the emplacement of unit 4. If truncation did not take place this probably implies that subsidence took place and that the relative sea-level rose prior to the eruption of unit 4. Within unit 6 the fronts of a large number of prograding subunits have been identified and mapped (Fig. 4) and this shows that the direction of progradation of unit 6 is variable in that the main part of the flows prograded from the SE towards the NW, although flows originating from south, SSE and SW are also identified. The outer parts of unit 6 gradually build out as a lava delta in the basin and interfinger laterally and vertically with unit 5. It is not possible from the data available to this study to judge if units 5 and 6 are of the same origin. The basin between units 5

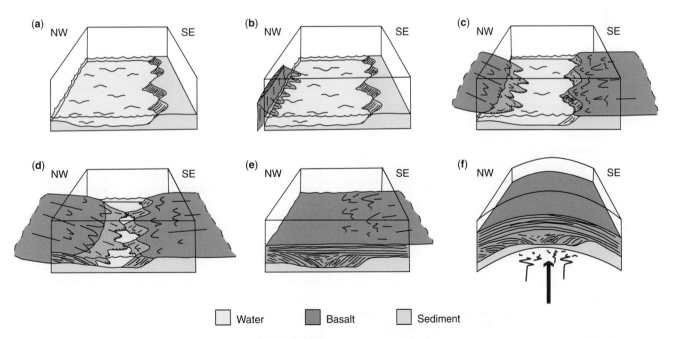

Fig. 5. Geological model for the study area. (**a**) Seismic mapping indicates that there is a basin underneath the volcanic succession in the study area. (**b-d**) At the beginning of volcanism lava flowed into the basin forming two prograding units – one from the NW (interpreted unit 5) and one from the SE (interpreted unit 6). (**e**) After filling up the basin (interpreted unit 4) the volcanism dominantly came from the SE, covering the lava filled basin with plane-parallel bedded basalts (interpreted units 3–5). (**f**) After the volcanism ceased, the area and the volcanic succession underwent differential subsidence and was subject to compression bending the volcanic succession. Blue lines indicate sill intrusions.

and 6 extends outside the data coverage. The time relation between units 5 and 6 cannot be unravelled for sure but, as the tops of units 5 and 6 are at a similar depth interval, it is suggested that units 5 and 6 are within the same time frame.

In the central part of the study area unit 4 is located geographically between units 5 and 6. The internal reflector pattern of unit 4 suggests that it consists of hyaloclastites deposited in a lava delta that entered the basin from various directions and gradually filled the basin.

(e) Unit 3 consists of thin plane-parallel bedded lava flows. The unit is most distinct and well developed in terms of internal configuration in the southeastern part of the studied area and the unit thins towards the NW. Detailed interpretation and visualization technique shows that small internal units prograde from SE to NW and that the lower part of unit 3 fills in minor lows. It is suggested that unit 3 is subaerially extruded onto a dry landscape and that the surface of the underlying unit has been exposed to weathering and erosion, indicating that a pause in eruption to the underlying unit is likely. The thickness variation of the unit and the direction of progradation of the small internal units strongly indicate that unit 3 originated from the SE outside the study area.

The internal reflector pattern in units 1 and 2 is plane-parallel bedded basalt flows and the units are very much alike. The units are separated by a pronounced reflector of positive reflection coefficient (hard kick). The internal configuration of the basalt unit indicates that the lava flows of units 1 and 2 are thicker than those in unit 3 and the thickness of both units decreases towards the SE. In the southeastern part of the study area it is seen that the lowermost part of unit 2 in places onlaps unit 3 towards the SE. It is found that units 1 and 2 are subaerial extruded basalt flows that originated from a (north) western direction and that there seems to have been a pause between the extrusions of the units, represented by the pronounced reflector.

(f) After the volcanism ceased, sedimentation took place first from the east and later on from the west, as is well known from a number of studies. The area and the volcanic succession underwent differential subsidence in the area and was subject to compression that bent the succession (e.g. Boldreel & Andersen 1993). From the use of 3D data it is found that a large number of faults offset the volcanic succession and this further complicates the interpretation of both the lower part of the volcanic succession and the sedimentary succession found below.

Sills are found to have intruded into units 4–6, that is, the lower part of the volcanic succession. The studied area is located close to the main sill complex of likely Early Eocene age where a sill has been dated as Early Eocene (Smallwood & Harding 2009), and as the behaviour of that finding is similar to the sills in the present study, it is suggested that the sills identified in the present study are of Early Eocene age. From the present seismic interpretation it is suggested that the sills were intruded either at a similar time to the eruption of units 4–6 or by later tectonic activity, but according to Smallwood & Harding (2009) it is most likely that the sills were intruded in Early Eocene, which corresponds to the uppermost part of the Beinsvørð Formation or the younger basalt formations (Ellis et al. 2009).

From the seismic interpretation it is found that the internal reflector patterns and identified minor escarpments in units 6 and 3 indicate that these two basaltic units originated from a (south) eastern direction within the study area whereas units 5, 4, 2 and 1 mainly originated from a (north) western direction as expected and illustrated from various papers (e.g. Ellis et al. 2009). It may seem that units 6 and 3 are sourced from a different area from the remaining basaltic units within the data coverage area as unit 4 fills in a basin area that extends outside the study area. The two units may have been emplaced from fissures or a volcano located further to the east of the study area. From the work of Ellis et al. (2009, fig. 3) it is indicated that lava is present on the eastern side of the Faroe–Shetland area but the published literature does not mention fissures or volcanoes located close to the eastern part of the studied area. This seems to indicate that the basin that was filled by unit 4 may have been an embayment and thus the basalt flows originated from a western to northwestern source and part of the flows were directed towards the basin area and part flowed on land to the other side of the embayment, from where the flows thus entered the basin from an easterly direction. From the data used in the present study, the validity of the two scenarios cannot be judged.

Conclusion

A detailed interpretation on a recently acquired 3D reflection seismic survey in a basalt-covered area SE of the Faroe Islands has been carried out. Based on seismic stratigraphic principles, the use of flattening of original horizontally embedded lava flows and 3D visualization software, a new approach to interpreting seismic data in basalt-covered regions is proposed (Fig. 2). In addition, knowledge from onshore exposed volcanic successions from west Greenland and information from the interpretation of wireline log data in basalt covered areas west of Shetland and on the Faroe Islands gives valuable geological information that can be evaluated in view of seismic resolution.

We divided the volcanic succession into six seismic facies units. The lowermost two facies units represent two individual successions building out from the NW and SE, respectively, on a landmass into a water-covered basin where the two volcanic successions form the flanks of the basin. The outer parts of the individual flows within the units interfinger with each other in the basin.

Following the emplacement of the two units the basin was gradually filled up and the following unit was clearly subaerially parallel bedded basalt extruded from the SE. The youngest two units consist of parallel bedded basalt flows that were subaerially extruded from the NE.

It is proposed that the 'base volcanic reflector' is not located at the base of the parallel bedded basalts, but is found at a greater depth where it reflects the base of lava deltas (hyaloclastites) formed in a basin. This implies that only in regions where lavas flowed onto dry land is the base basalt reflector located at the lower boundary of the parallel bedded basalt succession.

Sills have been identified by careful interpretation of the seismic data using the possibility of choosing optimal orientated profiles and inspecting the data in a 3D visualization software. In this manner it was found that the sills are saucer-shaped and often part of a common feeder system.

Below the volcanic succession a sedimentary succession is suggested to be present with indications of highs and basins offset by faults. It is suggested that the proposed interpretation technique is a valuable tool for interpreting reflection seismic data east of the Faroe Islands and in other basalt-covered regions.

We are grateful for the support from 006 Licence partners Statoil, Atlantic Petroleum and Dong Energy for putting the seismic data at our disposal and for the permission to publish part of the study. The manuscript benefited from critical reviews by Simon Passey and an anonymous referee. The interpretations and conclusions expressed, however, are the sole responsibility of the authors and do not necessarily reflect the views of all the Partners.

References

Andersen, M. S., Egerton, P. D., Hitchen, K. & Boldreel, L. O. 2007. Seismic facies analysis of volcanic sequences offshore the British Isles and the Faroes. *Evolution of Basaltic Provinces, 1st Jóannes Rasmussen Conference on the Faroe Islands*, August 2007, Extended Abstract.

Andersen, M. S., Boldreel, L. O. & SeiFaBa Group. 2009. Log responses in basalt successions in 8 wells from the Faroe–Shetland Channel – a classification scheme for interpretation of geophysical logs and case studies. *In*: Varming, T. & Ziska, H. (eds) *Faroe Island Exploration Conference Proceedings of the 2nd Conference*, 364–391. Føroya Fróðskaparefelag, The Faroese Academy of Sciences. *Annales Societatis Scientiarum Færoensis Supplementum*, **50**, Tórshavn.

Boldreel, L. O. 2006. Wire-line log-based stratigraphy of floodbasalts from the Lopra-1/1A well, Faroe Islands. *In*: Chalmers, J. A. & Waagstein, R. (eds) *Scientific Results from the Deepened Lopra-1 Borehole, Faroe Islands*. Geological Survey of Denmark and Greenland Bulletin, **9**, 7–22.

Boldreel, L. O. & Andersen, M. S. 1993. Late Paleocene to Miocene compression in the Faroe-Rockall Area. *In*: Parker, P. J. R. (ed.) *Petroleum Geology of Northwest Europe: Proceedings of the 4th Conference*. Geological Society, London, 1025–1034; doi: 10.1144/0041025.

Boldreel, L. O. & Andersen, M. S. 1994. Tertiary development of the Faroe–Rockall Plateau based on reflection seismic data. *Bulletin of the Geological Society of Denmark*, **41**, 162–180.

Boldreel, L. O., Neish, J. K. & Ziska, H. 1996. Exploration of the Faroes region: seismic images beneath a basalt province. *American Association of Petroleum Geologists Bulletin*, **80**. Abstract; doi: 10.1306/64EDA596-1724-11D7-8645000102C1865D.

Eccles, J. D., White, R. S., Christie, P. A. F., Roberts, A. W. & iSIMM Team. 2009. Insight into sub-basalt lithology from wide-angle converted shear wave analysis. *In*: Varming, T. & Ziska, H. (eds) *Faroe Islands Exploration Conference Proceedings of the 2nd Conference*, **226**, 43–60. Føroya Fróðskaparefelag, The Faroese Academy of Sciences. *Annales Societatis Scientiarum Færoensis Supplementum*, **50**, Tórshavn.

Ellis, D., Bell, B. R., Jolley, D. W. & O'Callaghan, M. 2002. The stratigraphy, environment of eruption and age of the Faroes Lava Group, NE Atlantic Ocean. *In*: Jolley, D. W. & Bell, B. R. (eds) *The North Atlantic Igneous Province: Stratigraphy, Tectonic, Volcanic and Magmatic Processes*. Geological Society, London, Special Publications, **197**, 253–269.

Ellis, D., Passey, S. R., Jolley, D. W. & Bell, B. R. 2009. Transfer zones. The application of new geological information from the Faroe Islands applied to the offshore exploration of intra and sub-basalt strata. *In*: Varming, T. & Ziska, H. (eds) *Faroe Islands Exploration Conference, Proceedings of the 2nd Conference*, 205–226. Føroya Fróðskaparefelag, The Faroese Academy of Sciences. *Annales Societatis Scientiarum Færoensis Supplementum*, **50**, Tórshavn.

Gallagher, J. W. & Dromgoole, P. W. 2007. Exploring below the basalt, offshore Faroes: a case history of sub-basalt imaging. *Petroleum Geoscience*, **13**, 213–225.

Gallagher, J. W. & Dromgoole, P. W. 2009. Sub-basalt seismic imaging-offshore Faroes. *In*: Varming, T. & Ziska, H. (eds) *Faroe Islands Exploration Conference Proceedings of the 2nd Conference*, 319–332. Føroya Fróðskaparefelag, The Faroese Academy of Sciences. *Annales Societatis Scientiarum Færoensis Supplementum*, **50**, Tórshavn.

Gatliff, R. W., Hitchen, K., Ritchie, J. D. & Smythe, D. K. 1984. Internal structure of the Erland Tertiary volcanic complex north of Shetland, revealed by seismic reflection. *Journal of the Geological Society*, **141**, 555–562.

Hald, N. & Waagstein, R. 1984. Lithology and geochemistry of a 2-km sequence of lower Tertiary tholeiitic lavas drilled on Suduroy, Faroe Islands (Lopra-1). *In*: Bertelsen, O., Noe-Nygaard, A. & Rasmussen, J. (eds) *The Deep Drilling Project 1980–1981 in the Faroe Islands*. Føroya Frodskaparfelag, Torshavn, 13–38.

Hansen, D. M. 2006. The morphology of intrusion-related vent structures and their implications for constraining the timing and intrusive events along the NE Atlantic margin. *Journal of the Geophysical Society, London*, **163**, 789–800.

Hansen, T. J. A. 2007. *Investigation of basalts east of the Faroe Islands, using geophysical logs and seismic methods*. Unpublished MSc thesis in Danish (abstract in English), University of Copenhagen, Institute of Geology.

Hansen, D. M. & Cartwright, J. A. 2006. Saucer-shaped sill with lobate morphology revealed by 3D seismic data: implications for resolving a shallow level sill emplacement mechanism. *Journal of the Geological Society*, **163**, 509–523.

Hansen, D. M., Cartwright, J. A. & Thomas, D. 2004. 3D seismic analysis of the geometry of igneous sills and sill junction relationships. *In*: Davis, R. J., Stewart, S. A., Cartwright, J. A., Lappin, M. & Underhill, J. R. (eds) *3D Seismic Technology: Application to the Exploration of Sedimentary Basins*. Geological Society, London, Memoirs, **29**, 199–208.

Heinesen, M. V. 1987. *Nedre tertiære basalt breccier og undervandslavastrømme, sydlige Disko, Vestgrønland. Strukturelle, petrografiske og mineralogiske studier*. Unpublished MSc thesis in Danish. University of Copenhagen, Institute of Geology.

Hopper, J. R., Dahl-Jensen, T. *et al.* 2003. Structure of the SE Greenland margin from seismic reflection and refraction data: implications for nascent spreading center subsidence and asymmetric crustal accretion during North Atlantic opening. *Journal of Geophysical Research*, **108**, 2269.

Jerram, D. A. 2002. Volcanology and facies architecture of flood basalts. *In*: Menzies, M. A., Klemperer, S. L., Ebinger, C. J. & Baker, J. (eds) *Volcanic Rifted Margins*. Geological Society America, Boulder, CO, Special Papers, **362**, 119–132.

Jones, J. G. & Nelson, P. H. H. 1970. The flow of basalt lava from air into water – its structural expression and stratigraphic significance. *Geological Magazine*, **107**, 13–19.

Kiørboe, L. & Petersen, S. A. 1995. *Seismic Investigation of the Faeroe Basalts and their Substratum*. Geological Society, London, Special Publications **90**, 111–123.

Larsen, L. M., Waagstein, R., Pedersen, A. K. & Storey, M. 1999. Trans-Atlantic correlation of the Paleogene volcanic successions in the Faroe Islands and East Greenland. *Journal of the Geological Society, London*, **156**, 1081–1095.

Leathard, M., McHugo, S. & Hoare, R. 2009. Imaging below Faroese Basalts using over/under acquisition. *In*: Varming, T. & Ziska, H. (eds) *Faroe Islands Exploration Conference, Proceedings of the 2nd Conference*, 61–66. Føroya Fróðskaparefelag, The Faroese Academy of Sciences. *Annales Societatis Scientiarum Færoensis Supplementum*, **50**, Tórshavn.

Naylor, P. H., Bell, B. R., Jolley, D. W., Durnall, P. & Fredsted, R. 1999. Paleogene magmatism in the Faroe–Shetland Basin: influences on uplift history and sedimentation. *In*: Fleet, A. J. & Boldy, S. A. R. (eds) *Petroleum Geology of North West Europe: Proceedings of the 5th Conference*, 545–558; doi: 10.1144/0050545.

Nelson, C. E., Jerram, D. A., Single, R. T. & Hobbs, R. W. 2009. Understanding the facies architecture of flood basalts and volcanic rifted margins. *In*: Varming, T. & Ziska, H. (eds) *Faroe Islands Exploration Conference, Proceedings of the 2nd Conference*, 84–103. Føroya Fróðskaparefelag, The Faroese Academy of Sciences. *Annales Societatis Scientiarum Færoensis Supplementum*, **50**, Tórshavn.

Nielsen, L., Boldreel, L. O. & Surlyk, F. 2004. Ground-penetrating radar imaging of carbonate mound structures and implications for interpretation of marine seismic data. *American Association of Petroleum Geologists Bulletin*, **88**, 1069–1082.

Passey, S. R. 2004. *The volcanic and sedimentary evolution of the Faroe Plateau Lava Group, Faroe Islands and the Faroe–Shetland basin, NE Atlantic*. PhD thesis, University of Glasgow.

Passey, S. R. 2007. Geological field guide to the central and north-eastern Islands, Faroe Islands. *Basaltic Plains Volcanism and Sedimentation. 1st Jóannes Rasmussen Conference: Evolution of Basaltic Provinces*, 28 August 2007, Jarðfeingi.

Passey, S. R. 2009. Recognition of a faulted basalt lava flow sequence through the correlation of stratigraphic marker units, Skopunarfjørdur, Faroe Islands. *In*: Varming, T. & Ziska, H. (eds) *Faroe Islands Exploration Conference Proceedings of the 2nd Conference*, 174–204. Føroya Fróðskaparefelag, The Faroese Academy of Sciences. *Annales Societatis Scientiarum Færoensis Supplementum*, **50**, Tórshavn.

Passey, S. R. & Bell, B. R. 2007. Morphologies and emplacement mechanisms of the lava flows of the Faroe Islands Basalt Group, Faroe Islands, NE Atlantic Ocean. *Bulletin of Volcanology*, **70**, 139–156.

Passey, S. R., Jolley, D. W. & Bell, B. R. 2006. Lithostratigraphic framework for the Faroe Islands Basalt Group, NE Atlantic. *A George P. L.*

Walker Symposium on Advances in Volcanology, 12–17 June, Reykholt, Iceland.

Peate, I. U., Larsen, M. & Lesher, C. E. 2003. The transition from sedimentation to flood volcanism in the Kangerlussuaq Basin, East Greenland: basaltic pyroclastic volcanism during initial Paleogene continental break-up. *Journal of the Geological Society, London*, **160**, 759–772.

Pedersen, A. K. & Dueholm, K. S. 1992. New methods for the geological analysis of Tertiary volcanic formations on Nuussuaq and Disko, central West Greenland, using multi-model photogrammetry. *In*: Dueholm, K. S. & Pedersen, A. K. (eds) *Geological Analysis and Mapping Using Multi-model Photogrammetry*. Rapport Grønlands Geologiske Undersøgelse, **156**, 19–34.

Pedersen, A. K., Larsen, L. M., Pedersen, G. K. & Dueholm, K. S. 2006. Five slices through the Nuussuaq Basin, West Greenland. *Geological Survey of Denmark and Greenland Bulletin*, **10**, 53–56.

Planke, S. 1994. Geophysical response of flood basalts from analysis of wire-line logs. Ocean Drilling Program Site 642, Vøring volcanic margin. *Journal of Geophysical Research*, **99**, 9279–9296.

Planke, S. & Eldholm, O. 1994. Seismic response and construction of seaward dipping wedges of flood basalts: Vørring volcanic margin. *Journal of Geophysical Research*, **99**, 9263–9278.

Planke, S., Alvestad, E. & Eldholm, O. 1999. Seismic characteristics of basaltic extrusive and intrusive rocks. *The Leading Edge*, **18**, 342–348.

Planke, S., Symmonds, P. A., Alvestad, E. & Skogseid, J. 2000. Seismic volcanostratigraphy of large-volume basaltic extrusive complexes on rifted margins. *Journal of Geophysical Research*, **105**, 19335–19351.

Planke, S., Rasmussen, T., Rey, S. S. & Myklebust, R. 2005. Seismic characteristics and distribution of volcanic intrusions and hydrothermal vent complexes in the Vøring and Møre basins. *In*: Doré, A. G. & Vining, B. A. (eds) *Petroleum Geology: North-West Europe and Global Perspectives. Proceedings of the 6th Petroleum Geology Conference*, 883–844; doi: 10.1144/0060883.

Rasmussen, J. & Noe-Nygaards, A. 1969. Beskrivelse til geologiske Kort over Færøerne. *Geological Survey of Denmark* I, series 24.

Rasmussen, J. & Noe-Nygaard, A. 1970. Geology of the Faroe Islands. *Danmarks Geologiske Undersøgelser* I, serie 25.

Rasmussen, J. & Noe-Nygaard, A. 1990. The origin of the Faroe Islands. Ministry of the Environment, Geological Survey of Denmark.

Ritchie, J. D., Gatliff, R. W. & Richards, P. C. 1999. Early Tertiary magmatism in the offshore NW UK margin & surrounds. *In*: Fleet, A. J. & Boldy, S. A. R. (eds) *Petroleum Geology of North-West Europe: Proceedings of the 5th Conference*, 573–584; doi: 10.1144/0050573.

Saunders, A. D., Fitton, J. G., Kerr, A. C., Norry, M. J. & Kent, R. W. 1997. The North Atlantic igneous province. *In*: Mahoney, J. J. & Coffin, M. L. (eds) *Large Igneous Provinces: Continental, Oceanic and Planetary Flood Volcanism*. American Geophysical Union, Geophysical Monographs, **100**, 45–93.

Smallwood, J. R. & Harding, A. 2009. New seismic imaging methods, dating, intrusion style and effects of sills: a drilled example from the Faroe-Shetland Basin. *In*: Varming, T. & Ziska, H. (eds) *Faroe Islands Exploration Conference Proceedings of the 2nd Conference*, 104–123. Føroya Fróðskaparefelag, The Faroese Academy of Sciences. *Annales Societatis Scientiarum Færoensis Supplementum*, **50**, Tórshavn.

Varming, T. 2009. Results from drilling of the 1st license round wells in the Faroese part of Judd Basin. *In*: Varming, T. & Ziska, H. (eds) *Faroe Islands Exploration Conference Proceedings of the 2nd Conference*, 346–363. Føroya Fróðskaparefelag, The Faroese Academy of Sciences. *Annales Societatis Scientiarum Færoensis Supplementum*, **50**, Tórshavn.

Waagstein, R. 1988. Structure, composition and age of the Faroe basalt plateau. *In*: Morton, A. C. & Parson, L. M. (eds) *Early Tertiary Volcanism and the Opening of the NE Atlantic*. Geological Society, London, Special Publications, **39**, 225–238.

Waagstein, R. 1996. Early Tertiary volcanism in the Faroe region and the initiation of the North Atlantic mantle plume. *International Lithosphere Programme*, August–September, Tórshavn, Faroe Islands.

Exploring for gas: the future for Angola

C. A. FIGUEIREDO,[1] L. BINGA,[1] J. CASTELHANO[1] and B. A. VINING[2]

[1]*Sonangol Gas Natural, Av. Lenin no. 58, 5/6 Andar, Luanda, Angola*
(e-mail: carlos.a.figueiredo@sonangol.co.ao)
[2]*Gaffney, Cline & Associates, Bentley Hall, Blacknest, Alton GU34 4PU, UK*

Abstract: Offshore Angola is a world-class petroleum province with oil production in excess of 2 million barrels per day in 2008. Exploration activity in the offshore commenced in earnest in the early 1990s, with the contractual terms permitting only the production of oil. There is no provision for the production of gas. However, there have been several discoveries of associated and non-associated gas, by accident, in the search for oil. Gas will play a significant role in the future of Angola. A new gas business is emerging. Demand from a new liquefied natural gas (LNG) plant, the emerging Soyo industrial complex and the growing domestic and industrial sectors have created a business environment whereby near-term gas exploration, development and production is a critical need for the future of Angola. What is the gas resource potential of Angola: from mature basins to new frontiers? Is there potential for new giant gas fields? Conceivably, there are multiple petroleum system opportunities. This paper provides some insights into the petroleum geoscience exploration efforts being made by Angola's national gas company, Sonangol Gas Natural. The focus of this paper will be on the exploration for non-associated gas. The associated gas potential of Angola is relatively well known and will not be discussed. Sonangol Gas Natural has an extensive offshore seismic and well database. This is characterized by a series of merged, high-quality 3D seismic surveys that facilitate the application of the most advanced technologies. In excess of 130 offshore wells have encountered gas at multiple stratigraphic levels. However, the interior basins are a truly frontier province and data are limited to a regional gravity and magnetic dataset and surface rock outcrops. The conventional thermogenic petroleum systems of the offshore and the adjacent onshore Lower Congo, Kwanza/Benguela and Namibe basins are relatively well known. These comprise a Late Cretaceous and Cenozoic post-salt petroleum system, and an Early Cretaceous and Late Jurassic pre-salt petroleum system. The primary emphasis of the exploration effort in these basins is to better understand and to delineate the gas kitchens for the known source horizons, for example Malembo, Iabe (post-Aptian salt) and Bucomazi (pre-Aptian salt) formations. Biogenic gas plays constitute an unconventional petroleum system that requires different exploration concepts from those of conventional thermogenic petroleum systems. The principal challenges are to identify palaeo-gas hydrates, reservoired in Miocene clastics, to define the biogenic gas fairway and to evaluate the play elements of reservoir, trap and seal. The geological conditions necessary to establish a biogenic gas play in the deepwater and ultra-deepwater will be examined. The interior Owambo, Okawango and Kassanje Basins have no proven petroleum systems. However, some new play concepts of Neoproterozoic age will be presented using possible analogues worldwide. The future demand for gas will increase globally. New developments in LNG will change the world of gas. Angola will be a key player.

Keywords: gas, biogenic, petroleum systems, Angola

Exploration for gas in Angola is only now commencing in earnest. However, Angola has long been known as a hydrocarbon province. In 1767, 59 barrels of viscous oil were exported to Lisbon. The first exploration well, Dande-1, was drilled in 1915 to a total depth of 602 m, but was a dry hole. There then followed a period of 40 years when 30 exploration wells were drilled in the onshore Kwanza Basin, before the first commercial oil discovery, the Benfica field, close to the capital Luanda, was made in 1955 by Missao de Pesquisa de Petroleos, an affiliate of the Belgian Petrofina group. Subsequently, exploration activity proceeded principally onshore and focused in the Cabinda area. The first offshore well, 96-1X, was drilled in 1966, resulting in the discovery of the Limba field.

The state oil company, Sonangol, was formed in 1978 and actively participated in 2D seismic acquisition and later drilling and oil production. The award of offshore licence blocks commenced in 1980 in the shallow waters (less than 200 m water depth) in the Lower Congo Basin. The contractual terms permitted only the production of oil. There was no provision for the production of gas. Exploration activity in shallow waters continued to increase in pace throughout the 1980s with discoveries in the Cretaceous Albian, Pinda carbonates and Cenomanian clastics.

Deepwater exploration (between 200 and 2000 m water depth) commenced in 1993 with the award of Blocks 16 and 17 in the Lower Congo Basin followed by the award of other deepwater blocks in the mid-1990s. The first well in deepwater, Bengo-1 in 1994, resulted in an oil and associated gas discovery that was sub-commercial at that time. However, it unveiled the presence of a Cenozoic petroleum system that is characterized by what were known as direct hydrocarbon indicators (DHIs) on seismic data. The discovery of the giant Girassol field in 1996, in a water depth of 1400 m, confirmed Angola as an emerging world-class petroleum province. Other exciting discoveries in the Lower Congo Basin soon followed, including: Kuito (Block 14), Dalia (Block 17) in 1997; Kissanje (Block 15) in 1998; and Platina and Plutonio (Block 18) in 1999. These discoveries enabled Angola to exceed production of 1 million barrels per day of oil (MBPD) in 2001.

Exploration moved into progressively deeper water depths, the first ultra-deepwater (between 2000 and 3000 m water depth) licence awards being made in 1998.

Today, Angola is well recognized as a world-class oil province with oil production exceeding 2 MBPD. It is well positioned to supply gas to countries surrounding the Atlantic basin, notably in Europe and North America. What does the future hold for gas?

Associated gas and some non-associated gas has been discovered, by accident, in the search for oil. Undoubtedly, Angola will continue to develop as a world-class oil province. However, what is the potential for gas? From a global perspective, demand for energy worldwide continues to increase. World consumption of gas is approximately 100×10^{12} SCF/year. Recent advances in liquefied natural gas (LNG) technologies are changing the world of gas, for the first time providing flexibility of supply. It is predicted that world LNG production may double by 2016. Angola will be a key player in the future world gas business. As mentioned above, the current Petroleum Sharing Contracts (PSC) are for oil; there is no provision for gas production. A contractual framework for gas has to be established in the near term. First production of LNG in Angola is scheduled for 2012, supplied solely by associated gas.

This paper provides some insights into the petroleum geoscience efforts being made by Angola's national gas company, Sonangol Gas Natural (Sonagas), in exploring for gas. Recently founded in 2004, Sonagas is charged to explore, develop and produce gas to address the growing demand. This demand comes from a new LNG plant and the emerging Soyo industrial complex in the northern coastal fringe, and the growing number of industrial and domestic consumers.

The focus of this paper is on the challenges facing the effective and efficient exploration for non-associated gas in a world-class prolific oil province. The associated gas potential of Angola is relatively well known and will not be discussed.

The potential for gas throughout Angola, both offshore and onshore, will be reviewed. Following a discussion of the regional tectonostratigraphic framework that underpins play definition, the potential for gas in the conventional thermogenic Pre-Aptian salt and Post-Aptian salt plays will be investigated. These are relatively emerging to mature plays. The possibility of an unconventional biogenic petroleum system, with multiple TCF resource potential in frontier plays, will then be examined in the deep and ultra-deepwater. Finally, the potential for gas in the truly frontier interior basins of Angola, from highly conceptual Neoproterozoic plays, will also be addressed, in the first instance using a global analogue approach.

Regional framework

A location map of the sedimentary basins of Angola is shown in Figure 1. These comprise the predominantly offshore Lower Congo Basin, the Kwanza/Benguela Basin and the Namibe Basin. The Lower Congo Basin is currently considered to be the most prospective. The modern South Atlantic passive margins started their evolution as part of a continental rift system that developed in the southern parts of the Gondwana super-continent at the Jurassic–Cretaceous boundary (Brice *et al.* 1982; Ala & Selley 1997). The interior basins of Angola are also shown in Figure 1. The location of offshore blocks in shallow, deep and ultra-deepwater and onshore blocks is shown.

A Mesozoic and Cenozoic stratigraphic summary of potential gas-prone petroleum systems is shown in Figure 2. The principal gas plays are highlighted by the red gas symbol. The source rocks of the Lower Congo Basin and the Kwanza/Benguela Basin are discussed in Burwood (1999). Thermogenic petroleum systems are proven in the Pre-Aptian salt plays and the Post-Aptian salt plays. An unconventional biogenic gas petroleum system is postulated in deepwater, comprising Middle to Late Miocene and Plio-Pleistocene plays. In addition, a highly conceptual, unproven Neoproterozoic petroleum system is being investigated in the interior basins, principally the Okavango and Owambo Basins.

Sonagas has access to an extensive offshore seismic and well database. This is summarized in Figure 3. Digital seismic databases comprise in excess of 17 000 km of 2D seismic, shown in the central panel of Figure 3, and in excess of 58 000 km² of merged

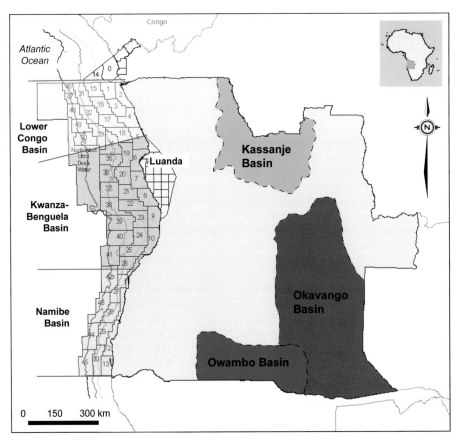

Fig. 1. Location map of the sedimentary basins of Angola.

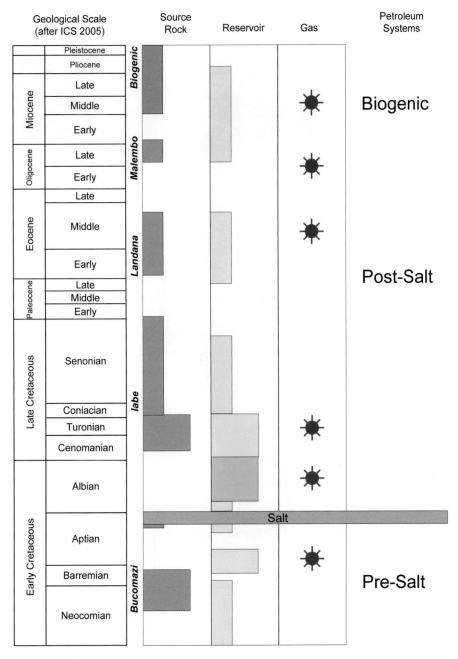

Fig. 2. Stratigraphic summary of potential gas-prone petroleum systems.

3D seismic surveys as shown in the panel on the right of Figure 3. The majority of the merged 3D seismic surveys are located in the Lower Congo Basin and provide an excellent quality dataset, both for regional play synthesis and prospect definition. The seismic datasets are of varying vintages and quality. Overall, seismic data quality in the Post-Aptian salt section is good to excellent. However, seismic data quality in the Pre-Aptian salt section is poor to fair and is a current limitation to prospect generation and maturation efforts. There is access to in excess of 450 wells of which 130 have gas shows, the latter shown in red in Figure 3. The number of gas well discoveries by block and by play are shown in Figure 4. Gas has been encountered in all the principal plays. The shallow water Albian Pinda carbonate play shows the highest number of gas discoveries. These gas wells lie principally in the Lower Congo Basin. Their geographic distribution, in the context of the interpreted limits of their respective plays, is shown in Figure 5. However, it must be remembered that all drilling activity has intentionally avoided the discovery of gas by focusing on predicted oil-prone areas. The discovery of gas has occurred by accident.

The deliberate search for gas, as mentioned above, is focused on three distinctly different types of petroleum system. Firstly, conventional thermogenic petroleum systems, where all the gas discoveries have been made to date, will be reviewed.

Thermogenic petroleum systems

Pre-Aptian salt plays

An understanding of the basin configuration for the Pre-Aptian salt plays, in particular the depth to basement, is somewhat tenuous due to poor seismic imaging. Gravity and magnetic studies are currently underway to better define basin architecture and the depth to crystalline basement. However, the main structural style is believed to be rotated fault block geometries, reflecting a syn-rift lower section of Late Triassic to Jurassic age and an Early Cretaceous sag phase

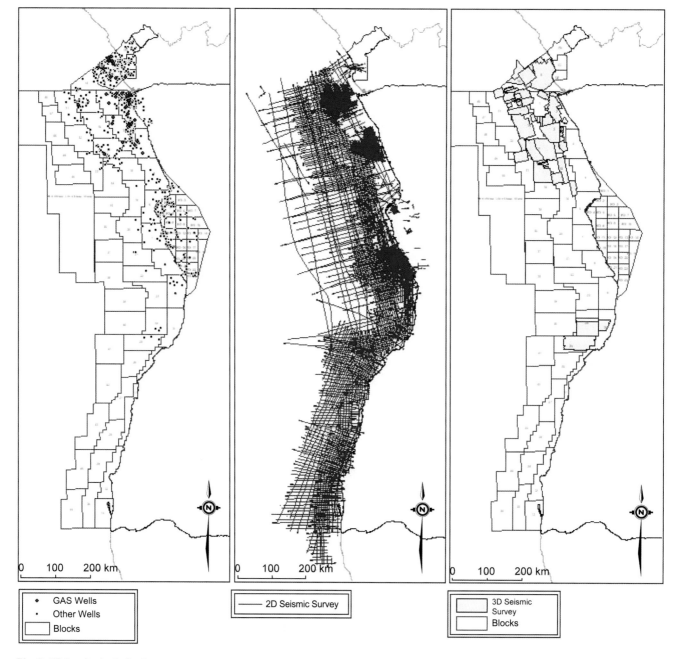

Fig. 3. Well and seismic database.

represented as a sedimentary wedge. The active rifting has been divided into three separate phases: Berriasian, Hauterivian and Late Barremian to Early Aptian (Karner & Driscall 1999). The key facies associations characterizing the syn-rift areas are continental-fluviatile and lacustrine. The sag phase sedimentary wedge has an increasing marine influence with some carbonate deposition. A schematic play cross-section of the Pre-Aptian salt plays in the Kwanza/Benguela Basin is shown in Figure 6.

The principal source rock for the Pre-Aptian salt plays is the Neocomian–Barremian Bucomazi Formation. This is a prolific super source rock that is believed to be areally extensive. However, additional source horizons exist in the syn-rift section that are predominantly lacustrine in facies.

Extensive hydrocarbon maturation studies have been performed, calibrated where possible to well data. A generalized present-day basal heat flow map is shown in Figure 7. Two areas of high heat flow are shown in red. Maturation modelling of the Bucomazi Formation present day delineates gas kitchens corresponding to the two areas of high heat flow (Fig. 7) and a further conjectural area offshore Luanda. These interpreted gas kitchen areas for the Bucomazi Formation present day are shown in Figure 8.

Migration of hydrocarbons from pre-salt source rocks is driven by buoyancy and believed to take place vertically up fault planes and laterally updip where there is an effective stratigraphic conduit. These hydrocarbon migration pathways are shown schematically in Figure 6.

Exploration for pre-salt gas plays is in progress, focused in the vicinity of the predicted gas kitchens (Fig. 8), and encouraged by the exciting recent discoveries in the conjugate margin offshore Brazil. The latter discoveries are used as analogues to aid the interpretation of structural style and reservoir distribution in the Angolan margin.

The leads and prospects identified to date have a variety of structural styles, for example antiformal features, fault-related closures and combination structural and stratigraphic traps. Many of the large leads lie below the Aptian salt. As with sub-salt plays in

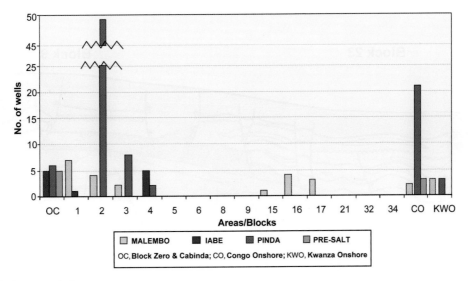

Fig. 4. Gas discoveries by area and play.

other areas of the world, such as the Gulf of Mexico, seismic imaging is challenging. In many of the seismic datasets offshore Angola there is poor seismic imaging beneath salt. Detailed velocity analyses have revealed that, in areas of massive salt, large features identified in the seismic time domain are not present in the depth domain. The effect of salt pullups is critical in lead generation. In this regard, it is imperative to clearly differentiate between massive salt and thin salt sills, below which there is a clastic stratigraphic section, albeit masked by poor seismic data quality. In addition, leads are being generated in the inter-salt structurally high areas, the areas beneath the salt affording the structural lows for trap definition.

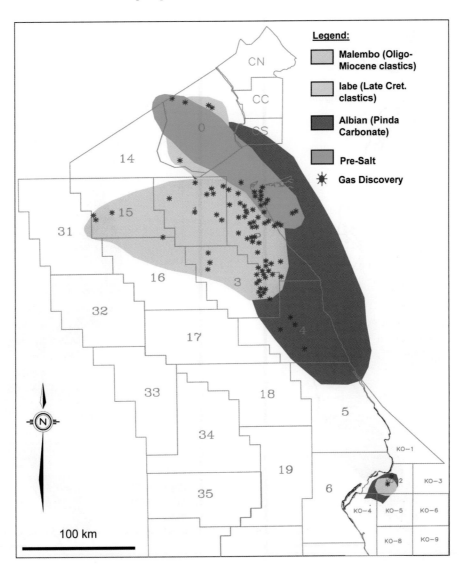

Fig. 5. Lower Congo Basin gas discoveries by play.

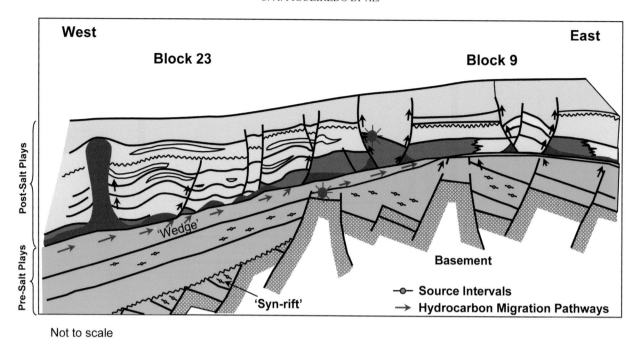

Fig. 6. Schematic play cross-section Kwanza/Benguela basin.

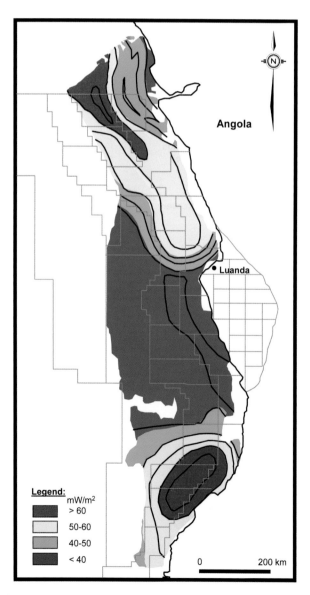

Fig. 7. Generalized basal heat flow.

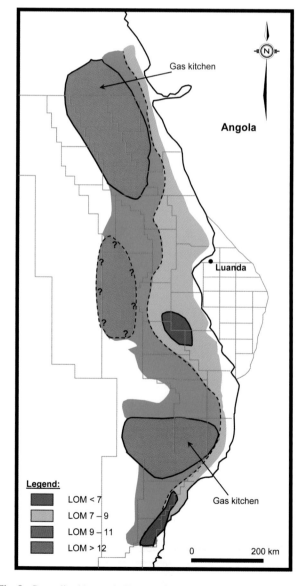

Fig. 8. Generalized base salt (Bucomazi Fm.) maturity present day.

The salt provides a regional super seal. Invariably, a sedimentary wedge, mentioned earlier, directly underlies the salt and is the primary exploration objective. In these areas, a syn-rift section is also sometimes present. An intra-formational seal(s) is therefore required to form stacked objectives. The presence of intra-formational seals may be problematic, particularly in areas of lacustrine facies, where the potential sealing horizon is predicted to be thin and discontinuous. A detailed knowledge of the nature of the depositional systems and facies associations is therefore required to facilitate the distribution of intra-formational seals.

The nature and distribution of the depositional systems for the pre-salt sedimentary wedge and syn-rift sequences, as it relates to seal and reservoir, is highly conceptual and is, in large part, driven by an interpretation of the mega-regional tectono-stratigraphic framework in the Mesozoic at the time of continental breakup. The principal marine influence is from the SW. The reservoir in the syn-rift Jurassic lacustrine facies is believed to be discontinuous and exploration is focused on the prediction of possible fluviatile systems, not identifiable on the current seismic datasets. Reservoir distribution in the Early Cretaceous sedimentary wedge is believed to have a more marine influence and the reservoir intervals are predicted to be better developed and more widespread. Well control in the pre-salt section is limited, primarily to the Lower Congo Basin. Long distance extrapolation of the depositional models is therefore required for pre-salt exploration in the Kwanza/Benguela and Namibe Basins.

Post-Aptian salt plays

Exploration for gas in thermogenic Post-Aptian salt plays is focused on identifying windows in the salt through which gas, derived from Pre-Aptian salt source rocks, enters the Post-Aptian salt petroleum system. Predicted gas migration conduits, as in the pre-salt section, are vertically up faults and laterally updip contingent on a stratigraphic conduit being present (Fig. 6).

Several post-salt source rocks have been identified: the Cenomanian–Senonian Iabe Formation; the Eocene Landana Formation and the Late Oligocene Malembo Formation as shown in Figure 2. A generalized present-day level of maturation map for the Iabe Formation source rocks is shown in Figure 9. A further gas kitchen exists derived from source rocks of the Malembo Formation. This is of limited areal extent and coincident in area with that shown in red in Figure 9. These stacked, areally constrained Post-Aptian salt gas kitchens immediately overlie that of the Pre-Aptian salt source rocks in the Lower Congo Basin (Fig. 8), thereby providing an increased focus area for gas exploration in this region.

Structural trap geometries are largely influenced by salt tectonics. The latter vary in style and intensity in the offshore areas, from low relief salt swells resulting in anticlinal closures in the shallow water areas, to diapirs and sills in the deeper water areas. Raft tectonics, a term first introduced by Burollet (1975) for the Kwanza Basin, are prevalent in the more proximal, listric fault areas (Duval et al. 1992; Lundin 1992; Valle et al. 2001), extension being taken up in the more distal areas by compressional, thrust features. Examples of the 3D geometry and displacement configuration of fault arrays from a raft system in the Lower Congo Basin are discussed in detail in Dutton et al. (2004). The controls on Miocene sand deposition during gravity-driven extension in Block 4 are well documented in Anderson et al. (2000). Combination structural/stratigraphic and stratigraphic traps are present; their identification and maturation relying on the application of advanced seismic technologies.

The Post-Aptian salt stratigraphic section, in the most part, possesses the physical properties conducive to the seismic identification of gas. Seismic amplitude anomalies may indicate the

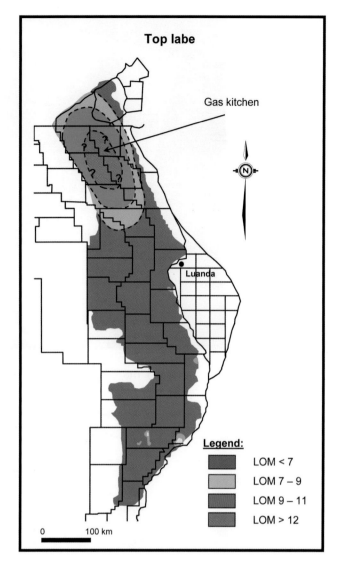

Fig. 9. Generalized present day level of maturity top Tabe formation.

presence of gas in the immediate vicinity of fault planes and in amalgamated channel complexes. For example, this is illustrated in Block 15 by a prospect, defined by high seismic amplitudes, shown on the seismic line A–A′ in Figure 10. A seismic flat spot, indicating a possible gas–water contact and conformable to the structural spill point, may be present. The distribution of this seismic anomaly is delineated by the high amplitude response shown in the time slice in the upper part of Figure 10. The reservoir is interpreted as an amalgamated Miocene channel complex. Gas charge is predicted by migration up a bounding fault from a pre-salt gas kitchen in a structurally high area. There is crestal faulting and possible seismic evidence of a gas migration trail to the seabed resulting in pock marks. Similar seismic indications of gas are observed adjacent to fault planes and as amalgamated channel complexes in Block Zero, as shown within the green ellipses in Figure 11.

Exploration for gas in the Post-Aptian salt section is further focused in areas where there is evidence of the presence of pock marks on the seabed. These latter areas are readily identifiable on the extensive quilt of merged 3D seismic surveys (Fig. 3) Pock marks are primarily the result of escape of gas to the seabed. However, they may also be formed by the escape of fluids, principally water.

A combination of multiple seismic and geological indicators for gas reduces uncertainty and risk in the exploration effort. Current

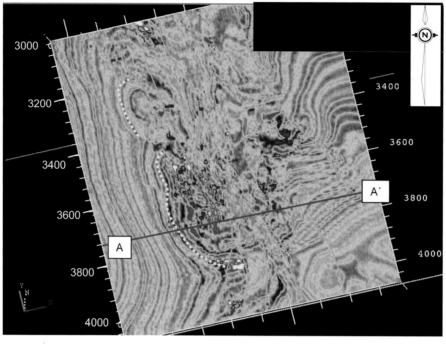

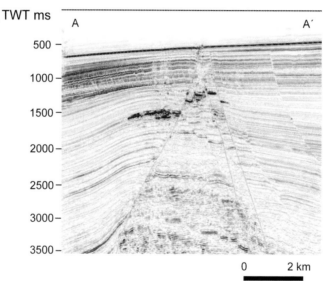

Fig. 10. Block 15 post-salt gas plays.

technologies such as neural networks are currently being applied to the identification and subsequent maturation of Post-Aptian salt thermogenic prospects and unconventional biogenic gas prospects. These technologies are predicated on seismic attribute responses, such as seismic amplitude, being associated with gas-charged sands in a particular area. The neural network technique then searches for similar responses throughout the seismic dataset. Numerous gas migration trails have been identified.

Although exploration focus is in the gas kitchen areas, exploration methodologies use the play as the fundamental unit of analysis. The Post-Aptian salt traps are structural, stratigraphic and combination traps. Reservoir prediction in deepwater depositional systems leverages seismic identification of reservoir facies. Exploration techniques include the reconstruction of seismic geomorphologies at times of large sea-level lowstands, for example Middle Miocene and Middle Oligocene, when large influxes of coarse clastics, principally transported by the Congo and Kwanza rivers, are predicted to dominate the depositional systems. Seismic RMS amplitude and variance displays, derived from 3D seismic datasets, are successful in defining these amalgamated channel complexes and predicting reservoir distribution, architecture and quality. The major amalgamated channel complexes of Middle Miocene and Middle Oligocene age are invariably located off-structure, infilling topographic lows adjacent to the lowside of listric faults, and possess a major stratigraphic trapping component. The geometry and evolution of deepwater submarine channels, particularly in offshore Angola, are described extensively in Kolla et al. (1998, 2001), Anderson et al. (2000), Mayall & Stewart (2000), Sikkema & Wojcik (2000), Abreu et al. (2003), Deptuck et al. (2003), Fonnesu (2003) and Gee & Gawthorpe (2007). Further insights into the distribution of these types of depositional systems on the Angolan and Brazilian margins may be obtained from their gravity signatures (Dickson et al. 2003). These channel complexes are illustrated by a prospect in Block 2 shown in Figure 12. The dark blue area is a fault plane, juxtaposed to the lowside of which are the

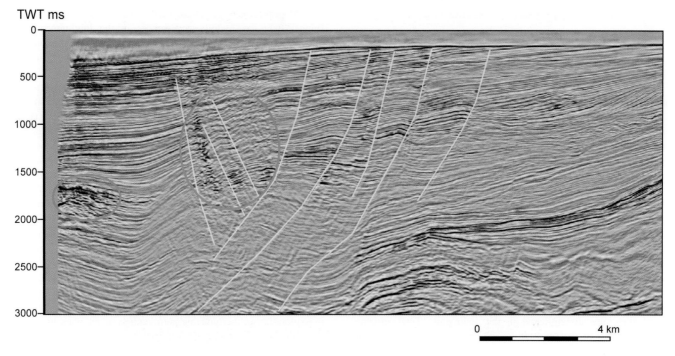

Fig. 11. Block zero seismic amplitude gas indications.

interpreted amalgamated channel complexes of Middle Miocene and Middle Oligocene age, shown as yellow and red seismic amplitudes.

Prospect generation and maturation efforts are therefore focused on placing leads in their regional play context, in particular with respect to the environment of deposition and in proximity to gas kitchens. Risk mitigation in the prospect maturation process comprises the application of amplitude versus offset (AVO) and electromagnetic (EM) techniques.

These exploration techniques for trap definition, reservoir presence, quality and pore-fill are not wholly confined to the search for gas in thermogenic petroleum systems. They can be applied to the search for biogenic gas prospects. However, the search for biogenic gas requires a different exploration approach.

Fig. 12. Block 2 thermogenic petroleum systems: post-salt.

Biogenic gas unconventional petroleum system

The presence of an unconventional biogenic gas petroleum system in deepwater offshore Angola is currently being investigated. A possible analogue is the Reliance Industries Limited giant D1–D3 gas field in the Krishna Godavari Basin, offshore east coast of India, which commenced first production in 2009.

The relative volumes of gas produced by organic source rocks, as an integral of time and temperature, is shown schematically in red in Figure 13 (Rice & Claypool 1981, p. 6, fig. 1). Thermogenic gas, as discussed above, is generated from source rocks in the late mature oil window and in the overmature stage. Biogenic gas can be generated in large quantities in the shallow stratigraphic section in areas immature for thermogenic gas. Exploration for biogenic gas thereby significantly increases the prospective area for gas.

Biogenic gas generation requires a source rock of microbial origin. The microbes produce methane through metabolic activities in the sediment. At low temperatures in deepwater, the resultant methane forms gas hydrates. Seafloor microbes may produce biosurfactants that are concentrated on bubbles which, with upward fluid flow, are disseminated in the sediments. Total organic carbon (TOC) of the biogenic gas source rocks may be low, at 2–3%, disseminated through a shale succession. Alternatively, the TOC may be concentrated in thin, centimetre-scale, condensed zones. Figure 14 (Rice & Claypool 1981, p. 7, fig. 2) shows the vertical microbial ecosystem succession for methane generation. Biogenic methane generation requires anaerobic conditions, immediately below the sulphate reduction zone and only a few metres below the sediment–water interface.

Exploration for biogenic gas in Angola is predicated on identifying palaeo-gas hydrate occurrences in the low temperature conditions found in deepwater. These exploration methodologies are similarly being investigated, as mentioned above, in the Krishna–Godavari Basin, India (Kundu *et al.* 2008).

The present-day stability field for gas hydrates is temperature- and pressure-dependent. The relationship between methane solubility, depth of burial and water depth is illustrated in Figure 15 (Rice & Claypool 1981, p. 10, fig. 6). The line of gas hydrate stability is also shown (Fig. 15) below which gas hydrates break down to release biogenic methane into the surrounding sediments. The base of the gas hydrate zone and the possible top of the free gas zone is characterized seismically by a bottom simulating reflector (BSR). Shallow BSRs on the Angola margin and their relationship to gas and gas hydrates are discussed in detail in Nouze & Baltzer (2003). This biogenic methane free gas is therefore available for migration and subsequent entrapment, contingent on migration conduits transecting this gas hydrate-free gas transition zone. Entrapment is envisaged primarily as stratigraphic traps where an effective updip, base and top seals are present. Reservoir is therefore considered to be the key play element to be investigated. As in exploration in thermogenic petroleum systems, pock marks on the seabed are direct indicators of fluid escape features and particularly gas. Figure 16 graphically shows the dimensions and widespread occurrence of pock marks in Block 31. In this example, it is recognized that the gas may be biogenic, thermogenic or a combination of both. Pock marks, although indicating the presence of gas on the seabed, represent gas leakage from the subsurface. This may indicate a gas accumulation at depth, but could also indicate a gas leakage risk whereby the seismic amplitude response is a

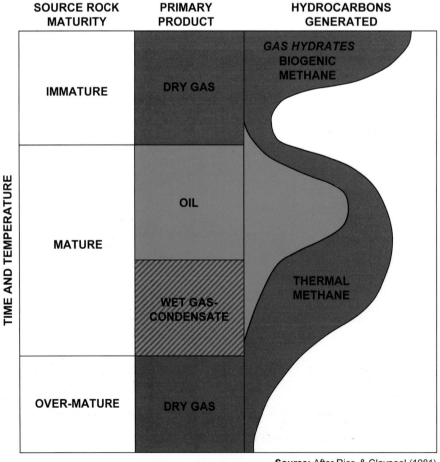

Fig. 13. Gas generation schematic.

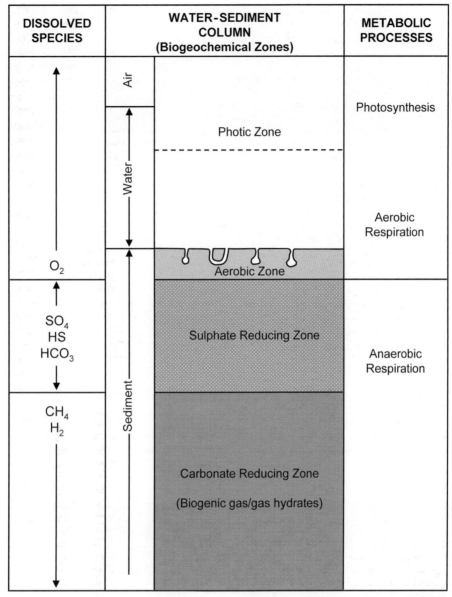

Fig. 14. Microbial ecosystem succession for methane generation.

consequence of residual gas saturations. The seismic line C–C', highlighted in Figure 16, is shown in cross-section view in Figure 17. The presence of two salt features is shown. The high seismic amplitude events in the shallow section are interpreted to be biogenic gas. Pock marks are seen as perturbations on the seabed.

In this overall context present day, the most favourable conditions for the entrapment of biogenic gas occur in 2000–4000 m water depth at approximately 1000 m below the seabed. In deepwater passive margin settings, such as offshore Angola, similar bathymetric and depth of burial relationships are predicted to have occurred from the Middle Miocene to the present day. Exploration in this unconventional petroleum system is focused in the ultra-deepwater blocks (2000–3000 m water depth) where the Middle to Late Miocene and Plio-Pleistocene sequences are at burial depths in excess of 2000 m. The depth of burial of the trap, as it relates to reservoir pressure, is of paramount importance in controlling the producibility characteristics, flow rates and recovery. The deeper targets, at higher reservoir pressures, will exhibit better flow rates and recovery factors. The sustained long-term production characteristics of these deepwater, shallow depth of burial, gas reservoirs is unknown. Reservoir depressurization resulting in the need for installation of compression facilities in the early life of a field can have a serious adverse effect on the commercial viability of the field.

A biogenic gas play fairway can then be defined both in geographic areal extent and within a vertical depth window. Exploration efforts are focused on the development of a sequence stratigraphic framework for the Middle to Late Miocene and Plio-Pleistocene sequences to determine, through the environment of deposition maps, the distribution of reservoir and seal in each time-stratigraphic equivalent unit. The principal reservoir targets are amalgamated, stacked channel sand complexes.

The identification of leads and prospects is facilitated by the seismic attribute response of some of these gas sands. Simple 2D seismic wedge models and fluid (gas) substitution studies indicate the seismic amplitude response to gas reservoirs of varying thickness and porosity characteristics within the play. Furthermore, seismic AVO and EM methodologies are used as risk mitigation techniques. The optimal methodology for trap definition is through mapping high-amplitude response.

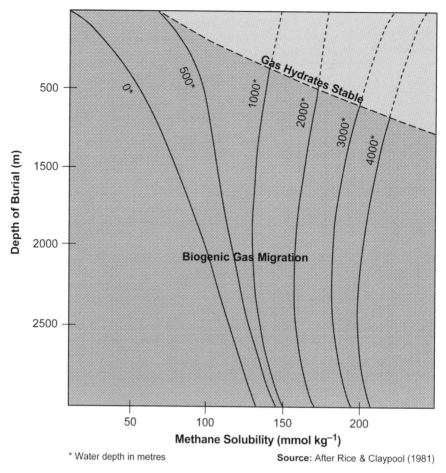

Fig. 15. Estimated methane solubility.

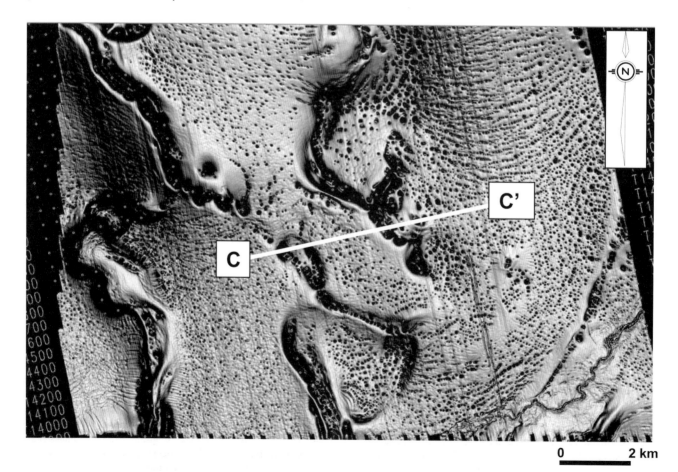

Fig. 16. Direct gas indicators: Block 31.

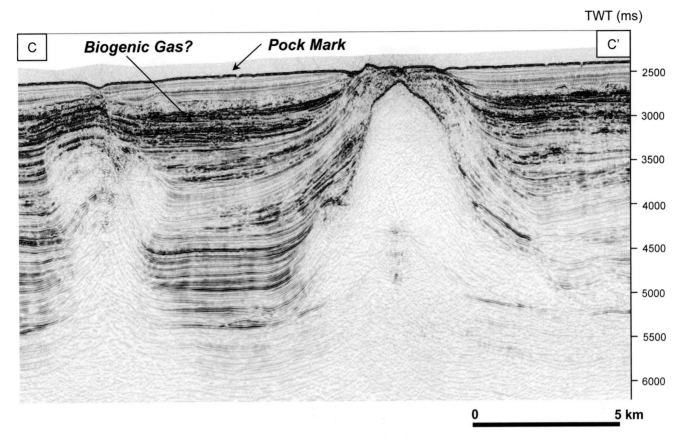

Fig. 17. Possible biogenic gas play: Block 31.

It is envisaged that biogenic gas prospects will be predominantly stratigraphic traps. The field size distribution for biogenic gas stratigraphic traps is expected to be larger than that of the post-salt plays, with structural, stratigraphic and combination traps, in the thermogenic petroleum system.

A third and final petroleum system to be discussed is a conceptual Neoproterozoic petroleum system that may be present in the interior basins of Angola.

Conceptual Neoproterozoic petroleum system

The interior basins of Angola, the Kassanje Basin, Okavango Basin and Owambo Basin cover a significant geographic area, as shown in Figure 1. The geology, as it pertains to possible petroleum systems, is poorly understood. There are no proven petroleum systems. There is a high level of uncertainty and it is high risk. The database to date is limited to gravity and magnetic data and surface geology. There are no seismic or well data.

The exploration approach currently in progress is two-fold. Firstly, an investigation of the plate tectonic setting in the Neoproterozoic, examining the terranes and possible depositional environments in a mega-regional Rodinia super continent context. Rodinia is shown schematically in the lower part of Figure 18. The interior basins of Angola are highlighted by the red star, juxtaposed to similar terranes in the Central Basin of the Democratic Republic of Congo and the Sao Francisco Basin in Brazil. The Neoproterozoic–Early Palaeozoic tectonostratigraphy and palaeogeography of the peri-Gondwanan terranes of West Africa and Brazil are discussed in detail in Murphy *et al.* (2004) and Nance *et al.* (2008). The second approach is the use of possible analogues worldwide. For example, these include the proven glaciogenic petroleum systems of the Infra-cambrian of North Africa, notably Algeria and Mauritania; the Middle East, in particular the Ghaba Salt Basin of Oman; the Volga–Urals and east Siberia Basins in Russia; the Sichuan Basin in China; the Rajasthan Basin in India; and the Amadeus, Officer and McArthur Basins in Australia.

A preliminary sediment thickness map, based on gravity and magnetic data, is shown in the upper part of Figure 18. A more detailed gravity and magnetic study, integrated with crustal structure, remote sensing data and surface geology data is in progress. At the time of publication, it is believed that the definition of the basin boundaries and depth to crystalline basement may change significantly. However, the southern part of the Okavango Basin and the eastern part of the Owambo Basin, located in the SE of Angola, are interpreted to have sediment thicknesses in excess of 8 km. Contingent on the presence of source rocks, this is sufficient sediment thickness to generate hydrocarbons. A schematic play cross-section, located A–A' on Figure 18, is shown in Figure 19. The crystalline basement is believed to be composed of Proterozoic metasediments. Neoproterozoic depositional systems infill and are influenced by this pre-existing topography. Source rocks are the Neoproterozoic Mulden Black Shales or the equivalent Lagoa de Jacare Formation in the Sao Francisco Basin in Brazil. Basin modelling in the Neoproterozoic is recognized as being problematic. Technological capability and input data are severely limited. For example, source rock kinetics are unknown. A chronostratigraphic framework, as it pertains to burial analysis, is highly conceptual. Age dating relies on geochronology isotope analyses. However, preliminary indications suggest that hydrocarbon charge may have occurred several hundred million years ago. Hydrocarbon retention is therefore considered a major risk. Reservoirs are the glaciogenic sandstones of the Marinoan and Sturtian glaciations, analogous to those cited from the Infra-cambrian above, and the Sansikwa carbonates of the intervening greenhouse periods. This latter carbonate play may comprise reservoirs analogous to the Neoproterozoic microbial–metazoan reefs similar to those found in the Nama Basin of southern Namibia (Wood 2005). These pinnacle reefs and platforms are over 300 m thick, in

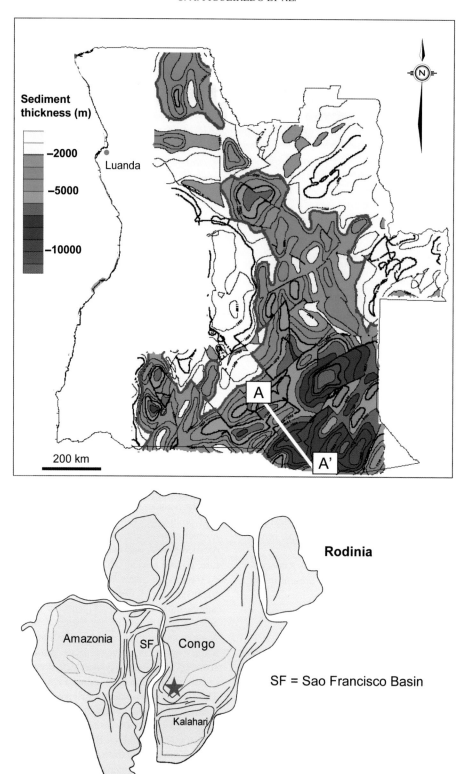

Fig. 18. A conceptual Neoproterozoic petroleum system?

excess of 7 km long and have the potential to be possible future giant fields.

A series of collaborative studies is in progress with NamCor in Namibia to investigate potential analogue petroleum systems and to determine their possible extension northwards into Angola.

Conclusions

In conclusion, energy demand, particularly for gas, is increasing globally. The advent of new technological developments in LNG will, for the first time, provide flexibility in supply. Angola will be a key player supplying this demand.

The exploration and exploitation of gas in Angola is in its infancy. From a petroleum geoscience perspective, there are multiple petroleum system opportunities. These comprise the emerging and mature conventional thermogenic Pre-Aptian salt and Post-Aptian salt petroleum systems of the offshore, renowned as a world-class oil province. Exploration for gas in these areas has the benefit of extensive, high quality seismic and well datasets. An unconventional biogenic gas petroleum system is postulated

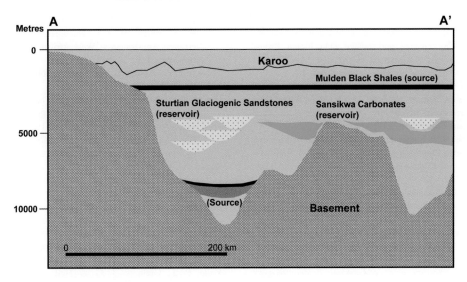

Fig. 19. Schematic Neoproterozoic play cross-section.

in deepwater that may extend beyond the boundaries of the thermogenic gas kitchens. Although unproven, the potential for gas in this frontier is large, principally in giant stratigraphic traps. Finally, the onshore interior Okavango and Owambo basins are vast – a true frontier. Data and knowledge are limited. However, conceptual Neoproterozoic petroleum systems are being developed using possible analogue settings from proven Neoproterozoic petroleum systems worldwide. As in the theme of the conference: '*from Mature Basins to New Frontiers*', Angola has both.

Sonagas has an ambitious discovery targets. This will require exploration for giant fields. A gas resource assessment of Angola is in progress suggesting potential in the tens to hundreds of TCF. The single most important issue to attain this potential is field size; do giant fields exist? Sonagas is confident that success exploring for gas in Angola will be unlocked by a combination of creativity and technology in petroleum geoscience.

The authors wish to thank the following current and former members of the Sonangol Gas Natural Exploration Department: S. Abiodun, A. Lasam, S. Dabeer, J. Gu, R. Rocca, S. Bastiot and A. Setiawan. We thank K. Black (GCA) for the graphics.

Delegate discussion

Question from John Argent (BG Group)

Is the current associated gas volume from the existing oil and serendipitous gas discoveries sufficient for multiple LNG trains? If so, when should we consider the deliberate exploration for gas in Angola?

Answer

The current associated gas volume from existing oil and gas discoveries is sufficient for the first train of LNG. Subsequent trains will rely on the exploration, development and production of new gas discoveries. The Soyo LNG facility is under construction and is due to be commissioned in 2012. It is therefore imperative to fast track the discovery, development and production of new gas volumes in preparation for a second train of LNG. In this regard, Sonangol Gas Natural and its partners are currently pursuing the deliberate exploration for gas in Angola.

References

Abreu, V., Sullivan, M., Pirmez, C. & Mohrig, D. 2003. Lateral accretion packages (LAPS): an important reservoir element in deep-water sinuous channels. *Marine and Petroleum Geology*, **20**, 631–648.

Ala, M. A. & Selley, R. C. 1997. *The West African Coastal Basins*. SADCC Energy Sector Technical and Administrative Unit, TAU, 129–150.

Anderson, J. E., Cartwright, J. A., Drysdall, S. J. & Vivian, N. 2000. Controls on turbidite sand deposition during gravity-driven extension of a passive margin: examples from Miocene sediments in Block 4, Angola. *Marine and Petroleum Geology*, **17**, 1165–1203.

Brice, S. E., Cochran, M. D., Pardo, G. & Edwards, A. D. 1982. Tectonics and sedimentation of the South Atlantic rift sequence; Cabinda, Angola. *Studies in Continental Margin Geology*, **34**, 5–18.

Burollet, P. F. 1975. Tectonique en radeaux en Angola. *Bulletin de la Societe Geologique de France*, **17**, 503–504; Les marge continentales et leur interet economique.

Burwood, R. 1999. Angola: source rock control for Lower Congo Coastal and Kwanza Basin petroleum systems. *In*: Cameron, N. R., Bate, R. H. & Clure, V. S. (eds) *The Oil and Gas Habitats of the South Atlantic*. Geological Society, London, Special Publications, **153**, 181–194.

Deptuck, M. E., Steffens, G. S., Barton, M. D. & Pirmez, C. 2003. Architecture and evolution of upper fan channel belts on the Niger Delta slope and in the Arabian Sea. *Marine and Petroleum Geology*, **20**, 649–676.

Dickson, W. G., Danforth, A. & Odegard, M. 2003. Gravity signatures of sediment systems: predicting reservoir distribution in Angolan and Brazilian basins. *In*: Arthur, T. J., MacGregor, D. S. & Cameron, N. R. (eds) *Petroleum Geology of Africa: New Themes and Developing Technologies*. Geological Society, London, Special Publications, **207**, 241–256.

Dutton, D. M., Lister, D., Trudgill, B. D. & Pedro, K. 2004. Three-dimensional geometry and displacement configuration of a fault array from a raft system, Lower Congo Basin, Offshore Angola: implications for the Neogene turbidite play. *In*: Davies, R. J., Cartwright, J. A., Stewart, S. A., Lappin, M. & Underhill, J. R. (eds) *3D Seismic Technology: Application to the Exploration of Sedimentary Basins*. Geological Society, London, Memoirs, **29**, 133–142.

Duval, B. C., Cramez, C. & Jackson, M. P. A. 1992. Raft tectonics in the Kwanza Basin, Angola. *In*: Jackson, M. P. A. (ed.) *Special Issue; Salt Tectonics. Marine and Petroleum Geology*. Butterworth/Geological Society, London, **9**, 389–404.

Fonnesu, F. 2003. 3D seismic images of a low sinuosity slope channel and related depositional lobe (West Africa deep-offshore). *Marine and Petroleum Geology*, **20**, 615–629.

Gee, M. J. R. & Gawthorpe, R. L. 2007. Early evolution of submarine channels offshore Angola revealed by three-dimensional seismic data. *In*: Davies, R. J., Posamentier, H. W., Wood, L. J. & Cartwright, J. A. (eds) *Seismic Geomorphology: Applications to Hydrocarbon Exploration and Production*. Geological Society, London, Special Publications, **277**, 223–235.

Karner, G. D. & Driscoll, N. W. 1999. Tectonic and stratigraphic development of the west African and eastern Brazilian margins: insights from quantitative basin modelling. *In*: Cameron, N. R., Bate, R. H. & Clure, V. S. (eds) *The Oil and Gas Habitats of the South Atlantic*. Geological Society, London, Special Publications, **153**, 11–40.

Kolla, V., Bourges, P., Urruty, J. M., Claude, D., Morice, M., Durand, J. & Kenyon, N. H. 1998. Reservoir architecture in recent and subsurface deep-water meandering channel and related depositional forms. *European Association of Geoscientists and Engineers/AAPG Third Research Symposium Extended Abstracts*, unpaginated.

Kolla, V., Bourges, P., Urruty, J. M. & Safa, P. 2001. Evolution of deep-water Tertiary sinuous channels offshore Angola (West Africa) and implications for reservoir architecture. *American Association of Petroleum Geologists Bulletin*, **85**, 1373–1405.

Kundu, N., Pal, N., Sinha, N. & Budhiraja, I. L. 2008. Palaeo-hydrate and its role in deep-water Plio-Pleistocene gas reservoirs in Krishna–Godavari Basin, India. *Proceedings of the 6th International Conference on Gas Hydrates (ICGH 2008)*, Vancouver, BC.

Lundin, E. R. 1992. Thin-skinned extensional tectonics on a salt detachment, northern Kwanza Basin, Angola. *In*: Jackson, M. P. A. (ed.) *Special Issue; Salt Tectonics. Marine and Petroleum Geology*. Butterworth/Geological Society, London, **9**, 405–411.

Mayall, M. & Stewart, I. 2000. The architecture of turbidite slope channels. *In*: Weimer, P., Slatt, R. M. *et al.* (eds) *Deep-water Reservoirs of the World: Gulf Coast Section Society of Economic Palaeontologists and Mineralogists Foundation 20th Annual Research Conference*, Houston, TX, 578–586.

Murphy, J. B., Pisarevsky, S. A., Nance, R. D. & Keppie, J. D. 2004. Neoproterozoic–Early Palaeozoic evolution of peri-Gondwanan terranes: implications for Laurentia–Gondwana connections. *International Journal of Earth Sciences*, **93**, 659–682.

Nance, R. D., Murphy, J. B. *et al.* 2008. Neoproterozoic–early Palaeozoic tectonostratigraphy and palaeogeography of the peri-Gondwanan terranes: Amazonian v. West African connections. *In*: Ennih, N. & Liegeois, J.-P. (eds) *The Boundaries of the West African Craton*. Geological Society, London, Special Publications, **297**, 345–383.

Nouze, H. & Baltzer, A. 2003. Shallow bottom-simulating reflectors on the Angola margin, in relation with gas and gas hydrate in the sediments. *In*: Van Rensbergen, P., Hillis, R. R., Maltman, A. J. & Morley, C. K. (eds) *Subsurface Sediment Mobilisation*. Geological Society, London, Special Publications, **216**, 191–206.

Rice, D. D. & Claypool, G. E. 1981. Generation, accumulation and resource potential of biogenic gas. *American Association of Petroleum Geologists Bulletin*, **65**, 5–28.

Sikkema, W. & Wojcik, K. 2000. 3D visualisation of turbidite systems, Lower Congo Basin, offshore Angola. *In*: Weimer, P., Slatt, R. M. *et al.* (eds) *Deep-water Reservoirs of the World: Gulf Coast Section Society of Economic Palaeontologists and Mineralogists Foundation 20th Annual Research Conference*, Houston, TX, 928–939.

Valle, P. J., Gielberg, J. G. & Helland-Hansen, W. 2001. Tectono-stratigraphic development in the eastern Lower Congo Basin, offshore Angola, West Africa. *Marine and Petroleum Geology*, **18**, 909–927.

Wood, R. A. 2005. Neoproterozoic microbial-metazoan reefs, Nama Basin, Namibia. *Geological Society of America Abstracts with Programmes*, **37**, No. 7, 484.

Session: Unconventional Hydrocarbons Resources

Unconventional oil and gas resources and the geological storage of carbon dioxide: overview

H. JOHNSON[1] and A. G. DORÉ[2]

[1]*British Geological Survey, Murchison House, West Mains Road, Edinburgh EH9 3LA, UK (e-mail: hj@bgs.ac.uk)*
[2]*Statoil, 2103 CityWest Boulevard, Suite 800, Houston, TX 77042, USA*

Abstract: The 'Unconventional oil and gas resources and the geological storage of carbon dioxide' section of the *Proceedings* is designed to provide new insights from recent research and exploitation within these major growth areas in applied geology. Research and innovation on unconventional oil and gas have been driven by market needs – specifically concerns over oil and gas supply – as well as technological development. A cross-section of this work is highlighted in the current set of papers, covering ultra-heavy oil, shale oil, shale gas, basin-centred gas, tight gas and clathrates. Unsurprisingly, much of this research has been pioneered in North America. This work is complemented herein by some initial studies from Europe. Similarly, the geological storage of CO_2 is not simply a story of technological advance but also a response to an urgent societal imperative. Carbon dioxide sequestration is recognized as an important method for reducing greenhouse gas emissions in the near future, and is expected to have growing relevance to the oil and gas industry and the energy sector. Recent findings from current commercial carbon capture and storage projects in the North Sea and North Africa are presented and are complemented by an overview of major research programmes established in North America.

Keywords: unconventional hydrocarbons, heavy oil, shale gas, shale oil, gas hydrates, CO_2, carbon capture and storage

The Seventh Petroleum Geology Conference, held in London in 2009, reflected the significant interest in the geology of unconventional oil and gas resources that has developed over the past 5–10 years. Market drivers, specifically the concerns over supply, which culminated in wildly inflated oil and gas prices during 2008, probably provided the main impetus for this research. Given this perception, it is no surprise that much research and innovation regarding the exploration and development of unconventional oil and gas resources has been pioneered in North America, where these resources already have a long history of exploitation, contribute significantly to current production, and represent very large future energy sources (e.g. Jarvie *et al.* 2007; Ross & Bustin 2008). Nine of the papers in this section provide a good cross-section of research into the commercial exploration and exploitation of unconventional resources in North America and internationally, including ultra-heavy oil, shale oil, shale gas, basin-centred gas, tight gas and clathrates.

Although there is some history of commercial carbon dioxide (CO_2) injection or reinjection as a means of enhancing oil recovery, the current growth of interest in carbon capture and storage (CCS) is primarily driven by societal imperatives, with concerns about climate change raising the need to stabilize atmospheric concentrations of CO_2. The geological storage of CO_2 is gaining recognition as an important large-scale option for reducing greenhouse gas emissions in the near and foreseeable future. Four papers in this section describe very different approaches to CCS in the North Sea, North Africa and the USA.

Unconventional oil and gas resources

Unconventional oil and gas resources can be characterized by either intractable rock with very low permeability or intractable fluids (or in the case of clathrates, solids) where some form of stimulation is typically required for commercial production. These resources overturn many of the exploration and production paradigms that are applied to conventional hydrocarbons. For example, some unconventional gas resources are found in synclines rather than anticlines. In some cases, oil and gas is reservoired in rocks previously considered sources or seals. In other cases water is encountered updip of gas. In fact, many unconventional hydrocarbon resources lack fluid contacts. Whereas conventional gas resources are buoyancy-driven deposits, which form discrete accumulations in structural and/or stratigraphic traps, unconventional gas resources do not share these characteristics (e.g. Law & Curtis 2002). Unconventional hydrocarbons are commonly described as continuous or regionally pervasive in nature. Although the total in-place resource volume may be large, the overall recovery factor is commonly relatively low (e.g. Schmoker 2002; Sonneberg & Pramudito 2009). Completely dry holes cannot be drilled within the boundaries of such accumulations, but it is very possible to drill wells that are not economic. In other words, these resources are heterogeneous with commercial 'sweet spots', which are controlled by geological parameters such as structure, stratigraphy and diagenesis (e.g. Law 2002; Sonneberg & Pramudito 2009). Economically speaking, unconventional hydrocarbons are usually (but not always) characterized by high break-even product prices and, as such, become the focus of attention as conventional resource supplies diminish or become unavailable to the multinational companies.

Tight oil/shale oil

North America has a long history of producing oil from shales or very tight reservoirs, and considerable interest currently focuses on the Devonian–Mississippian Bakken petroleum system of the Williston Basin in central North America. This formation is generally regarded as a prime example of a continuous, unconventional tight oil exploration play, that is, a petroleum system where the source, seal and reservoir are the same unit, and prospectivity is not constrained by conventional trapping mechanisms. The play is commonly described as self-contained, with short migration distances from mature source rock of the Bakken Formation (Sonneberg & Pramudito 2009). High pressure within the formation has been attributed to the generation of hydrocarbons within the Bakken and only limited fluid expulsion. However, there is some debate regarding the degree of migration of the hydrocarbons generated. Insight into this question is provided by Kuhn *et al.* through careful examination of geochemical parameters. The authors conclude that there is probably more migration within the Bakken Formation than previously assumed, and that the system can in fact be described as partially 'open'.

Shale gas

During the Sixth Petroleum Geology Conference held in 2004, a single paper by Richard Selley drew attention to the use of shale gas in parts of the USA and to the potential for exploitation of this resource in the UK (Selley 2005). Since then there has been an astonishing growth in research on the exploration and exploitation of shale gas plays. Harnessing this resource has converted a local curiosity into a multi-billion dollar international business, and has helped transform the North American market from gas starvation to guaranteed supply for 20 years or more. It is a classic example of market-driven research, whereby both innovation and pre-existing technology are brought together by economic necessity. This significant change was reflected by the high interest in shale gas shown at the Seventh Conference.

As with shale oil, shale gas systems are considered discrete, self-enclosed systems in which the source, seal and reservoir are one and the same. Shale gas is a very important exploration target in North America, where the resource falls into two distinct types: biogenic and thermogenic, although there can also be mixtures of the two gas types (Jarvie *et al.* 2007 and references therein). Known shale gas systems have been characterized by a number of parameters including total organic carbon content, thermal maturity, kerogen transformation, the efficiency of the source rock to retain its generated hydrocarbon products and the nature of the storage system. However, examination of productive shale gas systems indicates that the current parameters used to assess shale gas prospectivity vary greatly and may not provide a strong predictive model. Consequently, additional criteria, such as the clay and mineral content of the shales, the burial history and the precise nature of the gas storage and retention systems are fertile grounds for further research (e.g. Ross & Bustin 2008).

No major economic shale-gas enterprises are currently known from outside of North America, but many parts of Europe contain targets for shale gas exploration and commercial production is considered purely a matter of time. In this volume, Schulz *et al.* provide an overview of black shale successions that may be attractive for shale gas exploration in European basins, where conventional production is declining, underutilized gas gathering infrastructure exists and markets are accessible. Smith *et al.* follow Selley's pioneering work in providing a summary of organic-rich shale successions in the UK that may offer potential for shale gas resources. While no bonanzas equivalent to the Barnett or Marcellus Shales of the USA are envisaged, particular attention is drawn to the potential regional importance of the Mississippian Bowland Shales of northern England.

Basin-centred and tight gas

The basin-centred gas exploration play within the Western Canada Foreland Basin is characterized by a regionally pervasive gas saturated accumulation in low permeability, abnormally pressured reservoirs, with a lack of downdip water contact. Within this unconventional gas play, zones with commercially favourable reservoir producibility characteristics are controlled by a number of geological factors. For commercial success, Boettcher *et al.* highlight the importance of fully understanding the stratigraphy, sedimentology, rock properties and structure. They show that a process of long-term regional migration of undersaturated gas from source rocks into adjacent reservoir rocks, and the associated expulsion of water, is particularly important.

Kiraly *et al.* describe a basin-centred gas exploration play within the Pannonian Basin, Hungary, where thick Upper Miocene shales provide excellent pressure sealing above the significantly overpressured Upper and Middle Miocene formations. Within the overpressured zone, thick and poorly consolidated turbidite sandstones, as well as the shales, form unconventional reservoirs for basin-centred hydrocarbon accumulations.

Gas within unconventional fine-grained reservoirs is commonly extracted using hydraulic fracture treatments. Horizontal wells with multiple fracture stimulations are currently the most economic means of producing gas from tight reservoirs. Such wells are very expensive and typically suffer from a steep early production decline. Natural fractures, which are present in most shale units, act as weak planes that reactivate during hydraulic fracturing. Gale and Holder explain how knowledge of the geometry and intensity of the natural fracture system is therefore necessary for effective hydraulic fracture treatment design. Commonly occurring fracture types are reviewed and evaluated as to whether they are relevant to gas production.

Ultra-heavy oil

Ultra-heavy oil is too heavy or viscous to be produced by conventional flow from a reservoir. Biodegradation is generally responsible for the specific gravity and viscosity. Consequently, such accumulations are often found shallowly buried or even at surface. Again, there has been a resurgence of interest in exploiting ultra-heavy oil driven by the price rises of the mid-2000s. This enterprise is, however, technically and logistically demanding and is characterized by high break-even prices. Concerns also exist over the environmental challenges of exploiting this resource, since such enterprises can be seen as environmentally disruptive and CO_2-intensive. The heavy oil deposits of Canada are a case in point. With an estimated 1.7×10^{12} barrels in place, some assessments rank Canada second behind Saudi Arabia in terms of oil resources. Heavy oil makes up 98% of this resource, with the Athabasca Oil Sands of the Lower Cretaceous McMurray Formation in northern Alberta containing approximately 900×10^9 barrels in place. Depending upon the amount of overburden, the oil sands are developed through either surface mining or thermal *in situ* techniques, including steam-assisted gravity drainage. Like most unconventional resources, the accurate prediction and location of 'sweet spots' is critical. In this volume, Peacock shows how the development of an understanding of the regional distribution of fluvial–estuarine point bars, which comprise the principal reservoir rocks, is critical to maximizing commercial production from the Athabasca Oil Sands. For heavy oil thermal developments, the understanding and prediction of reservoir characteristics is particularly important because mudstone units within the reservoir can have a dramatic effect on steam chamber growth and the subsequent oil recovery.

Gas hydrates (clathrates)

Geoscientists have long speculated that gas hydrate accumulations could eventually be a commercially producible energy resource. The gas hydrates are naturally occurring 'ice-like' combinations of natural gas and water that have the potential to provide an immense resource of natural gas from the world's oceans and polar regions. The amount of gas in the world's gas hydrate accumulations is thought to exceed the volume of known conventional gas resources. However, until recently significant technical and economic hurdles made gas hydrate development seem a distant goal. This view began to change in recent years with the realization that this unconventional resource could possibly be developed with existing conventional oil and gas production technology. In this volume, Collett summarizes the most significant recent developments in this regard, such as gas hydrate production testing conducted during 2002 at the Mallik site in Canada and, more recently, the testing of the 'Mount Elbert Prospect' gas hydrate accumulation on the North Slope of Alaska.

Coal gasification

Coal has commonly been seen as a potential fall-back fossil fuel resource as hydrocarbons become depleted. Younger *et al.* describe how deep-seated coal resources may be exploited by means of underground coal gasification (UCG) with the possibility for coupling the technique to CCS. Significantly for the petroleum industry, the technological requirements for UCG, using directionally drilled boreholes from ground level, are far more akin to those of oil and gas production than they are to those of deep mining.

Geological storage of carbon dioxide

There is a range of options regarding the geological storage of CO_2. One possibility is the use of CO_2 in enhanced oil recovery and this is already commercial in some circumstances. The other main options are storage within depleted oil and gas fields or saline aquifers, while other techniques, such as storage in coals with enhanced coal bed methane, remain at an early stage. A key challenge for wider deployment of CO_2 geological storage is the requirement for the technique to be accepted as safe and verifiably effective. Senior *et al.* describe a commercial project that has been operating at In Salah in Algeria since 2004. CO_2 recovered from the natural gas production of the Krechba Field amounts to 1×10^6 tonnes per year, and this is injected into the water leg of the gas reservoir for the purpose of geological storage.

Injection of CO_2 commenced at the Sleipner Field in the Norwegian sector of the North Sea in 1996, and over 10×10^6 tonnes is currently stored in the late Cenozoic Utsira Sand saline aquifer. As a means of verifying its effectiveness, a comprehensive seismic monitoring programme of the storage site has been implemented, with repeat time-lapse 3D surveys that image the progressive development of the CO_2 plume. The total amount of CO_2 injected is known, and Chadwick and Noy show that independent verification of this amount can be achieved from the seismic data. Encouragingly, their observations of the laterally spreading CO_2 plume at Sleipner seem to confirm the integrity of the topseal to the Utsira Sand. Of equal significance, however, is their observation that history-matching techniques do not adequately predict the flow pathways indicated on the time-lapse 3D seismic, suggesting that our ability to model fluid behaviour in complex reservoirs remains at a simplistic level. Hermanrud *et al.* describe how monitoring of the CO_2 plume at Sleipner provides insights into the processes of subsurface fluid migration, especially with regard to vertical migration.

Esser *et al.* provide an insight into research by one of the seven major Regional Carbon Sequestration Partnerships in the USA and Canada, which aim to establish the best practices for permanently storing CO_2 in different geological formations. The research will help develop regulations and infrastructure requirements for future sequestration deployment and also includes a deep saline test in a structural dome 120 miles south of Salt Lake City, Utah, where deep saline formations are major targets for commercial-scale sequestration associated with the coal-fired power plants currently planned for the area.

References

Jarvie, D. M., Hill, R. J., Ruble, T. E. & Pollastro, R. M. 2007. Unconventional shale-gas systems: the Mississippian Barnett Shale of north-central Texas as one model for thermogenic shale-gas assessment. *American Association of Petroleum Geologists, Bulletin*, **91**, 475–499.

Law, B. E. 2002. Basin-centred gas systems. *American Association of Petroleum Geologists, Bulletin*, **86**, 1891–1919.

Law, B. E. & Curtis, J. B. 2002. Introduction to unconventional petroleum systems. *American Association of Petroleum Geologists, Bulletin*, **86**, 1851–1852.

Ross, D. J. K. & Bustin, R. M. 2008. Characterizing the shale gas resource potential of Devonian-Mississippian strata in the Western Canada sedimentary basin: application of an integrated formation evaluation. *American Association of Petroleum Geologists, Bulletin*, **92**, 87–125.

Schmoker, J. W. 2002. Resource-assessment perspectives for unconventional gas systems. *American Association of Petroleum Geologists, Bulletin*, **86**, 1993–1999.

Selley, R. C. 2005. UK shale-gas resources. *In*: Doré, A. G. & Vining, B. A. (eds) *Petroleum Geology: North-West Europe and Global Perspectives: Proceedings of the 6th Petroleum Geology Conference.* Geological Society, London, 707–714; doi: 10.1144/0060707.

Sonneberg, S. A. & Pramudito, A. 2009. Petroleum geology of the giant Elm Coulee field, Williston Basin. *American Association of Petroleum Geologists, Bulletin*, **93**, 1127–1153.

Bulk composition and phase behaviour of petroleum sourced by the Bakken Formation of the Williston Basin

P. KUHN, R. DI PRIMIO and B. HORSFIELD

Helmholtz-Centre Potsdam, German Research Centre for Geosciences (GFZ), Section 4.3, Organic Geochemistry, Telegrafenberg, D-14473 Potsdam, Germany (e-mail: kuhn@gfz-potsdam.de)

Abstract: The Bakken Formation is currently regarded primarily as a self-contained, unconventional petroleum system. While previously viewed as a source for oil occurring in overlying formations, it is now predicted that resources of more than 3.5 billion barrels of oil are trapped intraformationally. New insights into the formation's open v. closed nature are presented here using the physical properties of natural petroleum, source rock characteristics and the numerical modelling of phase behaviour. In the mature western part of the basin petroleum accumulations have been postulated to be continuous in nature, characterized by very short migration distances of indigenous hydrocarbons. This necessitates that the composition and therefore physical properties of the generated hydrocarbons must be controlled by the maturity of the source rock in the immediate vicinity. This assumption is not supported by the clustering of higher gas–oil ratios and lighter oil gravities along the locations of the anticlines in the basin. We have used open and closed system pyrolysis techniques to predict the bulk composition of the petroleum generated at different transformation stages, both cumulatively and instantaneously. Based on these predictions the Bakken would contain dominantly undersaturated fluids throughout the basin. Differences in predicted GORs of cumulative and instantaneous models support the conclusion that the reported hydrocarbon compositions cannot completely be explained by a tight self-contained petroleum system. The observed variability of in-place hydrocarbon compositions is readily explained by lateral migration of petroleum in the main middle Bakken carrier, and vertical leakage of emplaced hydrocarbons from the fractured reservoir at anticline locations. This has resulted in the loss of the early generated petroleum, and led to a present-day dominance of late generation products. These results reveal that the Bakken Formation is a partly open petroleum system, at least along the major anticlines of the Williston Basin.

Keywords: unconventional resources, shale oil play, North America, geochemistry, phase kinetics, phase behaviour, play assessment

The Devonian–Mississippian Bakken Formation of the Williston Basin in central North America (Fig. 1), recently described as a continuous, unconventional source rock–reservoir formation (Pitman *et al.* 2001; Pollastro *et al.* 2008; Sonnenberg & Pramudito 2009), is experiencing another renaissance as a major hydrocarbon resource. Earliest exploration for Bakken resources was initiated about half a century ago, three years after an unsuccessful well, targeting the overlying Madison Formation, recovered oil on a drill stem test in 1953 (LeFever 2005*b*). In succeeding decades, exploration in the 'Bakken Fairway' along the Nesson anticline proved to be successful. Unfortunately, the volumetric expectations for large amounts of oil based on its high hydrocarbon generation potential were not realized and interest in the Bakken as a reservoir decreased. With improvement of drilling technologies (Fritz 1990) and introduction of horizontal drilling techniques to the Williston Basin, interest returned and the Bakken reservoir potential was again targeted. Focusing on the middle member promised higher yield per well and greater success for this play, but high cost per well and low oil prices left the Bakken in disregard again. It took another 15 years and successful horizontal drilling in the middle member of the Bakken Formation in Elm Coulee and Parshall Field to kick off a new era in Bakken Formation drilling. With a new assessment of undiscovered oil resources (Pollastro *et al.* 2008) and higher oil prices, the Bakken play is back to prominence as an attractive onshore exploration target.

Despite the vast amount of work already performed on the Bakken Formation of the Williston Basin, questions regarding the dynamics of the system with respect to hydrocarbon generation and migration still remain. Here we combine natural data, including petroleum physical properties and geochemistry, with experimental results obtained on a multitude of Bakken source rock samples to provide an integrated understanding of the Bakken petroleum system.

Geological overview

The intracratonic, symmetrical Williston Basin (Hansen 1972) contains sediments of all geological periods from the Cambrian to the Cenozoic (Carlson & Anderson 1965) that amount to a maximum thickness of more than 4800 m sediment (Pitman *et al.* 2001) and multiple oil plays. Prominent source rocks were identified in the three members of the Mississippian Madison Group, the Mississippian–Devonian Bakken Formation, the Devonian Duperow Formation, and the Ordovician Red River and Winnipeg Formations (Jarvie 2001). Additionally, there has been some effort to produce gas from Cretaceous sediments and glacial drifts. The overall structural development of the basin is simple and defined mainly by the Nesson, Billings and Cedar Creek anticlines. While earlier publications described inversion of less than 100 m (Webster 1984), both Sweeney *et al.* (1992) and Osadetz *et al.* (2002) agree on greater inversion of around 550–700 m during the Cenozoic for the US and the Canadian parts of the basin. Additionally, based on apatite fission-track thermochronology, another inversion of *c.* 700 m during the Permian is proposed (Osadetz *et al.* 2002) for at least the Canadian part of the basin.

The low-permeability Bakken petroleum system, which can be found only in the subsurface of the basin, has a maximum thickness of 49 m (Murphy *et al.* 2009). The black shale source rocks have maximum thicknesses of 7 m for the upper and 15.2 m for the lower member (Webster 1984). The lithology in both shale members throughout their extent is invariant and homogenous,

Fig. 1. Location of the Williston Basin at the border of the USA and Canada; important structural features are indicated.

consisting of evenly distributed organic material of Type II kerogen, which is mixed with lesser amounts of clay, silt and dolomite grains (Meissner 1978; Webster 1984). Each of the three successively higher members of the Bakken Formation reaches across a wider area than the preceding one as they all pinch out against the older Upper Devonian Three Forks Formation that they overlay unconformably (Fig. 2; Meissner 1978; Webster 1984; LeFever 2008). The Bakken Formation, which extends blanket-like in the subsurface of the Williston Basin, consists of the middle reservoir layer encased by the upper and lower source rocks that are easily correlated through the basin by their high gamma ray signals (Murray 1968; Dembicki & Pirkle 1985; LeFever 2005a). Additionally, in some locations in the deeper part of the basin, the Sanish Sands of the Three Forks Formation are hydrologically connected to the Bakken Formation and need to be included when discussing the Bakken Petroleum system (Murray 1968). The total organic carbon (TOC) of the Bakken shales falls between 10 and 15 wt% and is seen petrologically as amorphous sapropelic kerogen (Schmoker & Hester 1983; Price et al. 1984; Webster 1984). Hydrogen indices are typically >500 mg HC g^{-1} TOC at a maturity below 0.4% vitrinite reflectance (Muscio & Horsfield 1996). Deposition occurred in a marine, euxinic environment in the photic zone (Requejo et al. 1992) during sea-level rise (Smith & Bustin 1998) in Devonian to Mississippian times based on conodont dating (Hayes & Holland 1983).

The lithology of the middle Bakken is more variable and changes laterally from dolomitic fine-grained silt to very silty fine-crystalline dolomite (Meissner 1978; Webster 1984). Porosity varies over a wide range from 1 to 16% but is generally low, averaging around 5% (Pitman et al. 2001) to 3.7% (Ropertz 1994). While the vertical permeability of the Bakken Formation ranges from 0.01 to 0.001 nd (Burrus et al. 1996a), the permeability of the middle Bakken is much higher, varying between 0 and 20 mD, and averaging around 0.04 mD (Pitman et al. 2001). These variations in the porosity and permeability together with the variability of the lithology of the middle Bakken should be kept in mind in the following, due to their first-order control on Bakken petroleum migration. The Lodgepole Formation conformably overlies the Bakken Formation, and is the basal unit of the Madison Group (Meissner 1978; Webster 1984).

Geochemistry has played a growing role in assessing the vertical and lateral extent of the Bakken petroleum system by addressing one of the most controversial issues, namely the contribution of Bakken sourced petroleum to other reservoir formations, that is, migration from the Bakken into the Madison Group. Williams (1974) proposed a vertical migration of petroleum from the Bakken based on carbon-isotope ratio measurements, gas chromatography, optical-rotation measurements and infrared spectroscopy. This was supported by Dow (1974), who indicated that oils expelled from the Bakken migrated vertically and laterally into Madison Formation traps or would eventually be retained by an evaporite seal of the Charles Salt of the Madison Group. In disagreement with his earlier publication (Price et al. 1984), Price & LeFever (1992, 1994) found support for the idea of a closed system without any major migration into overlying formations based on interpretation of reservoir rock extracts analysed by liquid and gas chromatography and mass spectrometry. Based on biomarker

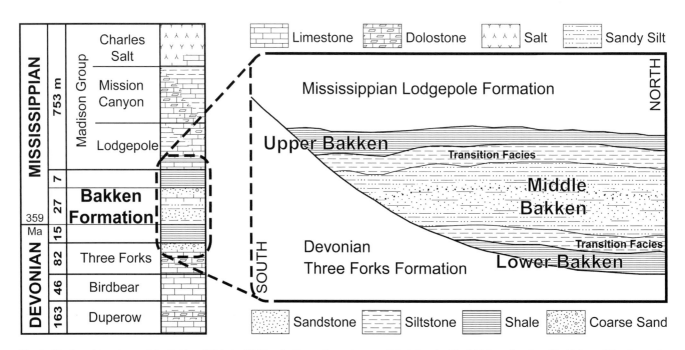

Fig. 2. Stratigraphic column of the Bakken (Webster 1984) and adjacent formations modified after the North Dakota Stratigraphic Column (Murphy et al. 2009); schematic profile of the Bakken Formation.

analysis Osadetz et al. (1994) agree that the mid-Madison accumulations were not sourced by Bakken-generated petroleum as they, in fact, had a carbonate biomarker signature inconsistent with Bakken biomarkers. This assessment was supported by Jarvie (2001), who used light hydrocarbon data to complement higher molecular weight biomarker data and discounted vertical migration in the US part of the Williston Basin except along major faults. However, Jiang & Li (2002) interpreted their quantitative biomarker analysis to support oil accumulations in the Canadian part of the basin, the Madison Group have significant contribution from Bakken-generated hydrocarbon.

An argument for a limited contribution from other sources to the Bakken petroleum system is provided by high overpressure within the Bakken system (Murray 1968; Finch 1969; Meissner 1978), which is attributed to the generation of hydrocarbons (Spencer 1987; Burrus et al. 1996b). Further, Pitman et al. (2001) discuss the development of small fracture systems in the Bakken Formation as related to hydrocarbon generation. Fractured areas in which the Bakken was exposed to greater stress by either external (structural movement) or internal (overpressure) processes are important, since exploration in these areas resulted in greater economical success, for example, the Bakken fairway along the Nesson anticline (LeFever 2005b).

The main result of the intense analysis and discussion of the Bakken Formation over the last four decades is a better understanding of the play associated with a continuous increase in the predicted in-place volume of hydrocarbons (Dow 1974; Williams 1974; Schmoker & Hester 1983; Price et al. 1984; Webster 1984; Flannery & Kraus 2006), which currently culminates in estimates of undiscovered resources of 3.65×10^9 BBL of oil and 1.84×10^{12} SCF of associated/dissolved natural gas (Pollastro et al. 2008). Further, the Bakken is currently described as an unconventional, continuous tight oil formation that has to be approached with new exploration and production techniques. The in situ generated petroleum in this continuous play is commonly found in close proximity to the mature source rock (Sonnenberg & Pramudito 2009), underlining the low impact of intraformation migration (Muscio et al. 1994), while stratigraphic (Johnson 2009) and maturity related traps (Coskey & Leonard 2009) are also described.

Goals and approach

Very little has been published on the bulk composition of Bakken petroleums. Therefore, this study integrates production data characterizing the physical properties of Bakken petroleums with geochemical analyses of the source rock including phase-predictive compositional kinetic models of hydrocarbon generation. By comparing these with the petroleum that is actually produced from the Bakken across North Dakota, further information regarding the effectiveness and extent of petroleum migration can be derived.

Data sources

Results of new Rock-Eval measurements were combined with data published by Muscio (1995), Sweeney et al. (1992), Webster (1984) and Price et al. (1984) to produce newly compiled maps. All newly measured data and model calculations are based on a sample set of more than 250 samples including all members of the Bakken and adjacent formations from 22 wells. The new dataset covers a maturity range from 405 to 453 °C T_{max}, which corresponds to an equivalent vitrinite reflectance of about 0.35–1.10% (Dembicki & Pirkle 1985).

Gas–oil ratios (GOR) and oil gravity data (production API°) were taken from drilling reports and the database of the North Dakota Geological Survey and the Department of Mineral Resources (Oil and Gas Division) of the North Dakota Industrial Commission.

Analytical techniques

TOC was determined using a LECO SC632 analyser. Further, the petroleum generation potential was quantified by Rock-Eval pyrolysis (Espitalié et al. 1977) using a Rock-Eval 6 instrument following procedures of *The Norwegian Industry Guide to Organic Geochemical Analyses* (NIGOGA, Weiss et al. 2000). For analytical details of published data, please refer to the references cited above.

Open system pyrolysis gas chromatography (Horsfield et al. 1989) was used for kerogen-type characterization as well as determination of petroleum-type organofacies. Samples were heated from 300 to 600 °C using a temperature ramp of 50 °C min^{-1}. Pyrolysis products were collected in a cryogenic trap (liquid nitrogen) from which they were liberated by ballistic heating (300 °C held for 10 min). Gas chromatography analysis was made using an Agilent 6890 GC with a HP-Ultra 1, 50 m × 0.32 mm internal diameter, dimethylpolysiloxane-coated (0.52 μm film thickness) column programmed from 30 to 320 °C at 5 °C/min (held for 35 min). Identification of prominent peaks of the chromatogram traces was carried out via retention time correlations with a standard sample and literature data (Horsfield & Dueppenbecker 1991) to determine the composition of generated hydrocarbons at different maturities.

Bulk kinetic parameters for Bakken source rocks were determined using an SR analyser (Humble Instruments). Open system programmed-temperature pyrolysis was performed at four different heating rates (0.7, 2, 5 and 15 °C min^{-1}). The ensuing petroleum formation rate curves were analysed assuming a maximum of 25 first-order parallel reactions and a single frequency factor using Kinetics 2000 software (Lawrence Livermore National Laboratory). The kinetic model thus obtained allows extrapolation to hydrocarbon generation in nature (Braun & Burnham 1990; Ungerer 1990; Schenk et al. 1997).

Based on the results from the 0.7 °C min^{-1} heating rate experiments, laboratory temperatures for conversion stages representing 10, 30, 50, 70 and 90% transformation were determined. In preparation for closed system non-isothermal pyrolysis (MSSV, Horsfield et al. 1989), source rock material was sealed in small glass tube aliquots and heated to the five determined temperatures at a heating rate of 0.7 °C min^{-1}. Analysis of the generated products was performed using the same analytical technique as for the open system pyrolysis experiments. This approach enabled insight into the compositional evolution of the generated hydrocarbons and calculated the GOR at different stages of source rock transformation. These results were further tested against natural conditions employing the compositional mass balance approach (Santamaría-Orozco & Horsfield 2003).

The evolving petroleum composition derived from MSSV pyrolysis was further used as input to phase kinetic modelling (di Primio & Horsfield 2006). Based on the composition of generated hydrocarbons and after gas phase compositional tuning, the behaviour of the generated fluid at different pressure and temperature conditions could be determined using PVT simulation software (PVTSim, Calsep, Denmark) to predict phase behaviour, physical properties and single or multiple phase compositions in the subsurface.

Finally, new and available well data were compiled to give an overview of the maturity, petroleum generation potential and physical production parameter distributions across the Bakken Formation. These maps were prepared utilizing a geographical information system (ArcGIS, ESRI Inc.).

Results

Bulk geochemical signature

During the first phase of this evaluation, the individual datasets were checked for consistency. It emerged that vitrinite reflectance data from two different datasets (Dembicki & Pirkle 1985; Muscio 1995) were inconsistent. In lieu of using vitrinite reflectance from the vitrinite-poor Bakken Formation, the Rock-Eval parameter T_{max} was used as a maturity parameter. Data from earlier publications (Price et al. 1984; Webster 1984; Sweeney et al. 1992; Muscio 1995) and newly derived data were compiled in a maturity map (Fig. 3a). In general the maturity was positively correlated with current burial depth, disregarding differentiation of upper or lower Bakken members. Highest values around 450 °C were measured on samples taken from wells within the deepest area of the basin, while the T_{max} decreased with shallower burial depth, for example, in an eastern direction. Furthermore, the higher T_{max} values in the northern part of the depocentre indicated a higher maturity, while in the southern part the source rock was less mature at similar depth. Therefore slightly increasing maturity from south to north in the western part of North Dakota is evident. Currently, it is not possible to further relate these changes to an increased heat flow or other variations in the temperature history, but the earlier mentioned higher inversion proposed for the Canadian part in contrast to the US part of the basin could have had an impact on this maturity trend.

The distribution of hydrogen index (HI) values, which describe the remaining petroleum generation potential of the Type II kerogen source rocks, conformed to the maturity distribution given by T_{max}. The lowest HI values in western North Dakota clustered around the deeper buried areas, in which the highest T_{max} values were encountered. Also, the shift to higher maturities from south to north in shallower buried areas away from the deeper parts of the basin was evident (Fig. 3b). In general, HI increased almost radially from deeper to shallower depths. While the negative relationship between HI and maturity was consistent with a uniform kerogen type across the area prior to maturation, inconsistencies for HI and T_{max} correlations were nevertheless found. In the shallower, immature part of the Bakken located east of 101°40'W longitude, a decrease of HI was not always accompanied by an increasing T_{max}. These low HI values in the counties of Renville, Ward and further east, which were in a similar range to the HI values in the deeper area, were not related to the catagenetic transformation of the source rocks but instead to either slightly different depositional environments in locations closer to the edge of the accumulation area or effects related to decarboxylation at low mature levels. Concerning the latter, a maximum HI of the Bakken source rocks may not be present in the original immature samples but rather at the end of diagenesis (Boudou et al. 1994; Killops et al. 1998). In Figure 4 the low maturity samples, which were sampled from a well in Rolette County, North Dakota, have the highest oxygen index (OI) values of all displayed samples, while the HI of more mature samples that are still in the immature range (Baskin 1997) are c. 100 mg HC g^{-1} TOC higher. This would indicate that the Bakken kerogen in the eastern part of the basin is also very similar to the composition in the west, but needs to further mature to fall into line with the original trends as they were described for the different kerogen types in the pseudo van Krevelen diagram (Tissot & Welte 1984).

Since neither explanation for the lower HI in the low mature area can be verified on the available sample set, samples from these areas will not be used as a corresponding immature counterpart for the samples in the deeper western part of the basin.

The TOC content of the Bakken Formation in North Dakota supports the thesis of a slight but recognizable facies variation of the source rocks. The reason for the decline of the percentage of organic carbon content in the east of the research area (Fig. 3c) can possibly be related to a slightly different depositional environment with poorer preservation potential due to its location closer to the edge of the basin. In this area changes of TOC related to hydrocarbon generation can be disregarded based on low T_{max} values of the rock samples, indicating immaturity. In contrast to the east, the lower TOC in the western part of the research area can be explained reasonably well by reduction due to hydrocarbon generation.

Quantification of generation and expulsion

Using the algebraic scheme developed by Cooles et al. (1986) and underlying assumptions (e.g. fixed amount of inert carbon, which is the part of the kerogen, i.e. not converted to petroleum during maturation), the original TOC in samples of elevated maturity was calculated. Based on the discussion above, samples from shallow areas in the east with lowest maturities were excluded. A sample whose T_{max} (427 °C) is still outside the threshold of hydrocarbon generation (Baskin 1997) with a production index less than 0.1 was selected as a standard representative of the immature source rock of the Bakken.

Following the Cooles et al. (1986) approach, the petroleum generation index (PGI) and petroleum expulsion efficiency (PEE) for the natural maturity sample set were calculated. The results show (Fig. 5) that the source rock starts to expel almost half of the generated hydrocarbons at a PGI of less than 0.15, where '0' is immature and '1' indicates the maximum amount of generated hydrocarbons. Further, the PEE (0 = no expulsion, 1 = expulsion of all generated hydrocarbons) increases to 0.9 at a PGI greater 0.4, which shows that 90% of the hydrocarbons generated in the source rock so far do not remain in the rock, but have been expelled into adjacent sediments. This indicates weak hydrocarbon retention in the Bakken source rocks, especially in the higher mature areas.

While Cooles et al. (1986) define the inert kerogen as constant throughout the main phase of petroleum generation, during which it undergoes internal rearrangement, publications by Horsfield (1997), Mann et al. (1997) and Jarvie et al. (2007) consider the formation of additional amounts of 'dead carbon' during petroleum generation. Dead carbon formation during hydrocarbon generation has been described for the Bakken source rocks (Muscio & Horsfield 1996); therefore, all results based on the Cooles approach have to be considered as maxima, since an increasing amount of dead carbon would lead to an overestimation of original immature TOC.

Further, when reviewing the PEE, the light composition of petroleum generated by Bakken source rocks has to be considered. Since some of the core samples were recovered nearly 30 years ago, part of the volatile components could have been lost during sample handling and storage. This would elevate the PEE for individual samples. Still, the majority of heavier hydrocarbons (C_{5+}) should have remained in the low-permeability and -porosity source rock samples that were used to conduct the measurements. The general trend will be the same, although slightly lower PEE values are possible.

Pyrolysis gas chromatography

After assessment of the bulk potential and behaviour of the petroleum generated by the Bakken, more detailed analysis of the hydrocarbon composition was performed. Pyrolysis gas chromatography was conducted on more than 60 source rock samples, representing the entire maturity range of the Bakken up to the end of the oil window between 450 and 460 °C T_{max} (Baskin 1997). Parts of the results are displayed in the petroleum-type organofacies diagram (Horsfield et al. 1993) and the kerogen-type sulphur

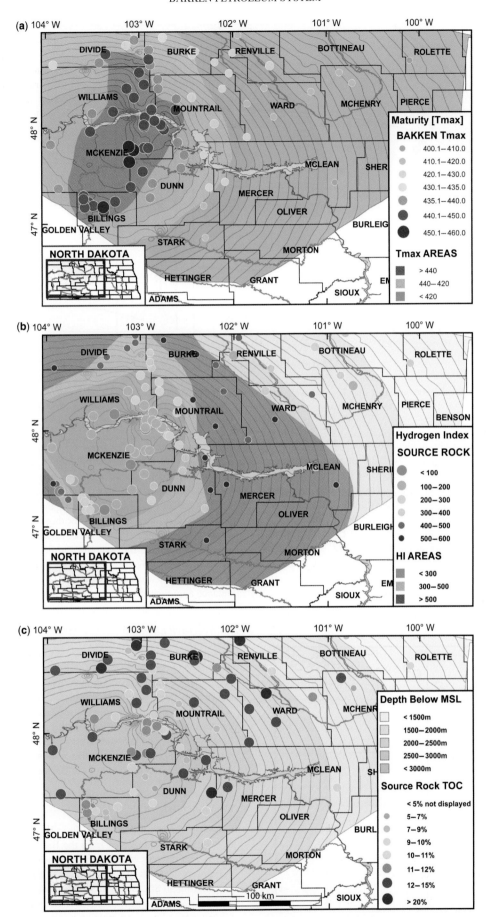

Fig. 3. (a) Bakken Formation maturity map based on T_{max} value of the S2 peak from Rock-Eval pyrolysis; data based on own measurements and publications referred to in the text. (b) HI map of the Bakken source rock, disregarding differences in the upper and lower shale; lowest values cluster in the deeper western area, while scattered lower values in the east indicate variability in the original source rock potential in this area. (c) Map of the weight percentage of TOC measured on source rock samples of several wells; in the background the depth of the top of the Bakken Formation below sea-level in the North Dakota part of the Williston Basin based on LeFever (2008) is displayed.

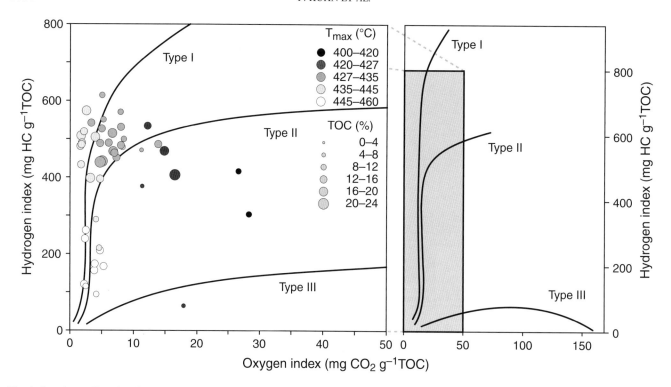

Fig. 4. Pseudo van Krevelen diagram (modified after Tissot & Welte 1984) displaying the HI and OI of samples from the upper Bakken member; maturity (T_{max}) and TOC (weight %) are greyscale- and size-coded.

diagram (Eglinton *et al.* 1990) in Figure 6, which shows clustering of the samples in both diagrams, with no distinction between the upper or lower Bakken member. Based on these results, a low variability of the composition of generated hydrocarbons and therefore a narrow range of variability in the original Bakken kerogen across the North Dakota part of the basin can be inferred.

Further, when comparing the percentage of C_{1-5}/C_{1+} of the bulk pyrolysis to the maturity of the samples (Fig. 7), it is clear that the change in composition is very linear, even including

the low mature samples from the eastern part of the basin at $T_{max} < 415$ °C. Especially at the higher mature range, which represents the deeper part of the basin, the petroleum generated by the Bakken has a very low variability and a predictable GOR. Greater variability is only found in the immature interval from 415–428 °C T_{max}. The reasons for this remain unclear, but might be related to some source rock variability in these immature areas from which the samples originate. The significance of these laboratory-based results was further tested against a natural Bakken maturity sequence employing the compositional mass balance approach (Santamaría-Orozco & Horsfield 2003).

Source rock kinetics

Kinetic analyses of 12 low-maturity source rocks of five wells were performed to determine the variability of the petroleum generation behaviour of the Bakken kerogen. The majority of the resulting conversion rates that were calculated for a constant geological heating rate of 3 °C/Ma (Fig. 8) indicate a very narrow range of generation temperatures, disregarding any difference of upper or lower member. The low variability of 13 °C at 50% transformation ratio (TR) implies a very similar response to maturation, which underlines the homogeneity of the Bakken kerogen.

Phase behaviour

Based on analytical procedures described above, the composition of hydrocarbons generated from the Bakken Formation's kerogen at different degrees of transformation was assessed. As displayed in Figure 9 the GOR that is generated at different generation stages ranges from 30 Sm^3 Sm^{-3} at a TR of 10% to c. 220 Sm^3 Sm^{-3} at a TR of 90%. The higher rate of GOR increase at higher TR stages is related to the almost exponential increase in gas (C_{1-5}) generation. Based on T_{max} the maturity of the Bakken source rocks in the deepest area of the basin did not pass the oil window; there the highest TR was at 96%. This limits the cumulative naturally generated GOR in the Bakken to around 240 Sm^3 Sm^{-3}.

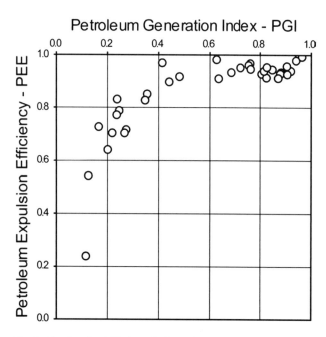

Fig. 5. Based on Rock-Eval pyrolysis results and the mass balance approach, the PGI and the PEE were calculated. The Bakken source rocks have an expulsion efficiency of greater than 60 and 80% at a generation of greater than 20 and 40%, respectively.

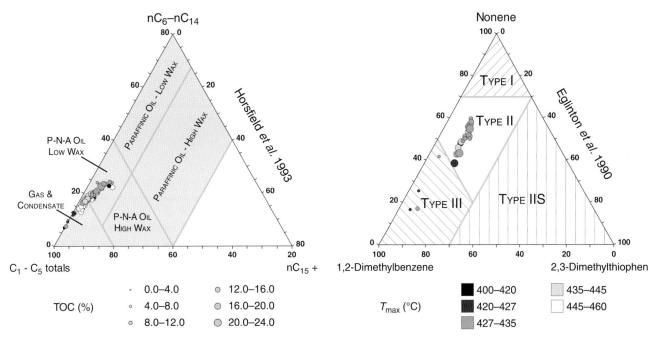

Fig. 6. Ternary plots show chain length distribution (left) and relative abundance of 2,3-dimethylthiophen, n-non-1-ene and 1,2-dimethylbenzene based on pyrolysis–gas chromatography; the majority of the samples from the upper and lower Bakken member cluster tightly together, implying a low kerogen variability; maturity (T_{max}) and TOC (weight %) are greyscale- and size-coded; plots are based on Horsfield et al. (1993) and Eglinton et al. (1990).

Using the phase kinetic approach, the phase behaviour of the generated hydrocarbons at variable pressure and temperature combinations could be defined for Bakken-generated hydrocarbons. Figure 9 displays the phase envelopes of petroleums generated from a selected source rock sample at five transformation stages. The pressure–temperature (PT) conditions within the phase envelopes describe conditions at which the hydrocarbons in the reservoir will occur in a two-phase state. All PT combinations outside the phase envelopes will result in an undersaturated phase of either liquid or gas (di Primio & Horsfield 1996). The temperature and pressure conditions are in general dependent on depth, while in some areas pressure significantly above hydrostatic conditions has been described for the Bakken petroleum system (Murray 1968; Finch 1969; Meissner 1978). Based on regular PT gradients of the area, temperature and minimum pressure conditions for the variable depth in which the Bakken can be found were identified. With these simple assumptions it is possible to determine if a single or a two-phase fluid can be expected at different locations of the Bakken. The phase distribution map (Fig. 10) is based on a Bakken Formation depth map (LeFever 2008), log tops of several wells provided by the North Dakota Geological Survey and average PT gradients. This map indicates that all regions in which the Bakken is currently under exploration and production have PT conditions that result in an undersaturated fluid. Even the elevated areas of the anticlines are not in or near the transition zone where PT conditions are suitable for phase separation.

Gas–oil ratio

After detailed laboratory analysis of the source rocks of the Bakken petroleum system, natural petroleum production data (predominantly surface produced GOR and API gravity) were assessed to be incorporated into a consistent description of the petroleum

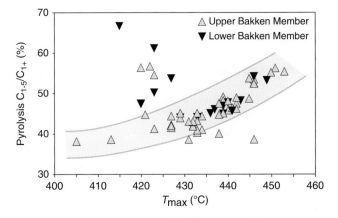

Fig. 7. Percentage of the C_{1-5}-bulk of the total C_{1+}-bulk yield based on pyrolysis–gas chromatography plotted against maturity in T_{max} (°C); in the range above 427 °C, which includes the oil window of the Bakken, a close relation between maturity and hydrocarbon yield composition is visible, irrespective of upper or lower source rock member; a continuation of this trend is also identified for a maturity interval below 415 °C. Outliers are visible for a few samples that have not yet reached the oil window, the reasons for which remain uncertain.

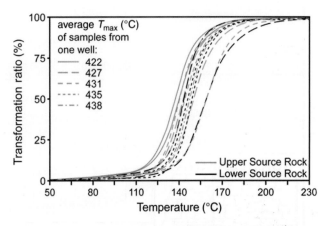

Fig. 8. Computed transformation ratios of 12 different source rock samples from five different wells and both source rock members are displayed for a constant heating rate of 3 °C/Ma; the majority of the curves cluster in a narrow range, which spans across a difference of 13 °C at a transformation of 50%.

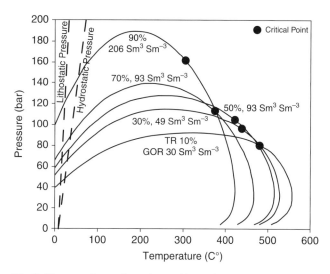

Fig. 9. Phase envelopes of petroleum of increasing maturity generated by MSSV from a Bakken Formation sample. Lithostatic and hydrostatic pressure gradients that can be expected in the Williston Basin are indicated by the dashed lines. GOR are included for the individual transformation stages in $Sm^3 Sm^{-3}$.

system. As displayed for around 500 wells, which all tested Bakken pools, in Figure 11a, the GOR ranges from <50 to c. 900 $Sm^3 Sm^{-3}$ (<280 to c. 5050 ft^3/BBL; in the following to convert to ft^3/BBL multiply values in $Sm^3 Sm^{-3}$ by a factor of 5.61), with only eight wells showing GORs in excess of 500 $Sm^3 Sm^{-3}$. While a limited number of values show the presence of gas condensates, the gross production from the Bakken has a GOR below 350 $Sm^3 Sm^{-3}$, which characterizes the Bakken as a black oil play (McCain 1990).

When GORs are plotted by well location, a correlation with the major structures is visible. In Figure 12a, the highest GORs are situated in the areas of the major anticlines, while oil-dominated fluids are distributed in all other mature areas, regardless of depth. A relationship of GOR to depth could not be identified in the total dataset as displayed in Figure 11b.

Oil density

Similar observations as for the GOR arise from the distribution of oil gravities measured on Bakken pool-producing wells in North Dakota. In the following, the API° gravity scale is used to describe the oil gravity [API° at 15.6 °C = 141.5/density (g cm^{-3}) – 131.5]. The API° histogram in Figure 11c has a unimodal distribution with a mean around 42° API, implying that more than 600 wells across the state produce a non-degraded high-quality oil. Again, a correlation of API° gravity to areas of structural deformation is evident (Fig. 12b). Furthermore, plotting the oil gravity against depth (Fig. 11d) does not generate any trend and therefore implies a limited API° variability which is independent of source rock maturity.

Even though both parameters – light API° gravity and high GOR – cluster along the same locations, an expected statistical correlation cannot be reported (Fig. 13). This independence becomes obvious when compared with a dataset of the Devonian Exshaw Formation of the neighbouring Western Canadian Basin (Evans et al. 1971), which has been added for orientation. Even though the values are in a comparable range in some parts of the datasets, a similar trend is not identifiable.

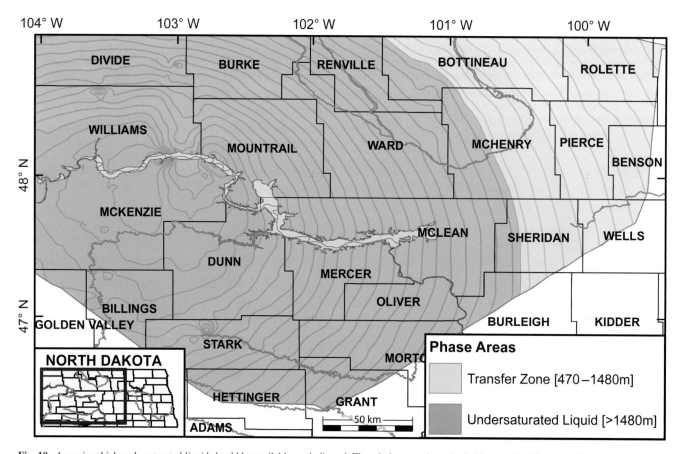

Fig. 10. Areas in which undersaturated liquid should be available are indicated. The whole area where the Bakken reached the threshold of petroleum generation is included in this area; only parts in the still immature shallower east are in the transfer zone in which first evidence of phase separation can be expected. This map is based on a depth map by LeFever (2008).

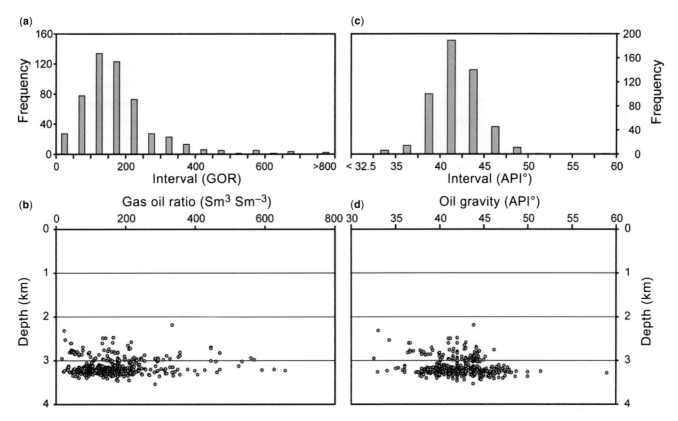

Fig. 11. GOR and oil gravity data of wells testing Bakken hydrocarbon pools from the database of the North Dakota Geological Survey is displayed: the upper left histogram (**a**) shows the GOR measured on c. 500 wells in the Williston Basin; most measured GORs are in the range from 50 to 250 $Sm^3\ Sm^{-3}$, while higher values are less frequent; the lower left graph (**b**) displays the same dataset against the depth of the Bakken Formation for their respective location; the right upper histogram (**c**) shows the oil gravity measured on c. 600 wells, most measured GORs are in the range from 37.5 to 45 API°, indicating only small variability in the Bakken Formation. In the lower right graph (**d**) the API° dataset is displayed against the depth of the Bakken Formation in their respective locations.

Further, from pyrolysis-gas chromatography and in agreement with Muscio & Horsfield (1996), petroleum compositions generated from the Bakken Formation are considerably lighter as compared with other Type II kerogen-containing source rocks like the Kimmeridge Clay or the Posidonia Formation. However, there are other marine source rocks that yield fairly light oils, for example, the Barnett shale (Hill *et al.* 2007).

Discussion

The abundance of source rock samples as well as production data allows a comprehensive investigation of the processes and mechanisms controlling petroleum generation, migration and accumulation for the Bakken Formation. Based on a collection of geochemical and production data, the variability of the Bakken source rocks is very narrow, concerning appearance, kerogen composition and general composition of the generated hydrocarbons, especially in the deeper buried western part of the Williston Basin in North Dakota. This is deduced based on the observed systematic changes in source rock properties as a function of increasing maturity as exemplified in the mass balance calculations and Figures 3a, b and 6–8. The eastern immature part of the source rock appears more variable with respect to the TOC and HI in the mature areas. The possible inhomogeneities require further investigation but might be related to their position closer to the edge of the basin that could facilitate variability in the source rock's depositional environment. Alternatively very low maturity of these samples may result in the remaining decarboxylation potential being absent in the samples that have gone through the early diagenetic stage. However, for this study their anomalous characteristics eliminate them from the list of potential immature reference samples representative of the mature western Bakken areas.

The upper and lower Bakken are indistinguishable from each other based on appearance and geochemical characteristics; accordingly they most likely both contribute to the hydrocarbon accumulations in the Bakken Formation and associated reservoirs.

The current understanding of the Bakken Formation as expressed by Pollastro *et al.* (2008) is that it is a continuous, unconventional, tight oil play where migration distances are short and accumulations are present throughout the formation, at least where the source rock is mature. Based on this description it should be expected that the in-place hydrocarbons should predominantly reflect the original source rock maturity or degree of transformation in any given location. This statement can be tested based on the data available from our own sampling as well as from the database of the North Dakota Geological Survey.

It is observed that the highest GORs and lightest oil gravities cluster along the locations of the anticlines. Further, the general lack of a depth trend for both GOR and API gravity indicate that source rock maturity, which is dependent on burial depth, does not control the petroleum properties encountered. The results from the mass balance calculations following the methodology of Cooles *et al.* (1986) indicate that, even at low transformation ratios (PGI c. 20%), the PEE reaches values above 60%, while PGIs higher than 40% result in PEEs above 90%. These values indicate that the Bakken source rock is an excellent expeller, as observed for a multitude of high-quality source rocks worldwide. In other words, there is no evidence from the available data that the Bakken source rocks have been subject to delayed expulsion due to a low-permeability surrounding environment, where

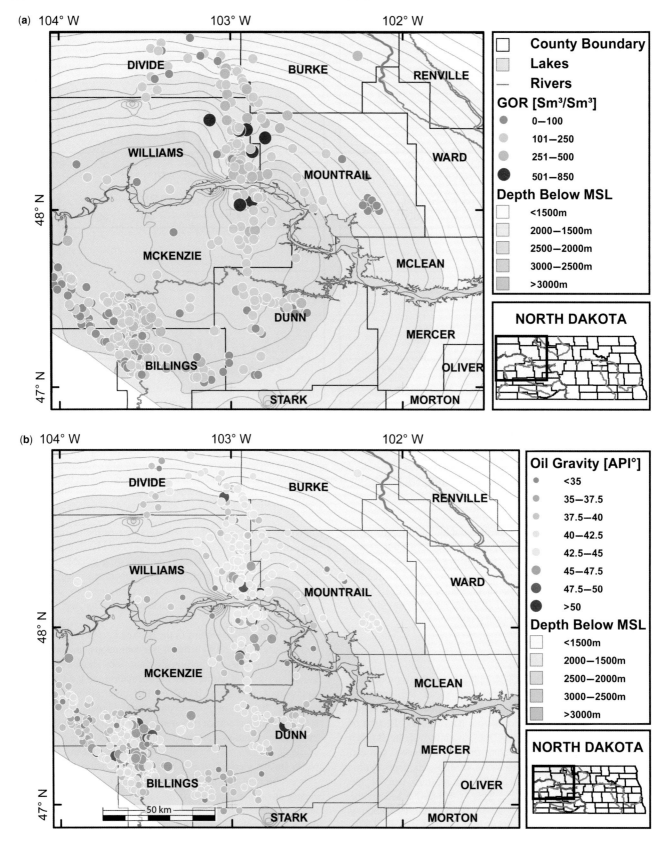

Fig. 12. Maps show the distribution of the wells with GOR (**a**) and oil gravity (**b**) measurements for the Bakken Formation; colours and size of the circles indicate the specific value at a well location.

secondary migration distances are short and a build-up of saturation during expulsion would be expected to have influenced the primary migration from the source. From these observations, the Bakken appears to be a petroleum system where generated hydrocarbons migrate to structural highs or stratigraphic traps and accumulate there. This is, of course, only a simplistic explanation for the distribution of GORs and oil gravities and contradicts current opinion on the Bakken Formation. The concept of an open Bakken petroleum system is, however, not supported by the persistent overpressure observed in the formation, which indicates a relatively tight formation where significant fluid loss has been hindered over geological periods.

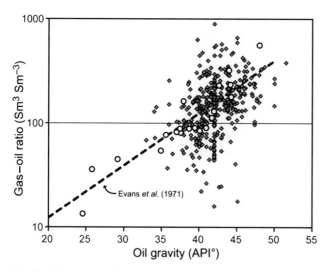

Fig. 13. GOR of the Bakken Formation plotted against the API°. For comparison, data from Evans et al. (1971) from the Devonian Exshaw from the neighbouring Western Canada Basin is included.

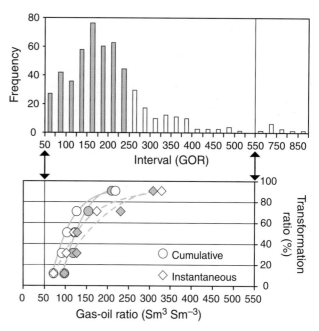

Fig. 14. Comparison of the actual measured GORs from the Bakken Formation with the laboratory based results from the PhaseKinetic approach. In the top histogram GOR values below 250 $Sm^3\ Sm^{-3}$ in grey, while higher values are white. The higher values correspond to the orange and red coloured GOR values in Figure 12a, which cluster primarily along the areas of the anticlines. In the lower graph the cumulative GOR that is generated at different transformation stages from two different immature source rock samples is displayed with circles; hydrocarbons generated at higher thermal maturity are lighter, leading to higher GORs at higher transformation stages. The cumulative generation of GORs can explain the majority of the production GORs. The diamond-shaped data points indicate the change of the GOR of the instantaneous generated petroleum. These reach higher GOR values and can therefore explain the higher GORs in the areas of the anticlines.

Laboratory simulation of hydrocarbon generation following the phase kinetic approach (di Primio & Horsfield 2006) indicates that the span of petroleum GORs generated reaches from roughly 50 $Sm^3\ Sm^{-3}$ at lowest stages of conversion to around 200 $Sm^3\ Sm^{-3}$ at complete conversion for a cumulative system. In a petroleum system characterized by short migration routes and no losses, it would be expected that reservoired petroleum should reflect the cumulative charge and, hence, be directly comparable to the laboratory predictions. This is in fact the case for the majority of the fluids tested in the basin, as is evident from the GOR histogram (Fig. 14). Fluid GORs from 50 to 250 $Sm^3\ Sm^{-3}$ can thus easily be explained. The lack of correlation between GOR and depth (i.e. maturity) indicates that the middle Bakken member petroleum is able to facilitate migration at least over moderate distances, thus erasing a clear relationship of petroleum maturity (as expressed by GOR or API gravity) and depth. The higher GORs observed, however, indicate a different situation. Higher GORs than those predicted experimentally can be explained by:

(1) variation in the source rock facies, which results in more gas-prone products; this is excluded based on the low kerogen variability of the source rock;
(2) influence of secondary cracking, which is unlikely to be dominant based on the present-day maturity of the Bakken; or
(3) dominance of a late stage generation phase, that is, an instantaneous as opposed to a cumulative composition.

The most likely explanation for the observed high GORs in the anticlines appears therefore to be leakage of the early generation stage and accumulation of late stage generation products. This process seems to have dominated only in the structurally influenced, and probably fractured anticlines, since GORs and API gravities are lower in both deeper and shallower structurally less disturbed areas of the Bakken. This observation underlines the relevance of the structural influence on the composition of accumulated hydrocarbons as opposed to the reservoir depth.

Based on the composition of Bakken-generated petroleum, which is inferred from closed system experiments and observed in the natural fluids, the phase behaviour at different pressure and temperature conditions was calculated. The applied average pressure and temperature gradients, which are very conservative since we are dealing with an overpressured formation (Murray 1968; Finch 1969; Meissner 1978), did not reach values that would allow a phase separation at depths at which mature Bakken source rocks are present. Thus, the hydrocarbons that were generated in the Bakken Formation should only exist as a single-phase undersaturated liquid, thus reducing the probability of phase separation and differential migration of oil and gas phases in the basin to a large extent.

Conclusion

Analysis of the geochemical data of a source rock maturity sequence, the GOR and gravity data of more than 600 wells and earlier published rock-evaluation data provided a detailed insight into the development and behaviour of source rock potential and generated petroleum in the Bakken Formation.

The upper and lower members of the Bakken Formation represent excellent source rocks which generate and very efficiently expel high API gravity petroleum. Source rock maturity within the basin is constrained to the oil window, reaching almost total kerogen conversion in the deepest regions of the Williston Basin. Secondary cracking of residual oil to gas, based on source rock maturity, is irrelevant in the Bakken Formation.

The middle Bakken member is a low-permeability carrier in which the generated and expelled petroleum is able to migrate laterally and accumulate in structural or stratigraphic traps.

The continuous nature of the middle Bakken accumulations is not apparent in our data, where fluid properties and source rock maturity do not correlate. In areas where structural deformation has increased the permeability of the Bakken system, hydrocarbon losses to surrounding formations through vertical leakage are assumed to have occurred, resulting in the isolation of a late mature petroleum phase in the reservoir.

We would like to thank the North Dakota Geological Survey, the Wilson M. Laird Core and Sample Library and staff for providing the core samples of the Bakken Formation and database access. Special thanks to J. A. LeFever for assistance and advice with the samples. Further, we would like to acknowledge D. M. Jarvie and an additional reviewer for their constructive and helpful comments and the effort they invested to improve the quality of the manuscript. The research on the Bakken Formation is part of the GFZ Industry Partner Project 'Predicting Petroleum Quality' that is sponsored by Chevron, ConocoPhillips, Devon Energy, Petrobras, and Statoil. This publication is part of a PhD study on the organic geochemistry of the Bakken Formation (P.P.K.).

References

Baskin, D. K. 1997. Atomic H/C ratio of kerogen as an estimate of thermal maturity and organic matter conversion. *American Association of Petroleum Geologists Bulletin*, **81**, 1437–1450.

Boudou, J. P., Espitalie, J., Bimer, J. & Salbut, P. D. 1994. Oxygen groups and oil suppression during coal pyrolysis. *Energy & Fuels*, **8**, 972–977.

Braun, R. L. & Burnham, A. K. 1990. Mathematical model of oil generation, degradation, and expulsion. *Energy & Fuels*, **4**, 132–146.

Burrus, J., Osadetz, K., Wolf, S., Doligez, B., Visser, K. & Dearborn, D. 1996a. A two-dimensional regional basin model of Williston Basin hydrocarbon systems. *American Association of Petroleum Geologists Bulletin*, **80**, 265–291.

Burrus, J., Wolf, S., Osadetz, K. & Visser, K. 1996b. Physical and numerical modelling constraints on oil expulsion and accumulation in the Bakken and Lodgepole petroleum systems of the Williston Basin (Canada-USA). *Bulletin of Canadian Petroleum Geology*, **44**, 429–445.

Carlson, C. G. & Anderson, S. B. 1965. Sedimentary and tectonic history of North Dakota part of Williston Basin. *American Association of Petroleum Geologists Bulletin*, **49**, 1833–1846.

Cooles, G. P., Mackenzie, A. S. & Quigley, T. M. 1986. Calculation of petroleum masses generated and expelled from source rocks; advances in organic geochemistry, 1985; Part I, Petroleum geochemistry. *Organic Geochemistry*, **10**, 235–245.

Coskey, R. J. & Leonard, J. E. 2009. Bakken oil accumulations – what's the trap? *In*: *American Association of Petroleum Geologists Annual Convention & Exhibition*. American Association of Petroleum Geologists, Denver, CO, 48.

Dembicki, H., Jr. & Pirkle, F. L. 1985. Regional source rock mapping using a source potential rating index. *American Association of Petroleum Geologists Bulletin*, **69**, 567–581.

di Primio, R. & Horsfield, B. 1996. Predicting the generation of heavy oils in carbonate/evaporitic environments using pyrolysis methods. Proceedings of the 17th International Meeting on Organic Chemistry; Part III, Origin of natural gases; petroleum geochemistry, impact of organic geochemistry on exploration; migration and expulsion of oil and gas. *Organic Geochemistry*, **24**, 999–1016.

di Primio, R. & Horsfield, B. 2006. From petroleum-type organofacies to hydrocarbon phase prediction. *American Association of Petroleum Geologists Bulletin*, **90**, 1031–1058.

Dow, W. G. 1974. Application of oil-correlation and source-rock data to exploration in Williston basin. New ideas; origin, migration and entrapment of oil, symposium. *American Association of Petroleum Geologists Bulletin*, **58**, 1253–1262.

Eglinton, T. I, Damste, J. S. S., Kohnen, M. E. L. & Deleeuw, J. W. 1990. Rapid estimation of the organic sulfur-content of kerogens, coals and asphaltenes by pyrolysis-gas chromatography. *Fuel*, **69**, 1394–1404.

Espitalié, J., Laporte, J. L. *et al.* 1977. Methode rapide de caracterisation des roches meres de leur potentiel petrolier et de leur degre d'evolution. *Revue de l'Institute Francais du Petrole*, **32**, 23–42.

Evans, C. R., Rogers, M. A. & Bailey, N. J. L. 1971. Evolution and alteration of petroleum in western Canada. *Chemical Geology*, **8**, 147–170.

Finch, W. C. 1969. Abnormal pressure in the Antelope field, North Dakota. *Journal of Petroleum Technology*, **21**, 821–826.

Flannery, J. & Kraus, J. 2006. *Integrated Analysis of the Bakken Petroleum System, U.S. Williston Basin*. American Association of Petroleum Geologists Annual Convention, 10–12 April 2006. Search and Discovery.

Fritz, M. 1990. Promising play endures. *American Association of Petroleum Geologists Explorer*, **11**, 14–15.

Hansen, A. R. 1972. Petroleum and natural gas; the Williston Basin. *In*: *Geological Atlas of the Rocky Mountain Region*. Rocky Mountain Association of Geologists, Denver, CO.

Hayes, M. D. & Holland, F. D., Jr. 1983. Conodonts of Bakken Formation (Devonian and Mississippian), Williston Basin, North Dakota. *In*: *American Association of Petroleum Geologists Rocky Mountain Section Meeting, 67*, Billings, MT. American Association of Petroleum Geologists, Tulsa, OK, Bulletins, 1341–1342.

Hill, R. J., Jarvie, D. M., Zumberge, J., Henry, M. & Pollastro, R. M. 2007. Oil and gas geochemistry and petroleum systems of the Fort Worth Basin. *American Association of Petroleum Geologists Bulletin*, **91**, 445–473.

Horsfield, B. 1997. The bulk composition of first-formed petroleum in source rocks. *In*: Welte, D. H., Horsfield, B. & Baker, D. R. (eds), *Petroleum and Basin Evolution*. Springer, Berlin, 335–402.

Horsfield, B. & Dueppenbecker, S. J. 1991. The decomposition of Posidonia shale and green river shale kerogens using microscale sealed vessel (MSSV) pyrolysis. *Journal of Analytical and Applied Pyrolysis*, **20**, 107–123.

Horsfield, B., Disko, U. & Leistner, F. 1989. The micro-scale simulation of maturation; outline of a new technique and its potential applications; geological modelling; aspects of integrated basin analysis and numerical simulation. *Geologische Rundschau*, **78**, 361–374.

Horsfield, B., Dueppenbecker, S. J., Schenk, H. J. & Schaefer, R. G. 1993. Kerogen typing concepts designed for the quantitative geochemical evaluation of petroleum potential; basin modelling; advances and applications. *Proceedings of the Norwegian Petroleum Society Conference*. Norwegian Petroleum Society, Trondheim, Special Publications, **3**, 243–249.

Jarvie, D. M. 2001. Williston Basin petroleum systems; inferences from oil geochemistry and geology. *The Mountain Geologist*, **38**, 19–41.

Jarvie, D. M., Hill, R. J., Ruble, T. E. & Pollastro, R. M. 2007. Unconventional shale-gas systems; the Mississippian Barnett Shale of North-Central Texas as one model for thermogenic shale-gas assessment; Special issue; Barnett Shale. *American Association of Petroleum Geologists Bulletin*, **91**, 475–499.

Jiang, C. & Li, M. 2002. Bakken/Madison petroleum systems in the Canadian Williston Basin; Part 3, Geochemical evidence for significant Bakken-derived oils in Madison Group reservoirs. *Organic Geochemistry*, **33**, 761–787.

Johnson, M. S. 2009. Parshall field, North Dakota, discovery of the year for the Rockies and beyond. *In*: *American Association of Petroleum Geologists Annual Convention & Exhibition*, Denver, CO, 109.

Killops, S. D., Funnell, R. H. *et al.* 1998. Predicting generation and expulsion of paraffinic oil from vitrinite-rich coals. *Organic Geochemistry*, **29**, 1–21.

LeFever, J. A. 2005a. The Bakken play of Montana and North Dakota. *The Bulletin of the Houston Geological Society*, **47**, 45.

LeFever, J. A. 2005b. Oil production from the Bakken Formation; a short history. *NDGS Newsletter*, **32**, 5–10.

LeFever, J. A. 2008. Bakken formation map series. *In*: LeFever, J. A. (ed.) *Geological Investigations*. North Dakota Geological Survey, Bismarck, ND.

Mann, U., Hantschel, T. *et al.* 1997. Petroleum migration; mechanisms, pathways, efficiencies and numerical simulation. *In*: Welte, D. H., Horsfield, B. & Baker, D. R. (eds) *Petroleum and Basin Evolution*. Springer, Berlin, 403–520.

McCain, W. D. 1990. *The Properties of Petroleum Fluids*. Penn Well Books, Tulsa, OK.

Meissner, F. F. 1978. Petroleum geology of the Bakken Formation, Williston Basin, North Dakota and Montana. *In*: *Williston Basin Symposium; 24th Annual Conference*. Montana Geological Society, Billings, MT.

Murphy, E. C., Nordeng, S. H., Juenker, B. J. & Hoganson, J. W. 2009. *In*: Helms, L. D. (ed.) *North Dakota Stratigraphic Column*. Miscellaneous Series 91. North Dakota Geological Survey, Bismarck, ND.

Murray, G. H., Jr. 1968. Quantitative fracture study; Sanish pool, McKenzie County, North Dakota. *American Association of Petroleum Geologists Bulletin*, **52**, 57–65.

Muscio, G. 1995. *The fate of oil and gas in a constrained natural system – implications from the Bakken petroleum system.* PhD thesis, Institut für Chemie und Dynamik der Geosphäre 4: Erdöl und Geochemie, RWTH Aachen University, Jülich.

Muscio, G. P. A. & Horsfield, B. 1996. Neoformation of inert carbon during the natural maturation of a marine source rock: Bakken Shale, Williston Basin. *Energy & Fuels*, **10**, 10–18.

Muscio, G. P. A., Horsfield, B. & Welte, D. H. 1994. Occurrence of thermogenic gas in the immature zone – implications from the Bakken in-source reservoir system. *Organic Geochemistry*, **22**, 461–476.

Osadetz, K. G., Snowdon, L. R. & Brooks, P. W. 1994. Oil families in Canadian Williston Basin (southwestern Saskatchewan). *Bulletin of Canadian Petroleum Geology*, **42**, 155–177.

Osadetz, K. G., Kohn, B. P., Feinstein, S. & O'Sullivan, P. B. 2002. Thermal history of Canadian Williston basin from apatite fission-track thermochronology; implications for petroleum systems and geodynamic history; low temperature thermochronology; from tectonics to landscape evolution. *Tectonophysics*, **349**, 221–249.

Pitman, J. K., Price Leigh, C. & LeFever Julie, A. 2001. *Diagenesis and Fracture Development in the Bakken Formation, Williston Basin; Implications for Reservoir Quality in the Middle Member.* US Geological Survey Professional Paper, **1653**, 19.

Pollastro, R. M., Cook, T. A. *et al.* 2008. *National Assessment of Oil and Gas Fact Sheet: Assessment of Undiscovered Oil Resources in the Devonian–Mississippian Bakken Formation, Williston Basin Province, Montana and North Dakota, 2008.* US Department of the Interior, US Geological Survey, 2.

Price, L. C. & LeFever, J. A. 1992. Does Bakken horizontal drilling imply a huge oil-resource base in fractured shales? *In*: Schmoker, J. W. C., Brown, E. B. & & Charles, A. (eds) *Geological Studies Relevant to Horizontal Drilling; Examples from Western North America.* Rocky Mountain Association of Geologists, Denver, CO, 199–214.

Price, L. C. & LeFever, J. 1994. Dysfunctionalism in the Williston Basin; the Bakken/mid-Madison petroleum system. *Bulletin of Canadian Petroleum Geology*, **42**, 187–218.

Price, L. C., Ging, T., Daws, T. A., Love, A., Pawlewicz, M. J. & Anders, D. E. 1984. Organic metamorphism in the Mississippian-Devonian Bakken Shale, North Dakota portion of the Williston Basin. *In*: Woodward, J., Meissner, F. F. & Clayton, J. L. (eds) *Rocky Mountain Association of Geologists 1984 Symposium; Hydrocarbon Source Rocks of the Greater Rocky Mountain Region.* Rocky Mountain Association of Geologists, Denver, CO, 83–133.

Requejo, A. G., Allan, J., Creaney, S., Gray, N. R. & Cole, K. S. 1992. Aryl isoprenoids and diaromatic carotenoids in Palaeozoic source rocks and oils from the Western Canada and Williston Basins. *Organic Geochemistry*, **19**, 245–264.

Ropertz, B. 1994. *Wege der primären Migration: Eine Untersuchung über Porennetze, Klüfte und Kerogennetzwerke als Leitbahnen für den Kohlenwasserstoff-Transport.* PhD thesis, Institut für Chemie und Dynamik der Geosphäre 4: Erdöl und Geochemie, RWTH Aachen University, Jülich.

Santamaría-Orozco, D. & Horsfield, B. 2003. *Gas Generation Potential of Upper Jurassic (Tithonian) Source Rocks in the Sonda de Campeche, Mexico. In*: Bartolini, C., Buffler, R. T. & Blickwede, J. F. (eds) American Association of Petroleum Geologists Memoirs, **79**. American Association of Petroleum Geologists, Tulsa, OK, 349–363.

Schenk, H. J., Di Primio, R. & Horsfield, B. 1997. The conversion of oil into gas in petroleum reservoirs; Part 1, Comparative kinetic investigation of gas generation from crude oils of lacustrine, marine and fluviodeltaic origin by programmed-temperature closed-system pyrolysis. *Organic Geochemistry*, **26**, 467–481.

Schmoker, J. W. & Hester, T. C. 1983. Organic carbon in Bakken Formation, United States portion of Williston Basin. *American Association of Petroleum Geologists Bulletin*, **67**, 2165–2174.

Smith, M. G. & Bustin, R. M. 1998. Production and preservation of organic matter during deposition of the Bakken Formation (Late Devonian and Early Mississippian), Williston Basin. *Palaeogeography, Palaeoclimatology, Palaeoecology*, **142**, 185–200.

Sonnenberg, S. A. & Pramudito, A. 2009. Petroleum geology of the giant Elm Coulee Field, Williston Basin. *American Association of Petroleum Geologists Bulletin*, **93**, 1127–1153.

Spencer, C. W. 1987. Hydrocarbon generation as a mechanism for overpressuring in Rocky Mountain region. *American Association of Petroleum Geologists Bulletin*, **71**, 368–388.

Sweeney, J. J., Gosnold, W. D., Braun, R. L. & Burnham, A. K. 1992. *A Chemical Kinetic Model of Hydrocarbon Generation from the Bakken Formation, Williston Basin North Dakota.* Lawrence Livermore National Laboratory, Livermore, CA, 60.

Tissot, B. P. & Welte, D. H. 1984. *Petroleum Formation and Occurrence.* Springer, Heidelberg.

Ungerer, P. 1990. State of the art of research in kinetic modelling of oil formation and expulsion. *Organic Geochemistry*, **16**, 1–25.

Webster, R. L. 1984. Petroleum source rocks and stratigraphy of the Bakken Formation in North Dakota. *In*: Woodward, J., Meissner, F. F. & Clayton, J. L. (eds) *Rocky Mountain Association of Geologists 1984 Symposium; Hydrocarbon Source Rocks of the Greater Rocky Mountain region.* Rocky Mountain Association of Geologists, Denver, CO, 57–81.

Weiss, H. M., Wilhelms, A. *et al.* 2000. *The Norwegian Industry Guide to Organic Geochemical Analyses.* Norsk Hydro, Statoil, Geolab Nor, SINTEF Petroleum Research, Norwegian Petroleum Directorate.

Williams, J. A. 1974. Characterization of oil types in Williston Basin; new ideas; origin, migration and entrapment of oil; symposium. *American Association of Petroleum Geologists Bulletin*, **58**, 1243–1252.

Shale gas in Europe: a regional overview and current research activities

H.-M. SCHULZ,[1] B. HORSFIELD[1] and R. F. SACHSENHOFER[2]

[1]*GFZ German Research Centre for Geosciences, Telegrafenberg, D-14473 Potsdam, Germany*
(e-mail: schulzhm@gfz-potsdam.de)
[2]*University of Leoben, Chair of Petroleum Geology, Peter-Tunner-Straße 5,*
A-8700 Leoben, Austria

Abstract: Shale gas is produced from fine-grained siliciclastic sediments that are typically rich in organic carbon. Nearly all shales contain thermal gas generated *in situ* at mature to overmature levels of thermal alteration, although gas of biogenic origin is also produced from some shales. While shale gas production in the USA began in 1821, it is only in the last few years that it has become widely significant (currently about 8% of the domestic gas). In contrast, European shale gas exploration is still in its infancy. In general, European sedimentary basins offer the best potential for shale gas occurrence because thick, organic matter-rich sediments occur in nearly all Phanerozoic strata. Even so, there is little knowledge about the factors controlling shale gas generation and, more importantly, shale gas production in European basins. These factors are not necessarily the same as those that control commercial shale gas production in the USA. Palaeozoic sediments of Cambrian to Ordovician age are currently being tested for their shale gas potential and productivity in Sweden, as are those of Silurian age in Poland. Moreover, Lower and Upper Carboniferous sedimentary successions from England in the west to Poland in the east probably contain shale gas, but their depth, thickness and thermal maturity may be limiting factors for exploration in continental regions. Lower Carboniferous black shales in the Dniepr–Donets Basin of the Ukraine may also hold a significant potential. Moreover, organic-rich sediments of Oligocene/Miocene age in the Paratethyan Basin may offer shale gas potential, for example in the Pannonian Basin. At present, Upper Jurassic black shales are currently being tested for their shale gas potential in the Vienna Basin. European analogues of known biogenic shale gas systems may occur locally in organic-rich Lower Cretaceous sediments in the North German Basin with gas generation being related to Pleistocene glaciation/deglaciation cycles.

Keywords: Shale gas, Europe, sedimentary basin, maturity, organic carbon, geological controls, resource potential

A great many formations with inherent shale gas potential occur throughout the world. Whether these potential targets emerge as economically viable depends not only upon geological characteristics, such as gas-in-place and rock properties, but also upon whether a distribution infrastructure exists and markets are readily accessible. Commercial production of shale gas occurs exclusively in North America to date. Here, shale gas is amongst the most active plays in recent years as a result of relatively high gas prices and the remarkable technological advances that have enabled the shales to be fractured and their gas potential exploited, beginning with the Mississippian Barnett Shale of the Fort Worth Basin (Curtis 2002; Frantz *et al.* 2006).

Annual natural gas production from shale gas reservoirs in the USA is approximately 1.0×10^{12} SCF and comes from more than 40 000 shale gas wells of which nearly half have been drilled in the last 10 years (around 6–10% of total natural gas production in the USA today; numbers vary according to reference). The Energy Information Agency (EIA) forecasts that shale gas production in the USA will more than double by 2020 if advanced exploration technologies are applied, and possibly reach 50% all told. The Haynesville Formation (Upper Jurassic, East Texas–North Louisiana, USA) is the most recent and promising addition to the existing US shale gas plays. Although production data are only preliminary because the basin is underexplored, the potential of the Haynesville is probably many times larger than that of the Barnett Shale. Very promising initial production rates, but steeper decline rates, have been reported (about 80% first year projected declines; Phasis 2008). A very high potential for shale gas has been determined in the Western Canada Basin (estimates are up to 1000×10^{12} SCF; Phasis 2008) and for the Devonian Muskwa Shale of the Horn River Basin, British Columbia, which is showing outstanding initial production rates (D. Jarvie pers. comm.).

A first estimation of global shale gas resources was published by Rogner (1997; $16\,112 \times 10^{12}$ SCF) with North America and China as the regions with the largest potential (both around 3000–4000 $\times 10^{12}$ SCF).

According to this study, Western Europe is a region with a lower potential shale gas resource ($\sim 510 \times 10^{12}$ SCF; Rogner 1997). It is true that Europe's geological setting is strongly compartmentalized and prospective basins of the size of the Fort Worth Basin, in other words roughly the area of Belgium, are therefore unlikely to occur. However, as black shales with gas window thermal maturity occur in almost all European Phanerozoic formations, a significant potential for exploration success exists and Europe is well positioned to buffer local energy demands.

European shale gas exploration is still in its infancy, although the first ideas to search for this unconventional gas resource were published by Selley (1987) about 20 years ago. Today, many companies have included this topic in their portfolios and actively explore for shale gas in Europe. For example, applications for licences to explore and develop shale gas were submitted during the 13th Onshore Round (launched in 2007) by the UK Department for Business Enterprise and Regulatory Reform (BERR). Moreover, the significance of shale gas is visible as, for example, shale gas prospects are available throughout Europe in large acreage positions, e.g. in France.

In recent years high energy prices and advances in fracturing technology and directional drilling have enabled shale gas production to become a lucrative reality in the USA, also because infrastructure is already in place. It is the predictions ahead of drilling of gas-in-place and rock properties that are of paramount importance for reducing risk and identifying 'sweet spots' or fairways. In other words, there is still a lot to learn scientifically about how shale gas systems actually work. Significant advances have been made over the last five years, but there are great opportunities for major advances in science and technology.

The well known shale gas systems in the USA differ from one another as regards the key factors which controlled shale gas formation, and therefore exhibit different shale properties and gas compositions. Thus, the gas shale systems may contain thermogenic (Mississipian Barnett Shale) or biogenic gas (Devonian Antrim Shale), or mixtures of both (Cretaceous Lewis Shale; Fishman & Hill 2008). It is therefore difficult to transfer a single genetic model to potential European systems.

In this short contribution, general geological and shale gas-relevant features of selected onshore European black shales are introduced and analysed for significance. Moreover, known industry activities are presented alongside major basic research activities.

Stratigraphic overview of selected potential gas shales and case studies

Cambro-Ordovician black shales in northern Europe

A prominent example from northern Europe is the Scandinavian Alum Shale (Upper Cambrian to Lower Ordovician), exploited over the last century and more for heavy metals like uranium and vanadium. Total organic carbon (TOC) contents can exceed 20% (Schovsbo 2003; Nielsen & Schovsbo 2007, and references therein), and the organic matter, albeit of marine origin, has a high intrinsic potential for generating and storing gas at all maturity levels (Horsfield et al. 1992). A significant gas potential has also been found by hydrous pyrolysis experiments (Buchardt 1999). In this region the Alum Shale displays organic matter content and type, maturity and thicknesses that are typical of productive shale gas areas (Dyni 2006). Because of deeper burial in a southward direction the Alum Shale exhibits maturities of more than 2% vitrinite reflectance (random reflectance R_r) in the southern Baltic Sea region. Currently, the Alum Shale is being evaluated by Shell for its shale gas potential in southern Sweden, where the sediments are about 100 m thick.

Silurian black shales in Poland

Organic-rich sediments of the Silurian are widespread in Europe and contain intervals with TOC contents of more than 20%. In the Danish–Polish Trough organic-rich sediments can be more than 200 m thick and are proven source rocks for oils in the Baltic region. In general, the maturity of the sedimentary organic matter increases from the NE part of the Baltic Syneclise towards the Teisseyre–Tornquist Zone (Zdanaviciute & Lazauskiene 2004, 2007; Fig. 1). It is not clear yet whether these sedimentary successions contain shale gas systems. 3Legs Resources plc has secured licences covering over 1 000 000 acres in the Baltic region of Poland to investigate shale gas potential.

Of the Silurian formations outside Poland that may contain shale gas potential, an excellent example is afforded by the Barrandian Basin of the Czech Republic where shales of Llandovery to Ludlow age have TOC contents of up to 8% and graptolite reflectances of up to 1.53% R_r (Volk et al. 2002).

Carboniferous black shales of the Variscan Foreland Basin

Investigations of Lower Carboniferous sediments from Ireland in the west to Poland in the east indicate an extended, northern

Fig. 1. Maturity, thickness and TOC contents of the Silurian dark-grey argillaceous complex (modified after Zdanaviciute & Lazauskiene 2004).

calcareous shelf ('Carboniferous limestone facies') grading into deep basins north of the uprising Variscan orogen. The deep basins were filled by siliciclastic material sourced from the southern mountain chains. Fine-grained sediments of the basinal 'Culm' facies were enriched with organic carbon during periods of high productivity during the deposition of the Tournaisian Alum Shale ('Lower Alum Shale'; Gursky 2006). These sediments have TOC contents of more than 10% in the Rhenish Massif, for example (Siegmund et al. 2002). Interestingly, they are also regarded as good magnetotelluric conductors in the North German Basin (Al Duba et al. 1988; Hoffmann et al. 2001). The palaeogeographic situation during the Upper Visean and Serpukhovian (Early Namurian) is comparable to the Tournaisian situation with black shale deposition in deep basins (e.g. the Serpukhovian 'Upper Alum Shale' with TOC contents up to 3% in Germany; Zimmerle & Stribrny 1992). The belt with similar black shales extends to the west as the Bowland Shale facies (Hoffmann et al. 2001).

Locally high maturities (3–7% R_o) along outcrops of Lower Carboniferous sediments in the Rhenish Massif may exclude shale gas prospectivity. On the other hand, maturities of Lower Carboniferous shales below 3% R_o in central Germany (e.g. Harz mountains) as well as in boreholes in NE Germany indicate a potential for the occurrence of productive gas shales (Franke & Neumann 1999).

Gas-prone Westphalian black shales have gas window maturities along the Baltic coast of eastern Germany and Poland. Thicknesses of more than 1 km with frequent shaly intervals represent promising shale gas systems with ongoing gas generation (Hoth 1997). Towards the SW, increasing overburden coupled with increasing maturity and local intrusions during the Carboniferous/Permian transition may restrict shale gas occurrence. Locally high nitrogen contents of conventional gas shows in Namurian intervals might also point to reduced shale gas calorific value.

Both a thermogenic and biogenic shale gas potential is assumed for basinal Lower Carboniferous sediments across England and Wales, and for the Oil Shale Group in the Midland Valley of Scotland (Selley 2005). Details about UK data and their ramifications for shale gas prospectivity are discussed by Smith et al. (2010). Moreover, Lower Carboniferous black shales in the Northwest Carboniferous Basin in Ireland (Bundoran Formation; Clayton

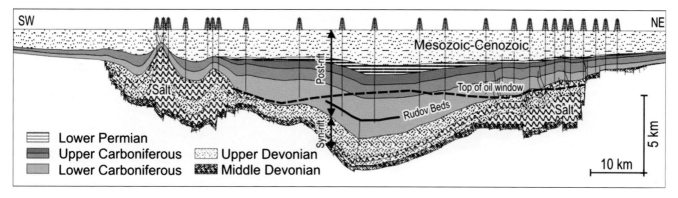

Fig. 2. Cross-section through the northwestern part of the Dniepr–Donets Basin (modified after Kivshik *et al.* 1993). Position of the Rudov Beds at the base of the Upper Visean section is indicated.

& Baily 1999) and in the Pomerian Basin in NW Poland (Lobzonka Shale Formation, Sapolno Calcareous Shale Formation; Matyja 2008) also are promising for shale gas.

Carboniferous black shales of the Dniepr–Donets Basin

The Dniepr–Donets Basin (DDB), a Late Devonian rift structure located within the East European Craton in Ukraine, hosts a large number of conventional gas and oil fields (Kabyshev *et al.* 1998; Ulmishek 2001). The basin fill includes about 4 km of syn-rift deposits and a Carboniferous to Cenozoic post-rift succession (Fig. 2) whose thickness increases towards the SE and reaches 15 km near the transition to the coal-bearing Donbas Foldbelt. Salt activity, tectonic reactivations and erosional events in the Permian and Late Cretaceous/Paleocene occurred during the post-rift stage (Stephenson *et al.* 2006 and references cited therein).

The Lower Carboniferous succession in the DDB is characterized by a high number of transgressive–regressive cycles controlled by tectonics and eustatic sea-level fluctuations (Dvorjanin *et al.* 1996). These cycles include shallow marine organic-rich shaly horizons several tens of metres thick. Typical TOC contents of these horizons range from 2 to 4%, though TOC contents averaging 6% and values as high as 12% occur in the 'Rudov Beds' defining the base of the Upper Visean section. This black shale, up to 70 m thick, is considered to be the main petroleum source rock in the DDB (Gavrish *et al.* 1994; Machulina & Babko 2004). According to Rock-Eval data, the organic matter in Upper Visean shales is mainly of kerogen type III–II (Kabyshev *et al.* 1999). Oil-prone coals with a Serpukhovian age occur mainly along the southern basin margin, whereas highly oil-prone shales with TOC contents above 10% occur locally in the northwestern part of the basin. Pelitic rocks in Bashkirian and Moscovian horizons typically have low TOC contents, but coal seams are widespread in the transition zone to the Donbas Foldbelt.

According to Shpak (1989) and Shymanovskyy *et al.* (2004), vitrinite reflectance at 5 km depth ranges from less than 1% R_o to 2% R_o in the northwestern DDB. Values above 2% R_o occur in the southeastern part of the DDB and the transition zone to the Donbas Foldbelt (Shpak 1989). At least in its northwestern part, the main oil and gas generation in Lower Carboniferous rocks occurred during Late Carboniferous/Permian time (Shymanovskyy & Sachsenhofer 2002).

Based on maturity and TOC contents, Upper Visean rocks (especially Rudov Beds) are considered the most interesting from a shale gas perspective.

Lower Jurassic black shales

Early Toarcian black shales are widespread in Europe (Farrimond *et al.* 1989). Occasionally, the sediments have TOC contents up to 19% and the marine organic matter enrichment is attributed to oceanic anoxic events (Schouten *et al.* 2000). These sediments are proven hydrocarbon source rocks in the South Central Graben in the North Sea, the Paris Basin and the Lower Saxony Basin (for an overview see Dill *et al.* 2008).

Early Toarcian black shales such as the Posidonia Shale display highly variable and locally very high maturity in the areas of Bramsche, Uchte and Vlotho (so-called massifs) in the Lower Saxony Basin. Maturation history is controlled by either deep burial at low, moderate or elevated heat flows or of shallow burial at very high heat flows such as in the vicinity of magmatic intrusions or volcanoes (Petmecky *et al.* 1999). The maturity of the Posidonia Shale with a thickness of several tens of metres increases constantly towards the centre of the massifs and can reach values of more than 4% R_o (Rullkötter *et al.* 1988).

Upper Jurassic shale gas in the Vienna Basin

The Vienna Basin is about 200 km long and 55 km wide, and extends from the Czech and Slovak Republics in the north to Austria in the south. It is of Miocene age, and lies superimposed on the outer allochthonous nappes of the Alpine–Carpathian thrustbelt (Kováč *et al.* 2004). Since conventional gas production started during the 1930s, more than 3000 wells have been drilled. The Vienna Basin is a mature oil and gas province with at least 46 fields. However, there is additional potential for 15×10^{12} SCF of shale gas in the Upper Jurassic Mikulov Marls (Langanger 2008), which is the main source rock for oil and gas in the basin (Ladwein 1988; Picha & Peters 1998; Adámek 2005). The TOC content of the more than 1500 m thick interval can exceed 10% and is composed of a kerogen type II–III (Ladwein *et al.* 1991). Mature gas-containing rocks are found below 5000 m, and cover an area of 500–800 km^2 (Fisher 2008).

Fractured intervals with gas kicks in the Mikulov Marls (Kimmeridgian to Upper Tithonian) have been observed at great depths (7500–8500 m in well Zistersdorf UT 2a; Wessely 1990; Fig. 3). This fracturing is argued to be the result of a pressure increase due to hydrocarbon generation.

The shale gas potential of the Mikulov Marls is currently under investigation by OMV (Vienna), utilizing horizontal drilling, completion and fracture stimulation and using an existing deep well.

Jurassic black shales are locally present also in the Carpathian foreland in Poland and Ukraine, but with less effective shale gas properties (Lafargue *et al.* 1994).

Neogene shale (basin centred) gas in the Makó Trough (Pannonian Basin)

The Makó Trough, located in SE Hungary and extending into Romania, is a 75 km long, NW–SE trending sub-basin of the

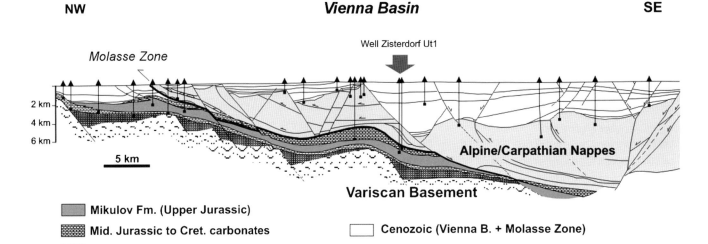

Fig. 3. Geological cross-section of the Vienna Basin highlighting the Mikulov Marl as a potential shale gas system (modified and simplified after Arzmüller et al. 2006).

Pannonian Basin System (Fig. 4). The total sediment thickness of Neogene–Quaternary rocks amounts to about 7000 m. Subsidence commenced during late Early Miocene times. Marine transgressions reached the Makó Trough during the Middle Miocene when black shales were deposited in oxygen-depleted deep marine settings (Haas 2001). Maximum water depth was attained during the Early Pannonian. This event is documented by the presence of pelagic fine-grained rocks of the Tótkomlós Member and the overlying turbidite-rich Szolnok Formation. At the end of the Pannonian, subsidence slowed down and the Makó Trough was filled by delta-slope and delta-plain deposits (post-rift phase).

Subsidence rates increased again during Late Pliocene times as a result of east–west directed compression (cf. Horváth & Cloetingh 1996).

Middle Miocene basinal rocks (e.g. Makó Formation) and the Late Miocene Tótkomlós Member are considered the main sources of oil and gas (Dolton 2006). TOC contents range up to 5% (Middle Miocene) and 2% (Late Miocene), respectively, and are composed mainly of kerogen type III (and II) (Clayton et al. 1994). Minor source rocks occur in the basinal marls of the Szolnok Formation, whereas the delta-plain deposits include excellent source rocks, but are probably immature.

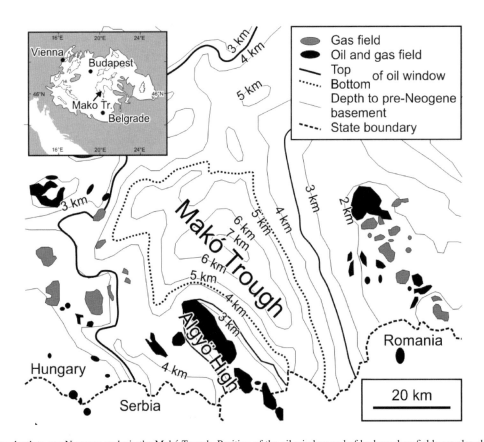

Fig. 4. Map showing depth to pre-Neogene rocks in the Makó Trough. Position of the oil window and of hydrocarbon fields are also shown (after Korosi 2007). Insert illustrates position of the Makó Trough in the Pannonian Basin system.

The oil window in SE Hungary is located in the depth range 2.4–4.3 km (Horváth et al. 1988). High heat flows (80–120 mW m^{-2}) and geothermal gradients (4–8 °C/100 m) are associated with lithospheric thinning. Hydrocarbon generation started during the latest Miocene and is still in progress (Magyar et al. 2006). The generated hydrocarbons filled major oil and gas fields above the surrounding basement highs (Horváth & Tari 1999; Tari & Horváth 2006), including the giant Algyö field, which produced 31×10^6 tonnes of oil and 70×10^9 m^3 of gas (Magyar et al. 2006). However, major resources estimated at about 600×10^9 m^3 of potential recoverable unconventional gas (shale gas and basin centred gas) remain within the Makó Trough (Korosi 2007) and are currently being evaluated by ExxonMobil, MOL and Falcon Oil & Gas.

Is there potential to find Antrim Shale-analogues with biogenic methane in Europe?

Before the Barnett became *the* shale gas system in the USA, shale gas was predominantly produced from the Upper Devonian Antrim Shale along the margins of the Michigan Basin during the 1990s (Curtis 2002). The gas in these areas is predominantly of biogenic origin (δ^{13}C as low as $-56‰$) and contains significant amounts of CO_2 (Martini et al. 1998; Jenkins & Boyer II 2008). Post-Pleistocene deglaciation resulted in a decrease of hydrostatic pressure. Meteoric water circulating along newly created weak zones diluted the highly saline pore water and enabled microbes to produce biogenic methane (Formolo et al. 2008). Is such a scenario applicable to other parts of northern Europe?

Organic matter-rich intervals of Lower Cretaceous age (Wealden, Barremian–Aptian boundary; Littke et al. 1998) may have undergone similar geological pathways, e.g. in the Lower Saxony Basin (Germany). This region was glaciated during the Pleistocene and experienced a similar geological history to the Michigan Basin. Intervals of Wealden sediments and along the Barremian/Aptian transition may have a biogenic shale gas potential in this basin. Moreover, some prerequisites for shale gas formation may also exist in the Puchkirchen Formation (Upper Oligocene to Early Miocene) of the Alpine Foreland Basin, which can be more than 1000 m thick (de Ruig 2003). Prominent bacterial methane gas accumulations occur in these clastic deep-water sediments. The gas is sourced from clayey horizons of the formation which contain immature kerogen type III (to II). A detrimental element may be the low level of compaction and therefore ductility of the Puchkirchen Formation.

Shale gas research activities in Europe

There is an urgent need for research concerning different aspects of shale gas formation in Europe. Firstly it must be stated that Europe still lacks a systematic database in which the occurrence of European black shales and their properties (thickness, TOC contents, maturity, brittleness, etc.) are compiled. This situation needs to be rectified. Secondly, the fraccability, gas contents and adsorption/desorption properties have to be analysed for individual gas shales. These characteristics are an indirect result of the complex geological history of the sediments in time and space (palaeogeography and depositional environment, diagenesis, basin history, etc.). In particular, tectonic, geomechanical and basin modelling approaches must be integrated to predict relevant gas shale properties.

Numerous small shale gas research projects have already started in Austria, England, France and Germany. For example, GeoEnergie focuses on the German federal state of Brandenburg and adjacent federal states. In the UK Lower Namurian shale gas prospectivity is being investigated by the University of Leicester. Furthermore, research at the University of Vienna focuses on the Mikulov Marl (Malmian gas shales in the Vienna Basin) and its diagenetic and clay mineralogical changes with depth to enable an optimized hydraulic fracturing. By far the largest effort is the 'Gas Shales in Europe' (GASH) project, coordinated by the German National Research Centre GFZ. This industry-sponsored project is interdisciplinary and run by a multinational expert task force drawn from leading research institutions, geological surveys, universities and consultants. Its overall goal is to predict shale gas formation and occurrence in time and space, and consists of reservoir-scale and regional-scale projects incorporating geophysics, rock mechanics, organic geochemistry, mineralogy and basin modelling. The research consortium focuses on potential gas shales of Europe, but it also integrates proven US gas shales (e.g. Barnett Shale) for calibration of key variables.

Outlook

Shale gas is an unconventional resource in Western Europe with predicted reserves of 510×10^{12} SCF gas (Rogner 1997). The international E&P industry has started to investigate single potential plays. To date, it is not clear whether these plays, e.g. in Poland, Sweden or Austria, can be exploited economically. However, the numerous Phanerozoic black shales in Europe offer promising targets for future exploration campaigns. Basic scientific work conducted as part of GASH and other projects will help to better understand European sedimentary basins in terms of shale gas formation over the coming three years. One major effort will be to compile a black shale database containing key properties of shale gas on a European scale. Moreover, research projects on both regional and reservoir scale aspects (e.g. ability to stimulate, fraccability, gas isotherms) for selected European gas shales will deliver insight into specific processes and properties. In this context, the available knowledge from North American shale gas systems can help to initially categorize potential targets, but detailed research due to different basin histories is required.

European offshore regions like the North or Baltic Seas are not considered here as no current activities are known from these areas. However, the existing infrastructure in such mature offshore hydrocarbon provinces may prove to be an advantage.

The authors thank reviewers N. Smith and D. Jarvie for their constructive reviews that helped to improve the manuscript. Editorial help by H. Johnson and S. Pickering is gratefully acknowledged.

References

Adámek, J. 2005. The Jurassic floor of the Bohemian Massif in Moravia – geology and paleogeography. *Bulletin of Geosciences*, **80**, 291–305.

Al Duba, Huenges, E., Nover, G., Will, G. & Jödicke, H. 1988. Impedance of black shale from Munsterland 1 borehole: an anomalously good conductor? *Geophysical Journal*, **94**, 413–419.

Arzmüller, G., Buchta, S., Ralbovsky, E. & Wessely, G. 2006. The Vienna basin. *In*: Golonka, J. & Picha, F. J. (eds) *The Carpathians and their Foreland: Geology and Hydrocarbon Resources*. AAPG Memoir, **84**, 191–204.

Buchardt, B. 1999. Gas potential of the Cambo-Ordovician Alum Shale in Southern Scandinavia and the Baltic Region. *In*: Whiticar, M. J. & Faber, E. (eds) *The Search for Deep Gas Selected*. Papers presented at the I.E.A./BMFT International Deep Gas Workshop, Hannover. Geologisches Jahrbuch Reihe D, **107**, 9–24.

Clayton, G. & Baily, H. 1999. Organic maturation levels of pre-Westphalian carboniferous rocks in Ireland, and in the Irish offshore. *In*: Whiticar, M. J. & Faber, E. (eds) *The Search for Deep Gas Selected*. Papers presented at the I.E.A./BMFT International Deep Gas Workshop, Hannover. Geologisches Jahrbuch Reihe D, **107**, 25–42.

Clayton, J. L., Koncz, I., King, J. D. & Tatar, E. 1994. Organic geochemistry of crude oils and source rocks, Bekes basin. *In*: Teleki, P. G., Mattick, R. E. & Kokay, J. (eds) *Basin Analysis in Petroleum Exploration. A Case Study from the Bekes Basin, Hungary*. Kluwer Academic, Dordrecht, 161–185.

Curtis, J. B. 2002. Fractured shale–gas systems. *American Association of Petroleum Geologists Bulletin*, **86**, 1921–1938.

De Ruig, M. 2003. Deep marine sedimentation and gas reservoir distribution in Upper Austria. *Oil Gas European Magazine*, **2**, 1–7.

Dill, H. G., Sachsenhofer, R. et al. 2008. The origin of mineral and energy resources of Central Europe II. *In*: McCann, T. (ed.) *Geology of Central Europe*. Geological Society, London, Special Publications, 1341–1449.

Dolton, G. L. 2006. *Pannonian Basin Province, Central Europe (Province 4808) – Petroleum Geology, Total Petroleum Systems, and Petroleum Resource Assessment*. US Geological Survey Bulletin, **2204–B**.

Dvorjanin, E. S., Samoyluk, A. P., Egurnova, M. G., Zaykovsky, N. Ya., Podladchikov, Yu. Yu., van den Belt, F. J. G. & de Boer, P. L. 1996. Sedimentary cycles and paleogeography of the Dnieper Donets Basin during the late Visean-Serpukhovian based on multiscale analysis of well logs. *Tectonophysics*, **268**, 169–187.

Dyni, J. R. 2006. Geology and resources of some world oil-shale deposits. US Geological Survey Scientific Investigations Report 2005–5294. World Wide Web Address: http://pubs.usgs.gov/sir/2005/5294/pdf/sir5294_508.pdf.

Farrimond, P., Eglinton, G. & Brassell, S. C. 1989. Toarcian anoxic event in Europe: an organic geochemical study. *Marine and Petroleum Geology*, **6**, 136–147.

Fisher, T. 2008. Austria turns to unconventional, shale gas. *International Gas Report*, **596**, 7–8.

Fishman, N. & Hill, R. 2008. Organic-rich shales as important, potentially prolific unconventional petroleum reservoirs – the US experience. GEP-17 Unconventional gas? Coalbed, shale, and tight gas-sands gases. 33rd International Geol. Congress Oslo, 6–14 August 2008. World Wide Web Address: http://www.cprm.gov.br/33IGC/1351784.html.

Formolo, M. J., Petsch, S. T., Martini, A. M. & Nüsslein, K. 2008. A new model linking atmospheric methane sources to Pleistocene glaciation via methanogenesis in sedimentary basins. *Geology*, **36**, 139–142.

Franke, D. & Neumann, E. 1999. Geology and hydrocarbon of the pre-Westphalian in the deep underground of the NE German Basin. *In*: Whiticar, M. J. & Faber, E. (eds) *The Search for Deep Gas*. Selected papers presented at the I.E.A./BMFT International Deep Gas Workshop, Hannover. Geologisches Jahrbuch Reihe D, **107**, 43–54.

Frantz Jr, J. H., Jochen, V. & Contributors. 2006. Shale gas. *Schlumberger White Paper*. World Wide Web Address: www.oilfield.slb.com/whitepaper/shalegas.

Gavrish, V. K., Machulina, S. A. & Kurilenko, V. S. 1994. Visean oil-source formation of the Dnieper-Donets basin. *Doklady Akademii Nauk Ukrainy*, **7**, 92–95.

Gursky, H.-J. 2006. Paläogeography, Paläoozeanographie und Fazies. *In*: Deutsche Stratigraphische Kommission (ed.) *Stratigraphie von Deutschland VI. Unterkarbon (Mississippium)*. Schriftenreihe der Deutschen Gesellschaft für Geowissenschaften, **41**, 51–68.

Haas, J. (ed.) 2001. *Geology of Hungary*. Eötvös University Press, Budapest.

Hoffmann, N., Jödicke, H. & Gerling, P. 2001. The distribution of pre-Westphalian source rocks in the North German Basin – evidence from magnetotelluric and geochemical data. *Geologie en Mijnbouw, The Netherlands Journal of Geosciences*, **80**, 71–84.

Horsfield, B., Bharati, S., Larter, S. R., Leistner, F., Littke, R., Schenk, H. J. & Dypvik, H. 1992. On the atypical petroleum-generating characteristics of alginite in the Cambrian Alum Shale. *In*: Schidlowski, M. et al. (eds) *Early Organic Evolution: Implications for Mineral and Energy Resources*. Springer, Berlin, 257–266.

Horváth, F. & Cloetingh, S. 1996. Stress-induced late-stage subsidence anomalies in the Pannonian Basin. *Tectonophysics*, **266**, 287–300.

Horváth, F. & Tari, G. 1999. IBS Pannonian Basin project: a review of the main results and their bearings on hydrocarbon exploration. *In*: Durand, B., Jolivet, L., Horváth, F. & Seranne, M. (eds) *The Mediterranean Basins: Tertiary Extension within the Alpine Orogen*. Geological Society, London, Special Publications, **156**, 195–213.

Horváth, F., Dovenyi, P., Szalay, A. & Royden, L. H. 1988. Subsidence, thermal, and maturation history of the Great Hungarian Plain. *In*: Royden, L. & Horváth, F. (eds) *The Pannonian Basin: A Study in Basin Evolution*. American Association of Petroleum Geologists Memoirs, **45**, 355–372.

Hoth, P. 1997. Fazies und Diagenese von Präperm-Sedimenten der Geotraverse Harz – Rügen. *Schriftenreihe für Geowissenschaften*, **4**, 1–139.

Jenkins, C. D. & Boyer II, C. M. 2008. Coalbed- and shale-gas reservoirs. *Journal of Petroleum Technology*, 92–99. SPE paper 103514.

Kabyshev, B., Krivchenkov, B., Stovba, S. & Ziegler, P. A. 1998. Hydrocarbon habitat of the Dniepr–Donets Depression. *Marine and Petroleum Geology*, **15**, 177–190.

Kabyshev, B. P., Kabyshev, Yu. B., Krivosheev, V. T., Prigarina, T. M. & Ulmishek, G. F. 1999. Oil-gas-generating characteristics of Paleozoic rocks of the Dniepr–Donets Basin based on Rock-Eval pyrolysis (in Russian). *Dopovidi Natsionalnoi Akademii Nauk Ukrainy*, 112–117.

Kivshik, M. K., Stovba, S. M., Turchanenko, M. T. & Redkolis, V. A. 1993. *The Regional Seismostratigraphic Investigations in the Dniepr–Donets Depression* (in Russian). Ukrgeofisika, Kiev.

Korosi, T. 2007. *Natural Gas Overview Hungary 2006*. 17th Session of Working Party on Gas, United Nations, Geneva, 23–24 January 2007. World Wide Web Address: http://www.unece.org/energy/se/pdfs/wpgas/session/17_countr/korosi.pdf.

Kováč, M., Baráth, I., Harzhauser, M., Hlavatý, I. & Hudáčková, N. 2004. Miocene depositional systems and sequence stratigraphy of the Vienna Basin. *Courier Forschungsintitut Senckenberg*, **246**, 187–212.

Ladwein, W. 1988. Organic geochemistry of Vienna Basin: model for hydrocarbon generation in overthrust belts. *American Association of Petroleum Geologists Bulletin*, **72**, 586–599.

Ladwein, W., Schmidt, F., Seifert, P. & Wessely, G. 1991. Geodynamics and generation of hydrocarbons in the region of the Vienna basin, Austria. *In*: Spencer, A. M. (ed.) *Generation, Accumulation, and Production of Europe's Hydrocarbons*. Oxford University Press, Oxford, 289–305.

Lafargue, E., Elliouz, N. & Roure, F. 1994. Thrust-controlled exploration plays in the outer Carpathians and their foreland (Poland, Ukraine and Romania). *First Break*, **12**, 69–79.

Langanger, H. 2008. *Exploration and Production*. OMV Capital Markets Day 2008, London. World Wide Web Address: http://www.omv.com/SecurityServlet/secure?cid=1225986876259&lang=de&swa_site=wps.vp.com&swa_nav=OMV+Holding%7C%7CInvestor+Relations%7C%7CEvents%7C%7CCapital+Markets+Day+2008%7C%7C&swa_pid=6_B_QFO+%5BCONTENT_NODE%3A3096224743844344%5D&swa_lang=de.

Littke, R., Jendrzejewski, L., Lokay, P., Shuangqing, W. & Rullkötter, J. 1998. Organic geochemistry and depositional history of the Barremian–Aptian boundary interval in the Lower Saxony Basin, northern Germany. *Cretaceous Research*, **19**, 581–614.

Machulina, S. A. & Babko, A. K. 2004. On the geology of Visean Domanik-type rocks in the Dniepr-Donetsk Depression (in Ukrainian). *Naftova i gazova promyslovist*, **5**, 3–8.

Magyar, I., Fogarasi, A., Vakarcs, G., Bukó, L. & Tari, G. C. 2006. The largest hydrocarbon field discovered to date in Hungary: Algyo. *In*: Golonka, J. & Picha, F. J. (eds) *The Carpathians and their Foreland: Geology and Hydrocarbon Resources*. American Association of Petroleum Geologists Memoirs, **84**, 619–632.

Martini, A. M., Walter, L. M., Budai, J. M., Ku, T. C. W., Kaiser, C. J. & Schoell, M. 1998. Genetic and temporal relations between formation waters and biogenic methane: Upper Devonian Antrim Shale, Michigan Basin, USA – CO_2 reduction vs. acetate fermentation-isotopic evidence. *Geochimica et Cosmochimica Acta*, **62**, 1699–1720.

Matyja, H. 2008. Pomeranian basin (NW Poland) and its sedimentary evolution during Mississippian times. *Geological Journal*, **43**, 123–150.

Nielsen, A. T. & Schovsbo, N. H. 2007. Cambrian to basal Ordovician lithostratigraphy in southern Scandinavia. *Bulletin of the Geological Society of Denmark*, **53**, 47–92.

Petmecky, S., Meier, L., Reiser, H. & Littke, R. 1999. High thermal maturity in the Lower Saxony Basin: intrusion or deep burial? *Tectonophysics*, **304**, 317–344.

Phasis, 2008. *US Shale Gas Brief September 2008*. World Wide Web Address: http://www.phasis.ca/files/pdf/Phasis_Shale_Gas_Study_Web.pdf.

Picha, F. J. & Peters, K. E. 1998. Biomarker oil-to-source rock correlation in the Western Carpathians and their foreland, Czech Republic. *Petroleum Geoscience*, **4**, 289–302.

Rogner, H.-H. 1997. An assessment of world hydrocarbon resources. *Annual Review of Energy and the Environment*, **22**, 217–262.

Rullkötter, J., Leythaeuser, D. *et al.* 1988. Organic matter maturation under the influence of a deep intrusive heat source: a natural experiment for quantitation of hydrocarbon generation and expulsion from a petroleum source rock. (Toarcian Shale, Northern Germany). *In*: Mattavelli, L. & Novelli, L. (eds) *Advances in Organic Geochemistry 1987*. Pergamon Press, Oxford, 847–856.

Schouten, S., van Kaam-Peters, H. M. E., Rijpstra, W. I. C., Schoell, M. & Sinninghe Damsté, J. S. 2000. Effects of an oceanic anoxic event on the stable carbon isotopic composition of early Toarcian carbon. *American Journal of Science*, **300**, 1–22.

Schovsbo, N. H. 2003. The geochemistry of Lower Palaeozoic sediments deposited on the margins of Baltica. *Bulletin of the Geological Society of Denmark*, **50**, 11–27.

Selley, R. C. 1987. British shale gas potential scrutinized. *Oil & Gas Journal*, **85**, 62–64.

Selley, R. C. 2005. UK shale-gas resources. *In*: Doré, A. G. & Vining, B. A. (eds) *Petroleum Geology: North-West Europe and Global Perspectives: Proceedings of the 6th Petroleum Geology Conference*. Geological Society, London, 707–714. doi: 10.1144/0060707.

Shpak, P.F. (ed.), 1989. *Geology and Petroleum Productivity of the Dniepr-Donetsk Depression – Petroleum Productivity* (in Russian). Naukova Dumka, Kiev.

Shymanovskyy, V. A. & Sachsenhofer, R. F. 2002. Hydrocarbon generation in the NW Dniepr–Donets Basin, Ukraine. *63rd Conference, European Association of Geoscientists & Engineers*, Extended Abstracts, paper 121.

Shymanovskyy, V. A., Sachsenhofer, R. F., Izart, A. & Li, Y. 2004. Numerical modelling of the thermal evolution of the northwestern Dniepr–Donets Basin (Ukraine). *Tectonophysics*, **381**, 61–79.

Siegmund, H., Trappe, J. & Oschmann, W. 2002. Sequence stratigraphic and genetic aspects of the Tournaisian 'Liegender Alaunschiefer' and adjacent beds. *International Journal of Earth Sciences*, **91**, 934–949.

Smith, N., Turner, P. & Williams, G. 2010. UK data and analysis for shale gas prospectivity. *In*: Vining, B. A. & Pickering, S. C. (eds) *Petroleum Geology: From Mature Basins to New Frontiers – Proceedings of the 7th Petroleum Geology Conference*. Geological Society, London, 1087–1098; doi: 10.1144/0071087.

Stephenson, R. A., Yegorova, T. *et al.* 2006. Late Paleozoic intra- and pericratonic basins on the East European Craton and its margins. *In*: Gee, D. G. & Stephenson, R. A. (eds) *European Lithosphere Dynamics*. Geological Society, London, Memoirs, **32**, 463–479.

Tari, G. C. & Horvath, F. 2006. Alpine evolution and hydrocarbon geology of the Pannonian Basin: an overview. *In*: Golonka, J. & Picha, F. J. (eds) *The Carpathians and their Foreland: Geology and Hydrocarbon Resources*. American Association of Petroleum Geologists Memoirs, **84**, 605–618.

Ulmishek, G. F. 2001. *Petroleum Geology and Resources of the Dnieper–Donets Basin, Ukraine and Russia*. U.S. Geological Survey Bulletin 2201-E, US Department of the Interior U.S. Geological Survey. World Wide Web Address: http://geology.cr.usgs.gov/pub/bulletins/b2201-e/.

Volk, H., Horsfield, B., Mann, U. & Suchý, V. 2002. Variability of petroleum inclusions in vein, fossil and vug cements – a geochemical study in the Barrandian Basin (Lower Palaeozoic, Czech Republic). *Organic Geochemistry*, **33**, 1319–1341.

Wessely, G. 1990. Geological results of deep exploration in the Vienna basin. *Geologische Rundschau*, **79**, 513–520.

Zdanaviciutė, O. & Lazauskiene, J. 2004. Hydrocarbon migration and entrapment in the Baltic Syneclise. *Organic Geochemistry*, **35**, 517–527.

Zdanaviciutė, O. & Lazauskiene, J. 2007. The petroleum potential of the Silurian succession in Lithuania. *Journal of Petroleum Geology*, **30**, 325–337.

Zimmerle, W. & Stribrny, B. 1992. Organic carbon rich pelitic sediments in the Federal Republic of Germany. *Courier Senckenbergsche Naturforschende Gesellschaft*, **152**, 1–142.

UK data and analysis for shale gas prospectivity

N. SMITH, P. TURNER and G. WILLIAMS

British Geological Survey, Kingsley Dunham Centre, Nicker Hill, Keyworth NG12 5GG, UK
(e-mail: njps@bgs.ac.uk)

Abstract: Organic-rich shale contains significant amounts of gas held within fractures and micropores and adsorbed onto organic matter. In the USA shale gas extracted from regionally extensive units such as the Barnett Shale currently accounts for 6% of gas production and is likely to reach 30% by 2015. Shale gas prospectivity is controlled by the amount and type of organic matter held in the shale, its thermal maturity, burial history, microporosity and fracture spacing and orientation. Potential targets range in age from Cambrian to the late Jurassic, within the main UK organic-rich black shales: younger shales have been excluded because they have not reached the gas window, but they may possess a biogenic gas play. A geographic information system, showing the distribution of potential reservoir units, has been created combining information on hydrocarbon shows, thermal maturity, fracture orientation, gas composition, and isotope data to identify potentially prospective areas for shale gas. Some of these data are shown as graphs and maps, but crucial data is lacking because earlier exploration concentrated on conventional reservoirs. The prospects include Lower Palaeozoic shale basins on the Midland Microcraton (a high risk because no conventional gas has been proved in this play), Mississippian shales in the Pennine Basin (the best prospect associated with conventional fields and high maturity), Pennsylvanian shales in the Stainmore and Northumberland Basin system (high risk because no conventional gas discoveries exist) and Jurassic shales in Wessex and Weald basins (small conventional fields signify potential here).

Keywords: shale, gas, maturity, thickness, source rock and reservoir

Organic-rich shale contains significant amounts of gas held within fractures and micropores and adsorbed onto organic matter. In the USA, shale gas extracted from regionally extensive units such as the Barnett Shale accounts for c. 6% of gas production. The success of US shale gas exploitation (over 28 000 wells producing c. 380×10^9 SCF per year) has stimulated significant interest in identifying potential reservoirs throughout the world. The depth range of the US shale gas plays extends down to 4500 m at present. Selley (1987, 2005), farsightedly, advocated shale gas exploration in the UK, based on Upper Devonian gas fields of the Appalachian Basin, which have been producing since 1821. However, in the past decade the Mississippian Barnett Shale of the Fort Worth Basin has become the most productive shale gas reservoir in the USA. US shales generally, and the Barnett Shale in particular, provide good analogues for potential shale gas plays in the UK, which has thick Mississippian shales both on and offshore; therefore the geology and geochemistry of the Barnett Shale are discussed briefly below.

The Barnett Shale

The Barnett Shale Formation (354–323 Ma) of the Fort Worth Basin is up to 300 m thick and underlies an area of c. 13 000 km² (Fig. 1). It contains c. 2.5×10^{12} SCF of proven gas reserves held in a low porosity and very low permeability shale matrix. Permeability is in the micro- to nanodarcy range and porosity rarely exceeds 6% (Bowker 2007): consequently the Barnett Shale Formation is slightly overpressured, with formation pressures in excess of 4000 psi (Frantz et al. 2005).

The Fort Worth Basin shows a northward-thickening, half-graben-like structure, but is bounded by reverse faults against the Lower Palaeozoic Muenster Arch (part of the Southern Oklahoma Aulacogen; Fig. 1). The Mississippian Barnett Shale unconformably overlies Ordovician strata (Viola Formation limit in Fig. 1) and can be subdivided into two units, the Upper and Lower Barnett, separated by the Forestburg Limestone. The upper shale unit is overlain by the Marble Falls Limestone, also of Mississippian age, which is conformable with a thick succession of overlying Pennsylvanian sediments. Sedimentary structures suggest that the main shale units were deposited by distal turbidity flows in a sediment-starved anoxic basin environment. The Carboniferous sequence is truncated by a Cretaceous supercrop above the Variscan unconformity. The Ouachita (Variscan) fold belt lies at right angles to the Muenster Arch and impinges on the Fort Worth Basin in the SE.

The three main factors controlling prospectivity of the Barnett are the thermal maturity, thickness and total organic carbon (TOC) content of the shale (Zhao et al. 2007). Local and regional structures such as joints, folds and faults control fracture porosity and thus influence production potential at a variety of scales. Most natural fractures are sealed but these can potentially be exploited by artificial fracturing techniques (Bowker 2007) to improve flow rates around a well. Siltstone bands and chert nodules can also affect prospectivity locally.

The Newark East shale gas field lies updip west of the depocentre of the Barnett Shale in the wedge between the Muenster Arch and the Ouachita (Variscan) fold belt (Fig. 1; Pollastro et al. 2004). The rest of the depocentre lies under the Fort Worth–Dallas conurbation. The Barnett Shale also produces oil from the area to the NW (Fig. 1), where overlying conventional reservoirs are also present.

In a similar tectonic position the Big Sandy shale gas field in eastern Kentucky lies over the thickest part of the Devonian Brown Shale in a foreland basin, which is bounded to the SE by the Pine Mountain thrust (Ray 1976) of the Appalachian (Variscan) fold belt.

The UK's closest tectonic analogue

Satellite basins of the Worcester Graben (Fig. 2) beneath the Bristol Coalfield and in Berkshire to the east (Mississippian strata are thin in the latter area) provide the closest UK tectonic

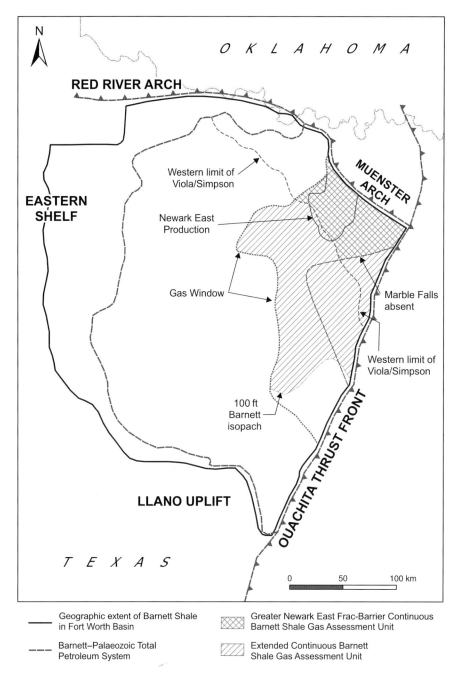

Fig. 1. Barnett Shale of the Fort Worth Basin, Texas, USA. The Muenster Arch and Ouachita Thrust Front form the northern and eastern boundaries of the foreland-type basin (barbed lines). The Barnett Shale lies unconformably on Ordovician strata. The Barnett Shale gas system, the Newark East Gasfield outline, the western limit of the gas window and other geological features relevant to completion fracturing are shown (Montgomery et al. 2005).

analogues to the Fort Worth Basin. They are small foreland basins preserved between the Variscides and the Worcester Graben. The latter formed in response to extensional faulting during the Lower Palaeozoic and inverted during the Variscan Orogeny comparable with the Southern Oklahoma Aulacogen (Smith 1993; Smith & Rushton 1993). The Bristol depocentre lacks the acreage of the Fort Worth Basin but contains the Lower Limestone Shale Group, conformably overlying Devonian and of roughly equivalent age (359–327 Ma) to the Barnett Shale Formation. Few wells and no maturity or geochemical data are available though, and seismic is sparse.

It is important to note, however, that there are significant differences between shale gas plays developed in different basins in the USA (age, maturity, tectonic position). So many black shale plays are currently being developed there, that whilst the three main factors (above) are paramount, exact comparisons with US shales are not deemed necessary to confirm or refute potential. Consequently this contribution considers the main generic controls on shale gas prospectivity identified in key US plays (see above) in a regional UK context, before identifying potential exploration targets.

General characteristics of any potential source rocks

Organic-rich sediments have a lower density, lower sonic velocity, higher porosity, higher resistivity and higher gamma ray (GR) values compared with sedimentary units of equal compaction and comparable mineralogy. Potential source rocks can thus be readily identified on well logs, provided they are rich in organic matter. For example, the Arnsbergian Sabden Shale typically exhibits anomalously low sonic velocity and high GR values in

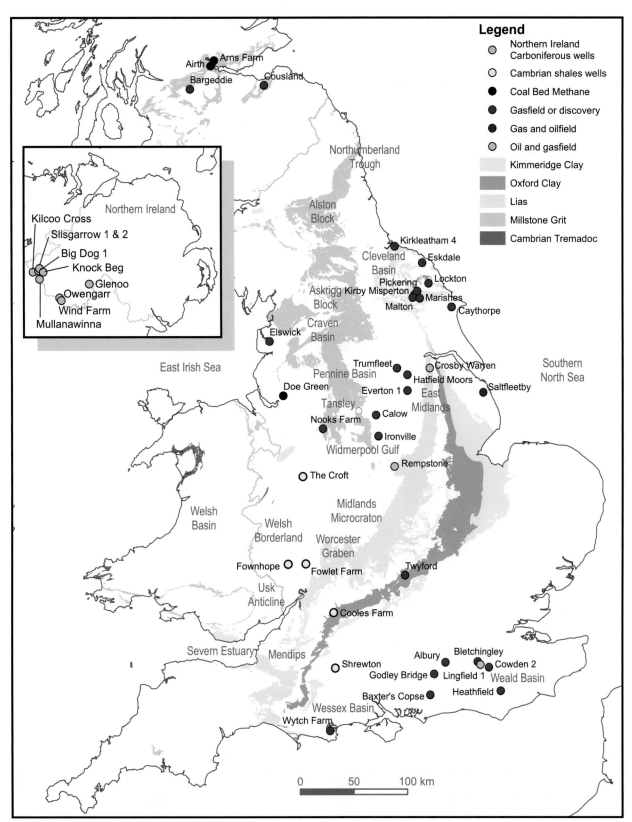

Fig. 2. Outcrops of main potential hydrocarbon source rocks (Lower Carboniferous strata beneath Millstone Grit are not shown). Wells reaching Tremadoc to Cambrian in England indicate the few provings at this level. Gas fields and discoveries indicate that gas window maturity has been reached. Hydrocarbon wells, plotted in inset of Northern Ireland, tested a tight gas Carboniferous play in several phases (Griffith 1983) as recently as 2001. Coalbed methane fields are also shown.

the Pennine Basin (Smith *et al.* 2005). Classification schemes based on wireline logs break down at high water saturations because water and organic material have comparable densities: the specific gravity of organic matter is about $0.95-1.05 \text{ g cm}^{-3}$ (Stocks & Lawrence 1990).

The outcrop distribution of key organic-rich shale formations identified on well logs is summarized in Figure 2. In the UK, organic-rich shales can be subdivided into three broad stratigraphic groups belonging to the Lower Palaeozoic (principally Cambrian), Mississippian and Jurassic (of Liassic, Oxfordian and

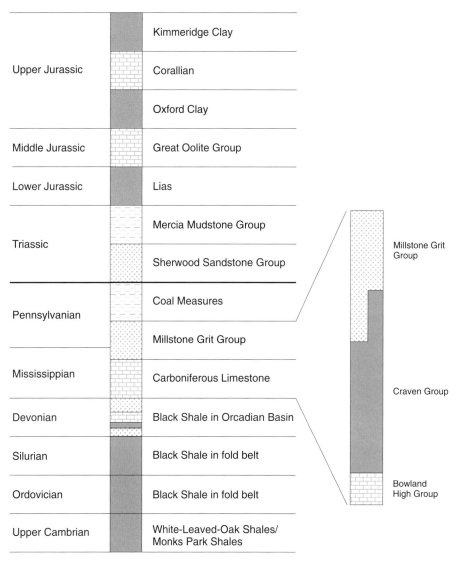

Fig. 3. Generalized stratigraphic column for UK onshore shows the main black shale formations (grey shading) and other representative lithologies in a typical sequence. The Sherwood Sandstone (stipple) lies unconformably on Coal Measures. Some black shales are only found in the Caledonian and Variscan fold belts (e.g. in the Orcadian Basin, Ordovician and Silurian shales in the Caledonides), which have been largely excluded from this study.

Kimmeridgian age), although such shales occur in other periods and groups (Fig. 3).

Pre-Carboniferous black shales

Black shales are developed in the Upper Cambrian, Upper Ordovician and Lower Silurian strata of the UK (Leggett 1980). Ordovician and Silurian shales are largely restricted to deeper basins and deformed in fold belts, while Cambrian black shale deposits are also found on the Midland Microcraton. Unexplained gas shows in the Welsh Borderland and Welsh Caledonides may have been sourced from concealed Lower Palaeozoic black shales. In most US basins conventional gas fields are present above the shale gas being developed. However this is not the case with the Cambrian Conasauga Shale in Alabama, where exploration and some production has been achieved. So-called overmature shales in conventional hydrocarbon terminology, often located in tectonic positions where conventional exploration has therefore been deemed unfavourable, may retain potential for shale gas.

Upper Cambrian black shales outcrop in small inliers close to zones of uplift, but seismic data suggest that they are widespread in the subsurface (Smith *et al.* 2005). Shales of this age were proved by the Fowlet Farm and Fownhope boreholes SW of the Malverns (Fig. 2; Earp & Hains 1971). In these boreholes, c. 68 m of black shale has been intruded by dolerite sills; the shales have high GR values and represent a potential source rock. Outcrop measurements of TOC support this interpretation, with values of c. 5% recorded by Parnell (1983).

In Shropshire c. 5 m of black shales with bituminous limestone concretions are underlain by 20 m of dark grey shale. To the NE at Lilleshall a small outcrop and the Croft borehole (Fig. 2) revealed c. 53 m of black, pyrite-rich shale lying unconformably on Lower Cambrian strata (Rushton *et al.* 1988). Unfortunately TOC and thermal maturity analyses were not made in this borehole. The Lilleshall deposits may be the erosional remnants of a tilted fault-block high (Middle Cambrian strata are missing; Smith & Rushton 1993), with a half-graben basin deepening eastwards (Smith *et al.* 2005).

Merevale borehole No. 2 (located on Fig. 5) encountered c. 60 m of condensed black mudstone (Monk's Park Shales; Taylor & Rushton 1971). TOC values exceeded 2% and sulphur content reached 10%; phosphate was present in the form of scattered nodules. These deposits may provide some shale gas potential. Conversely, the Cooles Farm borehole tested part of the Worcester Graben (Fig. 2). It was drilled into Cambrian strata but did not record the presence of any Upper Cambrian high gamma shale units. Seismic hereabouts reveals many reflectors beneath the well terminal depth. This play derives encouragement from the Cambrian Conasauga Shale in the Appalachians, where production has been achieved.

Ordovician and Silurian shale units seem to have less potential as a shale gas play, based on limited drilling. In the Fownhope well only a short section of Llandovery-age clay approached GR values indicative of a potential source rock. A borehole drilled at Usk (east of the Usk Anticline in Fig. 2) penetrated no potential source rocks in the Lower Palaeozoic, although minor gas shows occurred in some sandstone units. However this well was drilled on a structural high missing a large section of Llandovery or older sequence where black shales might have formed in the depocentre beneath the Usk Anticline. A black shale facies of Llandovery age is known in the Welsh Caledonides to the west (Leggett 1980).

UK Lower Palaeozoic black shales often show preferential enrichment of various heavy elements including As, Co, Cr, Cu, Mo, Ni, Pb, S, U and V (Leggett 1980; Jones & Plant 1989). This is particularly true of the Upper Cambrian interval and appears to be associated with organic matter. The presence of these minerals may be useful proxies for high TOC shales, where mining exploration data is available.

A few samples from the Withycombe Farm borehole (between Cooles Farm and Twyford; Fig. 2) suggest a porosity of 5–10% survives down to a depth of 1000 m in Lower Cambrian (Poole 1978). In many places these rocks have been subjected to significant compaction, intrusion and tectonism and primary porosity would be expected to be very low. There is, however, scope for induced fracture permeability in these rocks. These few porosity values are comparable with US shales being developed currently.

In contrast to the Appalachian Basin in the USA, there are no Devonian black shale sequences in the Variscan foreland of the UK. Lacustrine shale deposits are present in the Middle Devonian of the Orcadian Basin in Scotland: TOC values typically average c. 1.4%. Thermal maturity is variable, but these rocks largely fall within the oil window or are overmature (in the vicinity of underlying intrusions), and offshore they probably helped charge the Beatrice Oilfield (Hillier & Marshall 1992). Two boreholes drilled at Dounreay proved up to 150 m of interbedded high gamma shale beds.

Carboniferous black shales

Mississippian shale units (Fig. 3) offer the best potential for shale gas exploration in the UK because they have sourced hydrocarbons, have high TOCs (Armstrong et al. 1997) and were deposited in deep half-graben, within the Pennine Basin in northern England (Fraser et al. 1990). The younger (Namurian) of these shales are more widely deposited as the extensional half-graben subsidence was transforming to a more regional thermal subsidence.

Gas content and kerogen type

The gas content of Coal Measures strata was measured by the National Coal Board (Creedy 1989). Carbonaceous mudstones (including marine bands) contained 0.18–0.63 m^3/t (6.3–22.2 SCF/t) gas; whereas non-carbonaceous mudstones contained 0.009–0.1 m^3/t (0.3–3.5 SCF/t). Budge (1932) reported methane adsorption of 1.8 m^3/t (63.6 SCF/t) in carbon-rich Coal Measures mudstone and 0.25 m^3/t (8.8 SCF/t) in black shale from the South Wales Coalfield. These values in Westphalian strata are comparable with the reported values in the Cretaceous Lewis Shale in the USA, which has the lowest values of the US explored shales.

Ferguson (1984) measured methane values of up to 46 000 ppb in the Carboniferous Limestone of northern England, near to mineralized areas on the Alston Block: there also appeared to be an association between base metal vein mineralization in Derbyshire and the adjacent shale-rich Pennine Basin half-graben (Plant & Jones 1989). Ewbank et al. (1993) noted that Alportian–Pendleian mudstones exposed at outcrop in the northern part of the Pennine Basin predominantly contained gas-prone type III kerogen, whereas mudstone samples from a Widmerpool Gulf well (Arnsbergian age) and various Visean mudstones were dominated by type II kerogen (gas- and oil-prone).

Total organic carbon

Namurian marine shales have generally higher TOC values (over 4%) compared with non-marine shales (Spears & Amin 1981), which have an average value of around 2% (Fig. 4). Maynard et al. (1991) found that two Namurian black shale marine bands had a TOC content of between 10–13%, whereas values within interbedded strata ranged between 2 and 3%. The Namurian Holywell Shale, source rock for the southern East Irish Sea and Formby hydrocarbon fields, has TOC values ranging from 0.7–5%, with an average of 2.1% (Armstrong et al. 1997). The lower part of this formation gave an average of 3% TOC and pyrolysate yields of 7 kg/t^{-1}. These values are comparable with US shales.

Thermal maturity

Vitrinite reflectance (%R_o) measurements at outcrop and in boreholes provide a widely accepted proxy for thermal maturity and hydrocarbon generation. These are shown for the nearest value to the basal Namurian (Fig. 5; at Knutsford it is probably within the Westphalian D). Maturity zones can be divided into immature for oil, within the oil window, within the gas window and overmature for hydrocarbons. The oil window is normally considered to fall between 0.6 and 1.3% R_o, although it varies according to kerogen type. There is no agreement on the lower limit of the gas window, which has variously been put at 2% R_o (Landes 1967), 3% R_o (Dow 1977), 3.2% R_o (Dow & Connor 1982) and 5% R_o (Hood et al. 1975). In this study a value of 1.1% R_o has been used to differentiate maturity levels above and below the gas window (Fig. 5), because this value defines the limit of the Newark East gas field in the Barnett Shale (Pollastro et al. 2004 and see Fig. 1).

Porosity, permeability and fracture porosity

Limited porosity data from BGS boreholes in the southern Midlands suggest that porosities of 5–10% survive to a depth of 1000 m in Upper Palaeozoic shales (Poole 1977, 1978). Coal Measures mudstones, seat-earths and siltstones have measured permeability values in the range 4.34×10^{-6} to 7.1×10^{-3} mD (Oldroyd et al. 1971).

Joints, developed in the limestone platforms of Derbyshire and the Askrigg Block (Moseley & Ahmed 1967), are predominantly sub-vertical and perpendicular to bedding. Joint development preceded the main phase of movement on the main fault zones and they are frequently mineralized. Well bedded marine shales tend to have a high joint density, in contrast to mudstones and sand-rich shale units. Minor joints are more difficult to distinguish in argillaceous rocks, but master joints persist strongly with a high fracture density (0.6–3 m spacing).

The predominant NW trend of the main Carboniferous joint sets is coincident with the present-day maximum horizontal stress

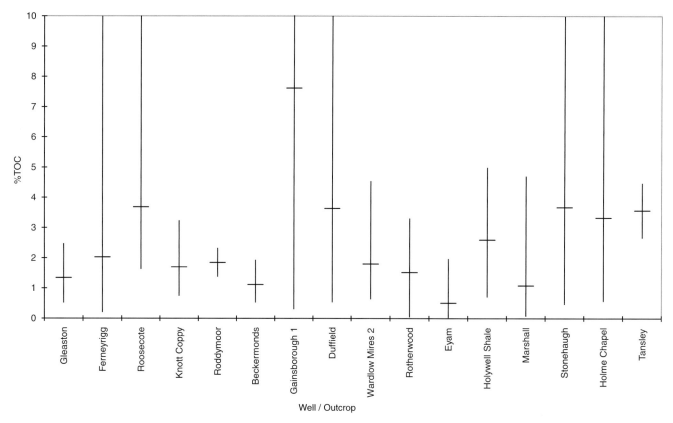

Fig. 4. Variation of TOC contents in the Carboniferous of northern England. Gleaston borehole is in south Cumbria; Ferneyrigg, Stonehaugh and Marshall Meadows are in the Northumberland Trough; Roddymoor is on the Alston Block; and Beckermonds is on the Askrigg Block. The other boreholes are in the Pennine Basin.

direction (145/325°; Evans & Brereton 1990). Rogers (2003) predicted that the stress-induced fracture permeability is likely to be higher along this trend in the current UK stress field.

Mineralogy

The Al_2O_3 content of UK Carboniferous mudstones ranges between 12 and 38%, with an average of around 25% (Ramsbottom et al. 1981). Marine and non-marine shales in the Tansley borehole (Fig. 2) have average Al_2O_3 contents of 20.6 and 17.9% respectively, with clay minerals making up 59 and 56% of the shale matrix (Spears & Amin 1981). The Barnett Shale has a relatively low clay content (c. 27%; Jarvie et al. 2004) compared with other shales. The presence of significant amounts of quartz (45%) and carbonate (10%) in the shale matrix imparts brittleness to the rock, facilitating artificial fracturing.

Gas fields and discoveries

The presence of conventional gas fields in Carboniferous and Jurassic basins (Fig. 2) demonstrates that gas has been generated. In a few cases there is unambiguous evidence of the source being Namurian shales (e.g. the Elswick Gasfield; Fig. 2), because Coal Measures are absent in the subcrop (Fig. 5; Smith 1985). In some gas fields migration could have occurred from either a Coal Measures source or from Namurian strata uplifted in an inverted basin (e.g. Nooks Farm; Fig. 2). Along the southern margin of the Cleveland Basin (Marishes to Malton gas fields; Fig. 2) a third migration direction, along east–west faults, from the Southern North Sea, is also a possibility.

Gas wetness (percentage of non-methane gas) values (Fig. 6) are high when gas is associated with oil in the source rock and lower (<5%) when the gas has been exclusively derived from coal-rich strata, is biogenic or overmature. Jarvie et al. (2004) typified two gas samples in the Barnett Shale as overmature dry gas, with gas wetness values of around 5%: in contrast, oil-associated gas had values over 12%. In the Carboniferous of the UK, high values measured at Welbeck Colliery are associated with oil shows, but at nearby Thoresby Colliery (also associated with oil) the gas is rich in methane and was probably sourced from the Coal Measures (Challinor 1990).

Insufficient gas composition and carbon isotope data were available to resolve the migration directions in most gas fields. Carbon and hydrogen isotope data from methane samples accompanying water flows within shallow tunnels in Namurian strata (Fig. 7) suggest that the methane either had a biogenic (modern) origin or formed through CO_2 reduction (Bath et al. 1988). The US Antrim Shale and New Albany Shale (Fig. 7; Illinois Quaternary-hosted gas) have late generation biogenic gas systems related to glacial meltwater ingress via fractures (Shurr & Ridgley 2002). US exploration has used stable isotopes to confirm source, maturity and hydrocarbon generation zone (Fig. 7).

Principal source rocks

Two lithostratigraphic units from the Mississippian represent potential shale gas plays in the UK: these are the Craven Group found in the Pennine Basin (Fig. 3) and the Lower Limestone Shale around the Mendips (Fig. 2). The Craven Group comprises a thick sequence of interbedded limestone and shale. The Pennine Basin formed to the north of the Wales–Brabant Massif: this geological and tectonic setting is not a direct tectonic analogue to the Barnett Shale. Conversely the Lower Limestone Shale was deposited in a similar tectonostratigraphic setting to the Barnett,

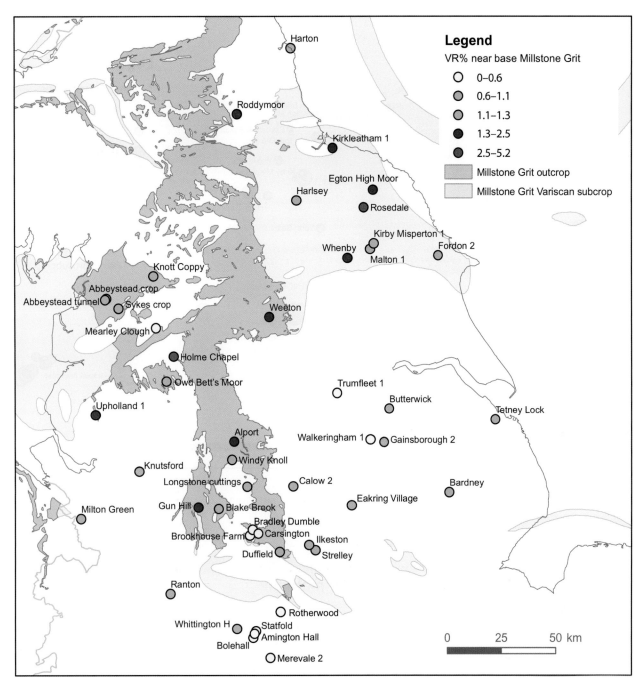

Fig. 5. Vitrinite reflectance percentage values obtained nearest to the base Namurian. Yellow spots are immature for hydrocarbons, green represents early part of oil window, orange late part of oil window (cf. Barnett Shale top of gas window), red the gas window and grey overmature (conventional hydrocarbon zones). The overmature areas should not be ruled unprospective for shale gas. Most existing hydrocarbon wells were drilled on structural highs and are likely to possess lower maturity than the potential source rock.

in the foreland of the Variscan Front, west of the Worcester Graben. It thickens southward from c. 35 m north of the Severn Estuary to c. 170 m in the Mendips (Kellaway & Welch 1948) and is also present south of the Bristol Channel.

The Pennine Basin has been the focus of extensive coal and hydrocarbon exploration in the past, producing a comprehensive database, available to evaluate shale gas potential in the region. The Craven Group (Mississippian) ranges in thickness from >1450 m in the Craven sub-basin to >5000 m in the Widmerpool Gulf (Waters et al. 2007; Fig. 2). The mudstones were deposited in distal slope turbidite and hemi-pelagic environments in relatively narrow, deep depocentres within the Pennine Basin (e.g. Craven and Widmerpool; Fig 2). Individual shale units that might represent potential source rocks include (Table 1) the Worston Shale Group, the Long Eaton Formation and the Widmerpool Formation. The latter has a high TOC content in Duffield borehole (Figs 5 & 8), but the GR values are low (Aitkenhead 1977).

Younger shale gas targets include the Bowland (Edale) Shales and the Sabden Shales (both Namurian). The condensed black Bowland Shale sequence outcropping in the Derbyshire Peak District is thin and immature for hydrocarbon generation, but these units thicken to the east and represent the main source rocks that charged the East Midlands hydrocarbon system. In the Tansley borehole (Fig. 2), TOC content averaged 4.48% in marine bands and 2.66% in non-marine shale sequences (Fig. 4). A GR peak is found close to the base of the Arnsbergian sequence, extending down into the Pendleian. In the nearby Uppertown borehole this interval coincides with phosphatic nodules and collophane, in a sequence of dark grey shale, containing abundant fossilized plants. Ponsford (1955) recorded higher than average

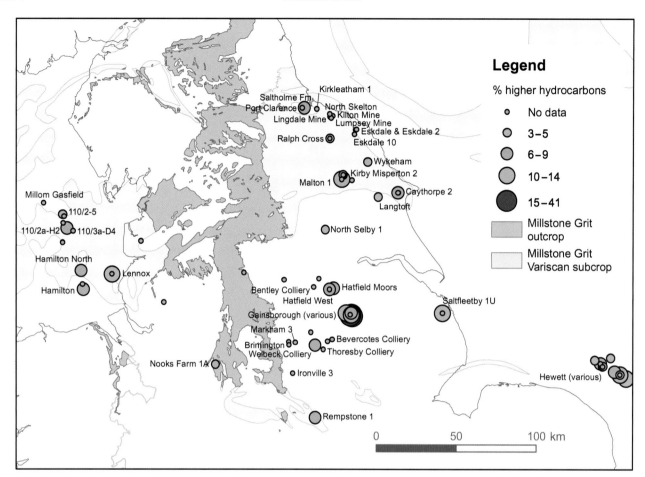

Fig. 6. Gas wetness ratio of analysed gases. The gases associated with oil fields have higher gas wetness values (a greater percentage of higher hydrocarbons). Welbeck Colliery has a high gas wetness, suggesting perhaps some gas from sources other than Coal Measures. These values suggest that source rocks, which have sourced the gas, exist in strata other than Coal Measures. However potential shale gas is likely to be nearer to dry methane within deeper parts of the half-graben.

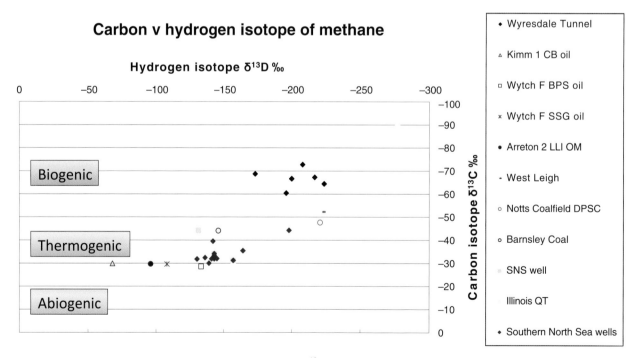

Fig. 7. Carbon and hydrogen isotope ranges in various methane gases. The $\delta^{13}C$ values 0 to $-20‰$ are probably abiogenically derived (e.g. Fischer–Tropsch reactions with hydrogen). The $\delta^{13}C$ values of -20 to about $-55‰$ represent thermogenic methane. The gas field methanes (Lokhurst 1998) are migrated gases. Coal field gases (Hitchman et al. 1989) might be more depleted than the migrated gases, with biogenic and CO_2-reduced gases (Bath et al. 1988) even more depleted. BC, Barnsley Coal; BPS, Bridport Sandstone; CB, Cornbrash; SSG, Sherwood Sandstone Group; LIAS, lower Lias organic matter; Arreton NC, Nottinghamshire Coalfield deep soft coal; SNS, Southern North Sea; WL, West Leigh; WT, Wyresdale Tunnel, Illinois; QT, Quaternary.

Table 1. Names of main black shale formations and their equivalents

Formation	Equivalents	Stages	Basin
Kimmeridge Clay		Kimmeridgian	Weald
Oxford Clay		Callovian–Oxfordian	Weald
Lias		Hettangian–Toarcian	Weald
Sabden Shales		Arnsbergian	Pennine
Upper Bowland Shale	Edale Shale	Pendleian	Pennine
Lower Bowland Shale	Widmerpool Formation	Brigantian	Pennine
Worston Shale	Hodder Mudstone	Chadian–Holkerian	Craven
Long Eaton Formation		Chadian–Asbian	Widmerpool
Other names			
Holywell Shale	Pentre Chert, Bowland Shale	Namurian	NE Wales
Craven Group	All local named formations	Mississippian–Namurian	Pennine

The Craven Group is a new term, extending from Courceyan to Yeadonian stages which has rendered a lot of the local names in the Carboniferous obsolete for stratigraphic purposes (Waters *et al.* 2007).

radioactivity in these shale units in this area. The gross thickness of the Upper Bowland Shale (Namurian), together with the net thickness of high-gamma shale layers, is shown for most of the hydrocarbon wells (Fig. 9). The Upper Bowland Shale typically has a GR profile which shows a higher value basal section, overlying Mississippian limestone: this type of profile is also seen in the Haynesville Shale of Louisiana, USA. The presence of underlying limestone is also similar to the Barnett Shale. In the latter the Ordovician Viola–Simpson provides an impermeable barrier. This is eroded southwards where the Ellenburger Limestone, an aquifer, subcrops (Fig. 1): this latter configuration makes artificial fracturing of the Barnett more difficult.

The Sabden Shales are characterized by anomalously low density and velocity on wireline logs, which might be indicative of high organic matter content. Overpressure caused by disequilibrium compaction or volume expansion related to hydrocarbon generation (Osborne & Swarbrick 1997) can also cause anomalous velocity–density trends. Core samples indicate that the unit predominantly comprises hard, dark grey to black micaceous mudstone with abundant plant fossils. The presence of interbedded coal

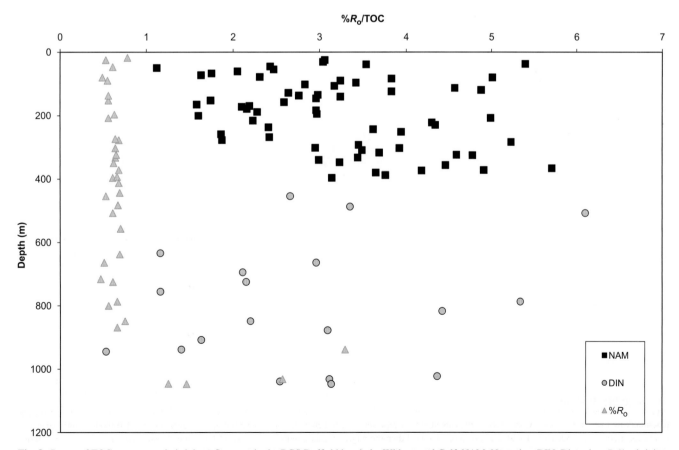

Fig. 8. Range of TOC contents and vitrinite reflectance in the BGS Duffield borehole, Widmerpool Gulf. NAM, Namurian; DIN, Dinantian; $R_o\%$, vitrinite reflectance. The high $R_o\%$ near the base is caused by igneous sills but may lead to gas window maturity near or below terminal depth.

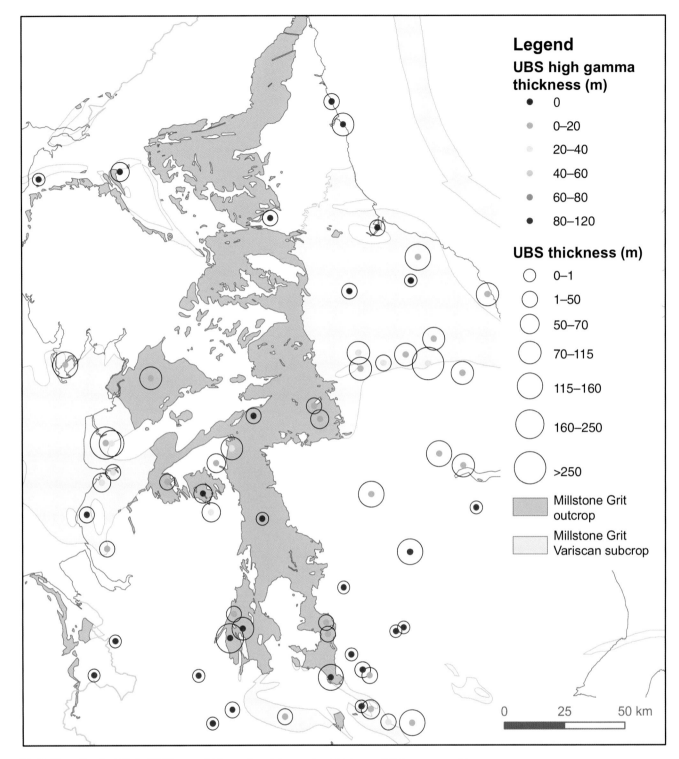

Fig. 9. Upper Bowland Shale (UBS): gross thickness of mudstones and net thickness of high gamma mudstones.

seams in some places suggests that at least part of the sequence may be gas-prone. TOC measurements in the Knott Coppy borehole (Figs 4 & 5) are in the range 1–3%, making these shale units potential hydrocarbon source rocks; however in this borehole TOC-rich shales do not always correspond to high gamma intervals.

Jurassic black shales

Black shales are found in the three main Jurassic mudstone formations: the Lias, Oxford Clay and Kimmeridge Clay (Figs 2 & 3). Liassic shales form the main source rocks for the Wessex and Weald oil fields (Ebukanson & Kinghorn 1986). These shales are mature for gas generation and small gas accumulations (dubbed 'enigmatic' by Butler & Pullan 1990) occur on the margins of the Weald Basin and are associated with oil at Wytch Farm in the Wessex Basin (Fig. 2). In northern England the Lias is within the oil window in the Cleveland Basin. The Oxford and Kimmeridge Clays contain organic-rich formations and might also prove to be potential source rocks in onshore basins, but not for thermogenic gas.

Porosity and permeability values are likely to be higher in Jurassic shale units because they have not been subject to as much compaction as Palaeozoic shale formations. Jurassic mudstones encountered in shallow (<30 m) engineering boreholes have porosities in the range 30–40%. Organic-rich shale units are preferentially enriched in Cu, Ni, V, Ag and Mo (Gad et al. 1969) in common with shales of Palaeozoic age. Exploration of a number of Cretaceous and Jurassic shales is underway in the USA, for example the Haynesville Shale of Texas and Louisiana. This formation lies above the concealed Ouachita fold belt in a similar relationship to the Weald Basin Jurassic shales, which overlie the Variscan fold belt.

Conclusions

At the regional scale the Mississippian Bowland Shale Formation provides the most prospective shale gas play. The Bowland Shale is a proven source rock mature for gas production. Those shales, which have not sourced conventional hydrocarbons, need to be carefully evaluated to see if they can make a contribution. Locally, shales of Lower Palaeozoic (principally Upper Cambrian) and Jurassic age might prove suitable targets. The small size of basins and the lack of overpressure, relative to the USA, make it unlikely that a large potential exists onshore UK, unless these older shales are proved to be prospective. However, additional data are required to minimize exploration risk. In particular, isotope studies of gas source provenance and more information on joint and fracture densities are required. Not all the organic-rich shale sections occur where there are high GR values. Shale gas is most likely to occur in the sub-basin depocentres of the Pennine and Weald basins. Previous conventional hydrocarbon exploration wells tested structural highs and fault traps, with the result that some entirely missed source rocks known from other wells or interpreted from seismic. Drilling wells in the depocentres and the greater attention to shales required for this unconventional exploration are likely to lead to a new surge in subsurface geological knowledge of the UK.

This paper is published with the permission of the Executive Director of the British Geological Survey (NERC). We thank S. Holloway, M. Stephenson, H. Johnson and A. Doré for improving drafts and D. Entwisle for providing porosity data.

References

Aitkenhead, N. 1977. The Institute of Geological Sciences Borehole at Duffield, Derbyshire. *Bulletin of the Geological Survey of Great Britain*, **59**, 1–59.

Armstrong, J. P., Smith, J., D'Elia, V. A. A. & Trueblood, S. P. 1997. The occurrence and correlation of oils and Namurian source rocks in the Liverpool Bay–North Wales area. *In*: Meadows, N. S., Trueblood, S. P., Hardman, M. & Cowan, G. (eds) *Petroleum Geology of the Irish Sea and Adjacent Areas*. Geological Society, London, Special Publications, **124**, 195–211.

Bath, A., Darling, W. G. et al. 1988. *Chemical and Stable Isotopic Analyses of Dissolved Gases and Groundwater Seepages collected from Wyresdale Tunnel, November 1987*. WE/88/1C.

Bowker, K. A. 2007. Barnett shale gas production, Fort Worth Basin: issues and discussion. *American Association of Petroleum Geologists*, **91**, 523–533.

Budge, G. D. 1932. Methane storage in strata. *Proceedings of the South Wales Institute of Engineers*, **48**, 177–200.

Butler, M. & Pullan, C. P. 1990. Tertiary structures and hydrocarbon entrapment in the Weald Basin of southern England. *In*: Hardman, R. F. P. & Brooks, J. (eds) *Tectonic Events Responsible for Britain's Oil and Gas Reserves*. Geological Society, London, Special Publications, **55**, 371–391.

Challinor, P. J. 1990. Oil ingress into mine workings. *The Mining Engineer*, August, 68–74.

Creedy, D. P. 1989. Geological sources of methane in relation to surface and underground hazards. *Paper 1.4 Methane – Facing the Problems*. Symposium, 26–28 September, Nottingham.

Dow, W. G. 1977. Kerogen studies and geological interpretations. *Journal of Geochemical Exploration*, **7**, 77–79.

Dow, W. G. & O'Connor, D. E. 1982. Kerogen maturity and type by reflected light microscopy applied to petroleum exploration. *In*: Staplin, F. L., Dow, W. G. et al. (eds) *How to Assess Maturation and Paleotemperatures*. Short Course No. 7. Society of Economic Paleontologists and Mineralogists, Tulsa, OK, 133–157.

Earp, J. R. & Hains, B. A. 1971. *The Welsh Borderland*. British Geological Survey, London, British Regional Geology Memoirs.

Ebukanson, E. J. & Kinghorn, R. R. F. 1986. Oil and gas accumulations and their possible source rocks in southern England. *Journal of Petroleum Geology*, **9**, 413–428.

Evans, C. J. & Brereton, N. R. 1990. In situ crustal stress in the United Kingdom from borehole breakouts. *In*: Hurst, A., Lovell, M. A. & Morton, A. C. (eds) *Geological Applications of Wireline Logs*. Geological Society, London, Special Publications, **48**, 327–338.

Ewbank, G., Manning, D. A. C. & Abbott, G. D. 1993. An organic geochemical study of bitumens and their potential source rocks from the South Pennine Orefield, central England. *Organic Geochemistry*, **20**, 579–598.

Ferguson, J. 1984. The methane content of some Carboniferous limestones from the northern Pennines and its relationship to mineralisation. *Proceedings of the Yorkshire Geological Society*, **45**, 67–69.

Frantz, J. H., Waters, G. A. & Jochen, V. A. 2005. Operators re-discover shale gas value. *E&P*, 1 October 2005.

Fraser, A. J., Nash, A. J., Steele, R. P. & Ebdon, C. C. 1990. A regional assessment of the intra-Carboniferous play of northern England. *In*: Brooks, J. (ed.) *Classic Petroleum Provinces*. Geological Society, London, Special Publications, **50**, 417–440.

Gad, M. A., Gatt, J. A. & Le Riche, H. H. 1969. Geochemistry of the Whitbian (Upper Lias) sediments of the Yorkshire coast. *Proceedings of the Yorkshire Geological Society*, **37**, 105–139.

Griffith, A. E. 1983. The search for petroleum in Northern Ireland. *In*: Brooks, J. (ed.) *Petroleum Geochemistry and Exploration of Europe*. Blackwell Scientific Publications, 213–222.

Hillier, S. & Marshall, J. E. A. 1992. Organic maturation, thermal history and hydrocarbon generation in the Orcadian Basin, Scotland. *Journal of the Geological Society, London*, **149**, 491–502.

Hitchman, S. P., Darling, W. G. & Williams, G. M. 1989. *Stable Isotope Ratios in Methane Containing Gases in the United Kingdom*. BGS Technical Report, WE/89/30.

Hood, A., Gutjahr, C. C. M. & Heacock, R. L. 1975. Organic metamorphism and the generation of petroleum. *American Association of Petroleum Geologists*, **59**, 986–996.

Jarvie, D., Pollastro, R., Hill, R., Bowker, K., Claxton, B. & Burgess, J. 2004. *Evaluation of Hydrocarbon Generation and Storage in the Barnett Shale, Fort Worth Basin, Texas*. World Wide Web Address: www.humble-inc.com.

Jones, D. G. & Plant, J. A. 1989. Geochemistry of shales. *In*: Plant, J. A. & Jones, D. G. (eds) *Metallogenic Models and Exploration Criteria for Buried Carbonate-hosted Ore Deposits – a Multidisciplinary Study in Eastern England*. British Geological Survey, Keyworth, 65–94.

Kellaway, G. A. & Welch, F. B. A. 1948. Bristol and Gloucester district. *British Regional Geology* (2nd edn). British Geological Survey, London.

Landes, K. K. 1967. Eometamorphism, and oil and gas in time and space. *Bulletin of the American Association of Petroleum Geologists*, **51**, 828–841.

Leggett, J. K. 1980. British Lower Palaeozoic black shales and their palaeo-oceanographic significance. *Journal of the Geological Society*, **137**, 139–156.

Lokhurst, A. (ed.) 1998. *The Northwest European Gas Atlas – Composition and Isotope Ratios of Natural Gases in Northwest European Gasfields*. NITG-TNO, Haarlem (CD ROM).

Maynard, J. R., Wignall, P. B. & Varker, W. G. 1991. A hot new shale facies from the Upper Carboniferous of Northern England. *Journal of the Geological Society*, **148**, 805–808.

Montgomery, S. L., Jarvie, D. M., Bowker, K. A. & Pollastro, R. M. 2005. Mississippian Barnett Shale, Fort Worth Basin, north central Texas: gas shale play with multi-trillion cubic foot potential. *American Association of Petroleum Geologists Bulletin*, **89**, 155–175.

Moseley, F. & Ahmed, S. M. 1967. Carboniferous joints in the north of England and their relation to earlier and later structures. *Proceedings of the Yorkshire Geological Society*, **36**, 61–90.

Oldroyd, G. C., McPherson, M. J. & Morris, L. H. 1971. Investigations into sudden abnormal emissions of firedamp from the floor strata of the Silkstone seam at Cortonwood Colliery. *The Mining Engineer*, **130**, 577–593.

Osborne, M. J. & Swarbrick, R. E. 1997. Mechanisms for generating overpressure in sedimentary basins: a reevaluation. *Bulletin of the American Association of Petroleum Geologists*, **81**, 1023–1041.

Parnell, J. 1983. The distribution of hydrocarbon minerals in the Welsh Borderlands and adjacent areas. *Geological Journal*, **18**, 129–139.

Plant, J. A. & Jones, D. G. 1989. *Metallogenic Models and Exploration Criteria for Buried Carbonate-hosted Ore Deposits – a Multidisciplinary Study in Eastern England*. British Geological Survey and the Institution of Mining and Metallurgy, London.

Pollastro, R. M., Hill, R. J. et al. 2004. *Assessment of Undiscovered Oil and Gas Resources of the Bend Arch–Fort Worth Basin Province of North-Central Texas and Southwestern Oklahoma, 2003: U.S. Geological Survey Fact Sheet 2004–3022*. World Wide Web Address: http://pubs.usgs.gov/fs/2004/3022/

Ponsford, D. R. A. 1955. Radioactivity studies of some British sedimentary rocks. *Bulletin of the Geological Survey*, **10**, 24–44.

Poole, E. G. 1977. Stratigraphy of the Steeple Aston Borehole, Oxfordshire. *Bulletin of the British Geological Survey*, **57**.

Poole, E. G. 1978. Stratigraphy of the Withycombe Farm Borehole, near Banbury, Oxfordshire. *Bulletin of the British Geological Survey*, **68**.

Ramsbottom, W. H. C., Sabine, P. A., Dangerfield, J. & Sabine, P. W. 1981. Mudrocks in the Carboniferous of Britain. *Quarterly Journal of Engineering Geology*, **14**, 257–262.

Ray, E. O. 1976. Devonian shale development in eastern Kentucky. *In*: *Natural Gas from Unconventional Geologic Sources*. Board on Mineral Resources and Commission on Natural Resources. National Academy of Sciences, Washington, DC, 100–111.

Rogers, S. F. 2003. Critical stress-related permeability in fractured rocks. *In*: Ameen, M. (ed.) *Fracture and In-situ Stress Characterization of Hydrocarbon Reservoirs*. Geological Society, London, Special Publications, **209**, 7–16.

Rushton, A. W. A., Hamblin, R. J. O. & Strong, G. E. 1988. The Croft Borehole in the Lilleshall Inlier of north Shropshire. *Reports of the British Geological Survey*, **19**, 1–14.

Selley, R. C. 1987. British shale gas potential scrutinized. *Oil and Gas Journal*, 15 June, 62–64.

Selley, R. C. 2005. UK shale-gas resources. *In*: Doré, A. G. & Vining, B. A. (eds) *Petroleum Geology: North-west Europe and Global Perspectives: Proceedings of the 6th Petroleum Geology Conference*. Geological Society, London, 707–714; doi: 10.1144/0060707.

Shurr, G. W. & Ridgley, J. L. 2002. Unconventional shallow biogenic gas systems. *Bulletin of the American Association of Petroleum Geologists*, **86**, 1939–1969.

Smith, N. J. P. (Compiler) 1985. *Map 1: Pre-Permian Geology of the United Kingdom (South)*. 1:1,000,000 scale. British Geological Survey.

Smith, N. J. P. 1993. The case for exploration of deep plays in the Variscan fold belt and its foreland. *In*: Parker, J. R. (ed.) *Petroleum Geology of Northwest Europe: Proceedings of the 4th Conference*. Geological Society, London, 667–675; doi: 10.1144/0040667.

Smith, N. J. P. & Rushton, A. W. A. 1993. Cambrian and Ordovician stratigraphy related to structure and seismic profiles in the western part of the English Midlands. *Geological Magazine*, **130**, 665–671.

Smith, N. J. P., Kirby, G. A. & Pharaoh, T. C. 2005. Structure and evolution of the south-west Pennine Basin and adjacent area. British Geological Survey, London, Subsurface Memoirs.

Spears, D. A. & Amin, M. A. 1981. Geochemistry and mineralogy of marine and non-marine Namurian black shales from the Tansley borehole. *Sedimentology*, **28**, 407–417.

Stocks, A. E. & Lawrence, S. R. 1990. Identification of source rocks from wireline logs. *In*: Hurst, A., Lovell, M. A. & Morton, A. C. (eds) *Geological Applications of Wireline Logs*. Geological Society, London, Special Publications, **48**, 241–252.

Taylor, K. & Rushton, A. W. A. 1971. The pre-Westphalian geology of the Warwickshire Coalfield. *Bulletin of the Geological Survey of Great Britain*, **35**, 1–150.

Waters, C. N., Browne, M. A. E., Dean, M. T. & Powell, J. H. 2007. *Lithostratigraphical Framework for Carboniferous Successions of Great Britain (Onshore)*. British Geological Survey, London, **RR/07/01**.

Zhao, H., Givens, N. B. & Curtis, B. 2007. Thermal maturity of the Barnett Shale determined from well-log analysis. *American Association of Petroleum Geologists Bulletin*, **91**, 535–549.

The Western Canada Foreland Basin: a basin-centred gas system

D. J. BOETTCHER,[1] M. THOMAS,[1] M. G. HRUDEY,[2] D. J. LEWIS,[2] C. O'BRIEN,[2] B. OZ,[2] D. REPOL[2] and R. YUAN[2]

[1]*Shell Canada Ltd (Retired)*
[2]*Shell Canada Ltd, 400 4th Avenue SW, PO Box 100, Station M, Calgary, Alberta, Canada T2P 2H5 (e-mail: michael.hrudey@shell.com)*

Abstract: Enormous volumes of gas (>30 Tcf) are contained within the deepest portions of the Western Canada Foreland Basin, where tight gas-saturated Cretaceous sandstones grade updip into porous water-saturated sandstones. Production has occurred from coarse-grained shoreline sands both near the updip gas–water interface, such as those found in the Elmworth Field, and from low-porosity–permeability reservoirs found deeper in the basin. These basin-centred gas (BCG) reservoirs are characterized by regionally pervasive gas-saturated lithologies, abnormal pressures and no downdip water contact, and occur in low-permeability reservoirs. The keys to Shell's exploration success were an understanding of the stratigraphy, sedimentology and rock properties of the basin, the development of structural, petrophysical and geomechanical models, development of an understanding of the desiccation or dewatering process, the distribution of water within the basin and how the pressure regime evolved, interpretation of 3D seismic, and an aggressive land strategy. The evaluation of structural leads was aided when seismic and geomechanical modelling were combined, thereby aiding in the prediction of zones with a higher probability of encountering favourable reservoir producibility characteristics, that is, areas where a well developed, well connected open fracture network is expected. This multidisciplinary approach has resulted in economic success in regions once thought to be non-productive, and where it was once said, 'People go broke chasing the Nikanassin'.

Keywords: basin-centred gas, tight gas sandstones, Canada, Cretaceous, Nikanassin, Cadomin, Falher, Cadotte, desiccated reservoir, fracture modelling, permeability enhancement, geomechanical modelling, updip water

Basin-centred gas systems, as defined by Law (2002), are those characterized by being: regionally gas pervasive, abnormally pressured (high or low), lacking a downdip water contact and occurring in low-permeability reservoirs. Bennion *et al.* (2002) stated that, in order to produce economic volumes of gas from these low-permeability (<0.05 mD) reservoirs, the initial water saturation must be significantly lower than what would be expected from a system in capillary equilibrium where the matrix is in dynamic contact with a free water level. In the past these have been variably referred to as tight gas sands or 'Deep Basin' gas with the latter first used by Masters (1979) to describe the thick, low-permeability, gas-bearing, Cretaceous section of west-central Alberta.

From the initial phases of exploration in this portion of Western Alberta to the mid-1970s, industry concentrated on either deep Palaeozoic objectives or conventional Cretaceous zones where reservoir properties were good and production rates were high. Beginning with the Elmworth discovery in 1976, the focus began to shift to the 'upside down trapping conditions of the Deep Basin' (Masters 1984). Initially the exploration and development concentrated on low-permeability (<0.4 mD) but well sorted shoreline sandstones of the Lower Cretaceous. Prolific drilling continued in the Deep Basin throughout the following decades, such that within the central focus area nearly 18 000 wells have been located and drilled. Of these, nearly 10 000 wells are currently producing a total of 4.92×10^9 SCF/day with a total cumulative gas volume of 23.8×10^{12} SCF. At present, the average well deliverability is approximately 0.5 mmcf/day.

As a relative latecomer to the area, in January 2003, Shell Canada began to investigate opportunities in the Deep Basin of west-central Alberta. The initial area examined (Fig. 1a) ranged from west of Calgary Alberta in the SE, to the NE corner of British Columbia, covering an area of 100 000 km^2. The western boundary was the edge of the Rocky Mountain disturbed belt (Fig. 1) while the eastern boundary was the updip water edge associated with each formation. The formations examined ranged from the Upper Cretaceous Cardium Formation to the Lower Cretaceous/Jurassic Nikanassin Formation (Fig. 3). From this analysis, an area of interest ranging from Township 45 Range 20W5 in the SE to 93-P in northeastern British Columbia (Fig. 2), covering 50 000 km^2, was selected. This region was chosen as all reservoirs of interest are present, principally the Cadotte, Falher, Cadomin and Nikanassin formations (Fig. 3), and pressures (along the basinal axis) are for the most part above hydrostatic (except the Cardium Formation). Within these units, the thickest sandstones occur in the upper Nikanassin Formation, where net reservoir can exceed 100 m. Available petroleum and natural gas rights directed Shell's attention into the deepest portions of the basin, corresponding to the region of lowest matrix permeability, where pressures were highest, where sandstone thickness were highest and where multiple thrust cored structures were evident on Shell's existing 2D seismic data base.

Shell's understanding of data within the study area favoured an exploration strategy for basin-centred gas (BCG) and, in the latter part of 2004, Shell drilled and completed its first BCG well at 15-20-65-13W6. Gas was encountered in the Cadotte (absolute open flow potential, AOFP, of 3.5 mmcf/day from 18.7 m of 5.6%phi), Cadomin (AOFP of 4.8 mmcf/day, 10.0 m at 5.0%phi) and Nikanassin formations (three intervals were tested; two were shut in while still cleaning up but the uppermost zone had a calculated AOFP of 9.0 mmcf/day, 93 m at 4.2%phi) and flowed from sandstones and conglomerates whose matrix permeability is commonly less than 100 μD. Continued exploration and subsequent exploitation has resulted in an understanding of the regional depositional history, rock properties, stratigraphic and

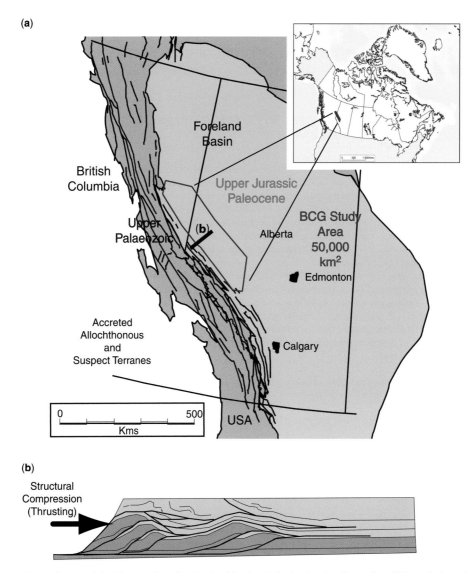

Fig. 1. Schematic regional tectonic map of the Western Canadian Foreland Basin. (**a**) Basin showing the outline of Upper Palaeozoic, Upper Jurassic/Paleocene sediments and faulting. (**b**) Generalized cross-section east of Palaeozoic outcrop.

structural geophysics, reservoir petrophysics (including water saturations, S_w, and updip water line complexities), reservoir desiccation and pressure distributions. These different facets are used in conjunction with geomechanical models to aid in the targeting of wells into features where the probabilities of encountering an effective fracture network and associated enhanced gas productivity have been maximized.

Deep basin stratigraphy, sedimentology and rock properties

During Late Jurassic and Early Cretaceous time, uplift and exhumation associated with the Columbian orogeny (Mossop & Shetsen 1994) provided abundant sediment to the foredeep trough, defined by Leckie & Smith (1992) as a succession of sedimentary rocks deposited in a cratonic region adjacent to an active orogenic belt. Sediments were derived mainly from the orogenic zone and thinned away from it. In Western Canada, this foreland basin was an elongate trough that developed between the eastern flanks of the ancestral Rocky Mountains and the stable interior platform, represented by the North American craton and extending across most of western Alberta and NE British Columbia.

In the Western Canada Foreland Basin (WCFB, also referred to as the Alberta Basin), the amount and rate of subsidence was greatest adjacent to the advancing thrust sheets and it was here that the greatest amount of sediment accumulated, resulting in stacked, eastward-thinning clastic wedges. The asymmetric subsidence of the basin also affected drainage patterns, creating a prevailing drainage system that was largely parallel to the basin axis (i.e. NW–SE).

Leckie & Smith (1992) subdivided the depositional history of the WCFB into five cycles of differing ages, each representing strata bounded by major unconformities or lithological changes. Prospective BCG sandstones are contained within the first two cycles, the first of which extends from Oxfordian (earliest Upper Jurassic) to late Valanginian time (mid Lower Cretaceous) and includes sediments from the base of the Fernie Formation shales to the top of the Nikanassin Formation (Fig. 3). The base of Cycle 1 (Leckie & Smith 1992), in the grey or green beds of the Late Jurassic Fernie Formation, marks the first widespread appearance of sediments that were derived from the rising Cordillera to the west. The top of Cycle 1 is represented by a regional unconformity that separates the Nikanassin Formation or Kootenay Group sediments from the overlying Lower Cretaceous Cadomin Formation (Fig. 4). This unconformity represents a hiatus that may have been in excess of 27 Ma (Leckie & Smith 1992). The second cycle incorporates the majority of the remaining zones of interest ranging from the Cadomin Formation to the Cadotte Formation.

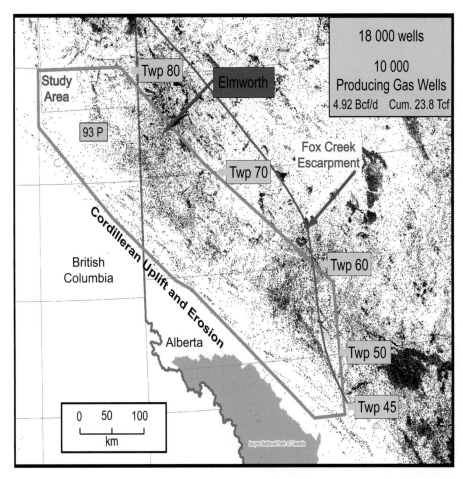

Fig. 2. Well density across study area.

Jurassic to Eocene tectonics produced a series (Fig. 5) of anticlinal structures of decreasing magnitude to the NE across the basinal axis, from structures of high amplitude to updip regions where no structures are evident and the reservoir units are essentially flat lying, dipping at angles of 1–2° to the SW.

Allen (1980) stated that, within the WCFB, hydrocarbons have been found in 20 rock units ranging in age from Permian to Late Cretaceous with the majority of reserves contained within the Lower Cretaceous Spirit River Formation (mainly Falher Member) and in the tight fluvial sandstones of the Upper Jurassic Nikanassin Formation. Each of the principal objective formations are discussed below.

Nikanassin Formation

The name 'Nikanassin Formation' (Poulton et al. 1990) extends northwards from outcrops into the subsurface in west-central Alberta, although no type section has been formally designated for the Formation. In the outcrop belt of the Rocky Mountain Foothills of northeastern British Columbia the equivalent Upper Jurassic sandstones and associated sediments have been included in the Minnes Group (Fig. 3). Stott (1993) correlated the Minnes Group eastward into the subsurface where the upper zones are absent below the sub-Cadomin Formation unconformity and only the Monteith Formation and parts of the Beattie Peaks Formation are preserved (Fig. 3). These two formations are not easily distinguishable in well logs and together comprise what is more commonly referred to as the Nikanassin Formation.

The Fernie/Nikanassin Formation assemblages form a clastic wedge that represents the earliest orogenic detritus in the foreland basin. The lower units, including the Green Beds, Fernie and Passage Beds, consist of marine deposits that initially filled the foredeep. These marine sediments grade upward into continental clastic deposits of the Kootenay Group and Nikanassin Formation (Barclay & Smith 1992), with basin filling occurring in a northerly direction along the axis of the basin. While the Passage Beds (siltstone and shale) represent marine deposition during the initial stages of basin-fill, sandstones of the Morrissey Formation or basal unit of the Kootenay Group (Fig. 3) were deposited along a northerly to northwesterly prograding high-energy shoreline. In southwestern Alberta, the Morrissey is overlain by fluviodeltaic quartz and chert-rich sandstones, siltstones, shales and coals of the Mist Mountain Formation. North of the North Saskatchewan River (off the south edge of Fig. 2), the coeval Nikanassin Formation is dominated by flood plain deposits in the west, grading to shoreline and estuarine sandstones near its eastern erosional limits. The equivalent Monteith Formation in northeastern British Columbia contains fluviodeltaic lithic sandstones, shales and coals in the west and shoreline to estuarine quartz arenites in the east, suggesting that both western and eastern provenances are represented in preserved strata in this early narrow foredeep trough.

The Minnes Group is interpreted to have been deposited in an entirely fluvial system in northeastern British Columbia, south of the Sukunka River (at the north edge of Fig. 2). Similar to the Kootenay Group of southwestern Alberta, it contains flood plain shales, fluvial channel sandstones and coal in its lower part, which grade upward into upland fluvial sandstones and conglomerates. During the late Jurassic, basin fill apparently occurred from south to north, suggesting that shoreline sequences became progressively younger in the same direction (Mossop & Shetsen 1994). Detailed log evaluation of the Nikanassin Formation, especially from wells in northeastern British Columbia, suggests

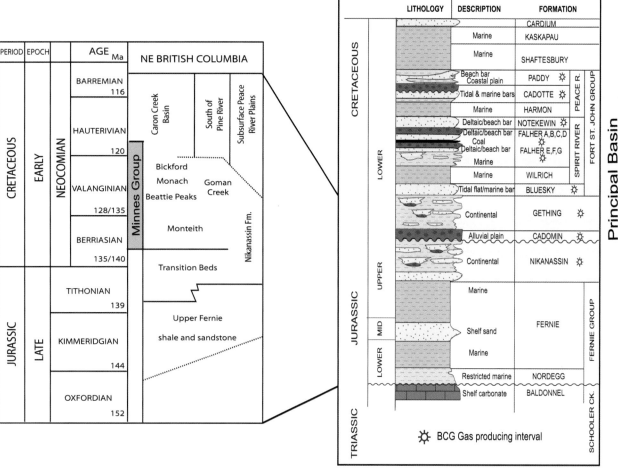

Fig. 3. BCG stratigraphic summary for NE British Columbia and west-central Alberta. Surface and subsurface.

that the lower portions of the formation consist of marine sediments characterized by thin, relatively 'dirty' coarsening upward beach sequences. These grade upward into units composed predominantly of continental (flood plain?) sediments including numerous coals and fluvial sandstones.

Depositional model

As previously indicated, those zones that are correlative with the Nikanassin Formation to the SE in the Kootenay Group and to the NW in the Minnes Group are primarily fluvial in nature. In the central portion of the BCG study area, the Nikanassin Formation can be subdivided into two main units, referred to simply as the Upper and Lower Nikanassin (Fig. 6). Net sand in the Upper Nikanassin can exceed 130 m in thickness along the axis of the trough. This depocentre is approximately 50 km long and can range in width from 10 to 20 km. To the east and NE these sands thin to a zero edge. The basal portion of the Upper Nikanassin (Fig. 6b) consists of discrete point bar and channel sand deposits (as determined from log signatures) separated by siltstones and shales. Moving up in the section, the sands become more continuous, both vertically and laterally, although some irregularly distributed shales are present. Stacked channel sandstones in the Upper Nikanassin can exceed 130 m in thickness (Fig. 6a). In cross-section (Fig. 7) the Upper Nikanassin thins to the east, with the more continuous sands near the top over much of the area.

The Lower Nikanassin consists predominantly of siltstones and shales with some scattered sandstones. Resistivity at the boundary between the two units shows a distinctive shift (Fig. 6). Zaitlin *et al.* (2002) indicated that changes in Basal Quartz deposition (early Cretaceous–southeastern Alberta) from low accommodation braided deposits to higher accommodation meandering deposits were associated with local syn-depositional faulting. Lukie *et al.* (2002) indicated that mudstone-dominated deposits consisting of terrestrial overbank sediments with included isolated channel sandstones are interpreted to reflect a progressive increase in accommodation space that led to successively less amalgamation of the channel deposits.

In the Western Canada Foreland Basin, the amount and rate of subsidence was greatest adjacent to the advancing thrust sheets and it was here that the greatest amount of sediment accumulated (Leckie & Smith 1992). Arnott *et al.* (2002) stated that in basins with increasing accommodation space meander deposits predominate in which irregularly distributed fluvial sands are preserved in thick overbank deposits. Thus, it is interpreted that the continental sediments of the Lower Nikanassin as well as those in the basal portions of the Upper Nikanassin were deposited in meander systems that formed in response to increasing accommodation space resulting from subsidence in the foreland basin. In contrast, it is postulated that the more regionally continuous sheet-like sandstones of the Upper Nikanassin were deposited in a braid plain complex where little overbank sediment is preserved. These braid plain deposits formed when accommodation space decreased (or sediment supply increased) and the sands build laterally as opposed to vertically (as noted for the Cadomin Formation by Johnson & Dalrymple 2005). In northeastern British Columbia the more laterally continuous

Fig. 4. Outcrop photograph of the Cadomin to Top Nikanassin interval showing the top Nikanassin unconformity. Note the difference in fracture density and patterns in the coarse-grained Nikanassin sandstones v. the conglomerates of the Cadomin.

braid plain sandstones are not evident. Instead the Nikanassin Formation consists predominantly of thick sequences of siltstones, coals and shales with some interbedded channel and point bar sands indicative of deposition in a meander system.

The distribution of the braid plain and meander systems in the Nikanassin Formation can be explained through examination of deeper features in the basin as seen on a seismic strike line (Fig. 8), flattened on the 4th coal in the Falher Member (Spirit River Formation), extending from Township 63 into NE British Columbia. Moving from SE to NW it is apparent that accommodation space is increasing from the 4th coal to the Triassic level. The break in slope where the thickness starts to increase can also be correlated to a Devonian reef edge deeper in the section and a hinge line in the Basement. The braid plain system was developed across the region to the SE, where the seismic 4th coal to Triassic isotime is relatively constant. Northwest of the break in slope, accommodation space increases and the Nikanassin Formation is characterized by sandstones deposited primarily in meander systems. Overall, the Nikanassin Formation represents a significant period of basin fill as the gross isopach ranges from approximately 200 m in the NE to over 1500 m in the centre of the foreland basin adjacent to the disturbed belt in the NW portions of the study area.

Rock properties

The upper portion of the Nikanassin Formation consists primarily of medium- to coarse-grained lithic sandstones in which the principal framework grains are quartz, chert and argillaceous rock fragments. Primary porosity was destroyed by the combined effects of mechanical compaction (physical rearrangement of grains into a closest packing in response to increasing overburden pressure), chemical compaction (suturing or welding of grain boundaries), quartz overgrowth development and the precipitation of kaolinite and late stage ankerite (iron-carbonate) cement (Fig. 9). The current porosity in these sandstones is mainly secondary and resulted from the partial to complete dissolution of chemically unstable argillaceous clasts and chert grains.

In a cross-plot of ambient porosity v. air permeability (Fig. 10a), the dark blue rhombs represent core averages (thickness weighted porosity and geometrically averaged permeability) for individual wells. The magenta squares are ambient whole-core values, from the Nikanassin Formation in Shell's 11-21-65-13W6 well. The green triangles represent ambient values on drill plugs from the same core. All of these values fall within the same general trend except for a few core averages (dark blue rhombs) where the permeability values are near to or greater than 1 mD, the result of an effective fractures network. From the individual well analysis (dark blue rhombs) average ambient porosity and permeability values are 0.05 and 0.14 mD, respectively. When conducting the regional scoping work to assess the thickness and distribution of the Upper Nikanassin Formation sandstones, it was noted that the sonic response is very consistent throughout the interval, averaging about 200 $\mu s\ m^{-1}$. Using a sonic-porosity transform of ([interval transit time $-$ 180]/400 $\mu s\ m^{-1}$) results in a porosity of 5%. Core analysis is consistent with this, giving porosity values in the Upper Nikanassin Formation sandstones of 4–6%.

In situ permeability values tend to be one order of magnitude lower than corresponding ambient values, but the actual reduction

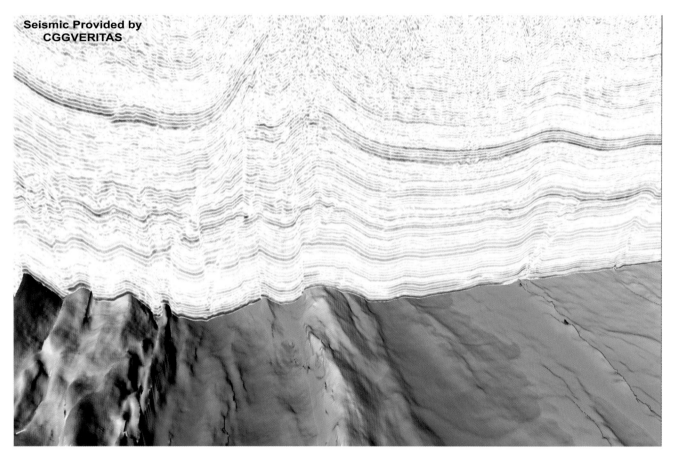

Fig. 5. Seismic and structure example showing a series of anticlinal structures related to a number of fault propagation folds. Seismic runs from SW to NE across the deepest portion of the BCG play area. Displayed horizon is in the lowermost Falher; line is 18 km long, displayed depth 3000 m. Viewed to NW.

is controlled by the fabric, composition and local stress values of the sandstones. More quartz-rich sandstones tend to be less pressure-sensitive than more argillaceous samples. For example, analysis of two core plugs from the Nikanassin Formation (Fig. 10b) in the 11-21-65-13W6 well shows the reduction in permeability is less than one order of magnitude and is more gradual than three samples shown from the low-permeability, illitic Rotliegendes sandstones.

Laboratory measurements conducted on whole-core and drill plugs, measuring matrix properties, showed that small-scale features such as argillaceous laminae, grain size variations and bedding have an impact on permeability. Reservoir performance, however, is controlled by larger scale facies and rock properties variations as well as natural and mechanically induced fractures. In the central portion of the BCG study area understanding such properties as concentration, size and distribution of fractures is key to optimizing reservoir performance, with fracture distribution being controlled by lithology, structural location, degree and effectiveness of the fracture network and tectonic history. Pressure transient analysis (Fig. 10a) shows the effect of fractures, as system permeabilities with values of hundreds of microDarcies are not uncommon.

Cadomin Formation

The Cadomin Formation in west-central Alberta has been documented extensively by Jackson (1984), Gies (1984), and Smith *et al.* (1984). The formation was deposited upon a widespread unconformity at the top of the Nikanassin Formation which formed as a result of uplift associated with the docking of allochthonous terrains along the western margin of the North American craton. Cadomin Formation strata are bounded to the NE by the Fox Creek Escarpment and by the Cut Bank Valley edge in the SE (Hayes *et al.* 1994). East of the topographically high Fox Creek escarpment the formation is absent. Lack of well control results in a poorly defined NW limit and to the SW the regime is breached with moderate to high amplitude foothill structuring.

Cadomin gas production is modest but widespread in the Deep Basin. Low-porosity–permeability sandstones and conglomerates forced Canadian Hunter to write off hundreds of BCF of Cadomin reserves at Elmworth in the late 1990s, as the company could not demonstrate that the gas could be recovered economically.

Depositional model

The conglomerates of the Cadomin Formation are considered to have formed in an alluvial plain environment as sediments were shed in an easterly direction from the rising Cordillera. At various places along the Cordilleran uplift rivers debouched coarse-grained sediment onto the interior plains. Upon emerging from the confines of mountain canyon systems these streams abruptly deposited their bed-load material to build large alluvial fans. With continued water flow and redistribution of sediment, these fans expanded away from their source area and ultimately merged to form a single widespread alluvial plain covering much of Western Alberta, with the eastern limit marked by the Fox Creek Escarpment (Fig. 2). Between the escarpment and the alluvial plain a broad braid-plain was formed by a north-flowing river system called the Spirit River channel. The alluvial plain deposits consist of poorly sorted, sandy, chert pebble conglomerates, while the Spirit River braid plain conglomerates are better sorted and consist of paler coloured finer grained chert pebbles. The main Deep Basin gas accumulation in the Cadomin Formation occurs in a downdip location, in low-permeability

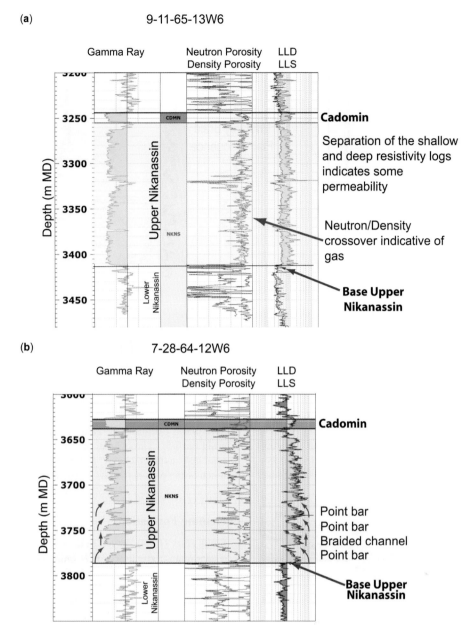

Fig. 6. Well log characteristics of the upper portion of the Nikanassin Formation.

alluvial plain sediments, while the more permeable braided river deposits predominate throughout the updip region near the gas–water contact.

Johnson & Dalrymple (2005) suggest that during Cadomin Formation time there was little accommodation space, which resulted in widespread deposition of the relatively thin Cadomin Formation conglomerates. This is apparent through much of northwestern Alberta where the Cadomin Formation averages about 5–10 m thick (Fig. 11a). Rapid changes in thickness and an irregular distribution of contours suggest deposition on either a peneplained surface or river reworking. In British Columbia (Fig. 11b) a regionally definable 10 m contour exists between Shell's Thunder and Bennett acreage. East of this contour the distribution of Cadomin Formation conglomerates is very similar to that observed in Alberta. Northwest of this contour, however, the Cadomin Formation can thicken to over 40 m with contours following the basin axis, suggesting some degree of subsidence and increased accommodation space in this area. Several relatively narrow zones of increased thickness are observed extending eastward from the disturbed belt, interpreted as alluvial fans feeding sediments into the basin.

Rock properties

In contrast to the Nikanassin Formation (Fig. 10a), there is no correlation between porosity and permeability in the Cadomin Formation (Fig. 12) within the BCG area. This lack of correlation is attributed to changes in rock fabric and the presence of fractures within the analysed core. The overall average ambient porosity (thickness weighted) and permeability (geometric average) values for cores are 5% and 2 mD, respectively. As with the Nikanassin Formation, *in situ* permeabilities are generally an order of magnitude lower.

In the conglomerates, reservoir properties are dependent upon the amount and quality of matrix. Chert pebbles contain variable amounts of microporosity and generally do not contribute to the overall properties of the reservoir. The coarser pebble-supported conglomerates tend to have poorer reservoir quality compared with those rocks where the larger grains are suspended in a 'finer' grained matrix (i.e. matrix supported). The chert pebbles can be very clean, consisting essentially of very finely crystalline silica, while others can contain variable amounts of impurities such as clays and organic matter. In some cases, impurities have

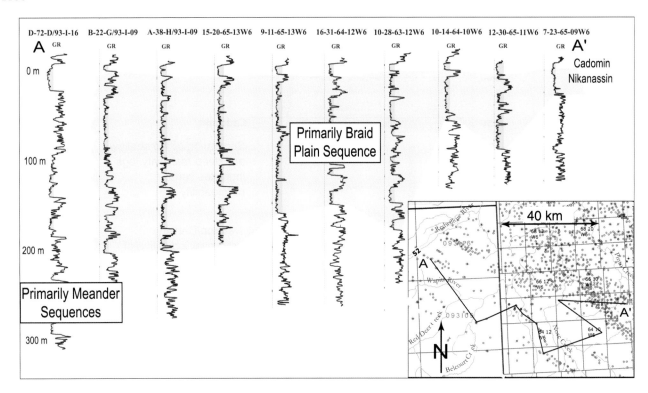

Fig. 7. Cross-section of upper Nikanassin Formation showing distribution of sandstones across the central BCG region.

been leached out, resulting in chert grains containing variable amounts of ineffective microporosity. As chert grains are composed of numerous small silica crystals, each with their own crystallographic c-axis, they generally do not develop large equant pore-filling overgrowths; rather a quartz druse will form on the grain surface. Throughout the area of interest, kaolinite is the principal authigenic clay and can infill isolated pores and/or much of the pore network.

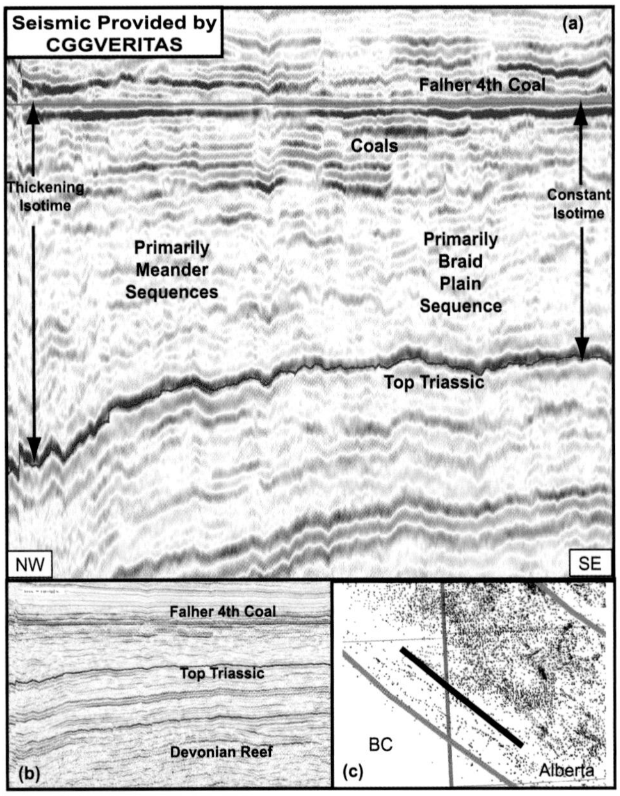

Gething Formation

Smith et al. (1984) suggest that the Gething Formation is a terrestrial sequence composed of interbedded fine-grained sandstones, siltstones, mudstones and coals. The Gething Formation can best be described as having been deposited on a low relief interior drainage plain on the eastern flanks of the Cordillera. The physiography of this drainage basin was a low-lying swampy plain with numerous lakes and forests (Smith et al. 1984). Sandstones in the Gething Formation were deposited in fluvial systems with the main drainage pattern oriented NE from the Cordillera and then switching to the NW along the Spirit River channel system, the same system that affected Cadomin Formation deposition. Johnson & Dalrymple (2005) found that Gething Formation sequence boundaries display little incision, indicating a high rate of subsidence during deposition. Where thin (<5–7 m) vertically isolated Gething Formation sandstones are observed in logs through the study area, they typically have a point bar or channel signature. Given the physiography of the basin at this time and the observed distribution of facies, it is assumed that the sandstones were deposited in meander systems resulting in a lower degree of reservoir predictability from well and/or seismic data.

Spirit River Formation

Falher Member

Sediments associated with the Falher Member of the Spirit River Formation were deposited in an overall regressive cycle. Five distinct cycles have been identified in the Falher Member with each cycle representing a rapid transgression followed by a slow regression as the coastline oscillated (Smith et al. 1984).

Fig. 8. NW to SE strike seismic line running across the British Columbia–Alberta border from Township 63 to south of Bearhole Lake. Both scales display the degree of changes in accommodation space from the SE to NW in the Nikanassin interval. Both displays (**a** and **b**), different scales, are flattened on the 4th Falher coal and are 70 km (**c**) in length.

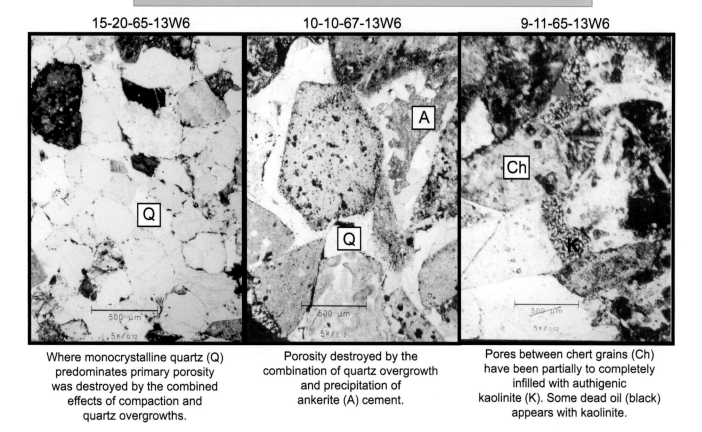

Fig. 9. Upper Nikanassin thin sections showing primarily medium- to coarse-grained lithic quartz rich sandstones. Secondary porosity is blue, and results from the dissolution of unstable argillaceous clasts and chert grains.

Coastal/deltaic sandstones and high-productivity conglomerates are common through the Elmworth Field (Fig. 2). To the south the Falher Member sediments represent deposition in a coastal plain environment and are characterized by shales, siltstones and coals, thus making the subdivision into cycles more difficult. The irregularly distributed Falher Member sandstones in this environment were deposited primarily in fluvial meander systems and consist of channel and point bar deposits.

Continuous coals (some up to 14 m thick) can be mapped over several townships. The thickest coal seam, referred to as the 4th

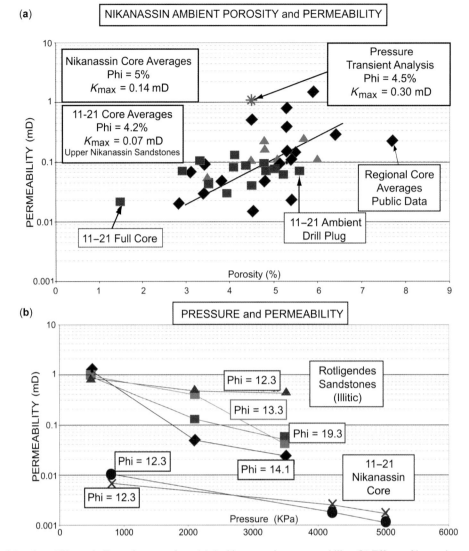

Fig. 10. Cross-plots of data from Nikanassin Formation core plugs. (**a**) Ambient porosity v. permeability. (**b**) Effects of increasing pressure on permeability for Nikanassin Formation and Rotliegendes sandstones.

coal, is found near the bottom of the Falher Member and contains approximately 10×10^{12} SCF of gas-in-place (Wyman 1984). Geochemical analysis of the Elmworth coal seams (Wyman 1984) indicates that the coals and carbonaceous shales were the most prolific source beds for the gas, which is still being expelled today (Michael & Bachu 2001).

Notikewin Member

Schmidt & Pemberton (2004) state that the Notikewin Member is the youngest member of the Spirit River Formation, and is underlain by the Falher and Wilrich Members, respectively. During deposition, the Western Canada Foreland Basin was inundated by the Boreal Sea, resulting in the formation of a shallow epieric seaway. The Notikewin Member represents a conglomerate-rich shoreline, formed within that seaway, which the aforementioned authors subdivided into four allomembers. The Notikewin Member had previously been described as a barrier island complex that had prograded as much as 150 km, although modern barrier islands generally form in transgressive environments. Schmidt & Pemberton's (2004) work demonstrated, however, that the barrier deposits formed in Notikewin time did not prograde a great distance as a barrier island/lagoon complex; instead sandstones and conglomerates of limited lateral extent were deposited in an orientation that was approximately parallel to the shoreline. The most prolific Notikewin Member wells occur in a narrow east–west trend through Township 67 from ranges 8 to 13W6. In this trend Schmidt & Pemberton (2004) defined a barrier spit that was formed as sediments were introduced into a shoreline as a result of channel avulsion. This barrier had little progradational extent, instead growing longshore, creating a large estuarine/lagoonal complex. The greatest concentration of net sand in the Notikewin Member occurs north of Township 65. To the south, the Notikewin Member consists primarily of coastal plain sediments, including silts, shales and coal.

Cadotte Member

As was observed with the Cadomin Formation, the Cadotte Member of the Peace River Formation can be found throughout the BCG study area in west-central Alberta and northeastern British Columbia. Early in middle Albian time, the Boreal Sea advanced across much of what is now known as the Western Canada Foreland Basin depositing marine shales of the Harmon Formation. This transgressive event occurred during a global rise in sea-level and was the beginning of a major transgression, which intensified during Late Albian time with the spread of the Shaftesbury/Colorado sea across much of western North America (Smith et al. 1984). A pause in this overall transgressive event occurred with deposition of the Paddy/Cadotte Members during

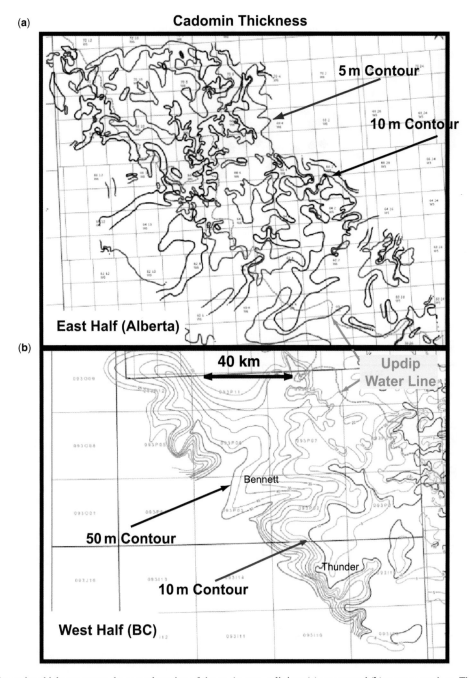

Fig. 11. Cadomin Formation thickness across the central portion of the study area, split into (**a**) eastern and (**b**) western portions. The east shows widespread thin conglomerates and uniformly low accommodation, whereas the west shows thickening from an increase in accommodation space, and the impact of alluvial fans forming from the west.

a regressive phase. Lettley *et al.* (2006) stated that the Cadotte Member forms an extensive shallow-marine complex that spans about 300 km along depositional strike and 150 km alongdip. A sharp, but conformable facies contact is typically seen in the middle of the Cadotte Member and is interpreted as a scour surface, separating the underlying very fine- to fine-grained sandstones of the lower to middle shoreface from the overlying chert-bearing sandstones and conglomerates of the upper shoreface. Lettley *et al.* (2006) observed lateral variations in texture and mineralogy throughout the upper shoreface sequence, which they attributed to the influence of deltaic sources.

To observe reservoir distribution, the thickness of the upper shoreface sands and conglomerates were mapped. Throughout the central portion (Fig. 13) of the study area the thickness of the upper shoreface is generally less than 10 m, with the thicker areas generally having a relatively well defined east–west orientation. This also matches the deviations seen on the updip water edge (blue line, Fig. 13). The exception to this is in the NW corner of the study area in British Columbia where the upper shoreface sediments follow a regular NW–SE oriented contour pattern and are interpreted to be filling a low coincident with the basinal axis. This pattern of basin fill is similar to that observed in the Cadomin Formation (Fig. 11a), although the area covered by the Cadotte Member is smaller, suggesting that between Cadomin and Cadotte time, the basin had progressively filled toward the NW.

Rock properties. Throughout the study area, both in British Columbia and Alberta, a significant scatter occurs on cross-plots of Cadotte Member permeability and porosity (Fig. 14a). The scatter on these data is most likely due to facies variations (i.e. sandstone v. conglomerate). Variability in grain size as well as porosity and permeability occurs in upper shoreface sandstones, where the finer grained rocks have a higher porosity but lower permeability than the coarser grained conglomerates where the pore

1110

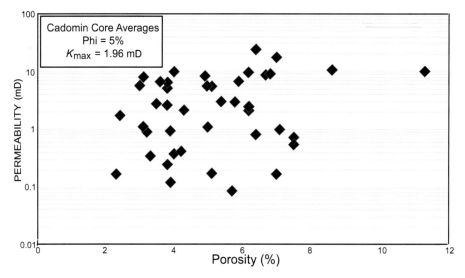

Fig. 12. Cadomin Formation porosity–permeability cross-plot.

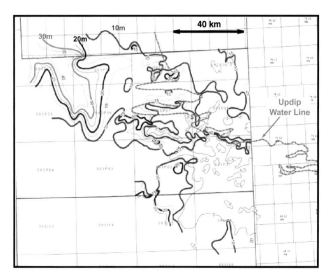

Fig. 13. Isopach of upper shoreface sandstones and conglomerates of the Cadotte Member across the western portion of the study area. Thicknesses are generally less than 10 m, except in the far NW, where up to 40+ m is encountered. Updip water edge outlined in blue.

throats are larger. When comparing the upper shoreface sandstones to those found in the very fine- to fine-grained middle to lower shoreface sandstones (Fig. 14b), a relatively good trend between core measured porosity and permeability is encountered, showing the impact of facies on these parameters. If the entire core is averaged, irrespective of facies, the average porosity and ambient permeability values are 6.5% and 0.68 mD, respectively. When evaluating just the upper shoreface the average porosity is 9.2% while the permeability is 2.32 mD.

In the unstructured regions of the study area, where the reservoir units are relatively flat, production is dependent upon matrix permeability, which in turn is dependent on facies and grain size. To the west in the structured areas, matrix permeability is enhanced by natural fractures.

Western Canadian BCG basin model

An idealized cross-section through the central BCG region (Fig. 15) shows that the deepest part of the basin occurs immediately in front of the disturbed zone with each unit rising gradually to the NE where the updip water contact is encountered. The Shaftesbury Formation shales (top seal) and Fernie Formation shales (bottom seal) encase the main reservoir units that extend from the Cadotte Member to the Nikanassin Formation. Also contained in this interval is an abundance of coals, which Wyman stated (for the Elmworth area, in 1984) were and continue to be the most prolific source beds for gas.

Since maximum burial, the Deep Basin of Western Canada has been uplifted with the loss of from 1500 to 3500 m of overburden (Issler *et al.* 1999). Over geological time, the process of hydrocarbon generation, the expulsion of dry gas, updip migration of these gases and basin exhumation have produced a pressure system in a state of pseudo-equilibrium. Overpressures (Fig. 16) are still maintained in Cretaceous strata in the deepest and most compacted portion of the basin, maintained by the very low-permeability strata. To the west of this pressure high, within the highly deformed structure belt, tectonic complications (blind fault zones) have provided leak points and/or impaired gas migration within structural units. Throughout this region, pressure lows are thought to be produced by breaches in reservoir units and as such some zones are totally water-wet.

To the east and near the updip water line, the BCG portion of the basin is unstructured with only a gentle 1 or 2° western dip. Gas migration continues from deeper regions towards the updip water edge, where permeability and porosities increase. At this point, hydrocarbon loss increases above the migration rate and formation pressures fall. As gas continues to escape from the system, water slumps back into the basin (Fig. 17) along higher permeability streaks, driven by hydrostatic forces. This slumping occurs until the relative permeability inhibits the movement of water and capillary forces become dominant. Further downdip from this, water continues to move deeper into the basin, drawn by capillary pressure, until gas migration and desiccation within the reservoir prevent further slumping. The transition region from gas to water across this interface is observed to have subnormal pressures, whereas further updip, pressures are supported by the hydrostatic gradient. Across this complex horizontal transition zone, variations in the formation permeability produce water-saturated and undersaturated intervals directly next to each other within the same reservoir unit. In the deepest portions of the basin, the pore pressure increases to values above the hydrostatic gradient. High gas gradients across

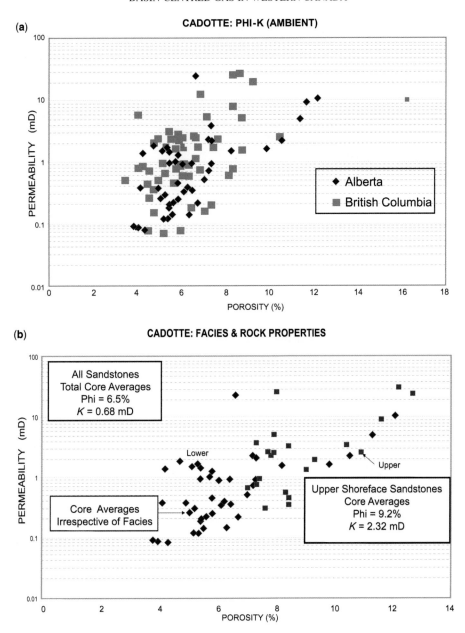

Fig. 14. Cadotte permeability v. porosity cross-plots. (**a**) BC v. Alberta data shows similar scatter of sandstone v. conglomerate. (**b**) Impact of facies on porosity and permeability values.

this area are the result of the burial, maturation, overpressuring, unroofing and expulsion history of the basin and where low leak-off rates maintain dynamic pressures above hydrostatic values.

It has also been found that vertical pressures can be highly variable, depending on the formation permeability. The overall high K_v/K_h results in the location of the recharging aquifer being at a significant distance from and at much lower elevations than the well location, thereby producing further under pressure. The location of the observed water edge is controlled by the magnitude of the capillary pressure and the concurrent hydrodynamic drive. Overall, all of these factors produce a complex and abnormal pressure regime that can range from significantly underpressure to significantly overpressure.

Early in the geological development of the Western Canada Foreland Basin, sedimentation, subsidence and thermal maturation resulted in overpressured source rocks expelling dry gas into 'tight' but higher porosity/permeability lithologies. Local gas pools would have been formed at this time. As time progressed, continued dry gas expulsion removed connate water and, in the process, desiccated the column through which the gas was migrating, although variations in sweep efficiency would have created regions of higher and lower water-saturated zones. The gas with dissolved water migrated updip, to regions where the relative permeability for water allowed the mobile phase to be water, and the situation switched from gas-carrying water to gas in solution in water. With time and unroofing, gas migration rates slowed and capillary and hydrostatic pressures collapsed water back into the basin. This occurred where porosity, permeability and the wetting phase permitted this process to happen. The present state of pseudo-equilibrium is the result of a balance between these pressures, gas expulsion and migration from the basin's source rocks and loss of gas from the system.

Water saturation

In order to produce economic volumes of gas from low-permeability reservoirs, the initial or irreducible water saturation must be less than what would be expected in a reservoir that is in capillary equilibrium with free water. In such a water-wet, tight gas sandstone, with very small pore throats and 'normal' water

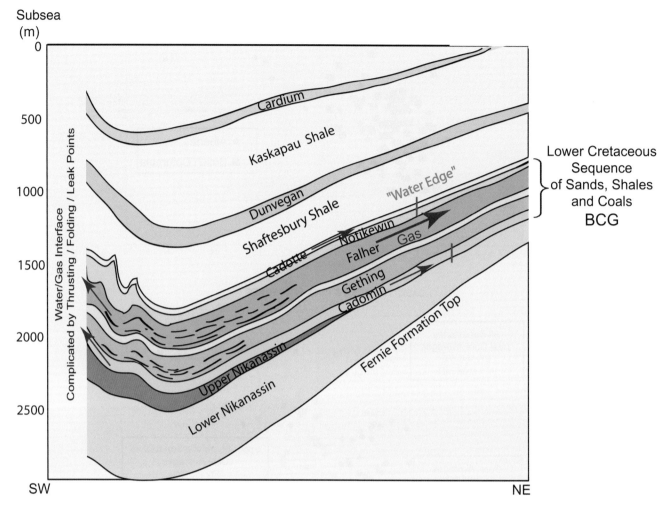

Fig. 15. BCG model of gas expulsion from coals into low porosity and permeability units and migration updip toward higher quality reservoir sandstones with water.

saturation, as determined from capillary pressure (with a transition zone of 250 m, S_w would be 45%, Fig. 18) gas-in-place would be significantly reduced and relative permeabilities would inhibit the flow of gas. However, Bennion et al. (2002) point out that economic volumes of gas are produced from these reservoirs on a worldwide basis, with the only water production being fresh water of condensation, generally in the range of 1–2 BBL/MMscf of gas. This is identical to the area within and surrounding the principal BCG region, where the average water production is 1.5–2 BBL/MMscf of gas (Fig. 19). This very low water production and corresponding low water saturation are the result of the removal of water, caused by long-term regional migration of undersaturated gas from source rocks adjacent to the producing interval. Newsham et al. (2002) stated that the key element required to remove water effectively is the process of vaporizing connate water into the gas phase. They further suggest that, based on laboratory data, at pressures and temperatures exceeding 10 000 psi and 350 °C, respectively, significant volumes of water vapour may be dissolved in the hydrocarbon gas. Thus, as long as hydrocarbon gas continues to migrate into and through the reservoir, a mechanism exists to continuously vaporize and effectively remove connate water.

In the Bossier sandstones of the East Texas Basin, Bennion et al. (2002) found that measured water saturations in the reservoir ($k < 0.01$ mD) averaged 5%. In Nikanassin Formation core from the 11-21-65-13W6 well, where average porosity was 4.5%, averaged core-measured water saturations of 15% and values as low as 6% were recorded. Overall, the desiccation process results in an increase in salinity in the remaining water phase, thereby suppressing the true in situ R_w (resistivity of the formation brine). Within the BCG region of Western Canada, the calculation of water saturation is difficult, as no representative sample of formation water can be produced from a desiccated reservoir. As such, R_w cannot been determined, but values for the Archie equation must be low to balance the extremely low water saturations, and as such, caution must be exercised when evaluating S_w in water-desiccated (BCG) systems.

Reservoir and petrophysical summary

The Deep Basin of Western Canada displays all the characteristics of BCG as defined by Law (2002):

- Regionally pervasive gas-saturated accumulations – gas is present in all prospective reservoir intervals examined and deliverability and commerciality is dependent upon reservoir quality (mainly 'system' permeability).
- Lack of a downdip water contact – no free water level has been identified in any of the zones of interest. The only water produced is water of condensation at rates of approximately 1.5–2 BBL/MMCF.
- Situated in low-permeability reservoirs – ambient matrix permeability values generally range from 100 to 500 μD with

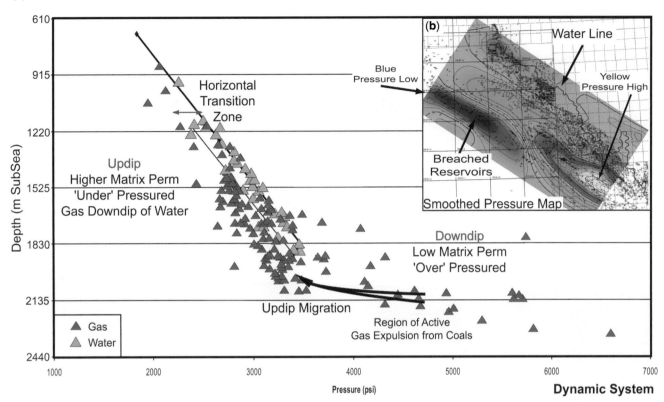

Fig. 16. (a) Cadomin Formation pressure–depth plot across the central BCG region showing the continuum of pressure with respect to depth from downdip, low-permeability, high pressure to updip, higher permeability, low pressure. Throughout the area, gas is expelled from coals and migrates updip to more permeable sandstones until, near the water level, relative permeability has the gas escaping into the water zones, lower pressures, but not allowing the water to slide further downdip. (b) The smoothed pressure map also shows the low pressures along the southwestern side of the study area, where regional tectonics has produced pressure leaks across faults and along folds.

in situ values about one order of magnitude less. Grain size does influence permeability with conglomerates being more permeable than fine-grained sandstones. Permeability increases updip towards the gas–water contact. In the non-structured areas reservoir performance is dependent primarily upon matrix permeability, while in the structured areas natural fracturing can enhance permeability. In structured fields, 'system' permeability values in excess of 1 mD have been determined from pressure transient analysis.

- Reservoirs are abnormally pressured (high or low) – pressures can range from greater than 40 MPa in the Cadomin/Nikanassin Formation intervals in the deeper portions of the basin to less than 5 MPa in the Cadotte Formation in northeastern British Columbia, just downdip of the water edge. A depth–pressure plot for the Cadomin Formation across the central BCG study area (Fig. 16) shows that, in the deepest portions of the basin, overpressuring is evident as charging occurs from the coals into the low-permeability sandstones. Gas then migrates updip, through these sandstones, to areas where permeability increases and the leaky gas–water interface is encountered and pressures fall as the relative permeability of gas v. water comes into play.

In summary, 'BCG basins' will display a continuum between conventional tight gas and 'true' BCG reservoirs. As dry gas is expelled from source units into low-porosity–permeability reservoir units, water is absorbed and carried updip to the leaky water interface. As geological time progresses, and as continued migration occurs, water saturation drops, until eventually the column is gas-saturated and desiccated of water. At the updip limits of the BCG region, higher porosity–permeability streaks will exhibit a slumping into the basin of higher S_w values, both vertically and laterally (see Fig. 13, Cadotte Formation water level).

Structural style and strain

Reservoir performance in non-structured areas is dependent primarily upon *in situ* matrix permeability which, as stated above, is often less than 50 μD. Pressure-transient analysis, on the other hand, indicates 'system' permeability in the hundreds of milliDarcies, well above matrix values. Many of these 'sweet spots' occur along structures where natural fracture permeability greatly influences the reservoir performance, and locating the best of these sweet spots was a challenge in the early phases of Shell Canada's exploration efforts. As such, much effort has been devoted both to understanding the detailed architecture of the thrust and fold front, and modelling these areas appropriately for the prediction of areas of enhanced fracture permeability.

The Western Canada Foreland Basin forms a shallowly west-dipping panel, tilted toward the Foothills and Foreland Belt (Rocky Mountains) of the Canadian Cordillera, lying between the conventional Foothills gas fields and the unstructured SW-dipping Deep Basin Belt. The deepest portion of this low-permeability BCG system is telescoped into the thrust front wedge or 'Triangle Zone', forming relatively low-displacement fault-transported fold structures (Fig. 5). The folding and thrust faulting of the deepest portion of the BCG system provided an opportunity for Shell Canada to target natural permeability pathways along strain-related fracture sets. Along this structured belt, understanding the geometry and kinematics of the structures was a necessary part of the process of well location and wellbore design. Further to the east, very little structure is evident (Fig. 5) and the reservoir units are

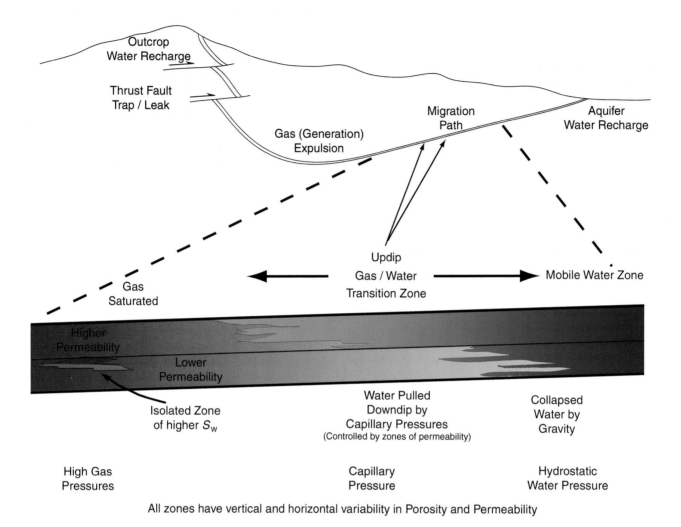

Fig. 17. Schematic of BCG basin pressures and gas–water transition zone.

essentially flat lying. In these areas reservoir performance is dependent primarily on matrix permeability, making facies analysis a more critical part of the evaluation process.

The structural style in the Foothills fold-thrust belt in west-central Alberta to northeastern British Columbia is strongly influenced by the mechanical stratigraphy of the Palaeozoic/Mesozoic succession. The thick Palaeozoic carbonates form thick thrust plates, folded around relatively low-angle thrust ramps. Opposed to this, the Jurassic–Cretaceous section in the study area, comprising relatively thin bedded sandstone–siltstone–shale–coal

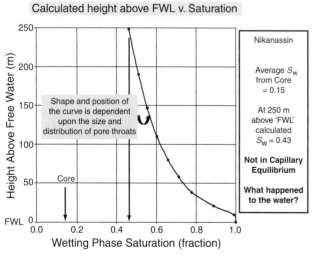

Fig. 18. Normal water saturation, as determined from capillary pressure, with a transition zone of 250 m (2.5% porosity sandstone) would be 45%. At these values, gas will not flow, but in fact, reservoirs do flow at economic rates with only minor volumes of water.

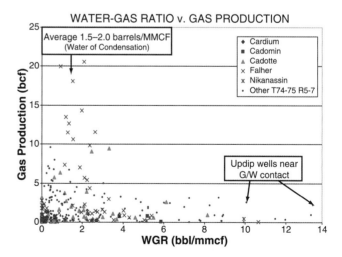

Fig. 19. Distribution of water to gas production across basin. Updip wells show a marked increase in water with respect to gas, whereas downdip wells only produce water at rates consistent with water of condensation.

intervals, appears to be almost completely decoupled from the underlying panels and is deformed more commonly by folding (see also Cooper et al. 2004). The detachment between the Palaeozoic thrust panels and overlying Mesozoic fold panels apparently occurs in the lowermost Nikanassin and/or Fernie formations. The thin-skinned fold-and-thrust belt is therefore decoupled into several mechano-stratigraphic wedges (see also Cooper et al. 2004). In general the Palaeozoic succession is detached from the Triassic succession, and the Triassic succession is detached from the Jurassic–Cretaceous wedge. This study focuses on the tectono-stratigraphic panel above the detachment in the lowermost Nikanassin and/or Fernie formations, and will be referred to here as the Cretaceous wedge. The Cretaceous wedge is then further internally decoupled on detachments of varying regional significance in the Gething, Spirit River and Shaftesbury formations (see also McMechan 1999). Interpretations of regional 3D seismic surveys aided the development of an understanding of the geometry and kinematics of mapped structures, with relevance to the exploitation of high permeability fracture networks and the stress–strain relationships in the deformed portion of the BCG study area.

Megascale structure

The interval between the Dunvegan Formation and Nikanassin Formation, within the Cretaceous wedge, is deformed into fault-bend folds, fault-propagation folds, detachment folds and variations between these end members (e.g. Fig. 5). The structures that define the thrust front, and various BCG target areas, are typically broad anticlines with 1000–1500 m wavelength, and up to approximately 500 m amplitude. These structures vary from ones having more typically planar limbs with tight hinges and flat crests, to ones that are arcuate in shape. The anticlines are generally about 10 km in strike length, with down plunge terminations commonly transferring laterally into en echelon folds.

Folds with tight hinges and relatively steeply dipping limbs are characterized seismically by steeply dipping 'wash-out' zones of poor data quality (e.g. Fig. 20a). These kink bands deflect at overlying or underlying fault ramps (Fig. 20b, d), and define fold hinge surfaces through tight hinge zones. These kink band zones could alternately be interpreted as steep fault zones; however, since they fold overlying thrust fault ramps (Fig. 20b, d) rather than offset them, they are generally thought to comprise hinge surfaces rather than fault zones.

The geometry of these structures has been strongly influenced by the mechanical stratigraphy. The highly layered Mesozoic succession, juxtaposing and repeating relatively competent sandstone and siltstone with incompetent shale and coal, leads to structures that have many internal complications produced through flexural slip between the competent beds, and complex fault linkages (Fig. 20; for a photograph see Fig. 21).

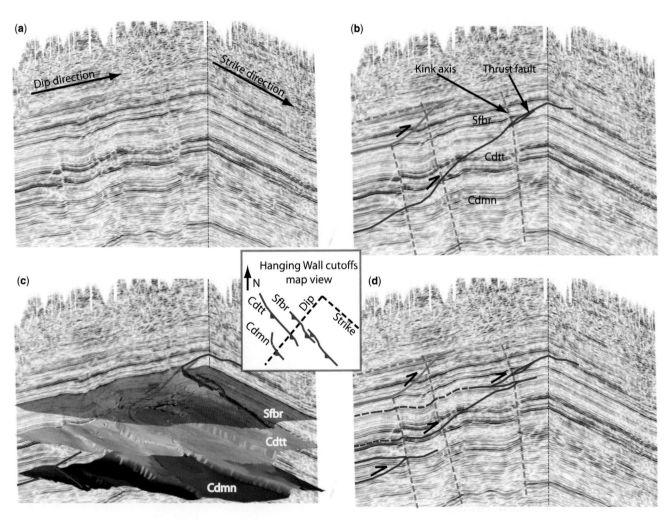

Fig. 20. Dip and strike seismic profile through frontal portion of the thrust front in the BCG system, NE British Columbia. (**a**) Uninterpreted. (**b**) One thrust ramp interpretation, and 'wash-out' at kink axes. (**c**) Surfaces added to provide map view of thrust linkages (inset also shows map view of hanging wall cutoffs) showing difficulty in linking thrust ramps. (**d**) Preferred interpretation with several thrust ramps linked by bedding-parallel décollement, and slightly folded at 'kink axis' fold hinges. Cdmn, Cadomin Formation; Cdtt, Cadotte member; Sfbr, Shaftesbury Formation.

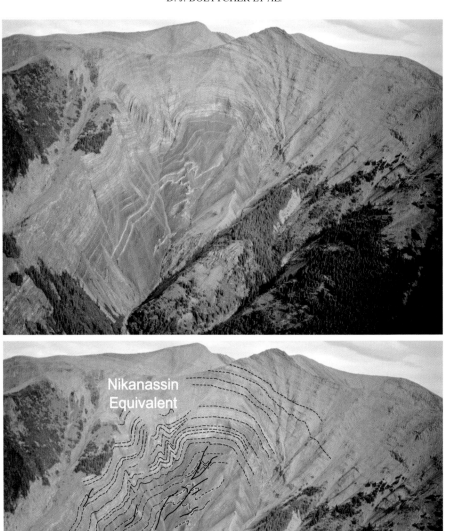

Fig. 21. Interbedded sandstones, siltstones, shales and coals of the Lower Fernie and Nikanassin equivalent formations. The anticlinal structure shows the typical internal complications from flexural slip between competent beds. Photo is looking NW up Davey Creek to the Davey Creek anticline formed in the immediate hanging wall of the Mount Russell thrust. Mountain is unnamed; photo by M. McMechan.

Interpreting internal fold structure

The Cretaceous wedge generally is deformed into folds across the central and northern portions of the study area; however, in detail, stacked low-angle folded thrust faults are visible within these anticlines. A high quality 3D seismic 'dip' profile shows what appears, at first glance, to be an apparent fault ramp between two horizons (e.g. Fig. 20b). Detailed analysis shows this apparent single ramp comprises several stacked fault ramps, linked by bed-parallel décollements. While hanging wall to footwall separation between décollements may be apparent on dip seismic data, imaging commonly precludes precise measurements of fault separation. However, when viewed at map orientation, hanging wall cutoffs with disparate strike lengths and different cutoff orientations between horizons (e.g. Fig. 20c) make it difficult to link the horizontal fault segments that form the hanging wall cutoffs into one surface. If fault surfaces are interpreted by linking segments along only 'dip' seismic profiles, it is possible that one erroneous fault surface will be interpreted where there are actually several linked fault surfaces. Such a mispicked fault surface will have inaccurate cutoff locations, angles and displacement vectors across ramps and, as such, will lead to inaccurate kinematic and geomechanical models and potentially mistargeted well paths.

These low-angle, low-displacement thrust faults commonly define a complex duplex network between stacked bedding-parallel décollements within the fold structures (e.g. Fig. 20d). Generally, the faults are linked only to the floor and roof décollement (lower Fernie or above outcrop in Fig. 21), and not laterally to each other. They do not form horses that would compartmentalize the reservoirs, but rather form en echelon duplexes (Fermor & Price 1987), with distinct lateral fault tips (Figs 22 & 23). These en echelon duplex structures (between décollements) are folded by the broad anticlines of interest, and developed in-sequence (either pre- or syn-folding).

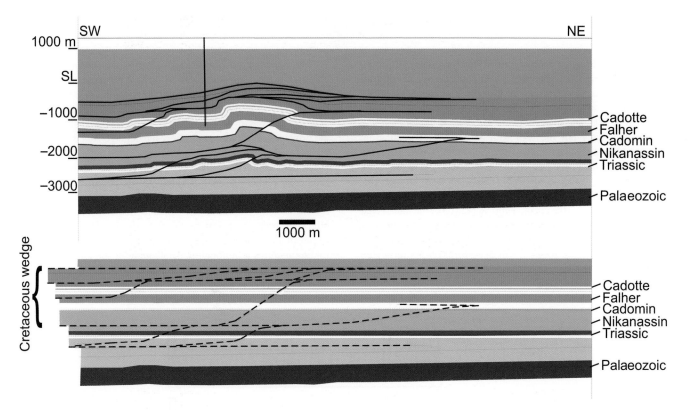

Fig. 22. Typical structural geometry (and restored) at thrust front, in the deepest portion of the BCG system. The thrust front is segmented by bedding-parallel décollement, defining the 'Mesozoic wedge'.

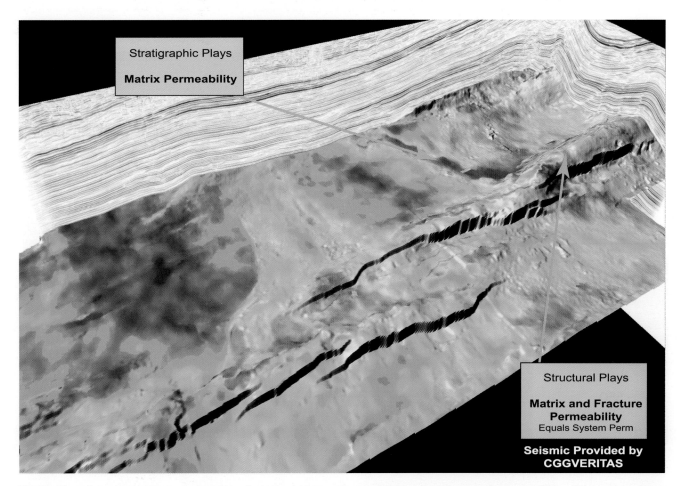

Fig. 23. Three-dimensional structure, stratigraphic and seismic view across the central BCG region. View includes an area of 20 × 40 km, with north to the right.

Fractures, mechanical modelling and structural sweet spot prediction

Unlike the identification of stratigraphic leads (e.g. facies mapping from seismic, Fig. 23), structural sweet spots do not necessarily need closely spaced well data. This especially applies to the structured portion along the west side of the BCG play area where the inherent risks and high costs associated with exploration and development have left the area with a sparse well population. Moreover, since the beginning of the 20th century the search for gas has been mainly focused on the exploration and exploitation of conventional hydrocarbon plays (Palaeozoic and Triassic carbonates). Overall this contributes toward a more sparse sampling of data to characterize the unconventional portion of the intersected gas-prone reservoir sequences (Jurassic/Cretaceous clastics) in this structured domain. In this context, regions where the probability of encountering zones of enhanced permeability through a well developed and/or well connected fracture network need to be identified through a combination of 2D and 3D seismic-based structural modelling and stratigraphic reservoir mapping techniques combined with geometric, kinematic and geomechanical modelling.

There are two types of quantitative approaches in structural modelling, geometric and numerical. In the first, geometric methods are based on either the principle of fixed hinge folding (De Sitter 1956; Biot 1964), or active-hinge-folding-related deformation (Suppe et al. 1992; Salvini & Storti 1997, 2001, 2004). Historically, in the principle of fixed hinge folding, the damage zone is calculated by means of layer curvature (Decker et al. 1989; Schultz-Ela & Yeh 1992; Lemiszki et al. 1994; Lisle 1994; Hennings & Olson 1997; Fischer & Wilkerson 2000; Bergbauer et al. 2003) to approximate the strain distribution across a structure. However, while useful in certain tectonic regimes, this method neglects the cumulative effects of strain during progressive deformation. The correct application of the principles of fixed-hinge or active-hinge-migration folding requires a thorough understanding of the kinematic evolution that characterizes each structure. In active-hinge-folding, high-strain areas or damage zones on the deformation panels are

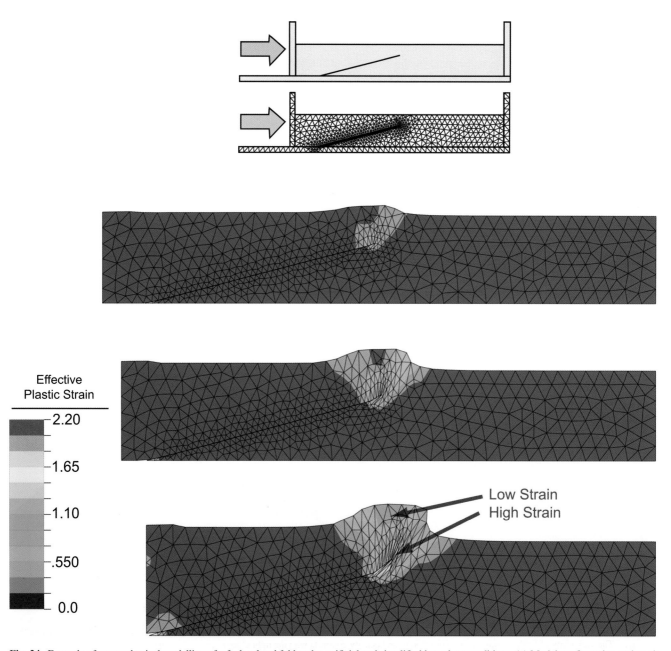

Fig. 24. Example of geomechanical modelling of a fault-related fold under artificial and simplified boundary conditions. (**a**) Model configuration and mesh discretization. (**b**) Evolution of plastic strain during fault-related folding.

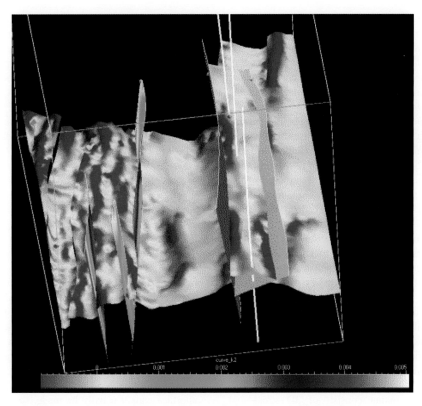

Fig. 25. Three-dimensional output from geomechanical modelling package displays regions of higher (red)/lower (green) probabilities of encountering favourable reservoir producibility characteristics, open fractures. It is in these locations where stress is predicted to be lower and drill targets located.

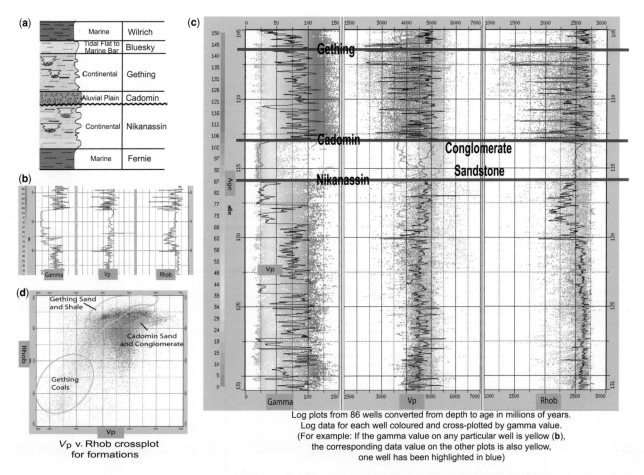

Fig. 26. Acoustic properties across the Gething to Upper Nikanassin (**a**) intervals from (**b**) a single well and (**c**) 86 wells near the BC Alberta border of the study area. The variability of Gething and Nikanassin velocity and density contrasts with the tight cluster of values from the Cadomin and demonstrates the variability of the total seismic response. (**d**) The V_p v. Rhob cross-plot also demonstrates the same issue, the overlapping of P-wave velocity and density of the Gething and Cadomin formations.

the consequence of tectonic interference between these domains, defined by the rock volumes as they migrated through the active axial surface(s) (Mosar & Suppe 1991; Salvini & Storti 1997, 2001, 2004). Overall, this method constitutes a better approach to replicating the consequences of strain accumulation in fault-related folding and/or fold-related faulting which occurred during tectonic transport in thin-skinned fold-and-thrust belts.

A further step in the prediction of damage zones in compressional tectonic settings is obtained through the use of numerical models (Erickson & Jamison 1995; Strayer & Hudleston 1997; Erickson et al. 1999; Salvini & Storti 2004), incorporating the evolution of contractional fold-and-thrust structures into the complex process of stress redistribution, strain localization and evolving mechanical properties of the deformed rocks. The use of this class of mechanical analysis requires the introduction of a complete set of equations that include stress equilibrium, strain compatibility, constitutive relations of rock units and appropriate numerical schemes, such as finite element analysis (FEA) or discrete element (DE) methods to solve boundary value problems with complex geometry, initial and boundary conditions, and non-linear mechanical behaviour. The application of such modelling computes the spatial and temporal distribution of damage zones associated with folding and faulting in these contractional regimes. Results show, for example (Fig. 24), that frictional slip along a pre-existing planar fault on a generic fault propagation fold concentrates inelastic strain along limbs, away from the crest.

In a total 3D analysis, output consists of 3D stress and strain envelopes, fault strength distribution, unbalanced strain tensors (equates to plastic deformation) and active deformation boundaries. From this, a non-linear, multiscale point source integration computes the local stress and strain for each point in the reservoir. The final product is a 3D representation of a structure (Fig. 25) with zones of higher and lower probabilities of encountering favourable reservoir producibility characteristics, that is, areas where a well developed and/or well connected open fracture network is expected. Unknown, or poorly understood in the analysis, is the 'detailed' tectonic history, including fault sequencing, the local stress field and the specific lithological failure conditions, all of which complicate the predictions.

Seismic stratigraphic mapping

An analysis of rock physics for 86 wells, both above and below the Cadomin Formation, demonstrates the difficulty in geophysical interpretation of many Cretaceous tight gas sandstones (Fig. 26). The well logs have been aligned based on formation top age and using linear interpolation between picks. The consistency of acoustic and elastic rock properties of the Cadomin Formation can be readily observed (Fig. 26) where little difference occurs between conglomerate and sandstone lithology. The over- and underlying formations, on the other hand, comprise siltstones, shales, coals and sandstones with a large variability in acoustic and elastic impedance. As the geophysical response of the Cadomin Formation is a tuned combination of thickness and a contribution from the surrounding rocks, the seismic reflection from this specific geological interface cannot usually be made, so a nearby tuned reflector is picked, usually caused by the nearest high impedance contrast rocks, often coals. Any stratigraphic interpretation or quantitative

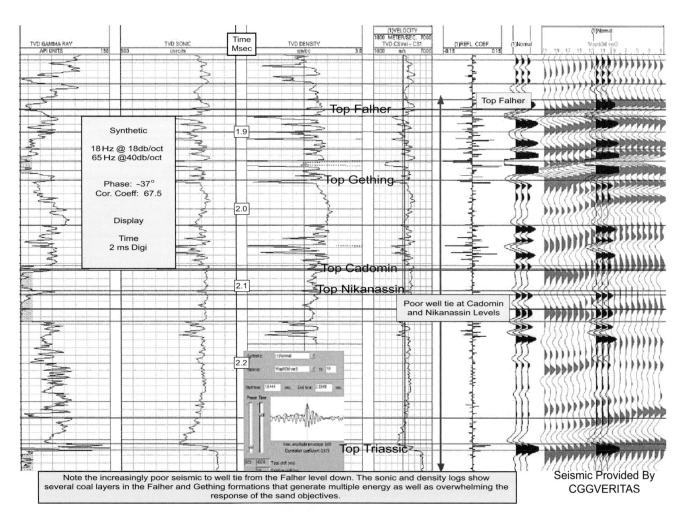

Fig. 27. Seismic to synthetic tie for a typical well.

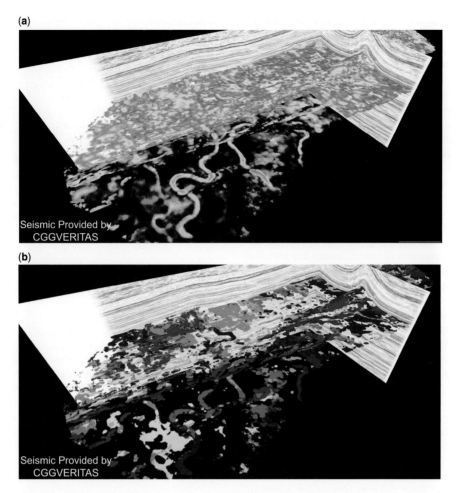

Fig. 28. Mid Cretaceous stratigraphic interpretation from events over an seismic time interval equivalent to 0.8 Ma (60 ms). (**a**) Events coloured by amplitude. (**b**) Events subdivided and coloured by relative age, revealing the stacking of channel systems.

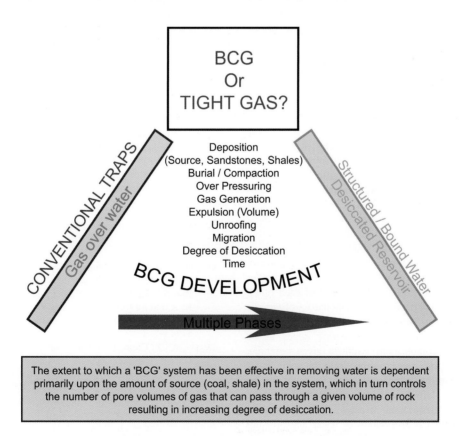

Fig. 29. 'Tight gas' to 'BCG' occurs over time involving amount and quality of source rocks, burial history, gas generation and degree of (over)pressuring, expulsion timing and rate, degree of unroofing, quality of reservoir and gas flow-through.

geophysical measurements are as such not accurate, as it is not known what formation and/or lithology/thickness variations are contributing to the measured response. Gas effects are also not observed or expected, with the low porosities of the sandstones and conglomerates.

A seismic to well tie (Fig. 27) demonstrates an additional problem in events below the coals of the Falher, where intra-bed multiples become dominant. At and above the Falher interval, good ties are evident, but below this level (where the coals begin) ties degrade severely into the Cadomin–Nikanassin formations. Across these deeper BCG units, attribute analysis is meaningless, although structural correlations may still be inferred.

In spite of the above complications, volume-based visualization techniques have still been very useful. Within the Falher Member, regional scale stratigraphic variations have mapped (Fig. 23) channels, valley systems and coaly regions. Above this interval, chronostratigraphic interpretation of seismic events has provided additional insight into depositional environments. In a younger event (Fig. 28), an amplitude extraction over a seismic volume has been age sculpted across an equivalent interval of 0.8 million years (60 ms). The fluvial channels are clearly visible and depositional environments can be interpreted (Fig. 28a). Coloured by relative age (Fig. 28b), the separation and stacking of channel systems is clearly interpretable.

Discussion

Law (2002) defined basin-centred gas regions as those that display four characteristics: regionally pervasive gas, abnormal pressure (high or low), common lack of a downdip water contact and located in low-permeability reservoirs. In general, 'BCG basins' will display all of these characteristics to one degree or another, and lie somewhere between a 'true' conventional tight gas or a 'true' BCG basin (Fig. 29). The complexities or degree of BCG development will be dependent on the basin's history, including lithology, burial, compaction, amount of overpressuring, gas maturation, expulsion, migration, degree of desiccation and gas loss out of the system. As gas is expelled from source units into a basin's low-porosity–permeability reservoir units, water is absorbed by the gas and carried updip, lowering the Deep Basin's water saturation (and R_w). That water is removed over time until eventually the entire objective column may be gas-saturated and 'fully' desiccated. Enough gas must flow through the system to remove this volume of water. Eventually, depending on local conditions, a dynamic updip gas–water interface will be established, controlled by a complex relationship between free, capillary and collapsed water, porosity, permeability and hydrostatic and capillary pressures regimes. Both the collapsed and capillary water will be present across each reservoir zone, producing a different geographical 'water line' across formations, often mimicking depositional environments. Capillary pressures acting to wick water along internal reservoir permeability and porosity variations will also produce a complex interfingering of higher and lower water saturations, often within metres of each other. This will in turn lead to a vertical intraformational set of various S_w lines, established along lines of common permeability. It is not unexpected for water-wet and gas-bearing zones to be encountered in various combinations within a given borehole, especially near to but also at some distance from the updip water transition zone. Some of the downdip wetter zones could also be by-passed water pools left behind and produced during the basin migration and desiccation process. Basin evolution will dictate the geographical distribution of water and pressure highs and lows across the basin, within the dynamics of an active system.

Conclusions

Western Canada, a mature BCG basin, displays all of Law's (2002) defining points, variable and complex pressures and water saturations within very tight gas-saturated sandstones and conglomerates, downdip of free water. With very low matrix permeabilities, commonly less than 100 μD, structural and geomechanical models were developed by Shell Canada to aid in the prediction of regions where producibility probabilities were highest, thereby increasing the likelihood of encountering areas of enhanced and effective natural fractures within the depositional and structural framework. With the integration of geological, petrophysical, geophysical and geomechanical models, Shell Canada was able to mount a successful exploration, land acquisition and eventual developmental programme. Both structural and stratigraphic leads were mapped and from the initiation of the unconventional gas plays in 2003, through organic and inorganic growth, more than 1000 drill spacing units (square miles) were acquired and more than 100 wells were drilled. In summary, a multidisciplinary approach resulted in economic success in regions once thought to be non-prospective.

M. Minchau is thanked for his directorship throughout the project, from a small study group to a full business unit. A. Cortis is thanked for his insightful review and technical guidance, J. Noad for his generous assistance, and C. Spencer for her assistance in the preparation of many of the figures and diagrams. Thanks are also given to Shell Canada for permission to publish this manuscript. Special thanks go to CGGVERITAS for permission to publish examples of their 3D seismic data.

References

Allen, F. H. 1980. Deep Basin, Alberta. *Annual AAPG-SEPM-EMD Convention*, Denver, 8–11 June 1980, Abstract.

Arnott, R. W. C., Zaitlin, B. A. & Potocki, D. J. 2002. Stratigraphic response to sedimentation in a net-accommodation-limited setting, Lower Cretaceous Basal Quartz, south-central Alberta. *Bulletin of Canadian Petroleum Geology*, **50**, 92–104.

Barclay, J. E. & Smith, D. G. 1992. Western Canada foreland basin oil and gas plays. *In*: Macqueen, R. W. & Leckie, D. A. (eds) *Foreland Basins and Fold Belts*. American Association of Petroleum Geologists, Tulsa, OK, Memoirs, **55**, 191–227.

Bennion, D. B., Thomas, F. B., Schulmeister, B. E. & Rushing, J. 2002. Laboratory and field validation of the mechanism of establishment of very low initial water saturations in ultra-low permeability porous media. *Canadian Institute of Mining, Metallurgy and Petroleum; Petroleum Society's Canadian International Petroleum Conference*, Calgary, Alberta, 11–13 June 2002, paper 2002–063.

Bergbauer, S., Mukerji, T. & Hennings, P. 2003. Improving curvature analyses of deformed horizons using scale-dependent filtering techniques. *American Association of Petroleum Geologist Bulletin*, **87**, 1255–1272.

Biot, M. A. 1964. Theory of internal buckling of a confined multi-layered structure. *Geological Society of America Bulletin*, **75**, 563–568.

Cooper, M., Brealey, C., Fermor, P., Green, R. & Morrison, M. 2004. Structural models of subsurface thrust-related folds in the foothills of British Columbia; case studies of sidetracked gas wells. *In*: McClay, K. R. (ed.) *Thrust Tectonics and Hydrocarbon Systems*. American Association of Petroleum Geologists, Tulsa, OK, Memoirs, **82**, 579–597.

Decker, A. D., Close, J. C. & McBane, R. A. 1989. The use of remote sensing, curvature analysis, and coal petrology as indicators of higher coal reservoir permeability. *International Coalbed Methane Symposium Proceedings*, 325–340.

De Sitter, L. U. 1956. Elastic or plastic buckling of the earth's crust. *Leide Geologische Mededelingen*, 171–176.

Erickson, G., Strayer, L. & Suppe, J. 1999. Mechanical modelling of deformation mechanism in folds. *Thrust Tectonics 99*. Royal Holloway University of London, Abstracts, 188.

Erickson, S. G. & Jamison, W. R. 1995. Viscous-plastic finite-element models of fault-bend folds. *Journal Structural Geology*, **17**, 561–573.

Fermor, P. R. & Price, R. A. 1987. Multiduplex structure along the base of the Lewis thrust sheet in the southern Canadian Rockies. *Bulletin of Canadian Petroleum Geology*, **35**, 159–185.

Fischer, M. P. & Wilkerson, M. S. 2000. Predicting the orientation of joints from fold shape, Results of pseudo three-dimensional modeling and curvature analysis. *Geology*, **28**, 15–18.

Gies, R. M. 1984. Case history for a major Alberta Deep Basin gas trap: the Cadomin Formation. *In*: Masters, J. A. (ed.) *Elmworth – Case Study of a Deep Basin Gas Field*. American Association of Petroleum Geologists, Tulsa, OK, Memoirs, **38**, 115–140.

Hayes, B. J. R., Christopher, J. E., Rosenthal, L., Los, G., McKercher, B., Minken, D. F., Tremblay, Y. M., Fennell, J. W. & Smith, D. G. 1994. Cretaceous Manville Group of the western Canada sedimentary basin. *In*: Mossop, G. D. & Shetsen, I. (compilers) *Geological Atlas of the Western Canada Sedimentary Basin*. Canadian Society of Petroleum Geology and Alberta Research Council, 317–334.

Hennings, P. H. & Olson, J. E. 1997. The relationship between bed curvature and fracture occurrence in a fault-propagation fold. AAPG/SEPM Annual Meeting, Abstracts, **6**, 49.

Issler, D. R., Willett, S. D., Beaumont, C., Donelick, R. A. & Griet, A. M. 1999. Paleotemperature history of two transects across the Western Canada Sedimentary Basin: constraints from apatite fission track analysis. *Bulletin of Canadian Petroleum Geology*, **47**, 475–486.

Jackson, P. C. 1984. Paleogeography of the Lower Cretaceous Mannville Group of Western Canada. *In*: Masters, J. A. (ed.) *Elmworth – Case Study of a Deep Basin Gas Field*. American Association of Petroleum Geologists, Tulsa, OK, Memoirs, **38**, 49–77.

Johnson, M. F. & Dalrymple, R. W. 2005. The Early Cretaceous Cadomin and Gething formations of the Deep Basin area, Western Canada: Cordilleran tectonics and alluvial sedimentation in a retroarc foreland basin. *American Association of Petroleum Geologists Annual Convention*, Calgary, Alberta, 22–22 June, Abstract.

Law, B. E. 2002. Basin-centred gas systems. *American Association of Petroleum Geologists Bulletin*, **86**, 1891–1919.

Leckie, D. A. & Smith, D. G. 1992. Regional setting, evolution, and depositional cycles of the Western Canada foreland basin. *In*: Macqueen, R. W. & Leckie, D. A. (eds) *Foreland Basins and Fold Belts*. American Association of Petroleum Geologists, Tulsa, OK, Memoirs, **55**, 9–46.

Lemiszki, P. J., Landes, J. D. & Hatcher, R. D. Jr. 1994. Controls on hinge-parallel extension fracturing in single-layer tangential-longitudinal strain fold. *Journal of Geophysical Research*, **99**, 22027–22041.

Lettley, C. D., Pemberton, S. G., Gingras, M. K. & Saunders, T. 2006. Depositional control of lateral reservoir variability within an extensive shoreline complex: example from the Lower Cretaceous Cadotte Member of Western Canada. *American Association of Petroleum Geologists Bulletin*, **90**, Program Abstracts.

Lisle, R. J. 1994. Detection of zones of abnormal strains in structures using Gaussian curvature analysis. *American Association of Petroleum Geologists Bulletin*, **78**, 1811–1819.

Lukie, T. D., Ardies, G. W., Dalrymple, R. W. & Zaitlin, B. A. 2002. Alluvial architecture of the Horsefly unit (Basal Quartz) in southern Alberta and northern Montana: influence of accommodation changes and contemporaneous faulting. *Bulletin of Canadian Petroleum Geology*, **50**, 73–91.

Masters, J. A. 1979. Deep Basin gas trap, western Canada. *American Association of Petroleum Geologists Bulletin*, **63**, 152–181.

Masters, J. A. 1984. Lower Cretaceous oil and gas in Western Canada. *In*: Masters, J. A. (ed.) *Elmworth – Case Study of a Deep Basin Gas Field*. American Association of Petroleum Geologists, Tulsa, OK, Memoirs, **38**, 1–33.

McMechan, M. E. 1999. Geometry of the structural front in the Kakwa area, foothills of northern Alberta. *Bulletin of Canadian Petroleum Geology*, **47**, 31–42.

Michael, K. & Bachu, S. 2001. Fluids and pressure distributions in the foreland-basin succession in the west-central part of the Alberta basin, Canada: evidence for permeability barriers and hydrocarbon generation and migration. *American Association of Petroleum Geologists Bulletin*, **85**, 1231–1252.

Mosar, J. & Suppe, J. 1992. Role of shear in fault-propagation folds. *In*: McClay, K. R. (ed.) *Thrust Tectonics*. Chapman & Hall, London, 123–132.

Mossop, G. & Shetsen, I. 1994. *Geological Atlas of the Western Canada Sedimentary Basin*. Alberta Geological Survey.

Newsham, K. E., Rushing, J. A., Chaouche, A. & Bennion, D. B. 2002. *Laboratory and Field Observations of an Apparent Sub-capillary Equilibrium Water Saturation Distribution in a Tight Gas Sand Reservoir*. SPE **75710**.

Poulton, T. P., Tittemore, J. & Dolby, G. 1990. Jurassic strata of northwestern Alberta and northeastern British Columbia. *Bulletin of Canadian Petroleum Geology*, **38A**, 159–175.

Salvini, F. & Storti, F. 1997. Spatial and temporal distribution of fractured rock panels from geometric and kinematic models of thrust-related folding (abstract). *American Association of Petroleum Geologists Bulletin*, **81**, 1409.

Salvini, F. & Storti, F. 2001. The distribution of deformation in parallel fault-related folds with migrating axial surfaces: comparison between fault-propagation and fault-bend folding. *Journal of Structural Geology*, **23**, 25–32.

Salvini, F. & Storti, F. 2004. Active-hinge-folding-related deformation and its role in hydrocarbon exploration and development – insights from HCA modeling. *In*: McClay, K. R. (ed.) *Thrust Tectonics and Hydrocarbon Systems*. American Association of Petroleum Geologists, Tulsa, OK, Memoirs, **82**, 453–472.

Schmidt, G. A. & Pemberton, S. G. 2004. Stratigraphy and paleogeography of a conglomeratic shoreline: Notikewin Member of the Spirit River Formation in the Wapiti area of west-central Alberta. *Bulletin of Canadian Petroleum Geology*, **52**, 57–76.

Schultz-Ela, D. D. & Yeh, J. 1992. Predicting fracture permeability from bed curvature. *In*: Tillerson, J. R. & Wawersik, W. R. (eds) *Rock Mechanics; Proceedings of the 33rd US Symposium*, **33**, 579–589.

Smith, D. G., Zorn, C. E. & Sneider, R. M. 1984. The paleogeography of the Lower Cretaceous of western Alberta and northeastern British Columbia in and adjacent to the Deep Basin of the Elmworth area. *In*: Masters, J. A. (ed.) *Elmworth – Case Study of a Deep Basin Gas Field*. American Association of Petroleum Geologists, Tulsa, OK, Memoirs, **38**, 79–114.

Stott, D. F. 1993. *Evolution of Cretaceous foredeeps: a comparative analysis along the length of the Canadian Rocky Mountains*. Geological Association of Canada, Boulder, CO, Special Papers, **39**, 131–150.

Strayer, L. M. & Hudleston, P. J. 1997. Numerical modeling of fold initiation at thrust ramps. *Journal of Structural Geology*, **19**, 551–556.

Suppe, J., Chou, G. T. & Hook, S. C. 1992. Rates of folding and faulting determined from growth strata. *In*: McClay, K. R. (ed.) *Thrust Tectonics*. Chapman & Hall, London, 105–121.

Wyman, R. E. 1984. Gas resources in Elmworth coal seams. *In*: Masters, J. A. (ed.) *Elmworth – Case Study of a Deep Basin Gas Field*. American Association of Petroleum Geologists, Tulsa, OK, Memoirs, **38**, 173–187.

Zaitlin, B. A., Warren, M. J., Potocki, D., Rosenthal, L. & Boyd, R. 2002. Depositional styles in a low accommodation foreland basin setting: an example from the Basal Quartz (Lower Cretaceous), southern Alberta. *Bulletin of Canadian Petroleum Geology*, **50**, 31–72.

Tight gas exploration in the Pannonian Basin

A. KIRÁLY, K. MILOTA, I. MAGYAR and K. KISS

MOL Hungarian Oil and Gas Plc, Budapest, Október 23. u. 18, H-1117, Hungary
(e-mail: Andras.Kiraly@ina.hr)

Abstract: The Pannonian Basin is a mature exploration area where most of the obvious structural features have been tested and the largest conventional accumulations probably have been found. The history of hydrocarbon exploration and production in the central Pannonian Basin, Hungary, is more than 90 years old. The evaluation of unconventional hydrocarbon reserves offers an opportunity to rejuvenate exploration in this young and hot basin.

Keywords: unconventional hydrocarbon, tight gas, basin-centred gas, Pannonian Basin, hydrocarbon system, overpressure

Structural and sedimentary evolution of the Pannonian Basin

Structural evolution

The Pannonian Basin is a Neogene extensional basin located in the east Central European segment of the convergence zone between the European and African plates (Fig. 1). The basin has a complex deformation history with distinct structural episodes (Horváth *et al.* 2006). Extensional tectonics started in the Early Miocene and culminated in the Middle Miocene. Because of strain localization along zones of crustal weakness inherited from the Cretaceous, that is, rejuvenation of thrust planes as low-angle normal faults, the extension was heterogeneous and resulted in a highly irregular basement morphology. Sub-basins of the Pannonian Basin system are thus separated by prominent basement highs, and are often treated as individual basins (e.g. Nagymarosy 2008).

Thinning of the lithosphere caused subsidence starting from the late Middle Miocene or early Late Miocene. This process affected almost the entire area between the Alps, Carpathians and Dinarides, and led to the formation of the geographically separate Pannonian Basin. Short-term local or regional (but not basinwide) compressional events also occurred during this period, the most significant of which caused a prominent unconformity between the Middle and Upper Miocene successions (Horváth 1995).

Subsidence anomalies at about the Miocene–Pliocene boundary mark the onset of structural inversion in the Pannonian Basin. Basin centres continued to subside whereas basin flanks suffered uplift and erosion at this advanced stage of basin evolution (Horváth & Cloetingh 1996; Magyar & Sztanó 2008).

Palaeogeographic evolution and sedimentary environments

Palaeogeographic evolution of the Pannonian Basin was intimately associated with the history of the Paratethys, the northern heir of the Tethys Ocean, which was more or less effectively separated from the Mediterranean Sea by the Alpine Orogen (Popov *et al.* 2004). Tectonic processes and eustasy controlled the palaeogeographic connections of the Pannonian Basin both towards the Mediterranean and towards the neighbouring Paratethyan basins (Vakarcs *et al.* 1999).

In the early, extensional phase of basin evolution, marine connections opened between the Pannonian Basin and the Mediterranean. Marine sedimentary environments first formed in the northern and western sub-basins and then became widespread across the entire intra-Carpathian area by the Middle Miocene. In the late Middle Miocene, however, the basin lost its direct connection to the sea, and became part of a large, closed or semi-closed inland sea, a uniform Paratethys. The marine environments were replaced by marginal marine (brackish to hypersaline) depositional settings and by euxinic environments in the basin centres.

A compressional tectonic phase at the Middle/Late Miocene boundary isolated the Pannonian Basin from the rest of the Paratethys, thus forming a brackish lake in the basin (Lake Pannon). The former archipelago of the Pannonian Basin gradually submerged as regional subsidence became dominant. In the Late Miocene through the Pliocene, sedimentation took place in the several-hundred-metre-deep lake and in the adjacent deltaic and fluvial environments (Magyar *et al.* 1999). Sediment input to the lacustrine basin was most intense from the NW and, in the eastern sub-basins, from the NE. The deep basin was gradually converted into fluvial plains as the shoreline prograded basinward. The lithostratigraphic boundaries within this shoaling-upward sequence are strongly diachronous.

Generalized stratigraphy of the basin fill

The area of the Pannonian Basin system extends over 10 countries. In this paper we focus on the central sub-basins which lie mostly within Hungary. In these basins, the Lower Miocene is dominated by terrestrial and limnic deposits, such as breccias, conglomerates, non-marine sandstones and siltstones, and freshwater clays. Following a transgression, the lower Middle Miocene (Badenian stage in the regional Paratethyan chronostratigraphy) is characterized by marine shales in the deep basins and corallinacean limestones and calcarenites on the highs. The upper Middle Miocene (Sarmatian stage) is represented by marls (often laminated, bituminous, with fish fossils) in the basins, and biogenic limestones, sandstones and evaporites above basement highs and in minor closed basins. There is a difference, however, in the development of these formations between the western and eastern sub-basins. In the west, the Lower and Middle Miocene succession is relatively thick, and generally each unit (Lower Miocene, Badenian, Sarmatian) is well represented in the sequence. In addition, traces of marine ingressions may occur in the Lower Miocene. In contrast, the Lower and Middle Miocene units are significantly thinner in the depocentres of the eastern sub-basins, there are no marine sediments older than Badenian and the separation of the Badenian,

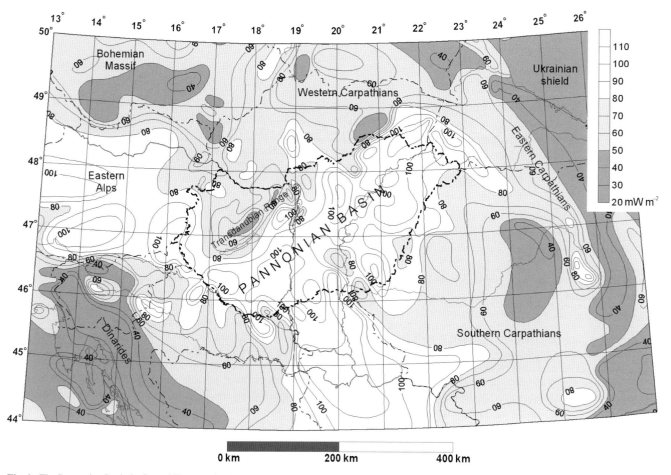

Fig. 1. The Pannonian Basin in Central Europe with the present-day heat flow (Lenkey et al. 2009). The borders of Hungary are indicated with a thicker line.

Sarmatian, and Upper Miocene Pannonian stages is often problematic due to condensed sedimentation, effective reworking of fossils and redeposition of sediments at various time intervals during the Neogene (Szuromi-Korecz et al. 2004). It remains unclear whether these differences between the western and eastern sub-basins are due to their different geographic positions relative to the sediment source, or to the different timing of the opening of the basins.

The Upper Miocene and Pliocene (together referred to as the 'Pannonian stage' in Hungary) are relatively uniform in the entire basin. Generally, they consist of five, strongly diachronous lithological units (Juhász 1991; Fig. 2). Shales deposited in the several-hundred-metre deep, giant brackish-water lacustrine basin comprise the Endrőd Formation. The Endrőd Formation is overlain by the turbidite-bearing Szolnok Formation. The distribution of this unit was closely controlled by the coeval basin morphology. The overlying shaly slope deposits comprise the Algyő Formation. The sequence is capped by the deltaic Újfalu Formation and by fluvial deposits (Zagyva and Nagyalföld Formations). Finally, much of the Pannonian Basin is covered by Quaternary deposits, which are missing from the inverted areas but can be as thick as 700 m in the subsiding sub-basins.

The Neogene petroleum system of the Pannonian Basin

Source rocks

Three main source units can be distinguished within the Neogene sequence based on their stratigraphic position, depositional environments and organic facies: transgressive marine shales of Badenien age; various Sarmatian sediments; and deep-water marls and deltaic–alluvial deposits of Pannonian age (Fig. 3). The deep water facies are not exposed in the surface, thus their description is based on cores and cutting samples from deep exploration wells.

The deep-water marine shales of the Badenian usually remain unexplored in the deepest parts of the sub-basins. Hydrocarbon shows and well test results, however, often indicate the presence of Badenian source rocks in the central parts of the sub-basins. The present source richness of the investigated samples reflects the maturity phase after migration of generated hydrocarbons: commonly low total organic carbon (TOC) content, low remaining potential and low hydrogen indices (HI) are characteristic. The thickness of the Badenien is highly variable, and 10–30% of the studied samples proved to be fair- to good-quality source rocks based on organic richness (Fig. 4). The kerogen type can hardly be estimated due to the high level of maturity. Low HI values and the relatively low rate of liptinites in the composition of the kerogen can be related to type III, but the biomarker composition of the extracts refers to organic matter of marine or mixed marine/terrigenous origin. The source maturity corresponds to the wet gas or dry gas window, but it is not surprising as all samples came from deeper than 3000 m.

The distribution, thickness and lithology of the Sarmatian sedimentary rocks within the Pannonian Basin reflect both a highly complex basin topography and a complex sedimentary system. The high variety of lithologies and facies compounded with uncertainties of stratigraphic correlation could be the reason why only a limited number of Sarmatian source rocks were geochemically studied. The organic facies of the Sarmatian deposits varies considerably with the depositional environment. It seems to be common that the higher organic content is related to anoxic

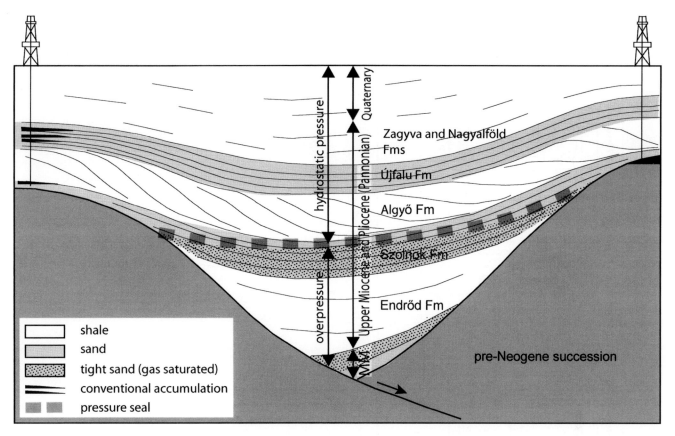

Fig. 2. Cartoon with the major lithostratigraphic units and pressure regimes in a deep sub-basin of the Pannonian basin. Conventional accumulations are known above the structural highs. MM, Middle Miocene.

or dysoxic conditions. NW from the largest Hungarian oil and gas field, Algyő, Sarmatian calcareous marl source rock was found with more than 12% average TOC content. The hydrocarbon yield is higher than 7 kg hydrocarbon/t rock. The kerogen content is dominated by type II. From shaly beds of some deep wells good-quality gas-prone sources of Sarmatian age were identified.

The transgressive basal marls and clays of Late Miocene Pannonian age are the main source rocks of both the conventional and unconventional accumulations. The average TOC is between 1 and 4% and goes up to 10%. The hydrocarbon yield shows great variability from less than 1 kg/t up to 12 kg hydrocarbon/t rock. Kerogen content varies, type II with HI up to 600 and more frequently a mixture of type III and II occur. The shaly beds interlayered between distal turbidites also contain fair-quality effective gas-prone source rocks (Fig. 3). Because of the deep burial and high maturity of these rocks (from onset of the oil window down to dry gas window) the original richness, hydrocarbon potential and kerogen type can only be estimated/calculated. Visual kerogen analysis indicates the organic matter being primarily of type III, terrigenous. In the upper part of the Pannonian sedimentary column coaly shale beds are known within the delta plain sediments; these might be good gas sources but are generally immature.

Maturity

The uppermost limit of the oil window (at 0.6% vitrinite reflectance) lies at about 2300 m or deeper across all sub-basins. The base of the oil or wet gas generation zone is variable and generally at an uncertain depth. Usually it can be characterized only by a calculated value as in most cases the maturity of the penetrated Neogene sequences remains within the oil or wet gas generation phase. In the deep sub-basins the depth zone of peak generation (0.9–1.1% vitrinite reflectance) is between 3900 and 4500 m.

Considering the present geothermal gradient of 44–55 °C/km, the onset of the oil generation is at a formation temperature of at least 100 °C, which is significantly higher than the world average. The oils of the largest Hungarian (conventional oil) field, Algyő, were generated at about 130 °C (Sajgó 1984). It seems to be that the Pannonian Basin was recently subjected to a relatively short but intensive heating event resulting in hydrocarbon generation. As a consequence of the high present-day heat flow (Fig. 1) and high geothermal gradient (Völgyi 1978), the source rocks are still actively generating oil, condensate and natural gas. This dynamic hydrocarbon system makes the Pannonian Basin exceptional.

Reservoirs and seals

Middle Miocene corallinacean limestones and calcarenites, sandstones of the Upper Miocene section, including deep-water turbidites and deltaic deposits (Szolnok and Újfalu Formations,

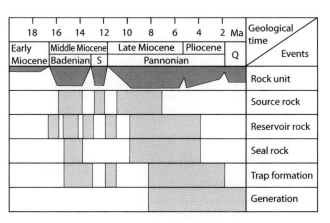

Fig. 3. Generalized events chart for the Neogene hydrocarbon system of the Pannonian Basin. S, Sarmatian.

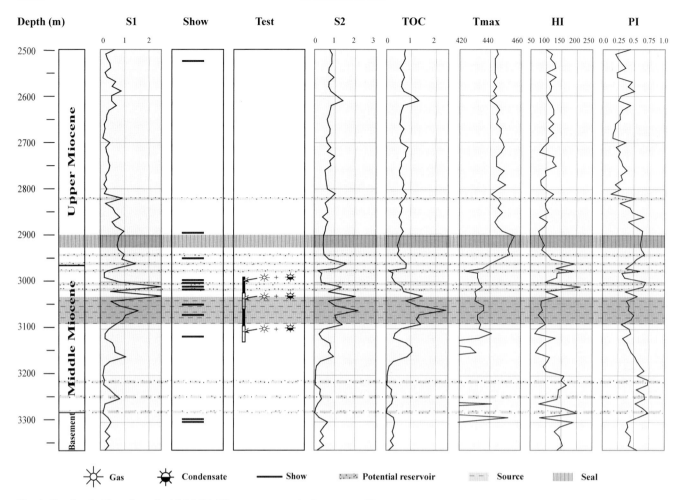

Fig. 4. Geochemical log of a well with Middle Miocene source rocks from eastern Hungary.

respectively), transgressive basal conglomerates of various age and even the weathered basement are confirmed reservoirs for conventional accumulations. These reservoirs are sealed by interbedded shales (Figs 3 & 4).

The thick shaly beds of the Upper Miocene Algyő Formation divide the sedimentary complex into two parts hydrodynamically (Fig. 2). These shales can provide excellent pressure sealing above the significantly overpressured basal shales, marls and turbidite sandstones (Endrőd and Szolnok Formations). The extension of the high pressure zone below the thick shale beds of the Algyő Formation exceeds thousands of square kilometres, and its thickness ranges between a few hundred metres and more than 1000 m. This thick overpressured succession that includes the turbidite sandstones of Szolnok Formation, the shales and marls of the Endrőd Formation and the transgressive middle Miocene sediments encompasses the unconventional reservoirs of the basin centre hydrocarbon accumulations (tight gas in the sand-dominated units and shale gas in the Endrőd Formation; Fig. 2). The sealing has been predicted to be capillary seal developed under high deposition rate and rapid compaction. On the flanks of the deep sub-basins producing conventional hydrocarbon fields related to structural highs were also discovered.

Timing

Thermal history simulations show imbalance between the present formation temperature and kerogen maturity expressed as vitrinite reflectance in the Pannonian Basin. The relatively low maturity values do not support an assumption of high palaeo-heat flow. It seems probable that the present-day formation temperature (the gradient is at least twice the world average) relates to a geologically very recent increase in heat flow. In other words, the actual heat flow (Fig. 1) is currently the highest since the beginning of the Neogene.

According to thermal history simulations, the earliest Neogene hydrocarbon generation started about 9–8 million years ago due to the significant subsidence and burial of the pre-Neogene succession (Fig. 3). Later the heat flow increased everywhere in the basin, resulting in intensive cracking of hydrocarbons. In the deep sub-basins the rising heat flow was partly balanced by the cooling effect of the huge sediment pile, which can explain the variability in the position of the oil window.

The development of capillary sealing happened within the last 5 Ma, resulting in high petroleum saturation within the fine-grained basal sediments and turbidite sandstones.

Tight gas plays in the Neogene of the Pannonian Basin

Two tight gas reservoirs are known in the Neogene sequence of Hungary: one in the Middle Miocene Badenian sandstones, and one in the Upper Miocene Szolnok Formation (Fig. 2).

Middle Miocene play

Exploratory wells in recent years proved the presence of a working hydrocarbon system in the Middle Miocene. The Badenian tight gas reservoirs are at great depth (deeper than 3000 m). They comprise mostly sandstones with 6–10% porosity and low permeability. The source rocks are probably Middle Miocene marine shales. The discoveries have particularly high pressure and high temperature (HPHT mainly HT) conditions, such as 50–60 MPa reservoir pressure and 180–200 °C. These wells produced mainly dry and/or good-quality wet gas.

The reservoirs showed typical tight gas behaviour under exploration tests and in the early production phase. Production started – without any stimulation – with higher daily rates and pressure, and then declined within some weeks. The original pressure was 10 times higher than the pressure during the tail production. After the pressure drop, production was stabilized at a lower level with limited water production and at a low well head pressure.

Although production has been continuous since the discovery, the commerciality of the accumulation has not been proven yet. Analysis of the reservoir system and fluid quality indicated that rates may be successfully enhanced with reservoir stimulation. The technical and economical analysis of two projects, i.e. drilling highly deviated wells and/or hydraulic fracturing of the reservoir, is in progress.

Upper Miocene play

Exploration for tight gas occurrences are focused on the Upper Miocene fill of the deep sub-basins, especially the turbidites of the Szolnok Formation (Fig. 2). Tight gas may form a continuous basin-centred accumulation in this reservoir. The major source rock is the underlying, highly overpressured Upper Miocene brackish-water shale (Endrőd Fm.), but the shaly beds of the Szolnok Fm. may have also contributed to gas generation. The porosity of the Szolnok sandstones goes up to 5–8%; their permeability is about 1.5 mD or less. The estimated (residual) water saturation varies widely (50–85%). The simulated/calculated (PetroFlow software) gas saturation varies with source richness in the Szolnok Formation and with the vertical distance from the Endrőd source. Reservoir quality also varies depending on the present depth (between 3000 and 4500 m) and on the depositional environment. Intensive exploration programmes are currently evaluating this play.

Summary

Tight gas occurrences in the sub-basins of the Pannonian Basin include a working system in the Middle Miocene (Badenian) and a more voluminous, continuous-type accumulation in the Upper Miocene. The deep Neogene sub-basins are characterized by high heat flow and high geothermal gradient, and the source rocks are actively generating hydrocarbons. The reservoirs are sandstones, characterized by high overpressures and extremely high temperatures. These conditions require special solutions in drilling, completion and stimulation technology.

The authors thank the two reviewers, Robert Gatliff (British Geological Survey, Edinburgh) and Hans-Martin Schulz (Helmholtz Centre, Potsdam) for their useful comments on the manuscript.

References

Horváth, F. 1995. Phases of compression during the evolution of the Pannonian basin and its bearing on hydrocarbon exploration. *Marine and Petroleum Geology*, **12**, 837–844.

Horváth, F. & Cloetingh, S. 1996. Stress-induced late-stage subsidence anomalies in the Pannonian Basin. *Tectonophysics*, **266**, 287–300.

Horváth, F., Bada, G., Szafián, P., Tari, G., Ádám, A. & Cloetingh, S. 2006. Formation and deformation of the Pannonian Basin: constraints from observational data. *In*: Gee, D. G. & Stephenson, R. A. (eds) *European Lithosphere Dynamics*. Geological Society, London, Memoirs, **32**, 191–206.

Juhász, Gy. 1991. Lithostratigraphical and sedimentological framework of the Pannonian (s.l.) sedimentary sequence in the Hungarian Plain (Alföld), Eastern Hungary. *Acta Geologica Hungarica*, **34**, 53–72.

Lenkey, L., Zsemle, F., Mádl-Szőnyi, J., Dövényi, P. & Rybach, L. 2009. Possibilities and limitations in the utilization of the Neogene geothermal reservoirs in the Great Hungarian Plain, Hungary. *Central European Geology*, **51**, 241–252.

Magyar, I. & Sztanó, O. 2008. Is there a Messinian unconformity in the Central Paratethys? *Stratigraphy*, **5**, 247–257.

Magyar, I., Geary, D. H. & Müller, P. 1999. Paleogeographic evolution of the Late Miocene Lake Pannon in Central Europe. *Palaeogeography, Palaeoclimatology, Palaeoecology*, **147**, 151–167.

Nagymarosy, A. 2008. Pannonian basins system. *In*: McCann, T. (ed.) *The Geology of Central Europe*. Geological Society, London, 1070–1074.

Popov, S. V., Rögl, F., Rozanov, A. Y., Steininger, F. R., Shcherba, I. G. & Kovač, M. (eds) 2004. *Lithological–Paleogeographic Maps of Paratethys. 10 Maps Late Eocene to Pliocene*. Courier Forschungsinstitut Senckenberg, **250**. E. Schweizerbart'sche Verlagsbuchhandlung, Stuttgart, 1–46.

Sajgó, Cs. 1984. Organic geochemistry of the crude oils from south-east Hungary. *Organic Geochemistry*, **6**, 569–578.

Szuromi-Korecz, A., Sütő-Szentai, M. & Magyar, I. 2004. Biostratigraphic revision of the Hód-I well: Hungary's deepest borehole failed to reach the base of the Upper Miocene Pannonian stage. *Geologica Carpathica*, **55**, 475–485.

Vakarcs, G., Hardenbol, J., Abreau, V. S., Vail, P. R., Tari, G. & Várnai, P. 1999. Correlation of the Oligocene–Middle Miocene regional stages with depositional sequences, a case study from the Pannonian Basin, Hungary. *In*: Degraciansky, P.-C., Hardenbol, J., Jacquin, T., Vail, P. R. & Farley, M. B. (eds) *Mesozoic–Cenozoic Sequence Stratigraphy of European Basins*. SEPM, Special Publications, **60**, 211–233.

Völgyi, L. 1978. Geothermal inhomogeneity in the Hungarian Great Plain (Pannonian Basin). *Acta Mineralogica–Petrographica*, **24**, 137–147.

Natural fractures in some US shales and their importance for gas production

JULIA F. W. GALE[1] and JON HOLDER[2]

[1]*Bureau of Economic Geology, Jackson School of Geosciences, The University of Texas at Austin, University Station Box X, Austin, TX 78713, USA (e-mail: julia.gale@beg.utexas.edu)*
[2]*Department of Petroleum and Geosystems Engineering, The University of Texas at Austin, 1 University Station C0300, Austin, TX 78712, USA*

Abstract: Shale gas reservoirs are commonly produced using hydraulic fracture treatments. Microseismic monitoring of hydraulically induced fracture growth shows that hydraulic fractures sometimes propagate away from the present-day maximum horizontal stress direction. One likely cause is that natural opening-mode fractures, which are present in most mudrocks, act as weak planes that reactivate during hydraulic fracturing. Knowledge of the geometry and intensity of the natural fracture system and the likelihood of reactivation is therefore necessary for effective hydraulic fracture treatment design. Changing effective stress and concomitant diagenetic evolution of the host-rock controls fracture initiation and key fracture attributes such as intensity, spatial distribution, openness and strength. Thus, a linked structural-diagenesis approach is needed to predict the fracture types likely to be present, their key attributes and an assessment of whether they will impact hydraulic fracture treatments significantly. Steep ($>75°$), narrow (<0.05 mm), calcite-sealed fractures are described in the Barnett Shale, north-central Texas, the Woodford Formation, west Texas and the New Albany Shale in the Illinois Basin. These fractures are weak because calcite cement grows mostly over non-carbonate grains and there is no crystal bond between cement and wall rock. In bending tests, samples containing natural fractures have half the tensile strength of those without and always break along the fracture plane. By contrast, samples with quartz-sealed fractures do not break along the fracture plane. The subcritical crack index of Barnett Shale is >100, indicating that the fractures are clustered. These fractures, especially where present in clusters, are likely to divert hydraulic fracture strands. Early, sealed, compacted fractures, fractures associated with deformation around concretions and sealed, bedding-parallel fractures also occur in many mudrocks but are unlikely to impact hydraulic fracture treatments significantly because they are not widely developed. There is no evidence of natural open microfractures in the samples studied.

Keywords: natural fractures, shale gas play, mudrock, hydraulic fracture treatment, reactivation

Shales and mudrocks form a substantial proportion of the sedimentary record and in petroleum exploration have previously been regarded as source or seal rocks. Recognition that mudrocks may also act as gas reservoirs has led to a rapid increase in production from basins across the USA (Fig. 1) and elsewhere. The Barnett Shale of the Fort Worth Basin (FWB; Table 1) in north-central Texas has been one of the most successful plays, where a combination of horizontal wells with staged, slick water hydraulic fracture treatments is the preferred completion strategy (Montgomery *et al.* 2005; Steward 2007). Friction reducers are added to the water (slick water) so that high rates can be pumped. Exploration for other shale gas reservoirs that may be exploited using a similar approach has proceeded rapidly, but it is now understood that most are sufficiently different in some key aspect to require a case-specific approach to drilling and completions. Emerging highly successful plays include the Jurassic (Kimmeridgian) Haynesville Formation in East Texas and Louisiana (Durham 2008), the Devonian Marcellus Shale in the Appalachian Basin (Curtis 2002) and the Devonian Horn River and Muskwa Shales in the Horn River Basin, NE British Columbia (Ross & Bustin 2008). The gas-generation mechanism must also be considered. Barnett Shale gas is thermogenic in origin (Hill *et al.* 2007; Jarvie *et al.* 2007), but some other successful plays, including the Antrim Shale in the Michigan Basin, produce biogenic gas and are not generally subject to hydraulic fracture stimulation (Fig. 1). Rather, the biogenic plays utilize horizontal drilling across open natural fractures in shallow parts of the basin (Curtis 2002). The purpose of this paper is to explain how natural fractures can be important in thermogenic shale gas reservoirs; biogenic gas plays are not considered further. Some commonly occurring fracture types will be reviewed and discussed as to whether they are relevant to gas production.

Montgomery *et al.* (2005) listed key attributes for a successful thermogenic shale gas reservoir on the basis of their knowledge of the Barnett Shale. These include high organic matter content (TOC), high thermal maturity, and sufficiently low clay content so that the rock is brittle enough to allow hydraulic fracture treatment. This last attribute gives rise to a problem with terminology. Many so-called 'shale gas' reservoirs, including the Barnett Shale, are not true shales but are better described as siliceous mudrocks with less than one-third clay minerals (Loucks & Ruppel 2007). Carbonate-rich mudrocks or marls can also fit the brittleness requirement; for example, the Cretaceous Niobrara Formation, a series of chalk-marl couplets, forms gas reservoirs in the Denver–Julesburg Basin (Fig. 1).

Relevance of natural fractures

Montgomery *et al.* (2005) concluded that natural fractures are not important to production in the Barnett Shale and at worst are detrimental. Their understanding was based on observations of opening-mode fractures sealed with calcite cement (Fig. 2), and they concluded, quite reasonably, that these sealed fractures could neither store nor transmit gas. The only observed fractures that are transmissive are large faults, some of which are associated with sag structures due to palaeokarst collapse in the underlying Ellenburger

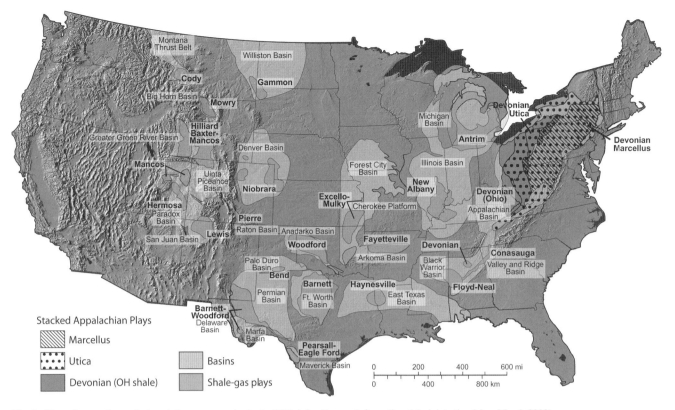

Fig. 1. Map of currently producing shale-gas reservoirs in the USA (after Energy Information Administration Map, March 2009).

Group (McDonnell et al. 2007). The Montgomery et al. (2005) conclusion that some fractures could be detrimental relates to these faults because they are thought to connect with the water-wet Ellenburger Group.

Warpinski & Teufel (1987) and Zhang et al. (2007) showed that mechanical discontinuities, including natural fractures and bedding, can affect hydraulic fracture propagation. On this basis, Gale et al. (2007) recognized that sealed fractures in the Barnett Shale might be weak planes that could reactivate in shear or opening-mode during hydraulic fracture treatment (Fig. 3). The natural fracture reactivates as the hydraulic fracture approaches, causing delamination of the cement from the wall rock. Hydraulic fracture fluid enters the opening and possibly causes further opening normal to the walls if treatment fluid pressure continues to increase (Fig. 3a, b). Once fracture reactivation has proceeded to the sealed fracture tip, the hydraulic fracture resumes growth parallel to S_{Hmax}.

Fracture origins

Processes giving rise to natural fracture sets in mudrocks include:

(1) local stress perturbations due to differential compaction and pore fluid during early burial;
(2) regional burial plus oil and gas generation and other diagenetic reactions, for example, smectite to illite conversion;
(3) regional, tectonic palaeostress;
(4) accommodation effects around major faults and folds;
(5) local structures, for example, in the FWB, fault-bounded sag features associated with underlying karst;
(6) stress release during uplift.

A mudrock reservoir may contain fractures generated by all or any subset of the mechanisms listed. Burial conditions and diagenetic alteration of the mudrock may be quite different at the time of

Table 1. Summary of cores and outcrops examined and fracture types observed

Basin	Formation	Age	No. of cores	Approx. total length (m)	Fracture types (opening-mode only)		
					Vertical planar	Vertical deformed	Horizontal sealed
Fort Worth (N)	Barnett Shale	Mississippian	5	187	Many	No	Few, local
Fort Worth (S)	Barnett Shale	Mississippian	2	4 + outcrop	Few	No	No
Delaware	Barnett Shale	Mississippian	1	91	Many	Few	Few
Fort Worth (S)	Smithwick Fm.	Pennsylvanian	2	58	Yes	No	Multiple swarms
Delaware	Woodford Fm.	Devonian	20	171	Some	Few	No
Illinois	New Albany Shale	Devonian	6	210 + outcrop	Many	Some	No
Denver–Julesberg	Niobrara Fm.	Cretaceous	2	105 + outcrop	Many	No	Few
Gulf of Mexico	Austin Chalk	Cretaceous	3	159 + outcrop	Many	Few	No

Fig. 2. Tall calcite-sealed fractures: (**a**) Barnett Shale, FWB; (**b**) Woodford Shale, Permian Basin; (**c**) New Albany Shale, Illinois Basin; and (**d**) Smithwick Formation, southern FWB. This type of fracture is prone to reactivation during hydraulic fracture treatments.

formation of each set. The interplay of mechanical and chemical processes during growth of a fracture set, termed *structural diagenesis*, governs propagation and sealing processes (Laubach *et al.* 2004; Olson *et al.* 2007). Difference in ambient conditions and rock properties at the time of formation of each set leads to differences in intensity, geometry and sealing attributes, and spatial arrangement.

Fracture types

Several shale gas reservoirs have been studied using cores and outcrops (Fig. 4; Table 1). The natural fractures observed are grouped into three categories: sub-vertical deformed, bedding parallel and sub-vertical planar. We confine our outcrop observations to cemented fractures.

Although we observed unmineralized joint sets in some outcrops, we did not include these as part of this study. Other authors have associated joint sets with fractures in the subsurface. For example, Lash & Engelder (2008) described joints in the Marcellus Shale, and Carr (1981) and Comer *et al.* (2006) described lineaments, joints and fractures in the New Albany Shale (Table 1). We are of the opinion that unmineralized joints are not representative of structures in the subsurface at depths sufficient for gas thermogenesis. Observations in core indicate that natural fractures at these depths have partial or complete mineralization on their walls, even if the mineralization constitutes only a very thin veneer of cement (Hooker *et al.* 2009). Moreover, where the natural fractures are parallel to present day S_{Hmax}, reactivation of natural fractures during drilling is common, such that the tips are unmineralized whereas the main body of the fracture contains significant cement. There are examples in thermogenic gas shales of microfractures with bituminous fill or coatings and otherwise no cement (Lash & Engelder 2005), but bitumen is known as a cement inhibitor (e.g. Marchand *et al.* 2000). Unmineralized opening-mode fractures (joints) may be important in shallow biogenic shale gas plays such as the Antrim or parts of the New Albany shale, where temperatures are too low for significant cement growth, but here we focus on thermogenic gas shales. Unmineralized opening-mode fractures could also be present in tectonically active regions, for example the Wabash Valley Fault System in the central part of the Illinois basin (New Albany Shale), where there would have been too little time for significant cementation in recently formed fractures. These fractures, if present, would require a separate systematic study to distinguish them from drilling-induced fractures. The fractures we observed in cores from wells near the Wabash Valley Fault System (Figs 2c & 4b) are partly cemented with quartz, pyrite and calcite.

Deformed fractures

Sealed fractures that have been subsequently deformed are present in several mudrock formations. Vertical shortening may be manifested by folding – for example, calcite-sealed fractures in the Woodford Formation in the Permian Basin (Table 1; Fig. 5a) or

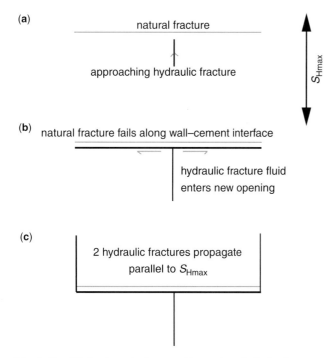

Fig. 3. Simplified, schematic diagram of interaction of a hydraulic fracture with a pre-existing natural fracture. (a) The natural fracture (grey) is shown normal to the maximum horizontal stress, S_{Hmax}. The hydraulic fracture (black) grows parallel to S_{Hmax}. (b) Natural fracture reactivation causes delamination of the cement from the wall rock. (c) Resumption of hydraulic fracture growth parallel to S_{Hmax}. The original single strand in diagram (a) is now two strands.

folding and faulting in dolomite veins in outcrops and cores of the lower part (Blocher Member) of the New Albany Shale in the Illinois Basin (Fig. 5b–d). It is likely that these fractures formed at an early stage of lithification. The rock was brittle enough to fracture, but subsequent compaction shortened the fracture and deformed the host-rock fabric around it. Fractures of this group have irregular walls because they are deformed and are therefore less likely to reactivate. They are commonly confined to mechanical layers on the order of a few centimetres to half a metre. Larger examples in the lower part of the New Albany Shale preserve centimetre-wide bitumen-filled pores, but they lie well below the main reservoir interval, and the bitumen has an origin different from that of the organic matter in the shales (Dumitrescu et al. 2004). This fracture type is not likely to be significant for gas production on a reservoir scale but could provide some local enhancement if the fractures coincided with the producing interval.

Bedding-parallel fractures

Bedding-parallel fractures include sealed macrofractures, which we describe here, and sealed microfractures that are observable in thin sections and have been reported in the literature. Speculation about the presence of open, bedding-parallel microfractures is reserved for the Discussion section because no such fractures were observed.

Sealed bedding-parallel macrofractures that can be seen without visual aid in core or outcrop samples are present in many mudrocks, but they vary considerably in intensity and in spatial distribution. They were not seen in the Barnett Shale cores from the FWB that were described by Gale et al. (2007). Examples do occur, however, in other FWB Barnett Shale cores, where they are most commonly associated with deformed laminae around the margins of carbonate concretions and short, vertical fractures between the concretions (Fig. 6a) (Gale, unpublished data). Local fractures associated with concretions are most likely formed by mechanism 1. Rare bedding-parallel fractures in the Barnett Shale in the Delaware Basin, west Texas, contain petroleum fluid inclusions in fibrous barite. In the Pennsylvanian Smithwick Formation from the southern FWB, swarms of bedding-parallel fractures are sealed with fibrous calcite (Fig. 6b). These most likely formed by mechanism 2. Cobbold & Rodrigues (2007) interpreted fibrous bedding-parallel veins in the Jurassic 'Shales-with-Beef' of SW England as being formed as a result of *seepage forces* that develop in response to an overpressure gradient.

Bedding-parallel fractures in outcrops of the Barnett Shale in central Texas are sealed with gypsum (Fig. 6c). Gypsum is not present as a fracture-sealing cement in the subsurface so these fractures are not considered to be representative of fractures in the reservoir, highlighting the potential danger in using outcrops as direct analogues to the subsurface. The gypsum-filled fractures probably developed during uplift and exposure of the black, pyrite-bearing mudrocks to oxidizing groundwater. When pyrite is oxidized, sulphuric acid is liberated, and this acid most likely reacted with calcite to produce gypsum. Calcite is present as concretions and discontinuous layers in the Barnett Shale at this location. Reactions are as follows:

$$4FeS_2 + 15O_2 + 14H_2O = 4Fe(OH)_3 + 8H_2SO_4 \quad (1)$$

and

$$H_2SO_4 + CaCO_3 + H_2O = CaSO_4 \cdot 2H_2O + CO_2 \quad (2)$$

Sealed bedding-parallel microfractures have been noted in several mudrocks. For example, Lash & Engelder (2005) observed bedding-parallel, bitumen-filled microfractures within clay laminae in the Upper Devonian Dunkirk Shale, western New York, which they interpreted as being due to overpressure during conversion of kerogen to bitumen (mechanism 2). With respect to their effect on production, because they are typically less abundant than sub-vertical sealed fractures, we contend that the sealed bedding-parallel fractures are of secondary importance.

Sub-vertical sealed planar fractures

The most common fractures observed in mudrock cores are the straight-sided, calcite-sealed planar fractures that were described in the Barnett Shale in the FWB by Gale et al. (2007). These are present in many other mudrocks, including the Woodford Formation and New Albany Shale and the Pennsylvanian Smithwick Formation (Fig. 2). These fractures are not deformed and most likely form as a result of mechanism 2 or 3, developing after compaction is complete.

Gale & Holder (2008) conducted bending tests on Barnett Shale samples and found that samples with calcite-sealed natural fractures have approximately half the tensile strength of the host rock. In these tests the sample always failed along the natural fracture. The reason for the weakness is that there is no physical bond between the calcite cement and the siliceous mudrock wall rock; the calcite cement simply grows over the wall rock. Calcite precipitates easily in this manner, unlike quartz and dolomite, which tend to require an existing grain of that mineral on which to precipitate (Lander et al. 2008; Gale et al. 2009). Although these fractures are dominantly sealed with calcite, some are sealed with other minerals, including quartz, albite, dolomite, barite and pyrite (Gale et al. 2007). In the example shown, quartz cement has grown over quartz grains in the wall rock, forming incomplete mineral bridges (Fig. 7). Albite cement makes up most of the rest of the fracture fill. Our hypothesis is that this part of the fracture

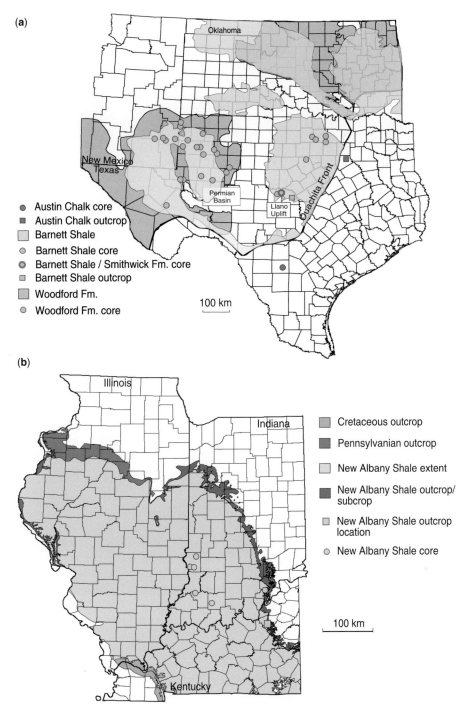

Fig. 4. Maps showing locations of cores and outcrops examined. (**a**) Texas locations; (**b**) New Albany Shale locations.

is stronger than the calcite-filled fractures tested by Gale & Holder (2008) and that, more generally, fractures sealed with cements that template onto grains in the wall rock will be stronger than those that simply precipitate over a surface regardless of composition. We could not test the fracture shown (Fig. 7) because of its small extent, but tests on other samples so far support the hypothesis. For example, in a chert–siliceous mudrock interbedded sequence (Woodford Formation), where the mineral fill is quartz, tested samples did not break along natural fracture planes but broke along planes away from the fracture. By contrast, calcite-sealed fractures in siliceous mudrock are consistently weak. In addition to quantitative bending tests conducted by Gale & Holder (2008), observation of calcite-sealed fractures in cores commonly reveals the cement–wall rock interface to be a weakness, along which the core breaks. In extreme cases the cement may part from both walls, leaving it as a completely detached thin sliver (Fig. 8).

Subcritical crack growth

Subcritical fracturing occurs when fractures propagate, even though the stress intensity at the crack tip is well below the fracture toughness of the material. They propagate at velocities several orders of magnitude slower than the dynamic rupture velocity of critical cracks. The process is assisted by chemical weakening at the crack tip (Lawn 1975; Atkinson & Meredith 1981; Atkinson 1984). Fracture patterns can be predicted from geomechanical models of subcritical fracture growth (Olson 2004). Fracture development and the geometry of the patterns that form are sensitive to mechanical layer thickness, fracture toughness of the rock, Young's

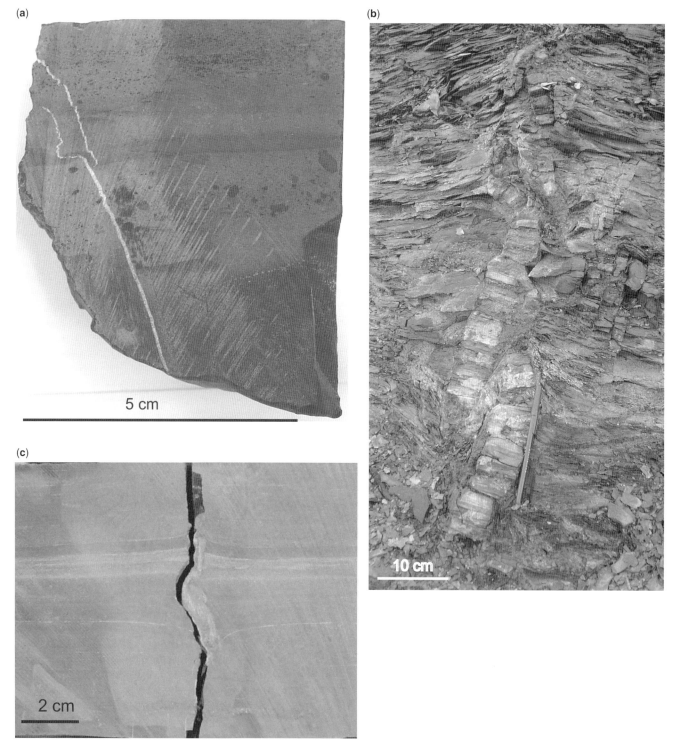

Fig. 5. (a) Folded calcite-sealed fracture in core from the Woodford Formation in the Permian Basin. (b) Folded dolomite veins in an outcrop of the lower part (Blocher Member) of the New Albany Shale. (c) Folded and faulted dolomite vein also with surrounding bedding laminae compaction in New Albany Shale core, Indiana (855 m depth).

modulus, Poisson's ratio, *in situ* stress anisotropy and the subcritical crack index. Values of these parameters at the time of fracturing are the main input to the model. Holder *et al.* (2001) adapted a dual-torsion-beam test apparatus and procedure for measuring subcritical crack index in sedimentary rocks. The subcritical index, *n*, is the slope of a curve on a plot of log load applied in the test v. log crack propagation velocity. We used their test procedure to measure subcritical indices for several shale samples from two Barnett Shale cores. Samples from the Mitchell Energy #2 T. P. Sims core were not distinguished by lithology; they are all dark, laminated mudrock, but there are no compositional data for the samples. Samples from a second core, core A, were selected across a wide range of lithologies, including light, carbonate-rich layers and dark layers (Fig. 2a). Subcritical crack indices for samples from both these Barnett Shale cores are mostly >100 and the range of values for the two cores is similar. The highest values occur in the more carbonate-rich facies in core A (Fig. 9) (Gale *et al.* 2007; Gale & Holder 2008). The range of subcritical index values for the Mitchell Energy #2 T. P. Sims core samples suggests these are of variable composition.

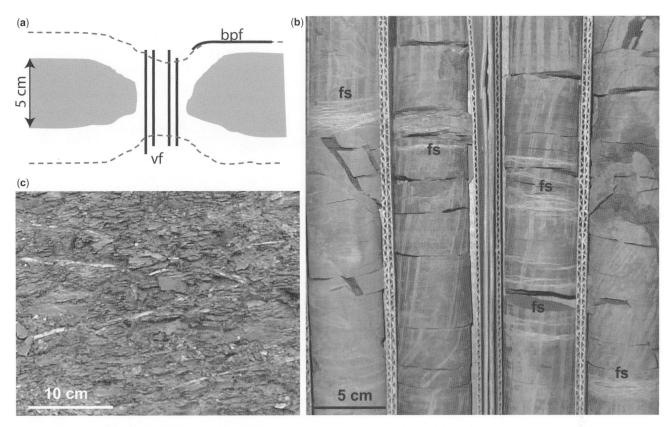

Fig. 6. (a) Sketch of bedding-parallel fractures (bpf) associated with carbonate concretions (grey) and sub-vertical fractures (vf) in the zone between the concretions. (b) Bedding-parallel fracture swarms (fs) sealed with fibrous calcite in core from the Smithwick Formation, central Texas. (c) Bedding-parallel gypsum-filled fractures in Barnett Shale in a quarry in San Saba County, central Texas.

The subcritical index measured from a present-day sample is a fair representation of the subcritical index at the time of formation of tall, planar, calcite-sealed fractures. Because these fractures show no evidence of compaction, they probably formed after the shale was fully compacted, and the host rock has probably not altered much from the time of fracture growth to present day. Geomechanical modelling, using a subcritical index that is >100 as an input parameter, predicts that fractures are clustered, with spaces between large fracture clusters on the order of several times the mechanical layer thickness (Olson 2004). Gale et al. (2007) concluded that if the total Barnett Shale thickness of 200 m is taken as the upper boundary to mechanical layer thickness, the distance between large clusters may be several hundred metres. The narrow, sealed fractures observed in core are distributed between clusters, possibly reflecting smaller internal mechanical layers or growth within mechanical layer boundaries. For shales with a

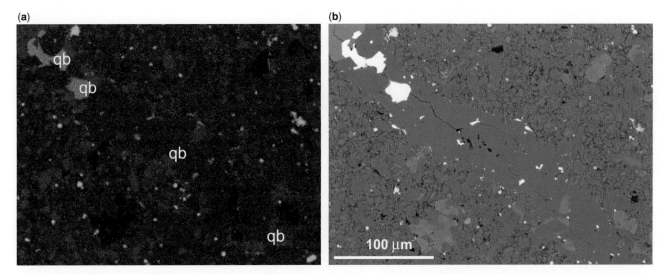

Fig. 7. Quartz and albite fracture fill in a Barnett Shale mudrock sample. (a) False colour combination of three greyscale element maps. Quartz is bright red, albite is dull red, barite is green and calcite is blue. Growths of quartz, interpreted as incomplete cement bridges (qb), overgrow groups of quartz grains in the wall rock. (b) The fracture wall is best differentiated in the backscattered electron image of the same area shown in (a). The quartz bridges in the fracture are overgrowing areas of quartz grains in the wall rock.

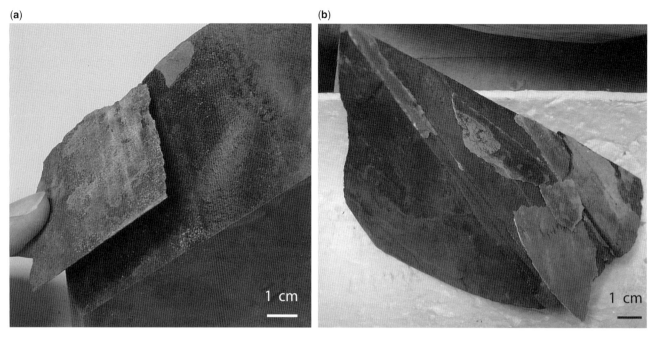

Fig. 8. (a, b) Calcite-sealed fractures in a New Albany Shale core where the cement fill has detached from both fracture walls during core handling.

moderate subcritical index, a less strongly clustered pattern would be expected, with dominant fractures spaced closer to the mechanical layer thickness.

Fracture size distributions and open fractures

Although the narrow, steep fractures of the most common type are sealed and probably do not enhance reservoir permeability, their population may follow a power-law size distribution. By analogy with fractured chalks we infer that the largest fractures in this distribution would be open and present in widely spaced clusters, several hundred feet apart. This inference is made on the basis of observed power distributions of opening-mode fractures in the Austin Chalk (Gale 2002), where the largest fractures (>11 mm) are partly open, whereas the smaller fractures (<11 mm wide) are sealed. A similar relationship is seen in the Niobrara Formation, although the crossover size from sealed to partly open is smaller (~3 mm) (Fig. 10). The crossover size of a given population, termed *emergent threshold* by Laubach (2003), depends partly on the composition of the cement. For a fracture to seal, rate of cement growth is in competition with fracture opening rate (Laubach *et al.* 2004). That is, for identical temperatures and opening rates, fracture populations sealed with fast-growing calcite have a larger emergent threshold than those sealing with quartz. Indeed, fractures as narrow as 4 mm in the Barnett Shale in the Delaware Basin retain porosity, being only partly sealed with quartz (Gale 2008), whereas calcite-sealed fractures of this width in the Barnett Shale in the FWB are completely sealed.

Discussion

Effect of natural fractures on hydraulic fracture treatment efficiency

Gale *et al.* (2007) concluded that the effect of natural fracture reactivation would be beneficial to gas production because it would

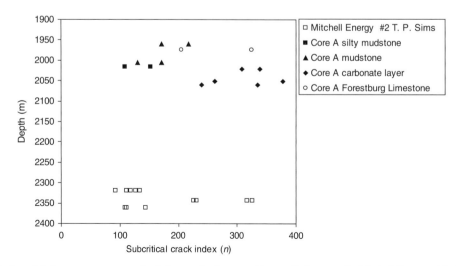

Fig. 9. Plot of subcritical crack index measurements v. depth for samples from two Barnett Shale cores, the Mitchell Energy #2 T. P. Sims and a second core, labelled core A, from the FWB, north-central Texas.

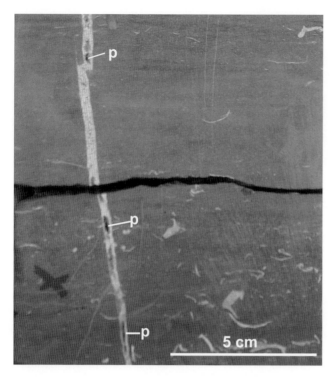

Fig. 10. Fracture porosity (p) in Niobrara Formation core from the Denver–Julesberg Basin (920 m depth). The fracture is otherwise sealed with calcite.

serve to multiply the number of hydraulic fracture strands and, thus, the total surface area, connected to the host rock (Fig. 3). This might not be the case. Doubling of the number of hydraulic fracture strands will not lead to doubling of surface area because increase in the number of strands will be offset by decrease in individual strand length as the fracture treatment energy is dissipated through more fractures. The movement of proppant through such a tortuous fracture network is also problematic. Figure 3 is simplistic because it only considers the case where the natural fracture is opened. Whether hydraulic fractures arrest at, cross over, divert at or reactivate natural fractures is a function of many factors, including relative strength of natural fracture planes and orientation with respect to the stress field (Olson 2008). Spreading the treated zone normal to S_{Hmax} might not be desirable if a staged hydraulic fracture treatment is planned because the rock volume expected to be treated by each successive stage could already have been fractured. New fractures might then link up to old ones, yielding less freshly broken shale surface area than anticipated. Moreover, the volumes of rock between the diverted fracture strands (Fig. 3c) remain unstimulated and represent lost pay. In any case, prediction of the hydraulic fracture geometry resulting from interaction with natural fractures is desirable.

Open microfractures

On the basis of production rates and volumes, some speculate that storage and permeability in mudrocks must be due to an open microfracture network (e.g. Bowker 2003). Most commonly this network is thought to be bedding parallel because of the known bedding-parallel weakness of mudrock. The only open microfractures we have seen are in thin sections and are due to section damage during preparation. Reed & Loucks (2007) examined many samples of Barnett and Woodford Shale using SEM-based techniques, some on argon-ion-milled surfaces, in search of microfractures and/or pores, but they saw no evidence of an open microfracture network in these formations. Rather, gas storage is accommodated partly by micro- to nanosized pores (free gas) and partly by sorption of methane and other molecules in organic matter and clay particles (Ruppel & Loucks 2008; Ross & Bustin 2009). The mechanism by which free or sorbed gas can move from its storage location into a hydraulic fracture network remains a challenging problem. Bedding-parallel weaknesses could allow transient openings, but unless cement or bitumen remains along the fracture, there would be no record of its presence (e.g. Lash & Engelder 2005). Of course it is well known during drilling, especially in the presence of reactive fluids, that shales and mudrocks can fracture, but this fracturing is due to the drilling process rather than a natural event. Drilling and sample preparation commonly cause fractures to form parallel to bedding, and these may be wrongly interpreted as fractures open in the subsurface.

Conclusions

Host-rock and fracture characteristics in shale gas reservoirs are diverse, and the role that natural fractures play must be considered case by case. Some common relationships can, however, be stated as general principles:

(1) Planar, sub-vertical, calcite-sealed fractures in siliceous mudrocks are weak and may reactivate during hydraulic fracture treatments. These form the most common type of natural fracture in many reservoirs. Interaction with hydraulic fracture networks may be beneficial or detrimental.
(2) Planar, sub-vertical fractures completely sealed with quartz can produce strong planes. Hydraulic fractures could arrest at these planes, causing premature screen-out of the fracture treatment.
(3) Early formed fractures are commonly deformed during host-rock compaction. They have irregular shapes and are commonly confined to mechanical layers of the order of a few centimetres to half a metre. These fractures are not likely to be significant for gas production on a reservoir scale, but they could provide some local enhancement if they are open and coincide with the producing interval.
(4) Horizontal fractures are preserved in some mudrocks. They are now sealed with bitumen or mineral cements, some of which have petroleum fluid inclusions in them. These fractures are evidence of fracturing during the hydrocarbon cracking process. Other mudrocks such as the Barnett may have had similar transient fractures during cracking, but they are not preserved.
(5) Large opening-mode natural fractures may be open above the emergent threshold for that fracture set. If the subcritical index is high, which is typical in siliceous mudrocks, then the large fractures are likely to be clustered, with spaces between clusters being several times the mechanical layer thickness.
(6) There is no evidence of open microfracture networks in the mudrocks examined in this study. In thin sections, open microfractures with no cement or hydrocarbon along them, if present, are likely to have been caused by drilling or sample damage and are not representative of structures in the subsurface.

J. F. W. Gale was supported in part by a GDL Foundation Fellowship. The principal funding groups for this work are the Fracture Research and Application Consortium at The University of Texas at Austin, the State of Texas Advanced Resource Recovery program, and the Permian Basin Geological Synthesis Project, the latter two being Bureau of Economic Geology projects. Funding for work on the New Albany Shale is through the Research Partnership to Secure Energy for America. R. Reed provided images for Figure 7. J. Olson and S. Laubach provided insight and discussion of fracture processes and helpful reviews were given by Robert Gatliff and an anonymous reviewer. Publication is authorized by the Director, Bureau of Economic Geology.

References

Atkinson, B. K. 1984. Subcritical crack growth in geologic materials. *Journal of Geophysical Research*, **89**, 4077–4114.

Atkinson, B. K. & Meredith, P. G. 1981. Stress corrosion cracking of quartz. A note on the influence on chemical environment. *Tectonophysics*, **77**, T1–T11.

Bowker, K. A. 2003. Recent development of the Barnett Shale play, Fort Worth Basin. *West Texas Geological Society Bulletin*, **42**, 4–11.

Carr, D. D. 1981. Lineament analysis. *In*: Hasenmueller, N. R. & Woodard, G. S. (eds) *Studies of the New Albany Shale (Devonian and Mississippian) and Equivalent Strata in Indiana*. US Department of Energy, contract number DE-AC 21-76MC 05204, 62–69.

Cobbold, P. R. & Rodrigues, N. 2007. Seepage forces, important factors in the formation of horizontal hydraulic fractures and bedding-parallel fibrous veins ('beef' and 'cone-in-cone'). *Geofluids*, **7**, 313–322.

Comer, J. B., Hasenmueller, N. R., Mastalerz, M. D., Rupp, J. A., Shaffer, N. R. & Zuppann, C. W. 2006. The New Albany Shale gas play in southern Indiana. Program with Abstracts, 2006 Eastern Section. *American Association of Petroleum Geologists 35th Annual Meeting, Buffalo, NY*, 17.

Curtis, J. B. 2002. Fractured shale–gas systems. *American Association of Petroleum Geologists Bulletin*, **86**, 1921–1938.

Dumitrescu, M., Finkelstein, D. B., Lazar, R., Schieber, J. & Brassell, S. C. 2004. Origin and history of bitumen in geodes of the New Albany Shale. *In*: Schieber, J. & Lazar, R. (eds) *Devonian Black Shales of the Eastern U.S.* Field Guide for the 2004 Annual Field Conference of the Great Lakes Section of the Society for Sedimentary Geology, Indiana Geological Survey, 61–67.

Durham, L. 2008. Haynesville, lower Tertiary the latest allures; Gulf Coast still revealing its charms. *American Association of Petroleum Geologists Explorer*, **29**, 9.

Gale, J. F. W. 2002. Specifying lengths of horizontal wells in fractured reservoirs. *Society of Petroleum Engineers Reservoir Evaluation and Engineering*, **5**, 266–272.

Gale, J. F. W. 2008. Natural fractures in the Barnett Shale in the Delaware Basin, Pecos Co., West Texas: comparison with the Barnett Shale in the Fort Worth Basin. *West Texas Geological Society Fall 2008 Symposium, Proceedings Abstracts*. WTGS digital publication 08-120, 13.

Gale, J. F. W. & Holder, J. 2008. Natural fractures in the Barnett Shale: constraints on spatial organization and tensile strength with implications for hydraulic fracture treatment in shale-gas reservoirs. *Proceedings of the 42nd US Rock Mechanics Symposium*. Paper no. ARMA 08-96.

Gale, J. F. W., Reed, R. M. & Holder, J. 2007. Natural fractures in the Barnett Shale and their importance for hydraulic fracture treatments. *American Association of Petroleum Geologists Bulletin*, **91**, 603–622.

Gale, J. F. W., Lander, R. H., Reed, R. M. & Laubach, S. E. 2009. Modeling fracture porosity evolution in dolostone. *Journal of Structural Geology*; doi: 10.1016j.jsg.2009.04.018.

Hill, R. J., Jarvie, D. M., Pollastro, R., Henry, M. & King, J. P. 2007. Geochemistry of oils and gases from the Fort Worth Basin, USA. *American Association of Petroleum Geologists Bulletin*, **91**, 445–473.

Holder, J., Olson, J. E. & Philip, Z. 2001. Experimental determination of subcritical crack growth parameters in sedimentary rock. *Geophysical Research Letters*, **28**, 599–602.

Hooker, J. N., Gale, J. F. W., Gomez, L. A., Laubach, S. E., Marrett, R. & Reed, R. M. 2009. Aperture-size scaling variations in a low-strain opening-mode fracture set, Cozzette Sandstone, Colorado. *Journal of Structural Geology*, **31**, 707–718.

Jarvie, D. M., Hill, R. J., Ruble, T. E. & Pollastro, R. 2007. Unconventional shale-gas systems: the Mississippian Barnett Shale of north-central Texas as one model for thermogenic shale-gas assessment. *American Association of Petroleum Geologists Bulletin*, **91**, 475–499.

Lander, R. H., Larese, R. E. & Bonnell, L. M. 2008. Toward more accurate quartz cement models – the importance of euhedral vs. non-euhedral growth rates. *American Association of Petroleum Geologists Bulletin*, **92**, 1537–1563.

Lash, G. G. & Engelder, T. 2005. An analysis of horizontal microcracking during catagenesis: example from the Catskill delta complex. *American Association of Petroleum Geologists Bulletin*, **89**, 1433–1449.

Lash, G. G. & Engelder, T. 2008. Tracking the burial and tectonic history of Devonian shale of the Appalachian Basin by analysis of joint intersection style. *Geological Society of America Bulletin*, **121**, 265–277.

Laubach, S. E. 2003. Practical approaches to identifying sealed and open fractures. *American Association of Petroleum Geologists Bulletin*, **87**, 561–579.

Laubach, S. E., Reed, R. M., Olson, J. E., Lander, R. H. & Bonnell, L. M. 2004. Coevolution of crack-seal texture and fracture porosity in sedimentary rocks: cathodoluminescence observations of regional fractures. *Journal of Structural Geology*, **26**, 967–982.

Lawn, B. R. 1975. An atomistic model of kinetic crack growth in brittle solids. *Journal of Materials Science*, **10**, 469–480.

Loucks, R. G. & Ruppel, S. C. 2007. Depositional setting and lithofacies of the Mississippian Deepwater Barnett Shale in the Forth Worth Basin, Texas. *American Association of Petroleum Geologists Bulletin*, **91**, 579–601.

Marchand, A. M. E., Haszeldine, R. S., Macaulay, C. I., Swennen, R. & Fallick, A. E. 2000. Quartz cementation inhibited by crestal oil charge; Miller deep water sandstone, UK North Sea. *Clay Minerals*, **35**, 201–210.

McDonnell, A., Loucks, R. G. & Dooley, T. 2007. Quantifying the origin and geometry of circular sag structures in northern Fort Worth Basin, Texas: paleocave collapse, pull-apart fault systems, or hydrothermal alteration? *American Association of Petroleum Geologists Bulletin*, **91**, 1295–1318.

Montgomery, S. L., Jarvie, D. M., Bowker, K. A. & Pollastro, R. M. 2005. Mississippian Barnett Shale, Fort Worth basin, north-central Texas: gas–shale play with multi-trillion cubic foot potential. *American Association of Petroleum Geologists Bulletin*, **89**, 155–175.

Olson, J. E. 2004. Predicting fracture swarms – the influence of subcritical crack growth and the crack-tip process zone on joint spacing in rock. *In*: Cosgrove, J. W. & Engelder, T. (eds) *The Initiation, Propagation, and Arrest of Joints and Other Fractures*. Geological Society, London, Special Publications, **231**, 73–87.

Olson, J. E. 2008. Multi-fracture propagation modeling: applications to hydraulic fracturing in shales and tight gas sands. *Proceedings of the 42nd US Rock Mechanics Symposium*. Paper no. ARMA 08-327.

Olson, J. E., Laubach, S. E. & Lander, R. H. 2007. Combining diagenesis and mechanics to quantify fracture aperture distributions and fracture pattern permeability. *In*: Lonergan, L., Rawnsley, K. & Sanderson, D. (eds) *Fractured Reservoirs*. Geological Society, London, Special Publications, **270**, 97–112.

Reed, R. M. & Loucks, R. G. 2007. Imaging nanoscale pores in the Mississippian Barnett Shale of the northern Fort Worth Basin (abstract). *Abstracts Volume, American Association of Petroleum Geologists Annual Convention*, **16**, 115.

Ross, D. J. K. & Bustin, R. M. 2008. Characterizing the shale gas resource potential of Devonian–Mississippian strata in the Western Canada sedimentary basin: application of an integrated formation evaluation. *American Association of Petroleum Geologists Bulletin*, **92**, 87–125.

Ross, D. J. K. & Bustin, R. M. 2009. The importance of shale composition and pore structure upon gas storage potential of shale gas reservoirs. *Marine and Petroleum Geology*, **26**, 916–927.

Ruppel, S. C. & Loucks, R. G. 2008. Black mudrocks: lessons and questions from the Mississippian Barnett Shale in the southern Midcontinent. *The Sedimentary Record*, **6**, 4–8.

Steward, D. B. 2007. *The Barnett Shale Play, Phoenix of the Fort Worth Basin: A History*. The Fort Worth Geological Society & The North Texas Geological Society.

Warpinski, N. R. & Teufel, L. W. 1987. Influence of geologic discontinuities on hydraulic fracture propagation. *Journal of Petroleum Technology*, **39**, 209–220.

Zhang, X., Jeffrey, R. G. & Thiercelin, M. 2007. Effects of frictional geological discontinuities on hydraulic fracture propagation. *Proceedings of the Society of Petroleum Engineers Hydraulic Fracturing Technology Conference*, Paper SPE 106111.

Athabasca oil sands: reservoir characterization and its impact on thermal and mining opportunities

M. J. PEACOCK

Imperial Oil Resources Limited, 237 4 Avenue SW, Calgary, Alberta, T2P 3M9, Canada (e-mail: mike.j.peacock@esso.ca)

Abstract: The heavy oil deposits of Canada are a large resource with an estimated 1.7×10^{12} barrels of bitumen in place. The Lower Cretaceous McMurray Formation in the Athabasca area of northern Alberta contains about 900×10^9 barrels of bitumen in place. This resource can be developed through surface mining and thermal *in situ* techniques. This paper examines the size of this resource in a global context and highlights its position relative to the North American market. The regional geology of the Western Canada Sedimentary Basin and the Athabasca area will also be reviewed. Understanding the regional reservoir distribution of the McMurray Formation is critical to understanding oil sands opportunities. Fluvial estuarine point bar reservoirs are a large portion of the resource that can be developed. Examples will be shown from the type outcrop location, where the stratigraphy can be organized into a hierarchy that subdivides channel-fills into bedsets, storeys, bars and barsets. Inclined heterolithic strata (IHS) surfaces can be identified. Considerable resource delineation drilling has occurred in the basin. The regulator for the Athabasca area specifies minimum drilling densities before project approvals are granted. Regional 2D seismic lines and project-specific 3D and 4D seismic datasets have also been acquired, to reduce the reservoir uncertainty and improve resource definition in this complex depositional environment. These techniques provide a unique opportunity to analyse a complex depositional system with abundant well and core control, outcrop data and seismic information.

To determine preliminary deposition environments, software techniques have been successfully used to interpret large datasets quickly. Laser imaging of mine faces has also been used to record stratigraphy and determine the mined volume of ore. The importance of detailed reservoir characterization studies and their impact on thermal *in situ* recovery mechanisms will also be discussed. Understanding reservoir facies distributions and lateral relationships affects any recovery process, but has an even greater significance in a heavy oil reservoir.

Keywords: Athabasca, oil sands, thermal, *in situ*, mining, 2D, 3D, 4D seismic, reservoir, characterization

The heavy oil deposits of Canada are predominantly located in northern Alberta (Fig. 1) and range in age from Upper Devonian to Lower Cretaceous (Fig. 2). This large resource contains c. 1.7×10^{12} barrels of bitumen in place. The Energy Resources Conservation Board (ERCB), the regulator for Alberta, quotes a recoverable resource of 175×10^9 barrels from these deposits (Energy Resources Conservation Board 2008). Canada ranks second behind Saudi Arabia in oil resources, with heavy oil contributing 98% of this resource (Fig. 3). The proximity of these resources to the North American market highlights the strategic importance they will have in supplying the future energy requirements of the North American continent.

Currently, industry extracts about 1.1×10^6 barrels of bitumen each day, which represents about one-third of the province's total crude oil production. This rate is expected to rise to 2.7×10^6 barrels a day by 2015 (Energy Resources Conservation Board 2008).

The Lower Cretaceous McMurray Formation in the Athabasca area of northern Alberta contains about 900×10^6 barrels of bitumen in place. This resource can be developed through surface mining and thermal *in situ* techniques. Generally, where the overburden is less than 70 m thick, the resource is extracted through open-pit mining, with all land being reclaimed at the end of pit life. New regulatory standards are accelerating the reclamation timing of inactive pit areas. In areas where the overburden is greater than 70 m, thermal *in situ* recovery techniques are used. One of the thermal *in situ* recovery techniques used in the Athabasca area is a process called steam-assisted gravity drainage, commonly referred to as SAGD (Fig. 4). In this process, two horizontal wellbores are drilled about 5 m apart. A typical SAGD well length is 1 km. Steam is then continuously injected into the upper wellbore, which reduces the viscosity of the bitumen, allowing continuous production from the lower wellbore through gravity drainage of the bitumen (Fig. 4). These well pairs are typically drilled from a common well pad in a set pattern or configuration. The SAGD process was originally patented by Imperial Oil Limited (Imperial) in 1982. For a detailed description of the process, see Butler (1982). Because the SAGD process relies on gravity drainage to recover the bitumen, the wells must be located in higher energy depositional units with minimal shale interbeds that could act as barriers or baffles to steam movement.

Regional setting

The Western Canada Sedimentary Basin comprises a northeastward thinning wedge of supracrustal rocks overlapping the North American craton (Price 1994). The thickest and stratigraphically the most complete section is in the Rocky Mountain foreland fold-and-thrust belt. Two main stages of development of the basin have been identified (Bally *et al.* 1966; Price and Mountjoy 1970). The first stage is represented by a Late Proterozoic to Late Jurassic miogeosyncline platform, dominated by carbonate, evaporate and shale deposition. This stage has been correlated to the continental rifting and drifting that created the initial continental margin of the North American craton and its adjacent ocean basin. The subsequent continental terrace wedge (miogeocline) prograded outboard from this passive margin. The second stage consisted of a Late Jurassic to Early Eocene foreland basin, bounded to the west by the emerging Cordilleran mountain belt,

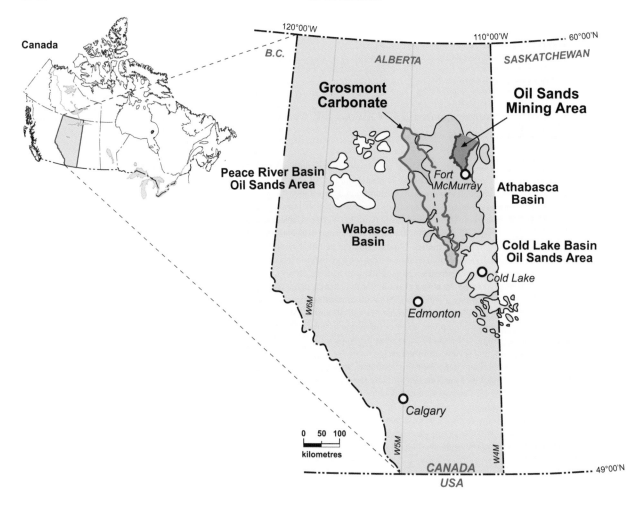

Fig. 1. General location map showing the heavy oil deposits of Alberta.

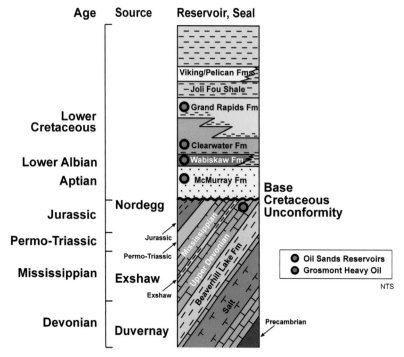

Fig. 2. Generalized stratigraphic column of the Western Canada Sedimentary Basin.

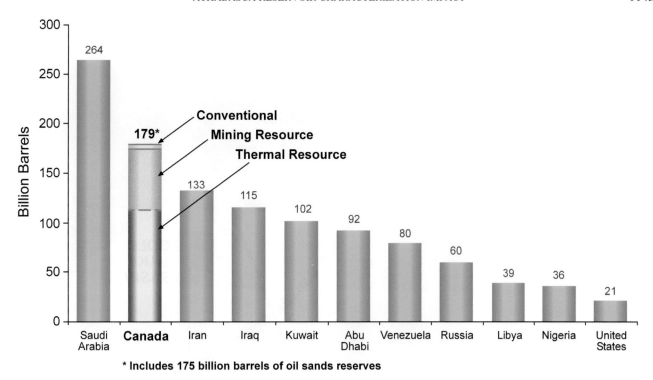

Fig. 3. Global oil resources. *Source*: Oil & Gas Journal, December 2005.

and to the east by the Precambrian (Canadian) Shield. The depositional system was mixed, varying between fluvial and marine. The Cordilleran mountain belt eventually resulted in detachment, thickening and compression of the fold belt, with tectonic thickening and subsidence of the foreland basin (Fig. 5). This compression continued until the Middle Eocene, when it was terminated by a period of crustal extension in the central part of the Cordillera.

In northern Alberta, the large, areally extensive Athabasca Anticline forms the regional trap for the heavy oil accumulations, and was the regional focus for migration. The anticline was the result of one of the final Laramide compressive events associated with the formation of the Cordillera.

Several petroleum source rocks (Fig. 2), ranging in age from Middle Devonian to Lower Cretaceous, have been identified in

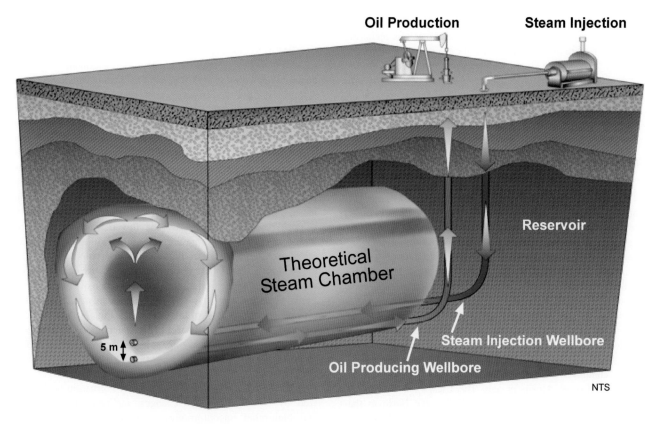

Fig. 4. Steam-assisted gravity drainage (SAGD) process.

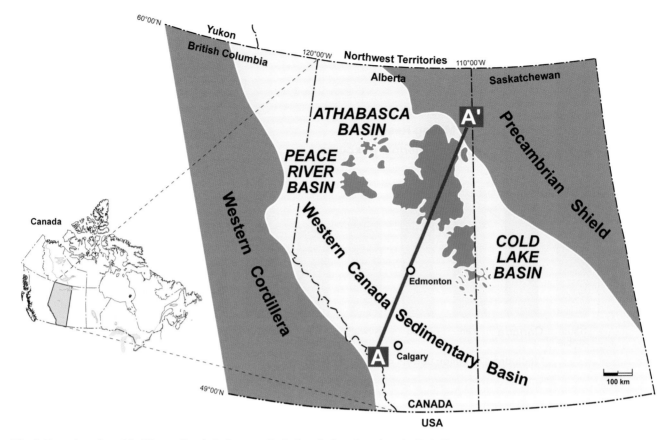

Fig. 5. Tectonic setting of the Western Canada Sedimentary Basin from the Late Jurassic to the Early Eocene.

the Western Canada Sedimentary Basin (Creaney and Allan 1990). A number of these source rocks have world-class source characteristics of high total organic content and hydrogen indices. The development of an effective regional top seal (Joli Fou Shale), combined with simple synclinal foreland basin geometry, enabled a highly efficient migration system to develop (Fig. 6), with migration focused on the Athabasca Anticline. Freshwater meteoric recharge on the eastern edge of the basin, where the major heavy oil reservoirs onlap the Canadian Shield, resulted in conventional oil biodegrading to heavy oil, that is, less than 25° API gravity. The Western Canada Sedimentary Basin contains about 1.75×10^{12} barrels of oil in place, of which an estimated 98% (1.7×10^{12} barrels) is heavy oil (Energy Resources Conservation Board 2008).

Stratigraphy

The Athabasca Basin is located on the eastern margin of the larger Western Canada Sedimentary Basin and, in some places, onlaps the Canadian Shield. During Barremian time, a major drop in relative sea-level occurred, resulting in substantial erosion across the entire foreland basin. The earliest post-unconformity sediments in the Athabasca Basin are the fluvial estuarine deposits of the McMurray Formation (Fig. 2). The McMurray Formation is equivalent to the Aptian age Lower Mannville Group that contains prolific conventional reservoirs in other parts of the basin. A marine transgression followed, resulting in the deposition of the shallow marine sands of the Lower Albian Wabiskaw Formation. These sands formed a secondary reservoir in the Athabasca Basin. Another third-order sea-level drop resulted in the return to fluvial estuarine conditions and the deposition of the Clearwater Formation, also Lower Albian in age. The Clearwater Formation is the main reservoir objective in the Cold Lake Basin to the south of the Athabasca Basin (Fig. 1). However, this formation becomes shaley to the north and is not a reservoir unit in the Athabasca Basin. The transgressive event that followed then deposited the Grand Rapids Formation, which consists of a series of shoreface sands cut by several incised valley-fill units. The Grand Rapids Formation is present in the southern part of the Athabasca Basin and is considered a secondary reservoir target.

The primary heavy oil reservoir target in the Athabasca Basin for both SAGD and mining exploitation is the Aptian McMurray Formation. Regionally, the McMurray Formation is a deepening-upward fluvial estuarine complex deposited within an Early Cretaceous palaeovalley system. The establishment of this palaeovalley system was controlled in part by structure on the Base Cretaceous unconformity and proximity to the Canadian Shield to the east (Fig. 5). The main valley trended north to NW toward the Boreal Sea. The isopach map of the McMurray Formation (Fig. 7) illustrates the size of the depositional system, which extends about 320 km northwards, and in places is up to 160 km wide. The thickness of the McMurray Formation can be greater than 120 m. There is a strong correlation between isopach thickness and a high net-to-gross ratio. The tributary systems are evident on the isopach map, as indicated by the fluvial transport directions. These tributary systems are normally filled with thick deposits of quartz-rich sands that form high-quality reservoirs. As a result of the shallow burial of these reservoirs, the sands are unconsolidated, which allows for open-pit mining in areas of thin overburden cover. Porosity ranges from 30 to 35%, with multidarcy permeability and high oil saturation. East of this water line (Fig. 7), the reservoir is structurally low and water-wet. This water line is controlled by the dissolution of underlying salt in the Palaeozoic section (Fig. 2). Figure 7 shows the location of several of the largest SAGD projects and the major mining developments.

In the Athabasca Basin, the McMurray Formation is characterized by abrupt lithofacies changes, inclined stratal geometry and high-relief unconformities in a complex environment of deposition. To reduce uncertainty in resource description and

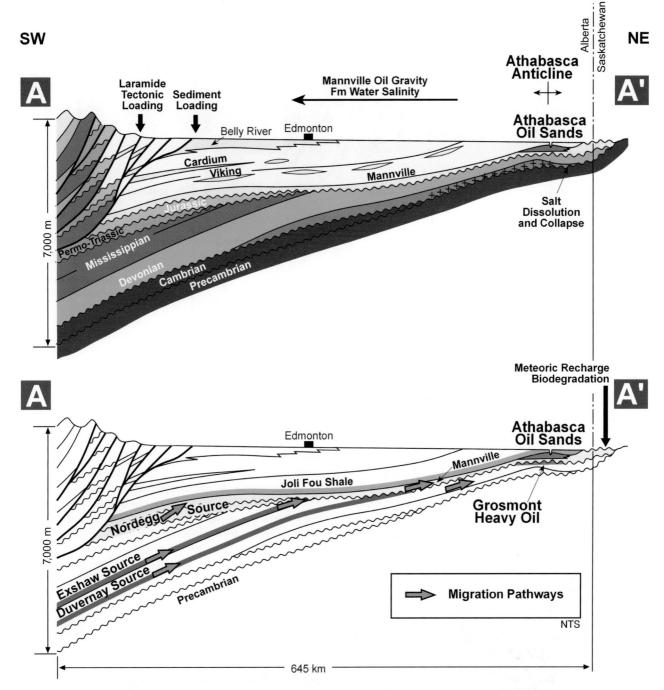

Fig. 6. Regional cross-section through the Western Canada Sedimentary Basin showing location of the Athabasca heavy oil deposits.

assessment, the ERCB stipulates a minimum well spacing before it will approve a project for development. The required well spacing is reduced to 100 m in certain areas. These wells have to be cored and logged. Dipmeter data is critical to understanding palaeodepositional trends. In addition, in the thermal or SAGD areas, baseline 3D seismic is required for 4D monitoring of steam flooding. This, combined with regional 2D seismic type outcrop exposures and mine face exposures, provides world-class datasets for detailed reservoir characterization studies. Fluvial estuarine point bar reservoirs represent a large fraction of the resource that can be developed. Examples will be given of the type of outcrop location where the stratigraphy can be organized into a hierarchy that subdivides channel-fills into bedsets, storeys, bars and barsets. Inclined heterolithic surfaces can be identified.

This complex stratigraphic environment of deposition can be organized within a sequence stratigraphic framework (Nardin et al. 2005). Several sequence-keyed facies are identifiable, including thick, amalgamated, braided stream facies, fluvial floodplain and estuarine facies, and large-scale fluvial estuarine point bars. The fluvial estuarine point bar facies are the major reservoir targets for Imperial's mining and SAGD developments. The characteristics of the fluvial estuarine point bar facies will be described in more detail, along with the effect of the facies on thermal reservoir performance.

The fluvial estuarine point bar facies is an upward-fining or thinning succession of stacked beds, bedsets and storeys. An example of a stacked reservoir hierarchy is the Steepbank River outcrop (Fig. 8). The main features of a typical point bar, as described by Thomas et al. (1987), includes basal channel sands, inclined heterolithic strata (IHS) and sand-prone inclined strata (IS).

The basal channel sands are typically amalgamated and characterized by trough and tabular cross-bedding (Fig. 9). These are the

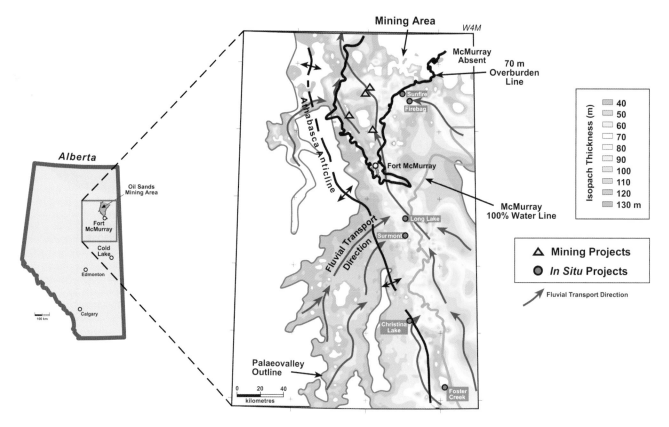

Fig. 7. Isopach map of the Lower Cretaceous McMurray Formation.

highest quality reservoir sands. They are generally overlain in an upward-fining succession by IS sandbodies and IHS sequences. Both IS and IHS are formed by laterally accreting point bars, with IS formed in higher energy, higher discharge depositional periods. Inclined heterolithic strata are dominated by mud, and were deposited in periods of lower energy conditions. In core, they can be recognized by an increase in mudstone bed thickness and frequency, degree of bioturbation and association with mudstone clast facies. The thickness and lateral extent of the mudstone facies can have a considerable impact on thermal reservoir performance. The estuarine conditions can be recognized in core by the brackish water trace fossil assemblage, including *Palaeophycus*, *Cylindrichnus*, *Planolites* and *Skolithos*. Matrix-supported angular mudstone clasts are common in all facies and are derived

Fig. 8. Steepbank river outcrop, lower McMurray Formation.

Fig. 9. Core example of a fluvial estuarine point bar sequence of the lower McMurray Formation.

from the erosion associated with the lateral accretion of the point bars (Nardin et al. 2005). Being able to differentiate between clasts and more laterally continuous IHS beds is important. On a gamma-ray log, clasts and IHS beds can have the same characteristics, so core is critical for making the ultimate identification. In the thermal recovery process, steam will still be able to move, and a steam chamber will grow within a reservoir section that contains clasts. However, thin mudstone beds can substantially slow and impede steam chamber growth.

The dipmeter is another log that is critical for identifying and interpreting the environment of deposition. Inclined heterolithic strata and IS dips range from 3 to 20°, with an average of about 10°. Typically, azimuths vary, but are generally within 90°. Even with a well spacing of 100 m in places, dips in the range of 10° will have a substantial impact on the ability to correlate surfaces between wells. Imperial has used regional 2D seismic lines to assist in identifying IHS sequences. Inclined heterolithic strata can be recognized on 2D seismic logs, with characteristics of moderate- to high-amplitude, discontinuous reflectors that appear to downlap the underlying sequence boundaries (Nardin et al. 2005).

Laser imaging of mine faces is also used for recording stratigraphy and determining the mined volume of ore. Laser images are collected frequently. For a particular mine face, multiple images are collected, capturing the changes in stratigraphy as the mine face migrates through the deposit.

Impact of reservoir heterogeneity on thermal recovery and performance

Thermal recovery efficiency depends on establishing an effective steam chamber within a reservoir unit. Two horizontal wells are drilled 5 m apart, with steam injected into the upper wellbore, and bitumen produced from the lower wellbore. Thermal communication between the two wellbores generally occurs within the first three months of steaming. Once communication has been established, the steam chamber grows like a cone (Fig. 10). The wellbores are positioned to reduce the effect of potential thief zones for the steam. Thief zones, which include top and bottom water and top gas, are zones that significantly reduce the efficiency of establishing a steam chamber. The theoretical recovery efficiency of the process is very high. Typically, theoretical recovery efficiencies are modelled and simulated from reservoirs with minimal mudstone interbeds. However, actual steam chamber growth is much more irregular than the models (Fig. 10), and is heavily affected by reservoir heterogeneity and, in particular, mudstone or shale interbeds. This can be demonstrated by repeatedly running temperature logs in observation wells and acquiring 4D or repeat seismic surveys.

Temperature logs can be used to monitor heat in the reservoir. When integrated with core coverage and detailed facies descriptions, a more accurate interpretation of steam chamber growth is possible. As previously mentioned, identifying clasts v. mudstone or shale interbeds is critical to understanding thermal reservoir performance. In the example from the Athabasca area, shown in Figure 11, the steam injector was located just below the base of the core. Steaming began and the steam chamber rose 4 m in the first 18 months through a section dominated by clasts. Steam chamber growth was then halted for six months by a thin, 1–2 cm mudstone interbed. Eventually, the steam broke through this thin interbed and the growth of the steam chamber continued. The steam chamber then took only six months to grow a further 3 m. The core in Figure 11 shows that this growth occurred in an interval with lower clast content. The steam chamber was then permanently halted by an interval dominated by IHS and more laterally continuous mudstone or shale interbeds.

Seismic monitoring at Imperial's Cold Lake development began in 1985 with the first seismic imaging pilot study. This pilot has evolved into the development of important seismic imaging tools for reservoir monitoring and production optimization (Eastwood et al. 1994, 1996), which are currently being used at Imperial's in situ heavy oil properties. The ERCB also mandates a baseline 3D seismic survey before approving the development of any SAGD project.

Seismic monitoring of in situ heavy oil production relies on the acoustic contrast that results from the introduction of gas within the pore space of the reservoir from both the injection of

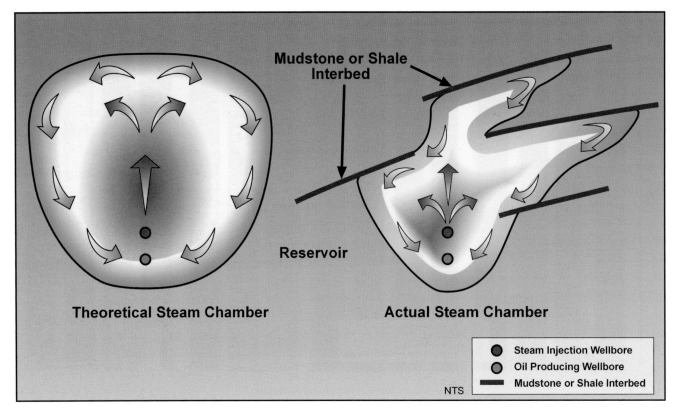

Fig. 10. Steam chamber growth in SAGD process.

steam and the release of gas associated with heating the bitumen. The seismic response associated with the introduction of this gas is a high-amplitude reflection at the top of the gas chamber, a time delay of reflections below the gas chamber because of the decreased seismic velocity within the chamber, and a loss of frequency resulting from the attenuation of the seismic wave through the gas chamber. These attributes can be used to construct areal conformance maps and vertical conformance volumes to monitor the behaviour of the heated region through time (Fig. 12). Conformance maps indicate the efficiency of steam

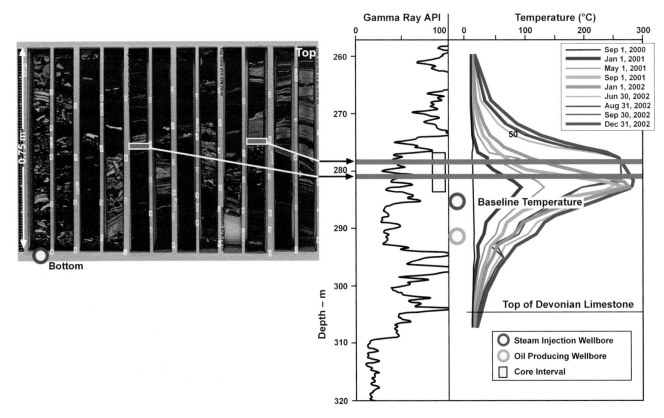

Fig. 11. Integration of core and thermal logging data to monitor steam chamber growth in a vertical observation well.

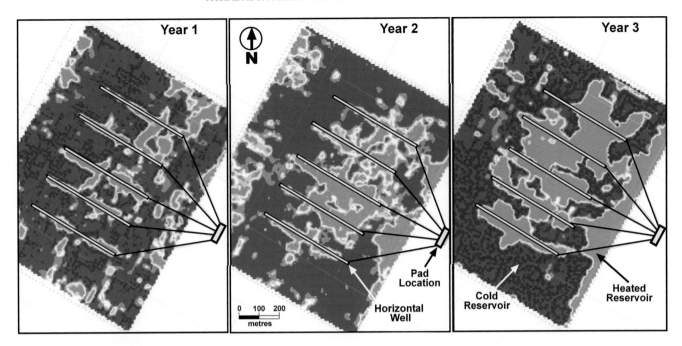

Fig. 12. Steam conformance maps generated from 4D seismic at the Cold Lake Field, Alberta.

chamber growth and areas that have been accessed by the steam injection. The integration of seismic monitoring surveys with the detailed sequence stratigraphy of the reservoir from logs, core and baseline seismic surveys can be used to show the influence of the stratigraphy on the distribution of steam within the reservoir. This understanding allows future well placement and infill drilling to be planned in a way that optimizes bitumen recovery.

The Cold Lake example (Fig. 12) shows the steam conformance mapping of a single five-horizontal-well pad. Through time, the steam front can be mapped. The steam front does not move or build up regularly, but is heavily influenced by the heterogeneity of the reservoir in which that particular well is positioned. The complex environment of deposition of these fluvial estuarine deposits and the relatively steep dips of the beds, averaging 10°, result in high variability of the reservoir along any particular wellbore. During the pre-development phase, vertical pilot holes are used to help determine the best placement of the horizontal wellbores.

Managing large well datasets

In the Athabasca Basin areas where bitumen is recovered through open-pit mining, well density is high (Fig. 13). In places, the well spacing is reduced to 100 m in compliance with regulatory requirements. All wells are cored and logged, providing a large and detailed database to analyse. In a typical mining development, a single pit might have as many as 10 000 core holes drilled before the end of pit development. This provides a large amount of data that conventional 3D geological modelling software is unable to accommodate. Consequently, mining software is used because it can manage the large datasets. However, mining software is not

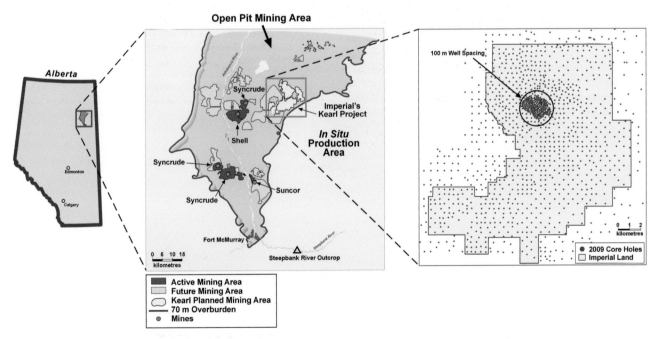

Fig. 13. Map of the Athabasca Basin that can be developed through surface mining.

as sophisticated as geological modelling software in its manipulative capabilities.

Conclusions

The Athabasca oil sands represent a substantial hydrocarbon resource located in a stable, accessible environment. The ERCB, the regulator for Alberta, estimates a recoverable resource of 175×10^9 barrels from these deposits (Energy Resources Conservation Board 2008). The proximity of these resources to the North American market highlights the strategic importance they will have in supplying the future energy requirements of the continent. Thermal processes are still in the early stages of development, but will play an increasing role in Canadian oil production. Lessons learned about recovery performance from new projects will greatly enhance the optimization of these projects. The integration of reservoir characterization studies will be a critical component of interpreting performance. Currently, industry extracts around 1.1×10^6 barrels of bitumen each day, a rate expected to rise to 2.7×10^6 barrels a day by 2015 (Energy Resources Conservation Board 2008).

Detailed reservoir characterization studies are important for any type of reservoir, be it a conventional or heavy oil field development. For heavy oil thermal developments, understanding reservoir characteristics is especially important, because shale interbeds have such a dramatic effect on steam chamber growth and, ultimately, on reservoir performance. Constructing a predictive stratigraphic model will reduce the risk associated with well placement. Horizontal well pairs, typically 1 km in length, will encounter varying stratigraphy along the wellbore.

Monitoring the effective steam chamber growth is critical to understanding reservoir performance and, ultimately, to optimizing the reservoir development plan. The reservoir development plan can be optimized by integrating reservoir description with temperature monitoring techniques, including temperature logs and 4D seismic conformance mapping. The detailed integration of these tools results in an increased improvement in field development management. In Alberta, the ERCB mandates baseline 3D seismic surveys before a project is approved. The incremental cost of repeat surveys is competitive when compared with the cost of misplacing an infill well, drilled to optimize production and improve reservoir conformance.

Reservoir characterization is also essential for mining developments. Mining facilities are designed to process the average parameters of the deposit. To achieve this, the different facies are required to be blended. Depending upon the facies being mined, lower quality facies (facies with an increased content of IHS beds) may be mixed with higher quality facies to produce an average blend for the facilities to process. This is managed on a daily basis, so a comprehensive understanding of the volume of a particular facies to be mined allows an improved predictive capability and increased facility efficiency.

In the Athabasca Basin, the unique combination of abundant well control, core and log data, 2D, 3D and 4D seismic data, mine exposures and outcrops, provides world-class datasets available for detailed reservoir characterization studies of a complex fluvial estuarine depositional environment. Laser images of mine faces are collected frequently to record stratigraphy and to determine the mined volume of ore. Multiple images are collected for each mine face to capture the stratigraphic changes as the mine face migrates through the deposit.

The author wishes to thank Imperial Oil Resources Limited for permission to publish this paper.

References

Bally, A. W., Gordy, P. L. & Stewart, G. A. 1966. Structure, seismic data and orogenic evolution of the southern Canadian Rockies. *Bulletin of Canadian Petroleum Geology*, **14**, 337–381.

Butler, R. M. 1982. *Method for continuously producing viscous hydrocarbons by gravity drainage while injecting heated fluids*. Canadian Patent CA1130201.

Creaney, S. & Allan, J. 1990. Hydrocarbon generation and migration in the Western Canada Sedimentary Basin. *In*: Brooks, J. (ed.) *Classic Petroleum Provinces*. Geological Society, London, Special Publications, **50**, 189–202.

Eastwood, J., Lebel, P., Dilay, A. & Blakeslee, S. 1994. Seismic monitoring of steam-based recovery of bitumen. *The Leading Edge*, **13**, 242–251.

Eastwood, J., Anderson, D. & Boone, T. 1996. 3D seismic monitoring for enhancing thermal recovery. *Canadian Society of Exploration Geophysicists, Recorder*, **21**, 3–5.

Energy Resources Conservation Board. 2008. *Alberta's Energy Reserves 2007 and Supply/Demand Outlook 2008–2017*. Energy Resources Conservation Board, Edmonton, **ST98-2008**.

Nardin, T. R., Carter, B. J. *et al.* 2005. Sequence stratigraphic and depositional facies framework of the Lower Cretaceous McMurray Formation, Kearl Oil Sands Project, Alberta. *Core Workshop 2005*. Canadian Society of Petroleum Geologists, Calgary.

Price, R. A. 1994. Cordilleran tectonics and the evolution of the Western Canada Basin. *In*: Mossop, G. D. & Shetsen, I. (comps.) *Geological Atlas of the Western Canada Sedimentary Basin*. Canadian Society of Petroleum Geologists and Alberta Research Council, Calgary, 13–24.

Price, R. A. & Mountjoy, E. W. 1970. Geologic structure of the Canadian Rocky Mountains between the Bow and Athabasca Rivers: a progress report. *In*: Wheeler, J. O. (ed.) *Structure of the Southern Canadian Cordillera*. Geological Association of Canada, St John's, Special Publications, **6**, 7–25.

Thomas, R. G., Smith, D. G., Wood, J. M., Visser, J., Calverly-Range, E. A. & Koster, E. H. 1987. Inclined heterolithic stratification – terminology, description, interpretation and significance. *Sedimentary Geology*, **53**, 123–179.

Resource potential of gas hydrates: recent contributions from international research and development projects

T. S. COLLETT

US Geological Survey, Denver Federal Center, MS-939, Box 25046, Denver, CO 80225, USA
(e-mail: tcollett@usgs.gov)

Abstract: It is generally accepted that the amount of gas in the world's gas hydrate accumulations exceeds the volume of known conventional gas resources. Researchers have long speculated that gas hydrates could eventually be a commercial producible energy resource yet technical and economic hurdles have historically made gas hydrate development a distant goal rather than a near-term possibility. This view began to change in recent years with the realization that this unconventional resource could possibly be developed with existing conventional oil and gas production technology. The most significant development has been gas hydrate production testing conducted at the Mallik site in Canada's Mackenzie Delta. The Mallik Gas Hydrate Production Research Well Program has yielded the first modern, fully integrated field study and production test of a natural gas hydrate accumulation. More recently, BP Exploration (Alaska) Inc. with the US Department of Energy and the US Geological Survey have successfully cored, logged and tested a gas hydrate accumulation on the North Slope of Alaska known as the Mount Elbert Prospect. The Mallik project along with the Mount Elbert effort has for the first time allowed the rational assessment of the production response of a gas hydrate accumulation. In addition to the gas hydrate production tests in Canada and the USA, marine gas hydrate research drilling, coring and logging expeditions launched by the national gas hydrate programmes in Japan, India, China and South Korea have also contributed significantly to our understanding of how gas hydrates occur in nature and have provided a much deeper appreciation of the geological controls on the occurrence of gas hydrates. With an increasing number of highly successful gas hydrate field studies, significant progress has been made in addressing some of the key issues on the formation, occurrence and stability of gas hydrates in nature.

Keywords: gas hydrate, natural gas, resources, energy, Alaska, Mackenzie Delta

Gas hydrates are naturally occurring 'ice-like' combinations of natural gas and water that have the potential to provide an immense resource of natural gas from the world's oceans and polar regions. Gas hydrates are known to be widespread in permafrost regions and beneath the sea in sediments of outer continental margins. It is generally accepted that the volume of natural gas contained in the world's gas hydrate accumulations greatly exceeds that of known gas reserves. There is also growing evidence that natural gas can be produced from gas hydrates with existing conventional oil and gas production technology (Dallimore *et al.* 2008; Moridis *et al.* 2008; Yamamoto & Dallimore 2008). This review of natural gas hydrates is intended to provide an up-to-date analysis of the geological controls on the occurrence of gas hydrates in nature with a focus on understanding the energy resource potential of gas hydrates. The results of some of the more important international gas hydrate research projects are discussed.

Occurrence of gas hydrates

As shown in Figure 1, gas hydrates have been recovered at about 40 locations throughout the world. However, only a limited number of gas hydrate accumulations have been examined and delineated with data collected by deep scientific drilling operations. Included in the following discussion are descriptions of several of the best known drilled marine and onshore permafrost-associated gas hydrate accumulations in the world.

Gulf of Mexico, USA

The occurrence of gas hydrates in the Gulf of Mexico was confirmed during DSDP Leg 96 when numerous gas hydrate samples were recovered from sub-bottom depths ranging from 20 to 40 mbsf in the Orca Basin (Sites 618 and 618A), which is located about 300 km south of Louisiana beneath about 2000 m of water. Near-surface (0–5 m) marine sediment coring has also recovered numerous gas hydrate samples on the Louisiana continental slope.

In 2005, the Chevron-led Gulf of Mexico Gas Hydrate Joint Industry Project (JIP) conducted scientific drilling, coring and downhole logging to assess hydrate-related hazards in fine-grained sediments with low concentrations of gas hydrate (Claypool 2006). This expedition targeted two deep-water locations in the Atwater Valley and Keathley Canyon areas of the Gulf of Mexico. Although gas hydrate was not physically recovered from the Keathley Canyon core hole, other indicators of gas hydrate, such as elevated downhole measured electrical resistivities, suggest the probable occurrence of gas hydrate in the KC151-2 well. Analysis of downhole measured resistivities and resistivity-at-the-bit (RAB) images from the KC151-2 hole revealed the occurrence of fracture filling gas hydrate at relatively high concentrations. The analysis of downhole well log data from the two JIP Atwater Valley wells shows little evidence of significant gas hydrate occurrences, other than several thin, possibly stratigraphically controlled gas-hydrate-bearing intervals. The next phase of the Gulf of Mexico JIP is being extended to coarser grained sediments with much higher expected gas hydrate concentrations. In the spring of 2009, the Gulf of Mexico JIP expects to conduct exploratory drilling and logging to better understand gas hydrate-bearing sands in the deep-water Gulf of Mexico.

Also in the Gulf of Mexico, at the Alaminos Canyon 818 site (AC818), gas hydrate is interpreted to occur within the Oligocene Frio volcaniclastic sand at the crest of a fold that is shallow enough to be in the hydrate stability zone (Smith *et al.* 2006). Examination of the well log data obtained from the Frio section of the Chevron Tiger Shark well drilled in AC818 indicates

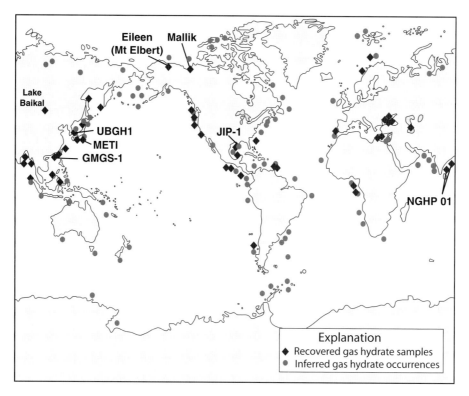

Fig. 1. Locations of sampled and inferred gas hydrate occurrences in oceanic sediment of outer continental margins and permafrost regions. Most of the recovered gas hydrate samples have been obtained during deep coring projects or shallow seabed coring operations. Most of the inferred gas hydrate occurrences are sites at which bottom-simulating reflectors (BSRs) have been observed on available seismic profiles. The gas hydrate occurrences reviewed in this report have also been highlighted on this map.

approximately 18 m of sand (3209–3227 m drilling depth) with porosity of about 30% and downhole measured resistivity in the range 30–40 Ω m. Initial volumetrics derived from downhole log data show very high gas hydrate saturations (up to 80% of available pore volume).

Alaska North Slope, Alaska, USA

Before the recently completed coring and downhole-logging operations in the BP Exploration (Alaska) Mount Elbert well in Milne Point, the only direct confirmation of gas hydrate on the North Slope was obtained in 1972 with data from the Northwest Eileen State-2 well located in the northwestern part of the Prudhoe Bay Field. Gas hydrates are also inferred to occur in an additional 50 exploratory and production wells in northern Alaska based on downhole log responses calibrated to the known gas hydrate occurrences in the Northwest Eileen State-2 well. Most of the well-log inferred gas hydrates occur in six laterally continuous sandstone units; all are geographically restricted to the area overlying the eastern part of the Kuparuk River Field and the western part of the Prudhoe Bay Field. The volume of gas within the gas hydrates of the Prudhoe Bay–Kuparuk River area, which has come to be known as the Eileen Gas Hydrate Accumulation, is estimated to be about $1.0–1.2 \times 10^{12}$ m^3

Under the Methane Hydrate Research and Development Act of 2000 (renewed in 2005), the Department of Energy (DOE) has funded field research on both Arctic and marine gas hydrates. Among the current Arctic studies, BP Exploration (Alaska) Inc. (BPXA) and the DOE have undertaken a project to characterize the commercial viability of gas hydrate resources in the Prudhoe Bay, Kuparuk River and Milne Point field areas on the Alaska North Slope. As part of this effort, the Mount Elbert Gas Hydrate Stratigraphic Test Well was completed in February 2007 and yielded one of the most comprehensive datasets yet compiled on naturally occurring gas hydrates (Boswell *et al.* 2008). In 2005, extensive analysis of BPXA's proprietary 3D seismic data and the integration of that data with existing well log data (enabled by collaborations with the US Geological Survey (USGS) and the Bureau of Land Management), resulted in the identification of more than a dozen discrete and mappable gas hydrate prospects within the Milne Point area. Because the most favourable of those targets was a previously undrilled, fault-bounded accumulation, BPXA and the DOE decided to drill a vertical stratigraphic test well at that location (named the 'Mount Elbert' prospect) to acquire critical reservoir data needed to develop a longer term production testing programme. Gas hydrates were expected and found in two stratigraphic sections. An upper zone, (Unit D) contained *c.* 14 m of gas hydrate-bearing reservoir-quality sandstone. A lower zone (Unit C), contained *c.* 16 m of gas hydrate-bearing reservoir. Both zones displayed gas hydrate saturations that varied with reservoir quality as expected, with typical values between 60 and 75%. The Mount Elbert gas hydrate stratigraphic test well project included the acquisition of pressure transient data from four short-duration pressure-drawdown tests with Schlumberger's wireline MDT (Boswell *et al.* 2008). These tests were conducted in open-hole, and were designed to build upon the knowledge gained from cased-hole MDT tests conducted during the Mallik 2002 testing programme. The MDT and NMR log data from the Mount Elbert well also confirmed the presence of a mobile pore-water phase even in the most highly gas hydrate-saturated intervals. Gas hydrate dissociation and production was confirmed in the later stages of each test in which the pressure was drawn down below gas hydrate equilibrium conditions. Additional work anticipated within this effort includes a long-term production testing programme designed to determine reservoir deliverability under a variety of production/completion scenarios.

Mackenzie River Delta, Mallik, Canada

The JAPEX/JNOC/GSC Mallik 2L-38 gas hydrate research well, drilled in 1998 near the site of the Mallik L-38 well, included extensive scientific studies designed to investigate the occurrence of *in situ* natural gas hydrate in the Mallik field area (Dallimore *et al.* 1999). Approximately 37 m of core was recovered from the gas hydrate interval (878–944 m) in the Mallik 2L-38 well. Pore-space gas hydrate and several forms of visible gas hydrate were observed in a variety of unconsolidated sands and gravels interbedded with non-hydrate-bearing silts. Because of the success of the 1998 Mallik 2L-38 gas hydrate research well programme, the Mallik site was elevated as an important gas hydrate production test site with the execution of two additional gas hydrate production research programmes: (1) the Mallik 2002 Gas Hydrate Production Research Well Program; and (2) the 2006–2008 JOGMEC/NRCan Mallik Gas Hydrate Production Research Program.

In June of 2005, the partners in the Mallik 2002 Gas Hydrate Production Research Well Program publicly released the results of the first modern, fully integrated field study and production test of a natural gas hydrate accumulation (Dallimore & Collett 2005). During the Mallik 2002 testing programme, the response of gas hydrates to heating and depressurization was evaluated. The results of three short-duration gas hydrate tests demonstrate that gas can be produced from gas hydrates exclusively through pressure stimulation. Thermal stimulation experiments were designed to destabilize gas hydrates by using circulated hot water to increase the *in situ* temperature. Gas was continuously produced throughout the test at varying rates with maximum flow rate reaching 360 m^3 per day (Dallimore & Collett 2005). The total volume of gas flowed was small, reflecting that the test was a controlled production experiment rather than a long-duration well test. It also demonstrated the difficulty of heating a relatively large rock mass by conductive heat flow alone.

As described by Dallimore *et al.* (2008) and Yamamoto & Dallimore (2008), the 2006–2008 JOGMEC/NRCan Mallik Gas Hydrate Production Research Program was designed to build on the results of the Mallik 2002 project with the main goal of monitoring long-term production behaviour of gas hydrates. The primary objective of the winter 2006–2007 field activities was to install equipment and instruments to allow for long-term production gas hydrate testing during the winter of 2007–2008. After completing drilling operations, a short pressure drawdown test was conducted to evaluate equipment performance and assess the short-term 'producibility' of the gas hydrate-bearing section. During the most successful 12.5 h of the test, at least 830 m^3 of gas were produced. The test results verified the effectiveness of the depressurization method even for such a short duration. The following winter (2007/2008), the team returned to the site to undertake a longer term production test with the implementation of countermeasures to overcome the problems encountered in the previous year's programme. The 2007/2008 field operations consisted of a six-day pressure drawdown test, during which 'stable' gas flow was measured at the surface. The 2007/2008 testing programme at Mallik established a continuous gas flow ranging from 2000 to 4000 m^3/day.

Nankai Trough, Japan

The presence of extensive bottom-simulating reflectors (BSRs) in the Nankai Trough was confirmed with seismic surveys carried out as a part of METI's domestic geophysical survey programme. The 1999/2000 Nankai Trough drilling and coring programme targeted an area of a prominent BSR located about 50 km from the mouth of the Tenryu River in central Japan at a water depth of 945 m (Takahashi & Tsuji 2005). This drilling project, consisting of a pilot well and three post-survey wells, confirmed the existence of gas hydrate in the intergranular pores of turbiditic sands based on the analysis of downhole-logging data, and observations from both conventional and pressure cores. Gas hydrate was determined to fill the pore spaces in these deposits, reaching saturations up to 80% in some layers. Individual hydrate-bearing sand layers were less than 1 m thick, with the cumulative thickness of the hydrate-bearing sands totalling about 12–14 m.

A multi-well drilling programme titled 'METI Toaki-oki to Kumano-nada' was successfully carried out in early 2004 (Takahashi & Tsuji 2005). A total of 16 sites were drilled at water depths ranging from 720 to 2030 m. Based on the analysis of both the available downhole log data and core observations, three different types of gas hydrate occurrences were identified: (1) sand with pore-filling hydrate; (2) silt with pore-filling hydrate; and (3) nodular or fracture-filling massive hydrate in fine-grain sediments. Analysis of pressure cores and downhole log data indicates that average gas hydrate saturations in the cored sand layers ranged from 55 to 68%, with the average sediment porosities ranging from 39 to 41%.

NGHP Expedition 01, India

NGHP Expedition 01 was designed to study the occurrence of gas hydrate off the Indian Peninsula and along the Andaman convergent margin with special emphasis on understanding the geological and geochemical controls on the occurrence of gas hydrate in these two diverse settings. During its 113.5-day voyage (28 April to 19 August 2006), the research drill ship *JOIDES Resolution* (*JR*) cored or drilled 39 holes at 21 sites (one site in Kerala-Konkan, 15 sites in Krishna–Godavari, four sites in Mahanadi and one site in Andaman deep offshore areas), penetrated more than 9250 m of sedimentary section and recovered nearly 2850 m of core. Twelve holes were logged with logging-while-drilling tools and an additional 13 holes were wireline logged. NGHP Expedition 01 was among the most complex and comprehensive methane hydrates field ventures yet conducted. All of the primary data collected during NGHP Expedition 01 are included in either the NGHP Expedition 01 Initial Reports (Collett *et al.* 2008*a*) or the NGHP Expedition 01 Downhole Log Data Report (Collett *et al.* 2008*b*), which were prepared by the USGS and published by the DGH on behalf of the MOP&NG.

NGHP Expedition 01 established the presence of gas hydrates in Krishna–Godavari, Mahanadi and Andaman basins. The expedition discovered one of the richest gas hydrate accumulations yet documented (Site 10 in the Krishna-Godavari Basin), documented the thickest and deepest gas hydrate stability zone yet known (Site 17 in Andaman Sea), and established the existence of a fully developed gas hydrate system in the Mahanadi Basin (Site 19). For the most part, the interpretation of downhole-logging data and linked imaging of recovered cores, analyses of interstitial water from cores and pressure core imaging from the sites drilled during NGHP Expedition 01 indicate that the occurrence of gas hydrate is mostly controlled by the presence of fractures and/or coarser grained (mostly sand-rich) sediments (Collett *et al.* 2008*a*).

Drilling Expedition GMGS-1, China

In June of 2007, a deep-water gas hydrate drilling and coring programme was successfully completed by the Guangzhou Marine Geological Survey (GMGS), China Geological Survey (CGS) and the Ministry of Land and Resources of P. R. China (Wu *et al.* 2008). Drilling expedition GMGS-1 was carried out from April to June 2007 in the Shenhu Area on the north slope of the South China Sea. During Expedition GMGS-1, eight sites were drilled in water depths of up to 1500 m. Each site was wireline logged to depths of up to 300 mbsf with a set of high-resolution slim wireline tools. Five of the eight sites occupied during the expedition were cored. Gas hydrate was detected at three of the five core sites. The sediments were predominantly clay, with a variable amount of silt-sized particles including foraminifera. The sediment layers rich in gas hydrate were about 10–25 m thick and were found just above the base of the predicted gas hydrate stability zone (BGHSZ) at all three sites. Analysis of pressure cores confirmed that the gas hydrate occurred within fine-grained foraminifera-rich clay sediments, with gas hydrate saturations ranging from 20 to 40%.

Drilling Expedition UBGH1, South Korea

In November of 2007 South Korea completed its first large-scale gas hydrate exploration and drilling expedition in the East Sea: Ulleung Basin Gas Hydrate Expedition 1 (UBGH1). Leg 1 of UBGH1 included the drilling of five logging-while-drilling holes in the Ulleung Basin, which was used to select a sub-set of three sites that were more likely to contain gas hydrate for Leg 2 drilling and coring operations. Coring during Leg 2, at water depths between 1800 and 2100 m, confirmed the presence of gas hydrate-bearing reservoirs up to 150 mbsf (Park 2008). Gas hydrate was recovered at all three core sites, occurring as veins and layers in clay-rich sediments, and as pore-filling material within the silty/sandy layers. At one site, a 130 m thick hydrate-bearing sedimentary section of interbedded sands and clays was penetrated. Analysis of pore-water freshening revealed average gas hydrate saturations of about 30% for the hydrate-bearing sand layers.

Summary

There are numerous research projects under way to investigate the geological origin of gas hydrates, their natural occurrence, the factors that affect their stability and the possibility of using this vast resource in the world energy mix. We have seen highly successful cooperative research projects, such as various phases of the Mallik effort that has for the first time tested the technology needed to produce gas hydrates. We have also seen other highly successful co-operative gas hydrate research studies in India, northern Alaska and the Gulf of Mexico. In most cases, the co-operative nature of these efforts directly contributed to their success. It is also not surprising that the most aggressive and well funded gas hydrate research programmes are in countries highly dependent on imported energy resources, such as Japan and India.

References

Boswell, R., Hunter, R., Collett, T. S., Digert, S., Hancock, S., Weeks, M. & Mount Elbert Science Team 2008. Investigation of gas hydrate bearing sandstone reservoirs at the Mount Elbert stratigraphic test well, Milne Point, Alaska. *Proceedings of the 6th International Conference on Gas Hydrates* (ICGH 2008), Vancouver, British Columbia, 6–10 July.

Claypool, G. E. 2006. The Gulf of Mexico Gas Hydrate Joint Industry Project; covering the cruise of the Drilling Vessel Uncle John; Mobile, Alabama to Galveston, Texas; Atwater Valley Blocks 13/14 and Keathley Canyon Block 151; 17 April to 22 May, 2005. World Wide Web Address: http://www.netl.doe.gov/technologies/oil-gas/publications/Hydrates/reports/GOMJIPCruise05.pdf.

Collett, T., Riedel, M. *et al.* & The NGHP Expedition 01 Scientific Party, 2008a. *Indian National Gas Hydrate Program Expedition 01 Initial Reports*. US Geological Survey/Directorate General of Hydrocarbons. Ministry of Petroleum & Natural Gas (India), 1 DVD.

Collett, T., Riedel, M. *et al.* & The NGHP Expedition 01 Scientific Party, 2008b. *Indian National Gas Hydrate Program Expedition 01 Downhole Log Data Report*. US Geological Survey/Directorate General of Hydrocarbons. Ministry of Petroleum & Natural Gas (India), 2 DVD set.

Dallimore, S. R. & Collett, T. S. (eds) 2005. *Scientific results from the Mallik 2002 gas hydrate production research well program*. Mackenzie Delta, Northwest Territories, Canada. Geological Survey of Canada Bulletin, **585**, two CD-ROM set.

Dallimore, S. R., Uchida, T. & Collett, T. S. 1999. Scientific results from JAPEX/JNOC/GSC Mallik 2L-38 gas hydrate research well. Mackenzie Delta, Northwest Territories, Canada. *Geological Survey of Canada Bulletin*, **544**.

Dallimore, S. R., Wright, J. F. *et al.* 2008. Geologic and porous media factors affecting the 2007 production response characteristics of the JOGMEC/NRCAN/AURORA Mallik Gas Hydrate Production Research Well. *Proceedings of the 6th International Conference on Gas Hydrates* (ICGH 2008), Vancouver, British Columbia, 6–10 July.

Moridis, G. J., Collett, T. S. *et al.* 2008. Toward production from gas hydrates: assessment of resources and technology and the role of numerical simulation. *Proceedings of the 2008 SPE Unconventional Reservoirs Conference, Keystone, Colorado, 10–12 February*, SPE 114163.

Park, K.-P. 2008. Gas hydrate exploration activities in Korea. *Proceedings of the 6th International Conference on Gas Hydrates* (ICGH 2008), Vancouver, British Columbia, 6–10 July.

Smith, S., Boswell, R., Collett, T. S., Lee, M. W. & Jones, E., 2006. Alaminos Canyon Block 818: documented example of gas hydrate saturated sand in the Gulf of Mexico: Fire in the ice. *Methane Hydrate Newsletter*. US Department of Energy, Office of Fossil Energy, National Energy Technology Laboratory, Fall Issue, 12–13.

Takahashi, H. & Tsuji, Y. 2005. Multi-Well Exploration Program in 2004 for natural hydrate in the Nankai-Trough Offshore Japan. *Proceedings of 2005 Offshore Technology Conference, Houston, Texas, 2–3 May* (OTC17162).

Wu, N., Yang, S., Zhang, H., Liang, J., Wang, H., Su, X. & Fu, S. 2008. Preliminary discussion on gas hydrate reservoir system of Shenhu area, North Slope of South China Sea. *Proceedings of the 6th International Conference on Gas Hydrates* (ICGH 2008), 6–10 July.

Yamamoto, K. & Dallimore, S. 2008. Aurora-JOGMEC-NRCan Mallik 2006–2008. Gas Hydrate Research Project progress. DOE-NETL Fire In the Ice. *Methane Hydrate Newsletter*, Summer, 1–5. World Wide Web Address: http://www.netl.doe.gov/technologies/oil-gas/publications/Hydrates/Newsletter/HMNewsSummer08.pdf#Page=1.

King coal: restoring the monarchy by underground gasification coupled to CCS

P. L. YOUNGER, D. J. RODDY and G. GONZÁLEZ

Sir Joseph Swan Institute for Energy Research, Newcastle University, NE1 7RU, UK
(email: p.l.younger@ncl.ac.uk)

Abstract: Coal has hitherto been seen as a potential fall-back resource as hydrocarbons become depleted. However, amidst anxieties over peak oil and gas, some recent studies have painted a picture almost as gloomy about the longer term prospects for coal. Such evaluations are misleading, as they identify as reserves only that coal accessible at reasonable cost by means of conventional mining, which are increasingly modest compared with deeper-seated coal reserves amenable to underground coal gasification (UCG) using directionally drilled boreholes from the surface. Significantly for the petroleum industry, the technological requirements for UCG are far more akin to those of oil and gas production than they are to those of deep mining. A number of projects around the world are revealing the feasibility of UCG. We highlight preliminary findings from a recent investigation of the potential for UCG in NE England, which has the longest history of conventional coal mining at industrial scale anywhere in the world. Despite this history, fully 75% of the coal resources in NE England remain in place. A significant proportion of these is likely to move to the 'reserves' register as underground gasification technology begins to be deployed. A particular attraction of UCG lies in its suitability for coupling to CCS: we can use our long-standing knowledge of the response of incumbent strata to longwall coal mining to predict substantial increases in permeability in and immediately above the voids created by gasification. As these engineered zones of high permeability will already be connected to surface power plants by the wells and pipelines used to produce synthesis gas during gasification, they represent ideal prospects for permanent sequestration of a large proportion of the carbon dioxide arising. Stored CO_2 will be kept in place by cap rocks higher up in the sequence.

Keywords: carbon dioxide, CCS, coal, gasification, *in situ*, reserves, UK, underground, wells

Most forecasters are now confident that 'Hubbert's Peak' – the global peak in oil production – will occur some time in the five years from 2017 to 2021 (e.g. Strahan 2008). The global peak in natural gas will follow a few decades thereafter. As Hubbert's Peak looms into view, the economics of petroleum production are finally shifting in favour of long-known but as yet little-exploited unconventional resources, such as the Athabascan tar sands, oil shales, shale gas and its close relative coalbed methane. Exploitation of coal itself, rather than the methane it contains, has often been identified as the best long-term prospect for maintaining global energy supplies beyond the maximum 50-year horizon offered by nuclear (Trainer 2006). However, atmospheric CO_2 emissions associated with conventional use of coal are the worst of all the fossil fuel categories, so that (outside of the developing world at least) strong resistance to increased coal use is now being experienced, due to concerns over runaway climate change. Opponents of increased use of coal rightly point to the ultimate need to move to a wholly renewable energy economy. Even in the absence of a need to combat climate change, this would still be the only option eventually, once all energy minerals have been exhausted. However, as Trainer (2006) has shown, all available renewable energy resources, exploited using all known or reasonably anticipatable conversion technologies, are insufficient to meet the extravagant demands of a modern consumer society, and will thus be very far from sufficient for a larger global population, especially if the developing countries aspire to levels of consumption already typical of the Global North. Furthermore, no-one has yet devised a renewable energy technology that does not require industrial-scale manufacturing to produce its component parts – and many of these manufacturing processes (e.g. production of steel) rely on continued use of coal. It is therefore likely that coal use will continue, and indeed expand worldwide in coming decades, not only while we develop better renewable technologies, but also while macro-economic adjustments take place (by design or disaster) to lower energy demand per capita in wasteful consumer societies.

Many commentators have complacently assumed that coal reserves are sufficient to support several centuries of exploitation without exhaustion. More recently, concerns have been expressed about some of the assumptions on which the more optimistic projections have been based (e.g. Strahan 2008). However, the pessimistic downgrading of likely reserve estimates is based on a fundamental assumption which cannot be left unchallenged: namely, that coal will only be exploited using conventional mining, which has depth limits of about 1500 m (because of severe floor heave issues), and is usually not economic below depths of a few hundred metres. If the possibility of underground coal gasification (UCG) is taken into account, the world's usable coal reserves are increased by a factor of at least three (McCracken 2008), and this increase will be achieved without the risk to life and limb posed by deep mining. Coupled with carbon capture and storage (CCS), UCG also offers a highly attractive carbon management option whereby most of the CO_2 arising can be permanently sequestered back in deep subsurface voids produced by the UCG process itself. If implemented at commercial scale, UCG with CCS (UCG–CCS) offers mankind's last, best 'bridging technology' to the inevitable renewable energy economy of the future.

A brief history of UCG

Although briefly postulated by Lord Kelvin in the late 19th century, the first UCG experiments were actually carried out by Sir William Ramsay (the discoverer of the noble gases), in 1912 in County Durham, NE England. Although these experiments were successful, further progress was halted by the First World War. Relatively low prices for oil and gas meant that for many years there was limited interest in further development of the technology. Nevertheless, a number of countries around the world have developed UCG operations. Most notably these have included the former Soviet Union and China, where commercial-scale operations have been conducted. Feasibility studies or trial operations have also been conducted in Australia, the USA, Spain, South Africa, India and the UK. The Underground Coal Gasification Partnership (UCGP) has estimated that around 20×10^9 m^3 of syngas have been produced to date from UCG activities across the world, equivalent to about 15×10^6 tonnes of coal. To date, the largest power generation plant based on UCG is a 100 MW steam turbine plant at Angren in Uzbekistan.

The focus of early trials such as those carried out in the former Soviet Union from 1930 and in China was on finding ways of controlling the underground gasification process, developing effective ways of drilling the injection and production wells, and of linking the two wells as a precursor to gasifying the coal. Most of this work can only be done at a significant scale in real coal seams, so trials tended to be expensive. In time the trials became bigger and longer in duration up to a point where it became sensible to build power generation facilities to run off the syngas being produced from an ongoing programme of coal seam gasification. Europe started to look at UCG during the 1950s and 1960s. Meanwhile, trials in the USA developed new variants on UCG technology and also explored more carefully the potential environmental impacts (Burton 2007). During the 1970s and 1980s, 32 separate tests were carried out along with a large supporting development programme. The main centre of interest was the Powder River Basin in Wyoming. One outcome from all of this was the development of a new control technique called 'moveable injection' (Friedmann 2008). Some work was also done on routes to chemicals production from syngas as an alternative to power generation. India started looking at UCG in the 1980s.

By the early 1990s, UCG was considered to be largely technically proven, but unfortunately that coincided with the start of the era of low-cost natural gas. Consequently, interest in UCG diminished and much of the development activity stalled except in China and also in Europe, where UCG came to be seen as an alternative to mining in respect of deeper, thinner coal seams. Between 1992 and 1998, trials were carried out at depths in excess of 500 m by a European consortium (UK, Spain and Belgium) on a site at Teruel in Spain (DTI 1999). These trials demonstrated the efficacy of a new technology termed 'CRIP' (controlled retractable injection point) in which the nozzle releasing the steam and oxygen into the coal is gradually drawn back out of a horizontal stretch of borehole as the coal surrounding it is gasified. The trials demonstrated that UCG in deep seams is feasible with minimal environmental impact at surface level. They also found that the gas produced had a calorific value similar to that achievable with surface-level gasification of coal, and showed that, at the higher operating pressure involved, there were significant volumes of methane produced in addition to the normal syngas components. Following on from these trials, a review of the feasibility of UCG in the UK was carried out, leading to proposals for a trial under the Firth of Forth in Scotland (DTI 2006).

The costs of performing UCG using a CRIP are dominated by the costs of geological exploration and drilling. However, parallel developments in the oil and gas industry over the last few decades have led to the opportunity to use effective forms of directional drilling (first used in the Spanish trial) to access coal for UCG. Many countries with indigenous coal resources are now re-examining the opportunities offered by UCG, including South Africa and Australia. The driving forces include security of energy supply, the cost of syngas relative to natural gas and crude oil, and the option of using the syngas for power generation in a combined-cycle power plant.

To date there are no examples of integrated UCG–CCS projects being constructed anywhere in the world. This is because storage of CO_2 as a supercritical fluid (i.e. a fluid with the density of a liquid but the compressibility, viscosity and diffusivity of a gas) requires depths of at least 750 m (and probably more than 800 m for impure CO_2 recovered from flue gas), whereas UCG projects around the world to date have targeted coal seams which are shallower than this (≤ 600 m).

The UCG process

UCG involves gasifying coal *in situ* by means of directionally drilled wells, using drilling technology developed by the oil and gas industry. The sequence of events involved in UCG is shown in Figure 1. Gasification means 'partial oxidation', and in the case of coal, about 80% of the original calorific value of the solid coal will be present in the resultant gas. Gasification is achieved by an exothermic reaction, which is initiated by reaction with hot steam and oxygen introduced via injection boreholes. As the operator controls the availability of oxygen, so the degree of oxidation is under the operator's control. The resultant hot gas mixture – known as synthesis gas or 'syngas' – contains hydrogen, methane and carbon monoxide, all of which have significant calorific value. Depending on precise gasification conditions, varying proportions of CO_2 and hydrogen sulphide may also be present in the syngas, although hydrogen sulphide is mobilized to a far lesser degree than in conventional coal combustion (NCC 2008). The precise proportions of the various component gases in any particular syngas mixture is a function of depth (since gasification is more efficient at high pressure), oxygen injection rate and coal seam quality. Examples of typical UCG syngas compositions from a variety of coals are reported by Galli *et al.* (1983), Pirard *et al.* (2000), Perkins & Sahajwalla (2006), Khadse *et al.* (2007) and Yang (2008). These sources reveal component gas fractions in the following ranges:

- H_2, 11–35%;
- CO, 2–16%;
- CH_4, 1–8%;
- CO_2, 12–28%;
- H_2S, 0.03–3.5%.

The syngas is drawn to the surface via neighbouring production boreholes, whence it can be transported by pipeline for use in a wide range of applications, such as driving turbines to generate electricity or for manufacturing products ranging from plastics to gas and liquid transport fuels. Pre- and/or post-combustion cleanup to minimize emissions of SO_x and NO_x is typically not required for UCG applications, due to the paucity of H_2S and NH_3 in the raw syngas (NH_3 is usually entirely absent because of the strongly exothermic nature of the nitrogen oxidation reaction, which at high temperatures and pressures favours the persistence of nitrogen gas). Gaseous emissions of toxic metals are also generally negligible, as the ash present in the coal remains below ground, and largely avoids fusion (NCC 2008). Given that most UCG processes are oxygen fuelled, CO_2 and water vapour are the only gaseous exhaust streams produced after gasification, thus making separation and capture of the CO_2 relatively simple and cheap. The process is, therefore, particularly compatible with CCS.

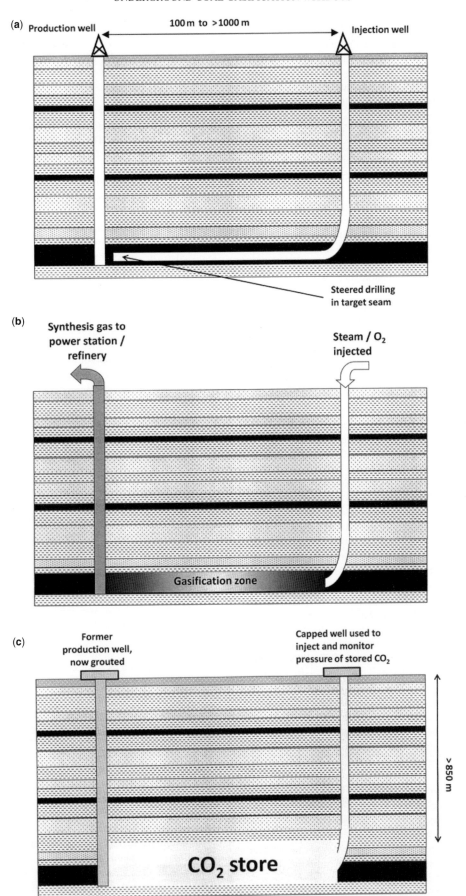

Fig. 1. Three stages in UCG–CCS. (**a**) Directional drilling of injection and production wells; the two wells do not coalesce, but a small pillar of coal is left between them which is amongst the first coal to gasify. (**b**) Creation of void by gasification to produce syngas; the gasification zone develops from left to right, progressively consuming coal closer and closer to the injection well as the controlled injection point is retracted from the end of the lateral bore. (**c**) Sealing of injected CO_2 in goaf produced by collapse of void in former gasification zone.

Prospects for CO_2 storage in UCG goaf

The UCG process creates voids deep underground following gasification of the coal. These voids will inevitably collapse, just as voids produced by longwall coal mining do, leaving high-permeability zones of artificial breccias – known as 'goaf' (from the Welsh word *ogof*, meaning a cave) – which are almost invariably isolated from the surface by low-permeability superincumbent strata (Younger *et al.* 2002). Where UCG has taken place at depths in excess of about 700–800 m, storage of CO_2 in these artificial high-permeability zones is a very attractive proposition. A combined UCG–CCS project could achieve a reduction in CO_2 emissions of as much as 85% compared with conventional coal-fired power generation. Such a project therefore offers a very attractive solution and is the only process yet devised that offers integrated energy recovery from coal and storage of CO_2 at the same site. In principle, UCG–CCS can also sit happily alongside some other CCS approaches: where CO_2 collection and transmission pipelines can be linked together, new degrees of freedom for carbon management emerge (Roddy 2008).

Subsurface injection of gases is being successfully accomplished worldwide for different purposes and in different scenarios. This includes oil and gas operations, temporary storage and permanent disposal. As some examples of this, since the 1970s, the oil industry has been practising enhanced oil recovery (EOR), which involves the injection of CO_2 into the oil reservoir, and more recently enhanced gas recovery (EGR) for gas reservoirs, including coalbed methane systems (e.g. Ross *et al.* 2009). For almost 100 years natural gas storage in salt caverns has been practised to allow supply flexibility against a fluctuating demand, and acid gas has been injected underground since the 1990s as waste in Canada.

With regard to CO_2 geological sequestration, the Intergovernmental Panel on Climate Change (Metz *et al.* 2005) proposed the following main scenarios for underground storage of CO_2: active and depleted oil and gas fields, deep saline aquifers, deep unmineable coal seams and (marginally) caverns or basalts. Based on the expected storage capacity and current experience, most of the efforts in research and all of the commercial-scale operations have been directed to storage in oil and gas operations, depleted hydrocarbon fields and associated deep saline aquifers. That is the case with Sleipner, Weyburn, In-Salah and more recently, Snohvit. Their individual annual injection rates are in a range of $0.7–2 \times 10^6$ tonnes of CO_2 and their total storage will amount to $17–20 \times 10^6$ tonnes of CO_2 each. Injection into deep unmineable coal seams has been tested in laboratory and field, with disparate results. The Recopol project in Poland found major problems in the injection of the CO_2 due to the plasticization and swelling of coal when the CO_2 is adsorbed in the coal matrix and displaces the methane. However, one option that has not been widely considered yet and could be of great interest due to a combination of economic and technical aspects is the storage of the CO_2 in the voids created by UCG.

The prospects for carbon sequestration in a UCG operation arise from a serendipitous association of a source of CO_2 and a viable long-term storage site. As with the other major CCS options, UCG–CCS takes place in a sedimentary basin with specific geological features that are particularly appropriate for geological storage. The general requirements of a site for carbon geological storage are (Metz *et al.* 2005):

- proximity to a source of CO_2, to guarantee the supply of CO_2 and improve the economics of the operation by avoiding long transportation routes;
- injectivity – the formation needs a high enough permeability to allow the injection of the fluid;
- storage capacity – sufficient to store the CO_2 produced during the plant lifetime;
- containment – some trapping mechanism has to guarantee the permanence of the CO_2 store for a considerable amount of time, *c.* 1000 years.

In addition to the generic site requirements, it is important to note the effect of the characteristics of the CO_2 stream in the constraints set on the storage site. The first of the four requirements is fully achieved by the UCG–CCS configuration. The plant and CO_2 injection infrastructures, geological and geophysical studies will have already been developed for the UCG operation when the time comes for CCS. Although capture is the main component of the cost of CCS (70–80%), the cost reduction in the remaining 20–30% is very significant.

For the other three requirements, extensive experience and knowledge in underground coal mining, especially caving methods of mining (longwall and its derivatives), provides the insights for the assessment and preliminary prediction of CO_2 storage capability. In a longwall panel, all of the coal is progressively removed from a rectangular area, and the roof is allowed to collapse, forming goaf. Typical longwall panels are usually about 1 km long, 150–250 m wide and 1–3 m high. Although there are different possible layouts for a UCG operation, one configuration is a chamber with a length of 500–600 m, 30–40 m wide and with the height equal to the thickness of the coal seam. A longitudinal pillar would separate the gasification chambers. A gasification chamber of the above dimensions corresponds to the 'shortwall' variant of collapse-based coal mining, a configuration which is well understood, as it is widely used in conventional mining to achieve rapid face movement with minimal disturbance of overlying aquifers.

Regarding the second requirement, injectivity, the absolute rock permeability is the most significant parameter to consider, the injection pressure is the second most significant parameter, and porosity is not as relevant. Thus, the injected amount varies almost linearly with permeability, while a 20% increase in injection pressure results in a 50% increase in injected CO_2 and a 100% increase in porosity shows only a 7% increase in injected CO_2. It is also important to note that the existence of a high-permeability area (or 'sweet zone') close to the injection point and a high contrast with the regional permeability acts as an enhancing factor for injectivity (Law & Bachu 1996).

Permeability changes in the surrounding strata of underground coal mining operations have been thoroughly studied in the past to inform engineering design for safe operations which avoid inducing large water inflows or gas outbursts (e.g. Younger *et al.* 2002; Esterhuizen & Karacan 2005). The process of goaf formation behind a longwall shearer track induces the formation of four distinct 'layers' above the cavity (Younger & Adams 1999):

- A caved zone, with broken blocks that have come off the roof – this is the broken material referred to as 'goaf'. The zone extends vertically to between three and six times the coal seam thickness. The final permeability of this zone will depend on the grade of re-compaction of the goaf. In the case of longitudinal pillars along the cavity, these would help to decrease the compaction, resulting in a higher permeability of the goaf. Direct measurements of saturated goaf are rather rare, but reported values are in the range of 1–20 Da (Younger *et al.* 2002), while values inferred from the hydrological behaviour of large systems of flooded panels range up to several hundred Da.
- A fractured zone with continuous fractures, joint opening and low stress. It extends to between 15 and 60 times the extraction height. Water and gas can drain directly to the void, as

Table 1. Comparison of typical permeabilities (in Da) of oil/gas reservoirs (and thus of deep saline aquifers) (after Levorsen 1967) and goaf formed by total collapse of extracted voids in coal (from literature sources and modelling results collated by Younger & Adams 1999)

Permeabilities (Da) of oil and gas reservoirs (and of deep saline aquifers)		Permeabilities (Da) of goaf from longwall mining (analogues for those formed by UCG)	
Poor reservoir	0.001–0.01	With mudstone roof strata	1–10
Good reservoir	0.01–0.1	With thinly interbedded silt–sand–mud–stone roof strata	10–50
Excellent reservoir	0.1–1	With strong sandstone/limestone roof strata	20–500

permeability in this zone can be up to 40 times the original permeability.
- A bending zone where horizontal bed separation and joint opening takes place, increasing horizontal hydraulic conductivity. This can extend to 60 m ahead of the longwall face.
- A zone of intact rock, often subject to compression beneath a final carapace of mildly extensionally disturbed rock, at or below the ground surface.

In longwall mining under the North Sea, changes in permeability of three orders of magnitude due to mining have been reported in the fractured zones above goaf (Neate & Whittaker 1979). The usual figures assumed are an increase of 100 mDa in vertical permeability and 50 mDa in horizontal. Predictions with improved modelling techniques show increases of up to 35 times in vertical permeability and 1000–2000 in horizontal permeability up to 50 m above the mining void (Guo et al. 2009).

In contrast, the permeabilities of oil and gas reservoirs (and thus of deep saline aquifers generally) have long been known to range from about 0.001 up to 1 Da (Levorsen 1967). As summarized in Table 1, UCG goaf and the relaxed roof strata above this will typically have permeabilities one to three orders of magnitude greater than the high end of this range. Therefore, it can be concluded that the UCG goaf and the zones of enhanced permeability above them represent a highly attractive prospect for CO_2 injection. UCG voids can be expected to cool rapidly due to goaf formation and groundwater ingress, so that even rapid subsequent CO_2 injection would be into pre-quenched zones. However, there are still some gaps in knowledge on the thermal effects on the overlying strata of the cavity, the presence of ashes and coal which can swell, and the effect of the injection pressure if it takes place before collapse.

The third requirement to consider is the storage capacity. Obviously, a remarkable advantage of UCG is the creation of a void that was not previously present. However, a rough estimation shows that the volume needed at 800 m depth to store the CO_2 produced from the syngas can be four or five times the volume occupied by the extracted coal. As with depleted hydrocarbon fields and deep saline aquifers, the storage capacity will depend on the specific storage, that is, the compressibility of the strata without exceeding the fracturing limit of the rock. Regulations for subsurface injection of waste gases in Alberta set the injection pressure limit at 90% of the rock fracturing pressure (Law & Bachu 1996).

The last requirement of a site for CO_2 geological sequestration is that it can provide containment for a considerable period of (say) 1000 years. The trapping mechanisms have been described in detail for deep saline aquifers and depleted hydrocarbon fields (Hitchon et al. 1999; Metz et al. 2005; Bradshaw et al. 2007). First of all a caprock is required which acts as a structural seal for the buoyant supercritical CO_2. Then, with increasing temporal and spatial scale, the rest of the mechanisms start to work: hydrodynamic trapping, residual gas dissolution in formation water and mineral precipitation. All these processes will occur according to the physical and chemical characteristics of the site and the CO_2 stream. In addition, in the case of UCG, the organic and inorganic by-products will certainly be mobilized with the CO_2 due to its solvent power, which is a key consideration in planning to avoid any pollution of adjoining aquifers in the event of contaminant migration. This will affect the chemical reactions in the water and the rock. It is also possible that the CO_2 is adsorbed in the coal matrix due to its higher affinity compared with other elements. In such a case, any resulting methane emissions would need to be factored into the overall greenhouse gas sequestration calculation.

As with permeability issues, the experience of underground coal mining is helpful in the quantification of containment in a coal basin. For more than a century, underground coal mining has been carried out under the sea in NE England using longwall methods, with workings extending as far as 14 km from shore; yet no incident of sea water in-rush ever occurred (Bičer 1987). Coal mining regulations in Britain state that the minimum distance from the seabed to mine works is 105 m, 60 m of which has to be within Coal Measures strata. Given that the viscosity of water is higher than that of supercritical CO_2, further comfort might be sought. In a more closely analogous case, Whittles et al. (2006) simulated the leakage of methane (with a viscosity of 1.75×10^{-5} Ns m^{-2}) into the mine workings in a UK colliery. They found that potential sources beyond 20 m of the working face would not leak into the roadway due to the reduced permeability resulting from increased confining stress.

In conclusion, it can be expected that the trapping mechanisms described in the literature will work in the case of UCG, and that specific sites which meet the requirements can provide the necessary confinement. However, caution has to be exercised, as there is no experience of the large scale that CO_2 storage would imply. The cumulative effects of multiseam extraction could also be important.

Another critical aspect which influences the mechanisms and requirements for the CO_2 storage site is the characteristic of the CO_2 stream to be injected. Anthropogenic CO_2 contains impurities which depend on the combustion process and the capture method. Some of these impurities are H_2O, SO_2, NO, H_2S, O_2, CH_4, HCN, Ar, N_2, H_2 and particulates (Anheden et al. 2005) and they will affect the thermodynamics (density, viscosity, critical point) compared with pure CO_2 (Li et al. 2009). In general, the presence of impurities decreases the critical temperature and increases the critical pressure (Seevam et al. 2008) at which CO_2 enters its supercritical state – which is essential for geological storage without further reaction. A stream emanating from a post-combustion process shows the smallest difference compared with pure CO_2, but in the case of pre-combustion or oxyfuel processes, the supercritical pressure can reach 83 or 93 bar while the critical temperature decreases to 29 or 27 °C respectively (Seevam et al. 2008). For typical northern Europe conditions, this would imply minimum depths in the order of 800 m for CCS to work (it could be deeper than 900 m if an oxyfuel combustion flue gas is to be stored).

Remaining technical challenges for UCG–CCS implementation

UCG–CCS faces the same challenges as any other comparable industrial operation, in relation to issues such as pollution prevention (e.g. WS Atkins Consultants Ltd 2004; Liu *et al.* 2007), landscape protection, nature conservation, etc. As UCG involves total extraction of coal in certain areas, it has the potential to induce land subsidence in a manner similar to longwall coal mining, albeit at the depths appropriate for UCG–CCS, surface lowering would be barely detectable. Long-established procedures for dealing with coal mining subsidence (e.g. NCB 1975) can be readily transferred across to UCG–CCS operations.

Nevertheless, a number of open questions remain, which will only be settled as UCG–CCS is more widely implemented at full scale and refined in the light of experience. These include the following issues:

- optimal layouts for injection and production boreholes, and strategies for managing simultaneous multiple-seam UCG–CCS development;
- borehole integrity, especially in view of the extremes of heat which are experienced by production boreholes, and the extremes of chilling which can be expected during CO_2 injection;
- the feasibility of adapting existing offshore platform technology to accommodate the full range of high-temperature operations required for UCG–CCS implementation;
- lack of precise process control in comparison to conventional (surface) gasifiers – it is not possible to fully predict or control rates of groundwater ingress, heterogeneity in reactant availability within gasification zones, or the precise limits and rates of expansion of voids and the concomitant formation of goaf and, because of this, and because of the lack of any equivalent of stockpiling fuel, it is difficult to get UCG to proceed as a steady-state process (in terms of both gas yields and heating values) unless a large number of auxiliary boreholes are installed (cf. NCC 2008);
- the response of strata to coal removal under non-isothermal conditions – in conventional coal mining, thermal effects can be ignored in geomechanical modelling, but it is unlikely that this approach will remain valid in situations of extreme temperature gradients as expected during UCG.

Active research on all of these topics is currently under way at a number of centres worldwide, and early indications are that all of these difficulties are surmountable.

The economic case for UCG–CCS

As with all commodities, the economic case for UCG–CCS amounts to a balance of credits and debits. On the credit side, UCG–CCS offers a low-cost route to emissions reduction; the cost is lower than for surface gasification plants because there is no need to mine, store or transport coal, there are no solid residues to dispose of, and there is no need to purchase a gasifier; it converts an abundant natural resource into a secure, economic supply of gas; it enables stranded coal resources (e.g. deep or offshore) to be converted into commercial reserves; there is a range of potential end uses and markets, for example, power generation, heating, synthetic fuels, chemicals and hydrogen; it is largely immune to crude oil price swings (unlike conventional coal mining which relies on diesel-fuelled equipment and transportation); it is cheaper than natural gas for power generation; and finally, as has been explained in greater detail above, UCG creates conditions for deep geological storage of CO_2 which are orders of magnitude more favourable than natural saline aquifers or depleted hydrocarbon reservoirs.

On the debit side of the balance sheet for UCG–CCS are: technical and commercial uncertainties (e.g. lack of economies of scale) since the technology has not yet been widely deployed; syngas production rates and composition are variable compared with pipeline-delivered natural gas; open-cast coal mining (where acceptable) is cheaper; ground subsidence must be managed, and there is some risk of aquifer contamination; trials and prospective site evaluation are expensive; there can be significant costs in transporting the syngas to the point of use; carbon capture technology for high-temperature pre-combustion applications is not yet a commercial reality (though capture post-combustion is); and planning approval processes (not only for UCG but also for CCS) are still under development in the majority of countries.

Wide-scale proliferation of commercial UCG projects has thus far been inhibited by the availability of comparatively cheap supplies of crude oil and natural gas. Looking at 2008 data, against a natural gas price (in the USA) of $9 per million Btu, raw syngas can be produced via UCG in the onshore USA for $1.8 per million Btu based on air gasification (Green 2008). Using oxygen-blown UCG in onshore Europe, the cost of syngas becomes $3.8 per million Btu. These figures are now sufficiently low for UCG to look commercially attractive whenever oil and gas prices are reasonably high.

The economic case for UCG syngas displacing natural gas or coal for power generation is relatively straightforward. Alternative uses such as conversion of syngas into liquid fuels, chemical intermediates or hydrogen are more difficult because, whilst the added value is well known (and much higher than for power generation), there is a tighter requirement for syngas cleanup. Technologies for cleaning up UCG syngas to chemical feedstock standard are still under development and so the costs are less well known. There are several such projects under way at present, which should help elucidate the figures in due course.

Other factors beyond straight economics then come into play to tip the balance. The main considerations are: CO_2 emissions and climate change; air quality and power station emissions; a desire for some protection against volatile and rising oil and gas prices; and considerations of national energy and fuel security in a politically unstable world – after all, the distribution of coal around the world is different from the distribution of oil and gas.

There is an additional factor to consider – and one that is difficult to place a value on in economic terms. More than 5000 deaths a year occur in the coal mines of China: four deaths for every million tonnes of coal mined. In Ukraine, the death rate is even worse: seven deaths per million tonnes. To put these figures into perspective, the last time death rates in UK coal mines were as high as they currently are in China was back in the 1920s; in the case of Ukraine the parallel figures occur way back in the 1880s. Much of this mining is linked to energy provision in support of the manufacture of goods for export to Western countries. UCG could provide an ethically acceptable way of enabling this economically driven low-cost manufacturing activity to continue.

Approaches to environmental risk assessment for UCG–CCS

An Environmental Risk Assessment (ERA) should answer four questions: what can occur that causes adverse consequences; what is the probability of occurrence of these consequences; how severe can they be; and how can they be reduced? This last issue is often referred to as risk management, though it differs from the more general risk management which takes into account economic and social considerations. The steps in performing an ERA to address these questions are:

- hazard identification, to reveal the contaminants and adverse situations that can be expected;
- exposure assessment, to describe the intensity, frequency and duration of exposure, routes of exposure and the nature of the population exposed;
- effect assessment, to describe the response of the receptors;
- risk characterization, to provide an estimate of the likelihood for adverse impacts, with endpoints definition and a qualitative or quantitative approach;
- risk management, which includes monitoring and mitigation options.

The ERA is a critical aspect in the development of both UCG and CCS and so far the two processes have been addressed separately. However, they share substantial common ground. Usually, when the ERAs are undertaken for UCG and for CCS, each is split into surface operations and what happens underground. The handling of gases like syngas from UCG or CO_2 is common practice in industry, and environmental, health and safety and other standards and regulations are well established. Also the engineering design is controlled, resulting in low failure rates. For example, CO_2 pipeline failures in the USA in the period of 1990–2001 had a frequency of 3.2×10^{-4} incidents per kilometre per year and the frequency of oil well blowouts in the Gulf of Mexico and the North Sea was of 10^{-4} incidents per well per year (IEA 2004). However, when it comes to the underground dimension, lack of previous experience and large uncertainties in geology and in hydrogeological, chemical and geomechanical behaviour appear.

The environmental risks associated with UCG were one of the reasons why UCG was not further developed in the 1970s and 1980s in the USA. Despite its indisputable benefits, initial trials at shallow depth (less than 200 m below surface), which produced groundwater contamination and even severe surface subsidence (as in Hoe Creek), discouraged pursuit of new experiments. Nevertheless, more recent projects (like Chinchilla in Australia) have proved that with good site selection and operation control, groundwater contamination can be avoided.

The main risks to be considered in UCG are groundwater depletion, groundwater contamination, gas leakage and subsidence. Set against these, its environmental advantages compared with conventional coal mining and surface combustion include the elimination of coal transport, stockpiles and much of the disturbance at surface, low dust and noise levels, the absence of health and safety concerns relating to underground workers, the avoidance of ash handling at power stations, and the virtual elimination of SO_x and NO_x emissions. Most of the contaminants produced in coal gasification are included in List I of the Water Framework Directive (2000/60/EC), which forbids release into a water body. Consequently, for a UCG operation to be permitted in the EU, any potential water contamination would almost certainly have to be restricted to water which had been previously classified as Permanently Unusable. The potential contaminants have been described and are well known (Humenick & Mattox 1978; Liu et al. 2007). They include organic compounds (phenols, benzene, PAHs and heterocyclics) and inorganics (calcium, sodium, sulphate, bicarbonate, aluminium, arsenic, boron, iron, zinc, selenium, hydroxide and uranium).

The approach that is proposed for environmental assessment of UCG is a risk-based approach with a source–pathway–receptor scheme (WS Atkins Consultants 2004). For CO_2 storage, the main risks identified are divided into three groups (Chadwick et al. 2008): leakage, dissolution in formation water and displacement. At a local scale, leakage into the atmosphere or the shallow subsurface can cause asphyxiation to animals or humans, or affect plants and underground ecosystems. If the leakage is offshore, it can affect the living organisms in the water column and the seabed and interfere with other legitimate uses of the sea. It is also important to make a distinction between sudden large releases and continuous small ones. Large releases from the storage site, however, are not expected unless there is a secondary accumulation close to the surface. The CO_2 injection could also initiate the mobilization of methane which is potentially explosive. On a global scale, the leakage of CO_2 or methane would hinder the ultimate aim of the sequestration, which is to reduce the concentration of greenhouse gases in the atmosphere. If a leak contains some other contaminants, they can pose an additional threat if exposure times and concentrations are toxic or carcinogenic. Regarding the risks resulting from dissolution in other fluids, the variation in pH of water caused by CO_2 can lead to the mobilization of metals. As CO_2 is a very good solvent, it can also transport other organic contaminants and contaminate potable water. The displacement of the CO_2 plume can induce seismicity or ground heave or subsidence. The brines pushed away can contact and contaminate potable aquifers or damage other mineral or energetic resources. It is also important to note that, although a single specific site should not pose a high risk, there is a cumulative effect as the number of storage sites increases in response to the large-scale opportunity for global warming mitigation.

The coupling of UCG–CCS alters the hazards and the risks inherent in UCG or CCS on its own (Burton 2007). On the one hand, the operation takes place at a much greater depth than conventional UCG (so that the conditions for CO_2 in its supercritical state are met). This certainly decreases the risk of potable aquifer contamination and of subsidence effects on the surface. In addition, the physical response of the surrounding coal to the CO_2 injection can help reduce the migration pathways by swelling. On the other hand, the pressurization of the cavity with the injection of the CO_2 can increase the risk of fracture propagation. Under these circumstances, the organic and inorganic by-products of gasification are forced out of the reaction chamber as the CO_2 is injected and pressurized in the void. The transport of organic and inorganic contaminants dissolved in the CO_2 through a fractured and porous medium is an area that has not been studied yet. Their concentration in the CO_2 plume and the changes in flow and chemical reactions between the CO_2, tars, ash, coal, brine and the formation rocks are unknown.

Although it is desirable to have a quantitative risk assessment, this has not always been possible in the ERAs performed for CO_2 storage projects. The reason for this is the lack of available data at this stage. Consequently, some studies have been performed using a deterministic analysis based on well proven numerical simulation tools and examining different scenarios (Espie et al. 2005). There have been various projects around the world to perform the ERA of CO_2 geological storage using several methodologies to characterize risk (Damen et al. 2006). From these projects, it seems likely that the most effective approach for UCG–CCS risk characterization in the future will have to combine deterministic numerical models with probabilistic analysis. These models will have to be able to couple thermal, geomechanical, transport and chemical processes.

Monitoring is especially important at the local scale, to be able to detect and mitigate any threat to safety. On a global scale, monitoring is important for dealing with issues related to credits for emissions reduction. Monitoring of the UCG operation is based on measurement of temperature, pressure and mass balance, as well as water chemistry and water level in monitoring wells. In the CO_2 injection phase, CO_2 stream injection rates, composition, temperature and pressure have to be monitored. The migration of the CO_2 can be checked with periodic sampling of air, water and soil, with pressure and logs in wells, with CO_2 flux chambers or using eddy covariance and indirect techniques such as geophysics or remote sensing. Since sealed wells represent the preferential pathway for leakage, their integrity has to be assured. That can be done with cement bond logs (Metz *et al.* 2005; Burton 2007; Chadwick *et al.* 2008). The mitigation options in the event of leakage are: recapping of leaking wells, reducing injection pressure, stopping injection, sealing the fracture or transferring the CO_2 (Anonymous 2006).

Northeast England case study

Despite the long history of industrial coal mining in the region (starting in 1585), huge reserves of coal remain in NE England. Only about 25% of the total coal resources have been extracted. There is coal in abandoned mines that is technically mineable – perhaps in excess of 500×10^6 tonnes. Some of this coal lies under land: most under the sea. However, because the mines have not been maintained there would be expensive problems to overcome. Typically it is found that roadways deteriorate, and electrical and mechanical equipment is destroyed as a result of flooding once water pumping operations cease. There is also the possibility of trapped water leading to sudden in-rushes, which would present a risk to personnel in the event of underground mining being re-established (Younger *et al.* 2002).

Then there is the coal that lies at depths considered uneconomic for conventional mining – both under the land and under the sea. This is particularly attractive for UCG when linked to CCS for the reasons given above, and could easily exceed another 500×10^6 tonnes. Previous estimates for the UK have suggested that between 7 and 16×10^9 tonnes of coal suitable for UCG could be available – and that ignores all coal below a depth of 1200 m (Green 2008).

Project Ramsay (named after Sir William Ramsay) was established to assess the opportunity for UCG–CCS in NE England. As part of the project, specialists were commissioned to undertake a thorough review of all available data to determine the quantity and accessibility of coal suitable for UCG and for UCG–CCS in nearshore areas of the NE coast. The study examined available data from a number of sources including the Coal Authority, the British Geological Survey and BERR as well as data held by others, including Newcastle University.

Suitability of the area was considered for UCG and UCG–CCS taking into account coal seam thickness, depth of cover between the top of target coal seams and the seabed where relevant, permeability of the pertinent strata and, where relevant, stand-off distances from old workings. For UCG a depth of 100 m or greater was used, and for UCG–CCS the minimum depth was increased to 800 m to achieve the storage pressures necessary for CO_2 in its supercritical state.

The project considered both nearshore coal seams (<2 km) and offshore coal seams (up to 10 km) at a few locations. The primary difference in approach between nearshore areas and further offshore is in the ability to reach the coal reserves from a wholly shore-based enterprise using directional drilling v. the need to utilize offshore rigs. Cost analysis has shown that there is not a significant cost advantage in one approach over the other for the coal reserves under consideration. The initial high cost of offshore rigs is broadly offset by the more expensive long-reach drilling costs associated with a nearshore project. The project found some very interesting coal seams and concluded that previous estimates of UCG-compatible coal resource had been conservative.

However, generating syngas from coal is only part of Project Ramsay: the region also provides ready energy and chemicals markets for syngas and its derivatives, and therefore offers a genuine prospect for a commercial UCG–CCS operation. Geography is important: these markets need to be sufficiently close to the chosen UCG base to be serviceable economically. The siting of a UCG production operation in NE England allows ready access to the process industry markets on Teesside for syngas and for derived gas products of methane and hydrogen. Equally, there are a number of existing power users and potential new investments in power generation plant at a scale that could make syngas a viable fuel. These options were all reviewed as part of the feasibility study.

From its inception, Project Ramsay always considered CCS as being an essential element of a successful UCG project. Consequently, detailed consideration has been given to those coal targets that are at sufficient depth to provide the option for CO_2 storage and where significant revenues can be generated by providing a long-term storage site for CO_2. Note, however, that CO_2 is also generated in large quantities by the same process and power industries that provide a potential market for the syngas and its derivatives. Increasingly there is a business opportunity in CO_2 collection, transmission and storage. There is therefore the option of extending the envelope to take in CO_2 from other industrial sources and offer additional storage capacity. The voids created through the UCG process in deep coal seams provide a storage option for CO_2 whether that CO_2 was produced through use of UCG syngas or from other industrial activities.

The study took account of specific local factors such as: the location of the most suitable coal seams relative to existing power plants and potential new power plants; the existence of pipeline corridors; the location of the most suitable coal seams relative to large industrial users of syngas and hydrogen; the potential for linking into other sources of CO_2 and CO_2 collection systems; the potential for connecting the UCG facility to the proposed new CO_2 pipeline linking the Eston Grange IGCC/CCS plant to storage locations under the North Sea, and so on.

The broad conclusions were that: previous estimates for UCG-compatible coal had been conservative; there are coal seams that appear to be usable for CO_2 storage following UCG; and some of the end uses for syngas are potentially attractive. The most attractive options in financial terms are (1) to sell syngas, take back captured CO_2 and store it for a fee, and (2) to sell decarbonized hydrogen and methane. It was concluded that a project could be done in three phases, ramping up the scale over time in order to minimize technical risk and investor exposure. Such a project could deliver a profit before year 10 and therefore might warrant follow-on discussion on a more commercial basis with interested parties. If developed on a broader scale, it could act as a source of investment funds for the renewable energy sector and thereby go beyond the ambition of being a bridging technology on the road to a sustainable energy future.

Conclusion

Estimates of the total global coal resource are in the order of thousands of billions of tonnes, whereas figures usually quoted for accessible coal reserves are typically tens of billions of tonnes. There is thus a huge gap between reserves and resources. UCG offers the tantalizing prospect of closing that gap quite considerably. If the UCG opportunity can be linked successfully to emerging CCS technology, then the implications for addressing the twin challenges of climate change and finite fossil fuel reserves is truly game-changing. There are particular attractions in developing a 'self-contained' solution whereby clean use of coal and CO_2 sequestration are combined in the same location without a need for material transfer. From a different perspective, there is an attraction in extending the envelope to include syngas export and CO_2 import/export. The former opens up the prospect of linking into lucrative opportunities beyond the power generation sector: the latter offers contingency plans on a number of fronts. The main environmental challenges lie in guarding against (1) aquifer contamination which can impact on potable water supplies and (2) land subsidence. Well established techniques exist for dealing with both of these. The main economic challenges relate to the up-front costs associated with evaluating specific sites from a commercial perspective and from an environmental perspective, largely because of the drilling costs associated with characterizing deep coal seams. The pace is tending to be set by those countries and regions that are blessed with significant coal resources and are concerned about the greenhouse gas emissions agenda.

We are grateful for funding from One NorthEast, HSBC Partnership for Environmental Innovation and Newcastle University. Project management by Stephen Price is greatly appreciated. Principal contractors on the North East case study were IMC Geophysics, PB Power and UCG Engineering Ltd.

References

Anheden, M., Andersson, A. *et al.* 2005. CO_2 quality requirement for a system with CO_2 capture, transport and storage. *In*: *Greenhouse Gas Control Technologies 7*. Elsevier Science, Oxford, 2559–2564.

Anonymous. 2006. Risk assessment and management framework for CO_2 sequestration in sub-seabed geological structures. *28th Consultative Meeting of Contracting Parties under the London Convention and 1st Meeting of Contracting Parties under the London Protocol*.

Bičer, N. 1987. *The inflow of water into mine workings*. PhD Thesis, Department of Mining Engineering, Newcastle University.

Bradshaw, J., Bachu, S. *et al.* 2007. CO_2 storage capacity estimation: issues and development of standards. *International Journal of Greenhouse Gas Control*, **1**, 62–68.

Burton, E. 2007. *Best Practice in Underground Coal Gasification*. Lawrence Livermore National Laboratories, contract no. W-7405-Eng-48.

Chadwick, A., Arts, R. *et al.* 2008. *Best Practice for the Storage of CO_2 in Saline Aquifers: Observations and Guidelines from the SACS and CO2STORE Projects*. British Geological Survey, Keyworth.

Damen, K., Faaij, A. *et al.* 2006. Health, safety and environmental risks of underground CO2 storage – overview of mechanisms and current knowledge. *Climatic Change*, **74**, 289–318.

DTI. 1999. *Underground Coal Gasification – Joint European Field Trial in Spain*. Project Summary, DTI Publication no. 017-1999.

DTI. 2006. *The feasibility of UCG under the Firth of Forth*. Project Summary 382, DTI Publication, URN 06/885.

Espie, T., Rubin, E. S. *et al.* 2005. Understanding risk for the long term storage of CO_2 in geologic formations. *In*: *Greenhouse Gas Control Technologies 7*. Elsevier Science, Oxford, 1277–1282.

Esterhuizen, G. S. & Karacan, C. O. 2005. Development of numerical models to investigate permeability changes and gas emissions around longwall mining panels. *40th US Rock Mechanics Symposium*, 25–29 June 2005, Anchorage.

Friedmann, J. 2008. North American prospects for UCG in a carbon constrained world. *Proceedings of the International Coal Conference*, October, Pittsburgh, PA.

Galli, R. D., Jones, G. E. & Kiviat, F. E. 1983. Simple field test model for underground coal gasification. *Industrial Engineering and Chemical Process Design and Development*, **22**, 538–544.

Green, M. 2008. The future prospects for underground coal gasification. *Proceedings of the International Coal Conference*, October, Pittsburgh, PA.

Guo, H., Adhikary, D. P. *et al.* 2009. Simulation of mine water inflow and gas emission during longwall mining. *Rock Mechanics and Rock Engineering* **42**, 25–51.

Hitchon, B., Gunter, W. D. *et al.* 1999. Sedimentary basins and greenhouse gases: a serendipitous association. *Energy Conversion and Management*, **40**, 825–843.

Humenick, M. J. & Mattox, C. F. 1978. Groundwater pollutants from underground coal gasification. *Water Research*, **12**, 463–469.

IEA. 2004. *Risk Assessment Workshop*. Report PH4/31. London, IEA GHG R&D Programme.

Khadse, A., Qayyumi, M., Mahajani, S. & Aghalayam, P. 2007. Underground coal gasification: a new clean coal utilization technique for India. *Energy*, **32**, 2061–2071.

Law, D. H. S. & Bachu, S. 1996. Hydrogeological and numerical analysis of CO_2 disposal in deep aquifers in the Alberta sedimentary basin. *Energy Conversion and Management*, **37**, 1167–1174.

Levorsen, A. I. 1967. *Geology of Petroleum*. W H Freeman, New York.

Li, H., Yan, J. *et al.* 2009. Impurity impacts on the purification process in oxy-fuel combustion based CO_2 capture and storage system. *Applied Energy*, **86**, 202–213.

Liu, S.-Q., Li, J.-G. *et al.* 2007. Groundwater pollution from underground coal gasification. *Journal of China University of Mining and Technology*, **17**, 467–472.

McCracken, R. 2008. Mining without mines: UCG. *Energy Economist*, **31**, 216.

Metz, B., Davidson, O., de Coninck, H. C., Loos, M. & Meyer, L. A. (eds) 2005. *IPCC Special Report on Carbon Dioxide Capture and Storage*. Prepared by Working Group III of the Intergovernmental Panel on Climate Change. Cambridge University Press, Cambridge.

NCB. 1975. *The Subsidence Engineers' Handbook*. National Coal Board, London.

NCC. 2008. *Underground Coal Gasification*. National Coal Council, Washington, DC. World Wide Web Address: www.nationalcoalcouncil.org/Documents/Underground_Coal_Gasification.pdf (accessed 13 January 2010).

Neate, C. J. & Whittaker, B. N. 1979. Influence of proximity of longwall mining on strata permeability and ground water. *U.S. Symposium on Rock Mechanics*. Austin, TX, 3rd June, 217–224.

Perkins, G. & Sahajwalla, V. 2006. A numerical study of the effects of operating conditions and coal properties on cavity growth in underground coal gasification. *Energy & Fuels*, **20**, 596–608.

Pirard, J. P., Brasseur, A., Coëme, A., Mostade, M. & Pirlot, P. 2000. Results of the tracer tests during the El Tremedal underground coal gasification at great depth. *Fuel*, **79**, 471–478.

Roddy, D. J. 2008. Linking underground coal gasification to CCS and the downstream opportunities. *Proceedings of the International Coal Conference*, October, Pittsburgh, PA.

Ross, H. E., Hagin, P. & Zoback, M. D. 2009. CO_2 storage and enhanced coalbed methane recovery: reservoir characterization and fluid flow simulations of the Big George Coal, Powder River Basin, Wyoming, USA. *International Journal of Greenhouse Gas Control*, **3**, 773–786.

Seevam, P. N., Race, J. M. *et al.* 2008. Transporting the next generation of CO_2 for carbon capture and storage: the impact of impurities on supercritical CO_2 pipelines. *7th International Pipeline Conference*, Calgary, Alberta. ASME, New York.

Strahan, D. 2008. Coaled comfort. *Geoscientist*, March, 18–21.

Trainer, T. 2006. *Renewable Energy Cannot Sustain A Consumer Society*. Springer, Heidelberg.

Whittles, D. N., Lowndes, I. S. *et al.* 2006. Influence of geotechnical factors on gas flow experienced in a UK longwall coal mine panel. *International Journal of Rock Mechanics and Mining Sciences*, **43**, 369–387.

WS Atkins Consultants Ltd. 2004. *Review of Environmental Issues of Underground Coal Gasification – Best Practice Guide*. DTI Cleaner Coal Technology Transfer Programme. DTI, Birmingham.

Yang, L. 2008. Coal properties and system operating parameters for underground coal gasification. *Energy Sources, Part A: Recovery, Utilization, and Environmental Effects*, **30**, 516–528.

Younger, P. L. & Adams, R. 1999. *Predicting Mine Water Rebound*. R&D Technical Report W179. Environment Agency, Bristol, UK.

Younger, P. L., Banwart, S. A. & Hedin, R. S. 2002. *Mine Water: Hydrology, Pollution, Remediation*. Kluwer Academic Publishers, Dordrecht.

Geological storage of carbon dioxide: an emerging opportunity

W. J. SENIOR,[1] J. D. KANTOROWICZ[2] and I. W. WRIGHT[2]

[1]*Senior CCS Solutions Ltd, Mt Pleasant Cottage, Sly Corner, Bucks HP16 9LD, UK (e-mail: bill@senior-ccs.co.uk)*
[2]*BP Alternative Energy, Chertsey Road, Sunbury-on-Thames, Middlesex TW16 7LN, UK*

Abstract: Concerns about climate change and the need to stabilize atmospheric CO_2 concentrations are driving the development of a lower carbon future. Within this context, carbon dioxide capture and storage (CCS) is gaining momentum as a large-scale option to reduce greenhouse gas emissions. This paper reviews the rationale and potential scale of CCS, the status of geological storage options and lessons from the operating In Salah project. CCS is expected to have applications in the oil and gas industry, and other industries, particularly the coal and power sectors. CO_2-enhanced oil recovery, depleted oil and gas fields and saline formations are considered the most important geological storage options. Experience with geological storage is being gained at the In Salah project in Algeria. Operating since 2004, it is the world's first industrial-scale project storing CO_2 in the water leg of a gas reservoir. A key challenge for wider deployment is for geological storage to be accepted as a safe and effective option, providing long-term CO_2 containment, with high integrity. This has several associated technical and regulatory challenges, including site characterization and selection, geological and well integrity risk assessment, performance prediction, the design of appropriate monitoring schemes and handling the closure and post-closure phases. The petroleum industry has the capabilities and know-how to deploy CCS and to manage the associated risks. This lends confidence that CCS will be a viable option and that deployment will help enable a low-carbon future.

Keywords: carbon dioxide, capture, storage, sequestration

Carbon dioxide capture and storage (CCS) consists of the separation, capture and compression of CO_2 from industrial and power sources, transportation to storage sites and injection of CO_2 into geological formations enabling long-term isolation from the atmosphere (IPCC 2005). This paper reviews the rationale for CCS, the major geological storage options and the operating In Salah CO_2 storage project in Algeria, and summarizes the main challenges for geological storage.

Rationale for carbon capture and storage

Concerns about climate change are driving the need to stabilize atmospheric CO_2 concentrations and to develop a low-carbon energy future. Within this context, CCS is gaining momentum as an important large-scale option to reduce greenhouse gas emissions. By enabling 80–90% reductions in CO_2 emissions from fossil fuels from large industrial sources, CCS can be considered a win–win solution to enable global energy supply, energy security and mitigation of climate change.

Many components of the technologies have been demonstrated at scale and are ready for commercial deployment. Further innovation and technology development should broaden application and reduce costs. CCS can be economically competitive with other low-carbon options such as renewables, especially where used with enhanced oil recovery (EOR). A key challenge for CCS is the urgent need to move ahead with large-scale integrated demonstration projects and commercial deployment, which will require policy and financial support. This requirement is similar to other sustainable low-carbon energy options such as wind power. CCS implementation will also involve creation of value chains which are technically and commercially complex, although analogous to the complex value chains that are involved in the Liquefied Natural Gas business.

The oil and gas industry will have a key role to play in CCS development and deployment. Given existing capabilities, stakeholders are looking to the industry to take a lead on transportation and storage. At the strategic level CCS can be an enabler for cleaner fossil fuels in a low-carbon world. The technology will also be required to manage oil and gas industry emissions. Early applications are expected to be high-CO_2 gas clean-up, concentrated emissions from hydrogen plants, and feedstock conversion including applications in refining, heavy oil, gas-to-liquids and coal-to-liquids, particularly to reduce the carbon footprint of heavier feed stocks. The oil and gas industry could play a leading role in geological storage, and new business models are emerging, for example providing storage solutions to third parties such as power utilities, and for decarbonized energy providers. CCS will be vital in other energy sectors, notably coal and power. Coal is the most carbon-intensive fuel, responsible for 39% of global energy-related CO_2 emissions. It is a low-cost, secure primary energy source that dominates the energy system in several leading economies, notably China and the USA. Resource holders are unlikely to forego the use of coal and thus CCS is vital to control emissions. CCS is also recognized as a key option for the power sector, which generates about 40% of world energy-related CO_2 emissions. Linked to hybrid and electric vehicles, decarbonization of power could also help address transport sector emissions.

Potential scale of CCS

Future energy and emissions models can be used to predict the timing and scale of potential CCS deployment. The scale is linked to emissions trajectories, which are dependent on the climate change objectives and requirement for stabilization of atmospheric greenhouse gas concentrations. The potential for global deployment of CCS could be up to 9000×10^6 tonnes/year of CO_2 by 2050, around 19% of projected business-as-usual CO_2 emissions (International Energy Authority 2008). These figures equate to equipping 15–17 GW of coal-fired power plant with CCS each year, implying that it will become the technology standard, and to an industry handling c. 82.5×10^6 BOE/day equivalent of CO_2, comparable to today's oil industry.

Geological storage options

There are a range of options for geological storage of CO_2 in the subsurface. It is generally assumed that CO_2 will be injected and stored in supercritical phase at depths greater than 800 m, assuming normal geothermal gradients. The single most important challenge for wider deployment is for geological storage to be accepted as safe and effective in containing CO_2 underground. The main technical issues relate to the geological integrity of the site: the long-term integrity of any wells and potential pathways creating risks of leakage to atmosphere. There are several associated technical and regulatory challenges, including site characterization and selection, geological and well integrity risk assessment, performance prediction, the design of fit-for-purpose monitoring schemes and handling the closure and post-closure phases, including the transfer of liabilities to the nation.

We consider three large-scale storage options: CO_2 EOR, depleted oil and gas fields, and saline aquifer formations, each of which is reviewed below. These are already being deployed at demonstration scale for CCS, and offer potential for commercial-scale storage. We consider other options to be lower priority. The potential for storage in coals during enhanced coal bed methane production has been piloted in USA, Canada, Poland and China; however technical and economic viability at larger scale is to be confirmed. Research stage options include basalts, salt caverns, disused mines and carbonation of ultramafic rocks.

Storage with CO_2 EOR

CO_2 has been used for EOR since the 1980s (Jarrell et al. 2002). Over 120 reservoirs use CO_2 EOR, with total worldwide production of nearly 1×10^6 BOPD, and an average 11% increase in recovery. Many current projects are in the USA, where about 45×10^6 tonnes per year of mostly natural CO_2 is transported by pipeline and injected. The technology is sufficiently well understood for full-field development to be implemented with confidence. Conceptually, the use of CO_2 to enhance recovery could be expanded into other areas of conventional oil and gas development. Firstly, CO_2 could replace natural gas in miscible gas floods. Secondly, under the right circumstances, CO_2 flooding could enhance recovery from condensate reservoirs. Finally, it is theoretically possible to use CO_2 to displace and so enhance recovery of natural gas, which is being piloted at two locations.

Deploying CCS with CO_2 EOR is an important option given the potential benefit of increased revenues. In existing CO_2 EOR projects in North America, the field operators usually purchase CO_2 at prices equivalent to $11–32/tonne CO_2 at oil prices of $15–50/barrel (Jarrell et al. 2002). Geological integrity is not a concern, although well integrity risks should be addressed. An advantage over other storage options is that CO_2 EOR is usually permissible within the petroleum licensing agreements. There is commercial experience of CCS with EOR at the Weyburn field in Canada, using CO_2 captured from lignite gasification in North Dakota and which has operated a storage demonstration and monitoring project since 2000 (Wilson & Monea 2004). EOR is being, or has been, considered in CCS project proposals in the North Sea, USA, Canada and the Middle East.

Challenges. If fiscal policy encourages storage of greenhouse gases, then CO_2 EOR projects may also benefit from carbon credits for storage. This opens up the possibility that field development may be optimized to store CO_2 rather than to maximize oil recovery. This will be a function of the relative price of carbon credits and hydrocarbons, and regulatory developments. There are no major obstacles to CCS with EOR, although a number of issues may constrain it from becoming volumetrically significant for storage. These include geography (transportation cost), impact on existing infrastructure (cost), volumetrics, timing, well integrity, storage assurance and individual field viability, which are discussed in turn below.

Geography is important because some major oil provinces are remote from sources of CO_2. Economics may not justify the pipeline cost. The potentially corrosive impact of CO_2 is a consideration for fields that have been developed with metallurgy that will be affected by produced CO_2. This has not been an issue onshore, or where associated gas contains CO_2; however it was cited by Statoil and Shell as a contributing factor at the Draugen field, offshore Norway (Provan 1992), where an offshore CO_2 EOR scheme integrated as part of a value chain to capture carbon at an onshore gas power plant was evaluated. The high cost and one-year shut-down requirement for major modifications to the offshore platform were cited as issues contributing to the unfavourable economics of EOR along with the low extra oil volumes. As a consequence the integrated project and value chain was considered uneconomic (Statoil 2007).

Volumetrics are a consideration given that the amount of CO_2 to be injected may be considerably greater than the available pore space. Larger current EOR projects such as the Rangely field in Wyoming inject up to 3×10^6 tonnes per year CO_2, only equivalent to the emissions of small coal power plants. Time-scales are important in two respects. Firstly, many EOR developments will not require a continuous stream of CO_2, rather after an initial peak, a declining rate is required as the produced CO_2 is recycled. Secondly, whilst the cessation of production is a function of oil price, some oil fields may not continue producing long enough to provide a sustainable storage option. In view of these issues, CO_2 EOR may need to be used in conjunction with other storage options. Well integrity is a key issue, which will be field-specific and which needs to consider longer term risks in a CO_2-rich environment. This is linked to storage assurance: if carbon credits require monitoring and verification to assure storage, then the associated costs need to be offset by the value of the credits to provide sufficient incentive for EOR with storage. The CO_2 produced from additional oil production v. baseline scenario and other options will also need to be factored in. Finally the viability of CO_2 EOR is field specific, and is a key consideration for integrated projects.

Storage in depleted oil and gas fields

Oil and gas fields at or close to the end of production are potentially attractive sites for CO_2 storage. They benefit from data and knowledge from the hydrocarbon life cycle, which should provide high confidence in geological storage integrity and potential capacity, together with opportunities to re-use wells and facilities. This method is being demonstrated in the Dutch sector of the North Sea, where Gaz de France is injecting 20,000/tonnes per year into a mature Rotliegendes gas field in the K12 block. The potential for enhanced gas recovery is still under investigation (Van der Meer et al. 2006). Other demonstrations are taking place in Australia (Otway) and Germany (Altmark) and new projects are planned in several areas. In addition, the In Salah project described below is ultimately intended to store CO_2 in a depleted gas reservoir.

Storage in depleted fields will not necessarily involve any attempt to enhance recovery in late field life. It is principally an opportunity to take advantage of a well understood reservoir with facilities and wells that could have their life extended, deferring decommissioning. Clearly, this option will have easier application in depleted fields, as opposed to fields with strong natural aquifers, or that have been water-flooded to maintain pressure.

Challenges. There are no insurmountable technical obstacles to storage in depleted fields. Geographic constraints are the limitations of field distribution, in a similar way to EOR, although more widespread. Perhaps the simplest is size: the 200 000 SCF/day CO_2 output from a 500 MW coal fired power station equates to $>1 \times 10^{12}$ SCF over 20 years. Larger fields may be able to store such volumes; smaller fields could be part of a portfolio. Timing may be a constraint as the optimal timing would involve cessation of production coinciding with the start of injection for storage. If not, the value of early curtailment (reserve loss) would have to be outweighed by the value of carbon credits. Conversely, if cessation of production is too early, then costs may be incurred to maintain facilities to await CO_2 supply. The availability of the fields for storage may also conflict with other possible uses. In some areas depleted gas fields have commercial value as natural gas storage sites.

Well integrity is a key issue for both depleted fields and EOR projects. This needs to take account of the risk of leakage in a CO_2-rich environment over the longer term from both pre-existing and newly drilled wells. With older fields that are approaching the end of design life there may also be costs associated with maintenance and integrity issues. It is also possible that the metallurgy used on old infrastructure and wells may be incompatible with impurities in the CO_2 stream, or not resistant to the corrosion that would result if water entered the system. The field history should also allow inferences regarding direct and indirect communication between the reservoir formations and the surrounding formations, upon which to base any risk assessment on the impact of injection on other users of pore space in the area. The latter is important as the surrounding aquifer system may have repressurized a field over time, reducing the amount of pore space that can readily be used for storage. With higher pressure, injection will cause a pressure pulse and may displace fluids out of the field. Compartmentalization of the reservoir and artificial pressure maintenance of the reservoir during production will limit the amount of CO_2 that can be injected without increasing the pressure to unacceptable levels.

Storage in saline formations

Formations where the pore water is unsuitable for human use or agriculture are expected to be a major geological storage option because of their occurrence in all the world's sedimentary basins. Their existence outside oil and gas provinces will be important in providing storage options in regions such as Europe, China and NE USA. Worldwide capacity is expected to be large and the Intergovernmental Panel on Climate Change (IPCC) provided a lower estimate of 1000 Gtonnes CO_2, with an upper estimate being an order of magnitude greater (IPCC 2005). However the methodology and validity of the estimates has been the subject of considerable debate (Bradshaw *et al.* 2007). This will require further development as storage is commercialized with similarities to oil and gas reserves reporting.

The option is at the early commercial stage overall. There is operating experience of CO_2 injection and storage in a saline aquifer at the commercial Sleipner project in the Norwegian sector of the North Sea since 1996 at a scale of 1×10^6 tonnes CO_2 per year. This scheme involves injection into saline aquifers of the Tertiary Utsira Formation at a depth of around 800 m subsea (Torp & Gale 2003). There has been extensive data acquisition, monitoring and modelling of the project as a demonstration of CO_2 storage. Seismic imaging from repeat 3D seismic surveys has been particularly effective here in understanding plume migration and demonstrates effective underground containment (Holloway *et al.* 2004). Additional experience has been gained at In Salah in Algeria (see below) and from other smaller projects in USA, Germany, Japan and Australia. Scale-up is planned for the Gorgon project in Australia to c. 3×10^6 tonnes CO_2 per year.

Challenges. The single most important issue for saline formations is storage integrity, a prerequisite for stakeholders' acceptance that this is a safe option. Confidence that secure sites are selected and that injection performance and integrity can be monitored and managed over the life cycle of the project and after injection ceases will be vital. The main underlying geological considerations are the trapping mechanisms for CO_2 within the aquifers, potential leakage paths and the risk of leakage to surface. At the pore scale a range of trapping mechanisms are expected to secure the CO_2 over differing time-scales: residual saturation trapping, dissolution, buoyancy trapping, mineralization and adsorption. At the field scale different trap types are being considered: structural trapping, stratigraphic trapping and 'migration' traps. These are closely related to potential leakage mechanisms and risk, which are not detailed in this paper. The initial industry focus is on sites involving structural traps, which are considered lowest risk by analogy with oil and gas fields. Recent work has highlighted concerns about the pressure building up in the storage site, together with its potential impacts such as fracturing of the seal and reduction of storage capacity, and how best to manage pressure regimes. Another key research area is how to optimize storage which will need to bring together both subsurface processes and injection strategy in a similar manner to reservoir management. Researchers are reviewing different strategies to manage the tradeoffs with integrity, i.e. whether to inject CO_2, or alternate water and CO_2, or even to dissolve CO_2 in water at the surface. These strategies may provide opportunities to enhance trapping, and gain wider acceptance and thereby reduce monitoring requirements and duration.

In general there is less seismic and well data available for saline formations at the outset of assessment. Therefore resources and lead time may be required for seismic and drilling data acquisition. However, even after these efforts, it is also likely that there will be greater uncertainties in predicting integrity and storage performance than for hydrocarbon fields, because there may be no proof of trap integrity or production data prior to the start of injection. One of the key challenges is that there may be no proof of integrity of aquifer sites until after the injection has ceased and CO_2 has stabilized in the subsurface. This is a risk that will need to be managed, through appropriate monitoring and performance assessment.

The licensing and regulatory frameworks are being developed to provide assurance around these issues. The key elements of these include proposals for site characterization protocols, risk assessment, performance prediction, licensing frameworks, monitoring programmes, remediation options, closure and transfer of longer term liabilities to the nation. The case study presented for In Salah illustrates the importance of demonstration projects to inform these issues.

Storage experience at the In Salah project

In Salah Gas is a BP, Sonatrach and Statoil Joint Venture that operates a number of gas fields in Algeria (Fig. 1). The gas contains up to 10% CO_2, higher than the sales specification, and so 1×10^6 tonnes per year CO_2 is separated from the gas stream. Rather than venting, it is injected into the Krechba C-10 reservoir via three (1500 m long) horizontal wells at a depth of 1900 m subsurface. CO_2 injection started in August 2004, and to date 3×10^6 tonnes have been injected (of the planned 17×10^6 tonnes that will be injected over the project life). This is an industrial-scale

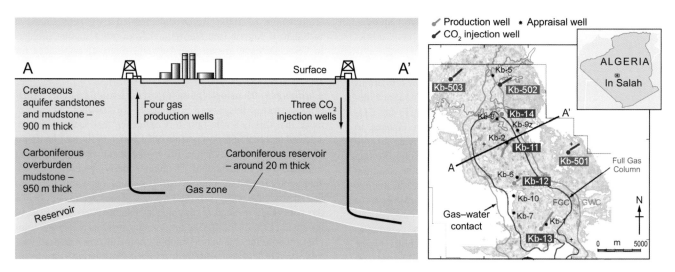

Fig. 1. Schematic cross-section through the Krechba field at In Salah, showing the general geology and well positions. Inset is a map of the field and location within Algeria.

demonstration of CO_2 geological storage in a deep saline formation and is the first such project in the world to store CO_2 in the water leg of a gas reservoir.

The C-10 Carboniferous storage reservoir in Krechba is a sandstone unit which is interpreted as a tidally influenced estuarine deposit of Tournaisian age, within an elongate NW–SE trending anticline (Riddiford et al. 2003). The reservoir is 20 m thick, naturally fractured, with porosities of 11–20% and permeability around 10 millidarcies. The reservoir sandstones are sealed by 950 m of Carboniferous mudstones, overlain unconformably by a further 900 m of Cretaceous clastics that contain a regional potable water aquifer (Fig. 1). The field is being produced by NE–SW horizontal wells along the crest of the structure, designed to intersect the expected natural fracture system.

Monitoring demonstration

Separately from the production operation, a joint industry project (JIP) was formed in 2004 to evaluate technologies for long-term monitoring of CO_2 geological storage. The project is a 5 year $30 million JIP involving the European Commission and US Department of Energy. The objective is to identify appropriate monitoring technology to deploy during 15 years of injection that will provide assurance of secure storage for the next 1000 years. This monitoring is also intended to show that underground CO_2 storage is a viable greenhouse gas mitigation option, and to support development of frameworks for the regulation of injection and monitoring for storage.

Prior to injection, a risk assessment concluded that the key risks to permanent storage were well bore integrity (within pre-existing wells) and the direction of CO_2 movement in the subsurface (depth conversion uncertainty). Available monitoring techniques were then assessed to identify specific tools to deploy to monitor the movement of CO_2 in the subsurface.

The technologies were assessed in terms of potential benefit and information gained v. cost (Fig. 2a). In some cases assessment was limited by the immaturity of the technology or not being proven in the operating conditions. These technologies had to be evaluated and tested to confirm deployment or rejection. For example, the water table is 100 m below the surface, and there is sparse surface vegetation, potentially diminishing the value of soil gas monitoring. The current and planned programme following this assessment is shown in Figure 2b. The aim of the programme is to end up with a cost-effective portfolio of technologies in the bottom left 'Deploy' quadrant. Note that technologies such as repeat 3D (4D) seismic (which are higher cost), may not be required. Satellite imagery has moved into the bottom left since the programme started as it has returned unexpectedly useful results. Technologies shown in blue are in the programme; those in pink are subject to further evaluation.

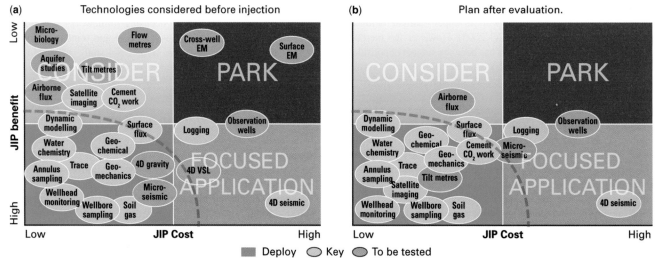

Fig. 2. A comparison of technologies considered by In Salah. (**a**) Technologies considered before injection. (**b**) Plan after evaluation.

The In Salah project receives no commercial credit for the CO_2 storage, but if the site was being regulated, then the monitoring programme would have to be agreed with the regulators and verification agencies. The dashed red line on Figure 2 depicts an envelope that should provide fit-for-purpose site-specific and cost-effective monitoring.

Satellite imagery

Calibrated INSAR satellite images after four years of injection show deformation patterns around producers and injectors, and at natural features (Mathieson et al. 2008). These are consistent with the (CO_2) re-pressurization and (natural gas) depletion history of the reservoir as well as surface erosion (Fig. 3). Natural features provide calibration that the induced deformation is being imaged correctly. These images suggest that the injected CO_2 is moving away from the wells along NW–SE trends, inferred to be the natural fracture systems, consistent with the original reservoir models. The surface is moving up to 5 mm per year in response to the reservoir level injection. Pressure changes caused by CO_2 movement have been tracked by satellite since injection commenced. The satellite imagery was recently calibrated by the arrival of CO_2 at a monitoring well 1.5 km from the KB502 injector. The sampled CO_2 contained a tracer, which was added to the CO_2 injected at KB502.

Further work is planned, including images from now higher frequency satellite passes, as well as micro-seismic and tilt-metres to provide accurate ground-level calibration, to interpret these results. This work will address how the reservoir injection pressure is transmitted to the overburden and provide a basis for determining how effectively satellite imagery can be used as a monitoring tool at this and other locations.

Conclusions

Carbon capture and geological storage is emerging as a major option to reduce CO_2 emissions from fossil fuels as part of a lower-carbon future. Progressing geological storage is part of the urgent need to move ahead with CCS. Implementing a wide range of projects is an urgent priority to gain operating experience and deeper understanding for risk and performance optimization. This will meet the key challenge for CCS: gaining public acceptance that geological storage is a safe and effective mitigation option.

The three major geological storage options, EOR, storage in depleted fields and saline aquifer formations, are ready for wider deployment at commercial scale, although learning from demonstration and deployment and further research and development will be needed in parallel.

CO_2 EOR will be an important early option as it can provide economic benefits and can reduce the cost of CCS demonstration and deployment. In future there may be a gradual transition from commercial EOR to CO_2 storage with EOR, with increasing value for carbon mitigation and the goal of maximizing CO_2 storage. Important issues to be addressed are how to maximize CO_2 storage, safe re-use of wells and infrastructure and well integrity in CO_2-rich conditions over the longer term. Similar challenges will exist for re-use of depleted oil and gas fields for CO_2 storage.

Aquifers are seen as a major long-term option, with a greater need for research and development than oil and gas fields. Further work is required in many areas, including characterization, resource assessment methodology, storage mechanisms and capacities, together with injection strategies to optimize storage efficiency and monitoring approaches.

The In Salah project shows the viability of storage, and the importance of practical experience in developing the storage option. It is a test bed for CO_2 storage which will assist in designing fit-for-purpose monitoring programmes. Projects like this will be essential to develop appropriate regulatory frameworks and policy instruments that will be essential for wider commercial deployment.

This paper is based on work by the authors at BP Alternative Energy, work with BP's partners on In Salah (Sonatrach and Statoil) and independent work by the lead author since leaving BP. We acknowledge these companies and colleagues within them.

References

Bradshaw, J., Bachu, S., Bonijoly, D., Burruss, R., Holloway, S., Christensen, N. P. & Mathiassen, O. M. 2007. CO_2 storage capacity estimation: issues and development of standards. *International Journal of Greenhouse Gas Control*, **1**, 62–68.

Holloway, S. et al. 2004, *Saline Aquifer CO_2 Storage Project (SACS) – Best Practice Manual*. IEA Greenhouse Gas & D Programme.

Intergovernmental Panel on Climate Change (IPCC). 2005. *Carbon Dioxide Capture and Storage*. IPCC Special Report. Cambridge University Press, New York.

International Energy Authority (IEA). 2008. *CO_2 Capture and Storage – a Key Carbon Abatement Option*. IEA Report. IEA, Paris.

Jarrell, P. M., Fox, C. E., Stein, M. H. & Webb, S. E. 2002. *Practical Aspects of CO_2 Flooding*. SPE Monograph **22**.

Mathieson, A., Wright, I. W., Roberts, D. M. & Ringrose, P. 2008. Satellite imaging to monitor CO_2 movement at Krechba, Algeria. In: *Proceedings of the 9th International Conference on Greenhouse Gas Control Technologies*, November, Washington, DC.

Provan, D. 1992. Draugen Oil Field, Haltenbanken Province, Offshore Norway. In: *Giant Oil and Gas Fields of the Decade 1978–1988*. American Association of Petroleum Geologists, Memoirs, **54**, 371–381.

Riddiford, F. A., Tourqui, A., Bishop, C. D., Taylor, B. & Smith, M. 2003. A cleaner development: the In Salah gas project, Algeria. In: *Proceedings of the 6th International Conference on Greenhouse Gas Control Technologies*, October, Kyoto.

Statoil, 2007. Halten CO_2 value chain: technically feasible, but not commercially viable. Press release.

Torp, T. A. & Gale, J. 2003. Demonstrating storage of CO_2 in geological reservoirs: the Sleipner and SACS projects. In: *Proceedings of the 6th International Conference on Greenhouse Gas Control Technologies*, October, Kyoto.

Van der Meer, L. G. H., Kreft, E., Geel, C. R., D'Hoore, D. & Hartman, J. 2006. Enhanced gas recovery testing in the K12-B reservoir by CO2 injection, a reservoir engineering study. In: *Proceedings of the 8th International Conference on Greenhouse Gas Control Technologies*, June, Trondheim.

Wilson, M & Monea, M. (eds) 2004. IEA GHG Weyburn CO_2 monitoring and storage project summary report 2000–2004. In: *Proceedings of the 7th International Conference on Greenhouse Gas Control Technologies*, September, Vancouver.

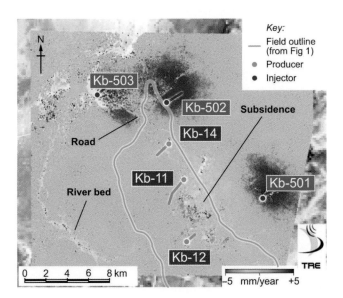

Fig. 3. INSAR satellite image of the northern end of the field showing changes in surface elevation.

History-matching flow simulations and time-lapse seismic data from the Sleipner CO_2 plume

R. A. CHADWICK and D. J. NOY

British Geological Survey, Keyworth, Nottingham NG12 5GG, UK (e-mail: rach@bgs.ac.uk)

Abstract: Since its inception in 1996, the CO_2 injection operation at Sleipner has been monitored by 3D time-lapse seismic surveys. Striking images of the CO_2 plume have been obtained showing a multi-tier feature of high reflectivity. In the medium to longer term, the topmost layer of CO_2, accumulating and migrating directly beneath the topseal, is the main determinant of storage site performance. Fortunately it is this topmost layer that can be most accurately characterized, its rate of growth quantified, and CO_2 flux arriving at the reservoir top estimated. The latter is mostly controlled by pathway flow through thin intra-reservoir mudstones. This has increased steadily with time, suggesting either that pathway transmissivities are increasing with time and/or the pathways are becoming more numerous. Detailed 3D history-matching of the topmost layer cannot easily match the observed rate of spreading. Isotropic permeabilities result in a stronger radial component than observed and a degree of anisotropic permeability, higher in a north–south direction, is possible. The main contributor to the mismatch, however, is likely to be small but significant uncertainty in the depth conversion. Irrespective of uncertainty, the observed rate of lateral migration seems to require very high permeabilities, and is, moreover, suggestive of a topseal which behaves like a 'hard' impermeable flow barrier. Detailed studies such as this will provide important constraints on longer term predictive models of plume evolution and storage performance which are key regulatory requirements.

Keywords: Sleipner, Utsira Sand, CO_2 plume, seismic monitoring, time-lapse seismic, reservoir flow simulation, history-matching

Large-scale underground storage of industrially produced carbon dioxide has the potential to play a key role in reducing man-made emissions of greenhouse gases (IPCC 2005). The CO_2 injection operation at Sleipner, in the central North Sea between the UK and Norway, commenced in 1996 and is the world's longest running industrial-scale storage project. It is also, so far, the only example of underground CO_2 storage arising as a direct response to environmental legislation (Baklid *et al.* 1996).

Carbon dioxide separated from natural gas produced at Sleipner is injected into the Utsira Sand, a major saline aquifer of late Cenozoic age (Fig. 1a). Well logs (Fig. 1b) show the Utsira Sand to have a sharply defined top and base, and gamma-ray (GR) values typically around 25 API units compared with 90 API units in the overlying caprock. GR and resistivity peaks within the sand are interpreted as thin intra-reservoir mudstones, mostly around 1 m thick (Zweigel *et al.* 2004).

The injection point is at a depth of about 1012 m below sea-level, some 200 m below the reservoir top, with around 12 million tonnes (Mt) of CO_2 currently stored. A comprehensive monitoring programme has been carried out, with multiple time-lapse 3D seismic surveys, augmented by high-resolution 2D seismic, sea-bottom gravity, controlled-source electromagnetic and seabed imaging surveys. This paper describes analysis of the time-lapse 3D seismic datasets acquired up to 2006, comprising the 1994 (baseline), 1999 (cumulative 2.35 Mt of CO_2 injected), 2001 (cumulative 4.26 Mt injected), 2002 (cumulative 4.97 Mt injected), 2004 (cumulative 6.84 Mt injected) and 2006 (cumulative 8.4 Mt injected) surveys.

The time-lapse seismic data clearly image the progressive development of the CO_2 plume, around 200 m in height and elliptical in plan view (Fig. 2). By 2006 the NNE-trending long axis of the plume spanned 3.6 km with a short axis of around 1 km. The plume forms a prominent multi-tier feature comprising a number of bright sub-horizontal reflections, interpreted as arising from discrete layers of CO_2, each up to a few metres thick, and mostly accumulating beneath the intra-reservoir mudstones (Fig. 1b). The CO_2 layers formed early in plume evolution (by 1999) and have remained identifiable ever since. The upper layers continue to spread laterally and generally increase in brightness, whereas the lower layers have stabilized in size and are growing progressively dimmer. Vertical linear features within the plume characterized by reduced reflection amplitudes and localized velocity pushdown are interpreted as 'chimneys' of moderate or high CO_2 saturation (Chadwick *et al.* 2004). The most prominent of these is located roughly above the injection point (Fig. 2), and is interpreted as the main conduit for CO_2 upward transport through the reservoir and the main feeder of the laterally spreading thin layers. Within the reservoir overburden, there is no evidence of systematic changes in seismic signature, indicating that CO_2 is being contained within the storage reservoir.

Quantitative analysis (Chadwick *et al.* 2004, 2005) has shown that, while the seismic images are consistent with the known injected amounts of CO_2, they do not provide a unique verification of injected mass. Significant uncertainties remain, particularly regarding seismic properties of CO_2-saturated rock, and the fine-scale distribution of dispersed CO_2 in between the reflective layers. Similarly, reservoir flow simulations (Lindeberg *et al.* 2001; Chadwick *et al.* 2008) have reproduced the current observed development of the CO_2 plume as a multi-tier layered structure. However, because the structural geometry of the sealing intra-reservoir mudstones is not precisely known, simulated layer thicknesses are not tightly constrained and have not yet been robustly matched to thicknesses obtained directly from the seismic data.

The objective of this paper is to look at key aspects of the seismic data that constrain models of CO_2 migration through the reservoir, and to assess whether flow processes in the reservoir are understood to the extent that predictions and simulations of future, longer-term, plume behaviour are likely to be robust.

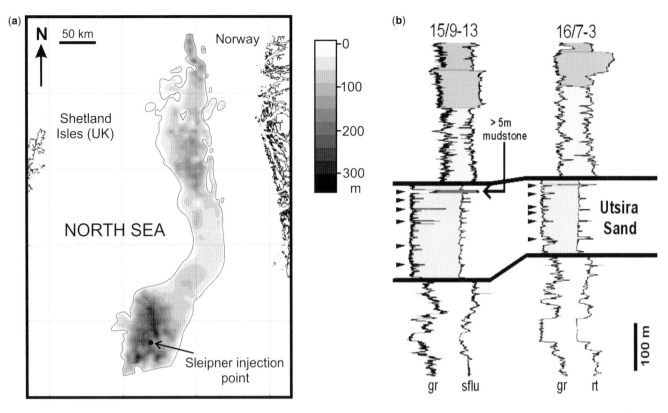

Fig. 1. (**a**) Location of the Sleipner injection operation showing distribution and thickness of the Utsira Sand. (**b**) Geophysical logs through the Utsira Sand showing GR and resistivity peaks corresponding to thin intra-reservoir mudstones. Note thicker (>5 m) mudstone near reservoir top. Wells 15/9-13 and 16/7-3 lie around 1 km WSW and 12 km ENE of the injection point respectively. Log types: gr, gamma ray; rt, deep resistivity; sflu, spherical-focused resistivity.

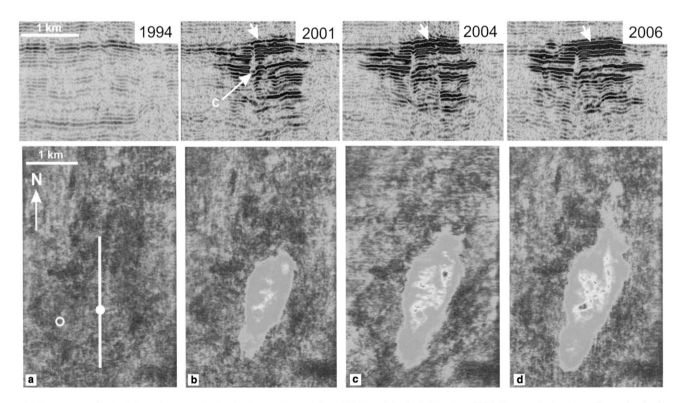

Fig. 2. Seismic images of the Sleipner plume showing its development from 1994 (pre-injection) through to 2006. Top panels show a north–south seismic section through the plume; bottom panels show plan views of the plume displayed as total reflection amplitude. C (on 2001 section) denotes the main feeder chimney in the plume. White arrow denotes topmost CO_2 layer. Location of north–south seismic section, injection point (solid circle) and well 15/9-13 (open circle) also shown.

Analysis of the topmost CO_2 layer

Although up to nine individual CO_2 layers have been identified in the plume (Chadwick et al. 2004), from the viewpoint of determining medium- to long-term storage site performance, understanding the topmost layer is of paramount importance. Long-term flow simulations of the Sleipner plume (Erik Lindeberg, pers. comm. 2007) show that within a few decades of ceasing injection most of the CO_2 will have migrated to the top of the reservoir as a buoyant fluid phase. Here, trapped beneath the undulating caprock, it will migrate laterally over significant distances. Over the succeeding centuries, dissolution of CO_2 at the base of the migrating layer, and convective sinking of CO_2 in solution, will gradually 'erode' the CO_2 layer and progressively reduce its mobility. Ultimately, after a few thousand years, most of the CO_2 will be fixed in a gravitationally stable dissolved form. Risk assessment therefore requires robust prediction of how the CO_2 will migrate in the free phase for a protracted period after site closure.

Fortunately, of all the layers in the plume, the topmost is imaged most clearly by the time-lapse monitoring data. Currently its lateral spread can be tracked in detail and its progressive increase in thickness and volume quantified with a degree of accuracy. The latter parameters allow the total upward flux of CO_2 through the reservoir to be calculated, which provides insights into reservoir flow behaviour and how this changes with time.

Quantitative analysis of the 2006 time-lapse processing ensemble, comprising the 1994, 2001, 2004 and 2006 surveys, forms the basis of this paper. Reflection amplitude changes at the top of the Utsira Sand correspond to the development of the topmost CO_2 layer and show how it has grown through time (Fig. 3). A north-trending linear prolongation of the layer is particularly prominent and corresponds to CO_2 migrating northwards along a linear ridge at the reservoir top (Fig. 3d). The CO_2–water contacts (CWC) correspond to the outer edges of the CO_2 layers, defined as the outer limit of detectable amplitude change.

Amplitude maps of the topmost layer from the full time-lapse dataset (Fig. 4) show how CO_2 first reached the reservoir top in 1999 (just prior to the first time-lapse repeat survey), as two small separate accumulations. By 2001 these had coalesced into a single accumulation which continued to expand thereafter.

Layer thicknesses from seismic amplitudes

Earlier work (Arts et al. 2003; Chadwick et al. 2005) has shown that layer reflectivity for the most part follows a thin-layer tuning relationship, with reflection amplitudes directly related to layer thicknesses. Rock physics properties were calculated for a range of CO_2 saturations in the Utsira Sand using the Gassman fluid substitution equation (Mavko et al. 2003), calibrated by P- and S-wave velocity and density data from well logs (Fig. 5a). For uniform fluid mixing (likely in the high saturation CO_2 layers), substitution of water by CO_2 reduces V_p from 2050 to around 1420 m s^{-1} for CO_2 saturations above 0.2. A synthetic wedge model was constructed using the calculated seismic properties and from this an amplitude–thickness relationship was derived (Fig. 5b). The amplitude–thickness relationship for the observed data was then obtained by scaling the maximum tuning amplitude of the synthetic data to the maximum seismic amplitudes observed in the plume. Two simplifying assumptions were made in this analysis. Firstly the CO_2 wedge model incorporated a uniformly high CO_2 saturation. In terms of the calculated seismic response this is acceptable given the characteristic Gassman fluid response with rapid velocity decrease even at low CO_2 saturations and the fact that low CO_2 saturations are likely to be restricted to the very lowermost part of the layer (Fig. 5c). Secondly, rock properties were assumed to be laterally uniform. This is also reasonable given the absence of data to the contrary and the very large changes induced by the CO_2 compared with likely lithological effects.

The amplitude–thickness relationship was used to transform observed amplitudes to layer thicknesses (Fig. 6a–c). CO_2 saturations within the layer were then calculated from a capillary pressure–saturation relationship determined by centrifuge experiments on core material from the Utsira Sand (Erik Lindeberg, pers. comm.). The capillary pressure, p_c, between the formation brine and the injected CO_2 will cause the CO_2 saturation, S_{CO_2}, to vary with height, h, in each CO_2 layer. The saturation can be computed by balancing the buoyancy, $\Delta \rho \cdot g \cdot h$, with the capillary pressure.

In SI units:

$$\Delta \rho \cdot g \cdot h = p_c = 810.35(1 - S_{CO_2})^{-0.948}. \tag{1}$$

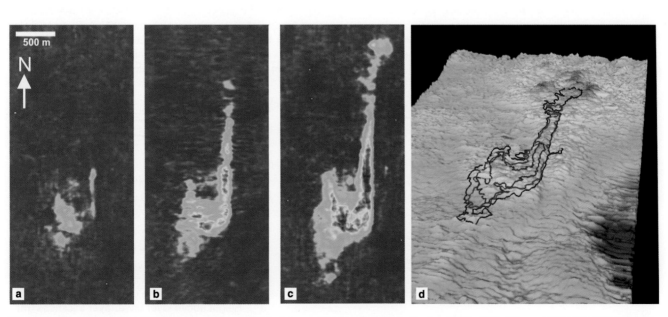

Fig. 3. Growth of the topmost CO_2 layer mapped as amplitude changes at the top of the Utsira Sand: (**a**) 2001–1994 difference; (**b**) 2004–1994 difference; (**c**) 2006–1994 difference; (**d**) 3D view (looking north) of the top Utsira Sand surface (mapped on the baseline 1994 dataset) showing the CWC in 2001 (red), 2004 (purple) and 2006 (blue). RMS amplitudes were extracted from a 4 ms window centred on the trough event corresponding to the negative acoustic impedance change at the top of the CO_2 layer.

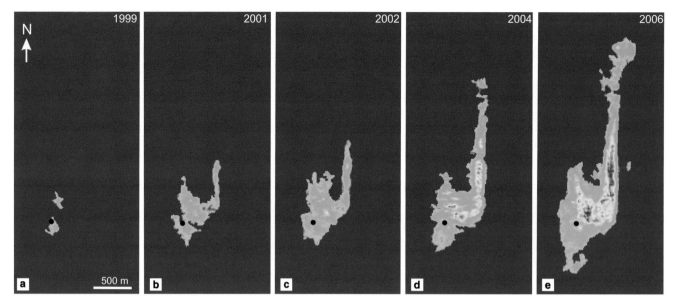

Fig. 4. Amplitude maps of the topmost layer through time. Black spot denotes the location of the injection point for reference. (Note however that the injection point lies roughly 200 m beneath the topmost layer so does not directly control its growth.)

The variation of S_{CO_2} with h was thereby computed and also the average value of S_{CO_2} for a range of layer thicknesses (Fig. 5c). This relationship was used to compute average saturations for the topmost layer (Fig. 6d–f).

Layer thicknesses from structural analysis

An alternative, wholly independent way of obtaining layer thicknesses is by topographic analysis of the reservoir top (Fig. 7). In map view (Fig. 3d), the outer limit of CO_2 reflectivity at this level corresponds to the CWC. The 3D form of the fluid contact at the base of the topmost layer was constructed by fitting a smooth sub-horizontal surface through the elevations of the CWC (note this surface is not truly horizontal because it is constructed in two-way time, not depth, and also because of significant fluid dynamic effects – see below). The two-way time thickness of the topmost layer was then calculated by subtracting the elevations of the reservoir top from the corresponding elevations of the CWC. These were then converted to true (depth) thicknesses (Fig. 8) using a simple velocity model for the overburden, based on well velocity data. Finally, as above, layer saturations were calculated (equation 1).

Evaluation of the thickness maps produced by the two methods shows that structurally derived layer thicknesses have a clear correlation with layer reflectivity (Fig. 3), supporting the general assumption of thin layer tuning. However, it is also clear that the structurally derived thicknesses are in places significantly larger than the thicknesses derived from reflectivity. The main reason for this is probably the non-linearity of the amplitude–thickness relationship whereby amplitude change is quite small for layer thicknesses above about 5 m (Fig. 5b). Application of the tuning assumption will generally tend to progressively underestimate layer thicknesses around and, especially, above the tuning thickness. Other uncertainties in the reflectivity-derived thicknesses include estimation of the maximum tuning amplitude and also the layer velocity. The structurally derived thicknesses are therefore considered to be generally the more reliable. On the other hand, it is clear that layer reflectivity extends into localized areas where the structurally derived layer does not, for example in the southern part of the layer (compare Figs 6 & 8). This is because the structural analysis does not allow CO_2 to be present wherever the constructed CWC is shallower than the topseal (Fig. 7). In reality the CO_2 layer is a dynamic entity with significant horizontal flow and perhaps supplied from below by a number of feeders, some of which may lie

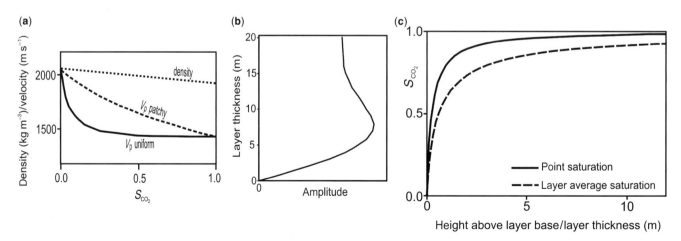

Fig. 5. (a) Variation of seismic velocity in the Utsira Sand with CO_2 saturation, calculated for uniform (Reuss bound) and patchy (Hill bound) fluid mixing. (b) Amplitude–thickness relationship from a synthetic wedge model of a CO_2-saturated layer within water-saturated sand. (c) Layer thickness–saturation function based on capillary pressure data from Utsira core.

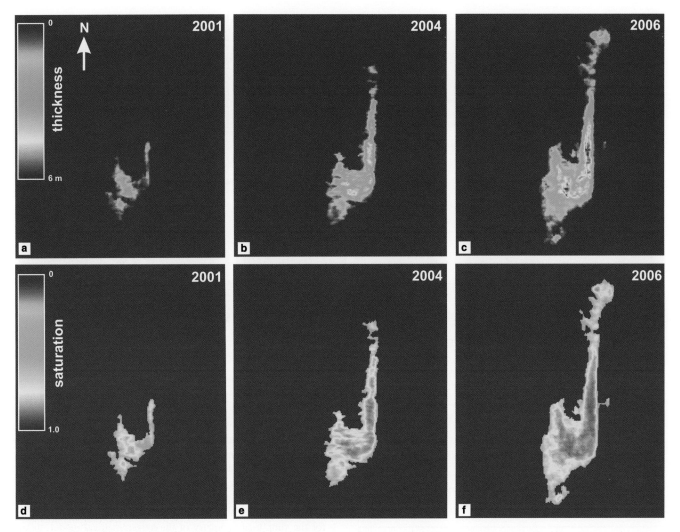

Fig. 6. Topmost CO_2 layer properties: (a–c) thicknesses derived from reflection amplitudes; (d–f) average saturations determined from capillary pressure data.

beneath topographic depressions in the topseal. These processes will lead to general layer thickening and allow locally 'overdeepened' areas of caprock topography to be underlain by CO_2.

The volume of CO_2 within the topmost layer was computed for the two methods of thickness determination using calculated average saturations and an assumed mean sand porosity of 0.38 (Table 1). As discussed above, volumes derived from the structural analysis are rather higher than those derived from the reflectivity, and, overall are considered to be more reliable.

From the topmost layer volumes, the rate at which CO_2 has arrived at the top of the reservoir can be estimated, bearing in mind that CO_2 first arrived at the reservoir top in 1999, just prior to the first seismic repeat survey when just two small patches of CO_2 had accumulated (Fig. 4a). Taking the structurally derived volumes, an estimated 1.07×10^5 m^3 of CO_2 had arrived at the reservoir top between 1999 and the time of the 2001 survey, an average flux of c. 135 m^3 per day. Between the 2001 and 2004 surveys c. 3.22×10^5 m^3 of CO_2 arrived at the reservoir top, an average flux of c. 316 m^3 per day. Between the 2004 and 2006 surveys a further c. 3.76×10^5 m^3 of CO_2 arrived at the reservoir top, averaging c. 540 m^3 per day. This marked increase in flux arriving at the reservoir top occurred despite a relatively uniform rate of CO_2 input at the injection point.

2D flow simulation of the whole plume flux

A simple TOUGH2 flow model was set up to simulate the observed growth of the CO_2 plume (Fig. 9). The modelled reservoir geometry was axisymmetrical (radial), with a flat reservoir top and horizontal intra-reservoir mudstones. The latter were all 1 m thick except for the uppermost, which was 5 m thick (Fig. 1b). These deliberately minimal assumptions are justified by the lack of detailed information on internal reservoir structure.

The vertical spacing of the mudstones was set to match the relative spacing of the seismic reflections (Fig. 2). The temperature of the reservoir top was set at 29 °C and the injection point at 36 °C, values in line with available information (see discussion below). Sand properties were based on laboratory core measurements. Mudstones were assumed to be uniformly semi-permeable with relative permeability to CO_2 increasing with CO_2 saturation. The flow properties of the mudstones, permeability and capillary

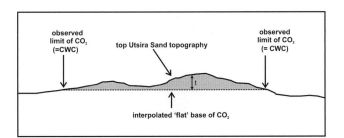

Fig. 7. Schematic cross-section through the topmost CO_2 layer, showing top reservoir topography and a nominally flat CWC. Thickness of the CO_2 layer (t) is the difference in elevation between the top reservoir and the CWC.

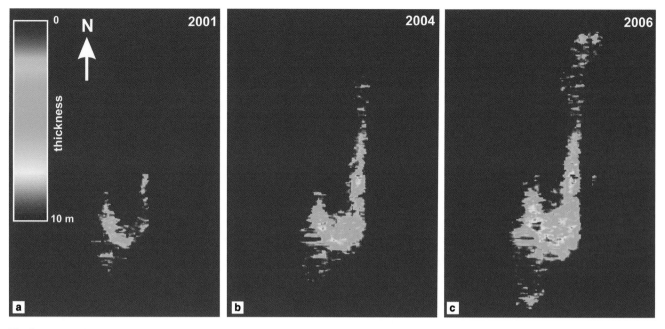

Fig. 8. Topmost CO$_2$ layer thicknesses derived from structural analysis.

Table 1. Observed volumes of CO$_2$ in the topmost layer computed from the two different methods

Date	Volume from amplitudes (m^3)	Volume from structural analysis (m^3)
1999	Very small	Very small
2002	66 684	107 380
2004	320 219	429 341
2006	615 876	805 409

entry pressure (Table 2) are the dominant factors in determining the rate of upward migration through the reservoir, and were adjusted such that the simulation matched the observed arrival of CO$_2$ at the top of the reservoir in 1999. This occurred just prior to the first repeat survey (Fig. 4a) and is a very well constrained key calibrating observation.

The flow simulation (Fig. 9) reproduces the key properties of the CO$_2$ plume, in particular the characteristic tiered 'Christmas-tree' profile caused by the accumulation of thin layers of CO$_2$ spreading laterally beneath each intra-reservoir mudstone. It also predicts that the uppermost CO$_2$ layer shows very rapid lateral spread (Fig. 9). This is driven by a progressive increase in the modelled rate of CO$_2$ arriving at the topmost layer from 1999 onwards (Fig. 10). This is despite the relatively uniform CO$_2$ input at the injection point and is largely due to increased effective permeabilities within the central part of the plume as the intra-reservoir mudstones become progressively more saturated with CO$_2$ and relative permeability to CO$_2$ increases.

Fluxes predicted by the flow simulation are, therefore, consistent with observed fluxes derived from the seismic data, showing a steady increase in the rate of CO$_2$ flowing into the topmost layer. The simulation, however, assumes Darcy flow through semi-permeable intra-reservoir mudstones with capillary entry pressures much lower than measured values from the caprock (Table 2), which laboratory testing shows to be a very low permeability capillary seal (Harrington *et al.* 2010). These modified values are required for the CO$_2$ to reach the top of the reservoir in the observed time-span (immediately prior to the 1999 survey). Even allowing for the simplified geometry of the axisymmetrical model, this is paradoxical, because the well logs (Fig. 1b) suggest that the thin mudstones are lithologically similar to the caprock and would, at least when intact, form capillary seals. It seems likely therefore that CO$_2$ migration through these mudstones is actually not by Darcy flow but via some form of pathway flow, the pathways becoming more effective, or more numerous with time. This is also consistent with measured pressure distributions in laboratory samples during long-term flow testing, which suggest pathway rather than Darcy flow (Harrington *et al.* 2010). The nature of these pathways is uncertain however. Small faults have been identified in the Utsira Sand, probably related to differential compaction during burial. More radically, Hermanrud *et al.* (2007) have suggested that fluidization effects within the main feeder chimney may have induced minor collapse and faulting of the layers within the reservoir. Whether natural or induced, small faults displacing the intra-reservoir mudstones are plausible candidates for providing the permeability enhancement required to explain the observed upward flow. A number of alternative explanations may be considered. Zweigel *et al.* (2004) showed that most of the individual

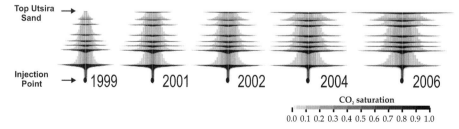

Fig. 9. Simulated growth of the CO$_2$ plume from 1999 to 2006 from the TOUGH2 axisymmetrical flow model.

Table 2. Flow properties of intra-reservoir mudstones in the flow model, compared with core measurements from caprock (0.9869×10^{12} m^2 equates to 1 Darcy)

	Intra-reservoir mudstones in TOUGH2 model	Laboratory caprock core
Capillary entry pressure (Pa)	1.6×10^4	$1.6 - 1.9 \times 10^6$
Permeability (m^2)	9×10^{-14}	4×10^{-19}

intra-reservoir mudstones are not correlatable from well to well, indicating lateral impersistence and/or the presence of 'holes'. Holes in the mudstones could be due to erosive removal, possibly by channelling, or soft sedimentary processes such as sand mobilization. Alternatively, geochemical processes may play a part. In the short-term, evaporation of residual water into the CO_2 could cause dehydration and possible shrinkage of the mudstones. Mineralogical changes are thought unlikely to occur significantly in the short time-scales considered here. Moreover experimental work on the reactivity of CO_2–caprock systems at reservoir conditions (summarized in Chadwick et al. 2008) indicates that mineralogical changes would tend to reduce porosity (and by implication permeability) in mudstone lithologies. Finally, recent experimental work has suggested that in CO_2–water systems in argillaceous rocks (i.e. sealing strata), CO_2 may show intermediate wetting properties, thereby reducing capillary entry pressures to CO_2 (e.g. Chiquet et al. 2005). However, by itself this process cannot explain the observed rate of migration of CO_2 to the top of the reservoir because the intrinsic permeability of the mudstones is still much too low.

3D flow simulation of the topmost layer spreading

The growth of the topmost layer between 1999 and 2006 is quite striking (Fig. 4), and has involved rapid lateral spreading of free CO_2 controlled by top reservoir topography. The CO_2 initially impacted the topseal within a local topographic dome, then spilled northwards along a prominent north-trending linear ridge before entering a more vaguely defined northerly topographic high (Fig. 3). Lateral migration was particularly rapid along the linear ridge where the CO_2 front advanced northwards at about 1 m per day between 2001 and 2004. In order to examine layer evolution more closely, a 3D flow model was set up. The top reservoir surface was mapped from the baseline (1994) seismic dataset and depth-converted using a layer-cake, laterally uniform velocity model (available wells constrain regional velocities but are not sufficiently closely spaced to constrain local velocity variation). The nature of the CO_2 supply to the topmost layer depends on the transport properties of the relatively thick (>5 m) mudstone immediately beneath (Fig. 1b). As discussed above, pathway flow is considered most likely, but the number of individual pathways is uncertain. In 1999, initial development of the topmost layer was as two seemingly distinct small accumulations (Fig. 4a), of which the southerly appears to have been fed directly by the main feeder chimney (Fig. 2). The simplest interpretation of the northerly accumulation is that it was supplied laterally from the southerly one, but a second smaller feeder, supplying from directly below, cannot be ruled out. For simplicity, the flow modelling assumed a single feeder, located between the two initial accumulations, with modelled CO_2 flux into the layer matched to the structurally derived volumes (Table 1). A minor adjustment to the model was also made to allow for the fact that the CO_2 injected at Sleipner contains up to about 2% methane, lowering density and increasing buoyancy. The TOUGH2 version used for modelling does not have an option for CO_2–methane mixtures, so to account for this, the top reservoir temperature in the model was raised to 31.5 °C, giving a pure CO_2 density comparable with that of a CO_2–methane (2%) mixture at 29 °C.

Permeability effects

Permeability values for the Utsira Sand range from laboratory determinations of 2–3 Darcy (with no evidence of horizontal/vertical anisotropy), to more regional estimates of up to 8 Darcy from large-scale water production (Zweigel et al. 2004).

An initial flow simulation was run using an isotropic reservoir permeability of 3 Darcy (Fig. 11), in accordance with the core measurements. The spread and lateral migration of the simulated layer are much slower than that of the real layer, the modelled CO_2 front barely reaching the southern end of the north-trending ridge. The modelled migration is also more obviously radial from

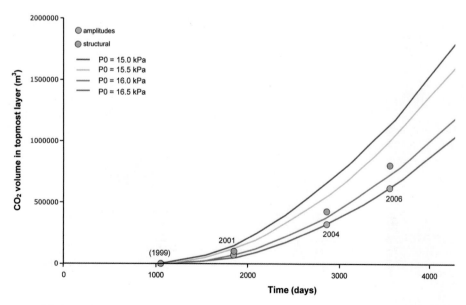

Fig. 10. Growth of the topmost CO_2 layer. Green and orange circles denote observed volumes derived from the seismic amplitudes and from structural analysis of the CWC respectively (Table 1). Solid lines show TOUGH2 flow simulation with semi-permeable intra-reservoir mudstones and a range of capillary entry pressure.

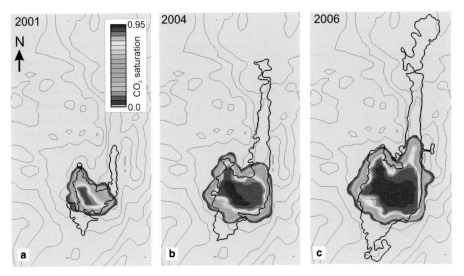

Fig. 11. Modelling topmost layer spreading from 2001 to 2006 assuming a reservoir permeability of 3 Darcy. Polygons denote observed CWC.

the feeder point than the observed spreading, with a greater tendency to spread east–west.

In order to increase the slow simulated spreading rates, a number of flow simulations were run with higher permeability values (Fig. 12). Permeabilities of 6 Darcy (Fig. 12b) and also 10 Darcy, produced more pronounced lateral spreading, but were still insufficient to match the observations. The modelled migration also remained more radial about the feeder point. In order to address this, permeability anisotropy was introduced: 3 Darcy east–west and 10 Darcy north–south. Specific geological justification of this is not obvious, though the assumption is qualitatively consistent with the generally north-trending depositional geometry of the Utsira Sand (Gregersen et al. 1997). With anisotropic permeability the match is improved somewhat (Fig. 12c), but migration along the north-trending ridge is still much slower than actually observed. It is clearly difficult to match the observed spreading of the topmost layer by adjusting permeability alone. Even simulations with improbably extreme values of anisotropic permeability, 40 Darcy north–south and 10 Darcy east–west, failed to match the observed northward spreading.

Buoyancy effects

Lateral migration of the topmost layer is driven by buoyancy forces arising from the relatively low density of the injected CO_2 relative to the reservoir pore-water. To test whether enhanced buoyancy could contribute to better history-matching, the top reservoir temperature in the model was increased to 36 °C, reducing the density of the CO_2 by about 300 kg m^{-3} and also its viscosity. A simulation was then run at this higher temperature but with the same injected volumes and with 3/10 Darcy anisotropic permeabilities (Fig. 12d). This simulation shows a better match than the lower temperature models but still fails to reproduce both the northward and southward extents of the observed layer. The justification for reducing plume density via increased temperatures is in any case questionable. The reservoir temperature at Sleipner is believed to

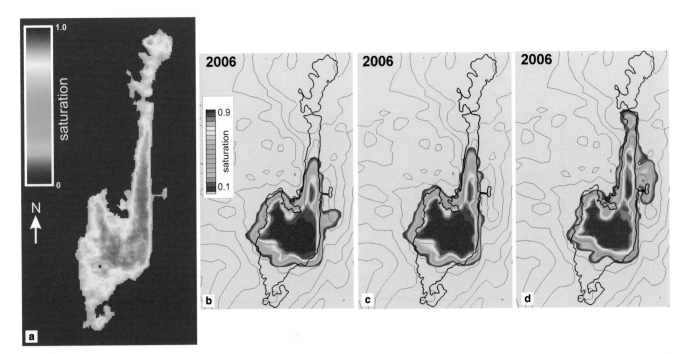

Fig. 12. The topmost layer in 2006. (**a**) Observed extents; (**b**) TOUGH2 simulation with $k = 6$ Darcy; (**c**) TOUGH2 simulation with $k = 3$ Darcy east–west and 10 Darcy north–south; (**d**) TOUGH2 simulation with $k = 3$ Darcy east–west, 10 Darcy north–south and higher reservoir temperature.

be well constrained: a temperature measurement from a Sleipner well has been recently confirmed by downhole measurements from the nearby Volve Field where water is being produced from the Utsira Sand (Ola Eiken, pers. comm.). Nevertheless some doubt remains: interpolation between the seabed and measured temperatures in the much deeper Sleipner East gas field (Christian Hermanrud, pers. comm. 2007) would suggest temperatures in the Utsira Sand up to a few degrees higher than the measured values.

Topographic effects

It is clear that neither adjusting reservoir permeability nor injected fluid properties can satisfactorily match the simulations with the observed spread of the topmost CO_2 layer, in terms of either its northerly or southerly extents. The lack of southerly spreading can partly be explained by the model simplification of a single CO_2 feeder located between the two 1999 patches. Placing a feeder directly beneath the southerly patch would produce much more southward spread. Additional feeders to the south may also be developing as the layer grows. The lack of northward spread in the model is more problematical. Additional feeders beneath the north–south trending tongue of CO_2 can probably be discounted due to the lack of a CO_2 source layer beneath. In addition, specific details of migration which also do not match the observations, such as the simulated spillage eastward from the north-trending linear ridge (Fig. 12d), require explanation. It is likely that slight inaccuracies in the depth topography of the top reservoir surface may explain these discrepancies.

A detailed cross-section through the topmost layer (Fig. 13a) shows that its observed northward progression involved rapid migration along an apparently near-horizontal ridge crest. Top reservoir depths were calculated using a laterally uniform 'layer-cake' overburden velocity model based on average velocities from a number of wells in the vicinity; depths are, in consequence, a simple transform of seismic two-way travel-time. Apparently paradoxical migration trends could therefore be explained by errors in the topography of the topseal produced by lateral velocity variations which have not been accounted for in the depth-conversion.

An initial test of this hypothesis is to examine the overall dip of the topseal. The feeder chimney impacts the topseal at point A1 at a two-way travel-time of 873 ms (Fig. 13). A local topographic culmination north of the linear ridge, point A2, lies at a two-way travel-time of 856 ms. The travel-time difference between A1 and A2 converts to a depth difference of 18.5 m, which over a horizontal spacing of 2249 m gives an average gradient of 0.0082. A more conservative estimate of topseal dip may be obtained by comparing the relative elevations of the southernmost and northernmost extremities of the layer, points B1 and B2 respectively. These two points have an elevation difference of 17.4 m over a distance of 2895 m, giving an average gradient of 0.0058.

In order to test potential CO_2 migration rates along dips of this order, a simple synthetic surface model was constructed. This has two circular domes whose relief and separation match the observed topseal surface, joined by a linear ridge of similar geometry to that observed. Two surface options were examined, one with a uniform

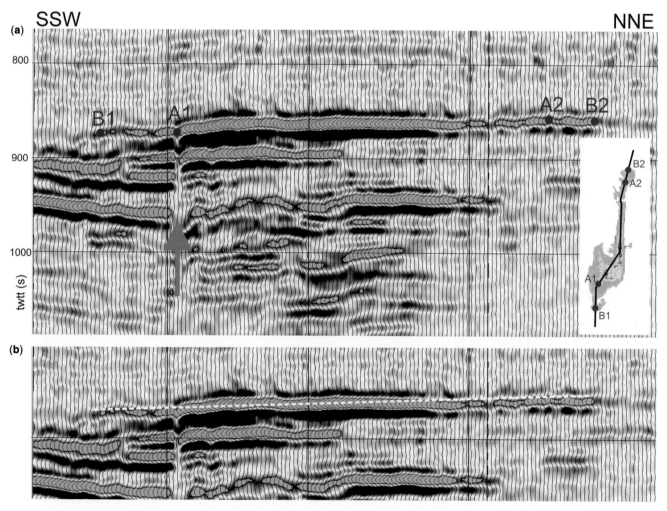

Fig. 13. Seismic line through the CO_2 plume. (**a**) Detailed geometry of the topmost layer in two-way travel-time. The main feeder chimney (arrowed) supplies the layer at point A1. (**b**) Deviation of the topmost layer from planar geometry (white dots). Reflective CO_2 layers in green.

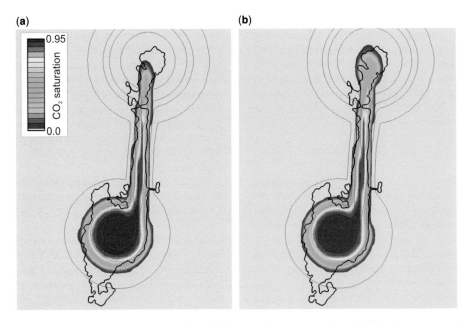

Fig. 14. Flow simulation of synthetic surface with sand permeability of 10 Darcy: (**a**) average gradient of 0.0058; (**b**) average gradient of 0.0082.

gradient along the ridge of 0.0058 and one with a gradient of 0.0082 (Fig. 14).

The simulations show that a synthetic surface with a smooth, uninterrupted dip in the observed range can produce the required rate of lateral migration, albeit requiring a high modelled permeability. A key property of the synthetic surface is that a uniform dip is maintained along the linear ridge. Close inspection of the real data (Fig. 13b) shows local perturbations of up to 5 ms two-way travel-time along the ridge with respect to a uniform gradient. The downwarp towards the northern end of the ridge effectively retards northward migration of the CO_2 in the flow simulations. If, however, this downwarp were not real, but just a localized time shift (pushdown) due to lateral velocity variation, then the discrepancy between the observed and simulated migration patterns could be explained.

There are a number of wells in the vicinity with velocity data, but these are too spatially scattered (typically more than 2 km apart), to explicitly map local velocity variations. Average velocities to the top of the Utsira Sand were calculated for these wells. Well pairs were selected, their spacing noted, and the difference in velocity between them calculated. Velocity difference against well spacing was then plotted for all well pairs (Fig. 15). Significant velocity variation is evident, with differences up to more than 50 m s^{-1}, but these are not clearly related to well spacing with velocity differences up to 30 m s^{-1} between wells only 2–3 km apart. This suggests that the velocity field is characterized by rapid local velocity changes which are not being properly spatially sampled by the widely spaced wells. Velocity difference is readily converted to two-way travel-time difference. Two-way travel-time differences of 10 ms or so are common at well spacings of 2–3 km. The linear best fit, albeit very poorly correlated, suggests that similar time differences could well occur over even smaller distances. It is clear therefore that the postulated time shifts noted on the cross-section (Fig. 13b) – up to 5 ms over a distances of about 1 km – are fully compatible with observed lateral velocity changes between the wells.

Accurate mapping of local velocity variation is required to resolve this issue but this is not possible with the widely spaced well data. Seismic stacking velocities are currently being assessed, but do not appear to be sufficiently precise. Another possibility is interpretive use of the seismic data itself. Bright spots in the overburden are clustered at certain stratigraphical levels and are probably related to natural accumulations of shallow gas which would lower seismic velocity. By mapping integrated bright-spot amplitudes it may be possible to scale these to relative velocity-induced time shifts. This is currently under investigation.

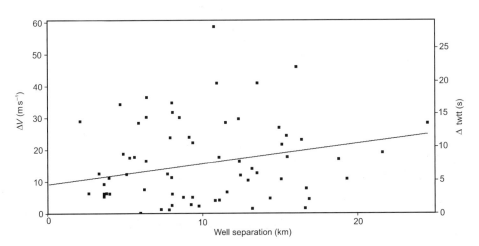

Fig. 15. Average velocity variation to top Utsira Sand showing velocity difference between pairs of wells and corresponding time shifts. Linear best fit also shown.

Fig. 16. Flow simulations (after six years) beneath a planar dipping topseal of variable flow properties: (**a**) $k = 4 \times 10^{-19}$ m^2; $P_c = 2$ MPa; (**b**) $k = 9 \times 10^{-14}$ m^2; $P_c = 40$ kPa; (**c**) $k = 9 \times 10^{-14}$ m^2; $P_c = 17$ kPa. Note figure has vertical exaggeration.

To summarize, the well data show that local velocity variations are present which induce significant localized time shifts. Time shift corrections applied to the topseal surface prior to depth conversion could produce a smoother depth surface which would allow simulated layer spreading to more closely match the observed data. Explicit mapping of these time shifts has not currently been achieved, however.

Topseal properties

The flow simulations have all assumed an impermeable reservoir topseal. This is consistent with laboratory testing on caprock core samples which measured capillary entry pressures of around 2 MPa, considerably higher than any credible reservoir pressure increase. However, although they seem to have similar geophysical log properties to the caprock, the thin intra-reservoir mudstones are required to be effectively semi-permeable for the whole plume reservoir simulation to match the observed arrival of CO_2 at the reservoir top (Fig. 4a). The question arises therefore as to whether the basal layers of the caprock might show similar flow properties to the intra-reservoir mudstones. There is no evidence of changes in the seismic signature of the caprock that might indicate migration of CO_2 through the topseal, but this is subject to finite detection capability and so does not unequivocally prove seal integrity, particularly as diffuse migration into the caprock might be difficult to detect seismically.

In order to investigate this, a simple 3D flow model was set up, with a planar dipping topseal of gradient 0.0082 (comparable to the real dips discussed above) and a reservoir permeability of 3 Darcy. CO_2 was input beneath the topseal, at the volumes calculated above, for a range of topseal flow properties and the simulation run for six years (Fig. 16). With topseal properties identical to the core measurements (Harrington *et al.* 2010) the CO_2 migrates updip more than 600 m (Fig. 16a). Modifying the topseal permeability to the modelled permeability of the intra-reservoir mudstones, but maintaining a higher capillary entry pressure, allows limited migration into the caprock and reduces the updip migration to about 550 m (Fig. 16b). With topseal flow properties set identical to the modelled properties of the intra-reservoir mudstones (Fig. 16c), there is major ingress into the topseal with negligible updip migration. It is clear that migration into the caprock would significantly retard lateral migration and is not consistent with the observed very rapid lateral migration of the topmost layer. The high observed migration rates, and the very high reservoir permeabilities that are required to simulate these, suggest therefore that the topseal is behaving as a very 'hard' flow barrier exerting negligible 'drag' on the flowing layer beneath.

Discussion

The type of analysis described above provides an interesting pointer to how the regulatory requirements of future large-scale underground storage may be addressed. For example, the European Storage Directive has three key requirements which have to be satisfied before liability for the storage site can revert to the licensing authority:

- actual behaviour of the injected CO_2 conforms with the modelled behaviour;
- there is no detectable leakage;
- the storage site is evolving towards a situation of long-term stability.

Because the topmost CO_2 layer represents the ultimate trapping level for mobile free CO_2 in the reservoir, understanding its behaviour is key to demonstrating both conformance and constraining future plume behaviour. Results so far suggest that the main processes governing lateral spreading of the topmost layer are quite well understood and its actual behaviour conforms reasonably well with the modelling. Regarding leakage detection, the lack of seismic changes in the caprock, together with the apparent lack of fluid transport into the caprock as deduced above, together make a strong case for no leakage. The third requirement clearly will involve much longer term predictive modelling than the short-term history matching described here; nevertheless, the latter does provide an essential starting point on which to base longer term prediction.

Conclusions

The topmost CO_2 layer in the Sleipner plume represents the ultimate trapping level for mobile free CO_2 in the reservoir. As such it can provide useful insights into key elements of reservoir performance, both in the short and longer term.

Volumetric analysis of the layer closely constrains total upward CO_2 flux through the reservoir. This allows assessment of bulk reservoir flow properties and how these change with time. The thin intra-reservoir mudstones exert the main control on reservoir flow. Laboratory measurements suggest that, in intact form, these should act as capillary seals, so it is inferred that transport is via pathway flow rather than Darcy flow, possibly associated with networks of small faults or other 'holes' perhaps of sedimentary origin. Calculated CO_2 fluxes to the topmost layer have steadily increased with time, suggesting that the feeder pathways are evolving, becoming either more transmissive with time and/or increasing in number.

Detailed 3D history-matching of the topmost layer is challenging, most likely because of uncertainties in exact topseal topography. It nevertheless provides insights into the flow properties of both the topmost reservoir sand and also the integrity of the seal. High observed lateral spreading rates are consistent with very high permeabilities, possibly with some anisotropy. They also suggest that the topseal presents a very 'hard' flow barrier which exerts minimal 'drag' on the lateral flow beneath.

Studies such as this, concentrating on defining the process and performance of the storage system, within and immediately around the reservoir, can provide a robust basis for meeting the regulatory requirements for the large-scale deployment of CCS.

We thank the CO2ReMoVe consortium for permission to publish this work. CO2ReMoVe is funded by the EU 6th Framework Programme and by industry partners BP, ConocoPhillips, ExxonMobil, Statoil, Schlumberger, Total, Vattenfall and Wintershall. R&D partners are BGR, BGS, BRGM, CMI, DNV, ECN, GFZ, GEUS, IFP, IMPERIAL, OGS, SINTEF, TNO and URS. This paper is published with permission of the Executive Director, BGS (NERC).

References

Arts, R., Eiken, O., Chadwick, R. A., Zweigel, P., van Der Meer, L. & Zinszner, B. 2003. Monitoring of CO_2 injected at Sleipner using time lapse seismic data. *In*: Gale, J. & Kaya, Y. (eds) *Greenhouse Gas Control Technologies*. Elsevier Science, Oxford, 347–52.

Baklid, A., Korbol, R. & Owren, G.. 1996. Sleipner vest CO_2 disposal, CO_2 injection into a shallow underground aquifer. *SPE Annual Technical Conference and Exhibition*, Denver, CO, SPE Paper 36600, 1–9.

Chadwick, R. A., Arts, R., Eiken, O., Kirby, G. A., Lindeberg, E. & Zweigel, P. 2004. 4D seismic imaging of a CO_2 bubble at the Sleipner Field, central North Sea. *In*: Davies, R. J., Cartwright, J. A., Stewart, S. A., Lappin, M. & Underhill, J. R. (eds) *3-D Seismic Technology: Application to the Exploration of Sedimentary Basins*. Geological Society, London, Memoirs, **29**, 311–320.

Chadwick, R. A., Arts, R. & Eiken, O. 2005. 4D seismic quantification of a CO_2 plume at Sleipner, North Sea. *In*: Doré, A. G. & Vining, B. A. (eds) *Petroleum Geology: North-West Europe and Global Perspectives: Proceedings of the 6th Petroleum Geology Conference*. Geological Society, London, 1385–1399; doi: 10.1144/0061385.

Chadwick, R. A., Arts, R., Bernstone, C., May, F., Thibeau, S. & Zweigel, P. 2008. *Best Practice for the Storage of CO_2 in Saline Aquifers*. British Geological Survey, Keyworth, Occasional Publications, **14**.

Chiquet, P., Broseta, D. & Thibeau, S. 2005. *Capillary Alteration of Shaly Caprocks by Carbon Dioxide*. Society of Petroleum Engineers Paper SPE 94183.

Gregersen, U., Michelsen, O. & Sorensen, J. C. 1997. Stratigraphy and facies distribution of the Utsira formation and the Pliocene sequences in the northern North Sea. *Marine and Petroleum Geology*, **14**, 893–914.

Harrington, J. F., Noy, D. J., Horseman, S. T., Birchall, D. J. & Chadwick, R. A. 2010. Laboratory study of gas and water flow in the Nordland Shale, Sleipner, North Sea. *In*: Grobe, M., Pashin, J. & Dodge, R. (eds) *Carbon Dioxide Sequestration in Geological Media – State of the Science*. American Association of Petroleum Geologists, Tulsa, OK, Studies in Geology, **59**, 521–543.

Hermanrud, C., Zweigel, P., Eiken, O., Lippard, J. & Andresen, T. 2007. CO_2 flow in the Utsira formation: inferences made from 4D seismic analysis in the Sleipner area. *European Conference of the American Association of Petroleum Geologists*, Athens, Abstract.

IPCC, 2005. Metz, B., Davidson, O., de Coninck, H. C., Loos, M., Meyer, L. A. (eds) *IPCC Special Report on Carbon Dioxide Capture and Storage. Prepared by Working Group III of the Intergovernmental Panel on Climate Change*. Cambridge University Press, Cambridge.

Lindeberg, E., Zweigel, P., Bergmo, P., Ghaderi, A. & Lothe, A. 2001. Prediction of CO_2 distribution pattern by geology and reservoir simulation and verified by time-lapse seismic. *In*: Williams, D. J., Durie, R. A., McMullen, P., Paulson, C. A. J., Mavko, G., Mukerji, T. & Dvorkin, J. 2003. *The Rock Physics Handbook*, 2nd edn. Cambridge University Press, Cambridge, 329.

Mavko, G., Mukerji, T. & Dvorkin, J. 2003. *The Rock Physics Handbook: Tools for Seismic Analysis in Porous Media*. Cambridge University Press, Cambridge.

Zweigel, P., Arts, R., Lothe, A. E. & Lindeberg, E. 2004. Reservoir geology of the Utsira Formation at the first industrial-scale underground CO_2 storage site (Sleipner area, North Sea). *In*: Baines, S. & Worden, R. H. (eds) *Geological Storage for CO_2 Emissions Reduction*. Geological Society, London, Special Publications, **233**, 165–180.

Differences between flow of injected CO_2 and hydrocarbon migration

CHRISTIAN HERMANRUD,[1,2] GUNN MARI GRIMSMO TEIGE,[1] MARTIN IDING,[1] OLA EIKEN,[1] LARS RENNAN[1] and SVEND ØSTMO[1]

[1]*Statoil ASA, 7005 Trondheim, Norway (e-mail: che@statoil.com)*
[2]*University of Bergen, Faculty of Mathematics and Natural Sciences, 2020 Bergen, Norway*

Abstract: Knowledge of fluid flow processes in the subsurface is important for CO_2 storage operations as well as for hydrocarbon exploration. Repeated seismic surveys for more than 10 years of CO_2 injection into the Utsira Formation, in the Sleipner area, offer a unique dataset. This dataset holds information on fluid migration processes that can be analysed for the benefit of hydrocarbon exploration and CO_2 storage considerations alike. Thorough analyses of these datasets reveal several features that give useful information of subsurface fluid flow processes. The CO_2 in the Utsira Formation has flowed laterally beneath thin, intra-formational shales. At the same time, CO_2 has flowed vertically through shaly horizons that would normally be considered as barriers to fluid flow. This flow has apparently taken place through vertically stacked flow conduits through the shales. These conduits may to some extent have existed prior to the start of CO_2 injection, but may also have been augmented by the CO_2 injection process. The calculated pushdown of seismic reflectors below the CO_2 plume is less than that observed, which may point to the presence of hitherto unrecognized flow paths for the CO_2. Hydrocarbon migration pathways are in general not recognizable in seismic data. This implies that such avenues are significantly thinner than those of the CO_2 migration in the Utsira Formation. This result points to the presence of mixed-wet migration pathways, in which capillary flow resistance may not control the (sub-horizontal) flow path thickness. A circular depression at the top of the Utsira Formation that existed prior to the injection may be interpreted as a result of a deeper seated sand remobilization feature. Such features will also promote vertical hydrocarbon migration where they are present. A more widespread occurrence of such features may explain why hydrocarbons are generally found beneath thick shales, but are less likely to be found below thin intra-formational shales below the structural spillpoint of the top seal. These observations suggest that seal thickness is an important parameter, even if the capillary entry pressure of the sealing rock is sufficiently high to preserve significant hydrocarbon columns.

Keywords: CO_2 injection, CO_2 flow, hydrocarbon migration, Utsira Formation, top seal, intra-formational shales

Knowledge of hydrocarbon migration processes is important for predictions of oil and gas occurrence in the subsurface. Direct field observations of hydrocarbon migration are rare and seismic identification of migration routes carries significant uncertainties. As a result, important aspects of hydrocarbon migration remain elusive.

The CO_2 injection that has been carried out in the Utsira Formation above the Sleipner West Field in the Norwegian North Sea has been monitored by a set of repeated seismic surveys (Arts *et al.* 2003; Chadwick *et al.* 2005). These surveys reveal how CO_2 has migrated in the Utsira Formation from 1996 to 2008, and give unique observations of fluid migration in a porous medium.

The conditions of hydrocarbon migration and CO_2 injection differ in many respects. Such differences are partly due to different fluid properties between hydrocarbons and CO_2, including interfacial tensions, solubilities, densities, viscosities, and interactions with pore water (and consequences for the pH). Also, significant differences exist between the injection rate of CO_2 and the much slower hydrocarbon migration velocity. These differences would be expected to significantly influence the flow paths of the migrating fluids.

Nevertheless, the physical parameters that control hydrocarbon migration and the flow of injected CO_2 in the subsurface are the same. Thus, knowledge of hydrocarbon migration processes should be helpful in predicting the short- and medium-term movements of injected CO_2. Likewise, the insight that has been gained from monitoring of the CO_2 flow in the Utsira Formation can give knowledge of fluid dynamics in the subsurface that will aid the understanding of petroleum migration.

The purpose of this paper is to examine the main characteristics of the CO_2 flow in the Utsira Formation, and compare it with the behaviour of hydrocarbon migration. Emphasis will be put on understanding the processes (or realization of the lack of such understanding) that emerges from investigations of the CO_2 flow, and the implication of these findings for analyses of hydrocarbon migration.

CO_2 injection in the Utsira Formation

The Sleipner West Field is situated in the northern North Sea. The field produces gas from a Jurassic structural trap. CO_2 is separated from this gas offshore, and is then reinjected into the Utsira Formation of late Middle Miocene to earliest Late Pliocene age. The CO_2 injector was drilled as a very shallowly inclined well from the Sleipner A platform. This well does not penetrate the caprocks above the Utsira Formation at the position where the CO_2 is being injected.

Only a limited section (7 m) of core material has been retrieved from the Utsira Formation. This core consists of clean, unconsolidated sand with 3% carbonate content (Holloway *et al.* 2004). Also wireline log data suggest that the Utsira Formation in this region (Fig. 1) consists of clean sands. At the injection site, the average porosity is 35–40%, the vertical and horizontal permeabilities are around 2 D, and the net/gross ratio is 0.90–0.97 (Zweigel *et al.* 2004). The formation has been described as a lower shoreface deposit, heavily influenced by longshore currents after deposition (Galloway 2002). The Cenozoic strata below the Utsira Formation contain abundant sand injection features at the Tampen Spur, 300 km north of the injection site (Rodrigues *et al.* 2009), and sand injection phenomena may have affected rocks of similar age as the Utsira Formation in this area (H. Løseth, pers. comm.). Injected sands in the Utsira Formation in the Sleipner area have not been documented.

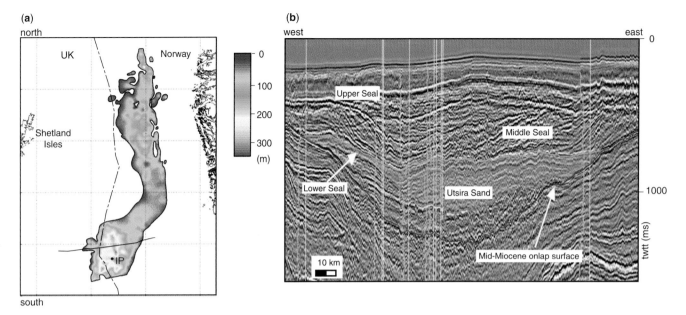

Fig. 1. (a) Location and extent of the Utsira Formation in the North Sea. Colours refer to formation thickness (metres); (b) a West–East cross-section of the Sleipner area. The location of the line is marked with a black solid line in (a). IP, injection point; twtt (ms), two-way travel time in milliseconds. From Holloway et al. (2004), and Schlumberger (courtesy of Schlumberger).

More than 12×10^6 tons of CO_2 have been injected into the Utsira Formation at Sleipner Field since 1996. A seismic survey was shot prior to the injection (1994), and six repeated seismic surveys have been acquired since then to monitor the CO_2 flow in the Utsira Formation. The repeated seismic surveys show how the CO_2 plume has grown with time (Fig. 2). The seismic data also reveal that the CO_2 has been distributed in nine individual layers (Fig. 3), that several of these have grown laterally with time, and that at the same time the CO_2 has moved through (as opposed to around) the intra-formational shaly layers on its way towards the top seal (Fig. 4). The shale beds that cap the individual CO_2 layers are typically 1–1.5 m thick, judging from well data in neighbouring wells (Zweigel et al. 2004). By assuming that the brightest amplitudes correspond to the maximum constructive interference between top and bottom of a thin CO_2-filled layer, a corresponding thickness of about 7–8 m has been suggested (Arts et al. 2003). Figure 5 shows how the CO_2 is envisaged to be distributed in the Utsira Formation from this assumption. The eight deepest CO_2 layers are trapped by intra-reservoir shales. Only the shallowest of these can be identified seismically; the others are inferred from the trapping of CO_2 and from log responses within the Utsira Formation in nearby wells (Zweigel et al. 2004).

CO_2 flow characteristics inferred from seismic data

The seismic data reveal that the injected CO_2 has followed flow paths that differ from those that would be inferred from knowledge of hydrocarbon migration processes alone:

(1) CO_2 is moving laterally in layers that are clearly visible on seismic data.
(2) The pushdown of the seismic reflectors below the CO_2 plume is larger than what would be expected from CO_2 in the nine

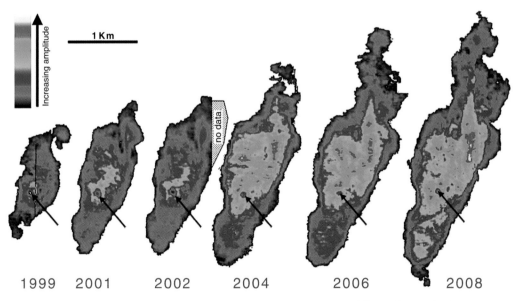

Fig. 2. Cumulative total reflection strength for all nine layers of the Sleipner CO_2 plume as seen on different vintage seismic surveys. Arrows show seismic chimney.

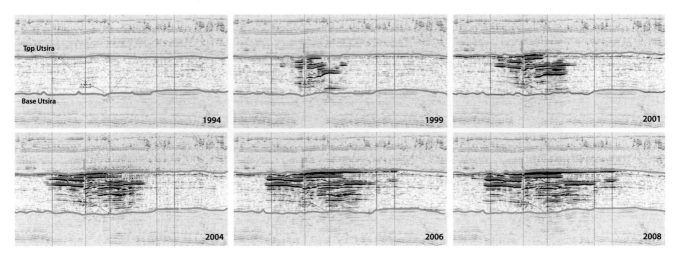

Fig. 3. Seismic signatures of the CO_2 plume in the Utsira Formation in different time-lapse seismic data. All lines are displayed at the same location, oriented north–south, and pass through the injection point. Arrows show seismic chimney.

layers alone. Chadwick *et al.* (2005) noted that a diffuse (low saturation) distribution of about 20% of the CO_2 is required to derive a satisfactory match between seismic amplitudes and pushdown, provided that the influence of gas saturation on seismic data follows the Gassman equation (Gassmann 1951). No flow model that explains how such diffuse gas saturation could have developed has to our knowledge been proposed.

(3) The vertical flow paths that have been exploited by the CO_2 are intriguing. The CO_2 in the individual layers is trapped by seals, yet the CO_2 flowed through all of these and reached the top layer within a three-year period (from injection started in 1996 to shooting of the first post-injection seismic survey in 1999). The vertical permeability of clays is typically at least six orders of magnitude below the horizontal permeability of the Utsira Formation sands (Katsube & Coyner 1994; Schlömer & Krooss 1997), which implies that the flow of CO_2 in the pore network would give a vertical CO_2 displacement of 1 mm or less during the time the CO_2 migrated 1 km horizontally. Clearly, the CO_2 did not flow vertically through continuous and unbroken shale horizons.

The seismic imaging of the CO_2 reveals a central area (chimney) with no coherent signals. This area emerges as a circular feature on the amplitude maps for all layers combined (Fig. 2), but does not show up in the amplitude maps of the uppermost layer. This circular feature is positioned approximately above the position of the CO_2 injection, and is included in the 95% confidence ellipsoid of this position.

The coincidence of the position of the central chimney approximately above the injector and the circular expression of the chimney on the amplitude maps renders the possibility that the chimney displays an image of a vertical flow path that was created by the injection process. A laboratory experiment was carried out to evaluate this hypothesis. In this experiment, a Plexiglas cylinder was filled with water. First sand then kaolin were added to the water and were allowed to settle for several weeks. Thereafter, the cylinder was tilted, and then shaken to promote sand compaction (Fig. 6). The tilting was made to ensure that the water flow velocity was sufficiently large to result in fluidization as the compact-derived excess water rose to the surface along the upper slope of the cylinder. The left picture was taken during fluidized water flow, when compaction-derived water broke the sand layering along the upper

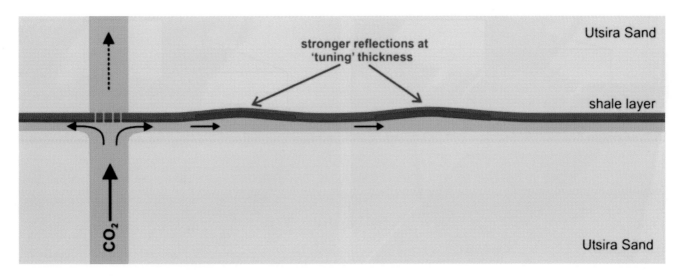

Fig. 4. Flow model of CO_2 from the injection point towards the top of the Utsira Formation and into discrete layers. From Arts *et al.* (2003), reproduced courtesy of Elsevier.

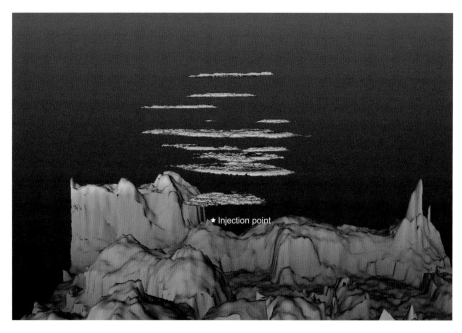

Fig. 5. Modelled 3D CO_2 distribution in the Utsira Formation, with layer thicknesses calculated from tuning thicknesses of seismic amplitudes. Flat CO_2–water contacts were applied for the construction of this figure. Courtesy of Permedia Inc.

slope of the tilted cylinder and erupted into the kaolin (Fig. 6, inserted picture). The right picture was taken after the fluidized flow had ceased, approximately half a minute after the first picture. At this stage, the porosity in the flow path is being reduced, and kaolin sinks into the vacant space at the top of the sand layer.

A vertical flow path through the shaly beds of the Utsira Formation might have resulted from fluidization of the unconsolidated sand, and could have been triggered by localized sand matrix collapse resulting, for example, from the dissolution of carbonate material in coquina beds. Calcite dissolution would be expected to take place as CO_2-saturated water is slightly acidic (pH = 5). An increasingly large portion of the injected CO_2 has found its way to the shallowest layer as the injection has progressed (Chadwick et al. 2009). This observation is consistent with an injection-made flow path that becomes more effective as injection proceeds.

However, a comparison of the seismic signatures from the (pre-injection) 1994 survey with that of later surveys demonstrates that a circular depression existed above layer 8 even before the injection started (Fig. 7). The seismic data quality precludes identification of possible circular features at deeper levels. The circular feature was thus not caused by the injection, although it could have been augmented by it. One could speculate that this feature is a collapse feature from a sand injection deeper in the Utsira Formation, created by processes similar to that of the laboratory experiment displayed in Figure 6.

Implications for the understanding of hydrocarbon migration

The seismic data acquired before and during the CO_2 injection at Sleipner Field give unique images of fluid migration within the subsurface. As such they also reveal information on hydrocarbon migration processes. We see several aspects of the migrating CO_2 that give such information, although the interpretation of this information may be ambiguous.

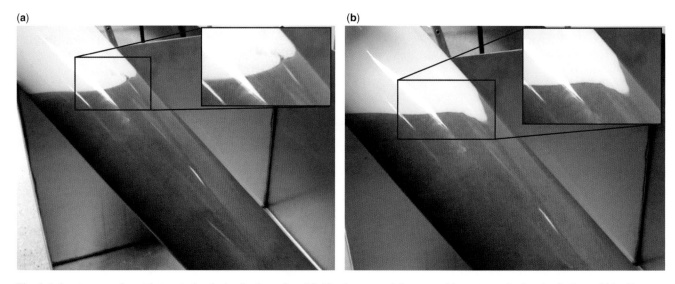

Fig. 6. Laboratory experiment demonstrating the implications of sand fluidization on overlying strata: (**a**) water eruption into kaolin layer; (**b**) kaolin sinks into sand layer. See text for further explanations.

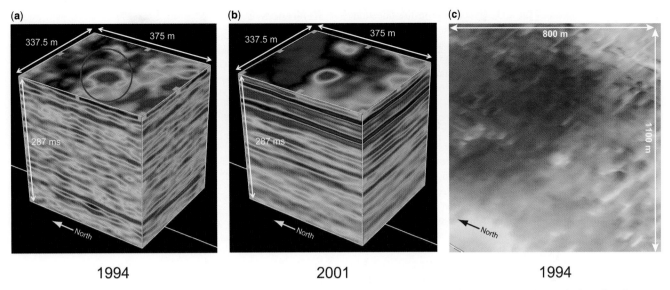

Fig. 7. Seismic signatures of the base of the top seal above CO_2 layer 8. Note the circular feature that is present above the chimney in time slices in both the (**a**) 1994 and (**b**) 2001 data. This circular feature emerges as a depression when the reflector that defines the base of the top seal is mapped (**c**).

The calculations of reflector pushdown by the Gassmann equation with all the CO_2 in layers of high CO_2 saturation give less pushdown than that observed in the seismic data. These calculations involve several parameters that carry some uncertainty. Also, documentations of the accuracy of Gassman-calculated pushdown variations due to fluid substitutions in the subsurface are sparse. Nevertheless, the mismatch between observed and calculated pushdown may hold information on CO_2 distributions and flow paths that are presently unrecognized. Resolving this would be beneficial for both CO_2 monitoring and for hydrocarbon migration and production studies.

Only trapping below the top seals would be considered in hydrocarbon exploration in reservoirs with properties similar to those of the Utsira Formation in the Sleipner area. Intra-reservoir shales of metre thickness would not be expected from exploration experience to result in multiple reservoirs. However, even thin but laterally continuous shale beds should in theory trap hydrocarbons at deeper levels and prevent them from reaching the top seal. That hydrocarbons in most cases find pathways through sequences with interfingering shale beds demonstrates that vertical flow paths are more prevalent than one would infer if thin intra-reservoir shales were thought of as unbroken and laterally continuous. Vertical seal bypass systems have recently attained significant attention (Cartwright et al. 2007). The effective vertical seal bypass apparent at Sleipner Field is the first dataset that we are aware of that images vertical flow through heterogeneous sandstones in (almost) real time. The observation of an effective seal bypass system in these data opens the possibility that such bypass systems are relatively common, and that their presence may explain why hydrocarbons so frequently find pathways to the shallowest positions in clastic reservoir traps. Whatever causes the migration to bypass the internal seals (such as water escape structures, sand injections, erosional features), they are probably more likely to disrupt thin than thick seal rock intervals. The practical implication of this assumption is that seal rock thickness is an important parameter in hydrocarbon sealing, even if mercury injection measurements should conclude that the capillary entry pressure of the seal has not been overcome by the buoyancy of the hydrocarbon column.

Migrating CO_2 is clearly visible in the seismic images of the Utsira Formation in the Sleipner area. With these clear images in mind, why do we not observe hydrocarbon migration pathways in seismic data?

The thickness of a sub-horizontally migrating fluid layer can in general be calculated by assuming that the buoyancy in this layer must overcome the capillary entry pressure for *lateral* flow in the top of the migrating layer. This entry pressure is overcome when

$$(\rho_w - \rho_g)gh > 2\gamma/r_t \quad (1)$$

where ρ_w is the density of formation water, ρ_g is the density of the migrating, non-wetting fluid, g is the acceleration of gravity, h is the column height of the migrating fluid, γ is the interfacial tension and r_t is the largest pore throat radius in contact with the non-wetting fluid. The minimum thickness of a migrating body of a non-wetting fluid should thus be controlled by

$$h > 2\gamma/r_t(\rho_w - \rho_g)g \quad (2)$$

The density difference between water and that of migrating CH_4 and CO_2, respectively, may be about 2:1 (Nordgård Bolås et al. 2005; Nooner et al. 2007). On the other hand, the interfacial tension of CH_4–water is approximately twice that of CO_2–water in the Utsira Formation in the Sleipner area (Nordgård Bolås et al. 2005; Chalbaud et al. 2009). Thus, the minimum thickness of laterally migrating CO_2 should be approximately equal to the thickness of migrating CH_4.

One could calculate this minimum thickness if the critical pore-throat size of the Utsira Formation sands was known. Such knowledge has not been published. The critical pore-throats can nevertheless be calculated from the rock permeability. Unfortunately, such calculations carry significant uncertainties, and the deviation between calculations by the methods of Pittman (1992) and Berg & Avery (1995) give an order-of-magnitude difference in the calculated height (from 7 to 70 cm by assuming a density contrast between migrating CO_2 and water of 0.3 g cm^{-3}, and an interfacial tension of 30 mN between CO_2 and water; Chalbaud et al. 2009).

The highest seismic amplitudes are assumed to correspond to the maximum tuning of the CO_2-saturated layers. The amplitude of a CO_2 layer at 1 m thickness would be expected to be 15% of this, which approximately corresponds to the detection limit on the time-lapse difference cubes. The thickness of migrating CO_2 is at or possibly below the seismic detection limit at saddle points (such as to the north in the 2004 data of Fig. 2). This observation may suggest that migration pathways of less than a metre are in fact what one should expect in both CO_2 flow and CH_4 migration.

The greater thicknesses of the CO_2 layers close to the injection point are mainly a result of the large injection rate, and thus the build-up of slightly inclined gas–water contacts, below the intraformational shale beds (Bickle et al. 2007).

The above calculations and observations suggest that capillary trapping may well determine the retention of both CO_2 and CH_4 in the subsurface, and yet result in migration paths that are too thin for seismic detection in rocks with permeability of 2 D. However, the thickness of migrating CH_4 and CO_2 stringers alike should be ten times the thickness observed in the Utsira Formation in less permeable carrier beds with permeabilities of 20 mD instead of 2 D. This follows from the permeability v. pore throat relationships of both Pittman (1992) and Berg & Avery (1995). Gas migration in water-wet, low-permeable carrier beds should thus be visible in seismic data, due to their high capillary entry pressures. The observation that gas migration routes have not been identified as continuous stringers of bright seismic amplitudes may suggest that such migration routes are not controlled by capillary trapping over geological time, which would imply that the carrier bed rocks have a mixed wettability. Mixed-wetted rocks would result in less trapping of residual hydrocarbons in the carrier beds, and would also result in less storage capacity of CO_2 below or outside structural spill points, especially since wettabilities may change with time in the presence of both hydrocarbons (Teige 2008) and CO_2 (Chiquet et al. 2005). Further examinations of the CO_2 flow paths in the Utsira Formation can possibly result in more accurate knowledge of the role of capillary trapping, and thereby of the subsurface wettability distribution.

Summary and conclusions

The CO_2 that has been injected into the Utsira Formation in the Sleipner area has been distributed in nine different layers, most of which are capped by metre-thick shales. The injected CO_2 penetrated these layers and migrated to the top layer during the three year period between the start of injection and the first repeat seismic survey. The CO_2 has migrated several kilometres laterally in layers of c. 8 m thickness.

Inferences of hydrocarbon migration processes are based on indirect evidence. As migration pathways are not in general identified in seismic data, they are by inference significantly thinner than what is observed as typical thicknesses of flowing CO_2 in the Sleipner Field data. Observations of minimum CO_2 migration thicknesses close to the c. 1 m detection limit at spill points are consistent with calculated values, and so do not preclude that capillary trapping plays an important role in controlling the subsurface flow of CO_2 and CH_4. The importance of such trapping depends on the wettability of the carrier beds and possible change of wetting characteristics with time, a topic that needs further investigations.

In theory, hydrocarbons would not be expected to migrate vertically through shale beds, yet vertical migration of hydrocarbons is an efficient process in clastic rocks, which in the large majority of cases contain shale layers of significant lateral extents. It is suggested that vertical migration of hydrocarbons is frequently enhanced by the presence of vertical sand conduits, created by dewatering and/or injection of deeper sands. Wettability changes in the presence of migrating hydrocarbons may also play a significant role in explaining the efficiency of vertical hydrocarbon migration.

Thanks are due to Andrew Cavanagh for preparing Figure 5 and to Elin Storsten for drafting the figures. M. C. Akhurst and an anonymous referee are thanked for constructive comments to an earlier version of the manuscript. We are thankful for Statoil's permission to publish this paper.

References

Arts, R., Eiken, O., Chadwick, A., Zweigel, P., van der Meer, B. & Zinszner, B. 2003. Monitoring of CO_2 injected at Sleipner using time-lapse seismic data. In: Gale, J. & Kaya, Y. (eds) Greenhouse Gas Control Technologies. Elsevier, Oxford, 347–352.

Berg, R. R. & Avery, A. H. 1995. Sealing properties of Tertiary growth faults, Texas Gulf Coast. American Association of Petroleum Geologists Bulletin, 79, 375–393.

Bickle, M., Chadwick, A., Huppert, H. E., Hallworth, M. & Lyle, S. 2007. Modelling carbon dipoxide accumulation at Sleipner: implications for underground carbon storage. Earth and Planetary Science Letters, 225, 164–176.

Cartwright, J., Huuse, M. & Aplin, A. 2007. Seal bypass systems. American Association of Petroleum Geologists Bulletin, 91, 1141–1166.

Chadwick, A., Arts, R. & Eiken, O. 2005. 4D seismic quantification of a growing CO_2 plume at Sleipner, North Sea. In: Dore, A. G. & Vining, B. A. (eds) Petroleum Geology: North-west Europe and Global Perspectives: Proceedings of the 6th Petroleum Geology Conference. Geological Society, London, 1–15. doi: 10.11440060001.

Chadwick, A., Noy, D., Arts, R. & Eiken, O. 2009. Latest time-lapse seismic data from Sleipner yield new insights into CO2 plume development. Energy Procedia 1, 2103–2110.

Chalbaud, C., Robin, M., Lombard, J-M., Martin, J-M., Egermann, P. & Bertin, H. 2009. Interfacial tension measurements and wettability evaluation for geological storage. Advances in Water Resources, 32, 98–109.

Chiquet, P., Broseta, D. & Thibeau, S. 2005. Capillary Alteration of Shaly Caprocks by Carbon Dioxide. Presented at the 14th Europec Biennial Conference, Madrid, 13–16 June 2005. SPE 94183.

Galloway, W. E. 2002. Paleogeographic setting and depositional architecture of a sand-dominated shelf depositional system, Miocene Utsira Formation, North Sea Basin. Journal of Sedimentary Research, 72, 476–490.

Gassmann, F. 1951. Über die Elastizität poröser Medien. Vierteljahrsschrift der Naturforschenden Gesellschaft, 96, 1–23.

Holloway, S., Chadwick, A., Lindeberg, E., Czernichowski-Lauriol, I. & Arts, R. 2004. Best Practice Manual. Saline Aquifer CO_2 storage (SACS) project.

Katsube, T. J. & Coyner, K. 1994. Determination of permeability–compaction relationship from interpretation of permeability-stress data for shales from eastern and northern Canada. Geological Survey of Canada, Current Research, 1994-D, 169–177.

Nooner, S. L., Eiken, O., Hermanrud, C., Sasagawa, G. S., Stenvold, T. & Zumberge, M. A. 2007. Constraints on the in situ density of CO_2- within the Utsira formation from time-lapse seafloor gravity measurements. International Journal of Greenhouse Gas Control, 1, 198–214.

Nordgård Bolås, H., Hermanrud, C. & Teige, G. M. G. 2005. Seal capacity estimation from subsurface pore pressures. Basin Research, 17, 583–599.

Pittman, E. D. 1992. Relationship of porosity and permeability to various parameters derived from mercury injection-capillary pressure curves for sandstone. American Association of Petroleum Geologists Bulletin, 76, 191–198.

Rodrigues, N., Cobbold, P. R. & Løseth, H. 2009. Physical modelling of sand injectites. Tectonophysics, 474, 610–632.

Schlömer, S. & Krooss, B. M. 1997. Experimental characterization of the hydrocarbon sealing efficiency of cap rocks. Marine and Petroleum Geology, 4, 565–80.

Teige, G. M. G. 2008. Sealing mechanisms at pore scale, and consequences for hydrocarbon exploration. Ph.D. thesis, University of Oslo, Norway.

Zweigel, P., Arts, R., Lothe, A. E. & Lindeberg, E. B. G. 2004. Reservoir geology of the Utsira Formation at the first industrial-scale underground CO_2 storage site (Sleipner area, North Sea). In: Baines, S. J. & Worden, R. H. (eds) Geological Storage of Carbon Dioxide. Geological Society, London, Special Publications, 23, 165–180.

Preparing for a carbon constrained world; overview of the United States regional carbon sequestration partnerships programme and its Southwest Regional Partnership

R. ESSER,[1] R. LEVEY,[1] B. McPHERSON,[1,2] W. O'DOWD,[3] J. LITYNSKI[3] and S. PLASYNSKI[3]

[1]*Energy & Geoscience Institute, 423 Wakara Way, Suite 300, University of Utah, Salt Lake City, UT 84108, USA (e-mail: resser@egi.utah.edu)*
[2]*Department of Civil and Environmental Engineering, University of Utah, Salt Lake City, UT 84112, USA*
[3]*National Energy Technology Laboratory, PO Box 1094, US Department of Energy, Pittsburgh, PA 15236, USA*

Abstract: The Southwest Carbon Partnership (SWP), one of seven United States Department of Energy-funded Regional Carbon Sequestration Partnerships, has been tasked with assessing the CO_2 sequestration potential within the southwestern United States. Carbon dioxide is considered a 'greenhouse' gas and is emitted, in large volumes, by the burning of fossil fuels and other industrial processes. CO_2 capture from point source emitters and subsequent geological sequestration is being considered as a viable short- to intermediate-range mitigation option to combat the phenomena of global warming. Significant fossil fuel reserves and consumers exist within the seven member states of the SWP and, as such, the Partnership is dedicating a large amount of resources to the challenges posed by large-scale CO_2 sequestration. Three distinct phases of work have been or will be performed by the SWP: a Characterization Phase to identify carbon capture and sequestration potential; a Validation Phase to test small-scale field injection of CO_2; and a Deployment Phase to test commercial-scale field injection of CO_2. Each phase presents challenges and opportunities to the refinement of the best approach to safe and efficient geological storage of CO_2 within the SW region of the United States.

Keywords: CO_2, sequestration, EOR, ECBM, geological storage, saline aquifer, fossil fuel, global warming, greenhouse gas

Preliminary reports by the United State Federal Energy Information Administration (EIA) indicate that total US energy-related carbon dioxide emissions have grown by 19.4% since 1990. More than 80% of US greenhouse gas emissions are energy-related carbon dioxide emissions. In 2007 US carbon dioxide emissions from fossil fuel consumption increased by 1.6%, from 5888×10^6 metric tons of carbon dioxide ($MMTCO_2$) in 2006 to 5984 $MMTCO_2$ in 2007 (United States EIA 2008).

The United States Department of Energy designated an initiative of partnerships throughout the USA and Canada to help develop the best practices for permanently storing carbon dioxide in geological formations. The seven Regional Carbon Sequestration Partnerships (RCSPs) include over 350 organizations in 42 states and four Canadian provinces (Litynski *et al.* 2008). It is anticipated that the outcomes of studies and tests will help develop regulations and infrastructure requirements for future sequestration deployment. These seven RCSPs represent regions with elements that are geologically representative of all the sequestration site variability in the USA. They encompass 97% of coal-fired CO_2 emissions, and 97% of industrial CO_2 emissions. The RCSPs initiative is being implemented in three phases:

- Phase I – the Characterization Phase (2003–2005), which characterized opportunities for carbon capture and sequestration;
- Phase II – the Validation Phase (2005–2010), in which small-scale field tests are being concluded;
- Phase III – the Deployment Phase (2010–2017), which will conduct large-volume CO_2 storage tests.

The Southwest Regional Partnership (SWP) includes all or part of eight states (Arizona, Colorado, Kansas, Nevada, New Mexico, Oklahoma, Texas, Utah and Wyoming) that represent vast opportunities for carbon storage, spanning enhanced oil recovery, saline reservoir sequestration, enhanced coalbed methane production and terrestrial carbon sequestration.

In the SW region, over 95% of anthropogenic CO_2 emissions result from fossil fuel combustion, with approximately half of these emissions from power plants (National Energy Technology Laboratory 2008). Geological storage options include coal seams (1×10^9 metric tons of minimum storage potential), natural gas and depleted and marginal oil fields (23×10^9 metric tons minimum storage potential), and deep saline formations (82×10^9 metric tons minimum storage potential). The SWP is also exploring the viability of supplanting the CO_2 currently produced from natural CO_2 reservoirs, used for enhanced oil and natural gas recovery, with anthropogenic CO_2 from power plants. The presence of CO_2 pipelines that bridge CO_2 sources and potential CO_2 sinks improves the viability of this possibility. Although terrestrial CO_2 sequestration appears to be a viable alternative in several parts of the SW region, low rainfall in some areas decreases the relative terrestrial storage capacity, limiting the applications of this option.

This paper summarizes the breadth of the RCSPs and focuses on the current results of field injection tests conducted by the SWP. The SWP has completed one pilot injection test, is in the active injection phase of two additional tests, and the results of all three are providing the basis for a commercial-scale test (exceeding 1×10^6 tons injection). This large-scale injection deployment is planned for the Gordon Creek Field, near Price, Utah, 120 miles south of Salt Lake City. The target storage reservoirs are Jurassic sandstones, present throughout the region from Wyoming to Northern New Mexico. We will inject over 1×10^6 tons of CO_2 over 4 years (up to 500 000 tons per year) from a natural CO_2 deposit and monitor migration and storage efficacy. The project will evaluate state-of-the-art monitoring techniques to determine the impact of injection operations and the integrity of the storage reservoirs. The area provides an excellent deployment test opportunity for analysis of high injection rates and high-resolution monitoring of CO_2 in multiple geological horizons. These deep saline formations are major targets for commercial-scale CO_2 sequestration associated with future coal-fired power plants planned for the area.

Overview of completed and ongoing validation tests

During 2010, the SWP will conclude five separate field tests, including the three geological tests mentioned previously as well as two terrestrial analyses, each designed to validate the most promising carbon sequestration technologies and infrastructure concepts. The field tests represent a variety of carbon sink targets, including enhanced oil recovery (EOR) with carbon sequestration, enhanced coal bed methane (ECBM) production with carbon sequestration and geological sequestration combined with terrestrial (surface) sequestration.

The primary project goal of these tests is to develop an optimum sequestration strategy for the southwestern United States, subject to the constraints unique to the region (such as water availability). Development of this sequestration 'framework' will incorporate best practices for (1) permitting and regulatory requirements, (2) comprehensive geological characterization, (3) risk mitigation, (4) monitoring, verification and accounting (MVA) protocols and (5) effective outreach and communication to stakeholders and the general public.

The SW region contains a large number and volume of geological sinks, including depleted oil and natural gas fields, saline formations and coal beds. EOR using CO_2 has been conducted in the region since the early 1970s. Several hundred miles of CO_2 pipelines through the region provide access to CO_2 near many candidate geological sequestration sites. McElmo Dome, Colorado is the major source of CO_2 for many of the EOR/ECBM operations located within the SWP region. Ideally, these sources would be supplanted by the abundant anthropogenic CO_2 sources in the region, such as coal-fired power plants.

Paradox Basin, Utah: Aneth EOR-sequestration test

Since August 2007, the SWP has conducted EOR and concomitant CO_2 sequestration in the Aneth oil field in San Juan County, Utah. Discovered in 1956, the Aneth Field is currently owned and operated by Resolute Natural Resources Company and the Navajo Nation Oil and Gas Company, and has produced greater than 149×10^6 barrels of oil (of an estimated 421×10^6 barrels of oil in place). The pilot test site is centrally located within the Paradox Basin in the Aneth mound complex (Fig. 1). The primary oil-producing zones and CO_2 sequestration targets are the Desert Creek and overlying Ismay carbonate members of the Pennsylvanian Paradox Formation. The oil and injected CO_2 are accommodated within a stratigraphic trap that is sealed by the overlying Gothic Shale and underlying Chimney Rock Shale. The approximate depth of the Desert Creek/Ismay Members is 5600 ft with a cumulative thickness of less than 200 ft. Average porosity and permeability for the CO_2 injection zone are 10% and 20 mD, respectively. Porosity and permeability measurements for the overlying Gothic Shale (thickness c. 15 ft) yield c. 3% and less than 0.01 mD, respectively.

Beginning in late 2005, we conducted pre-injection geological characterization, model development and simulations. The majority of data used for the baseline characterization of the Aneth site was obtained from geophysical well logs made available by the Utah Division of Oil, Gas and Mining (DOGM). Approximately 460 digitized well logs from 1077 Aneth area wells yielded direct and indirect data pertaining to formation depth, thickness, temperature, porosity, permeability and lithology. The digitized well logs were used to construct a 3D model of the injection site (Petrel® software by Schlumberger) from which various simulations were conducted to assess CO_2 capacity, migration of reservoir fluids and forces (CO_2, oil, groundwater, tracers, pressure, stress), as well as short- and long-term CO_2 trapping phenomena (structural, residual, solubility and mineral).

Between August 2007 and January 2010, an average of 29 000 MCF (1700 tons) per day was injected into 16 wells within the Aneth Field for a total of c. 1.5×10^6 tons of injected CO_2. Monitoring activities at Aneth are ongoing, and include periodic surface CO_2 flux surveys, downhole sampling (water, pressure, temperature, natural and artificial tracers), vertical seismic profiling, time-lapse 2D seismic, self-potential and geodetic surveys (tilt, GPS, InSar). Final analyses of each monitoring technique are anticipated by mid-2010. To date, no unanticipated migration of the injected CO_2 at Aneth, either within or beyond the injection zone, has been detected.

Permian Basin, Texas: SACROC EOR sequestration test

The SACROC Unit in the Texas Permian Basin is the oldest CO_2-EOR operation in the USA, with CO_2 injection dating from 1972; only the CO_2-EOR project in Hungary is older, dating from c. 1969 (Kane 1979). The Kinder–Morgan-operated SACROC Field/Unit lies along a trend of fields described by Galloway et al. (1983) as the Horseshoe Atoll Play (Fig. 2). The major

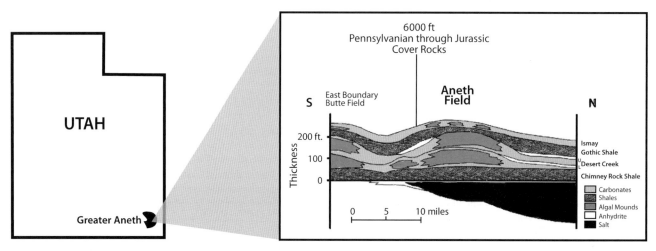

Fig. 1. Location map of Aneth Field and pilot site (left) and representative cross-section (right).

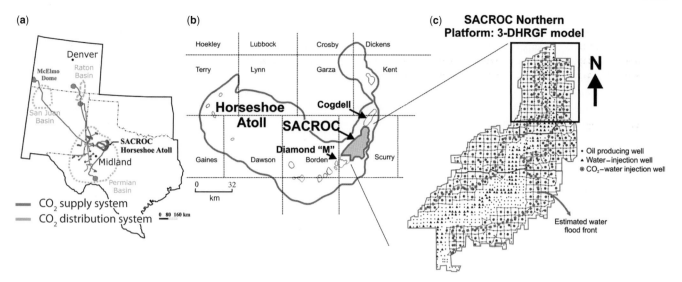

Fig. 2. Regional location map for the SACROC EOR Field, Texas.

oil-bearing and CO_2 injection zones are carbonates of the Pennsylvanian Cisco and Canyon Group formations. Lithologies of the Cisco and Canyon Group formations vary in observed facies, from interbedded pellet, crinoid, algal, to intraclast grainstones and boundstones with an average porosity and permeability of 8% and 20 mD. The oil-bearing zones are overlain by the low-permeability, Permian Wolfcamp Formation (shale). The depth of the CO_2 injection zone ranges from 6300 to 7100 ft.

The SACROC Field has an extensive dataset for evaluating virtually all aspects of the oil production and CO_2 injection since operations began in the 1950s. The SWP has gained access to much of this historical data to build comprehensive geological models and perform detailed simulations to evaluate capacity, injectivity, migration and storage of the CO_2, not only for current field tests but also to evaluate potential impacts of long-term EOR operations. The SACROC pilot represents an opportunity to match, in tandem with large-scale monitoring operations, injection of greater than 1.5×10^6 tons of CO_2. Early, fully parameterized simulations of historical CO_2 injection and multiphase flow of oil, CO_2 and brine at SACROC focused on determining the various trapping mechanisms, short and long term (Fig. 3; Han et al. 2010). An important outcome of this project is the efficiency with which CO_2 is trapped within saline aquifers that are immediately overlain by oil.

The total volume of CO_2 injected at the SACROC oil field between 1972 and January 2010 is greater than 175×10^6 metric tons (Fig. 3). At the end of the current injection test, CO_2 injection will exceed 18×10^6 metric tons per year total. Like Aneth, the monitoring activities at SACROC are ongoing and include periodic surface CO_2 flux surveys, downhole sampling (water, pressure, temperature, natural and artificial tracers), vertical seismic profiling, time-lapse 2D/3D seismic, self-potential and geodetic surveys (tilt, GPS, InSar). Some early monitoring analyses resulted in the identification of shallow groundwater flow paths, regional trends in groundwater chemistry and impacts to shallow groundwater from pre-1975 produced water evaporation pits. Additional monitoring results are not expected until mid-2010. Currently, no unanticipated migration of the injected CO_2 at SACROC, either within or beyond the injection zone, has been detected.

Because of the relatively large dataset at SACROC, encompassing a greater number of years of CO_2 injection, the pilot will be an initial high-resolution analysis of the potential for CO_2 storage in the broader carbonate Horseshoe Atoll system. Given that most of the western side of the atoll is below the oil–water contact, it is particularly agreeable for large-scale sequestration.

San Juan Basin, New Mexico: ECBM sequestration test

The San Juan basin (SJB) is one of the top ranked basins in the world for CO_2 coalbed sequestration because it has: (1) advantageous geology and coals with high methane content; (2) abundant anthropogenic CO_2 from nearby power plants; (3) low capital and operating costs; (4) well developed natural gas and CO_2 pipeline systems; and (5) local companies, for example, ConocoPhillips, with coalbed methane (CBM) and ECBM expertise. The geology consists of the thick (>200 ft) coal-bearing Fruitland Formation with an average porosity and permeability of >10% and >200 mD, respectively. The Fruitland coal is saturated with methane (and variable quantities of low total dissolved solids (TDS) groundwater) that is contained structurally and stratigraphically by the low-permeability Kirtland Shale.

ConocoPhillips has operated the SJB project in collaboration with the SWP, specifically to examine ECBM efficacy with CO_2 sequestration. Pilot CO_2 injection began in July 2008, and

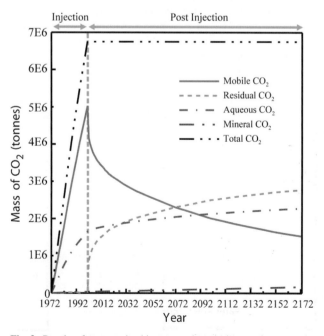

Fig. 3. Results of an extensive history-match analysis to evaluate trapping mechanisms at SACROC for 100 years starting in 1972.

continued for c. 1 year; the total amount of CO_2 injected is c. 18 000 tons. The SJB pilot has tested a suite of monitoring approaches tailored for its unique geology (e.g. increased geodetic surveys to assess swelling of coalbeds) and value-added benefit (methane production). Early monitoring results suggest differential transport rates between natural (N_2) and man-made (perfluoroocarbon) tracers and the injected CO_2, although a full analysis awaits. Likewise, tilt-meter and GPS stations indicate no significant swelling of the CO_2-saturated coals. No unanticipated migration of the injected CO_2 at the SJB has been detected.

SWP terrestrial pilot test and terrestrial Riparian Restoration Project

Terrestrial carbon capacity in the SW region is limited by low average annual precipitation and yearly variability in precipitation. Even in systems managed for carbon storage, wet years followed by a series of dry years may result in a net carbon flux out of the system. Limited opportunities exist for increasing carbon storage on rangelands because most areas are at a relatively stable equilibrium with respect to land use history and management. Much of the desert grassland and shrubland areas with less than 12 inches of annual precipitation are subject to loss of cover and exposure to wind and water erosion. Retaining soil carbon levels in these ecosystems will require active restoration practices that are challenging given current technologies. Two demonstration projects are nearing completion: a regional-scale terrestrial analysis of land-use trends and a local-scale arid-land restoration project.

The regional analysis includes a detailed reporting and monitoring system that functions consistently across hierarchical scales and is compatible with the existing technology underlying the DOE's Energy Information Administration Voluntary Reporting of Greenhouse Gases (1605b) Program. Within this system, the project has: (1) developed improved technologies and systems for direct measurement of soil and vegetation carbon at reference sites selected within the region; (2) developed remote sensing and classification protocols to improve mesoscale (km^2) soil and vegetation carbon estimates; (3) constructed ecological process (state and transition) models that reflect soil/vegetation changes resulting from current land use and land use associated with implementation of programs to sequester carbon or reduce carbon losses; and (4) developed a regional inventory and decision

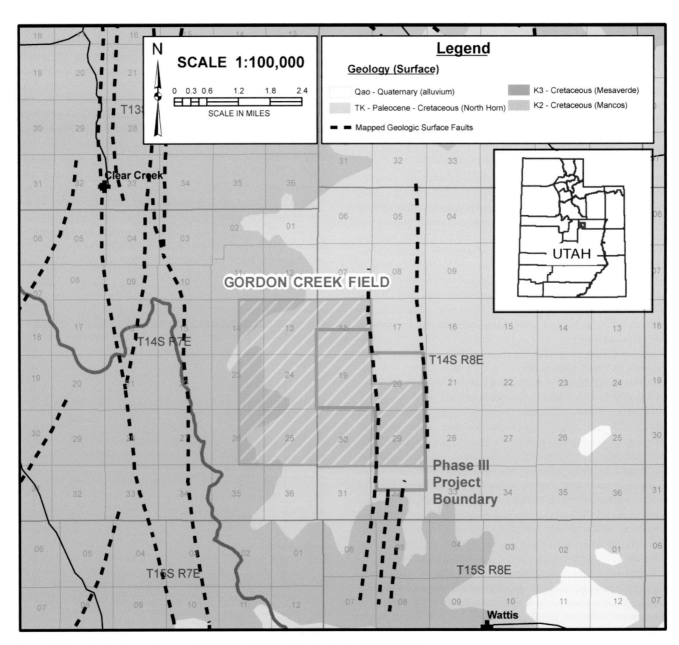

Fig. 4. Site location/surface geology map for the Phase III Site at the Gordon Creek Field, Utah.

support tool. The value-added products of the test include new carbon credits and increased land productivity.

The local-scale arid-land restoration project entailed desalination of small quantities of produced water from the SJB ECBM pilot (described previously in this paper). This water was utilized for irrigating a tract of arid land, with carbon soil content measured before and after irrigation. Preliminary results suggest over 10% increases in soil carbon uptake resulting from this irrigation. The test site is the San Juan Basin Coal Fairway, near Navajo City, New Mexico, and involves such value-added products as new carbon credits, improved water quality and improved ecological conditions.

Medium-scale validation project impacts

The SWP medium-scale validation tests will impact the USA by providing a comprehensive assessment of the sources and potential sinks for CO_2 in the SW region. These data are being integrated with the data from other partnerships to provide a database covering the entire nation. The data generated by these field tests are providing information to evaluate potential commercial-scale sequestration projects in the southwestern USA and other regions of the country. Some value-added benefits of the project include enhanced recovery of oil, natural gas and coalbed methane. Part of the value-added benefits for oil, natural gas and methane recovery is that revenue from sale of the recovered hydrocarbons mitigates some of the cost of CO_2 storage. Currently, all such enhanced resource recovery operations use CO_2 drawn from natural CO_2 reservoirs. If all enhanced recovery operations in the southwestern USA were to use powerplant-generated CO_2 (from fossil fuels) rather than natural CO_2, it is estimated that the region would see at least a 10% reduction in greenhouse gas intensity.

Commercial-scale deployment project overview

Gordon Creek Field, Utah: commercial-scale injection test

The SWP will accomplish a major sequestration deployment in the Gordon Creek Field, located 120 miles SE of Salt Lake City in central Utah (Fig. 4). This test will follow an injection schedule over four years, leading up to 500 000 tonnes (551 000 US tons) of CO_2 per year, and a total injection amount exceeding 1 000 000 tonnes. The target formations for this deployment are several deep Jurassic sandstones. These formations are also targets of potential commercial sequestration throughout the western USA. The SWP plans include a possible 'dual completion' with injection in two different formations at the same time. By carrying out two tests in two different formations within the same stratigraphy, portability of science and engineering results can begin to be evaluated.

Gordon Creek sits on an elongated surface anticline located along the northern plunge of the San Rafael uplift. Drilling at Gordon Creek in the 1950s–1970s resulted in the discovery of significant deposits of CO_2 in the Jurassic–Triassic Sinbad Member of the Moenkopi Formation and Permian White Rim Sandstone. Also, CO_2 shows have occurred in a number of other Triassic, Permian and Pennsylvanian reservoirs. The most significant and consistent shows have been in the White Rim Sandstone. The area provides an excellent deployment test opportunity for analysis of high injection rates and high-resolution monitoring of CO_2 in multiple reservoirs. These deep saline formations are major targets for commercial-scale sequestration associated with future coal-fired power plants and CBM (coalbed methane) development planned for the area. For this research project, no profits will be earned, and no CO_2 will leave the site. Much of the Gordon Creek site falls under the jurisdiction of the state government of Utah.

For the sequestration deployment test at the Gordon Creek Field, the primary target formation is the Jurassic Navajo Formation Sandstone. The Navajo Formation lies within a true 'stacked' system of alternating reservoirs and seals (Fig. 5), and possesses ample storage capacity for a commercial-scale sequestration demonstration (>1 000 000 tons of CO_2).

The Gordon Creek Field is a methane-producing field located on the eastern flank of the Wasatch Plateau, which has produced c. 4 MMcf of methane from the Cretaceous Ferron Sandstone since 2002 (Utah DoGM, pers. comm.). Additionally, the Permian White Rim Sandstone hosts an untapped source of >98% pure CO_2 (Morgan & Chidsey 1991). A 9 mile long by 5 mile wide anticline serves as the structural trap for the Ferron Sandstone methane and possibly the White Rim Sandstone CO_2 (Fig. 5; estimated from existing data by T. C. Morgan). The anticline itself is obliquely divided by north–south normal faults that terminate in the middle Jurassic evaporite deposits of the Carmel Formation. The entire sedimentary package at the Gordon Creek Field is in excess of 15 000 ft thick, although wells within the field only penetrate the Permian White Rim Sandstone (Fig. 5). The thick (>4000 ft) Cretaceous Mancos Shale is exposed at the surface of the Gordon Creek Field. Beneath the Mancos Shale, from the

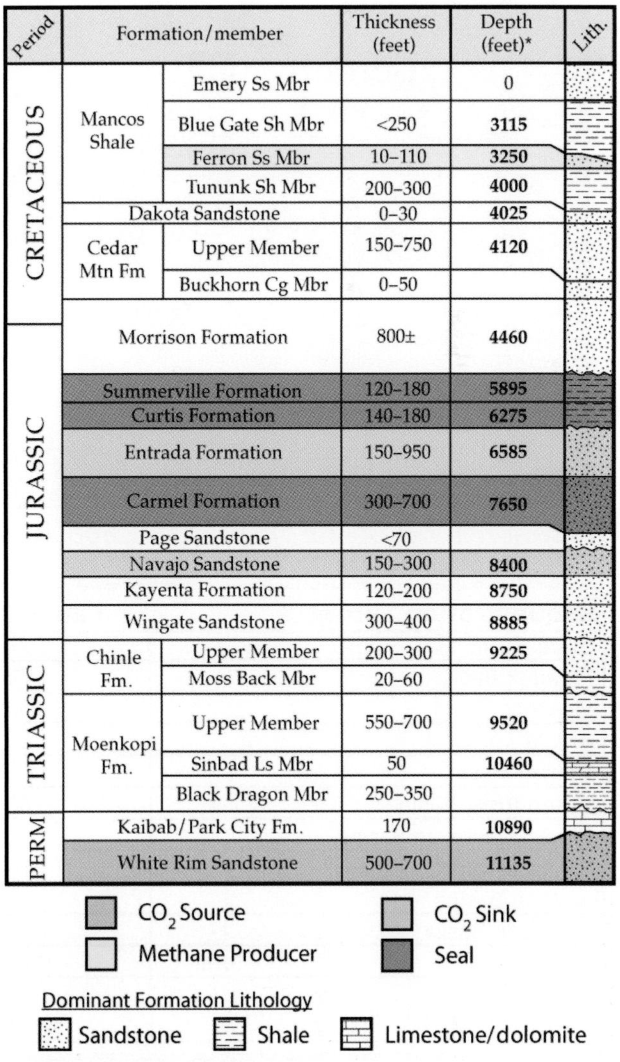

Period	Formation/member		Thickness (feet)	Depth (feet)*
CRETACEOUS	Mancos Shale	Emery Ss Mbr		0
		Blue Gate Sh Mbr	<250	3115
		Ferron Ss Mbr	10–110	3250
		Tununk Sh Mbr	200–300	4000
	Dakota Sandstone		0–30	4025
	Cedar Mtn Fm	Upper Member	150–750	4120
		Buckhorn Cg Mbr	0–50	
JURASSIC	Morrison Formation		800±	4460
	Summerville Formation		120–180	5895
	Curtis Formation		140–180	6275
	Entrada Formation		150–950	6585
	Carmel Formation		300–700	7650
	Page Sandstone		<70	
	Navajo Sandstone		150–300	8400
	Kayenta Formation		120–200	8750
	Wingate Sandstone		300–400	8885
TRIASSIC	Chinle Fm.	Upper Member	200–300	9225
		Moss Back Mbr	20–60	
	Moenkopi Fm.	Upper Member	550–700	9520
		Sinbad Ls Mbr	50	10460
		Black Dragon Mbr	250–350	
PERM	Kaibab/Park City Fm.		170	10890
	White Rim Sandstone		500–700	11135

■ CO_2 Source ■ CO_2 Sink
□ Methane Producer ■ Seal

Dominant Formation Lithology
Sandstone Shale Limestone/dolomite

Fig. 5. Stratigraphy of the Gorden Creek area (modified from Hintze 1992); Ss, sandstone; Sh, shale; Ls, limestone; Cg, conglomerate; Fm., formation; Mbr, member.

mid Cretaceous to the Mississippian, are a series of interbedded sandstones, shales and limestones. Potential CO_2 sinks in the section include the Dakota, Morrison, Entrada, Navajo, Wingate and White Rim sandstones, although the Navajo Sandstone is the preferred candidate due to its intermediate depth and proven injectivity; the Navajo is currently being utilized as a salt water disposal zone for waste waters from the methane production in the Ferron Sandstone. Assuming an average depth of 8400 ft and an average thickness of 215 ft, the Navajo Sandstone within the 45 square mile anticlinal structure of the Gordon Creek Field has a CO_2 capacity of 5.9 MMtons (10% porosity with only 1% of the porosity occupied with CO_2). The Entrada Sandstone may have additional capacity of about three times this volume. The primary seal stratigraphically above the Navajo Sandstone is the Carmel Formation, a series of interbedded, low-porosity, low-permeability limestones, shales and sandstones. Additional stratigraphic seals include the Curtis and Summerville formations (shales and low permeability sandstones) and the thick Mancos Shale.

The project will require extensive monitoring and simulation to determine if storage operations are effective with respect to

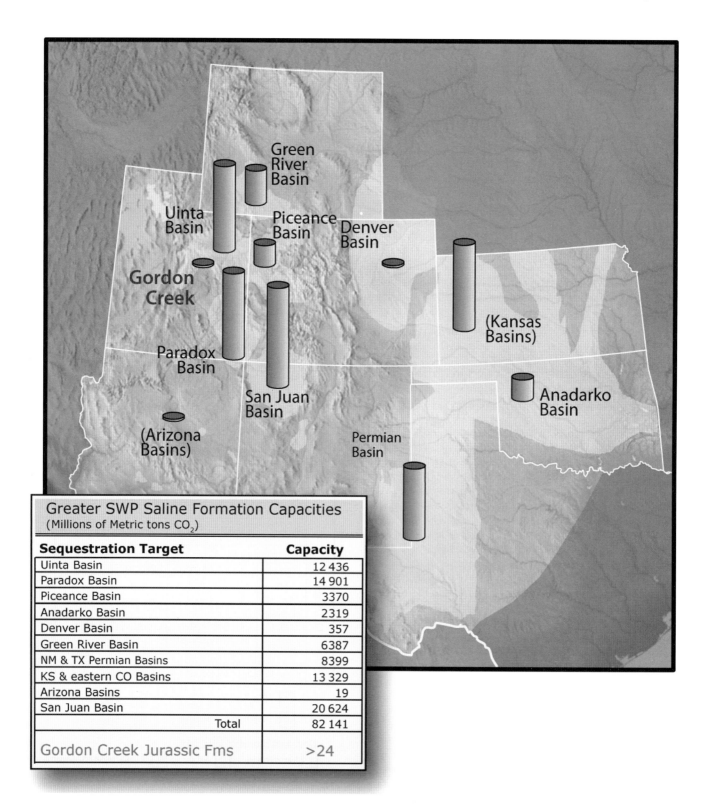

Fig. 6. SWP CO_2 sequestration capacities for deep saline formations. Size of column for each site represents relative CO_2 storage capacity.

trapping injected CO_2. Vertical seismic profiling and other geophysical methods will be particularly utilized, given their proven ability to resolve CO_2 plumes (Hovorka *et al.* 2006). Monitoring techniques that will be used include repeat 3D (4D) seismic surveys, pressure monitoring, groundwater chemistry monitoring, pressure and fluid sample monitoring from other locations, soil gas sampling, and other methods. A variety of software code created and modified by SWP organizations as well as commercial/public simulation model tools are in use for this project.

Anticipated carbon storage capacity

The SWP determined that the region's point sources emit $c. 335 \times 10^6$ tonnes (369×10^6 US tons) of CO_2 per year (National Energy Technology Laboratory 2008); over 100 years, assuming no change in emissions rate, these emissions translate to $c. 3.4 \times 10^9$ tonnes ($c. 3.7 \times 10^9$ short tons) of total storage capacity needed. Preliminary geological characterization by the SWP provides an initial estimate of capacity of the deep Jurassic and older saline formations in the SWP region to exceed 82×10^9 tonnes ($c. 90 \times 10^9$ short tons; National Energy Technology Laboratory 2008). The Gordon Creek Field site, in comparison, is projected to store at least 24×10^6 metric tons of CO_2 in formations that are Jurassic in age (Fig. 6). During continued project activities, the SWP and collaborators at the National Energy Technology Laboratory will continue to refine capacity estimates and evaluate injectivity and other critical factors relevant to regional storage goals.

Conclusions

The Southwest Regional Partnership has been charged by the US Department of Energy, along with six other regional partnerships, with providing a framework for the safe and efficient storage of CO_2 within geological sinks. To that end, the SWP has successfully completed two of the three phases of the DOE mandate: (1) assess the major point sources of anthropogenic CO_2 and potential sinks within the southwestern USA; and (2) utilize small pilot CO_2 injection operations to evaluate the steps necessary for larger scale ($>1 \times 10^6$ metric tons of CO_2 injected per year). The third phase of the SWP operations, commercial-scale CO_2 injection, will be commencing in mid to late 2010.

Within the SWP, annual emissions from fossil fuel burning electrical generation as well as cement and fertilizer plants exceed 335×10^6 metric tons of CO_2. With a conservative estimate of 82×10^9 metric tons of storage in deep saline aquifers alone, the SWP can potentially sequester over 100 years of anthropogenic CO_2, at current emission rates. Additional geological storage potential within depleted oil reservoirs (EOR) and unmineable coal seams (ECBM) has been successfully assessed by SWP organizations at sites in Texas, Utah and New Mexico. Detailed regulatory observance, geological characterization, simulations and surface/subsurface monitoring have been utilized at each site in order to determine the best approach to CO_2 sequestration elsewhere.

The third phase of the SWP project is the large-scale injection of CO_2 within a deep saline aquifer in the Gordon Creek Field of central Utah. Up to 1×10^6 metric tons of CO_2 will be injected into the Jurassic Navajo Formation over a four year period beginning in 2010. The lessons learned from the pilot-scale injection operations in Texas, Utah and New Mexico will be applied to the project at the Gordon Creek site, although it is anticipated that deep saline CO_2 sequestration will have a unique set of challenges when compared with previous EOR or ECBM studies. Ultimately, the lessons learned from all phases of the SWP and other regional partnership programs will populate DOE/NETL databases so that future CO_2 sequestration operators can achieve maximum efficiency while minimizing risk.

References

Galloway, W. E., Ewing, T. E., Garrett, C. M., Tyler, N. & Bebout, D. G. 1983. *Atlas of Major Texas Oil Reservoirs*. The University of Texas at Austin, Bureau of Economic Geology, Austin, TX.

Han, W. S., McPherson, B. J., Lichtner, P. C. & Wang, F. P. 2010. Evaluation of trapping mechanisms in geologic CO_2 sequestration: case study of SACROC northern platform, a 35-year CO_2 injection site. *American Journal of Science*, **310**, 282–324; doi: 10.2475/04.2010.03.

Hintze, L. F. 1992. *Geologic History of Utah*. Brigham Young University, Provo, UT. Geology Studies Special Publications, **7**.

Hovorka, S. D., Benson, S. M. *et al.* 2006. Measuring permanence of CO_2 storage in saline formations: the Frio experiment. *Environmental Geosciences*, **13**, 1–17.

Kane, A. 1979. Performance review of a large-scale CO_2-WAG Enhanced Recovery Project, SACROC Unit Kelly–Snyder Field. *Journal of Petroleum Technology*, **31**, 217–231.

Litynski, J. T., Plasynski, S., McIlvried, H. G., Mahoney, C. & Srivastava, R. D. 2008. The United States Department of Energy's Regional Carbon Sequestration Partnerships Program Validation Phase. *Environment International*, **34**, 127–138; doi: 10.1016/j.envint.2007.07.005

Morgan, T. C. & Chidsey, T. 1991. Gordon Creek, Farnham Dome and Woodside fields, Carbon and Emery Counties, Utah. *In*: Chidsey, T. C., Jr (ed.) *Geology of East-central Utah*. Utah Geological Association, Salt Lake City, UT, **19**, 301–309.

National Energy Technology Laboratory. 2008. *Carbon Sequestration Atlas of United States and Canada*. NETL, Washington, DC.

United States EIA. 2008. *US energy-related carbon dioxide emissions rose by 1.6 percent in 2007*. Press Release. Retrieved May 20, 2008. World Wide Web Address: http://www.eia.doe.gov/neic/press/press298.html.

Index

Page numbers in *italic* denote figures. Page numbers in **bold** denote tables.

AAPG. *See* American Association of Petroleum Geologists (AAPG)
å1 Basin area 139
Abolag-1 well
   stromatolitic carbonates *680*
   Taoudenni Basin *679*
Abu Zenima 772
Acacus Formation 686
acoustic impedance (AI) 510
   Chalk units *543*, 547
   juxtaposition difference value of zero 513
   PHIT relationship 547
   South Arne 527, *528*
   v. total porosity *543*
acoustic pull-ups
   MMU 84
acquisition and processing. *See* Four dimensional acquisition and processing
acquisition plan
   ISROCK-96 seismic survey *556*
   Rockall Trough *556*
active rifting
   Angola 1046
Adriatic basin
   geological sections *117*
   hydrocarbon occurrences *117*
   Porto Corsini East gas field *122*
Adriatic Sea 121
aeolian sandstones 826
aeromagnetic compilations
   average density *570–571*
   basement rocks magnetic properties **570**
   basement structure map *575*
   bathymetry and topography map *561*
   datasets 559–569
      aeromagnetic 560
      bathymetric 559–560
      gravity 560–565
      petrophysical 565–569
      topographic 559–560
   Greenland *580–581*
   magnetic data *579*
   methods and applications 559–583
   modelling 571
   Norway 559–583
   Norwegian–Greenland Sea *579*
   results 569–580
      basement structures 569–573
      North Atlantic oceanic spreading 578–580
      quaternary sand channels 575–576
      salt-related magnetic anomalies 576–578
      sedimentary subcrop pattern 573–575
   rocks magnetic properties **572**
   surveys **562**, *579*
   susceptibility log *577*
aeromagnetic data acquisitions
   NUG 33

aeromagnetic modelling
   NUG 33
Africa. *See also* specific areas e.g. North Africa
   erosion 679
   exploring older and deeper hydrocarbon plays 681–682
   heritage palaeohighs 718
   orogen 708
AFTA. *See* apatite fission track analysis (AFTA)
aggradational prism 89
aggrading channels
   architecture 84
   central North Sea basin Late Neogene below Horizon 6, 83–85
   geological controls 89–90
Ægir ridges
   spreading rates 579
Ahara High 712–714
   Algeria 712–714
      interpretation 713–714
      stratigraphic architecture 712–713
   Cambro-Ordovician succession stratigraphic pinchout *712*
   chronostratigraphic chart *720*
   Devonian succession *713*
   interpretation 713–714
   shape 713
   stratigraphic architecture 712–713, *713*
Ahara uplift
   Silurian 713
Ahnet Basin 686, 729, 733, 738
   Azzel Matti Ridge *720*
   chronostratigraphic chart *720*
   Fammenian sediments 738
   Palaeozoic petroleum system 686
AI. *See* acoustic impedance (AI)
airborne laser scanner 20
Airy–Heiskanen root 565
Ajjer Tassil outcrops
   Palaeozoic succession stratigraphic column *709*
Alaminos Canyon area
   roof-edge thrust *904*
Alaminos Canyon 818 site 1151–1152
Alaska
   Amethyst State-1 well *637*, *638*, 639, *640*
      palaeotemperatures *639*
   Brookian sequence 828
   exhumation episodes 642
   Mount Elbert Prospect 1151
   North Slope
      burial/uplift history *639*
      gas hydrates occurrence 1152
      individual episode magnitude 643
      location map *637*
      wells *640*
   sedimentary rocks magnetic properties **572**

Alba Field 337–342
   Alba drilling technology 340–342
   cartoon geological models *340*
   4D data showing water cones *341*
   4D seismic results 346
   4D seismic showing OWC movement *341*
   EOR mechanisms 346
   fault baffling 340
   fault interpretation *339*
   field edge seismic tuning *342*
   first step-change 337
   injected sands 340
   lithology distribution 342
   MPD 346
   pilot hole-attic oil capture *343*
   production *338*
   reservoir level
      faulting 337
   seismic data history **339**
   seismic datasets 340
   seismic technology 337–340
   shear impedance inversion 337
   TTD 346
   UKNS maximizing offshore heavy oils fields 337–347
   field history 337–342, 337–347
   viscous oil 339
Alba Northern Platform (ANP) 340
   drilling slots 342
Al Basin
   wells VR trends *100*
Alberta
   BCG stratigraphic summary *1102*
   British Colombia border *1106*
   Cold Lake Field 7
   4D seismic steam conformance maps *1149*
   Cretaceous section 1099
   Early Cretaceous 1102
   heavy oil deposits *1142*
   Lower Cretaceous McMurray Formation 1062
   sandstone *v.* conglomerate *1111*
Albian age
   Early
      Wabiskaw Formation 1144
Albian–Cenomanian *120*
   foreland 124
   hydrocarbon occurrences 124
Albian Pinda carbonate play 1045
Albian Wabiskaw Formation 1144
Albo-Aptian Kopervik Sandstone 53
algal-derived fluorescent amorphous organic matter (AOM) 98
Algeria 707, 708, 709, 718, 721, 729, 731, 733
   Ahara High, Illizi, and Berkine basins 712–714

Algeria (*Continued*)
  Early Silurian Tannezuft hot shale thermal maturity *690*
  field map *1168*
  Neoproterozoic geology *677*
  platform sub-basins 707–708
  reservoirs *753*
  tectonic events *689*
  Tihemboka High, Illizi and Murzuq basins 715–716
Allen, Mark 589
allochthonous biofacies
  North Sea Eldfisk Field 481
Allochthonous Complex 119
Almada Basin 875
  conjugate margin evolution 879–880
  data 868–869
  2D gravity modelling outline 870
  margin setting 869–874
  nature of crust transitional domain 874–877
  salt diapirs 871
  total rift setting 877–879
Almada–Gabon margin 868
Alpin Anticline 967
  seismic profile *969, 972*
Alpine
  collision 265
  compression 975
  orogen 113
Alpine Foredeep 743
Alpine foreland
  petroleum geoscientists virtual fieldtrips 25
Alportian–Pendleian mudstones 1091
Alps 113
  Villafortuna–Trecate oil field *119*
Alston Block 1091, *1092*
Altay mountains 660
amalgamated turbidites 279
Amerasia Basin 589, *590*
American Association of Petroleum Geologists (AAPG) 13
Amethyst State-1 well, Alaska *637, 638,* 639, 640
  palaeotemperatures 639
amplitude variation with offset (AVO) 245, 279, 448
  applied time-lapse seismic methods 528–533
    compaction monitoring on northern crest 532–533
    fault control of injected water 530–532
    reservoir depletion of northern crest 528–530
  character 443
  distribution 441
  Donan Field *439, 441, 442*
    signature *439*
  Forties response with gas v. oil *220*
  gas sand calculation *291*
  gas sand thickness uncertainty *295*
  Laggan *283, 285, 289, 291, 295*
  Lochranza field development *449*
  net gas sand map *285*
  rock physics analysis 437
  Seismofacies *289*
    traces *294*
  structural map *283*
    Tormore *283,* 296

time-lapse rock physics 528–533
  Tormore *283*
analogue models 19
Andrew fan axis 436
Andrew Formation
  interbedded turbidites 431
Aneth Field
  location map *1190*
Angola. *See also* specific location
  active rifting 1046
  biogenic source rocks TOC 1052
  block 12
    thermogenic petroleum systems post-salt *1051*
  block 15
    post-salt gas plays *1050*
  block 31, *1014*
  block zero seismic amplitude gas indications *1051*
  database *1046*
  deepwater 1013, 1014
  direct gas indicators *1054*
  estimated methane solubility *1054*
  future gas exploration 1043–1057
    biogenic gas unconventional petroleum system 1052–1056
    conceptual Neoproterozoic petroleum system 1055–1056
    regional framework 1044–1045
    thermogenic petroleum systems 1045–1052
      Post-Aptian salt plays 1049–1052
      Pre-Aptian salt plays 1045–1049
  gas discoveries *1047*
  gas generation schematic *1052*
  generalized basal heat flow *1048*
  hydrocarbon maturation studies 1046
  methane generation microbial ecosystem succession *1053*
  oil exportation 1043
  oil production 1043
  possible biogenic gas play *1055*
  pre-salt gas plays 1046
  sedimentary basins 1044, *1044*
  syn-rift areas 1046
ANP. *See* Alba Northern Platform (ANP)
Anti-Atlas 729
Antrim Shale-analogues with biogenic methane in Europe 1083
AOM. *See* algal-derived fluorescent amorphous organic matter (AOM)
apatite fission track analysis (AFTA) 633, 980
  Amethyst State-1 well *638,* 639, 640
  Arctic *641*
  burial history reconstruction *635*
  Carboniferous sandstones 996
  Denmark 642
  detailed analysis 639
  Eocene palaeogeothermal gradients *636*
  Gro-3 well in Nuussuaq Basin 640
  Norwegian wells *989*
  palaeotemperatures characterizing 636
  Svalbard 640
  Trail Island 641
Apennines 113
  Eocene to Pliocene times 113
  frontal thrusts *121*
  geological sections *117*
  hydrocarbon occurrences *117*

mountain chains 113
  Pliocene times 113
  Porto Corsini East gas field *122*
  Settala gas field *123*
  Val d'Agri oil field *121*
  Villafortuna–Trecate oil field *119*
Appalachian Basin
  US 1090
applied time-lapse seismic methods 523–534
  reservoir management 523–534
  South Arne 523–524
  time-lapse rock physics AVO inversion 528–533
    compaction monitoring on northern crest 532–533
    fault control of injected water on SW flank 530–532
    reservoir depletion of northern crest 528–530
  time-lapse seismic data acquisition 524–528, *525*
    AVO inversion 526–527
    time shifts 525–527
  time-lapse time shift observations 528–533
Aptian 662
  coal-bearing muddy sediments 598
Apulian Platform 119
  Cretaceous age *120*
  geological sections *117*
  hydrocarbon occurrences *117*
  Val d'Agri oil field *121*
aquifer
  saline 1191, 1195
  seal capacity
    determined 498
    v. holes 502
Arab Field
  Jurassic GDE map *795*
Arab Formation
  carbonate grainstone 795
  grainstones 796
Arabian Palaeozoic petroleum system 692
Arabian Plate 684, 812
  geology 792
  Middle East exploration potential 794–799
Ara Group 674
Arctic *641. See also* Russian Arctic shelves (RAS) geology
  assessment of undiscovered petroleum resources 621–631
  frontier margins 823
  offset and curvature of Novaya Zemlya 645–656
  sedimentary basin *590*
Areoligera gippingensis biomarker 435
Argana Basin *930*
  location map *922*
Argana Valley *927*
Arkitsa fault, Greece
  petroleum geoscientists virtual fieldtrips 24–25
Arne–Elin Graben 129
Arnsbergian Sabden Shale
  general characteristics 1088–1089
A sandstones
  Dooish 950
  Dooish area 950

Rockall Basin *946*, 950
Rockall Basin palaeodrainage 944, 946
Ashgill aged glacial valleys 717
Askrigg Block 67, *1092*
Asl Sandstone 778, *778*
Asmari carbonate reservoir
   Cenozoic age *791*
Asterosoma 133
   facies core photos *135*
Asti Formation
   foreland ramp 123
Atar Group 679
Athabasca
   Lower Cretaceous McMurray
      Formation 1141
Athabasca Basin
   heavy oil deposits *1145*
   sediments 1144
   surface mining map *1149*
Athabasca Oil Sands
   Lower Cretaceous McMurray
      Formation 1062
   reservoir characterization 1141–1150
Athabasca reservoir characterization impact 1141–1150
   managing large well datasets 1149–1150
   regional setting 1141–1144
   reservoir heterogeneity 1147–1149
   stratigraphy 1144–1147
Atlantic. *See also* specific areas e.g. North Atlantic
   borderland
      location map *922*
      crustal thickness map *864*
      lead space plots *948*
      pre-drift reconstruction *1031*
Atlantic basins
   base basalts *1028*
   Mesozoic 938–940
   seismic section *1028*
Atlantic Margin
   basin inversion 971
   bathymetric map *964*
   Cenozoic unconformities *974*
   geological and structural features *844*
   imaged sub-basalt strata 1029
   Ireland thick sediments 924
   Mesozoic 938–940
   Permo-Triassic Basin evolution 921–934
      Central High Atlas Basin, Morocco 931–934
      developing analogue depositional 928–929
      Fundy Basin, Nova Scotia, Canada 929–931
      geological framework 921–927, 922–925, 925–927
      integrated regional study 921
      provenance studies 927–928
      structural and terrane map *923*
Atlantic sub-basalt hydrocarbon prospectivity 1025–1031
   Faroe–Shetland and Møre basins 1025–1028
   Rockall 1028–1029
   seismic data acquisition and processing 1025
Atlantis Field
   OBS nodes 1016

Atlas Basin
   Morocco 928
Atwater Valley 1151
Auðhumla Basin
   bathymetric gorge *973*
Auðhumla Basin Syncline 969, 973
   seismic profile *973*
Auk–Argyll–Flora area 60–62
Auk–Argyll region 58, 59
Auk–Flora Ridge 66
   seismic profile
      along crest *61*
      crossing northern flank *62*
      Dogger Granite *63*
      southern flank *60*
Auk Terrace 200
   remnant topography *196*
Australia. *See also* specific location
   Jackson Field 355
   petroleum systems 828
Austria. *See also* specific location
   shale 1081
Austrian Unconformity 742
   recognition 749
AVO. *See* amplitude variation with offset (AVO)
Avon shale 206
Awanat Wanin I formation
   incised valleys *712*
Axial Structural Domain
   Novaya Zemlya 651
axis-perpendicular sub-regional correlation
   Dagny area 173
   Sigrun–Gudrun area 168–173
Azienda Generale Italiana Petroli (AGIP) 113–114
AZIP. *See* Azienda Generale Italiana Petroli (AGIP)
Azzel Matti Ridge 716
   chronostratigraphic chart *720*
   Silurian sediments 717

back-barrier sediments
   Gert-1 well *148*
   Upper Jurassic barrier island system of Freja discovery 149–150
Badenian
   deepwater marine shales 1126
Baghewala Oil Field
   Rajasthan, India 676
Bai Hassan
   wells 804
Bakken Fairway 1065
Bakken Formation 1066, *1075*
   API *1075*
   bulk kinetic parameters 1067
   computed transformation ratios *1071*
   Devonian age 1065
   gas chromatography analysis 1067
   GOR 1067, *1074*, *1075*
   maturity map *1069*
   MSSV *1072*
   oil gravity data 1067, *1074*
   PEE *1070*, 1073
   petroleum generation threshold *1072*
   petroleum migration 1066
   PGI *1070*, 1073
   phase distribution map 1071
   Rock-Eval pyrolysis 1067, *1069*, *1070*

stratigraphic column *1066*
TOC 1067, 1073
upper *v.* lower 1073
Williston Basin 1065–1076
Bakken hydrocarbon pools
   GOR *1073*
   oil gravity data *1073*
Bakken member
   maturity *1070*
   pseudo van Krevelen diagram *1070*
   TOC *1070*
Bakken petroleum system 1065–1076
   analytical techniques 1067
   data sources 1067
   geological overview 1065–1067
   goals 1067
   results 1068–1073
      bulk geochemical signature 1068–1073
      gas–oil ratio 1071–1072
      oil density 1072–1073
      phase behaviour 1070–1071
      pyrolysis gas chromatography 1068–1069
      source rock kinetics 1070
Bakken shales
   TOC 1066
Balder Field 46
   hydrocarbons *48*
Balder Formation
   Eocene age 45
   maturity map *662*
Balder Sand Formation 49
Balmoral. *See also* Lower Balmoral
   sandstone system 441
Baltica Craton 604
Baltica Timanide sources 595
banked oil *356*
Barents Basin 604
   geological cross-sections *600*
   plate-tectonic setting *603*
Barents–Kara region *592*, *593*
   areal occurrence *602*
   depth-to-basement map *599*
   petroleum play elements chart *601*
Barents Megabasin 598
Barents–Pechora Shelf 601
Barents Province 598–603, *600*, *601*
Barents region 598
   gas trap formation 603
Barents Sea *564*, *565*, 591
   basement thickness 893
   Cenozoic exhumation episodes *637*
   exhumation episodes 642, 643
   exhumation histories 634–637
   free-air gravity field *595*
   geology *634*
   high-resolution aeromagnetic surveys 576
   individual contributions to basement depth map 889
   individual episode magnitude 643
   Moho map 892
   Paleocene cooling 637
   remapped 560
Barents Shelf 593, 649
   basinal connections 656
   fold-and-thrust belt 646, *653*
   palaeogeography maps *655*

Barents Shelf (*Continued*)
   pre-Permian subsidence events 647
   structural curvature 654–656
   structural pattern 651
   subsidence history 654–656
Barnett Shale 7, 1131
   bedding-parallel fractures *1137*
   Fort Worth Basin 1087, *1088*
   fracture *1137*
   gas production 1087
   mudrock cores 1134
   mudrock sample *1137*
   subcritical crack index measurements *1138*
   tall calcite-sealed fractures *1133*
   thermal maturity 1087
   UK shale gas 1087
barrier island sandstones
   Freja discovery area 146
   palaeotopography 150
   reservoir architecture 145–154
basal heat flow
   Angola *1048*
basal Silurian hot shale source rock
   Murzuq Basin 689
basal Silurian succession 712
basal Triassic Bunter Shale 38
basalts
   escarpments *969, 970*
   NE Atlantic basins *1028*
   Palaeogene age 1025
   Rockall–Faroe area *969, 970*
basal Volgian–Ryazanian megasequence unconformity (BVRU) 196, *204*
   broad log character motifs 199
   formation 200
   tectonic reconfiguration events *209*
   well correlation *198, 202*
Base Balder unconformity
   Faroe–Shetland Basin 975
base basalts
   NE Atlantic basins *1028*
Base Buzzard 373
Base Chalk
   Central North Sea Chalk role 502
   hydrocarbon discoveries *v.* aquifer seal capacity *505*
Base Cretaceous 38
   clastics *433*
   prospect map *433*
   surface *44*
Base Cretaceous Unconformity (BCU) 177, 186, 419, 505
   aquifer gradient 502
   hydrocarbon discoveries *v.* aquifer seal capacity *504*
basement
   Caledonian or Precambrian 893–894
   definition 893
Base Permian Saalian Unconformity 37
base salt maturity present day
   Bucomazi Formation *1048*
base volcanic reflector 1038
Base Zechstein
   fault trends *43*
Basilicata–Calabria accretionary arc
   Val d'Agri oil field *121*
basin architecture
   deepwater reservoirs Palaeogene *v.* Late Jurassic 50

basin-centred gas (BCG) *1104*, 1112
   distribution *1106*
   3D structure *1117*
   gas expulsion *1112*
   gas–water transition zone schematic *1114*
   reservoirs 1099
   seismic profile *1115*
   structural geometry at thrust front *1117*
   systems 1099
   tight gas *1121*
   Western Canada 1099–1122
      Cadomin Formation 1104–1106
         depositional model 1104–1105
         rock properties 1105–1106
      deep basin stratigraphy 1100–1101
      fractures 1118–1120
      Gething Formation 1106
      interpreting internal fold structure 1116
      megascale structure 1115
      Nikanassin Formation 1101–1104
         depositional model 1102–1103
         rock properties 1103–1104
      reservoir 1112–1113
      seismic stratigraphic mapping 1120–1122
      Spirit River Formation 1106–1110
         Cadotte Member 1108–1110
         Falher Member 1106–1108
         Notikewin Member 1108
      structural style 1113–1114
      water saturation 1111–1112
   Western Canadian BCG basin model 1110–1111
basin modelling 78
basin setting/sea-level response
   deepwater reservoirs 49–50
Bastrup delta *983*
bathymetric gorge
   Auðhumla Basin *973*
   North Ymir Ridge anticlines *973*
Batnfjordsøra *918*
Bayesian Belief Networks
   global petroleum systems 6–7
bay-fill *164*
   wireline-log cross-plot *165*
Bay Fill Facies Association *163*
   Hugin Formation sedimentology 159–162
      description 159
      interpretation 159–161
      sequence stratigraphy 161
      wireline-log character 161–162
bay-fill mudstones
   gamma ray logs 161
   transgression 161
Bazhenov Formation
   gas yield *665*
   source rocks 661
BCG. *See* basin-centred gas (BCG)
BCU. *See* Base Cretaceous Unconformity (BCU)
beach ridges 148
Beatrice Oilfield 1090
Beavington-Penny 589
Bechar Basin 725, 733
bedding-parallel fractures
   carbonate concretions *1137*
Belayim Land Field 778
Bennett Island 597–598, *598*, 610

benthic biofacies
   Eldfisk Field *480*
Bergman, Steve 589
Berkine Basin 686, 688, 712–714, *712*
   burial history *689*
   Cambro-Ordovician configuration *714*
   chronostratigraphic chart *720*
   Devonian succession *714*
   Frasnian source rocks 689
   interpretation 713–714
   isopach map *714*
   Palaeozoic succession stratigraphic column *709*
   stratigraphic architecture 712–713, *713*
   tectonic events *689*
   Triassic reservoirs 689
Berry, John 589
Beryl Embayment area 46
Besano Shales 118
   foreland 124
   hydrocarbon occurrences 124
BGS. *See* British Geological Survey (BGS)
Bighorn Basin 5
   hydrocarbons 6
Big Sandy shale gas field 1087
Bill Bailey's Bank 966–967
Billund delta *983*
biofacies
   allochthonous 481
   benthic *480*
   Early Maastrichtian Tor Formation *484*
   Eldfisk Field *480*
   Late Campanian Magne Formation *483*
   Late Danian Middle Ekofisk Formation *488*
   Late Danian Upper Ekofisk Formation *488*
   Late Maastrichtian Tor Formation Subzone *485, 486*
   Mid-early Danian Lower Ekofisk Formation *487*
   North Sea Eldfisk Field 479–481
   Upper Ekofisk Formation 488
biogenic gas
   accumulations 122
   Angola *1055*
   Angola future gas exploration 1052–1056
   conceptual Neoproterozoic petroleum system 1055–1056
   fairway 1053
   Netherlands petroleum exploration 264
   unconventional petroleum system 1052–1056
biostratigraphy 376
   A1-NC198 exploration well in Kufra Basin, SE Libya 761–769
   Balmoral sandstone system 441
   Buzzard Field *375*
   Central Graben 77
   Eldfisk Field 474, *490*
   Hugin Formation 174
   Laggan wells 288
   Lower Balmoral sandstone reservoir *440*
   Middle Ekofisk Formation 476
   Pentland succession *193*
   turbidite lobe *381*
bioturbated shale-prone interval
   Kristin Field 425
bitumen 1141
Bivrost Lineament 571, 893

black shales **1095**
  Cambro-Ordovician 1080
  Carboniferous age 1092
    Dniepr-Donets Basin 1081
    mineralogy 1092
    principal source rocks 1092–1096
    UK shale gas 1091–1096
  Early Jurassic age 1081
  Europe 1081
  Infracambrian age 676
  Jurassic age 1096–1097
  Kimmeridge Clay 1096
  Lower Bowland **1095**
  Lower Palaeozoic age 1091
  Poland 1080
  Silurian age *762*, 1080
  UK shale gas 1090–1091
Bled el Mass area
  Djebel Tamamat 716–718
block 1
  dynamic modelling study 397
block 12
  post-salt thermogenic petroleum systems *1051*
block 15
  post-salt gas plays *1050*
block 31
  extrasalt *1015*
  location 1013, *1014*
  petroleum geology 1013
  RMS amplitude extractions *1015*
  salt coverage *1014*
  salt geometries *1014*
  subsalt areas *1015*
  WATS survey acquisition area *1023*
block zero seismic amplitude gas indications *1051*
Blodøks Formation 540
Blomidon Formation 929, 931, *932*
Bo Member
  depth range *97*
  hopanes *107*
  hopane:sterane ratio 103
  oil-prone 98
  oil-prone shales 103
  sapropelic kerogen photomicrographs *102*
  source rock potential *104*
  thermal maturity map *101*
  wells GR *99*
Book Cliffs of Utah 24, 52
Boreal Ocean 420
borehole Leak Off Test
  Central North Sea 496
Bossier sandstones
  East Texas Basin 1112
bottom-simulating reflectors (BSR)
  gas hydrate *1152*
  Nankai Trough 1153
bottom-water currents
  Cenozoic compressional structures 963
  Faroe Bank Anticline 967
  Hatton–Rockall Basin 965
  North Rockall Trough 973
  Rockall–Faroe 972–973, *973*
    compression 973
  Wyville–Thomson Ridge 969
Bouguer anomaly
  Norwegian continental shelf *887*

Bouguer gravity 565
  data *567*
  profiles *574*
Bovanenkovskoye fields 664, *664*
Bovanenkovskoye gas charge 666
Bowland Shale
  black shale **1095**
  Lower **1095**
  mudstones thickness *1096*
  thickness 1095
Brazil. *See also* Almada Basin; Camamu Basin; Santos Basin
  margins *879*
    Bouguer-corrected gravity anomaly field *870*, *873*
    characteristics 873
    COB *870*, *873*, *873*, *877*, *878*
    conjugate margins *876*
    conjugate transects *877*
    continental lithosphere thinning 865
    crustal extension 869
    crustal-scale gravity modelling 868
    geological map *869*
    gravity modelled transect *872*
    gravity modelling 869, 871
    gravity modelling transect *871*
    hinge-lines *878*
    MCS profile *874*, *875*
    plate reconstructions *870*
    POC 874
    rifted *856*, 859, *877*
    subsidence analysis using flexural backstripping 859
    tectonomagmatic evolution 879
Bream Field
  hydrocarbons 987
Brennand 38
Brent Delta systems
  Jurassic age 174
Brent Field 43
Brent Formation sandstones 565
Bridge Anticline 969–970
  Cenozoic sediments 969
  seismic profile *972*
British Columbia
  Alberta border *1106*
  Alberta sandstone v. conglomerate *1111*
  BCG stratigraphic summary *1102*
  Horn River Basin 1079
British Geological Survey (BGS) 27
British Sedimentological Research Group (BSRG) 27
Broad Fourteens Basin 269
Brookian sequence
  Alaska 828
Brooks, Jim 31
Brown Shale 1087
B sandstones
  Rockall Basin *946*
    palaeodrainage 944–946, *947*
BSR. *See* bottom-simulating reflectors (BSR)
BSRG. *See* British Sedimentological Research Group (BSRG)
Buah Formation 674, 679
Buchan Graben 369, *371*
Bucomazi Formation
  generalized base salt maturity present day *1048*

Bunter Shale
  Triassic age 262, *270*
  basal 38
burial anomaly 987
burial history modelling
  coal-measure source rocks 71
Burns reservoir
  LKO 345
Bush Formation petroleum system 674
Buzzard discovery 48
Buzzard Field 53, 377
  appraisal well correlation *373*
  biostratigraphy markers *375*
  bottom hole pressure *378*
  B4 reservoir *381*
  Central Panel wells *384*
  chemostratigraphic sand *384*
  data acquisition 372
  depositional sequence compensational stacking *385*
  development area location map *370*
  downhole gauge data *379*
  end 2007, 377–379
  exploration *371*
  geochemical data interpretation 380
  interference data plot *378*
  late 2005 to end 2006, 375–377
    consistent biostratigraphic interpretation 376–377
    pre-development drilling results 375–376
    pre-development geological model 377
  location 369
  mid 2003, 372–374
    geological model at field sanction 373–374
    reservoir stratigraphy 372–373
    seismic interpretation 373
  mid 2003 reservoir stratigraphy 372–373
  planar laminated shales 375
  pressure measurements 377–379
  reservoir from appraisal to production 369–384
  Southern Panel wells *384*
  stratigraphy *372*
  today mid 2009, 379–384
    Buzzard chemostratigraphy integration 380–381
    reservoir connectivity current understanding 382–384
    revised geological model 381–382
  top reservoir depth map *371*
  wells *374*
Buzzard Field Development Plan 372
  appraisal drilling 373
  Donan 448
Buzzard reservoir 380
  geological model 384
  model 380
Buzzard Sandstone Member
  palaeogeography *370*
BVRU. *See* basal Volgian–Ryazanian megasequence unconformity (BVRU)
bypassed oil 349
  defining 349
  locating 349, 351
  Nelson Field *365*
  T70 *366*
  T70 unit *366*

bypassed pay
  case histories 456–461
    Type 1 Scenario A 456–457
    Type 1 Scenario B 457–459
    Type 1 Scenario C 459
    Type 2 Scenario A 459
    Type 2 Scenario B 459–460
    Type 2 Scenario C 460–461
  classification **454**, **455**
  evolution 455–456
    commerciality 456
    identification 455–456
    resources estimation 456
  identification 453–461
  net pay 453
  pay 453–454
  types 454–455
    Type 1 Scenario A 454
    Type 1 Scenario B 454
    Type 1 Scenario C 454
    Type 2 Scenario A 454
    Type 2 Scenario B 455
    Type 2 Scenario C 455

Cadomin Formation 1099–1104, 1122
  basin-centred gas in Western Canada
      1104–1106
    depositional model 1104–1105
    rock properties 1105–1106
  dip and strike seismic profile *1115*
  Lower Cretaceous age 1100
  outcrop photograph *1103*
  porosity–permeability cross-plot *1110*
  pressure–depth plots *1113*
  thickness *1109*
Cadotte Member
  basin-centred gas in Western Canada
      1108–1110
    dip and strike seismic profile *1115*
    permeability v. porosity cross-plots *1111*
    upper shoreface sandstones isopach *1110*
calcispheres 481
calcite-sealed fractures
  Barnett Shale *1133*
  hydraulic fracture treatments *1133*
  New Albany Shale *1138*
  reactivation *1133*
Caledonian basement 888, 889, 893–894
  mid-Norwegian margin 895
  Vøring margin 890
Caledonian erosion 754
  Temis 3D model *757*
Caledonian event 686
Caledonian gravity collapse 893
Caledonian Høybakken detachment 890
Caledonian nappes 889, 890
Caledonian orogeny 265
Caledonian unconformity
  Upper Silurian age 711
Caledonoid granite core
  potential field modelling 61
calibrated difference reflectivity (CDR) 517
  4D acquisition and processing
    additional 4D noise reduction 521
    processing flow improvements 520
  improvements **522**
  map *519*
  processing and acquisition *521*
  reductions *520*

calibrated INSAR satellite images 1169
  surface elevation changes *1169*
Callovian age
  Kimmeridgian basin breakup 205
  Kimmeridgian erosion 209
  Kimmeridgian megasequence 192, *203*
    Humber Group 186–192
    regional sequence stratigraphic
        correlation of wells *183*
    well correlation *204*
  Late 186
  Ryazanian interval
    sequence stratigraphic schemes *179*
    unconformities 185
  tectonostratigraphic succession 186
Caltanissetta accretionary arc
  Gela oil field *120*
Camamu Basin
  conjugate margin evolution 879–880
  data 868–869
  margin setting 869–874
    continent–ocean boundary 872–874
    crustal architecture 870–872
    Gabon 872
    Northeast Brazil 870–871
    structural inheritance 869–870
  M-reflector 875
  nature of crust transitional domain
      874–877
    crustal extension mode 876–877
    exhumed mantle 875–876
    S-reflector resemblance 876
  total rift setting 877–879
  transect construction 868–869
  2D gravity modelling outline 870
Camamu margin 868
Cambrian age
  basin remnant *681*
  Novaya Zemlya *652*
  plays in North Africa *682*
  succession 707–710
    fluvial sandstones 717
    North Africa 684
Cambro-Ordovician age
  Ahara High *712*
  black shales in Europe 1080
  configuration and Berkine Basin *714*
  North Africa stratigraphy
      682–686, *684*
  succession stratigraphic pinchout *712*
Campanian age
  Magne Formation biofacies map *483*
  sediments *482*
Canada. *See also* specific location
  BCG 1099–1122
  gas hydrates occurrence in Mallik 1153
  heavy oil deposits 1141
  Mackenzie River Delta 1153
  ultra-heavy oil 1062
  Williston Basin *1066*
canopy-margin thrust systems
  salt tectonics emerging concepts 904–905
Captain Field 342–346
  attic oil map *346*
  Captain pilot holes 344–345
  cross section showing reservoirs *344*
  discovery 342
  drainage schematic *345*
  drilling technology 343–344

3D seismic data acquisition 343
dual-lateral wells 346
geosteering LWD tools 345
phased development areas *345*
pilot well example 345
reservoirs 343
showing main reservoir intervals *344*
UKNS maximizing production 337–347
  Alba Field development 340–342
  drilling technology 343–344
  field history 340–342
  pilot holes 344–345
  pilot well example 345
  Well A3 pilot holes *346*
Captain Sand
  Lower 343
Caradocian Qasim Formation 695
carbonate
  Albian Pinda play 1045
  Alpine foreland 25
  buildups 733
  facies 826
  grainstone 795
  methane gases *1094*
  petroleum geoscientists virtual fieldtrips 25
  platforms 823
  slump Eldfisk Field *479*
carbon capture and storage (CCS) 1061, 1155
  defined 1165
  rationale 1165–1166
  scale 1165–1166
  technology 17
  underground gasification 1155–1162
carbon constrained world
  preparation 1189–1195
    anticipated carbon storage capacity 1195
    commercial-scale deployment project
        overview 1193–1195
    completed and ongoing validation
        tests 1190
    Gordon Creek Field 1193–1195
    medium-scale validation project impacts
        1193
    Paradox Basin 1190
    Permian Basin 1190–1191
    San Juan Basin 1191–1192
    SWP terrestrial pilot test 1192–1193
    terrestrial Riparian Restoration Project
        1192–1193
carbon dioxide 1155. *See also* Sleipner carbon
    dioxide plume
  emissions 1189
  flow characteristics 1184–1186
  geological storage 1165–1169
    challenges 1167
    geography 1166
    options 1166–1167
    storage 1166
      depleted oil and gas fields 1166–1167
      experience at In Salah project
          1167–1169
      monitoring demonstration 1168–1169
      saline formations 1167
      satellite imagery 1169
      unconventional oil and gas resources
          1063
      volumetrics 1166
  history-matching flow simulations
      1171–1182

injected hydrocarbon migration
1183–1188
carbon dioxide flow characteristics
1184–1186
implications 1186–1188
Utsira Formation 1183–1184
Sleipner Field 1063
Sleipner plume 1171–1182
store 16
UCG with CCS 1158–1159
Utsira Formation *1185*, 1187
well integrity 1167
carbon dioxide water contacts (CWC)
1173–1174
structural analysis *1177*
Carboniferous age 320, 321
basin configuration *72*
black shales 1092
Dniepr–Donets Basin 1081
mineralogy 1092
principal source rocks 1092–1096
UK shale gas 1091–1096
block and basin structure *71*
Central North Sea petroleum system 57–74
geophysical methods synthesis 67–69
hydrocarbon prospectivity implications
69–74
play types and traps 71–74
source rocks 69–71
potential field modelling 63–67
input model construction data 64–65
sub-Permian basin architecture 65–67
three-dimensional forward 63–64
three-dimensional inversion
solution 65
seismic interpretation 59–63
Auk–Argyll–Flora area 60–61
Jaeren high area 62
Mid North Sea High area 61
sub-Zechstein relationships 60–63
West Central Shelf 62–63
source rocks 69–71
sub-Zechstein well sections 57–59
Chiswick Field UK SNS reservoir
317–319, *319*
Early 730
Krechba reservoir 1168
Late
Netherlands petroleum exploration 263
palaeogeography maps *655*
North Africa
outcrops *726*, *727*
plays *731*
sequence frameworks *728*
petroleum system 74
sandstone 641, 996
sediments 316
shallow marine 596
Shannon Basin 23
source rock 71, 73
source rocks elevated structural
locations 71
stratigraphic traps 725–733
Early Bashkirian regression
729–730
Early to Middle Tournaisian
transgression 729
global transgressive–regressive cycles
730–731

Late Carboniferous 730
Late Devonian regression 725–729
Latest Viséan to Serpukhovian
transgression 729
Late Tournaisian regression 729
Late Tournaisian to earliest Viséan
transgression 729
potential stratigraphic trapping types
731–733
Carbonate buildups 733
depositional pinchout 731–733
incised valley-fills 732
Serpukhovian–Bashkirian incised
valleys 733
Strunian to Tournaisian incised valleys
733
truncation traps 731
Viséan incised valleys 733
Viséan transgressive–regressive cycles
729
subcrop 318, 319, 322
total organic content (TOC), *1092*
Unayzah formations 697
Variscan Foreland Basin 1080–1081
Catskill Mountains
gas-generated hydraulic fractures 505
CCS. *See* carbon capture and storage (CCS)
CDR. *See* calibrated difference reflectivity
(CDR)
Celtic Sea basins 924
location map *922*
Cenomanian age
deltaic coals 663
fields 660
fluvial-deltaic deposits 662
gas charge 660, 662, 663
gas fields 667
gas volumes *664*
Iabe Formation 825
play 659, 662, 663
Pokur Formation 659, 662, 663
reservoirs *664*
traps 664, 666
Cenozoic age 46. *See also* Early Cenozoic age
Arctic synchronous exhumation 633–643
Asmari carbonate reservoir *791*
Barents Sea *637*
Bridge Anticline 969
clastic input *273*
compression 118, 963
compressional structures 963
deep Oligocene play 698–700
deepwater reservoirs Palaeogene *v.*
Late Jurassic 45–46
Early 847, 849
East Greenland continental margin 975
exhumation *637*
intraplate inversional swells areal
occurrence *602*
Late 971
Middle East exploring potential of
hydrocarbon plays 673–702
Mordor Anticline 967
NE Atlantic Margin unconformities *974*
Netherlands petroleum exploration with
stratigraphic plays 273
North Africa exploring potential of
hydrocarbon plays 673–702
North Africa volcanic centers 673

rifted continental margin seismic
profiles 839
Rockall–Faroe area 963, *974*
Scandinavia 982
Slyne Basin 1005
structural elements *265*
thrust-and-fold belts 113
uplift 916
Val d'Agri oil field *121*
Viera Anticline 971
Central Europe
Pannonian Basin 1126, *1126*
Central Graben 226. *See also* Danish Central
Graben (DCG)
biostratigraphic records 77
chalk lithostratigraphic nomenclature 473
Dutch 265
East 180, *189*
high-amplitude reflector 185
shoreface deposition *190*
horizon between four wells 82
horizon characteristics **81**
Humber Group 177–210
seismic profile 80
seismic profile perpendicular to basin axis
80, *81*
sequence stratigraphic schemes *179*
tectonic reconfiguration evidence
200–201, 200–205
Central High Atlas Basin, Morocco 929
Atlantic margin Permo-Triassic Basin
evolution 931–934
location map *922*
Permian section *930*
Central North Sea (CNS) 70, 178
aquifer seal capacity 505
borehole Leak Off Test 496
Carboniferous basin configuration *72*
Carboniferous petroleum system
57–74, 69
crystalline basement surface *73*
diapir flank structure 518
direct pressures in chalk *498*
discoveries analysis *503*
dry holes 497, *503*
Gannet Area 46
Huntington discoveries 213–223
hydrocarbon discoveries 497
v. aquifer seal capacity *504*, *505*
v. aquifer seal capacity histogram *502*
hydrocarbon leakage 493
hydrocarbon seal capacity calculations
499–501
Jasmine discover 225–242
lithostatic stress 496
location *350*
long-offset seismic data 58
pressure-depth plot *496*
retention capacity 496
stratigraphic thicknesses control *504*
Tay formation fan system 46
Triassic *407*
Triassic strata truncation *41*
wells 58
Central North Sea Basin 85
aggradational channels *91*
channel troughs traces *89*
depositional model *91*
erosive trace along channel base 90

Central North Sea Basin (*Continued*)
  Late Neogene development contour currents 77–92
    aggrading channels geological controls 89–90
    channels and salt diapir leakage zones 85–86
    contour current depositional setting evidence 86–89
    geophysical data and methods 78
    Late Neogene basin development implications 90–92
    regional background 77–78
    results 78–85
      aggrading channels below Horizon 6, 83–85
      channel features detailed mapping 83
      Late Neogene succession seismic-stratigraphy 78–83
  Merganser leakage zone *88*
  three channel pathways *88*
  time-sliced map *89*
Central North Sea Chalk
  aquifer seal capacity 496–497
    methods comparison 497
  deep overpressure
    comparisons of dry holes and discoveries 497–502
      base Chalk 502
      base Cretaceous unconformity 502
      top reservoir/base seal 498–502
  low porosity and low permeability *497*
  pressure transition zone 493–495
    mud pressure profiles 494–495
    non-reservoir characteristics 494
    role as regional seal 495
    Upper Cretaceous pressures 494
  role in development of deep overpressure 493–506
  seal breach analysis 495–496
Central Panels 369
  Buzzard Field *384*
  high-density turbidites 372
  injection well *382*
  OWC *372*
Central Structural Domain
  Novaya Zemlya 651
Central Taimyr Fold Belt
  crustal tectonic *594*
Central Trough 43, 48, 50
  deepwater reservoirs Palaeogene *v.* Late Jurassic 49
  salts regional traverses *45*
  wells 51
central Utah deltaic reservoir analogues
  petroleum geoscientists virtual fieldtrips 24
central West Siberian basin 661, *662*
Central Ymir Ridge Anticline
  seismic profile *971*
Cervia Mare
  Porto Corsini East gas field *122*
CGGVeritas 249
Chalk
  acoustic impedance *v.* total porosity *543*
  biostratigraphic analysis 474
  Central Graben 473
  characterized 539
  composite seismic section *543*
  composition 463

Ekofisk Formation 464
Eldfisk Field *475*
facies *495*
holostratigraphic approach 473–491
holostratigraphic study 473
hydrocarbon seal 505
inversion results *545*
JCR 473
Late Cretaceous age
  Inoceramid bivalves 480
  Netherlands petroleum exploration with stratigraphic plays 269–273
lithostratigraphic nomenclature 473
North Sea *538*
  basin 479
  well correlation diagram *540*
PHIE *v.* true vertical depth *541*
PHIT *546*
reflector 4D seismic monitoring 343
represents 539
reservoir and non-reservoir *495*
Tor Formation 464
units *541*, *546*
Chalk Group
  AI-PHIT relationship 547
  hydrocarbon producing 273
  petrophysical log analysis 547
  time isochore map *539*
  well log correlation 547
Chalk Sea 88
channel axis drainage cells 365
channel features detailed mapping
  central North Sea basin Late Neogene 83
channelized turbidites
  geological characteristics 351
channel troughs traces
  Central North Sea basin *89*
chaotic hummocky basin floor facies 1035
chaotic hummocky mounded facies 1035
charging giant gas fields
  Siberia Basin 659–667
chemostratigraphy 374
  A1-NC198 exploration well, Kufra Basin, Libya 767
    material and methods 765
    results 765–766
    study objective 765
  Buzzard Field *384*
Chimney Rock Shale 1190
China
  Drilling Expedition GMGS-1, 1153–1154
Chiswick Field 316
  appraisal well results *317*
  BP drilled appraisal well 315
  Carboniferous depth structure map *317*
  Carboniferous rocks 316
  Carboniferous subcrop 318, 319
  gas producers 321
  geological well section *322*
  hydraulic fracture locations *322*
  intra-Carboniferous reflectors 319
  LASMO drilled well 315
  location 315
  location map *316*
  seismic section *320*
  stratigraphic column *318*
  UK SNS overcoming multiple uncertainties 315–323
    Carboniferous reservoir 317–319

Westphalian 319
Westphalian B 319
field history 315–316
phase I field development 321–323
  development plan 321
  drilling results 321
  Gamma well 321
phase II field development 322–323
Rotliegend reservoir 319
seismic interpretation 319–320
  database 319
  depth conversion 320
  interpretation 319–320
stratigraphy 316–317
structural history and stratigraphy 316–317
structure 316
Westphalian 319
Venture Production's involvement 315
wells penetration 317
chlorite
  coatings 292
  Laggan 279, *284*, *289*
  porosity and permeability preservation 288
Chondrites facies core photos *135*
Chukchi Basin 613
  HC play elements 614
  internal structure *613*
Chukchi Peninsula 597
Chukchi Sea 591, 598, *608*
  RAS sedimentary basins and petroleum geology 611–614
Chukchi Shelf
  structural elements *612*
Cimmerian age
  Late
    rift architecture and stratal geometries 44–45
    tectonics and fault geometries 39
  North Sea hydrocarbon systems 37–45
  reservoir distribution controls 38–39
  rifting seismic line *44*
Clair Field 32, 299–313
  core segment *311*
  data acquisition packages 306
  depletion lack 310
  development 299–313
  discovery 299
  fault juxtaposition *310*
  field pressure decline graph 308
  first water injector well data *308*
  formation pressure tests 307
  4D difference map *311*
  4D seismic data 311
  fractured oil reservoir 299–313
  fractures 300
  fractures azimuth establishment 305
  Graben segments 301
  granulation seams 310
  horst segments 301
  injector positioning and timing 307–310
  location 299
  location and seismic data coverage *300*
  LoFS 311
  managing field start-up 303–310
    data acquisition strategy 304–306
    early well results 306–307
    producer-led start-up 303–304

OBC dataset 301
oil and water production data
    309, *311*
original field development plan 303
petrographic analysis 310
play in Yell Basin *960*
pre-development appraisal 301–302
reservoir communication 306
reservoir pressure 309
reservoir quality 300
seismic line and crossline *302*
start-up, oil production profile *312*
streamer dataset *303*
structure and stratigraphy 299–301
3D seismic data 303
Top Unit V 301
Top Unit V map *303*
uninterpreted seismic data
    comparison *304*
water breakthrough 310–312
water injector well data *309*
well data *306, 307*
well interference testing 309
Witch's Hat segment 309, *310*
Clair Group
    stratigraphy and matrix properties *300*
Clair Unitization and Field Operating
    Agreement 299
Clare Shale 23
clastic basin fill
    stratigraphic traps 269
clastics 262
    Jurassic–Cretaceous 1118
Clavering Island 641
Clavering Ø
    thermal history solutions *998*
clay mineral
    Shiranish Formation 816
Claymore Field 391
claystone packages
    Buzzard Field *384*
Clearwater Formation 1144
Cleo-1 wells 138
Clyde Field
    subdivided 182
CNS. *See* Central North Sea (CNS)
coal
    Cenomanian deltaic 663
    gas content 1091
    gasification 1063
    gasifying *in situ* 1156
    Pennine Basin 1093
    source rocks 71
    underground gasification 1155–1163
coal mines
    deaths 1160
coastal plain *164*
    facies *163*, 167
        description 167
        Hugin Formation sedimentology 167
        interpretation 167
        sequence stratigraphy 167
        wireline-log character 167
    sequence stratigraphic surfaces 185
Cockburn Basin
    geoseismic profile *926*
Coffee Soil Fault 537
Cold Lake development
    Imperial 1147

Cold Lake Field, Alberta 7
    4D seismic steam conformance maps *1149*
    4D seismic surveys *1149*
Colin Oakman core workshop 27–28
commercial-scale injection test
    Gordon Creek Field, Utah 1193–1195
common risk segment (CRS) mapping
    process 791
    defined 794
    play fairway 794
compressional structures
    Rockall–Faroe area compression
        Hatton Bank Anticline 964–965
conceptual play model
    Lower Cretaceous early syn-rift stage 962
    Lower Cretaceous rift initiation 962
    Middle Jurassic pre-rift sandstone 962
    West Shetland Platform *961*
conceptual stratigraphic architecture
    Judy Member *409*
conductivity–temperature–depth (CTD)
    probe
    Rockall Trough 550, *554*
Congo 827
    craton 876
Congo Basin
    Lower *1044*
        3D seismic survey 1045
        gas discoveries by play *1047*
conjugate continental margins
    bathymetry contours *868*
    crust deformations 880
    rifting and crustal thinning 877
continental crust 825
continental epeiric sea basins 828
continental margin
    deeper structure 824
    petroleum plays 975
    post-rift evolution 824
    subsidence and heat flow history 840
continental red beds 934
continental shelf
    Norway sedimentary rocks properties **572**
continental uplift 827
continent–ocean transition (COT) 6, 868
    Faroe Islands *838, 840*
    HVLC P-wave velocities 831
controlled retractable injection point (CRIP)
    1156
controlled-source electromagnetic (CSEM)
    survey 254, 258, 828
    Northern Judd Basin T38 sequence
        prospectivity 254–258
    Suilven *257*
    T38 anomaly *257*
Cooles Farm borehole 1090
cooling episodes 996
core intervals **28**
Corrib gas field 938
    Dooish K-feldspars *950*
    sandstones *949*
Corsica
    Pleistocene age Golo turbidite system 382
COT. *See* continent–ocean transition (COT)
counter-depositional dip 46
Crag Formation 90
Craven Basin 62
    southern margins analogue *64*
    thickness 1093

Craven Group
    black shale **1095**
    Pennine Basin 1092–1093
    shale 1093
Creaney, Steve 589
Cretaceous age 420. *See also* Early
    Cretaceous age; Late Cretaceous age
    Alberta 1099
    Alpine orogen 113
    Apulian Platform succession *120*
    breakup 855
    chalk Tor Formation 225
    clastics 1118
    depocentres *846*
    Early 1100, 1102
    Faroe–Shetland Basin syn-rift play
        953–962
    gas yield *665*
    Khaz Dumi and Garau major play
        systems *793*
    Late 811–818
    Lewis Shale 1091
    Middle
        early syn-rift stage mature 962
        Faroe–Shetland syn-rift play 957–958
        relay ramps 962
        stratigraphic interpretation *1121*
    Møre Basin 420
    oil 808
    petroleum system 119–120
    reservoir 803, 805
    Faroe–Shetland syn-rift play 958–962
        early syn-rift 961
        Greenland analogues 958–959
        Jurassic rift stage 960
        rift initiation 960–961
        structural control on sedimentation
            959–961
        syn-rift plays 961–962
    Rockall Basin sandstones *1031*
    stratigraphic interpretation *1121*
    syn-rift play models 962
    tight gas sandstones 1120
    Vøring Basin *1031*
    wedge 1114–1116
crew change 1
CRIP. *See* controlled retractable injection
    point (CRIP)
Cromer Knoll 495
    mudrock 505
Cromer Knoll Group 186, 495
cross-bedded sandstone
    Garn Formation *423*
CRS. *See* common risk segment (CRS)
    mapping process
crustal scale deformation
    North Sea hydrocarbon systems 39–41
crustal thinning 6
Cruziana ichnofacies 164
Cryogenian age
    Dar Cheikh Group 677
crystalline basement
    depth *67*
    Devil's Hole Granite *68*
    Devono-Carboniferous basin 72
    Ireland *948*
    lead space plots *948*
    output *69*
    overlain with gas chimneys *73*

crystalline basement (*Continued*)
   sub-Zechstein composite layer *72*
   three dimensional model *70*
C sandstones
   Rockall Basin *946*
   Rockall Basin palaeodrainage *946, 947*
CSEM. *See* controlled-source electromagnetic (CSEM) survey
CTD. *See* conductivity–temperature–depth (CTD) probe
Curlew area
   well correlation *198*
Curlew Embayment *191*
   deltaic conditions *190*
CWC. *See* carbon dioxide water contacts (CWC)
Cylindrichnus 133
Czech Republic
   shale 1081

Dagmar Field 544, 547
Dagny area
   axis-perpendicular sub-regional correlation *173*
   well-log correlation panel *171*
Dakhla source rock 788
Dan Field 543
Danian age
   Chalk Group deposition 538
   Ekofisk Formation 539, 541
      biofacies map *487, 488*
   Late *488*
Danish Basin 569, 576
   individual contributions to basement depth map *891–892*
   potential field *892*
   Precambrian basement *895*
Danish Central Graben (DCG) 95
   barrier island sandstone reservoir architecture 145–154
      Freja discovery 148–153
         back-barrier sediments 149–150
         depositional environments 149–150
         shoreface sandstones 150
         stratigraphy 149
         thick stacked 150–153
      Holocene–Recent Rømø 146–148
         methods 146
         morphology and depositional environments 146
         palaeogeographic reconstruction 148
         vertical section trough Rømø barrier island 147–148
   biomarker ratios for oils and source rocks **108–109**
   chronostratigraphic scheme *130*
   cores sedimentology 129
   facies distributions and basin evolution *134*
   Late Kimmeridgian palaeogeographic map *152*
   Late Kimmeridgian to Ryazanian *134*
   log correlation panel *131, 132, 133*
   Lower Cretaceous succession *128*
   map *96*
   oil types 95–111
   potential kitchen areas *110*
   reactive kerogen wells **106**
   reservoir sandstones 127–142
      depositional environments 129–133
      distribution 129–141
      Lower Volgian shoreface sandstones 138–139
         depositional environment 139
         palaeogeography 139
         reservoir characteristics 139
      methods 129
      palaeogeography 135–136
      reservoir characteristics 133–134
      structural setting 127–129
      Upper Kimmeridgian shallow marine sandstones 136–137
         depositional environment 136–137
         reservoir characteristics 137
      Upper Kimmeridgian to Lower Volgian sandstones 137–138
         depositional environment 137–138
         palaeogeography 138
         reservoir characteristics 138
      Upper Middle Volgian to Ryazanian sandstones 139–141
         depositional environment 139–141
         palaeogeography 141
         reservoir characteristics 141
   sandstone play area 95–111
      methods 95–96
      oil composition and source 103
      petroleum evaluation 95–111
      regional setting 96
   sedimentological and biostratigraphical data 129
   sedimentological core log *140*
   simplified structural profile *129*
   source rock maturity and quality 96–103
      generative kerogen estimation 101–103
      kerogen composition 96–100
   source rock potential *104*
   steranes in oils and source rocks *110*
   structural elements *146*
   structural map *128*
   structural section *98*
   syn-rift sedimentary fill 96
   time stratigraphic chart *128*
   Upper Jurassic
      barrier island sandstone 145–154
      isochore map *128*
      reservoir sandstones 127–142
      sandstone play area 95–111
      succession *128*
   wells *131, 132, 133, 146*
Danish North Sea
   location map *538*
   time isochore map *539*
Danish Wadden Sea 148, *149, 151*
Darag sub-basin
   seismic profile *780*
Dar Cheikh Group
   Late Cryogenian age 677
Dawn Anticline 967–968
   bathymetric gorge *973*
DCG. *See* Danish Central Graben (DCG)
DDB. *See* Dniepr–Donets Basin (DDB)
Dead Sea fault system 812
deaths
   coal mines 1160
debates
   GeoControversies 11–17
Deep Joanne model 412–414
   gas and oil recovery factors 412
   model data **412**
   parameter sensitivities **411**
   porosity–depth realization 412
   reservoir depth 412
   results 412–414
Deep Judy model results 412
   initial gas rate *v.* CGR 412
   model data **412**
   parameter sensitivities **411**
   porosity-depth realization 412
   reservoir depth 412
   RF *v.* CGR/GOR 412
   sensitivity *414*
deep Triassic reservoirs
   porosity–depth relationships 410
   production behaviour 410
deepwater 48
   Angola 1013, 1014
   Badenian marine shales 1126
   far-outboard areas 823
   fold belts 823
   Gulf of Mexico 1014
   Ormen Lange system *917, 918*, 920
deepwater reservoirs
   basin architecture 50
   ESB and Jameson Land 49
   Late Jurassic *v.* Palaeogene age 50
   North Sea hydrocarbon systems 45–50, *50*
      basin setting/sea-level response 49–50
      Cenozoic examples and play considerations 45–46
      Central Trough Upper Jurassic comparison 49
      provenance/hinterland 49
      sandstone injectites 46
      Upper Jurassic deepwater depositional systems 46–48
   Palaeogene *v.* Late Jurassic 45–50
   provenance/hinterland 49
   salt withdrawal role 50
Default Thermal History 634, *635, 636*, 639
Delaware Formation *456*
De Long Massif 597
deltaics 826
   central Utah reservoir analogues 24
   depocentres 823
DEM. *See* digital elevation model (DEM)
Denmark *563*. *See also* Danish Central Graben (DCG)
   basin 569, 576, 891–892, 895
   Danish–German basin 891
   Felicia-1 wells AFTA data 642
   North Sea *538, 539*
   Norwegian Basin 891
   petrophysical samples locations *568*
   potential field anomalies 891
   sedimentary rocks magnetic properties **572**
   Wadden Sea 148, *149, 151*
depositional architecture
   tectonic and climactic influence *934*
depositional shoreline break (DSB) 82
depth velocity plots
   compaction depth–velocity *750*
Desana Formation
   Settala gas field *123*
desiccated reservoir 1112

detrital potassium feldspars
    lead 940
        isotope analysis 827
        isotopic composition 950
detrital zircon
    geochronology 6
    global database 6
    plate reconstruction 6
Devil's Hole Granite
    crystalline basement output 68
    depth contours 68
    3D model showing granite morphologies 68
Devil's Hole High wells 59
Devonian age *711*. *See also* Early Devonian age
    Ahara High *713*
    argillaceous *750*
    Bakken Formation 1065
    basin development 57
    Berkine Basin *714*
    Brown Shale 1087
    Carboniferous basins 66
        3D display of crystalline basement surface *73*
        locations 69
    chronostratigraphic diagram *747*
    depth velocity plots *750*
    Exshaw Western Canada Basin *1075*
    Ghadames Basin *711*
    Ghadames–Illizi Basin *756*
    Horn River Basin, British Columbia 1079
    hot shale *756*
    hydrocarbon prospectivity 686–688
    Jaurf Formation 692
    Kyle Limestone 59
    Late 725–729, 745
        GDE map *746*
    Mississippian Bakken Formation 1065
    Muskwa Shale 1079
    Netherlands petroleum exploration 266
    North Africa stratigraphy 686–688
    palaeogeography maps *655*
    regional-scale synthesis 57
    sandstones 569
    stratigraphic succession 710
    stratigraphic wedges 715, 721
    succession 707–710, *713*, *714*
    Tihemboka 715
    unconformities types 711
    Williston Basin 1065
diagenetic trapping 262
diapiric salt
    Gulf of Mexico 902
    intrude downwards 911
    Niendorf II diapir *903*
diapirs 899. *See also* Salt diapirs
    Central North Sea 518
    flank structure 518
    Gulf of Mexico *900*, *901*
    Hildesheimer Wald 902, *903*
    intrude salt plumes *900*
    Niendorf II 902
    rise *905*
    stages *901*
digital elevation model (DEM) 20
digital outcrop model (DOM) 20
Dinantian palaeogeography United Kingdom 69

Diplocraterion 161
dipmeter 1147
direct gas indicators
    Angola *1054*
Disko *981*
Djebel Tamamat area 716–718
    interpretation 717–718
    stratigraphic architecture 716–717
    wedge geometry *717*
Djebel Tebaga
    Austrian Unconformity *743*
Dniepr–Donets Basin (DDB)
    Carboniferous black shales 1081
    cross-section *1081*
    shale 1081
Dogger Granite 65
    crystalline basement output 68
    depth contours 68
    3D model showing granite morphologies 68
DOM. *See* digital outcrop model (DOM)
Donan Field 443, *446*
    appraisal well location *434*, *441*
    AVO *439*, *441*, *442*
    data summary 450
    discovery and appraisal of well results *434*
    drilling efficiencies 449
    FDP 448
    FPSO 443
    full stack seismic data *440*
    geological model 432
    high-resolution biostratigraphy 437
    high-resolution dinocyst-based biostratigraphy study 434
    historical behaviour analysis 442
    infill locations *437*, *448*
    lithology impedance volume *447*
    location map *432*
    oil price trends chart *437*
    overburden canyons *445*
    OWC behaviour *444*
    polarity flip *440*
    post-well analysis 441
    production history 443
    production plot *435*
    redevelopment 437
    renaissance old field in new landscape 431–450
        beyond second oil Donan 2007–?, 447–449
        2008 drilling campaign 448–449
        infill well performance 449
        preparing for infill drilling 448
        well performance 447
    first life Donan 1987–1997 431–437
        abandonment 437
        Donan changing geological view 435–437
        Donan discovery and appraisal 431–432
        Donan field development 432
        early field performance 432–434
        infill drilling 434–435
        location and block history 431
    future challenges 449
        identifying remaining potential 449
        Lochranza birth 449
        new beginning prospects Donan 2001–2007 437–447
        development drilling 443

Dumbarton Project birth 441–442
    field redevelopment plan 443
    leveraging technology 443–447
    reappraisal 441–442
    reservoir modelling 442–443
    technologies influencing appraisal 437–441
    residuals map *445*
    rock physics model *439*
    structure map 441
    time slice *445*
    wedge modelling *442*
Donan model
    layer-cake *435*
    revised *436*
Donan reservoir
    pressure plot showing depletion *443*
Dooish A sandstones 950
Dooish gas condensate discovery, Eastern Rockall Basin 938
    palaeodrainage 940–942
Dooish potassium feldspars *950*
    Corrib gas field *950*
    Porcupine Basin *950*
    Rockall Basin margin *950*
    Slyne Basin *950*
Dooish sandstones
    deposition 948
    lead space plots *947*
    potassium feldspar *947*
Dooish well
    Ireland *939*
    lead concentration **942–943**
Downey, Marlan 16
    motion National Oil Companies petroleum industry future 13
Drachev, Sergey 589, 590
drainage cells 350, *350*
    creation 361
    number and location 367
Drauppe Formation 100
Drilling Expedition
    GMGS-1, China 1153–1154
    UBGH1, South Korea 1154
drilling technology 340–342
drill stem tests (DST) 372, 374
dry holes *v.* aquifer seal capacity 502
DSB. *See* depositional shoreline break (DSB)
DSDP Leg 96, 1151
DST. *See* drill stem tests (DST)
Duffield borehole
    TOC *1095*
Dumbarton Project 431, 442, 443, 447, 449
    birth 441–442
    development wells *446*
    production performance *447*
Dutch Central Graben 265

earliest Volgian *204*
Early Albian age
    Wabiskaw Formation 1144
Early Cambrian age
    basin remnant *681*
    North Africa plays *682*
Early Carboniferous age
    North Africa 730
Early Cenozoic age
    Laramide Orogeny genetic basin analysis 5
    magmatic underplating 847, 849

Early Cenozoic age (*Continued*)
  Middle East exploring hydrocarbon plays 673–702
  North Africa exploring hydrocarbon plays 673–702
Early Cretaceous age
  Alberta 1102
  Cadomin Formation 1100
  conceptual play model 962
  DCG *128*
  Faroe–Shetland syn-rift play 957
  faults analytic signal shaded relief map *576*
  McMurray Formation
    Alberta 1062
    Athabasca 1141
    Athabasca Oil Sands 1062
    fluvial estuarine point bar sequence *1147*
    isopach map *1146*
    steepbank river outcrop *1146*
  Netherlands petroleum exploration 268–269
  Nikanassin Formation 1099
  Rauwerd stratigraphic trapping *272*
  rifting in Ghadames–Illizi Basin 741–742
  rifting in mid-Norwegian margin 849
  rift initiation 962
  sandstone *958*
  sediment 1100
  Spirit River Formation 1101
  syn-rift stage 962
  thickening in Ghadames–Illizi Basin 742
  West Shetland Platform *956*
Early Devonian age
  Caledonian unconformity 711
  fluvial sediments 714
  isopach map *714*
  MPath modelling *758*
  palaeogeography maps *655*
  Tadrart formation outcrop *711*
  Tadrart play *758*
  Tadrart reservoir GDE map *748*
Early Eocene age
  burial *635*
Early Jurassic age
  black shales 1081
  Netherlands petroleum exploration with stratigraphic plays 268
  paralic sandstone units 419
  petroleum system 118–119
  Posidonia shale 263–264
  source rock *271*
Early Maastrichtian age
  porosity and bulk volume water *477*
  Tor Formation biofacies map *484*
Early Miocene age *635*
  Pannonian Basin 1126
Early Neogene age
  exhumation 983–984
Early Oxfordian age
  sequence J46 186
Early Palaeozoic age
  black shales 1091
  genetic basin analysis 5
  stratigraphic architecture *718*
Early Permian age
  basin development 57
  Rotliegend Group 325
  sediments deposition 739

Early Pleistocene age
  progradational system *83*
early post-rift 961
Early Silurian age
  Tannezuft hot shale *690*
  Tanzuft 743–745
  Tanzuft Source Rock *744*
early syn-rift 961
  conceptual play model 962
Early Tournaisian age 729
Early Triassic age
  North Atlantic region *949*
Early Volgian age
  basin-scale palaeogeographic reconfiguration 205
  palaeovalley model 205–206
  prospect-scale models 206–207
  reactivation 205
  reconfiguration for exploration models 205–207
  sequence J64 199
  sequence J66a *195*, 199
  shoreface sandstones 138–139
    depositional environment 139
    palaeogeography 139
    reservoir characteristics 139
  turbidites 206
  uplift 205
Earth model
  channel complex details *1018*
  density mode *1018*
  model-based band-limited reflectivity *1018*
  NATS data *1018*
  velocity mode *1018*
East Central Graben (ECG) 180, *189*
  high-amplitude reflector 185
  shoreface deposition *190*
Eastern Structural Domain
  Novaya Zemlya 651
East Shetland Basin (ESB) 43, 44, 48
  deepwater reservoirs 49
East Siberian Sea 593, 611
East Siberian Sea Province (ESSP)
  crustal domain 609
  petroleum geology 611
  RAS sedimentary basins 609–611
  tectonic history 609
East Texas Basin
  Bossier sandstones 1112
Ebbing, Jorg 589
EBN. *See* Energie Beheer Nederland B.V. (EBN)
Eburnean Orogeny 677
ECBM. *See* enhanced coal bed methane (ECBM)
ECG. *See* East Central Graben (ECG)
Edoras–Hatton margin
  LCB 849
EGR. *See* enhanced gas recovery (EGR)
Egret Field
  high production performance 405
Egypt. *See also* specific locations
  geological section *699*
  Nile Delta *699, 700, 701*
  North Red Sea basin hydrocarbon prospectivity 783–788
    gravity/magnetics 787–788
    reservoir and seal 786
    seismic acquisition and processing 786

  seismic mapping 786–787
  source 783
  stratigraphy and tectonic framework 783
  Western Desert *697*
EIA. *See* Energy Information Administration (EIA)
Eikesdalen *918*
Eileen Gas Hydrate Accumulation 1152
Ekofisk Field
  Lower 475–476
  porosity and bulk volume water *477*
Ekofisk Formation 463, 473, 478, 523, *542*. *See also* Lower Ekofisk Formation
  air–mercury capillary pressure 469
  biofacies map *487, 488*
  cementation factor 466
  Chalk 463, 464
  Chalk well micrographs *467*
  core data *468*
  depth trend *468*
  effective stress *532*
  Hg-capillary pressure *467*
  Middle *488*
  porosity, water saturation and PWFT *469*
  reservoir *525*
  shale to gas-bearing chalk 465
  time-lapse time shift *532*
  Upper *488*
  water saturation 465
Eldfisk Alpha
  porosity v. permeability plots *479*
Eldfisk Bravo
  planktonic foraminifera 486
Eldfisk Field
  benthic biofacies groups *480*
  biostratigraphical database 474
  biostratigraphic and tectonostratigraphic summary *490*
  carbonate slump *479*
  Chalk 475
  Chalk sequences unconformities *480*
  Cretaceous reworking 478
  depth map *474*
  environmental factors 479
  fine-grained nannofossil component 481
  holostratigraphic approach to chalk 473–491
  holostratigraphic study 474
  hydrodynamic sorting pattern *479*
  intra Danian reworking 478
  location 473
  location map *474*
  planktonic bathymetric foraminifera biofacies groups *480*
  planktonic morphotypes 479
  porosity and bulk volume water *477*
  reworked horizons 478
  stratigraphic summary *476*
  tectonic and eustatic events *489*
El Gassi Shale 684
Elgin–Franklin area *189, 193*
Elgin–Franklin–Shearwater–Erskine trend
  productive reservoirs 187
Elm Coulee 1065
Emgyet Shale 686
Emsian flooding surface *713*
  stratigraphic pinchout *716*
endorheic dryland system 406

Endrõd Formation 1126
Energie Beheer Nederland B.V. (EBN) 261
Energy Information Administration (EIA) 1079, 1189
Energy Resources Conservation Board (ERCB) 1141, 1145, 1147
England. *See also* specific location
 biogenic shale gas 1080
 Carboniferous TOC *1092*
 CCS 1162
 thermogenic shale gas 1080
 UCG case study 1162
 wells *1089*
enhanced coal bed methane (ECBM) 1190
enhanced gas recovery (EGR) 1158
enhanced oil recovery (EOR) 1158, 1165–1166, 1190
Ensign Field
 appraisal Phase I 329–332
  well 48/14a-5 evaluation 330–332
   illite cementation 330
  well 48/14-2 evaluation 329–330
   hydraulic fracture stimulation parameters 330
 appraisal phase II 332–335
  well 48/14a-6 evaluation 332–335
   drilling induced fractures 333
   faults 333
   fractures 333–334
 appraisal wells *329*
 core porosity–permeability *330*
 discovery 325
 field history 325
 hydraulic fracture proppant size and volumes **331**
 improving well deliverability 325–336
 Leman Sandstone depth map *326*
 Leman Sandstone reservoir 327
 location 325
 map showing location *326*
 overburden interval 329
 productivity in tight gas reservoirs 335–336
 reservoir sedimentology 327–328
 reservoir structure 328
 reservoir zones 328
 Schlumberger FMI 332
 seismic interpretation 328–329
 seismic section showing overburden *328*
 stratigraphic sequence *327*
 stratigraphy 327
 structure and tectonic setting 325
 Venture's appraisal campaign 335
 well data *327*
Eocene age
 Apennines 113
 Balder Formation 45
 Gannet Area *46*
 hemipelagic, deep-marine deposits 984
 North Atlantic breakup transition 986
 palaeogeothermal gradients *636*, *640*
 plateau basalts *957*
 Tay formation fan system *46*
EON Ruhrgas UK Exploration and Production Ltd 223
EOR. *See* enhanced oil recovery (EOR)
ERCB. *See* Energy Resources Conservation Board (ERCB)

ERD. *See* extended reach drilling (ERD) wells
Erris Basin 1009–1012
 data reprocessing 1009
 exhumation estimates from seismic velocity data 1010–1012
 location map *1006*
 reprocessing example *1010*
 seismic interpretation 1009–1012
 Slyne Basin 1009–1010
Erskine area
 shoreface environments *191*
ESB. *See* East Shetland Basin (ESB)
Essaouira Basin 926
ESSP. *See* East Siberian Sea Province (ESSP)
Eugene Island Block 338 Field
 fluid saturation crossplot *461*
 production profile *460*
 seismic velocity *461*
Euphrates Graben
 geological cross-section *812*
 Shiranish Formation 811–818
 stratigraphy *813*
 Syria sedimentology 811–818
EUR. *See* expected ultimate recovery (EUR)
Eurasian–African plate collision
 tectonic cycle 90
Eurasian Arctic
 onshore fold belt 593
 tectonic events 593
Eurasian Basin 589, *590*
Eurasian Continental Margin
 free-air gravity field *595*
Europe 31–33. *See also* specific location
 Antrim Shale-analogues with biogenic methane 1083
 basin architectures regional-scale polyphase rifting *42*
 black shale 1081
 exploration 32
 exploration and exploitation techniques 33
 field development and production 32–33
 gas shales case studies 1080–1083
 Pannonian Basin 1126, *1126*
 Petroleum Geology Conference series history 31
 proceedings overview 31–32
 shale gas 1079–1083
  regional overview 1079–1083
European Margin
 lead potassium-feldspar data 937
EUROSTRATAFORM project 914
exotic fragments
 originate 910
 travel 911
expected ultimate recovery (EUR)
 Joanne 415
 Judy 415
 Judy Field 412
expendable conductivity–temperature–depth (XCTD) probe
 seismic acquisition 557
expendable sound speed probe (XSP)
 seismic acquisition 557
Exshaw Western Canada Basin *1075*
extended reach drilling (ERD) wells 340
extensional margin evolution 867

facies. *See also* shoreface
 associations *164*, 933
 carbonate 826
 chaotic hummocky basin floor 1035
 chaotic hummocky mounded 1035
 coastal plain *163*, 167
 core photos *135*
 DCG *134*
 Hugin Formation **160–161**
 Ile Formation *424*
 Joanne Member 407, 410
 Lagoa Feia lacustrine 862
 map *361*
 Mississippian *267*
 Shiranish Formation 815–816
 sigmoid aggradational 1035
 Tofte Formation *426*
 units description 1035–1038, *1037–1038*
facies 5
 movement direction *1038*
facies 6
 movement direction *1038*
Faleide, Jan Inge 589
Falher Member 1103, *1104*
 basin-centred gas 1106–1108
Fammenian sediments
 Ahnet Basin 738
Fangst Group
 shallow marine 419
Farne Granite 65
 crystalline basement output *68*
 depth contours *68*
 3D model showing granite morphologies *68*
Faroe Bank Anticline 967
 bottom-water current activity 967
Faroe Bank Channel 973
Faroe Bank Channel Syncline 967
 seismic profile *968*
Faroe Island Basalt Group (FIBC) 1033
Faroe Islands *832*, *836*, *837*, 865
 conceptual breakup models *863*
 COT *838*, *840*
 description 1033
 effect of LVZ *834*
 geological structures *1034*
 geophone component *835*
 intra-basalt units 1033–1040
 iSIMM seismic profile *838*
 map *1026*
 North Atlantic Ocean 1033
 profile constraints 839
 P-wave velocity structure 836
 rifted continental margins 836
 sub-basalt LVZ *839*
 TD 1033
 3D seismic data 1033–1040
 wide-angle data *835*
Faroe–Shetland Basin 245, 246, 252, 827, 837
 Base Balder unconformity 975
 complete crustal structure *1027*
 conceptual breakup models *863*
 crustal thickness map *864*
 Faroe–Shetland syn-rift play 953
 future exploration 962
 hydrocarbon provinces 1025
 Mesozoic–Palaeozoic structural play 954
 Paleocene age *281*
 structural features *300*

Faroe–Shetland Basin (*Continued*)
  structural setting 959
  sub-basalt hydrocarbon prospectivity
      1025–1031
  syn-rift play 953–962
  UKCS location map *280*
Faroe–Shetland Channel
  sub-basalt imaging 1034
Faroe–Shetland region 953–962
  Cretaceous reservoir formations 958–962
    early syn-rift 961
    Greenland analogues 958–959
    Jurassic rift stage 960
    late syn-rift and early post-rift 961
    rift climax 961
    rift initiation 960–961
    structural control on sedimentation
        959–961
    syn-rift plays 961–962
  Early Cretaceous 957
  exploration history 953
  Late Cretaceous 958
  methods 953–954
  Mid-Cretaceous 957–958
  Mid-Late Cretaceous 958
  onshore analogues 955
  Pre-Early Cretaceous 957
  reflection seismic data 953–954
  structural map *954*
  structural observations 955–958
  wells 954–955
Farsund Formation 139
  dead carbon average proportion *106*
  depth map *98*
  depth range *97*
  HI v. depth plot *102*
  hopane:sterane ratio 103
  immature to marginally mature
      samples *103*
  kerogen 98
  non-bioturbated claystones 139
  oil window 96, *102*
  PI v. depth plot *102*
  sandstones and siltstones 141
  sapropelic kerogen photomicrographs and
      oil immersion *102*
  shales 98
  source rock potential *104*
  source rock quality 96
  thermal maturity map *101*
  TOC contest *105*
fault. *See also* specific name
  analytic signal shaded relief map *576*
  Arkitsa 24–25
  48/14a-6 well evaluation 333
  Base Zechstein *43*
  control 532
  displacement *806*
  fold *1118*
  imaging *512*
  inner and outer seismic distortion zones *510*
  Jurassic age *576*
  juxtaposition difference 515
  network 513, 515
    AI volume-rendered view *513*
    juxtaposition difference values *513*
  Puffin horst area *43*
  relative seismic volume analysis 509–515
    case study 513–515
    caveats and discussion 515
    defining damage zone 510
    defining fault network 509–510
    fault relative analysis associated with
        damage zones 510–513
    fault-sealing potential 509–515
    inner zone 509
    outer zone 509
  reservoir depletion 532
  rift oblique lineaments 961
  sealing potential 509–515
  zones 958, 961
fault damage zone (FDZ) 509
  index volumes *512*
FDZ. *See* fault damage zone (FDZ)
Feda Graben 82, 103, 136, 150
  wells VR trends *100*
feldspars. *See also* detrital potassium
      feldspars; potassium feldspars
  concentration **942–943**
  detrital lead isotope analysis 827
  Dooish potassium *950*
  Dooish well **942–943**
  European Margin 937
  provenance technique 934
Felicia-1 wells, Offshore Denmark
  AFTA data 642
Fell Sandstone Group 59
Fennoscandian grid 565
Fennoscandian Shield 598
Fernie Formation
  interbedded sandstones *1116*
Ferron Sandstone 24
  lidar studies 24
FFZ. *See* Florianopolis Fracture Zone (FFZ)
FIBC. *See* Faroe Island Basalt Group (FIBC)
fieldtrips
  petroleum geoscientists virtual 19–25
Finland *563*
Finnmark 565
Fisher Bank Basin
  Jurassic age 214
  wells *218*
five-tile WATS *1021*
Fjellanger, Erik 589
Fladen Ground Spur 48
flank well
  example of interpretation *1008*
Flett Basin
  gas-prone 252
Flett Sub-Basin
  location map *280*
floating production storage and
      offloading (FPSO)
  Donan Field 443
  Maersk Oil's Leadon Field 443
Flora Field 59
Florianopolis Fracture Zone (FFZ) 855
flowing tubing head pressure (FTHP) 315
fluvial distributary system *408*
fluvial erosion
  Ormen Lange system *918*
fluvial sandstones
  Cambrian succession 717
fluvio-deltaic distributary system 86
fluvio-tidal channel-fill *164*
  facies association *163*
  facies association D 166–167
    description 166
    interpretation 167
    sequence stratigraphy 167
    wireline-log character 167
  Hugin Formation sedimentology 166–167
  wireline-log cross-plot *165*
FMH. *See* Forties-Montrose High (FMH)
FMI. *See* Schlumberger Full-bore
    Micro-Imager log data (FMI)
Foinaven Field 248
foreland 124
  Asti Formation 123
  basin 113
  Italy hydrocarbon occurrences 113–125
  Mesozoic carbonate substratum 119
  Po Plain *119*
  ramp 122, 123
  thrust propagation folds 122
Foreland Basin 1 Megasequence 738
  Hercynian Orogeny 739
  Pre-Hercynian burial 738
  Pre-Hercynian structures 738–739
Foreland Basin 2 megasequence 743
Former Soviet Union (FSU) 13
Forties Field
  bypassed oil target 365
  depth structure *221*
  sandstone reservoir prediction *221*
  well log section *222*
Forties-Montrose High (FMH) *203*, 350
  Early Volgian reactivation 205
  palaeogeographic maps 205
  recording reactivation *204*
  seismic line *203*
Fort Worth Basin (FWB) 1131
  Barnett Shale Formation 1087, *1088*
  Mississippian Barnett Shale 1087
forward and inverse model *8*
Fosen Peninsula 565
fossil fuel 1189, 1193, 1195
Foula Formation *930*
four dimensional acquisition and processing
    517–522
  acquisition repeatability *521*
  analysis method 517
  case study details 517–518
  CDR calculation 518–521
    additional 4D noise reduction 521
    improved acquisition repeatability
        520–521
    processing flow improvements 520
  noise levels 517
  noise reduction 521
  North Sea case study 517–522
  repeatability measurements 517
  seismic CDR *521*
  surveys *521*
four dimensional model
  Alba data showing water cones *341*
  Alba seismic showing OWC movement *341*
  Clair field difference map *311*
  Cold Lake Field seismic conformance maps
      1149
  response characterized 312
  seismic data
    Clair field 311
    monitor 311
    Nelson Field areal flow detection
        controls methods 360–361
    OWC 342

seismic monitoring Chalk reflector 343
seismic response T70 sequence *362*
seismic surveys 366, 1147
    Cold Lake Field *1149*
stem flooding monitoring 1145
four-tile WATS *1021*
FPSO. *See* floating production storage and offloading (FPSO)
fractures. *See also* Hydraulic fractures
  bedding-parallel carbonate concretions *1137*
  calcite-sealed *1133*, *1138*
  Ensign enigma 335–336
  gas-generated hydraulic 505
  horizontal well 48/14a-6 *335*
  model 808, 809, *809*
    strike-slip element 804
  oil reservoir 299–313
  reservoir models 23
  stimulation process 325
    well 48/14-2 332, 336
    well 48/14a-6 336
  Taq Taq field appraisal and development 807–808
  well 48/14a-6 evaluation 333–334
France. *See also* specific location
  petroleum geoscientists virtual fieldtrips 24
Franz Josef Land Archipelago 598
Fraser, Al 589
Frasnian age
  charge 754
  'hot' shale 754
  source rocks deposition 722
  source rocks kitchen maps *757*
  unconformity 686
Freja Field 145, *152*
  barrier island sandstones 146
  palaeotopography 145, 152
  well log panel *154*
Freja–Mjølner Field *146*
Froan Basin 925
  geoseismic profile *924*
  location map *922*
FSU. *See* Former Soviet Union (FSU)
FTHP. *See* flowing tubing head pressure (FTHP)
Fugelli, Edith 589
Fugloy Ridge 837
  effect of LVZ *834*
  seismic data *834*
Fulmar–Clyde Terrace 190, *191*
  turbidite deposition *195*
Fulmar Field
  correlation model *208*
  defined 216
  exploration 216
  fault-crest seismic interpretation 207
  Jurassic age *214*
  pinchout prospects 207
  reservoir 495, *498*
  sandstone 178, *207*
    aggradational packages 186
    BVRU 205
    deposited in narrow palaeovalleys 206
    deposition tectonic reconfiguration 206
    deposits sequences J52-J54b *193*
    Humber Group thickness 207
    initial distribution 180
    isopach *218*

    sequences J52 and J54 186
    shallow-marine 177
    subdivided 182
    well 22/30b-11 189
  well log section *222*
Fundy Basin 928
  Atlantic margin evolution and fill 929–931
  detailed analysis 929
  location map *922*
  Permian and Triassic sections *930*
FWB. *See* Fort Worth Basin (FWB)

Gabon
  conjugate margin evaporitic deposits 874
  crustal architecture 872
  margin 868, 872, *878*
    hinge-lines 872
    tectonosedimentary *879*
Gaffney, Cline & Associates (GCA) 13
Gaffney, Peter 13–14
Gaggiano Field
  Villafortuna–Trecate oil field *119*
Gamma well
  challenging gas development 321
Ganges–Brahmaputra 827
Gannet Area
  Eocene Tay formation fan system *46*
Garau major play systems *793*
Gargaff Arch 710–712
  chronostratigraphic chart *720*
  geological map *710*
  incised valleys *712*
  interpretation 711–712
  Siluro-Devonian succession stratigraphic architecture *711*
  stratigraphic architecture 710–711
Garn Formation 421, 422–423
  chlorite 422
  cross-bedded sandstone *423*
  depositional synthesis 423
  description 422–423
  petrographic porosity *421*
  pressure profiles *422*
  rapid pressure decline explanation 423
gas 123
  Angola discoveries *1047*
  Angola exploration 1043–1057
  Angola generation schematic *1052*
  Bakken Formation 1067
  Barnett Shale 1087
  Bazhenov Formation *665*
  BCG model expulsion *1112*
  biogenic 122, 264, 1052–1056
    England 1080
    Scotland 1080
    Wales 1080
  Cenomanian *664*, 667
  charge
    Cenomanian fields 660
    Cenomanian play 662, 663
    Urengoy field 667
  charging
    giant 659–667
  chimneys
    long-offset seismic data *58*
    sub-Zechstein composite layer *72*
  Chiswick Field 321

  chromatography
    analysis 1067
    Bakken petroleum system 1068–1069
    chain length distribution *1071*
  Coal Measures 1091
  condensate
    Joanne Sandstone Member 228
    Triassic Skagerrak Formation 228
    West Limb core area 233
  Corrib 938, *949*, *950*
  Cretaceous sources 665
  Deep Judy model results 412
  direct gas indicators in Angola *1054*
  enhanced recovery 1158
  Europe case studies 1080–1083
  generated hydraulic fractures 505
  greenhouse 1189, 1192–1193
  initial rate *v.* CGR 412
  Italy *115*
  Judy model 414
  Jurassic sources 665
  Leningradskoye field drainage area *663*, *664*, 665
  liquefied natural 827, 1044
  Lower Congo Basin discoveries by play *1047*
  North Africa occurrences 762
  Pokur Formation *663*, *664*
  recovery 412
  recovery factor 414
  Settala field 122, 123, *123*
  shales 1080–1083
  South Arne rock physics model *528*
  Tanopcha Formation *663*, *664*, 665
  thermogenic shale gas 1080
  trap formation East Barents region 603
  typical permeability reservoirs **1159**
  UK shale gas 1091
  Urengoy field drainage 665
  wetness ratio *1094*
GASH. *See* Gas Shales in Europe (GASH) project
gas hydrates
  BSR *1152*
  Gas Hydrate Production Research Program 1153
  India occurrence 1153
  JAPEX/JNOC/GSC Mallik 2L-38 1153
  JIP Gulf of Mexico 1151
  location *1152*
  occurrences 1151–1154
    Alaska North Slope, Alaska, USA 1152
    drilling expedition GMGS-1, China 1153–1154
    Drilling Expedition UBGH1, South Korea 1154
    Gulf of Mexico, USA 1151–1152
    Mackenzie River Delta 1153
    Nankai Trough, Japan 1153
    NGHP Expedition 01, India 1153
    North Africa 762
  resource potential 1151–1154
  unconventional oil and gas resources 1062–1063
gas–oil ratios (GOR)
  Bakken Formation 1067, *1074*, *1075*
  Bakken hydrocarbon pools *1073*
  Bakken petroleum system 1071–1072
  Williston Basin *1072*

gas plays
    block 15, *1050*
    Cenomanian 659
    Cenomanian Pokur Formation 659
    post-salt *1050*
    salt 1046
gas-prone petroleum systems
    stratigraphic summary *1045*
Gas Shales in Europe (GASH) project 1083
Gassman equation 1185, 1187
GCA. *See* Gaffney, Cline & Associates (GCA)
GEBCO. *See* General Bathymetric Chart of Oceans (GEBCO)
Gebel Duwi North Red Sea Hills 786
    field photograph *790*
    palaeotopography of palaeostructuring *790*
    pre-rift Cretaceous Nubia sandstone *790*
Gebel Zeit
    post-rift phase 776
Gedinnian fluvial unit *713*
Gela Field
    age *120*
    Caltanissetta accretionary arc *120*
    discovery 118
    Italy hydrocarbon occurrences 118–119
Gela Formation *120*
Gemsa Basin
    footwall prospectivity 775–777
    half graben cross-section *775*
    landsat 7742 geological image *776*
General Bathymetric Chart of Oceans (GEBCO) 560
Generic Mapping Tools (GMT) *554*
genetic analysis summary chart *4*
genetic basin analysis
    global petroleum systems 2–7
    processes hierarchy *3*
genetic models
    predictions *7*
Gent Ridge
    wells VR trends *100*
geochemistry 1065–1066, 1076
    oil 360
    water 357–358, 360
GeoControversies debates 11–17, *16*
    Debate 1 peak oil no longer concern 11–12
        for motion David Jenkins 11–12
        against motion Jeremy Leggett 12
    Debate 2 National Oil Companies
        petroleum industry future 12–14
        for motion Marlan W. Downey 13
        against motion Peter Gaffney 13–14
    Debate 3 North Sea finished 14–15
        against motion Jim Hannon 15
        for motion Richard Hardman 14–15
    summary Julian Rush 15–17
geographic information system (GIS)
    applications 20
geological conceptual model *350*
geological sidetrack 30/6-6Z 230
Geological Survey of Norway (NGU)
    aeromagnetic and gravity data 559
    aeromagnetic data acquisitions and modelling 33
Geology of Petroleum (Levorsen) 14
geomechanical modelling 1118
    example *1118*
    three-dimensional output *1119*

geoscientists virtual fieldtrips 19–25
geosteering 342
    Captain Field 345
    LWD tools 342
German squeezed diapirs *903*
Gert Member
    porosity and oil saturation plot *136*
Gertrud Graben 103
    wells VR trends *100*
Gert-1 well 130, 145
    back barrier sediments *148*
    HTHP environment 134
    interpreted washover fans photographs *153*
    porosity–permeability plot *148*
    sedimentological core log *147*
Gething Formation
    acoustic properties *1119*
    basin-centred gas in Western Canada 1106
    shale 264
Ghadames Basin 686, 689, 710–712, 725
    chronostratigraphic chart *720*
    interpretation 711–712
    Siluro-Devonian succession stratigraphic architecture *711*
    stratigraphic architecture 710–711
Ghadames–Illizi Basin *736*
    basin modelling approach and workflow *749*
    chronostratigraphic diagram *747*
    compaction depth–velocity relationship *751*
    depocentres *755*
    depth velocity plots *750*, *751*
    Devonian 'hot' shale *756*
    Early Cretaceous rifting 741–742
    Early Cretaceous thickening 742
    genesis burial plots *754*, *755*
    Hercynian Unconformity 739
    integrated petroleum systems 735–759
    outline *740*
    petroleum resources by country *753*
    petroleum systems and play fairway analysis 735
    reservoir-seal combinations 745
    seismic profiles *741*
    Silurian 'hot' shale *756*
    stratigraphic framework *737*
    Tanzuft source rock *758*
    temperature *v.* depth *754*
    velocity analysis 745
Gharib
    Zeit evaporite section *787*
GIS. *See* geographic information system (GIS) applications
glacial terrain
    Norway *918*
global database
    detrital zircon dates 6
global oil resources *1143*
global petroleum systems 1–8
    Bayesian Belief Networks 6–7
    concepts 6
    detrital zircon geochronology 6
    distributions predictions *3*
    extensional systems 6
    forward modelling 7
    genetic basin analysis 2–7

example 5–6
    global data plate motion 4–5
    overview 2–4
    methods 6
    opportunities 7–8
    space and time 1–8
    technology 6
    tools 6–7
global plate
    motion models 4
    reconstruction 4, *5*
global positioning system (GPS) 20
global transgressive–regressive cycles and North Africa 730–731
global warming 1189
Glossifungites ichnofacies 139
GMT. *See* Generic Mapping Tools (GMT)
goaf formation 1158–1159
    caved zone 1158
    fractured zone 1158–1159
Goldsmith, Phil 38
Gondwana
    craton 708
    supercontinent 674, 718
Google Earth 20, 25
GOR. *See* gas–oil ratios (GOR)
Gordon Creek Field
    carbon constrained world preparation 1193–1195
    commercial-scale injection test 1193–1195
    site location/surface geology map *1192*
    stratigraphy *1193*
Gorm Field 547
Gorm-Lola Ridge 547
Gothic Shale 1190
GPS. *See* global positioning system (GPS)
Graben Fault 309
Graben segments
    Clair field 301
graben system
    tectonic reconfiguration 178
grainstones
    Arab Formation 796
    carbonate 795
Grand Coyer sub-basin 24
granite
    Caledonoid potential field modelling 61
    Devil's Hole Granite 3D model *68*
    Dogger Granite 65
    Dogger Granite 3D model *68*
    Farne Granite 65
    Farne Granite 3D model *68*
Grantz, Art 589
gravitational loading
    Neogene depositional units 92
gravity
    Bouguer gravity data *567*
    compilations 559–583
    flow sandstones *138*
    gradiometry *579*
    Iris-1 location *138*
    Norway 559–583
    oil data 1067, *1073*, *1074*
    residual *567*
    surveys compilation *566*
    Svane-1 location *138*
    Tail End Graben *138*
Greater Britannia area
    fluid pressure indications 502

Greece
  Arkitsa fault 24–25
  petroleum geoscientists virtual fieldtrips 24–25
Green, Paul 589
Green Canyon area
  Marco Polo minibasin 909
  roof-edge thrust 904
  Suprasalt Miocene strata 909
greenhouse gas 1189, 1192–1193
Greenland 643, 955
  analogues 958–959
  case C episodic uplift and exhumation
    AFTA data defining cooling episodes 990
    thermochronological and geomorphological integration 990–992
  case D nature of Oligocene hiatus
    episodic uplift and exhumation 992–994
  case E Ordovician source rocks
    episodic uplift and exhumation 994–995
    Ordovician outlier implications 995
  case F post-rift burial episodic uplift and exhumation 995–999
    Cenozoic burial and exhumation 996–997
    East Greenland margin development 998–999
    mismatch between apatite dating and AFT results 997–998
    Palaeogene intrusive bodies thermal effects 996–997
  continental margin
    Cenozoic compressional features 975
    compression and inversion of underlying basin 975
  exhumation episodes 642
  hydrocarbons generation 999, 1001
  landscape 981
  margin
    aeromagnetic data reconstruction 580–581
    geological map 997
    uplifting and erosion in Norway 998
  Mesozoic basins 980
  Oligocene 993
  Palaeogene deposits 998
  plate reconstruction 582
  post-basalt heating 1000
  regional exhumation 1000
  regional structures 573
  seismic profile 996
  thermal history 641
  thermal history solutions 998
Greenland Sea
  aeromagnetic surveys 579
  embayment 654
Grès d'Annot Formation, SE France
  petroleum geoscientists virtual fieldtrips 24
Grouw and Leeuwarden Field area
  stratigraphic trapping 272
Gro-3 well
  Nuussuaq Basin 640
Gryphon complex 46
Gulf of Cadiz 555
Gulf of Evia 24
Gulf of Mexico 456, 458
  allochthonous fragments 908

allochthonous minibasin 910
bathymetric profile 906
bulge of salt 907
buried salt canopies 900
compressed, uplifted area 906
deepwater 1014
density profiles 906
diapiric salt 902
extruding salt stock 908
Gas Hydrate JIP 1151
gas hydrates resource potential 1151–1152
gravity and regional shortening 905
Marco Polo minibasin 909
prograding sedimentary wedge 907
salt tectonics 902
salt tectonics concepts 899–911
source-to-sink systems 914
spreading salt canopies 900, 910
squeezed diapirs intrude salt plumes 900
squeezed diapir stages 901
subsalt Mad Dog Field 1015
thin roof disruption 908
thrust faults offset 907
thrusts 904
WATS experience 1023
Gulf of Suez 783
  3D OBC seismic data 780
  hydrocarbon play 774
  isotope analysis 786
  location map and structural elements 772
  Nukul syncline 24
  petroleum system 788
  pre-rift beds 771
  pre-rift reservoirs 777
  sand pinchouts 780
  sandstone distribution 779
  seismic datasets 774
  source rock biomarker 786
  stratigraphic column 773
  subtle exploration plays 771–781
    exploration targets 774–775
    Gemsa Basin footwall prospectivity 775–777
    Gulf of Suez structural setting 771
    post-rift reservoirs 774
    pre-rift reservoirs 773
    Ras Abu Darag Concession 778–780
    regional stratigraphy 771–773
    reservoirs 773–774
    Rudeis Formation 777–780
    seismic data quality 774
    source rocks 773
    South October Concession 777–778
    syn-rift reservoirs 773–774
    types 774
    Upper Rudeis prospectivity 777
  syn-rift beds 771
  syn-rift sedimentation 774, 780
  watershed 787
Gull Island Formation 23
Guo, Li 590

Haldan Field 537, 547
  Jurassic GDE map 795
Haltenbanken
  stratigraphic column 420
Halten Terrace 420, 422
  hydrocarbon generation 419
  overpressure development 419

Halten-Trøndelag Basin 422
Hammerfest Basin 634
Hamra Quartzites 685
  isopach map 714
  Ordovician 685
Hanifa Field
  source rock development 795
Hanifa Formation
  Jurassic age 793
  source rocks 799
Hannon, Jim 14, 16, 16
  against motion North Sea finished 15
Hannon Westwood 14, 15
Hans-1, Offshore Denmark
  AFTA data 642
Hardangerfjord Shear Zone 893
  satellite image 916
Harding/Gryphon complexes 46
Hardman, Richard 16
  for motion North Sea finished 14–15
Hareelv Formation 49
Hassi Messaoud Field 684, 707
Hassi R'Mel fields 707
Hatton Bank
  COT 836
  crustal thickness 837
  igneous intrusions 837
  P-wave velocity 836, 837
  rifted continental margins 836
  seismic velocity structure of crust 836
Hatton Bank Anticline 964–965
  seismic profile 967
  seismic profile of unconformities 967
Hatton–Rockall Basin
  bottom-water current activity 965
Heather Formation
  Ling Sandstone Member 174
heavy oil deposits
  Alberta 1142
  Athabasca Basin 1145
  Canada 1141
heavy oil production
  seismic monitoring 1147–1148
  in situ 1147–1148
Hejre-1 well 127
  HTHP environment 134
  oil discovery 99
Helgeland Basin
  top basement 890
Helgeson's model 44
Helland–Hansen Arch
  basin inversion 971
Heno Formation 149
  porosity–permeability relationship 136
  sandstones 127
Heno Plateau 129, 136
Henrietta Island 598
Hercynian age
  burial 738
  erosion 679, 689
  foredeep 738
  orogeny 708, 711, 739
  structures 738–739
  unconformity 731
    Geoprobe image 742
    Palaeozoic geology 694
    subcrop map 740
Hermod system 46, 49
  hydrocarbons 48

Heron cluster
   Skagerrak reservoirs 242
Heron Field
   high production performance 405
HI. *See* hydrogen index (HI)
hidden pay
   example 458
Hidra Formation 473, 540
high-amplitude reflector
   ECG 185
High Atlas 926
high-density turbidites
   Central and Southern Panels 372
high pressure–high temperature (HPHT) field 127, 1128–1129
   appraisal and development 33
   Hejre-1 and Gert-1 wells 134
   Kristin Field sedimentology 419–428
      diagenetic controls on reservoir quality 421–422
      Garn formation 422–423
         depositional synthesis 423
         description 422–423
         rapid pressure decline explanation 423
      Ile formation 424–425
         depositional synthesis 424–425
         description 424
         rapid pressure decline explanation 425
      Not formation 423–424
      palaeogeography 420
      reservoir sedimentology 422–426
      tectonics and burial history 419–420
      Tofte Formation 425–426
         depositional synthesis 425
         description 425
         rapid pressure decline explanation 426
      Upper Ror Formation 425
   Norwegian shelf 420
   Skagerrak Formation 405
   Svane-1 well 138
high-resolution biostratigraphy
   Donan Field 437
high-resolution dinocyst-based biostratigraphy study
   Donan Field 434
high-velocity lower-crust (HVLC)
   P-wave velocities 831
Hildesheimer Wald
   diapirs 902, *903*
Hith Formation
   evaporites 795
   Upper Jurassic GDE map *797*
Hod Formation 473, 478, 539, 540
   listric faults and channel cuts *543*
   porosity *v.* permeability plots *479*
   time isochore maps *542*
Hoggar region 725
   observations 745
Holocene–Recent Rømø barrier island system
   analogue study 145
   DCG 146–148
   methods 146
   morphology and depositional environments 146
   palaeogeographic reconstruction 148
   reservoir architecture 146–148
   vertical section trough 147–148

Holywell Shale
   black shale **1095**
Horda Platform 852
   location map *922*
Horizon 4
   coherency map *89*
   TWT structure map *86*
Horizon 6
   TWT structure 83, *87*
horizon interval *82*
horizontal pilot holes 342
Horn River Basin
   Devonian age 1079
   Muskwa Shale 1079
horst segments
   Clair field 301
hot shale
   Algeria *690*
   basal Silurian source rock 689
   Devonian age *756*
   Frasnian age 754
   Ghadames–Illizi Basin *756*
   Murzuq Basin 689
   Silurian age *756*
Howe area *189*
Howells, Dvid 31
Høybakken detachment 569–571
HPHT. *See* high pressure–high temperature (HPHT) field
Hubbert, M. King 11
Hubbert's Peak 11, 1155
Hugin Formation
   biostratigraphic data 174
   core descriptions and GR log characteristics *163*
   cored wells distribution *162*
   facies associations *164*
   facies summary **160–161**
   Jurassic age 157, 168
   Jurassic strata 168
   MFS 173–174
   Norwegian sector Quadrant 15, 157–175
      axis-perpendicular sub-regional correlation 168–173
      bay-fill facies association A 159–162
         description 159
         interpretation 159–161
         sequence stratigraphy 161
         wireline-log character 161–162
      coastal plain facies association E 167
         description 167
         interpretation 167
         sequence stratigraphy 167
         wireline-log character 167
      exploration implications 174–175
      fluvio-tidal channel-fill facies association D 166–167
         description 166
         interpretation 167
         sequence stratigraphy 167
         wireline-log character 167
      mouth bar facies association C 166
         cross-bedding 166
         interpretation 166
         sequence stratigraphy 166
         wireline-log character 166
      offshore open-marine facies association F 167–168
         description 167

         interpretation 167
         sequence stratigraphy 167
         wireline-log character 168
      previous studies 157–158
      sedimentology 159–174
      sequence stratigraphic framework 168–174
         Dagny area 173
         regional correlation along basin axis 168
         Sigrun–Gudrun area 168–173
         sub-MFS A regression 173–174
         sub-MFS J44 regression 174
      shoreface facies association B 162–166
         description 162–164
         interpretation 164–166
         sequence stratigraphy 166
         wireline-log character 166
      tectonic setting 159
   palaeogeographic reconstructions of deposition *172*, 173–174
   present-day structural setting *158*
   regional maximum flooding surfaces 168
   sedimentology 159–169
   sequence stratigraphy 173–174
   shallow-marine and marginal-marine deposits 157
   shallow-marine deposition 168
   transgression 157, 168
   well-log correlation panel *169, 170, 171*
   wireline-log cross-plot *165*
Humber Group
   sequence stratigraphy
      comparison of interpretations *181*
      interpretation 205
      regional sequence stratigraphic correlation of wells *184*
   succession
      seismic line *203*
   turbidite deposits 189
   UK Central Graben sequence stratigraphy reappraisal 177–210
      background 177–178
      Callovian–Kimmeridgian megasequence 186–192
      sequence J44 186–192
      sequence J46 186
      sequence J52 186–187
      sequence J54a 187–190
      sequence J54(b) 190
      sequence J56 190–192
      sequence J62a 192
      sequence J62(b) 192
      sequence J63 192
   Central Graben unconformities 185–186
   Early Volgian reconfiguration for exploration models implications 205–207
      basin-scale palaeogeographic reconfiguration 205
      palaeovalley model modification 205–206
      prospect-scale models 206–207
      Volgian turbidites 206
   geological setting 180
   previous sequence stratigraphic studies 180–182
   revised sequence stratigraphic model 182–185

study methodology 178–180
tectonic reconfiguration 200–205
  basement uplift evidence 200–201
  block rotation 200
  events timing and style 201–205
  salt redistribution 200
Volgian–Ryazanian megasequence
  192–200
  sequence J64 199
  sequence J66a 199
  sequence J66b 199
  sequence J71-J72 199
  sequence J73-J76 199
  sequence K10 199–200
Humble Source Rock Analyzer 95
Hungary
  geochemical log *1128*
  oil window 1083
  Pannonian Basin 1062, 1126
  source rocks *1128*
Huntington discoveries *219*
  Central North Sea exploration 213–223
    drilling results 221–223
    market perceptions 219–221
    pre-discovery exploration 213–214
    pre-drill predictions 219
    technical evaluation 214–219
  channel axis development *217*
  comparison of results **223**
  reservoir potential *215*
  stratigraphic framework *215*
Huntington Forties depth structure map *220*
Huntington Fulmar well location *219*
HVLC. See high-velocity lower-crust
  (HVLC)
HWC. See hydrocarbon–water contract
  (HWC)
hyaloclastites 1033–1039
Hyblean platform
  Gela oil field *120*
  geological sections *117*
  hydrocarbon occurrences *117*
hydraulic fractures 321, 323, 330, 1062,
  1131–1132
  48/14a-5 well 332
  48/14a-6 well *335*
  Chiswick Field *322*
  Ensign Field **331**
  Leman Sandstone interval *333*
  pre-existing natural fracture diagram *1134*
  well 48/14a-6 *333*
  48/14-2 well evaluation 330
hydrocarbon(s) 2, 124, 591, 598
  Adriatic basin *117*
  Angola 1046
  Apennines *117*
  Apulian platform *117*
  v. aquifer seal capacity 502
  Bakken *1073*
  Balder *48*
  Bighorn Basin 6
  Bream Field 987
  v. carbon dioxide 1183–1188
  Carboniferous petroleum system
    play types and traps 71–74
    search beneath Central North Sea 69–71
  Cenozoic plays 698–700
    exploration challenges 700
    stratigraphy 698–700

Central North Sea 493
Chalk 505
Chalk Group 273
charge and maturation 824–826
  major marine source rocks 824
  oceanic crust exploration 826
deepwater equivalents 49
East Greenland 999, 1001
entrepreneurial exploration 113
exploration 1, 51
exploring older and deeper plays
  673–702
Faroe–Shetland and Møre basins 1025
geochemical characteristics 106
Gulf of Suez 774
Halten Terrace 419
Hermod *48*
Hyblean platform *117*
Infracambrian age 674
intervals and well completion 454
inter-well volumes 455
Istrian platform *117*
Italy 113–125, *114, 116*
Jameson Land Basin 997
Judd Basin 247
Lagoa Feia lacustrine facies 862
Laptev Sea Province 606
Lombard basin and ridges *117*
Lower Balmoral reservoir *439*
Mesozoic plays 697–698
Middle East plays 674–676, 691–697
migration 1186–1188
modelling 825
Neoproterozoic petroleum system
  677–681
Neoproterozoic plays 676–682
  hydrocarbon plays 677–679
  stratigraphy and tectonic evolution
    676–677
  thermal evolution and hydrocarbon
    generation 679–681
North Africa plays 688–691
  Cambro-Ordovician stratigraphy
    682–686
  Silurian and Devonian stratigraphy
    686–688
  thermal evolution 688–691
North Atlantic passive margins 986
Northern Ireland wells *1089*
Northern Judd Basin 251–253
North Kara Sea 605
Oman plays 674–676
Palaeozoic plays 682–684, 697
Pennine Basin 1093
plays 673–702
potential 4
predictive capabilities 51
RAS 603
reservoir 27, 95
Saudi Arabia plays 691–697
  exploration challenges 695–697
  stratigraphy and tectonic evolution
    692–695
seismic reflection datasets 549
Sicilian–Maghrebian chain *117*
SKB 604
source rock 818
Taoudenni Basin *678*, 679–681
3D seismic 51

Triassic Skagerrak Formation 225
untapped accumulations 455
Val d'Agri giant oil field *120*
West African basins 7
West Africa plays 676
West Emila accretionary arc *117*
hydrocarbon–water contract (HWC) 230
hydrogen index (HI) 1068

Iapetus closure 265
Iberia margin
  drilling results 867
iceberg keel erosion *83*
IEA. See International Energy
  Agency (IEA)
IGMAS. See Interactive Gravity and
  Magnetic Application System (IGMAS)
Ile Formation 421, 424–425
  depositional synthesis 424–425
  description 424
  facies components *424*
  modern analogues *424*
  pressure profiles *422*
  rapid pressure decline explanation 425
Illizi Basin 712–716, 725, 1145. See also
  Ghadames–Illizi Basin
  burial and uplift models tested *752*
  chronostratigraphic chart *720*
  interpretation 713–716
  stratigraphic architecture 712–713,
    713, 715
  stratigraphic pinchout *716*
imaged sub-basalt strata
  NE Atlantic margin 1029
Imperial
  Cold Lake development 1147
inboard areas 825
inclined stocks squeezing
  stages *901*
inclined strata (IS) 1145
  sand-prone 1145
India. See also specific location
  gas hydrates occurrence 1153
Indonesia. See also specific location
  Kalimantan *458*, 459, *459*
industry
  challenges 1
  plate tectonic theory 2
Industry Taskforce on Peak Oil and Energy
  Security (ITPOES)
  members 12
  report 12
infill wells 349, *350*, 355, 362
  drilling in mature fields offshore 349
  use of predictive models to plan
    349–350
Infracambrian age 679
  black oil shales 676
  hydrocarbon plays 674
Inge High 139
injected carbon dioxide
  hydrocarbon migration 1183–1188
    flow characteristics 1184–1186
    implications 1186–1188
    Utsira Formation 1183–1184
injected sands and field edge 340
injection and production wells
  directional drilling *1157*
Inner Moray Firth 38

In Salah Field
  carbon dioxide geological storage
      1168–1169
  satellite imagery 1169
  technologies *1168*
In Salah Krechba Field
  cross-section *1168*
in situ coal gasifying 1156
in situ heavy oil production
  seismic monitoring 1147–1148
integrated petroleum systems
  Ghadames–Illizi Basin, North Africa
      735–759
integrated Seismic Imaging and Modelling of
    Margins (iSIMM) 831
  airgun source 832
  Faroe Islands *838*
  profile 833, *838*
Interactive Gravity and Magnetic Application
    System (IGMAS) 888
interbedded turbidites
  Andrew Formation 431
Intergovernmental Panel on Climate Change
    (IPCC) 15, 1167
International Energy Agency (IEA)
  World Energy Outlook (2005) 11
interpretation method 1035
interpreted seismic reflectors
  3D visualization environment 1035
interpreted unit boundaries
  two-dimensional perspective *1036*
Intra-B4 shale
  field-wide correlation and map *377*
  holes 377
  recognition 376
intra-Buzzard 373
intra-chalk unit porosity variation 537–548
  basin development 540–541
  database 537–538
  geological setting 538–539
  inversion results along key profiles
      543–547
    key profile 1, 543–544
    key profile 2, 544–547
    key profile 3, 547
    key profile 4, 547
  seismic inversion 541
  2D seismic inversion work flow
      541–543
intracontinental deformation 824
intra-Pennsylvanian stratigraphic traps 266
Intra-Pliocene Unconformity 642
inverse genetic scheme 185
Ionian Ocean formation 741–742
IPCC. *See* Intergovernmental Panel on
    Climate Change (IPCC)
iPhone 25
Iraq. *See also* specific location
  Kurdistan region 801–809
  location map *801*
  Taq Taq field 801–809
Ireland. *See also* specific location
  Atlantic margin region 924
  basins offshore *939*
  crystalline basement *948*
  Dooish well *939*
  hydrocarbon wells *1089*
  lead space plots *948*
  Mesozoic sandstones 951

passive margins case histories 1005–1012
Shannon Basin 23–24
Iris-1 location
  gravity-flow sandstones *138*
irreversible compaction
  grain-to-grain bonds failing 534
IS. *See* inclined strata (IS)
iSIMM. *See* integrated Seismic Imaging and
    Modelling of Margins (iSIMM)
ISIS software 541
  two dimensional inversion
      workflow *544*
isoprobe multicollector 6
ISROCK
  acquisition plan *556*
  seismic survey *554, 555, 556*
  water layer reflectivity *551*
  water layer structure *553*
Istrian platform
  geological sections *117*
  hydrocarbon occurrences *117*
  Porto Corsini East gas field *122*
Italy 116–123
  cumulative discovered reserves *125*
  gas provinces and oil provinces *115*
  geological framework and hydrocarbon
      occurrences *114*
  hydrocarbon occurrences 116–123, *116*
    Cretaceous petroleum system 119–120
    foredeep wedges biogenic gas 121–123
    foredeep wedges thermogenic gas 121
    geological framework 113
    historical perspective 113–116
    petroleum system 118–119
    thermogenic gas 116–120
    thrust-and-fold belt to foreland
        113–125
  petroleum
    acreage *114*
    endowment 114
    permits 114
    systems *124*
  seismic surveys and wells *115*
  tectono-stratigraphic cycles *116*
ITPOES. *See* Industry Taskforce on Peak Oil
    and Energy Security (ITPOES)

Jabal Hasawnah 710
Jackson Field
  Australia 355
Jacuípe Volcanic Complex 870
Jade Field 225, 407, 412
  high production performance 405
  models 410
  predictive history-matched model 405
  reservoir depth 410
  well 30/02c-4 242
Jaeren High
  remnant topography *196*
  seismic profile *64*
  sub-Zechstein seismic stratigraphy 62
Jameson Land Basin
  deepwater reservoirs *49*
  erratic variation of fission track age 996
  hydrocarbons generation 997
  time-equivalent strata 420
Jan Mayen Fracture Zone (JMFZ) 579–580
Jan Mayen microcontinent
  plate reconstruction *582*

Japan. *See also* specific location
  Nankai Trough 1153
Jasmine discovery 225–242
  analogues 228
  appraisal drilling locations *233*
  appraisal phase 230–234
    drilling results 231–234
    30/06-07, 234
    30/06-07 and -07Z 231–233
    30/06-07Y 234
    programme 230–231
    secondary reservoirs summary 234
  appraisal sealing conundrum *237*
  exploration drilling phase summary *230*
  exploration phase 227–230
    discovery well results 228–230
    pre-drill prospect evaluation 227–228
  licence and exploitation history 225–226
  location *226*
  location and aims 225
  LWD 234
  pressure data *236*
  reservoir characterization programme
      234–242
    compartmentalization 242
    depositional model 236–240
    drill stem tests 242
    intra-reservoir correlation 240
    reservoir quality 242
    reservoir zonation scheme 240–242
      Top Joanne 240–241
      Top Julius Mudstone 241–242
      Top Lower Joanne 241
      Top Middle Joanne 241
    seismic reprocessing 234–236
    structural model 236
  reservoir zonation 240
  sealing conundrum *232, 242*
  seismic depth section wells *233*
  stratigraphic and structural setting
      226–227
  wireline log data 30/06-6 *231*
Jasmine Field 225
Jasmine–Julia structural cross section *229*
Jasmine Northern Terrace area 227
Jasmine West Limb 231
  faulted anticline 227
Jaurf Formation
  Devonian age 692
J Block 225
  Joanne Member 410
  seismic reprocessing strategy *238*
  shoreface deposition *191*
  survey acquisition area *238*
JCR. *See* Joint Chalk Research (JCR)
Jenkins, David 15
  for motion peak oil no longer concern
      11–12
JIP. *See* Joint Industry Project (JIP)
JMFZ. *See* Jan Mayen Fracture Zone (JMFZ)
Joanne Field 225, 234
Joanne Formation
  porosity *v.* permeability plot *413*
Joanne Member 231, 405, 406, *408*
  channel-dominated 416
  characteristic decline 407
  characteristic radial composite model *409*
  conceptual stratigraphic architecture *409*
  core differences 407

dynamic modelling workflow *411*
EUR resources *415*
facies 407
gross porosity plotted with depth by facies *410*
heterogeneous 417
J Block 410
linear flow behavior 409
porosity-depth trends 416
production 411
production performance 405, 407
production profile v. CGR *415*
reservoir 226, 410
   architecture 408
   production performance 412
RF. v. CGR/GOR *415*
shalier interval 240
SIS 410
static modelling workflow *411*
vertical pressure barriers 407
well test *409*
Joanne Sandstone Member 226, 240
   depositional environment 406
   erosion 241
   rich gas condensate 228
   sandstone 406
Joint Chalk Research (JCR) 473
Joint Industry Project (JIP) 1168
   Gas Hydrate Gulf of Mexico 1151
Jonathan Mudstone 226, 241
   climatic reinterpretation 236
Josephine Sandstone Member 240, 406
J Ridge area 39, *40*, 180, 190, 200, *203*, 225
   structural evolution 226
   Top Triassic Unconformity 227
Judd Basin
   basin inversion *249*
   hydrocarbon occurrences *247*
   oil-prone basin 252
   seismic amplitude anomaly 249
   stratigraphic column *247*
   T38 seismic anomaly *249*
Judy Field 225, 407
   description 406
   EUR 412
   gas production *408*
   high production performance 405
   production profiles 412
   reservoir properties with depth 414
Judy Formation
   reservoir architecture 409
   vertical communication 409
Judy Member 406, *408*
   conceptual stratigraphic architecture *409*
   EUR resources *415*
   gas recovery factor 414
   gross porosity plotted with depth by facies *410*
   heterogeneous 417
   porosity-depth trends 416
   production profile v. CGR *415*
   radial flow *409*
   reservoir architecture 408
   reservoirs 405, 410
   RF v. CGR/GOR *414*
   sandstones contrast 407
   sheetflood-dominated 409, 416
   Skagerrak Formation 410

static and dynamic modelling workflow *411*
variogram-controlled SIS simulation 411
vertical pressure barriers 407
well 407, 410
Judy Sandstone Member 240
   depositional environment 406
Julia well 228
   boundary *232*
   pressure data *236*
   structural cross section *229*
Julius Mudstone Member 226
   climatic reinterpretation 236
Jurassic age. *See also* Late Jurassic age
   analytic signal shaded relief map *576*
   Arabian Plate *794–799*
   barrier island system 150
   basins *216*
   black shales 1096–1097
   Brent Delta systems 174
   Buzzard Field reservoir 369–371
   clastics 1118
   conceptual play model 962
   Early
      Netherlands petroleum exploration 268
      petroleum system 118–119
      Posidonia shale 263–264
      source rock *271*
   faults *576*
   fields *178*
   Fisher Bank Basin 214
   Fulmar Field *214*
   gas yield *665*
   Hanifa *793*
   Hugin Formation 157, 168
   Jura 437, *438*
   lithostratigraphy *159*
   Middle
      conceptual play model 962
      North Atlantic region *949*
      palaeogeography *420*
      pre-rift sandstone 962
   Middle East exploration potential 794–799
   Navajo Formation Sandstone 1193
   North Sea rift complex 127
   palaeogeography *420*
   palaeotopography 150
   Pentland Formation reservoir 231
   pre-rift sandstone 962
   pre-rift succession *957*
   rift stage 960
   sedimentation *216*
   stratigraphy 369–371
   Tyumen Formation 666
   wedge 1114–1115
   West Siberia Basin depth map *660*
Jutland 569
   lithological and susceptibility log *579*
juxtaposition difference 510
Jylland *149*, *150*

Kalak Nappe Complex 565
Kalimantan, Indonesia *458*, *459*, *459*
Kara Massif (KM) 595. *See also* Barents–Kara region
   tectonic basement 596
Kara Province 600

Kara Sea 591, 595, 598, 605, *661*
   fold-and-thrust belt 649
   free-air gravity field *595*
Kara Shelf
   fold-and-thrust belt 646
Kareem Formation 772, 773, 775
   seismic character 776
Karoo Group
   South Africa 52
Kasimovian age
   palaeogeography maps *655*
Kassanje Basin *1044*, 1055
Kattegat Platform 891
   individual contributions to basement depth map 891–892
KCF. *See* Kimmeridge Clay Formation (KCF)
Keathley Canyon 1151
kerogen
   composition and source rock quality 96–100
   DCG 96–100, **106**
   Farsund Formation 98, *102*
   NW Danish Central Graben source rock 101–103
   oil-prone through Farsund Formation 96
   photomicrographs and oil immersion *102*
   reactive wells **106**
   sapropelic *102*
   Stellar default kinetic parameters 663
Kettla Tuff 245, 246
Keyhole Markup Language (KML) 20
KG. *See* Krishna Godavari (K-G) Basin
Khatanga Saddle 622, 624
   assessment unit 624–627
   densities and median sizes *625*
   distributions and AUs probabilities *629*
   field densities, field sizes, and exploration maturities **623–624**
   Monte Carlo simulation 624
Khaz Dumi
   major play systems *793*
Khuff Formation
   Saudi Arabia 692
Kimmeridge Clay Formation (KCF) 51, 178, 186, 214, 495, 825
   black shale **1095**, 1096
   sequence J64 *179*, *181*, 192–200
   subcrop *207*
   well correlation *197*
   stratigraphic thicknesses control *504*
   subordinate turbidite 192
   turbidite flows 369
   turbidite sands pressures *503*
Kimmeridge Fulmar sandstones
   distribution 205
Kimmeridge–Piper Transition unit (KPT) 390
Kimmeridgian age 264
   Late
      DCG *134*
      sequence J63 192
      time stratigraphic scheme *152*
      well correlation *197*
   sequences J62a 192
   sequences J62b 192
Kingston, Dave 2
Kintail fan 246
Kirchhoff time migration
   South Arne *526*

kitchen maps 754
Klett, Timothy 589
Klinkenberg permeability
   Chalk empirical relation 469
Klubben Formation 565
KM. *See* Kara Massif (KM)
KML. *See* Keyhole Markup Language (KML)
Kollstraumen detachment 569, 571
Kometan Formation 805
   description *803*
   lithological model *807*
   seismic analysis 807
Kong Oscar Fjord 996
Kongsberg-Bamble Complex 569
   Bouguer gravity profiles *574*
   NRM 565
Kootenay Group 1101
Kotel'nyi Terrane 597
KPT. *See* Kimmeridge–Piper Transition unit (KPT)
Kraka Extension Survey 539
   location map *538*
Kraka Field 543
Kraka structure 547
Krechba
   Carboniferous storage reservoir 1168
Krishna Godavari (K-G) Basin 825, 1052
Kristiansand 569
Kristin Field
   bioturbated shale-prone interval 425
   chemostratigraphy 423
   depletion from 12 wells 421
   footwall stratigraphy 428
   growth across faults 419
   HPHT sedimentology 419–428
   location map *420*
   major fault overstep *428*
   marine transgression 425
   permeability distribution *427*
   quartz cement growth 422
   structural model *421*
   unexpected pressure decline 419–428
Kufra Basin 761–769, *761, 764*
   burial history *691*
   Cambro-Ordovician sandstone *767*
   chemostratigraphy 763–766
      material and methods 765
      results 765–766
      study objective 765
   depth-to-basement map *762*
   fluid inclusions 766–767, *767*
      material and methods 767
      results 767
      study objective 766
   palynology 763
      material and methods 763
      results 763
      study objective 763
   petroleum source rock evaluation 767–768
      material and methods 768
      results 768
      study objective 767–768
   regional hydrocarbon prospectivity implications 768–769
   seismic section *681*
   spore colouration 768
      material and methods 768
      results 768
      study objectivity 768

Spore Colour Index against depth *769*
tectonic events *691*
vitrinite reflectance 768, *768*
well A1-NC198 exploration 761–769
wells locations *762*
Kupfershiefer
   Permian age 264
Kurdistan region
   Taq Taq field 801–809
Kwanza-Benguela Basin *1044*
   schematic play cross-section *1048*
Kyle Group 57
Kyle Limestone 59

laboratory caprock core **1177**
Labrador Sea conjugate
   exhumed mantle 848
Laggan Field
   appraisal wells 279
   AVO *283*
      gas sand calculation *291*
      gas sand thickness uncertainty *295*
      net gas sand map *285*
      seismofacies 289
   biostratigraphic zones 288
   cataclastic bands *293*
   chlorite 279, *284*, 289
   compartments and segments definition *293*
   depositional model *286*
   exploration wells 279
   fault juxtaposition analysis *293*
   feeder system 282
   gas condensate discoveries 279
   leaking segments *294*
   location map *280*
   Low Reflectivity Package 279
   mature understanding of undeveloped discovery 279–297
      AVO analysis 289
      compartmentalization 289–292
      depositional model 279–282
      3D geological models 294–296
      net pay estimation 289
      petrographic overview 288
      petrophysical interpretation 288
      reservoir correlation 288
      seismic interpretation 288–289
      seismic inversion 289
      static uncertainty study 292–294
      Tormore 296
   net pay *v.* net gas sand map *296*
   Paleocene gas condensate discoveries 279
   petrochemical analysis uncertainty 288
   Poisson ratio 289
   pressure data 289, *292*
   pressure *v.* depth plot *292*
   reservoir
      correlation *287*
      depth structure map *282*
      properties *284*
      subdivision 282
      thickness *290*
   sand isochore map *285*
   sands sediment supply 282
   seismic impedance 282
   seismic interpretation 288
   seismic IP and well data *286*

   static uncertainty *291*
   stratigraphy *281*
   structural map *283*
   traps 279
   undeveloped discovery 282
   water saturation uncertainty 288
Lagoa Feia Formation 856
   facies 862
   hydrocarbon source rock 862
La Marcouline Formation 25
Lamba Formation 245
Lapland Granulite Belt 569
   Bouguer gravity profiles *574*
Laptev Rift System (LRS) 605
   structure *607*
Laptev Sea 591, 593, 597, 598, 611
   hydrocarbon 606
   RAS sedimentary basins 605–609, *606–609*
Laptev Shelf
   oil generation 607
lasers
   ablation system 6
   airborne scanner 20
   scanning technology 25
Last Chance Sand 359
Last Chance Shale 359, *359, 360*
   shut-in flowmeter pass *358*
Late Callovian age
   sequence J46 186
Late Campanian age
   Magne Formation biofacies map *483*
Late Carboniferous age
   Netherlands petroleum exploration 263
   palaeogeography maps 655
Late Cenozoic age
   Viera Anticline 971
Late Cimmerian age
   fault geometries 39
   North Sea hydrocarbon systems 39, 44–45
   rift architecture 44–45
   tectonics 39
Late Cretaceous age *488*, 1101
   Cardium Formation 1099
   Chalk 463, 464
      Central North Sea 493
      Inoceramid bivalves 480
      limestone porosity *465–466*
      natural gamma ray log *465–466*
      Netherlands petroleum exploration 269–273
      pressures 494
   Danian Chalk Group deposition 538
   Euphrates Graben sedimentology 811–818
   Faroe–Shetland syn-rift play 958
   genetic basin analysis 5
   global genetic analysis 4
   Nubian sandstone 783
   Shiranish Formation sedimentology 811–818
Late Cryogenian age
   Dar Cheikh Group 677
Late Danian age
   Middle Ekofisk Formation biofacies map *488*
   Upper Ekofisk Formation biofacies map *488*
Late Devonian age 725–729, 745
   Antrim Shale 1083

GDE map *746*
incised valleys *712*
Late Jurassic age 795
  Arab play 795
    CRS map *799*
    formation 799
    for oil *798*, 799
    for oil charge *796*
    play cartoon *798*
    for reservoir *796*
    for seal *797*
  barrier island sandstones 145–154
    Freja discovery 148–153
      back-barrier sediments 149–150
      depositional environments 149–150
      shoreface sandstones 150
      stratigraphy 149
      thick stacked 150–153
  basin architecture 50
  Bazhenov Formation 666
  DCG 95–111, 127–142, 129–141
  deepwater depositional systems 46–48
  deposits 203
  extension multiple phases interplay *42*
  Fulmar reservoir potential *215*
  Fulmar Sandstone 215, *218*
  global plate reconstruction *5*
  isochore map *128*
  isopach *218*
  Kimmeridge Clay *215*
  Lower Volgian gravity-flow sandstones 137–138
    depositional environment 137–138
    palaeogeography 138
    reservoir characteristics 138
  Lower Volgian shoreface sandstones 138–139
    depositional environment 139
    palaeogeography 139
    reservoir characteristics 139
  Netherlands petroleum exploration 268–269
  Nikanssin Formation 1101
  v. Palaeogene age deepwater reservoirs 50
  platform area 369
  play map *271*
  reservoir architecture 145–154
  reservoir sandstones 129–141
  sandstone play area 95–111
  shale gas in Europe 1081
  shale gas in Vienna Basin 1081
  shallow marine sandstones 136–137
    depositional environment 136–137
    reservoir characteristics 137
  shoreface sandstones 129–136
  succession *128*
  Upper Kimmeridgian paralic 129–138
    depositional environments 129–133
    palaeogeography 135–136
    reservoir characteristics 133–134
  Upper Middle Volgian 139–141
    depositional environment 139–141
    distribution and stratigraphy 139
    palaeogeography 141
    reservoir characteristics 141
Late Kimmeridgian age
  DCG *134*
  sequence J63 192, *197*

time stratigraphic scheme *152*
well correlation *197*
Late Maastrichtian age
  Tor Formation Subzone
    biofacies map *485*, *486*
Late Miocene age
  Algyõ Formation 1128
  brackish-water shale 1129
  North Sea Basin palaeogeography *985*
  Pannonian Basin 1126, 1127–1128
  shales 1062
Late Neogene age
  basin development implications 90–92
  depocentres *82*, *89*
  development contour currents 77–92
  exhumation 983–984
Late Neoproterozoic age
  Peri-Gondwana Margin *675*
Late Ordovician age 686
  Lower Silurian shales 715
  Mamouniyat Formation *685*
  Sarah Formation
    striated glacial surface *695*
Late Oxfordian age
  Callovian–Kimmeridgian megasequence (J44-J63) 190–192
  sequence J54a 187–190
    depositional environments *190*
    regressive shoreface sandstones 187
    turbidite sandstones *188*
  sequence J54b 190
    depositional environments *191*
    well correlation *197*
Late Palaeozoic age
  genetic basin analysis *5*
Late Paleocene age 45
Late Permian age
  North Atlantic region *949*
lateral horizon density variations *64*
Late Ryazanian age
  sequence K10 199–200
Late Silurian age
  Caledonian unconformity 711
  closure 265
late syn-rift *961*
  Faroe–Shetland syn-rift play *958*
Late Tournaisian age 729
  regression 729
Late Triassic age
  Dutch Central Graben 265
  Italy hydrocarbon occurrences 118–119
Late Viséan age 729
Late Volgian age *202*
layer density values input
  3D inversion model *66*
LCB. *See* lower crustal body (LCB)
LCS. *See* Lower Captain Sand (LCS)
lead
  basement source 938
  concentration **942–943**
  Dooish well **942–943**
  isotopic data 948
  North Atlantic region 938
  potassium feldspar
    European Margin *937*
    isotopic composition *950*
    palaeodrainage *938*
    Porcupine Bank *940*

provenance technique 934
Rockall Basin 938, *941*
provenance analysis 942–944
source ranges 948
space plots *930*
  crystalline basement from onshore Ireland 948
  Dooish sandstones K-feldspar *947*
  Lewisian Complex Rocks *948*
  Nagssugtoqidian Mobile Belt *948*
  offshore crystalline basement blocks from Atlantic *948*
Leadon Field
  FPSO 443
Leggett, Jeremy 11, 15
  against motion peak oil no longer concern 12
Leiosphaerida morphology 98
Leman Sandstone 325, 327
  correlation panel *329*
  depth map *326*
  Ensign Field *326*, 327
  hydraulic fracture-stimulated intervals *333*
  illite cementation 328
  reservoir 327
Lena-Anabar Basin 624
  assessment units 627, *629*
  densities and median sizes *625*
  field densities, field sizes, and exploration maturities **623–624**
Lena Delta
  crustal tectonic *594*
Leningradskoye field drainage area *664*
  gas yield 663, *664*, *665*
lessons learnt 347
  drilling pilot holes 347
  seismic 347
Levantine Basin
  exploring older and deeper hydrocarbon plays 697–698
  Mesozoic 697–698
Levorsen, A. I. 14
Lewisian Complex Rocks
  lead space plots *948*
Lewis Shale
  Cretaceous age 1091
Lias
  black shale **1095**, 1096
Libya 729. *See also* specific location
  basins 730
  Gargaff Arch 710–712
    chronostratigraphic chart *720*
    geological map *710*
  Ghadames basin 710–712
  Illizi basin 715–716
  Kufra Basin 761–769
  Murzuq basin 710–712, 715–716
  outcrops 729
  platform 707–708
  reservoirs *753*
  sub-basins and arches 707–708
  tectonic events *691*
  Tihemboka High 715–716
lidars 20, 23
  Ferron Sandstone 24
  Nukhul syncline dataset *22*
  Panther Tongue sandstone 24
  scan data 24

Life of Field Seismic system
  4D data 312
  Valhall 311
Limestone
  Devonian Kyle Limestone 59
  Mendips 1092–1093
  Top Palaeogene Chalk 465–466
  Upper Cretaceous Chalk 465–466
Limits to Growth 15
Ling Sandstone Member
  Heather Formation 174
liquefied natural gas (LNG) 1044
  globalization 827
lithofacies prediction
  Taq Taq field 806
Little, Alice 590
LKO. *See* Lowest Known Oil (LKO)
LNG. *See* liquefied natural gas (LNG)
Lochkovian age
  palaeogeography maps 655
Lochranza Field 449
  AVO and structure 449
Lodgepole Formation 1066
Lofoten–Vesterålen margin 847
  geological and structural
    features 844
  individual contributions to basement depth
    map 889–890
  NRM 565
  petroleum geoscientists virtual
    fieldtrips 24
log facies map
  Nelson Field 361
logging while drilling (LWD) 306, 342
  advances 446
  geosteering 342, 347
  Jasmine accumulation 234
  well locations 347
Lola Formation 130
  shoreface sandstones 135
Lombard Basin
  geological sections 117
  hydrocarbon occurrences 117
  Settala gas field 123
  Villafortuna–Trecate oil field 119
Lomonosov Ridge 590
Longa Basin 610
Long Eaton Formation
  black shale **1095**
  shale 1093
Lopra Formation 1038
Louisiana
  offshore 460, 461
  water saturation data 457
Lousy Bank Anticline 965–966
  seismic profile 968
Lower Balmoral
  net pay thickness 434
  overlying Lista roof shale 448
  reservoir 439
  sandstone 431, 437
    biostratigraphic subdivision 440
  vertical seismic profile 441
Lower Bowland
  black shale **1095**
Lower Captain Sand (LCS) 343
Lower Congo Basin 1044
  3D seismic survey 1045
  gas discoveries by play 1047

Lower Cretaceous age. *See* Early Cretaceous
  age
lower crustal body (LCB) 843
  Edoras–Hatton margin 849
  mid-Norwegian margin 843–853, 845, 852
    3D models compilation 844
    isostatic considerations 844–846
    process orientated approach
      846–847
    regional considerations 847–850
  Norwegian model Moho horizons 844
  Vøring margin 847
Lower Devonian age. *See* Early Devonian age
Lower Ekofisk Field 475–476
  porosity and bulk volume water 477
Lower Ekofisk Formation 486–487
  age 486
  biofacies 486
  biofacies map 487
  lithology/petrophysics 486
  tectonostratigraphy 486–487
Lower Fernie Formation
  interbedded sandstones 1116
Lower Limestone Shale
  Mendips 1092–1093
Lower Magne Formation 482–483
  age 482
  biofacies 482
  correlation 483
  lithology/petrophysics 482
  tectonostratigraphy 482–483
Lower Miocene age. *See* Early Miocene age
Lower Palaeozoic age. *See* Early Palaeozoic
  age
Lower Piper sections 393
Lower Rotliegend 59, 62
  subcrop traps 74
  units 61
Lower Rudeis Formation 777
Lower Scott Sands 393
  J35 post-drill correlation 394
Lower Silurian shales
  Upper Ordovician surface 715
Lower Tor Formation 484
  age 484
  biofacies 484
  lithology/petrophysics 484
  tectonostratigraphy 484
Lower Volgian gravity-flow sandstones
  137–138
Lowest Known Oil (LKO)
  Ross and minor Burns reservoir 345
low velocity zone (LVZ)
  sub-basalt 833
LRS. *See* Laptev Rift System (LRS)
Ludlov Recess 653
Lutetian compressional event 975
LVZ. *See* low velocity zone (LVZ)
LWD. *See* logging while drilling (LWD)

Maastrichtian age
  Early 477, 484
  Late 485, 486
  Tor Formation 539, 541, 547
Macaronichnus 139
Mackenzie River Delta 827, 1151
  gas hydrates occurrence 1153
Macuspana Basin 456, 458
Mad Dog Field

pre-stack depth-migrated data 1015
  subsalt 1015
  WATS survey 1016
Madison Formation 1065, 1066
magmatic underplating 843
Magne Field
  porosity and bulk volume water 477
  tectonic and eustatic events 489
Magne Formation 473, 477, 478. *See also*
    Lower Magne Formation
  Late Campanian age 483
  porosity *v.* permeability plots 479
  sequences 481
magnetic basement
  Norwegian margin 888
Magnus basin
  location map 922
Magnus Fields 43
  bypassed oil 356
  water overrun 356
Maider Basin 731, 732
major deltaic depocentres 823
Makó Trough 1083
  pre-Neogene rocks 1082
  shale gas in Europe 1081–1083
Malaysia 459, 460
Mali
  Neoproterozoic geology 677
Mallik, Canada
  gas hydrates occurrence 1153
Mallik Gas Hydrate Production Research
    Well Program 1151, 1153
Malta 741
Mamouniyat Formation
  Late Ordovician glaciogenic 685
Marcellus Shale 1133
Marco Polo minibasin
  Green Canyon area 909
  Gulf of Mexico 909
margins. *See also* specific name
  3D isostatic behaviour 824
  evolution 2D 824
Margins Programme 913
Marinated fold belt 679
marine shales
  Tofte Formation 426
Masjid-i-Sulaiman wells 791
Massif Ancient
  Pre-Cambrian rocks 925
maturity 1079–1081, 1083, 1087–1091
Mauritania
  Neoproterozoic geology 677
  Taoudenni Basin 676
Maximum flooding surfaces (MFS) 180
  J44 transgression 173–174
  palaeogeographic reconstruction
    173–174
  T75 358
    bridge plug 357
    vertical flow barrier 359
  transgression 173
maximum regressive surfaces (MRS) 180
McMurray Formation 1144
  Lower Cretaceous age
    Alberta 1062
MDT. *See* modular dynamics tester (MDT)
  data points
measurement while drilling (MWD) sensors
  446

Median Volcanic Ridge
  continental crust 848
Mediterranean Basin
  tectonic elements *698*
Mediterranean passive margin rifting
  739–741
Megion Formation 661
Meharez High 733
Mendips
  Lower Limestone Shale 1092–1093
mercury capillary pressure 464
Merganser leakage zone
  Central North Sea basin *88*
Meride Limestone 118
  foreland 124
  hydrocarbon occurrences 124
Mesozoic age *119*
  Atlantic margin basins 938–940
  basins regional 3D data 41
  carbonate cores *121*
  depocentres Geoprobe image *742*
  evolving drainage patterns 937
  exploring older and deeper hydrocarbon
    plays 697–698
  extension 118
  Faroe–Shetland Basin 954
  foreland 119
  Greenland basins 980
  hydrocarbon plays 697–698
  intraplate inversional swells *602*
  Ireland sandstones 951
  Levantine Basin 697–698
  Møre basin magmatic underplating 853
  petroleum systems 698
  plays 697–698
  rifting events 265
  Rockall Basin palaeodrainage
    938–940
  Rockall Basin stratigraphy 938
  Rockall Trough strata 1031
  sediment dispersal into NE Atlantic margin
    basins 938–940
  Sirte Basin 697–698
  structural elements *265*
  structural play 954
  Val d'Agri oil field *121*
  Villafortuna–Trecate oil field *119*
  Vøring basin magmatic underplating 853
  Western Desert 697–698
  West Siberian basin
    chronostratigraphy *662*
methane
  Angola *1053*
  carbon and hydrogen isotope
    ranges *1094*
  enhanced coal bed 1190
  generation microbial ecosystem succession
    1053
  northern West Siberia *665*
  solubility *1054*
metric tons of carbon dioxide 1189
MFS. *See* Maximum flooding surfaces (MFS)
Michigan Basin 1083
microscale sealed vessel (MSSV) pyrolysis
  1067
  Bakken Formation *1072*
Mid-Clysmic Unconformity 777
Middle Akakus Shale
  Silurian age GDE map *748*

Middle Cimmerian age
  reservoir distribution controls 38–39
Middle Cretaceous age
  early syn-rift stage mature 962
  Faroe–Shetland syn-rift play 957–958
  relay ramps 962
  stratigraphic interpretation *1121*
Middle East
  case study of exploration potential
    791–799
  composite play fairway map *799*
  description 792
  exploration potential 791–799
    Arabian plate Jurassic 794–799
    exploration process 794
    future potential 799
    history 791
    plays 792–793
  exploring hydrocarbon plays 691–697
    challenges 695–697
    stratigraphy 692–695
  exploring older and deeper hydrocarbon
    plays 674–676, 682–684, 697
  hydrocarbons development 695
  Khuff Formation 692
  major play systems *793*
  Palaeozoic basins *674*
  Palaeozoic geology *694*
  regional geology *693*
  Silurian Tannezuft Formation *687*
  striated glacial surface *695*
  tectonostratigraphic chart *793*
  Unayzah complex 692
Middle Ekofisk Formation 476, 487
  age 487
  biofacies 487
  biostratigraphic data 476
  Late Danian age
    biofacies map *488*
  lithology/petrophysics 487
Middle Eldfisk Field
  porosity and bulk volume water *477*
Middle Joanne shalier interval 242
Middle Jurassic age
  conceptual play model 962
  North Atlantic region *949*
  palaeogeography *420*
  pre-rift sandstone 962
  Tyumen Formation 666
Middle Miocene age
  North Sea Basin palaeogeography *985*
  Pannonian Basin 1126, 1127–1128
  Pannonian Basin tight gas plays
    1128–1129
  shales 1062
  source rocks geochemical log *1128*
Middle Norwegian margin
  LCB 843–853
    3D models compilation 844
    isostatic considerations 844–846
    process orientated approach 846–847
    regional considerations 847–850
Middle Oxfordian age
  sequence J52 186–187, *189*
Middle Tor Formation 484
Middle Tournaisian age 729
Middle Volgian age
  sequence J66b 199
  sequence J71-J72 199

mid-early Danian age
  Lower Ekofisk Formation biofacies
    map *487*
Mid-Late Cretaceous age
  Faroe–Shetland syn-rift play 958
Mid Miocene Unconformity (MMU) 78
  acoustic pull-ups 84
  3D horizon maps *79*
  TWT structure map *79*
Mid North Sea High
  structural and stratigraphic relationships *70*
mid-Norwegian body
  formed 843
mid-Norwegian margin
  Caledonian basement 895
  Early Cretaceous rifting 849
  geological and structural features *844*
  individual contributions to basement depth
    map 889–890
  isostatic flexural base 845
  isostatic flexural Moho *848*
  LCB *845*
  LCB properties *852*
  magmatic underplate 846
  Moho configuration *849*
  profiles *851*
  Rockall Trough 849
Mikulov Marls 1081, *1082*
Minas Basin 929
  location *931*
  sedimentary logs through Triassic
    succession *932*
  Triassic outcrops *931*
minibasins
  formation 911
  salt tectonics emerging concepts
    905–909
  Sigsbee Escarpment 908
  triggers 905–909
Minnes Group 1101
Miocene age. *See also* Late Miocene age
  Early *635*
    Pannonian Basin 1126
  Middle
    North Sea Basin palaeogeography *985*
    Pannonian Basin 1126, 1127–1128
    Pannonian Basin tight gas plays
      1128–1129
    shales 1062
    source rocks geochemical log *1128*
  North Sea Basin palaeogeography *985*
  palaeogeothermal gradients *640*
  Pannonian Basin 1127–1128
  Pannonian Basin tight gas plays
    1128–1129
  shales 1062
  source rocks geochemical log *1128*
Mirren salt dome
  seismic cross-section of channel features *84*
Mississippian age
  facies map *267*
  Netherlands petroleum exploration 266
  reef leads 266
  sedimentological model *266*
  shale 1090
  UK 1090
Mississippian Barnett Shale
  Fort Worth Basin 1087
Mjølner Field 145

MMU. *See* Mid Miocene Unconformity
    (MMU)
Moab fault, Utah
    petroleum geoscientists virtual fieldtrips 25
modular dynamics tester (MDT) data points
    Judy Sandstone 228
Mohns ridges
    spreading rates 579
Moho boundary
    Norwegian model 844
Moho configuration
    mid-Norwegian margin *849*
Moho difference
    Vøring margin *850*
monitoring, verification and accounting
    (MVA) 1190
monitoring wells 1161
Monk's Park Shale 1090
Moray Firth 43, 44, 48, 53
Mordor Anticline 967
    Cenozoic sediments 967
    seismic profile *970, 971*
Møre Basin
    continental crystalline crust *846*
    Cretaceous age 420
    hydrocarbon provinces 1025
    magmatic underplating 853
    rifted basins *1029*
    seismic section *1029*
    sub-basalt hydrocarbon prospectivity
        1025–1031
    sub-basalt imaging 1025
Møre coastal area 852
Møre margin
    basement thickness *892*
    geological and structural features *844*
    individual contributions to basement depth
        map *890*–*891*
    lateral volcanic distribution *846*
    Moho map *892*
    negative Moho difference *846*
    profile *846*
    satellite image *916*
Møre–Trøndelag Fault Zone *917*
    Palaeozoic-age *916*
Morgan Accommodation Zone 775
Morocco 733. *See also* specific location
    Atlas Basin 928
    Central High Atlas Basin, Morocco *929*
    High Atlas *926*
    Maider Basin 731, *732*
    palaeogeographic map at Triassic
        time *927*
    Permo-Triassic outcrops *934*
    present-day structure summary map *927*
    rifting and continental breakup *921*
    structural style and evolution evidence
        from outcrop 925–927
    Tafilalt Basin 731
    tectonic and climactic influence on
        depositional architecture *934*
Morrissey Formation 1101
mounted laser scanner system *21*
Mount Elbert Prospect 1151
    Alaska 1151
    gas hydrate stratigraphic test well *1152*
Mouthbar Facies Association *163, 164*
    cross-bedding *166*
    Hugin Formation sedimentology *166*
    interpretation *166*
    sequence stratigraphy *166*
    wireline-log character *166*
    wireline-log cross-plot *165*
MRS. *See* maximum regressive
    surfaces (MRS)
MSSV. *See* microscale sealed vessel (MSSV)
    pyrolysis
Muckle Basin
    structural evolution and sediment dispersal
        960
    structural reconstruction *956*
mud and silt turbidites 282
mudrock 1131–1136, 1139
    Barnett Shale 1134, *1137*
    Cromer Knoll 505
mudstones
    Alportian–Pendleian 1091
    bay-fill 161
    climatic reinterpretation 236
    Jonathan Mudstone Member 226, 241
    Pokur Formation 663
Muenster Arch *1088*
multiscale model context
    petroleum geoscientists virtual fieldtrips
        22–23
Mungo Central North Sea
    CDR frequency spectrum *519*
    seismic time difference volumes *518*
Mungo diapir
    seismic time section *518*
Mungo four dimensional Noise Case
    Study 517
Mungo monitor surveys *521*
Murzuq Basin 710–712, 715–716
    burial history *691*
    interpretation 711–712, 715–716
    Silurian hot shale source rock 689
    stratigraphic architecture 710–711, 715
    tectonic events *691*
Muskwa Shale 1079
MVA. *See* monitoring, verification and
    accounting (MVA)
MWD. *See* measurement while drilling
    (MWD) sensors

Nafun Group 674
Nagssugtoqidian Mobile Belt
    lead space plots *948*
Namibe Basin *1044*
Namurian age
    marine shales TOC 1091
    onshore UK 69
    vitrinite reflectance percentage
        values *1093*
Nana-1XP well 543, 544
    quality-control plot *544*
Nankai Trough
    BSR 1153
    gas hydrates occurrence 1153
Naokelekan Formation 802, 808
narrow-azimuth towed steamer
    (NATS) 1014
    acquisition configuration *1016*
    associated generic costs *1022*
    data *1018, 1020*
    data quality *v.* tile number *1022*
    earth model *1018*
    v. WATS tile number comparison *1021*
Narve Field
    porosity and bulk volume water *477*
    tectonic and eustatic events *489*
Narve Formation 473, 478, 481
    age 481
    biofacies 481
    lithology/petrophysics 481
    regional correlation 481
    tectonostratigraphy 481
National Oil Companies petroleum industry
    future
    GeoControversies debates 12–14
National Science Foundation (NSF)
    source-to-sink studies 913
NATS. *See* narrow-azimuth towed
    steamer (NATS)
natural fractures 1062
    effect on hydraulic fracture treatment
        efficiency 1138–1139
    gas shales 1131–1139
    open 1138
    open microfractures 1139
    origins 1132–1133
    reactivation *1134*, 1138–1139
    relevance 1131–1132
    size distributions 1138
    subcritical crack growth 1135–1138
    types 1133–1137
        bedding-parallel fractures 1134
        deformed fractures 1133–1134
        sub-vertical sealed planar fractures
            1133–1134
    US shales 1131–1139
natural gas. *See also* gas hydrates
    liquefied 1044
        globalization 827
    production and shale gas reservoirs 1079
natural remanent magnetization (NRM)
    Norway 565
Navajo Formation Sandstone
    Jurassic age 1193
Nelson Field *53*
    bypassed oil configuration *365*
    bypassed oil volumes 361
    drainage cells 361
    hydraulic units *359*
    lateral compartmentalization 357
    locating remaining oil in 349–367
        analogue information 351–352
            channelized turbidite reservoirs
                production patterns 351–352
            channel margins bypassed oil
                351–352
            mud drapes 351
            sand complexes amalgamation 351
            shale blankets 351
        background information 350–351
            central channel complex 351
            eastern channel complex 351
            western channel complex 351
        data integration 352–358, 354
        data integration vertical flow 354–358
            oil–water contact movement
                354–355
            post-production formation tester logs
                355
            shut-in crossflow 355
            water geochemistry 357–358
            water shutoff performance 355

flow detection controls methods
    359–361
  areal flow barriers and baffles 361
  4D seismic data 360–361
  water and oil geochemistry 360
  water cut maps used in conjunction
    with log facies maps 359–360
  locating remaining oil workflow 350
  predictive models to plan infill wells
    349–350
  screening tools 361–362
  target oil volumes 364–366
    oil bypassed in channel margins
      365–366
    significant attic oil volumes 364–365
    slow oil 365
  uneconomic oil volumes 362–364
    channel margin architecture 364
    heterogeneous reservoir intervals 363
    local sand pinchouts 364
    minor attic oil volumes 363
    thin beds 362–363
    thin beds partially disconnected by
      faulting 363–364
  vertical flow barriers 359
location 350
log facies map 361
open hole gamma-ray logs 357
principal reservoir intervals 353
resistivity logs 357
top structure map 352
turbidite reservoirs 356
well 22/11-7 351
Neocomian–Barremian Bucomazi Formation
  1046
Neogene age. See also Late Neogene age
  basin 77, 78
    unconformity 91
  depositional units 92
  Early
    exhumation 983–984
  exhumation 983–984
  hydrocarbon system events chart 1127
  Makó Trough rocks 1082
  Norwegian wells **989**
  petroleum system 1126–1128
  Scandinavia uplift 1000
  shale gas in Europe 1081–1083
  succession seismic-stratigraphy 78–83
Neoproterozoic petroleum system 679, 682
  Algeria 677
  conceptual 1056
  exploring older and deeper hydrocarbon
    plays 676–681
  Mali 677
  Mauritania 677
  Taoudenni Basin 676–681, 677, 679
    hydrocarbon plays 677–679
    stratigraphy and tectonic evolution
      676–677
    thermal evolution and hydrocarbon
      generation 679–681
Neoproterozoic plays 673–702
  basin remnant 681
  cross-section schematic 1057
  exploring older and deeper hydrocarbon
    plays 681–682
  Middle East hydrocarbon plays 674
  North Africa 681–682, 682

stromatolitic carbonate reservoir 679
succession 679
timescale 675
West Africa 681–682
Neotethys 742–743
Nesna detachment 569
Nesna Shear Zone 893
Netherlands 261–274
  exploration drilling 261
  index map 262
  petroleum exploration with stratigraphic
    plays 261–274
    based on basin evolution and
      depositional model 265–266
    basin evolution structural framework
      264–265
    biogenic gas 264
    Cenozoic 273
    Devonian and Mississippian 266
    Early Jurassic 268
    Early Jurassic Posidonia shale 263–264
    Geverik shale 264
    Late Carboniferous gas 263
    Late Cretaceous Chalk 269–273
    Late Jurassic and Early Cretaceous
      268–269
    Pennsylvanian 266
    plays 263–264
      biogenic gas 264
      Early Jurassic Posidonia shale
        263–264
      Geverik shale 264
      Late Carboniferous gas 263
    Rotliegend 266–267
    structural v. stratigraphic traps 261–262
    Triassic 267–268
    Zechstein 267
  stratigraphical log 262
  stratigraphic traps 266, 273, 274
  unconformities 262
net to gross (NTG) character
  Skagerrak Formation 410
  well production performance 410
New Albany Shale 1134
  calcite-sealed fractures 1138
  cores and outcrops 1135
  tall calcite-sealed fractures 1133
Newark East shale gas field 1087
Newfoundland margin
  drilling results 867
New Mexico
  ECBM sequestration test 1191–1192
New Siberian Islands
  crustal tectonic 594
New Siberian-Wrangel Foreland Basin
  609–610, 612
New York
  Catskill Mountains 505
Nezzazat Group 773, 774
NGHP Expedition 01, India 1153
NGU. See Geological Survey of Norway
  (NGU)
Niendorf II
  diapiric salt 903
  diapirs 902
Nikanassin Formation 1099–1100,
  1102–1105, 1112–1113, 1122
  ambient porosity v. permeability 1108
  basin-centred gas 1101–1104

  depositional model 1102–1103
  rock properties 1103–1104
  cross-section 1106
  data cross-plots 1108
  interbedded sandstones 1116
  Lower Cretaceous age 1099
  outcrop photograph 1103
  shales 1110
Nile Delta
  burial history 701
  Cenozoic plays
    deep Oligocene play 698–700
  depositional system 701
  geological section 699
  lithology 699
  Oligocene sequence 700
  petroleum systems 699
  sedimentary characteristics 700
  sedimentary stacking patterns 699
  stratigraphy 700
  tectonic events 699, 701
Niobrara Formation
  fracture porosity 1139
non-bioturbated claystones
  Farsund Formation 139
non-reservoir chalk facies
  coccolith fragments 495
Nordfjord
  satellite image 916
Nordfjord–Sogn detachment 569–571
Nordkapp Basin Aeromagnetic
  Survey 564
Nordland
  petrophysical samples locations 568
  susceptibility log 577
normalized root mean square (NRMS)
  517, 525
North Africa
  Cambrian succession 684
  Cambro-Ordovician stratigraphy 682–686,
    684
  Carboniferous outcrop sections 726, 727
  Carboniferous plays 731
  Carboniferous stratigraphic traps
    725–733
  Cenozoic volcanic centres 673
  Devonian stratigraphy 686–688
  Early Carboniferous cycle 730
  exploring older and deeper hydrocarbon
    plays 676, 682–691
  gas occurrences 762
  Ghadames–Illizi Basin, North Africa
    integrated petroleum systems
      735–759
  and global transgressive–regressive cycles
    730–731
  Hercynian unconformity 683
  hydrocarbon generation 688–691
  hydrocarbon prospectivity 682–688
  Late Ordovician glaciogenic 685
  Neoproterozoic and Neoproterozoic–Early
    Cambrian plays 682
  oil occurrences 762
  palaeohigh influence on plays 707–723
  Palaeozoic basins 674, 688, 722
  Palaeozoic hydrocarbon source rocks 691
  Palaeozoic platform geological setting
    707–708
  Palaeozoic times 719

North Africa (*Continued*)
  plays
    Palaeohigh influence 721–722
      structural style Pan-African heritage 718
      uplifts timing 718–721
  Precambrian and Palaeozoic plays *692*
  Silurian stratigraphy 686–688
  Silurian Tannezuft Formation *687*
  stratigraphic architecture *718*
  stratigraphic trapping types *731*
  tectonic evolution 682–688
  thermal evolution 688–691
  uprising palaeohigh *719*
North Africa Carboniferous sequence framework *728*
  stratigraphic traps North Africa 725–731
    Early to Middle Tournaisian transgression 729
    global transgressive–regressive cycles 730–731
    Late Devonian (Strunian) regression 725–729
    Latest Viséan to Serpukhovian transgression 729
    Late Tournaisian regression 729
    Late Tournaisian to earliest Viséan transgression 729
    Strunian regression 725–729
    Viséan transgressive–regressive cycles 729
North America 1065. *See also* specific areas
  shale gas 1062
  shales and tight reservoirs 1061
North Atlantic *550*
  AFTA and VR data *641*
  areas displaying Cenozoic exhumation *642*
  conceptual breakup models *863*
  continental breakup 831, 986
  Faroe Islands 1033
  flood basalt province 838
  gravity compilations 578–580
  hot mantle 839
  Late Permian–Early Triassic and Middle Jurassic *949*
  lead basement source 938
  Norway aeromagnetic compilations 578–580
  oceanic spreading 578–580
  palaeogeographic reconstructions *949*
  Paleocene–Eocene transition 986
  pre-quaternary geology *980*
  rifted continental margins *832*
  schematic Triassic palaeogeographic reconstruction *929*
  seafloor spreading 834
  seismic profile *832*
North Atlantic margins
  basalt flows 833
  P-wave velocities 839
  seismic profile *832*
  vertical seismic profiles *832*
North Atlantic passive margins
  erosion surfaces 981
  geological map *886*
  hydrocarbons preservation 986
  Permo-Triassic 921–934
  post-rift uplift 979
  rifting 893
  seismic sections *983*
  uplift and erosion 981
  uplift and exhumation 979–1001
North Chukchi Basin 613
  internal structure *613*
Northeast Atlantic
  base basalts *1028*
  basins *1028*
  bathymetry and topography map *561*
  Bouguer gravity *566*
  CHAMP satellite magnetic anomaly *564*
  gravity surveys compilation *566*
  magnetic compilation *564*
  magnetic surveys compilation *563*
  pre-drift reconstruction *1031*
  seismic section *1028*
  sub-basalt hydrocarbon prospectivity 1025–1031
    Faroe–Shetland and Møre basins 1025–1028
    Rockall 1028–1029
    seismic data acquisition and processing 1025
Northeast Atlantic Margin
  basin inversion 971
  bathymetric map *964*
  Cenozoic unconformities *974*
  geological and structural features *844*
  imaged sub-basalt strata *1029*
Northeast England
  Carboniferous TOC *1092*
  case study 1162
  underground coal gasification with CCS 1162
Northeast North Sea
  individual contributions to basement depth map 891–892
Northern Judd Basin
  CSEM survey 258
  depositional concept *251*
  model far-stack *256*
  OMV05-2 *255*
  OMV05-2 far-stack *256*
  optimized thickness model *257*
  rock physics 253
  rock physics evaluation 258
  seismic amplitude anomaly 251, 253
  T38 sequence prospectivity 245–258
    CSEM survey 254–258
    hydrocarbon sourcing 251–253
    within Lamba formation 249–258
      CSEM survey 254–258
      hydrocarbon sourcing 251–253
      reservoir 249–251
      rock physics analysis 253–254
      seal 253
      trap 251
    main prospectivity 247–249
    stratigraphy 245–246
    structure 246–247
northern Norway
  Precambrian units 889
Northern Panel 369
Northern Priverkhoyansk Foredeep
  assessment unit 624, 627–629
  densities and median sizes *625*
  distributions and assessment unit probabilities *629*
  field densities, field sizes, and exploration maturities **623–624**
Northern Terrace 231
  seismic reprocessing results *239*
Northern Yamal Peninsula *661*
North German Basin 1080
North Ireland
  hydrocarbon wells *1089*
North Kara Province
  petroleum geology 605
  RAS sedimentary basins 605
  shelves 593
North Kara Sea
  HC systems 605
North Red Sea *786*
  basin hydrocarbon prospectivity Egypt 783–788
North Rockall Trough 973
north Russia
  topography *646*
North Sea 11, *16*, 27, **28**, 135, 138, *148*, *150*, 463, 1183. *See also* Central North Sea (CNS)
  adjacent petroliferous basins 32
  CDR frequency spectrum *519*
  Chalk *538*
  debate 16
  deepwater reservoirs *50*
  erosion removal 988
  exploration 14, 17, 31, 37
  four dimensional acquisition and processing case study 517–522
  Gannet Area
    Eocene Tay formation fan system *46*
  GeoControversies debates 14–15
  geological evolution and exploration history 37
  individual contributions to basement depth map 891–892
  Jurassic rift complex 127
  largest fields 32
  location map *538*
  Mungo *518*, *519*
  oil field 443
  polyphase rifting 43
  porosity–permeability relationship *141*
  pressure distribution *52*
  rift basin architecture 37–38
  Scott Field 387
  seismic technique datasets 33
  seismic time difference volumes *518*
  sub-Zechstein seismo-stratigraphic relationship 61
  tectonic model evolution *40*
  3D seismic mapping *52*
  time isochore map *539*
  2D seismic mapping *52*
  UK seismic and drilling technology 337–347
  unconformities 185
  Utsira Formation *1184*
  Volgian sandstone intervals *139*
  well data *41*
North Sea Basin
  case A tectonic control
    Cenozoic stratigraphy and palaeogeography 982–986
    Early and Late Neogene exhumation 983–984
    episodic uplift and exhumation 982–986

Southern Scandinavia post-breakup
    uplift phases 986
  Chalk deposition 479
  Late Miocene palaeogeography *985*
  maximum burial *982*
  Middle Miocene palaeogeography *985*
  mid-Oligocene palaeogeography *984*
  palaeogeography *984*
  Pliocene palaeogeography *985*
  Scandinavia 1001
North Sea Central Graben 463
  wells location *494*
North Sea Dome
  North Sea hydrocarbon systems
    37–4537–38
North Sea Eldfisk Field
  holostratigraphic approach to Chalk
      473–491
    biofacies methodology 479–481
    Ekofisk Formation 478
    geophysical data 477–481, 479–481
    Magne Formation 478
    methodology and integration of results
      474–478
    results 481–488
    stratigraphic background 473
    study objectives 474
  Lower Ekofisk Formation 475–476,
      486–487
    age 486
    biofacies 486
    lithology/petrophysics 486
    tectonostratigraphy 486–487
  Lower Magne Formation 482–483
    age 482
    biofacies 482
    correlation 483
    lithology/petrophysics 482
    tectonostratigraphy 482–483
  Lower Tor Formation 484
    age 484
    biofacies 484
    lithology/petrophysics 484
    tectonostratigraphy 484
  Magne Formation 477
  Middle Ekofisk Formation 476, 487
    age 487
    biofacies 487
    lithology/petrophysics 487
  Middle Tor Formation
    age 484
    biofacies 484
    lithology/petrophysics 484
    tectonostratigraphy 484
  Narve Formation 478, 481
    age 481
    biofacies 481
    lithology/petrophysics 481
    regional correlation 481
    tectonostratigraphy 481
  Thud Formation 478, 481–482
    age 481
    biofacies 482
    correlation 482
    lithology/petrophysics 481–482
    tectonostratigraphy 482
  Tor Formation 476–477, 478
  Upper Ekofisk Formation 475, 487–488
    age 487

  biofacies 488
  lithology/petrophysics 487
  Upper Magne Formation 483–484
    age 483
    biofacies 483
    correlation 483
    formation 483
    lithology/petrophysics 483
    tectonostratigraphy 483
  Upper Tor Formation 484–486
    age 484
    biofacies 484–485
    lithology/petrophysics 484
    tectonostratigraphy 486
North Sea Exploration Manager of Amerada
    Hess 14
North Sea Group 273
North Sea High
  structural and stratigraphic relationships 70
North Sea hydrocarbon systems 37–53
  deepwater reservoirs Palaeogene *v.* Late
      Jurassic 45–50
    basin architecture 50
    basin setting/sea-level response 49–50
    Cenozoic examples 45–46
    Central Trough Upper Jurassic
      comparison 49
    play considerations 45–46
    provenance/hinterland 49
    reservoir fairway development
      48–49
    salt withdrawal role 50
    sandstone injectites 46
    sandstone provenance 48–49
    Upper Jurassic deepwater depositional
      systems 46–48
  historical milestones 37–45
  petroleum geology retrospective view
      51–53
  regional setting 37
  tectonics 37–45
    basin forming mechanisms 39–41
    Cimmerian events 37–38
    crustal scale deformation 39–41
    fault geometries 39
    Late Cimmerian 39
    Late Cimmerian rift architecture 44–45
    Mid Cimmerian reservoir distribution
      controls 38–39
    North Sea Dome 37–38
    polyphase rifting 41–43
    shear 39–41
    stratal geometries fundamental change
      44–45
    structural processes 41
    trap scale salt-related geometries 43–44
    Triassic reservoir distribution controls
      38–39
North Sea Rift
  3D seismic data blanket 39
  generalized stratigraphy *39*
  structural elements *38*
  Triassic stratigraphy *40*
North Siberian Arch
  crustal tectonic *594*
North Slope of Alaska
  location map *637*
  wells *640*
Northumberland Trough 59

Northwest Danish Central Graben
    95–111
Northwest European Atlantic margin
  structural and terrane map *923*
North West Europe and Global Perspectives
    31
Northwest Siberia 667
  basin sensitivity parameters 661 **664**
  isotopic composition of methane *665*
North Ymir Ridge Anticline
  bathymetric gorge *973*
  seismic profile *971*
Norway 559–583, *563. See also*
    mid-Norwegian margin
  aeromagnetic compilations 559–583
  analytic signal shaded relief map *576*
  bathymetry and topography map *561*
  body formed 843
  case B early Neogene exhumation 987–990
    episodic uplift and exhumation 987–990
      basic concepts 987
      burial anomalies 987
      timing and magnitude 987–988
  continental shelf sedimentary rocks **572**
  Danish Basin potential field anomalies 891
  Eldfisk Field holostratigraphic approach to
    Chalk 479–488
  Hugin Formation 157–175
  maximum burial *982*
  middle margin 843–853
  North Atlantic passive margin
    rifting 893
  North Sea 479–488, 1183
  NRM 565
  onlapping marine shales 419
  petroleum geoscientists virtual fieldtrips 24
  petrophysical data *581*
  petrophysical samples locations *568*
  Precambrian units 889
  pre-glacial terrain *918*
  profile data 560–561
  regional structures *573*
  reservoired hydrocarbons 989
  rifting commenced 925
  uplifting and erosion 998
Norwegian Barents Shelf 599
Norwegian Central Graben Chalk fields 481
Norwegian continental margin
  geological features *917*
  onshore geomorphology 916
  pre-glacial terrain *918*
Norwegian Continental Shelf 885–895
  basement thickness 885–895
  basement thickness maps 892–893
  Bouguer anomaly *887*
  database presentation 887–888
    gravity anomaly map 888
    magnetic anomaly map 887–888
    regional seismic data 888
  data integration 888
  geological setting 885–887
  individual contributions to basement depth
    map 889–892
    Barents Sea 889
    Kattegat 891–892
    Lofoten margin 889–890
    Mid-Norwegian margin 889–891
    Møre margin 890–891
    Northeastern North Sea 891–892

Norwegian Continental Shelf (*Continued*)
    Norwegian–Danish Basin 891–892
        Skagerrak 891–892
        Viking Graben 891
        Vøring margin 890
    Moho depth 892–893
    new compilation 885–895
    passive margin system 885–895
    petroleum industry 885
    potential field data 888
    segmentation of passive margin system 885–895
    seismic profiles *887*
    top basement 885–895
        compilation **890**
        map *889*
    total magnetic field anomaly *887*
Norwegian–Danish Basin 569, 576
    individual contributions to basement depth map 891–892
    potential field 892
    Precambrian basement 895
Norwegian–Greenland Sea
    aeromagnetic surveys 579
    embayment 654
Norwegian Margin *847*
    basin inversion 971
    Caledonian basement 895
    distribution of lower crustal bodies 843–853
    flexural isostasy 843
    formation 887
    individual contributions to basement depth map 889–890
    magmatic underplating 852
    magnetic basement 888
    physiographically 887
    plate reconstruction *582*
    properties of lower crustal bodies 843–853
Norwegian model
    LCB and Moho horizons 844
    Moho boundary 844
    potential field models 844
    volcanic distribution 844
Norwegian Sea Overflow
    compression 973
Norwegian shelf
    HP/HT field 420
Norwegian wells
    AFTA and FR data *989*
    burial anomaly **989**
    Neogene phase **989**
    palaeogeothermal gradient values *990*
    sonic logs *988*
    thermochronological constraints *989*
    3/7-6 well 138
    2/12-1 well 130
Not formation 423–424
Notikewin Member
    basin-centred gas in Western Canada 1108
Noto Formation 118
    foreland 124
    hydrocarbon occurrences 124
Nova Scotia, Canada
    Fundy Basin 928, 929–931
    Minas Basin 929, *931*
    Permo-Triassic 934
    Permo-Triassic outcrops 934

Novaya Zemlya 660
    archipelago 649
    archipelago dimensions *650*
    arcuate fold-and-thrust belts *648*
    Axial Structural Domain 651
    Central Structural Domain 651
    classification schemes 647–649
    curvature 647–649
    Eastern Structural Domain 651
    fold-and-thrust belt 645–656, *653*
    foreland deformation 603
    offset 645–656
    palaeogeography maps *655*
    Precambrian–Cambrian exposures 652
    structural trend lines *652*
    structure *652*, *653*
        curvature 654–656
        curvature cause 651–654
        domain 651
        interpretation 651–654
        offset cause 654
        patterns 649–651, *649*
        trend lines *649*, *650*
    thrust-sheet transport 654
    topography *646*, *652*
    Western Structural Domain 651
Novaya Zemlya Fold Belt (NZFB) 594, 598, *603*
NRM. *See* natural remanent magnetization (NRM)
NRMS. *See* normalized root mean square (NRMS)
NSF. *See* National Science Foundation (NSF)
NTG. *See* net to gross (NTG) character
Nubia Sandstone 773, 774, 786
    structural traps 788
Nukhul Formation 772
    basal conglomerates 772
    lidar-derived DEM *22*
    location map *22*
    marine to brackish depositional conditions 777
    petroleum geoscientists virtual fieldtrips 24
    Quickbird Satellite image *22*
    syncline dataset *22*, 24
    syn-rift sedimentation 772
    syn-rift stratigraphy 777
    VRGS software *22*
Nuussuaq Basin
    Gro-3 well 640
NZFB. *See* Novaya Zemlya Fold Belt (NZFB)

OBC. *See* Ocean Bottom Cable (OBC)
oblique progradational facies 1035
OBM. *See* oil-based mud (OBM)
O'Brien, Mike 14
OBS. *See* ocean bottom seismometers (OBS)
Ocean Bottom Cable (OBC)
    acquisition programme 477
    data 390
    survey 337
ocean bottom seismometers (OBS) 831, 833
    Atlantis Field 1016
    component 834
    nodes 1016

oceanic deepwater basins
    free-air gravity field *595*
oceanic sound speed variations 549
October Concession Rudeis Formation 778
October Revolution Island 594, 596
offshore aeromagnetic surveys **562**
offshore crystalline basement blocks from Atlantic
    lead space plots *948*
offshore deposition
    indicator and predictor 919
Offshore Equatorial Guinea 827
offshore open-marine facies association
    Hugin Formation sedimentology 167–168
        description 167
        interpretation 167
        sequence stratigraphy 167
        wireline-log character 168
offshore three dimensional seismic data
    interpretation 1033–1040
    analogues 1038–1039
    base volcanic effect and subvolcanic unit 1038
    data and methods 1034–1035
    facies units description 1035–1038
        facies unit 1 1037
        facies unit 2 1037
        facies unit 3 1037
        facies unit 4 1037
        facies unit 5 1037
        facies unit 6 1037–1038
    geological development 1039–1040
    geological model *1039*
    intrusions 1038
    study background 1034
offshore West Greenland
    seismic profile *996*
oil
    Angola exploration 1043
    Angola production 1043
    Bakken Formation 1067, *1073*, *1074*
    composition 95
    Cretaceous age 808
    Danish Central Graben 95–111
    density 1072–1073
    East Barents region 603
    enhanced recovery 1158, 1165–1166, 1190
    Farsund Formation 96, 105
    generation potential 105, 607
    geochemistry 360
    geology-led approach 366
    gravity data 1067, *1074*
    heavy 1141, *1142*, *1145*, 1147–1148
    hydrocarbon pools *1073*
    Italy provinces *115*
    kerogens 96, 100, 103, 104, 110
    Laptev Shelf 607
    locating methods 366
    mature fields 366
    migration model *253*
    North Africa occurrences *762*
    permeabilities **1159**
    predicting production behaviour 405–417
    price trends chart *437*
    recovery 412
    reservoir 299–313
    shales 103
    source rock 98

T38 anomaly *253*
  Westray Fault system *253*
  window 96, 105, 110
oil-based mud (OBM) 95
oil–water contact (OWC) 337, 351, 355
  Donan Field *444*
  4D seismic and reservoir model calibration 342
  injection strategy 391
  Scott Field *389*
  Tyra Field 465, 471
Okavango Basin *1044*, 1055
  thickness 1055
Old Red Sandstone 299
Oligocene age 113, 689
  depositional features *701*
  Nile Delta sequence *700*
  succession *701*
  West Greenland hiatus offshore *993*
  West Greenland unconformity *994*
Oman
  exploring hydrocarbon plays 674–676
Onika Anticline 970
  seismic profile *972*
onlapping marine shales 419
onshore United Kingdom 69
open fractures
  geomechanical modelling 3D output *1119*
Open Geospatial Consortium 25
Ophelia-1 well 136
  sedimentological core logs *137*
Ophiomorpha 133, 166
  burrows 130
O.P.V. Seillean 432
Orca Basin 1151
Orcadian Basin 1090, *1090*
Ordovician age
  Hamra Quartzites 685
  Late *685*, 686, *695*
  North Africa stratigraphy 682–686, *684*
  Sarah Formation 695, 697
  shale 1091
  shallow water clastic sediments 596
  Unayzah formation 697
organic carbon 813, 814, 818
organic facies
  Shiranish Formation 815–816
Origin of Sediment-Related AeroMagnetics (OSRAM) Project 565
Ormen Lange system
  fluvial erosion *918*
  Maastrichtian–Paleocene age *917*
  segment-style analysis 920
Oslo Rift 569
OSRAM. *See* Origin of Sediment-Related AeroMagnetics (OSRAM) Project
Ouachita Thrust Front *1088*
Ouan Kasa Formation 686
Oued Mya Basin
  thermal maturity of Early Silurian Tannezuft hot shale *690*
Ougarta Arch 725
Oukaimeden Sandstone Formation 932
  distribution of facies associations *933*
Oukaimeden Valley 925, *927*
  alluvial fan breccias *928*
  sedimentation *928*
  stereonet plot *928*
  structural cross-section across *928*

Ouled Cheb formation 688
Ourika Valley 925, *927*
outboard areas 825
outcrops
  analogue studies 23
  fracture types **1132**
  West Greenland 1034
Outer Moray Firth 387
Outer Rough Basin 97
  VR gradients 96
  wells VR trends *100*
Outer Rough Sand 138, 139
  facies core photos *135*
  porosity–permeability relationship *141*
  sedimentological core log *140*
Owambo Basin *1044*, 1055
  thickness 1055
OWC. *See* oil–water contact (OWC)
Oxford Clay
  black shale **1095**, 1096
Oxfordian age
  Early
    sequence J46 186
  fairways 49
  Kimmeridgian basin fabric 201
  Kimmeridgian turbidite sandstone reservoirs 205
  Late 190–192
    sequence J54a 187–190
      depositional environments *190*
      regressive shoreface sandstones 187
      turbidite sandstones *188*
  Middle
    sequence J52 186–187, *189*

Pabdeh in Cenozoic
  major play systems *793*
Pai Khoi region
  major strike-slip fault systems 654
Pakistan
  Sindh and Punjab regions 676
palaeoburial
  data 980
  landscape 982
Palaeogene age 463
  basalts 1025
  v. Late Jurassic age deepwater reservoirs 50
  volcanic distribution *1026*
palaeogeographic reconstruction 4
palaeogeothermal gradients
  derived from AFTA and VR data *636*, *640*
palaeohigh 722
  influence on North African plays 707–723
    Ahara High, Illizi, and Berkine basins, Algeria 712–714
    Cambrian to Devonian succession 707–710
    Djebel Tamamat in Bled el Mass area 716–718
    Gargaff Arch, Ghadames and Murzuq basins, Libya 710–712
    geological setting 707–710
    Palaeozoic platform 707–710
    Palaeozoic succession around syn-tectonic highs 710–718
    petroleum system implications 722
      implications on reservoir quality 722
      potential tectonostratigraphic traps 722

    pre-Hercynian migration process 722
    source rock deposition 722
    stratigraphic architecture 722
    stratigraphic architecture 707–718, 721–722
    structural style Pan-African heritage 718
    sub-basins and arches 707–708
    synthesis tectonostratigraphic evolution 718–722
    Tihemboka High, Illizi and Murzuq basins 715–716
    uplifts timing 718–721
  Palaeozoic petroleum systems 707–723
  Pan-African heritage 718
Palaeotethys Ocean Drift 738
palaeothermal 980
Palaeozoic age
  Ahnet Basin 686, 738
  Ajjer Tassil outcrops *709*
  Berkine Basin *709*
  depocentres Geoprobe image *742*
  Early 5, *718*, 1091
  exploring older and deeper hydrocarbon plays 682–684, 697
  Faroe–Shetland Basin structural play 954
  Hercynian unconformity *694*
  hydrocarbon plays 682
  integrated petroleum systems 735–759
  Late 5
  Middle East *694*
    basins *674*
    plays 697
  Møre–Trøndelag Fault Zone 916
  North Africa 707–723, *719*
    basins *674*, 688, *761*
    location 735
    plays *692*
    tectonic elements *736*
  petroleum system 688, 691, 697, 707–723
  Reggane Basin 686, 738
  Rub Al Khali Basin 695
  sedimentation 708
  structural elements *264*
  succession 710–718
    stratigraphic column *709*
  Taoudenni Basin 738
  Tindouf Basin 738
Paleocene age
  Andrew sands 225
  Barents Sea cooling 637
  Faroe–Shetland Basin *281*
  gas condensate discoveries 279
  Laggan 279
  Late 45
  Lower Balmoral sandstones 437
  Ormen Lange system 916–919
  palaeogeothermal gradients *636*, *640*
  Pladda prospects 437, *438*
  reservoir 431, 432
  sandstones *214*
  shelfal Hermod sequences 45
  Top Lower Balmoral Structure *438*
  Tormore 279
  transition to North Atlantic breakup 986
  Vøring Basin strata *1031*
Paleocene Forties Sandstone Member *214*, 215
  exploration 213

Paleocene Forties Sandstone Member
  (*Continued*)
    isopach map *217*
    reservoir interval *351*
    reservoir potential *215*
Pannonian Basin 1082, *1082*
  Central Europe 1126, *1126*
  dynamic hydrocarbon system 1127
  Hungary 1062, 1126
  lithostratigraphic units and pressure
      regimes *1127*
  Lower Miocene 1126
  Middle Miocene 1126–1129
  Neogene hydrocarbon system events chart
      1127
  Neogene petroleum system 1126–1128
    maturity 1127
    reservoirs and seals 1127–1128
    source rocks 1126–1127
    timing 1128
  overpressured basal shales 1128
  overpressured Upper Miocene
      brackish-water shale 1129
  Pliocene 1126
  present-day heat flow *1126*
  Sarmatian sedimentary rocks 1126
  shale gas in Europe 1081–1083
  structural and sedimentary evolution
      1125–1126
    palaeogeographic evolution 1125
    sedimentary environment 1125
    stratigraphy of basin fill 1125–1126
    structural evolution 1125
  tight gas exploration 1125–1129
  tight gas plays 1128–1129
  TOC 1126
  unconventional hydrocarbon 1128
  Upper Miocene 1126, 1127–1128
Panther Tongue sandstone 24
  lidar studies 24
paper theme 372
Paradox Basin sequestration test 1190
parallel bedded platform facies 1035
Paratethys 1126
Parshall Field 1065
Passage Beds 1101
passive margins 823–828
  evolution 1005–1012
  exploration models 828
  exploration technology 828
  frontiers 828
    arctic 828
    palaeo-passive margins 828
  hydrocarbon charge and maturation
      824–826
    heat flow models 825–826
    hydrates and biogenic gas 825
    local source rocks 824–825
    marine source rocks 824
    oceanic anoxic events 824
    oceanic crust exploration 826
  hydrocarbon prospectivity 825
  late structural evolution 827
  magmatic addition 824
  megasequence 738
  new development concepts 827–828
  Norwegian continental shelf 885–895
  passive margin evolution 823–824
  reservoirs 826–827

large drainage basin systems 826
  remaining deepwater plays 827
  reservoirs and basement in multiphase
      rifts 826
  reservoirs and outer margins 826
  shelf edge deltas and plays 826–827
  retention 827
  series of case histories from Ireland
      1005–1012
  subtle migration paths 827
PCG7 audience *12*
peak oil no longer concern
  GeoControversies debates 11–12
Pechora Sea 598
PEE. *See* petroleum expulsion
    efficiency (PEE)
Pelagian Basin 741
Pelion Formation
  transgressive sandstones 422
Pennine Basin *1092*
  coal and hydrocarbon
      exploration 1093
  Craven Group 1092–1093
Pennsylvanian
  Netherlands petroleum exploration with
      stratigraphic plays 266
  play map *268*
Pentland basin
  marine environments *191*
Pentland Formation 185
  reservoir 231
  tectono-stratigraphic model 234
Pentland Hills Inlier 67
Pentland succession
  biostratigraphic data *193*
Peri-Gondwana Margin palaeogeography *675*
Periscope 15
  planned *v.* actual geosteered wellpath *446*
permeability preservation 288
Permian age
  architecture 65–67
  Argana Basin *930*
  basins 325, *456*
    location map *922*
    morphology 57
    preparing for carbon constrained world
        1190–1191
  Central High Atlas Basin *930*
  Early 57, 739
  Fundy Basin *930*
  hopanes and steranes in oil sample *107*
  Kupfershiefer 264
  lithostratigraphy *58*
  plate-tectonic setting and depositional
      environments *603*
  red beds 926
  Rotliegend Groningen gas field 261
  Scandinavia *572*
  seismic data for wells 57
  shallow marine 596
  traps 262
  Upper Rotliegendes Group 262, *269*
  volcanic rocks magnetic properties *572*
Permian White Rim Sandstone 1193
Permo-Carboniferous
  Unayzah formations 697
Permo-Triassic Basin 921–934, *923*
  Central High Atlas and Argana basins 925
  evolution and fill 921–934

implications for future exploration
    921–934
  Morocco and Nova Scotia 934
  North Atlantic passive margins 921–934
  outcrops 934
PESGB. *See* Petroleum Exploration Society
    of Great Britain (PESGB)
petroleum
  acreage in Italy *114*
  Bakken Formation 1066
  endowment in Italy 114
  migration 1066
  permits in Italy 114
  play fairway analysis 735–759
  plays in continental margins 975
  Shiranish Formation 818
Petroleum Exploration Society of Great
    Britain (PESGB) 27
petroleum expulsion efficiency (PEE) 1068
  Bakken Formation *1070*, 1073
petroleum generation 96, 98
  Bakken Formation *1072*
  potential 101
  threshold *1072*
petroleum generation index (PGI) 1068
  Bakken Formation *1070*, 1073
petroleum geology
  North Sea hydrocarbon systems 51–53
  and reservoir production of South Arne
      523–524
Petroleum Geology Conference
  history from 1974 to date from NW Europe
      to international 31
petroleum geoscientists virtual fieldtrips
    19–25
  applications 20–23
  DVD examples 23–25
    Arkitsa fault, Greece 24–25
    carbonates, Alpine foreland 25
    deltaic reservoir analogues from central
        Utah 24
    Grès d'Annot Formation, SE France 24
    Lofoten, Norway 24
    Moab fault, Utah 25
    Nukul syncline, Gulf of Suez 24
    Shannon Basin, West of Ireland
        23–24
  future trends 25
  virtual fieldtrips creation 19–20
petroleum industry
  Norwegian continental shelf 885
  petroleum geoscientists virtual fieldtrips
      20–23
    additional field time 20–22
    discontinuity dimensions 23
    geometry constraints 23
    multiscale model context 22–23
    traditional fieldtrip training tool 20
petroleum resources
  Siberian Craton 621–631
Petroleum Sharing Contracts (PSC) 1044
petroleum systems *675*, *676*, *688*, *700*, *824*,
    826, *827*
  advances 2
  analysis 2
  Arabian Palaeozoic 692
  Bush Formation 674
  characterized 673
  Cretaceous age 119–120

Devonian–Mississippian Bakken
    Formation 1061–1062
distribution and nature 6
Early Jurassic age 118–119
Early Silurian Tanzuft 743–745
gas-prone *1045*
Ghadames–Illizi Basin 735
Gulf of Suez 788
hydrocarbon distribution 752–754
implications on reservoir quality 722
integrated 735–759
Italy 118–119, *124*
Late Devonian Frasnian 745
Late Triassic age 118–119
megasequence framework 735–743
    Foreland Basin 1 megasequence 738
    Foreland Basin 2 megasequence 743
    passive margin 1 megasequence 738
    passive margin 2 megasequence 738
    post-rift 2 megasequence 741
    post-rift 3 megasequence 742–743
    syn-rift 1 megasequence 738
    syn-rift 2 megasequence 739–742
Mesozoic 698
migration modelling 754–755
missing section map and erosion maps
    745–752
model 791
Nile Delta, Egypt *699*
North African Palaeozoic basins 722
palaeohigh influence on North African
    plays 722
petroleum source rocks 743–745
phases 674
play fairway analysis 735–759
post-salt thermogenic *1051*
potential tectonostratigraphic traps 722
pre-Hercynian migration process 722
reservoir-seal combinations 745
Rub Al Khali Basin, Saudi Arabia *696*
source access 754
source rock deposition 722
source rock pressure 754
South Atlantic Ocean breakup 862–865
stratigraphic architecture 722
between Timor and Australia 828
viability 4
petrophysical data
    compilation *581*
    NGU 559
    Norwegian mainland *581*
    rock bodies 578
PGI. *See* petroleum generation index (PGI)
PGS Central North Sea MegaSurvey 78
PGS North Sea Digital Atlas time
    horizons 64
phase kinetics 1065
photomosaic and lidar data
    stratigraphy 24
pick-and-mix forward models 826
Piedmont glacier *648*
Piggvar Terrace
    wells VR trends *100*
Pila Spi Formation 801
pilot holes 342, *343*
    Captain Field 344–345, *346*
    drilling 347
    horizontal 342
    well locations 347

Piper Field 391
    sandstones 389, 398, 437
    Upper and Lower 393
Pladda prospects
    Paleocene 437, *438*
planktonic foraminifera
    biofacies groups *480*
    Eldfisk Bravo 486
    Eldfisk Field *480*
plate reconstruction 2
    constrain 4
    detrital zircons 6
    Greenland *582*
    Norwegian margin *582*
plate tectonics 51
play fairway
    analysis 791, 794
    assessment 1065
    CRS description 794
    distribution 4
    frontier basin distribution 2
    locate new oil and gas resources 791
    lowest risk segments 794
    quality 6
    risk mapping techniques 791–799
play systems 793. *See also* specific location
    syn-rift models 962
Pleistocene age
    Corsica 382
    dip map *904*
    Early *83*
    glaciations *918*
    Golo turbidite system 382
Pliocene age
    Apennines 113
    North Sea Basin palaeogeography *985*
    Pannonian Basin 1126
Plio-Pleistocene
    foredeep wedges biogenic gas 121–123
    Italy hydrocarbon occurrences 121–123
    Porto Corsini East gas field *122*
    Settala gas field *123*
    terrigenous units *122*, *123*
PMR. *See* Porcupine Median Ridge (PMR)
POC. *See* proto-oceanic crust (POC)
pod/interpod terminology 44
Pokur Formation 659, 662, 664, 667
    Cenomanian deltaic coals and
        mudstones 663
    Cenomanian fluvial-deltaic deposits 662
    Cenomanian gas play 659
    gas yield *663*, *664*
    temperature map *666*
    thermal yields 663
Poland. *See also* specific location
    black shales 1080
polyphase faulting 824
polyphase rifting
    European basin architectures *42*
    hydrocarbon systems 41–43
    North Sea 41–43
ponded turbidites 282, 288
Ponta Grossa dyke 861
Po Plain 113, 123
    foreland *119*
    Villafortuna–Trecate oil field *119*
Porcupine Bank 972
    lead data from detrital K-feldspar
        grains 940

seabed slope *552*
Porcupine Basin 44, 848, 940, *949*,
    1005–1009
    database 1005–1007
    Dooish K-feldspars *950*
    interpretation grid basemap *1006*
    location map *1006*
    seismic interpretation 1005–1009
Porcupine Median Ridge (PMR) 1008
    original and reprocessed section *1007*
porosity
    and permeability preservation 288
    uncertainty 294
Porto Corsini East Field 122–123, *122*
Posidonia shale
    Early Jurassic age 263–264
Post-Aptian salt plays 1049–1052
post-basalt heating
    East Greenland 1000
post-rift
    early 961
post-rift uplift
    along passive margins 1000
    megasequence 741, 742–743
    North Atlantic passive margins 979
post-salt gas plays *1050*
post-salt thermogenic petroleum
    systems *1051*
potassium feldspars
    detrital 940, 950
    Dooish *947*, *950*
    lead 934, 938
    lead isotopic composition 950
    Rockall Basin 951
potential field modelling
    Caledonoid granite core 61
    Carboniferous petroleum system
        65–67
potential stratigraphic trapping types
    Carboniferous stratigraphic traps North
        Africa 731–733
Powder River Basin in Wyoming 1156
Pre-Aptian salt plays 1045–1049
Precambrian age
    basement 888, 890, 891, 893–894
    gneisses 889
    Massif Ancient rocks 925
    North Africa plays *692*
    northern Norway 889
    Norwegian–Danish Basin 895
    Novaya Zemlya 652
    Taoudenni Basin stratigraphy *678*
predicting production behaviour
    from deep HPHT Triassic reservoirs
        405–417
    Deep Joanne model results
        412–414
    Deep Judy model results 412
    field production and well test behaviour
        407
    methods 410
    modelling parameters 410–412
    results 412–414
    sedimentological interpretation
        408–409
    Skagerrak diagenesis 409–410
    Skagerrak Formation geology 405–407
    sedimentary architecture impact on
        recovery 405–417

pressure distribution
  North Sea 52
pressure-temperature (PT) conditions 1071
pre-stack depth migration (PSDM)
  processing 236, 390
  reprocessed 1011
  Top Zeit rugosity 787
  velocity models 787
pre-stack time migration (PSTM)
  data 236, 437
  three dimensional AVO products 216, 217
production behaviour
  predicting 405–417
Project Ramsay 1162
prospect evaluation 261
proto-oceanic crust (POC)
  West Africa and Brazilian margins 874
Provenance analysis 934
  categories 937
  deepwater reservoirs Palaeogene v. Late Jurassic 49
  Rockall Basin palaeodrainage 942–944, 948–950
PSC. See Petroleum Sharing Contracts (PSC)
PSDM. See pre-stack depth migration (PSDM)
PSTM. See pre-stack time migration (PSTM)
PT. See pressure–temperature (PT) conditions
Puffin horst area
  fault trends 43
Puffin Terrace 191, 200
  turbidite deposition 195
  well correlations 193
Punjab region
  Pakistan 676
Purple Pick 373
Pyrenean compression 975
pyrolysis gas chromatography
  Bakken petroleum system 1068–1069
  chain length distribution 1071

Qamchuqa Formation 807
  description 804
  lithological model 807
  porosity 806
Qamchuqa matrix 809
Qara Chauq 804
Qasaiba in Silurian
  major play systems 793
Q-Marine survey 343
quartz cement growth
  Kristin Field 422
Quaternary sand channels
  analytical signal 578
Quickbird Satellite image
  Nukhul syncline dataset 22
Qulleq-1 well
  unconformity below Neogene succession 994
  VR data v. depth 994

radon demultiple 1009
Rajasthan, India
  Baghewala Oil Field 676
Ramsay, William 1155–1156
RAS. See Russian Arctic shelves (RAS) geology

Ras Abu Darag well
  Nebwi 81-1 779
  seismic amplitude anomalies 780
Rås Basin 420
Ras Hamia formation 688
Rattray Volcanics 397
Rauwerd
  Lower Cretaceous age 272
Rauwerd-1-Sneek-1
  cross-section 272
Ravn Member 136
  sedimentological core logs 137
RCSP. See Regional Carbon Sequestration Partnerships (RCSP)
reactivation 1131–1133
  natural fractures 1134, 1138–1139
reactive kerogen wells **106**
recovery factor (RF)
  v. CGR/GOR 415
  Deep Judy model results 412
red bed deposits 928
Red Sand Formation 90
Red Sea
  location map 784
  North 783–788
  prospects and leads 787
  source rock biomarker and isotope analysis 786
  stratigraphy and tectonic framework 785
  tectonic setting 785
Red Sea basin
  aeromagnetic data 788
  3D SRME processing 786
  gravity data 788
  onshore field mapping 784
  pre- and syn-rift sequences 786–787
  PSDM, foldout
  PSTM, foldout
  salt flood velocity model, foldout
  seismic data coverage 784
  subsalt events 787
  subsalt illumination 786
Red Sea Rift 771
Reggane Basin 686, 729, 733, 738
Regional Carbon Sequestration Partnerships (RCSP) 1189
regressive shoreface sandstones
  sequence J54a 187
regressive–transgressive scheme 185
relay ramps
  fault zones 958, 961
  Mid-Cretaceous early syn-rift stage mature 962
  rift oblique lineaments 961
  rift topography 959
  West Shetland Platform 957
reservoirs
  Algeria 753
  architecture
    Joanne Member 408
    Judy Formation 409
    Athabasca Oil Sands 1141–1150
    behaviour implications for sedimentological interpretation 408–409
  Buzzard Field 369–384, 381
  Captain Field 344
  Carboniferous 317–319, 1168
  Cenomanian 664

  Cenozoic Asmari carbonate 791
  central Utah deltaic analogues 24
  Chalk facies
    coccolith fragments 495
    dissolution seams 495
    localized stylolites 495
  characterization programme 234–242
  compaction
    observed and predicted 533
    South Arne 524, 526
  Cretaceous 803, 805, 958–959, 958–962
  deep Joanne model 412
  deep Judy model 412
  deep Triassic 410
  deepwater 45–50, 49, 50
  depletion fault control 532
  depth 410, 412, 414
  depth structure map 282
  desiccated 1112
  Donan 443
  Egyptian North Red Sea basin 786
  Ekofisk 525
  exploring subtle exploration plays in Gulf of Suez 773–774
    post-rift reservoirs 774
    pre-rift reservoirs 773
    syn-rift reservoirs 773–774
  flow simulation 1171
  fractures 23
  Fulmar Field 495, 498
  geomechanical model 524, 533
  hydrocarbon(s) 95, 786
  Jade Field 410
  Jasmine discovery 234–242
    depositional model 236–240
    intra-reservoir correlation 240
    seismic reprocessing 234–236
    structural model 236
    zonation scheme 240–242
      Top Joanne 240–241
      Top Julius Mudstone 241–242
      Top Lower Joanne 241
      Top Middle Joanne 241
  Joanne Member 226, 410
  Judy Field 414
  Judy Member 405, 410
  Laggan Field 282, 284, 287, 288, 290
  Late Jurassic age 127–142
  Leman Sandstone 327
  Libya 753
  Lower Balmoral 439, 440
  Norway hydrocarbons 989
  oil 299–313
  passive margins 826–827
    large drainage basin systems 826
    remaining deepwater plays 827
    reservoirs and basement in multiphase rifts 826
    reservoirs and outer margins 826
  predicting production behaviour 408–409
  Ross and minor Burns 345
  sandstones
    characteristics 134
    defined 138
    Forties Field 221
    Lower Balmoral 440

Scott Field 387
shale gas 1131
    annual natural gas production 1079
    map *1132*
Shiranish Formation 802
Skagerrak Formation 226
    stratigraphy 372–373
Taq Taq field development 802
tight in North America 1061
Triassic age 410
Upper Leman interval **334**
well 48/14-2 **334**
well 48/14a-5 **334**
well 48/14a-6 **334**
reservoir sedimentology
    Garn formation 422–423
        depositional synthesis 423
        description 422–423
        rapid pressure decline explanation 423
    HP/HT Kristin Field 422–425
    Ile formation 424–425
        depositional synthesis 424–425
        description 424
        rapid pressure decline explanation 425
    Not formation 423–424
    Tofte Formation 425–426
        depositional synthesis 425
        description 425
        rapid pressure decline explanation 426
    Upper Ror Formation 425
RF. *See* recovery factor (RF)
Rheic Ocean Drift 738
Rheic Ocean rifting 738
Rhuddanian Sandstones 692
Ribble Turbidites
    erosions 208
ridge-trough sedimentary structures 78
rifting 827, 999–1000
    basin architecture 37–38
    climax 961
    and continental breakup 921
    events 265, 824
    heat flow anomaly 663
    hot/wet spot 824
    initiation 962
    Jurassic age 960
    Late Triassic age 265
    Norwegian North Atlantic passive margin 893
    Norwegian region 925
    phase 825
    Rheic Ocean 738
    Rockall Trough 1029
    shear 824
    Solan Bank High 923
    topography relay ramps 959
    Yell Basin margin *960*
Rijswijk oil province 269
Rio Grande Rise
    location *868*
Rio Muni transform
    gravity modelling 875
Rita-1 well 130, 136
    sedimentological core logs *137*
RMS. *See* root mean square (RMS)
robust depositional model development 387
Rockall Basin *930*
    B sandstones *946*
    Cretaceous sandstones *1031*
    C sandstones *946*
    Dooish gas condensate discovery 938, 940–942
    Dooish K-feldspars *950*
    geoseismic profile and seismic profile *925*
    geoseismic section *940*
    Hatton–Rockall Basin 965
    K-feldspars bulk *951*
    lead-in-potassium-feldspar provenance tool 938
    location map *922, 1026*
    margin
        sandstones 937–951
        sedimentology 937–951
    Mesozoic sediment dispersal 938–940
    Mesozoic stratigraphy 938
    methodology 942–944
        core logging and sampling 942
        imaging and EMPA analysis 942
        lead analysis 942–944
        provenance analysis 942–944
    NE Atlantic sub-basalt hydrocarbon prospectivity 1028–1029
    palaeodrainage 937–952
        results 944–948
            analytical results 947–948
            B sandstones 944–946, *947*
            core description 944
            C sandstones 946, *947*
            depositional environments 944–946
            sandstone petrography 946–947
            A sandstones 944, 946, *946*, 950
            well stratigraphy 944
        rotated faults *1030*
        sedimentological logs of cores *945*
        strike line *1030*
        variation of lead K-feldspar populations *941*
        wireline profiles *941*
    provenance interpretation 948–950
    regional sand-sourcing implications 950
    sub-basalt hydrocarbon prospectivity 1025–1031
Rockall–Faroe area
    age and characteristics **965**
    Atlantic Margin timing, controls and consequences 963–975
    basalt escarpments *969, 970*
    bottom-water currents *973*
    Cenozoic compressional structures 963
    Cenozoic unconformities *974*
    compression 963–975
        basin control on orientation 971–972
        bottom-water current activity 972–973
            Norwegian Sea Overflow 973
            subsidence 972–973
        driving mechanisms 973–975
        prospectivity implications 975
        structures 964–971, *968*
            Alpin Anticline 967
            Auðhumla Basin Syncline 969
            Bill Bailey's Bank 966–967
            Bridge Anticline 969–970
            Dawn Anticline 967–968
            Faroe Bank Anticlines 967
            Faroe Bank Channel Syncline 967
            Hatton Bank Anticline 964–965
            location *966*
            Lousy Bank 965–966
            Mordor Anticline 967
            Onika Anticline 970
            Viera Anticline 971
            West Lewis Ridge reverse faults 970–971
            Wyville–Thomson Ridge Anticline 968–969
            Ymir Ridge Complex 968
        unconformities 963–964
    top-basalt isochron map *966*
    well/borehole data location *964*
    Ypresian isochore *970*
Rockall–Faroe continental margin
    volcanism constraints 831–841
Rockall–Faroe continental margins
    magmatism 831–841
    compressional structures 837
    crustal structure 837
    crustal velocity models 835–837
    geological interpretation 837–839
    igneous intrusion 838
    igneous intrusion and extrusion 837
    lithological constraints from P- and S-wave velocities 839–849
    source and receiver design 832–833
    wide-angle seismic data 833–835
Rockall Trough
    acquisition plan *556*
    age of rifting 1029
    bottom-water currents 973
    CTD probe 550
    location *550*
    Mesozoic strata 1031
    mid-Norwegian margin 849
    oceanic mixing 549, 557
    reflection strength 555
    seismic reflectivity 555
    sound speed
        CTD probe casts *554*
        profiles *555*
    water layer
        reflectivity *551*
        sound sections *552*
        stratigraphy *554*
        structure *553*
Rock-Eval pyrolysis
    Bakken Formation 1067, *1069*
rock physics
    amplitude anomaly 258
    inversion 532
    model 527, 534
Rocky Mountain foreland fold-and-thrust belt 1141
Rodinia Supercontinent 677
Rolf Field 537
Romagna accretionary arc
    Porto Corsini East gas field *122*
Rømø barrier island. *See also*
    Holocene–Recent Rømø barrier island system
    GPR profile *150, 151, 152*
    modern analogue 152
    palaeogeographic reconstruction 148
    reservoir analogue 146
    tidal inlet sands *152*
    vertical section trough 147–148
    wells position *150*
    west–east cross-sections *150*

Rona Fault 960
Rona Ridge 245
roof disruption
  Gulf of Mexico *908*
roof-edge thrust
  Alaminos Canyon area *904*
root mean square (RMS) 517
  amplitude extraction *1015*, 1019
Ross Formation 23
Ross reservoir
  LKO 345
rotary steerable systems (RSS) 340
rotated faults
  Rockall Basin *1030*
Rotliegend Group 61, 66, 328
  Chiswick Field 319
  desert sandstones 73
  Lower 59, 74
  Lower Permian age 325
  Netherlands petroleum exploration 266–267
  reservoir 319
  sequence 329
  strata 62, 67, 69
  thickness *69*
  two-way time and depth maps *320*
  Upper and Lower 61
Rotliegend Slochteren sand 266
Rover Boys 2
RSS. *See* rotary steerable systems (RSS)
Rub Al Khali Basin 695
  burial history *696*
  Palaeozoic petroleum system 695
  petroleum system 695
  petroleum systems *696*
  structural elements *693*
  tectonic elements *693*
  tectonic events *696*
Rudeis Formation 772, 774
  exploring subtle exploration plays 777–778
  Lower 777
  South October Concession 778
  syn-rift 775, 777
Rush, Julian 16
  GeoControversies debates summary 15–17
Russia. *See also* specific location
  CSP 611–614
  FSU and Circum-Arctic final frontier 589–590
  topography *646*
Russian Arctic shelves (RAS) geology 591–615
  consolidated basement main characteristics 593–598
    eastern sector 596–598
    western sector 593–596
  crustal tectonic *594*
  free-air gravity field *595*
  HC resources 603
  heterogeneous folded tectonic basement 598
  sedimentary basins 598–614
    CSP Russian sector 611–614
    east Barents Province 598–603
    ESSP 609–611
    Laptev Sea Province 605–609
    North Kara Province 605

petroleum geology 604–609, 611, 614
South Kara Province 603–605
sparsely explored 614
surveys and offshore wells location *592*
tectonic history and petroleum geology 591–615
tectonic setting 593
Ryazanian gravity-flow sandstones
  Upper Jurassic reservoir sandstones 139–141

Sabden Shales
  black shale **1095**
SACROC EOR
  history-match analysis *1191*
  sequestration test 1190–1191
  Texas 1190
SAGD. *See* steam-assisted gravity drainage (SAGD)
Saharn Platform
  highs *708*
saline aquifer 1191, 1195
salt
  basin architecture 180
  Block 31, offshore Angola *1014*
  canopies *904*, 911
    advancing *903*
    Gulf of Mexico *900*, *910*
  deepwater reservoirs 50
  diapiric *902*, *903*, 911
  gas plays 1046
  Gulf of Mexico *907*, *908*
  plays 1045–1049
  sheets *903*
  withdrawal minibasins 24
  withdrawal role 50
salt diapirs 78, 83, *86*, *87*, 92
  chimneys *82*
  fluid leakage 90
  leakage zones 85–86
  structure 79
Saltire Formation *394*
Saltire Shale 397
salt plumes
  intrusive 899–904
salt tectonics 899
  emerging concepts 899–911, *900*
    Allochthonous fragments 909–911
    canopy-margin thrust systems 904–905
    intrusive salt plumes 899–904
    minibasin triggers 905–909
    state of knowledge 911
  Gulf of Mexico 902
Salvador–N'Komi transfer zones 877
sand fluidization
  laboratory experiment demonstrating implications *1186*
SandGEM 6–7
sandstones 240, *778*, *778*
  aeolian 826
  Albo-Aptian Kopervik 53
  Balmoral 441
  barrier island 146
  basin-centred, distal lowstand *732*
  Bossier 1112
  Buzzard Sandstone Member *370*
  Carboniferous 641, 996

v. conglomerate *1111*
Corrib gas field *949*
Cretaceous *1031*
cross-bedded *423*
C sandstones *946*
deepwater reservoirs 46, 48–49
deposition tectonic reconfiguration 206
deposits sequences J52-J54b *193*
Devonian sandstones 569
Dooish *947*, *948*
Early Jurassic age 419
Early Volgian age 138–139
Farsund Formation 141
Fell Sandstone Group 59
fluvial 717
Fulmar Field 178, *207*
  aggradational packages 186
  BVRU 205
  deposited in narrow palaeovalleys 206
  Humber Group thickness 207
  initial distribution 180
  J52 and J54 sequences 186
  shallow-marine 177
  subdivided 182
  well 22/30b-11 189
gravity-flow *138*
Gulf of Suez *779*
Heno Formation 127
Ireland Mesozoic 951
isopach *218*
Joanne Sandstone Member 226, 406
  erosion 241
  rich gas condensate 228
Judy Member 407
Judy Sandstone Member 240
  depositional environment 406
Kimmeridge Fulmar 205
Kimmeridge Fulmar sandstones 205
Late Jurassic age 127–142
Leman Sandstone 325, 327
  correlation panel *329*
  depth map *326*
  hydraulic fracture-stimulated intervals *333*
  illite cementation 328
  reservoir 327
Lower Balmoral 431
Lower Cretaceous age *958*
Navajo Formation Sandstone 1193
  reservoir *440*
reservoir prediction *221*
Rockall Basin *946*
  margin 937–951
  palaeodrainage 944–946
Sgiath Sandstone 369
Sherwood Sandstone 938, *1090*
shoreface 174
  Lola Formation 135
  porosity–permeability relationship *141*
  Upper Jurassic barrier island system of Freja discovery 150
Skagerrak Formation 39, 406, 409
Strunian–Tournaisian 731
turbidites *196*
Upper Jurassic
  barrier island 145–154
  reservoir 127–142
Upper Rudeis 777

San Juan Basin (SJB) 1191–1192
San Paulo Plateau (SPP) 855
   bedded evaporite deposition age 857
   drilling 856
   subsidence analysis 860
   Tupi oil discovery *859*
San Rafael Swell 24
Santerno Formation 123
Santonian Thud Formation
   sediments *482*
Santos Basin 824, 826
   air gravity data *856*
   bathymetry *856*
   beta factor maps *860*
   Bouguer gravity anomaly map *856*
   conceptual breakup models *863*
   continental lithosphere thinning and crustal thickness 859
   crustal cross-section *862*
   crustal thickness map *861*
   gravity inversion 859, 860, *862*
   incipient breakup 861
   margin breakup 857
   rifted margin 861
   rifting 855
   seismic cross-sections *858*
   South Atlantic Ocean breakup 855–865
      regional geology 855–857
      regional structure 858–859
   stratigraphy *857*
   subsidence analysis 859
   subsidence association 865
   subsidence modelling 860
Sao Francisco craton 870, 876
Sao Paulo Plateau 824
   beta factor maps *860*
   breakup of South Atlantic Ocean 855–859
   conceptual breakup models *863*
   continental lithosphere thinning 859
   crustal cross-section *862*
   crustal thickness 859
   gravity inversion 859, 860, *862*
   rifting 855
   seismic cross-sections *858*
   South Atlantic Ocean breakup 859–860, 860–862
   subsidence analysis from 2D backstripping 859
Sao Paulo Ridge
   location *868*
sapropelic kerogen
   Farsund Formation *102*
Sarah Formation
   Late Ordovician age *695*
   Ordovician 695, 697
Sarawak
   offshore 459, *459*, *460*
Sargelu Formation 802
Sarmatian sedimentary rocks
   Pannonian Basin 1126
Sartirana Formation
   Settala gas field *123*
Saudi Arabia. *See also* specific location
   exploring hydrocarbon plays 691–697
      challenges 695–697
      stratigraphy 692–695
      hydrocarbons development 695

Khuff Formation 692
Rub Al Khali Basin 693–696
striated glacial surface *695*
Unayzah complex 692
Saxo-1 well
   sedimentological core log *140*
   Volgian sandstone intervals *139*
Scandes
   Cenozoic uplift 916
Scandinavia 986
   basement area
      palaeogeography *984*
   Cenozoic development 982
   eastern North Sea Basin 1001
   erosion basement 986
   gravity surveys compilation *566*
   hemipelagic, deep-marine deposits of Eocene age 984
   landscape 981
   Neogene uplift 1000
   onshore–offshore geological map *886*
   Permian volcanic rocks magnetic properties *572*
   post-breakup uplift phases 986
   regional exhumation 1000
   tectonic pulses 982
   uplift phase
      progradation *984*, *985*
   volcanic rocks magnetic properties *572*
Scandinavian Alum Shale 1080
scanning electron microscope (SEM) analysis 330
Schiehallion Field 248
   amplitude shutoff *250*
Schlumberger Full-bore Micro-Imager log data (FMI)
   Ensign Field 332
Sciacca Formation
   porosity origin 118
Scoresby Sund area 998
   erosional unloading 999
Scotland 387
   biogenic gas 1080
   Orcadian Basin 1090
Scott, Robert 589, 590
Scott Field
   asset 403
   asset matures 387
   continuous drilling programme 391
   depositional perspective 393
   drilled infill wells 403
   drilling campaign 391, 402, *402*
   dynamic modelling 393
   infill drilling targets 387
   initial development drilling 390
   injection strategy 391
   integrated analysis 402
   interpreted open hole results *395*, *396*
      J32Z *398*
      J33 *396*
      J35 *395*
   IW, SW–NE cross section *399*
   location map *388*
   mature field revitalization 387–403
      bypassed sweep targets 393–400
      downdip injector 402
      drilling campaign impact on field production 402–403

   geological description 387–390
   infill campaign (2005–2007) 391–402
   production overview 391
   seismic database 390
   seismic interpretation 390–391
   updip attic targets 391–393
   MDT logs 393
   North Sea 387
   OBC data with 1999 PSDM data *390*
   PBU derivative J16 *400*
   Piper depletion J16 *400*
   post-drill correlation J32 *401*
   production *402*
   production history 403
   recent drilling 387
   reservoirs 387
   sands 398
   simulation modelling 402
   Skene 391
   structuration *389*
   structure map *388*
   target
      J31 *399*
      J32Z *397*
      J35 *392*, *393*
   tectonostratigraphic development 390
   wells
      J4 393
      J30Z *402*
      J32Z 396
      J34 396
   zonation *389*
Scott Formation 397
Scott Sands
   Lower 393, *394*
Scott–Telford field complex 32
Scremerston Coal Group 59, 69
sealed wells 1162
sea water injection 351
sedimentology 27, 827
   architecture 406
      production behaviour with depth 416
      production performance 405
   basins 591, 1079–1080
      Angola 1044, *1044*
   fill patterns 2
   Gulf of Mexico *907*
   Rockall Basin margin 937–951
   Shiranish Formation 811–818
segment-style analysis
   Ormen Lange system 920
seismic data
   acquisition
      XCTD probe 557
      XSP 557
   acquisition parameters **519**
   CDR 4D surveys *521*
   Cimmerian rifting 44
   cold-water statics 557
   crossline profile orientated NW–SE *1036*
   cross-section of channel features *84*
   database 1005–1007
   exhumation estimates 1010–1012
   facies analysis 805–806
   facies types 1035
   imaging of variable layer sound speed 549–557

seismic data (*Continued*)
  interpretation 1009
    Carboniferous petroleum system 59–63
      sub-Zechstein seismo-stratigraphic
        relationships
          Auk–Argyll–Flora area 60–61
          Jaeren High area 62
          Mid North Sea High area 61
          West Central Shelf 62–63
    new methods of improving 1005–1012
  inversion 538
  monitoring 1171
  new methods for improving 1005–1012
  Norwegian Continental Shelf basement
    map 888
  passive margin evolution case histories
    from Ireland 1005–1012
  profile
    Auk–Flora Ridge *61*, *62*, *63*
    Central Graben *80*, *81*
    Jaeren High *64*
    Norwegian continental shelf *887*
    West Central Shelf *65*, *66*
  project-specific four dimensional 1141
  project-specific three dimensional 1141
  reflection imaging
    hydrocarbon exploration 549
    oceanic variability 550–557
      longer term 557
      short-term 555–556
      spatial variation 553–555
      temporal variation 555–557
    oceanography implications 557
    sound speed variation 549
    surveying implications 557
    water layer variability 549–557
  regional two dimensional 1141, 1147
  repeat four dimensional 1168
  repeat three dimensional 1168
  reprocessing 1007–1009
  sequence stratigraphy 52
  stratigraphy 39, 51
  survey **519**, 557
  type outcrop exposures 1145
  velocity data 1010–1012
  well log expressions *47*
seismic fault distortion zone (SFDZ) 509
  expressions *512*
  fault network 510
  geobody *511*
  juxtaposition across damage zones 510
  juxtaposition analysis 513
  juxtaposition-based sealing 515
  juxtaposition difference *511*, 514
  mapped *512*
  reflectivity volumetric fault attributes *512*
  skin geometries *511*
  thickness 510
  volume render of single fault *514*
seismic technology 337–340
Selandian Våle Formation 487
Sele formation *518*
Sele/Hermod shelf
  seismic expressions *47*
Selkirk Field
  discovery 49
  turbidites *202*
SEM. *See* scanning electron microscope
  (SEM) analysis

Semberah Field *458*, 459
  geological model *458*, *459*
Senja
  NRM 565
Sentusidinium 149
sequence
  J44
    Callovian age 186–192
  J46
    depositional environments *187*
    Early Oxfordian age 186
    Late Callovian age 186
    well correlation *188*
  J52 *203*
    Fulmar Sandstone deposits *193*
    Middle Oxfordian age 186–187, *189*
  J54 *194*
    depositional environments *191*
    Fulmar Sandstone deposits *193*
    Late Oxfordian age 187–190, *190*
      depositional environments *190*
      regressive shoreface
        sandstones 187
      turbidite sandstones *188*
    well correlation *188*, *197*
  J56 *194*, *203*
    depositional environments *191*
    turbidite deposition *191*
    well correlations *193*
  J62
    depositional environments *195*
    Kimmeridgian 192
    well correlation *197*, *198*
  J63
    late Kimmeridgian age 192
    well correlation *197*
  J64
    Early Volgian age 199
    shoreface *196*
    turbidite sandstones *196*
  J66
    Early Volgian age *195*, 199
    Middle Volgian age 199
    turbidite sandstones *196*
  J71
    Middle Volgian age 199
  J72
    Middle Volgian age 199
  J73 199
  J76 199
  K10
    Late Ryazanian age 199–200
  stratigraphy 51
    application 45
    correlation of wells *183*, *184*
    framework 168
    model *47*, 182–185
    schemes *179*
    studies 180–182
sequential Gaussian simulation (SGS) 296
sequential indicator simulation (SIS)
  Joanne model 410
sequestration 1165
Serpukhovian–Bashkirian incised
  valleys 733
Settala gas field 122, 123, *123*
SFDZ. *See* seismic fault distortion
  zone (SFDZ)
Sgiath Sandstone 369

SGS. *See* sequential Gaussian
  simulation (SGS)
Shaftsbury Formation *1115*
shale 377
  Arnsbergian Sabden Shale 1088–1089
  Austria 1081
  Avon 206
  Bakken 1066
  Besano Shales 118, 124
  Big Sandy shale gas field 1087
  bioturbated shale-prone interval 425
  black 676, *762*, 1081, 1091, **1095**,
    1096–1097
  Chimney Rock Shale 1190
  Clare Shale 23
  Craven Group 1093
  Cretaceous Lewis Shale 1091
  Czech Republic 1081
  DDB 1081
  deepwater marine shales 1126
  Devonian 'hot' shale *756*
  Ekofisk Formation 465
  El Gassi Shale 684
  Emgyet Shale 686
  Europe 1083
  Farsund Formation 98
  Frasnian 'hot' shale 754
  gas 1080–1083
  gas-bearing chalk 465
  GDE map *748*
  geometry 722
  Gething Formation 264
  Ghadames-Illizi Basin *756*
  Gothic Shale 1190
  Holywell Shale **1095**
  Intra-B4 shale 376, 377
  Lewis Shale 1091
  Lias **1095**, 1096
  Long Eaton Formation 1093
  Lower Bowland **1095**
  Lower Limestone Shale 1092–1093
  Middle Miocene 1062
  Miocene 1062
  Mississippian 1090
  New Albany Shale *1133*, *1135*, *1134*, *1138*
  Nikanssin Formation 1110
  North America 1061
  Ordovician 1091
  Poland 1080
  Posidonia shale 263–264
  Scandinavian Alum Shale 1080
  Silurian age 715, 1091
  Slovak Republic 1081
  stratigraphic groups 1089–1090, *1090*
  thickness 1087
  Tofte Formation *426*
  Ukraine 1081
  Widmerpool Formation 1093
  Worston Shale Group 1093
shale gas 1062
  case studies 1080–1083
    Cambro-Ordovician black shales 1080
    Carboniferous black shales 1080–1081,
      1081
    Lower Jurassic black shales 1081
    Neogene shale (basin centred) gas
      1081–1083
    Silurian black shales 1080
    Upper Jurassic shale gas 1081

Europe 1079–1083
  biogenic methane 1083
  current research activities 1079–1083
  outlook 1083
  research activities 1083
Newark East 1087
North America 1062
play 1133
reservoirs 1131
  annual natural gas production 1079
  map *1132*
UK 1062, 1087
United States 1079–1180
Western Canada Basin 1079
Western Europe 1079, 1083
shale oil
  play 1065
  unconventional oil and gas resources 1061–1062
shallow marine
  exhibit 157
  Fangst Group 419
  Fulmar Sandstone 177
  Fulmar sandstone or turbidite sandstones 178
  Hugin deposition 168
  Sigrun–Gudrun area 173
Shannon Basin
  Carboniferous age 23
  petroleum geoscientists virtual fieldtrips 23–24
shear rifting 824
Shearwater area
  shoreface environments *191*
shelf margin wedge 45
Sherwood Sandstone 938, *1090*
Shetland. *See* Clair field
Shetland area *1031*
  structural elements *246*
Shetland Basin
  location map *922*
Shetland Isles 245, 953
Shiranish Formation 802
  activation energy *817*
  description *803*
  Euphrates Graben 811–818
    bulk petroleum generation kinetics 817–818
    hydrocarbon source rock potential 813–815
    mineralogy 816–817
    organic facies 815–816
    regional geology 811–813
    results 813
    samples 813
    trace element palaeoredox proxies 817
  fault movement *805*
  high clay mineral content 816
  kaolinite 816
  lithology *814*
  mineralogical composition *817*
  open pyrolysis gas chromatograms *815*
  organic geochemical characterization *814*
  petroleum formation 818
  petroleum geology 811
  sedimentology 811–818
  seismic analysis 807
  tectonic map *812*
  TOC *816*
  total nitrogen content *816*
  vitrinite reflectance *817*
Shiranish model
  fault displacement *806*
shoreface *164*
  deposits 166
  facies association B 162–166
  Hugin Formation sedimentology 162–166
    description 162–164
    interpretation 164–166
    sequence stratigraphy 166
    wireline-log character 166
  sandstones 174
    Freja discovery 150
    Lola Formation 135
    porosity–permeability relationship *141*
  sequences J64 *196*
  wireline-log cross-plot *165*
Shoreface Facies Association *163*
Shuttle Radar Topography Mission (SRTM) 20
Siberian Arch
  crustal tectonic *594*
Siberian Arctic Shelf
  petroleum play elements chart *608*
  sedimentary basin *606*
  tectonostratigraphy *608*
Siberian Basin 667
  basin sensitivity parameters 661, **664**
  charging gas fields 659–667
    analyses 662–663
    geology 660–662
    results 663–667
    stratigraphy 661–662
    structure 660–661
  isotopic composition of methane *665*
Siberian–Chukchi Fold Belt 597
Siberian Craton 604
  consists 621
  field densities **623–624**
  geological provinces *622*
  geological provinces assessment results **630**
  location 621
  north of Arctic Circle 621–631
  petroleum resources 621–631, 624
    assessment results 631
    assessment units 622
    geological analogues for assessment 622–631
      Khatanga Saddle assessment unit 624–627
      Lena-Anabar Basin assessment unit 627
      Northern Priverkhoyansk Foredeep assessment unit 627–629
      Yenisey–Khatanga Basin assessment unit 629–631
    geological provinces 621
    petroleum occurrence geological models 621–622
    resource quantities 622
Siberian Sea 591, 598
  free-air gravity field *595*
Siberian Sea Basin 609, *611*
Siberian Sea Province 609–611
Siberian Sea Shelf *610*
Siberian shelves 596
Sicilian-Maghrebian chain
  Gela oil field *120*
  geological sections *117*
  hydrocarbon occurrences *117*
Sicily 741
sigmoid aggradational facies 1035
sigmoidal clinoform package 83
Sigrun-Gudrun area
  axis-perpendicular sub-regional correlation 168–173
  flooding 173
  shallow-marine sedimentation 173
  well-log correlation panel *170*
Sigsbee Escarpment 904
  minibasins 908
  thrusting 905
siliciclastic rocks
  depth v. average density *906*
siltstones
  Farsund Formation 141
Silurian age *711*
  agrilleceous depth velocity plots *751*
  Ahara uplift 713
  basal 712
  Berkine Basin *714*
  black shales *762*, 1080
  chronostratigraphic diagram *747*
  dark-grey agrillaceous complex *1080*
  Early *690*, 743–745, *744*
  flooding *712*, 716
  Ghadames Basin *711*
  hot shale Ghadames–Illizi Basin *756*
  hydrocarbon prospectivity 686–688
  Late 265
  Middle Akakus Shale *748*
  Middle East *687*
  North Africa 686–688, *687*
  sediments Azzel Matti Ridge 717
  shale 715, 1091
  succession 686
  Tanezzuft Formation *765*
  Tannezuft Formation *687*
  Tihemboka 715
  well A1-NC198 *767*
Sindh region
  Pakistan 676
Single Well Oil Production System (SWOPS) 432
Sirte Basin
  exploring hydrocarbon plays 697–698
SIS. *See* sequential indicator simulation (SIS)
six-tile WATS *1021*
SJB. *See* San Juan Basin (SJB)
SJU600 SB
  tectonic reconfiguration event 196
Skagerrak Field
  exploration 216
Skagerrak Formation *214*, 226, 234, 236, 240, 405
  basement depth map 891–892
  channel and overbank processes *408*
  degrade with depth 405
  depositional facies 409
  gas condensate 228
  geology 405–407
  HPHT 405
  hydrocarbon fields 225
  Judy Member well 410
  locations producing *406*

Skagerrak Formation (*Continued*)
  NTG 410
  pods as basin infilled 406
  porosity with depth 409–410
    predicting production behaviour
      409–410
  production performance 407, 416
  proximal to distal trends *408*
  quad 30 area *406*
  reservoirs 226, 407
    Heron cluster 242
  sandstones 39, 409
    sequences 406
  sedimentation *240*
  stratigraphic column *407*
  stratigraphy changes 416
  structural tilt 416
  succession 226
  VClay log 411
Skagerrak Platform 891
Skallingen peninsula
  washover fan *151*
SKB. *See* South Kara Basin (SKB)
Skjold Flank-1 well 547
Skolithos ichnofacies 139, 164
Skrubbe Fault 487
Skua Field
  high production performance 405
Sleipner carbon dioxide plume 1171–1182
  cumulative total reflection
    strength *1184*
  feeder chimney *1179*
  flow simulation of synthetic surface *1180*
  flow simulations *1181*
  schematic cross-section *1175*
  seismic images *1172*
  seismic line through *1179*
  time-lapse monitoring 1173
  time-lapse seismic data 1171–1182
  topmost carbon dioxide layer
    analysis 1173–1175, *1178*
    buoyancy effects 1178–1179
    growth *1177*
    observed volume **1176**
    permeability effects 1177–1178
    properties *1175*
    spreading 3D flow simulation
      1177–1181
    thickness derived from structural
      analysis *1176*
    thicknesses from seismic amplitudes
      1173–1174
    thicknesses from structural analysis
      1174–1175
    topographic effects 1179–1181
    topseal properties 1181
  TOUGH2 axisymmetrical flow model *1176*
  whole plume flux 2D flow simulation
    1175–1177
Sleipner Delta
  shallow-marine 174
Sleipner Field
  carbon dioxide injection 1063
  injection operation *1172*
Sleipner Formation 168
Sleipner West Field 158, 1183
slope by-pass systems 823
Slovak Republic
  shale 1081

Slyne Basin 938, 1009–1010
  Cenozoic exhumation history 1005
  Dooish K-feldspars *950*
  exhumation 1009, 1010
  exhumation estimates from seismic
    velocities *1011*
  exhumation values 1012
  location map *922*, *1006*
  seismic data 1005
smart mobile devices 25
Smith, John 31
Smith Bank Formation 38
Smith Bank Shale 39
Smithwick Formation
  tall calcite-sealed fractures *1133*
Society of Petroleum Engineers
  (SPE) 13
Søgne Basin 138
Sognefjord
  satellite image *916*
Solan Bank High
  classical rifting 923
  geoseismic profile *924*
  location map *922*
Solarcentury 11
Sole Pit area
  map showing location *326*
Sole Pit Basin 325
source-to-sink systems
  drainage basin–shelf–slope basin floor
    profile *914*
  elements *914*
  exploration value 919
  Gulf of Mexico 914
  Norwegian continental shelf 916–919
  NSF 913
  passive margins 913–914, 913–920
    ancient systems analysis 915
    linked segment-style approach 915–916
    methods 914–916
    modelling 915
    Norwegian continental margin 913–920
    Norwegian continental shelf
      916–919
      drainage system characteristics 919
      fan volume *v.* onshore source area size
        916
    significance 913–914
    uncertainties 919
  segments *915*
South Africa
  Karoo Group 52
South Anyui
  ophiolitic suture 597
  suture *607*
South Arne Field 487
  acoustic impedances 526, 527, *527*
  base survey *526*
  changes in water saturation *531*
  compaction 523
  depositional environment 527
  fluid flow 534
  Kirchhoff time migration *526*
  monitor survey *526*
  petroleum geology
    applied time-lapse seismic methods
      523–524
  Poisson's ratio *527*, *528*
  reservoir compaction 524, *526*

reservoir geomechanical model 524
rock physics model 527
  change in gas saturation *528*
  change in porosity *528*
  change in water saturation *528*
surveys 526
three-dimensional view *525*, *530*
time-lapse AVO inversion 527
time-lapse rock physics inversion *529*
time-lapse shift *526*
top-reservoir structure *524*
South Atlantic Ocean
  bathymetry contours *868*
  breakup 855–865
    continental lithosphere thinning
      859–860
    crustal thickness determination
      859–860
    failed breakup model 860
    history 855, 860–862
    petroleum systems 862–865
    rifted margin formation 865
    subsidence analysis 859
    thinned continental crust i 855–865
  conceptual breakup models *863*
  Santos Basin 855–859
    regional geology 855–857
    regional structure 858–859
  Sao Paulo Plateau 855–859
  sedimentary basins evolution 878
South Bróna Basin
  geoseismic profile *925*
South Chukchi Basin 613
  HC play elements 614
  internal structure *613*
Southern North Sea
  Chiswick Field 315–323
Southern Panels 369
  Buzzard Field *384*
  high-density turbidites 372
  Lower B4 unit *380*
Southern Permian Basin 325
Southern Scandinavia post-breakup uplift
  phases 986
South Kara Basin (SKB) 604,
  646, 654
  HC systems 604
  MCS data 604
  petroleum potential 604
  plate-tectonic setting and depositional
    environments 603
  structural mechanisms proposed 604
South Kara Province
  petroleum geology 604
  RAS sedimentary basins 603–605
South Kara Sea 660
South Korea
  Drilling Expedition UBGH1, 1154
South October Concession
  exploring subtle exploration plays in Gulf
    of Suez 777–778
  Rudeis Formation 778
South Taimyr Fold Belt
  crustal tectonic *594*
South Viking Graben
  Hugin Formation 157–175
  sedimentology and sequence
    stratigraphy 173
  well-log correlation panel *169*

Southwest Carbon Partnership
     (SWP) 1189
   carbon dioxide sequestration
      capacities *1194*
   terrestrial Riparian Restoration
      Project 1192–1193
South Ymir Ridge Anticline 970
   seismic profile *970*
Soviet Union
   Former 13
SPE. *See* Society of Petroleum
     Engineers (SPE)
Spirit River Formation
   basin-centred gas in Western Canada
      1106–1110
      Cadotte Member 1108–1110
      Falher Member 1106–1108
      Notikewin Member 1108
   Lower Cretaceous age 1101
SPP. *See* San Paulo Plateau (SPP)
SRME. *See* surface related multiple
     elimination (SRME)
SROCK 557
SRTM. *See* Shuttle Radar Topography
     Mission (SRTM)
STAR
   application 517, 519
   methodology *520*
steam-assisted gravity drainage (SAGD)
     1141, *1143*, 1145, 1147
   steam chambers *1148*
steam chambers 1147
   logging data *1148*
   SAGD *1148*
Stellar
   default kinetic parameters
      kerogen transformation 663
   three dimensional basin modelling software
      662, 667
   three dimensional basin models 663
stem flooding
   four dimensional monitoring 1145
Stendal, Henrik 590
sterane maturity parameters 103
Storelv formation 565
STRATAFORM project 914
stratigraphic architecture
   Ahara High 712–713, *713*
   Berkine Basin *713*
   Djebel Tamamat 716–717
   Gargaff Arch 710–711
   Illizi Basin 712–713
   Jasmine discovery 226–227
   North Africa *718*
   North African plays 710–718
      Algeria 712–716
      Bled el Mass area 716–718
      Libya 710–712
   North Sea Eldfisk Field,
      Norway 473
   Palaeozoic succession *718*
   Taq Taq field *805*
   Tihemboka High 715
stratigraphic column *263*
   Chiswick Field *318*
   CNS Triassic *407*
   Gulf of Suez *773*
   Judd Basin *247*
   Skagerrak Formation *407*

stratigraphic framework
   Ghadames–Illizi Basin *737*
   Huntington discoveries *215*
stratigraphic pinchout
   Emsian flooding surface *716*
   Illizi Basin *716*
   Silurian flooding *716*
stratigraphic trap 262
   clastic basin fill 269
   defined 261
   Grouw *272*
   Lower Cretaceous Rauwerd *272*
   Netherlands 266, 273
   North Africa 725–733
   types 731–733, *733*
      Carbonate buildups (Waulsortian reefs)
         733
      depositional pinchout 731–733
      incised valley-fills 732
      North Africa *731*
      Serpukhovian–Bashkirian incised
         valleys 733
      Strunian to Tournaisian incised
         valleys 733
      Viséan incised valleys 733
stratigraphy
   Buzzard Field *372*
   charging gas fields 661–662
   Chiswick Field 316–317
   Chukchi basins *613*
   Egyptian North Red Sea basin 783
   Euphrates Graben *813*
   Faroe–Shetland Paleocene *281*
   gas-prone petroleum systems *1045*
   Laggan sands *281*
   Netherlands 262
   Nile Delta *700*
      play 698–700
   Northern Judd Basin 245–246
   North Sea Rift *39*
   photomosaic data 24
   Red Sea *785*
   Santos Basin *857*
   Saudi Arabia 692–695
   Skagerrak Formation 416
   Taoudenni Basin 676–677
   Taq Taq *802*
   thicknesses control *504*
   Vaila Formation *281*
   West of Shetlands Area *281*
Streppenosa Formation 118
   foreland 124
   hydrocarbon occurrences 124
stress-fields
   tectonic 90
strike-slip faulting 41
stromatolitic carbonates
   Abolag-1, *680*
   Taoudenni Basin *680*
structural architecture
   Almada Basin 867–880
   Camamu Basin 867–880
   North Sea hydrocarbon
      systems 41
structural diagenesis 1133
Strunian Sands 738
Strunian–Tournaisian
   incised valleys 733
   sandstones 731

sub-basalt imaging
   Faroe–Shetland Channel 1034
   Møre Basin 1025
subsalt
   Block 31, *1015*
   imaging challenges 1013–1015
   Mad Dog Field 1015
subsidence
   calculating 831
   modelling programmes 839
sub-Zechstein
   composite layer 72
   seismo-stratigraphic relationship 60–61
   well sections 57–59
Suez Rift 771
Suilven
   CSEM 257
   gas discovery 252
   oil discovery 252
Sukkertoppen Iskappe *981*
   idealized profile *993*
   landscape development 992, *993*
   uplift history *995*
Suprasalt Miocene strata
   Green Canyon area *909*
surface related multiple elimination (SRME)
     236, 1007
   three dimensional Red Sea basin 786
   visualization quality control *1007*
suspension-transport by contour
     currents 90
Svalbard
   AFTA data 640
Svane-1 well 138
   formation pressures **136**
   gravity-flow sandstones *138*
   HTHP environment 138
Svecokarelian greenstone belts 569
Sverdrup Basin 643
Swab–Tanf Formation 813
Sweden *563*
   exhumation 983
SWOPS. *See* Single Well Oil Production
     System (SWOPS)
SWP. *See* Southwest Carbon
     Partnership (SWP)
synchronous exhumation 633–643
   Alaska North Slope 633–643
   Barents Sea 633–643
   exhumation histories from North Slope
      Alaska 637–640
   exhumation histories in Barents Sea
      634–637
   implications for hydrocarbon prospectivity
      643
   regional extent 643
   regional synchroneity 642–643
   thermal history reconstruction 634
syn-rift 823. *See also* Faroe–Shetland Basin,
     syn-rift play
   Angola 1046
   early 961
   late 958, 961
   Lower Cretaceous age 962
   plays 961–962
   poorly dated 824
   Rudeis 775
   sedimentary fill 96
   stratigraphy 777

syn-rift 1 megasequence 738
syn-rift 2 megasequence 739–742
Syria. *See also* specific location
   Euphrates Graben 811–818
   tectonic map *812*
Szolnok Formation 1126, 1128

Tabe Formation
   generalized present day level of
      maturity *1049*
Tadrart fields 754
Tadrart Formation
   Lower Devonian age
      GDE map *748*
      MPath modelling *758*
      outcrop photo *711*
Tafilalt Basin 731, 732
Tail End Graben 129
   gravity-flow sandstones *138*
Taimyr Fold Belt
   crustal tectonic *594*
Tannezuft Formation
   Silurian age *765*
      Middle East *687*
      North Africa *687*
Tanopcha Formation 659, 662, 667
   coal-bearing muddy sediments 598
   gas yield *663, 664, 665*
   temperature map *666*
   thermal yields 663
Tanzuft 'hot' shale 754
Tanzuft source rocks
   kitchen maps *757*
Taoudenni Basin 676, 681, 682, 688
   Abolag-1 well *679*
   burial histories *680*
   depositional model for stromatolitic
      carbonates *680*
   exploring older and deeper hydrocarbon
      plays 676–681
   geological cross-section *678*
   Neoproterozoic geology *677*
   Neoproterozoic petroleum system
      676–681
   Neoproterozoic succession *679*
   petroleum system *679*
   Precambrian stratigraphy *678*
   regional tectonic events *680*
   tectonic events and hydrocarbon
      play *678*
Taq Taq field 801–809
   appraisal and development 801–809
      current plans 809
      fluids 808–809
      pore system 806–807
      reservoirs 802
      seismic facies analysis 805–806
      Shiranish Formation 802
      structure 802–805
   appraisal and development of 801–809
   lithofacies prediction 806
   location map *801*
   log, core, mud gas and image interpretation
      808
   matrix-fracture interaction under
      water *807*
   seismic dip line *805*
   stratigraphy *802, 805*
   structural interpretation *805*

target oil volumes
   locating remaining oil in Nelson Field
      365–366
   slow oil 365
Target Remaining Oil 350
Tasmanites morphology 98
Tassili N'Ager outcrops
   Palaeocurrents 745
Tayarat Limestone 813
Tay formation fan system
   Eocene age *46*
TD. *See* total depth (TD)
TDS. *See* total dissolved solids (TDS)
Teal embayment *191*
Telford fields 387
Temis three dimensional model 754
   Caledonian erosion *757*
Tempa Rossa oil fields
   Val d'Agri oil field *121*
temperature logs 1147
Tern horst 44
   evolution *43*
   Late Jurassic extension multiple phases
      interplay *42*
terrestrial laser scanning 25
   workflow *24*
Terroir, Didier 589
Tertzakian, Peter 14
Tethyan rim of Arabia 828
Tethys Ocean 420, 1126
Texas
   Bossier sandstones 1112
   cores and outcrops *1135*
   Fort Worth Basin *1087, 1088,* 1131
   SACROC EOR Field, Texas 1190
      history-match analysis *1191*
      regional location map *1191*
      sequestration test 1190–1191
Thebes Formation 773
thermal history algorithms 826
thermal mining *in situ*
   techniques 1141
thermogenic gas shales
   microfractures 1133
thermogenic petroleum systems
   Angola future gas exploration 1045–1052
      Post-Aptian salt plays 1049–1052
      Pre-Aptian salt plays 1045–1049
Thousand Barrels a Second
   (Tertzakian) 14
three channel pathways
   Central North Sea basin *88*
three-component geophones
   S-waves 833
three dimensional horizon maps
   MMU *79*
three dimensional maps
   West Greenland *991*
three dimensional modelling
   basin project *660*
   compilation 844
   crystalline basement surface *70*
   finite-difference
      area *1019*
      generation *1019*
      NATS *v.* WATS *1020*
      WATS oil field subsalt offshore Angola
         image 1016–1017
   forward and inverse 63–64

   geological 294–296
   geological software 1149
   inversion
      free air gravity computation 65
      layer density values input *66*
      stages 63
   isostatic behaviour 824
   petroleum systems 686, 689
   showing granite morphologies *68*
   visualization environment 1035
three dimensional seismic data
   Captain Field 343
   Clair field 303
   hydrocarbon systems 51
   and sequence stratigraphy of North Sea *53*
three dimensional seismic mapping
   North Sea *52*
   and porosity variation of intra-chalk
      units 537–548
Three Forks Formation 1066
thrust-and-fold belts
   Cenozoic 113
   Italy hydrocarbon occurrences 113–125
thrust faults offset
   Walker Ridge area *907*
thrusting *904*
   Gulf of Mexico *904*
   propagation folds 122
   Sigsbee Escarpment 905
   types *903*
Thud Field
   tectonic and eustatic events *489*
Thud Formation 473, 478, 481–482
   age 481
   biofacies 482
   correlation 482
   lithology/petrophysics 481–482
   tectonostratigraphy 482
TIB. *See* Transscandinavian Igneous
   Belt (TIB)
tidal inlet sands
   Rømø barrier island *152*
tight gas sandstones 1099, 1120
tight oil
   unconventional oil and gas resources
      1061–1062
tight reservoirs
   North America 1061
Tihemboka area
   chronostratigraphic chart *720*
   geological map *715*
   Siluro-Devonian stratigraphic architecture
      715
Tihemboka High 715–716
   interpretation 715–716
   stratigraphic architecture 715
Tilje Formation 425
Timan–Pechora Basin
   Palaeozoic oil-prone source rocks 602
time-lapse seismic data 532
   acquisition, processing and
      inversion 525
   AVO inversions 526–527
   South Arne 527
   volumetric strain *533*
time-lapse time shift
   Ekofisk 532
   South Arne *526*
   Valhall 532

INDEX

time-lapse time shift observations
  applied time-lapse seismic methods 528–533
Timor
  petroleum systems 828
Tindouf Basin 738
TOC. See total organic carbon (TOC)
Tofte Formation 421, 425–426
  depositional synthesis 425
  description 425
  facies elements 426
  marine shales 426
  pressure profiles 422
  rapid pressure decline explanation 426
tomographic inversions
  wide-angle arrival times **835**
Top Buzzard 373
Top Chalk Group
  time-structure map 538
Top Chalk horizon
  depth maps 1009
top crystalline basement depths
  3D inversion of free air gravity computation 65
Top Eocene
  depth maps 1009
Top Hermod surface 48
Top Joanne
  reservoir zonation scheme 240–241
Top Julius Mudstone
  reservoir zonation scheme 241–242
Top Jurassic depth map
  West Siberia Basin 660
Top Lower Joanne
  reservoir zonation scheme 241
Top Middle Joanne
  reservoir zonation scheme 241
Top Palaeogene Chalk
  limestone porosity 465–466
  natural gamma ray log 465–466
Top Reservoir
  Base Cretaceous and Base Chalk horizon **499–501**
  dry holes and discoveries analysis 503
  hydrocarbon discoveries v. aquifer seal capacity histogram 502
  South Arne structure 524
top seal 1183–1184, 1187
  seismic signatures 1187
  trapping below 1187
Top Shiranish
  depth map 804
Top Triassic Unconformity 227
Top Zeit rugosity
  PSDM velocity models 787
Tor Field
  hydrodynamic sorting pattern 479
  porosity and bulk volume water 477
  tectonic and eustatic events 489
Tor Formation 473, 476–477, 478, 481, 523, 539. See also Lower Tor Formation
  age 484
  air–mercury capillary pressure 469
  biofacies 484
  Chalk 464
  Chalk well micrographs 467
  core data 468
  Cretaceous age 225
  depth trends 468
  effective stress 532
  lithology/petrophysics 484
  mercury capillary pressure 467
  penecontemporaneous allochthonous sediments 476
  porosity, water saturation and PWFT 469
  reservoir 525
  tectonostratigraphy 484
Tormore
  AVO 283, 296
  chlorite 296
  gas condensate discoveries 279
  Laggan mature understanding 296
  location map 280
  Paleocene gas condensate discoveries 279
  structural map 283, 296
Tornquist Fan
  Bouguer gravity profiles 574
total depth (TD)
  Faroe Islands 1033
total dissolved solids (TDS) 1190
total magnetic field anomaly
  Norwegian continental shelf 887
total organic carbon (TOC) 811
  Bakken Formation 1073
  Bakken member 1070
  Bakken shales 1066
  Carboniferous 1092
  England 1092
  Namurian 1091
  Northeast England 1092
  Pannonian Basin 1126
  upper Bakken member 1070
TOUGH2 model 1178
  flow simulation 1177
  intra-reservoir mudstones **1177**
Tournaisian age
  Early 729
  Late 729
  Middle 729
Tournaisian Alum Shale 1080
T38 prospectivity
  Lamba formation 249–258
  Northern Judd Basin sequence 249–259
    CSEM survey 254–258
    hydrocarbon sourcing 251–253
    reservoir 249–251
    rock physics analysis 253–254
    seal 253
    seismic anomaly 252
    seismic model 254
    trap 251
    reducing uncertainty level 249–258
traditional fieldtrip training tool
  petroleum geoscientists virtual fieldtrips 20
Traill Island
  AFTA data 641
Traill–Vøring igneous complex 580–581
Trans-Saharan Seaway 742–743
Transscandinavian Igneous Belt (TIB) 569
traps
  scale salt-related geometries implications 43–44
  stratigraphy role 262
  style 350
Trecate oil field 119

Triassic age 663
  Argana Basin 930
  basins location map 922
  basins of Morocco 928
  Berkine Basin 689
  Bunter fields 262, 270
  Bunter Shale 38
  Central High Atlas Basin 930
  CNS 407
  Early 949
  Fundy Basin 930
  Late 118–119, 265
  Main Buntsandstein Formation 267
  Morocco palaeogeographic map 927
  Netherlands petroleum exploration with stratigraphic plays 267–268
  outcrops 931
  palaeogeographic reconstruction 929
  plate-tectonic setting and depositional environments 603
  pods 201, 410
  pre-rift succession 957
  red beds 926
  reservoirs 38–39, 405–417
  salt ridges 78
  sand dispersal on NW European margin 927–928
  sedimentary architecture 405–417
  Skagerrak Formation 214, 214, 405
  strata 41, 191
  stratigraphy 40, 228
  succession 118, 932
  traps 262
Trinity three dimensional basin model 754
Troll delta 45
Trøndelag Platform 890, 925
  amplitude magnetic anomalies 574
  offshore top basement map 894
  onshore tectonostratigraphic map 894
true vertical depth subsea (TVDSS) 355
Tunguska Basin 621
Tunisia reservoirs 753
Tupi oil discovery
  SPP 859
turbidites 282
  amalgamated 279
  Andrew Formation 431
  biostratigraphy 381
  channel complexes 351
  chemostratigraphy integration 381
  deposition
    Puffin Terrace 195
    Puffin Terrace and Fulmar–Clyde Terrace 195
    sequence J56 191
  high-density 372
  Humber Group 189
  KCF 192, 369
  mud and silt 282
  ponded 282, 288
  reservoirs 356
  sand deposition phases 369
  sands pressures 503
  sandstones sequences J64 196
  Selkirk 202
  slope 826
TVDSS. See true vertical depth subsea (TVDSS)

two dimensional modelling
  gravity 868, 870
  margin evolution 824
  perspective *1036*
  seismic data 57
two dimensional seismic inversion
  intra-chalk unit porosity variation 541–543
  well log correlation 537
two dimensional seismic mapping
  North Sea *52*
two way travel time (TWT) structure
  Horizon 4, *86*
  Horizon 6, *87*
  MMU *79*
TWT. *See* two way travel time (TWT) structure
Tyra Field 463–464, *465–466*
  bulk density and water saturation *466*
  calculated water saturation *465*
  capillary pressure 471
  capillary pressure curve *469*
  Chalk oil–water contact tiling 463–471
    log- and core data analysis 464–471
    results 471
  conceptual cross section *463*
  conceptual water saturation *464*
  depth structure map *464*
  GOC *464*
  hydrodynamic action 471
  lithostratigraphic and fluid contacts in wells **468**
  OWC 465, 471
  porosity, water saturation and PWFT *470*
  water saturation calculation 471
  water-saturation-depth curve *469*
  well logs analysis **468**
Tyumen Formation
  Jurassic age 666
  source rocks 661

UBGH1. *See* Ulleung Basin Gas Hydrate Expedition 1 (UBGH1)
UCG. *See* underground coal gasification (UCG)
UCGP. *See* Underground Coal Gasification Partnership (UCGP)
UKCS. *See* United Kingdom Continental Shelf (UKCS)
Ukraine
  shale 1081
Ulleung Basin Gas Hydrate Expedition 1 (UBGH1) 1154
ultra-heavy oil
  biodegradation 1062
  Canada 1062
  unconventional oil and gas resources 1062
Unayzah complex
  Saudi Arabia 692
Unayzah formations
  Ordovician 697
unconformities
  nature and significance 824
  Neogene basin 91
  traps 262
unconventional oil and gas resources 1061–1063, 1065
  basin-centred and tight gas 1062
  carbon dioxide geological storage 1063
  coal gasification 1063
  gas hydrates 1062–1063
  geological storage of carbon dioxide 1061–1063
  shale gas 1062
  shale oil 1061–1062
  tight oil 1061–1062
  ultra-heavy oil 1062
underground coal gasification (UCG) 1063, 1155
  CCS 1155–1163
    economic case 1160
    ERA 1160–1162
    history 1155–1156
    implementation technical challenges 1160
    Northeast England case study 1162
    process 1156–1157
    prospects for carbon dioxide storage 1158–1159
    stages *1157*
  UK feasibility 1156
Underground Coal Gasification Partnership (UCGP) 1156
uneconomic oil volumes
  locating remaining oil in Nelson Field 362–364
    heterogeneous reservoir intervals 363
    minor attic oil volumes 363
    thin beds 362–363
United Kingdom *47*, 1155–1156, 1159–1160, 1162. *See also* specific location
  Central Graben sequence stratigraphy reappraisal 177–210
  Central North Sea 213–223, *219*
  data and analysis for shale gas prospectivity 1087–1097
  hydrocarbon self-sufficiency 31
  Lower Palaeozoic black shales 1091
  Mississippian shale 1090
  North Sea 337–347
  Norway Quad 26 area *47*
  onshore 69
  Sele/Hermod shelf *47*
  sequence stratigraphic models *47*
  shale gas 1062, 1087–1097
    Barnett Shale 1087
    Carboniferous black shales 1091–1096
    gas content and kerogen type 1091
    gas fields and discoveries 1092
    Jurassic black shales 1096–1097
    mineralogy 1092
    porosity, permeability and fracture porosity 1091–1092
    potential source rocks general characteristics 1088–1090
    pre-Carboniferous black shales 1090–1091
    principal source rocks 1092–1096
    thermal maturity 1091
    TOC 1091
    UK's closest tectonic analogue 1087–1088
  shale gas exploration 1087
  shales stratigraphic groups 1089–1090, *1090*
  Southern North Sea 315–323
  UCG feasibility 1156
  wells 505
    direct pressures in Chalk *498*
    fluid pressures *497*
    pressure-depth plot *496*
United Kingdom Continental Shelf (UKCS) 15, 225–242
  exploration 213
  Faroe–Shetland Basin location map *280*
United States. *See also* specific location
  NSF source-to-sink studies 913
  regional carbon sequestration partnerships programme 1189–1195
  shale gas 1079–1180
  Williston Basin *1066*
untapped hydrocarbon accumulations 455
updip water 1099–1100, 1109–1110, 1122
uplift and erosion
  North Atlantic passive margins 981
  shelves of Greenland 1000
  subsidence 982
  tectonic 92
uplift and exhumation 979–1001
  case A tectonic control 982–986
  case B early Neogene exhumation 987–990
  case C West Greenland 990–992
  case D Oligocene hiatus 992–994
  case E Ordovician source rocks 994–995
  case F post-rift burial 995–999
  elevated rift margins 999–1000
  estimating timing and magnitude 981–982
  North Atlantic passive margins 979–1001
  origin of post-rift uplift along passive margins 1000
upper Bakken member
  maturity *1070*
  pseudo van Krevelen diagram *1070*
  TOC *1070*
Upper Bowland Shale
  black shale **1095**
  mudstones thickness *1096*
  thickness 1095
Upper Captain Sand (USC) 343
Upper Cretaceous age. *See* Late Cretaceous age
Upper Devonian age. *See* Late Devonian age
Upper Ekofisk 475, 487–488
  age 487
  biofacies 488
  biostratigraphy 475
  lithology/petrophysics 487
Upper Ekofisk Formation *488*
Upper Fars Formation 804
Upper Jurassic age. *See* Late Jurassic age
Upper Kimmeridgian 137–138
  depositional environment 137–138
  palaeogeography 138
  paralic and shoreface sandstones 129–136
    depositional environments 129–133
    palaeogeography 135–136
    reservoir characteristics 133–134
  reservoir characteristics 138
  shallow marine sandstones 136–137
    depositional environment 136–137
    reservoir characteristics 137
Upper Leman interval
  reservoir properties **334**
Upper Magne Formation 483–484
  age 483
  biofacies 483

correlation 483
formation 483
lithology/petrophysics 483
tectonostratigraphy 483
Upper Middle Volgian 139–141
  depositional environment 139–141
  distribution and stratigraphy 139
  palaeogeography 141
  reservoir characteristics 141
Upper Miocene age. *See* Late
    Miocene age
Upper Nikanssin age
  acoustic properties *1119*
  secondary porosity *1107*
  sections *1107*
Upper Ordovician age
  Lower Silurian shales 715
Upper Piper sections 393
Upper Ror Formation 425
Upper Rotliegendes Group
  Permian 262
  play map *269*
  strata 62
  units 61
Upper Rudeis Formation
  characterization 777
  sandstone distribution 777, *779*
Upper Rudeis prospectivity
  exploring plays in Gulf of Suez 777
Upper Silurian age
  Caledonian unconformity 711
Upper Tor Formation 484–486
  age 484
  biofacies 484–485
  lithology/petrophysics 484
  tectonostratigraphy 486
Uralian deformation 646
Uralian Ocean 656
Uralian Orogeny 645, 654
Ural mountains 660
Urengoy field 664, *664*
  gas charge 667
  gas yield *665*
USC. *See* Upper Captain Sand (USC)
U-shaped minibasin
  Walker Ridge area *905*
Ust' Lena Rift 605
Utah
  Book Cliffs of Utah 24, 52
  deltaic reservoir analogues 24
  EOR-sequestration test 1190
  Gordon Creek Field 1193–1195
    commercial-scale injection test
      1193–1195
    site location/surface geology map *1192*
    stratigraphy *1193*
Utsira Formation
  carbon dioxide
    flow model *1185*
    flow v. hydrocarbon migration 1183
    injection 1183–1184
    migration 1187
    plume seismic signatures *1185*
  location and extent *1184*
  modelled 3D carbon dioxide
    distribution *1186*
  North Sea *1184*
Utsira Sand 1171
  amplitude changes *1173*

amplitude–thickness
    relationship *1174*
average velocities variation to top
    1179, *1179*
carbon dioxide saturations 1173
Cenozoic saline aquifer 1063
core material centrifuge experiments 1173
CWC *1173*
distribution and thickness *1172*
faults 1176
geophysical logs *1172*
layer thickness-saturation
    function *1174*
north-trending depositional geometry 1178
permeability values 1177
reflection amplitude changes 1173
seismic velocity *1174*
water production 1179

Vaila Formation 245
  stratigraphy *281*
Val d'Agri oil field *121*
  discovery 119
  hydrocarbons *120*
  Italy hydrocarbon occurrences 119–120
  petrophysical characteristics *120*
Våle Formation 487, *488*
Valhall
  LoFS system 311
  time-lapse time shift 532
valid exploration targets
  Carboniferous petroleum system 74
Variscan basement 924
Variscan Foreland Basin 1080–1081
Vasyugan Formation 661
Venture appraisal well 332
Venture-operated Chiswick Field 332
Verkhoyansk Fold Belt 596
Vernet Field 458
vertical seismic profiling (VSP) 234
Verzhbitsky, Vladimir 589
Vienna Basin
  geological cross-section *1082*
Viera Anticline 971
  Late Cenozoic compression 971
Viking Graben 38, 43, 48, 50, 893
  individual contributions to basement depth
    map 891
  sedimentology and sequence stratigraphy
    173–174
  well-log correlation panel *169*
Vil'kitski Rift Basin 610, 611
Villafortuna Field *119*
  discovered 118
virtual fieldtrips
  creation 19–20
  petroleum geoscientists 19–25
  workflow 20
virtual fieldwork
  basis 20
  workflow *21*
virtual outcrop (VO) 19, 20
Virtual Reality Geological Studio (VRG)
  lidar scan data 24
Viséan age
  Late 729
Viséan incised valleys 733
Viséan transgressive–regressive cycles 729
vitrinite reflectance (VR) 633, 980

data v. depth *994*
gradients 96
measurements 95
percentage values *1093*
software 22
Vlieland section
  cross section *272*
VO. *See* virtual outcrop (VO)
volcanic rocks magnetic properties *572*
volcanism onset 1026
Volgian age *204*. *See also* Early Volgian age
  depocentre
    development 201
    downdip *203*
  Late *202*
  sandstone intervals of North Sea *139*
  sequence J64 199
  sequence J66a *195*, 199
  turbidites 206
    sediment sources *207*
  well correlation *202*
Volgian–Ryazanian
  broad log character motifs 199
  environments *196*
  formation process *200*
  Humber Group sequence stratigraphy
    reappraisal 199
  log profiles 209
  megasequence 199
    sequence J64 199
    sequence J66a 199
    sequence J66b 199
    sequence J71-J72 199
    sequence K10 199–200
    sequences J73-J76 199
    succession *184*, 199
  sequence age *203*
  tectonic reconfiguration 199, 209
Volve Field 1179
Vøring Basin *847*
  Cretaceous strata *1031*
  Eocene volcanic rocks 568
  magmatic underplating 853
  modelled strains *583*
  Paleocene strata *1031*
Vøring margin
  basement thickness *892*
  Caledonian basement 890
  crustal thickness *847*
  geological and structural features *844*
  individual contributions to basement depth
    map 890
  lateral volcanic distribution 846
  LCB distribution *847*
  magmatic rocks *847*
  Moho difference *850*
  Moho map *892*
  negative Moho difference 846
VR. *See* vitrinite reflectance (VR)
VRG. *See* Virtual Reality Geological Studio
    (VRG)
VSP. *See* vertical seismic profiling (VSP)

Wabiskaw Formation 1144
Wadden Sea 145–154, *148*, *149*, *151*
Wadi Shati
  incised valleys *712*
Wales
  biogenic and thermogenic shale gas 1080

Walker Ridge area
  thrust faults offset 907
  U-shaped minibasin 905
Walvis Ridge
  location 868
water
  distribution to gas production across basin 1114
  layer stratigraphy Rockall Trough 554
  saturation capillary pressure 1114
water cones 341
WATS. See wide-azimuth towed streamer (WATS)
Waulsortian reef 733
WCG. See West Central Graben (WCG)
Welbeck Colliery
  gas wetness ratio 1094
wells 1155–1156. See also Infill wells; specific name
  20/06-2, 371
  20/06-2(a-B1A)
    geochemical profiles and binary diagrams 383
  20/06-5 372
  20/60(a-B)-8 380
  20/60(a-B1A,-B5,-B6) 379
  20/60(a-B14Z)
    pressure depletion 378
    updip production 378
  22/11 360
  30/6-07 234
  30/06-6 228, 230
  30/06-7
    migrated VSP image 235
  48/14-2
    discovery well 332
    evaluation 329–330
    fracture stimulation 332
    fracture stimulation process 336
    future orientation 335
    hydraulic fracture stimulation parameters 330
    reservoir properties **334**
    scanning electron photomicrograph of fibrous illite 331
  48/14-5
    rose diagram showing strike orientation 333
  48/14(a)-5 330–332, 332
    reservoir properties **334**
    scanning electron photomicrograph of fibrous illite 331
  48/14(a)-6 332–335
    FMI log 334
    horizontal well 335
    hydraulic fracture-stimulated intervals 333
    reservoir properties **334**
  204/18-1 251
    reservoir analogue 250
  A3
    pressure-tester formation 358
  Amalie-1 wells 138
  Amethyst State-1 well, Alaska 637
    AFTA and VR data 638, 639, 640
    palaeotemperatures 639
  A1-NC43 764
  A1-NC198 761–769
    buried hill structure 763
    chemostratigraphy 767
    Kufra Basin 761–769
    lithology and GR characteristics 765
    palynomorphs 767
    Silurian to Carboniferous stratigraphy 767
    Spore Colour Index against depth 769
    well log 766
  business justification of targeting blind test 1022
  Captain Field, Well A3 346
  Central North Sea 58
  Central Panel injection well 382
  Central Trough 51
  Cleo-1 wells 138
  correlation panel 241
  Dooish well 939, **942–943**
  England 1089
  Felicia-1 and Hans-1 wells, Offshore Denmark 642
  Gert-1 well 130, 134, 145
    back barrier sediments 148
    interpreted washover fans photographs 153
    porosity-permeability plot 148
    sedimentological core log 147
  Gro-3 well 640
  Hejre-1 134
  HH84-1, GG85-1, GS216-1 778
  Huntington Fulmar well location 219
  hydrocarbon(s) 1089
    injection and production 1157
  J4 393
  J30 393
  J30(Z) 402
  J32(Z) 396, 397, 398
  J34 393, 396, 402
  J66(a SB) 198
  Jade Field 242
  JAPEX/JNOC/GSC Mallik 2L-38 gas hydrate research well 1153
  Judy Member 410
  Julia horst well 228, 232
  Julia well 236
  kerogen **106**
  leaking 1162
  log correlation 2D seismic inversion 537
  monitoring 1161
  N15
    shale baffle 355
  N20(Z) 357
  N27 364
  North Sea data 41
  North Slope of Alaska 640
  pressure and logs 1162
  Qara Chauq 804
  Rockall Basin palaeodrainage stratigraphy 944
  Saxo-1 139
  Scott zonation 389
  sealed 1162
  seismic to synthetic tie 1120
  Svane-1 138
    formation pressures **136**
  T35 246
  T36 246
  T38
    prospectivity 249–258
  T70
    bypassed oil 366
    bypassed oil volumes 366
    4D seismic response 362
    macroforms and drainage cells 354
  T75
    fill and spill geometry 355
    fuel tank display 363
    macroforms and drainage cells 353
    western channel baffles 359
  Wessel-1 139
  Western Canadian Foreland Basin density 1101
Wernigerode Phase tectonics 482
  Peine Phase tectonics 482
Wessel-1 well
  Volgian sandstone intervals 139
West Africa
  exploring older and deeper hydrocarbon plays 676–681
  hydrocarbons 7
  POC 874
  Taoudenni Basin 676–681
West Canada
  BCG basin model 1110–1111
  Nikanassin Formation 1101–1104
West Canada Basin
  centred gas system 1062, 1099–1122
  Devonian Exshaw 1075
  shale gas 1079
West Canada Sedimentary Basin
  regional cross-section 1145
  regional setting 1141–1144
  stratigraphic column 1142
  tectonic setting from Late Jurassic to Early Eocene 1144
West Canadian Foreland Basin
  schematic regional tectonic map 1100
  well density 1101
West Central Graben (WCG) 180
  high-amplitude reflector 185
  marine conditions 187
  seismic line 200, 201, 203
  shoreface deposition 190
  well correlation 188, 194, 197
West Central Shelf 69
  seismic profile 65, 66
  seismo-stratigraphic relationships 62–63
West Emila accretionary arc
  geological sections 117
  hydrocarbon occurrences 117
  Settala gas field 123
  Villafortuna-Trecate oil field 119
Western Channel wells
  22/11-N24 and 22/11-N28Z 357
  N15 bridge plug 358
Western Desert
  burial history 697
  exploring older and deeper hydrocarbon plays 697–698
  tectonic events 697
Western Platform 191
Western Structural Domain
  Novaya Zemlya 651
West Europe
  shale gas 1079, 1083
West Forties basin
  tectonic reconfiguration 201

West Gondwana
 reconstructions 876
West Greenland *981*
 burial and uplift history *995*
 3D maps *991*
 erosional unconformity 1001
 evaluating palaeotemperature profiles 982
 events chronology *992*
 geomorphological and geological
  observations 990
 landscape 1001
 landscape and detailed
  thermochronological dataset 999
 Oligocene hiatus offshore *993*
 Oligocene unconformity *994*
 outcrops 1034
 seismic profile *994*, *996*
 unconformity below Neogene
  succession *994*
 uplift and erosion of offshore
  shelves 1000
West Lewis Basin
 seismic profile *972*
West Lewis Ridge
 reverse faults 970–971
  inversion 970
 seismic profile *973*
West Limb
 Boundary Fault *232*
 core area gas condensate *233*
 exploration drilling phase summary *230*
 seismic reprocessing results *239*
West Lutong Field 459
 by-passed hydrocarbons identification *460*
 geological models *459*
West Orkney Basin
 classical rifting 923
Westphalian
 black shale 1080
 Chiswick Field 319
 gas development 319
Westray Fault 245, 252
 oil *253*
 oil migration model *253*
West Shetland Area
 aeromagnetic map *248*
 stratigraphy *281*
West Shetland Platform
 breached relay ramps 957
 conceptual model for structural
  evolution *959*
 conceptual play model *961*
 Early Cretaceous *956*
 seismic acquisitions/processing *955*
West Siberia 667
 Cenomanian gas fields 667
 Cenomanian reservoirs 667
 isotopic composition of methane *665*
West Siberia Basin 660, 661, 662
 cross-section of petroleum systems *661*
 3D basin modelling project *660*
 Mesozoic chronostratigraphy *662*
 northern and central 661, *662*
 source rocks 659
 Top Jurassic depth map *660*
White Zone 249

wide-azimuth towed streamer (WATS)
 acquisition configuration *1016*
 acquisition details *1017*
 acquisition parameters *1017*
 data *1020*
 five-tile *1021*
 four-tile *1021*
 Gulf of Mexico 1023
 oil field subsalt offshore Angola image
  1013–1024
  business case support 1023
  3D finite-difference modelling
   1016–1017
  model evaluation 1018–1022
  recommended WATS survey 1023
  subsalt imaging challenges 1013–1015
  wide-azimuth solution 1015–1016
 six-tile *1021*
 survey
  Block 31 acquisition area *1023*
  Mad Dog Field 1016
  subsalt offshore Angola image 1023
 three-tile *1021*
 tile number comparison
  v. NATS *1021*
 tiles 1019
wide sag-type basins 824
Widmerpool Formation
 shale 1093
Widmerpool Gulf *1095*
Williston Basin *1066*
 Bakken Formation 1065–1076
 Devonian age 1061–1062, 1065
 GOR *1072*, *1073*
 lithostatic and hydrostatic pressure
  gradients *1072*
 location *1066*
 Mississippian Bakken Formation 1065
 petroleum system 1061–1062
Witch's Hat segment
 Clair field 309, 310
Withycombe Farm borehole 1091
Wolfville fluvial succession 931
Wolfville Formation 929, 931, *932*
 summary correlation *933*
Woodford Formation
 folded calcite-sealed fracture *1136*
 folded dolomite veins *1136*
Woodford Shale
 tall calcite-sealed fractures *1133*
Worcester Graben satellite basins 1087–2088
World Energy Outlook (2005)
 International Energy Agency (IEA) 11
Worston Shale Group
 black shale **1095**
 shale 1093
Wrangel Foreland Basin 609–610, 612
Wrangel-Herald Arch 612, 613
Wrangel Island 597
 crustal tectonic *594*
Wyoming 5, 6
Wyoming Craton 5
Wyville–Thomson Ridge Anticline
 968–969
 bottom-water current activity 969
 seismic profile *971*

XCTD. *See* expendable
 conductivity-temperature-depth
 (XCTD) probe
XSP. *See* expendable sound speed
 probe (XSP)

Yamal Peninsula *661*
Yell Basin
 Clair play *960*
 rifted margin *960*
 seismic acquisitions/processing *955*
 structural evolution and sediment
  dispersal *960*
 structural reconstruction *956*
Yenisei-Khatanga Basin
 Assessment Unit
  densities and median sizes *628*
  distributions and assessment unit
   probabilities *629*
Yenisei-Khatanga region 593
Yenisey-Khatanga Basin 622
 assessment unit 629–631
 exploration maturities **626–627**
 field densities and sizes **626–627**
Ylvingen Fault 571
Ymir Ridge Anticline 970
 North *973*
 seismic profile *971*
Ymir Ridge Complex 968
Yowlumne Field 352
Ypresian isochore
 Rockall–Faroe area *970*

Zagros mountain-building
 folding and faulting 803
Zechstein Group *59*, 66
 composite layer
  crystalline basement surface *72*
  gas chimneys *72*
  isopach map *72*
  isopachs *69*
 evaporite seal 71
 faults 891
 isochore two-way time and depth
  maps *320*
 isopachs 65
 Netherlands petroleum exploration
  267
 play map *270*
 salts 320
  BCU depth structure map *219*
  halokinetic activity 269
  underpinning 44
 seismo-stratigraphic relationship
  Auk–Argyll–Flora area 60–61
  Jaeren high area 62
  Mid North Sea High area 61
  West Central Shelf 62–63
 topsoil 73
 Upper Palaeozoic 59
 well sections 57–59
Ziegler, Peter 38, 39, 51
zircon
 age spectra 6
 detrital 6
 geochronology 6